List of Elements with Their Symbols and Atomic Masses

Element	Symbol	Atomic Number	Atomic Weight	Element	Symbol	Atomic Number	Atomic Weight
Actinium	Ac	89	227.03[a]	Molybdenum	Mo	42	95.94
Aluminum	Al	13	26.98	Neodymium	Nd	60	144.24
Americium	Am	95	243.06[a]	Neon	Ne	10	20.18
Antimony	Sb	51	121.76	Neptunium	Np	93	237.05[a]
Argon	Ar	18	39.95	Nickel	Ni	28	58.69
Arsenic	As	33	74.92	Niobium	Nb	41	92.91
Astatine	At	85	209.99[a]	Nitrogen	N	7	14.01
Barium	Ba	56	137.33	Nobelium	No	102	259.10[a]
Berkelium	Bk	97	247.07[a]	Osmium	Os	76	190.23
Beryllium	Be	4	9.012	Oxygen	O	8	16.00
Bismuth	Bi	83	208.98	Palladium	Pd	46	106.42
Bohrium	Bh	107	264.12[a]	Phosphorus	P	15	30.97
Boron	B	5	10.81	Platinum	Pt	78	195.08
Bromine	Br	35	79.90	Plutonium	Pu	94	244.06[a]
Cadmium	Cd	48	112.41	Polonium	Po	84	208.98[a]
Calcium	Ca	20	40.08	Potassium	K	19	39.10
Californium	Cf	98	251.08[a]	Praseodymium	Pr	59	140.91
Carbon	C	6	12.01	Promethium	Pm	61	145[a]
Cerium	Ce	58	140.12	Protactinium	Pa	91	231.04
Cesium	Cs	55	132.91	Radium	Ra	88	226.03[a]
Chlorine	Cl	17	35.45	Radon	Rn	86	222.02[a]
Chromium	Cr	24	52.00	Rhenium	Re	75	186.21
Cobalt	Co	27	58.93	Rhodium	Rh	45	102.91
Copper	Cu	29	63.55	Roentgenium	Rg	111	272[a]
Curium	Cm	96	247.07[a]	Rubidium	Rb	37	85.47
Darmstadtium	Ds	110	271[a]	Ruthenium	Ru	44	101.07
Dubnium	Db	105	262.11[a]	Rutherfordium	Rf	104	261.11[a]
Dysprosium	Dy	66	162.50	Samarium	Sm	62	150.36
Einsteinium	Es	99	252.08[a]	Scandium	Sc	21	44.96
Erbium	Er	68	167.26	Seaborgium	Sg	106	266.12[a]
Europium	Eu	63	151.96	Selenium	Se	34	78.96
Fermium	Fm	100	257.10[a]	Silicon	Si	14	28.09
Fluorine	F	9	19.00	Silver	Ag	47	107.87
Francium	Fr	87	223.02[a]	Sodium	Na	11	22.99
Gadolinium	Gd	64	157.25	Strontium	Sr	38	87.62
Gallium	Ga	31	69.72	Sulfur	S	16	32.07
Germanium	Ge	32	72.64	Tantalum	Ta	73	180.95
Gold	Au	79	196.97	Technetium	Tc	43	98[a]
Hafnium	Hf	72	178.49	Tellurium	Te	52	127.60
Hassium	Hs	108	269.13[a]	Terbium	Tb	65	158.93
Helium	He	2	4.003	Thallium	Tl	81	204.38
Holmium	Ho	67	164.93	Thorium	Th	90	232.04
Hydrogen	H	1	1.008	Thulium	Tm	69	168.93
Indium	In	49	114.82	Tin	Sn	50	118.71
Iodine	I	53	126.90	Titanium	Ti	22	47.87
Iridium	Ir	77	192.22	Tungsten	W	74	183.84
Iron	Fe	26	55.85	Uranium	U	92	238.03
Krypton	Kr	36	83.80	Vanadium	V	23	50.94
Lanthanum	La	57	138.91	Xenon	Xe	54	131.293
Lawrencium	Lr	103	262.11[a]	Ytterbium	Yb	70	173.04
Lead	Pb	82	207.2	Yttrium	Y	39	88.91
Lithium	Li	3	6.941	Zinc	Zn	30	65.41
Lutetium	Lu	71	174.97	Zirconium	Zr	40	91.22
Magnesium	Mg	12	24.31	*[c]		112	277[a]
Manganese	Mn	25	54.94	*[b]		113	284[a]
Meitnerium	Mt	109	268.14[a]	*[b]		114	289[a]
Mendelevium	Md	101	258.10[a]	*[b]		115	288[a]
Mercury	Hg	80	200.59	*[b]		116	292[a]

[a]Mass of longest-lived or most important isotope.
[b]The names of elements have not yet been decided.
[c]During the time of printing, Element 112 has a proposed name of Copernicium, which is currently in the review process by IUPAC.

Tenth Edition

INTRODUCTION TO
CHEMICAL PRINCIPLES

Tenth Edition

INTRODUCTION TO CHEMICAL PRINCIPLES

H. Stephen Stoker
Weber State University

Prentice Hall

Boston Columbus Indianapolis New York San Francisco Upper Saddle River
Amsterdam Cape Town Dubai London Madrid Milan Munich Paris Montréal Toronto
Delhi Mexico City São Paulo Sydney Hong Kong Seoul Singapore Taipei Tokyo

Editor in Chief, Chemistry and Geosciences: Nicole Folchetti
Acquisitions Editor: Terry Haugen
Marketing Manager: Erin Gardner
Assistant Editors: Laurie Hoffman and Carol DuPont
Editorial Assistant: Lisa Tarabokjia
Managing Editor, Chemistry and Geosciences: Gina M. Cheselka
Project Manager: Wendy A. Perez
Senior Operations Supervisor: Nick Sklitsis
Operations Specialist: Amanda A. Smith
Composition/Full Service: GEX Publishing Services
Art Editor: Connie Long
Art Studio: Stacy B. Smith
Art Director/Cover Design: Jayne Conte
Photo Researcher: Eric Schrader

Cover Photo Credit: Digital Art/Corbis

Library of Congress Cataloging-in-Publication Data
Stoker, H. Stephen (Howard Stephen), 1939-
    Introduction to chemical principles/H. Stephen Stoker.--10th ed.
      p. cm.
    Includes bibliographical references and index.
    ISBN 978-0-321-66604-8 (pbk. : alk. paper)
    1. Chemistry.   I. Title.
  QD33.2.S76   2011
  540--dc22   2009041052

Printed in the United States
10 9 8 7 6 5 4 3 2 1

Prentice Hall
is an imprint of

www.pearsonhighered.com

ISBN-13: 978-0-321-66604-8
ISBN-10: 0-321-66604-6

CONTENTS

PREFACE

Introduction to Chemical Principles is a text for students who have had little or no previous instruction in chemistry or whose instruction was so long ago that a thorough review is needed. The text's purpose is to give students the background (and confidence) needed for a subsequent successful encounter with a main sequence, college-level, general chemistry course.

Many texts written for preparatory chemistry courses are simply watered-down versions of general chemistry texts: They treat almost all topics found in the general chemistry course, but at a superficial level. *Introduction to Chemical Principles* does not fit this mold. My philosophy is that it is better to treat fewer topics extensively and have the student understand those topics in greater depth. I resisted the very real temptation to include lots of additional concepts in this new edition. Instead, my focus for this edition was on rewriting selected portions to improve the clarity of presentation.

NEW FEATURES OF THE TENTH EDITION

- **New page design.** A completely new page design is present throughout the tenth edition. Many of the line drawings also have a new look.
- **Increased number of worked-out example problems.** The number of in-text worked-out examples has been increased by 17 and now totals 240. These worked-out-in-detail problems with their extensive commentary constitute one of the greatest strengths of the text.
- **Extensive revision of "End-of-Chapter Problem Sets."** Although the total number of end-of-chapter problems, which already exceeds that of most other similar texts, has not increased significantly, many of the previous edition's problems (35%) have been replaced with new problems. A special effort was made to create new problems that address specifically the "core concepts" associated with a given chapter section's subject matter. In most chapters several of the newly added problems involve presentation of data in a "visual form" rather than in a "sentence form." Many of the "visual problems" involve situations where reasoning, with little or no calculation, is needed to test a student's grasp of a key concept.

Content changes to individual chapters. After nine successful editions of *Introduction to Chemical Principles*, the need for drastic alterations in chapter ordering and chapter content does not exist. Changes that have been made relate to "fine tuning" of the presentation of the subject matter. Among the most important changes to this edition are the following:

- *Chapter 4:* Material concerning atoms, molecules, and chemical formulas, which previously was the starting material for Chapter 5, has been moved to the end of Chapter 4. This transferred material is a logical extension of the material already present in Chapter 4 on the classifications of matter (mixtures, pure substances, elements, and compounds).
- *Chapter 5:* The previously mentioned shift of some Chapter 5 material to Chapter 4 allows for the inclusion into Chapter 5 of material concerning nuclear stability and instability (radioactivity). This material, which was previously in a separate chapter on radioactivity, is now discussed immediately after the discussion about an atom having a nucleus.
- *Chapter 11:* A new section has been added on hydrogen bonding and how the properties of water relate to this concept. This new section replaces a deleted section that dealt with types of solids.
- *Chapter 12:* Within this chapter, the order in which the various gas laws are introduced has been altered. All discussion of STP conditions is now presented in a section entitled "molar volume." Discussion of gas laws as they apply to chemical reactions now occurs after all of the fundamental gas laws have been presented.
- *Chapters 14–15:* The topic of net ionic equations has been moved from Chapter 14 to Chapter 15 where it is discussed in conjunction with considerations of electrolytes and stoichiometric calculations involving ions. Both of these latter topics are new additions to Chapter 15.

- **Expanded "Multiple-Choice Practice Test" feature.** The emphasis in this text has always been and still is on working problems from scratch. Some, but certainly not all instructors, use this same approach when giving class examinations. A multiple-choice question examination is another common type of examination given. To aid students whose examinations involve multiple-choice questions, in each chapter the feature called "multiple-choice practice test" has been expanded. Each test now contains 20 questions rather than the previous 12 questions. It is intended that students use this feature as an aid in reviewing subject matter for an upcoming multiple-choice examination.

IMPORTANT CONTINUING FEATURES IN THE TENTH EDITION

1. **Development of each topic starts out at ground level.** Because of the varied degrees of understanding of chemical principles possessed by students taking a preparatory chemistry course, each topic is developed step by step from ground level until the level of sophistication required for a further chemistry course is attained.

2. **Problem-solving pedagogy is based on dimensional analysis.** Forty-two years of teaching experience suggest to me that student "troubles" in general chemistry courses are almost always centered on the inability to set up and solve problems. Whenever possible, I use dimensional analysis in problem solving. This method, which requires no mathematics beyond arithmetic and elementary algebra, is a powerful and widely applicable problem-solving tool. Most important, it is a method that an average student can master with an average amount of diligence. Mastering dimensional analysis also helps build the confidence that is so valuable for future chemistry courses.

3. **Detailed commentary accompanies all worked-out example problems.** In all chapters, one or more worked-out example problems follow the presentation of key concepts. These examples walk students through the thought processes involved in solving the particular type of problem. Detailed commentary accompanies all of the steps involved in solving a problem. In addition, an unworked practice exercise is coupled to each worked-out example. It is intended that students work this exercise immediately after examining the worked-out example. Such action gives a student immediate feedback on whether he or she actually understands the worked-out example. Inability to correctly work the practice exercise, whose answer is provided, is indicative that more study of the worked-out example is needed. In total, the number of worked-out examples is significantly greater than that found in most texts.

4. **"Answer Double Check" feature.** Over half (60%) of the text's worked-out examples are enhanced by the feature called "answer double check." The purpose of this feature, which is appended to the end of the worked-out example discussion, is to encourage students to consider whether the answer they obtain in working a problem is a reasonable answer in terms of items such as numerical magnitude, number of significant figures present, sign convention (plus or minus), and direction of change (increase or descrease). An unreasonable answer is often a sign that a calculator error has been made.

5. **Significant-figure concepts are emphasized in all problem-solving situations.** Routinely, electronic calculators display answers that contain more digits than are needed or acceptable. In all worked-out examples, students are reminded about these unneeded digits by the appearance of two answers to the example: the calculator answer (which does not take into account significant figures) and, in color, the correct answer (which is the calculator answer adjusted to the correct number of significant figures).

6. **Operation rules for standardizing uncertainty in numbers are used.** Students often experience a relatively high degree of frustration when they correctly solve a problem and yet obtain an answer that differs *slightly* from the one given in the answer section at the back of the book. They want to get the exact number shown in the answer section. Most often the discrepancy is due to differing

degrees of uncertainty in the input numbers used for the calculation, for example, in molecular mass values. To minimize such frustration, operational rules have been introduced for standardizing uncertainty in input numbers. The standard mode of operation is always (1) to round all atomic masses to hundredths before using them in molecular mass calculations, and (2) to specify frequently used numbers, such as Avogadro's number, molar volume, and the ideal gas constant to four significant figures. Using these operational rules for input numbers, student answers will match the back-of-the-book answers *to the last significant digit.*

7. **Defined terms always appear in self-standing complete sentences.** All definitions are highlighted in the text when they are first presented, using boldface and italic type. Each defined term appears as a complete sentence; students are never required to deduce a definition from context. In addition, the definitions of all terms appear in a separate glossary found at the end of the text. All defined terms have been reexamined to see if they could be stated with greater clarity. The result is a rewording of many defined terms. In addition, the number of defined terms has been increased. There are 29 new or modified definitions in this new edition of the text.

8. **All end-of-chapter exercises occur in matched pairs.** In essence, each chapter has two independent, but similar, problem sets. Counting subparts to problems, there are over 5000 questions and problems available for students to use in their journey to proficient problem solving. Answers to all of the odd-numbered problems are found at the end of the text. Thus, two problem sets exist, one with answers and one without.

9. **Each end-of-chapter problem set, except for Chapters 1 and 2, is divided into four sections:** (1) Practice Problems, (2) Additional Problems, (3) Cumulative Problems, and (4) Multiple Choice Practice Test. The practice problems are categorized by topic and are arranged in the same sequence as the chapter's textual material. These problems, which are always single-concept, are drill problems that most students will find routine. The additional problems section contains problems that involve more than one concept from the chapter and are usually more difficult than the practice problems. The cumulative-skills section draws not only on materials from the current chapter, but also on concepts discussed in previous chapters. The working of problems in this third group allows students to continue to use, rather than forget, problem-solving techniques presented earlier.

10. **Historical vignettes are used to address some of the "people aspects" of chemistry.** These vignettes, entitled "The Human Side of Chemistry," are brief biographies of scientists who helped develop the foundations of modern chemistry. In courses such as the one for which this text is written, it is very easy for students to completely lose any feeling for the people involved in the development of the subject matter they are considering. If it were not for the contributions of these people, many of whom worked under adverse conditions, chemistry would not be the central science that it is today.

11. **Marginal notes are used extensively.** The two main functions of the marginal notes are (1) to summarize key concepts and often give help for remembering concepts or distinguishing between similar concepts, and (2) to provide additional details, links between concepts, or historical information about the concepts under discussion.

SUPPLEMENTS

For the Instructor

Instructor Solutions Manual: (ISBN 0-321-67620-3) by Nancy J. Gardner, California State University–Long Beach. Contains full solutions to all of the end-of-chapter problems in the text.

Printed Test Bank: (ISBN 0-321-67618-1) by Pamela Kerrigan, Mount Saint Vincent. Contains approximately 1000 multiple-choice and short-answer questions, all referenced to the text.

Instructor Images: (ISBN 0-321-67621-1) All images, tables, and photos from the text have been made available as JPEG files on the Instructor Resource Catalog. Instructors may download these and enhance their lecture presentation notes to suit the needs of their class.

CourseSmart: (ISBN 0-321-67708-0) Access your college textbook in online format at www.coursesmart.com.

For the Student

Student Solutions Manual: (ISBN 0-321-67619-X) by Nancy J. Gardner, California State University–Long Beach. Includes full solutions to all odd-numbered end-of-chapter problems in the text.

ACKNOWLEDGMENTS

I'd like to gratefully acknowledge the valuable contributions of my accuracy reviewer Andreas Lippert of Weber State University.

Every effort has been made to rid this text of any typographical errors. I encourage my readers who notice anything suspicious, or who have other questions or comments, to e-mail me at the address below.

H. Stephen Stoker
e-mail: hstoker1@weber.edu

Reviewers of the Tenth Edition of Introduction to Chemical Principles, *Stoker*

John M. Allen,
Indiana State University

Ashton T. Griffin,
Wayne Community College

Lisa DeVane,
Bladen Community College

Charles Spillner,
Solano Community College

Todd M. Johnson,
Weber State University

Ann van Heerden,
Lonestar College–CyFair

Virginia Lea Miller,
Montgomery College

Douglas S. Cody,
Nassau Community College

Andreas Lippert,
Weber State University

Reviewers of the Ninth Edition of Introduction to Chemical Principals, *Stoker*

Doris Eckey
University of Iowa

James Falender,
Central Michigan University

William D. Lumbley,
Indiana University Bloomington

James Murphy,
Santa Monica College

Kristen Murphy,
University of Wisconsin-Milwaukee

Anthony P. Toste,
Missouri State University

Carrie Woodcock,
Eastern Michigan University

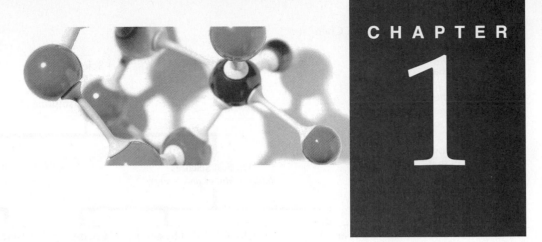

The Science of Chemistry

1.1 CHEMISTRY—A SCIENTIFIC DISCIPLINE

Chemistry is part of a larger body of knowledge called *science*. **Science** *is the study in which humans attempt to organize and explain, in a systematic and logical manner, knowledge about themselves and their surroundings.*

Because of the enormous scope of science, the sheer amount of accumulated knowledge, and the limitations of human mental capacity to master such a large and diverse body of knowledge, science is divided into smaller subdivisions called *scientific disciplines*. A **scientific discipline** *is a branch of science limited in size and scope to make it more manageable.* Examples of scientific disciplines are *chemistry*, astronomy, botany, geology, physics, and zoology.

Figure 1.1 shows an organizational chart, with emphasis on chemistry, for the various scientific disciplines. These disciplines can be grouped into *physical sciences* (the study of matter and energy) and *biological sciences* (the study of living organisms). Chemistry is a physical science.

Rigid boundaries between scientific disciplines *do not exist*. All scientific disciplines borrow information and methods from each other. No scientific discipline is totally independent. Environmental problems that scientists have encountered in the last two decades particularly show the interdependence of the various scientific disciplines. For example, chemists attempting to solve the problems of chemical contamination of the environment find that they need some knowledge of geology, zoology, and botany. It is now common to talk not only of chemists, but also of geochemists, biochemists, chemical physicists, and so on. The middle portion of Figure 1.1 shows the overlap of the other scientific disciplines with chemistry.

Discipline overlap requires that scientists, in addition to having in-depth knowledge of a selected discipline, also have limited knowledge of other disciplines. Discipline overlap also explains why a great many college students are required to study chemistry. One or more chemistry courses are required because of their applicability to the disciplines in which the student has more specific interest.

FIGURE 1.1 An organizational chart showing the relationship of the scientific discipline called chemistry to other scientific disciplines and also the sub-structuring that occurs within the discipline of chemistry.

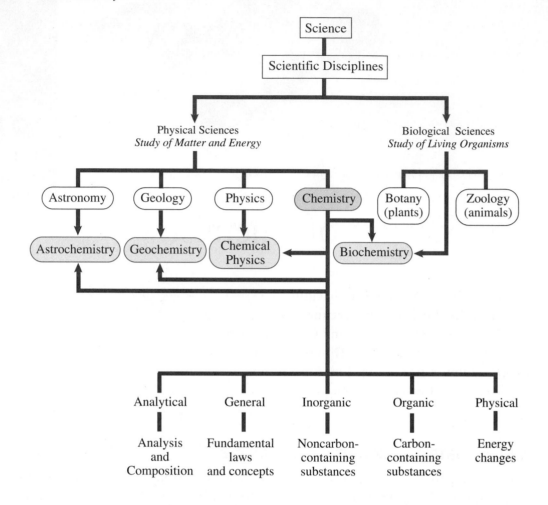

The body of knowledge found within the scientific discipline of chemistry is itself vast. No one can hope to master completely all aspects of chemical knowledge. However, the fundamental concepts of chemistry can be learned in a relatively short period of time.

The vastness of chemistry is sufficiently large that it, like most scientific disciplines, is partitioned into *subdisciplines*. The lower portion of Figure 1.1 shows the five fundamental branches of chemistry: analytical, general, inorganic, organic, and physical. Most of the subject matter of this textbook falls within the realm of *general chemistry*, the fundamental laws and concepts of chemistry.

The American Chemical Society (ACS) is the largest scientific organization in the world. Examination of the names of the 33 subdivisions of ACS (see Table 1.1) further illustrates the wide diversity of subject matter and activities encompassed within the discipline of chemistry.

1.2 SCIENTIFIC RESEARCH AND TECHNOLOGY

The basic activity through which new knowledge is added to the various scientific disciplines, including chemistry, is that of *scientific research*. **Scientific research** *is the process of methodical investigation into a subject in order to discover new information about the subject.* There are two general types of scientific research—*basic* and *applied*. **Basic scientific research** *is research whose major focus is the discovery of new fundamental information about humans and other living organisms and the universe in which they live.*

TABLE 1.1 Names of the Divisions of the American Chemical Society	
Agricultural and Food Chemistry	Fluorine Chemistry
Agrochemicals	Fuel Chemistry
Analytical Chemistry	Geochemistry
Biochemical Technology	History of Chemistry
Biological Chemistry	Industrial and Engineering Chemistry
Business Development and Management	Inorganic Chemistry
Carbohydrate Chemistry	Medicinal Chemistry
Cellulose and Renewable Materials	Nuclear Chemistry and Technology
Chemical Education	Organic Chemistry
Chemical Health and Safety	Petroleum Chemistry
Chemical Information	Physical Chemistry
Chemical Technicians	Polymer Chemistry
Chemical Toxicology	Polymeric Materials: Science and Engineering
Chemistry and the Law	Professional Relations
Colloid and Surface Science	Rubber
Computers in Chemistry	Small Chemical Businesses
Environmental Chemistry	

Numerous scientists, but far from the majority, are involved with basic scientific research. Most scientists function in the arena of applied scientific research. **Applied scientific research** *is research with the major focus of discovery of useful products and processes that can be used to benefit humankind.* Almost always, basic scientific research is the lifeline that supplies applied scientific research with new ideas on which to work.

Closely associated with the phrase *applied scientific research* is the word *technology*. **Technology** *is the application of scientific knowledge to the production of new products to improve human survival, comfort, and quality of life.* No change in our conditions results from basic scientific research endeavors unless something is done with the body of information that accumulates. That is the role of applied scientific research and the ensuing technology. Technology manipulates nature for the advantage of humankind. Technological advances began affecting our society more than 200 years ago, and new advances still continue at an accelerating pace to have a major impact on human society. Section 1.3 considers numerous contributions of chemical technology to human well-being.

Both benefits and detriments can be obtained from the same piece of scientific knowledge, depending on the technology used to put it to work. For example, knowledge concerning the closely related structures of the naturally-occurring substances morphine and codeine, obtained through basic research, has led to the development of several important codeine-derivatives currently used in modern medicine as prescription painkillers (hydrocodone and oxycodone) as well as the synthesis of the illegal drug heroin, whose structure parallels closely that of morphine.

Whether or not a given piece of scientific knowledge is technologically used for beneficial or detrimental purposes depends on the motives of those men and women, whether in industry or government, who have the decision-making authority. In democratic societies, citizens (the voters) can influence many technological decisions. It is important for citizens to become informed about scientific and technological issues.

1.3 THE SCOPE OF CHEMISTRY AND CHEMICAL TECHNOLOGY

Although chemistry is concerned with only a part of the scientific knowledge that has been accumulated, it is in itself an enormous and broad field. Chemistry touches all parts of our lives.

Many of the clothes we wear are made from synthetic fibers produced by chemical processes. Even natural fibers, such as cotton or wool, are the products of naturally occurring chemical reactions within living systems. Our transportation usually involves vehicles powered with energy obtained by burning chemical mixtures such as gasoline or diesel and jet fuels. The drugs used to cure many of our illnesses are the result of chemical research. The paper on which this textbook is printed was produced through a chemical process, and the ink used in printing the words and illustrations is a mixture of many chemicals. Almost all of our recreational pursuits involve objects made of materials produced by chemical industries. Skis, boats, basketballs, bowling balls, musical instruments, and television sets all contain materials that do not occur naturally, but are products of human technological expertise.

Our bodies are a complex mixture of chemicals. The principles of chemistry are fundamental to an understanding of all processes of the living state. Chemical secretions (hormones) produced within our bodies help determine our outward physical characteristics such as height, weight, and appearance. Digestion of food involves a complex series of chemical reactions. Food itself is an extremely complicated array of chemical substances. Chemical reactions govern our thought processes and how knowledge is stored in and retrieved from our brains. In short, chemistry runs our lives.

A formal course in chemistry can be a fascinating experience because it helps us understand ourselves and our surroundings. We cannot truly understand or even know very much about the world we live in or about our own bodies without being conversant with the fundamental ideas of chemistry.

1.4 HOW CHEMISTS DISCOVER THINGS—THE SCIENTIFIC METHOD

There is no one single correct way to do scientific research, be it *basic* or *applied* (Sec. 1.2). Different scientists often have different approaches to solving the same problem. However, the various approaches used always have embodied within them a number of common characteristics that constitute the problem-solving approach known as the *scientific method.*

The **scientific method** *is a set of general procedures based on experimentation and observation used to acquire scientific knowledge and explain natural phenomena.* The procedural steps in the scientific method are as follows:

1. Identify the problem, break it into small parts, and carefully plan procedures to obtain information about all aspects of this problem.
2. Collect data concerning the problem through observation and experimentation (see Figure 1.2).
3. Analyze and organize the data in terms of general statements (generalizations) that summarize the experimental observations.
4. Suggest probable explanations for the generalizations.
5. Experiment further to prove or disprove the proposed explanations.

Although two different scientists rarely approach the same problem in exactly the same way, there are always similarities in their approaches. These similarities are the procedures associated with the scientific method.

On occasion, a great discovery is made by accident, but the majority of scientific discoveries are the result of the application of these five steps over long periods of time. There are no instantaneous steps in the scientific method: applying them requires considerable amounts of time. Even in those situations where luck is involved, it must be remembered that chance favors the prepared mind. To take full advantage of an accidental discovery, a person must be well trained in the procedures of the scientific method.

FIGURE 1.2 Chemistry is an experimental science. Most discoveries in chemistry are made through analysis of data obtained from experiments carried out in laboratories. *(iStockphoto)*

The imagination, creativity, and mental attitude of a scientist using the scientific method are always major factors in scientific success. The procedures of the scientific method must always be enhanced with the abilities of a thinking scientist.

There are special vocabulary terms associated with the scientific method and its use. This vocabulary includes the terms *experiment, scientific fact, scientific law, scientific hypothesis*, and *scientific theory*. An understanding of the relationships among these terms is the key to a real understanding of how to obtain chemical knowledge.

Experiments, Observations, and Data

The beginning step in the search for chemical knowledge is the identification of a problem concerning some chemical system that needs study. After determining what other chemists have already learned about the selected problem, a chemist sets up *experiments* for obtaining more information. An **experiment** *is a well-defined, controlled procedure for obtaining information about a system under study.*

Performing an experiment involves making careful observations about a system under study. An **observation** *is a statement that describes something we see, hear, smell, taste, or feel.* Instrumentation is most often used as an aid in making observations. Observations obtained while performing an experiment are called *data*. Such data may be *qualitative* or *quantitative*, with the latter being preferred. **Qualitative data** *is non-numerical data consisting of general observations about a system under study.* The observation that ice is less dense than liquid water is an example of qualitative information about a system. **Quantitative data** *is numerical data obtained by various measurements on a system under study.* The information that ice has a density of 0.9170 grams per cubic centimeter at 0°C whereas liquid water has a density of 0.9999 grams per cubic centimeter at the same temperature represents quantitative data. Quantitative observations are more useful than qualitative ones because they can be compared with each other and trends or patterns in information can be seen.

An experiment typically involves study of at least two quantities, that is, variables that have changing values. Usually, the effect of change in one variable on another variable, with all other variables held constant, is measured. For example, the effect that temperature change has on the density of a fixed quantity of a gas, with pressure held constant, can be measured.

A well-designed experiment is always performed under controlled conditions, that is, the values of all variables are always noted, not just those that are changing. When such is the case, the experimental data can be reproduced, if needed, by repeating the experiment.

EXAMPLE 1.1 **Distinguishing between Qualitative and Quantitative Data**

Classify each of the following pieces of information as *qualitative data* or *quantitative data*.

   a. The patient's high fever has reached 105.3°F.
   b. The cricket is chirping more loudly tonight than last night.
   c. The package of candy contains about 200 gummi bears.

SOLUTION

   a. Quantitative data—the temperature was measured with a thermometer.
   b. Qualitative data—no measurement was made.
   c. Qualitative data—even though a number is specified, it is an estimated number rather than a measured number.

▶ **Practice Exercise 1.1** Classify each of the following pieces of information as *qualitative data* or *quantitative data*.

   a. The moon is in its last phase.
   b. John has a very large foot size, wearing size 15 shoes.
   c. Sixteen of the runners completed the race in less than 6 minutes.

Answers to all practice exercises in this chapter are found in the back-of-the-book answer section.

Scientific Facts

The individual pieces of new information (data) about a system under study, obtained by carrying out experimental procedures, are called *scientific facts*. A **scientific fact** *is a reproducible piece of data about some natural phenomenon that is obtained from experimentation.* Note the word *reproducible* in this definition. If a given experiment is repeated under exactly the same conditions, the same results (scientific facts) should be obtained. To be acceptable, all scientific facts must be verifiable by anyone who has the time, means, and knowledge needed to repeat the experiments that led to their discovery.

It is important that scientific data be published so that other scientists have the opportunity to critique and double-check both the data and experimental design. The most common publication avenue is that of articles in scientific journals. Other communication avenues include papers presented at scientific meetings and specialized textbooks. Communication of research results to other scientists is as important as the actual obtaining of the results in the laboratory.

Scientific Laws

After obtaining scientific facts through experimentation, a scientist then makes an effort to determine ways in which the scientific facts about a given system relate both to each other and to scientific facts known about similar systems. Repeating patterns often emerge among the collected scientific facts. These patterns that describe the behavior of chemical systems under specific conditions are called *scientific laws*. A **scientific law** *is a generalization that summarizes scientific facts about a natural phenomenon.*

Do not assume that scientific laws are easy to discover. Often, many years of work and thousands of facts are needed before the true relationships among variables in the area under study emerge.

A scientific law is a description of what happens in a given type of experiment. No new understanding of nature results from simply stating a scientific law. A scientific law merely summarizes already known observations (scientific facts).

A scientific law can be expressed either as a verbal statement or as a mathematical equation. An example of a verbally stated scientific law is "If hot and cold pieces of metal are placed in contact with each other, the temperature of the hot piece always decreases and the temperature of the cold piece always increases."

It is important to distinguish between the use of the word *law* in science and its use in a societal context. Scientific laws are *discovered* by research (see Fig. 1.2), and researchers have *no control* over what the laws turn out to be. Societal laws, which are designed to control aspects of human behavior, are *arbitrary conventions* agreed upon (in a democracy) by the majority of those to whom the laws apply. These laws *can be* and *are changed* when necessary. For example, the speed limit for a particular highway (a societal law) can be decreased or increased for various safety or political reasons.

Scientific Hypotheses

There is no mention in a scientific law about why the occurrence described happens. The scientific law simply summarizes experimental observations without attempting to clarify the reasons for the occurrence. Chemists and other scientists are not content with such a situation. They want to know *why* a certain type of observation is always made. Thus, after a scientific law is discovered, scientists work out *plausible, tentative* explanations of the behavior encompassed by the scientific law. These explanations are called *scientific hypotheses*. A **scientific hypothesis** *is a model or statement that can be tested by experiment, which offers an explanation for a scientific law.* Note the inclusion of the concept of *testability* in this definition. A scientific hypothesis is different from other kinds of hypotheses because of the testability requirement. In other academic disciplines, such as philosophy, hypotheses concerning the meaning of life are often considered. Such hypotheses are not *scientific* hypotheses because they cannot be tested by experiment.

Once a scientific hypothesis has been proposed, experimentation begins again. Scientists run more experiments under varied, but controlled, conditions to test the reliability of the proposed explanation. The scientific hypothesis must be able to predict the outcome of as-yet-untried experiments. The validity of the scientific hypothesis depends upon its predictions being true.

In practice, scientists usually start with a number of alternative scientific hypotheses for a given scientific law. Evaluation proceeds by demonstrating that certain proposals are *not* valid. A successful experiment is one in which one or more of the alternative scientific hypotheses are demonstrated to be inconsistent with experimental observation and are thus rejected. Scientific progress is made in the same way a marble statue is: Unwanted bits of marble are chipped away. Example 1.2 contains a simple illustration of this "chipping away" principle in a scientific context.

A contrast exists between the ways in which scientific facts and the results of technology (Sec. 1.2) are shared. Scientists publish their observations (scientific facts) as widely, openly, and quickly as possible. Technological breakthroughs, on the other hand, are usually kept secret by an individual or company until patent rights for the new process or product are obtained. Even then, only limited information is released.

| **EXAMPLE 1.2** | **Relating Scientific Hypotheses to Experimental Information** |

Suppose you encounter a situation involving two unopened books with no identification on their covers and four alternative scientific hypotheses about these books, which are (1) the thinner book is a chemistry textbook, (2) the thicker book is a chemistry textbook, (3) both books are chemistry textbooks, and (4) neither book is a chemistry textbook. What evaluative information about these scientific hypotheses can be obtained by opening the thicker book and determining that it is a chemistry textbook?

As a problem-solving approach, the scientific method is used by many people who do not call themselves scientists. An automobile mechanic uses the scientific approach when working on a car. First, based on tests (experimental observations), the automobile mechanic deduces a probable cause of the problem (a hypothesis). Then parts are adjusted or replaced, and the car is checked (more experimental observations) to see if the problem has been corrected (testing of the hypothesis). An experienced automobile mechanic learns that certain observations almost always indicate a specific problem (a validated hypothesis).

SOLUTION

This experiment (opening the thicker book) proves scientific hypothesis 2 and disproves scientific hypothesis 4; it does not, however, prove that *only one* of the scientific hypotheses is true. The fact that the thicker book is a chemistry textbook does not rule out the possibility that the thinner book is also a chemistry textbook.

▶ **Practice Exercise 1.2**   Based on the same "two-book, four-hypothesis" situation stated in Example 1.2, what evaluative information about the scientific hypotheses is obtained from the single observation that the thinner book is *not* a chemistry textbook?

EXAMPLE 1.3 **Differentiating among Terminologies Associated with the Scientific Method**

Classify each of the following as a *scientific fact*, *scientific law*, or *scientific hypothesis*.

a. The burning candle generated both heat and light.
b. As a candle burns, its wax gradually disappears.
c. All burning candles generate heat and light.
d. Burning candles generate heat as the result of the decomposition of melted wax.

SOLUTION

a. This is a single observation, that is, a *scientific fact*.
b. This is a single observation, that is, a *scientific fact*.
c. This is a generalization based on many observations, that is, a *scientific law*.
d. This is a tentative testable explanation for observations, that is, a *scientific hypothesis*.

▶ **Practice Exercise 1.3**   Classify each of the following as a *scientific fact*, *scientific hypothesis*, or *scientific law*.

a. Your car will not start.
b. Cars are more difficult to start in sub-zero weather than in warm weather.
c. It is probable that your car will not start because it is out of gas.
d. It is probable that your car will not start because it has a frozen fuel line.

Scientific Theories

The term *theory* is often misused by nonscientists in everyday contexts. "I have a theory that such and such is the case" is a frequently heard comment. In this case, *theory* means a "speculative guess," which is not what a theory is. The terminology *unvalidated hypothesis* would be closer to what is meant.

As further experimentation continues to validate a particular scientific hypothesis, its acceptance in scientific circles increases. If after extensive testing, the reliability of a scientific hypothesis is still very high, confidence in it increases to the extent that it is accepted by the scientific community at large. After more time has elapsed and more positive support has accumulated, the scientific hypothesis assumes the status of a *scientific theory*. A **scientific theory** *is a scientific hypothesis that has been tested and validated over a long period of time*. The dividing line between a scientific hypothesis and a scientific theory is arbitrary and cannot be precisely defined. There is no set number of supporting experiments that must be performed to give scientific theory status to a scientific hypothesis.

Scientific theories serve two important purposes: (1) they allow scientists to predict what will happen in experiments that have not yet been run, and (2) they simplify the very real problem of being able to remember all the scientific facts that have already been discovered. Figure 1.3 shows the interplay that must occur between scientific hypotheses and experimentation before an acceptable scientific theory is obtained.

Scientific theories must often undergo modification. As scientific tools, particularly instrumentation, become more accurate, there is an increasing probability that some experimental observations will not be consistent with all aspects of a given scientific theory. A scientific theory inconsistent with new observations must either be modified to accommodate the new results or be restated in such a way that scientists know where it is

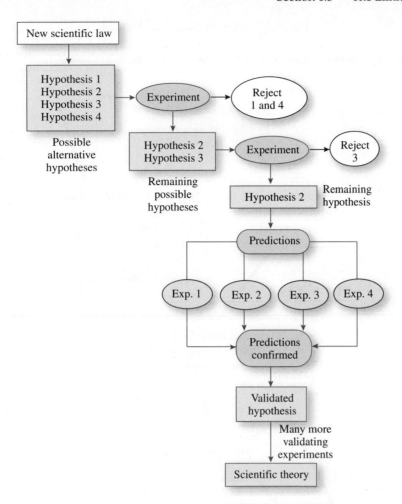

FIGURE 1.3 A possible sequence of events experienced by scientists doing experimental research based on a newly discovered scientific law. (1) A number of alternative scientific hypotheses are proposed to explain the new scientific law; (2) experiments are carried out to eliminate invalid scientific hypotheses; (3) predictions are made based on the surviving scientific hypothesis, and further experiments are carried out to test these predictions; (4) confirmed predictions produce a validated scientific hypothesis; and (5) many more validating experiments give the scientific hypothesis the status of a scientific theory.

useful and where it is not. Most scientific theories in use have known limitations. These imperfect scientific theories are simply the best ideas anyone has found *so far* to describe, explain, and predict what happens in the world in which we live. Scientific theories with limitations are generally not abandoned until a better scientific theory is developed.

Scientific facts that have been verified by repeated experiments will never be changed, but the scientific theories that were invented to explain these scientific facts are subject to change. In this sense, scientific facts are more important than the scientific theories devised to explain them. It is a mistake to believe that, by knowing all the scientific laws and scientific theories that are derived from experimental observations, the experimental facts are not needed. New scientific theories can be developed only by people who have a wide knowledge of the scientific facts relating to a particular field, especially those scientific facts that have not been satisfactorily accounted for by existing scientific theories.

Figure 1.4 highlights the central role that experimentation plays in the scientific method and also summarizes the general steps, procedures, and terminology associated with this most important pattern of action for acquiring scientific knowledge.

1.5 THE LIMITATIONS OF THE SCIENTIFIC METHOD

Scientists do not view scientific theories or their precursors—scientific hypotheses—as absolute truth. All scientific theories are considered provisional—subject to change in the light of new experimental observations. There is always the chance that some future, yet-unthought-of experiment using instrumentation that has not yet been

FIGURE 1.4 A diagram showing the central role that experimentation plays in the scientific method for obtaining new scientific information.

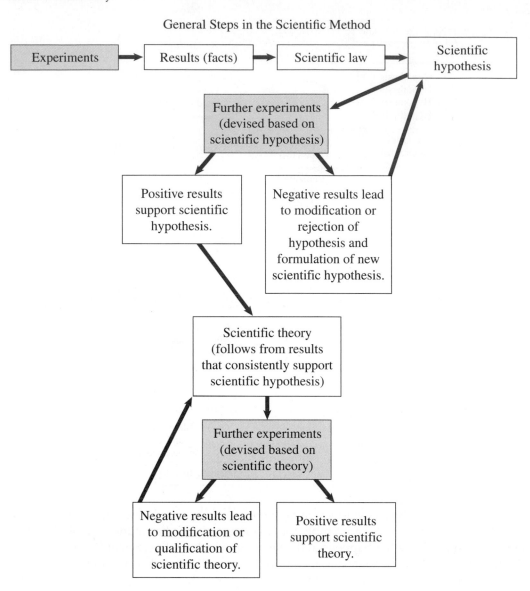

General Steps in the Scientific Method

invented will be devised that disproves or places limitations on a currently well-established scientific theory. If such were the case, the scientific theory would then be modified to take into account the new experimental developments. In many ways, science is like a living organism: It continues to grow and change.

Thus, inherent in the use of the scientific method is the not-often-stated concept that "theories can be proven false or not proven false, but they can never be proven true." This concept that a thing (idea or concept or theory) cannot be proven to be true, but can only be proven not to be true results from the following line of reasoning:

1. Experiments designed to test a theory have the purpose of trying to prove that the theory cannot explain the experimental observations obtained from the experiment.
2. If such experiments do not prove the theory wrong, new experiments are designed and conducted with the purpose of still trying to disprove the theory.
3. Experimentation to "disprove" continues in a theoretically never-to-be completed process. (In practice, after many, many failed attempts to "disprove," experimentation attempts become infrequent and the scientific theory becomes one of the foundations of the scientific discipline.)

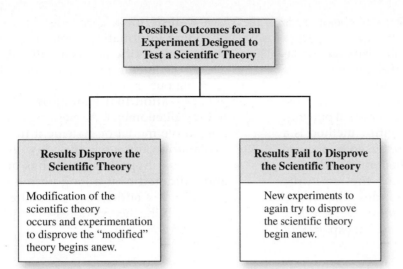

FIGURE 1.5 Possible results from attempting to disprove a scientific theory.

Figure 1.5 summarizes diagrammatically the preceding line of reasoning relative to disproving a scientific theory.

1.6 THE LIMITATIONS OF SCIENCE

The testability requirement associated with scientific theories and scientific hypotheses also has the effect of limiting the application of these theories and hypotheses to the realm of the natural world. The methods of science are not applicable to supernatural phenomenon. The dictionary definition for *supernatural* is "not of the natural world," that is, something that cannot be explained using scientific laws. Thus, science is not equipped to deal with philosophical or religious questions such as "What is the purpose of life?" or "Does the human body contain a spirit that gives it life?" Questions such as these fall outside the realm of science because of their lack of testability using scientific methods. This does not mean that such questions are not valid questions nor that they are not important questions. It simply means that answers to such questions cannot be obtained using the methodology of science.

Summary

1. **Science and Scientific Disciplines** Science is the study in which humans attempt to organize and explain, in a systematic manner, knowledge about themselves and their surroundings. Scientific disciplines, of which chemistry is one, are branches of science limited in size and scope to make them more manageable.

2. **Scientific Research** Scientific research is the process of methodical investigation into a subject in order to discover new information about the subject. Such research is further classified as *basic* scientific research or *applied* scientific research. The major focus of basic scientific research is the discovery of new fundamental information, and the major focus of applied scientific research is the discovery of useful products and processes that can be used to benefit humankind.

3. **Technology** Scientific disciplines represent abstract bodies of knowledge. Technology is the physical application of scientific knowledge to the production of new products to improve human survival, comfort, and quality of life.

4. **Scope of Chemistry** The scope of chemistry is extremely broad, and it touches every aspect of our lives. The principles of chemistry are fundamental to an understanding of all processes of the living state. Chemical processes produce the products needed for our clothes, housing, transportation, medications, and recreational pursuits.

5. **Scientific Method** The scientific method is a set of systematic procedures used to acquire knowledge and explain phenomena. The terminology associated with the scientific method and its use includes the terms *experiment, scientific fact, scientific law, scientific hypothesis*, and *scientific theory*. An experiment is a well-defined, controlled procedure for obtaining information about a system under study. A scientific fact is a valid observation about some natural phenomenon obtained by carrying out experiments. A scientific law is a generalization that summarizes facts about natural phenomena. A scientific hypothesis is a tentative model or statement that offers an explanation for a scientific law. A scientific theory is a hypothesis that has been tested and validated over a long period of time. A scientific law addresses how matter behaves, and a scientific theory addresses why it behaves that way.

Key Terms

The new terms defined in this chapter are

applied scientific research *Sec 1.2*
basic scientific research *Sec 1.2*
experiment *Sec 1.4*
observation *Sec 1.4*
qualitative data *Sec 1.4*
quantitative data *Sec 1.4*
science *Sec 1.1*
scientific discipline *Sec 1.1*
scientific fact *Sec 1.4*
scientific hypothesis *Sec 1.4*
scientific law *Sec 1.4*
scientific method *Sec 1.4*
scientific research *Sec 1.1*
scientific theory *Sec 1.4*
technology *Sec 1.2*

Practice Problems

SCIENTIFIC DISCIPLINES (SEC. 1.1)

1.1 Indicate whether each of the following statements is true or false.
a. Science is the study in which humans attempt to organize and explain, in a systematic and logical manner, knowledge about themselves and their surroundings.
b. Scientific disciplines are limited in size and scope to the extent that subdividing of subject matter within a discipline is not necessary.
c. Boundaries between scientific disciplines are very rigid.
d. Collectively, the knowledge in scientific disciplines constitutes the whole of scientific knowledge currently known.

1.2 Indicate whether each of the following statements is true or false.
a. Scientific disciplines are branches of scientific knowledge limited in size and scope to make them more manageable.
b. Scientific disciplines are defined in such a manner that each discipline is totally independent of other disciplines.
c. Complete mastery of all concepts within a scientific discipline is difficult (but possible) because of limited size and scope for the discipline.
d. The scientific discipline of chemistry has some overlap with other physical sciences but no overlap with biological sciences.

1.3 Based on information found in Figure 1.1, indicate whether each of the following statements is true or false.
a. Chemistry and botany are both classified as physical sciences.
b. Inorganic chemistry is the chemistry subdiscipline that focuses on carbon-containing substances.
c. Physical chemistry is the chemistry subdiscipline that focuses on energy changes in substances.
d. Physical science disciplines primarily study various aspects of living organisms.

1.4 Based on information found in Figure 1.1, indicate whether each of the following statements is true or false.
a. Chemistry and zoology are both classified as biological sciences.
b. Organic chemistry is the chemistry subdiscipline that focuses on non-carbon-containing substances.

c. General chemistry is the chemistry subdiscipline that focuses on the fundamental laws and concepts of chemistry.
d. Biological science disciplines primarily study various aspects of matter and energy.

SCIENTIFIC RESEARCH AND TECHNOLOGY (SEC 1.2)

1.5 Classify each of the following scientific research endeavors relating to a newly synthesized material as involving *basic* research or *applied* research.
a. determination of the melting and freezing points of the material
b. determination of its properties relative to use as a motor oil additive
c. study of its decomposition behavior at a high temperature
d. study of its properties relative to use as a flame retardant for household draperies

1.6 Classify each of the following scientific research endeavors relating to a newly synthesized material as involving *basic* research or *applied* research.
a. determination of the electrical conductivity of the material at low temperatures
b. determination of its properties relative to use as a bleaching agent in manufacture of paper from wood pulp
c. study of its toxicity relative to its use as a food additive
d. study of its antioxidant properties relative to its use as a food additive

THE SCIENTIFIC METHOD (SEC. 1.4)

1.7 Arrange the following steps in the scientific method in the sequence in which they normally occur.
a. Suggest probable explanations for generalizations obtained from data.
b. Collect data concerning a problem through observation and experimentation.
c. Identify a problem, and carefully plan procedures to obtain information about all aspects of this problem.
d. Experiment further to prove or disprove proposed explanations.
e. Analyze and organize data in terms of general statements that summarize experimental observations.

1.8 Arrange the following terms associated with the scientific method in the order in which they are normally encountered as the scientific method is applied to a problem.
a. scientific law
b. scientific fact
c. scientific theory
d. experiment
e. scientific hypothesis

1.9 Classify each of the following hypotheses as *scientific* hypotheses or *nonscientific* hypotheses.
a. In northern climates cars rust faster during the winter months than during the summer months because of the use of salt on roads.
b. The purpose of life for human beings is to collect as many material possessions as possible.
c. Birds can perch on electrical transmission lines without consequences because their feet contain a shock-resistant material.
d. John Stockton is the greatest point guard in NBA history because he took a chemistry class during each year in high school.

1.10 Classify each of the following hypotheses as *scientific* hypotheses or *nonscientific* hypotheses.
a. Karl Malone is the greatest power forward in NBA history because he was born in Louisiana.
b. The author of this textbook is bald because he chewed his food too fast as a child.
c. The force of gravity exerted upon an object is related to the color of the object.
d. Copper and gold do not possess the characteristic silver-gray color of other metals because they always contain color-changing impurities.

1.11 Classify each of the following statements as a *scientific fact*, *scientific law*, or *scientific hypothesis*.
a. Ice floats on liquid water because its density is less than that of liquid water.
b. The boiling point of water increases as atmospheric pressure increases.
c. The diameter of the moon is 3476 kilometers.
d. The extent of pain-relieving effects of an aspirin tablet are dependent on the shape of the aspirin tablet.

1.12 Classify each of the following statements as a *scientific fact*, *scientific law*, or *scientific hypothesis*.
a. A 12 in. thick block of lead will stop gamma radiation.
b. The freezing point of water is both 32°F and 0°C.
c. Positive and negative charges repel each other because of sound waves that they emit.
d. All brands of peanut butter when heated to a sufficiently high temperature will decompose.

1.13 Indicate whether each of the following statements is true or false.
a. A scientific theory is a summary of experimental observations.
b. A scientific hypothesis is a summary of experimental facts.
c. A scientific theory is subject to modification in light of new experimental observations.
d. An experiment is a well-defined, controlled procedure for obtaining facts.

1.14 Indicate whether each of the following statements is true or false.
a. A scientific theory is a scientific hypothesis that has not yet been subjected to experimental testing.

b. It is much easier to disprove a false scientific hypothesis than it is to prove a valid one.

c. Established scientific theories eventually become scientific laws.

d. A scientific law is an explanation of why a particular phenomenon occurs.

1.15 Constructively criticize the statement "You needn't take it too seriously; after all, it's only a theory."

1.16 Constructively criticize the statement "The results of the experiment do not agree with the theory. Something must be wrong with the experiment."

1.17 Assume that you have four pennies with unknown mint dates and four hypotheses concerning these dates: (1) all dates are the same; (2) two different dates are present; (3) three different dates are present; and (4) all dates are different. Which of the listed scientific hypotheses could be eliminated by determining that

a. two pennies have the same date?

b. two pennies have different dates?

c. two of three pennies have the same date?

d. three pennies have different dates?

1.18 Assume that you have four red balls of equal size and four hypotheses concerning the masses of the balls: (1) each ball has a different mass; (2) there are balls of two masses; (3) balls of three different masses are present; and (4) all balls have the same mass. Which of the listed scientific hypotheses could be eliminated by determining that

a. two balls have the same mass?

b. three balls have the same mass?

c. there are two masses among three balls?

d. there are two masses among four balls?

1.19 Indicate whether each of the following statements represents *qualitative data* or *quantitative data*.

a. The sun sets in the west.

b. The automobile tire pressure is 32 pounds per square inch.

c. The length of the copper rod is 6.37 meters.

d. The colorless liquid has an alcohol-like odor.

1.20 Indicate whether each of the following statements represents *qualitative data* or *quantitative data*.

a. The empty vial weighs 54.2 grams.

b. The density of oxygen gas increases as its temperature increases.

c. The density of gold at 20°C is 19.3 grams per cubic centimeter.

d. The odorless liquid has a reddish-brown color.

1.21 A researcher studies the behavior of a fixed amount of a gas under constant temperature conditions with the following results:

a. At a pressure of 4.0 atmospheres the gas occupies a volume of 2.0 liters.

b. At a pressure of 1.0 atmosphere the gas occupies a volume of 8.0 liters.

c. At a pressure of 2.0 atmospheres the gas occupies a volume of 4.0 liters.

d. At a pressure of 8.0 atmospheres the gas occupies a volume of 1.0 liter.

What generalization (scientific law) concerning the relationship between volume and pressure, under the conditions of the experiments, can be obtained from these data?

1.22 A researcher studies the behavior of a gas under constant temperature and constant volume conditions with the following results:

a. 10.0 grams of gas exerted a pressure of 4.0 atmospheres.

b. 40.0 grams of gas exerted a pressure of 16.0 atmospheres.

c. 5.0 grams of gas exerted a pressure of 2.0 atmospheres.

d. 20.0 grams of gas exerted a pressure of 8.0 atmospheres.

What generalization (scientific law) concerning the relationship between amount of gas and pressure, under the conditions of the experiments, can be obtained from these data?

1.23 What are the differences between a scientific law and a societal law?

1.24 What is the reason for repeating experiments several times before developing a scientific law based on the experiments?

1.25 Why is it important that scientific data be published?

1.26 The phrase "It has been proved scientifically" is rarely used by scientists. Explain why.

1.27 Why is it useless to conduct an experiment under uncontrolled conditions?

1.28 What are the two important purposes that scientific theories serve?

1.29 What is the difference between a qualitative observation and a quantitative observation?

1.30 If a scientific theory is false, how will the scientific method, applied over time, reveal that such is the case?

Multiple-Choice Practice Test

Use this bank of 10 multiple-choice questions as a review of key concepts presented in this chapter. For some of the questions, there may be more than one correct answer (choice d) or no correct answer (choice e).

1.31 The scientific discipline called *chemistry* is a discipline that

a. is totally independent of all other scientific disciplines.

b. has some overlap with other physical sciences but no overlap with biological sciences.
c. has totally rigid boundaries.
d. more than one correct response
e. no correct response

1.32 Which of the following is a correct pairing of concepts?
a. general chemistry; fundamental laws and concepts of chemistry
b. inorganic chemistry; chemistry of carbon-containing substances
c. organic chemistry; chemistry of non-carbon-containing substances
d. more than one correct response
e. no correct response

1.33 Which of the following scientific research endeavors would be classified as *basic* research rather than *applied* research?
a. research on a substance's solubility in various solvents
b. research on a substance's toxicity relative to its use as a food additive
c. research on a substance's magnetic properties at low temperatures
d. more than one correct response
e. no correct response

1.34 Which of the following is a correct ordering of terminology as it is normally encountered in the application of the scientific method to a research problem?
a. scientific law, scientific hypothesis, scientific theory
b. scientific fact, scientific law, scientific hypothesis
c. scientific law, scientific fact, scientific theory
d. more than one correct response
e. no correct response

1.35 Which of the following are examples of *quantitative* data?
a. The sun sets in the west.
b. The odorless liquid had a reddish-brown color.
c. The metal iron has a high melting point.
d. more than one correct answer
e. no correct response

1.36 Which of the following pairings of terms are correct?
a. scientific theory and validated scientific hypothesis
b. scientific law and unvalidated scientific hypothesis
c. fact and valid observation
d. more than one correct response
e. no correct response

1.37 Which of the following statements concerning a *scientific theory* is correct?
a. It is a scientific law that has been tested and validated over a long period of time.
b. It allows scientists to predict what will happen in yet-to-be-run experiments.
c. It is an explanation that is considered to be absolute truth.
d. more than one correct response
e. no correct response

1.38 Which of the following statements concerning a *scientific law* is correct?
a. It is an explanation for why a given occurrence happens.
b. It is a generalized summary of facts about a natural phenomenon.
c. It is always formulated in terms of a mathematical equation.
d. more than one correct response
e. no correct response

1.39 The statement "The boiling point of water is 100°C at 1 atmosphere pressure" is an example of
a. a scientific law.
b. a scientific hypothesis.
c. a scientific fact.
d. more than one correct response
e. no correct response

1.40 Which of the following concepts applies to both *scientific* hypotheses and *non-scientific* hypotheses?
a. can be tested using experimental methods
b. pertains only to substances that can be seen
c. offers an explanation for several observations
d. more than one correct response
e. no correct response

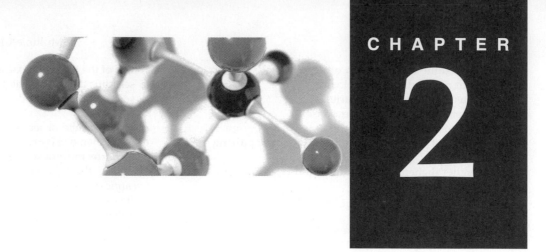

Numbers from Measurements

2.1 THE IMPORTANCE OF MEASUREMENT

It would be extremely difficult for a carpenter to build cabinets without being able to use tools such as hammers, saws, drills, rulers, straight edges, and T squares. They are a carpenter's "tools of the trade." Chemists also have "tools of the trade." Their most used tool is the one called *measurement*. **Measurement** *is the determination of the dimensions, capacity, quantity, or extent of something*. In chemical laboratories, the most common types of measurements are those of mass, volume, length, time, temperature, pressure, and concentration.

Understanding measurement is indispensable in the study of chemistry. Questions such as "how much ...?", "how long ...?", and "how many ...?" simply cannot be answered without resorting to measurements. It is the purpose of this chapter and the next to help students acquire the necessary background to deal properly with measurement. Almost all of the material in these two chapters is mathematical. An understanding of mathematics is a necessity for students of chemistry who want their encounters with the subject to be successful. The following analogy is appropriate for the situation. Physical exertion in sports can be fun, relaxing, and challenging for those in good physical shape. But for those not in good physical condition, such exertion is not satisfying and may even be downright painful (especially the day after). Being "in good shape" mathematically has the same effect on the study of chemistry. It can cause that study to be a very satisfying and enjoyable experience. On the other hand, a lack of the necessary mathematical skills can cause "chemical exercise" to be somewhat painful. The message should be clear. The contents of this chapter (and Chapter 3) must be taken very seriously. Skimming over this material is a sure invitation to frustration and struggle with the chemical topics that follow.

2.2 EXACT AND INEXACT NUMBERS

In scientific work, numbers are grouped in two categories: *exact numbers* and *inexact numbers*. An **exact number** *is a number that has a value with no uncertainty in it; that is, it is known exactly.* Exact numbers occur in definitions (for example, there are exactly 12 objects in a dozen, not 12.01 or 12.02), in counting (for example, there can be 7 people in a room, but never 6.99 or 7.02), and in simple fractions (for example, 1/3, 3/5, and 5/9).

An **inexact number** *is a number that has a value with a degree of uncertainty in it.* Inexact numbers result any time a measurement is made. It is impossible to make an *exact* measurement; some uncertainty will always be present. Flaws in measuring-device construction, improper calibration of an instrument, and the skills (or lack of skills) possessed by a person using a measuring device all contribute to error (uncertainty).

2.3 ACCURACY, PRECISION, AND ERROR

Two important terms relating to the uncertainties associated with measurement values are *precision* and *accuracy*. Although these terms are used somewhat interchangeably in nonscientific discussions, they have distinctly different meanings in science.

Precision *is an indicator of how close a series of measurements on the same object are to each other.* Note that precision determination involves a series of measurements; it is not proper to speak of the precision of a single measurement made on an object. **Accuracy** *is an indicator of how close a measurement (or the average of multiple measurements) comes to a true or accepted value.* The activity of throwing darts at a target illustrates nicely the difference between these two terms (see Fig. 2.1). Accuracy refers to how close the darts are to the center (bull's-eye) of the target. Precision refers to how close the darts are to each other.

Both the precision and accuracy of a series of measurements usually relate directly to the actual physical measuring device used. You would expect, and it is most often the case, that the precision and accuracy of temperature readings obtained from a thermometer with a scale marked in tenths of a degree would be better than readings obtained from a thermometer whose scale has only degree marks. A series of readings obtained from a stopwatch whose dial shows tenths of a second will usually be more precise and accurate than readings from a stopwatch that shows only seconds.

Both precision and accuracy depend not only on the measuring device used but also on the technical skill of the person making the measurement. How well can that person read the numerical scale of the instrument? How well can that person calibrate the instrument before its use?

Normally, high accuracy accompanies high precision. However, high precision and low accuracy are also possible. For example, results obtained using a poorly calibrated instrument could give high precision but low accuracy. All measurements would be off by a constant amount as a result of the improper calibration.

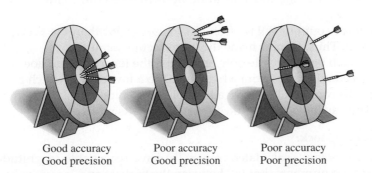

Good accuracy   Poor accuracy   Poor accuracy
Good precision   Good precision   Poor precision

FIGURE 2.1 The difference between precision and accuracy.

Errors in measurement can be classified as either *random errors* or *systematic errors*. A **random error** *is an error originating from uncontrollable variables in an experiment.* Such errors result in experimental values that fluctuate about the true value. A variation in the angle from which a measurement scale is viewed will cause random error. Momentary changes in air currents, atmospheric pressure, or temperature near a sensitive balance for weighing would cause random errors. The net result of random errors, which can never be completely eliminated, is a decrease in the precision of a series of measurements.

A **systematic error** *is an error originating from controllable variables in an experiment.* They are "constant" errors that occur again and again. A flaw in a piece of equipment, such as a chipped weight in a balance, would cause systematic error. All readings would be off by a specific amount because of that flaw. Systematic errors affect the accuracy of measurements. Results are consistently either too high or too low compared to the true value.

EXAMPLE 2.1   **Determining the Accuracy and Precision of a Series of Measurements**

Two teams of three students each count the number of people entering the main gate of a football stadium in the 5-minute period just prior to kickoff. Their individual and team counts are

Team A	Team B
Miranda: 577 people	Spencer: 574 people
Brinley: 579 people	Taylor: 585 people
Brianna: 581 people	Brandi: 593 people

An "electronic counter" indicates that 581 people passed through the gate in the designated time period.

  a. Which person's count is the most accurate? *Brianna*
  b. Which team's count is the most accurate? *team A*
  c. Which person's count is the most precise?
  d. Which team's count is the most precise?

SOLUTION

  a. Briana's count is the most accurate because her count is the same as the electronic count.
  b. The average count of 579 for team A is closer to 581 than the average count of 584 for team B. Thus, team A made the more accurate count.
  c. Because each person made only one count, the term *precision* does not apply.
  d. Team A's counts range from a high of 581 to a low of 577, which gives a spread of 4. Team B's counts range from a high of 593 to a low of 574, which gives a spread of 19. The smaller spread makes Team A's counts more precise.

Answer Double Check:

Do the "team averages" calculated in part (b) have reasonable magnitudes? Yes. Both team averages are numbers that fall between the highest and lowest values in the individual counts, a requirement for a correct average.

▶ **Practice Exercise 2.1**   Two teams of three students each count the number of birds sitting on a power line on a cold wintry day. Their individual and team counts are

Answers to all practice exercises in this chapter are found in the back-of-the-book answer section.

Team A	Team B
Marjorie: 71 birds	Howard: 76 birds
Marlene: 74 birds	Alice: 76 birds
Kathleen: 71 birds	Sharon: 67 birds

Detailed analysis of a photograph of the "wire" birds shows that 76 birds are actually present.

 a. Which person's count is the most accurate?
 b. Which team's count is the most accurate?
 c. Which person's count is the most precise?
 d. Which team's count is the most precise?

2.4 UNCERTAINTY IN MEASUREMENTS

As noted in Section 2.2, every measurement carries a degree of uncertainty or error. Even when very elaborate and expensive measuring devices are used, some degree of uncertainty will be present in the measurement.

   To illustrate how measurement uncertainty arises, let us consider how two different thermometer scales, illustrated in Figure 2.2, are used to measure a given temperature. Determining the temperature involves determining the height of the mercury column in the thermometer. The scale on the left in Figure 2.2 is marked off in one-degree intervals. Using this scale, we can say with certainty that the temperature is between 29 and 30 degrees. We can further say that the actual temperature is closer to 29 degrees than to 30 and estimate it to be 29.2 degrees. The scale on the right has more

Every measurement has some degree of uncertainty associated with it.

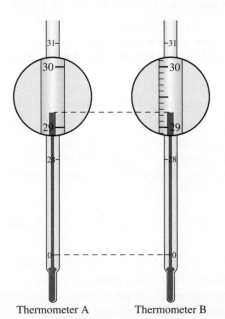

FIGURE 2.2
Measuring a temperature. A portion of the degree scale on each of the two differently scaled thermometers has been magnified.

Thermometer A   Thermometer B

subdivisions, being marked off in tenths of a degree rather than in degrees. Using this scale, we can definitely say that the temperature is between 29.2 and 29.3 degrees and can estimate it to be 29.25 degrees. Note how both temperature readings contain some digits (all those except the last one) that are exactly known and one digit (the last one) that is estimated. It is this last digit, the estimated one, that reflects uncertainty in the measurement. Note also that the uncertainty in the second temperature reading is less than that in the first reading—an uncertainty in the hundredths place compared to an uncertainty in the tenths place.

In reading a measurement scale, all digits known for certain are recorded plus one estimated digit. It is wrong to record more than one estimated digit.

Only one estimated digit is ever recorded as part of a measurement. It would be incorrect for a scientist to report that the height of the mercury column in Figure 2.2, as read on the scale on the right, corresponds to a temperature of 29.247 degrees. The value 29.247 contains two estimated digits (the 4 and the 7) and would indicate a measurement with smaller uncertainty than is actually obtainable with that particular measuring device.

The magnitude of the uncertainty in the last recorded digit in a measurement (the estimated digit) may be indicated using a "plus–minus" notation. The following three mass measurements illustrate this notation.

$$15 \pm 1 \text{ grams}$$
$$15.3 \pm 0.1 \text{ grams}$$
$$15.34 \pm 0.03 \text{ grams}$$

The measurement 15 ± 1 gram designates a mass that lies between 14 grams and 16 grams; where the mass lies within this range is not known. The measurement 15.34 ± 0.03 grams designates a mass between 15.31 and 15.37 grams; where the mass lies within this range is again not known.

Most often the uncertainty in the last recorded digit is one unit (as in the first two mass measurements), but it may be larger (as in the third mass measurement). In this text we will follow the almost universal practice of dropping the "plus–minus" notation if the magnitude of the uncertainty is one unit. Thus, in the absence of "plus–minus" notation you will be expected to assume that there is an uncertainty of one unit in the last recorded digit. A measurement reported simply as 27.3 inches means 27.3 ± 0.1 inch. Only in the situation where the uncertainty is greater than one unit in the last recorded digit will the amount of the uncertainty be explicitly shown.

Example 2.2 relates measurement uncertainty to actual measuring-device scales.

EXAMPLE 2.2 **Recording Measurements to the Proper Uncertainty Level**

What should the recorded uncertainty be (± 0.1 unit, ± 0.01 unit, etc.) for measurements made using the following measuring-device scales.

- **a.** a thermometer scale with markings in 1 degree intervals
- **b.** a measuring cup scale with markings in 100 fluid ounce intervals
- **c.** a ruler scale with markings in 10 centimeter intervals
- **d.** a barometer scale with markings in 0.1 of a millimeter intervals

SOLUTION

In each case the recorded uncertainty should have one estimated digit beyond the markings on the measuring device.

- **a.** For a scale graduated in units of one, readings should be made to the closest 0.1 of a unit; thus, the recorded uncertainty should be ± 0.1 degree.
- **b.** Since the scale markings are in 100 unit intervals, the reading should be estimated to the closest tens unit (± 10 fluid ounces).

c. For a scale graduated in units of ten, readings should be made to the closest ones unit (±1 centimeter).

d. This time, the measurement should be made to the closest 0.01 of a unit because the scale intervals are 0.1 unit in size (±0.01 millimeter).

Answer Double Check:

Are the answers reasonable in terms of uncertainty? Yes. In each case the recorded uncertainty is smaller, by a factor of 10, than the scale unit interval.

▶ **Practice Exercise 2.2** What should the recorded uncertainty be (±0.1 unit, ±0.01 unit, etc.) for measurements made using the following measuring-device scales.

a. a measuring cup scale with markings in 0.1 quart intervals
b. a meter stick scale with markings in 10 millimeter intervals
c. a protractor scale with markings in 1 degree intervals
d. a balance scale with markings in 100 gram intervals

Scientists, when making measurements, always attempt to minimize the uncertainty associated with the measurement. Uncertainty, however, cannot be completely eliminated.

The two most common origins for uncertainty in a measurement are the following:

1. *Human error*: A person's hands have limits as to how well they can manipulate laboratory equipment and a person's eyes can see (read) only so well. There are limits to how well a person can calibrate an instrument.

2. *Instrument error*: Instruments themselves can have imperfections (flaws) associated with their manufacture. Proper calibration of an instrument does not always negate such manufacturing defects. Corrosion problems as well as "wear and tear" on a heavily used instrument can also contribute to instrument error.

2.5 SIGNIFICANT FIGURES

Because measurements are never exact (Sec. 2.4), two types of information must be conveyed whenever a numerical value for a measurement is recorded: (1) the magnitude of the measurement and (2) the uncertainty of the measurement. The magnitude is indicated by the digit values. Uncertainty is indicated by the number of *significant figures* recorded. **Significant figures** *are the digits in any measurement that are known with certainty plus one digit that is uncertain*. To summarize, in equation form,

> The term *significant figures* is often verbalized in shortened form as "sig figs."

Number of significant figures = All certain digits + One uncertain digit

Determining the number of significant figures in a measurement is not always as straightforward as Example 2.2 infers. In this example, you knew the type of instrument used for each measurement and its limitations because you made the measurement. Quite often when someone else makes a measurement, such information is not available. All that is known is the reported final result—the numerical value of the measured quantity. In this situation questions often arise about the "significance" of various digits in the measurement. For example, consider the published value of the distance from the earth to the sun, which is 93,000,000 miles. Intuition tells you that it is highly improbable that this distance is known to the closest mile. You suspect that this is an estimated distance. To what digit has this number been estimated? Is it to the nearest million miles, the closest hundred thousand miles, the nearest ten thousand miles, or what?

A set of guidelines has been developed to aid scientists in interpreting the significance of reported measurements or results calculated from measurements. Four rules constitute the guidelines, one rule for the digits 1 through 9 and three rules for the digit 0. A zero in a measurement may or may not be significant depending on its location in the sequence of digits forming the numerical value for the measurement. There is a rule for each of three classes of zeros—leading zeros, confined zeros, and trailing zeros.

Nonzero digits are always significant.

Rule 1 The digits 1 through 9 inclusive (all of the nonzero digits) always count as significant figures.

14.232	five significant figures
3.11	three significant figures
244.6	four significant figures

Leading zeros are never significant.

Rule 2 *Leading zeros* are zeros that occur at the start of a number, that is, zeros that precede all nonzero digits. Located between the decimal point and the first nonzero digit, such zeros do not count as significant figures. Their function is to determine the position of the decimal point. (The single zero often written to the left of the decimal point is also never significant; its function is simply to draw attention to where the decimal point is located. It is not considered to be a leading zero as it does not help determine decimal point location.)

0.00045	two significant figures
0.0113	three significant figures
0.000000072	two significant figures

Leading zeros are always to the left of the first nonzero digit and to the right of the decimal place.

Confined zeros are always significant.

Rule 3 *Confined zeros* are zeros between nonzero digits. Such zeros always count as significant figures.

2.075	four significant figures
6007	four significant figures
0.03007	four significant figures

Trailing zeros are sometimes significant.

Rule 4 *Trailing zeros* are zeros at the end of a number. They are significant if (a) there is a decimal point present in the number or (b) they carry overbars. Otherwise trailing zeros are not significant.

The following numbers, all containing decimal points, illustrate condition (a) of rule 4.

62.00	four significant figures
24.70	four significant figures
0.02000	four significant figures
4300.00	six significant figures

By condition (b), trailing zeros in numbers lacking an explicitly shown decimal point become significant when marked with a bar above the zero(s).

$$36,\overline{000} \quad \text{five significant figures}$$

$$36,0\overline{00} \quad \text{four significant figures}$$

$$36,00\overline{0} \quad \text{three significant figures}$$

$$10,02\overline{0} \quad \text{five significant figures}$$

In cases involving trailing zeros where neither a decimal point nor over-bar(s) is present, "confusion exists" because these zeros may or may not be significant. For example, it is not possible to definitely know how many significant figures are present in the measurement 5600 grams (no decimal place explicitly shown). There are three possible interpretations—two, three, or four significant figures—depending on the uncertainty associated with the measurement (5600 ±100, 5600 ±10, and 5600 ±1). Standard operating procedure in such cases, where no other information about the measurement is available, is to assume the largest of the uncertainties possible for the measurement. Thus, the measurement 5600 grams contains two significant figures. Generally, then, in cases involving trailing zeros where neither a decimal point nor overbar(s) is present, the trailing zeros are assumed to be nonsignificant.

Another method, more convenient than rule 4, for dealing with the significance of trailing zeros involves expressing the number in scientific notation. In this notation, to be presented in Section 2.7, only significant digits are shown.

93,000,000	two significant figures
360,000	two significant figures
330,300	four significant figures
6310	three significant figures

EXAMPLE 2.3 **Determining the Number of Significant Figures in a Numerical Value**

Determine the number of significant figures in the numerical value in each of the following statements.

a. A wire has a diameter of 0.05082 inch.
b. The mass of the earth is 6,600,000,000,000,000,000,000 tons.
c. A hospital patient's blood glucose level was determined to be 4850 micrograms per milliliter of blood.
d. Normal body temperature for a chickadee is 41.0°C.

SOLUTION

a. There are four significant figures. The leading zero is not significant (rule 2) and the confined zero is significant (rule 3).
b. There are two significant figures. The trailing zeros are not significant because no decimal point or overbar notation is present (rule 4).
c. There are three significant figures. The trailing zero is not significant (rule 4).
d. There are three significant figures. The trailing zero is significant because a decimal point is present (rule 4).

▶ **Practice Exercise 2.3**   Determine the number of significant figures in the numerical value in each of the following statements.

   a. The uncut diamond has a mass of 0.4021 g.
   b. The beehive contains 14,000 bees.
   c. The maximum speed of a three-toed sloth is 0.15 miles per hour.
   d. A sheet of paper is 0.0042 inch thick.

It is important to remember what is "significant" about significant figures. The number of significant figures in a measurement conveys information about the uncertainty associated with the measurement. The "location" of the last significant digit in the numerical value for a measurement specifies the measurement's uncertainty: Is the last significant digit located in the hundredths, tenths, ones, or tens position, etc.? Consider the following measurement values (with the last significant digit in color for emphasis).

   4620.0 has five significant figures and an uncertainty of tenths.

   4620 has three significant figures and an uncertainty in the tens place.

   462,000 has three significant figures and an uncertainty in the thousands place.

All numbers obtained by measurement have uncertainty. By convention, this uncertainty is assumed to be in the number's final significant digit.

EXAMPLE 2.4   Determining the Number of Significant Figures and Magnitude of Uncertainty in a Numerical Value

The number of carbon monoxide molecules in a sample of automobile exhaust is verbally reported as two hundred and five thousand. What meaning, in terms of significant figures and magnitude of uncertainty, is conveyed by each of the following written notations for this number?

   a. 205,000   **b.** 205,$\overline{0}$00   **c.** 205,000.   **d.** 205,$\overline{000}$

SOLUTION

   a. Three significant figures are present in this number; the confined zero is significant but the trailing zeros are not. Since the last significant digit, the 5, is located in the fourth place to the left of the understood decimal point (the thousands place), the uncertainty is ±1000.
   b. This number has four significant figures. The overbar above the first of the three trailing zeros makes this zero significant. The last significant digit, the zero with the overbar, occupies the hundreds place in the number. Thus, the uncertainty is ±100.
   c. There are six significant figures present. Explicitly placing a decimal point at the end of the number makes all the trailing zeros significant. The uncertainty is ±1 since the last of the trailing zeros is in the ones position.
   d. With the overbar notation present on all trailing zeros, all six digits present are significant. The uncertainty is ±1 since the last of the trailing zeros is in the ones position.

▶ Practice Exercise 2.4   The mass of a whale is verbally reported to be two hundred and fifty thousand kilograms. What meaning, in terms of significant figures and magnitude of uncertainty is conveyed by each of the following written notations for this number?

a. 250,000   **b.** 250,000   **c.** 250,000   **d.** 250,000.

EXAMPLE 2.5 **Uncertainty and the Number of Significant Figures in a Measurement**

A particular type of chemical balance is designed to give masses with an uncertainty of ±0.001 gram. How many significant figures should be reported when weighing each of the following samples on this balance?

  a. a sample with a mass of approximately 25 grams
  b. a sample with a mass of approximately 2 grams
  c. a sample with a mass of approximately 1 gram
  d. a sample with a mass of approximately 2/10 gram

SOLUTION

  a. Five significant figures—there will be two digits to the left of the decimal point (24 or 25) and three digits after the decimal point. 25.035 gram would be a possible mass value.
  b. Four significant figures—there will be one digit to the left of the decimal point (1 or 2) and three digits after the decimal point. 1.993 gram would be a possible mass value.
  c. Three or four significant figures depending upon whether the actual mass is greater or less than one. 1.014 gram and 0.991 gram would be possible mass values.
  d. Three significant figures—the digit to the left of the decimal point is a zero that is not significant. 0.211 gram would be a possible mass value.

Answer Double Check:

Are the answers reasonable in terms of the number of recorded digits? Yes. Each of the masses has three digits after the decimal point, a requirement for a measurement where the uncertainty is ±0.001.

▶ Practice Exercise 2.5   A centigram balance is designed to give masses with an uncertainty of ±0.01 gram. How many significant figures should be reported when weighing each of the following samples on this balance?

  a. a sample with a mass of approximately 120 grams
  b. a sample with a mass of approximately 100 grams
  c. a sample with a mass of approximately 1 gram
  d. a sample with a mass of approximately 1/4 gram

Figure 2.3 summarizes, in diagram form, the four rules used in determining how many significant figures a measured number possesses.

FIGURE 2.3 The rules for determining how many significant figures are present in a measured number.

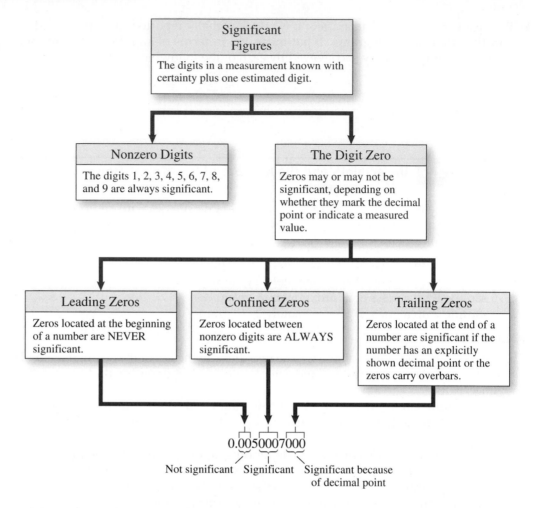

2.6 SIGNIFICANT FIGURES AND MATHEMATICAL OPERATIONS

When measurements are added, subtracted, multiplied, or divided, consideration must be given to the number of significant figures in the computed result. *Mathematical operations should not increase (or decrease) the uncertainty of experimental measurements.*

Concern about the number of significant figures in a calculated number is particularly critical when an electronic calculator is used to do the arithmetic of the calculation. Hand calculators now in common use are not programmed to take significant figures into account. Consequently, the digital readouts on them more often than not display more digits than are needed. It is a mistake to record these extra digits, since they have no significance; that is, they are not significant figures (see Fig. 2.4).

To record correctly the numbers obtained through calculations, students must be able to (1) round off numbers to a specified number of significant figures and (2) determine the allowable number of significant figures in a calculated result based on a set of operational rules. We will consider these skills in the order listed.

Rounding Off Numbers

When calculator answers are obtained that contain too many digits, it is necessary to drop the unneeded (nonsignificant) digits, a process that is called *rounding off*. **Rounding off** *is the process of deleting unwanted (nonsignificant) digits from a calculated number.* Three simple rules govern the process.

FIGURE 2.4 The digital readout on an electronic calculator usually shows more digits than are needed or justified. Electronic calculators are not programmed to take significant figures into account. *(Texas Instruments Incorporated)*

Rule 1 If the first digit to be dropped is less than 5, that digit and all digits that follow it are simply dropped.

Thus, 62.312 rounded off to three significant figures becomes 62.3.

Rule 2 If the first digit to be dropped is a digit greater than 5, or a 5 followed by digits other than all zeros, the excess digits are all dropped and the last retained digit is increased in value by one unit.

Thus, 62.782 and 62.558 rounded off to three significant figures become, respectively, 62.8 and 62.6.

Rule 3 If the first digit to be dropped is a 5 not followed by any other digit or a 5 followed only by zeros, an odd–even rule applies. Drop the 5 and any zeros that follow it and then

a. increase the last retained digit by one unit if it is *odd*, or
b. leave the last retained digit the same if it is *even*.

Thus, 62.650 and 62.350 rounded to three significant figures become, respectively, 62.6 (even rule) and 62.4 (odd rule). The number zero as a last retained digit is always considered an even number; thus, 62.050 rounded to three significant figures becomes 62.0.

These rounding rules must be modified slightly when digits to the left of the decimal point are to be dropped. To maintain the inferred position of the decimal point in such situations, nonsignificant zeros must replace all of the dropped digits that are to the left of the inferred decimal point. Parts (b) and (d) of Example 2.6 illustrate this point.

An additional point about rounding off is that numbers are not rounded *sequentially*. In rounding the number 4.548 to two significant figures we do not round it first to 4.55 and then to 4.6. When rounding we look at the first number to the right of the last retained digit and base the rounding operation on its value. The number 4.548 rounds to 4.5 because the first number to the right of the last retained digit, 4, has a value of less than 5 (rule 1).

EXAMPLE 2.6   **Rounding Numbers to a Specified Number of Significant Figures**

Round off each of the following numbers to two significant figures.

  a. 25.7   **b.** 432,117   **c.** 0.435   **d.** 15,500

SOLUTION

a. Rule 2 applies. The last retained digit (the 5) is increased in value by one unit.

25.7   becomes   26

b. Since the first digit to be dropped is a 2, rule 1 applies.

432,117   becomes   430,000

Note that to maintain the position of the inferred decimal point, nonsignificant zeros must replace all of the "dropped" digits. This will always be the case when digits to the left of the inferred decimal place are "dropped."

c. Rule 3 applies. The first and only digit to be dropped is a 5. The last retained digit (the 3) is an odd number, so, using the odd–even rule, its value is increased by one unit.

0.435   becomes   0.44

d. This is a rule 3 situation again. Since an odd digit (a 5) occupies the second significant figure place, its value is increased by one.

15,500   becomes   16,000

Note again that nonsignificant zeros must take the place of all digits that are "dropped" to the left of the inferred decimal place.

▶ **Practice Exercise 2.6**   Round off each of the following numbers to three significant figures.

a. 25.07   b. 25.70   c. 0.0003125   d. 33,333

Operational Rules for Mathematical Operations

Calculations should not increase or decrease the uncertainty in measurements. Two operational rules exist to help ensure that this is the case. One rule covers the operations of multiplication and division and the other the operations of addition and subtraction.

Rule 1   *Multiplication and Division.* In multiplication and division, the number of significant figures in the product or quotient is the same as in the number in the calculation that contains the fewest significant figures.

$$6.038 \times \mathbf{2.57} = 15.51766 \quad \text{(calculator answer)}$$
$$= 15.5 \quad \text{(\textbf{correct answer})}$$

This number limits the answer
to three significant figures.

Rule 2   *Addition and Subtraction.* In addition and subtraction of a series of measurements, the uncertainty in the answer should be the same as that of the measurement in the series that has the greatest uncertainty.

		Position of uncertainty
347	⟵	Ones position
+ 2.03	⟵	Hundredths position
+ 23.6	⟵	Tenths position
372.63	(calculator answer)	
373	(**correct answer**)	

The measurement with the greatest uncertainty in this series of numbers is 347, which is uncertain to the ones position. Thus, the calculator answer must be rounded to the ones position, the position of greatest uncertainty.

Note that for multiplication and division (rule 1), significant figures are counted, and that for addition and subtraction (rule 2), uncertainties are considered. The answers from either addition or subtraction can have more or fewer significant figures than any of the numbers that have been added or subtracted, as is shown in Example 2.9.

EXAMPLE 2.7 **Predicting the Number of Significant Figures That Should Be Present in the Answer to a Multiplication Problem**

Without actually doing any multiplications, indicate the number of significant figures that should be present in the answer to each of the multiplications. Assume that all numbers are measured quantities.

   a. $6.00 \times 6.00 \times 6.00$
   b. $6.00 \times 0.600 \times 60.60$
   c. $0.006 \times 0.060 \times 0.600$
   d. $60,600 \times 6060 \times 606$

SOLUTION

   a. Each number to be multiplied contains three significant figures. Thus, the answer should also contain three significant figures.
   b. The first two input numbers contain three significant figures and the third number contains four significant figures. Only three significant figures should be present in the answer.
   c. The input numbers contain, respectively, one, two, and three significant figures. The one significant figure input number limits the answer to one significant figure.
   d. All three input numbers contain three significant figures. Thus, the answer should contain three significant figures.

Answer Double Check:

Are the answers reasonable in terms of the number of significant figures present? Yes. Each answer contains no more significant figures than the least number found in the numbers used to generate the answer.

▶ **Practice Exercise 2.7**   Without actually doing any multiplications, indicate the number of significant figures that should be present in the answer to each of the multiplications. Assume that all numbers are measured quantities.

   a. $2.70 \times 27.0 \times 270$
   b. $0.10 \times 0.11 \times 0.12$
   c. $25 \times 30 \times 35$
   d. $3,303 \times 3,033 \times 3,330$

EXAMPLE 2.8 **Expressing Multiplication/Division Answers to the Proper Number of Significant Figures**

Perform the following computations, all of which involve multiplication and/or division. Express your answers to the proper number of significant figures. Assume that all numbers are measured quantities.

   a. 3.751×0.42

   b. $\dfrac{1,810,000}{3.1453}$

   c. $\dfrac{1800.0}{6.0000}$

   d. $\dfrac{3.130 \times 3.140}{3.15}$

SOLUTION

a. The calculator answer to this problem is

$$3.751 \times 0.42 = 1.57542$$

The input number with the least number of significant figures is 0.42, which has two significant figures. Thus the calculator answer must be rounded off to two significant figures.

1.57542	becomes	1.6
(calculator answer)		(**correct answer**)

b. The calculator answer to this problem is

$$\frac{1,810,000}{3.1453} = 575,461.8$$

The input number 1,810,000 contains three significant figures and the input number 3.1453 has five significant figures. Thus the correct answer is limited to three significant figures and is obtained by rounding the calculator answer to three significant figures.

575,461.8	becomes	575,000
(calculator answer)		(**correct answer**)

A decimal point is not explicitly shown in the number 575,000 since doing so would make the trailing zeros significant.

c. The calculator answer to this problem is

$$\frac{1800.0}{6.0000} = 300$$

Both input numbers contain five significant figures. Thus the correct answer must also contain five significant figures.

300	becomes	300.00
(calculator answer)		(**correct answer**)

Note here how the calculator answer had too few significant figures. Most calculators cut off zeros after the decimal point even if they are significant. *Using too few significant figures in an answer is just as wrong as using too many.*

d. This problem involves both multiplication and division. The calculator answer is

$$\frac{3.130 \times 3.140}{3.15} = 3.1200634$$

The input number with the least number of significant figures is 3.15, which contains three significant figures. Thus, the calculator answer must be rounded off to three significant figures.

3.1200634	becomes	3.12
(calculator answer)		(**correct answer**)

In a calculation that involves more than one multiplication–division operation, such as part (d) of this example, carry all of the digits that show on your calculator until you arrive at the final answer and then round off using the rounding rules. Do not round off at each step in the calculation.

Answer Double Check:

Are the answers reasonable in terms of the number of significant figures present? Yes. Each answer contains no more significant figures than the least number found in the numbers used to generate the answer.

▶ **Practice Exercise 2.8**   Perform the following computations, all of which involve multiplication and/or division. Express your answers to the proper number of significant figures. Assume that all numbers are measured quantities.

a. 0.3722×3410 **b.** $\dfrac{32.14}{0.033}$

c. $\dfrac{0.3200}{0.0044}$ **d.** $\dfrac{67 \times 300 \times 2.000}{2.33 \times 2500}$

EXAMPLE 2.9	**Expressing Addition/Subtraction Answers to the Proper Number of Significant Figures**

Perform the following computations, all of which involve addition or subtraction. Express your answers to the proper number of significant figures. Assume that all numbers are measured quantities.

a. $13.01 + 13.001 + 13.010$
b. $10.2 + 3.4 + 6.01$
c. $0.6700 - 0.6644$
d. $34.7 + 0.0007$

SOLUTION

a. The calculator answer to this problem is

$$13.01 + 13.001 + 13.010 = 39.021$$

Since this is an addition problem, in going from the calculator answer to the correct answer we must consider uncertainties rather than significant figures. (The multiplication–division rule is based on significant figures and the addition–subtraction rule is based on uncertainties.)

The uncertainties in the input numbers are

13.01   hundredths

13.001   thousandths

13.010   thousandths

The number 13.01 has the greatest uncertainty (hundredths) and so the last retained digit in the correct answer should reflect this uncertainty. Hence, the calculator answer is rounded off to hundredths.

39.021   becomes   39.02
(calculator answer)   (**correct answer**)

b. The calculator answer to this problem is

$$10.2 + 3.4 + 6.01 = 19.61$$

The uncertainty in the first two input numbers is tenths, and the third input number involves an uncertainty of hundredths. Thus, the last retained digit in the correct answer will be in the tenths place, the largest uncertainty among the input numbers.

19.61   becomes   19.6
(calculator answer)   (**correct answer**)

Note that the input number 3.4 possesses two significant figures and yet the correct answer contains three significant figures. Why? The number of significant figures is not the determining factor in addition and subtraction (rule 2) as it is in multiplication and division (rule 1).

c. The calculator answer to this problem is

$$0.6700 - 0.6644 = 0.0056$$

Both input numbers are known to the ten-thousandths place. Thus, the answer should also have an uncertainty involving the ten-thousandths place.

In this particular problem the calculator answer and the correct answer are the same, a situation that does not occur very often. The correct answer is 0.0056. Note that two significant figures were "lost" in the subtraction. The answer has two significant figures. The two input numbers each have four significant figures. For addition and subtraction this is allowable; for multiplication and division it would not be allowable.

d. The calculator answer to this problem is

$$34.7 + 0.0007 = 34.7007$$

The uncertainty in the input number 34.7 is tenths and the uncertainty in the input number 0.0007 is ten-thousandths. Thus, the calculator answer must be rounded off to the tenths place.

34.7007	becomes	34.7
(calculator answer)		**(correct answer)**

The correct answer, 34.7, is the same as one of the input numbers. The message of this situation is that the number 0.0007 is negligible when added to the number 34.7.

Answer Double Check:

Were the answers rounded to the correct number of significant figures using the proper rule? Yes. The rounding of calculator answers was done using uncertainty considerations rather than the number of significant figures in input numbers. The addition–subtraction rule, which involves uncertainty considerations, was used rather than the multiplication–division rule, which involves input-number significant figure considerations.

▶ **Practice Exercise 2.9**   Perform the following computations, all of which involve addition or subtraction. Express your answers to the proper number of significant figures. Assume that all numbers are measured quantities.

a. 34.6 + 3.20 + 109
b. 8.13 + 7.51 + 0.02
c. 0.4500 − 0.4300
d. 430 + 4.33

Example 2.8 was an exercise that involved only multiplication/division operations and the significant figure rule for these operations was applied in each part. Similarly, Example 2.9 was an exercise that involved only addition/subtraction operations and the significant figure rule for such operations was always used. What about the occasionally encountered situation where both multiplication/division and addition/subtraction are encountered in the same problem, such as in

$$23.77 \times (1.3 + 2.58 + 6.671)$$

For such a problem, both calculational significant-figure rules must be applied by using a two-step approach. First we add the three numbers in parentheses, obtaining an intermediate answer that is rounded based on the addition–subtraction rule. This intermediate answer is then multiplied by 23.77 to generate a final answer that is then adjusted for significant figures using the multiplication–division rule.

This "undesirable" situation of having to round twice within the same problem, thus introducing two rounding errors, is minimized to some degree by using the following operational rules for the intermediate answer rounding:

1. The intermediate answer is rounded such that one extra digit is maintained (the first insignificant digit), which is highlighted in a way to denote that it is not significant.
2. This "incorrectly" rounded intermediate answer (one too many digits) is then carried into further steps of the calculation and "correct" rounding occurs with the final answer.

Following this plan, we solve the preceding problem in this manner.

Addition: (1.3 + 2.58 + 6.671) = 10.551 (calculator answer)

= 10.5⬚5⬚ (intermediate answer, with one extra digit retained when rounding using the addition/subtraction rule)

Multiplication: (23.77 × 10.5⬚5⬚) = 250.7735 (calculator answer)

= 251 (correct answer, rounded remembering that 23.77 has four significant figures and 10.5⬚5⬚ has only three significant figures)

Note, again, that the just employed "rounding procedures" are used only when both significant figure rules (addition/subtraction and multiplication/division) are needed in the same problem. The reason for doing this is to minimize the rounding errors associated with having to round twice.

Significant Figures and Exact Numbers

In Section 2.2 we noted that some numbers are exact. Counted numbers, defined numbers, and simple fractions all fall into this category. The conventions of significant figures do not apply to exact numbers because there is no uncertainty associated with them. Therefore, such numbers, when present in a calculation, will never limit the number of significant figures in the computational answer; that is, they will have no effect on the number of significant figures in a calculated result.

An alternative approach to the topic of significant figures and exact numbers is to view exact numbers as containing an infinite number of significant figures. From this viewpoint, such numbers will never be the limiting factors in significant-figure considerations. Thus, when we say that there are 3 feet in 1 yard (a definition), both the numbers 3 and 1 are considered to have an infinite number of significant figures.

Multiplication by a Small Whole Number

An interesting significant figure situation arises when a measured quantity is multiplied by a small whole number (an exact number). There are two approaches to solving this type of problem with the two approaches often giving a different number of significant figures in the answer. These two approaches are the following:

1. Treating the problem as a simple multiplication problem and using the multiplication/division significant figure rule when rounding the calculator answer.
2. Treating the problem as an addition problem in which the measured quantity is added to itself several times and then rounding the calculator answer using the addition/subtraction rule.

The simple problem of obtaining the mass of three identical coins, each of which has a mass of 6.32 grams, nicely illustrates this "dual-approach" situation.

Using the multiplication approach generates 19.0 grams as the correct answer.

$$3 \times 6.32 \text{ grams} = 18.96 \text{ grams} \quad \text{(calculator answer)}$$

$$\left(\begin{array}{c}\text{exact} \\ \text{number}\end{array}\right)\left(\begin{array}{c}\text{three significant} \\ \text{figures}\end{array}\right) = 19.0 \text{ grams} \quad \textbf{(correct answer)}$$

The correct answer has three significant figures, the same number present in the input number 6.32.

Using the addition approach generates 18.96 grams as the correct answer.

6.32 grams

6.32 grams

6.32 grams

18.96 grams (calculator and **correct answer**)

The correct answer should have an uncertainty in the hundredths place since each of the input numbers has this uncertainty.

In situations such as the preceding, the operational rule followed in this text will be to use the addition approach since it often, but not always, allows for an additional significant figure in the answer. Example 2.10 further explores the situation just discussed.

| **EXAMPLE 2.10** | **Significant Figures in Calculations Involving Multiplication by a Small Whole Number** |

For a collection of identical 19.13 centimeter-long pencils, what would be the total combined length of the pencils if

a. six pencils are present and the problem is solved using multiplication?
b. six pencils are present and the problem is solved using addition?
c. three pencils are present and the problem is solved using multiplication?
d. three pencils are present and the problem is solved using addition?

SOLUTION

An exact number has no effect on the number of significant figures in an answer obtained by calculation.

a. Using multiplication, the set-up is

$$(6 \times 19.13) \text{ centimeters} = 114.78 \text{ centimeters} \quad \text{(calculator answer)}$$
$$= 114.8 \text{ centimeters} \quad \textbf{(correct answer)}$$

The correct answer is limited to four significant figures based on the multiplication/division significant figure rule because the input number 19.13 contains four significant figures.

b. Using the addition approach, the set-up is

19.13 centimeters

19.13 centimeters

19.13 centimeters

19.13 centimeters

19.13 centimeters

19.13 centimeters

114.78 centimeters (calculator and **correct answer**)

The final answer has an uncertainty of hundredths, the same as that of all of the input numbers, based on the addition/subtraction rule for significant figures.

Note that the addition approach answer (114.78) contains one more significant figure than the multiplication approach answer (114.8) obtained in part (a). The addition approach is the preferred approach because of this.

c. Using the multiplication approach (3×19.13) generates the answer 57.39 centimeters as both the calculator and correct answer.

d. The addition approach ($19.13 + 19.13 + 19.13$) also generates the answer 57.39 centimeters as both the calculator and correct answer.

Thus, for three pencils, the two approaches give the same answer.

▶ **Practice Exercise 2.10**   What is the total combined volume, in liters, of seven identical 0.65 liter beverage containers when the problem is solved as

a. a multiplication problem.
b. an addition problem.

2.7 SCIENTIFIC NOTATION

Up to this point in the chapter, we have expressed all numbers in decimal notation, the everyday method for expressing numbers. Such notation becomes cumbersome for very large and very small numbers (which occur frequently in scientific work). For example, in one drop of blood, which is 92% water by mass, there are approximately

$$1,600,000,000,000,000,000,000$$

molecules (Sec 5.2) of water, each of which has a mass of

$$0.000000000000000000000030 \text{ gram}$$

Recording such large and small numbers is not only time consuming but also open to error; often, too many or too few zeros are recorded. Also, it is impossible to enter such numbers into most calculators because they cannot accept that many digits. (Most calculators accept eight or ten digits.)

A method called *scientific notation* exists for expressing in compact form multi-digit numbers that involve many zeros. **Scientific notation** *is a numerical system in which numbers are expressed in the form* $A \times 10^n$, *where A is a number with a single nonzero digit to the left of the decimal point and* n *is a whole number.* The number *A* is called the *coefficient*. The number 10^n is called the *exponential term*. The coefficient is always multiplied by the exponential term. The scientific notation form of the number 703 is

Scientific notation is also called *exponential notation.*

The two previously cited numbers that deal with molecules of water are expressed in scientific notation as

$$1.6 \times 10^{21} \text{ molecules} \quad \text{and} \quad 3.0 \times 10^{-23} \text{ gram}$$

Such scientific notation is compatible with most calculators.

Exponents

A brief review of exponents and their use is in order before we consider the rules for converting numbers from ordinary decimal notation to scientific notation and vice versa. An **exponent** *is a number written as a superscript following another number and indicates how many times the first number, the base, is to be multiplied by itself.* The following examples illustrate the use of exponents.

$$6^2 = 6 \times 6 = 36$$
$$3^5 = 3 \times 3 \times 3 \times 3 \times 3 = 243$$
$$10^3 = 10 \times 10 \times 10 = 1000$$

Exponents are also frequently referred to as *powers* of numbers. Thus, 6^2 may be verbally read as "six to the second power" and 3^5 as "three to the fifth power." Raising a number to the second power is often called "squaring," and raising it to the third power "cubing."

Scientific notation exclusively uses powers of ten. *When 10 is raised to a positive power, its decimal equivalent is the number 1 followed by as many zeros as the power.* This one-to-one correlation between power magnitude and number of zeros is shown in color in the following examples.

$$10^2 = 100 \qquad \text{(two zeros and a power of 2)}$$
$$10^4 = 10,000 \qquad \text{(four zeros and a power of 4)}$$
$$10^6 = 1,000,000 \qquad \text{(six zeros and a power of 6)}$$

The notation 10^0 is a defined quantity.

$$10^0 = 1$$

The preceding generalization easily explains why. Ten to the zero power is the number 1 followed by no zeros, which is simply 1.

All of the examples of exponential notation presented so far have had positive exponents. This is because each example represented a number of magnitude greater than one. Negative exponents are also possible. They are associated with numbers of magnitude less than one.

A negative sign in front of an exponent is interpreted to mean that the base and the power to which it is raised are in the denominator of a fraction in which 1 is the numerator. The following examples illustrate this interpretation.

$$10^{-1} = \frac{1}{10^1} = \frac{1}{10} = 0.1$$

$$10^{-2} = \frac{1}{10^2} = \frac{1}{10 \times 10} = \frac{1}{100} = 0.01$$

$$10^{-3} = \frac{1}{10^3} = \frac{1}{10 \times 10 \times 10} = \frac{1}{1000} = 0.001$$

When the number 10 is raised to a negative power, the absolute value of the power (the value ignoring the minus sign) is always one more than the number of zeros between the decimal point and the one. This correlation between power magnitude and number of zeros is shown in color in the following examples.

$$10^{-2} = 0.01 \qquad \text{(one zero and a power of } -2)$$
$$10^{-4} = 0.0001 \qquad \text{(three zeros and a power of } -4)$$
$$10^{-6} = 0.000001 \qquad \text{(five zeros and a power of } -6)$$

Differences in magnitude between numbers that are powers of 10 are often described with the phrase *orders of magnitude*. An **order of magnitude** *is a single exponential value of the number 10.* Thus, 10^6 is four orders of magnitude larger than 10^2, and 10^7 is three orders of magnitude larger than 10^4.

Converting from Decimal to Scientific Notation

The procedure for converting a number from decimal notation to scientific notation involves two operational rules:

Rule 1 The coefficient must be a number between 1 and 10 that contains the same number of significant figures as are present in the original decimal number. The coefficient is obtained by rewriting the decimal number with a decimal point after the first nonzero digit and deleting all *nonsignificant* zeros.

> The decimal and scientific notation forms of a number *always* contain the same number of significant figures.

For 233,000, the coefficient is 2.33

For 0.00557, the coefficient is 5.57

For 0.35500, the coefficient is 3.5500

Rule 2 The value of the exponent for the power of ten is obtained by counting the number of places the decimal point in the coefficient must be moved to give back the original decimal number. If the decimal point movement is to the *right*, the exponent has a *positive* value and if the decimal point movement is to the *left*, the exponent has a *negative* value.

Numerous applications of these two rules are found in Example 2.11.

EXAMPLE 2.11   **Expressing Decimal Numbers in Scientific Notation**

Without use of a calculator, express in scientific notation the number in each of the following statements.

  a. Light travels at a speed of 186,000 miles per second.
  b. A person exhales approximately 320,000,000,000,000,000,000 molecules of carbon dioxide in one breath.
  c. The diameter of a human hair is 0.0016 inch.
  d. The maximum allowable amount of chromium in drinking water (EPA standard) is 0.00000010 gram per milliliter of water.

SOLUTION

  a. For the number 186,000 the scientific notation coefficient is 1.86. This coefficient meets the requirement that it contain the same number of significant figures as the original decimal number.

  The value of the exponent in the exponential term of the scientific notation is +5 since the decimal point in the coefficient must be moved five places to the right to generate the original decimal number.

1.86000

five-place movement

Note that zeros are added to the coefficient as the decimal point is moved in order to obtain the original decimal point position.

Multiplying the coefficient by the exponential term 10^5 gives the scientific notation form of the number, which is

$$1.86 \times 10^5$$

b. The coefficient for the number 320,000,000,000,000,000,000 is 3.2. It, like the original number, contains two significant figures.

The value of the exponent for the exponential term is +20 since the decimal point in the coefficient must be moved 20 places to the right to generate the original decimal number.

$$3.20000000000000000000$$
20-place movement

Movement of the decimal point in the coefficient to the *right* always results in a *positive* exponent.

The scientific notation form of the number, obtained by multiplying the coefficient and exponential term, is

$$3.2 \times 10^{20}$$

c. The scientific notation coefficient for the number 0.0016 is 1.6. The exponent for the power of ten is –3, since moving the decimal point in the coefficient three places to the left generates the original decimal number.

$$0.001.6$$
three-place movement

Movement of the decimal point in the coefficient to the *left* always means that the exponent will be *negative*. The scientific notation form of the number is, thus, 1.6×10^{-3}.

d. The scientific notation coefficient for the number 0.00000010 is 1.0. The coefficient is 1.0 rather than 1 because the original number has two significant figures. The exponent for the power of ten is –7.

$$0.0000001.0$$
seven-place movement

The number in scientific notation is, thus, 1.0×10^{-7}.

Some numbers, such as 2.4, 0.911, and 57, are simpler in their original form than in scientific notation. Such numbers are usually left in decimal notation. Although there are no fixed rules as to when scientific notation should be used, it is generally not used for numbers between 0.1 and 1000.

Answer Double Check:

Are the answers reasonable in terms of the sign (plus or minus) for the exponent? Yes. When converting from decimal notation to scientific notation, numbers greater than one will always have *positive* exponents and numbers less than one will always have *negative* exponents. Such is the case for the answers here.

▶ **Practice Exercise 2.11** Without the use of a calculator, express in scientific notation the number in each of the following statements.

a. The earth has an estimated mass of 6,600,000,000,000,000,000,000 tons.

b. The diameter of a red blood cell is 0.000006 meters.

c. A blood analysis report gives a cholesterol level of 226 milligrams per deciliter of blood.
d. The white blood cell concentration in normal blood is approximately 12,000,000,000 cells per milliliter of blood.

Significant Figures and Scientific Notation

When a number is expressed in scientific notation, *only significant digits become part of the coefficient.* Because of this there is never any confusion (ambiguity) in determining the number of significant figures in a number expressed in scientific notation. There are five possible precision interpretations for the number 10,000 (one, two, three, four, or five significant figures). In scientific notation each of these interpretations assumes a different form.

1×10^4 (10,000 with one significant figure)
1.0×10^4 (1$\bar{0}$,000 with two significant figures)
1.00×10^4 (1$\overline{0,0}$00 with three significant figures)
1.000×10^4 (1$\overline{0,00}$0 with four significant figures)
1.0000×10^4 (1$\overline{0,000}$ with five significant figures)

Most scientists use the preceding scientific notation method for designating significant figures in an "ambiguous" number instead of the overbar notation discussed in Section 2.5. The overbar notation is used only in those situations where there is reason for not expressing the number in scientific notation.

Converting from Scientific to Decimal Notation

To convert a number in scientific notation, such as 6.02×10^{23}, into a regular decimal number, we start by examining the exponent. The value of the exponent tells how many places the decimal point must be moved. If the exponent is positive, movement is to the right to give a number greater than one; if it is negative, movement is to the left to give a number less than one. Zeros may have to be added to the number as the decimal point is moved.

EXAMPLE 2.12   Expressing Scientific Notation Numbers in Decimal Notation

Without use of a calculator, convert the scientific notation number in each of the following statements to a decimal number.

a. The announced attendance at a football game was 5.3127×10^4.
b. The circumference of the earth is 2.5×10^4 miles.
c. The concentration of gold in seawater is 1.1×10^{-8} gram per liter.
d. The mass of a hydrogen atom is 1.67×10^{-24} gram.

SOLUTION

a. The exponent +4 tells us the decimal is to be located four places to the right of where it is in 5.3127.

$$5.3127$$
Decimal point shift

The decimal number is 53,127.

b. The exponent +4 tells us the decimal is to be located four places to the right of where it is in 2.5. Trailing zeros will have to be added to accommodate the decimal point change.

$$2.\overbrace{5000}^{\text{Added zeros}}$$

Decimal point shift

These added "trailing zeros" are not significant zeros. Thus, the number of significant digits remains at two. The decimal form of the number is

$$25,000$$

c. The exponent –8 tells us the decimal is to be located eight places to the left of where it is in 1.1. Leading zeros will have to be added to accommodate the decimal point change.

$$0.\underbrace{00000001}.1$$

Added zeros

Decimal point shift

These added "leading zeros" are not significant zeros. Thus, the number of significant digits remains at two. The decimal form of the number is

$$0.000000011$$

d. The exponent –24 tells us the decimal is to be located 24 places to the left of where it is in 1.67. This will produce an extremely small number.

$$0.00000000000000000000001.67$$

Decimal point shift

Twenty-three leading zeros were needed to mark the new decimal place. (In numbers with negative exponents, the number of added leading zeros will always be one less than the value of the exponent.) The decimal form of this number is, thus,

$$0.00000000000000000000000167$$

The number of significant figures present is three, the same number as in the original scientific notation form of the number.

Answer Double Check:

Are the answers reasonable in terms of significant figures? Yes. The number of significant figures in the decimal form of the number must be the same as the number of significant figures in the coefficient of the scientific notation form of the number. Such is the case in each part of this problem.

▶ Practice Exercise 2.12   Without use of a calculator, convert the scientific notation number in each of the following statements to a decimal number.

a. The mass percent lead in a soil sample is found to be 1.2×10^{-3} percent.
b. A person exercising vigorously can lose up to 1.6×10^{7} microliters of water per day through perspiration.
c. A 1-ounce silver coin contains 1.60×10^{23} atoms of silver.
d. The radius of an atom is approximately 1.3×10^{-13} kilometer.

Uncertainty and Scientific Notation

The uncertainty associated with a measurement whose value is expressed in scientific notation cannot be obtained directly from the coefficient in the scientific notation. The coefficient decimal point location is not the true location for the decimal point. The value of the exponential term must be taken into account in determining the uncertainty.

    The uncertainty associated with a scientific notation number is obtained by determining the uncertainty associated with the coefficient and then multiplying this value by the exponential term. For the number 3.753×10^2, we have

$$
\underset{\substack{\text{uncertainty} \\ \text{in coefficient}}}{10^{-3}} \quad \times \quad \underset{\substack{\text{exponential} \\ \text{term}}}{10^{2}} \quad = \quad \underset{\substack{\text{uncertainty} \\ \text{in value}}}{10^{-1}}
$$

That the uncertainty is, indeed, 10^{-1} can be readily seen by rewriting the number in decimal notation.

$$3.753 \times 10^2 = 375.3$$

The uncertainty for the decimal number is in the tenths place (10^{-1}). Example 2.13 further illustrates the process of determining the uncertainty associated with numbers that are expressed in scientific notation.

EXAMPLE 2.13 **Determining the Uncertainty Associated with Numbers Expressed in Scientific Notation**

What is the uncertainty associated with each of the following measured values?

    a. 4.200×10^4 **b.** 4.2×10^5 **c.** 4.2×10^{-1} **d.** 4.200×10^{-3}

SOLUTION

Each part will be worked the same way. The uncertainty in the coefficient is multiplied by the exponential term to obtain the overall uncertainty for the value.

    a. The uncertainty in the coefficient is ± 0.001

$$10^{-3} \times 10^4 = 10^1$$

        The overall uncertainty is ± 10.

    b. $10^{-1} \times 10^5 = 10^4$, which gives an uncertainty of $\pm 10{,}000$. That the uncertainty is this large becomes readily apparent when the number is written in decimal notation.

$$4.2 \times 10^5 = 420{,}000$$

    c. $10^{-1} \times 10^{-1} = 10^{-2}$, which gives an uncertainty of ± 0.01
    d. $10^{-3} \times 10^{-3} = 10^{-6}$, which gives an uncertainty of ± 0.000001

▶ **Practice Exercise 2.13** What is the uncertainty associated with each of the following measured values?

    a. 3.72×10^4 **b.** 3.72×10^2 **c.** 3.7200×10^2 **d.** 3.72×10^{-2}

2.8 MATHEMATICAL OPERATIONS IN SCIENTIFIC NOTATION

A major advantage of writing numbers in scientific notation is that it greatly simplifies the mathematical operations of multiplication and division.

Multiplication in Scientific Notation

Multiplication of two or more numbers expressed in scientific notation involves two separate operations or steps.

An electronic calculator combines steps 1 and 2 into one operation.

Step 1 Multiply the coefficients (the decimal numbers between 1 and 10) together in the usual manner.

Step 2 *Add* algebraically the exponents of the powers of ten to obtain a new exponent.

In general terms, we can represent the multiplication of two scientific notation numbers as follows.

$$(a \times 10^x) \times (b \times 10^y) = ab \times 10^{x+y}$$

EXAMPLE 2.14   Multiplication of Scientific Notation Numbers

Carry out the following multiplications involving scientific notation. Be sure to take into account significant figures in obtaining your final answer.

a. $(2.05 \times 10^{-3}) \times (1.19 \times 10^{-7})$

b. $(7.92 \times 10^{10}) \times (2.3 \times 10^{-4})$

SOLUTION

a. Multiplying the two coefficients together gives

$$2.05 \times 1.19 = 2.4395$$

Since both input numbers for the multiplication contain three significant figures, the calculator answer must be rounded to three significant figures.

2.4395	becomes	2.44
(calculator answer)		**(correct answer)**

An electronic calculator is not programmed to take into account significant figures. It cannot completely substitute for your brain.

Next, the exponents of the powers of ten are added to generate the new power of ten.

$$10^{-3} \times 10^{-7} = 10^{(-3)+(-7)} = 10^{-10}$$

(To add two numbers of the same sign, either positive or negative, just add the numbers and place the common sign in front of the sum.) Combining the coefficient and the exponential term gives the answer of

$$2.44 \times 10^{-10}$$

b. Multiplying the coefficients gives

$$7.92 \times 2.3 = 18.216$$

The answer must be rounded to two significant figures because the input number 2.3 has only two significant figures.

18.216	becomes	18
(calculator answer)		**(correct answer)**

In combining the exponential terms, we have exponents of different signs. The smaller exponent (4) is subtracted from the larger exponent (10), and the sign of the larger exponent (+) is used. Thus,

$$(+10) + (-4) = (+6)$$

and

$$10^{10} \times 10^{-4} = 10^6$$

Combining the coefficient and the exponential term gives

$$18 \times 10^6$$

This answer has something wrong with it; it is not in correct scientific notation form. The coefficient should be a number between 1 and 10. This problem is corrected by recognizing that 18 is equal to 1.8×10^1, making this a substitution for 18, and then combining exponential terms.

$$18 \times 10^6 = 1.8 \times 10^① \times 10^⑥ = 1.8 \times 10^⑦$$

An electronic calculator automatically makes this coefficient–exponent adjustment for you.

The correct answer is 1.8×10^7.

▶ **Practice Exercise 2.14**   Carry out the following multiplications involving scientific notation. Be sure to take into account significant figures in obtaining your final answer.

a. $(2.543 \times 10^3) \times (2.003 \times 10^6)$
b. $(5.329 \times 10^{-1}) \times (3.11 \times 10^9)$

In Example 2.14, each multiplication was considered in two parts—a coefficient part and an exponential term part. Such an approach is helpful for understanding the role that exponents play in the calculations. In practice, with an electronic calculator, multiplications like those in Example 2.14 are usually done in one step with the numbers being entered directly into the calculator in scientific notation form. Example 2.15 considers this one-step approach with several comments added about how to enter scientific notation numbers into calculators in that form.

EXAMPLE 2.15   Multiplication of Numbers in Scientific Notation Form

Carry out the following multiplication, directly entering the numbers into a calculator in scientific notation.

$$(4.72 \times 10^{-6}) \times (8.21 \times 10^{23})$$

SOLUTION

We first enter the number 4.72×10^{-6} into the calculator. This involves the following operations.

1. Enter 4.72 into the calculator.
2. Press the $\boxed{\text{EE}}$ or $\boxed{\text{EXP}}$ key (or something similar). This enters the "× 10" portion of the notation.
3. Enter 6 (the magnitude of the exponent) and then press the $\boxed{+/-;}$ key to change the exponent from +6 to –6.

The number 4.72×10^{-6} is now entered into the calculator. How it is displayed varies with the brand of calculator. Variations include

$$\boxed{4.72 \quad -06} \qquad \boxed{4.72^{-06}} \qquad \boxed{0.00000472}$$

The multiplication key is then pressed and the second number is entered into the calculator in a similar manner to that just explained. The $\boxed{=}$ key is pressed to generate the answer.

$$(4.72 \times 10^{-6}) \times (8.21 \times 10^{23}) = 3.87512 \times 10^{18} \qquad \text{(calculator answer)}$$

$$= 3.88 \times 10^{18} \qquad \textbf{(correct answer)}$$

A calculator entry mistake that students often make is to enter the coefficient (4.72), then enter "times" and "10," then press $\boxed{\text{EE}}$ or $\boxed{\text{EXP}}$ and enter –6. The calculator result is 4.72×10^{-5} a number 10 times larger than wanted. Entering "× 10" multiplies the number by an extra ten. Pressing the $\boxed{\text{EE}}$ or $\boxed{\text{EXP}}$ key stands for "× 10."

▶ **Practice Exercise 2.15**   Carry out the following multiplication, directly entering the numbers into a calculator in scientific notation.

$$(8.27 \times 10^{14}) \times (3.11 \times 10^{-17})$$

Division in Scientific Notation

Division of two numbers expressed in scientific notation involves two separate operations or steps.

Step 1   Divide the coefficients (the decimal numbers between 1 and 10) in the usual manner.

Step 2   *Subtract* algebraically the exponent in the denominator (bottom) from the exponent in the numerator (top) to give the exponent of the new power of ten.

Note that in multiplication we add exponents, and in division we subtract exponents.

| **EXAMPLE 2.16** | **Division of Scientific Notation Numbers** |

Carry out the following divisions in scientific notation. Be sure to take significant figures into account in obtaining your final answer.

a. $\dfrac{2.05 \times 10^5}{1.19 \times 10^3}$ b. $\dfrac{3.92 \times 10^{10}}{9.1 \times 10^{-4}}$

SOLUTION

a.   Performing the indicated division involving the coefficients gives

$$\frac{2.05}{1.19} = 1.722689$$

Since both input numbers for the division have three significant figures, the calculator answer must be rounded off to three significant figures.

$$1.722689 \qquad \text{becomes} \qquad 1.72$$
$$\text{(calculator answer)} \qquad \textbf{(correct answer)}$$

Dividing exponential terms involves the algebraic subtraction of exponents.

$$\frac{10^5}{10^3} = 10^{(5)-(+3)} = 10^2$$

Algebraic subtraction involves changing the sign of the number to be subtracted and then following the rules for addition (as outlined in Example 2.12). In this problem the number to be subtracted (+3) becomes, upon changing the sign, (–3). Then we add (+5) and (–3). The answer is (+2), as shown in the preceding equation. Combining the coefficient and the exponential term gives

$$1.72 \times 10^2$$

b. The new coefficient, obtained by dividing 3.92 by 9.1, should contain two significant figures, the same number as in the input number 9.1.

$$\frac{3.92}{9.1} = 0.43076923$$

0.43076923	becomes	0.43
(calculator answer)		**(correct answer)**

Performing the exponential term division by subtracting the powers of the exponential terms gives

$$\frac{10^{10}}{10^{-4}} = 10^{(+10)-(-4)} = 10^{14}$$

Combining the coefficient and the exponential term gives

$$0.43 \times 10^{14}$$

which is not in correct scientific notation form because the coefficient is a number less than one. This problem is remedied by recognizing that 0.43 is equal to 4.3×10^{-1}, making this substitution for 0.43, and then combining the two exponential terms.

$$0.43 \times 10^{14} = (4.3 \times 10^{-1}) \times 10^{14} = 4.3 \times 10^{13}$$

An electronic calculator automatically makes this coefficient–exponent adjustment for you.

The correct answer is, thus, 4.3×10^{13}.

▶ **Practice Exercise 2.16** Carry out the following divisions in scientific notation. Be sure to take significant figures into account in obtaining your final answer.

a. $\dfrac{9.98 \times 10^{-3}}{2.341 \times 10^{-7}}$ **b.** $\dfrac{5.1 \times 10^{-10}}{8.76 \times 10^{7}}$

Addition and Subtraction in Scientific Notation

To add or subtract numbers written in scientific notation, *the power of ten for all numbers must be the same.* More often than not, one or more exponents must be adjusted. Adjusting the exponent requires rewriting the number in a form where the coefficient is a number greater than 10 or less than 1. With exponents all the same, *the coefficients are then added or subtracted and the exponent is maintained at its now common value.* Although any of the exponents may be changed, changing the smaller exponent to a larger one will usually produce a coefficient in the answer that is a number between 1 and 10.

EXAMPLE 2.17　**Addition and Subtraction of Scientific Notation Numbers**

Perform the following additions or subtractions with all numbers expressed in scientific notation. Answers will need to be checked for the correct number of significant figures and for the correct scientific notation form (coefficient is a number between 1 and 10).

　　a. $(2.661 \times 10^3) + (3.011 \times 10^3)$　　　　**b.** $(2.66 \times 10^4) - (1.03 \times 10^3)$
　　c. $(9.98 \times 10^{-3}) + (8.04 \times 10^{-5})$

SOLUTION

a. The exponents are the same to begin with. Therefore, we can proceed with the addition immediately.

$$
\begin{aligned}
2.661 &\times 10^3 \\
\underline{3.011} &\times 10^3 \\
5.672 &\times 10^3 \text{ (calculator and \textbf{correct answer})}
\end{aligned}
$$

Both input numbers have uncertainties in the thousandths place. The calculator answer has the same uncertainty. Thus, the calculator answer and correct answer are the same. Note that the exponent values are not added. They are maintained at their common value.

b. The exponents are different, so before subtracting we must change one of the exponents. Let us change 10^3 to 10^4:

$$10^3 \quad \text{can be written as} \quad 10^{-1} \times 10^4$$

Then, by substitution, we have

$$1.03 \times 10^3 = 1.03 \times 10^{-1} \times 10^4$$

The coefficient and the first exponent are then combined to give a new coefficient.

$$1.03 \times 10^{-1} \times 10^4 = 0.103 \times 10^4$$

We are now ready to make the subtraction called for in the original statement of the problem.

$$
\begin{aligned}
2.66 \times 10^4 =&\ \ 2.66\ \ \times 10^4 \qquad \text{common exponent}\\
-1.03 \times 10^3 =&\ -0.103 \times 10^4 \\
\hline
&\ \ 2.557 \times 10^4 \ \ \text{(calculator answer)}
\end{aligned}
$$

The calculator answer must be adjusted for significant figures. On the common exponent basis of 10^4, the uncertainty in 2.66 lies in the hundredths place and that in 0.103 lies in the thousandths place. The correct answer, therefore, is limited to an uncertainty of hundredths. (Recall, from Example 2.9, the rules on addition and significant figures.) Thus,

$$2.557 \times 10^4 \text{ becomes } 2.56 \times 10^4 \text{ (\textbf{correct answer})}$$

c. The exponents are 10^{-3} and 10^{-5}. Since –5 is smaller than –3, let us have 10^{-3} as the common exponent. (Always use the larger of the two exponents as the common exponent.)

$$10^{-5} \quad \text{can be rewritten as} \quad 10^{-2} \times 10^{-3}$$

Then by substitution we have

$$8.04 \times \underbrace{10^{-5}} = 8.04 \times \underbrace{10^{-2} \times 10^{-3}}$$

The coefficient and the first exponent are then combined to give a new coefficient.

$$\underbrace{8.04 \times 10^{-2}} \times 10^{-3} = \underbrace{0.0804} \times 10^{-3}$$

We are now ready to make the addition called for in the original statement of the problem.

$$
\begin{aligned}
9.98 \times 10^{-3} &= 9.98 \quad \times 10^{-3} \quad \text{common exponent} \\
8.04 \times 10^{-5} &= 0.0804 \times 10^{-3} \\
\hline
&\ \ 10.0604 \times 10^{-3} \quad \text{(calculator answer)}
\end{aligned}
$$

The calculator answer must be adjusted for significant figures and also changed into correct scientific notation since the coefficient has a value greater than 10. The significant figure adjustment rounds the answer to hundredths, giving

$$10.06 \times 10^{-3}$$

The correct scientific notation adjustment involves rewriting 10.06 as a power of ten and then simplifying the resulting expression.

$$\underbrace{10.06 \times 10^{-3}} = \underbrace{1.006 \times 10^{1}} \times 10^{-3} = 1.006 \times 10^{-2} \ \textbf{(correct answer)}$$

▶ **Practice Exercise 2.17**   Perform the following additions and subtractions with all numbers expressed in scientific notation. Besides containing the correct number of significant figures, answers should also be in standard scientific notation (coefficient is a number between 1 and 10).

a. $(2.333 \times 10^3) + (1.22 \times 10^3)$ **b.** $(2.333 \times 10^3) - (2.33 \times 10^2)$

c. $(9.9 \times 10^8) + (9.9 \times 10^7)$

Summary

1. Exact and Inexact Numbers Numbers are of two kinds: exact and inexact. An exact number has a value that has no uncertainty associated with it. Exact numbers occur in definitions, in counting, and in simple fractions. An inexact number has a value that has a degree of uncertainty associated with it. Inexact numbers result any time a measurement is made.

2. **Precision and Accuracy of Measurements** Precision refers to how close a series of measurements on the same object are to each other. Accuracy refers to how close a measurement (or the average of multiple measurements) comes to a true or accepted value. The precision and accuracy of a measurement depend not only on the measuring device used but also on the technical skill of the person making the measurement.

3. **Uncertainty in Measurements** Every measurement has a degree of uncertainty associated with it. In reading a measurement scale, all digits known for certain are recorded plus one estimated digit (uncertainty). It is wrong to record more than one estimated digit in a measurement.

4. **Significant Figures** Significant figures in a measurement are those digits that are certain, plus a last digit that has been estimated. The maximum number of significant figures possible in a measurement is determined by the design of the measuring device. All nonzero digits in a measurement value are always significant. Zeros in a measurement value may or may not be significant depending on their placement within the measurement value.

5. **Significant Figures and Mathematical Operations** Calculations should never increase (or decrease) the degree of uncertainty in measurements. In multiplication and division, the number of significant figures in the answer is the same as that in the input number containing the fewest significant figures. In addition and subtraction of a series of measurements, the result can be no more certain than the least-certain measurement in the series.

6. **Scientific Notation** Scientific notation is a system for writing decimal numbers in a more compact form that greatly simplifies the mathematical operations of multiplication and division. In this system, numbers are expressed in the form $A \times 10^n$, where A is a number with a single nonzero digit to the left of the decimal point and n is a whole number.

Key Terms

The new terms defined in this chapter are

accuracy *Sec. 2.3*
exact number *Sec. 2.2*
exponent *Sec. 2.7*

inexact number *Sec. 2.2*
measurement *Sec. 2.1*
order of magnitude *Sec. 2.7*

precision *Sec. 2.3*
random error *Sec. 2.3*
rounding off *Sec. 2.6*

scientific notation *Sec. 2.7*
significant figures *Sec. 2.5*
systematic error *Sec. 2.3*

Practice Problems

EXACT AND INEXACT NUMBERS (SEC. 2.2)

2.1 Indicate whether the number in each of the following statements is an *exact* or an *inexact* number.
   a. A classroom contains 24 chairs.
   b. There are 60 seconds in a minute.
   c. A bag of cherries weighs 4.1 pounds.
   d. A newspaper article contains 421 words.

2.2 Indicate whether the number in each of the following statements is an *exact* or an *inexact* number.
   a. A classroom contains 44 students.
   b. The car is traveling at a speed of 63 miles per hour.
   c. The temperature on the beach is 93°F.
   d. The child is 6 years old.

2.3 Classify each of the following as an *exact* or an *inexact* number.
   a. 7 railroad cars   **b.** 14 gallons of gasoline
   c. 547 marbles   **d.** $23.54

2.4 Classify each of the following as an *exact* or an *inexact* number.
   a. 25 pounds of sugar   **b.** 12 dozen apples
   c. $23.00   **d.** 25 watermelon seeds

2.5 A person is told that there are 60 seconds in a minute and also that a garden hose is 60 feet long. What is the fundamental difference between the value of 60 in these two pieces of information?

2.6 A person is told that there are 12 students in a classroom and also that a train will pass by in 12 minutes. What is the fundamental difference between the value of 12 in these two pieces of information?

ACCURACY AND PRECISION (SEC. 2.3)

2.7 With a high-grade volumetric measuring device, the volume of a liquid sample is determined to be 6.321 L (liters). Three students are asked to determine the volume of the same liquid sample using a lower-grade measuring device. How do you evaluate the following work of the three students with regard to precision and accuracy?

	Students		
Trials	**A**	**B**	**C**
1	6.35 L	6.31 L	6.36 L
2	6.31 L	6.32 L	6.36 L
3	6.38 L	6.33 L	6.35 L
4	6.32 L	6.32 L	6.36 L

2.8   With a high-grade measuring device, the length of an object is determined to be 13.452 mm (millimeters). Three students are asked to determine the length of the same object using a lower-grade measuring device. How do you evaluate the following work of the three students with regard to precision and accuracy?

	Students		
Trials	**A**	**B**	**C**
1	13.6 mm	13.4 mm	13.9 mm
2	13.9 mm	13.5 mm	13.9 mm
3	13.3 mm	13.5 mm	13.3 mm
4	13.6 mm	13.4 mm	14.3 mm

UNCERTAINTY IN MEASUREMENTS (SEC. 2.4)

2.9   Consider the following rulers as instruments for the measurement of length in centimeters (cm)?

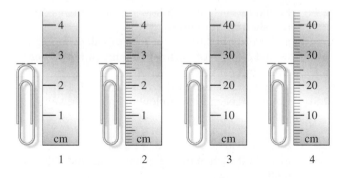

What would the uncertainty be in measurements made using the following?
a. Ruler 1   **b.** Ruler 2

2.10   Using the rulers given in Problem 2.9, what would the uncertainty be in measurements made using the following?
a. Ruler 3   **b.** Ruler 4

2.11   Using the rulers given in Problem 2.9, what is the length of the paper clip shown by the side of the following?
a. Ruler 3   **b.** Ruler 4

2.12   Using the rulers given in Problem 2.9, what is the length of the paper clip shown by the side of the following?
a. Ruler 1   **b.** Ruler 2

2.13   With which of the rulers in Problem 2.9 was each of the following measurements made, assuming that you cannot use a ruler multiple times in making a measurement? (It is possible that there may be more than one correct answer.)
a. 21.2 cm   **b.** 3.2 cm   **c.** 3.65 cm   **d.** 27 cm

2.14   With which of the rulers in Problem 2.9 was each of the following measurements made, assuming that you cannot use a ruler multiple times in making a measurement? (It is possible that there may be more than one correct answer.)
a. 2.3 cm   **b.** 2.33 cm   **c.** 17 cm   **d.** 4 cm

2.15   What is the magnitude of the uncertainty associated with each of the following measured values?
a. 3.147   **b.** 23,275   **c.** 0.1111   **d.** 27.9

2.16   What is the magnitude of the uncertainty associated with each of the following measured values?
a. 27.72   **b.** 363   **c.** 0.277   **d.** 231.2

2.17   What is the difference in meaning between the times 3.3 seconds and 3.30 seconds?

2.18   What is the difference in meaning between the lengths 0.54 inch and 0.540 inch?

2.19   The number of people present at a parade was estimated by police to be 50,000. How many people were present at the parade if you assume that this estimate has each of the following uncertainties?
a. 10,000   **b.** 1000   **c.** 100   **d.** 10

2.20   The number of people present at a protest rally was estimated by police to be 18,000. How many people were present at the rally if you assume that this estimate has each of the following uncertainties?
a. 10,000   **b.** 1000   **c.** 100   **d.** 10

SIGNIFICANT FIGURES (SEC. 2.5)

2.21   Determine the number of significant figures in each of the following measured values.
a. 0.43571   **b.** 0.00621   **c.** 0.505023   **d.** 0.0000003

2.22   Determine the number of significant figures in each of the following measured values.
a. 0.54382   **b.** 0.23100   **c.** 0.00004   **d.** 0.1302120

2.23   Determine the number of significant figures in each of the following measured values.
a. 3300   **b.** 3300.00   **c.** 635,730   **d.** 635.730

2.24   Determine the number of significant figures in each of the following measured values.
a. 4700   **b.** 4700.0   **c.** 23,700   **d.** 23.700

2.25   Determine the number of significant figures in each of the following measured values.
a. 3010.20   **b.** 0.00300300
c. 40,400   **d.** 33,000,000

2.26 Determine the number of significant figures in each of the following measured values.
    a. 250.00   **b.** 4004
    c. 0.0505050   **d.** 2,375,000

2.27 For each of the numbers in Problem 2.25, tell how many of the zeros present are
    a. confined zeros.
    b. leading zeros.
    c. trailing zeros that are significant.
    d. trailing zeros that are not significant.

2.28 For each of the numbers in Problem 2.26, tell how many of the zeros present are
    a. confined zeros.
    b. leading zeros.
    c. trailing zeros that are significant.
    d. trailing zeros that are not significant.

2.29 Identify the *estimated digit* in each of the measured values in Problem 2.25.

2.30 Identify the *estimated digit* in each of the measured values in Problem 2.26.

2.31 What is the magnitude of the uncertainty (± 10, ± 0.1, etc.) associated with each of the measured values in Problem 2.25?

2.32 What is the magnitude of the uncertainty (± 10, ± 0.1, etc.) associated with each of the measured values in Problem 2.26?

2.33 In the following pairs of numbers, tell whether both members of the pair contain the same number of significant figures.
    a. 11.01 and 11.00   **b.** 2002 and 2020
    c. 0.05700 and 0.05070   **d.** 0.000066 and 660,000

2.34 In the following pairs of numbers, tell whether both members of the pair contain the same number of significant figures.
    a. 2305 and 2350   **b.** 0.6600 and 0.0066
    c. 23,000 and 23,001   **d.** 936,000 and 0.000936

2.35 In the pairs of numbers of Problem 2.33, tell whether both members of the pair have the same uncertainty.

2.36 In the pairs of numbers of Problem 2.34, tell whether both members of the pair have the same uncertainty.

2.37 Using standard arithmetic notation and overbars (if needed), write the number twenty-three thousand in a manner such that it has the following numbers of significant figures.
    a. two   **b.** four   **c.** six   **d.** eight

2.38 Using standard arithmetic notation and overbars (if needed), write the number six hundred thousand in a manner such that it has the following numbers of significant figures.
    a. one   **b.** three   **c.** five   **d.** seven

2.39 Determine the number of significant figures in each of the following measured values.
    a. 2600 ± 10 **b.** 1.375 ± 0.001
    c. 42 ± 1 **d.** $73,000 \pm 1$

2.40 Determine the number of significant figures in each of the following measured values.
    a. 700 ± 100 **b.** 700 ± 10
    c. 43.57 ± 0.01 **d.** $64,000 \pm 1$

2.41 The uncertainty associated with a particular electronic balance is ± 0.0001 gram. A sample that weighs approximately 1 gram is weighed on the balance. How many significant figures should the value for this measurement contain?

2.42 The uncertainty associated with a particular electronic balance is ± 0.0001 gram. A sample that weighs approximately 100 grams is weighed on the balance. How many significant figures should the value for this measurement contain?

ROUNDING OFF (SEC. 2.6)

2.43 Round off each of the following measured values to four significant figures.
    a. 431.2071 **b.** 31.2071 **c.** 8.2071   **d.** 1.02071

2.44 Round off each of the following measured values to four significant figures.
    a. 233.3723 **b.** 33.3723 **c.** 6.3723   **d.** 1.03723

2.45 Round off each of the following measured values to three significant figures.
    a. 42.55   **b.** 42.65   **c.** 42.75   **d.** 42.85

2.46 Round off each of the following measured values to three significant figures.
    a. 67.25   **b.** 67.35   **c.** 67.45   **d.** 67.55

2.47 Round off each of the following measured values to three significant figures.
    a. 42,303   **b.** 42,360   **c.** 42,549   **d.** 42,601

2.48 Round off each of the following measured values to three significant figures.
    a. 67,222   **b.** 67,666   **c.** 67,495   **d.** 67,501

2.49 Round off each of the following measured values to two significant figures.
    a. 0.000333 **b.** 0.01234 **c.** 0.200007 **d.** 0.356303

2.50 Round off each of the following measured values to two significant figures.
    a. 0.000567 **b.** 0.07546 **c.** 0.400006 **d.** 0.454699

2.51 Round off each of the following measured values to tenths.
    a. 42.3337   **b.** 42.003
    c. 42.570056   **d.** 42.2822

2.52 Round off each of the following measured values to tenths.
    a. 55.5555   **b.** 55.007
    c. 55.073039   **d.** 55.9998

2.53 Round off each of the following numbers to the number of significant figures indicated in parentheses.
    a. 0.350763 (three)   **b.** 653.899 (four)
    c. 22.55555 (five)   **d.** 0.277654 (four)

2.54 Round off each of the following numbers to the number of significant figures indicated in parentheses.
    a. 3883 (two)   **b.** 0.00003011 (two)
    c. 4.4050 (three)   **d.** 2.1000 (three)

2.55 Round off the number 30,427.29 to the indicated number of significant figures.
    a. six   **b.** five   **c.** four   **d.** two

2.56 Round off the number 50,125.09 to the indicated number of significant figures.
    a. six   **b.** five   **c.** four   **d.** two

2.57 Using proper rounding techniques, decrease by two the number of significant figures in each of the following numbers.
    a. 0.03455   **b.** 2.5003
    c. 1,456,000   **d.** 100.0

2.58 Using proper rounding techniques, decrease by two the number of significant figures in each of the following numbers.
    a. 0.50505   **b.** 2,000,567
    c. 2.335   **d.** 1234.5

2.59 Rewrite each of the following numbers so that it contains two significant figures.
    a. 0.123   **b.** 123,000   **c.** 12.3   **d.** 0.000123

2.60 Rewrite each of the following numbers so that it contains two significant figures.
    a. 21.000   **b.** 210,000   **c.** 0.0210   **d.** 2.100

SIGNIFICANT FIGURES IN MULTIPLICATION AND DIVISION (SEC. 2.6)

2.61 Without actually solving the problems, indicate the number of significant figures that should be present in the answers to the following multiplications and divisions. Assume that all numbers are measured quantities.
    a. $4.5 \times 4.05 \times 4.50$ **b.** $0.100 \times 0.001 \times 0.010$
    c. $\dfrac{655,000}{6.5500}$ **d.** $\dfrac{6.00}{33.000}$

2.62 Without actually solving the problems, indicate the number of significant figures that should be present in the answers to the following multiplications and divisions. Assume that all numbers are measured quantities.
    a. $3.33 \times 3.03 \times 0.0333$
    b. $300,003 \times 20,200 \times 1.33333$
    c. $\dfrac{333,000}{3.33000}$ **d.** $\dfrac{0.0666}{1.3457}$

2.63 How many significant figures must the number Q possess, in each case, to make the following mathematical equations valid from a significant-figure standpoint?
    a. $7.312 \times Q = 4.13$ **b.** $7.312 \times Q = 0.0022$
    c. $7.312 \times Q = 20.44$ **d.** $7.312 \times Q = 0.1100$

2.64 How many significant figures must the number Q possess, in each case, to make the following mathematical equations valid from a significant-figure standpoint?
    a. $94,461 \times Q = 33,003$
    b. $94,461 \times Q = 1.03$
    c. $94,461 \times Q = 0.6200$
    d. $94,461 \times Q = 233,620,000$

2.65 Carry out the following multiplications and divisions, expressing your answers to the correct number of significant figures. Assume that all numbers are measured quantities.
    a. 4.2337×0.00706 **b.** 3700×37.00
    c. $\dfrac{5671}{4.44}$ **d.** $\dfrac{5.01}{5.07}$

2.66 Carry out the following multiplications and divisions, expressing your answers to the correct number of significant figures. Assume that all numbers are measured quantities.
    a. 350.00×0.00072 **b.** $620,000 \times 620.000$
    c. $\dfrac{3554}{2.22}$ **d.** $\dfrac{0.000623}{0.000632}$

2.67 Carry out the following mathematical operations, expressing your answers to the correct number of significant figures. Assume that all numbers are measured quantities.
    a. $\dfrac{4.5 \times 6.3}{7.22}$ **b.** $\dfrac{5.567 \times 3.0001}{3.45}$
    c. $\dfrac{37 \times 43}{4.2 \times 6.0}$ **d.** $\dfrac{112 \times 20}{30 \times 63}$

2.68 Carry out the following mathematical operations, expressing your answers to the correct number of significant figures. Assume that all numbers are measured quantities.
    a. $\dfrac{2.322 \times 4.00}{3.200 \times 6.73}$ **b.** $\dfrac{7.403}{3.220 \times 5.000}$
    c. $\dfrac{11.2 \times 11.2}{3.3 \times 6.5}$ **d.** $\dfrac{5600 \times 300}{22 \times 97.1}$

SIGNIFICANT FIGURES IN ADDITION AND SUBTRACTION (SEC. 2.6)

2.69 Without actually solving the problems, indicate the uncertainty (tenths, hundredths, etc.) that should be present in the answers to the following additions and subtractions. Assume that all numbers are measured quantities.
    a. $12.1 + 23.1 + 127.01$ **b.** $43.65 - 23.7$
    c. $1237.6 + 23 + 0.12$ **d.** $4650 + 25 + 200$

2.70 Without actually solving the problems, indicate the uncertainty (tenths, hundredths, etc.) that should be present in the answers to the following additions and subtractions. Assume that all numbers are measured quantities.
    a. $0.06 + 1.32 + 7.901$ **b.** $4.72 - 3.908$
    c. $23.6 + 33 + 17.21$ **d.** $46,230 + 325 + 45$

2.71 Perform the following additions or subtractions. Report your results to the proper number of significant figures. Assume that all numbers are measured quantities.
    a. $12 + 23 + 127$ **b.** $3.111 + 3.11 + 3.1$
    c. $1237.6 + 23 + 0.12$ **d.** $43.65 - 23.7$

2.72 Perform the following additions or subtractions. Report your results to the proper number of significant figures. Assume that all numbers are measured quantities.
  a. 237 + 37 + 7   **b.** 4.000 + 4.002 + 4.20
  c. 235.45 + 37 + 36.4   **d.** 4.111 − 3.07

2.73 Perform the following additions or subtractions. Report your results to the proper number of significant figures. Assume that all numbers are measured quantities.
  a. 999.0 + 1.7 − 43.7   **b.** 345 − 6.7 + 4.33
  c. 1200 + 43 + 7   **d.** 132 − 0.0073

2.74 Perform the following additions or subtractions. Report your results to the proper number of significant figures. Assume that all numbers are measured quantities.
  a. 1237.6 + 1237.4   **b.** 1237.6 − 1237.4
  c. 23,000 + 457 + 23   **d.** 3.12 − 0.00007

CALCULATIONS INVOLVING BOTH SIGNIFICANT FIGURE RULES (SEC. 2.6)

2.75 Perform the following mathematical operations, expressing your answers to the correct number of significant figures. Assume all numbers are measured quantities.
  a. 0.400 × (2.33 + 4.5)
  b. 0.300 × (3.73 + 3.0 + 2.777)
  c. 0.300 × (3.73 − 3.0)
  d. 0.650 × 7.23 × (6.772 + 3.10)

2.76 Perform the following mathematical operations, expressing your answers to the correct number of significant figures. Assume all numbers are measured quantities.
  a. 1.20 × (3.45 + 4.5)
  b. 0.40 × (3.03 + 3.50 + 7.123)
  c. 2.10 × (5.73 − 2.00)
  d. 0.500 × 0.1723 × (0.6772 + 0.31)

SIGNIFICANT FIGURES AND EXACT NUMBERS (SEC. 2.6)

2.77 Each of the following calculations involves the numbers 4.3, 230, 20, and 13.00. The numbers 4.3 and 13.00 are measurements; 230 and 20 are exact numbers. Express each answer to the proper number of significant figures.
  a. 4.3 + 230 + 20 + 13.00
  b. 4.3 × 230 × 20 × 13.00
  c. 4.3 + 230 − 20 − 13.00
  d. $\dfrac{4.3 \times 230}{20 \times 13.00}$

2.78 Each of the following calculations involves the numbers 200, 17, 24, and 40. The numbers 200 and 17 are exact; 24 and 40 are measurements. Express each answer to the proper number of significant figures.
  a. 200 + 17 + 24 + 40   **b.** 200 × 17 × 24 × 40
  c. 200 − 17 − 24 − 40   **d.** $\dfrac{200 \times 17}{24 \times 40}$

2.79 Based on the concept of multiplication and its associated significant figure rule, carry out the following mathematical operations.
  a. 24 (measured) × 7 (exact)
  b. 13.7 (measured) × 9 (exact)
  c. 43.73 (measured) × 3 (exact)
  d. 0.037 (measured) × 5 (exact)

2.80 Based on the concept of multiplication and its associated significant figure rule, carry out the following mathematical operations.
  a. 33 (measured) × 5 (exact)
  b. 23.7 (measured) × 5 (exact)
  c. 53.73 (measured) × 3 (exact)
  d. 0.0037 (measured) × 4 (exact)

2.81 Based on the concept of addition and its associated significant figure rule, carry out each of the mathematical operations given in Problem 2.79.

2.82 Based on the concept of addition and its associated significant figure rule, carry out each of the mathematical operations given in Problem 2.80.

SCIENTIFIC NOTATION (SEC. 2.7)

2.83 For each of the following numbers, will the exponent be positive, negative, or zero when the number is expressed in scientific notation?
  a. 0.0320   **b.** 321.7   **c.** 6.87   **d.** 63,002

2.84 For each of the following numbers, will the exponent be positive, negative, or zero when the number is expressed in scientific notation?
  a. 0.323   **b.** 10.23   **c.** 623,000   **d.** 9.003

2.85 For each of the following numbers, by how many places does the decimal point have to be moved in order to express the number in scientific notation?
  a. 0.0315   **b.** 0.00013   **c.** 3500   **d.** 63,003

2.86 For each of the following numbers, by how many places does the decimal point have to be moved in order to express the number in scientific notation?
  a. 375.3   **b.** 2200   **c.** 0.00112   **d.** 0.00000057

2.87 For each of the following numbers, how many significant figures should be present in the scientific notation form of the number?
  a. 27,520   **b.** 27.520   **c.** 0.001210   **d.** 1.37500

2.88 For each of the following numbers, how many significant figures should be present in the scientific notation form of the number?
  a. 88,200   **b.** 47.200   **c.** 0.00330   **d.** 0.27540

2.89 How many digits will there be in the coefficient when each of the following numbers is expressed in scientific notation?
  a. 23   **b.** 23.000   **c.** 2300   **d.** 2300.000

2.90 How many digits will there be in the coefficient when each of the following numbers is expressed in scientific notation?
  a. 47   **b.** 47.00   **c.** 47,000   **d.** 470.00

2.91 Express the following numbers in scientific notation.
  a. 473.2   **b.** 0.001234   **c.** 231.00   **d.** 231,000,000

2.92 Express the following numbers in scientific notation.
a. 787.6   **b.** 0.01798   **c.** 40.0   **d.** 675,000

2.93 Express each of the following numbers, to three significant figures, in scientific notation.
a. 0.00300300   **b.** 936,020
c. 25.5003   **d.** 450,000,300

2.94 Express each of the following numbers, to three significant figures, in scientific notation.
a. 0.50022247   **b.** 234,300
c. 37.5600   **d.** 23,593,000

2.95 Using scientific notation, express the number sixty-seven thousand to the following number of significant figures.
a. one   **b.** three   **c.** five   **d.** seven

2.96 Using scientific notation, express the number six-hundred-seventy-four thousand to the following number of significant figures.
a. two   **b.** four   **c.** six   **d.** eight

2.97 Express the following numbers in decimal notation.
a. 2.30×10^{-3} **b.** 4.35×10^3
c. 6.6500×10^{-2} **d.** 1.11×10^8

2.98 Express the following numbers in decimal notation.
a. 4.63×10^{-5} **b.** 3.327×10^2
c. 3.47700×10^{-1} **d.** 2.4010×10^7

2.99 Each of the following numbers is expressed in non-standard (incorrect) scientific notation. Convert each number to standard (correct) scientific notation.
a. 342×10^4 **b.** 23.6×10^{-4}
c. 0.0032×10^5 **d.** 0.12×10^{-3}

2.100 Each of the following numbers is expressed in nonstandard (incorrect) scientific notation. Convert each number to standard (correct) scientific notation.
a. 47.23×10^2 **b.** 23.60×10^{-2}
c. 0.100×10^5 **d.** 0.023×10^{-3}

2.101 For each of the following pairs of measured numbers, indicate whether the first member of the pair is larger or smaller than the second member of the pair.
a. 4.02×10^{-2} and 4.02×10^2
b. 2.0×10^2 and 2.0×10^4
c. 2.023×10^{-2} and 2.023×10^{-4}
d. 4.30×10^6 and 4.03×10^6

2.102 For each of the following pairs of measured numbers, indicate whether the first member of the pair is larger or smaller than the second member of the pair.
a. 3.0×10^{-1} and 3.0×10^{-2}
b. 2.0×10^6 and 2.0×10^7
c. 4.11×10^4 and 3.99×10^{-4}
d. 7.25×10^0 and 7.50×10^{-1}

UNCERTAINTY AND SCIENTIFIC NOTATION (SEC. 2.7)

2.103 What is the uncertainty, in terms of a power of ten, associated with each of the following measured values?
a. 3.700×10^4 **b.** 3.700×10^6
c. 3.70×10^5 **d.** 3.7×10^{-2}

2.104 What is the uncertainty, in terms of a power of ten, associated with each of the following measured values?
a. 5.70×10^{-2} **b.** 5.700×10^{-1}
c. 5.7×10^3 **d.** 5.700×10^4

2.105 Rewrite the number 365,000 in scientific notation so that it fits each of the following specifications.
a. has an uncertainty in the thousands place
b. has an uncertainty in the tens place
c. has an uncertainty in the tenths place
d. has an uncertainty in the thousandths place

2.106 Rewrite the number 725,000 in scientific notation so that it fits each of the following specifications.
a. has an uncertainty in the ten thousands place
b. has an uncertainty in the hundreds place
c. has an uncertainty in the tenths place
d. has an uncertainty in the ten-thousandths place

2.107 For each of the following pairs of measured numbers, indicate whether the uncertainty in the first member of the pair is greater than, the same as, or less than the uncertainty in the second member of the pair.
a. 2.50×10^2 and 2.5×10^2
b. 2.50×10^{-4} and 2.5×10^{-5}
c. 2.500×10^{-2} and 2.5×10^{-4}
d. 5.6×10^6 and 5.60×10^5

2.108 For each of the following pairs of measured numbers, indicate whether the uncertainty in the first member of the pair is greater than, the same as, or less than the uncertainty in the second member of the pair.
a. 4.612×10^2 and 4.609×10^2
b. 1.03×10^9 and 1.03×10^{-9}
c. 3.00×10^2 and 3.000×10^1
d. 4.5×10^4 and 4.500×10^6

MULTIPLICATION AND DIVISION IN SCIENTIFIC NOTATION (SEC. 2.8)

2.109 Without using a calculator, carry out the following multiplications of exponential terms.
a. $10^5 \times 10^3$ **b.** $10^{-5} \times 10^{-3}$
c. $10^5 \times 10^{-3}$ **d.** $10^{-5} \times 10^3$

2.110 Without using a calculator, carry out the following multiplications of exponential terms.
a. $10^7 \times 10^4$ **b.** $10^{-7} \times 10^{-4}$
c. $10^7 \times 10^{-4}$ **d.** $10^{-7} \times 10^4$

2.111 Carry out the following multiplications, making sure that your answer is expressed in correct scientific notation form and to the correct number of significant figures.
a. $(1.171 \times 10^6) \times (2.555 \times 10^2)$
b. $(5.37 \times 10^{-3}) \times (1.7 \times 10^5)$
c. $(9.0 \times 10^{-5}) \times (3.000 \times 10^{-5})$
d. $(3.0 \times 10^5) \times (9.000 \times 10^5)$

2.112 Carry out the following multiplications, making sure that your answer is expressed in correct scientific notation form and to the correct number of significant figures.
a. $(2.340 \times 10^{-3}) \times (2.60 \times 10^6)$

b. $(1.110 \times 10^5) \times (3.333 \times 10^{-7})$
c. $(9.8 \times 10^2) \times (7.00 \times 10^2)$
d. $(8.77 \times 10^{-6}) \times (5.030 \times 10^2)$

2.113 Without using a calculator, carry out the following divisions of exponential terms.
a. $\dfrac{10^5}{10^3}$ **b.** $\dfrac{10^5}{10^{-3}}$ **c.** $\dfrac{10^{-5}}{10^3}$ **d.** $\dfrac{10^{-5}}{10^{-3}}$

2.114 Without using a calculator, carry out the following divisions of exponential terms.
a. $\dfrac{10^2}{10^3}$ **b.** $\dfrac{10^2}{10^{-3}}$ **c.** $\dfrac{10^{-2}}{10^3}$ **d.** $\dfrac{10^{-2}}{10^{-3}}$

2.115 Carry out the following divisions, making sure that your answer is expressed in correct scientific notation form and to the correct number of significant figures.
a. $\dfrac{9.51167 \times 10^{-2}}{3.32 \times 10^{-3}}$ **b.** $\dfrac{4.500 \times 10^{10}}{5.0005 \times 10^{-8}}$
c. $\dfrac{3.32 \times 10^{-3}}{9.51167 \times 10^{-2}}$ **d.** $\dfrac{5.0005 \times 10^{-8}}{4.500 \times 10^{10}}$

2.116 Carry out the following divisions, making sure that your answer is expressed in correct scientific notation form and to the correct number of significant figures.
a. $\dfrac{3.5608 \times 10^3}{5.71 \times 10^5}$ **b.** $\dfrac{3.300 \times 10^{-5}}{4.0003 \times 10^2}$
c. $\dfrac{5.71 \times 10^5}{3.5608 \times 10^3}$ **d.** $\dfrac{4.003 \times 10^2}{3.300 \times 10^{-5}}$

2.117 Without using a calculator, carry out the following mathematical operations involving exponential terms.
a. $\dfrac{10^2 \times 10^3}{10^4}$ **b.** $\dfrac{10^{-2} \times 10^{-3}}{10^{-4}}$
c. $\dfrac{10^6}{10^{-5} \times 10^{-9}}$ **d.** $\dfrac{10^{-3} \times 10^2 \times 10^5}{10^{-6} \times 10^8}$

2.118 Without using a calculator, carry out the following mathematical operations involving exponential terms.
a. $\dfrac{10^4 \times 10^5}{10^6 \times 10^3}$ **b.** $\dfrac{10^{-3} \times 10^{-3} \times 10^{-3}}{10^{-6}}$
c. $\dfrac{10^2 \times 10^3 \times 10^4}{10^{-2} \times 10^{-3} \times 10^{-4}}$ **d.** $\dfrac{10^{-6} \times 10^4}{10^3 \times 10^{-5}}$

2.119 Perform the following mathematical operations. Be sure your answer contains the correct number of significant figures and that it is in correct scientific notation form.
a. $\dfrac{(6.0 \times 10^3) \times (5.0 \times 10^3)}{2.0 \times 10^7}$
b. $\dfrac{2.0 \times 10^7}{(6.0 \times 10^3) \times (5.0 \times 10^3)}$
c. $\dfrac{(3.571 \times 10^{-5}) \times (4.5113 \times 10^{-9})}{(5.10 \times 10^{-6}) \times (3.71300 \times 10^{10})}$
d. $\dfrac{(5 \times 10^{10}) \times (6.0 \times 10^7) \times (3.111 \times 10^{-5})}{(3 \times 10^3) \times (4.00 \times 10^{-6})}$

2.120 Perform the following mathematical operations. Be sure your answer contains the correct number of

significant figures and that it is in correct scientific notation form.
a. $\dfrac{(3.00 \times 10^5) \times (6.00 \times 10^3) \times (5.00 \times 10^6)}{2.00 \times 10^7}$
b. $\dfrac{4.1111 \times 10^{-3}}{(3.003 \times 10^{-6}) \times (9.8760 \times 10^{-5})}$
c. $\dfrac{(6 \times 10^5) \times (6 \times 10^{-5})}{(3 \times 10^2) \times (1 \times 10^{-10})}$
d. $\dfrac{(3.00 \times 10^6) \times (2.7 \times 10^3) \times (8.50 \times 10^3)}{(2.22 \times 10^2) \times (8.504 \times 10^6)}$

ADDITION AND SUBTRACTION IN SCIENTIFIC NOTATION (SEC. 2.8)

2.121 Carry out the following additions and subtractions, expressing each answer in correct scientific notation form to the correct number of significant figures.
a. $(3.245 \times 10^3) + (1.17 \times 10^3)$
b. $(9.870 \times 10^{-2}) - (5.7 \times 10^{-3})$
c. $(9.356 \times 10^5) + (3.27 \times 10^4)$
d. $(2.030 \times 10^4) - (1.111 \times 10^3)$

2.122 Carry out the following additions and subtractions, expressing each answer in correct scientific notation form to the correct number of significant figures.
a. $(5.405 \times 10^6) + (3.09 \times 10^5)$
b. $(7.777 \times 10^{-1}) - (5.3 \times 10^{-1})$
c. $(8.219 \times 10^2) - (1.901 \times 10^1)$
d. $(3.45 \times 10^3) + (3.45 \times 10^2)$

2.123 Carry out the following subtractions, expressing each answer in correct scientific notation form to the correct number of significant figures.
a. $(8.313 \times 10^7) - (6.00 \times 10^6)$
b. $(8.313 \times 10^7) - (6.00 \times 10^5)$
c. $(8.313 \times 10^7) - (6.00 \times 10^4)$
d. $(8.313 \times 10^7) - (6.00 \times 10^2)$

2.124 Carry out the following subtractions, expressing each answer in correct scientific notation form to the correct number of significant figures.
a. $(7.431 \times 10^8) - (4.00 \times 10^7)$
b. $(7.431 \times 10^8) - (4.00 \times 10^6)$
c. $(7.431 \times 10^8) - (4.00 \times 10^5)$
d. $(7.431 \times 10^8) - (4.00 \times 10^3)$

ADDITIONAL PROBLEMS

2.125 Indicate whether or not both members of each of the following pairs of measured values, when rounded to hundredths, contain the same number of significant figures.
a. 24.736 and 24.766 **b.** 3.736 and 11.736
c. 0.06736 and 0.0111 **d.** 3.003 and 13.003

2.126 Indicate whether or not both members of each of the following pairs of measured values, when

rounded to hundredths, contain the same number of significant figures.
a. 24.736 and 124.766
b. 13.736 and 11.736
c. 0.006736 and 0.0111
d. 3.003 and 3.0030

2.127 How many significant figures does each of the following numbers have?
a. 7770   b. 7.770
c. 7.770×10^{-3} d. 7.770×10^{3}

2.128 How many significant figures does each of the following numbers have?
a. 0.4500   b. 4.500
c. 4.500×10^{-3} d. 4.500×10^{3}

2.129 In the following pairs of numbers, tell whether both members of the pair contain the same number of significant figures.
a. 11.0 and 11.00
b. 600.0 and 6.000×10^{3}
c. 6300 and 6.3×10^{3}
d. 0.300045 and 0.345000

2.130 In the following pairs of numbers, tell whether both members of the pair contain the same number of significant figures.
a. 0.000066 and 660,000
b. 1.500 and 1.500×10^{2}
c. 54,000 and 5400.0
d. 0.05700 and 0.0570

2.131 Using scientific notation, write the measurement 600 pounds with an uncertainty of
a. ±1 pound.   b. ±0.1 pound.
c. ±10 pounds.   d. ±0.001 pound.

2.132 Using scientific notation, write the measurement 1300 miles with an uncertainty of
a. ±1 mile.   b. ±0.1 mile.
c. ±100 miles.   d. ±0.01 mile.

2.133 In the following pairs of numbers, tell whether both members of the pair have an uncertainty of less than ±0.01.
a. 0.006 and 0.016
b. 2.700 and 2.700×10^{2}
c. 3300.00 and $33\overline{00}$
d. 5.750×10^{1} and 5.750×10^{-1}

2.134 In the following pairs of numbers, tell whether both members of the pair have an uncertainty of less than ±0.01.
a. 3.71 and 3.50
b. 4.500 and 4.500×10^{-2}
c. 270.0 and 0.27000
d. 4.31×10^{-2} and 4.31×10^{2}

2.135 Write each of the following numbers in scientific notation to the number of significant figures indicated in parentheses.
a. 632,567 (4)   b. 0.312546 (3)
c. 63,000,023 (4)   d. 0.500000 (4)

2.136 Write each of the following numbers in scientific notation to the number of significant figures indicated in parentheses.
a. 0.00300300 (3)   b. 936,000 (2)
c. 23.5003 (3)   d. 450,000,001 (6)

2.137 How many significant figures must the number Q possess, in each case, to make the following mathematical equations valid from a significant-figure standpoint?
a. $6.000 \times Q = 4.0$
b. $\dfrac{5.000}{Q} = 3.175$
c. $5.250 + Q = 7.03$
d. $0.7777 - Q = 0.011$

2.138 How many significant figures must the number Q possess, in each case, to make the following mathematical equations valid from a significant-figure standpoint?
a. $450.0 \times Q = 3.00$
b. $\dfrac{Q}{5.1256} = 1.703$
c. $9.13 + Q = 10.2$
d. $Q - 0.111 = 9.25$

2.139 What is wrong with the statement "The number of objects is 12.00 exactly"?

2.140 What is wrong with the statement "Through counting, it was determined that the basket contained 6.70×10^{1} peaches"?

2.141 Each of the following calculations contains the numbers 4.2, 5.30, 11, and 28. The numbers 4.2 and 5.30 are measured quantities, and 11 and 28 are exact numbers. Do each calculation and express each answer to the proper number of significant figures.
a. $(4.2 + 5.30) \times (28 + 11)$
b. $4.2 \times 5.30 \times (28 - 11)$
c. $\dfrac{28 - 4.2}{5.30 \times 11}$
d. $\dfrac{28 - 4.2}{11 - 5.30}$

2.142 Each of the following calculations contains the numbers 3.111, 5.03, 100, and 33. The numbers 3.111 and 5.03 are measured quantities, and 100 and 33 are exact numbers. Do each calculation and express each answer to the proper number of significant figures.
a. $(3.111 + 5.03) \times (100 + 33)$
b. $3.111 \times 5.03 \times (100 + 33)$
c. $\dfrac{3.111 + 5.03}{100 \times 33}$
d. $\dfrac{5.03 - 3.111}{100 + 33}$

2.143 Arrange the following sets of numbers in ascending order (from smallest to largest).
a. 2.07×10^{2}, 243, 1.03×10^{3}
b. 0.0023, 3.04×10^{-2}, 2.11×10^{-3}
c. 23,000, 2.30×10^{5}, 9.67×10^{4}
d. 0.00013, 0.000014, 1.5×10^{-4}

2.144 Arrange the following sets of numbers in ascending order (from smallest to largest).
   a. $350, 3.51 \times 10^2, 3.522 \times 10^1$
   b. $0.000234, 2.341 \times 10^{-3}, 2.3401 \times 10^{-4}$
   c. $965,000,000, 9.76 \times 10^8, 2.03 \times 10^8$
   d. $0.00010, 0.00023, 3.4 \times 10^{-2}$

Multiple-Choice Practice Test

Use this bank of 20 multiple-choice questions as a review of key concepts presented in this chapter. For many of the questions, there may be more than one correct response (choice d) or no correct response (choice e).

2.145 Which of the following statements contains an exact number?
   a. There are 63 apples in the box.
   b. The magazine has 24 pages.
   c. The paper dimensions are 8.5×11 inches.
   d. more than one correct response
   e. no correct response

2.146 A student measures the mass of a metal object three times with the following results: 3.056 g, 3.057 g, and 3.056 g. Based on this information it is correct to say that the student's results are
   a. accurate.
   b. precise.
   c. both accurate and precise.
   d. more than one correct response
   e. no correct response

2.147 Which of the following length measurements is consistent with a ruler that has subdivisions of 0.1 inch?
   a. 21.14 inches
   b. 21.1 inches
   c. 21.00 inches
   d. more than one correct response
   e. no correct response

2.148 In which of the following pairs of numbers does each member of the pair contain the *same number* of significant figures?
   a. 22.30 and 22.3
   b. 314 and 3.14
   c. 63,000 and 6,300
   d. more than one correct response
   e. no correct response

2.149 Which of the following statements concerning the "significance" of zeros in measured numbers is correct?
   a. Leading zeros are always significant.
   b. Confined zeros are always significant.
   c. Trailing zeros are never significant.
   d. more than one correct response
   e. no correct response

2.150 In which of the following measured numbers are *all* of the zeros significant?
   a. 0.000003140
   b. 360,031,010
   c. 3.0101
   d. more than one correct response
   e. no correct response

2.151 In which of the following cases is the given number correctly rounded to three significant figures?
   a. 241,000 becomes 241
   b. 0.3334 becomes 0.334
   c. 42.357 becomes 42.3
   d. more than one correct response
   e. no correct response

2.152 In which of the following pairs of numbers do the two members of the pair have the same uncertainty?
   a. 3.102 and 5.134
   b. 0.20 and 0.200
   c. 301,000 and 300,000
   d. more than one correct answer
   e. no correct response

2.153 The calculator answer obtained from multiplying the measurements 42.44, 3.41, and 7.00 is 1013.0428. This answer
   a. is correct as written.
   b. should be rounded to 1013.
   c. should be rounded to 1010.
   d. more than one correct response
   e. no correct response

2.154 For which of the following calculations does the correct answer contain three significant figures?
   a. $6.00 \times 6.00 \times 6.00$
   b. $6.00 \times 0.600 \times 60.60$
   c. $0.006 \times 0.060 \times 0.600$
   d. more than one correct response
   e. no correct response

2.155 The correct answer obtained from adding the measurements 8.1, 3.14, and 97.143 contains
   a. two significant figures.
   b. three significant figures.
   c. four significant figures.
   d. five significant figures.
   e. no correct response

2.156 For which of the following sets of mathematical operations does the correct answer contain only one significant figure?
   a. $4.0 \times (2.3 + 4.5)$
   b. $3.0 \times (3.7 - 3.4)$
   c. $7.02 \times (0.0001 + 0.011)$
   d. more than one correct response
   e. no correct response

2.157 Given that 7 is an exact number and 0.13 is a measured number, for which of the following problems does the answer contain three significant figures?
   a. 7 multiplied by 0.13
   b. 7 divided by 0.13

c. 7 added to 0.13
d. more than one correct response
e. no correct response

2.158 When the number 3009.1 is converted to scientific notation, the coefficient in the number
  a. is 3.01.
  b. is 10^3.
  c. contains five significant figures.
  d. more than one correct response
  e. no correct response

2.159 In which of the following cases is the given number correctly converted to scientific notation?
  a. 321,000 becomes 3.21×10^4
  b. 0.00300 becomes 3×10^{-3}
  c. 31.040 becomes 3.104×10^1
  d. more than one correct response
  e. no correct response

2.160 Which of the following is a correct conversion from scientific notation to decimal notation?
  a. 3.00×10^{-3} becomes 0.003
  b. 5.00×10^3 becomes 5000
  c. 3.21×10^6 becomes 321,000
  d. more than one correct response
  e. no correct response

2.161 In which of the following pairs of numbers do the two members of the pair have the same uncertainty?
  a. 4.80×10^5 and 4.80×10^6
  b. 3.21×10^{-4} and 3.2×10^{-3}

c. 3×10^{-5} and 3.2211×10^{-1}
d. more than one correct response
e. no correct response

2.162 Which of the following represent the value "twenty-hundredths" written to two significant figures?
  a. 0.020
  b. 0.20
  c. 2.0×10^2
  d. more than one correct response
  e. no correct response

2.163 In which of the following pairs of numbers are the two numbers *totally* equivalent?
  a. 3200 and 3.200×10^3
  b. 0.20 and 2.0×10^{-1}
  c. 4713.0 and 4.7130×10^3
  d. more than one correct response
  e. no correct response

2.164 In which of the following measured values is the uncertainty ± 0.1?
  a. 0.100
  b. 2.1×10^2
  c. 1.0×10^{-2}
  d. more than one correct response
  e. no correct response

Unit Systems and Dimensional Analysis

3.1 THE METRIC SYSTEM OF UNITS

All measurements consist of three parts: (1) a number that tells the amount of the quantity measured, (2) an error that produces an amount of uncertainty, and (3) a unit that tells the nature of the quantity being measured. Chapter 2 dealt with the interpretation and manipulation of the number and error parts of a measurement. We now turn our attention to units.

A unit is a "label" that describes (or identifies) what is being measured (or counted). It can be almost anything: quarts, dimes, frogs, bushels, inches, or pages, for example. Having units for a measurement is an absolute necessity. If you were to ask a neighbor to lend you six sugar, the immediate response would be "How much sugar?" You would then have to indicate that you wished to borrow six pounds, six ounces, six cups, six teaspoons, or whatever amount of sugar you needed.

Two formal systems of units of measurement are used in the United States today. Common measurements in commerce—in supermarkets, lumberyards, gas stations, and so on—are made in the *English system*. The units of this system include the familiar inch, foot, pound, quart, and gallon. A second system, the *metric system*, is used in scientific work. Units in this system include the meter and kilogram. The United States is one of only a very few countries that use different unit systems in commerce and scientific work. On a worldwide basis, almost universally, the metric system is used in both areas.

Metric system use has become more common in the United States in recent years. Metric units now appear on many consumer products. Road signs in some states display distances in both miles and

FIGURE 3.1 Metric measurements are becoming increasingly common in the United States as exemplified by highway mileage signs and consumer products. *(iStockphoto; Eric Schrader/Pearson Science)*

kilometers, and soft drinks are now sold in 1-, 2-, and 3-liter containers (see Fig. 3.1). Automobile engine sizes are often given in liters. Canned and packaged goods (cereals, mixes, fruits, etc.) on grocery store shelves now have the content masses listed in grams as well as ounces or pounds.

The metric system is superior to the English system. Its superiority lies in the area of interrelationships between units of the same type (volume, length, etc.). Metric unit interrelationships are less complicated than English unit interrelationships because the metric system is a decimal unit system. In the metric system, conversion from one unit size to another can be accomplished simply by moving the decimal point to the right or left an appropriate number of places.

SI Units

The metric measurement system, which dates back to the year 1791, was devised under the auspices of the French Academy of Sciences. The use of this French system quickly spread to other nations and it became the preferred system of measurement for scientists throughout the world.

In 1960, again in France, a *revised* metric system, with improved definitions for units, was developed by an international committee. This revised system, called the *SI system of units*, is now the unit system of preference for scientists. The **SI system of units** *is a particular choice of metric units that was adopted in 1960 as a standard for making metric system measurements.* (The acronym *SI* comes from the phrase "French *S*ysteme *I*nternational d'Unites.")

The SI system of units has *seven base units*, each of which relates to a fundamental physical quantity. An **SI base unit** *is one of seven SI units of measurement from which all other SI measurement units can be derived.* Table 3.1 lists these seven base units along with the symbols used to denote them.

Other SI units, called *derived units*, can be obtained from the SI base units. An example of such a derived unit is the SI unit of speed, which is a combination of the SI base units for distance and time.

$$\text{SI unit of speed} = \frac{\text{SI unit of length (m)}}{\text{SI unit of time (s)}}$$

TABLE 3.1 The Seven Base Units in the SI System of Units

Base Quantity	Unit Name	Unit Abbreviation
Length	meter	m
Mass	kilogram	kg
Time	second	s
Temperature	kelvin	K
Amount of substance	mole	mol
Electric current*	ampere	A
Luminous intensity*	candela	cd

This unit will not be further encountered in this textbook.

An **SI-derived unit** *is an SI unit derived by combining two or more SI base units.* Several additional SI-derived units will be encountered in later sections of this chapter.

Metric System Prefixes

In the metric system, basic or derived units for each type of measurement—length, mass, time, etc.—are multiplied by appropriate powers of ten to form smaller or larger units. The names of the larger or smaller units are constructed from the unprefixed unit name by attaching to it a prefix that tells which power of ten is involved. These prefixes are given in Table 3.2, along with their symbols or abbreviations and mathematical meanings. The prefixes in color are those most frequently used.

Students often ask if other metric system prefixes exist besides those in Table 3.2. The answer is yes. The other approved prefixes are

10^{15}	peta (P)
10^{18}	exa (E)
10^{21}	zetta (Z)
10^{24}	yotta (Y)
10^{-15}	femto (f)
10^{-18}	atto (a)
10^{-21}	zepto (z)
10^{-24}	yocto (y)

TABLE 3.2 Metric System Prefixes and Their Mathematical Meanings

Prefix	Symbol	Mathematical Meaning	Pronunciation	Word Meaning
Tera-	T	$1{,}000{,}000{,}000{,}000 = 10^{12}$	TER-uh	trillion
Giga-	G	$1{,}000{,}000{,}000 = 10^{9}$	GIG-uh	billion
Mega-	M	$1{,}000{,}000 = 10^{6}$	MEG-uh	million
Kilo-	k	$1000 = 10^{3}$	KIL-oh	thousand
Hecto-	h	$100 = 10^{2}$	HEK-toe	hundred
Deca-	da	$10 = 10^{1}$	DEK-uh	ten
		$1 = 10^{0}$		one
Deci-	d	$0.1 = 10^{-1}$	DES-ee	tenth
Centi-	c	$0.01 = 10^{-2}$	SEN-tee	hundredth
Milli-	m	$0.001 = 10^{-3}$	MIL-ee	thousandth
Micro-	μ*	$0.000\,001 = 10^{-6}$	MY-kro	millionth
Nano-	n	$0.000\,000\,001 = 10^{-9}$	NAN-oh	billionth
Pico-	p	$0.000\,000\,000\,001 = 10^{-12}$	PEE-koh	trillionth

This is the Greek letter mu *(pronounced "mew," rhymes with "you").*

The use of numerical prefixes should not be new to you. Consider the use of the prefix *tri-* in the following words: triangle, tricycle, trio, trinity, triple. Every one of these words conveys the idea of three of something. We will use the metric system prefixes in the same way.

The meaning of a prefix always remains constant; it is independent of the base unit it modifies. For example, a kilosecond is a thousand seconds; a kilowatt, a thousand watts; and a kilocalorie, a thousand calories. The prefix *kilo-* will always mean a thousand.

EXAMPLE 3.1 **Recognizing the Mathematical Meanings of Metric System Prefixes**

Write the name of the power of 10 associated with the listed metric system prefix or the metric system prefix associated with the listed power of 10.

   a. nano-   **b.** micro-   **c.** deci-   **d.** 10^{-2} **e.** 10^{6} **f.** 10^{9}

SOLUTION

   a. *nano-* denotes 10^{-9} (one-billionth)
   b. *micro-* denotes 10^{-6} (one-millionth)
   c. *deci-* denotes 10^{-1} (one-tenth)
   d. 10^{-2} (one-hundredth) is denoted by the prefix *centi-*
   e. 10^{6} (one million) is denoted by the prefix *mega-*
   f. 10^{9} (one billion) is denoted by the prefix *giga-*

▶ **Practice Exercise 3.1**   Write the name of the power of 10 associated with the listed metric system prefix or the metric system prefix associated with the listed power of 10.

   a. kilo-   **b.** centi-   **c.** giga-   **d.** 10^{-9} **e.** 10^{-3} **f.** 10^{-12}

Answers to all practice exercises in this chapter are found in the back-of-the-book answer section.

3.2 METRIC UNITS OF LENGTH

The **meter** (m) *is the SI system base unit of length.* (*Metre* has been adopted as the preferred international spelling for the unit, but *meter* is the spelling used in the United States and in this book.) A meter is about the same size as the English yard unit; 1 meter equals 1.09 yards (Fig. 3.2a). The prefixes listed in Table 3.2 enable us to derive other units of length from the meter. The kilometer (km) is 1000 times larger than the meter; the centimeter (cm) and millimeter (mm) are, respectively, one-hundredth and one-thousandth of a meter. Most laboratory measurements are made in centimeters or millimeters rather than meters because of the meter's relatively large size.

Length is measured by determining the distance between two points.

3.3 METRIC UNITS OF MASS

The **kilogram** (kg) *is the SI system base unit of mass.* It is the only SI base unit whose name includes a prefix (*kilo-* in this case). In chemistry, the smaller *gram* (g) unit is generally used instead of the larger *kilogram* unit. Note that in forming smaller and larger SI mass units, the prefix is added to the word *gram* to give units such as microgram or centigram. (A second prefix is never added to the SI base unit kilogram; a microkilogram is not an appropriate designation.)

A gram is a very small unit compared with the English pound and ounce (Fig. 3.2b). It takes approximately 28 grams to equal 1 ounce and nearly 454 grams to equal 1 pound. Both grams (g) and milligrams (mg) are commonly used in the laboratory, where the kilogram (kg) is generally too large.

FIGURE 3.2
Comparisons of the metric base units of length (meter), mass (gram), and volume (liter) with common objects. *(Rim light/Getty Images, Inc.— PhotoDisc; Getty Images, Inc.—Photodisc; Eric Schrader/Pearson Science)*

(a) Length
A meter is slightly longer than a yard.
1 meter = 1.09 yards
A baseball bat is about 1 meter long.

(b) Mass
A gram is a small unit compared to a pound.
1 gram = 1/454 pound
Two pennies have a mass of about 5 grams.

(c) Volume
A liter is slightly larger than a quart.
1 liter = 1.06 quarts
Most beverages are now sold by the liter rather than by the quart.

Mass is measured by determining the amount of matter in an object. The determination is made using a balance.

The terms *mass* and *weight* are frequently used interchangeably in measurement discussions. Although in most cases this practice does no harm, technically it is incorrect to interchange the terms. Mass and weight refer to different properties of matter, and their difference in meaning should be understood.

Mass *is a measure of the total quantity of matter in an object.* **Weight** *is a measure of the force exerted on an object by gravitational forces.* The mass of a substance is a constant; the weight of an object is a variable dependent upon the geographical location of that object.

Matter at the equator weighs less than it would at the North Pole because Earth is not a perfect sphere but bulges at the equator. As a result, an object at the equator is farther from the center of Earth. It therefore weighs less because the magnitude of gravitational attraction (the measure of weight) is inversely proportional to the distance between the centers of the attracting objects; that is, the gravitational attraction is larger when the objects' centers are closer together and smaller when the objects' centers are farther apart. Gravitational attraction also depends on the masses of the attracting bodies; the greater the masses, the greater the attraction. For this reason, an object would weigh much less on the moon than on Earth because of the smaller size of the moon and the correspondingly lower gravitational attraction. Quantitatively, a 22.0 pound object weighing 22.0 pounds at Earth's North Pole would weigh 21.9 pounds at Earth's equator and only 3.7 pounds on the moon. In outer space an astronaut may be weightless but never massless. In fact, he or she has the same mass in space as on Earth (see Fig. 3.3).

3.4 METRIC UNITS OF VOLUME

Before specific metric units of volume are considered, a brief review of how the quantities *area* and *volume* are calculated is in order.

Area *is a measure of the extent of a surface.* The units for area are squared units of length. Common area units include square inches (in.^{2}), square feet (ft^2), square meters (m^2), and square centimeters (cm^2). Note that a squared unit is just that unit multiplied by itself. The unit cm^2 means centimeter × centimeter in the same way that 3^2 means 3×3. Figure 3.4 lists the formulas needed for determining the areas of commonly encountered geometrical objects. Notice that in each case the formula involves taking the product of

FIGURE 3.3 An astronaut on a space walk is weightless but not massless. He or she has exactly the same mass as on Earth. *(NASA Headquarters)*

two lengths. Note also that the constant pi (π) is needed in calculating the area of a circle. Its value to three, four, and five significant figures, respectively, is 3.14, 3.142, and 3.1416.

Volume is measured by determining the amount of space occupied by a three-dimensional object.

    Volume *is a measure of the amount of space occupied by an object.* It is a three-dimensional measure and thus involves units that have been cubed: in.^{3}, ft^3, m^3, cm^3, and so on. Again, a cubed unit is just that unit multiplied by itself three times in the same manner that 3^3 is $3 \times 3 \times 3$. Figure 3.5 lists volume formulas for commonly encountered three-dimensional objects. Note particularly the manner in which the volume of a cube is calculated—side × side × side. This is the key to understanding metric units of volume whose definitions are cube related.

    The **cubic meter** (m^3) *is the SI system base unit of volume.* It is a derived SI unit rather than a basic SI unit (see Table 3.1). It is the volume associated with a cube with sides of 1 meter.

$$1 \text{ cubic meter} = \text{volume of a cube with sides of 1 m}$$
$$= 1 \text{ m} \times 1 \text{ m} \times 1 \text{ m}$$
$$= 1 \text{ m}^3$$

FIGURE 3.4 Formulas for calculating the areas of various two-dimensional objects.

Square	Area = side × side $A = s \times s$ $= s^2$	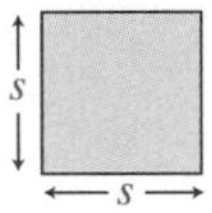
Rectangle	Area = length × width $A = l \times w$	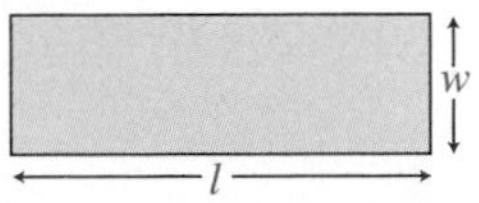
Circle	Area = π × (radius)^2 $A = \pi \times r^2$	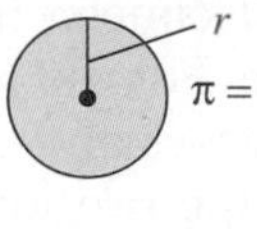 $\pi = 3.1416$
Triangle	Area = $\frac{1}{2}$ × base × height $A = \frac{1}{2} \times b \times h$	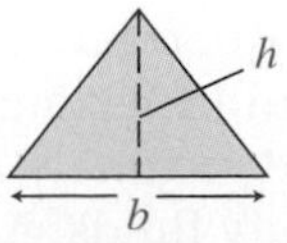

FIGURE 3.5
Formulas for
calculating the volumes
of various three-
dimensional objects.

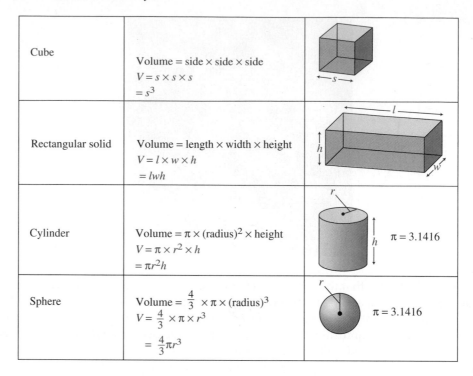

Cube	Volume = side × side × side $V = s \times s \times s$ $= s^3$	
Rectangular solid	Volume = length × width × height $V = l \times w \times h$ $= lwh$	
Cylinder	Volume = π × (radius)^2 × height $V = \pi \times r^2 \times h$ $= \pi r^2 h$	π = 3.1416
Sphere	Volume = $\frac{4}{3}$ × π × (radius)^3 $V = \frac{4}{3} \times \pi \times r^3$ $= \frac{4}{3}\pi r^3$	π = 3.1416

The cubic meter is too large a unit to be routinely used in the chemistry laboratory. The smaller units cubic decimeter (dm^3) and cubic centimeter (cm^3) are frequently used instead of m^3.

More frequently used than the preceding SI units of volume (m^3, dm^3, and cm^3) are the non-SI units *liter (L)* and *milliliter (mL)*. The abbreviation for liter is a capital L rather than a lowercase l because a lowercase l is easily confused with the number 1. (As with meter, we will use the U.S. spelling for liter rather than the international spelling, which is litre.)

Formal definitions for the *liter* and *milliliter* units of volume are based on cubic measurement. A **liter** *is a volume equal to that of a cube whose sides are 1 dm or 10 cm (1 dm = 10 cm) in length.*

$$1 \text{ liter} = 1 \text{ dm}^3 (1 \text{ dm} \times 1 \text{ dm} \times 1 \text{ dm})$$
$$1 \text{ liter} = 1000 \text{ cm}^3 (10 \text{ cm} \times 10 \text{ cm} \times 10 \text{ cm})$$

The cubic measurement equivalents for the smaller milliliter volume unit are obtained by dividing both sides of the preceding equations by 1000 since a milliliter is one-thousandth of a liter.

$$1 \text{ milliliter} = 0.001 \text{ dm}^3$$
$$1 \text{ milliliter} = 1 \text{ cm}^3$$

Thus, the liter and cubic decimeter units are the same, and the milliliter and cubic centimeter units are the same. For the latter case, in practice, mL is used when specifying volumes for liquids and gases and cm^3 when specifying volume for solids. Figure 3.6 shows diagrammatically the relationship between the units of liter and milliliter and cubic measurement.

A liter and a quart have approximately the same volume; 1 liter equals 1.06 quarts (Fig. 3.2c). The milliliter and deciliter (dL) are commonly used in the laboratory. Deciliter units are routinely encountered in clinical laboratory reports detailing the composition of body fluids. A deciliter is equal to 100 mL.

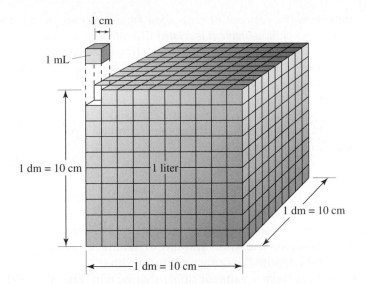

FIGURE 3.6 The relationship between the units liter and milliliter in terms of cubic units.

3.5 UNITS IN MATHEMATICAL OPERATIONS

Just as numbers can have exponents associated with them (3^2, 6^3, 10^2, 10^4, etc.), so also can units have exponents associated with them. In Section 3.4 in the discussion of volume, the cubic volume units were introduced. For example, cm^3 denotes the unit centimeter multiplied by itself three times ($cm \times cm \times cm$). Other commonly encountered units with exponents include

in.^{2}	square inches
ft^3	cubic feet
cm^2	square centimeters
m^2	square meters

To avoid confusion with the word *in*, the abbreviation for inches, in., includes a period. This is the only unit abbreviation in which a period appears.

In mathematical problems the powers on units are manipulated in the same manner as powers of ten (Sec. 2.8). Exponents are added during multiplication and subtracted during division. Just as

$$10^2 \times 10^1 = 10^3$$

so likewise

$$km^2 \times km = km^3$$

Similarly,

$$\frac{10^3}{10^2} = 10^1$$

and

$$\frac{km^3}{km^2} = km$$

3.6 CONVERSION FACTORS

With both the English and metric systems of units in common use in the United States, it is often necessary to change measurements from the English system to the metric system or vice versa. The mathematical tool used to accomplish this task is a general method of problem solving called *dimensional analysis* (Sec. 3.7). Central to the use of dimensional analysis is the concept of *conversion factors*. A **conversion factor** *is a ratio that specifies how one unit of measurement is related to another unit of measurement.*

Conversion factors are derived from equations (equalities) that relate units. Consider the quantities "1 minute" and "60 seconds," both of which describe the same amount of time. We may write an equation describing this fact:

$$1 \text{ min} = 60 \text{ sec}$$

This fixed relationship is the basis for the construction of a pair of conversion factors that relate seconds and minutes:

$$\frac{1 \text{ min}}{60 \text{ sec}} \quad \text{and} \quad \frac{60 \text{ sec}}{1 \text{ min}}$$

Note that conversion factors always come in pairs, one member of the pair being the reciprocal of the other. Also note that the numerator and the denominator of a conversion factor always describe the same amount of whatever we are considering. One minute and 60 seconds denote the same amount of time.

Conversion factors have a value of unity, that is, one. That this is the case can be seen by taking the equation

$$1 \text{ min} = 60 \text{ sec}$$

and dividing each side of it by the quantity "60 seconds," which gives

$$\frac{1 \text{ min}}{60 \text{ sec}} = \frac{60 \text{ sec}}{60 \text{ sec}}$$

Since the numerator and denominator of the fraction on the right are identical, we have

$$\frac{1 \text{ min}}{60 \text{ sec}} = 1$$

In general, we will always be able to construct a set of two conversion factors, each with a value of unity, from any two terms that describe the same "amount" of whatever we are considering. The two conversion factors will always be reciprocals of each other. (Conversion factors can also be constructed from two quantities that are equivalent rather than equal. We will consider this situation in Sec. 3.8.)

Three important categories of conversion factors are (1) English-to-English, (2) metric-to-metric, and (3) metric-to-English or English-to-metric. Further details concerning these types of conversion factors follow.

> A valid conversion factor requires a correct relationship between units. For example, even though the statement 1 minute = 45 seconds is false, you can foolishly make a conversion factor from it. However, any answers you obtain using this incorrect conversion factor will be incorrect. Correct answers require correct conversion factors, which require correct relationships between units.

English-to-English Conversion Factors

English-to-English conversion factors are used to change a measurement in one English system unit into another English unit. Both the numerator and the denominator of such conversion factors involve English units.

Most students are familiar with and have memorized numerous conversion factors within the English system of measurement. Some of these factors, with only one member of a conversion factor pair listed, are

$$\frac{12 \text{ in.}}{1 \text{ ft}} \quad \frac{1 \text{ yd}}{3 \text{ ft}} \quad \frac{1 \text{ gal}}{4 \text{ qt}} \quad \frac{16 \text{ oz}}{1 \text{ lb}}$$

Such conversion factors are exact; that is, they contain an unlimited number of significant figures because the numbers within them arise from definitions (Sec. 2.2).

Table 3.3 lists the unit relationships most commonly needed to derive English-to-English conversion factors.

TABLE 3.3 Relationships Needed to Derive Commonly Used English-to-English Conversion Factors

LENGTH	MASS	VOLUME
12 inches = 1 foot	16 ounces = 1 pound	32 fluid ounces = 1 quart
36 inches = 1 yard	2000 pounds = 1 ton	2 pints = 1 quart
3 feet = 1 yard		4 quarts = 1 gallon
5280 feet = 1 mile		

Metric-to-Metric Conversion Factors

Metric-to-metric conversion factors are used to change a measurement in one metric unit into another metric unit. Such conversion factors are similar to English-to-English conversion factors in that they arise from definitions and are thus exact.

Individual metric-to-metric conversion factors are derived from the meanings of the metric system prefixes (Table 3.2). For example, the set of conversion factors involving kilometer and meter come from the equality

$$1 \text{ kilometer} = 10^3 \text{ meters}$$

and those relating centigram and gram come from the equality

$$1 \text{ centigram} = 10^{-2} \text{ gram}$$

These two pairs of conversion factors, respectively, are

$$\frac{10^3 \text{ m}}{1 \text{ km}} \quad \text{and} \quad \frac{1 \text{ km}}{10^3 \text{ m}}$$
$$\frac{10^{-2} \text{ g}}{1 \text{ cg}} \quad \text{and} \quad \frac{1 \text{ cg}}{10^{-2} \text{ g}}$$

To write metric-to-metric conversion factors, you need to know the meaning of the metric system prefixes in terms of powers of ten (see Table 3.2).

A recommended general rule to follow in writing metric-to-metric conversion factors is that the numerical equivalent of the prefix (the power of ten) is always associated with the base (unprefixed) unit and the number one is always associated with the prefixed unit.

The number 1 always goes with the *prefixed* unit.

$$\frac{1 \text{ km}}{10^3 \text{ m}}$$

The power of 10 always goes with the *unprefixed* unit.

Metric-to-English and English-to-Metric Conversion Factors

Conversion factors that relate metric units to English units must be established by measurement since they involve two different unit systems. Such conversion factors are thus *not exact*. Table 3.4 lists commonly used "metric–English" conversion factors. These few conversion factors are sufficient to solve most of the problems that we will encounter.

Because they are not exact, metric-to-English conversion factors can be specified to differing numbers of significant figures. For example,

$$1.00 \text{ lb} = 454 \text{ g}$$
$$1.000 \text{ lb} = 453.6 \text{ g}$$
$$1.0000 \text{ lb} = 453.59 \text{ g}$$

To help avoid confusion about the accuracy of metric–English conversion factors, scientists have agreed on common experiment-based *definitions* for English units in terms of metric units. They have also agreed that these definitions should be taken to be *exact*. The three *exact* definitions are
1 in. = 2.540005 cm
1 lb = 453.59237 g
1 qt = 0.94633343 L
Conversion factors obtained from these exact definitions are almost always rounded to fewer digits. These rounded conversion factors are not exact.

TABLE 3.4 Conversion Factors That Relate the English and Metric Systems of Measurement to Each Other

Type of Measurement	Most Used Conversion Factor	Additional Conversion Factors		
Length	$\dfrac{2.540 \text{ cm}}{1 \text{ in.}}$	$\dfrac{1.609 \text{ km}}{1 \text{ mi}}$	$\dfrac{39.37 \text{ in.}}{1 \text{ m}}$	$\dfrac{1.094 \text{ yd}}{1 \text{ m}}$
Mass	$\dfrac{453.6 \text{ g}}{1 \text{ lb}}$		$\dfrac{2.205 \text{ g}}{1 \text{ kg}}$	
Volume	$\dfrac{0.9463 \text{ L}}{1 \text{ qt}}$		$\dfrac{3.785 \text{ L}}{1 \text{ gal}}$	

In a problem-solving context, which "version" of a conversion factor is used depends on how many significant figures there are in the other numbers of the problem. *Conversion factors should never limit the number of significant figures in the answer to a problem.* The conversion factors in Table 3.4 are given to four significant figures, which generally is sufficient for the applications we will make of them.

Figure 3.7 summarizes the major concepts about conversion factors considered in this section of the text.

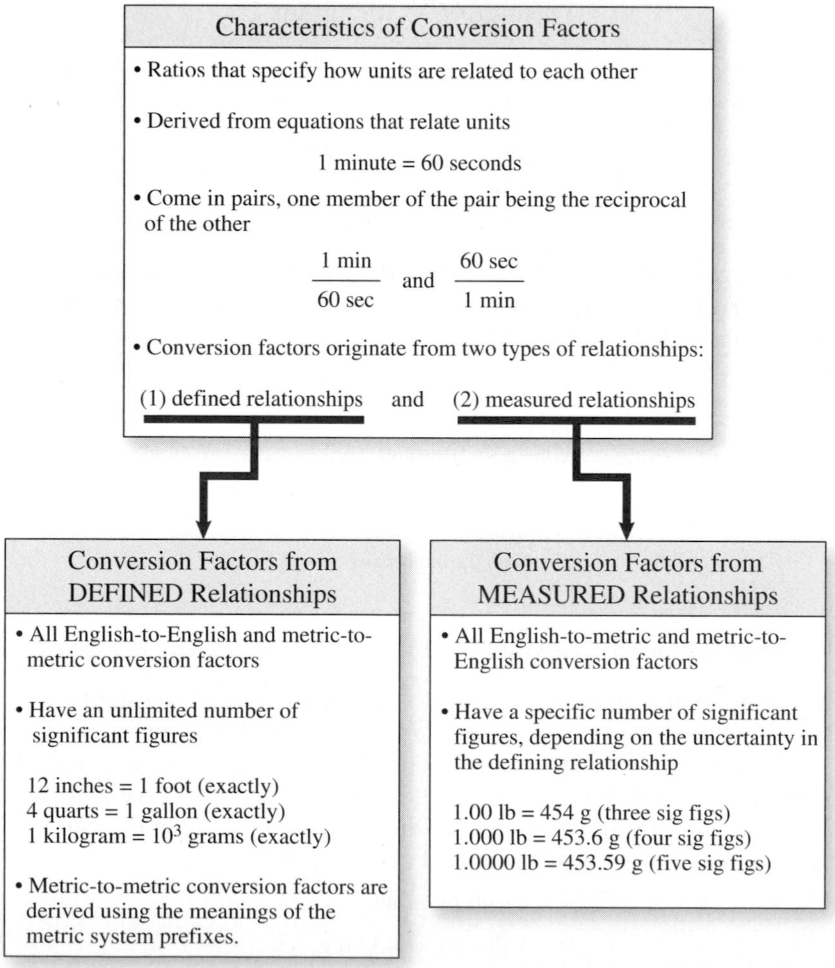

FIGURE 3.7 English-to-English and metric-to-metric conversion factors are based on defined relationships. Metric-to-English and English-to-metric conversion factors are based on measured relationships.

Characteristics of Conversion Factors

• Ratios that specify how units are related to each other

• Derived from equations that relate units

1 minute = 60 seconds

• Come in pairs, one member of the pair being the reciprocal of the other

$$\frac{1 \text{ min}}{60 \text{ sec}} \quad \text{and} \quad \frac{60 \text{ sec}}{1 \text{ min}}$$

• Conversion factors originate from two types of relationships:

(1) defined relationships   and   (2) measured relationships

Conversion Factors from DEFINED Relationships

• All English-to-English and metric-to-metric conversion factors

• Have an unlimited number of significant figures

12 inches = 1 foot (exactly)
4 quarts = 1 gallon (exactly)
1 kilogram = 10^3 grams (exactly)

• Metric-to-metric conversion factors are derived using the meanings of the metric system prefixes.

Conversion Factors from MEASURED Relationships

• All English-to-metric and metric-to-English conversion factors

• Have a specific number of significant figures, depending on the uncertainty in the defining relationship

1.00 lb = 454 g (three sig figs)
1.000 lb = 453.6 g (four sig figs)
1.0000 lb = 453.59 g (five sig figs)

3.7 DIMENSIONAL ANALYSIS

Dimensional analysis *is a general problem-solving method in which the units associated with numbers are used as a guide in setting up calculations.* In this method, units are treated in the same way as numbers; that is, they can be multiplied, divided, or cancelled. For example, just as

$$3 \times 3 = 3^2 \quad (3 \text{ squared})$$

we have

$$km \times km = km^2 \quad (km \text{ squared})$$

Also, just as the twos cancel in the expression

$$\frac{2 \times 3 \times 6}{2 \times 5}$$

the inches cancel in the expression

$$\frac{\cancel{in.} \times cm}{\cancel{in.}}$$

Like units found in the numerator and denominator of a fraction will always cancel, just as like numbers do.

The following steps indicate how to set up a problem using dimensional analysis.

Step 1 *Identify the known or given quantity (both a numerical value and units) and the units of the new quantity to be determined.*

This information will always be found in the statement of the problem. Write an equation with the given quantity on the left and the units of the desired quantity on the right.

Step 2 *Multiply the given quantity by one or more conversion factors in a manner such that the unwanted (original) units are cancelled out, leaving only the new desired unit.*

The general format for the multiplication is

information given × conversion factor(s) = information sought

The number of conversion factors used depends on the individual problem. Except in the simplest of problems, it is a good idea to predetermine formally the sequence of unit changes to be used. This sequence will be called the unit "pathway."

Step 3 *Perform the mathematical operations indicated by the conversion factor setup.*

In performing the calculation, you need to double check that all units except the desired set have cancelled out. You also need to check the numerical answer to see that it contains the *proper number of significant figures.*

Significant figures are not something we learn about in Chapter 2 and then forget about for the remainder of the course. Significant figures will be part of every calculation done in this course.

Now let us work a number of sample problems using dimensional analysis and the steps just outlined.

Metric-to-Metric Conversion Factor Use

Our first two examples involve only metric system units and thus will involve only metric-to-metric conversion factors.

EXAMPLE 3.2 | One-Step Metric-to-Metric Conversion Factor Problem

A vitamin-C tablet is found to contain 0.500 g of vitamin C. How many milligrams of vitamin C does this tablet contain?

SOLUTION

Step 1 The given quantity is 0.500 g, the mass of vitamin C in one tablet. The unit of the desired quantity is milligrams.

$$0.500 \text{ g} = ? \text{ mg}$$

Step 2 Only one conversion factor will be needed to convert from grams to milligrams—one that relates grams to milligrams. Two forms of this factor exist.

$$\frac{1 \text{ mg}}{10^{-3} \text{ g}} \quad \text{and} \quad \frac{10^{-3} \text{ g}}{1 \text{ mg}}$$

The first factor is used because it allows for the cancellation of the gram units, leaving us with milligrams as the new units.

$$0.500 \cancel{\text{ g}} \times \frac{1 \text{ mg}}{10^{-3} \cancel{\text{ g}}}$$

For cancellation, a unit must appear in both the numerator and the denominator. Because the given quantity (0.500 g) has grams in the numerator, the conversion factor used must be the one with grams in the denominator.

If the other conversion factor had been used, we would have

$$0.500 \text{ g} \times \frac{10^{-3} \text{ g}}{1 \text{ mg}}$$

No unit cancellation is possible in this setup. Multiplication gives g^2/mg as the final units, which is certainly not what we want. In all cases, only one of the two conversion factors of a reciprocal pair will correctly fit into a dimensional analysis setup.

Step 3 Step 2 takes care of the units. All that is left is to combine numerical terms to get the final answer; we still have to do the arithmetic. Collecting the numerical terms gives

$$\frac{0.500 \times 1}{10^{-3}} \text{ mg} = 500 \text{ mg} \qquad \text{(calculator answer)}$$

$$= 5.00 \times 10^2 \text{ mg} \quad \textbf{(correct answer)}$$

Since the conversion factor used in this problem is derived from a definition, it contains an unlimited number of significant figures and will not limit in any way the allowable number of significant figures in the answer. Therefore, the answer should have three significant figures, the same number as in the given quantity.

Answer Double Check:

Is the magnitude of the answer reasonable? Yes. In going from a large unit (g) to a smaller unit (mg), the numerical value should increase, which it has (0.500 to 500); there should be many "smalls" in a "large." A gram is equivalent to 1000 milligrams. Therefore, one-half gram is equivalent to 500 milligrams, which is our answer.

▶ **Practice Exercise 3.2** Capillaries, the microscopic vessels that carry blood from small arteries to small veins, are on the average only 1 mm long. What is the average length of a capillary in meters?

EXAMPLE 3.3 **Two-Step Metric-to-Metric Conversion Factor Problem**

The *ozone layer* is a region in the upper atmosphere, at altitudes between 25 and 35 km, where the concentration of ozone is several times higher than at ground level. Express the altitude 35 km (the upper limit of the ozone layer) in centimeter units.

SOLUTION

Step 1 The given quantity is 35 km, and the units of the desired quantity are centimeters.

$$35 \text{ km} = ? \text{ cm}$$

Step 2 In dealing with metric–metric unit changes where both the original and desired units carry prefixes (which is the case in this problem), it is recommended that you always channel units through the base unit (unprefixed unit). If you do that, you will not need to deal with any conversion factors other than those resulting from prefix definitions. Following this recommendation, the unit pathway for this problem is

$$\begin{array}{ccccc} \text{km} & \rightarrow & \text{m} & \rightarrow & \text{cm} \\ \text{prefixed} & & \text{base} & & \text{prefixed} \\ \text{unit} & & \text{unit} & & \text{unit} \end{array}$$

In the setup for this problem, we will need two conversion factors, one for the kilometer-to-meter change and one for the meter-to-centimeter change.

$$35 \text{ km} \times \frac{10^3 \text{ m}}{1 \text{ km}} \times \frac{1 \text{ cm}}{10^{-2} \text{ m}}$$

This conversion factor converts km to m.

This conversion factor converts m to cm.

The units cancel except for the desired centimeters.

Step 3 Carrying out the indicated numerical calculation gives

$$\frac{35 \times 10^3 \times 1}{1 \times 10^{-2}} \text{ cm} = 3.5 \times 10^6 \text{ cm} \quad \text{(calculator and } \textbf{correct answer})$$

Numbers from first conversion factor

Numbers from second conversion factor

The correct answer is the same as the calculator answer for this problem. The given quantity has two significant figures, and both conversion factors are exact. Thus, the correct answer should contain two significant figures.

Answer Double Check:

Is the magnitude of the answer reasonable? Yes. The answer is a large number (10^6), which it should be. There are many small units (cm) in a large unit (km).

▶ **Practice Exercise 3.3** A package of instant chocolate pudding contains 2750 mg of sodium. Express this amount of sodium in kilogram units.

The next example illustrates the use of cubic conversion factors in a dimensional analysis setting. The context is a volume measurement (Sec. 3.4) specified using cubic dimensions.

EXAMPLE 3.4 **Metric-to-Metric Conversion Factor Problem Involving Cubic Dimensions**

The volume of air in a room is determined to be 1.20×10^8 dm^3. What is this volume in cubic centimeters?

SOLUTION

Step 1 The given quantity is 1.20×10^8 dm^3, and the units of the desired quantity are cubic centimeters.

$$1.20 \times 10^8 \text{ dm}^3 = ? \text{ cm}^3$$

Step 2 The unit pathway will be

$$\text{dm}^3 \quad \rightarrow \quad \text{m}^3 \quad \rightarrow \quad \text{cm}^3$$
$$\text{prefixed} \qquad \text{base} \qquad \text{prefixed}$$
$$\text{unit} \qquad \text{unit} \qquad \text{unit}$$

As noted in Example 3.3, it is always recommended that in going from one prefixed metric unit to another prefixed metric unit, a pathway be used that goes through the base (unprefixed) unit.

The first conversion factor that is needed (dm^3 to m^3) is obtained by cubing the conversion factor, based on the relationship 10^{-1} m = 1 dm, that is,

$$\left(\frac{10^{-1} \text{ m}}{1 \text{ dm}}\right)^3$$

In a similar manner, the second conversion factor involves the relationship

$$\left(\frac{1 \text{ cm}}{10^{-2} \text{ m}}\right)^3$$

The dimensional analysis setup for the problem is

$$1.20 \times 10^8 \text{ dm}^3 \times \left(\frac{10^{-1} \text{ m}}{1 \text{ dm}}\right)^3 \times \left(\frac{1 \text{ cm}}{10^{-2} \text{ m}}\right)^3$$

Note that for cubed conversion factors, the exponent of 3 outside the parentheses affects everything within those parentheses, that is, both the number and the unit.

$$\left(\frac{10^{-1} \text{ m}}{1 \text{ dm}}\right)^3 = \frac{(10^{-1})^3 \text{ m}^3}{(1)^3 \text{ dm}^3} = \frac{10^{-3} \text{ m}^3}{1 \text{ dm}^3}$$

$$\left(\frac{1 \text{ cm}}{10^{-2} \text{ m}}\right)^3 = \frac{(1)^3 \text{ cm}^3}{(10^{-2})^3 \text{ m}^3} = \frac{1 \text{ cm}^3}{10^{-6} \text{ m}^3}$$

Rewriting the dimensional analysis setup using the preceding relationships enables us to easily see the cancellation of units that occurs

$$1.20 \times 10^8 \text{ dm}^3 \times \frac{10^{-3} \text{ m}^3}{1 \text{ dm}^3} \times \frac{1 \text{ cm}^3}{10^{-6} \text{ m}^3}$$

All units cancel except cm^3.

Step 3 Carrying out the indicated numerical calculation gives

$$\frac{1.20 \times 10^8 \times 10^{-3} \times 1}{1 \times 10^{-6}} \text{ cm}^3 = 1.2 \times 10^{11} \text{ cm}^3 \quad \text{(calculator answer)}$$

$$= 1.20 \times 10^{11} \text{ cm}^3 \text{ (\textbf{correct answer})}$$

The original data is given to three significant figures, and the conversion factors are exact (defined). Therefore, the answer can be specified to three significant figures.

Answer Double Check:

The most common mistake made in a problem of this type is forgetting to cube the numerical parts of cubed conversion factors. Such is not the case here. All numbers that should be cubed have been cubed. Note that the given number, 1.20×10^8, was not cubed; it was not part of a cubed conversion factor.

▶ Practice Exercise 3.4   The amount of water in the world's oceans is estimated to be $1.35 \times 10^9 \text{ km}^3$. What is this water volume in cubic centimeters?

As shown in the detailed solutions for Examples 3.2 through 3.4, when using dimensional analysis to set up and solve problems, three questions are always asked:

1. What data are given in the problem?
2. What quantity do we wish to obtain in the problem?
3. What conversion factors are needed to facilitate the transition from the given quantity to the desired one?

English-to-English Conversion Factor Use

Our next worked-out example problem involves the use of English-to-English conversion factors. Such conversion factors, which were previously considered in Section 3.6 (Table 3.3) should pose no problems for you. The example is intended to give you further insight into the dimensional analysis method for setting up the pathway for unit change.

EXAMPLE 3.5   **Multistep English-to-English Conversion Factor Problem**

If a person's stomach produces 87 fl oz of gastric juice in a day, what is the equivalent volume of the gastric juice in gallons?

SOLUTION

Step 1 The given quantity is 87 fl oz, and the unit of the desired quantity is gallons.

$$87 \text{ fl oz} = ? \text{ gal}$$

Step 2 The logical pathway to follow to accomplish the desired change is to convert fluid ounces to quarts (fluid ounces are subdivisions of a quart) and then convert the quarts to gallons (quarts are subdivisions of a gallon).

$$\text{fl oz} \rightarrow \text{qt} \rightarrow \text{gal}$$

Always use logical steps in setting up the pathway for a unit change. It does not have to be done in one big jump. Use smaller steps for which you know the conversion factors. Big steps usually get you involved with unfamiliar conversion factors. Most people do not carry around in their head the number of fluid ounces in a gallon, but they do know that there are 32 fl oz in a quart (see Table 3.3).

The setup for this problem will require two conversion factors: fluid ounces to quarts and quarts to gallons.

$$87 \ \cancel{\text{fl oz}} \times \frac{1 \ \cancel{\text{qt}}}{32 \ \cancel{\text{fl oz}}} \times \frac{1 \ \text{gal}}{4 \ \cancel{\text{qt}}}$$

$$\text{fl oz} \rightarrow \text{qt} \rightarrow \text{gal}$$

The units all cancel except for the desired gallons.

Step 3 Performing the indicated multiplications, we get

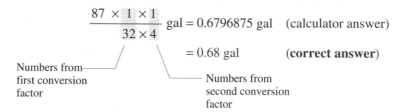

$$\frac{87 \times 1 \times 1}{32 \times 4} \ \text{gal} = 0.6796875 \ \text{gal} \quad \text{(calculator answer)}$$

$$= 0.68 \ \text{gal} \quad \textbf{(correct answer)}$$

Numbers from first conversion factor

Numbers from second conversion factor

The calculator answer contains too many significant figures. The correct answer should contain only two significant figures, the number in the given quantity. Again, both conversion factors involve exact definitions and will not enter into significant figure considerations.

▶ **Practice Exercise 3.5**   It is reported, somewhat humorously, that the distance from Los Angeles to New York City is 177,000,000 inches. What is this distance in terms of miles?

English-to-Metric and Metric-to-English Conversion Factor Use

We will now consider some sample problems that involve both the English and metric systems of units. As mentioned previously (Sec. 3.6), conversion factors between these two unit systems do not arise from definitions, but are determined experimentally. Hence they are not exact. Some of these experimentally determined conversion factors were given in Table 3.4.

Instead of trying to remember all of the conversion factors listed in Table 3.4, memorize only one factor for each type of measurement (mass, volume, length). Knowing only one factor of each type is sufficient information to work metric-to-English or English-to-metric conversion problems. The relationships that are the most useful for you to memorize are the following:

Length: 1 in. = 2.540 cm   (four significant figures)

Mass: 1 lb = 453.6 g   (four significant figures)

Volume: 1 qt = 0.9463 L   (four significant figures)

These three equalities can be considered "bridge relationships," connecting English and metric system measurement units of various types.

These bridge relationships are always applicable in problem solving. For example, no matter what mass units are given or asked for in a problem, we can convert to pounds or grams (bridge units), cross the bridge with our memorized conversion factor, and then convert to the desired final unit. The only advantage that would be gained by memorizing all the factors in Table 3.4 would be that you could work some problems with fewer conversion factors, usually only one factor fewer. The reduction in the number of conversion factors used is usually not worth the added complication of keeping track of the additional conversion factors.

EXAMPLE 3.6   English-to-Metric Conversion Factor Problem

Your feisty neighbor Landy Ippert weighs 244 lb. What is his mass in kilograms?

SOLUTION

Step 1 The given quantity is 244 lb, and the units of the desired quantity are kilograms.

$$244 \text{ lb} = ? \text{ kg}$$

Step 2 This is an English–metric mass-unit conversion problem. The mass-unit "measurement bridge" involves pounds and grams. Since pounds are the given units, we are at the bridge to start with. Pounds are converted to grams (crossing the bridge), and then the grams are converted to kilograms.

$$\text{lb} \rightarrow \text{g} \rightarrow \text{kg}$$

The conversion factor setup for this problem is

$$244 \text{ l\!b} \times \frac{453.6 \text{ g}}{1 \text{ l\!b}} \times \frac{1 \text{ kg}}{10^3 \text{ g}}$$

The units all cancel except for kilograms.

Step 3 Performing the indicated arithmetic gives

$$\frac{244 \times 453.6 \times 1}{1 \times 10^3} \text{ kg} = 110.6784 \text{ kg} \quad \text{(calculator answer)}$$

$$= 111 \text{ kg} \quad \textbf{(correct answer)}$$

The calculator answer must be rounded off to three significant figures, since the given quantity (244 lb) has only three significant figures. The first conversion factor has four significant figures, and the second one is exact.

An alternative pathway for working this problem makes use of the conversion factor in Table 3.4 involving pounds and kilograms. If we use this conversion factor, we can go directly from the given to the desired unit in one step. The setup is

$$244 \text{ l\!b} \times \frac{1 \text{ kg}}{2.205 \text{ l\!b}} = 110.65759 \text{ kg} \quad \text{(calculator answer)}$$

$$= 111 \text{ kg} \quad \textbf{(correct answer)}$$

As is the case here, for most problems there is more than one pathway that can be used to get the answer. When alternative pathways exist, it cannot be said that one way is more correct than another. The important concept is that you select a pathway, choose a correct set of conversion factors consistent with that pathway, and get the answer.

Although both pathways used in this problem are correct, we prefer the first solution because it uses our mass measurement bridge. Use of the measurement bridge system cuts down on the number of conversion factors you are required to know (or look up in a table).

> Conversion factors based on measurement, such as English–metric conversion factors, should never be the limiting quantity in determining the number of significant figures in a final answer. For this reason and also to minimize rounding error, we will always use four-significant-figure English–metric conversion factors (Table 3.4) in problem solving unless the given quantity dictates the use of conversion factors with even more significant figures.

Answer Double Check:

Using the second setup for solving the problem, it is easy to check that the magnitude of the answer is reasonable. Rounding the number 244 to 240 and dividing by 2 gives an estimated answer of 120. The correct answer, 111, is consistent with this estimate.

▶ **Practice Exercise 3.6** A precast concrete block has a mass of 0.250 tons. What is the mass of this concrete block in grams?

EXAMPLE 3.7 **Multistep Metric-to-English Conversion Factor Problem**

The average direct daily use of water in the United States per person is 3.0×10^5 mL. This includes water for drinking, cooking, bathing, dishwashing, laundry, flushing toilets, watering lawns, and so on. What is the volume, in gallons, of this amount of water?

SOLUTION

Step 1 The given quantity is 3.0×10^5 mL, and the units of the desired quantity are gallons.

$$3.0 \times 10^5 \text{ mL} = ? \text{ gal}$$

Step 2 The bridge relationship for volume involves liters and quarts. Thus, we need to convert the milliliters to liters, cross the bridge to quarts, and then convert quarts to gallons.

$$\text{mL} \rightarrow \text{L} \rightarrow \text{qt} \rightarrow \text{gal}$$

Following the pathway, the setup becomes

$$3.0 \times 10^5 \text{ mL} \times \frac{10^{-3} \text{ L}}{1 \text{ mL}} \times \frac{1 \text{ qt}}{0.9463 \text{ L}} \times \frac{1 \text{ gal}}{4 \text{ qt}}$$

Step 3 The numerical calculation involves the following collection of numbers.

$$\frac{3.0 \times 10^5 \times 10^{-3} \times 1 \times 1}{1 \times 0.9463 \times 4} \text{ gal} = 79.256049 \text{ gal} \quad \text{(calculator answer)}$$

$$= 79 \text{ gal} \quad \text{(\textbf{correct answer})}$$

The calculator answer must be rounded to two significant figures because the input number 3.0×10^5 has only two significant figures.

Again, as in Example 3.6, to illustrate that there is more than one way to set up almost any unit conversion problem, let us work this problem using the other volume conversion factor from Table 3.4.

To use the conversion factor based on 3.785 L = 1 gal requires a pathway of

$$\text{mL} \rightarrow \text{L} \rightarrow \text{gal}$$

The conversion factor setup is

$$3.0 \times 10^5 \text{ mL} \times \frac{10^{-3} \text{ L}}{1 \text{ mL}} \times \frac{1 \text{ gal}}{3.785 \text{ L}} = 79.260238 \text{ gal (calculator answer)}$$

$$= 79 \text{ gal (\textbf{correct answer})}$$

Note that the calculator answers obtained from this setup and the previous one for this problem are different—79.256049 and 79.260238—but each gives the same answer after rounding to two significant figures. The slight difference arises from the fact that the English-to-metric conversion factors used are not exact definitions. The two conversion factors used have different rounding-off errors in them.

Note that both methods give the same correct answer after rounding to the correct number of significant figures. Although each of the setups is correct, we still prefer the method of having a specific bridge relationship for volume, mass, and length to use in crossing the metric-to-English bridge.

▶ **Practice Exercise 3.7** The chain length for a 14 karat gold necklace is measured at 24.2 inches. What is this chain length in millimeters?

| **EXAMPLE 3.8** | **Metric-to-English Conversion Factor Problem Involving Cubic Units** |

A large flask contains 375 mL of strained (no seeds) watermelon juice. (Tests are to be run on the juice to see how much vitamin A it contains.) Specify the volume of this amount of juice in

   a. cubic centimeters.   **b.** cubic feet.

SOLUTION

   a. From Section 3.4, we note that

$$1 \text{ mL} = 1 \text{ cm}^3$$

Therefore,

$$375 \text{ mL} \times \frac{1 \text{ cm}^3}{1 \text{ mL}} = 375 \text{ cm}^3 \quad \text{(calculator and } \textbf{correct answer}\text{)}$$

   b. We will use the answer from part (a), 375 cm^3, as the starting point for the part (b) calculation.

Step 1 The given quantity is 375 cm^3, and the units for the desired quantity are ft^3.

$$375 \text{ cm}^3 = ? \text{ ft}^3$$

Step 2 If this problem were a problem involving just length rather than $(\text{length})^3$, the pathway would be

$$\text{cm} \rightarrow \text{in.} \rightarrow \text{ft}$$

The pathway for $(\text{length})^3$ is just an adaptation of this pathway.

$$\text{cm}^3 \rightarrow \text{in.}^3 \rightarrow \text{ft}^3$$

The conversion factors needed for this pathway are simply those for length raised to the third power; that is,

$$375 \text{ cm}^3 \times \left(\frac{1 \text{ in.}}{2.540 \text{ cm}} \right)^3 \times \left(\frac{1 \text{ ft}}{12 \text{ in.}} \right)^3$$

Note that the *entire* conversion factor, in each case, must be cubed, not just the units on the conversion factor. The notation $(\frac{1 \text{ ft}}{12 \text{ in.}})^3$ means that the conversion factor within the parentheses is multiplied by itself three times. Thus,

$$\left(\frac{1 \text{ ft}}{12 \text{ in.}} \right)^3 = \frac{1 \text{ ft}}{12 \text{ in.}} \times \frac{1 \text{ ft}}{12 \text{ in.}} \times \frac{1 \text{ ft}}{12 \text{ in.}}$$

$$= \frac{1 \text{ ft}^3}{1728 \text{ in.}^3}$$

The complete setup for this problem, showing the removal of parentheses and cancellation of units, is

$$375 \text{ cm}^3 \times \left(\frac{1 \text{ in.}}{2.540 \text{ cm}}\right)^3 \times \left(\frac{1 \text{ ft}}{12 \text{ in.}}\right)^3$$

$$= 375 \text{ cm}^3 \times \frac{1^3 \text{ in.}^3}{(2.540)^3 \text{ cm}^3} \times \frac{1^3 \text{ ft}^3}{(12)^3 \text{ in.}^3}$$

Note that the numbers as well as the units must be cubed with removal of parentheses.

Step 3 The numerical setup is

$$\frac{375 \times 1^3 \times 1^3}{(2.540)^3 \times (12)^3} \text{ ft}^3 = 0.013243 \text{ ft}^3 \quad \text{(calculator answer)}$$

$$= 0.0132 \text{ ft}^3 \quad \textbf{(correct answer)}$$

Answer Double Check:

The most common mistake made in solving a problem of this type is forgetting to cube numbers that need to be cubed. In the step 3 numerical setup for this problem, the denominator of the problem, in rewritten form, is

$$\underbrace{2.540 \times 2.540 \times 2.540}_{(2.540)^3} \times \underbrace{12 \times 12 \times 12}_{(12)^3}$$

▶ **Practice Exercise 3.8**   Loss of water through sweating for a human is about 450 mL per day. What is this water volume in

a. cubic centimeters?   **b.** cubic inches?

Units Involving More Than One Type of Measurement

As a final concept in our discussion of unit changes effected using dimensional analysis, "complex" measurement units are considered, that is, units that involve more than one type of measurement. Such "complex" units are frequently encountered in many settings. Two examples of "complex" units are the following:

    miles per hour (length and time): 60 miles/hr

    milligrams per milliliter (mass and volume): 3.0 mg salt/mL solution

Example 3.9 illustrates how such "complex" units are handled in a dimensional analysis unit-change setting.

EXAMPLE 3.9   **Conversion Factor Problem Involving Two Types of Units**

Worldwide emissions of carbon dioxide into the atmosphere are estimated at 2×10^9 tons per year. What is this emission rate in pounds per hour?

SOLUTION

Step 1 The given quantity is 2×10^9 ton/year, and the units of this quantity are to be changed to lb/hr.

$$2 \times 10^9 \frac{\text{ton}}{\text{yr}} = ? \frac{\text{lb}}{\text{hr}}$$

Step 2 This problem is more complex than previous examples because two different types of units are involved: mass and time. However, the approach we use to solve it is similar, other than that we must change two units instead of one. The tons must be converted to pounds and the years to hours.

The logical pathway for the mass change is the direct one-step path

$$\text{ton} \rightarrow \text{lb}$$

For time, it is logical to make the change in two steps.

$$\text{yr} \rightarrow \text{day} \rightarrow \text{hr}$$

It does not matter whether time or mass is handled first in the conversion-factor setup. We will arbitrarily choose to handle time first. The setup becomes

$$2 \times 10^9 \ \frac{\text{ton}}{\cancel{\text{yr}}} \times \frac{1 \cancel{\text{yr}}}{365 \cancel{\text{day}}} \times \frac{1 \cancel{\text{day}}}{24 \text{ hr}}$$

The units at this ──┘   └── The units at this
point are ton/day.   point are ton/hr.

Note that in the first conversion factor, years had to be in the numerator to cancel the years in the denominator of the given quantity.

We are not done yet. The time conversion from years to hours has been accomplished, but nothing has been done with mass. To take care of mass we do not start a new conversion factor setup. Rather, an additional conversion factor is tacked onto those we already have in place.

$$2 \times 10^9 \ \frac{\cancel{\text{ton}}}{\cancel{\text{yr}}} \times \frac{1 \cancel{\text{yr}}}{365 \cancel{\text{day}}} \times \frac{1 \cancel{\text{day}}}{24 \text{ hr}} \times \frac{2000 \text{ lb}}{1 \cancel{\text{ton}}}$$

The tons in the denominator of the last factor cancel the tons in the numerator of the given quantity. With this cancellation, the units now become lb/hr.

Step 3 Collecting the numerical factors and performing the indicated math gives

$$\frac{2 \times 10^9 \times 1 \times 1 \times 2000}{365 \times 24 \times 1} \frac{\text{lb}}{\text{hr}} = 4.56621 \times 10^8 \frac{\text{lb}}{\text{hr}} \qquad \text{(calculator answer)}$$

$$= 5 \times 10^8 \frac{\text{lb}}{\text{hr}} \qquad \textbf{(correct answer)}$$

The given quantity possesses only one significant figure. Rounding the calculator answer to one significant figure produces the correct answer of 5×10^8 lb/hr. Again, none of the conversion factors plays a role in significant-figure considerations since they all originated from definitions.

Answer Double Check:

In solving dimensional analysis problems involving compound units, a common error made is mistakenly writing one or more conversion factors in the inverted form. The double-check for this is to *always* formally cancel units using slash marks. If conversion factors are present in inverted form, the units will not cancel in the proper manner. Unit cancellation shows that there are no inverted conversion factors in the dimensional analysis setup for the preceding problem.

▶ **Practice Exercise 3.9**   Average gasoline mileage for a particular sports utility vehicle is 13 miles per gallon of gasoline. Express this miles/gallon value in the units feet/quart.

Although the emphasis in this section has been on using conversion factors to change units within the English or metric systems or from one to the other, the applications of conversion factors go far beyond this type of activity. We will resort to using conversion factors time and time again throughout this textbook in solving problems. What has been covered in this section is only the tip of the iceberg relative to dimensional analysis and conversion factors.

3.8 DENSITY

Density *is the ratio of the mass of an object to the volume occupied by that object; that is,*

$$\text{density } (d) = \frac{\text{mass}}{\text{volume}}$$

The SI-derived unit for density is kilograms per cubic meter (kg/m^3). This is too large a unit for almost all chemical applications that involve density use.

The most frequently encountered density units in chemistry are grams per cubic centimeter (g/cm^3) for solids, grams per milliliter (g/mL) for liquids, and grams per liter (g/L) for gases. Use of these units avoids the problem of having density values that are extremely small or extremely large numbers. Table 3.5 gives density values for a number of substances.

People often speak of one substance being heavier or lighter than another. For example, it is said that "lead is a heavier metal than aluminum." What is actually meant by this statement is that lead has a higher density than aluminum; that is, there is more mass in a specific volume of lead than there is in the same volume of aluminum. The density of an object is a measure of how tightly the object's mass is packed into a given volume. Even though the density of lead ($11.3\ g/cm^3$) is greater than that of aluminum ($2.70\ g/cm^3$), 1 g of lead weighs exactly the same as 1 g of aluminum—1 g is 1 g. Because the aluminum is less dense than the lead, the mass in the 1 g of aluminum will occupy a larger volume than the mass in the 1 g of lead. Said in another way, if equal-volume samples of lead and aluminum are weighed, the lead will have the greater mass. When we say lead is heavier than aluminum, we actually mean that lead is more dense than aluminum.

Metals are classified as light or heavy, based on density. Light metals are metals whose densities are less than $4.00\ g/cm^3$. Included among the light metals are

calcium	$1.55\ g/cm^3$
magnesium	$1.74\ g/cm^3$
aluminum	$2.70\ g/cm^3$

The lightest of the light metals is lithium, with a density of $0.53\ g/cm^3$.

The densities of solids and liquids are often compared to the density of water. Anything less dense (lighter) than water floats in it (see Fig. 3.8), and anything more

TABLE 3.5 Densities of Selected Solids, Liquids, and Gases

Solids	Density (g/cm^3 at 25°C)*	Liquids	Density (g/mL at 25°C)*	Gases	Density (g/L at 25°C, 1 atm)*
Gold	19.3	Mercury	13.55	Chlorine	3.17
Lead	11.3	Milk	1.028–1.035	Carbon dioxide	1.96
Copper	8.93	Blood plasma	1.027	Oxygen	1.42
Aluminum	2.70	Urine	1.003–1.030	Air (dry)	1.29
Table salt	2.16	Water	0.997	Nitrogen	1.25
Bone	1.7–2.0	Olive oil	0.92	Methane	0.66
Table sugar	1.59	Ethyl alcohol	0.79	Hydrogen	0.08
Wood, pine	0.30–0.50	Gasoline	0.56		

Density changes with temperature. (In most cases it decreases with increasing temperature, since almost all substances expand when heated.) Consequently, the temperature must be recorded along with a density value. In addition, the pressure of gases must be specified.

FIGURE 3.8 Icebergs float in water because ice (solid water) is less dense than liquid water. *(iStockphoto)*

dense (heavier) sinks. In a similar vein, densities of gases are compared to that of air. Any gas less dense (lighter) will rise in air, and anything more dense (heavier) will sink in air.

To obtain an object's density, we must make two measurements: One involves determining the object's mass, and the other, its volume; the mass is then divided by the volume.

EXAMPLE 3.10 **Using Mass and Volume to Calculate Density**

A student determines that the mass of a 20.0 mL sample of olive oil (to be used in oil-and-vinegar salad dressing) is 18.4 g.

  a. What is the density of the olive oil in grams per milliliter?
  b. Predict whether the olive oil layer will be on the top or bottom in unshaken oil-and-vinegar salad dressing.

SOLUTION

  a. Substituting the given mass and volume values into the formula

$$\text{density} = \frac{\text{mass}}{\text{volume}}$$

we have

$$\text{density} = \frac{18.4 \text{ g}}{20.0 \text{ mL}} = 0.92 \, \frac{\text{g}}{\text{mL}} \qquad \text{(calculator answer)}$$

$$= 0.920 \, \frac{\text{g}}{\text{mL}} \qquad \textbf{(correct answer)}$$

Because both input numbers contain three significant figures, the density is specified to three significant figures.

  b. Since vinegar is a water-based solution, its density will be slightly greater than that of water (1.00 g/mL). The olive oil will be the top layer because its density is less than that of water or vinegar.

Answer Double Check:

Is the magnitude of the numerical answer reasonable? Yes. The density should be expected to be close to 1.00 g/mL because the mass and volume values are approximately the same (18.4 and 20.0). Division of almost equal numbers gives a result close to 1.00.

▶ **Practice Exercise 3.10** Lithium is the lightest (least dense) of all metals. What is its density, in grams per cubic centimeter, if 24 g of the metal occupies a volume of 45 cm³?

Using Density as a Conversion Factor

Density may be used as a conversion factor to convert from mass to volume, or vice versa.

In a mathematical sense, density can be thought of as a conversion factor that relates the volume and mass of an object. Interpreting density in this manner enables us to calculate a substance's volume from its mass and density or its mass from its volume and density. Density is thus a bridge connecting mass and volume.

Density conversion factors, like all other conversion factors, have two reciprocal forms. For a density of 1.37 g/mL, the two conversion factors forms are

$$\frac{1.37 \text{ g}}{1 \text{ mL}} \quad \text{and} \quad \frac{1 \text{ mL}}{1.37 \text{ g}}$$

Note that the number 1 always goes in front of a "naked" unit in a conversion factor, that is, a density given as 4.3 g/mL means 4.3 grams per 1 mL.

EXAMPLE 3.11 **Using Density as a Conversion Factor to Relate Mass and Volume**

Methane, a gas that is naturally present in the earth's atmosphere in small amounts, has a density of 0.714 g/L at a particular temperature and pressure.

a. What is the mass, in grams, of 10.0 L of methane?
b. What is the volume, in liters, of 10.0 g of methane?

SOLUTION

Both parts of this problem can be solved using density as a conversion factor.

a.

Step 1 The given quantity is 10.0 L of methane. The unit of the desired quantity (mass) is grams. Thus,

$$10.0 \text{ L} = ? \text{ g}$$

Step 2 The pathway going from liters to grams involves a single step, since density, used as a conversion factor, directly relates grams and liters.

$$10.0 \text{ Ł} \times \frac{0.714 \text{ g}}{1 \text{ Ł}}$$

Step 3 Doing the indicated math gives the following answer.

$$\frac{10.0 \times 0.714}{1} \text{ g} = 7.14 \text{ g} \quad \text{(calculator and \textbf{correct answer})}$$

The calculator answer and correct answer turn out to be the same.

An alternative way of working this problem involves directly using the defining equation for density.

$$\text{density} = \frac{\text{mass}}{\text{volume}}$$

Rearranging this equation so that mass is isolated on the left side of the equation gives

$$\text{mass} = \text{density} \times \text{volume}$$

Since both density and volume are given in the problem statement, we have, by direct substitution,

$$\text{mass} = \frac{0.714 \text{ g}}{1 \text{ L}} \times 10.0 \text{ L}$$

$$= 7.14 \text{ g (calculator and \textbf{correct answer})}$$

b.

Step 1 The given quantity is 10.0 g of methane. The unit of the desired quantity (volume) is liters.

$$10.0 \text{ g} = ? \text{ L}$$

Step 2 The pathway going from grams to liters involves a single step, since density, used as a conversion factor, directly relates grams and liters.

$$10.0 \text{ g} \times \frac{1 \text{ L}}{0.714 \text{ g}}$$

Note that the conversion factor used is actually the reciprocal of the density. Use of the inverted form is necessary in order for the gram units to cancel.

Step 3 Doing the indicated math gives the following answer

$$\frac{10.0 \times 1}{0.714} \text{ L} = 14.005602 \text{ L} \quad \text{(calculator answer)}$$

$$= 14.0 \text{ L} \quad \text{(\textbf{correct answer})}$$

The correct answer is the calculator answer rounded to three significant figures, the number in each input number.

As in part (a), this problem can also be worked, using a rearranged form of the defining equation for density. Rearranging this equation to isolate volume on the left side gives

$$\text{volume} = \frac{\text{mass}}{\text{density}}$$

The given values for mass and density can then be directly substituted into this equation to produce the answer 14.0 L.

Answer Double Check:

Are the answers reasonable in terms of their magnitudes? Yes. In part (a), since the density is less than one, 10.0 L of methane will have a mass of less than 10.0 g, the mass it would have if the density were 1.00 g/mL. The answer 7.14 g is consistent with this reasoning. In part (b), 7.14 g of methane would have a volume of 10.0 L. Therefore, 10.0 g of methane will have a volume greater than 10.0 L. The answer 14.0 L is consistent with this reasoning.

▶ **Practice Exercise 3.11**   Ethylene glycol (automotive antifreeze) has a density of 1.11 g/mL.

   a. What is the mass, in grams, of 54.0 mL of this antifreeze?
   b. What is the volume, in milliliters, of 54.0 g of this antifreeze?

Density is a conversion factor of a different type from those previously used in this chapter. It involves a ratio whose numerator and denominator are *equivalent* rather than *equal*; that is, it is an *equivalence* conversion factor rather than the previously used *equality* conversion factors. An **equivalence conversion factor** *is a ratio that converts one* *type of measure to a different type of measure.* Density involves mass and volume, two different types of measure. This contrasts with equality conversion factors that have ratios in which the numerator and denominator involve the same measure. An **equality conversion factor** *is a ratio that converts one unit of a given measure to another unit of the same measure.* Twelve inches and one foot, which are both measures of length, generate an equality conversion factor.

Density is a property peculiar to a particular substance. The density of water applies only to water and to no other substance. Different substances have different densities at a given set of conditions.

A major difference between *equivalence* conversion factors and *equality* conversion factors is that the former have applicability only in the particular problem setting for which they were derived, whereas the latter are applicable in all problem-solving situations. Many different gram-to-cubic centimeter (mass-to-volume) relationships (densities) exist, but only one foot-to-inch relationship exists. Mathematically, equivalence conversion factors can be used the same way equality conversion factors are, and such conversion factors will be used often in later chapters.

3.9 EQUIVALENCE CONVERSION FACTORS OTHER THAN DENSITY

In the previous section we noted that density was an *equivalence* conversion factor rather than an *equality* conversion factor. Equivalence conversion factors have application only in a limited setting, the setting for which they were derived. Besides density, numerous other equivalence conversion factor types exist, three of which are considered in this section because they are often encountered in a chemical context. These additional equivalence conversion factor types are (1) concentration and dosage relationships, (2) rate relationships, and (3) cost relationships.

Concentration and Dosage Relationship Conversion Factors

If the concentration of salt in a salt water solution is 4.5 mg/mL, then we have

$$4.5 \text{ mg salt} = 1 \text{ mL of solution}$$

and the two conversion factors are

$$\frac{4.5 \text{ mg salt}}{1 \text{ mL solution}} \quad \text{and} \quad \frac{1 \text{ mL solution}}{4.5 \text{ mg salt}}$$

If the dosage for an antibiotic is 125 mg/kg of body weight, then we have

$$125 \text{ mg antibiotic} = 1 \text{ kg body weight}$$

and the two conversion factors are

$$\frac{125 \text{ mg antibiotic}}{1 \text{ kg body weight}} \quad \text{and} \quad \frac{1 \text{ kg body weight}}{125 \text{ mg anibiotic}}$$

Example 3.12 illustrates the usage of a concentration relationship in a problem-solving context.

EXAMPLE 3.12 Using Concentration Conversion Factors

A blood analysis shows a vitamin-C concentration of 0.2 mg/100 mL of blood. How many grams of vitamin C are there in 1.2 L of this blood?

SOLUTION

Step 1 The given quantity is 1.2 L of blood. The unit of the desired quantity is grams of vitamin C.

$$1.2 \text{ L blood} = ? \text{ g vitamin C}$$

Step 2 The pathway in going from liters of blood to grams of vitamin C is

$$\text{L blood} \rightarrow \text{mL blood} \rightarrow \text{mg vitamin C} \rightarrow \text{g vitamin C}$$

The conversion factors needed for this pathway are

$$1.2 \; \text{L blood} \times \frac{1 \; \text{mL blood}}{10^{-3} \; \text{L blood}} \times \frac{0.2 \; \text{mg vitamin C}}{100 \; \text{mL blood}} \times \frac{10^{-3} \; \text{g vitamin C}}{1 \; \text{mg vitamin C}}$$

Step 3 The numerical calculation involves the following collection of numbers.

$$\frac{1.2 \times 1 \times 0.2 \times 10^{-3}}{10^{-3} \times 100 \times 1} \; \text{g vitamin C} = 0.0024 \; \text{g vitamin C} \quad \text{(calculator answer)}$$

$$= 0.002 \; \text{g vitamin C} \quad \textbf{(correct answer)}$$

The given number of 0.2 mg vitamin C limits the answer to one significant figure.

Answer Double Check:

Is it reasonable that the answer should be as small as it is (0.002g)? Yes. Since the volume size of 1.2 L (1200 mL) is 12 times as large as a 100 mL volume, this larger volume should contain 12 times as much vitamin C as in a 100 mL volume. Twelve times 0.2 mg is 2.4 mg, which is equivalent to 0.0024 g. Therefore, the answer size is reasonable.

▶ **Practice Exercise 3.12** The concentration of a silver nitrate solution is 5.00 g silver nitrate per 100 mL of solution. How many grams of silver nitrate are present in 0.750 L of this solution?

Rate Relationship Conversion Factors

In a chemical setting, rate relationships involve the amount of a substance and a time unit. Industrially, the amount of a chemical substance that can be produced or is consumed in a given amount of time is an economically important consideration.

Examples of rate relationships for the production of a chemical substance could include

$$\frac{65,000 \; \text{lb substance}}{1 \; \text{day}} \qquad \frac{7200 \; \text{g substance}}{1 \; \text{min}} \qquad \frac{3.72 \; \text{tons substance}}{1 \; \text{hr}}$$

Example 3.13 illustrates how a rate conversion factor is incorporated into a dimensional analysis type problem.

EXAMPLE 3.13 | **Using Rate Conversion Factors**

A certain chemical process consumes water at the rate of 55 L/sec. How much water, in liters, is consumed in 3.0 hours?

SOLUTION

Step 1 The given quantity is 3.0 hours of time. The unit of the desired quantity is liters of water.

$$3.0 \; \text{hours} = ? \; \text{liters water}$$

Step 2 The pathway for going from hours to liters of water involves two time unit changes and the given rate conversion factor relationship.

$$\text{hours} \rightarrow \text{minutes} \rightarrow \text{seconds} \rightarrow \text{liters water}$$

The hours need to be changed to seconds because the rate conversion factor is stated in terms of seconds.

The conversion factor setup needed to effect the changes in this pathway is

$$3.0 \; \cancel{hr} \times \frac{60 \; \cancel{min}}{1 \; \cancel{hr}} \times \frac{60 \; \cancel{sec}}{1 \; \cancel{min}} \times \frac{55 \; L \; water}{1 \; \cancel{sec}}$$

Step 3 The numerical calculation, after cancellation of units, involves the following collection of numbers.

$$\frac{3.0 \times 60 \times 60 \times 55}{1 \times 1 \times 1} \; L \; water = 594,000 \; L \; water \; (\text{calculator answer})$$

$$= 590,000 \; L \; water \; (\textbf{correct answer})$$

Both the given time (3.0 hr) and the water amount (55 L) limit the correct answer to two significant figures.

Answer Double Check:

Is it reasonable that the answer is as large as it is? Yes. There are several thousand seconds in 3.0 hours and therefore the number 55 should increase by a factor of several thousand, which it has.

▶ **Practice Exercise 3.13**   How much time, in hours, is needed to produce 35,000 lb of salt if the salt production rate is 325 lb per minute?

Cost Relationship Conversion Factors

Cost relationships involve the amount of a substance and a monetary amount. There is a tremendous difference between having the price of copper at the following three levels.

$$\frac{13 \; cents}{1 \; lb \; copper} \qquad \frac{37 \; cents}{1 \; lb \; copper} \qquad \frac{82 \; cents}{1 \; lb \; copper}$$

The first cost relationship is so low that copper cannot be produced at a profit.

Example 3.14, which involves the cost of aluminum metal, illustrates how a cost conversion factor is used in a dimensional analysis setup.

EXAMPLE 3.14   **Using Cost Conversion Factors**

If aluminum metal is selling for 91.25 cents a pound, what would be the cost to a buyer, in dollars, who needs 2525 pounds of aluminum?

SOLUTION

Step 1 The given quantity is 2525 lb of aluminum. The unit of the desired quantity is dollars.

$$2525 \; lb \; aluminum = ? \; dollars$$

Step 2 Because the cost relationship is stated in terms of cents, the cents will need to be changed to dollars. The needed pathway is

$$lb \; aluminum \rightarrow cents \rightarrow dollars$$

The conversion factor sequence consistent with this pathway is

$$2525 \; \cancel{lb \; aluminum} \times \frac{91.25 \; \cancel{cents}}{1 \; \cancel{lb \; aluminum}} \times \frac{1 \; dollar}{100 \; \cancel{cents}}$$

Step 3 The numerical relationships that remain after cancellation of units are

$$\frac{2525 \times 91.25 \times 1}{1 \times 100} \text{ dollar} = 2304.0625 \text{ dollars (calculator answer)}$$

$$= 2304 \text{ dollars} \quad \textbf{(correct answer)}$$

Both 2525 and 91.25 are four-significant-figure quantities. Thus, the calculator answer is rounded to four significant figures.

▶ **Practice Exercise 3.14**   The commodity market price for wheat on a specific day is $5.13 per bushel. How many bushels of wheat can be purchased for $10,000.00 at that time?

3.10 PERCENTAGE AND PERCENT ERROR

Percent *is the number of items of a specified type in a group of 100 total items.* The quantity 45% means 45 items per 100 total items. In general, percent is parts per 100 total parts.

A mathematical statement of the percent concept is

$$\text{percent} = \frac{\text{number of items of interest}}{\text{total number of items}} \times 100$$

Example 3.15 shows a simple calculation of a percent.

EXAMPLE 3.15   **Percentage Calculation**

A professor proctoring an examination notices that 11 students out of a class of 80 students write with their left hand. Calculate the percentage, to three significant figures, of *right-handed* students in the class.

SOLUTION

We first calculate the number of right-handed students in the class.

right-handed students = 80 − 11 = 69   (calculator and **correct answer**)

The percentage of right-handed students is equal to the number of right-handed students divided by the total number of students times the factor 100.

$$\text{percentage right-handed students} = \frac{69}{80} \times 100$$

$$= 86.25\% \quad \text{(calculator answer)}$$

$$= 86.2\% \quad \textbf{(correct answer)}$$

▶ **Practice Exercise 3.15**   A customer gives a mixture of coins to a bank teller for deposit—17 nickels, 5 dimes, and 9 quarters. Calculate the percentage, to three significant figures, of the coins presented for deposit that are nickels.

Using Percentage as a Conversion Factor

Percent values find use as conversion factors in problem-solving situations where dimensional analysis is employed. Let us look at the mechanics of writing a percentage as a conversion factor, paying particular attention to the units involved. The percent value in the statement "A gold alloy is found upon analysis to contain 77% gold by mass" will be our focus point for the discussion. What are the mass units associated with the value 77%? The answer is that many mass units could be appropriate.

The following statements, written as fractions (conversion factors), are all consistent with our 77% gold analysis.

$$\frac{77 \text{ oz of gold}}{100 \text{ oz of gold alloy}} \qquad \frac{77 \text{ lb of gold}}{100 \text{ lb of gold alloy}}$$

$$\frac{77 \text{ g of gold}}{100 \text{ g of gold alloy}} \qquad \frac{77 \text{ kg of gold}}{100 \text{ kg of gold alloy}}$$

Thus, in writing a percent as a conversion factor, the choice of unit (for mass in this case) is arbitrary. In practice, the complete context of the problem in which the percentage is found usually makes obvious the appropriate unit choice.

In conversion factors derived from percentages, confusion about cancellation of units frequently arises. Both numerator and denominator contain the same units, for example, ounces. Yet the ounces do not cancel. Not considering the *complete unit* is what causes confusion in the minds of students. In the conversion factor

$$\frac{77 \text{ oz of gold}}{100 \text{ oz gold alloy}}$$

the numerator and denominator dimensions are not simply "ounces." The dimensions are, respectively, "ounces of gold" and "ounces of gold alloy." Ounces cannot be cancelled, because they are only a part of the complete dimension. Cancellation is possible only when *complete* units are identical.

To avoid the problem of mistakenly cancelling dimensions that are not the same, always write complete dimensions. Percent will always involve the same dimensional units (pounds, grams, meters) of different things. The identities of the different things, gold and gold alloy in our example, must always be included as part of the units.

EXAMPLE 3.16   **Using Percentage as a Conversion Factor**

A sample of clean, dry air is found to be 20.9% oxygen by volume. How many milliliters of oxygen are present in 375 mL of this air?

SOLUTION

Step 1 The given quantity is 375 mL of air, and the desired quantity is milliliters of oxygen.

$$375 \text{ mL air} = ? \text{ mL oxygen}$$

Step 2 This is a one-conversion-factor problem with the conversion factor, obtained from the given percentage, being

$$\frac{20.9 \text{ mL oxygen}}{100 \text{ mL air}}$$

The setup for the problem is

$$375 \text{ mL air} \times \frac{20.9 \text{ mL oxygen}}{100 \text{ mL air}}$$

Step 3 The numerical calculation involves the following arrangement of numbers.

$$\frac{375 \times 20.9}{100} \text{ mL oxygen} = 78.375 \text{ mL oxygen} \quad \text{(calculator answer)}$$

$$= 78.4 \text{ mL oxygen} \quad \textbf{(correct answer)}$$

The original percentage (20.9) and the air volume (375 mL) both have three significant figures. Therefore, the correct answer should also contain three significant figures. The number 100 in the setup is an exact number, since it originates from the definition of percent.

Answer Double Check:

Is the magnitude of the answer reasonable? Yes. Using rounded numbers, we note that 100 mL of air contains about 20 mL of oxygen and that 400 mL of air (4 times as much air) should contain about 80 mL of oxygen. A value of 78.4 mL for oxygen is, thus, appropriate for 375 mL of air.

▶ **Practice Exercise 3.16** A sample of a copper alloy is found to be 38.7% copper by mass. How many pounds of copper are present in a 30.2 lb sample of the alloy?

| **EXAMPLE 3.17** | **Multiple Use of Percentages in a Conversion Problem** |

A study of the 184 adult males living in a small town in Weber County in the state of Utah includes the following facts: (1) 20.1% of the adult males are baldheaded, (2) 62.1% of the baldheaded adult males are left-handed, and (3) 91.3% of the left-handed, bald-headed adult males are handsome. How many handsome, left-handed, baldheaded adult males live in the town?

SOLUTION

Step 1 The given quantity is 184 adult males, with the units of the desired quantity being handsome, left-handed, baldheaded adult males.

184 adult males = ? handsome, left-handed, baldheaded adult males

Step 2 When conversion factors, particularly unusual ones, are given in word form in the statement of a problem, it is suggested that you first extract them from the problem statement and write them down before starting to solve the problem. The given conversion factors in this problem are

$$\frac{20.1 \text{ baldheaded adult males}}{100 \text{ adult males}}$$

$$\frac{62.1 \text{ left-handed, baldheaded adult males}}{100 \text{ baldheaded adult males}}$$

$$\frac{91.3 \text{ handsome, left-handed, baldheaded adult males}}{100 \text{ left-handed, baldheaded adult males}}$$

The unit-conversion pathway for this problem will be

$$\text{adult males} \rightarrow \frac{\text{baldheaded}}{\text{adult males}} \rightarrow \frac{\text{left-handed,}}{\text{baldheaded}} \rightarrow \frac{\text{handsome,}}{\text{left-handed,}} \atop \text{baldheaded} \atop \text{adult males}$$

The dimensional analysis setup for this problem is

$$184 \text{ adult males} \times \frac{20.1 \text{ baldheaded}}{100 \text{ adult males}} \times \frac{62.1 \text{ left-handed}}{100 \text{ baldheaded}} \times \frac{91.3 \text{ handsome}}{100 \text{ left-handed}}$$

All units cancel except those in the numerator of the last conversion factor.

Step 3 Performing the indicated arithmetic gives

$$\frac{184 \times 20.1 \times 62.1 \times 91.3}{100 \times 100 \times 100} \text{ handsome, left-handed, baldheaded adult males}$$

= 20.968929 handsome, left-handed, baldheaded adult males (calculator answer)

= 21.0 handsome, left-handed, baldheaded adult males   (**correct answer**)

The calculator answer must be rounded off to three significant figures since all three of the given percentages contain only three significant figures.

▶ **Practice Exercise 3.17**   Examination of a collection of 325 marbles shows that 37.5% of them are of a smaller size than the rest, 35.2% of the smaller marbles are blue in color, and 16.3% of the smaller blue marbles glow in the dark. How many glow-in-the-dark, blue, small marbles are present in the collection?

Percent Error

Percent error *is the ratio of the difference between a measured value and the accepted value for the measurement and the accepted value itself, all multiplied by 100.* A mathematical statement of the percent error concept is

$$\text{percent error} = \frac{\text{measured value} - \text{accepted value}}{\text{accepted value}} \times 100$$

A percent error can be either positive or negative. If the measured value is greater than the accepted value, the difference (measured value – accepted value) will be positive, and the percent error will be positive. If the measured value is less than the accepted value, the difference will be negative and the percent error will be negative.

EXAMPLE 3.18   Calculation of a Percent Error

The accepted value for the density of copper is 8.93 g/cm^3. In a laboratory setting, three students are asked to experimentally determine the density of copper. Their results are

$$\begin{array}{ll}
\text{student 1:} & 8.91 \text{ g/cm}^3 \\
\text{student 2:} & 8.23 \text{ g/cm}^3 \\
\text{student 3:} & 8.99 \text{ g/cm}^3
\end{array}$$

Calculate the percent error associated with each student's reported density.

SOLUTION

We will use the equation

$$\text{percent error} = \frac{\text{measured value} - \text{accepted value}}{\text{accepted value}} \times 100$$

to calculate the percent error in each student's result.

STUDENT 1

$$\text{percent error} = \frac{(8.91 - 8.93) \; \cancel{\text{g/cm}^3}}{8.93 \; \cancel{\text{g/cm}^3}} \times 100$$

$$= \frac{-0.02}{8.93} \times 100$$

$$= -0.22396416 \quad \text{(calculator answer)}$$

$$= -0.2 \qquad \text{(\textbf{correct answer})}$$

The negative sign associated with the result indicates that the measured value was less than the accepted value.

STUDENT 2

$$\text{percent error} = \frac{(8.23 - 8.93) \; \cancel{g/cm^3}}{8.93 \; \cancel{g/cm^3}} \times 100$$

$$= \frac{-0.70}{8.93} \times 100$$

$$= -7.8387458 \quad \text{(calculator answer)}$$

$$= -7.8 \qquad \text{(\textbf{correct answer})}$$

The percent error is again negative, reflecting that the measured value was less than the accepted value.

STUDENT 3

This time we will have a positive percent error because the measured value is greater than the accepted value.

$$\text{percent error} = \frac{(8.99 - 8.93) \; \cancel{g/cm^3}}{8.93 \; \cancel{g/cm^3}} \times 100$$

$$= \frac{0.06}{8.93} \times 100$$

$$= 067189249 \quad \text{(calculator answer)}$$

$$= 0.7 \qquad \text{(\textbf{correct answer})}$$

Note that there are no units associated with percent error; g/cm^3, present in both numerator and denominator, cancel.

Answer Double Check:

Are the signs (plus or minus) associated with the percent error consistent with convention? Yes. A positive percent error arises when the measured value is greater than the accepted value (student 3), and a negative percent error results from a measured value that is less than the accepted value (students 1 and 2).

▶ **Practice Exercise 3.18**   The accepted experimental value for the solubility of oxygen gas in water at 20°C and atmospheric pressure is 0.044 grams per liter. In a laboratory setting, three students are asked to determine the solubility of oxygen gas in water at the same conditions of temperature and pressure. Their results are

student 1: 0.059 g/L

student 2: 0.041 g/L

student 3: 0.054 g/L

Calculate the percent error associated with each student's reported solubility.

3.11 TEMPERATURE SCALES

Temperature *is a measure of the hotness or coldness of an object*. The most common instrument for measuring temperature is the mercury-in-glass thermometer, which consists of a glass bulb containing mercury sealed to a slender glass capillary tube. The higher the temperature, the farther the mercury will rise in the capillary tube. Graduations on the capillary tube indicate the height of the mercury column in terms of defined units, usually called *degrees*. A tiny superscript circle (°) is used as the symbol for a degree.

Three different temperature scales are in common use—Celsius, Kelvin, and Fahrenheit. Both the Celsius and Kelvin scales are part of the metric measurement system, and the Fahrenheit scale belongs to the English measurement system. Different degree sizes and different reference points are what produce the various temperature scales.

The SI base unit for temperature is the kelvin, with symbol K (see Table 3.1). Note that there is no degree sign associated with the symbol K and that the k in the word kelvin is not capitalized when it denotes the SI unit kelvin. This contrasts with the Celsius and Fahrenheit scales where the notations °C and °F are used.

The Celsius Scale

Thirty is hot,
Twenty is pleasing;
Ten is quite cool,
And zero is freezing.

The Celsius scale, named after Anders Celsius (1701–1744), a Swedish astronomer, is the scale most commonly encountered in scientific work. On this scale the boiling and freezing points of water serve as reference points, with the former having a value of 100°C (degrees Celsius) and the latter, 0°C. Thus, there are 100 degree intervals between the two reference points. The Celsius scale was formerly called the centigrade scale.

The Kelvin scale is a close relative of the Celsius scale. The size of the temperature unit is the same on both scales, as is the interval between the reference points. The two scales differ only in the numerical values assigned to the reference points and the names of the units. On the Kelvin scale the boiling point of water is at 373.15 K (kelvins), and the freezing point of water is at 273.15 K. This scale, proposed by the British mathematician and physicist William Kelvin (1824–1907) in 1848, is particularly useful when working with relationships between temperature and pressure–volume behavior of gases (see Sec. 12.3). The degree sign is not used in specifying Kelvin temperatures (321 K instead of 321°K).

A unique feature of the Kelvin scale is that negative temperature readings never occur. The lowest possible temperature thought to be obtainable occurs at 0 on the Kelvin scale. This temperature, known as *absolute zero*, has never been produced experimentally, although scientists have come within a fraction of a degree of reaching it.

The Fahrenheit scale was designed by the German physicist Gabriel Fahrenheit (1686–1736) in the early 1700s. After proposing several scales, he finally adopted a system that used a salt–ice mixture and boiling mercury as the reference points. These were the two extremes in temperature available to him at that time. A reading of 0 was assigned to the salt–ice mixture and 600 to the boiling mercury. The distance between these two points was divided into 600 equal parts or degrees. On this scale, water freezes at 32°F (degrees Fahrenheit) and boils at 212°F. Thus, there are 180 degrees between the freezing and boiling points of water on this scale as contrasted to 100 degrees on the Celsius scale and 100 kelvins on the Kelvin scale. Figure 3.9 shows a comparison of the three temperature scales.

When changing a temperature reading on one scale to its equivalent on another scale, we must take two factors into consideration: (1) the size of the unit on the two scales may differ, and (2) the zero points on the two scales do not coincide.

Difference in unit size will be a factor any time the Fahrenheit scale is involved in a conversion process. The conversion factors necessary to relate the size of the Fahrenheit degree to the size of the Celsius degree or the kelvin are obtainable from the information in Figure 3.9. From that figure we see that 180 Fahrenheit degrees are

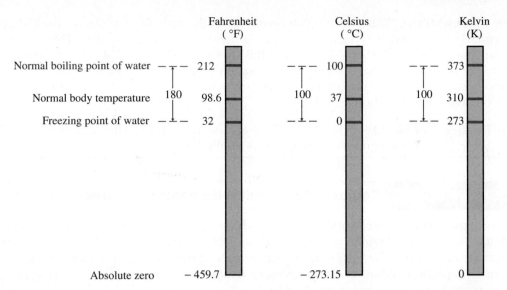

FIGURE 3.9 The relationships among the Fahrenheit, Celsius, and Kelvin temperature scales.

equivalent to 100 Celsius degrees or kelvins. Using this relationship and the fact that $\frac{180}{100} = \frac{9}{5}$, we obtain the following equalities.

$$5 \text{ Celsius degrees} = 9 \text{ Fahrenheit degrees}$$

$$5 \text{ kelvins} = 9 \text{ Fahrenheit degrees}$$

Conversion factors derived from these equalities will contain an infinite number of significant figures; that is, they are exact conversion factors.

Adjustment for differing zero-point locations is carried out by considering how many degrees above or below the freezing point of water (the ice point) the original temperature is. Examples 3.19 and 3.20 show how this zero-point adjustment is carried out, in addition to illustrating the use of temperature scale conversion factors.

EXAMPLE 3.19 **Fahrenheit-to-Celsius Temperature Conversion**

An oven for baking pizza operates at approximately 525°F. What is the equivalent temperature on the Celsius scale?

SOLUTION

First, we determine the number of degrees between the ice point (freezing point of water) and the given temperature on the original scale.

$$525°F - 32°F = 493°F \text{ above the ice point}$$

Second, we convert from Fahrenheit units to Celsius units.

$$493 \text{ Fahrenheit degrees} \times \frac{5 \text{ Celsius degrees}}{9 \text{ Fahrenheit degrees}} = 273.88888 \text{ Celsius degrees}$$
(calculator answer)
$$= 274 \text{ Celsius degrees}$$
(**correct answer**)

Third, taking into account the ice point on the new scale, we determine the new temperature. On the Celsius scale, the temperature will be 274 degrees above the ice point. Since the ice point is 0°C, the new temperature will be 274°C.

Answer Double Check:

Is the magnitude of the Celsius and Fahrenheit readings reasonable? Yes. At *high* temperatures the Celsius reading will always be less than the Fahrenheit reading, roughly

How hot is hot?

100°C	Boiling point of water
800°C	Campfire
875°C	Cigarette ember
1,600°C	Gas stove fire
2,300°C	Filament of lightbulb
7,500°C	Surface of the sun
30,000°C	Typical lightning bolt

by a factor of two. This relationship results from the fact that the size of the Celsius degree is almost twice as big as that of the Fahrenheit degree (9/5 = 1.8). This approximate factor-of-two relationship holds in this problem (525° and 274°C). Temperatures must be high enough that the zero-point adjustment (adding or subtracting 32) does not appreciably change the magnitude of the temperature readings in order for the factor-of-two relationship to hold.

▶ **Practice Exercise 3.19**   The temperature on a warm summer day in a desert area is 106°F. What is the equivalent temperature on the Celsius scale?

EXAMPLE 3.20   **Celsius-to-Fahrenheit Temperature Conversion**

The ozone layer over Antarctica thins dramatically every year during spring. Antarctica's unusual winter weather conditions of extreme cold (–85°C) and total darkness are necessary prerequisites for the occurrence of the chemical reactions that lead to ozone depletion. What is the equivalent temperature on the Fahrenheit scale of Antarctica's winter temperature of −85°C?

SOLUTION

First, we determine how many degrees there are between the original temperature and the ice point. On the Celsius scale this will always be equal numerically to the original temperature, since the ice point is 0°C.

$$-85°C = 85 \text{ Celsius degrees below the ice point}$$

Second, we change this number of degrees from Celsius units to Fahrenheit units.

$$85 \cancel{\text{ Celsius degrees}} \times \frac{9 \text{ Fahrenheit degrees}}{5 \cancel{\text{ Celsius degrees}}} = 153 \text{ Fahrenheit degrees}$$

Significant-figure considerations for changes in size of degree (Celsius to Fahrenheit or vice versa) represent an exception to the general rules for handling significant figures. Although size of degree changes involve a multiplication, position of uncertainty (addition–subtraction rules) is used to determine how the calculator answer should be modified to give the correct answer. If the original number of degrees is known to tenths, then the new number of degrees is also specified to tenths. Or, as in our specific problem here, if the original number of degrees is known to the closest degree (±1), then the new number of degrees is specified to the closest degree.

Third, taking into account the ice point on the new scale, we determine the new temperature. The new temperature will be 153 Fahrenheit degrees below the ice point on the Fahrenheit scale.

$$32°F − 153°F = −121°F \quad \text{(calculator and \textbf{correct answer})}$$

Answer Double Check:

Is the magnitude of the Celsius and Fahrenheit readings reasonable? Yes. Celsius and Fahrenheit scale readings are numerically equal at a temperature of –40°C (the only temperature at which they are numerically equal). At temperatures above –40°C, Fahrenheit readings are always higher than Celsius readings. At temperatures below −40°C, Fahrenheit readings are always lower than Celsius readings. Since this problem involves a temperature below −40°C, the Fahrenheit temperature should be numerically lower than the Celsius temperature, which is the case.

▶ **Practice Exercise 3.20**   The melting point of sodium chloride (table salt) is 755°C. What is the equivalent temperature on the Fahrenheit scale?

Examples 3.19 and 3.20 point out that Fahrenheit–Celsius temperature-scale conversions are more complicated than the unit conversions of the last three sections. Not only is multiplication by a conversion factor required, but also addition and subtraction.

The relationship between the Kelvin and Celsius scales is very simple because the sizes of the degree unit is the same. No conversion factors are needed. All that is required is an adjustment for the differing zero points. This adjustment involves the number 273, the number of units by which the two scales are offset from each other. The adjustment factor is specifically 273, 273.2, or 273.15 depending on the uncertainty in the temperature measurement. Since temperatures are most often stated in terms of a whole number of units (31°C, 45°C, etc.), 273 is the most used adjustment factor. However, the other two factors are needed when dealing with temperatures involving tenths or hundredths of a degree (31.5°C, 43.12°C, etc.).

To change a Celsius temperature to the Kelvin scale, we add the adjustment factor 273.

$$K = °C + 273$$

To change a Kelvin temperature to Celsius, we subtract this same adjustment factor.

$$°C = K - 273$$

Note that the symbol for the kelvin is K, not °K.

The relationship between the Fahrenheit scale and the Celsius scale can also be stated in an equation format.

$$°F = \frac{9}{5}(°C) + 32$$

$$°C = \frac{5}{9}(°F - 32)$$

Some students prefer to use these equations rather than the dimensional analysis approach used in Examples 3.19 and 3.20. The use of these equations is illustrated in Example 3.21.

EXAMPLE 3.21   **Temperature Scale Conversions**

A person suffering from heat stroke is found to have a body temperature of 41.1°C. What is this temperature on (a) the Fahrenheit scale? (b) the Kelvin scale?

SOLUTION

a. Substituting into the Celsius-to-Fahrenheit equation, we get

$$°F = \frac{9}{5}(°C) + 32 = \frac{9}{5}(41.1) + 32$$

$$= 74.0 + 32$$

$$= 106°F \quad \text{(calculator answer)}$$

$$= 106.0°F \quad \textbf{(correct answer)}$$

The multiplication $\frac{9}{5}(41.1)$ gives the calculator answer 73.98, which is rounded to 74.0, the closest tenth of a degree (see the significant-figure discussion in Example 3.20). Adding 32 (an exact number by definition) to 74.0 gives 106.0 as the correct answer.

b. Substituting into the Celsius-to-Kelvin equation, we get

$$K = °C + 273.2 = 41.1 + 273.2$$

$$= 314.3 \text{ K} \quad \text{(calculator and } \textbf{correct answer)}$$

Note that we used the adjustment factor 273.2 rather than simply 273 because the given temperature involved tenths of a degree.

▶ **Practice Exercise 3.21** The winter setting for a thermostat in a home is 70°F. What is this temperature on

a. the Celsius scale?   b. the Kelvin scale?

Temperature Readings and Significant Figures

Temperatures encountered in everyday situations are generally determined using a thermometer (or equivalent electronic device). Standard procedure in reading a thermometer is to estimate the temperature to the closest degree. Thus, thermometer-obtained temperature readings have an uncertainty in the "ones places," that is, to the closest degree.

When considering significant figures for temperature readings, the preceding operational procedure must be taken into account. Celsius or Fahrenheit temperatures of 10°, 20°, 30°, etc., are considered to have two significant figures even though no decimal point is explicitly shown after the zero (Sec. 2.5). A temperature reading of 100°C or 100°F is considered to possess three significant figures.

Note that the previous paragraph applies only to ordinary temperatures obtained using a thermometer (or equivalent electronic device). The temperature at the bottom of a blast furnace (used in steelmaking) often reaches 3600°F. Such a temperature is not a four-significant-figure temperature but rather would be assumed, without further information, to be a two-significant-figure temperature. Obviously, this temperature was not determined by inserting a standard thermometer into the bottom of the blast furnace.

Summary

1. **SI System of Units** The SI system of units, a metric unit system, is the preferred system of measurement in all scientific work. The SI system employs seven base units, from which other needed units are derived. It is a decimal system in which larger and smaller units of a quantity are related by factors of 10. Prefixes are used to designate relationships between the base unit and larger or smaller units.

2. **Conversion Factors** A conversion factor is a ratio that converts a measure expressed in one unit to a measure expressed in another unit. English-to-English and metric-to-metric conversion factors are defined quantities. Metric-to-English conversion factors are obtained by measurement and therefore have an uncertainty factor associated with them.

3. **Dimensional Analysis** Dimensional analysis is a general problem-solving method in which the units associated with numbers are used as a guide in setting up calculations. A given quantity is multiplied by one or more conversion factors in

such a manner that the unwanted (original) units are cancelled, leaving only the desired units.

4. **Density** Density is the ratio of the mass of an object to the volume occupied by that object. A correct density expression includes a numerical value, a mass unit, and a volume unit. Density can be used as a conversion factor to relate mass to volume or vice versa.

5. **Percentage and Percent Error** Percent means parts per hundred, that is, the number of items of a specified type in a group of 100 items. Percentage values find use as conversion factors in problem-solving situations where dimensional analysis is employed. Percent error compares a measured value with its accepted value.

6. **Temperature Scales** The three major temperature scales are the Celsius, Kelvin, and Fahrenheit scales. The size of the degree for the Celsius and Kelvin scales is the same. They differ only in the numerical values assigned to the reference points. The Fahrenheit scale has a smaller degree size than the other two temperature scales.

Key Terms

The new terms or concepts defined in this chapter are

area *Sec. 3.4*
conversion factor *Sec. 3.6*
cubic meter *Sec. 3.4*
density *Sec. 3.8*
dimensional analysis
    Sec. 3.7

equality conversion
    factor *Sec. 3.8*
equivalence conversion
    factor *Sec. 3.8*
kilogram *Sec. 3.3*
liter *Sec. 3.4*

mass *Sec. 3.3*
meter *Sec. 3.2*
percent *Sec. 3.10*
percent error *Sec. 3.10*
SI base unit *Sec. 3.1*
SI-derived unit *Sec. 3.1*

SI system of units *Sec. 3.1*
temperature *Sec. 3.11*
volume *Sec. 3.4*
weight *Sec. 3.3*

Practice Problems

METRIC SYSTEM UNITS (SECS. 3.1–3.4)

3.1 Indicate whether each of the following quantities is expressed in metric units.
   a. area of a field: 2.4 acres
   b. thickness of a sheet of paper: 0.0106 centimeter
   c. speed of light: 186,000 miles/second
   d. amount of an anti-inflammatory drug in a capsule: 0.500 gram

3.2 Indicate whether each of the following quantities is expressed in metric units.
   a. recommended daily intake of the vitamin thiamine: 1.4 milligrams
   b. highway speed limit: 80 kilometers/hour
   c. volume of a copper rod: 67.3 cubic inches
   d. amount of milk in a container: 2.33 deciliters

3.3 Identify the metric prefixes corresponding to each of the following powers of ten, or vice versa.
   a. 10^{-9} **b.** 10^{6} **c.** 10^{-3}
   d. kilo-   **e.** centi-   **f.** micro-

3.4 Identify the metric prefixes corresponding to each of the following powers of ten, or vice versa.
   a. 10^{-12} **b.** 10^{-6} **c.** 10^{-2}
   d. milli-   **e.** deci-   **f.** mega-

3.5 Write the symbol (abbreviation) for each of the following metric system units, or vice versa.
   a. microgram   **b.** kilometer
   c. centiliter   **d.** dm
   e. mL   **f.** pg

3.6 Write the symbol (abbreviation) for each of the following metric system units, or vice versa.
   a. megagram   **b.** microliter
   c. millimeter   **d.** cL
   e. nm   **f.** kg

3.7 Complete the following table by filling in the "blanks" in each line with the name of the metric system prefix, its abbreviation, and/or the power of ten to which it is equivalent. The first line is already completed as an example.

	Metric Prefix	Abbreviation for Prefix	Mathematical Meaning of Prefix
	milli-	m	10^{-3}
a.	tera-		
b.		n	
c.	giga-		
d.			10^{-6}

3.8 Complete the following table by filling in the "blanks" in each line with the name of metric prefix, its abbreviation, and/or the power of ten to which it is equivalent. The first line is already completed as an example.

	Metric Prefix	Abbreviation for Prefix	Mathematical Meaning of Prefix
	centi-	c	10^{-2}
a.	kilo-		
b.		M	
c.	pico-		
d.			10^{-1}

3.9 Complete the following table by filling in the "blanks" in each line with the name of the metric unit, the property being measured (mass, length, or volume), and/or the abbreviation for the metric unit. The first line is completed as an example.

	Metric Unit	Property being Measured	Abbreviation for Metric Unit
	microliter	volume	μL
a.	kilogram		
b.			Mm
c.	nanogram		
d.			mL

3.10 Complete the following table by filling in the "blanks" in each line with the name metric unit, the property being measured (mass, length, or volume), and/or the abbreviation for the metric unit. The first line is completed as an example.

	Metric Unit	Property being Measured	Abbreviation for Metric Unit
	gigagram	mass	Gg
a.	centimeter		
b.			dL
c.	picometer		
d.			Tg

3.11 Use the appropriate metric prefix abbreviation to replace the power of ten in each of the following values.
   a. 6.8×10^{-1} m   **b.** 3.2×10^{-12} L
   c. 7.23×10^{-2} L   **d.** 6.5×10^{6} g

3.12 Use the appropriate metric prefix abbreviation to replace the power of ten in each of the following values.
   a. 4.1×10^{-9} L   **b.** 9.9×10^{-6} g
   c. 8.721×10^{3} g   **d.** 4.4×10^{9} m

3.13 For each of the pairs of units listed, indicate whether the first unit is larger or smaller than the second unit, and then indicate how many times larger or smaller it is.
   a. centigram, gram   **b.** nanogram, microgram
   c. kilogram, decigram   **d.** milligram, megagram

3.14 For each of the pairs of units listed, indicate whether the first unit is larger or smaller than the second unit, and then indicate how many times larger or smaller it is.
   a. milliliter, liter   **b.** kiloliter, microliter
   c. nanoliter, deciliter   **d.** centiliter, megaliter

3.15 What type of quantity (length, mass, area, or volume) do each of the following units represent?
   a. cm   **b.** cm^2 **c.** cm^3 **d.** cL

3.16 What type of quantity (length, mass, area, or volume) do each of the following units represent?
   a. km^2 **b.** kL   **c.** km   **d.** km^3

3.17 For each of the pairs of units listed, indicate which quantity is larger.
   a. 1 centimeter, 1 inch   **b.** 1 meter, 1 yard
   c. 1 gram, 1 pound   **d.** 1 liter, 1 gallon

3.18 For each of the pairs of units listed, indicate which quantity is larger.
   a. 1 kilometer, 1 mile   **b.** 1 milliliter, 1 fluid ounce
   c. 1 kilogram, 1 pound   **d.** 1 liter, 1 quart

AREA AND VOLUME MEASUREMENTS (SECS. 3.4 AND 3.5)

3.19 Calculate the area of the following surfaces.
   a. a square surface whose side is 4.52 cm
   b. a rectangular surface whose width is 3.5 m and whose length is 9.2 m
   c. a circle whose radius is 4.579 mm
   d. a triangle whose height is 3.0 mm and whose base is 5.5 mm

3.20 Calculate the area of the following surfaces.
   a. a rectangular surface whose dimensions are 24.3 m and 32.1 m
   b. a circle of radius 2.7213 cm
   c. a triangular object whose base is 12.0 mm and whose height is 8.00 mm
   d. a square surface with sides of 6.7 cm

3.21 Calculate the volume of each of the following objects, each of which has a regular geometrical shape.
   a. a copper block 5.4 cm long, 0.52 cm high, and 3.4 cm wide
   b. a cylindrical piece of cheese that has a height of 7.5 cm and a radius of 2.4 cm
   c. a spherical piece of Styrofoam with a radius of 87 mm
   d. a piece of gold in the shape of a cube whose edge is 7.2 cm

3.22 Calculate the volume of each of the following objects, each of which has a regular geometrical shape.
   a. a cube of steel whose edge is 3.5175 mm
   b. a spherical marble with a radius of 1.212 cm
   c. a bar of iron 6.0 m long, 0.10 m wide, and 0.20 m high
   d. a cylindrical rod of copper whose length is 62 mm and whose radius is 3.2 mm

3.23 Indicate whether each of the following cubic volume measurements is equal to or not equal to a liter.
   a. 1 dm^3 **b.** 10 dm^3 **c.** 1 cm^3 **d.** 1000 cm^3

3.24 Indicate whether each of the following cubic volume measurements is equal to or not equal to a milliliter.
   a. 0.001 dm^3 **b.** 1 dm^3
   c. 1 cm^3 **d.** 10 cm^3

CONVERSION FACTORS (SEC. 3.6)

3.25 Write an equation that relates the members of each of the following pairs of time units and also write the two conversion factors associated with the equation.
   a. days and hours
   b. minutes and seconds
   c. decades and centuries
   d. days and years

3.26 Write an equation that relates the members of each of the following pairs of time units, and also write the two conversion factors associated with the equation.
a. days and weeks **b.** hours and minutes
c. months and years **d.** years and centuries

3.27 Based on each of the following descriptions, write an equation (similar to 60 seconds = 1 minute) that relates two English system units of measurement.
a. volume units, contains the number 2
b. length units, contains the number 3
c. mass units, contains the number 2000
d. length units, contains the number 36

3.28 Based on each of the following descriptions, write an equation (similar to 60 seconds = 1 minute) that relates two English system units of measurement.
a. volume units, contains the number 4
b. length units, contains the number 5280
c. mass units, contains the number 16
d. volume units, contains the number 32

3.29 Give the two forms of the conversion factor that relate each of the following pairs of units.
a. kL and L **b.** mg and g
c. m and cm **d.** μsec and sec

3.30 Give the two forms of the conversion factor that relate each of the following pairs of units.
a. ng and g **b.** dL and L
c. m and Mm **d.** psec and sec

3.31 Complete each of the following conversion factors by replacing the letter A found within the conversion factor with a numerical value.
a. $\dfrac{\text{A g}}{1\text{ kg}}$ **b.** $\dfrac{1\,\mu\text{m}}{\text{A m}}$ **c.** $\dfrac{\text{A g}}{1\text{ lb}}$ **d.** $\dfrac{1\text{ in.}}{\text{A cm}}$

3.32 Complete each of the following conversion factors by replacing the letter A found within the conversion factor with a numerical value.
a. $\dfrac{\text{A m}}{1\text{ Mm}}$ **b.** $\dfrac{1\text{ cg}}{\text{A g}}$ **c.** $\dfrac{\text{A L}}{1\text{ qt}}$ **d.** $\dfrac{\text{A cm}}{1\text{ in.}}$

3.33 Indicate how each of the following conversion factors should be interpreted in terms of significant figures present.
a. $\dfrac{1.609\text{ km}}{1\text{ mile}}$ **b.** $\dfrac{10^{-2}\text{ m}}{1\text{ cm}}$
c. $\dfrac{28.35\text{ g}}{1\text{ oz}}$ **d.** $\dfrac{12\text{ in.}}{1\text{ ft}}$

3.34 Indicate how each of the following conversion factors should be interpreted in terms of significant figures present.
a. $\dfrac{2.540\text{ cm}}{1\text{ in.}}$ **b.** $\dfrac{453.6\text{ g}}{1\text{ lb}}$
c. $\dfrac{2.113\text{ pt}}{1\text{ L}}$ **d.** $\dfrac{10^{-9}\text{ m}}{1\text{ nm}}$

3.35 Indicate whether each of the following equations relating units would generate an *exact* set of conversion factors or an *inexact* set of conversion factors relative to significant figures.
a. 1 dozen = 12 objects
b. 1 kilogram = 2.20 pounds
c. 1 minute = 60 seconds
d. 1 millimeter = 10^{-3} meter

3.36 Indicate whether each of the following equations relating units would generate an *exact* set of conversion factors or an *inexact* set of conversion factors relative to significant figures.
a. 1 gallon = 16 cups
b. 1 week = 7 days
c. 1 pint = 0.4732 liter
d. 1 mile = 5280 feet

3.37 Indicate whether or not each of the following is a valid relationship.
a. $1\text{ L} = 1\text{ dm}^3$ **b.** $1\text{ L} = 10\text{ cm}^3$
c. $1\text{ mL} = 1\text{ cm}^3$ **d.** $1\text{ mL} = 10\text{ dm}^3$

3.38 Indicate whether or not each of the following is a valid relationship.
a. $1\text{ L} = 0.001\text{ dm}^3$ **b.** $1\text{ L} = 1000\text{ cm}^3$
c. $1\text{ mL} = 10\text{ cm}^3$ **d.** $1\text{ mL} = 0.001\text{ dm}^3$

DIMENSIONAL ANALYSIS—METRIC–METRIC UNIT CONVERSIONS (SEC. 3.7)

3.39 Perform the following metric system conversions using dimensional analysis and one conversion factor.
a. 37 L to dL **b.** 37.0 mm to m
c. 0.37 pg to g **d.** 370 kL to L

3.40 Perform the following metric system conversions using dimensional analysis and one conversion factor.
a. 23 Mg to g **b.** 23.0 μm to m
c. 0.23 cL to L **d.** 230 g to ng

3.41 Perform the following metric system conversions using dimensional analysis and two conversion factors.
a. 47 Mg to mg **b.** 5.00 nL to cL
c. 6×10^{-2} μm to dm **d.** 37 pm to km

3.42 Perform the following metric system conversions using dimensional analysis and two conversion factors.
a. 3×10^{-3} Mm to mm **b.** 4.20 pg to dg
c. 45 kL to mL **d.** 0.31 cm to μm

3.43 A surface has an area of 365 m². What is this area in each of the following units?
a. km² **b.** cm² **c.** dm² **d.** Mm²

3.44 A surface has an area of 35 m². What is this area in each of the following units?
a. mm² **b.** pm² **c.** Gm² **d.** μm²

3.45 A object has a volume of 35 m^3. What is this volume in each of the following units?
   a. mm^3 **b.** pm^3 **c.** Gm^3 **d.** μm^3

3.46 An object has a volume of 365 m^3. What is this volume in each of the following units?
   a. km^3 **b.** cm^3 **c.** dm^3 **d.** Mm^3

3.47 Perform the following metric system conversions, using dimensional analysis.
   a. 6.0 cm^2 to m^2 **b.** 7.2 mm^3 to m^3
   c. 25 μm^2 to dm^2 **d.** 0.023 km^3 to nm^3

3.48 Perform the following metric system conversions, using dimensional analysis.
   a. 3.25 km^2 to m^2 **b.** 0.30 pm^3 to m^3
   c. 9.552 dm^2 to mm^2 **d.** 5.6 cm^3 to μm^3

DIMENSIONAL ANALYSIS—METRIC–ENGLISH UNIT CONVERSIONS (SEC. 3.7)

3.49 The length of a football field, between goal lines, is 100.0 yd. Express this length in the following units.
   a. meters   **b.** centimeters
   c. kilometers   **d.** inches

3.50 The length of a football field, between goal posts, is 120.0 yd. Express this length in the following units.
   a. meters   **b.** millimeters
   c. megameters   **d.** miles

3.51 A spray steam iron has a capacity of 75 mL of water. Express this water capacity in the following units.
   a. qt   **b.** gal   **c.** fl oz   **d.** cm^3

3.52 An automobile's gasoline tank has a capacity of 64 L. Express this capacity in the following units.
   a. qt   **b.** gal   **c.** fl oz   **d.** cm^3

3.53 The mass of the earth is estimated to be 6.6×10^{21} tons. Express the mass of the earth in the following units.
   a. g   **b.** kg   **c.** ng   **d.** oz

3.54 A defensive lineman on a professional football team has a mass of 295 lb. Express the mass of this football player in the following units.
   a. kg   **b.** Mg   **c.** mg   **d.** ton

3.55 An apple has a volume of 61 cm^3. Express this volume in the following units.
   a. cubic feet   **b.** cubic yards
   c. cubic inches   **d.** cubic miles

3.56 A large pond has a surface area of 1250 km^2. Express this volume in the following units.
   a. square yards   **b.** square inches
   c. square feet   **d.** square miles

3.57 A regular-issue U.S. postage stamp is 2.1 cm wide and 2.5 cm long. Express the surface area of this postage stamp in
   a. square centimeters.   **b.** square inches.

3.58 A rectangular piece of concrete has dimensions of 3.6 m and 1.2 m. Express its surface area in
   a. square meters.   **b.** square yards.

3.59 The luggage compartment of an automobile has the dimensions 95 cm × 105 cm × 145 cm. What is the volume of this compartment in cubic feet?

3.60 A copper block is 65 cm long, 3.0 cm high, and 4.0 cm wide. What is the volume of this block in cubic inches?

DIMENSIONAL ANALYSIS—UNITS INVOLVING TWO TYPES OF MEASUREMENT (SEC. 3.7)

3.61 A certain chemical process consumes water at a rate of 55 L/sec. Express this water consumption rate in the following units.
   a. L/hr   **b.** kL/sec
   c. dL/min   **d.** mL/day

3.62 A certain petroleum refinery operation uses hydrogen gas at the rate of 5×10^4 g/min. Express this hydrogen consumption rate in the following units.
   a. dg/min   **b.** g/sec
   c. kg/hr   **d.** ng/day

3.63 The amount of carbon monoxide, a common air pollutant, in a sample of air is measured at 0.0057 μg/mL air. Express this mass/volume ratio in the following units.
   a. μg/L   **b.** g/mL   **c.** mg/μL   **d.** kg/kL

3.64 The amount of mercury in a sample of water from a polluted lake is measured at 0.39 μg/mL water. Express this mass/volume ratio in the following units.
   a. μg/L   **b.** g/mL   **c.** ng/nL   **d.** cg/dL

DENSITY (SEC. 3.8)

3.65 Calculate the density, in grams per milliliter, for each of the following.
   a. 25.0 g of ethyl alcohol having a volume of 31.7 mL
   b. 25.0 g of chromium metal having a volume of 3.48 cm^3
   c. 25.0 mL of olive oil having a mass of 22.9 g
   d. 25.0 L of chloroform having a mass of 37,200 g

3.66 Calculate the density, in grams per milliliter, for each of the following.
   a. 15.0 g of seawater having a volume of 14.6 mL
   b. 15.0 g of cork having a volume of 60.0 cm^3
   c. 15.0 mL of kerosene having a mass of 12.3 g
   d. 15.0 L of helium gas having a mass of 2.67 g

3.67 Calculate the mass, in grams, for each of the following.
   a. 47.6 mL of acetone (d = 0.791 g/mL)
   b. 47.6 cm^3 of silver metal (d = 10.40 g/cm^3)
   c. 47.6 L of carbon monoxide gas (d = 1.25 g/L)
   d. 4.76 cm^3 of rock salt (d = 2.18 g/cm^3)

3.68 Calculate the mass, in grams, for each of the following.
   a. 63.8 mL of gasoline (d = 0.56 g/mL)
   b. 63.0 cm^3 of sodium metal (d = 0.93 g/cm^3)
   c. 63.0 L of ammonia gas (d = 0.759 g/L)
   d. 63.0 cm^3 of mercury (d = 13.6 g/mL)

3.69 Calculate the volume, in milliliters, for each of the following.
   a. 17.6 g of blood plasma (d = 1.027 g/mL)
   b. 17.6 g of gold metal (d = 19.3 g/cm^3)
   c. 17.6 g of dry air (d = 1.29 g/L)
   d. 17.6 g of urine (d = 1.008 g/mL)

3.70 Calculate the volume, in milliliters, for each of the following.
  a. 33.2 g of milk ($d = 1.03$ g/mL)
  b. 33.2 g of bone ($d = 1.8$ g/cm^3)
  c. 33.2 g of hydrogen gas ($d = 0.087$ g/L)
  d. 33.2 g of lead metal ($d = 11.3$ g/cm^3)

3.71 A small bottle contains 2.171 mL of a red liquid. The total mass of the bottle and liquid is 5.261 g. The empty bottle weighs 3.006 g. What is the density, in grams per milliliter, of the liquid?

3.72 A piece of metal weighing 187.6 g is placed in a graduated cylinder containing 225.2 mL of water. The combined volume of solid and liquid is 250.3 mL. What is the density, in grams per milliliter, of the metal?

3.73 An automobile gasoline tank holds 13.0 gal when full. How many pounds of gasoline will it hold, if the gasoline has a density of 0.56 g/mL?

3.74 Liquid sodium metal has a density of 0.93 g/cm^3. How many pounds of liquid sodium are needed to fill a container whose capacity is 15.0 L?

3.75 What mass of chromium (density = 7.18 g/cm^3) occupies the same volume as 100.0 g of aluminum (density = 2.70 g/cm^3)?

3.76 What volume of nickel (density = 8.90 g/cm^3) has the same mass as 100.0 cm^3 of lead (density = 11.3 g/cm^3)?

3.77 Water has a density of 1.0 g/mL at room temperature. State whether each of the following will sink or float when dropped in water.
  a. paraffin wax ($d = 0.90$ g/cm^3)
  b. limestone ($d = 2.8$ g/cm^3)

3.78 Air has a density of 1.29 g/L at room temperature. State whether each of the following will rise or sink in air.
  a. helium gas ($d = 0.18$ g/L)
  b. argon gas ($d = 1.78$ g/L)

3.79 What is the mass, in grams, of 16.0 gal of gasoline if the density of the gasoline is 1011 lb/yd^3?

3.80 What is the mass, in grams, of 12.0 fluid ounces of milk if the density of the milk is 0.0373 lb/in.^{3}?

EQUIVALENCE CONVERSION FACTORS (SEC. 3.9)

3.81 Classify each of the following as an *equality* conversion factor or an *equivalence* conversion factor.
  a. $\dfrac{1.28 \text{ g salt solution}}{1 \text{ mL salt solution}}$
  b. $\dfrac{10^{-3} \text{ L salt solution}}{1 \text{ mL salt solution}}$
  c. $\dfrac{4 \text{ qt salt solution}}{1 \text{ gal salt solution}}$
  d. $\dfrac{4.5 \text{ mg salt}}{1 \text{ g salt solution}}$

3.82 Classify each of the following as an *equality* conversion factor or an *equivalence* conversion factor.
  a. $\dfrac{1 \text{ mg sugar}}{10^{-3} \text{ g sugar}}$
  b. $\dfrac{1.20 \text{ mg sugar}}{10.0 \text{ kg sugar solution}}$
  c. $\dfrac{2 \text{ pt sugar solution}}{1 \text{ qt sugar solution}}$
  d. $\dfrac{1.20 \text{ mg sugar solution}}{10.0 \text{ mL sugar solution}}$

3.83 A pediatric dosage of a certain antibiotic is 32 mg/kg of body weight per day. How much antibiotic, in milligrams per day, should be administered to a child who weighs 15.9 kg?

3.84 A pediatric dosage of a certain analgesic is 225 mg/kg of body weight per day. How much analgesic, in milligrams per day, should be administered to a child who weighs 12.3 kg?

3.85 An arthritis medication dosage is 6.00 mg/kg of body weight. If you correctly give 375 mg of medication to a patient, what is the patient's weight in pounds?

3.86 A narcotic painkiller dosage is 3.00 mg/kg of body weight. If you correctly give 245 mg of medication to a patient, what is the patient's weight in pounds?

3.87 The amount of sugar in a sugar solution is found to be 2.30 μg/L. How much sugar, in grams, is present in the following amounts of the sugar solution?
  a. 2.30 L **b.** 2.30 mL **c.** 2.30 qt **d.** 2.30 fl oz

3.88 The amount of salt in a salt solution is found to be 4.5 mg/mL. How much salt, in grams, is present in the following amounts of the salt solution?
  a. 4.42 mL **b.** 4.42 L **c.** 4.42 qt **d.** 4.42 fl oz

3.89 At a gasoline pump, the gasoline is dispensed at the rate of 0.20 gallon per second. How much time, in minutes, is needed to dispense the following volumes of gasoline?
  a. 16.0 gal **b.** 13.0 qt **c.** 7.00 L **d.** 5355 mL

3.90 At a soft drink dispenser, liquid is dispensed at the rate of 2.00 fluid ounces per second. How much time is needed, in minutes, to obtain the following size soft drinks?
  a. 16.00 fl oz. **b.** 16.00 mL
  c. 2.00 qt **d.** 0.10 gal

3.91 How long, in minutes, will it take a car driven at 45 km/hr to traverse a distance of 4.2 miles?

3.92 A trip was completed, without stops, in 3 hours and 8 minutes. If the average car speed was 67.0 miles per hour, what was the length, in kilometers, of the trip?

3.93 The commodity market price for soybeans is $9.20 per bushel on a given day. Assume the cost is rounded to the closest 0.01 dollar.
  a. How many bushels of soybeans can be purchased for $1000.00?
  b. How much, in dollars, will 45 bushels of soybeans cost?

3.94 The commodity market price for corn is $3.72 per bushel on a given day. Assume the cost is rounded to the closest 0.01 dollar.
  a. How many bushels of corn can be purchased for $500.00?
  b. How much, in dollars, will 625 bushels of corn cost?

3.95 The cost of manufacturing a pain-killing drug is calculated to be 75 cents per gram. What is the cost, in dollars and cents, of the following amounts of the drug?
  a. 3.75 g **b.** 3.75 cg
  c. 6.20 lb **d.** 6.20 oz

3.96 The cost of manufacturing a cholesterol-lowering drug is calculated to be 62 cents per gram. What is the cost, in dollars and cents, of the following amounts of the drug?

    a. 42 g   **b.** 4.20 cg   **c.** 7.2 lb   **d.** 1.00 oz

PERCENTAGE AND PERCENT ERROR (SEC. 3.10)

3.97 An assortment of coins contains 15 pennies, 6 nickels, 3 dimes, and 10 quarters. Calculate, to three significant figures, the percentage of coins that
a. are nickels.
b. are quarters.
c. have a face value of 10 cents or less.
d. are smaller in diameter than a nickel.

3.98 An assortment of coins contains 7 pennies, 10 nickels, 6 dimes, and 5 quarters. Calculate, to three significant figures, the percentage of coins that
a. are dimes.
b. are pennies.
c. have a face value of 10 cents or more.
d. are larger in diameter than a dime.

3.99 A 1980 U.S. penny (a zinc–copper alloy) with a mass of 3.053 g contains 2.902 g of copper. What is the mass percentage in the penny of
a. copper?   **b.** zinc?

3.100 A 2004 U.S. penny (zinc plated with a thin layer of copper) with a mass of 2.552 g contains 2.488 g of zinc. What is the mass percentage in the penny of
a. copper?   **b.** zinc?

3.101 How many grams of water are contained in 65.3 g of a mixture of alcohol and water that is 34.2% water by mass?

3.102 How many grams of alcohol are contained in 467 g of a mixture of alcohol and water that is 23.0% alcohol by mass?

3.103 A solution of table salt in water contains 15.3% by mass of table salt. If 437 g of solution are evaporated to dryness, how many grams of table salt will remain?

3.104 A solution of table salt in water contains 15.3% by mass of table salt. In a 542 g sample of this solution, how many grams of water are present?

3.105 A vinegar solution (acetic acid and water) is 9.00% by mass acetic acid, and it has a density of 1.013 g/mL. How many grams of acetic acid are present in 1.00 gal of the vinegar solution?

3.106 A vinegar solution (acetic acid and water) is 18.00% by mass acetic acid, and it has a density of 1.025 g/mL. How many grams of acetic acid are present in 1.00 quart of the vinegar solution?

3.107 Consider the following facts about a candy mixture containing "Gummi bears" and "Gummi worms": (1) 30.9% of the 661 items present are Gummi bears, (2) 23.0% of the Gummi bears are orange, and (3) 6.4% of the orange Gummi bears have only one ear. How many one-eared, orange Gummi bears are present in the candy mixture?

3.108 An analysis of the makeup of a beginning chemistry class gives the following facts: (1) 47.1% of the 87 students are female; (2) 43.9% of the female students are married; and (3) 33.3% of the married female students are sophomores. How many students in the class are sophomore female students who are married?

3.109 The accepted value for the normal boiling point of ethyl alcohol is 78.5°C. In a laboratory setting, three students are asked to experimentally determine the normal boiling point of ethyl alcohol. Their results are
        student 1: 78.0°C
        student 2: 77.9°C
        student 3: 79.7°C
Calculate the percent error associated with each student's reported boiling point.

3.110 The accepted value for the normal boiling point of benzaldehyde, a substance used as an almond flavoring, is 178°C. In a laboratory setting, three students are asked to experimentally determine the normal boiling point of benzaldehyde. Their results are
        student 1: 175°C
        student 2: 190°C
        student 3: 181°C
Calculate the percent error associated with each student's reported boiling point.

3.111 The life expectancy for men in the United States rose from 53.6 years in 1920 to 72.6 years in 1995. What is the percentage increase in life expectancy for men during this time period?

3.112 The life expectancy for women in the United States rose from 54.6 years in 1920 to 79.0 years in 1995. What is the percentage increase in life expectancy for women during this time period?

TEMPERATURE SCALES (SEC. 3.11)

3.113 State the freezing point of water on each of the following temperature scales.
a. Fahrenheit
b. Celsius
c. Kelvin

3.114 State the boiling point of water on each of the following temperature scales.
a. Fahrenheit
b. Celsius
c. Kelvin

3.115 If the temperature increases by 20 degrees on the Celsius scale, how many degrees will it rise on the Fahrenheit scale?

3.116 If the temperature increases by 20 degrees on the Fahrenheit scale, how many degrees will it rise on the Celsius scale?

3.117 Convert each of the following Celsius temperatures to the Fahrenheit scale.
a. 1352°C   **b.** 37.6°C   **c.** −7°C   **d.** 295°C

3.118 Convert each of the following Celsius temperatures to the Fahrenheit scale.
   a. 652°C   **b.** 45.2°C   **c.** –17°C   **d.** 335°C

3.119 Convert each of the following Fahrenheit temperatures to the Celsius scale.
   a. 1352°F   **b.** 37.6°F   **c.** –7°F   **d.** 295°F

3.120 Convert each of the following Fahrenheit temperatures to the Celsius scale.
   a. 652°F   **b.** 45.2°F   **c.** –17°F   **d.** 335°F

3.121 Convert each of the following temperatures to the Kelvin scale.
   a. 275°C   **b.** 275.2°C
   c. 275.73°C   **d.** 275°F

3.122 Convert each of the following temperatures to the Kelvin scale.
   a. 169°C   **b.** 169.3°C
   c. 169.27°C   **d.** 169°F

3.123 Carry out the following temperature scale conversions.
   a. The temperature on a hot summer day is 101°F. What is this temperature in degrees Celsius?
   b. Oxygen, the gas necessary to sustain life, freezes to a solid at –218.4°C. What is this temperature in degrees Fahrenheit?
   c. The melting point of sodium chloride (table salt) is 804°C. What is this temperature in kelvins?
   d. Liquefied nitrogen boils at 77 K. What is this temperature in degrees Fahrenheit?

3.124 Carry out the following temperature scale conversions.
   a. Mercury freezes at 234.3 K. What is this temperature in degrees Celsius?
   b. Normal body temperature for a chickadee is 41.0°C. What is this temperature in degrees Fahrenheit?
   c. A recommended temperature setting for household hot water heaters is 140°F. What is this temperature in degrees Celsius?
   d. The metal aluminum melts at 934 K. What is this temperature in degrees Fahrenheit?

3.125 Which is the higher temperature, –10°C or 10°F?

3.126 Which is the higher temperature, –15°C or 4°F?

3.127 Which is the lower temperature, 223 K or –60°F?

3.128 Which is the lower temperature, 381 K or 98°C?

ADDITIONAL PROBLEMS

3.129 When each of the following measurements of length is converted to inches, using the conversion factor (12 inch/1 foot), how many significant figures should the answer have?
   a. 4.3 ft   **b.** 2.03 ft
   c. 0.33030 ft   **d.** 5.123 ft

3.130 When each of the following measurements of mass is converted to tons, using the conversion factor

(1 ton/2000 pound), how many significant figures should the answer have?
   a. 7 lb   **b.** 4.321 lb   **c.** 375 lb   **d.** 640 lb

3.131 By unit change within the metric system, write each of the following quantities in a form in which the numerical value is a number between 1 and 10.
   a. 6301 m   **b.** 1442 msec
   c. 1327 μg   **d.** 0.021 L

3.132 By unit change within the metric system, write each of the following quantities in a form in which the numerical value is a number between 1 and 10.
   a. 3333.0 km   **b.** 0.003 mg
   c. 1373.00 ML   **d.** 234 cL

3.133 What power of 10 should replace the question mark in each of the following equalities?
   a. 1 kL = ? L   **b.** 1 pL = ? L
   c. 1 nL = ? L   **d.** 1 dL = ? L

3.134 What power of 10 should replace the question mark in each of the following equalities?
   a. 1 μm = ? m   **b.** 1 Mm = ? m
   c. 1 cm = ? m   **d.** 1 mm = ? m

3.135 The heights of the starting five players on a basketball team are 20.9 dm, 2030 mm, 1.90 m, 0.00183 km, and 203 cm. What is the average height, in centimeters, of these five basketball players? (The average is the sum of the individual values divided by the number of values.)

3.136 The masses for the five heaviest defensive linemen on a football team are 141,000 g, 0.133 Mg, 1.28×10^8 mg, 126 kg, and 1.22×10^{11} μg. What is the average mass, in grams, of these defensive linemen? (The average is the sum of the individual values divided by the number of values.)

3.137 Using dimensional analysis, convert each of the following volume measurements to liters.
   a. 3.72 gal   **b.** 3.720 gal
   c. 3.7200 gal   **d.** 3.72000 gal

3.138 Using dimensional analysis, convert each of the following mass measurements to grams.
   a. 3.72 lb   **b.** 3.720 lb
   c. 3.7200 lb   **d.** 3.72000 lb

3.139 Using the dimensional analysis method of problem solving, set up and solve the following problem: "If your heart beats at a rate of 69 times per minute, how many times will your heart have beat by the time you reach your 8th birthday?" (Ignore the fact that two of the years were leap years.)

3.140 Using the dimensional analysis method of problem solving, set up and solve the following problem: "A chemistry course meets for 50-minute sessions four times a week for 15 weeks (a semester). How many seconds will a student with perfect attendance spend in class during the semester?"

3.141 If your blood has a density of 1.05 g/mL at 20°C, how many grams of blood would you lose if you donated 1.00 pint of blood?

3.142 If your urine has a density of 1.030 g/mL at 20°C, how many pounds of urine would you lose if you eliminated 0.500 pint of urine?

3.143 The density of a solution of sulfuric acid is 1.29 g/cm^3, and it is 38.1% acid by mass. What volume, in milliliters, of the sulfuric acid solution is needed to supply 325 g of sulfuric acid?

3.144 The density of a solution of sulfuric acid is 1.29 g/cm^3, and it is 38.1% acid by mass. What is the mass, in grams, of 25.0 mL of sulfuric acid?

3.145 Levels of blood glucose higher than 400 mg/dL are life-threatening. Is either of the following laboratory-measured glucose levels life-threatening?
a. 5000 μg/mL   b. 0.5 g/L

3.146 Levels of blood glucose lower than 40 mg/dL are life-threatening. Is either of the following laboratory-measured glucose levels life-threatening?
a. 2000 μg/mL   b. 20,000,000 ng/cL

3.147 The concentration of carbon monoxide, a common air pollutant, is measured at 5.7×10^{-3} μg/cm^3 inside a room. How many grams of carbon monoxide are present in the room if the room's dimensions are 3.5 m × 3.0 m × 3.2 m?

3.148 If the concentration of mercury in a polluted lake is 0.39 μg/mL, what is the total mass of mercury, in kilograms, present in the lake if it has a surface area of 125 mi^2 and an average depth of 35 ft?

3.149 A square piece of aluminum foil, 4.0 in. on a side, is found to weigh 0.466 g. What is the thickness of the foil, in millimeters, if the density of the foil is 269 cg/cm^3?

3.150 The density of osmium (the densest metal) is 2260 cg/cm^3. What is the mass, in grams, of a block of osmium with dimensions 5.00 in. × 4.00 in. × 0.25 ft?

3.151 The estimated earthworm population in a particular plot of ground is 22 worms per cubic meter of soil. How many worms would be expected to be found in the top 1.0 meter of soil if the plot dimensions are 2.00 km by 0.50 km?

3.152 A 60.3 g sample of a pulverized metal ore is placed in a container whose volume is 50.0 mL. The container is then completely filled with water and weighed. The ore and water have a combined mass of 102.3 g. Assuming that the density of the water is 1.00 g/cm^3, calculate the density, in g/cm^3, of the metal ore.

3.153 A scientist invented a new temperature scale called the Howard scale (°H) and assigned the boiling and freezing points of water the values 200°H and −200°H, respectively. What is a temperature of 50°F equivalent to on the Howard scale?

3.154 A scientist invented a new temperature scale called the Scott scale (°S) and assigned the boiling and freezing points of water the values 150°S and −50°S, respectively. What is a temperature of 50°C equivalent to on the Scott scale?

3.155 Given the following information about a bag of Gummi bears, calculate the number of Gummi bears in the bag: (1) 30.3% of the Gummi bears are orange; (2) 8.10% of the orange Gummi bears have only one ear; (3) there are three one-eared orange Gummi bears in the bag.

3.156 Given the following information about a class of students, calculate the number of students in the class: (1) 37.3% of the students have blue eyes; (2) 20.0% of the blue-eyed students are left handed; (3) there are five blue-eyed, left-handed students in the class.

CUMULATIVE PROBLEMS

3.157 A sample of a colorless liquid has a mass of two grams and a volume of four milliliters. Calculate the liquid's density using the following uncertainty specifications and express your answers in scientific notation.
a. 2.000 g and 4.000 mL
b. 2.00 g and 4.0 mL
c. 2.0000 g and 4.0000 mL
d. 2.000 g and 4.0000 mL

3.158 A one-gram sample of a powdery white solid is found to have a volume of two cubic centimeters. Calculate the solid's density using the following uncertainty specifications and express your answers in scientific notation.
a. 1.0 g and 2.0 cm^3
b. 1.000 g and 2.00 cm^3
c. 1.0000 g and 2.0000 cm^3
d. 1.000 g and 2.0000 cm^3

3.159 A rectangular box measures 10 cm wide, 200 cm long, and 4 cm high. Calculate the volume of the box, in cubic centimeters, given that all the dimensions are known to
a. the closest centimeter.
b. the closest tenth of a centimeter.
c. the closest hundredth of a centimeter.
d. two significant figures.

3.160 A rectangular room has a width of 9 m and a length of 21 m. Calculate the area of this room, in square meters, given that both dimensions are known to
a. the closest meter.
b. the closest tenth of a meter.
c. the closest hundredth of a meter.
d. three significant figures.

3.161 Indicate which measurement in each of the following sets of measurements has the least uncertainty.
a. 3.256×10^3 g, 3.256×10^4 g, 3.256×10^5 g
b. 3.34 g, 3.34 kg, 3.34 mg
c. 4.31 g, 4.31×10^{-3} kg, 4.31×10^3 mg
d. 325.0 cg, 3.2500 g, 0.00325 kg

3.162 Indicate which measurement in each of the following sets of measurements has the least uncertainty.
a. 2.53×10^{-3} m, 2.53×10^{-4} m, 2.53×10^{-5} m
b. 7.612 m, 7.612 km, 7.612 cm

c. 6.73 m, 6.73 × 10^2 cm, 6.73 × 10^6 μm
d. 35.300 mm, 3.530 cm, 0.0353 m

3.163 Indicate whether each of the following measurements is equivalent in all aspects to the measurement 1.2120 g.
a. 0.00121 kg
b. 121.20 cg
c. 12120 mg
d. 1212.0 μg

3.164 Indicate whether each of the following measurements is equivalent in all aspects to the measurement 3.4020 L.
a. 0.034020 kL
b. 340.20 cL
c. 3402 mL
d. 0.0000034020 ML

Multiple-Choice Practice Test

Use this bank of 20 multiple-choice questions as a review of key concepts presented in this chapter. For many of the questions, there may be more than one correct response (choice d) or no correct response (choice e).

3.165 Which of the following metric system prefixes is correctly paired with its mathematical meaning?
a. milli- and 10^{-2}
b. micro- and 10^{-6}
c. giga- and 10^{-9}
d. more than one correct response
e. no correct response

3.166 In which of the following sequences are the metric system prefixes listed in order of *decreasing* size?
a. mega-, giga-, kilo-
b. nano-, micro-, milli-
c. pico-, centi-, deci-
d. more than one correct response
e. no correct response

3.167 Which of the following metric system unit abbreviations denotes a unit of volume?
a. mL
b. kg
c. cm^3
d. more than one correct response
e. no correct response

3.168 Which of the following statements concerning conversion factors is correct?
a. English-to-English conversion factors come from measured relationships.
b. Metric-to-English conversion factors come from defined relationships.
c. Conversion factors always come in reciprocal pairs.
d. more than one correct response
e. no correct response

3.169 Which of the following is an *exact* conversion factor rather than an *inexact* conversion factor?
a. 1 mile = 5280 feet
b. 1 inch = 2.540 centimeters
c. 1 kg = 1 × 10^3 grams
d. more than one correct answer
e. no correct answer

3.170 Which of the following is a *correct* conversion factor?
a. 453.6 g/1 lb
b. 10^{-2} cm/1 in.
c. 0.9463 qt/1 L

d. more than one correct response
e. no correct response

3.171 Given that the pathway for solving a problem using dimensional analysis is weeks→days→hours→minutes, how many conversion factors will be needed?
a. one   b. two   c. three   d. four
e. no correct response

3.172 Which of the following conversion factor sequences (unit setups) would affect the change from milligrams to kilograms?
a. mg × (g/mg) × (kg/g)
b. mg × (g/kg) × (mg/g)
c. mg × (mg/g) × (kg/g)
d. mg × (g/mg) × (g/kg)
e. no correct response

3.173 Which of the following conversion factor sequences (unit setups) would affect the change from ft/hr to cm/sec?
a. ft/hr × (in./ft) × (min/hr) × (sec/min)
b. ft/hr × (cm/in.) × (hr/min) × (min/sec)
c. ft/hr × (in./ft) × (cm/in.) × (hr/min)
      × (min/sec)
d. ft/hr × (in./ft) × (in./cm) × (hr/min)
      × (sec/min)
e. no correct response

3.174 How many nanograms of liquid are present in a 2.5 megagram sample of the liquid?
a. 2.5 × 10^3 ng
b. 2.5 × 10^{-3} ng
c. 2.5 × 10^{15} ng
d. 2.5 × 10^{-15} ng
e. no correct response

3.175 Express the area measurement 24.0 cm^2 in the units of square yards.
a. 0.263 yd^2
b. 0.103 yd^2
c. 0.00729 yd^2
d. 0.00287 yd^2
e. no correct response

3.176 Express the measurement 4.22 quarts in the units of microliters.
a. 3.99 × 10^{-6} μL
b. 3.99 × 10^6 μL
c. 4.46 × 10^{-6} μL
d. 4.46 × 10^{-3} μL
e. no correct response

3.177 Which of the following would be a correct set of density units?
a. L/g   b. lb/gal   c. kg/cm
d. more than one correct response
e. no correct response

3.178 If object A has a mass of 24 g and a volume of 2.0 mL and object B has a mass of 36 g and a volume of 6.0 mL, which of the following is a correct density comparison?
   a. B is more dense than A.
   b. A is twice as dense as B.
   c. B has a density one-third that of A.
   d. A and B have equal densities because the given units are the same.
   e. no correct response

3.179 What mass, in grams, of a solution with a density of 1.13 g/mL is needed to provide 25.0 L of the solution?
   a. 0.0224 g   **b.** 22.4 g   **c.** 28.6 g
   d. 22,400 g   **e.** no correct response

3.180 Which of the following is an *equivalence* conversion factor rather than an *equality* conversion factor?
   a. 4.9 g salt solution/1 mL salt solution
   b. 5.2 g salt/1 mL salt solution
   c. 2.12 g salt/1 mL salt
   d. more than one correct response
   e. no correct response

3.181 A student experimentally determines that the boiling point of a particular compound is 178.5°C. What is the percent error in the student's measurement, given that the accepted value for the boiling point of the compounds is 165.9°C?
   a. +7.06%   **b.** –7.06%   **c.** +7.59%
   d. –7.59%   **e.** no correct response

3.182 Which of the following comparisons of the size of the degree on the major temperature scales are correct?
   a. A kelvin is larger than a Celsius degree.
   b. A Fahrenheit degree and a Celsius degree are the same size.
   c. A Fahrenheit degree is larger than a kelvin.
   d. more than one correct response
   e. no correct response

3.183 Which of the following statements concerning the major temperature scales are correct?
   a. Celsius temperatures can never have negative values.
   b. The addition of 273 to a Fahrenheit scale reading converts it to a Kelvin scale reading.
   c. The freezing point of water is the same on the Celsius and Kelvin scales.
   d. more than one correct response
   e. no correct response

3.184 A termperature change of 16.0 degrees Fahrenheit is equivalent to how many degrees of change on the Celsius scale?
   a. 8.89°   **b.** 14.8°   **c.** 18.7°
   d. 28.8°   **e.** no correct response

Basic Concepts About Matter

4.1 CHEMISTRY—THE STUDY OF MATTER

Chemistry *is the scientific discipline concerned with the characteristics, composition, and transformations of matter.* What is matter? What is it that chemists study? Intuitively, most people have a general feeling for the meaning of the word *matter*. They consider matter to be the materials of the physical universe—that is, the stuff from which the universe is made. Such an interpretation is a correct one.

Matter *is anything that has mass and occupies space.* Matter includes all things—both living and non-living—that can be seen (such as plants, soil, and rocks), as well as things that cannot be seen (such as air and bacteria). Various forms of energy, such as heat, light, and electricity, are not considered to be matter. However, chemists must be concerned with energy as well as with matter, because nearly all changes that matter undergoes involve the release or absorption of energy.

The scope of chemistry is extremely broad, and it touches every aspect of our lives. An iron gate rusting, a chocolate cake baking, the diagnosis and treatment of a heart attack, the propulsion of a jet airliner, and the digesting of food all fall within the realm of chemistry. The key to understanding such diverse processes is an understanding of how matter can be classified into a surprisingly small number of categories. The naturally occurring materials of the universe and the synthetic materials humans have fashioned from them are, indeed, much simpler in makeup than they outwardly appear.

4.2 PHYSICAL STATES OF MATTER

Three physical states exist for matter: *solid*, *liquid*, and *gas*. The classification of a given matter sample in terms of physical state is based on whether its shape and volume are definite or indefinite.

Most people think of rocks as solids, hence the sayings "solid as a rock" and "it is written in stone." But at the high temperatures and pressures within the earth, rock can turn to molten magma.

Solid *is the physical state characterized by a definite shape and a definite volume.* Sugar cubes have the same shape and volume whether they are placed in a large container or on a table top (Fig. 4.1a). For solids in powdered or granulated forms, such as sugar or salt, a quantity of the solid takes the shape of the portion of the container it occupies, but each individual particle has a definite shape and volume. **Liquid** *is the physical state characterized by an indefinite shape and a definite volume.* A liquid always takes the shape of its container to the extent that it fills the container (Fig. 4.1b). **Gas** *is the physical state characterized by an indefinite shape and an indefinite volume.* A gas always completely fills its container, adopting both the container's volume and shape (Fig. 4.1c).

The state of matter observed for a particular substance is always dependent on the temperature and pressure under which the observation is made. Because we live on a planet characterized by relatively narrow temperature extremes, we tend to fall into the error of believing that the commonly observed states of substances are the only states in which they occur. Under laboratory conditions, states other than the natural ones can be obtained for almost all substances. Oxygen, which is nearly always thought of as a gas, can be obtained in the liquid and solid states at very low temperatures. People seldom think of the metal iron as being a gas, its state at extremely high temperatures (above 3000°C). At intermediate temperatures (1535–3000°C), iron is a liquid. Water is one of the very few substances familiar to everyone in all three of its physical states: solid ice, liquid water, and gaseous steam (see Fig. 4.2).

Chapter 11 will consider in detail further properties of the different physical states of matter, changes from one state to another, and the question of why some substances decompose. Suffice it to say at present that physical state is one of the qualities by which matter can be classified.

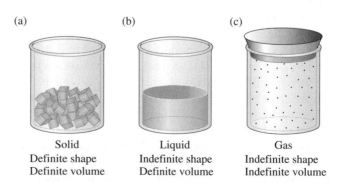

(a)	(b)	(c)
Solid	Liquid	Gas
Definite shape	Indefinite shape	Indefinite shape
Definite volume	Definite volume	Indefinite volume

FIGURE 4.1 (a) A solid has a definite shape and a definite volume. (b) A liquid has an indefinite shape—it takes the shape of its container—and a definite volume. (c) A gas has an indefinite shape and an indefinite volume—it assumes the shape and volume of its container.

FIGURE 4.2 Water can be found simultaneously in the solid, liquid, and vapor (gaseous) forms, as shown here at Yellowstone National Park. *(Shutterstock)*

4.3 PROPERTIES OF MATTER

How are the various kinds of matter differentiated from each other? The answer is simple—by their *properties*. **Properties** *are the distinguishing characteristics of a substance that are used in its identification and description.* Just as we recognize a friend by characteristics such as hair color, walk, tone of voice, or shape of nose, we recognize various chemical substances by how they look and behave. Each chemical substance has a unique set of properties that distinguishes it from all other substances. If two samples of matter have every property identical, they must necessarily be the same substance.

Knowledge of the properties of substances is useful in the following:

1. *Identifying an unknown substance.* Identifying a confiscated drug such as cocaine involves comparing the properties of the drug to those of known cocaine samples.
2. *Distinguishing between different substances.* A dentist can quickly tell the difference between a real tooth and a false tooth because of property differences.
3. *Characterizing a newly discovered substance.* Any new substance must have a unique set of properties different from those of any previously characterized substance.
4. *Predicting the usefulness of a substance for specific applications.* Water-soluble substances obviously should not be used in the manufacture of bathing suits.

Physical and Chemical Properties

There are two general categories of properties of matter: *physical* and *chemical*. A **physical property** *is a characteristic of a substance that can be observed without changing the substance into another substance.* Color, odor, taste, size, physical state, boiling point, melting point, and density are all examples of physical properties.

The physical appearance of a substance may change while a physical property is being determined, but the substance's identity will not. For example, the melting point of a solid cannot be measured without melting the solid, changing it to a liquid. Although the liquid's appearance is much different from that of the solid, the substance is still the same. Its chemical identity has not changed. Hence, melting point is a physical property.

A **chemical property** *is a characteristic of a substance that describes the way the substance undergoes or resists change to form a new substance.* When copper objects are exposed to moist air for long periods of time, they turn green; this is a chemical property of copper.

Chemical properties describe the ability of a substance to form new substances, either by reaction with other substances or by decomposition. Physical properties are associated with a substance's physical existence. They can be determined without reference to any other substance, and determining them causes no change in the identity of the substance.

TABLE 4.1 Selected Physical and Chemical Properties of Water

Physical Properties	Chemical Properties
1. Colorless 2. Odorless 3. Boiling point = 100°C 4. Freezing point = 0°C 5. Density = 1.000 g/mL at 4°C	1. Reacts with bromine to form a mixture of two acids. 2. Can be decomposed by means of electricity to form hydrogen and oxygen. 3. Reacts vigorously with the metal sodium to produce hydrogen. 4. Does not react with gold even at high temperatures. 5. Reacts with carbon monoxide at elevated temperatures to produce carbon dioxide and hydrogen.

The green coating formed on the copper is a new substance; it results from the reaction of copper metal with the oxygen, carbon dioxide, and water in air. The properties of this green coating are very different from those of metallic copper. On the other hand, gold objects resist change when exposed to air for long periods of time. The lack of reactivity of gold with air is a chemical property of gold.

Most often the changes associated with chemical properties result from the interaction (reaction) of a substance with one or more other substances. However, the presence of a second substance is not an absolute requirement. Sometimes the presence of energy (usually heat or light) can trigger the change called *decomposition*. The fact that hydrogen peroxide, in the presence of either heat or light, decomposes into the substances water and oxygen is a chemical property of hydrogen peroxide.

When we specify chemical properties, we usually give conditions such as temperature and pressure because they influence the interactions between substances. For example, the gases oxygen and hydrogen are unreactive toward each other at room temperature, but they interact explosively at a temperature of several hundred degrees Celsius.

Selected physical and chemical properties of water are contrasted in Table 4.1. Note how the chemical properties of water cannot be described without reference to other substances. It does not make sense to say simply that a substance reacts. The substance that it interacts with must be specified because it might interact with many different substances.

EXAMPLE 4.1 Classifying Properties as Physical or Chemical

Classify each of the following properties for selected metals as a *physical* property or a *chemical* property.

   a. Titanium metal can be drawn into thin wires.
   b. Silver metal shows no sign of reaction when placed in hydrochloric acid.
   c. Copper metal possesses a reddish-brown color.
   d. Potassium metal, which has a shiny appearance when freshly cut, turns dull gray when exposed to air.

SOLUTION

   a. Physical property. Pulling bulk metal into a thin wire does not change its composition.
   b. Chemical property. Reactivity with an acid as well as lack of reactivity with an acid are considered to be chemical properties.
   c. Physical property. The property of color is determined without changing the identity (composition) of the metal.
   d. Chemical property. A change in color indicates the formation of a new substance.

Answers to all practice exercises in this chapter are found in the back-of-the-book answer section.

▶ **Practice Exercise 4.1** Classify each of the following properties for selected metals as a *physical* property or a *chemical* property.

   a. Iron metal rusts in an atmosphere of moist air.

   b. Mercury metal is a liquid at room temperature.
   c. Nickel metal dissolves in acid to produce a light green solution.
   d. Beryllium metal, when inhaled in a finely divided form, can produce serious lung disease.

Intensive and Extensive Properties

The properties of a substance may be classified in a second manner—as *intensive* or *extensive* properties based on whether they depend on the *amount* of substance present. For example, two different-sized pieces of pure copper will have the same melting point and the same reddish-brown color (both intensive properties) but different masses (an extensive property).

An **intensive property** *is a property that is independent of the amount of substance present.* Temperature, color, melting point, and density are all intensive properties; they are the same for a small sample and for a large one of the same substance.

An **extensive property** *is a property that depends on the amount of substance present.* The mass, length, and volume of a substance are examples of extensive properties. Values of the same extensive property are additive. For example, two pieces of aluminum have a combined mass that is the sum of the individual masses. Intensive properties, such as melting point and density, which are the same for two samples of a substance, are not additive.

Because every sample of a given substance has the same intensive properties, such properties are more useful than extensive properties for characterizing (identifying) substances. Chemical properties of a substance, which are also intensive properties, are also key components of a substance-identification process.

Interestingly, taking the mathematical ratio of two numerical extensive properties produces an intensive property. In effect, the "amount factor" cancels out in such a calculation. The property of density (Sec. 3.8) is an example of such a situation. Density is the ratio of mass over volume. Both mass and volume are extensive properties; however, their ratio (density) is an intensive property.

Color is an intensive property that cannot always be used for identification. This property can fool you if particle size for a substance is extremely small. For example, silver is a silvery-white metal with a high luster; however, in the finely divided state, metallic silver appears black.

4.4 CHANGES IN MATTER

Changes in matter are common and familiar occurrences. Changes take place when food is digested, paper is burned, and iron rusts (see Fig. 4.3). Like properties of matter, changes in matter are classified into two categories: *physical* and *chemical*.

FIGURE 4.3 The rusting of an iron pipe is an example of a naturally occurring chemical change. *(Tobias Stinner/Fotolia)*

A **physical change** *is a process in which a substance changes its physical appearance but not its chemical composition.* A new substance is never formed as a result of a physical change.

A change in physical state is the most common type of physical change. The melting of ice, the freezing of liquid water, the conversion of liquid water into steam (evaporation), the condensation of steam to water, the sublimation of ice in cold weather, and the formation of snow crystals in clouds in the winter (deposition) all represent changes of state. The terminology used in describing changes of state, with the exception of the terms *sublimation* and *deposition*, should be familiar to almost everyone. Although the processes of sublimation and deposition—going from a solid directly to the gaseous state or vice versa—are not common, they are encountered in everyday life. Dry ice sublimes, as do mothballs placed in a clothing storage area. As mentioned previously, ice or snow forming in clouds is an example of deposition. Figure 4.4 summarizes the terminology used in describing changes of state.

In any change of state, the composition of the substance undergoing change remains the same even though its physical state and outward appearance have changed. The melting of ice does not produce a new substance. The substance is water before and after the change. Similarly, the steam produced from boiling water is still water. Changes such as these illustrate that matter can change in appearance without undergoing a change in chemical composition.

A **chemical change** *is a process in which a substance undergoes a change in chemical composition.* The creation of one or more new substances is always a characteristic of a chemical change. Carbon dioxide and water are two new substances produced when the chemical change associated with the burning of gasoline occurs. Ashes, carbon dioxide, and water are among the new substances produced when wood is burned. Chemical changes are often called *chemical reactions.* A **chemical reaction** *is a process in which at least one new substance is produced as a result of chemical change.* Table 4.2 classifies a number of changes for matter as being either physical or chemical.

Most changes for matter can easily be classified as either physical or chemical. However, not all changes are "black" or "white." There are some gray areas. For example, the formation of certain solutions falls in the gray area. Common salt dissolves easily in water to form a solution of saltwater. The salt can easily be recovered by the physical process of evaporating the water. When gaseous hydrogen chloride is dissolved in water, again a solution results, but in this case the starting materials cannot be easily recovered by evaporation. The formation of saltwater is considered a physical change because the original components can be recovered in an unchanged form using physical methods. The second solution presents classification problems because of the possibility that a chemical reaction took place.

The changes involved in the cooking of an egg also present classification problems. The cooked egg contains the same structural units as the uncooked egg. However, some changes in structural arrangement have taken place, so is the change physical or chemical? Despite

Frost on a windowpane is an example of deposition.

When roasted coffee is ground, the rich aroma is due to the sublimation of coffee components.

Not all physical changes involve a change of state. Pulverizing an aspirin tablet into a fine powder and cutting a piece of adhesive tape into small pieces are examples of physical changes that involve only the solid state.

FIGURE 4.4
Terminology associated with physical changes of state.

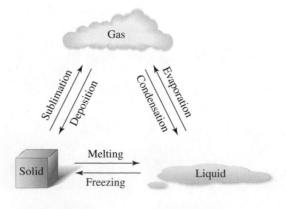

TABLE 4.2 Classification of Changes as Physical or Chemical

Change	Classification
Rusting of iron	chemical
Melting of snow	physical
Sharpening a pencil	physical
Digesting food	chemical
Taking a bite of food	physical
Burning gasoline	chemical
Slicing an onion	physical
Detonation of dynamite	chemical
Souring of milk	chemical
Breaking of glass	physical

the existence of gray areas, we shall continue to use the concepts of physical and chemical change because their usefulness far outweighs the problems created by a few exceptions.

Use of the Terms *Physical* and *Chemical*

Generalizations concerning the use of the terms *physical* and *chemical* are in order.

1. Whenever the term *physical* is used to modify another term, as in physical property or physical change, it always conveys the idea that the composition (identity) of the substance involved *did not change*.
2. Whenever the term *chemical* is used to modify another term, as in chemical property or chemical change, it always conveys the idea that the composition (identity) of the substance(s) involved *did change* or *successfully resisted change* as the result of an external challenge to identity.

The message of the "modifiers" *physical* and *chemical* is constant: *Physical* denotes no change in composition, and *chemical* denotes change in composition or resistance to such change.

Based on these generalizations, we note that techniques used to accomplish physical change are called physical methods or physical means. Chemical methods and chemical means are used to bring about chemical change. A physical separation would be a separation process in which none of the components experienced composition changes. Composition changes would be part of a chemical separation process.

EXAMPLE 4.2   **Correct Use of the Terms *Physical* and *Chemical***

Correctly complete each of the following sentences by placing the word *physical* or *chemical* in the blank.

   a. The fact that pure aspirin melts at 143°C is a _____ property of aspirin.
   b. The fact that sodium metal explosively interacts with water to produce hydrogen gas is a _____ property of sodium.
   c. Straightening a bent nail with a hammer is an example of a _____ change.
   d. Draining off the water from a water-and-hard-boiled-egg mixture is an example of a _____ separation.

SOLUTION

   a. Physical. Changing solid aspirin to liquid aspirin (melting) does not produce any new substances. We still have aspirin.

    b. Chemical. A new substance, hydrogen, is produced.

    c. Physical. The nail is still a nail.

    d. Physical. We started out with water and eggs, and after the separation we still have water and eggs.

▶ **Practice Exercise 4.2**   Correctly complete each of the following sentences by placing the word *physical* or *chemical* in the blank.

    a. The fact that the metal gold does not tarnish in the presence of oxygen is a _____ property of gold.

    b. A _____ change produces a different form of the same substance.

    c. The vaporization of dry ice (solid state carbon dioxide) is an example of a _____ change.

    d. The yellowish-green color of chlorine gas is a _____ property of this gas.

4.5 PURE SUBSTANCES AND MIXTURES

Substance is a general term used to denote any variety of matter. *Pure substance* is a specific term that applies only to matter that contains a single substance.

All samples of a pure substance, no matter what their source, have the same properties under the same conditions.

In addition to its classification by physical state (Sec. 4.2), matter can also be classified in terms of its composition as a *pure substance* or a *mixture*. A **pure substance** *is a single kind of matter that cannot be separated into other kinds of matter using physical means.* All samples of a pure substance contain only that substance and nothing else. Pure water is water and nothing else. Pure sucrose (table sugar) contains only that substance and nothing else.

    A pure substance always has a definite and constant composition. This invariant composition dictates that the properties of a pure substance are always the same under a given set of conditions. Collectively, these definite and constant physical and chemical properties constitute the means for identification of the pure substance.

    A **mixture** *is a physical combination of two or more pure substances in which each substance retains its own identity.* Components of a mixture retain their identity because they are physically mixed rather than chemically combined. Consider, for example, a mixture of salt and pepper (see Fig. 4.5). Close examination of such a

FIGURE 4.5 The individual particles of salt and pepper are easily recognizable in a mixture of the two substances because a mixture is a physical rather than chemical combination of substances. *(Ken Lax/ Pearson Education/ PH College)*

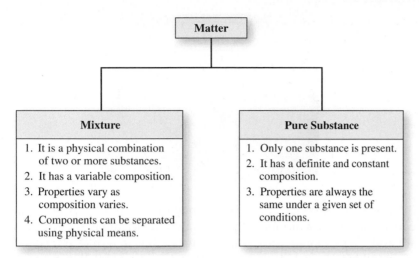

FIGURE 4.6 A comparison chart contrasting the differences between mixtures and pure substances.

mixture will show distinct particles of salt and pepper with no obvious interaction between them. The salt particles in the mixture are identical in properties and composition to the salt particles in the salt container, and the pepper particles in the mixture are no different from those in the pepper container.

Once a particular mixture is made up, its composition is constant. However, mixtures of the same components with different compositions can also be made up; thus, mixtures are considered to have variable compositions. Consider the large number of salt and pepper mixtures that could be produced by varying the amounts of the two substances present.

An additional characteristic of any mixture is that its components can often be retrieved intact from the mixture by physical means, that is, without a chemical change. In many cases, the differences in properties of the various components make the separation relatively easy. For example, in our salt–pepper mixture, if the pepper grains were large enough, the two components could be separated manually by picking out all the pepper grains. Alternatively, the separation could be carried out by dissolving the salt particles in water, removing the insoluble pepper particles, and then evaporating the water to recover the salt.

Most mixture separations are not as easy as that for a salt–pepper mixture; expensive instrumentation and numerous separation steps are required. Imagine the logistics involved in the separation of the components of blood, a water-based mixture of varying amounts of proteins, sugar (glucose), salt (sodium chloride), oxygen, carbon dioxide, and other components.

Figure 4.6 summarizes the differences between mixtures and pure substances. Further considerations about mixtures are found in Section 4.6, and Section 4.7 contains further details about pure substances.

Commercial table salt is not just the pure substance sodium chloride. It is a mixture containing small amounts of potassium iodide as a nutritional supplement, dextrose as a stabilizer, and calcium silicate as an anticaking agent (to keep it pouring when it rains) in addition to the sodium chloride.

Most naturally occurring samples of matter are mixtures. Gold and diamond are two of the few naturally occurring pure substances. Despite their scarcity in nature, numerous pure substances are known. They are obtained from natural mixtures by using various types of separation techniques or are synthesized in the laboratory from naturally occurring materials.

4.6 HETEROGENEOUS AND HOMOGENEOUS MIXTURES

Mixtures are subclassified as *heterogeneous* or *homogeneous*. This subclassification is based on visual recognition of the mixture's components. A **heterogeneous mixture** *is a mixture that contains two or more visually distinguishable phases (parts), each of which has different properties.* A heterogeneous mixture of sand and sugar is said to be composed of two phases: a sand phase and a sugar phase. Because two or more phases are always present, a nonuniform appearance is characteristic of all heterogeneous mixtures. Naturally occurring heterogeneous mixtures include rocks, soils, and wood (see Fig. 4.7).

Hetero is a prefix that means "different." Its use in the classification term *heterogeneous mixture* focuses on the different properties associated with different parts of the mixture.

FIGURE 4.7 Wood is an example of a heterogeneous mixture. *(Tim Mainiero/Shutterstock)*

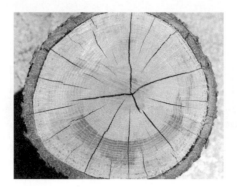

American dimes and quarters in current use are heterogeneous mixtures of metals. A copper phase is sandwiched between a copper–nickel phase. Older "silver" dimes and quarters are a homogeneous mixture of silver (90%) and copper (10%).

Homo is a prefix that means "the same." Its use in the classification term *homogeneous mixture* focuses on the same properties throughout the mixture.

Manufacturers add substances called emulsifying agents to products like chocolate bars and peanut butter to keep these products homogeneous. Without emulsifiers, the ingredients would slowly separate into phases and look unpalatable.

The phases in a heterogeneous mixture may or may not be in the same physical state. Set concrete contains a number of phases, all of which are in the solid state. A mixture of sand and water contains two phases, and each is in a different state (solid and liquid). It is possible to have heterogeneous mixtures in which all components are liquids. In order for these mixtures to occur, the mixed liquids must have limited solubility in each other. When this is the case, the mixed liquids form separate layers with the least dense liquid on top. An oil-and-vinegar salad dressing is an example of such a liquid–liquid mixture. Oil-and-vinegar dressing consists of two phases (oil and vinegar) regardless of whether the mixture consists of two separate layers or of oil droplets dispersed throughout the vinegar, a condition caused by shaking the mixture. All the oil droplets together are considered to be a single phase.

A **homogeneous mixture** *is a mixture that contains only one visually distinguishable phase (part), which has uniform properties throughout.* A sugar–water mixture in which all the sugar has dissolved is a one-phase (homogeneous) system with an appearance that cannot be distinguished from that of pure water. Air is a homogeneous mixture of gases, motor oil and gasoline are multicomponent homogeneous mixtures of liquids, and metal alloys such as 14-karat gold (a mixture of copper and gold) are examples of homogeneous mixtures of solids. The homogeneity present in solid metallic alloys is achieved by mixing the metals while they are in the molten state. Obviously, homogeneous mixtures are possible only when all components present are in the same physical state.

A thorough intermingling of the components in a homogeneous mixture is required in order for a single phase to exist. Sometimes this occurs almost instantaneously during the preparation of the mixture, as in the addition of alcohol to water. At other times, an extended period of mixing or stirring is required. For example, when a hard sugar cube is added to a container of water, it does not instantaneously dissolve to give a homogeneous solution. Only after much stirring does the sugar completely dissolve. Prior to that point, the mixture is heterogeneous, containing a solid phase (the undissolved sugar cube) and a liquid phase (sugar dissolved in water).

Figure 4.8 contrasts the properties common to all mixtures with those specific for heterogeneous and homogeneous mixtures.

Use of the Terms *Homogeneous* and *Heterogeneous*

A summary of the major concepts developed in both this section and Section 4.5 is presented in Figure 4.9. This summary is based on the interplay between the terms *heterogeneous* and *homogeneous* and the terms *chemical* and *physical*. From this interplay come the new expressions *chemically homogeneous, chemically heterogeneous, physically homogeneous,* and *physically heterogeneous.*

All pure substances are *chemically homogeneous.* Only one substance can be present in a chemically homogeneous material. Mixtures, which by definition must contain two or more substances, are always *chemically heterogeneous.* The term *physically homogeneous*

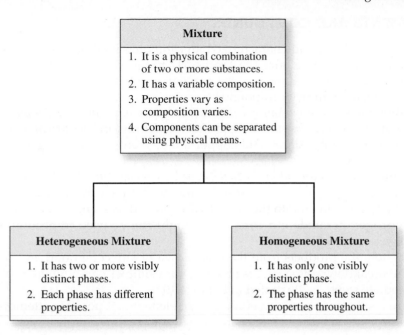

FIGURE 4.8 A comparison chart contrasting the properties common to all mixtures with those specific for heterogeneous and homogeneous mixtures.

describes materials consisting of only one phase. If two or more phases are present, then the term *physically heterogeneous* applies.

Pure water (Fig. 4.9a) is both chemically homogeneous and physically homogeneous because only one substance and one phase are present. Water with some sugar dissolved in it (Fig. 4.9b) is physically homogeneous with only one phase present, but is chemically heterogeneous because two substances, sugar and water, are present. A mixture of oil and water (Fig. 4.9c) is both chemically and physically heterogeneous because it contains two substances and two phases. An ice cube in liquid water (Fig. 4.9d) represents the somewhat unusual situation of chemical homogeneity and physical heterogeneity—one substance and two phases. This combination occurs only when a single substance is present in two or more physical states (solid, liquid, gas).

Blood appears homogeneous to the naked eye. But when looked at under a microscope, it can be seen to be a heterogeneous mixture of red and white blood cells and liquid called plasma.

Chemically homogeneous
Physically homogeneous

One substance and one phase
(a) Pure water

Chemically heterogeneous
Physically homogeneous

Two substances and one phase
(b) Sugar water

FIGURE 4.9
Describing a sample of matter using the terms *chemically heterogeneous, physically heterogeneous, chemically homogeneous,* and *physically homogeneous.*

Chemically heterogeneous
Physically heterogeneous

Two substances and two phases
(c) Oil and water

Chemically homogeneous
Physically heterogeneous

One substance and two phases
(d) Ice and water

4.7 ELEMENTS AND COMPOUNDS

Chemists have isolated and characterized an estimated 9 million pure substances. A very small number of these pure substances, 117 to be exact, are different from all the others. They are *elements*. All the rest, the remaining millions, are *compounds*. What distinguishes an element from a compound?

The definition for the term *element* given here will do for now. After considering the concept of atomic number (Sec. 5.2), we will present a more accurate definition.

An **element** *is a pure substance that cannot be broken down into simpler pure substances using ordinary chemical means such as a chemical reaction, an electric current, heat, or a beam of light.* The metals gold, silver, and copper are elements as are the gases hydrogen, oxygen, and nitrogen.

A **compound** *is a pure substance that can be broken down into two or more simpler pure substances using chemical means.* Water is a compound. By means of an electric current, water can be broken down into the gases hydrogen and oxygen, both of which are elements. Hydrogen peroxide is a compound. Light can be used to decompose it into water and gaseous oxygen.

Ultimately the products from the breakdown of any compound are elements. In practice the breakdown often occurs in steps, with simpler compounds resulting from the intermediate steps, as illustrated in Figure 4.10.

Every compound that exists is made up of some combination of two or more of the 117 known elements. Usually, four or fewer elements are present in a given compound.

Before a substance can be classified as an element, all possible attempts must be made chemically to subdivide it into simpler substances. If a sample of pure substance, S, is subjected to a decomposition process and two new substances, X and Y, are produced, S would be classified as a compound. If, on the other hand, a number of attempts made chemically to subdivide S proved unsuccessful, we might correctly call it an element, but until all possible reactions have proved unsuccessful, such a classification could be in error.

Figure 4.11 contrasts the properties general to all pure substances with those specific for the pure substance subclassifications of elements and compounds.

Even though two or more elements are obtained from decomposition of compounds, compounds are not mixtures. Why is this so? Remember, substances can be combined either physically or chemically. Physical combination of substances produces a mixture. Chemical combination of substances produces a compound, a substance in which combining entities are *bound together*. No such binding occurs during physical combination. Three important distinctions between compounds and mixtures are the following:

1. Compounds have properties distinctly different from those of the substances that combined to form the compound. The components of mixtures retain their individual properties.
2. Compounds have a definite composition. Mixtures have a variable composition.
3. Physical methods are sufficient to separate the components of a mixture. The components of a compound cannot be separated by physical methods; chemical methods are required.

FIGURE 4.10 Stepwise breakdown of a compound containing three elements (A, B, and C) to yield its constituent elements.

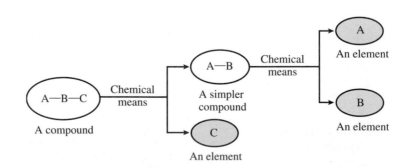

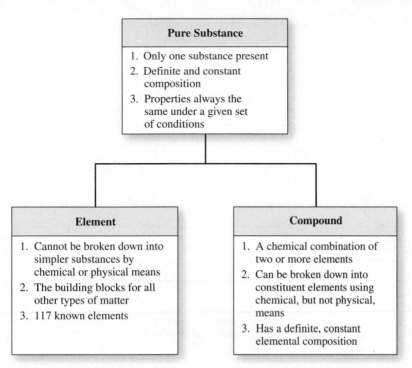

FIGURE 4.11 A comparison chart contrasting the properties common to all pure substances with those specific for elements and compounds.

Example 4.3, which involves two comparisons involving ballpoint pens and their caps, further illustrates the difference between compounds and mixtures.

EXAMPLE 4.3 **"Composition" Difference between a Mixture and a Compound**

Consider two boxes with the following contents: The first contains 25 ballpoint pens, each with its cap on; the second contains 25 ballpoint pens without caps and 25 ballpoint pen caps. Which box has contents that would be an analogy for a mixture and which box has contents that would be an analogy for a compound?

SOLUTION

The box containing the ballpoint pens with their caps on represents the compound. Two samples withdrawn from this box will always be the same; each will be a ballpoint pen with its cap on. Each item in the box has the same composition.

The box containing separated ballpoint pens and caps represents the mixture. Two samples withdrawn from this box need not be the same; results could be two ballpoint pens, two caps, or a cap and a ballpoint pen. All items in the box do not have the same composition.

▶ **Practice Exercise 4.3** Consider two boxes with the following contents: The first contains 10 paddle locks and 10 keys that fit the locks; the second contains 10 paddle locks with each lock's key inserted into the cylinder. Which box has contents that would be an analogy for a mixture and which box has contents that would be an analogy for a compound?

Figure 4.12 presents the thought processes that a chemist goes through in classifying a sample of matter as a heterogeneous mixture, a homogeneous mixture, a

FIGURE 4.12 The thought processes used in classifying matter into various categories.

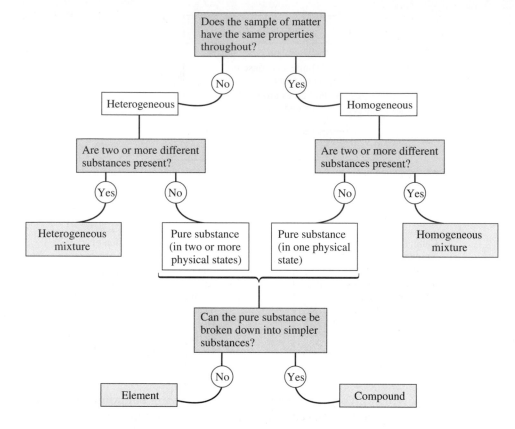

compound, or an element. This figure, which serves as a final summary of matter classification, is based on the following three questions about a sample of matter:

1. Does the sample of matter have the same properties throughout?
2. Are two or more different substances present?
3. Can the pure substance be broken down into simpler substances?

EXAMPLE 4.4   Classifying Substances as Element, Compound, or Mixture

Consider the following properties that involve the metal silver:

1. Silver metal cannot be decomposed into simpler substances using chemical techniques.
2. Silver metal reacts with gaseous hydrogen sulfide to form the black solid silver sulfide (silverware tarnish).
3. Sterling silver is a metal alloy that contains 92.5% silver and 7.5% copper by mass.
4. Argenite ore, a source for silver metal, is found in Mexico and Peru.

Based on these properties, classify each of the following silver-containing entities as an element, compound, homogeneous mixture, or heterogeneous mixture.

    a. sterling silver   **b.** argenite ore
    c. silverware tarnish   **d.** silver metal

SOLUTION

    a. Homogeneous mixture. Alloys are homogeneous physical combinations of substances obtained by mixing the substance while they are in a molten state.
    b. Heterogeneous mixture. Ores contain numerous visible components besides the metal to be extracted.

c. Compound. Silverware tarnish is produced by a chemical reaction rather than a physical mixing of substances.

d. Element. A substance that cannot be decomposed into simpler substances using chemical means is an element.

▶ **Practice Exercise 4.4** Consider the following properties that involve the metal copper:

1. Bronze is a metal alloy containing copper and up to 18% tin.
2. Patina, the green film that forms on weather-exposed copper objects contains both copper carbonate and copper hydroxide.
3. Copper metal cannot be broken down into simpler substances even when heated to extremely high temperatures.
4. Copper sulfide is the form in which copper is found in almost all copper ores.

Based on these properties, classify each of the following copper-containing entities as an element, compound, homogeneous mixture, or heterogeneous mixture.

a. patina
c. copper metal

b. bronze
d. copper sulfide

4.8 DISCOVERY AND ABUNDANCE OF THE ELEMENTS

The discovery and isolation of the 117 known elements, the building blocks for all matter, have taken place over a period of several centuries. Most of the discoveries have occurred since 1700, with the 1790s and 1880s being the most active decades.

Not all the elements are naturally occurring. Eighty-eight of the 117 elements occur naturally, and the remaining 29 have been synthesized in the laboratory by bombarding samples of naturally occurring elements with small particles (Sec. 5.10). The synthetic (laboratory-produced) elements are all unstable (radioactive) and usually quickly revert back to naturally occurring elements. Further details about the concept of radioactivity (unstable elements) are found in the next chapter (Secs 5.6 through 5.13).

The naturally occurring elements are not evenly distributed on Earth or in the universe as a whole. What is startling is the degree of inequality in the distribution. A very few elements account for the majority of elemental particles (atoms). (An atom is the smallest particle of an element that can exist; see Sec. 4.10.)

Studies of the radiation emitted by stars enable scientists to estimate the elemental composition of the universe. Results indicate that two elements, hydrogen and helium, are absolutely dominant (Fig. 4.13a). All other elements are mere "impurities" when their abundances are compared with those of these two dominant elements. In this big picture, in which Earth is but a tiny microdot, 91% of

A student who attended a university in the year 1700 would have been taught that 11 elements existed. In 1750, he or she would have learned about 17 elements, in 1800 about 43, in 1850 about 58, in 1900 about 83, and in 1950 about 96. Today's total of 117 elements was reached in 2006.

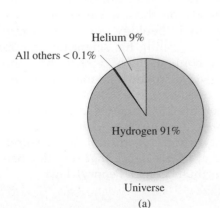

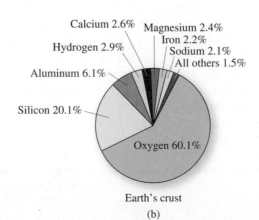

Helium 9%

All others < 0.1%

Hydrogen 91%

Universe

(a)

Calcium 2.6% Magnesium 2.4%
Iron 2.2%
Hydrogen 2.9% Sodium 2.1%
All others 1.5%
Aluminum 6.1%

Silicon 20.1%

Oxygen 60.1%

Earth's crust

(b)

FIGURE 4.13
Abundance of elements in the universe and in Earth's crust (in atom percent).

Any increase in the number of known elements from 117 will result from the production of additional synthetic elements. Current chemical theory strongly suggests that all naturally occurring elements have been identified. The isolation of the last of the known naturally occurring elements, rhenium, occurred in 1925.

all elemental particles (atoms) are hydrogen, and nearly all of the remaining 9% are helium.

If we narrow our view of elemental abundances to the chemical world of humans—the Earth's crust (its waters, atmosphere, and outer solid surfaces)—a different perspective emerges. Again, two elements dominate, but this time they are oxygen and silicon (Fig. 4.13b). The numbers in Figure 4.13 are atom percents, that is, the percentage of total atoms that are of a given type. Note that eight elements (the only elements with atom percents greater than 1%) account for over 98% of total atoms in Earth's crust. Furthermore, two elements (oxygen and silicon) account for 80% of the atoms that make up the chemical world of humans.

Oxygen, the most abundant element in Earth's crust, was isolated in pure form for the first time in 1774 by the English chemist and theologian Joseph Priestley (see "The Human Side of Chemistry 1"). Discovery years for the other "top five" elements of Earth's crust are 1824 (silicon), 1827 (aluminum), 1766 (hydrogen), and 1808 (calcium).

The Human Side of Chemistry 1

Joseph Priestley (1733–1804)

Oxygen, the most abundant element within Earth's crust, was isolated in gaseous form for the first time on Monday, August 7, 1774, by the English theologian and part-time chemist Joseph Priestley. The experiment leading to oxygen's isolation involved heating an oxygen-containing compound of mercury until it decomposed.

Born in Yorkshire, England, in 1733, Priestley was the eldest son of a nonconformist minister. As a youth, he studied languages, logic, and philosophy but never formally studied science. He developed strong religious beliefs of his own and became a Unitarian minister.

As a result of meeting Benjamin Franklin in London in 1766, when Franklin was attempting to settle the taxation dispute between the British government and the American colonists, Priestley became interested in science. At first his studies involved electricity (one of Franklin's interests), and then his interest turned to research on gases.

Coincident with his focus on gases was a move to Leeds, England, to take over a pastorate there. Here he lived next door to a brewery from which he could readily obtain carbon dioxide, the gas produced by fermenting grain products. His carbon dioxide studies led him to the idea of dissolving this gas under pressure in water. The resulting soda water became famous all over Europe.

Only three gases were known when Priestley began his gas studies: air, carbon dioxide, and the then recently discovered hydrogen. Numerous new gases were discovered by him, including ammonia, hydrogen chloride, sulfur dioxide, carbon monoxide, hydrogen sulfide, and his most notable discovery, oxygen.

During the period of his scientific studies, Priestley remained an outspoken man on religion and also openly supported the American colonists in their revolt against the British king. In 1791, while living in Birmingham, England, his church, home, and laboratory were burned by an angry mob upset because of his sympathetic attitude toward the American and French Revolutions. After fleeing the area in disguise, Priestley eventually emigrated to the United States, where he lived the remaining 10 years of his life in relative seclusion in Northumberland, Pennsylvania.

Priestley was the author of more than 150 books, mostly on theological subjects. He always considered theology more important than science. Concerning his scientific accomplishments, a contemporary wrote that "no single person ever discovered so many new and curious substances."

In 1874, on the 100th anniversary of the discovery of oxygen, the American Chemical Society, now the largest chemical organization in the world, was organized at the residence that Priestley had occupied in Northumberland, Pennsylvania.

4.9 NAMES AND CHEMICAL SYMBOLS OF THE ELEMENTS

Each element has a unique name, which in most cases was selected by its discoverer. A wide variety of rationales for choosing a name has been applied. Some elements bear geographical names. Germanium was named after the native country of its German discoverer. The elements francium and polonium acquired names in a similar manner. The elements mercury, uranium, neptunium, and plutonium are all named for planets. Helium gets its name from the Greek word *helios* for "sun," since it was first observed spectroscopically in the sun's corona during an eclipse. Some elements carry names that relate to specific properties of the element or compounds containing it. Chlorine's name is derived from the Greek *chloros* denoting "greenish yellow," the color of chlorine gas. Iridium gets its name from the Greek *iris* meaning "rainbow" because of the various colors of the compounds from which it was isolated.

In the early 1800s chemists adopted the practice of assigning *chemical symbols* to the elements. A **chemical symbol** *is a one- or two-letter designation for an element derived from the element's name.* In written communications, chemical symbols are used more frequently than names in referring to the elements. The system of chemical symbols now in use originated in 1814 with the Swedish chemist Jöns Jakob Berzelius (see "The Human Side of Chemistry 2").

> Learning the chemical symbols of the more common elements is an important key to having a successful experience in studying chemistry. Knowledge of chemical symbols is essential for writing chemical formulas (Sec. 4.13), naming compounds (Chapter 8), and writing chemical equations (Sec. 10.2).

The Human Side of Chemistry 2

Jöns Jakob Berzelius (1779–1848)

"These symbols are horrifying. A young student in chemistry might as well learn Hebrew as make himself acquainted with them." Such was the response of a contemporary when, in 1814, Jöns Jakob Berzelius first proposed the system of elemental symbols that forms the basis for the symbol system we use today. This initial opposition turned later to acceptance as the advantages of his symbol system became apparent, although acceptance was slow during his lifetime.

Born in Vaversunda, Sweden, in 1779, Berzelius was the son of a clergyman-schoolmaster. The death of both his parents, before he was 9, caused his youth to be a constant shuffling between relatives.

In 1796, at age 17, partly to get away from his unhappy home situation, he began the study of medicine in Uppsala, obtaining a degree six years later. His first position was physician to the poor in several neighborhoods in Stockholm. A university appointment in Stockholm as professor of Medicine and Pharmacy came in 1807. Berzelius's true interests were, however, in the field of chemistry, and he left medicine shortly thereafter. In 1815 he obtained an appointment as professor of chemistry.

Berzelius loved experimental (laboratory) work and was probably the best experimental chemist of his generation. His experimental contributions to chemistry are considered even more important than his chemical symbols. Working in a laboratory with facilities no more elaborate than those of a kitchen, he performed more than 2000 experiments over a 10-year period to determine accurate atomic masses (Sec. 5.4) for 43 of the known 48 elements. The values he obtained are remarkably accurate, as measured by today's standards, an amazing accomplishment considering the crudeness of his laboratory equipment. He reported atomic masses of 35.41 and 63.00 for chlorine and copper, respectively. Today's accepted values for these two elements are, respectively, 35.45 and 63.55. Many laboratory innovations, among them the wash bottle, filter paper, and rubber tubing, came from the work of Berzelius.

In addition to atomic symbols and atomic masses, Berzelius also discovered the elements cerium, thorium, selenium, and silicon. Silicon is the second most abundant element in Earth's crust, exceeded only by oxygen (Sec. 4.8). Silicon–oxygen compounds (silicates) make up most of Earth's rock, soil, and sand. In recent times, "silicon chips" have become the basis for the computer industry. Indeed, a region of California, near San Francisco, where many different computer-related industries are located, is called Silicon Valley.

A complete list of the known elements and their chemical symbols is given in Table 4.3. The chemical symbols and names of the more frequently encountered elements are shown in color in this table.

Fourteen elements have one-letter symbols, and the rest have two-letter symbols. If a symbol consists of a single letter, it is capitalized. In all two-letter symbols the first letter is always capitalized, and the second letter is always lowercase. Two-letter symbols usually, but not always, start with the first letter of the element's English name. The second letter of the symbol is frequently, but not always, the second letter of the name.

The distribution of elements in the human body and other living systems is very different from that found in the earth's crust. This results from living systems *selectively* taking up matter from their external environment rather than simply accumulating matter representative of their surroundings. The four most abundant elements in the human body, in terms of atom percent, are hydrogen (63%), oxygen (25.5%), carbon (9.5%), and nitrogen (1.4%). The high abundances of hydrogen and oxygen in the body reflect the high water content of the body. Hydrogen is over twice as abundant as oxygen, largely because water contains hydrogen and oxygen in a 2:1 atom ratio.

TABLE 4.3 The Chemical Symbols for the Elements

Ac	actinium	Ge	germanium	**Pt**	**platinum**
Ag	**silver***	**H**	**hydrogen**	Pu	plutonium
Al	**aluminum**	**He**	**helium**	Ra	radium
Am	americium	Hf	hafnium	Rb	rubidium
Ar	**argon**	**Hg**	**mercury***	Re	rhenium
As	arsenic	Ho	holmium	Rf	rutherfordium
At	astatine	Hs	hassium	Rg	roentgenium
Au	**gold***	**I**	**iodine**	Rh	rhodium
B	**boron**	In	indium	Rn	radon
Ba	**barium**	Ir	iridium	Ru	ruthenium
Be	**beryllium**	**K**	**potassium***	**S**	**sulfur**
Bh	bohrium	Kr	krypton	Sb	antimony*
Bi	bismuth	La	lanthanum	Sc	scandium
Bk	berkelium	**Li**	**lithium**	Se	selenium
Br	**bromine**	Lr	lawrencium	Sg	seaborgium
C	**carbon**	Lu	lutetium	**Si**	**silicon**
Ca	**calcium**	Md	mendelevium	Sm	samarium
Cd	cadmium	**Mg**	**magnesium**	**Sn**	**tin***
Ce	cerium	Mn	manganese	Sr	strontium
Cf	californium	Mo	molybdenum	Ta	tantalum
Cl	**chlorine**	Mt	meitnerium	Tb	terbium
Cm	curium	**N**	**nitrogen**	Tc	technetium
Co	**cobalt**	**Na**	**sodium***	Te	tellurium
Cr	**chromium**	Nb	niobium	Th	thorium
Cs	cesium	Nd	neodymium	Ti	titanium
Cu	**copper***	**Ne**	**neon**	Tl	thallium
Db	dubnium	**Ni**	**nickel**	Tm	thulium
Ds	darmstadtium	No	nobelium	**U**	**uranium**
Dy	dysprosium	Np	neptunium	V	vanadium
Er	erbium	**O**	**oxygen**	W	tungsten*
Es	einsteinium	Os	osmium	Xe	xenon
Eu	europium	**P**	**phosphorus**	Y	yttrium
F	**fluorine**	Pa	protactinium	Yb	ytterbium
Fe	**iron***	**Pb**	**lead***	**Zn**	**zinc**
Fm	fermium	Pd	palladium	Zr	zirconium
Fr	francium	Pm	promethium		
Ga	gallium	Po	polonium		
Gd	gadolinium	Pr	praseodymium		

Only 111 elements are listed in this table. Elements 112–116 and 118, discovered (synthesized) in the period 1996–2006, are yet to be named.

*These elements have symbols that were derived from non-English sources.

TABLE 4.4 Elements Whose Chemical Symbols Are Derived from a Non-English Name of the Element

English Name of Element	Non-English Name of Element	Chemical Symbol
Chemical Symbols From Latin		
Antimony	stibium	Sb
Copper	cuprum	Cu
Gold	aurum	Au
Iron	ferrum	Fe
Lead	plumbum	Pb
Mercury	hydragyrum	Hg
Potassium	kalium	K
Silver	argentum	Ag
Sodium	natrium	Na
Tin	stannum	Sn
Chemical Symbol From German		
Tungsten	wolfram	W

Gold gets its symbol from the Latin *aurum*, which means "shiny." The liquid metal mercury gets its symbol from the Latin *hydragyrum*, which means "runs like water." Lead gets its symbol from the Latin *plumbum*, from which we get the name "plumber," because pipes used to be made out of lead.

Consider the elements terbium, technetium, and tellurium, whose symbols are, respectively, Tb, Tc, and Te. Obviously, a variety of choices of second letters is necessary because the first two letters are the same in all three elements' names.

Eleven elements have chemical symbols that bear no relationship to the element's English-language name. In ten of these cases, the symbol is derived from the Latin name of the element; in the case of tungsten, a German name is the symbol source. Most of these elements have been known for hundreds of years and date back to the time when Latin was the language of scientists. Table 4.4 shows the relationship between the chemical symbol and the non-English name of these 11 elements.

The chemical symbols of the elements are also found on the inside front cover of this book. The chart of elements on the left side of the inside front cover is called a *periodic table*. More will be said about it in later chapters. Both cover listings also give other information about the elements. This additional information will be discussed in Chapter 5.

4.10 THE ATOM

Can a sample of an element, say gold, be divided endlessly into smaller and smaller pieces of gold, or is there a limit to the subdivision process whereby a "smallest possible piece" of gold is obtained? In other words, is matter "continuous" or "discontinuous"? This is a concept that was debated by philosophers for many centuries without a conclusion being reached, although the "continuous" concept tended to be favored.

A definitive answer to this "continuous–discontinuous" question is now available. It came in the nineteenth century and is based on scientific experimentation rather than philosophical speculation. In a series of papers published in the period 1803–1807, the English chemist John Dalton (1766–1844; see "The Human Side of Chemistry 3") proposed that matter is *discontinuous*; that is, there is a limit to the process of subdividing matter into smaller and smaller particles. Dalton's proposal was based on data he and other scientists had collected concerning the amounts of different substances that react with each other.

The Human Side of Chemistry 3

John Dalton (1766–1844)

Born in Cumberland, England, in 1766, John Dalton was the second son of a poverty-stricken Quaker weaver. His formal education at the village school lasted until age 11. At age 12, Dalton himself was teaching in the village school. Shortly thereafter, he made his first attempts at scientific investigations, recording weather observations.

Throughout his life, Dalton had a particular interest in the study of weather. In 1787 he made his first entry in a notebook entitled "Observations on the Weather." He continued to record temperature, barometric pressure, rainfall, dew point, and so on, for the next 57 years; the last of over 200,000 observations was made the evening before his death.

In 1793, Dalton moved to Manchester, England, where he remained the rest of his life. He supported himself by private tutoring, which left him time to pursue his scientific investigations on an almost full-time basis. Some of his investigations involved color blindness, a personal affliction of his. He was the first person to describe color blindness.

His interest in meteorology was responsible for his greatest contribution to chemistry, the atomic theory of matter. From "weather" he turned his attention to the nature of the atmosphere and then to the study of gases in general. Dalton's atomic theory, first published in 1808, was based on his observations of the behavior of gases. He is also the formulator of the gas law now called Dalton's law of partial pressures (Sec 12.16).

Dalton remained a devout Quaker all his life. He was a very poor speaker and was not well received as a lecturer. He shunned honors and never found time for marriage. In later years, honors did come to him, including honorary doctor's degrees from Oxford and Cambridge Universities.

In 1832, some of his colleagues sought to present him to King William IV. Dalton objected because he did not want to wear the court dress. He finally went in the scarlet robes of Oxford University. Quakers do not wear scarlet, but Dalton, being color-blind to red, saw scarlet as gray. So he appeared before the king in scarlet but in gray to himself.

Upon his death in 1844, he was accorded a public funeral in Manchester. Over 40,000 persons passed by his casket, an appropriate tribute to a man who was so instrumental in revolutionizing the science of chemistry.

The word *atom* is derived from the Greek *atmos* meaning "indivisible." The Greek philosopher Democritus (460–370 B.C.), a proponent of the discontinuous matter concept, was the first to use this term.

His data, inconsistent with the idea of infinitely divisible matter, were compatible with the concept that a limit to the process of physical subdivision of matter exists.

Dalton called these smallest particles of subdivision *atoms*. An **atom** *is the smallest particle of an element that can exist and still have the properties of the element.* Additional research, carried out by many scientists, has now validated Dalton's basic conclusion that the building blocks for all types of matter are atoms. Some of the details of Dalton's original proposals have had to be modified in light of later, more sophisticated experiments, but the basic concept of atoms remains.

Today, among scientists, the concept that atoms are the building blocks for matter is a foregone conclusion. The large accumulated amount of supporting evidence for atoms is impressive. Key concepts about atoms, in terms of current knowledge, are found in what is known as the *atomic theory of matter*. The **atomic theory of matter** *is a set of five statements that summarizes modern-day scientific thought about atoms.* These five statements are the following:

1. All matter is made up of small particles called atoms, of which 117 different "types" are known, with each type corresponding to a different element.
2. All atoms of a given type are similar to one another and significantly different from all other types.
3. The relative number and arrangement of different types of atoms contained in a pure substance (its composition and structure) determine its identity.

4. Chemical change is a union, separation, or rearrangement of atoms to give new substances.
5. Only whole atoms can participate in or result from any chemical change, since atoms are considered indestructible during such changes.

Just how small is an atom? Atomic dimensions and masses, although not directly measurable, are known quantities obtained by calculation. The data used for the calculations come from measurements made on macroscopic amounts of pure substances.

The diameter of an atom is about 10^{-10} m.

$$1 \text{ atom} \approx 10^{-10} \text{ m}$$

If one were to arrange atoms of this diameter in a straight line, it would take one million of them to extend across the dot that serves as a period at the end of this sentence.

The mass of an atom is about 10^{-23} g.

$$1 \text{ atom} \approx 10^{-23} \text{ g}$$

To produce a mass of 1 lb would require about 5×10^{25} such atoms. The number 5×10^{25} is so large that it is difficult to visualize its magnitude. The following comparison "hints" at this number's magnitude. If each of the 6.8 billion people on Earth were made a millionaire (receiving 1 million \$1 bills), we would still need 7 billion other worlds, each inhabited by the same number of millionaires to have 5×10^{25} dollar bills in circulation.

Atoms are incredibly small particles. No one has seen or ever will see an atom with the naked eye. The question may thus be asked: "How can you be absolutely sure that something as minute as an atom really exists?" The achievements of late twentieth-century and early twenty-first century scientific instrumentation have gone a long way toward removing any doubt about the existence of atoms. Electron microscopes, capable of producing magnification factors in the millions, have made it possible to photograph "images" of individual atoms. In 1976 physicists at the University of Chicago were successful in obtaining images of single atoms. One of these images is shown in Figure 4.14.

Atoms are very small.

- Imagine an apple enlarged to the size of Earth. The atoms in the apple would then be about the size of cherries.
- The diameter of an atom would have to be increased two hundred million times (2×10^8) to cause it to have the diameter of a penny.

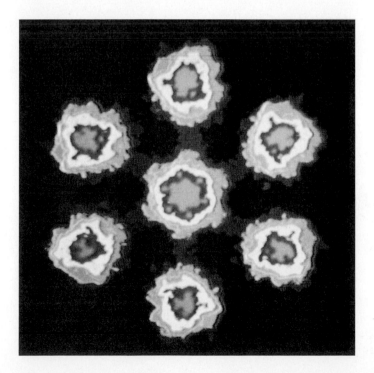

FIGURE 4.14 The bright spots in this photomicrograph are images of seven uranium atoms. The images were obtained using an electron microscope. *(Courtesy of M. Isaacson, Cornell University, and M. Ohtsuki, The University of Chicago)*

4.11 THE MOLECULE

Reasons for the tendency of atoms to collect together into molecules and information on the binding forces involved are considered in Chapter 7. The important point at this time is that a molecule is a collection of atoms that functions as a single composite unit.

Free isolated atoms are rarely encountered in nature. Instead, under normal conditions of temperature and pressure, atoms are almost always found together in aggregates or clusters ranging in size from two atoms to numbers too large to count. When the group or cluster of atoms is relatively small and bound together tightly, the resulting entity is called a *molecule*. A **molecule** *is a group of two or more atoms that functions as a unit because the atoms are tightly bound together.* This resultant "package" of atoms behaves in many ways as a single, distinct particle would.

A **diatomic molecule** *is a molecule that contains two atoms.* It is the simplest type of molecule that can exist. Next in complexity are *triatomic* molecules. A **triatomic molecule** *is a molecule that contains three atoms.* Continuing on numerically, we have *tetratomic* molecules, *pentatomic* molecules, and so on. The less specific term *polyatomic molecule* is often used to designate molecules containing three or more atoms.

The Latin word *mole* means "a mass." The word *molecule* denotes a "little mass."

The atoms contained in a molecule may all be of the same kind, or two or more kinds may be present. On the basis of this observation, molecules are classified into two categories: *homoatomic* and *heteroatomic*. A **homoatomic molecule** *is a molecule in which all atoms present are the same kind.* A pure substance containing homoatomic molecules must be an element. A **heteroatomic molecule** *is a molecule in which two or more different kinds of atoms are present.* Pure substances containing heteroatomic molecules must be compounds. Figure 4.15 shows general models for selected simple heteroatomic molecules.

The fact that homoatomic molecules exist indicates that individual atoms are not always the preferred structural unit for an element. The gaseous elements hydrogen, oxygen, nitrogen, fluorine, and chlorine exist in the form of diatomic molecules. There are four atoms present in a gaseous phosphorus molecule and eight atoms present in a gaseous sulfur molecule (see Fig 4.16). Some guidelines for determining which elements have individual atoms as their basic structural units and which exist in molecular form will be given in Section 7.2.

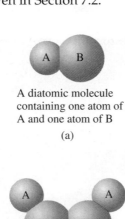

A diatomic molecule containing one atom of A and one atom of B

(a)

A triatomic molecule containing two atoms of A and one atom of B

(b)

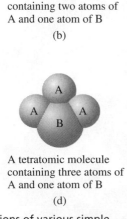

A tetratomic molecule containing two atoms of A and two atoms of B

(c)

A tetratomic molecule containing three atoms of A and one atom of B

(d)

FIGURE 4.15 General depictions of various simple heteroatomic molecules using models. Spheres of different sizes and colors represent different kinds of atoms.

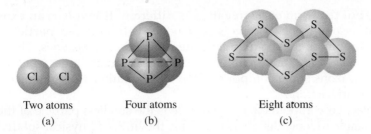

Two atoms
(a)

Four atoms
(b)

Eight atoms
(c)

FIGURE 4.16 Make-up of the homoatomic molecules present in the elements chlorine (a), phosphorus (b), and sulfur (c).

| **EXAMPLE 4.5** | **Classifying Molecules Based on Numbers of and Types of Atoms Present** |

Classify each of the following molecules as (1) *diatomic, triatomic*, etc., (2) *homoatomic* or *heteroatomic*, and (3) representing an *element* or a *compound*.

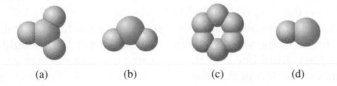

(a)   (b)   (c)   (d)

SOLUTION

  a. Tetratomic (four atoms); heteroatomic (two kinds of atoms); a compound (two kinds of atoms)
  b. Triatomic (three atoms); heteroatomic (two kinds of atoms); a compound (two kinds of atoms)
  c. Hexatomic (six atoms); homoatomic (one kind of atom); an element (one kind of atom)
  d. Diatomic (two atoms); heteroatomic (two kinds of atoms); a compound (two kinds of atoms)

Answer Double Check:

Are the answers reasonable? Yes. Heteroatomic molecule and compound are always companion terms, and homoatomic molecule and element are always companion terms. Such is the case for each of the answers. These term pairings are not affected by the number of atoms present in the molecule.

▶ **Practice Exercise 4.5** Classify each of the following molecules as (1) *diatomic, triatomic*, etc., (2) *homoatomic* or *heteroatomic*, and (3) representing an *element* or a *compound*.

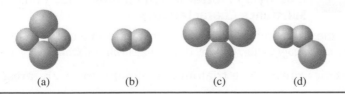

(a)   (b)   (c)   (d)

Ozone (as in ozone layer) is a form of oxygen different from the oxygen we breathe. In ozone the molecules are triatomic, and in the oxygen we breathe the molecules are diatomic. In both cases the molecules are homoatomic because oxygen is an element.

Many, but not all, compounds have heteroatomic molecules as their basic structural unit. Those compounds that do are called *molecular compounds*. Some compounds in the liquid and solid state, however, are not molecular; that is, the atoms present are not collected together into discrete heteroatomic molecules. These nonmolecular compounds still contain atoms of at least two kinds (a necessary requirement for a compound), but the form of aggregation is different. It involves an extended three-dimensional assembly of positively and negatively charged particles called *ions* (Sec. 7.4). Compounds that contain ions are called *ionic compounds*. The familiar substances sodium chloride (table salt) and calcium carbonate (limestone) are ionic compounds. The reasons some compounds have ionic rather than molecular structures are considered in Section 7.7.

For molecular compounds, the molecule is the smallest particle of the compound capable of a stable independent existence. It is the limit of physical subdivision for the compound. Consider the molecular compound sucrose (table sugar). Continued subdivision of a quantity of table sugar to yield smaller and smaller amounts would ultimately lead to the isolation of one single particle of table sugar—a molecule of table sugar. This molecule of sugar could not be broken down any further and still maintain the physical and chemical properties of table sugar. The sugar molecule could be broken down further by chemical (not physical) means to give atoms, but if that occurred, we would no longer have table sugar. The *molecule* is the limit of *physical* subdivision. The *atom* is the limit of *chemical* subdivision.

Every molecular compound has as its smallest characteristic unit a *unique* molecule. If two samples had the same molecule as a basic unit, both would have the same properties; thus, they would be one and the same compound. An alternative way of stating the same conclusion is that there is only one kind of molecule for any given molecular substance.

Since every molecule in a sample of a molecular compound is the same as every other molecule in the sample, it is commonly stated that molecular compounds are made up of a single kind of particle. Such terminology is correct as long as it is remembered that the particle referred to is the molecule. In a sample of a molecular compound, there are at least two kinds of atoms present but only one kind of molecule.

The properties of molecules are very different from the properties of the atoms that make up the molecules. Molecules do not maintain the properties of their constituent elements. Table sugar is a white crystalline molecular compound with a sweet taste. None of the three elements present in table sugar (carbon, hydrogen, and oxygen) is a white solid or has a sweet taste. Carbon is a black solid, and hydrogen and oxygen are colorless gases.

Figure 4.17 summarizes the relationships between hetero- and homoatomic molecules and elements, compounds, and pure substances. Example 4.6 deals with these same classifications as well as that of mixtures (heterogeneous and homogeneous; see Section 4.6).

EXAMPLE 4.6 **Classifying Matter Based on Molecular, Phase, and Substance Characteristics**

Assign each of the following descriptions of matter to one of the following categories: *element, compound, homogeneous mixture,* or *heterogeneous mixture.*

  a. All molecules present are triatomic; two substances are present; one phase is present.
  b. All molecules present are heteroatomic; one substance is present; two phases are present.
  c. All molecules present are identical and homoatomic; one phase is present.
  d. Both homoatomic and heteroatomic molecules are present; one phase is present.

SOLUTION

   a. This is a homogeneous (one phase) mixture (two substances). The molecular information (triatomic) does not differentiate between elements and compounds as mixture components.

   b. This is a compound (one substance, heteroatomic molecules). The compound is present in two physical states (two phases). An example of such a situation would be ice cubes in water.

   c. This is an element (one substance—all identical homoatomic molecules). The element is present in one physical state (one phase).

   d. This is a homogeneous (one phase) mixture (both homoatomic and heteroatomic molecules). The mixture contains at least one element (homoatomic) and at least one compound (heteroatomic).

▶ **Practice Exercise 4.6**   Assign each of the following descriptions of matter to one of the following categories: *element, compound, homogeneous mixture,* or *heterogeneous mixture.*

   a. One kind of homoatomic molecule is present; one phase is present.

   b. All molecules present are diatomic; two substances are present; one phase is present.

   c. Two kinds of heteroatomic molecules are present; two phases are present.

   d. All molecules present are identical, triatomic, and heteroatomic; one phase is present.

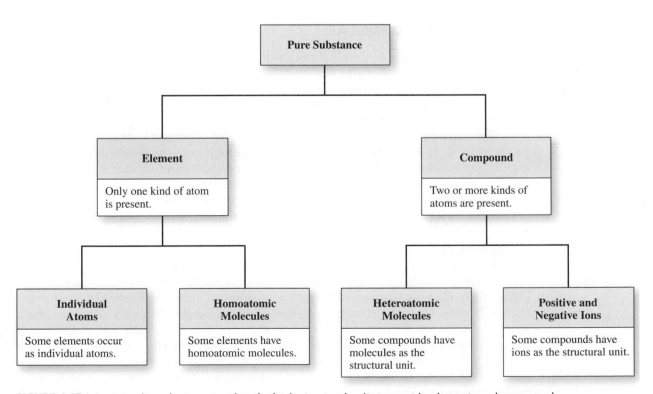

FIGURE 4.17   A comparison chart contrasting the basic structural units present in elements and compounds.

4.12 NATURAL AND SYNTHETIC COMPOUNDS

Approximately 9 million chemical compounds are now known, with more being characterized daily. No end appears to be in sight as to the number of compounds that can and will be prepared in the future. Approximately 10,000 new chemical substances are registered every week with Chemical Abstracts Service, a clearinghouse for new information concerning chemical substances.

Many compounds, perhaps the majority now known, are not naturally occurring substances. These synthetic (laboratory-produced) compounds are legitimate compounds and should not be considered second class or unimportant simply because they lack the distinction of being natural. Many of the plastics, synthetic fibers, and prescription drugs now in common use are synthetic materials produced through controlled chemical change carried out on an industrial scale.

We have noted that chemists can produce compounds not found in nature. The reverse is also true. Nature is capable of making many compounds, especially those found in living systems, that chemists are not yet able to prepare in the laboratory.

There is a middle ground also. Many compounds that exist in nature can also be produced in the laboratory. A fallacy exists in the thinking of some people concerning these compounds that have "dual origins." A belief still persists that there is a difference between compounds prepared in the laboratory and samples of the same compounds found in nature. This is not true for pure samples of a compound. All pure samples of a compound, regardless of their origin, have the same composition. Since compositions are the same, properties will also be the same. There is no difference, for example, between a laboratory-prepared vitamin and a "natural" vitamin if both are pure samples of the same vitamin, despite frequent claims to the contrary.

Two examples of compounds that were first discovered in nature and then later produced in a laboratory setting are the antibiotic penicillin G and the antirejection drug cyclosporin. The original source for penicillin G, the first antibiotic to be marketed, was a mold of the Penicillium family, from whence it got its name. Cyclosporin, a drug used to control rejection of a transplanted organ, such as a liver, by a patient's own immune system, was originally isolated from a type of soil fungus.

4.13 CHEMICAL FORMULAS

A most important piece of information about a compound is its composition. *Chemical formulas* represent a concise means of specifying compound compositions. A **chemical formula** *is a notation made up of the chemical symbols of the elements present in a compound and numerical subscripts (located to the right of each chemical symbol) that indicate the number of atoms of each element present in a structural unit of the compound.*

The chemical formula for the compound we call aspirin is $C_9H_8O_4$. This formula provides us with the following information about an aspirin molecule: Three elements are present—carbon (C), hydrogen (H), and oxygen (O)—and 21 atoms are present—9 carbon atoms, 8 hydrogen atoms, and 4 oxygen atoms.

When only one atom of a particular element is present in a molecule of a compound, the element's symbol is written without a numerical subscript in the formula of the compound. In the formula for isopropyl alcohol, C_3H_6O, for example, the subscript 1 for the element oxygen is not written.

A chemical symbol in a formula stands for one atom of the element. If more than one atom is to be indicated in a formula, a subscript number is used after the symbol.

To write formulas correctly, it is necessary to follow strictly the capitalization rules for elemental symbols (Sec. 4.9). Making the error of capitalizing the second letter of an element's symbol can dramatically alter the meaning of a chemical formula. The formulas $CoCl_2$ and $COCl_2$ illustrate this point; the symbol Co stands for the element cobalt, whereas CO stands for one atom of carbon and one atom of oxygen. The properties of the compounds $CoCl_2$ and $COCl_2$ are dramatically different. The compound $CoCl_2$ is a blue crystalline solid with a melting point of 724°C and a boiling point of 1029°C. The compound $COCl_2$ is a highly toxic colorless gas with a melting point of –118°C and a boiling point of 8°C.

For molecular compounds, chemical formulas give the composition of the molecules making up the compounds. For ionic compounds, which have no molecules, a chemical formula gives the ion ratio found in the compound. For example, the ionic

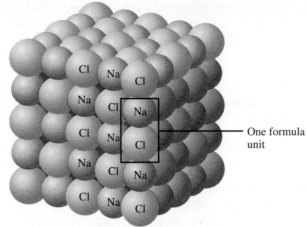

FIGURE 4.18   A comparison of a molecule of a molecular compound and a formula unit of an ionic compound. A molecule can exist as a separate unit, whereas a formula unit is simply two or more ions "plucked" from a much larger array of ions.

(a) A molecule of the molecular compound methane (CH_4)

(b) A formula unit of the ionic compound sodium chloride (NaCl)

One formula unit

compound sodium oxide contains sodium ions and oxygen ions in a two-to-one ratio— twice as many sodium ions as oxygen ions. The formula of this compound is Na_2O, which expresses the ratio between the two types of ions present. The term *formula unit* is used to describe this smallest ratio between ions. The distinction between the formula unit of an ionic compound and the molecule of a molecular compound is pictorially portrayed in Figure 4.18.

Sometimes chemical formulas contain parentheses, an example being $Al_2(SO_4)_3$. The interpretation of this formula is straightforward; in a formula unit there are present two aluminum (Al) ions and three SO_4 groups. The subscript following the parentheses always indicates the number of units in the formula of the polyatomic entity inside the parentheses. As another example, consider the compound $Pb(C_2H_5)_4$. Four units of C_2H_5 are present. In terms of atoms present, the formula $Pb(C_2H_5)_4$ represents 29 atoms: 1 lead (Pb) atom, $4 \times 2 = 8$ carbon (C) atoms, and $4 \times 5 = 20$ hydrogen (H) atoms. The formula could be (but is not) written as PbC_8H_{20}. Both versions of the formula convey the same information in terms of atoms present. However, $Pb(C_2H_5)_4$ gives the additional information that the C and H are present as C_2H_5 units and is therefore the preferred way of writing the formula. Further information concerning the use of parentheses (when and why) will be presented in Section 7.8. The important concern now is being able to interpret formulas that contain parentheses in terms of total atoms present. Example 4.7 deals with this skill in greater detail.

EXAMPLE 4.7   **Interpreting Chemical Formulas in Terms of Atoms and Elements Present**

Interpret each of the following chemical formulas in terms of how many atoms of each element are present in one structural unit of the compound.

   a. $C_8H_9O_2N$ (acetaminophen, the active ingredient in Tylenol)
   b. $(NH_4)_2C_2O_4$ (ammonium oxalate, used in the manufacture of explosives)
   c. $Ca_{10}(PO_4)_6(OH)_2$ (hydroxyapatite, present in tooth enamel)

SOLUTION

   a. We simply look at the subscripts following the symbols for the elements. This formula indicates that eight carbon atoms, nine hydrogen atoms, two oxygen atoms, and one nitrogen atom are present in one molecule of the compound.

b. The subscript following the parenthesis, 2, indicates that two NH_4 units are present. Collectively, in these two units, we have two nitrogen atoms and $2 \times 4 = 8$ hydrogen atoms. In addition, two carbon atoms and four oxygen atoms are present.

c. There are ten calcium atoms. The amounts of phosphorus, hydrogen, and oxygen are affected by the subscripts outside the parentheses. There are six phosphorus atoms and two hydrogen atoms present. Oxygen atoms are present in two locations in the formula. There are a total of 26 oxygen atoms: 24 from the PO_4 subunits (6×4) and two from the OH subunits (2×1).

▶ **Practice Exercise 4.7** Interpret each of the following chemical formulas in terms of how many atoms of each element are present in one structural unit of the compound.

a. H_2SO_4 (sulfuric acid, an industrial acid)
b. $C_{17}H_{20}N_4O_6$ (riboflavin, a B-vitamin)
c. $Ca(H_2PO_4)_2$ (calcium dihyrogenphosphate, an ingredient in baking powder)

Figure 4.19 pictorially relates chemical formulas to the matter classifications of element, compound, and mixture. Note in the top third of the diagram that the formulas for molecules of an element contain only one type of atom and thus only one elemental symbol. In the middle third of the diagram, we see that formulas for the compounds shown contain two types of atoms and thus two elemental symbols. Finally, in the bottom third of the diagram, we see that in mixtures, different types of molecules with different chemical formulas, must be present.

In addition to chemical formulas, compounds have names. Naming compounds is not as simple as naming elements. Although the nomenclature of elements (Sec. 4.9) has been largely left up to the imagination of their discoverers, extensive sets of systematic rules exist for naming compounds. Rules must be used because of the large number of compounds that exist. Chapter 8 is devoted to compound nomenclature, and we will not

FIGURE 4.19 A contrast between the chemical formulas for molecules of elements and molecules of compounds and their pictorial representations.

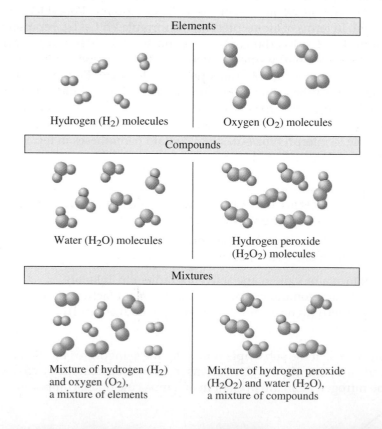

Elements

Hydrogen (H_2) molecules

Oxygen (O_2) molecules

Compounds

Water (H_2O) molecules

Hydrogen peroxide (H_2O_2) molecules

Mixtures

Mixture of hydrogen (H_2) and oxygen (O_2), a mixture of elements

Mixture of hydrogen peroxide (H_2O_2) and water (H_2O), a mixture of compounds

worry about naming rules until then. For the time being, our focus will be on the meaning and significance of chemical formulas; knowing how to name the compounds that the formulas represent is not a prerequisite for understanding the meaning of formulas.

Summary

1. **Chemistry and Matter** Chemistry is the branch of science concerned with the characterization, composition, and transformations of matter. Matter, the substances of the physical universe, is anything that has mass and occupies space.

2. **Physical States of Matter** Matter exists in three physical states: solid, liquid, and gas. Physical-state classification is based on whether a substance's shape and volume are definite or indefinite.

3. **Properties of Matter** Properties, the distinguishing characteristics of a substance used in its identification and description, are of two types: physical and chemical. Physical properties can be observed without changing a substance into another substance. Chemical properties are properties that matter exhibits as it undergoes or resists changes in composition. The failure of a substance to undergo change in the presence of another substance is considered a chemical property.

4. **Changes in Matter** Changes that can occur in matter are of two types: physical and chemical. A physical change is a process that does not alter the composition of the substance. No new substances are ever formed as a result of a physical change. A chemical change is a process that involves a change in the composition of the substance. Such changes always involve conversion of the material or materials under consideration into one or more new substances that have properties and composition distinctly different from those of the original materials. Chemical changes are also called chemical reactions.

5. **Pure Substances and Mixtures** All specimens of matter are either pure substances or mixtures. A pure substance is a form of matter that has a definite and constant composition. A mixture is a physical combination of two or more pure substances in which the pure substances retain their identity.

6. **Types of Mixtures** Mixtures can be classified as heterogeneous or homogeneous on the basis of the visual recognition of the components present. A heterogeneous mixture contains visibly different parts or phases, each of which has different properties. A homogeneous mixture contains only one phase, which has uniform properties throughout.

7. **Types of Pure Substances** A pure substance can be classified as an element or a compound on the basis of whether it can be broken down into two or more simpler substances by ordinary chemical means. Elements cannot be broken down into simpler substances. Compounds yield two or more simpler substances when broken down. There are 117 pure substances that qualify as elements. Not all of the elements are naturally occurring. There are millions of compounds.

8. **Chemical Symbols** Chemical symbols are a shorthand notation for the names of the elements. Most chemical symbols consist of two letters; a few involve a single letter. The first letter of a chemical symbol is always capitalized, and the second letter is always lowercase.

9. **Atoms and Molecules** An atom is the smallest particle of an element that can exist and still have the properties of the element. Free isolated atoms are rarely encountered in nature. Instead, atoms are almost always found together in aggregates or clusters. A molecule is a group of two or more atoms that functions as a unit because the atoms are tightly bound together.

10. **Types of Molecules** Molecules are of two types: homoatomic and heteroatomic. Homoatomic molecules are molecules in which all atoms present are of the same kind. A pure substance containing homoatomic molecules is an element. Heteroatomic molecules are molecules in which two or more different kinds of atoms are present. Pure substances that contain heteroatomic molecules must be compounds. The molecule is the limit of physical subdivision. The atom is the limit of chemical subdivision.

11. **Chemical Formulas** Chemical formulas are used to specify compound composition in a concise manner. They consist of the symbols of the elements present in the compound and numerical subscripts (located to the right of each symbol) that indicate the number of atoms of each element present in a molecule of the compound.

Key Terms

The new terms or concepts defined in this chapter are

atom *Sec. 4.10*
atomic theory of matter
    Sec. 4.10
chemical change *Sec. 4.4*
chemical formula *Sec. 4.13*
chemical property *Sec. 4.3*
chemical reaction *Sec. 4.4*
chemical symbol *Sec. 4.9*
chemistry *Sec. 4.1*

compound *Sec. 4.7*
diatomic molecule *Sec. 4.11*
element *Sec. 4.7*
extensive property *Sec. 4.3*
gas *Sec. 4.2*
heteroatomic molecule
    Sec. 4.11
heterogeneous mixture
    Sec. 4.6

homoatomic molecule
    Sec. 4.11
homogeneous mixture
    Sec. 4.6
intensive property *Sec. 4.3*
liquid *Sec. 4.2*
matter *Sec. 4.1*
mixture *Sec. 4.5*

molecule *Sec. 4.11*
physical change *Sec. 4.4*
physical property *Sec. 4.3*
properties *Sec. 4.3*
pure substance *Sec. 4.5*
solid *Sec. 4.2*
triatomic molecule
    Sec. 4.11

Practice Problems

PHYSICAL STATES OF MATTER (SEC. 4.2)

4.1   What physical characteristic
   a. distinguishes liquids from solids?
   b. is common to the gaseous and liquid states?

4.2   What physical characteristic
   a. distinguishes gases from liquids?
   b. is common to the liquid and solid states?

4.3   Indicate whether each of the following substances does or does not take the shape of its container and also whether it has a definite volume.
   a. copper wire   **b.** oxygen gas
   c. a hard sugar cube   **d.** bulk granulated sugar

4.4   Indicate whether each of the following substances does or does not take the shape of its container and also whether it has an indefinite volume.
   a. aluminum powder
   b. carbon dioxide gas
   c. a single grain of table salt
   d. bulk-granulated table salt

4.5   Ethyl alcohol, a liquid at room temperature, has a boiling point of 78.3°C and a melting point of –117°C. Specify the physical state of ethyl alcohol at each of the following temperatures.
   a. –120°C   **b.** –60°C
   c. 60°C   **d.** 120°C

4.6   Carbon tetrachloride, a liquid at room temperature, has a boiling point of 77°C and a melting point of –23°C. Specify the physical state of carbon tetrachloride at each of the following temperatures.
   a. –80°C   **b.** –30°C   **c.** 30°C   **d.** 80°C

PROPERTIES OF MATTER (SEC. 4.3)

4.7   The following are properties of the metal beryllium. Classify them as physical or chemical.
   a. In powdered form, it burns brilliantly on ignition.
   b. Bulk metal does not react with steam even when red hot.
   c. It has a density of 1.85 g/cm^3 at 20°C.
   d. It is a relatively soft silvery-white metal.

4.8   The following are properties of the metal aluminum. Classify them as physical or chemical.
   a. It generates a colorless, odorless gas when added to sulfuric acid.
   b. It can easily be formed into thin foils.
   c. It is a solid at room temperature.
   d. It is a good conductor of heat.

4.9   Indicate whether each of the following statements describes a physical or chemical property.
   a. Silver compounds discolor the skin by reacting with skin protein.
   b. Hemoglobin gives blood its red color.
   c. Lithium metal is light enough to float on water.
   d. Mercury is a liquid at room temperature.

4.10   Indicate whether each of the following statements describes a physical or chemical property.
   a. Charcoal lighter fluid can be ignited with a match.
   b. Magnesium metal does not react with cold water.
   c. Carbon monoxide is a colorless gas.
   d. Sodium metal is so soft that it can be cut with a sharp knife.

4.11   Classify each of the following as an intensive property or an extensive property.
   a. length   **b.** density   **c.** color   **d.** boiling point

4.12   Classify each of the following as an intensive property or an extensive property.
   a. mass   **b.** temperature
   c. volume   **d.** melting point

4.13   Classify each of the following pairs of characterizations as (1) differing in extensive properties, (2) differing in intensive properties, or (3) differing in both extensive and intensive properties.
   a. 20 g phosphorus at 25°C
        65 g phosphorus at 25°C
   b. 65 g liquid bromine at 25°C
        65 g liquid water at 25°C
   c. 35 g copper metal at 25°C
        35 g copper metal at 100°C
   d. 42 g gold metal at 25°C
        35 g gold metal at 15°C

4.14 Classify each of the following pairs of characterizations as (1) differing in extensive properties, (2) differing in intensive properties, or (3) differing in both extensive and intensive properties.
  a. 20 g sulfur at 25°C
  65 g phosphorus at 25°C
  b. 65 g liquid water at 35°C
  65 g liquid water at 25°C
  c. 65 g silver metal at 25°C
  65 g copper metal at 100°C
  d. 42 g gold metal at 25°C
  _____ 35 g gold metal at 25°C

CHANGES IN MATTER (SEC. 4.4)

4.15 Classify each of the following changes as physical or chemical.
  a. crushing a dry leaf
  b. hammering a metal into a thin sheet
  c. burning your chemistry textbook
  d. slicing a ham
4.16 Classify each of the following changes as physical or chemical.
  a. evaporation of water from a lake
  b. scabbing over of a skin cut
  c. cutting a string into several pieces
  _____ **d.** melting of candle wax
4.17 Classify each of the following changes as physical or chemical.
  a. a match burning
  b. "rubbing alcohol" evaporating
  c. a copper object turning green in color over time
  d. a pan of water boiling
4.18 Classify each of the following changes as physical or chemical.
  a. a newspaper page turning yellow over time
  b. a rubber band breaking
  c. a firecracker exploding
  _____ **d.** dry ice "disappearing" over time
4.19 Indicate whether each of the following methods for obtaining various substances involves physical or chemical change.
  a. Sodium chloride (salt) is obtained from saltwater by evaporation of the water.
  b. Nitrogen gas is obtained from air by letting the nitrogen boil off from liquid air.
  c. Oxygen gas is obtained by decomposition of the oxygen-containing compound potassium chlorate.
  d. Water is obtained by the high-temperature reaction of gaseous hydrogen with gaseous oxygen.
4.20 Indicate whether each of the following methods for obtaining various substances involves physical or chemical change.
  a. Mercury is obtained by decomposing a mercury–oxygen compound, liberating the oxygen and leaving the mercury behind.
  b. Sand is obtained from a sand–sugar mixture by adding water to the mixture and pouring off the resulting sugar–water solution.

c. Ammonia is obtained by the high-temperature, high-pressure reaction between hydrogen and nitrogen.
  d. Water is obtained from a sugar–water solution by evaporating off and then collecting the water.
4.21 Complete each of the following sentences by placing the word *chemical* or *physical* in the blank.
  a. The freezing over of a pond's surface is a _____ process.
  b. The crushing of some ice to make ice chips is a _____ procedure.
  c. The destruction of a newspaper through burning it is a _____ process.
  d. Pulverizing a hard sugar cube using a wooden mallet is a _____ procedure.
4.22 Complete each of the following sentences by placing the word *chemical* or *physical* in the blank.
  a. The reflection of light by a shiny metallic object is a _____ process.
  b. The decomposing of a blue powdered material to produce a white glasslike substance and a gas is a _____ procedure.
  c. A burning candle produces light by _____ means.
  _____ **d.** The grating of a piece of cheese is a _____ technique.
4.23 Give the name of the change of state associated with each of the following processes.
  a. Water is made into ice cubes.
  b. The inside of your car window fogs up.
  c. Mothballs in the clothes closet disappear with time.
  d. Perspiration dries.
4.24 Give the name of the change of state associated with each of the following processes.
  a. Dry ice disappears without melting.
  b. Snowflakes form.
  c. Dew on the lawn disappears when the sun comes out.
  _____ **d.** Ice cubes in a soft drink disappear with time.

PURE SUBSTANCES AND MIXTURES (SECS. 4.5 AND 4.6)

4.25 Consider the following classifications of matter: heterogeneous mixture, homogeneous mixture, and pure substance.
  a. In which of these classifications must two or more substances be present?
  b. In which of these classifications must the composition be uniform throughout?
4.26 Consider the following classifications of matter: heterogeneous mixture, homogeneous mixture, and pure substance.
  a. In which of these classifications must the composition be constant?
  b. In which of these classifications is separation into simpler substances using physical means possible?
4.27 Classify each of the following statements as true or false.
  a. Heterogeneous mixtures must contain three or more substances.

b. Pure substances cannot have a variable composition.

c. Substances maintain many of their properties in a heterogeneous mixture but not in a homogeneous mixture.

d. Pure substances are seldom encountered in the everyday world.

4.28 Classify each of the following statements as true or false.

a. Homogeneous mixtures must contain at least two substances.

b. Heterogeneous mixtures but not homogeneous mixtures can have a variable composition.

c. Pure substances cannot be separated into other kinds of matter using physical means.

d. The number of known pure substances is less than one hundred thousand.

4.29 Assign each of the following descriptions of matter to one of the following categories: heterogeneous mixture, homogeneous mixture, pure substance.

a. two substances present, two phases present

b. two substances present, one phase present

c. three substances present, one phase present

d. three substances present, three phases present

4.30 Assign each of the following descriptions of matter to one of the following categories: heterogeneous mixture, homogeneous mixture, pure substance.

a. one substance present, one phase present

b. one substance present, two phases present

c. one substance present, three phases present

d. three substances present, two phases present

4.31 Classify each of the following as a heterogeneous mixture, a homogeneous mixture, or a pure substance. Also indicate how many phases are present. (In each case the substances are present in the same container.)

a. water and dissolved salt

b. water and dissolved sugar

c. water and sand

d. water and oil

4.32 Classify each of the following as a heterogeneous mixture, a homogeneous mixture, or a pure substance. Also indicate how many phases are present. (In each case the substances are present in the same container.)

a. liquid water and ice

b. liquid water, oil, and ice

c. carbonated water (soda water) and ice

d. oil, ice, saltwater solution, sugar–water solution, and pieces of copper metal

4.33 In each of the following situations two of the four phrases *chemically homogeneous*, *chemically heterogeneous*, *physically homogeneous*, and *physically heterogeneous* apply. Select the two correct phrases for each situation.

a. pure water

b. tap water

c. oil-and-vinegar salad dressing with spices

d. oil-and-vinegar salad dressing without spices

4.34 In each of the following situations two of the four phrases *chemically homogeneous*, *chemically heterogeneous*, *physically homogeneous*, and *physically heterogeneous* apply. Select the two correct phrases for each situation.

a. water and dissolved table salt

b. water and white sand

c. carbonated beverage right after opening container

d. carbonated beverage that has gone flat

ELEMENTS AND COMPOUNDS (SEC. 4.7)

4.35 Based on the information given, classify each of the pure substances A through D as elements or compounds, or indicate that no such classification is possible because of insufficient information.

a. Analysis with an elaborate instrument indicates that substance A contains two elements.

b. Substance B decomposes upon heating.

c. Heating substance C to 1000°C causes no change in it.

d. Heating substance D to 500°C causes it to change from a solid to a liquid.

4.36 Based on the information given, classify each of the pure substances A through D as elements or compounds, or indicate that no such classification is possible because of insufficient information.

a. Substance A cannot be broken down into simpler substances by chemical means.

b. Substance B cannot be broken down into simpler substances by physical means.

c. Substance C readily dissolves in water.

d. Substance D readily reacts with the element chlorine.

4.37 Indicate whether each of the following statements is true or false.

a. Both elements and compounds are pure substances.

b. A compound results from the physical combination of two or more elements.

c. For matter to be heterogeneous, at least two compounds must be present.

d. Compounds, but not elements, can have a variable composition.

4.38 Indicate whether each of the following statements is true or false.

a. Compounds can be separated into their constituent elements using chemical means.

b. Elements can be separated into their constituent compounds using physical means.

c. A compound must contain at least two elements.

d. A compound is a physical mixture of different elements.

4.39 Based on the information given in the following equations, classify each of the pure substances A through G as elements or compounds, or indicate that no such classification is possible because of insufficient information.

a. $A + B \rightarrow C$ **b.** $D \rightarrow E + F + G$

4.40 Based on the information given in the following equations, classify each of the pure substances A through G as elements or compounds, or indicate that no such classification is possible because of insufficient information.

a. $A \rightarrow B + C$

b. $D + E \rightarrow F + G$

4.41 Consider two boxes with the following contents: The first contains 50 individual paper clips and 50 individual rubber bands; the second contains the same number of paper clips and rubber bands with the difference that each paper clip is interlocked with a rubber band. Which box has contents that would be an analogy for a mixture, and which has contents that would be an analogy for a compound?

4.42 Consider the characteristics of the two breakfast cereals Crispy Wheat 'N Raisins and Crispix. The first cereal contains wheat flakes and raisins. The second cereal contains a fused two-layered flake, one side of which is rice and the other side corn. Characterize the properties of these two cereals that make one an analogy for a mixture and the other an analogy for a compound.

4.43 Indicate whether or not each of the following characterizations applies to a *mixture* that contains the elements copper and sulfur. (The copper and sulfur have not reacted with each other.)

a. It has a variable composition.

b. Elements present maintain their individual properties.

c. Physical methods are sufficient to separate the elements present.

d. A chemical combination of the elements has occurred.

4.44 Indicate whether or not each of the following characterizations applies to a *compound* that contains the elements copper and sulfur.

a. It has a variable composition.

b. Elements present maintain their individual properties.

c. Physical methods are sufficient to separate the elements present.

d. A chemical combination of the elements has occurred.

DISCOVERY AND ABUNDANCE OF THE ELEMENTS (SEC. 4.8)

4.45 Indicate whether each of the following statements about elements is true or false.

a. All except three of the elements are naturally occurring.

b. New elements have been identified within the last 10 years.

c. Oxygen is the most abundant element in the universe as a whole.

d. Two elements account for over 75% of the elemental particles (atoms) in Earth's crust.

4.46 Indicate whether each of the following statements about elements is true or false.

a. The two most active decades for discovery of new elements were the 1790s and the 1840s.

b. The majority of the known elements have been discovered since 1900.

c. Aluminum is the third most abundant element in Earth's crust.

d. At present, 108 elements are known.

4.47 Indicate whether each of the following statements about elemental abundances is true or false.

a. Silicon is the second most abundant element in Earth's crust.

b. Hydrogen is the most abundant element in the universe but not in Earth's crust.

c. Oxygen and hydrogen are the two most abundant elements in the universe.

d. One element accounts for over one-half of the atoms in Earth's crust.

4.48 Indicate whether each of the following statements about elemental abundances is true or false.

a. Hydrogen is the most abundant element in both Earth's crust and the universe.

b. Oxygen and silicon are the two most abundant elements in the universe.

c. Helium is the second most abundant element in Earth's crust.

d. Two elements account for over three-fourths of the atoms in Earth's crust.

4.49 With the help of Figure 4.13, indicate whether the first listed element in each of the given pairs of elements is more abundant or less abundant in Earth's crust, in terms of atom percent, than the second listed element.

a. silicon and aluminum

b. calcium and hydrogen

c. iron and oxygen

d. sodium and potassium

4.50 With the help of Figure 4.13, indicate whether the first listed element in each of the given pairs of elements is more abundant or less abundant in Earth's crust, in terms of atom percent, than the second listed element.

a. oxygen and hydrogen

b. iron and aluminum

c. calcium and magnesium

d. copper and sodium

NAMES AND CHEMICAL SYMBOLS OF THE ELEMENTS (SEC. 4.9)

4.51 Give the name of the element associated with each of the following chemical symbols, or vice versa.

a. N   b. Ni

c. Pb   d. Sn

e. aluminum   f. neon

g. hydrogen   h. uranium

4.52 Give the name of the element associated with each of the following chemical symbols, or vice versa.

a. Li   b. He   c. F   d. Zn

e. mercury   f. chlorine   g. gold   h. selenium

4.53 Write the chemical symbol for each member of the following pairs of elements.
a. sodium and sulfur
b. magnesium and manganese
c. calcium and cadmium
d. arsenic and argon

4.54 Write the chemical symbol for each member of the following pairs of elements.
a. copper and cobalt
b. potassium and phosphorus
c. iron and iodine
d. silicon and silver

4.55 Several elements have chemical symbols that begin with the letter B. For each of the following chemical symbols, give the name of the corresponding element.
a. B   b. Ba   c. Be
d. Bi   e. Bk   f. Br

4.56 Several elements have chemical symbols that begin with the letter T. For each of the following chemical symbols, give the name of the corresponding element.
a. Ta   b. Tb   c. Tc
d. Te   e. Th   f. Tl

4.57 Each of the following names of elements is spelled incorrectly. Correct the misspellings.
a. flourine   b. zink
c. potasium   d. sulfer

4.58 Each of the following names of elements is spelled incorrectly. Correct the misspellings.
a. phosphorous   b. murcury
c. clorine   d. argone

4.59 Give the English name and symbol for each of the following elements, whose Latin name is
a. ferrum.   b. stannum.
c. natrium.   d. aurum.

4.60 Give the English name and symbol for each of the following elements, whose Latin name is
a. kalium.   b. argentum.
c. plumbum.   d. stibium.

4.61 Certain words can be viewed whimsically as sequential combinations of symbols of the elements. For example, the given name Stephen is made up of the following chemical symbol sequence: S-Te-P-He-N. Analyze each of the following given names in a similar manner.
a. Rebecca   b. Raymond
c. Nancy   d. Bruce
e. Sharon   f. Alice

4.62 Certain words can be viewed whimsically as sequential combinations of symbols of the elements. For example, the given name Stephen is made up of the following chemical symbol sequence: S-Te-P-He-N. Analyze each of the following given names in a similar manner.
a. Barbara   b. Eugene
c. Heather   d. Monica
e. Allan   f. Bryce

ATOMS AND MOLECULES (SECS. 4.10 AND 4.11)

4.63 Which of the following concepts are *not* consistent with the statements of modern-day atomic theory?
a. Atoms are the basic building blocks for all kinds of matter.
b. Different types of atoms exist.
c. All atoms of a given type are identical.
d. Atoms are considered indestructible during chemical change processes.

4.64 Which of the following concepts are *not* consistent with the statements of modern-day atomic theory?
a. Only whole atoms can participate in chemical reactions.
b. Atoms change identity during chemical change processes.
c. One-hundred twenty different "types" of atoms are known.
d. Chemical change is a union, separation, or rearrangement of atoms to give new substances.

4.65 Which of the terms *heteroatomic, homoatomic, diatomic, triatomic, element,* and *compound* apply to each of the following models for molecules? (More than one term may apply in a given situation.)

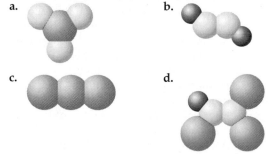

4.66 Which of the terms *heteroatomic, homoatomic, diatomic, triatomic, element,* and *compound* apply to each of the following models for molecules? (More than one term may apply in a given situation.)

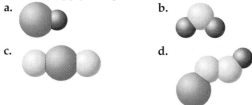

4.67 Indicate whether each of the following statements is *true* or *false*. If a statement is false, change it to make it true. (Such a rewriting should involve more than merely converting the statement to a negative one.)
a. Molecules must contain three or more atoms.
b. The atom is the limit of chemical subdivision for an element.
c. All compounds have molecules as their basic structural unit.
d. The diameter of an atom is approximately 10^{-8} meters.

4.68 Indicate whether each of the following statements is *true* or *false*. If a statement is false, change it to make it true. (Such a rewriting should involve more than merely converting the statement to a negative one.)

a. A molecule of an element may be homoatomic or heteroatomic depending on which element is involved.

b. Heteroatomic molecules do not maintain the properties of their constituent elements.

c. Only one kind of atom may be present in a homoatomic molecule.

d. The mass of an atom is approximately 10^{-23} gram.

4.69 Assign each of the following descriptions of matter to one of the following categories: *element, compound,* or *mixture.*

a. one substance present, two phases present, all molecules present are homoatomic

b. two substances present, one phase present, all molecules present are homoatomic

c. one phase present, two kinds of triatomic heteroatomic molecules present

d. triatomic homoatomic and diatomic heteroatomic molecules present

4.70 Assign each of the following descriptions of matter to one of the following categories: *element, compound,* or *mixture.*

a. one substance present, one phase present, one kind of heteroatomic molecule present

b. two substances present, two phases present, all molecules present are heteroatomic

c. one phase present, two kinds of diatomic heteroatomic molecules present

d. diatomic homoatomic and triatomic heteroatomic molecules present

CHEMICAL FORMULAS (SEC. 4.13)

4.71 Write chemical formulas, using the generalized symbols A, B, and C, for the substances represented by each of the following molecular models. List the symbols in alphabetical order in the chemical formulas.

a.

b.

c.

d.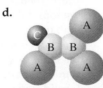

4.72 Write chemical formulas, using the generalized symbols A, B, and C, for the substances represented by each of the following molecular models. List the symbols in alphabetical order in the chemical formulas.

a.

b.

c.

d.

4.73 On the basis of its chemical formula, classify each of the following substances as an element or compound.
a. $NaClO_2$ b. CO c. S_8 d. Al

4.74 On the basis of its chemical formula, classify each of the following substances as an element or compound.
a. AlN b. CO_2 c. Co d. O_3

4.75 Write the chemical formulas for compounds in which the following numbers of atoms are present in a formula unit of the compound.

a. 20 carbon atoms, 30 hydrogen atoms

b. 2 hydrogen atoms, 1 sulfur atom, 4 oxygen atoms

c. 1 hydrogen atom, 1 carbon atom, 1 nitrogen atom

d. 1 potassium atom, 1 manganese atom, 4 oxygen atoms

4.76 Write the chemical formulas for compounds in which the following numbers of atoms are present in a formula unit of the compound.

a. 10 carbon atoms, 14 hydrogen atoms, 2 oxygen atoms

b. 2 carbon atoms, 6 hydrogen atoms, 1 oxygen atom

c. 1 sodium atom, 1 oxygen atom, 1 hydrogen atom

d. 1 potassium atom, 1 nitrogen atom, 3 oxygen atoms

4.77 In each of the following pairs of formulas, indicate whether the first listed formula denotes *more total atoms,* the *same number of total atoms,* or *fewer total atoms* than the second listed formula.

a. N_2O and NO_2

b. KNO_3 and $Ca(OH)_2$

c. $Ba(ClO)_2$ and $Ba(ClO_2)_2$

d. $Al_2(SO_4)_3$ and $Ba_3(PO_4)_2$

4.78 In each of the following pairs of formulas, indicate whether the first listed formula denotes *more total atoms,* the *same number of total atoms,* or *fewer total atoms* than the second listed formula.

a. HN_3 and NH_3

b. $NaClO_3$ and $Be(CN)_2$

c. $CaSO_4$ and $Mg(OH)_2$

d. $Be_3(PO_4)_2$ and $Be(C_2H_3O_2)_2$

4.79 What difference in meaning, if any, is there in the following pairs of notations?
a. NO and No b. Cs_2 and CS_2
c. $CoBr_2$ and $COBr_2$ d. H and H_2

4.80 What difference in meaning, if any, is there in the following pairs of notations?
a. Hf and HF b. TiN and Tin
c. $NICl_2$ and $NiCl_2$ d. O_2 and O_3

4.81 The following molecular formulas are *incorrectly* written. Rewrite each formula in the correct manner.

a. H3PO4

b. SICL₄ (a silicon–chlorine compound)

c. NOO

d. 2HO (two H atoms and two O atoms)

4.82 The following molecular formulas are *incorrectly* written. Rewrite each formula in the correct manner.
  a. H2CO3
  b. ALBR$_3$ (an aluminum–bromine compound)
  c. HSH
  d. 2NO$_2$ (two N atoms and four O atoms)

4.83 Write a chemical formula for each of the following substances based on the information given about a molecule of the substance.
  a. A molecule of hydrogen cyanide is triatomic and contains the elements hydrogen, carbon, and nitrogen.
  b. A molecule of sulfuric acid is heptaatomic and contains two atoms of hydrogen, one atom of sulfur, and the element oxygen.

4.84 Write a chemical formula for each of the following substances based on the information given about a molecule of the substance.
  a. A molecule of nitrous oxide contains twice as many atoms of nitrogen as of oxygen and is triatomic.
  b. A molecule of nitric acid is pentaatomic and contains three atoms of oxygen and the elements hydrogen and nitrogen.

ADDITIONAL PROBLEMS

4.85 Carbon monoxide is a colorless, odorless gas that is toxic to humans. It combines with the metal nickel to form nickel carbonyl, a colorless liquid that boils at 43°C.
  a. List all physical properties of substances found in the preceding narrative.
  b. List all chemical properties of substances found in the preceding narrative.

4.86 A hard sugar cube is pulverized, and the resulting granules are heated in air until they discolor and then finally burst into flame and burn.
  a. List all physical changes to substances found in the preceding narrative.
  b. List all chemical changes to substances found in the preceding narrative.

4.87 Assign each of the following descriptions of matter to one of the following categories: element, compound, mixture.
  a. One substance present, one phase present, substance can be decomposed by chemical means
  b. Two substances present, one phase present
  c. One substance present, two elements present
  d. Two elements present, composition is variable

4.88 Assign each of the following descriptions of matter to one of the following categories: element, compound, mixture.
  a. One substance present, one phase present, substance cannot be decomposed by chemical means
  b. One substance present, three elements present

  c. Two substances present, two phases present
  d. Two elements present, composition is definite and constant

4.89 Indicate whether each of the following samples of matter is a heterogeneous mixture, a homogeneous mixture, a compound, or an element.
  a. a colorless single-phase liquid that when boiled away (evaporated) leaves behind a solid white residue
  b. a uniform red liquid with a boiling point of 59°C that cannot be broken down into simpler substances using chemical means
  c. a nonuniform, white crystalline substance, part of which dissolves in water and part of which does not
  d. a colorless single-phase liquid that completely evaporates without decomposition when heated and produces a gas that can be separated into simpler components using physical means

4.90 Indicate whether each of the following samples of matter is a heterogeneous mixture, a homogeneous mixture, a compound, or an element.
  a. a colorless gas, only part of which reacts with hot iron
  b. a cloudy liquid that separates into two layers upon standing for two hours
  c. a green solid, all of which melts at the same temperature to produce a liquid that decomposes upon further heating
  d. a colorless gas that cannot be separated into simpler substances using physical means and that reacts with copper to produce both a copper–nitrogen compound and a copper–oxygen compound

4.91 Classify the following as (1) heterogeneous mixture, (2) heterogeneous, but not a mixture, (3) homogeneous mixture, or (4) homogeneous, but not a mixture.
  a. an undissolved sugar cube in water
  b. a partially dissolved sugar cube in water
  c. a completely dissolved sugar cube in water
  d. an ice cube in water

4.92 Classify the following as (1) heterogeneous mixture, (2) heterogeneous, but not a mixture, (3) homogeneous mixture, or (4) homogeneous, but not a mixture.
  a. molten iron metal
  b. solid iron metal
  c. mix of molten and solid iron
  d. solid iron in water

4.93 In which of the following sequences of elements do all of the elements have two-letter symbols?
  a. magnesium, nitrogen, phosphorus
  b. bromine, iron, calcium
  c. aluminum, copper, chlorine
  d. boron, barium, beryllium

4.94 In which of the following sequences of elements do all of the elements have symbols that start with a letter that is not the first letter of the element's English name?
 a. silver, gold, mercury
 b. copper, helium, neon
 c. cobalt, chromium, sodium
 d. potassium, iron, lead

4.95 The chemical symbols Co and Hf when split into two capital letters produce the symbols of other elements: Co gives C and O and Hf gives H and F. With the help of Table 4.3, identify the other two-letter chemical symbols that when split into two separate capital letters produce the symbols of two other elements.

4.96 Reversal of the letters in the chemical symbols Ni and Ca produce the chemical symbols of other elements (In and Ac). With the help of Table 4.3, identify the other two-letter symbols for which this reversal process produces the symbol of another element.

4.97 Based on the given information, determine the numerical value of the subscript x in each of the following chemical formulas.
 a. $Na_2S_xO_3$; formula unit contains 7 atoms.
 b. $Ba(ClO_x)_2$; formula unit contains 9 atoms.
 c. $Na_xP_xO_{10}$; formula unit contains 16 atoms.
 d. $C_xH_{2x}O_x$; formula unit contains 24 atoms.

4.98 Based on the given information, determine the numerical value of the subscript x in each of the following chemical formulas.
 a. BaS_2O_x; formula unit contains 6 atoms.
 b. $Al_2(SO_x)_3$; formula unit contains 17 atoms.
 c. SO_xCl_x; formula unit contains 5 atoms.
 d. $C_xH_{2x}Cl_x$; formula unit contains 8 atoms.

4.99 A mixture contains the following five pure substances: O_2, N_2O, H_2O, CCl_4, and CH_2Br_2.
 a. How many different kinds of heteroatomic molecules are present in the mixture?
 b. How many different kinds of pentaatomic molecules are present in the mixture?
 c. How many different compounds are present in the mixture?
 d. How many different kinds of atoms are present in the mixture?
 e. How many total atoms are present in a mixture sample containing three molecules of each component?

4.100 A mixture contains the following five pure substances: N_2, N_2H_4, NH_3, CH_4, and CH_3Cl.
 a. How many different kinds of molecules that contain four or fewer atoms are present in the mixture?
 b. How many different kinds of homoatomic molecules are present in the mixture?
 c. How many different kinds of atoms are present in the mixture?

 d. How many total atoms are present in a mixture sample containing five molecules of each component?
 e. How many total hydrogen atoms are present in a mixture sample containing four molecules of each component?

CUMULATIVE PROBLEMS

4.101 Specify the physical state of a pure substance at each of the following conditions, or indicate that the state determination is not possible from the information given.
 a. 10°C below its freezing point
 b. 30°C above its melting point
 c. after sublimation has taken place
 d. at its boiling point

4.102 Specify the physical state of a pure substance at each of the following conditions, or indicate that the state determination is not possible from the information given.
 a. 10°C below its melting point
 b. 30°C above its freezing point
 c. after decomposition has taken place
 d. after deposition has taken place

4.103 Calculate the following percents, expressing each percent to three significant figures.
 a. percent of the elements that are naturally occurring
 b. percent of the 111 chemical symbols that are one-letter symbols
 c. percent of the elements that have been discovered during the 1900s
 d. percent of elementary particles (atoms) in Earth's crust that are silicon, magnesium, or iron atoms

4.104 Calculate the following percents, expressing each percent to three significant figures.
 a. percent of the elements that are synthetic (laboratory produced)
 b. percent of the 111 chemical symbols in which the first two letters of the element's name are the symbol
 c. percent of the elements that were discovered during the 1800s
 d. percent of elementary particles (atoms) in Earth's crust that are aluminum, hydrogen, or calcium atoms

4.105 The following density determination data were obtained by three students analyzing unknown substances.
 student I: mass = 4.32 g   volume = 3.78 mL
 student II: mass = 5.73 g   volume = 5.02 mL
 student III: mass = 1.52 g   volume = 1.33 mL
 a. Is it likely that the students were working with different unknowns or that they were working with the same substance?
 b. Is it possible to tell from the given data whether the unknowns were elements or compounds?

4.106 The following density determination data were obtained by three students analyzing unknown substances.

student I: mass = 27.2 g   volume = 23.6 mL

student II: mass = 30.3 g   volume = 28.3 mL

student III: mass = 55.6 g   volume = 42.5 mL

a. Is it likely that the students were working with different unknowns or that they were working with the same substance?

b. Is it possible to tell from the given data whether the unknowns were elements or compounds?

4.107 Three samples of a substance were subjected to analysis with each sample analyzed by a different technique. The results were:

technique I: 34.1% of Q and 65.9% of X

technique II: 34.12% of Q and 65.88% of X

technique III: 34.12497% of Q and 65.87503% of X

Is the substance that was analyzed likely an element, a compound, or a mixture?

4.108 Three samples of a substance were subjected to analysis with each sample analyzed by a different technique. The results were:

technique I: 34.2% of Z and 65.8% of D

technique II: 36.32% of Z and 63.68% of D

technique III: 37.2111% of Z and 62.7889% of D

Is the substance that was analyzed likely an element, a compound, or a mixture?

4.109 The density of gold is 19.3 g/cm^3, and the mass of a single gold atom is 3.27×10^{-22} g. How many gold atoms are present in a piece of gold whose volume is 3.22 cm^3?

4.110 The density of copper is 8.93 g/cm^3, and the mass of a single copper atom is 1.06×10^{-22} g. How many copper atoms are present in a copper bar whose dimensions are 2.00 cm × 3.00 cm × 5.00 cm?

4.111 In 1.00 g of fluorine atoms there are 3.17×10^{22} fluorine atoms. If you lined these atoms up side by side, how many miles long would the line of fluorine atoms be? The diameter of a fluorine atom is 1.44×10^{-8} cm.

4.112 In 1.00 g of fluorine atoms there are 3.17×10^{22} fluorine atoms. If you started counting these atoms at the rate of 10 per second, how many years would it take to count all the atoms in the 1.00 g sample?

Multiple-Choice Practice Test

Use this bank of 20 multiple-choice questions as a review of key concepts presented in this chapter. For many of the questions, there may be more than one correct answer (choice d) or no correct answer (choice e).

4.113 Which of the following is a property of both the liquid state and the solid state?

a. a definite shape

b. an indefinite volume

c. an indefinite shape and a definite volume

d. more than one correct response

e. no correct response

4.114 In which of the following pairs of properties are both properties *chemical* properties?

a. melts at 73°C, blue in color

b. good reflector of light, is very hard

c. decomposes upon heating, reacts with oxygen

d. more than one correct response

e. no correct response

4.115 Which of the following changes is a *physical* change?

a. melting of ice

b. pulverizing of a hard sugar cube

c. tarnishing of a piece of silver jewelry

d. more than one correct response

e. no correct response

4.116 Which of the following statements concerning mixtures is correct?

a. All components of a heterogeneous mixture must be in the same physical state.

b. A homogeneous mixture can have components present in two physical states.

c. A heterogeneous mixture must contain at least two phases.

d. more than one correct response

e. no correct response

4.117 The description "two substances present, two phases present" is correct for

a. heterogeneous mixtures.

b. homogeneous mixtures.

c. pure substances.

d. more than one correct response

e. no correct response

4.118 Which of the following characterizations could represent a compound?

a. one substance present, one phase present, substance can be decomposed by chemical means

b. one substance present, three elements present

c. two elements present, composition is definite and constant

d. more than one correct response

e. no correct response

4.119 When substance A interacts with substance B, a new substance C is formed. Based on this information,

a. A must be an element.

b. B could be an element.

c. C must be a compound.

d. more than one correct response

e. no correct response

4.120 Which of the following statements concerning elements and compounds is correct?
a. Elements, but not compounds, are pure substances.
b. A compound is a physical combination of two or more elements.
c. Compounds, but not elements, can be broken down into simpler substances, using chemical means.
d. more than one correct response
e. no correct response

4.121 Which of the following statements concerning the known elements is correct?
a. All except three of the elements are naturally occurring.
b. New elements have been characterized within the last 20 years.
c. It is very probable that all naturally occurring elements have already been discovered.
d. more than one correct response
e. no correct response

4.122 Which of the following statements concerning elemental abundances (in atom percent) in Earth's crust is correct?
a. One element accounts for over one-half of all elemental particles (atoms).
b. Oxygen and iron are the two most abundant elements.
c. Elemental abundances for Earth's crust are very similar to those for the universe as a whole.
d. more than one correct response
e. no correct response

4.123 In which of the following sequences of elements do all members of the sequence have one-letter chemical symbols?
a. nitrogen, oxygen, carbon
b. fluorine, chlorine, iodine
c. helium, beryllium, lithium
d. more than one correct response
e. no correct response

4.124 In which of the following sequences of elements do all members of the sequence have chemical symbols starting with the same letter?
a. sulfur, silicon, sodium
b. gold, silver, aluminum
c. potassium, phosphorus, lead
d. more than one correct response
e. no correct response

4.125 Which of the following statements is *not* part of the atomic theory of matter?
a. Chemical change involves a union, separation, or rearrangement of atoms.
b. Atoms are considered indestructible during chemical change.
c. Only whole atoms can participate in or result from chemical change.
d. more than one correct response
e. no correct response

4.126 Which of the following statements about atoms is *correct?*
a. An atom is the limit of physical subdivision.
b. Free, isolated atoms are commonly found in nature.
c. An atom is the smallest "piece" of an element that can exist and still have the properties of the element.
d. more than one correct response
e. no correct response

4.127 Which of the following statements about atoms is *correct?*
a. The mass of an atom is of the order of 10^{-24} g.
b. The diameter of an atom is of the order of 10^{-5} m.
c. Different types of atoms are distinguished from each other by their color.
d. more than one correct response
e. no correct response

4.128 Which of the following statements about molecules is *correct?*
a. All compounds have molecules as their basic structural unit.
b. Heteroatomic molecules, upon chemical subdivision, always yield two or more kinds of atoms.
c. No two atoms in a heteroatomic molecule may be the same.
d. more than one correct response
e. no correct response

4.129 Which of the following is an impossibility for a molecule of a compound in which the molecules are triatomic?
a. Three different elements could be present.
b. Three different kinds of atoms could be present.
c. Two different kinds of atoms could be present.
d. more than one correct response
e. no correct response

4.130 What is the numerical value for the subscript x in the chemical formula H_xPO_4, given that molecules of the compound contain eight atoms?
a. one   b. two   c. three   d. four
e. no correct response

4.131 In which of the following sequences of chemical formulas do all members of the sequence fit the description "heteroatomic and triatomic"?
a. CO_2, HCN, O_3
b. N_2O, NO_2, NO
c. SO_2, S_2O, SO_3
d. more than one correct response
e. no correct response

4.132 In which of the following pairs of chemical formulas do both members of the pair have the same number of atoms per molecule?
a. HNO_3 and $Ca(OH)_2$
b. $CoCl_2$ and $COCl_2$
c. H_2SO_4 and H_2CO_3
d. more than one correct response
e. no correct response

CHAPTER 5

Subatomic Particles, Isotopes, and Nuclear Chemistry

5.1 SUBATOMIC PARTICLES: PROTONS, NEUTRONS, AND ELECTRONS

In Chapter 4 we learned that all matter is made up of small particles called atoms and that 117 different types of atoms are known, each type corresponding to a different element. Until the last two decades of the nineteenth century, scientists believed that atoms were solid, indivisible spheres without internal substructure. Today this concept is known to be incorrect. Evidence from a variety of sources, some of which will be discussed in Section 5.5, indicates that atoms themselves are made up of smaller, more fundamental particles called *subatomic particles*. In this chapter we consider the fundamental types of subatomic particles, how they arrange themselves within an atom, and the relationship between an atom's subatomic makeup and its chemical identity and its stability.

A **subatomic particle** *is a very small particle that is a building block for atoms.* Three major types of subatomic particles exist: the *electron*, the *proton*, and the *neutron*. The key properties of *electrical charge* and *mass* for these three types of subatomic particles are given in Table 5.1. An **electron** *is a subatomic particle that possesses a negative* (−) *electrical charge.* Electrons were characterized in 1897 by the English physicist Joseph John Thomson (1856–1940). A **proton** *is a subatomic particle that possesses a positive* (+) *electrical*

TABLE 5.1 Charges and Masses of the Major Subatomic Particles

	Electron	Proton	Neutron
Charge	−1	+1	0
Actual mass (g)	9.109×10^{-28}	1.673×10^{-24}	1.675×10^{-24}
Relative mass (based on the electron's being one unit)	1	1837	1839
Relative mass (based on the neutron's being one unit)	0 (1/1839)	1	1

charge. Protons were discovered in 1886 by the German physicist Eugen Goldstein (1850–1930). A **neutron** *is a subatomic particle that is neutral; that is, it has no charge.* The neutron, the last of the three subatomic particles to be identified, was characterized in 1932 by the English physicist James Chadwick (1891–1974).

Using the mass or relative mass values from Table 5.1, we see that the electron has the smallest mass of the three subatomic particles. Both protons and neutrons are very massive particles compared to the electron, being nearly 2000 times heavier. For most purposes the masses of protons and neutrons can be considered equal, although technically the neutron is slightly heavier (1839 versus 1837 on a relative mass scale where the electron has a value of 1; see Table 5.1).

Electrons and protons, the two types of *charged* subatomic particles, possess the same amount of electrical charge; the character of the charge is, however, opposite (negative versus positive). The fact that these subatomic particles are charged is most important because of the way in which charged particles interact. *Particles of opposite or unlike charge attract each other; particles of like charge repel each other.* This behavior of charged particles will be of major concern in many of the discussions in later portions of the text.

Atoms of all 117 elements contain the same three types of subatomic particles. Different kinds of atoms differ only in the number of the various subatomic particles they contain.

Arrangement of Subatomic Particles Within an Atom

The arrangement of subatomic particles within an atom is not haphazard. All protons and all neutrons are found at the center of an atom in a very small volume called the *nucleus* (Figure 5.1). A **nucleus** *is the small, dense, positively charged center of an atom; it contains an atom's protons and neutrons.* A nucleus always has a positive charge because of the positively charged protons that are present. Almost all of an atom's

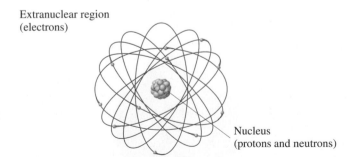

Extranuclear region
(electrons)

Nucleus
(protons and neutrons)

FIGURE 5.1 The protons and neutrons of an atom are found in the central nuclear region—the nucleus—and the electrons are found in an electron cloud outside the nucleus. Note that this figure is not drawn to scale; the correct scale would be comparable to a penny (the nucleus) in the center of a baseball field (the atom).

mass (over 99.9%) is found within its nucleus; all the heavy subatomic particles (protons and neutrons) are located there. The small size of the nucleus, coupled with its large amount of mass, causes nuclear material to be extremely dense.

Closely resembling the term *nucleus* is the term *nucleon*. A **nucleon** *is any subatomic particle found in the nucleus of an atom.* Thus, both protons and neutrons are nucleons and the nucleus can be regarded as containing a collection of nucleons (protons and neutrons).

The outer (extranuclear) region of an atom contains all of the electrons. It is an extremely large region compared to the nucleus. It is mostly empty space. It is a region in which the electrons move rapidly about the nucleus. The motion of the electrons in this extranuclear region determines the volume (size) of the atom in the same way as the blades of a fan determine a volume by their motion. The volume occupied by the electrons is sometimes referred to as the *electron cloud.* Since electrons are negatively charged, the electron cloud is said to be negatively charged.

Figure 5.1 contrasts the nuclear and extranuclear regions of an atom. The attractive force between the nucleus (positively charged) and the electrons (negatively charged) keeps the electrons within the extranuclear region of the atom. By analogy, the attractive force of gravity keeps the planets in their orbits about the sun.

Charge Neutrality of an Atom

An atom as a whole is neutral. How can it be that an entity possessing positive charge (the nuclear region) and negative charge (the extranuclear region, or electron cloud) can end up neutral overall? For this to occur, the same number of positive and negative charges must be present in the atom; equal numbers of positive and negative charges give a *net* electrical charge of zero. Atom neutrality thus requires that there be the same number of electrons and protons present in an atom, which is always the case for atoms.

number of protons = number of electrons

Size Relationships within an Atom

The diameter of the nucleus of an atom is approximately 10^{-15} m, which is about 1/100,000 the 10^{-10} m diameter of an atom (see Fig. 5.2). As a help in visualizing this size contrast, imagine enlarging (magnifying) the nucleus until it is the size of a baseball (about 2.9 inches in diameter). If the nucleus were this large, the whole atom would have a diameter of approximately 2.5 miles. The electrons would still be smaller than the periods used to end sentences in this text, and they would move about within that 2.5-mile region.

FIGURE 5.2 Key atomic dimensions (not drawn to scale).

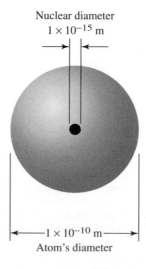

Nuclear diameter
1×10^{-15} m

1×10^{-10} m
Atom's diameter

The concentration of almost all the mass of an atom in the nucleus can also be illustrated by using our imagination. If a coin the same size as a copper penny contained copper nuclei (copper atoms stripped of their electrons) rather than copper atoms (which are mostly empty space), the coin would weigh 190,000,000 tons. Nuclei are indeed very dense matter.

Additional Subatomic Particles

Our just-completed discussion of the makeup of atoms in terms of subatomic particles is based on the existence of three types of subatomic particles: protons, neutrons, and electrons. This model of the atom is actually an oversimplification. In recent years, as the result of research carried out by nuclear physicists, the picture of the atom has lost its simplicity. Experimental evidence now available indicates that protons and neutrons themselves are made up of even smaller particles. Numerous other particles, with names such as leptons, mesons, and baryons, have been discovered. No *simple* theory is yet available that can explain all of these new discoveries relating to the complex nature of the nucleus.

Despite the existence of these smaller subatomic particles, further discussions in this chapter will be based on the three-subatomic-particle model of the atom. It readily explains almost all chemical observations about atoms. We will have no occasion to deal with any of the recently discovered types of subatomic (or sub-subatomic) particles. Protons, neutrons, and electrons will meet all our needs.

We will also continue to use the concept that atoms are the fundamental building blocks for all types of matter (Sec. 4.10) despite the existence of protons, neutrons, and electrons. This is because under normal conditions subatomic particles do not lead an independent existence for any appreciable length of time. The only way they gain stability is by joining together to form an atom.

A significant collection of evidence is consistent with and supports the existence, nature, and arrangement of subatomic particles as given in this section. Several pieces of such evidence will be considered in Section 5.5.

5.2 ATOMIC NUMBER AND MASS NUMBER

What determines whether a given atom is an atom of carbon or an atom of oxygen or an atom of gold? It is the number of protons present in the nucleus of the atom. The number of protons present determines atom identity. Every element has a characteristic number of protons associated with all of its atoms. For example:

All carbon atoms contain 6 protons.

All oxygen atoms contain 8 protons.

All gold atoms contain 79 protons.

If two atoms differ in the number of protons present, they must be atoms of two different elements. Conversely, if two atoms possess the same number of protons, they must be atoms of the same element.

Atomic Number

The characteristic number of protons associated with the atoms of a particular element is called the *atomic number* of the element. An **atomic number** *is the number of protons in the nucleus of an atom.*

Values of atomic numbers for the various elements, along with selected additional information, are printed on the inside front cover of this book. A check of the entries in the atomic number column of the data tabulation on the right shows that an entry exists

Atomic numbers must be *whole* numbers since they are obtained by counting whole objects (protons). You cannot have $3\frac{1}{2}$ or $5\frac{1}{4}$ protons in a nucleus.

for each of the numbers in the sequence 1 to 116 plus the number 118; an element with atomic number 117 is yet-to-be characterized (Sec. 5.10).

The existence of an element that corresponds to each of the atomic numbers 1 through 116 is an indication of the order existing in nature. Scientists also interpret this continuous atomic number sequence 1 through 116 as evidence that there are no "missing elements" yet to be discovered in nature. The highest-atomic-numbered element that is naturally occurring is element 92 (uranium); elements 93 through 116 and 118 are all synthetic (Secs. 4.8 and 5.10).

The atomic numbers of the elements are also found on the left side of the inside front cover. The diagram there is a *periodic table,* a graphical presentation of selected characteristics of the elements. (The periodic table will be considered in detail in Chapter 6.) We note at this time that each box in the periodic table designates an element, with that element's symbol in the center of the box. The number above the symbol is the element's atomic number. The elements are arranged in the periodic table in order of increasing atomic number.

Previously, in Section 4.7, an element was defined as a pure substance that cannot be broken down into simpler substances by ordinary chemical means. Although this is a good historical definition for an element, we can now give a more rigorous definition using the concept of atomic number. An **element** *is a pure substance in which all atoms present have the same atomic number; that is, all atoms have the same number of protons.*

An atomic number, besides giving information about the number of protons present in an atom, also gives information about the number of electrons present. Because of charge neutrality (Sec. 5.1) an atom has the same number of protons and electrons. Thus,

$$\text{atomic number} = \text{number of protons} = \text{number of electrons}$$

The symbol Z is used as a general designation for atomic number.

Mass Number

Information about the number of neutrons present in an atom is not obtainable solely from an atomic number. A second number, a *mass number,* is needed in addition to the atomic number. A **mass number** *is the sum of the number of protons and the number of neutrons in the nucleus of an atom.*

$$\text{mass number} = \text{number of protons} + \text{number of neutrons}$$

Mass numbers, like atomic numbers, must always be *whole* numbers.

The symbol A is used as a general designation for mass number. Since only protons and neutrons are present in a nucleus, the mass number gives the total number of subatomic particles present in the nucleus. The mass of an atom is almost totally accounted for by the protons and neutrons present (Sec. 5.1), hence the designation *mass* number.

Subatomic Particle Makeup of an Atom

The *sum* of the mass number and the atomic number for an atom (A + Z) has significance. It corresponds to the *total* number of subatomic particles present in the atom (protons + neutrons + electrons).

Knowing the atomic number and mass number of an atom uniquely specifies the atom's makeup in terms of subatomic particles. The following equations show the relationship between subatomic particles and the two numbers.

$$\text{number of protons} = \text{atomic number} = Z$$
$$\text{number of electrons} = \text{atomic number} = Z$$
$$\text{number of neutrons} = \text{mass number} - \text{atomic number} = A - Z$$

Note that the neutron count is obtained through subtraction of the atomic number from the mass number.

Mass numbers are not tabulated in a manner similar to atomic numbers because, as we learn in the next section, most elements lack a unique mass number.

The mass and atomic numbers of a given atom are often specified using the notation

$$^{A}_{Z}E$$

Here E represents the symbol of the element being considered. The atomic number is placed as a subscript in front of the elemental symbol. The mass number is placed as a superscript in front of the elemental symbol. Examples of such notation for actual atoms include

$$^{19}_{9}F \quad ^{23}_{11}Na \quad ^{197}_{79}Au$$

The first of these notations specifies a fluorine atom that has an atomic number of 9 and a mass number of 19.

Examples 5.1 and 5.2 are sample calculations showing the interrelationships among atomic number, mass number, and the subatomic particle composition of atoms.

EXAMPLE 5.1 **Determining the Subatomic Makeup of an Atom Given Its Atomic Number and Mass Number**

Determine the following information for an atom that has an atomic number of 11 and a mass number of 23.

a. The number of protons present
b. The number of neutrons present
c. The number of electrons present
d. The complete symbol ($^{A}_{Z}E$) for the atom

SOLUTION

a. There are 11 protons because the atomic number is always equal to the number of protons present.
b. There are 12 neutrons because the number of neutrons is always obtained by subtracting the atomic number from the mass number ($23 - 11 = 12$).

$$\underbrace{(\text{Protons} + \text{neutrons})}_{\text{Mass number}} - \underbrace{\text{protons}}_{\substack{\text{Atomic} \\ \text{number}}} = \text{neutrons}$$

c. There are 11 electrons because the number of protons and the number of electrons are always the same in an atom.
d. The superscript in the symbol notation (A) is the mass number; it is given as 23. The subscript in the symbol notation (Z) is the atomic number; it is given as 11. The identity of the atom is determined using the information found inside the front cover. Either tabulation there gives the information that the atomic number 11 belongs to the element sodium (Na). The complete symbol of the atom is, therefore, $^{23}_{11}Na$.

Answer Double Check:

Are the answers reasonable? Yes. The number of protons should always be the same as the number of electrons. Such is the case here. The mass number is always greater than the atomic number by a factor of 2 to 2.5. Such is the case here. The mass number, the larger of the two numbers, is always on the top in the elemental symbol. Such is the case here.

▶ **Practice Exercise 5.1** Determine the following information for an atom that has an atomic number of 34 and a mass number of 80.

a. The number of protons present
b. The number of electrons present
c. The number of neutrons present
d. The complete symbol ($^{A}_{Z}E$) for the atom

Answers to all practice exercises in this chapter are found in the back-of-the-book answer section.

EXAMPLE 5.2 **Determining the Subatomic Makeup of an Atom Given Its Complete Chemical Symbol**

Determine the following information for an atom whose complete chemical symbol is $^{63}_{29}Cu$.

   a. The total number of subatomic particles present in the atom
   b. The total number of subatomic particles present in the nucleus of the atom
   c. The total number of nucleons present in the atom
   d. The total charge (including sign) associated with the nucleus of the atom

SOLUTION

   a. The mass number gives the combined number of protons and neutrons present. The atomic number gives the number of electrons present. Adding these two numbers together gives the total number of subatomic particles present. The are 92 subatomic particles present (63 + 29).
   b. The nucleus contains all protons and all neutrons. The mass number (protons + neutrons), thus, gives the total number of subatomic particles present in the nucleus of an atom. There are 63 subatomic particles present in the nucleus.
   c. A nucleon is any subatomic particle in the nucleus. Thus, both protons and neutrons are nucleons. There are 63 nucleons present in the nucleus. Part (b) and part (c) of this problem ask the same question using different terminology.
   d. The charge associated with a nucleus originates from the protons present. It will always be positive because protons are positively charged particles. The atomic number, 29, indicates that 29 protons are present. Thus, the nuclear charge is +29.

Answer Double Check:

Are the answers reasonable? Yes. The number of subatomic particles present is always given by (A + Z). Such is the case here. The number of nucleons (subatomic particles in the nucleus) is always given by A. Such is the case here. The charge on the nucleus, which is always positive, is given by A. Such is the case here.

▶ **Practice Exercise 5.2**   Determine the following information for an atom whose complete symbol is $^{27}_{13}Al$.

   a. The total number of subatomic particles present in the atom
   b. The total number of subatomic particles present in the nucleus of the atom
   c. The total number of nucleons present in the atom
   d. The total charge (including sign) associated with the nucleus of the atom

---

5.3 ISOTOPES

Charge neutrality in an atom (Sec. 5.1) requires the presence of an equal number of protons and electrons. Because neutrons have no electrical charge, their numbers in atoms do not have to be the same as the number of protons or electrons. Most atoms contain more neutrons than either protons or electrons.

   Studies of atoms of various elements show that the number of neutrons present in atoms of an element is usually not constant; it varies over a small range. This means that not all atoms of an element have to be identical. They must have the same number of protons and electrons, but they can differ in the number of neutrons. For example, three kinds of naturally occurring oxygen atoms exist. All oxygen atoms have eight protons and eight electrons. Most oxygen atoms also contain eight neutrons. Some oxygen atoms exist, however, that contain nine neutrons, and a few exist that contain ten neutrons. Designations for these three kinds of oxygen atoms are

$$^{16}_{8}\text{O} \quad ^{17}_{8}\text{O} \quad ^{18}_{8}\text{O}$$

Atoms of an element that differ in neutron count are called *isotopes*. Three oxygen isotopes exist. **Isotopes** *are atoms of an element that have the same number of protons and electrons but different numbers of neutrons.* Isotopes always have the same atomic number and different mass numbers.

The presence of one or more additional neutrons in the tiny nucleus of an atom has essentially no effect on the way it behaves chemically. Thus, isotopes of an element have the same chemical properties. Isotopes have the same number of electrons, and it is electrons that determine chemical properties. When two atoms interact chemically, the outer part (electrons) of one interacts with the outer part (electrons) of the other. The small nuclear centers never come in contact with each other during a chemical interaction between atoms.

Isotopes of an element can have slightly different physical properties because they have different numbers of neutrons and therefore different masses. Physical property differences are greatest for elements of low atomic number. For such elements, differences in mass between isotopes are relatively large when compared to the masses of the isotopes themselves. For example, ^2_1H is twice as heavy as ^1_1H, and the density of ^2_1H is twice that of ^1_1H (0.18 g/L versus 0.090 g/L).

Most elements occurring naturally are mixtures of isotopes. The various isotopes of a given element are of varying abundance; usually one isotope is predominant. Typical of this situation is the element magnesium, which exists in nature in three isotopic forms: $^{24}_{12}\text{Mg}$, $^{25}_{12}\text{Mg}$, and $^{26}_{12}\text{Mg}$. The *percent abundances* for these three isotopes are, respectively, 78.70%, 10.13%, and 11.17%. A **percent abundance** *is the percent of atoms in a natural sample of a pure element that are a particular isotope of the element.* Percent abundances are number percents (number of atoms) rather than mass percents. A sample of 10,000 magnesium atoms would contain 7870 $^{24}_{12}\text{Mg}$ atoms, 1013 $^{25}_{12}\text{Mg}$ atoms, and 1117 $^{26}_{12}\text{Mg}$ atoms. Table 5.2 gives natural isotopic abundances and isotopic masses for the elements with atomic numbers 1 through 12. The unit used for specifying the mass of the various isotopes, amu, will be discussed in Section 5.4.

The percent abundances of the isotopes of an element may vary slightly in samples obtained from different locations, but such variations are ordinarily extremely small. We will assume in this text that the isotopic composition of an element is a constant.

Isotopic masses, although not whole numbers, have values that are very close to whole numbers. This fact can be verified by looking at the isotopic masses in Table 5.2. If an isotopic mass is rounded off to the closest whole number, this value is the same as the mass number of the isotope. This statement can be verified using the data in Table 5.2.

Twenty-three elements have only one naturally occurring form; that is, they are monoisotopic. For these elements, all atoms found in nature are identical to each other. Of the simpler elements (atomic numbers of 20 or less), those with only one form are

$$^{9}_{4}\text{Be} \quad ^{19}_{9}\text{F} \quad ^{23}_{11}\text{Na} \quad ^{27}_{13}\text{Al} \quad ^{31}_{15}\text{P}$$

The existence of isotopes adds clarification to the wording used in some of the statements of atomic theory (Sec. 4.10). Statement 1 reads "All matter is made up of small particles called atoms, of which 117 different 'types' are known." It should now be apparent why the word "types" was put in quotation marks. Because of the existence of isotopes, atoms of each type are similar, but not identical. Atoms of a given element are similar in that they have the same atomic number, but not identical since they may have different mass numbers.

The word *isotope* comes from the Greek *iso*, meaning "equal," and *topos*, meaning "place." Isotopes occupy an equal place (location) in listings of elements because all isotopes of an element have the same atomic number.

Hydrogen isotopes are unique among isotopes in that each isotope has a different name.

^1_1H protium (symbol H)

^2_1H deuterium (symbol D)

^3_1H tritium (symbol T)

TABLE 5.2 Isotopic Data for Elements with Atomic Numbers 1 through 12. Information given for each isotope includes mass number, isotopic mass in amu, and percent abundance.

1	Hydrogen	2	Helium	3	Lithium
$^1_1$H 1.008 amu 99.985% $^2_1$H 2.014 amu 0.015% $^3_1$H 3.016 amu trace		$^3_2$He 3.016 amu trace $^4_2$He 4.003 amu 100%		$^6_3$Li 6.015 amu 7.42% $^7_3$Li 7.016 amu 92.58%	

4	Beryllium	5	Boron	6	Carbon
$^9_4$Be 9.012 amu 100%		$^{10}_5$B 10.013 amu 19.6% $^{11}_5$B 11.009 amu 80.4%		$^{12}_6$C 12.000 amu 98.89% $^{13}_6$C 13.003 amu 1.11% $^{14}_6$C 14.003 amu trace	

7	Nitrogen	8	Oxygen	9	Fluorine
$^{14}_7$N 14.003 amu 99.63% $^{15}_7$N 15.000 amu 0.37%		$^{16}_8$O 15.995 amu 99.759% $^{17}_8$O 16.999 amu 0.037% $^{18}_8$O 17.999 amu 0.204%		$^{19}_9$F 18.998 amu 100%	

10	Neon	11	Sodium	12	Magnesium
$^{20}_{10}$Ne 19.992 amu 90.92% $^{21}_{10}$Ne 20.994 amu 0.26% $^{22}_{10}$Ne 21.991 amu 8.82%		$^{23}_{11}$Na 22.990 amu 100%		$^{24}_{12}$Mg 23.985 amu 78.70% $^{25}_{12}$Mg 24.986 amu 10.13% $^{26}_{12}$Mg 25.983 amu 11.17%	

Statement 2 reads "All atoms of a given type are similar to one another and significantly different from all other types." All atoms of an element are similar in chemical properties and differ significantly from atoms of other elements with different chemical properties.

EXAMPLE 5.3 **Identifying Characteristics of Isotopes of an Element**

Three isotopes exist for argon, element 18. One isotope has an isotopic mass of 35.967 amu and a percent abundance of 0.337. A second isotope has an isotopic mass of 37.963 amu and a percent abundance of 0.063. The third isotope has an isotopic mass of 39.962 amu.

   a. What are the mass numbers for the three argon isotopes?
   b. What is the percent abundance for the third argon isotope?

SOLUTION

   a. Isotopic masses, although not whole numbers, are numbers with values very close to that of whole numbers. Rounding an isotopic mass to the nearest whole number gives the mass number of that isotope. Thus, the mass numbers of the three argon isotopes are 36 (35.967 amu), 38 (37.963 amu), and 40 (39.962 amu). (The reason isotopic masses have values that are always very close to a whole number relates to protons and neutrons having masses on the amu scale that differ only slightly from the value 1.00 amu.)

b. The sum of the percents abundance for the three isotopes must add to 100%. Thus, by subtraction, we obtain a percent of 99.600% for the third isotope.

$$(100.000 - 0.337 - 0.063)\% = 99.600\% \text{ (calculator and \textbf{correct answer})}$$

▶ **Practice Exercise 5.3** Three isotopes exist for silicon, element 14. One isotope has an isotopic mass of 27.977 amu and a percent abundance of 92.21. A second isotope has an isotopic mass of 28.976 amu and a percent abundance of 4.70. The third isotope has an isotopic mass of 29.974 amu.

a. What are the mass numbers for the three silicon isotopes?
b. What is the percent abundance for the third silicon isotope?

It is possible for isotopes of two different elements to have the same mass number. For example, the element iron (atomic number 26) exists in nature in four isotopic forms, one of which is $^{58}_{26}$Fe. The element nickel, with an atomic number two units greater than that of iron, exists in nature in five isotopic forms, one of which is $^{58}_{28}$Ni. Thus, atoms of both iron and nickel exist with a mass number of 58. Thus, mass numbers are not unique for elements as are atomic numbers. Atoms of different elements that have the same mass number are called *isobars*; $^{58}_{26}$Fe and $^{58}_{28}$Ni are isobars. **Isobars** *are atoms that have the same mass number but different atomic numbers.* Even though atoms of two *different* elements can have the same mass number (isobars), they cannot have the same atomic number. All atoms of a given atomic number must necessarily be atoms of the same element.

There are 286 isotopes that occur naturally. In addition, over 2000 more isotopes have been synthesized in the laboratory (from the naturally occurring ones) using nuclear bombardment reactions. (Section 5.10 considers such nuclear reactions.) These unstable synthetic isotopes all have the common characteristic of being radioactive. Radioactive isotopes eventually revert back to naturally occurring isotopes. Many of these unstable isotopes, despite their instability, have important uses in chemical and biological research as well as in medicine.

EXAMPLE 5.4 **Distinguishing between Isotopes and Isobars**

Indicate whether the members of each of the following pairs are isotopes, isobars, or neither.

a. $^{42}_{20}$X and $^{43}_{20}$Q
b. $^{40}_{19}$X and $^{40}_{20}$Q
c. $^{44}_{20}$X and $^{45}_{21}$Q
d. an atom X with 20 protons and 21 neutrons and an atom Q with 19 protons and 21 neutrons

SOLUTION

a. These atoms are isotopes. Both atoms have the same atomic number of 20. Isotopes differ from each other in neutron count, which is the case here. Atom X has 22 neutrons, and atom Q has 23 neutrons.
b. These atoms are isobars. They have the same mass number (40) and different atomic numbers (19 and 20).

    c. These atoms are not isotopes or isobars. Isotopes must have the same atomic number and isobars must have the same mass number. Neither is the case here.

    d. These atoms are not isotopes or isobars. They are not isotopes, because a differing number of protons means differing atomic numbers. They are not isobars, because the mass numbers differ: 41 for atom X and 40 for atom Q. The two atoms contain the same number of neutrons. However, the definition for isobars is based on the same mass number rather than on the same neutron count.

▶ **Practice Exercise 5.4**   Indicate whether the members of each of the following pairs are isotopes, isobars, or neither.

    a. $^{69}_{31}X$ and $^{71}_{31}Q$

    b. $^{75}_{33}X$ and $^{75}_{32}Q$

    c. $^{78}_{34}X$ and $^{78}_{35}Q$

    d. An atom X with 30 protons and 34 neutrons and an atom Q with 31 protons and 33 neutrons

5.4 ATOMIC MASSES

An analogy involving isotopes and identical twins is helpful in understanding the fact that all atoms of an element need not have the same mass. Identical twins need not weigh the same even though they have identical "gene packages." They are identical twins by the gene criterion regardless of their masses. Likewise, isotopes, even though they have different masses, are atoms of the same element by atomic number criterion (same number of protons).

A detailed consideration of the periodic table, its value, significance, and use is the topic of a considerable portion of Chapter 6.

The existence of isotopes means that atoms of an element can have several different masses. For example, magnesium atoms can have any one of three masses because there are three magnesium isotopes. Which of these three magnesium isotopic masses is used in situations in which the mass of the element magnesium needs to be specified? The answer is none of them. Instead, a *weighted average mass* that takes into account the existence of isotopes and their relative abundances is used. These weighted average masses are called *atomic masses*.

Atomic mass values for the elements are found inside the front cover of this book. There are two such listings there. On the inside cover's right side (atomic number—atomic mass listing), atomic masses are given to the maximum number of significant figures possible, which varies from element to element. Typical atomic mass values include

$$14.0067 \quad \text{for the element nitrogen}$$
$$32.065 \quad \text{for the element sulfur}$$
$$126.90447 \text{ for the element iodine}$$

On the inside cover's left side (periodic table; see Sec. 5.2), the atomic mass values are the numbers underneath the symbols of the elements. (The number above each element's symbol is its atomic number; see Sec. 5.2.) In this periodic table listing, the atomic masses are given to the hundredths decimal place. It is this rounded form of the atomic mass that is most often used in chemical calculations (Chapters 9 and 10).

Atomic mass values are not *mass numbers*. They cannot possibly be mass numbers because they are not whole numbers. Mass numbers, which are *counts* of the number of protons and neutrons present in nuclei (Sec. 5.2), must be whole numbers. Atomic masses are calculated numbers obtained from data on isotopic masses and isotopic abundances.

The starting point for understanding the origins of atomic mass values is a consideration of the formal definition for an atomic mass. An **atomic mass** *is the relative mass of an average atom of an element on a scale using the* $^{12}_{6}C$ *atom as the reference.* The meaning of two terms found within this definition, *relative mass* and *average atom*, is crucial to understanding the definition as a whole.

Relative Mass

The usual standards of mass, such as grams or pounds, are not convenient for use with atoms, because very small numbers are always encountered. For example, the mass in grams of a $^{238}_{92}$U atom, one of the heaviest atoms known, is 3.95×10^{-22}. To avoid repeatedly encountering such small numbers scientists have chosen to work with relative rather than actual mass values.

A relative mass value for an atom is the mass of that atom relative to some standard rather than the actual mass value of the atom in grams. The term *relative* means "as compared to." The choice of the standard is arbitrary; this gives scientists control over the magnitude of the numbers on the relative scale, thus avoiding very small numbers.

For most purposes in chemistry, relative mass values serve just as well as actual mass values. Knowing how many times heavier one atom is than another, information obtainable from a relative mass scale, is just as useful as knowing the actual mass values of the atoms involved. Example 5.5 illustrates the procedures involved in constructing a relative mass scale and also points out some of the characteristics of such a scale.

EXAMPLE 5.5 **Constructing a Relative Mass Scale**

Construct a relative mass scale for the hypothetical atoms Q, X, and Z, given the following information about them.

1. Atoms of Q are four times heavier than those of X.
2. Atoms of X are three times heavier than those of Z.

SOLUTION

Atoms of Z are the lightest of the three types of atoms. We will arbitrarily assign atoms of Z a mass value of one unit. The unit name can be anything we wish, and we shall choose "snick." On this basis, one atom of Z has a mass value of 1 snick. Atoms of Z will be our scale reference point. Atoms of X will have a mass value of 3 snicks (three times as heavy as Z) and Q atoms a value of 12 snicks (four times as heavy as X).

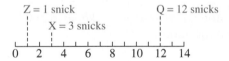

The name snick chosen for the mass unit was arbitrary. The assignment of the value 1 for the mass of Z, the reference point on the scale, was also arbitrary. What if we had chosen to call the unit a "smerge" and had chosen a value of 3 for the mass of an atom of Z? If this had been the case, the resulting relative scale would have appeared as

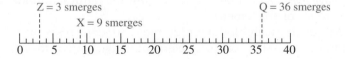

Which of the preceding relative scales is the "better" scale? The answer is that the scales are equivalent. The relationships between the masses of Q, X, and Z are the same on the two scales, even though the reference points and unit names differ. On the snick scale, for example, Q is four times heavier than X (12/3); on the smerge scale, Q is also four times heavier than X (36/9).

Notice that we did not need to know the actual masses of Q, X, and Z to set up either the snick or smerge scale. All that is needed to set up a relative scale is a set of interrelationships among quantities. One value—the reference point—is arbitrarily assigned, and all other values are determined by using the known interrelationships.

The information given at the start of this example is sufficient to set up an infinite number of relative mass scales. Each scale would differ from the others in choice of reference point and unit name. All the scales would, however, be equivalent to each other, and each scale would provide all of the mass relationships obtainable from an actual mass scale except for actual mass values.

▶ **Practice Exercise 5.5** Construct a relative mass scale for the hypothetical atoms Q, X, and Z, given the following information about them.

1. Atoms of Q are one-half as heavy as those of X.
2. Atoms of X are one-half as heavy as those of Z.
3. The reference point for the scale is X = 6.00 splats.

A relative scale of atomic masses has been set up in a manner similar to that used in Example 5.5. The unit is called the *atomic mass unit*, abbreviated amu. The arbitrary reference point involves a particular isotope of carbon, $^{12}_{6}C$. The mass of this isotope is set at 12.00000 amu. The masses of all other atoms are then determined relative to that of $^{12}_{6}C$. For example, if an atom is twice as heavy as a $^{12}_{6}C$ atom, its mass is 24.00000 amu on the scale, and if an atom weighs half as much as a $^{12}_{6}C$ atom, its scale mass is 6.00000 amu.

The masses of all atoms have been determined relative to each other experimentally. Actual values for the masses of selected isotopes on the $^{12}_{6}C$ scale are given in Figure 5.3.

Reread the formal definition of atomic mass given earlier in this section. Note how $^{12}_{6}C$ is mentioned explicitly in the definition because of the central role it plays in the setting up of the relative atomic mass scale.

On the basis of the values given in Figure 5.3, it is possible to state, for example, that $^{238}_{92}U$ is 4.256 times as heavy as $^{56}_{26}Fe$ (238.05 amu/55.93 amu = 4.256) and $^{56}_{26}Fe$ is 2.798 times as heavy as $^{20}_{10}Ne$ (55.93 amu/19.99 amu = 2.798). We do not need to know the actual masses of the atoms involved to make such statements; relative masses are sufficient to calculate the information.

Average Atom

Since isotopes exist, the mass of an atom of a specific element can have one of several values. For example, oxygen atoms can have any one of three masses, since three isotopes exist: $^{16}_{8}O$, $^{17}_{8}O$, and $^{18}_{8}O$. Despite mass variances among isotopes, the atoms of an element are treated as if they all had a single common mass. The common mass value used is a *weighted average mass*, which takes into account the natural abundances and atomic masses of the isotopes of an element.

The validity of the weighted average mass concept rests on two points. First, extensive studies of naturally occurring elements have shown that the percent abundance of the isotopes of a given element is generally constant. No matter where the element sample is obtained on Earth, it generally contains the same percentage of each isotope. Because of these constant isotopic ratios, the mass of an "average atom" does not vary. Second, chemical operations are always carried out with very large numbers of atoms. The tiniest piece of matter visible to the eye contains more atoms than can be counted by a person in a lifetime. The numbers are so great that any collection of atoms a chemist works with will be representative of naturally occurring isotopic ratios.

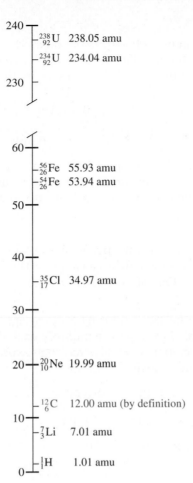

Weighted Averages

Atomic masses are weighted averages calculated from the following three pieces of information:

1. The *number* of isotopes that exist for the element
2. The *isotopic mass* for each isotope, that is, the relative mass of each isotope on the $^{12}_{6}C$ scale
3. The *percent abundance* of each isotope

Table 5.2 gives these data for selected elements.

Examples 5.6 and 5.7 illustrate the operations needed to calculate weighted averages. Example 5.6 is a general exercise concerning weighted averages, and Example 5.7 illustrates the calculation of an atomic mass by the method of weighted averages.

EXAMPLE 5.6 **Calculation of a Weighted Average**

Sulfur oxides are air pollutants that arise primarily from the burning of coal. A student measures the sulfur oxide concentration in the atmosphere on five successive days, in parts per billion (ppb), with the following results.

24 ppb   21 ppb   21 ppb   24 ppb   15 ppb

What is the average sulfur oxide concentration in the air over this time period, based on the student's measurements?

SOLUTION

Let us solve this problem using two different methods. The first method involves procedures familiar to you—the normal way of taking an average. By this method, the average is found by dividing the sum of the numbers by the number of values summed.

$$\frac{24 + 21 + 21 + 24 + 15}{5} \text{ ppb} = 21 \text{ ppb} \qquad \text{(calculator answer)}$$

$$= 21.0 \text{ ppb} \qquad \textbf{(correct answer)}$$

Now let us solve this same problem again, this time treating it as a weighted average problem. To do this, we organize the given information in a different way. In our list of pollutant concentrations we have three different concentration values: 24 ppb, 21 ppb, and 15 ppb.

Two of the five values (40.0%) are 24 ppb.
Two of the five values (40.0%) are 21 ppb.
One of the five values (20.0%) is 15 ppb.

We will use the data in this percent form for our weighted average calculation.

To find the weighted average, we multiply each distinct value (24, 21, and 15) by its fractional abundance, that is, by its percentage expressed in decimal form, and then we sum the products from the multiplications.

$$0.400 \times 24 \text{ ppb} = 9.6 \text{ ppb}$$

$$0.400 \times 21 \text{ ppb} = 8.4 \text{ ppb}$$

$$0.200 \times 15 \text{ ppb} = \underline{3.0 \text{ ppb}}$$

21.0 ppb (same average value as before)

This averaging method, although it appears somewhat more involved than the normal method, is the one that must be used in calculating atomic masses because of the form in which data about isotopes are obtained. The percent abundances of isotopes are experimentally determinable quantities. The total number of atoms of various isotopes present in nature, a prerequisite for using the normal average method, is not easily determined. Therefore, we use the percent method.

Answer Double Check:

Is the numerical value of the answer reasonable? Yes. In calculating an average, the result must always fall in the range defined by the highest (24 ppb) and lowest values (15 ppb) used in taking the average. Such is the case here. The average, 21.0 ppb, falls within the 15–24 ppb range.

▶ **Practice Exercise 5.6**   Ozone is an air pollutant that is a component of photochemical smog. It is formed by the action of sunlight on motor vehicle exhaust gases. The ozone concentration in the lower atmosphere on five successive days, in parts per million (ppm), was found to be at the following levels.

0.075 ppm   0.077 ppm   0.077 ppm   0.071 ppm   0.075 ppm

Using the weighted average method, calculate the average ozone concentration in the lower atmosphere over this time period.

EXAMPLE 5.7 Calculation of Atomic Mass from Isotopic Masses and Percent Abundances

Magnesium occurs in nature in three isotopic forms: $^{24}_{12}Mg$ (78.70% abundance), $^{25}_{12}Mg$ (10.13% abundance), and $^{26}_{12}Mg$ (11.17% abundance). The relative masses of these three isotopes, respectively, are 23.985, 24.986, and 25.983 amu. Calculate the atomic mass of magnesium from these data.

SOLUTION

The atomic mass of an element is calculated using the weighted average method illustrated in Example 5.6. Each of the isotopic masses is multiplied by the fractional abundance associated with that mass, and then the products are summed.

$$0.7870 \times 23.985 \text{ amu} = 18.876195 \text{ amu} = 18.88 \text{ amu}$$
$$0.1013 \times 24.986 \text{ amu} = 2.5310818 \text{ amu} = 2.531 \text{ amu}$$
$$0.1117 \times 25.983 \text{ amu} = 2.9023011 \text{ amu} = 2.902 \text{ amu}$$

Note that the method for converting percentages to fractional abundances is always the same. The decimal point in the percentage is moved two places to the left. For example,

$$78.70\% \text{ becomes } 0.7870$$

Significant figures are always an important part of an atomic mass calculation. Both isotopic masses and percent abundances are experimentally determined numbers.

Summing, to obtain the atomic mass, gives

$$(18.88 + 2.531 + 2.902) \text{ amu} = 24.313 \text{ amu} \quad \text{(calculator answer)}$$
$$= 24.31 \text{ amu} \quad \textbf{(correct answer)}$$

Since the number 18.88 is known only to the hundredths place, the answer can be expressed only to the hundredths place.

The above calculation involved an element that exists in three isotopic forms. An atomic mass calculation for an element having four isotopic forms would be carried out in an almost identical fashion. The only difference would be four products to calculate (instead of three) and four terms in the resulting sum.

Answer Double Check:

Is the numerical value of the atomic mass reasonable? Yes. The average mass should have a value near the mass of the isotope that is present in the greatest abundance. That is the case here. The mass of the most abundant isotope is 23.98 amu, and the calculated average is 24.31 amu. The other isotopic masses are 24.99 amu and 25.98 amu.

▶ **Practice Exercise 5.7** Chromium occurs in nature in four isotopic forms: $^{50}_{24}Cr$ (4.31% abundance), $^{52}_{24}Cr$ (83.76% abundance), $^{53}_{24}Cr$ (9.55% abundance), and $^{54}_{24}Cr$ (2.38% abundance). The relative masses of these four isotopes, respectively, are 49.9461 amu, 51.9405 amu, 52.9407 amu, and 53.9389 amu. Calculate the atomic mass for chromium from these data.

In Example 5.7 the atomic mass of magnesium was calculated to be 24.31 amu. How many magnesium atoms have a mass of 24.31 amu? The answer is none. Magnesium atoms have a mass of 23.985, 24.986, or 25.983 amu, depending on which isotope they are. The mass 24.31 amu is the mass of an average magnesium atom. It is this average mass that is used in calculations even though no magnesium atoms have masses equal to this average value. Only in the case where all atoms have the same mass will the isotopic mass and the atomic mass be the same.

The *fractional abundance* is the percent abundance divided by 100%. For a 92.301% abundance the fraction abundance is 0.92301. The decimal point is shifted two places to the left in converting percent abundance to a fractional abundance.

If a census bureau report gives the average number of children in a family in a given locality as 2.7, obviously no family within the area actually has 2.7 children. In a similar fashion, no atom of an element for which several isotopes exist has the average mass that is calculated using the methods of Example 5.7.

Atomic masses are subject to change, and they do change. Every two years an updated atomic mass listing is published by the International Union of Pure and Applied Chemistry (IUPAC). This update, produced by an international committee of chemists, takes into account all new research on isotopic abundances.

New atomic mass values often have less uncertainty than the older values they replace. This is a reflection of the increasingly sophisticated instrumentation available to current researchers, which enables them to make better measurements.

Illustrative of the atomic mass changes that do occur are those contained in the 2007 update report issued by the IUPAC. The atomic masses of three elements were changed:

element 30: zinc	65.409 becomes 65.38
element 42: molybdenum	95.94 becomes 95.96
element 70: ytterbium	173.04 becomes 173.054

Atomic mass revisions such as these explain why various textbooks (and periodic tables) often differ in a few atomic mass values. The differing values come from different IUPAC reports on atomic masses. Atomic mass values used in this text come from the IUPAC 2007 report on the atomic masses of the elements.

The uncertainty associated with atomic mass values varies from element to element. For example, we have

B	10.811 amu
F	18.9984032 amu
Si	28.0855 amu
Pb	207.2 amu

What causes such variance in uncertainty? The key factor is the constancy of isotopic percentage abundance measurements among various samples of an element. Although all elements have an essentially constant set of isotopic percentage abundances, there are slight variations among samples obtained from different sources. All variations are small; however, for some elements they are greater than for others. When only one form of an element occurs in nature, such as F, values with less uncertainty can be obtained for atomic mass.

An atomic mass cannot be calculated for all elements. Recall, from Sec. 4.8, that not all elements are naturally occurring substances. Twenty-nine of the known elements are synthetic, having been produced in the laboratory from naturally occurring elements. Obviously a weighted average atomic mass cannot be calculated for these laboratory-produced elements, since the amount of each isotope produced varies, depending on the laboratory experiment carried out.

Tabulations of atomic masses do contain entries for the synthetic elements. Such entries are the mass number of the most stable isotope of the synthetic element. (All isotopes of all synthetic elements are unstable.) Such mass numbers are always enclosed in parentheses to distinguish them from calculated atomic masses. Note the presence of such entries in both the atomic mass listing and the periodic table inside the front cover.

Mass Spectrometry Experiments

Before leaving the subject of atomic masses, we need to consider one additional question. How do scientists determine the abundances and masses of the various isotopes of an element? An instrument known as a *mass spectrometer* is the key to obtaining such information.

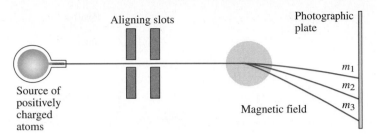

FIGURE 5.4 A simplified diagram of a mass spectrometer, an instrument used to obtain information about isotopic masses and abundances.

A schematic diagram of a mass spectrometer is given in Figure 5.4. The important components of this instrument are a source of charged particles of the element under investigation, an aligning system to create a narrow beam of the charged particles, a magnetic field to affect the path of the charged particles, and a means of detecting the charged particles (such as a photographic plate).

Suppose oxygen gas containing all three isotopes is admitted to the instrument. The gaseous molecules are bombarded with an energetic electron beam, producing positively charged oxygen species. The aligning system produces a narrow beam of these charged atoms, which then enters the magnetic field. The most massive particles (heaviest isotope) are not deflected by the magnetic field as much as the less massive ones, so the charged atoms are divided into separate beams that strike the photographic plate at different points depending on their masses. The more abundant isotopes will create more intense lines on the plate. The relative intensities of the lines correlate exactly with the relative abundances of the isotopes.

5.5 EVIDENCE SUPPORTING THE EXISTENCE AND ARRANGEMENT OF SUBATOMIC PARTICLES

A significant collection of evidence is consistent with and supports the existence, nature, and arrangement of subatomic particles as given in Section 5.1. Two historically important types of experiments illustrate some of the sources of this evidence. *Discharge tube experiments* resulted in the original concept that the atom contained negatively and positively charged particles. *Metal foil experiments* provided evidence for the existence of a nucleus within the atom.

Discharge Tube Experiments

Neon signs, fluorescent lights, and television tubes are all basic components of our modern technological society. The forerunner for all three of these developments was the *gas discharge tube*. Gas discharge tubes also provided some of the first evidence that an atom consisted of still smaller particles (subatomic particles).

The principle behind the operation of a gas discharge tube—that gases at low pressure conduct electricity—was discovered in 1821 by the English chemist Humphry Davy (1778–1829). Subsequently, gas discharge tube studies were carried out by many scientists.

A simplified diagram of a gas discharge tube is shown in Figure 5.5. The apparatus consists of a sealed glass tube containing two metal disks called *electrodes*. The glass tube also has a side arm for attachment to a vacuum pump. During operation, the electrodes are connected to a source of electrical power. (The electrode attached to the positive side of the electrical power source is called the *anode*; the one attached to the negative side is known as the *cathode*.) Use of the vacuum pump allows the amount of gas within the tube to be varied. The smaller the amount of gas present, the lower the pressure within the tube.

Early studies with gas discharge tubes showed that when the tube was almost evacuated (low pressure), electricity flowed from one electrode to the other and the residual gas became luminous (it glowed). Different gases in the tube gave different

FIGURE 5.5 A simplified version of a gas discharge tube.

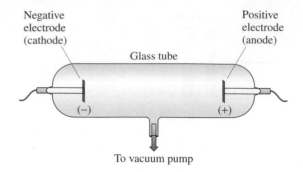

Negative electrode (cathode)

Positive electrode (anode)

Glass tube

(−)   (+)

To vacuum pump

colors to the glow. After the pressure in the tube was reduced to still lower levels (very little gas remaining), it was found that the luminosity disappeared but the electrical conductance continued, as shown by a greenish glow given off by the tube's glass walls. This glow was the initial discovery of what became known as *cathode rays*. Their discovery marked the beginning of nearly 40 years of discharge tube experimentation that ultimately led to the characterization of both the electron and the proton.

The term *cathode rays* comes from the observation that when an obstacle is placed between the negative electrode (cathode) and the opposite glass wall, a sharp shadow the shape of the obstacle is cast on that wall. This indicates that the rays are coming from the cathode.

Further studies showed that these cathode rays caused certain minerals such as sphalerite (zinc sulfide) to glow. Glass plates were coated with sphalerite and observed under high magnification while being bombarded with cathode rays. The light emitted by the sphalerite coating consisted of many pinpoint flashes. This observation suggested that cathode rays were in reality a stream of extremely small particles.

Joseph John Thomson (1856–1940), an English physicist, provided many facts about the nature of cathode rays. Using a variety of materials as cathodes, he showed that cathode ray production was a general property of matter. By using a specially designed cathode ray tube (see Fig. 5.6), he also found that cathode rays could be deflected by charged plates or a magnetic field. The rays were repelled by the north pole, or negative plate, and attracted to the south pole, or positive plate, thus indicating that they were negatively charged. In 1897, Thomson concluded that cathode rays were

Thomson's cathode ray tube experiments have evolved into today's televisions and computer monitors (still often called CRTs).

FIGURE 5.6 J. J. Thomson's cathode ray tube involved the use of both electrical and magnetic fields.

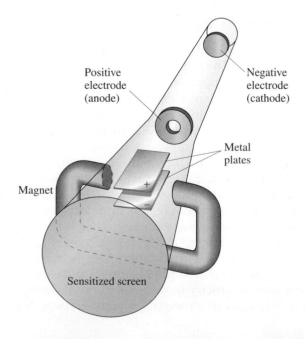

Positive electrode (anode)

Negative electrode (cathode)

Metal plates

Magnet

+

−

Sensitized screen

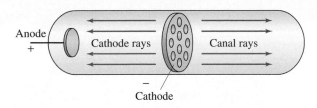

FIGURE 5.7 A gas discharge tube modified to detect canal rays.

streams of negatively charged particles, which today we call *electrons.* Further experiments by others proved that his conclusions were correct.

In 1886, a German physicist, Eugen Goldstein (1850–1930), showed that positive particles were also present in discharge tubes. He used a discharge tube in which the cathode was a metal plate with a large number of holes drilled in it. The usual cathode rays were observed to stream from cathode to anode. In addition, rays of light appeared to stream from each of the holes in the cathode in a direction opposite to that of the cathode rays (see Fig. 5.7). Because these rays were observed streaming through the holes or channels in the cathode, Goldstein called them *canal rays.*

Further research showed that canal rays were of many different types, in contrast to cathode rays, which are only one type, and that the particles making up canal rays were much heavier than those of cathode rays. The type of canal rays produced depended upon the gas in the tube. The simplest canal rays were eventually identified as the particles now called protons.

Canal rays are now known to be gas atoms that have lost one or more electrons. Their origin and behavior in a discharge tube can be understood as follows. Electrons (cathode rays) emitted from the cathode collide with residual gas molecules (air) on the way to the anode. Some of these electrons have enough energy to knock electrons away from the gas molecules, leaving behind a positive particle (the remainder of the gas molecule). These positive particles are attracted to the cathode, and some of them pass through the holes or channels. The fact that atoms, under certain conditions, can lose electrons will be discussed further in Section 7.4.

On the basis of discharge tube experiments, Thomson proposed in 1898 that the atom was composed of a sphere of positive electricity containing most of the mass, and that small negative electrons were attached to the surface of the positive sphere. He postulated that a high voltage could pull off surface electrons to produce cathode rays. Thomson's model of the atom, sometimes referred to as the "raisin muffin" or "plum pudding" model—with the electrons as the raisins or plums—is now known to be incorrect. Its significance is that it set the stage for an experiment, commonly called the gold foil experiment, that led to the currently accepted arrangement of protons and electrons in the atom.

Metal Foil Experiments

In 1911 Ernest Rutherford (1871–1937; see "The Human Side of Chemistry 4") designed an experiment to test the Thomson model of the atom. In this experiment thin sheets of metal foil were bombarded by alpha particles from a radioactive source. Alpha particles, which are positively charged, are ejected at high speeds from some radioactive materials. The phenomenon of radioactivity had been discovered in 1896 and gave further evidence that electrical charges existed within the atom. Gold was chosen as the target metal because it is easily hammered into very thin sheets. The experimental setup for Rutherford's experiment is shown in Figure 5.8. Alpha particles do not appreciably penetrate lead, so a lead plate with a slit was used to produce a narrow alpha particle beam. Each time an alpha particle hit the fluorescent screen, a flash of light was produced.

Rutherford expected that all the alpha particles, since they were so energetic, would pass straight through the thin gold foil. His reasoning was based on the

FIGURE 5.8
Rutherford's gold foil—alpha particle experiment. Most of the alpha particles went straight through the foil, but a few were deflected at large angles.

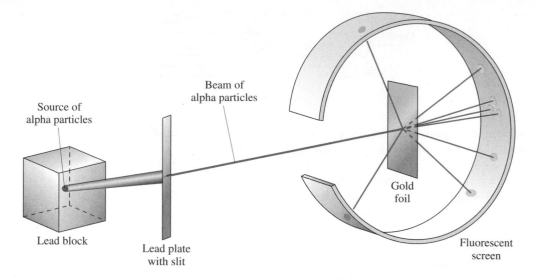

Source of alpha particles

Beam of alpha particles

Gold foil

Lead block

Lead plate with slit

Fluorescent screen

Thomson model, in which the mass and positive charge of the gold atoms were distributed uniformly through each atom. As each positive alpha particle neared the foil, Rutherford assumed that it would be confronted by a uniform positive charge. All particles would be affected the same way (no deflection), which would support the Thomson model of the atom.

The results from the experiment were very surprising. Most of the particles—more than 99%—went straight through as expected. A few, however, were appreciably deflected by something that had to be much heavier than the alpha particles themselves. A very few particles were deflected almost directly back toward the alpha particle source. Similar results were obtained when elements other than gold were used as targets.

Extensive study of the results of his experiments led Rutherford to propose the following explanation:

1. A very dense, small nucleus exists in the center of the atom. This nucleus contains most of the mass of the atom and all of the positive charge.
2. Electrons occupy most of the total volume of the atom and are located outside the nucleus.
3. When an alpha particle scores a direct hit on a nucleus, it is deflected back along the incoming path.
4. A near miss of a nucleus by an alpha particle results in repulsion and deflection.
5. Most of the alpha particles pass through without any interference, because most of the atomic volume is empty space.
6. Electrons have so little mass that they do not deflect the much larger alpha particles (an alpha particle is almost 8000 times heavier than an electron).

Many other experiments have since verified Rutherford's conclusion that at the center of an atom there is a nucleus that is very small and very dense.

5.6 NUCLEAR STABILITY AND RADIOACTIVITY

We now return to further discussion about the nucleus of an atom. In Section 5.1, we learned that an atom's nucleus, located at the center of an atom, is a small, dense region that contains all of the protons and all of the neutrons present in an atom. A nucleus is always positively charged because of the protons that are present within it.

Numerous studies concerning atomic nuclei show that they may be divided into two categories based on nuclear stability. Some nuclei are stable while others are not. A **stable**

The Human Side of Chemistry 4

Ernest Rutherford (1871–1937)

Ernest Rutherford, born in 1871 on a farm in New Zealand, was the fourth in a family of 12 children. He left New Zealand for England at the age of 25 after having won a scholarship to Cambridge University. Rutherford's gold foil experiment is only one of many contributions that he made to the sciences of chemistry and physics. Indeed, three years prior to carrying out this famous experiment, in 1908, he received the Nobel Prize in chemistry for his investigations into the nature of radioactivity.

While in graduate school at Cambridge, his professor, J. J. Thomson, encouraged him to study the newly discovered phenomenon of radioactivity. His research in this area led to the discovery of the alpha and beta radiation associated with radioactivity.

In 1899, he moved to Canada, spending nine years at McGill University doing further research on alpha and beta particles. At McGill he found that alpha particles are helium nuclei and that beta particles are electrons. For this work he was awarded his Nobel Prize.

In 1907, he returned to England (Manchester University) where further studies on alpha particles led to his famous gold foil experiment, considered to be Rutherford's greatest and most fruitful contribution to scientific knowledge. Many years later Rutherford described the unexpected results of this experiment as follows: "It was about as credible as if you had fired a 15-inch shell at a piece of tissue paper and it came back and hit you."

World War I brought an abrupt change in direction, a switch from atoms to submarines. He studied underwater acoustics, supplying the government with much information needed to advance the technology of submarine detection.

Following the war, in 1919, Rutherford moved to Cambridge University, assuming the position formerly held by J. J. Thomson, the professor who guided him as a graduate student. Here, he again was on the forefront of scientific advances, this time discovering nuclear transformations, the process in which an atom of an element changes into another element (Sec 5.8).

Many of the students Rutherford guided as a professor went on to make major scientific discoveries of their own. Among his graduate students were 10 future recipients of the Nobel Prize. He lived to see some of them receive their prizes.

Rutherford's research work was diverse—radioactivity at McGill, atomic physics at Manchester, and nuclear physics at Cambridge. His research was world-class at all three institutions. He was a very talented researcher. Element 104, rutherfordium, carries Rutherford's name.

nucleus *is a nucleus that does not easily undergo change.* Conversely, an **unstable nucleus** *is a nucleus that spontaneously undergoes change.* The spontaneous change that unstable nuclei undergo involves emission of radiation from the nucleus, a process by which an unstable nucleus can become more stable. The radiation emitted from unstable nuclei is called *radioactivity.* **Radioactivity** *is the radiation spontaneously emitted from an unstable nucleus.* Atoms that possess unstable nuclei are said to be *radioactive.* A **radioactive atom** *is an atom with an unstable nucleus from which radiation is spontaneously emitted.*

In discussions about radioactivity, the term *nuclide* is used as an alternate designation for an atom. A **nuclide** *is an atom with a specific atomic number and a specific mass number.* The term *isotopes* (Sec. 5.3) refers to different atomic forms of the same element; the term *nuclide* is used in describing atomic forms of different elements. The species $^{12}_{6}C$ and $^{13}_{6}C$ are isotopes of the element carbon. The species $^{12}_{6}C$, $^{15}_{7}N$, and $^{16}_{8}O$ are nuclides of different elements. The terms *radioactive atom* and *radioactive nuclide* are used interchangeably. The term *radioactive nuclide* is often shortened to *radionuclide.*

Also in radioactivity discussions there are two notation systems for designating a given nuclide. Consider a nuclide of nitrogen with seven protons and eight neutrons. This nuclide can be denoted as $^{15}_{7}N$ or nitrogen-15. In the first notation, which was previously introduced in Section 5.3, the superscript is the mass number and the subscript is the atomic number. In the second notation the mass number is appended to the name of the element with a hyphen. An advantage of the first notation is that the atomic number is shown; an advantage of the second notation is that superscripts and subscripts are not needed. Both types of notation will be used in the remainder of this chapter.

Of the 88 elements that are found in nature (Sec. 4.8), 29 have at least one naturally occurring radionuclide. Radionuclides, however, are known for all 117 elements, even though they occur naturally for only the aforementioned 29 elements. This is because laboratory procedures have been developed by which scientists convert nonradioactive nuclides (stable nucleus) into radioactive nuclides (unstable nucleus). Such procedures, called bombardment reactions, are considered in Section 5.10.

No simple rule exists for predicting whether a particular nucleus will be stable or unstable, that is, nonradioactive or radioactive. However, a consideration of some observations about those nuclei that are stable is helpful in understanding the stability–instability situation for nuclei. Two generalizations are readily apparent from a study of the properties of naturally occurring stable nuclei.

1. *There is a correlation between nuclear stability and the total number of nucleons (protons plus neutrons) found in a nucleus.* All nuclei with 84 or more protons present are unstable. The largest stable nucleus known is that of $^{209}_{83}$Bi, a nucleus that contains 209 nucleons. It thus appears that there is a limit to the number of nucleons that can be packed into a stable nucleus.

2. *There is a correlation between nuclear stability and neutron-to-proton ratio in a nucleus.* The number of neutrons present in a stable nucleus increases as the number of protons increases. For elements of low atomic number, neutron-to-proton ratios for stable nuclei are very close to 1. For heavier elements, stable nuclei have higher neutron-to-proton ratios, with the ratio reaching approximately 1.5 (3-to-2) for the heaviest stable elements. Figure 5.9 illustrates this changing neutron-to-proton ratio as a function of atomic number (the number of protons present). These

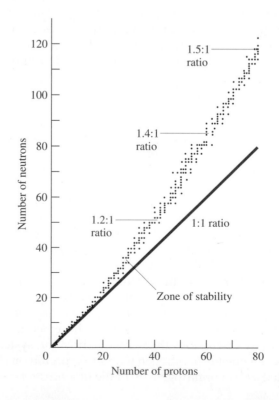

FIGURE 5.9 A graph of the number of neutrons versus the number of protons for stable nuclei.

neutron-to-proton ratio observations suggest that neutrons are at least partially responsible for the stability of the nucleus. It should be remembered that like charges repel each other and that most nuclei contain many protons (with identical positive charges) squeezed together into a very small volume. As the number of protons increase, the forces of repulsion between protons sharply increase. Therefore, a greater number of neutrons is necessary to "counteract" the increased repulsions. Finally, at element 84, the repulsive forces become so great that nuclei are unstable regardless of the number of neutrons present.

5.7 HALF-LIFE: A MEASURE OF NUCLEAR STABILITY

The term *radioactive decay* is used to denote the process of spontaneous radiation emission by an unstable nucleus. **Radioactive decay** *is the process whereby an unstable nucleus spontaneously gives off radiation.* Radioactive nuclides do not all undergo radioactive decay at the same rate. Some decay very rapidly; others undergo decay at very slow rates. This indicates that radioactive nuclides are not all equally unstable. The greater the radioactive decay rate, the lower the stability of the radionuclide.

The concept of *half-life* is used to quantitatively express nuclear stability. A **half-life** *is the time required for one half of any given quantity of a radioactive substance to undergo decay.* For example, if a radionuclide's half-life is 12 days and you have a 4.00 g sample of it, then after 12 days (one half-life) only 2.00 g of the sample (half the original amount) will remain undecayed; the other half will have decayed into some other substance.

Half-lives as long as billions of years and as short as a fraction of a second have been determined. Table 5.3 contains examples of the wide range of half-life values.

Most naturally occurring radionuclides have long half-lives. Some radionuclides with short half-lives, however, are also found in nature. Such short-lived species, because they decay rapidly, must be continually produced in order to be present. Processes that result in their production are (1) the decay of naturally occurring long-lived nuclides, (2) the decay of short-lived nuclides that have been produced in the previous manner, and (3) reactions involving cosmic rays, which take place naturally in the upper atmosphere. Examples of the second method of producing short-lived nuclides are presented in Section 5.13.

The decay rate (half-life) of a radionuclide is constant. It is independent of outward conditions such as temperature, pressure, and state of chemical combination. It is dependent only on the identity of the radionuclide. For example, radioactive sodium-24,

Half-life and rate of decay for a radionuclide are inversely related. The faster the rate of decay, the shorter the half-life.

TABLE 5.3 Range of Half-Lives Found for Naturally Occurring Radionuclides	
Element	**Half-life**
Vanadium-50	6×10^{15} yr
Platinum-190	6.9×10^{11} yr
Uranium-238	4.5×10^{9} yr
Uranium-235	7.1×10^{8} yr
Thorium-230	7.5×10^{4} yr
Lead-210	22 yr
Bismuth-214	19.7 min
Polonium-212	3.0×10^{-7} sec

FIGURE 5.10 Decay of 80.0 mg of $^{131}_{53}$I, which has a half-life of 8.0 days. After each half-life period, the quantity of original material present at the beginning of the period is reduced by half.

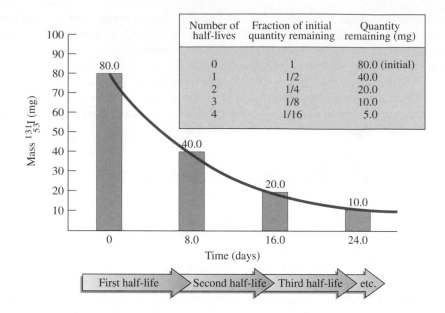

Number of half-lives	Fraction of initial quantity remaining	Quantity remaining (mg)
0	1	80.0 (initial)
1	1/2	40.0
2	1/4	20.0
3	1/8	10.0
4	1/16	5.0

whether incorporated into NaCl, NaBr, Na_2SO_4, or $NaC_2H_3O_2$, decays at the same rate. If a nuclide is radioactive, nothing will stop it from decaying and nothing will increase or decrease its decay rate.

Figure 5.10 shows graphically the meaning of half-life. After one half-life has passed, one half of the original atoms have decayed, so half remain. During the next half-life, one-half of the remaining half will decay, so one fourth of the original atoms remain undecayed. After three half-lives, $\frac{1}{2} \times \frac{1}{2} \times \frac{1}{2} = \frac{1}{8}$ of the original atoms remain undecayed, and so on. Note from Figure 5.10 that only a very small amount of original material (less than 1%) remains after seven half-lives have elapsed.

Calculations involving amounts of radioactive material decayed, amounts remaining undecayed, and time elapsed can be carried out by using the following equation.

$$\begin{pmatrix} \text{amount of radionuclide} \\ \text{undecayed after } n \text{ half-lives} \end{pmatrix} = \begin{pmatrix} \text{original amount} \\ \text{of radionuclide} \end{pmatrix} \times \frac{1}{2^n}$$

EXAMPLE 5.8 **Using Half-Life to Calculate the Amount of Radionuclide that Remains Undecayed after a Certain Time**

The half-life of cobalt-60 is 5.3 yr. If 2.0 g of cobalt-60 is allowed to decay for a period of 15.9 yr, how many grams of cobalt-60 remain?

SOLUTION

First, we must determine the number of half-lives that have elapsed.

$$15.9 \text{ yr} \times \frac{1 \text{ half-life}}{5.3 \text{ yr}} = 3.0 \text{ half-lives}$$

Knowing the number of elapsed half-lives and the original amount of radioactive cobalt present, we can use the equation

$$\begin{pmatrix} \text{amount of radionuclide} \\ \text{undecayed after } n \text{ half-lives} \end{pmatrix} = \begin{pmatrix} \text{original amount} \\ \text{of radionuclide} \end{pmatrix} \times \frac{1}{2^n}$$

to get our answer.

$$\left(\begin{array}{c} \text{amount of radionuclide} \\ \text{undecayed after } n \text{ half-lives} \end{array}\right) = 2.0 \text{ g} \times \frac{1}{2^3} \text{ Three half-lives}$$

$$= 2.0 \text{ g} \times \frac{1}{8}$$

$$= 0.25 \text{ g (calculator and \textbf{correct answer})}$$

Answer Double Check:

Is the magnitude of the answer reasonable? Yes. First, it must be less than the starting amount of 2.0 g, which it is. Second, the amount sequence for three half-lives is 2.0 g to 1.0 g to 0.50 g to 0.25 g. The latter is the calculated answer.

▶ **Practice Exercise 5.8**   The half-life of cesium-137, a radioisotope used in food irradiation, is 30.2 years. If 4.60 g of cesium-137 is allowed to decay for a period of 90.6 years, how many grams of cesium-137 remain?

| **EXAMPLE 5.9** | **Using Half-Life to Calculate the Time Needed to Reduce Radioactivity to a Specific Level** |

Iodine-135 is a nuclide found in radioactive fallout from nuclear weapon explosions. Its half-life is 6.70 hr. How long, in hours, will it take for 93.75% (15/16) of the iodine-135 atoms in a "fallout" sample to undergo decay?

SOLUTION

If 15/16 of the sample has decayed, then 1/16 of the sample remains undecayed. In terms of $1/2^n$, 1/16 is equal to $1/2^4$; that is,

$$\frac{1}{2} \times \frac{1}{2} \times \frac{1}{2} \times \frac{1}{2} = \frac{1}{2^4} = \frac{1}{16}$$

Thus four half-lives have elapsed in reducing the amount of iodine-135 to 1/16 of its original amount.

Since the half-life of iodine-135 is 6.70 hr, the total time elapsed will be

$$4 \text{ half-lives} \times \frac{6.70 \text{ hr}}{1 \text{ half-life}} = 26.8 \text{ hr}   \text{ (calculator and \textbf{correct answer})}$$

Answer Double Check:

Is the calculated number of elapsed half-lives reasonable? Yes. If 15/16 of the sample undergoes decay, only 1/16 of the sample remains undecayed. The fraction 1/16 is equal to $1/2^4$ and the exponent n of 2^n is the number of half-lives. Four half-lives is correct.

▶ **Practice Exercise 5.9**   Chromium-151 is a nuclide used in nuclear medicine for the assessment of kidney activity. Its half-life is 27.8 days. How long will it take, in days, for 87.5% (7/8) of the chromium-151 atoms in a nuclear medicine sample to undergo decay?

In both Examples 5.8 and 5.9 the time elapsed was equivalent to a whole number of half-lives. In order to work problems involving a fractional number of half-lives, equations involving logarithms must be used. Such equations will not be presented in this text; hence, only problems that involve a whole number of half-lives will be considered.

5.8 THE NATURE OF NATURAL RADIOACTIVE EMISSIONS

The radiation given off by radionuclides is of several different types, with the type given off by a specific radionuclide being dependent on the neutron-to-proton ratio of the nuclide.

The first information concerning the nature of the radiation emanating from naturally radioactive materials was obtained by Ernest Rutherford in the years 1898–1899. Using an apparatus similar to that shown in Figure 5.11, he found that if a beam of radiation is passed between electric plates, it is split into three components, indicating the presence of three different types of emissions from radioactive materials. A closer analysis of Rutherford's experiment reveals that one radiation component is positively charged (it is attracted to the negative plate), a second component is negatively charged (it is attracted to the positive plate), and the third component carries no charge (it is unaffected by either charged plate). Rutherford chose to call the three radiation components alpha rays (α rays) (the positive component), beta rays (β rays) (the negative component), and gamma rays (γ rays) (the uncharged component). (Alpha, beta, and gamma are the first three letters of the Greek alphabet.) We mention Rutherford's nomenclature system because it stuck; we still use these Greek letter designations. Today, we speak of alpha particles, beta particles, and gamma rays. Further research has shown that both alpha and beta radiation involve particles with mass and that gamma radiation has no mass; that is, it is a form of energy.

The complete characterization of the three types of natural radioactive emissions required many years. Early work in the field was hampered by the fact that many of the details concerning atomic structure were not yet known. For example, recall (Sec. 5.1) the neutron was not identified until 1932, 36 years after the discovery of radioactivity. In terms of modern day scientific knowledge, Rutherford's three types of radiation are characterized as follows.

An **alpha particle** *is a particle, in which two protons and two neutrons are present, that is emitted by certain radioactive nuclei.* The notation used to represent an alpha particle is $^4_2\alpha$. The numerical subscript indicates that the charge on the particle is + 2 (from the two protons). The numerical superscript indicates a mass of 4 amu. On the atomic mass scale (Sec. 5.4) protons and neutrons both have masses equal to 1.0 amu. Thus, the total mass of an alpha particle (two protons and two neutrons) is 4.0 amu. Alpha particles are identical with the nuclei of helium-4 ($^4_2$He) atoms.

A **beta particle** *is a particle, whose charge and mass are indentical to those of an electron, that is emitted by certain radioactive nuclei.* However, beta particles are not extranuclear electrons; they are particles that have been produced inside the nucleus and then ejected. More concerning this process will be given in Section 5.9. The symbol used to represent a beta particle is $^0_{-1}\beta$. The numerical subscript indicates that the charge on the beta particle is –1, that of an electron. The use of the superscript zero for the mass of a

FIGURE 5.11 Effect of an electric field on radiation emanating from a naturally radioactive substance. Gamma rays are unaffected. The lighter beta particles are deflected considerably more than the heavier alpha particles.

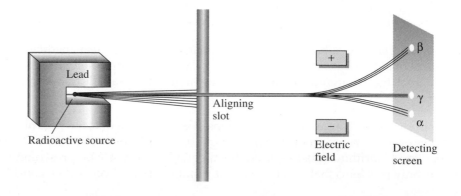

beta particle is not to be interpreted as meaning that a beta particle has no mass, but rather that its mass number (protons + neutrons) is zero. The actual mass of a beta particle on the atomic mass scale is 0.00055 amu.

A **gamma ray** *is a form of high-energy radiation without mass or charge that is emitted by certain radioactive nuclei.* Gamma rays are similar to X-rays, except they have higher energy. The symbol for gamma rays is $_0^0\gamma$.

5.9 EQUATIONS FOR RADIOACTIVE DECAY

Alpha, beta, and gamma emissions come from the nucleus of an atom. These spontaneous emissions alter nuclei; obviously, if a nucleus loses an alpha particle (two protons and two neutrons), it will not be the same as it was before the departure of the particle. In the case of alpha and beta emissions, the nuclear alteration causes atom identity to change; an atom of another element is formed.

Separate consideration of the production of alpha particles, beta particles, and gamma rays via radioactive decay provides further insights into the nature of radioactive processes and also introduces the concept of nuclear equations. In these considerations the terminology *parent nuclide* and *daughter nuclide* occurs. A **parent nuclide** *is the nuclide that undergoes decay in a radioactive decay process.* A **daughter nuclide** *is the nuclide that is produced as a result of a radioactive decay process.*

Alpha-Particle Decay

Alpha-particle decay *is the radioactive decay process in which an alpha particle is emitted from an unstable nucleus.* It always results in the formation of a nuclide of a different element. The daughter nuclide of such decay has an atomic number that is 2 fewer than that of the original nucleus and a mass number that is 4 fewer. We can represent alpha-particle decay in general terms by the equation

$$_Z^A X \longrightarrow \, _2^4\alpha + \, _{(Z-2)}^{(A-4)}Y$$

where X is the symbol for the nucleus of the original element undergoing decay and Y is the symbol for the nucleus of the element formed as a result of the decay.

Specific alpha-decay processes can be represented using *nuclear equations.* A **nuclear equation** *is an equation in which the chemical symbols used represent atomic nuclei rather than atoms.* Both $_{83}^{211}Bi$ and $_{92}^{238}U$ are radionuclides that undergo alpha-particle decay. The nuclear equations for these two decay processes are

$$_{83}^{211}Bi \longrightarrow \, _2^4\alpha + \, _{81}^{207}Tl$$

$$_{92}^{238}U \longrightarrow \, _2^4\alpha + \, _{90}^{234}Th$$

In the first equation, $_{83}^{211}Bi$ is the parent nuclide and $_{81}^{207}Tl$ is the daughter nuclide; in the second equation, $_{92}^{238}U$ is the parent nuclide and $_{90}^{234}Th$ is the daughter nuclide.

Both charge and mass must be conserved in a correctly written nuclear equation. That such is the case is ensured by application of the following two rules.

1. The sum of the subscripts (atomic number or particle charge) on both sides of the nuclear equation must be equal.
2. The sum of the superscripts (mass number) on both sides of the nuclear equation must be equal.

Both of our example equations are balanced. In the alpha decay of $_{83}^{211}Bi$, the subscripts on both sides total 83 and the superscripts total 211, For the decay of $_{92}^{238}U$, the subscripts total 92 on both sides and the superscripts total 238 on both sides.

> Loss of an alpha particle from an unstable nuclide *always* results in (1) a decrease of four units in the mass number (A) and (2) a decrease of two units in the atomic number (Z).

Beta-Particle Decay

Beta-particle decay *is the radioactive decay process in which a beta particle is emitted from an unstable nucleus.* Beta-particle decay also always results in the formation of a nuclide of a different element. The mass number of the new nuclide is the same as that of the original atom. The atomic number, however, has increased by one unit. The general equation for beta decay is

$$\underset{Z}{\overset{A}{X}} \longrightarrow \underset{-1}{\overset{0}{\beta}} + \underset{(Z+1)}{\overset{A}{Y}}$$

Specific examples of beta-particle decay are

$$\underset{4}{\overset{10}{Be}} \longrightarrow \underset{-1}{\overset{0}{\beta}} + \underset{5}{\overset{10}{B}}$$

$$\underset{90}{\overset{234}{Th}} \longrightarrow \underset{-1}{\overset{0}{\beta}} + \underset{91}{\overset{234}{Pa}}$$

Loss of a beta particle from an unstable nucleus results in (1) no change in the mass number (A) and (2) an increase of one unit in the atomic number (Z).

Both of these nuclear equations are balanced; superscripts and subscripts add to the same sums on both sides of the equation.

It is not immediately apparent how a nucleus, composed only of neutrons and protons, ejects a negative particle (beta particle) when no such particle is present in the nucleus. The accepted explanation is that through a complex series of steps a neutron in the nucleus is transformed into a proton and a beta particle; that is,

$$\underset{0}{\overset{1}{n}} \longrightarrow \underset{1}{\overset{1}{p}} + \underset{-1}{\overset{0}{\beta}}$$

Once formed within the nucleus, the beta particle is ejected with a high velocity. The net result of beta-particle formation is an increase by one in the number of protons present in the nucleus and a decrease by one in the number of neutrons present in the nucleus; the mass number is, however, constant, as the total number of subatomic particles in the nucleus (protons and neutrons) has not changed. Note in our two examples of beta emission that the daughter nuclide has one more proton than the parent as evidenced by the atomic number of the daughter being greater than that of the parent by one unit. Subtraction of the atomic number of the daughter nuclide from its mass number in each case—to get the number of neutrons—will reveal that the daughter nuclide contains one fewer neutron than the parent.

Gamma-Ray Emission

Among *synthetically* produced radionuclides (Sec. 5.10) pure "gamma emitters," radionuclides that give off gamma rays but no alpha or beta particles, occur. These radionuclides are important in diagnostic nuclear medicine. Pure "gamma emitters" are not found among naturally occurring radionuclides.

Gamma-ray emission *is the radioactive decay process in which gamma rays are emitted from an unstable nucleus.* For naturally occurring radionuclides, gamma-ray emission always occurs in conjunction with an alpha- or beta-decay process; it never occurs independently. Such gamma rays are most often not included in the nuclear equation, since they do not affect the balancing of the equation or the identity of the decay product.

EXAMPLE 5.10 **Writing Balanced Nuclear Equations Given the Parent Nuclide and Its Mode of Decay**

Write a balanced nuclear equation for the decay of each of the following radioactive nuclides. The mode of decay is indicated in parentheses.

    a. $\underset{54}{\overset{138}{Xe}}$ (beta emission)   **b.** $\underset{58}{\overset{142}{Ce}}$ (alpha emission)

    c. $\underset{78}{\overset{190}{Pt}}$ (alpha emission)   **d.** $\underset{35}{\overset{82}{Br}}$ (beta emission)

SOLUTION

In each case the atomic and mass numbers of the daughter nuclide are obtained by first writing the symbols of the parent nuclide and the particle emitted by the nucleus (alpha or beta particle) and then balancing the equation.

a. Let X represent the product of the radioactive decay, that is, the daughter nuclide. Then

$$^{138}_{54}\text{Xe} \longrightarrow ^{0}_{-1}\beta + \text{X}$$

Because the sums of the superscripts on both sides of the equation must be equal, the superscript for X must be 138. In order for the sums of the subscripts on both sides of the equation to be equal, the subscript for X must be 55. Then $54 = (-1) + (55)$. As soon as the subscript of X is determined, the identity of X can be determined from a periodic table (inside front cover). The element with an atomic number of 55 is cesium (Cs). Therefore,

$$^{138}_{54}\text{Xe} \longrightarrow ^{0}_{-1}\beta + ^{138}_{55}\text{Cs}$$

b. Similarly, letting X represent the product of the radioactive decay, we have for the alpha decay of $^{142}_{58}\text{Ce}$

$$^{142}_{58}\text{Ce} \longrightarrow ^{4}_{2}\alpha + \text{X}$$

Balancing the equation, making the superscripts on the right side of the equation total 142 and the subscripts total 58, we get

$$^{142}_{58}\text{Ce} \longrightarrow ^{4}_{2}\alpha + ^{138}_{56}\text{Ba}$$

c. Similarly, we write

$$^{190}_{78}\text{Pt} \longrightarrow ^{4}_{2}\alpha + \text{X}$$

Balancing superscripts and subscripts, we get

$$^{190}_{78}\text{Pt} \longrightarrow ^{4}_{2}\alpha + ^{186}_{76}\text{Os}$$

d. Finally, we write

$$^{82}_{35}\text{Br} \longrightarrow ^{0}_{-1}\beta + \text{X}$$

In beta emission the atomic number of the daughter nuclide is always greater by one and the mass number does not change from that of the parent. The balancing procedure gives us this result.

$$^{82}_{35}\text{Br} \longrightarrow ^{0}_{-1}\beta + ^{82}_{36}\text{Kr}$$

Answer Double Check:

The mathematical requirement for a balanced nuclear equation is that the subscript sum and superscript sum be the same on each side of the equation. Is such the case for each of these equations? Yes. For example, in part (a) the subscript sum is 54 on both sides of the equation and the superscript sum is 138 on both sides of the equation.

▶ **Practice Exercise 5.10**   Write a balanced nuclear equation for the decay of each of the following radioactive nuclides. The mode of decay is indicated in parentheses.

a. $^{212}_{85}\text{At}$ (alpha emission)   **b.** $^{72}_{31}\text{Ga}$ (beta emission)

5.10 TRANSMUTATION AND BOMBARDMENT REACTIONS

Radioactive decay, (Secs. 5.7–5.9), is an example of a *natural* transmutation reaction. A **transmutation reaction** *is a nuclear reaction in which a nuclide of one element is changed into a nuclide of another element.* It is also possible to cause transmutation to occur in a laboratory setting through use of *bombardment reactions.* A **bombardment reaction** *is a* *nuclear reaction brought about by bombarding stable nuclei with small particles traveling at very high speeds.* Bombardment reactions involve *artificial* transmutation, called artificial because the change does not occur naturally.

In bombardment reactions, there are always two reactants (the target nuclide and the small, high-energy bombarding particle) and also two products (the daughter nuclide and another small particle such as a neutron or proton).

The first successful bombardment reaction was carried out in 1919, 25 years after the discovery of radioactive decay, by Ernest Rutherford, the same Rutherford who earlier had investigated the nature of alpha, beta, and gamma rays (Sec. 5.8). Rutherford's initial successful bombardment experiment consisted of letting alpha particles from a natural source (radium) bombard nitrogen gas. In this process he found that a new stable nuclide was formed: oxygen-17. The nuclear equation for this transmutation is

$$\ce{^{14}_{7}N + ^{4}_{2}\alpha -> ^{17}_{8}O + ^{1}_{1}p}$$

Further research carried out by many investigators has shown that numerous nuclei experience change under the stress of bombardment by small, high-energy particles. In most cases, the new nuclide that is produced is radioactive (unstable). Two examples of bombardment reactions now carried out in laboratories in which the product nuclide is radioactive are

$$\ce{^{44}_{20}Ca + ^{1}_{1}p -> ^{44}_{21}Sc + ^{1}_{0}n}$$

$$\ce{^{23}_{11}Na + ^{2}_{1}H -> ^{21}_{10}Ne + ^{4}_{2}\alpha}$$

Radioactive nuclides produced by bombardment reactions, like naturally occurring radionuclides, undergo radioactive decay. In many cases, the previously discussed alpha- and beta-particle modes of decay (Sec. 5.9) occur. Additional modes of decay, to be discussed in Section 5.11 are also encountered.

Synthetic Elements

Production of the small, *high-energy* bombarding particles needed to effect a bombardment reaction requires use of a cyclotron or a linear accelerator (both very expensive, *large* pieces of equipment). Both use magnetic fields to accelerate charged particles to velocities at which the energy is sufficient to allow the particle to penetrate the nucleus and induce a nuclear change.

Over 2000 bombardment-produced radionuclides that do not occur naturally are now known. This number is seven times greater than the number of naturally occurring nuclides (Sec. 5.3). In this total is at least one radionuclide of every naturally occurring element. In addition, nuclides of 29 elements that do not occur in nature have been produced in small quantities as the result of bombardment reactions. Four of these "synthetic" elements, produced between 1937 and 1941, filled gaps in the periodic table for which no naturally occurring element had been found. These four elements are technetium (Tc, element 43), an element with numerous uses in nuclear medicine; promethium (Pm, element 61); astatine (At, element 85); and francium (Fr, element 87). The remainder of the synthetic elements, elements 93 to 116 and 118, are called the *transuranium elements* because of their occurrence immediately following uranium in the periodic table. (Uranium is the highest atomic-numbered, naturally occurring element.) All isotopes of all of the transuranium elements are radioactive. Table 5.4 gives information about the stability of the transuranium elements. Note the extremely short half-lives of the more recently produced elements.

Significant uses exist for some "synthetic" radionuclides, particularly in the field of medicine. For example, the synthetic radionuclides cobalt-60, yttrium-90, iodine-131, and gold-198 find use in radiotherapy treatment for cancer.

5.11 POSITRON EMISSION AND ELECTRON CAPTURE

Laboratory-produced radionuclides undergo radioactive decay just as do radionuclides from nature. Four modes of decay are encountered. They are alpha-particle emission and beta-particle emission (the same as for naturally occurring radionuclides—Sec. 5.9) and two new modes not found for naturally radioactive substances: positron emission and electron capture.

TABLE 5.4   Stability Characteristics of Transuranium Elements

Name	Symbol	Atomic Number	Mass Number of Most Stable Nuclide	Half-Life of Most Stable Nuclide	Discovery Year of First Isotope
Neptunium	Np	93	237	2.14×10^6 yr	1940
Plutonium	Pu	94	244	7.6×10^7 yr	1940
Americium	Am	95	243	8.0×10^3 yr	1944
Curium	Cm	96	247	1.6×10^7 yr	1944
Berkelium	Bk	97	247	1400 yr	1950
Californium	Cf	98	251	900 yr	1950
Einsteinium	Es	99	252	472 days	1952
Fermium	Fm	100	257	100 days	1953
Mendelevium	Md	101	258	52 days	1955
Nobelium	No	102	259	58 min	1958
Lawrencium	Lr	103	262	3.6 hr	1961
Rutherfordium	Rf	104	267	1.3 hr	1969
Dubnium	Db	105	268	1.2 days	1970
Seaborgium	Sg	106	271	1.9 min	1974
Bohrium	Bh	107	272	9.6 sec	1980
Hassium	Hs	108	270	3.6 sec	1984
Meitnerium	Mt	109	276	0.72 sec	1982
Darmstadtium	Ds	110	281	11.1 sec	1994
Roentgenium	Rg	111	280	3.6 sec	1994
Element 112	—	112	285	34 sec	1996
Element 113	—	113	284	0.48 sec	2004
Element 114	—	114	289	2.6 sec	1999
Element 115	—	115	288	87 msec	2004
Element 116	—	116	293	61 msec	2006
Element 118	—	118	294	0.89 msec	2006

Positron emission *is a radioactive decay process in which a positron is emitted from an unstable nucleus when a proton is converted to a neutron.* The particle involved in positron emission, the positron, is a particle we have not previously discussed. A **positron** *is a particle with the same mass as an electron or a beta particle, but with a positive charge.* The symbol for a positron is $^{0}_{1}\beta$. Its production in the nucleus is due to the conversion within the nucleus of a proton to a neutron.

Positron emission, although not found among naturally radioactive substances associated with Earth, is important in processes that occur in stars and in the sun.

$$^{1}_{1}\text{p} \longrightarrow {}^{1}_{0}\text{n} + {}^{0}_{1}\beta$$

This process is just the opposite of that occurring during beta-particle emission (Sec. 5.9). The general equation for position emission is

$$^A_Z X \longrightarrow {}^0_1\beta + {}_{(Z-1)}^A Y$$

The net effect of positron emission is thus to decrease the atomic number (number of protons), while the mass number remains constant. An example of a radioactive decay process involving positron emission is

$$^{30}_{15}P \longrightarrow {}^0_1\beta + {}^{30}_{14}Si$$

Electron capture *is a radioactive decay process in which an electron in a low-energy orbital, such as the 1s orbital, is pulled into an unstable nucleus, converting a proton to a neutron.*

$$^0_{-1}e + {}^1_1p \longrightarrow {}^1_0n$$

The general equation for electron capture is

$$^A_Z X + {}^0_{-1}e \longrightarrow {}_{(Z-1)}^A Y$$

The net effect of electron capture, a decrease of one in the atomic number, is the same as that of position emission. An example of electron capture is the reaction

$$^{87}_{37}Rb + {}^0_{-1}e \longrightarrow {}^{87}_{36}Kr$$

Why and when positron emission and electron capture occur is considered in Section 5.12.

EXAMPLE 5.11 **Writing Balanced Nuclear Equations Given the Parent Nuclide and Its Mode of Decay**

Write a balanced nuclear equation for the decay of each of the following radioactive nuclides. The mode of decay is indicated in parentheses.

   a. $^{62}_{29}$Cu (positron emission)   **b.** $^{118}_{52}$Te (electron capture)

   c. $^{105}_{47}$Ag (electron capture)   **d.** $^{82}_{37}$Rb (positron emission)

SOLUTION

In each case the atomic number and the mass number of the daughter nuclide are obtained by first writing the symbols of the parent nuclide and the particle emitted (positron) or absorbed (electron) and then balancing the equation.

   a. Let X represent the product of the radioactive decay, that is, the daughter nuclide. Then

$$^{62}_{29}Cu \longrightarrow {}^0_1\beta + X$$

   Note that the Greek letter β is used to denote not only a beta particle but also a positron. The difference between the two particles is that the former is negatively charged ($^0_{-1}\beta$) and the latter is positively charged ($^0_1\beta$).
   Since the sum of the superscripts on each side of the equation must be equal, the superscript for X must be 62. In order for the sums of the subscripts on both sides of the equation to be equal, at 29, the subscript for X must be 28. As soon as the subscript of X is determined, the identity of X is known. Looking at a periodic

table we determine that the element with an atomic number of 28 is nickel (Ni). Therefore,

$$^{62}_{29}\text{Cu} \longrightarrow ^{\ 0}_{1}\beta + ^{62}_{28}\text{Ni}$$

b. Similarly, letting X represent the product daughter nuclide of the radioactive decay, we have for $^{118}_{52}\text{Te}$ decaying by the electron capture mechanism

$$^{118}_{52}\text{Te} + ^{\ 0}_{-1}\text{e} \longrightarrow \text{X}$$

Note that in electron capture the electron appears on the reactant side of the equation. This makes equations for electron capture different from those for alpha, beta, and positron emissions where in each case the small particle involved is placed on the product side of the equation.

Balancing the above equation, making the superscripts on each side of the equation total 118 and the subscripts total 51, we get

$$^{118}_{52}\text{Te} + ^{\ 0}_{-1}\text{e} \longrightarrow ^{118}_{51}\text{Sb}$$

c. Similarly, for this electron capture we write

$$^{105}_{47}\text{Ag} + ^{\ 0}_{-1}\text{e} \longrightarrow \text{X}$$

Balancing superscripts and subscripts, we get

$$^{105}_{47}\text{Ag} + ^{\ 0}_{-1}\text{e} \longrightarrow ^{105}_{46}\text{Pd}$$

d. Finally, we write

$$^{82}_{37}\text{Rb} \longrightarrow ^{\ 0}_{1}\beta + \text{X}$$

In both positron emission and electron capture, the atomic number of the daughter nuclide decreases by 1 and the mass number does not change from that of the parent. The balancing process gives results consistent with this generalization in this part as well as in the previous three parts of this problem.

$$^{82}_{37}\text{Rb} \longrightarrow ^{\ 0}_{1}\beta + ^{82}_{36}\text{Kr}$$

Answer Double Check:

In both positron emission and electron capture, the mass number does not change and the atomic number decreases by 1. Is such the case for each of the answers? Yes.

▶ **Practice Exercise 5.11**   Write a balanced nuclear equation for the decay of each of the following radionuclides. The mode of decay is indicated in parentheses.

a. $^{91}_{42}\text{Mo}$ (positron emission)   **b.** $^{75}_{34}\text{Se}$ (electron capture)

Figure 5.12 is a summary diagram contrasting the various modes of decay for radioactive nuclides that we have considered; alpha-particle emission and beta-particle emission (Sec. 5.9) and positron emission and electron capture (Sec. 5.11). It relates these various decay modes to each other based on a common parent nuclide. Note from Figure 5.12 that for a parent nuclide (^A_ZE), the daughter nuclide has a lower atomic number for alpha-particle emission, positron emission, and electron capture and a higher atomic number only in the case of beta-particle emission.

FIGURE 5.12
Summary diagram for the various modes of radioactive decay. Alpha-particle emission, positron emission, and electron capture produce a daughter nuclide with an atomic number less than that of the parent nuclide. Beta-particle emission produces a daughter nuclide with an atomic number greater than that of the parent nuclide.

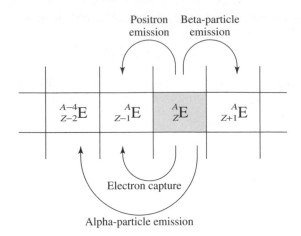

5.12 NEUTRON-TO-PROTON RATIO AND TYPE OF RADIOACTIVE DECAY

What determines whether a given radionuclide undergoes alpha particle decay, beta particle decay, positron emission, or electron capture? It is how the neutron-to-proton ratio for the radionuclide compares to that for stable nuclei containing approximately the same number of nucleons. Neutron-to-proton ratios for stable nuclides were previously discussed in Section 5.6; Figure 5.9 in that section shows this information graphically.

Based on position (location) in Figure 5.9, unstable nuclei can be classified into three categories:

1. *Unstable nuclei in which the neutron-to-proton ratio is too high.* Such nuclides, which can be considered proton-poor, lie to the left of the zone of stability in Figure 5.9.
2. *Unstable nuclei in which the neutron-to-proton ratio is too low.* Such nuclides, which can be considered proton-rich, lie to the right of the zone of stability in Figure 5.9.
3. *Unstable nuclei in which the total number of nucleons exceeds 209—the limit for a stable nucleus.* Such nuclei lie beyond the zone of stability in Figure 5.9.

Nuclei in each of these categories have a particular mode of decay that predominates over others.

Beta emission is the predominant decay mode for nuclides having neutron-to-proton ratios that are *too high* for stability. Such nuclides lie to the *left* of the zone of stability in Figure 5.9. In almost all cases, nuclei in this category have mass numbers greater than the atomic mass of the element. In the following example of beta emission, note that the mass number of the radioactive manganese nuclide (56) is greater than the atomic mass of manganese (54.94 amu).

$$\ce{^{56}_{25}Mn} \longrightarrow \ce{^{56}_{26}Fe} + \ce{^{0}_{-1}\beta}$$

As discussed previously (Sec. 5.9), beta emission involves the transformation of a neutron into a proton. This increases the number of protons, decreases the number of neutrons, and causes a decrease in the neutron-to-proton ratio—the desired result. Note that in the above reaction for the beta decay of manganese-56, the neutron-to-proton ratio of the parent nuclide is 1.24 and that of the daughter iron-56 is 1.15.

Radionuclides lying to the *right* of the zone of stability have *too low* a neutron-to-proton ratio. They decay by converting a proton into a neutron—just the opposite of the process that occurs in beta-particle emission. Radionuclei in this category generally have mass numbers that are lower than the atomic mass of the element. The conversion of a proton into a neutron can be accomplished by either of two ways: (1) by positron emission or (2) by electron capture. These decay modes were described in Section 5.11. The process of electron capture seems to be preferred over positron emission for nuclides of high atomic number. For lighter nuclides, numerous examples of both types of decay processes are known.

Alpha-particle emission is found primarily among elements in which the total number of nucleons exceeds 209. These are the nuclei that lie *beyond* the region of stability (Fig. 5.9). Alpha-particle emission is not, however, the only mode of decay for elements in this region; beta-particle emission is also common. Another characteristic of radionuclides of this type is that the decay process for reaching stability involves more than one step. This results in the formation of a *decay series* the topic of Section 5.13.

Before we leave this section, we should note that the preceding guidelines work in most instances; however, there are exceptions. For example, both $^{146}_{60}$Nd and $^{148}_{60}$Nd are stable and lie in the region of stability in Figure 5.9, but $^{147}_{60}$Nd, which is radioactive, also lies in the region of stability (between the two stable nuclides).

5.13 RADIOACTIVE DECAY SERIES

Radioactive nuclides with high atomic numbers cannot attain nuclear stability with a single emission; they are too far away from the region of stability (Fig. 5.9) for this to occur. Instead, a series of decay steps is required for such nuclei to reach stability. A large radionuclide undergoes decay to produce a product nucleus that is also radioactive. This product in turn produces a third radionuclide; this in turn decays to produce a fourth nuclide, and so forth, until ultimately a stable nucleus is produced. Such a sequence of decay products is called a *radioactive decay series*. A **radioactive decay series** *is a sequence of nuclear reactions in which one radioactive nuclide decays to a second, which then decays to a third, and so forth, until a stable nuclide is finally produced.*

Three naturally occurring decay series are known. Each starts with a long-lived radionuclide and ends with a stable nuclide. A fourth decay series was discovered after the synthesis of certain elements not found in nature (transuranium elements—Sec. 5.10). Because the parent of this fourth series, plutonium-241, does not occur in nature to a measurable extent, the series is not classified as naturally occurring. General characteristics of the four decay series are given in Table 5.5.

Figure 5.13 shows all of the members of the uranium-238 decay series. It is representative of the other three series. Note, from Figure 5.13, that both alpha and beta particles are part of the decay sequence and that there is no pattern relative to which type of particle is emitted when.

Many periodic tables give atomic masses rather than mass numbers for the elements Th, Pa, and U, even though all isotopes of these elements are radioactive. This is because small amounts of isotopes of these elements, with half-lives in millions of years, are found in Earth's crust.

TABLE 5.5   The Four Known Radioactive Decay Series

Parent	Number of Decay Steps	Final Product of Series
Uranium-238	14	lead-206
Thorium-232	10	lead-208
Uranium-235	11	lead-207
Plutonium-241	13	bismuth-209

FIGURE 5.13   In the uranium-238 decay series, each nuclide, except lead-206 (the final product), is radioactive; the successive transformations continue until this stable product is obtained.

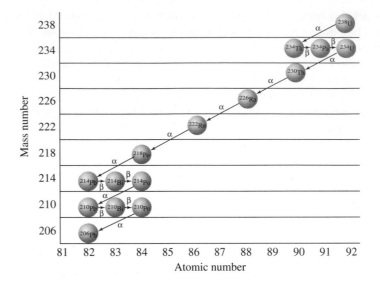

One of the intermediate products in the uranium-238 decay series is the radionuclide radon-222 (see Fig. 5.13). This substance is a gas at normal temperatures, and is therefore a very mobile species. Its presence has been detected in both aqueous and atmospheric environments.

Summary

1. **Subatomic Particles** Subatomic particles, the very small building blocks from which atoms are made, are of three major types: electrons, protons, and neutrons. Electrons are negatively charged, protons are positively charged, and neutrons have no charge. Protons and neutrons have much larger masses than electrons. All neutrons and protons are found at the center of the atom in the nucleus. The electrons occupy the region about (around) the nucleus. Discharge tube experiments played a key role in the discovery of electrons and protons. Metal foil experiments were instrumental in establishing that an atom had a nucleus.

2. **Atomic Number and Mass Number** Each atom has a characteristic atomic number (Z) and mass number (A). The atomic number is equal to the number of protons in the nucleus of the atom. The mass number is equal to the total number of protons and neutrons in the nucleus. In designating various types of atoms, the atomic number is specified as a subscript to the left of an atom's elemental symbol, and the mass number is specified as a superscript to the left of the atom's elemental symbol.

3. **Isotopes and Isobars** Isotopes are atoms that have the same number of protons and electrons but have different numbers of neutrons. The isotopes of an element always have the same atomic number and different mass numbers. Isotopes of an element have the same chemical properties. Isobars are atoms of different elements that have the same mass number.

4. **Atomic Mass** The atomic mass of an element is the relative mass of an average atom of the element on a scale using the $^{12}_{6}C$ atom as the reference. An atomic mass depends on the percent abundances and masses of the naturally occurring isotopes of an element.

5. **Nuclear Stability and Radioactivity** Some nuclides possess nuclei that are unstable because of an undesirable neutron-to-proton ratio. To achieve stability these unstable nuclides spontaneously emit radiation, which changes the neutron-to-proton ratio. Such nuclides are said to be radioactive.

6. **Half-Life** Radioactive nuclides are not all equally unstable. Some have fast decay rates and others decay slowly. The rate of decay for a radionuclide is given by its half-life. A half-life

is the time required for half of any given quantity of a radioactive substance to undergo decay.

7. **Radioactive Emissions** Three types of radiation are emitted by naturally occurring radioactive nuclides: alpha, beta, and gamma. These radiations are characterized by mass and charge values. Alpha radiation (alpha particle) has a mass of 4 amu and a +2 charge. Beta radiation (beta particle) has a very small mass (0.00055 amu) and a –1 charge. Gamma radiation, a form of energy, has no mass and no charge.

8. **Nuclear Equations** The chemical symbols in a nuclear equation represent nuclei rather than atoms. Atomic numbers and mass numbers are always shown when representing the nuclei. Balancing of a nuclear equation is based on both *charge* and *mass* conservation. Charge conservation means that the sum of the subscripts (number of protons, or positive charges, in the nucleus) must be equal on both sides of the equation. Similarly, the total number of nucleons (protons and neutrons) is conserved. This means that the sum of the superscripts (the mass numbers) must be the same on each side of the equation.

9. **Transmutation and Bombardment Reactions** A transmutation reaction is a nuclear reaction in which a nuclide of one element is changed into a nuclide of another element. A bombardment reaction is a transmutation process in which small particles traveling at very high speeds are collided with stable nuclei, with such collisions causing these nuclei to undergo nuclear change. Over 2000 laboratory-produced radionuclides have been produced in this manner including radionuclides of 29 elements not found in nature.

10. **Positron Emission and Electron Capture** Positron emission and electron capture are modes of radioactive decay found among laboratory-produced radionuclides. In positron emission a positron is ejected from an unstable nucleus when a proton is converted to a neutron. A positron has the characteristics of an electron except that it is positively charged. Electron capture involves an electron in a low-energy orbital being pulled into the nucleus, converting a proton to a neutron.

11. **Neutron-to-Proton Ratio and Type of Radioactive Decay** Neutron-to-proton ratio relative to stable nuclides with a similar number of nucleons determines the type of radioactive decay a radioactive nuclide undergoes.

12. **Radioactive Decay Series** The product of the radioactive decay of an unstable nuclide may be a stable nuclide or an unstable nuclide. If the product is an unstable nuclide, it will decay and produce still another nuclide. Further decay will continue until a stable nuclide is formed. Such a sequence of nuclear reactions is called a radioactive decay series.

Key Terms

The new terms defined in this chapter are

alpha particle *Sec. 5.8*
alpha-particle decay *Sec. 5.9*
atomic mass *Sec. 5.4*
atomic number *Sec. 5.2*
beta particle *Sec. 5.8*
beta-particle decay *Sec. 5.9*
bombardment reaction *Sec. 5.10*
daughter nuclide *Sec. 5.9*

electron *Sec. 5.1*
electron capture *Sec. 5.11*
element *Sec. 5.2*
gamma ray *Sec. 5.8*
gamma-ray emission *Sec. 5.9*
half-life *Sec. 5.7*
isobars *Sec. 5.3*
isotopes *Sec. 5.3*
mass number *Sec. 5.2*
neutron *Sec. 5.1*

nuclear equation *Sec. 5.9*
nucleon *Sec. 5.1*
nucleus *Sec. 5.1*
nuclide *Sec. 5.6*
parent nuclide *Sec. 5.9*
percent abundance *Sec. 5.3*
positron *Sec. 5.11*
positron emission *Sec. 5.11*
proton *Sec. 5.1*

radioactive atom *Sec. 5.6*
radioactive decay *Sec. 5.7*
radioactive decay series *Sec. 5.13*
radioactivity *Sec. 5.6*
stable nucleus *Sec. 5.6*
subatomic particle *Sec. 5.1*
transmutation reaction *Sec. 5.10*
unstable nucleus *Sec. 5.6*

Practice Problems

SUBATOMIC PARTICLES (SEC. 5.1)

5.1 Match the terms *proton*, *neutron*, and *electron* to each of the following subatomic particle descriptions. It is possible that more than one term may apply in a given situation.
a. possesses a negative charge
b. has a mass slightly less than that of a neutron
c. can be called a nucleon
d. is the heaviest of the three particles

5.2 Match the terms *proton*, *neutron*, and *electron* to each of the following subatomic particle descriptions. It is possible that more than one term may apply in a given situation.
a. has no charge
b. has a charge equal to but opposite in sign to that of an electron
c. is not found in the nucleus
d. has a positive charge

5.3 Indicate whether each of the following statements about the nucleus of an atom is *true* or *false*.
a. The nucleus of an atom is neutral.
b. The nucleus of an atom contains only neutrons.
c. The number of nucleons present in the nucleus is always equal to the number of electrons present outside the nucleus.
d. The nucleus accounts for almost all of the mass of an atom.

5.4 Indicate whether each of the following statements about the nucleus of an atom is *true* or *false*.
a. The nucleus accounts for almost all the volume of an atom.
b. The nucleus can be positively or negatively charged, depending on the identity of the atom.
c. The nucleus of an atom contains an equal number of protons, neutrons, and electrons.
d. The nucleus of an atom is always positively charged.

5.5 In terms of mass, how many electrons are needed to equal the mass of one proton?

5.6 In terms of mass, how many electrons are needed to equal the mass of one neutron?

ATOMIC NUMBER AND MASS NUMBER (SEC. 5.2)

5.7 Using the information inside the front cover, determine the atomic number associated with the listed element or the element name associated with the listed atomic number.
a. tin   b. silver
c. atomic number 28   d. atomic number 53

5.8 Using the information inside the front cover, determine the atomic number associated with the listed element or the element name associated with the listed atomic number.
a. lead
b. beryllium
c. atomic number 18
d. atomic number 56

5.9 For each of the following atoms, specify the atomic number and the mass number.
a. $^{53}_{24}Cr$ b. $^{103}_{44}Ru$ c. $^{256}_{101}Md$ d. $^{34}_{16}S$

5.10 For each of the following atoms, specify the atomic number and the mass number.
a. $^{67}_{30}Zn$ b. $^{9}_{4}Be$ c. $^{40}_{20}Ca$ d. $^{3}_{1}H$

5.11 What is the atomic number for atoms composed of the following sets of subatomic particles?
a. 5 protons, 5 electrons, and 6 neutrons
b. 8 protons, 8 electrons, and 8 neutrons
c. 13 protons, 13 electrons, and 14 neutrons
d. 18 protons, 18 electrons, and 22 neutrons

5.12 What is the atomic number for atoms composed of the following sets of subatomic particles?
a. 4 protons, 4 electrons, and 5 neutrons
b. 7 protons, 7 electrons, and 8 neutrons
c. 15 protons, 15 electrons, and 16 neutrons
d. 20 protons, 20 electrons, and 28 neutrons

5.13 What is the mass number for each of the atoms in Problem 5.11?

5.14 What is the mass number for each of the atoms in Problem 5.12?

5.15 What is the total number of nucleons present for each of the atoms in Problem 5.11?

5.16 What is the total number of nucleons present for each of the atoms in Problem 5.12?

5.17 What is the charge on the nucleus for each of the atoms in Problem 5.11?

5.18 What is the charge on the nucleus for each of the atoms in Problem 5.12?

5.19 What is the complete chemical symbol ($^A_Z E$) for each of the atoms in Problem 5.11?

5.20 What is the complete chemical symbol ($^A_Z E$) for each of the atoms in Problem 5.12?

5.21 Indicate whether (1) the atomic number, (2) the mass number, or (3) both the atomic number and the mass number are needed to determine the following.
a. number of protons in an atom
b. number of neutrons in an atom
c. number of nucleons in an atom
d. total number of subatomic particles in an atom

5.22 What information about the subatomic particles present in an atom is obtained from each of the following?
a. atomic number
b. mass number
c. mass number – atomic number
d. mass number + atomic number

5.23 Using the information inside the front cover, identify the element X based on the given complete symbol for X.
   a. $^{15}_{7}X$ **b.** $^{27}_{13}X$ **c.** $^{139}_{56}X$ **d.** $^{197}_{79}X$

5.24 Using the information inside the front cover, identify the element X based on the given complete symbol for X.
   a. $^{16}_{8}X$ **b.** $^{19}_{9}X$ **c.** $^{40}_{20}X$ **d.** $^{109}_{47}X$

5.25 Fill in the blanks in each line in the following table. The first line is already completed as an example.

	Symbol	Atomic number	Mass number	Number of protons	Number of neutrons
	$^{3}_{2}He$	2	3	2	1
a.			60	28	
b.		18	37		
c.			90		52
d.	$^{235}_{92}U$				

5.26 Fill in the blanks in each line in the following table. The first line is already completed as an example.

	Symbol	Atomic number	Mass number	Number of protons	Number of neutrons
	$^{37}_{17}Cl$	17	37	17	20
a.			232	94	
b.	$^{32}_{16}S$				
c.		26	56		
d.			40		20

5.27 Determine the number of protons, electrons, and neutrons present in atoms with the following characteristics.
   a. atomic number = 27 and mass number = 59
   b. mass number = 103 and Z = 45
   c. Z = 69 and A = 169
   d. atomic number = 9 and A = 19

5.28 Determine the number of protons, electrons, and neutrons present in atoms with the following characteristics.
   a. A = 103 and atomic number = 44
   b. Z = 41 and A = 93
   c. mass number = 59 and atomic number = 27
   d. Z = 59 and mass number = 141

5.29 Determine the following information for an atom whose complete symbol is $^{23}_{11}Na$.
   a. the total number of subatomic particles present
   b. the total number of subatomic particles present in the nucleus of the atom

c. the total number of nucleons present
   d. the total charge (including sign) associated with the nucleus of the atom

5.30 Determine the following information for an atom whose complete symbol is $^{37}_{17}Cl$.
   a. the total number of subatomic particles present
   b. the total number of subatomic particles present in the nucleus of the atom
   c. the total number of nucleons present
   d. the total charge (including sign) associated with the nucleus of the atom

5.31 Write complete symbols ($^{A}_{Z}E$) for atoms with the following characteristics.
   a. contains 15 electrons and 16 neutrons
   b. oxygen atom with 10 neutrons
   c. chromium atom with a mass number of 54
   d. gold atom that contains 276 subatomic particles

5.32 Write complete symbols ($^{A}_{Z}E$) for atoms with the following characteristics.
   a. contains 20 electrons and 24 neutrons
   b. radon atom with a mass number of 211
   c. silver atom that contains 157 subatomic particles
   d. beryllium atom that contains 9 nucleons

5.33 An atom with an atomic number of 12 contains 37 subatomic particles. How many of these subatomic particles have each of the following characteristics?
   a. carry a charge
   b. carry no charge
   c. located within the nucleus
   d. not found in the nucleus

5.34 An atom with an atomic number of 17 contains 54 subatomic particles. How many of these subatomic particles have each of the following characteristics?
   a. carry a charge
   b. carry no charge
   c. located within the nucleus
   d. not found in the nucleus

5.35 Calculate the mass number for atoms with the following characteristics.
   a. atomic number of 20; three more neutrons than protons
   b. atomic number of 20; three more neutrons than electrons
   c. atomic number of 30; equal number of neutrons and protons
   d. atomic number of 30; equal number of neutrons and electrons

5.36 Calculate the mass number for atoms with the following characteristics.
   a. atomic number of 17; three more neutrons than protons
   b. atomic number of 17; three more neutrons than electrons
   c. atomic number of 25; equal number of neutrons and protons
   d. atomic number of 25; equal number of neutrons and electrons

5.37 Characterize each of the following pairs of atoms as containing (1) the same number of neutrons, (2) the same number of electrons, or (3) the same total number of subatomic particles.
  a. $^{40}_{20}Ca$ and $^{41}_{19}K$ **b.** $^{30}_{14}Si$ and $^{32}_{16}S$
  c. $^{23}_{11}Na$ and $^{24}_{12}Mg$ **d.** $^{7}_{3}Li$ and $^{6}_{3}Li$

5.38 Characterize each of the following pairs of atoms as containing (1) the same number of neutrons, (2) the same number of electrons, or (3) the same total number of subatomic particles.
  a. $^{13}_{6}C$ and $^{14}_{7}N$ **b.** $^{18}_{8}O$ and $^{19}_{9}F$
  c. $^{37}_{17}Cl$ and $^{36}_{18}Ar$ **d.** $^{35}_{17}Cl$ and $^{37}_{17}Cl$

5.39 What are the names of the elements that each of the following atoms represents?
  a. mass number = 45; 24 neutrons are present
  b. Cl atom with two more neutrons than $^{35}_{17}Cl$
  c. atomic number = 38; 50 neutrons are present
  d. atom with two more protons and two more electrons than $^{70}_{31}Ga$

5.40 What are the names of the elements that each of the following atoms represents?
  a. mass number = 60; 32 neutrons are present
  b. Ca atom with three more neutrons than $^{40}_{20}Ca$
  c. atomic number = 14; 14 neutrons are present
  d. atom with one more proton and one more electron than $^{40}_{18}Ar$

5.41 Indicate whether each of the following pieces of information is sufficient to determine the element identity for an atom.
  a. number of neutrons
  b. atomic number
  c. number of nucleons
  d. nuclear charge

5.42 Indicate whether each of the following pieces of information is sufficient to determine the element identity for an atom.
  a. mass number
  b. number of protons
  c. total mass of the nucleus
  d. sum of the atomic number and mass number

ISOTOPES (SEC. 5.3)

5.43 Write complete symbols ($^{A}_{Z}E$) for the four naturally occurring isotopes of iron whose mass numbers are 54, 56, 57, and 58.

5.44 Write complete symbols ($^{A}_{Z}E$) for the four naturally occurring isotopes of chromium whose mass numbers are 50, 52, 53, and 54.

5.45 Five naturally occurring isotopes of the element zirconium exist. Knowing that the heaviest isotope has a mass number of 96 and that the other isotopes have, respectively, two, four, five, and six fewer neutrons, write the complete symbol ($^{A}_{Z}E$) for each of the five isotopes.

5.46 Four naturally occurring isotopes of the element strontium exist. Knowing that the lightest isotope has a mass number of 84 and that the other isotopes have, respectively, two, four, and five more neutrons, write the complete symbol ($^{A}_{Z}E$) for each of the four isotopes.

5.47 Four isotopes exist for iron, element 26. One isotope has an isotopic mass of 53.940 amu and a percent abundance of 5.82. A second isotope has an isotopic mass of 55.935 amu and a percent abundance of 91.66. The third isotope has an isotopic mass of 56.935 amu. The fourth isotope has an isotopic mass of 57.933 amu and a percent abundance of 0.33.
  a. What are the mass numbers for the four iron isotopes?
  b. What is the percent abundance for the third iron isotope?

5.48 Four isotopes exist for strontium, element 38. One isotope has an isotopic mass of 83.913 amu and a percent abundance of 0.56. A second isotope has an isotopic mass of 85.909 amu and a percent abundance of 9.86. The third isotope has an isotopic mass of 86.909 amu. The fourth isotope has an isotopic mass of 87.906 amu and a percent abundance of 82.56.
  a. What are the mass numbers for the four strontium isotopes?
  b. What is the percent abundance for the third strontium isotope?

5.49 Using the information given in the following table indicate whether each of the following pairs of atoms are isotopes.

	Protons	Neutrons	Electrons
Atom A	9	10	9
Atom B	10	9	10
Atom C	10	10	10
Atom D	9	9	9

  a. atom A and atom B
  b. atom C and atom D
  c. atom B and atom D

5.50 Using the information given in the table in Problem 5.49, indicate whether each of the following pairs of atoms are isotopes.
  a. atom A and atom C
  b. atom B and atom C
  c. atom A and atom D

5.51 The following are selected properties for the most abundant isotope of a particular element. Which of these properties would also be the same for the second-most-abundant isotope of the element?
  a. mass number is 70
  b. 31 electrons are present
  c. isotopic mass is 69.92 amu
  d. isotope reacts with chlorine to give a green compound

5.52 The following are selected properties for the most abundant isotope of a particular element. Which of these properties would also be the same for the second-most-abundant isotope of the element?
 a. atomic number is 31
 b. does not react with the element gold
 c. 40 neutrons are present
 d. density is 1.0301 g/mL

5.53 Indicate whether the members of each of the following pairs of atoms are isotopes, isobars, or neither.
 a. $^{24}_{12}X$ and $^{26}_{12}Q$ **b.** $^{71}_{31}X$ and $^{71}_{32}Q$
 c. $^{56}_{25}X$ and $^{54}_{24}Q$ **d.** $^{57}_{27}X$ and $^{60}_{27}Q$

5.54 Indicate whether the members of each of the following pairs of atoms are isotopes, isobars, or neither.
 a. $^{64}_{30}X$ and $^{64}_{29}Q$ **b.** $^{20}_{10}X$ and $^{22}_{10}Q$
 c. $^{36}_{16}X$ and $^{35}_{17}Q$ **d.** $^{60}_{28}X$ and $^{60}_{26}Q$

5.55 Indicate whether the members of each of the following pairs of atoms, specified in terms of subatomic particle composition, are isotopes, isobars, or neither.
 a. (24p, 24e, 26n) and (24p, 24e, 28n)
 b. (24p, 24e, 28n) and (25p, 25e, 27n)
 c. (24p, 24e, 26n) and (25p, 25e, 26n)
 d. (24p, 24e, 26n) and (24p, 24e, 24n)

5.56 Indicate whether the members of each of the following pairs of atoms, specified in terms of subatomic particle composition, are isotopes, isobars, or neither.
 a. (30p, 30e, 39n) and (31p, 31e, 38n)
 b. (30p, 30e, 35n) and (29p, 29e, 36n)
 c. (30p, 30e, 34n) and (30p, 30e, 36n)
 d. (30p, 30e, 38n) and (32p, 32e, 38n)

5.57 Write the complete chemical symbol ($^{A}_{Z}E$) for the three isotopes of the element phosphorus that have the following characteristics.
 a. 46, 48, and 50 subatomic particles are, respectively, present.
 b. 31, 32, and 33 nucleons are, respectively, present.
 c. 16, 17, and 19 neutrons are, respectively, present.
 d. 30, 32, and 34 subatomic particles are, respectively, present in the nucleus.

5.58 Write the complete chemical symbol ($^{A}_{Z}E$) for the three isotopes of the element silicon that have the following characteristics.
 a. 42, 44, and 46 subatomic particles are, respectively, present.
 b. 28, 29, and 31 nucleons are, respectively, present.
 c. 14, 15, and 16 neutrons are, respectively, present.
 d. 27, 28, and 30 subatomic particles are, respectively, present in the nucleus.

5.59 Isobars with a mass number of 40 exist for Ar, K, and Ca. Write the complete symbol ($^{A}_{Z}E$) for each of these isobars.

5.60 Isobars with a mass number of 50 exist for Ti, V, and Cr. Write the complete symbol ($^{A}_{Z}E$) for each of these isobars.

5.61 What is the mass number of each of the following atoms?
 a. an atom whose relative mass on the $^{12}_{6}C$ scale is 58.83201 amu
 b. an atom of element 20 that has an equal number of all three types of subatomic particles
 c. an isobar of $^{18}_{8}O$ that contains 9 neutrons
 d. an isotope of $^{16}_{8}O$ that contains 10 neutrons

5.62 What is the mass number of each of the following atoms?
 a. an atom whose relative mass on the $^{12}_{6}C$ scale is 30.97396 amu
 b. an atom of element 14 that has an equal number of all three types of subatomic particles
 c. an isobar of $^{14}_{7}N$ that contains 8 neutrons
 d. an isotope of $^{14}_{7}N$ that contains 8 neutrons

ATOMIC MASSES (SEC. 5.4)

5.63 Using the information inside the front cover, determine the atomic mass associated with the listed element or the element name associated with the listed atomic mass.
 a. iron
 b. nitrogen
 c. 40.08 amu
 d. 126.90 amu

5.64 Using the information inside the front cover, determine the atomic mass associated with the listed element or the element name associated with the listed atomic mass.
 a. phosphorus
 b. nickel
 c. 101.07 amu
 d. 20.18 amu

5.65 Construct a relative mass scale for the hypothetical atoms Q, X, and Z, given that atoms of Q are two times as heavy as those of X, atoms of X are two times as heavy as atoms of Z, and the reference point for the scale is X = 4.00 bebs.

5.66 Construct a relative mass scale for the hypothetical atoms Q, X, and Z, given that atoms of Q are three times as heavy as those of X, atoms of X are four times as heavy as atoms of Z, and the reference point for the scale is Z = 2.50 bobs.

5.67 The atoms of element Z each have an average mass three-fourths that of a $^{12}_{6}C$ atom. Another element, X, has atoms whose average mass is three times the mass of Z atoms. A third element, Q, has atoms with an average mass nine times that of $^{12}_{6}C$.
 a. Construct a relative mass scale, based on $^{12}_{6}C$ having a mass of 12 amu, for the elements Z, X, and Q.
 b. Based on atomic masses rounded off to whole numbers, what is the identity of elements Z, X, and Q?

5.68 The atoms of element Z each have an average mass one-third that of a $^{12}_{6}C$ atom. Another element, X, has atoms whose average mass is four times the mass of Z atoms. A third element, Q, has atoms whose average mass is twice the mass of X atoms.
   a. Construct a relative mass scale, based on $^{12}_{6}C$ having a mass of 12 amu, for the elements Z, X, and Q.
   b. Based on atomic masses rounded off to whole numbers, what is the identity of elements Z, X, and Q?

5.69 A football team has the following distribution of players: 22.0% are defensive linemen with an average mass of 271 lb, 19.0% are defensive backs with an average mass of 175 lb, 26.0% are offensive linemen with an average mass of 263 lb, 15.0% are offensive backs with an average mass of 182 lb, and 19.0% are specialty team members with an average mass of 191 lb. What is the average mass of a football player on this team?

5.70 The assortment of automobiles on a used-car lot is categorized as follows: 18.0% are 1 year old, 10.0% are 2 years old, 33.0% are 3 years old, 4.0% are 4 years old, 31.0% are 5 years old, and 4.0% are 6 years old. What is the average age of a car on this used-car lot?

5.71 Calculate the atomic mass of chlorine on the basis of the following percent composition and isotopic mass data for the naturally occurring isotopes.

Isotope	Mass (amu)	Abundance(%)
$^{35}_{17}Cl$	34.9689	75.53%
$^{37}_{17}Cl$	36.9659	24.47%

5.72 Calculate the atomic mass of copper on the basis of the following percent composition and isotopic mass data for the naturally occurring isotopes.

Isotope	Mass (amu)	Abundance(%)
$^{63}_{29}Cu$	62.9298	69.09
$^{65}_{29}Cu$	64.9278	30.91

5.73 Calculate the atomic mass of titanium on the basis of the following percent composition and isotopic mass data for the naturally occurring isotopes.

Isotope	Mass (amu)	Abundance(%)
$^{46}_{22}Ti$	45.95263	7.93
$^{47}_{22}Ti$	46.95176	7.28
$^{48}_{22}Ti$	47.94795	73.94
$^{49}_{22}Ti$	48.94787	5.51
$^{50}_{22}Ti$	49.94479	5.34

5.74 Calculate the atomic mass of sulfur on the basis of the following percent composition and isotopic mass data for the naturally occurring isotopes.

Isotope	Mass (amu)	Abundance(%)
$^{32}_{16}S$	31.9721	95.0
$^{33}_{16}S$	32.9715	0.76
$^{34}_{16}S$	33.9679	4.22
$^{36}_{16}S$	35.9671	0.014

5.75 Each of the following elements has only two naturally occurring isotopes. Determine, in each case, which isotope is more abundant, using only the atomic mass value for the element that is listed on the periodic table inside the front cover.
   a. $^{14}_{7}N$ and $^{15}_{7}N$ **b.** $^{50}_{23}V$ and $^{51}_{23}V$
   c. $^{121}_{51}Sb$ and $^{123}_{51}Sb$ **d.** $^{191}_{77}Ir$ and $^{193}_{77}Ir$

5.76 Each of the following elements has only two naturally occurring isotopes. Determine, in each case, which isotope is more abundant, using only the atomic mass value for the element that is listed on the periodic table inside the front cover.
   a. $^{10}_{5}B$ and $^{11}_{5}B$ **b.** $^{69}_{31}Ga$ and $^{71}_{31}Ga$
   c. $^{107}_{47}Ag$ and $^{109}_{47}Ag$ **d.** $^{203}_{81}Tl$ and $^{205}_{81}Tl$

5.77 If the mass number range for the isotopes of an element is from 54 to 59, which of the following calculated atomic mass values are impossible values for the atomic mass of the element?
   a. 54.23 amu
   b. 57.02 amu
   c. 58.37 amu
   d. 59.21 amu

5.78 If the mass number range for the isotopes of an element is from 75 to 82, which of the following calculated atomic mass values are impossible values for the atomic mass of the element?
   a. 74.23 amu
   b. 75.09 amu
   c. 80.37 amu
   d. 82.79 amu

5.79 Fill in the blanks in each line in the following table, based on reasoning rather than the use of a calculator. The first line is already completed as an example.

	Isotope A	Isotope B	Atomic Mass	Abundance % A	Abundance % B
	46 amu	48 amu	47 amu	50%	50%
a.	46 amu	50 amu	47 amu		
b.	46 amu	50 amu		25%	
c.	45 amu	50 amu	46 amu		
d.	45 amu	50 amu			80%

5.80 Fill in the blanks in each line in the following table, based on reasoning rather than the use of a calculator. The first line is already completed as an example.

	Isotope A	Isotope B	Atomic Mass	Abundance % A	Abundance % B
	36 amu	40 amu	39 amu	25%	75%
a.	36 amu	40 amu	37 amu		
b.	36 amu	40 amu		25%	
c.	35 amu	40 amu	37 amu		
d.	35 amu	40 amu			60%

5.81 Based on atomic mass values, what is the mass ratio (to three significant figures) of the following?
  a. 1 atom of Cl to 1 atom of Rb
  b. 2 atoms of Cl to 2 atoms of Rb
  c. 1 atom of Ag to 2 atoms of Cu
  d. 10 atoms of H to 1 atom of Be

5.82 Based on atomic mass values, what is the mass ratio (to three significant figures) of the following?
  a. 1 atom of Cr to 1 atom of Si
  b. 1 atom of Si to 1 atom of Cr
  c. 3 atoms of Ni to 3 atoms of Se
  d. 1 atom of Br to 5 atoms of Cs

5.83 What are the name and symbol of the element whose average atoms have a mass
  a. close to four times the mass of an average nitrogen atom?
  b. that is 81.2% of the mass of an average silver atom?
  c. that is three times the atomic number of lithium?
  d. close to one-fifth the mass of an average neon atom?

5.84 What are the name and symbol of the element whose average atoms have a mass
  a. close to three times the mass of an average beryllium atom?
  b. that is 37.1% of the mass of an average zinc atom?
  c. that is three times the atomic number of fluorine?
  d. close to one-half the mass of an average silicon atom?

5.85 The arbitrary standard for the atomic mass scale is the exact number 12 for the mass of $^{12}_{6}C$. Why, then, is the atomic mass of carbon listed as 12.01?

5.86 The atomic mass of fluorine is 19.00 amu and that of copper is 63.55 amu. All fluorine atoms have a mass of 19.00 amu, and not a single copper atom has a mass of 63.55 amu. Explain.

EVIDENCE SUPPORTING THE EXISTENCE AND ARRANGEMENT OF SUBATOMIC PARTICLES (SEC. 5.5)

5.87 Indicate whether each of the following statements about discharge tube and metal foil experiments is *true* or *false*.
  a. Information obtained from discharge tube experiments led to the characterization of both protons and electrons.
  b. Canal rays are negatively charged particles produced in a discharge tube.
  c. Metal foil experiments led to the discovery of neutrons.
  d. Many different types of cathode rays are known.

5.88 Indicate whether each of the following statements about discharge tube and metal foil experiments is *true* or *false*.
  a. In metal foil experiments almost all of the bombarding particles were stopped by the metal foil.
  b. Metal foil experiments led to the concept that an atom has a nucleus.
  c. Cathode rays and canal rays move in opposite directions in a discharge tube.
  d. Many different types of canal rays have been observed, but only one type of cathode ray is known.

NUCLEAR STABILITY AND RADIOACTIVITY (SEC. 5.6)

5.89 What physical manifestation indicates that an atom possesses an unstable nucleus?

5.90 What is radioactivity?

5.91 What is the limit for nuclear stability in terms of number of nucleons present in a nucleus?

5.92 How do the neutron-to-proton ratios compare for stable nuclei of low atomic number and stable nuclei of high atomic number?

5.93 What is the difference, if any, in meaning between the terms *radioactive atom* and *radioactive nuclide*?

5.94 What is the difference, if any, in meaning between the terms *nuclides* and *isotopes*?

5.95 Use two different notations to denote each of the following nuclides.
  a. contains 5 protons, 4 neutrons, and 5 electrons
  b. contains 19 protons, 25 neutrons, and 19 electrons
  c. contains 45 protons, 51 neutrons, and 45 electrons
  d. contains 73 protons, 109 neutrons, and 73 electrons

5.96 Use two different notations to denote each of the following nuclides.
  a. contains 7 protons, 9 neutrons, and 7 electrons
  b. contains 31 protons, 45 neutrons, and 31 electrons
  c. contains 56 protons, 84 neutrons, and 56 electrons
  d. contains 75 protons, 105 neutrons, and 75 electrons

5.97 Use a notation different from that given to designate each of the following nuclides.
　　a. nitrogen-14　　　**b.** gold-197
　　c. $^{92}_{37}$Rb　　　　**d.** $^{121}_{50}$Sn

5.98 Use a notation different from that given to designate each of the following nuclides.
　　a. oxygen-17　　　**b.** lead-212
　　c. $^{10}_{5}$B　　　　**d.** $^{209}_{83}$Bi

HALF-LIFE (SEC. 5.7)

5.99 Copper-67 has a half-life of 2.6 days. What fraction of atoms in a copper-67 sample will remain in an *undecayed* state after the following periods of time?
　　a. 5.2 days　　　　**b.** 13 days
　　c. 3 half-lives　　**d.** 6 half-lives

5.100 Germanium-69 has a half-life of 1.6 days. What fractions of atoms in a germanium-69 sample will remain in an *undecayed* state after the following periods of time?
　　a. 4.8 days　　　　**b.** 8.0 days
　　c. 4 half-lives　　**d.** 8 half-lives

5.101 How many half-lives have elapsed if after 8.0 years the fraction of *undecayed* nuclides present is
　　a. 1/16?　　　　**b.** 1/64?
　　c. 1/32?　　　　**d.** 1/512?

5.102 How many half-lives have elapsed if after 14 days the fraction of *undecayed* nuclides present is
　　a. 1/4?　　　　**b.** 1/8?
　　c. 1/256?　　　**d.** 1/1024?

5.103 The half-life of copper-60 is 2.4 minutes. How many grams of copper-60 in an 8.0 g sample will remain *undecayed* after the following periods of time?
　　a. 4.8 min　　　**b.** 9.6 min
　　c. 16.8 min　　**d.** 24.0 min

5.104 The half-life of arsenic-69 is 15 minutes. How many grams of arsenic-69 in a 6.0 g sample will remain *undecayed* after the following periods of time?
　　a. 45 min　　　**b.** 75 min
　　c. 105 min　　**d.** 165 min

5.105 The half-life of silver-112 is 3.2 hours. How many grams of this nuclide in a 10.0 g sample will have *decayed* after the following periods of time?
　　a. 6.4 hr　　　**b.** 19.2 hr
　　c. 28.8 hr　　**d.** 48 hr

5.106 The half-life of barium-129 is 2.5 hours. How many grams of this nuclide in a 4.00 g sample will have *decayed* after the following periods of time?
　　a. 5.0 hr　　　**b.** 15 hr
　　c. 17.5 hr　　**d.** 32.5 hr

5.107 Iron-52 has a half-life of 8.0 hours. How long will it take, in hours, for the following fractions of nuclides in an iron-52 sample to decay?
　　a. 3/4　　　　**b.** 15/16
　　c. 31/32　　　**d.** 127/128

5.108 Cobalt-55 has a half-life of 18 hours. How long will it take, in hours, for the following fractions of nuclides in a cobalt-55 sample to decay?
　　a. 7/8　　　　**b.** 31/32
　　c. 63/64　　　**d.** 127/128

THE NATURE OF NATURAL RADIOACTIVE EMISSIONS (SEC. 5.8)

5.109 Supply a complete symbol, with superscript and subscript, for each of the following types of radiation.
　　a. alpha particle　　**b.** beta particle
　　c. gamma ray

5.110 Give the charge and mass (in amu) of each of the following types of radiation.
　　a. alpha particle　　**b.** beta particle
　　c. gamma ray

5.111 State the composition of an alpha particle in terms of protons and neutrons.

5.112 What is the relationship between a beta particle and an electron?

EQUATIONS FOR RADIOACTIVE DECAY (SEC. 5.9)

5.113 Write balanced equations for the alpha decay of each of the following radionuclides.
　　a. $^{200}_{84}$Po　　　**b.** $^{244}_{96}$Cm
　　c. curium-240　　**d.** uranium-238

5.114 Write balanced equations for the alpha decay of each of the following radionuclides.
　　a. $^{147}_{62}$Sm　　　**b.** $^{192}_{78}$Pt
　　c. radon-217　　**d.** radium-224

5.115 Write balanced equations for the beta decay of each of the following radionuclides.
　　a. $^{10}_{4}$Be　　　**b.** $^{77}_{32}$Ge
　　c. iron-60　　　**d.** sodium-25

5.116 Write balanced equations for the beta decay of each of the following radionuclides.
　　a. $^{67}_{29}$Cu　　　**b.** $^{92}_{36}$Kr
　　c. silver-117　　**d.** rhenium-190

5.117 What is the effect on the mass number of the parent nuclide when each of the following occurs?
　　a. alpha particle decay　**b.** beta particle decay
　　c. gamma ray emission

5.118 What is the effect on the atomic number of the parent nuclide when each of the following occurs?
　　a. alpha particle decay
　　b. beta particle decay
　　c. gamma ray emission

5.119 Identify the mode of decay for each of the following radioactive decays, where the parent and daughter nuclide are given.
　　a. Parent nuclide = platinum-190 and daughter nuclide = osmium-186
　　b. Parent nuclide = oxygen-19 and daughter nuclide = fluorine-19

5.120 Identify the mode of decay for each of the following radioactive decays, where the parent and daughter nuclide are given.
  a. Parent nuclide = silicon-24 and daughter nuclide = phosphorus-24
  b. Parent nuclide = einsteinium-252 and daughter nuclide = berkelium-248

5.121 Supply the missing nuclear symbol in each of the following radioactive-decay processes.
  a. $^{12}_{5}B \longrightarrow {}^{12}_{6}C + X$
  b. $^{125}_{51}Sb \longrightarrow {}^{0}_{-1}\beta + X$
  c. $^{251}_{98}Cf \longrightarrow {}^{247}_{96}Cm + X$
  d. $^{233}_{92}U \longrightarrow {}^{4}_{2}\alpha + X$

5.122 Supply the missing nuclear symbol in each of the following radioactive-decay processes.
  a. $^{195}_{84}Po \longrightarrow {}^{4}_{2}\alpha + X$
  b. $^{75}_{31}Ga \longrightarrow {}^{75}_{32}Ge + X$
  c. $X \longrightarrow {}^{0}_{-1}\beta + {}^{88}_{36}Kr$
  d. $^{222}_{89}Ac \longrightarrow {}^{218}_{87}Fr + X$

5.123 Write nuclear equations for each of the following radioactive-decay processes.
  a. Mercury-199 is formed by beta emission.
  b. Cadmium-120 undergoes beta emission.
  c. Terbium-148 is formed by alpha emission.
  d. Radium-226 undergoes alpha emission.

5.124 Write nuclear equations for each of the following radioactive-decay processes.
  a. Thallium-206 is formed by beta emission.
  b. Palladium-109 undergoes beta emission.
  c. Plutonium-241 is formed by alpha emission.
  d. Fermium-249 undergoes alpha emission.

TRANSMUTATION AND BOMBARDMENT REACTIONS (SEC. 5.10)

5.125 Supply the missing nuclear symbol in each of the following equations for bombardment reactions.
  a. $^{15}_{7}N + {}^{1}_{1}H \longrightarrow X + {}^{12}_{6}C$
  b. $^{27}_{13}Al + X \longrightarrow {}^{4}_{2}\alpha + {}^{25}_{12}Mg$
  c. $^{80}_{34}Se + {}^{2}_{1}H \longrightarrow X + {}^{1}_{1}H$
  d. $X + {}^{2}_{1}H \longrightarrow {}^{9}_{3}Li + 2{}^{1}_{1}H$

5.126 Supply the missing nuclear symbol in each of the following equations for bombardment reactions.
  a. $^{14}_{7}N + X \longrightarrow {}^{14}_{6}C + {}^{1}_{1}H$
  b. $^{27}_{13}Al + {}^{4}_{2}\alpha \longrightarrow X + {}^{1}_{0}n$
  c. $^{56}_{26}Fe + {}^{2}_{1}H \longrightarrow {}^{54}_{25}Mn + X$
  d. $X + {}^{4}_{2}\alpha \longrightarrow {}^{109}_{47}Ag + {}^{1}_{1}H$

5.127 Write equations for the following nuclear bombardment processes.
  a. Beryllium-9 is bombarded with an alpha particle and emits a neutron.
  b. Nickel-58 is bombarded with a proton, and an alpha particle is emitted.
  c. Bombardment of cadmium-113 produces cadmium-114 and a gamma ray.
  d. Bombardment of a nuclide with and alpha particle produces phosphorus-30 and a neutron.

5.128 Write equations for the following nuclear bombardment processes.
  a. Bombardment of a radionuclide with an alpha particle produces curium-242 and one neutron.
  b. Bombardment of curium-246 with a small particle produces nobelium-254 and four neutrons.
  c. Aluminum-27 is bombarded with an alpha particle and produces a neutron.
  d. Bombardment of sodium-23 with hydrogen-2 produces neon-21.

5.129 Using Table 5.4 as your source of information, for how many of the transuranium elements does the most stable isotope have a half-life greater than 1 month?

5.130 Using Table 5.4 as your source of information, for how many of the transuranium elements does the most stable isotope have a half-life greater than 1 day?

POSITRON EMISSION AND ELECTRON CAPTURE (SEC. 5.11)

5.131 How does positron emission affect the mass number and atomic number of the parent nuclide undergoing decay?

5.132 How does electron capture affect the mass number and atomic number of a parent nuclide that undergoes electron capture?

5.133 What mode or modes of decay are associated with the conversion within the nucleus of a neutron to a proton?

5.134 What mode or modes of decay are associated with the conversion within the nucleus of a proton to a neutron?

5.135 Write balanced nuclear equations for the positron decay of the following radionuclides.
  a. $^{29}_{15}P$ **b.** antimony-112
  c. $^{46}_{23}V$ **d.** cerium-132

5.136 Write balanced nuclear equations for the positron decay of the following radionuclides.
  a. $^{33}_{17}Cl$ **b.** selenium-70
  c. $^{40}_{21}Sc$ **d.** tin-109

5.137 Write balanced nuclear equations for the electron-capture decay of the following radionuclides.
  a. $^{76}_{36}Kr$ **b.** xenon-122
  c. $^{100}_{46}Pd$ **d.** tantalum-175

5.138 Write balanced nuclear equations for the electron-capture decay of the following radionuclides.
  a. $^{80}_{38}Sr$
  b. rhenium-181
  c. $^{88}_{40}Zr$
  d. lead-196

5.139 Supply the missing nuclear symbol in each of the following equations for radioactive-decay processes.
  a. $^{17}_{9}F \longrightarrow X + ^{17}_{8}O$
  b. $^{88}_{37}Rb + X \longrightarrow ^{83}_{36}Kr$
  c. $X \longrightarrow ^{0}_{1}\beta + ^{103}_{46}Pd$
  d. $^{133}_{56}Ba + ^{0}_{-1}e \longrightarrow X$

5.140 Supply the missing nuclear symbol in each of the following equations for radioactive-decay process.
  a. $^{191}_{80}Hg + X \longrightarrow ^{191}_{79}Au$
  b. $X \longrightarrow ^{0}_{1}\beta + ^{172}_{72}Hf$
  c. $X + ^{0}_{-1}e \longrightarrow ^{108}_{49}In$
  d. $^{47}_{24}Cr \longrightarrow X + ^{47}_{23}V$

NEUTRON-TO-PROTON RATIO AND TYPE OF RADIOACTIVE DECAY (SEC. 5.12)

5.141 What is the dominant mode of decay for a radionuclide in which the neutron-to-proton ratio is too high for stability?

5.142 What is the dominant mode of decay for a radionuclide in which the neutron-to-proton ratio is too low for stability?

5.143 Calculate the neutron-to-proton ratio before and after each of the following processes takes place.
  a. beta decay of $^{65}_{28}Ni$
  b. alpha decay of $^{192}_{78}Pt$
  c. electron capture by $^{165}_{69}Tm$
  d. positron decay of $^{107}_{49}In$

5.144 Calculate the neutron-to-proton ratio before and after each of the following processes takes place.
  a. electron capture by $^{75}_{34}Se$
  b. beta decay of $^{78}_{33}As$
  c. positron decay of $^{37}_{19}K$
  d. alpha decay of $^{253}_{102}No$

5.145 One member of each of the following pairs of radionuclides decays by beta-particle emission and the other by positron emission. Which is which? Explain your reasoning.
  a. $^{74}_{36}Kr$ and $^{87}_{36}Kr$
  b. $^{68}_{33}As$ and $^{84}_{34}Se$
  c. $^{74}_{31}Ga$ and $^{64}_{31}Ga$
  d. $^{99}_{41}Nb$ and $^{99}_{46}Pd$

5.146 One member of each of the following pairs of radionuclides decays by beta-particle emission and the other by positron emission. Which is which? Explain your reasoning.
  a. $^{82}_{39}Y$ and $^{92}_{39}Y$
  b. $^{53}_{26}Fe$ and $^{68}_{29}Cu$
  c. $^{99}_{46}Pd$ and $^{115}_{46}Pd$
  d. $^{50}_{25}Mn$ and $^{50}_{21}Sc$

RADIOACTIVE DECAY SERIES (SEC. 5.13)

5.147 The plutonium-241 decay series terminates with bismuth-209. Would you expect bismuth-209 to be a stable or unstable nuclide? Explain your answer.

5.148 The thorium-232 decay series terminates with lead-208. Would you expect lead-208 to be a stable or unstable nuclide? Explain your answer.

5.149 The uranium-238 decay series begins with alpha-particle emission. The daughter nuclide undergoes beta-particle emission. What is the nuclide produced from this latter emission?

5.150 The thorium-232 decay series begins with alpha-particle emission. The daughter nuclide undergoes beta-particle emission. What is the nuclide produced from this latter emission?

5.151 In the thorium-232 natural decay series, the intermediate radon-220 undergoes alpha decay, the resulting daughter also undergoes alpha decay, and the succeeding two daughters both emit beta particles. Write four nuclear equations to represent these four steps of the thorium-232 natural decay series.

5.152 In the uranium-235 natural decay series, the intermediate thorium-231 undergoes beta decay, the resulting daughter undergoes alpha decay, and the two succeeding daughters emit a beta particle and an alpha particle in that order. Write four nuclear equations to represent these four steps of the uranium-235 natural decay series.

ADDITIONAL PROBLEMS

5.153 Indicate whether each of the following statements concerning sodium isotopes is *true* or *false*.
  a. $^{23}_{11}Na$ has one more electron than does $^{24}_{11}Na$.
  b. $^{23}_{11}Na$ and $^{24}_{11}Na$ contain the same number of neutrons.
  c. $^{23}_{11}Na$ has one fewer subatomic particle than $^{24}_{11}Na$.
  d. $^{23}_{11}Na$ and $^{24}_{11}Na$ have the same atomic number.

5.154 Indicate whether each of the following statements concerning magnesium isotopes is *true* or *false*.
  a. $^{24}_{12}Mg$ has one more proton than $^{25}_{12}Mg$.
  b. $^{24}_{12}Mg$ and $^{25}_{12}Mg$ contain the same number of subatomic particles in their nucleus.
  c. $^{24}_{12}Mg$ has one fewer neutron than $^{25}_{12}Mg$.
  d. $^{24}_{12}Mg$ and $^{25}_{12}Mg$ have different mass numbers.

5.155 Arrange the five atoms $^{42}_{20}Ca$, $^{39}_{19}K$, $^{44}_{21}Sc$, $^{37}_{18}Ar$, and $^{43}_{22}Ti$ in order of
  a. increasing number of electrons.
  b. decreasing number of neutrons.
  c. increasing number of protons.
  d. decreasing mass.

5.156 Arrange the five atoms $^{92}_{40}Zr$, $^{89}_{39}Y$, $^{94}_{41}Nb$, $^{87}_{38}Sr$, and $^{93}_{42}Mo$ in order of
  a. decreasing number of electrons.
  b. increasing number of neutrons.
  c. increasing number of nucleons.
  d. decreasing number of subatomic particles.

5.157 Write the complete symbol for the isotope of boron with each of the following characteristics.
    a. contains two fewer neutrons than $^{10}_{5}$B
    b. contains three more subatomic particles than $^{9}_{5}$B
    c. contains the same number of neutrons as $^{14}_{7}$N
    d. contains the same number of subatomic particles as $^{14}_{7}$N

5.158 Write the complete symbol for the isotope of chromium with each of the following characteristics.
    a. contains two more neutrons than $^{55}_{24}$Cr
    b. contains two fewer subatomic particles than $^{52}_{24}$Cr
    c. contains the same number of neutrons as $^{60}_{29}$Cu
    d. contains the same number of subatomic particles as $^{60}_{29}$Cu

5.159 Copper consists of two naturally occurring isotopes with masses of 62.9298 amu and 64.9278 amu.
    a. How many protons are in the nucleus of each isotope?
    b. How many electrons are in an atom of each isotope?
    c. How many neutrons are in the nucleus of each isotope?

5.160 Silver consists of two naturally occurring isotopes with masses of 106.9041 amu and 108.9047 amu.
    a. How many protons are in the nucleus of each isotope?
    b. How many electrons are in an atom of each isotope?
    c. How many neutrons are in the nucleus of each isotope?

5.161 Naturally occurring aluminum has a single isotope. Determine the following for naturally occurring atoms of aluminum.
    a. atomic number
    b. mass number
    c. number of neutrons in the nucleus
    d. isotopic mass, in amu, to three significant figures

5.162 Naturally occurring sodium has a single isotope. Determine the following for naturally occurring atoms of sodium.
    a. atomic number
    b. mass number
    c. number of neutrons in the nucleus
    d. isotopic mass, in amu, to three significant figures

5.163 Identify the element with each of the following characteristics.
    a. has an atomic mass 1.679 times that of phosphorus
    b. has three times as many electrons as zinc
    c. has 20 more protons than beryllium
    d. has twice as many charged subatomic particles as calcium

5.164 Identify the element with each of the following characteristics.
    a. has a nuclear charge twice that of magnesium
    b. has 30 more electrons than potassium
    c. has an atomic mass 2.381 times that of bromine
    d. has three times as many protons as sulfur

5.165 Suppose it was decided to redefine the atomic mass scale by choosing as an arbitrary reference point a value of 20.000 amu to represent the naturally occurring mixture of nickel isotopes. What would be the atomic mass of the following elements on the new atomic mass scale?
    a. silver
    b. gold

5.166 Suppose it was decided to redefine the atomic mass scale by choosing as an arbitrary reference point a value of 40.000 amu to represent the mass of fluorine atoms. (All fluorine atoms are identical; that is, fluorine is monoisotopic.) What would be the atomic mass of the following elements on the new atomic mass scale?
    a. sulfur
    b. platinum

5.167 What is the identity of a radionuclide X if it decays by
    a. alpha emission to give radon-217?
    b. beta emission to give calcium-47?
    c. positron emission to give bromine-78?
    d. electron capture to give chlorine-37?

5.168 What is the identity of a radionuclide X if it decays by
    a. alpha emission to give radon-218?
    b. beta emission to give magnesium-28?
    c. positron emission to give fluorine-17?
    d. electron capture to give cesium-129?

5.169 Fill in the blanks in the following radioactive decay series.
    a. $^{232}_{90}$Th $\xrightarrow{\alpha}$ ___ $\xrightarrow{\beta^-}$ ___ $\xrightarrow{\beta^-}$ $^{228}_{90}$Th
    b. ___ $\xrightarrow{\beta^-}$ ___ $\xrightarrow{\alpha}$ $^{224}_{88}$Ra $\xrightarrow{\alpha}$ ___

5.170 Fill in the blanks in the following radioactive decay series.
    a. $^{223}_{87}$Fr $\xrightarrow{\beta^-}$ ___ $\xrightarrow{\alpha}$ ___ $\xrightarrow{\alpha}$ $^{215}_{84}$Po
    b. ___ $\xrightarrow{\beta^-}$ ___ $\xrightarrow{\beta^-}$ $^{210}_{84}$Po $\xrightarrow{\alpha}$ ___

5.171 Phosphorus-31 is the only stable isotope of phosphorus. Predict how phosphorus-28 will decay and how phosphorus-34 will decay.

5.172 Fluorine-19 is the only stable isotope of fluorine. Predict how fluorine-16 will decay and how fluorine-21 will decay.

5.173 Consider the decay series

$$A \longrightarrow B \longrightarrow C \longrightarrow D$$

where A, B, and C are radioactive, with half-lives of 3.2 min, 25 days, and 9.0 sec, respectively, and D is nonradioactive, Starting with 1.00 mole of A, and none of B, C, and D, calculate the numbers of moles of A, B, C, and D present after 50 days.

5.174 Consider the decay series

$$E \longrightarrow F \longrightarrow G \longrightarrow H$$

where E, F, and G are radioactive, with half-lives of 5.0 min, 2.3 min, and 35 days, respectively, and H is nonradioactive. Starting with 1.00 mole of E, and none of F, G, and H, calculate the numbers of moles of E, F, G, and H present after 105 days.

5.175 A sample of radioactive ^{210}X initially weighed 4.000 g. After 35 days, 0.125 g of ^{210}X remained, the rest having decayed to the stable ^{206}Q. Calculate
 a. the half-life of ^{210}X.
 b. the mass of ^{206}Q formed.

5.176 A sample of radioactive ^{213}X initially weighed 2.000 g. After 15 days, 0.250 g of ^{213}X remained, the rest having decayed to the stable ^{209}Q. Calculate
 a. the half-life of ^{213}X.
 b. the mass of ^{209}Q formed.

5.177 For the following nuclear processes, characterize the relationship between parent and daughter nuclides as that of isobars, isotopes, or neither.
 a. positron emission
 b. alpha emission
 c. neutron absorption
 d. alpha decay followed by two beta decays

5.178 For the following nuclear processes, characterize the relationship between parent and daughter nuclides as that of isobars, isotopes, or neither.
 a. electron capture
 b. beta emission
 c. neutron emission
 d. beta emission followed by electron capture

CUMULATIVE PROBLEMS

5.179 What is the atomic number of the element of highest atomic mass whose symbol does not derive from the English name of the element?

5.180 What is the atomic number of the element of lowest atomic mass whose symbol does not derive from the English name of the element?

5.181 How many protons are present in seven molecules of the compound $C_6H_{12}O_6$ (glucose, blood sugar)?

5.182 How many electrons are present in nine molecules of the compound $C_{12}H_{22}O_{11}$ (sucrose, table sugar)?

5.183 The diameter of an atom is approximately 10^5 times as large as the diameter of the nucleus. If the nucleus were enlarged to the size of a Ping-Pong ball (1.5 in. in diameter), what would be the diameter of the atom, in miles?

5.184 The diameter of an atom is approximately 10^{-8} cm, and that of the nucleus is 10^{-13} cm. How many times larger is the volume of the atom than the volume of the nucleus? The formula for the volume of a sphere is $V = 0.524 \times d^3$ (where d is the diameter of the sphere).

Multiple-Choice Practice Test

Use this bank of 20 multiple choice questions as a review of key concepts presented in this chapter. For many of the questions, there may be more than one correct answer (choice d) or no correct answer (choice e).

5.185 Which of the following statements correctly describes the subatomic particle called an electron?
 a. has a mass slightly greater than that of a proton
 b. can also be called a nucleon
 c. has a positive charge
 d. more than one correct response
 e. no correct response

5.186 Which of the following statements concerning the nucleus of an atom is correct?
 a. It accounts for almost all of the mass of the atom.
 b. It accounts for only a small amount of the total volume of the atom.
 c. It contains only neutrons.
 d. more than one correct response
 e. no correct response

5.187 Atoms as a whole are neutral because
 a. an equal number of protons and neutrons are always present.
 b. the number of subatomic particles in the nucleus is always an even number.
 c. neutrons neutralize the charges found on the electrons and protons present.
 d. more than one correct response
 e. no correct response

5.188 The mass number and atomic number of an atom that contains 15 electrons, 15 protons, and 16 neutrons are, respectively,
 a. 46 and 15.
 b. 46 and 16.
 c. 31 and 15.
 d. 30 and 16.
 e. no correct answer

5.189 Which of the following statements is *correct* for $^{40}_{19}K$?
  a. contains more protons than neutrons
  b. contains more electrons than protons
  c. contains more neutrons than electrons
  d. more than one correct response
  e. no correct response

5.190 Phosphorus is element 15. This means that all atoms of phosphorus have
  a. an atomic number of 15.
  b. 15 protons in the nucleus.
  c. a total of 15 subatomic particles present.
  d. more than one correct response
  e. no correct response

5.191 Which of the following pieces of information is associated with the quantity $(A + Z)$?
  a. number of subatomic particles present in the nucleus of an atom
  b. number of charged subatomic particles present in an atom
  c. total number of subatomic particles present in an atom
  d. more than one correct response
  e. no correct response

5.192 Isotopes of a given element have
  a. the same mass number and different numbers of protons.
  b. the same atomic number and the same number of subatomic particles.
  c. the same atomic number and different mass numbers.
  d. more than one correct response
  e. no correct response

5.193 Which of the following properties for the most abundant isotope of an element would *not* be possessed by the second most abundant isotope of the same element?
  a. boiling point is 2403°C
  b. atomic number is 31
  c. reacts with chlorine to give the compound XCl_3
  d. more than one correct response
  e. no correct response

5.194 Which of the following pairs of atoms, specified in terms of subatomic particle composition, are isobars?
  a. (20p, 20e, 20n) and (21p, 21e, 21n)
  b. (20p, 20e, 20n) and (20p, 20e, 22n)
  c. (20p, 20e, 22n) and (21p, 21e, 21n)
  d. more than one correct response
  e. no correct response

5.195 Chlorine, which exists in nature in two isotopic forms, has an atomic mass of 35.45 amu. This means that
  a. all chlorine atoms have a mass of 35.45 amu.
  b. some, but not all, chlorine atoms have a mass of 35.45 amu.

  c. 35.45 amu is the upper limit for the mass of a chlorine atom.
  d. more than one correct response
  e. no correct response

5.196 What is the atomic mass of the hypothetical element supposium, given that 40.0% of supposium atoms have a mass of 16.0 amu and 60.0% of supposium atoms have a mass of 18.0 amu?
  a. 16.8 amu
  b. 17.0 amu
  c. 17.2 amu
  d. 17.4 amu
  e. no correct response

5.197 The notation "nitrogen-15" denotes a nitrogen nuclide that contains
  a. 15 protons.
  b. 15 neutrons.
  c. 15 subatomic particles in its nucleus.
  d. more than one correct response
  e. no correct response

5.198 A mass number increase is associated with which of the following modes of decay?
  a. alpha-particle emission
  b. beta-particle emission
  c. gamma-ray emission
  d. more than one correct response
  e. no correct response

5.199 In terms of subatomic particles, the composition of an alpha particle is
  a. three neutrons and one proton.
  b. two neutrons and two protons.
  c. one neutron and three protons.
  d. one neutron and two protons.
  e. no correct response

5.200 The explanation of how a beta particle is produced in the nucleus of a radionuclide and then emitted involves the conversion of a
  a. proton to a neutron.
  b. neutron to a proton.
  c. proton to an electron.
  d. neutron to an electron.
  e. no correct response

5.201 Which of the following sets of data is consistent with a half-life of 5.0 days for a radionuclide sample?
  a. time elapsed = 20.0 days;
       fraction of nuclides decayed = 7/8
  b. time elapsed = 15.0 days;
       fraction of nuclides undecayed = 1/16
  c. time elapsed = 10.0 days;
       fraction of nuclides decayed = 3/4
  d. more than one correct response
  e. no correct response

5.202 If 1.50 gram of a 2.00 gram radioactive sample undergoes decay in 60 minutes, then the half-life of the radioactive substance is
a. 15 minutes.
b. 30 minutes.
c. 120 minutes.
d. 240 minutes.
e. no correct response

5.203 Which of the following equations is an example of an incorrectly written nuclear equation?
a. $^{131}_{53}I \longrightarrow ^{\ 0}_{-1}\beta + ^{131}_{52}Te$
b. $^{238}_{92}U \longrightarrow ^{4}_{2}\alpha + ^{234}_{90}Th$
c. $^{87}_{37}Rb + ^{\ 0}_{-1}e \longrightarrow ^{87}_{36}Kr$
d. more than one correct response
e. no correct response

5.204 The bombardment reaction involving $^{23}_{11}Na$ and $^{2}_{1}H$ gives two products, one of which is $^{1}_{1}H$. The other product is
a. $^{24}_{11}Na$
b. $^{25}_{11}Na$
c. $^{25}_{12}Mg$
d. $^{22}_{12}Mg$
e. no correct response

Electronic Structure and Chemical Periodicity

6.1 THE PERIODIC LAW

During the early part of the nineteenth century, scientists began to look for order in the increasing amount of chemical information that had become available. They knew that certain elements had properties that were very similar to those of other elements, and they sought to use these similarities as a means for arranging or classifying the elements.

In 1869, these efforts culminated in the discovery of what is now called the *periodic law*. The **periodic law** *states that when elements are arranged in order of increasing atomic number, elements with similar chemical behavior occur at periodic (regularly recurring) intervals.* Proposed independently by both the Russian chemist Dmitri Ivanovich Mendeleev (1834–1907; see "The Human Side of Chemistry 5") and the German chemist Julius Lothar Meyer (1830–1895), the periodic law is one of the most important of all chemical laws.

The preceding statement of the periodic law is in modern-day language. It differs from the original 1869 statements in that the phrase *atomic number* has replaced "atom mass." The use of atom mass in the original statements reflected theories prevalent in 1869. According to these theories, the masses of atoms were their most important distinguishing properties—a knowledge of subatomic particles (Sec. 5.1) was still 30 years away. When the details of subatomic structure were finally discovered, it

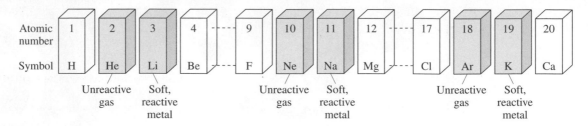

FIGURE 6.1 An illustration of the periodicity in properties that occurs when elements are arranged in order of increasing atomic number. He, Ne, and Ar have similar chemical properties as do Li, Na, and K.

became obvious that the properties of atoms were related not to their masses but to the number and arrangement of their electrons. Thus, the periodic law was modified to reflect this new knowledge.

Figure 6.1 shows parts of the repeating pattern for chemical properties (periodic law) for the sequence of elements with atomic numbers 1 through 20. The elements shown in the same color have similar chemical properties. For the sake of simplicity, only two of the periodic relationships (repeating patterns) are shown; for the elements listed, similar properties are found in every eighth element.

The concept of a periodic variation or pattern should not be new to you because numerous everyday examples of this exist. Most, but not all, involve time. Mondays occur at regular intervals of seven days. Office workers get paid at regular intervals, such as every two weeks. The red–green–yellow light sequence for a traffic signal repeats itself in a periodic fashion.

To be useful, the relationships generated by the periodic law must be easily visualized. This is the purpose of *periodic tables*, the subject of Section 6.2.

6.2 THE PERIODIC TABLE

A **periodic table** *is a tabular arrangement of the elements in order of increasing atomic number such that elements having similar chemical behavior are grouped in vertical columns.* It is a graphical portrayal of the periodic law.

The most commonly used form of the periodic table is the one shown in Figure 6.2 (and also inside the front cover of this book). Note the locations in this table for the elements He, Ne, and Ar and the elements Li, Na, and K—groups of elements with similar properties (Fig. 6.1). The elements He, Ne, and Ar are all found in the far right column of the table and the elements Li, Na, and K are all found in the far left column of the table. From further study of the far right column of this periodic table, we can deduce that the elements Kr, Xe, and Rn will have properties similar to those of He, Ne, and Ar since all of these elements are in the same column of the table.

Within the periodic table each element is represented by a square box. Within each box are given the symbol, atomic number, and atomic mass of the element, as shown in Figure 6.3. It should be noted that some periodic tables, but not the ones in this text, also give the element's name within the box.

Periods and Groups of Elements

Two terms integrally connected with the use of a periodic table are the terms *period* and *group*.

A **period** *is a horizontal row of elements in the periodic table.* For identification purposes, the periods are numbered sequentially, with Arabic numbers, starting at the top of the periodic table. (These period numbers are sometimes, but not usually, shown on the left-hand side of a periodic table.) Period 3 is the third row of elements, period 4 the

Using the information on a periodic table, you can quickly determine the number of protons and electrons for atoms of an element. However, no information concerning neutrons is available from a periodic table; mass numbers are not part of the information given because they are not unique for an element (Sec. 5.3).

The Human Side of Chemistry 5

Dmitri Ivanovich Mendeleev (1834–1907)

Dmitri Ivanovich Mendeleev, born in Siberia in 1834, the youngest of 17 children in his family, is Russia's most famous chemist. His periodic classification of the elements is heralded as his greatest achievement. Other areas of research for him included petroleum chemistry and pressure and temperature studies on gases.

Mendeleev's talents as a prolific writer and a popular teacher directly influenced his discovery of the periodic law. In 1867, two years prior to his great discovery, Mendeleev was appointed professor of general chemistry at the University of St. Petersburg. To help his students, he immediately began writing a new textbook. In preparation for writing some material about the elements and their properties, he wrote down the properties of each element on separate cards. While sorting through his "flash" cards to try to "organize things," he "discovered" the periodic repetition of properties of the elements. Within a month he had written a paper on the subject that was delivered before the Russian Chemical Society. The textbook he was writing, *General Chemistry*, became a classic, going through eight Russian editions and numerous English, French, and German editions.

Mendeleev's first periodic table listed the elements according to increasing atom mass and grouped them (in columns) according to chemical reactivity. In doing this he realized that several "holes" existed in the elemental arrangement, which he interpreted as evidence of missing (yet to be discovered) elements. Two of the vacant spaces were directly under aluminum and silicon. He named these undiscovered elements eka-aluminum and eka-silicon and predicted what their properties would be. Within 15 years, these elements—now known as gallium and germanium—were isolated from naturally occurring minerals. Their properties matched closely with those that Mendeleev had predicted.

Mendeleev remained a professor at St. Petersburg until 1890, when he resigned because of a controversy with the Ministry of Education concerning student political rights. He took the side of the students. In 1893, he became director of the Russian Bureau of Weights and Measures, a post he held until his death.

It should be pointed out that the German chemist Julius Lothar Meyer (1830–1895) discovered the periodic law simultaneously and independently of Mendeleev. Unknown to each other, both published papers on the subject in 1869. Because Mendeleev did more with the periodic law once it was published, he is generally given more credit for its development.

In 1882, when selected to receive the Davy medal of the British Royal Society for his work concerning systematic classification of the elements, Mendeleev insisted that Meyer be corecipient of the award, formally recognizing Meyer's independent conception of the periodic law.

Element 101, mendelevium, bears his name.

---

fourth row of elements, and so forth. The elements Na, Mg, Al, Si, P, S, Cl, and Ar are all members of period 3 (see Fig. 6.2). Period 1 has only two elements—H and He.

A **group** *is a vertical column of elements in the periodic table*. The elements within a group have similar chemical properties. Three different labeling schemes for groups are in common use, two of which are shown in Figure 6.2 (located above each group). The bottom set of group labels in Figure 6.2, which involve Roman numerals and the letters A and B, is widely used in North America. Europeans use a similar convention (not shown) that numbers the groups consecutively as IA through VIIIA and then IB through VIIIB. North American and European group labels match for only four groups (IA, IIA, IB, and IIB). In an effort to eliminate confusion between the two systems, the International Union of Pure and Applied Chemistry (IUPAC) has proposed a new convention that numbers the groups simply using the numbers 1 through 18 (no As or Bs and no Roman numerals), as is shown in the top set of labels in Figure 6.2. We will use the traditional North American convention because the IUPAC system is still not widely used.

FIGURE 6.2   The most commonly used form of the periodic table. Atomic masses are based on the 2007 IUPAC table of atomic masses.

1 Group IA	2 Group IIA	3 Group IIIB	4 Group IVB	5 Group VB	6 Group VIB	7 Group VIIB	8 Group VIIIB	9 Group VIIIB	10 Group	11 Group IB	12 Group IIB	13 Group IIIA	14 Group IVA	15 Group VA	16 Group VIA	17 Group VIIA	18 Group VIIIA
1 **H** 1.01																	2 **He** 4.00
3 **Li** 6.94	4 **Be** 9.01											5 **B** 10.81	6 **C** 12.01	7 **N** 14.01	8 **O** 16.00	9 **F** 19.00	10 **Ne** 20.18
11 **Na** 22.99	12 **Mg** 24.31											13 **Al** 26.98	14 **Si** 28.09	15 **P** 30.97	16 **S** 32.07	17 **Cl** 35.45	18 **Ar** 39.95
19 **K** 39.10	20 **Ca** 40.08	21 **Sc** 44.96	22 **Ti** 47.87	23 **V** 50.94	24 **Cr** 52.00	25 **Mn** 54.94	26 **Fe** 55.85	27 **Co** 58.93	28 **Ni** 58.69	29 **Cu** 63.55	30 **Zn** 65.38	31 **Ga** 69.72	32 **Ge** 72.64	33 **As** 74.92	34 **Se** 78.96	35 **Br** 79.90	36 **Kr** 83.80
37 **Rb** 85.47	38 **Sr** 87.62	39 **Y** 88.91	40 **Zr** 91.22	41 **Nb** 92.91	42 **Mo** 95.96	43 **Tc** (98)	44 **Ru** 101.07	45 **Rh** 102.91	46 **Pd** 106.42	47 **Ag** 107.87	48 **Cd** 112.41	49 **In** 114.82	50 **Sn** 118.71	51 **Sb** 121.76	52 **Te** 127.60	53 **I** 126.90	54 **Xe** 131.29
55 **Cs** 132.91	56 **Ba** 137.33	57 **La** 138.91	72 **Hf** 178.49	73 **Ta** 180.95	74 **W** 183.84	75 **Re** 186.21	76 **Os** 190.23	77 **Ir** 192.22	78 **Pt** 195.08	79 **Au** 196.97	80 **Hg** 200.59	81 **Tl** 204.38	82 **Pb** 207.2	83 **Bi** 208.98	84 **Po** (209)	85 **At** (210)	86 **Rn** (222)
87 **Fr** (223)	88 **Ra** (226)	89 **Ac** (227)	104 **Rf** (267)	105 **Db** (268)	106 **Sg** (271)	107 **Bh** (272)	108 **Hs** (270)	109 **Mt** (276)	110 **Ds** (281)	111 **Rg** (280)	112 (285)	113 – (284)	114 – (289)	115 – (288)	116 – (293)	117	118 – (294)

58 **Ce** 140.12	59 **Pr** 140.91	60 **Nd** 144.24	61 **Pm** (145)	62 **Sm** 150.36	63 **Eu** 151.96	64 **Gd** 157.25	65 **Tb** 158.93	66 **Dy** 162.50	67 **Ho** 164.93	68 **Er** 167.26	69 **Tm** 168.93	70 **Yb** 173.05	71 **Lu** 174.97
90 **Th** 232.04	91 **Pa** 231.04	92 **U** 238.03	93 **Np** (237)	94 **Pu** (244)	95 **Am** (243)	96 **Cm** (247)	97 **Bk** (247)	98 **Cf** (251)	99 **Es** (252)	100 **Fm** (257)	101 **Md** (258)	102 **No** (259)	103 **Lr** (262)

Non-metals

Metals

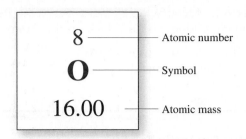

FIGURE 6.3
Arrangement of
information about
elements within boxes
of the periodic table.

Four groups of elements also have common (nonnumerical) names. On the extreme left side of the periodic table are found the *alkali metals* (Li, Na, K, Rb, Cs) and the *alkaline earth metals* (Be, Mg, Ca, Sr, Ba). **Alkali metal** *is a general name for any element in group IA of the periodic table, excluding hydrogen.* The alkali metals are soft, shiny metals that readily react with water. **Alkaline earth metal** *is a general name for any element in group IIA of the periodic table.* The alkaline earth metals are soft, shiny metals also, but are only moderately reactive with water. On the extreme right side of the periodic table are found the *halogens* (F, Cl, Br, I, At) and the *noble gases* (He, Ne, Ar, Kr, Xe, Rn). **Halogen** *is a general name for any element in group VIIA of the periodic table.* The halogens are very reactive colored substances that are gases at room temperature or become such at temperatures slightly above room temperature. **Noble gas** *is a general name for any element in group VIIIA of the periodic table.* Noble gases are unreactive gases that undergo few, if any, chemical reactions.

The location of any element in the periodic table is specified by giving its group number and its period number. The element gold (Au), with an atomic number of 79, belongs to group IB and is in period 6. Nitrogen (N), with an atomic number of 7, belongs to group VA and is in period 2.

The elements within a given periodic table group show numerous similarities in chemical properties, the degree of similarity varying from group to group. In no case are the group members clones of one another. Each element has some individual characteristics not found in other elements of the group. By analogy, the members of a human family often bear many resemblances to each other, but each member also has some (and often much) individuality.

EXAMPLE 6.1 **Identifying Groups, Periods, and Specially Named Families of Elements**

Identify the element, by giving its chemical symbol, that fits each of the following periodic table–related descriptions.

  a. located in both period 2 and group IIIA
  b. the period 2 noble gas
  c. the period 3 alkaline earth metal
  d. the period 4 alkali metal

SOLUTION

  a. Period 2 is the second row of elements, and group IIIA is the sixth column in from the right side of the periodic table. The element that has this column–row (period–group) location is B (boron).
  b. The noble gases are the elements of group VIIIA (the rightmost column in the periodic table). The period 2 (second row) noble gas is Ne (neon).
  c. The alkaline earth metals are the elements of group IIA (the second column in from the left side of the periodic table). The period 3 (third row) alkaline earth metal is Mg (magnesium).
  d. The alkali metals are the elements of the first column (extreme left) of the periodic table. The period 4 (fourth row) alkali metal is K (potassium).

▶ **Practice Exercise 6.1** Identify the element, by giving its chemical symbol, that fits each of the following periodic table–related descriptions.

Answers to all practice exercises in this chapter are found in the back-of-the-book answer section.

a. located in both period 3 and group VIA
b. the period 2 alkaline earth metal
c. the period 3 halogen
d. the period 5 noble gas

The Shape of the Periodic Table

There is one location in the periodic table of Figure 6.2 where the practice of arranging the elements according to increasing atomic number seems to be violated. This is the location of elements 57 and 89, which are both in group IIIB. Shown next to them, in group IVB, are elements 72 and 104, respectively. The missing elements, 58–71 and 90–103, are located in two rows at the bottom of the periodic table. Technically, these elements should be included in the body of the table, as shown in Figure 6.4a. However, to have a more compact table, they are placed in the position shown in Figure 6.4b. This arrangement should present no problems to the user of the periodic table as long as it is recognized for what it is—a space-saving device.

Students often assume that a corollary of the fact that the elements are arranged in the periodic table in order of increasing atomic number is that they are also arranged in order of increasing atomic mass. This latter conclusion is correct most of the time; however, there are exceptions.

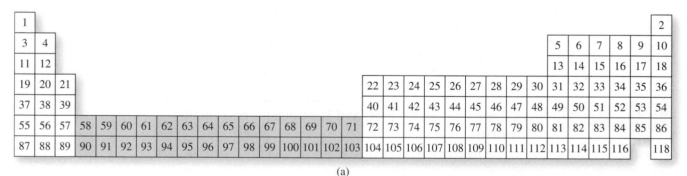

FIGURE 6.4 Long and short forms of the periodic table: (a) periodic table with elements 58–71 and 90–103 (in color) in their proper positions—the long form; (b) periodic table modified to conserve space by placing elements 58–71 and 90–103 (in color) below the rest of the elements—the short form.

The following element sequences are exceptions to the generalization that atomic mass increases with increasing atomic number. In each triad, the middle element has the smaller atomic mass.

18	19	20
Ar	**K**	**Ca**
39.95	39.10	40.08

27	28	29
Co	**Ni**	**Cu**
58.93	58.69	63.55

52	53	54
Te	**I**	**Xe**
127.60	126.90	131.29

These inversions in atomic mass result from isotopic abundances (Sec. 5.3). Let us consider such data for the Ar–K–Ca triad.

^{36}Ar	0.34%	^{39}K	93.10%	^{40}Ca	96.97%
^{38}Ar	0.07%	^{40}K	0.01%	^{42}Ca	0.64%
^{40}Ar	99.60%	^{41}K	6.88%	^{43}Ca	0.14%
				^{44}Ca	2.06%
				^{46}Ca	0.003%
				^{48}Ca	0.18%

For argon, the heaviest isotope ($A = 40$) is most abundant, and for potassium, the lightest isotope ($A = 39$) is most abundant. This results in potassium having a lower atomic mass than argon. The atomic masses of argon and calcium are almost equal (39.95 and 40.08 amu, respectively); the dominant isotope for both elements has a mass number of 40.

Later in this chapter we will find that a periodic table conveys much more information about the elements than the symbols, atomic numbers, and atomic masses that are printed on it. Information about the arrangement of electrons in an atom is coded into the table, as is information concerning some physical and chemical property trends. Indeed, the periodic table is considered to be the single most useful study aid available for organizing information about the elements.

For many years after the formulation of the periodic law and the periodic table, both were considered to be empirical. The law worked and the table was very useful, but there was no explanation available for the law or for why the periodic table had the shape it had. It is now known that the theoretical basis for both the periodic law and the periodic table involves the arrangement of electrons in atoms. The properties of the elements repeat themselves in a periodic manner because the arrangement of electrons about the nucleus of an atom follows a periodic pattern. Electron arrangements and an explanation of the periodic law and periodic table in terms of electronic theory are the subject matter for most of the remainder of this chapter.

6.3 THE ENERGY OF AN ELECTRON

In Section 5.1 we learned the following facts about electrons.

1. They are one of three fundamental subatomic particles.
2. They have very little mass in comparison to protons and neutrons.
3. They are located outside the nucleus of the atom.
4. They move rapidly about the nucleus in a volume that defines the size of the atom.

Much more must be known about electrons to understand the chemical behavior of the various elements, why some elements have similar chemical properties, and the theoretical basis for the periodic law.

More specific information concerning the behavior and arrangement of electrons within the extranuclear region of an atom is derived from a complex mathematical model for electron behavior called *quantum mechanics.* All the early work on this subject was done by physicists rather than by chemists.

During the early part of the twentieth century (1910–1930) a major revolution occurred in the field of physics, a revolution that profoundly affected chemistry. During this time it became clear that the established laws of physics, the laws that had been used for many years to predict the behavior of macroscopic objects, could not explain the behavior of extremely small objects such as atoms and electrons. The work of a number of European physicists led to this conclusion and also beyond it. These same physicists were able to formalize new laws that did apply to small objects. It is from these new laws, collectively called quantum mechanics, that information concerning the arrangement and behavior of electrons about a nucleus came.

A major force in the development of quantum mechanics was the Austrian physicist Erwin Schrödinger (1887–1961; see "The Human Side of Chemistry 6"). In 1926 he showed that the laws of quantum mechanics could be used to characterize the motion of electrons. Most of the concepts in the remainder of this chapter come from solutions to the equations developed by Schrödinger.

A formal discussion of quantum mechanics is beyond the scope of this course (and many other chemistry courses also) because of the rigorous mathematics involved. The answers obtained from quantum mechanics are, however, simple enough to be understood to a surprisingly large degree at the level of an introductory chemistry course. A consideration of these quantum-mechanical answers will enable us to develop a system for specifying electron arrangements around a nucleus and further to develop basic rules governing compound formation. This latter topic will occupy our attention in Chapter 7.

Present-day quantum-mechanical theory describes the arrangement of an atom's electrons in terms of their energies. Indeed, the energy of an electron is the property that is most important to any consideration of its behavior about the nucleus. The higher an electron's total energy, the larger the average distance of the electron from the nucleus with which it is associated.

A most significant characteristic of an electron's energy is that it is a *quantized property.* A **quantized property** *is a property that can have only certain values, that is, not all values are allowed.* Since an electron's energy is quantized, an electron can have only certain specific energies.

Quantization is a phenomenon not commonly encountered in the macroscopic world. Somewhat analogous is the process of a person climbing a flight of stairs. In Figure 6.5a you see six steps between ground level and the entrance level. As a person climbs these stairs there are only six permanent positions he or she can occupy (with both feet together). Thus the person's position (height above ground level) is quantized; only certain positions are allowed. The opposite of quantized is continuousness. A person climbing a ramp up to the entrance (Fig. 6.5b) would be able to assume a continuous set of heights above ground level; all values are allowed.

The energy of an electron determines its behavior about the nucleus. Since electron energies are quantized, only certain behavior patterns are allowed. Descriptions of electron

The floors of a building are quantized. An elevator will stop at the 5th or the 6th floor, but not at the 5.5 floor. The frets on the neck of a guitar make it quantized because the notes between the frets cannot be played. An instrument like a cello is continuous because it does not have frets; any note can be played.

FIGURE 6.5 A stairway with quantized position levels (the steps) versus a ramp with continuous position levels.

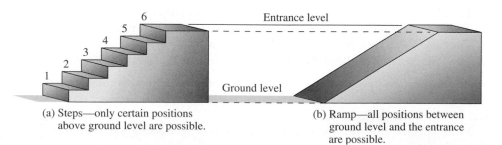

(a) Steps—only certain positions above ground level are possible.

(b) Ramp—all positions between ground level and the entrance are possible.

The Human Side of Chemistry 6

**Erwin Schrödinger
(1887–1961)**

Erwin Schrödinger, an only child, was born in 1887 in Vienna, Austria. He was taught at home until age 11, and science was always part of his environment—his father was a botanist, and his maternal grandfather a professor of chemistry. His formal education began in 1898 when he entered a gymnasium (high school), where he received a sound classical education. This early schooling in the classics influenced Schrödinger's actions throughout his life. The culmination of his formal education was a doctorate in theoretical physics in 1910 from the University of Vienna.

After service as an artillery officer during World War I, Schrödinger held professorships in Stuttgart and Zürich. During his years in Switzerland his pioneering papers on the equations of quantum theory were formulated and published.

To have more contact with notable physicists (including Albert Einstein), he moved in 1927 to Berlin. All went well until Hitler's new regime (which he could not accept) came into power. With the rise of Nazism, he accepted an invitation to be a guest professor at Oxford. In 1933, only a few weeks after leaving Germany for England, he received the Nobel Prize in physics for his quantum theory work.

Hitler's activities continued to affect his career. After a short stay in England, he returned to his native Austria despite an uncertain political situation there. When the Germans occupied Austria in 1938, Schrödinger fled to Italy and from there to the United States. Following a short stay at Princeton University, he became director of the School of Theoretical Physics in Dublin, Ireland. For the next 17 years he worked in Dublin. In 1956, he again returned to Austria, where he remained until his death.

In addition to his pioneering work in quantum theory, Schrödinger's research touched the areas of thermodynamics, specific heats of solids, theory of color, and physical aspects of the living cell.

Schrödinger was many-sided. He loved poetry and wrote verse of his own. He was a sculptor and also widely acquainted with modern and classical art. He quoted Greek philosophers as he lectured on problems of theoretical physics. A colleague's tribute upon his death was: "The breadth of his knowledge was as marvelous as the power of his thought."

behaviors involve the use of the terms *electron shell, electron subshell,* and *electron orbital.* Sections 6.4–6.6 consider the meaning of these terms and the mathematical interrelationships between them.

6.4 ELECTRON SHELLS

It was mentioned in Section 6.3 that electrons with higher energy will tend to be found farther from the nucleus than those with lower energy. Based on considerations of energy-distance from the nucleus, electrons can be grouped into *electron shells* or electron main energy levels. An **electron shell** *is a region of space about a nucleus that contains electrons that have approximately the same energy and that spend most of their time approximately the same distance from the nucleus.*

A shell number, n, is used to identify each electron shell. Electron shells are numbered 1, 2, 3, and so on, outward from the nucleus. Electron energy increases as the distance of the electron shell from the nucleus increases. An electron in shell 1 has the minimum amount of energy that an electron can have. No known atom has electrons beyond shell 7.

The maximum number of electrons possible in an electron shell varies; the higher the shell energy, the more electrons the shell can accommodate. The farther electrons are from the nucleus (a higher-energy shell), the greater the volume of space available for them; hence, the more electrons there can be in the shell. (Conceptually, electron shells

Electrons that occupy shell 1 are closer to the nucleus and have a lower energy than do electrons in shell 2.

TABLE 6.1 Important Characteristics of Electron Shells

Shell	Number Designation (n)	Electron Capacity ($2n^2$)
1st	1	$2 \times 1^2 = 2$
2nd	2	$2 \times 2^2 = 8$
3rd	3	$2 \times 3^2 = 18$
4th	4	$2 \times 4^2 = 32$
5th	5	$2 \times 5^2 = 50*$
6th	6	$2 \times 6^2 = 72*$
7th	7	$2 \times 7^2 = 98*$

The maximum number of electrons in this shell has never been attained in any element now known.

may be considered to be nested one inside another, somewhat like the layers of flavors inside a jawbreaker or similar type of candy.)

The lowest-energy shell ($n = 1$) accommodates a maximum of 2 electrons. In the second, third, and fourth shells, 8, 18, and 32 electrons, respectively, are allowed. A very simple mathematical equation can be used to calculate the maximum number of electrons allowed in any given shell:

$$\text{shell electron capacity} = 2n^2, \text{ where } n = \text{shell number}$$

For example, when $n = 4$, the value $2n^2 = 2(4^2) = 32$, which is the number previously given for the number of electrons allowed in the fourth shell. Although there is a maximum electron occupancy for each shell or main energy level, a shell may hold fewer than the allowable number of electrons in a given situation.

Table 6.1 summarizes concepts presented in this section concerning electron shells or electron main energy levels.

6.5 ELECTRON SUBSHELLS

The early development of quantum mechanics was closely linked to the study of light (spectral) emissions produced by atoms. The letters *s*, *p*, *d*, and *f*—used to denote subshells—are derived, respectively, from old spectroscopic terminology that describes spectral lines as *s*harp, *p*rincipal, *d*iffuse, and *f*undamental.

All (both) electrons in the first shell have the same energy, but in higher shells all the electrons do not have the same energy. Their energies are all close to each other in magnitude, but they are not identical. The range of energies for electrons in a shell is due to the existence within an electron shell of *electron subshells* or electron energy sublevels. An **electron subshell** *is a region of space within an electron shell that contains electrons that have the same energy.*

The number of subshells within a shell varies. A shell contains the same number of subshells as its own shell number; that is,

number of subshells in a shell = n,   where n = shell number

Thus, each successive shell has one more subshell than the previous one. Shell 3 contains three subshells, shell 4 contains four subshells, and shell 5 contains five subshells.

Subshells are identified by a number and a letter. The number indicates the shell to which the subshell belongs. The letters are *s*, *p*, *d*, or *f* (all lowercase letters), which, in that order, denote subshells of increasing energy within a shell. The lowest energy subshell within a shell is always the *s* subshell, the next higher the *p* subshell, then the *d*, and finally the *f*. Shell 1 has only one subshell, the 1*s*. Shell 2 has two subshells, the 2*s* and 2*p*. The 3*s*, 3*p*, and 3*d* subshells are found in shell 3. Shell 4 contains four subshells, the 4*s*, 4*p*, 4*d*, and 4*f* subshells. Figure 6.6 gives information concerning subshells for the first seven shells. Note in that figure that number-and-letter designations are not given

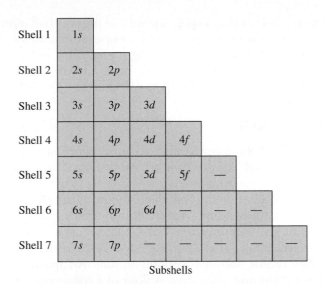

FIGURE 6.6 Subshells needed to describe the electron arrangements of the known elements. Note that not all subshells in shells 5, 6, and 7 are needed. Four of the five subshells in shell 5 are used, only three of the six subshells in shell 6 are used, and only two of the seven subshells in shell 7 are used.

for all the subshells in shells 5, 6, and 7; some of these subshells—those denoted with a dash—are not needed to describe the electron arrangements of the 117 known elements. Why they are not needed will become evident in Section 6.7. The important concept now is that we will never need subshell types other than s, p, d, and f in describing the electron arrangements of the known elements.

The maximum number of electrons that a subshell can hold varies from 2 to 14, depending on the type of subshell—s, p, d, or f. An s subshell can accommodate only 2 electrons. What shell the s subshell is located in does not affect the maximum electron occupancy figure; that is, the $1s$, $2s$, $3s$, $4s$, $5s$, $6s$, and $7s$ subshells all have a maximum electron occupancy of 2. Subshells of the p, d, and f types can accommodate maximums of 6, 10, and 14 electrons, respectively. Again the maximum numbers of electrons in these types of subshells depend only on the subshell types and are independent of shell number. Figure 6.7 summarizes the information presented in this section for subshells

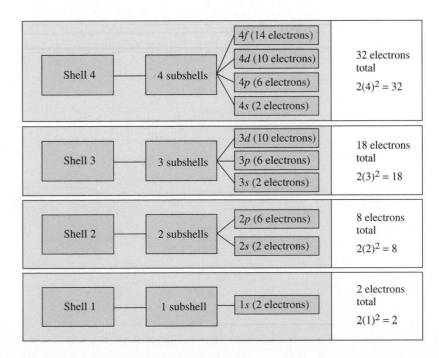

FIGURE 6.7 The distribution of electrons within shells and subshells for the first four electron shells.

found in shells 1 through 4. Within a shell, the sum of the subshell electron occupancies is the same as the shell electron occupancy ($2n^2$). For example, in shell 4, an *s* subshell containing 2 electrons, a *p* subshell containing 6 electrons, a *d* subshell containing 10 electrons, and an *f* subshell containing 14 electrons add up to a total of 32 electrons, which is the maximum occupancy of shell 4 as calculated by the $2n^2$ formula.

6.6 ELECTRON ORBITALS

The last and most basic of the three terms used in describing electron arrangements about nuclei is *electron orbital*. An **electron orbital** *is a region of space within an electron subshell where an electron with a specific energy is most likely to be found.*

An analogy for the relationship between shells, subshells, and orbitals can be found in the physical layout of a high-rise condominium complex. A shell is the counterpart of a floor of the condominium in our analogy. Just as each floor will contain apartments (of different sizes), a shell contains subshells (of different sizes). Further, just as apartments contain rooms, subshells contain orbitals. An apartment is a collection of rooms; a subshell is a collection of orbitals. A floor of a condominium building is a collection of apartments; a shell is a collection of subshells.

The characteristics of electron orbitals, the rooms in our "electron apartment house," include the following:

1. The number of orbitals in a subshell varies, being one for an *s* subshell, three for a *p* subshell, five for a *d* subshell, and seven for an *f* subshell.
2. The maximum number of electrons in an orbital does not vary. It is always two.
3. The notation used to designate orbitals is the same as that used for subshells. Thus, orbitals in the 4*f* subshell (there are seven of them) are called 4*f* orbitals.

According to the rules of quantum mechanics (Sec. 6.3), we cannot know the exact position of an electron within an orbital. It is considered to occupy the entire orbital in a manner somewhat analogous to rotating fan blades occupying a defined region of space.

We have already noted (Sec. 6.5) that all electrons in a subshell have the same energy. Thus all electrons in orbitals of the same subshell will have the same energy. This means that *shell and subshell designations are sufficient to specify the energy of an electron.* This statement will be of great importance in the discussions of Section 6.7.

Orbitals have a definite size and shape related to the type of subshell in which they are found. (Remember, an orbital is a region of space. We are not talking about the size and shape of an electron, but rather the size and shape of a region of space where an electron is found.) Typical *s*, *p*, *d*, and *f* orbital shapes are given in Figure 6.8. Notice that the shapes increase in complexity in the order *s*, *p*, *d*, and *f*. Some of the *d* and *f* orbitals have shapes related, but not identical, to those shown in Figure 6.8.

Orbitals within the same subshell differ mainly in orientation. For example, the three 2*p* orbitals look the same but are aligned in different directions—along the *x*, *y*, and *z* axes in a Cartesian coordinate system (see Fig. 6.9).

Orbitals of the same type but in different shells (for example, 1*s*, 2*s*, and 3*s*) have the same general shape but differ in size (volume) (see Fig. 6.10).

FIGURE 6.8 The shapes of atomic orbitals. To improve the perspective, the *f* orbital is shown within a cube with one lobe pointing to each corner of the cube. Only two electrons may occupy an orbital regardless of the number of lobes present in its shape. Some *d* and *f* orbitals have shapes related to, but not identical to, those shown.

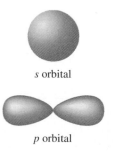

s orbital

p orbital

d orbital

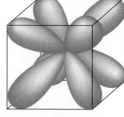

f orbital

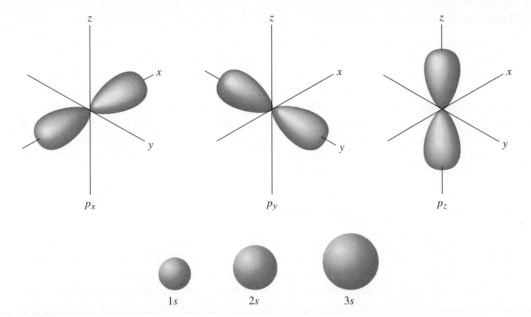

FIGURE 6.9 The orientation of the three orbitals in a p subshell.

p_x p_y p_z

FIGURE 6.10
Size relationships among s orbitals in different shells.

$1s$ $2s$ $3s$

EXAMPLE 6.2 | **Interrelationships among Electron Shells, Electron Subshells, and Electron Orbitals**

Determine the following information about electron shells, electron subshells, and electron orbitals.

a. The number of electron subshells in shell 4
b. The number of electron orbitals in a $3d$ subshell
c. The maximum number of electrons that could be contained in a $2p$ subshell
d. The maximum number of electrons that could be contained in a $2p$ orbital

SOLUTION

a. The number of subshells in a shell is the same as the shell number. Thus, shell 4 will contain four subshells.
b. The number of orbitals in a given type (s, p, d, f) subshell is independent of the shell number. Each s subshell ($1s$, $2s$, $3s$, etc.) contains one orbital, each p subshell contains three orbitals, each d subshell contains five orbitals, and each f subshell contains seven orbitals. Thus, a $3d$ subshell contains five orbitals.
c. The maximum number of electrons in a subshell depends on the number of orbitals the subshell contains, with each orbital holding two electrons. In a p subshell there are three orbitals. Thus, a p subshell can accommodate six electrons (two per orbital).
d. The maximum number of electrons in an orbital is two. It does not matter what type of orbital it is ($2s$, $2p$, $3d$, etc.). All orbitals hold a maximum of two electrons.

▶ **Practice Exercise 6.2**   Determine the following information about electron shells, electron subshells, and electron orbitals.

a. The number of electron subshells in shell 2
b. The number of electron orbitals in the $3s$ subshell
c. The maximum number of electrons that could be contained in a $3d$ subshell
d. The maximum number of electrons that could be contained in a $4p$ orbital

Electron Spin

A final feature of electron behavior is that electrons possess a property called *electron spin*. Electrons exhibit properties (magnetic) that can be explained by using the concept that electrons are spinning on their own axes in either a clockwise or counterclockwise direction. When two electrons are present in an orbital, they always have opposite spins; that is, one is considered to be spinning clockwise and the other, counterclockwise. This situation of opposite spins is the only state possible for two electrons in the same orbital. **Electron spin** *is a property of an electron associated with the concept that an electron is spinning on its own axis.* We will have more to say about electron spin when we discuss orbital diagrams in Section 6.8.

6.7 ELECTRON CONFIGURATIONS

An **electron configuration** *is a statement of how many electrons an atom has in each of its subshells.* Since subshells group electrons according to energy (Sec. 6.5), electron configurations indicate how many electrons an atom has of various energies.

Electron configurations are not written out in words; a shorthand system with symbols is used. Subshells containing electrons, listed in order of increasing energy, are designated using number–letter combinations (1s, 2s, 2p, etc.). A superscript following each subshell designation indicates the number of electrons in that subshell. The electron configuration for oxygen using this shorthand notation is

$$1s^2 2s^2 2p^4 \qquad \text{(read "one-}s\text{-two, two-}s\text{-two, two-}p\text{-four")}$$

An oxygen atom thus has an electron arrangement of two electrons in the 1s subshell, two electrons in the 2s subshell, and four electrons in the 2p subshell.

Aufbau Principle

To determine the electron configuration for an atom, a procedure called the *aufbau principle* (German *aufbauen*, to build) is used. The **aufbau principle** *states that electrons normally occupy electron subshells in an atom in order of increasing subshell energy.* This guideline brings order to what could be a very disorganized situation. Many subshells exist about the nucleus of any given atom. Electrons do not occupy these subshells in a random, haphazard fashion; a very predictable pattern, governed by the aufbau principle, exists for electron subshell occupancy. *Subshells are filled in order of increasing energy.*

Use of the aufbau principle requires knowledge concerning the electron capacities of orbitals and subshells (which we already have; see Sec. 6.6) and knowledge concerning the relative energies of subshells (which we now consider).

The ordering of electron subshells in terms of increasing energy, which is experimentally determined, is more complex than might be expected. This is because the energies of subshells in different shells often overlap, as shown in Figure 6.11. Beginning with shell 4, one or more lower-energy subshells of a specific shell have energies lower than the higher-energy subshells of a preceding shell and thus acquire electrons first. For example, the 4s subshell acquires electrons before the 3d subshell does (see Fig. 6.11). As another example, the s subshell of the sixth energy level fills before the d subshell of the fifth energy level or the f subshell of the fourth energy level (again, refer to Fig. 6.11).

Aufbau Diagram

The sequence in which subshells acquire electrons must be learned before electron configurations can be written. A useful mnemonic (memory) device, called an *aufbau diagram*, helps considerably with this learning process. As can be seen from Figure 6.12, an **aufbau**

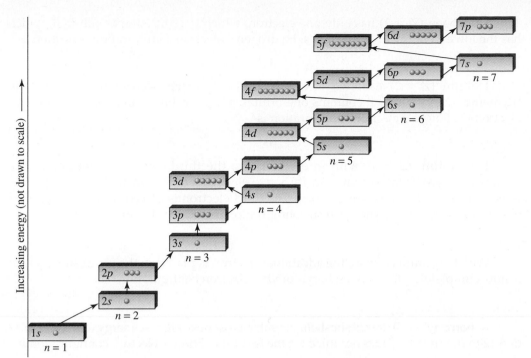

FIGURE 6.11
Relative energies and filling order for electron subshells.

diagram *is a listing of electron subshells in the order in which electrons occupy them.* All *s* sub-shells are located in column 1 of the diagram, all *p* subshells in column 2, and so on. Subshells belonging to the same shell are found in the same row. The order of subshell filling is given by following the diagonal arrows, starting with the top one. The 1*s* subshell fills first. The second arrow points to (goes through) the 2*s* subshell. It fills next. The third arrow points to both the 2*p* and 3*s* subshells. The 2*p* fills first, followed by the 3*s*. Any time a single arrow points to more than one subshell, start at the tail of the arrow and work to its head to determine the proper filling sequence. The 3*p* subshell fills next, and so on. An aufbau diagram is an easy way to catalog the information given in Figure 6.11.

Writing Electron Configurations

We are now ready to write electron configurations. Let us systematically consider electron configurations for the first few elements in the periodic table.

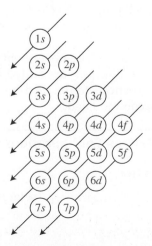

FIGURE 6.12 An aufbau diagram is an aid to remembering subshell filling order.

Hydrogen ($Z = 1$) has only one electron, which goes into the $1s$ subshell, which has the lowest energy of all subshells. Hydrogen's electron configuration is written as

$$1s^1$$

Helium ($Z = 2$) has two electrons, both of which occupy the $1s$ subshell. (Remember, an s subshell contains one orbital, and an orbital can accommodate two electrons.) Helium's electron configuration is

$$1s^2$$

For lithium ($Z = 3$), with three electrons, the third electron cannot enter the $1s$ subshell since its maximum capacity is two electrons. (All s subshells are completely filled with two electrons; see Sec. 6.6.) The third electron is placed in the next-highest energy subshell, the $2s$. The electron configuration for lithium is thus

$$1s^2 2s^1$$

With beryllium ($Z = 4$), the additional electron is placed in the $2s$ subshell, which is now completely filled, giving beryllium the electron configuration

$$1s^2 2s^2$$

In boron ($Z = 5$), the $2p$ subshell, the subshell of next-highest energy (check Fig. 6.11 or 6.12 to be sure), becomes occupied for the first time. Boron's electron configuration is

$$1s^2 2s^2 2p^1$$

An *electron configuration* is a shorthand notation designating the *subshells* in an atom that are occupied by electrons. The sum of the superscripts in an electron configuration equals the total number of electrons present and hence equals the atomic number of the element.

A p subshell can accommodate six electrons since there are three orbitals within it (Sec. 6.6). The $2p$ subshell can thus accommodate the additional electrons found in C, N, O, F, and Ne. The electron configurations of these elements are

$$\text{C } (Z = 6): \quad 1s^2 2s^2 2p^2$$

$$\text{N } (Z = 7): \quad 1s^2 2s^2 2p^3$$

$$\text{O } (Z = 8): \quad 1s^2 2s^2 2p^4$$

$$\text{F } (Z = 9): \quad 1s^2 2s^2 2p^5$$

$$\text{Ne } (Z = 10): \quad 1s^2 2s^2 2p^6$$

With sodium ($Z = 11$), the $3s$ subshell acquires an electron for the first time.

$$\text{Na } (Z = 11): \quad 1s^2 2s^2 2p^6 3s^1$$

Note the pattern that is developing in the electron configurations we have written so far. Each element has an electron configuration the same as the one just before it with the addition of one electron.

Electron configurations for other elements are obtained by simply extending the principles we have just illustrated. Electrons are added to subshells, always filling one of lower energy before adding electrons to the next-highest subshell, until the correct number of electrons has been accommodated.

EXAMPLE 6.3   **Writing Electron Configurations for Atoms**

Write the electron configurations for

    a. calcium ($Z = 20$)   **b.** selenium ($Z = 34$)

SOLUTION

a. The number of electrons in a calcium atom is 20. Remember, the atomic number (Z) gives the number of electrons (Sec. 5.2). We will need to fill subshells, in order of increasing energy, until 20 electrons have been accommodated.

   The 1s, 2s, and 2p subshells fill first, accommodating a total of 10 electrons among them.

$$1s^2 2s^2 2p^6 \ldots$$

Next, according to Figure 6.11 or 6.12, the 3s fills and then the 3p subshell.

$$1s^2 2s^2 2p^6 3s^2 3p^6 \ldots$$

We have accommodated 18 electrons at this point. We will need to add two more electrons to get our desired number of 20.

   These last two electrons are added to the 4s subshell.

$$1s^2 2s^2 2p^6 3s^2 3p^6 4s^2$$

These last two electrons completely fill the 4s subshell.

b. To write the electron configuration for selenium, we continue along the same lines as in part (a), remembering that the maximum electron subshell populations are $s = 2, p = 6,$ and $d = 10$.

   The first 18 electrons, as with calcium [part (a)], will fill the 1s, 2s, 2p, 3s, and 3p subshells.

$$1s^2 2s^2 2p^6 3s^2 3p^6 \ldots$$

The 4s subshell fills next, accommodating 2 electrons, followed by the 3d subshell, which accommodates 10 electrons.

$$1s^2 2s^2 2p^6 3s^2 3p^6 4s^2 3d^{10} \ldots$$

We now have a total of 30 electrons.

   Four more electrons are needed, which are added to the next-higher subshell in energy, the 4p.

$$1s^2 2s^2 2p^6 3s^2 3p^6 4s^2 3d^{10} 4p^4$$

The 4p subshell can accommodate 6 electrons, but we do not want it filled to capacity because that would give us too many electrons.

Answer Double Check:

Does each of the electron configurations contain the correct number of electrons? Yes. Since the atomic number of an element gives the number of electrons present, the sum of the superscripts in an electron configuration should add to the atomic number of the element. Such is the case for the answers to both parts of the problem. For example, in part (a), 2 + 2 + 6 + 2 + 6 + 2 = 20 (atomic number of Ca).

▶ **Practice Exercise 6.3** Write the electron configurations for

a. potassium $(Z = 19)$ b. nickel $(Z = 28)$.

It should be noted that for a few elements in the middle of the periodic table, the actual distribution of electrons within subshells differs slightly from that obtained using the aufbau principle and aufbau diagram. These exceptions are caused by very small energy differences between some subshells and are not important in the uses we shall make of electronic configurations.

Condensed Electron Configurations

The electrons in the outermost electron shell of an atom—the shell with the highest n value—are the electrons primarily responsible for the chemical behavior of an element (Sec. 7.2). It is these outer shell electrons that are "exposed" to other atoms when atoms interact with each other. The electrons in completed inner electron shells normally do not play a major role in determining the chemical behavior of an atom.

An abbreviated, or shorthand, method for writing electron configurations, which focuses on the outer shell electrons present, exists and is in common use. Such abbreviated electron configurations, which are called *condensed electron configurations*, separate the electrons present into two groups—*core electrons* (the inner shell electrons) and *outer electrons* (the outer shell electrons).

To illustrate how *condensed* electron configurations are written, let us consider the elements potassium and calcium. The *full* electron configurations for these two elements are

$$\text{K} \qquad 1s^2 2s^2 2p^6 3s^2 3p^6 4s^1$$

$$\text{Ca} \qquad 1s^2 2s^2 2p^6 3s^2 3p^6 4s^2$$

The outer electron shell for both of these elements is shell 4 (the highest numbered shell). Potassium's outer electron configuration is $4s^1$ and that for calcium is $4s^2$. The inner shell electrons for both potassium and calcium are the same: $1s^2 2s^2 2p^6 3s^2 3p^6$. This inner shell electron arrangement is the same as that for argon, the noble gas (Sec. 6.2) that precedes potassium and calcium in the periodic table.

To write condensed electron configurations for potassium and calcium, the inner core electrons are denoted by placing brackets about the chemical symbol for argon, the noble gas whose electron configuration is the same as that of the inner core electrons. Following this bracketed noble gas (argon) chemical symbol are the remaining electrons in the electron configuration, which are the outer electrons.

Thus the condensed electron configurations are

$$\text{K} \qquad [\text{Ar}]4s^1$$

$$\text{Ca} \qquad [\text{Ar}]4s^2$$

In all condensed electron configurations the chemical symbol of the noble gas is always enclosed in square brackets rather than parentheses. Two additional examples of condensed electron configurations are

$$\text{Zn} \qquad [\text{Ar}]4s^2 3d^{10}$$

$$\text{Sn} \qquad [\text{Kr}]5s^2 4d^{10} 5p^2$$

Thus, a **condensed electron configuration** *is an electron configuration in which the chemical symbol of the nearest noble gas element of lower atomic number is used to represent the electrons in the configuration up to that of the noble gas, and the remaining additional electrons are then appended to the chemical symbol of the noble gas.* The electrons within the condensed electron configuration denoted by the noble gas chemical symbol are called *core electrons* and the remaining electrons *outer electrons.* **Core electrons** *are the inner-shell electrons of an atom that are not normally involved in determining the chemical properties of the* *atom.* The number of core electrons present in an atom is the same as the atomic number of the noble gas whose chemical symbol is present in the condensed electron configuration. **Outer electrons** *are the electrons in a condensed electron configuration given after the noble-gas core electrons.* Among the outer electrons are those that determine the chemical properties of the atom. Example 6.4 gives additional details about the process of converting full electron configurations into condensed electron configurations.

EXAMPLE 6.4   **Writing Condensed Electron Configurations**

Convert each of the following electron configurations into condensed electron configurations.

a. $_7$N: $1s^2 2s^2 2p^3$

b. $_{12}$Mg: $1s^2 2s^2 2p^6 3s^2$

c. $_{34}$Se: $1s^2 2s^2 2p^6 3s^2 3p^6 4s^2 3d^{10} 4p^4$

d. $_{53}$I: $1s^2 2s^2 2p^6 3s^2 3p^6 4s^2 3d^{10} 4p^6 5s^2 4d^{10} 5p^5$

SOLUTION

a.   Helium, with the electron configuration $1s^2$, is the noble gas that precedes nitrogen in the periodic table. Replacing the $1s^2$ portion of nitrogen's electron configuration with the symbol [He] produces the abbreviated electron configuration for nitrogen: [He]$2s^2 2p^3$.

b.   Neon, with the electron configuration $1s^2 2s^2 2p^6$, is the noble gas that precedes magnesium in the periodic table. Using the symbol [Ne] to represent the $1s^2 2s^2 2p^6$ portion of magnesium's electron configuration gives [Ne]$3s^2$ as the condensed electron configuration.

c.   The preceding noble gas for selenium is argon, whose electron configuration is $1s^2 2s^2 2p^6 3s^2 3p^6$. The condensed electron configuration for selenium will, thus, be [Ar]$4s^2 3d^{10} 4p^4$.

d.   Iodine's position in the periodic table is immediately before that of the noble gas xenon (Xe). However, condensed electron configurations are based on the *preceding* noble gas, krypton in this case, rather than on the upcoming noble gas. Iodine's condensed electron configuration is thus based on krypton rather than xenon, and is [Kr]$5s^2 4d^{10} 5p^5$, where [Kr] represents the $1s^2 2s^2 2p^6 3s^2 3p^6 4s^2 3d^{10} 4p^6$ portion of iodine's electron configuration.

Answer Double Check:

The number of electrons to the left of the noble gas notation in a condensed electron configuration is always equal to the difference between the atomic number of the element and that of the noble gas. Such is the case in the answers to this problem. For example, in part (c) the atomic number of selenium is 34, that of argon is 18, and their difference is 16. The notation $4s^2 3d^{10} 4p^4$ involves 16 electrons.

▶ **Practice Exercise 6.4**   Convert each of the following electron configurations into condensed electron configurations.

a. $_7$F: $1s^2 2s^2 2p^5$

b. $_{14}$Si: $1s^2 2s^2 2p^6 3s^2 3p^2$

c. $_{20}$Ca: $1s^2 2s^2 2p^6 3s^2 3p^6 4s^2$

d. $_{48}$Cd: $1s^2 2s^2 2p^6 3s^2 3p^6 4s^2 3d^{10} 4p^6 5s^2 4d^{10}$

6.8 ORBITAL DIAGRAMS

The arrangement of electrons about a nucleus can be specified in terms of *subshell* occupancy or *orbital* occupancy. The notation for specifying subshell occupancy (electron configurations) was considered in the previous section. In this section we consider electron arrangements at the orbital level.

Two principles are needed to specify orbital occupancy for electrons, one we have already considered and one that is new. The principles are

1. The aufbau principle
2. Hund's rule

The aufbau principle was used extensively in Section 6.7 in writing electron configurations. Hund's rule has not been encountered previously. The namesake for Hund's rule is the German physicist Friedrich Hund (1896–1997), an early worker in the field of quantum mechanics.

Hund's rule *states that when electrons are placed in a set of orbitals of equal energy (the orbitals of a subshell), the order of filling for the orbitals is such that each orbital of the subshell receives an electron with the same spin before any orbital receives a second electron (of the opposite spin).* Such a pattern of orbital filling minimizes repulsions between electrons. Numerous applications of Hund's rule are required in stating electron arrangements in terms of orbital occupancies, that is, in drawing *orbital diagrams*.

An **orbital diagram** *is a diagram that shows how many electrons an atom has in each of its occupied electron orbitals.* In drawing orbital diagrams, we use circles to represent orbitals and arrows with a single barb to denote electrons. The orbital diagram for hydrogen, with its one electron, is

An *electron configuration* specifics *subshell* occupancy for electrons, and an *orbital diagram* specifies *orbital* occupancy for electrons.

$$\text{H:} \quad \textcircled{\uparrow}$$
$$1s$$

A helium atom contains two electrons, both of which occupy the $1s$ orbital. The orbital diagram for helium is

$$\text{He:} \quad \textcircled{\uparrow\downarrow}$$
$$1s$$

The two electrons present are of opposite spin (see Sec. 6.6). Note the notation used to indicate this—one arrow points up and the other points down. The two electrons are said to be *paired*. **Paired electrons** *are two electrons of opposite spin present in the same orbital.* When only one electron is present in an orbital, it is said to be *unpaired*. An **unpaired electron** *is a single electron in an orbital.* The electron configuration for hydrogen involves an unpaired electron.

Orbital diagrams for the next two elements, lithium and beryllium, are drawn according to reasoning similar to that followed for H and He. The two electrons in the $2s$ orbital of Be are paired.

$$\text{Li:} \quad \textcircled{\uparrow\downarrow} \quad \textcircled{\uparrow}$$
$$\quad 1s \qquad 2s$$

$$\text{Be:} \quad \textcircled{\uparrow\downarrow} \quad \textcircled{\uparrow\downarrow}$$
$$\quad 1s \qquad 2s$$

Boron has the electron configuration $1s^22s^22p^1$. The fifth electron in boron must enter a $2p$ orbital, since both the $1s$ and $2s$ orbitals are full. Boron's orbital diagram is

B: (1↓) (1↓) (1)( )( )
     1s   2s   2p

Note that it does not matter which one of the three $2p$ orbitals of boron is shown as containing an electron. All three orbitals have the same energy. Each of the following notations is equivalent.

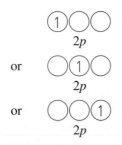

It does not matter whether the arrows denoting unpaired electrons are pointing up or down. Up and down are equivalent positions. What does matter is that all of the arrows for unpaired electrons are pointing in the same direction, either all up or all down.

With carbon, element 6, we encounter the use of Hund's rule for the first time. Carbon has two electrons in the $2p$ subshell ($1s^22s^22p^2$). Do the two electrons go into the same orbital (paired), or do they go into separate equivalent orbitals (unpaired)? Hund's rule indicates that the latter is the case. The orbital diagram for carbon is

C: (1↓) (1↓) (1)(1)( )
     1s   2s   2p

Again, it does not matter which two of the three $2p$ orbitals are shown as containing electrons since all three orbitals have the same energy. However, the two $2p$ electrons must have the same spin. It will always be the case, for energy reasons, that unpaired electrons in orbitals of equal energy will have the same spin.

Nitrogen, with the electron configuration $1s^22s^22p^3$, contains three unpaired electrons, all with the same spin.

N: (1↓) (1↓) (1)(1)(1)
     1s   2s   2p

With oxygen ($1s^22s^22p^4$), two of the four $2p$ electrons must pair up, leaving two unpaired electrons. Fluorine, the next element in the periodic table, has only one unpaired electron. Finally, with neon, all of the electrons are paired up.

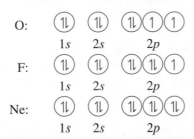

O: (1↓) (1↓) (1↓)(1)(1)
     1s   2s   2p

F: (1↓) (1↓) (1↓)(1↓)(1)
     1s   2s   2p

Ne: (1↓) (1↓) (1↓)(1↓)(1↓)
     1s   2s   2p

EXAMPLE 6.5 **Writing Orbital Diagrams for the Electrons in Atoms**

Draw an orbital diagram for the electrons in the element vanadium. The electron configuration for vanadium is

$$1s^2 2s^2 2p^6 3s^2 3p^6 4s^2 3d^3$$

SOLUTION

All the occupied subshells of vanadium are completely filled except for the $3d$. There are five orbitals in the $3d$ subshell, and there are three electrons present. Following Hund's rule, we will put one electron in each of three $3d$ orbitals, giving us three unpaired electrons. The three unpaired electrons will have the same spin.

Answer Double Check:

Do all of the unpaired electrons have the same spin? Yes. The arrows for the three unpaired electrons are all pointing in the same direction. It does not matter whether they are all pointing up or all pointing down, but they must all be pointing in the same direction, indicating that they all have the same spin.

▶ **Practice Exercise 6.5** Draw an orbital diagram for the electrons in the element silicon. The electron configuration for silicon is

$$1s^2 2s^2 2p^6 3s^2 3p^2$$

Atoms may be classified as *paramagnetic* or *diamagnetic* on the basis of unpaired electrons. A **paramagnetic atom** *is an atom that has an electron arrangement containing one or more unpaired electrons.* The presence of unpaired electrons causes paramagnetic substances to be slightly attracted to a magnet. Measurement of paramagnetism provides experimental verification of the presence of unpaired electrons. A **diamagnetic atom** *is an atom that has an electron arrangement in which all electrons are paired.*

6.9 ELECTRON CONFIGURATIONS AND THE PERIODIC LAW

A knowledge of electron configurations for the elements provides an explanation for the periodic law. Recall, from Section 6.1, that the periodic law points out that the properties of the elements repeat themselves in a regular manner when the elements are ordered in sequence of increasing atomic number. Those elements with similar chemical properties are placed one under another in vertical columns (groups) in a periodic table.

Groups of elements have similar chemical properties because of similarities that exist in the electron configurations of the elements of the group. *Chemical properties repeat themselves in a regular manner among the elements because electron configurations repeat themselves in a regular manner among the elements.*

To illustrate this correlation between similar chemical properties and similar electron configurations, let us look at the electron configurations of two groups of elements known to have similar chemical properties.

We begin with the elements lithium, sodium, potassium, and rubidium—all members of group IA of the periodic table. The electron configurations for these elements are

$$_3\text{Li:}\quad 1s^2\,\widehat{2s^1}$$
$$_{11}\text{Na:}\quad 1s^2 2s^2 2p^6\,\widehat{3s^1}$$
$$_{19}\text{K:}\quad 1s^2 2s^2 2p^6 3s^2 3p^6\,\widehat{4s^1}$$
$$_{37}\text{Rb:}\quad 1s^2 2s^2 2p^6 3s^2 3p^6 4s^2 3d^{10} 4p^6\,\widehat{5s^1}$$

We see that each of these elements has one outer s electron (shown in color), the last electron added by the aufbau principle. It is this similarity in outer shell electron arrangements that causes these elements to have similar chemical properties.

The elements fluorine, chlorine, bromine, and iodine of group VIIA of the periodic table have similar chemical properties. The electron configurations for these four elements are

$$_9\text{F:}\quad 1s^2\,\widehat{2s^2\,2p^5}$$
$$_{17}\text{Cl:}\quad 1s^2 2s^2 2p^6\,\widehat{3s^2\,3p^5}$$
$$_{35}\text{Br:}\quad 1s^2 2s^2 2p^6 3s^2 3p^6\,\widehat{4s^2}\,3d^{10}\,\widehat{4p^5}$$
$$_{53}\text{I:}\quad 1s^2 2s^2 2p^6 3s^2 3p^6 4s^2 3d^{10} 4p^6\,\widehat{5s^2}\,4d^{10}\,\widehat{5p^5}$$

Once again, similarities in electron configurations are readily apparent. This time the repeating pattern is the seven electrons (in color) in the outermost s and p subshells.

Section 7.2 will consider in depth the fact that the electrons most important in controlling chemical properties are those found in the outermost shell of an atom.

6.10 ELECTRON CONFIGURATIONS AND THE PERIODIC TABLE

One of the strongest pieces of supporting evidence for the assignment of electrons to shells, subshells, and orbitals is the periodic table itself. The basic shape and structure of this table, which were determined many years before electrons were even discovered, are consistent with and can be explained by electron configurations. Indeed, the specific location of an element in the periodic table can be used to obtain information about its electron configuration.

As the first step in linking electron configurations to the periodic table, let us analyze the general shape of the periodic table in terms of columns of elements. As shown in Figure 6.13, we have on the extreme left of the table 2 columns of elements, in the center an area containing 10 columns of elements, to the right a block of 6 columns of elements, and at the bottom of the table in two rows, 14 columns of elements.

The numbers of columns of elements in the various regions of the periodic table—2, 6, 10, and 14—is the same as the maximum numbers of electrons that the various types of subshells can accommodate. We will see shortly that this is a very significant observation; the number matchup is no coincidence. The various columnar regions of the periodic table are called the s area (2 columns), the p area (6 columns), the d area (10 columns), and the f area (14 columns), as shown in Figure 6.14.

The concept of *distinguishing electrons* is the key to obtaining electron configuration information from the periodic table. For an element, the **distinguishing electron** *is the last electron added to the element's electron configuration when the configuration is written according to the aufbau principle*. This last electron added is the one that causes

FIGURE 6.13 The periodic table can be divided into four areas—areas that are 2, 6, 10, and 14 columns wide.

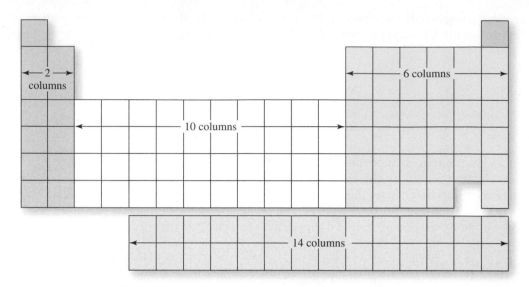

an element's electron configuration to differ from that of the element immediately preceding it in the periodic table, hence the term *distinguishing electron*.

For all elements located in the *s* area of the periodic table the distinguishing electron is always found in an *s* subshell. All *p* area elements have distinguishing electrons in *p* subshells. Similarly, elements in the *d* and *f* areas of the periodic table have, respectively, distinguishing electrons located in *d* and *f* subshells. Thus, the area location of an element in the periodic table can be used to determine the type of subshell that contains the distinguishing electron. Note that the element helium is considered to belong to the *s* rather than the *p* area of the periodic table even though its table position is on the right-hand side. (The reason for this placement of helium will be explained in Section 7.3.)

The extent of filling of the subshell containing an element's distinguishing electron can also be determined from the element's position in the periodic table. All elements in the first column of a specific area contain only one electron in the subshell, all elements in the second column contain two electrons in the subshell, and so on. Thus all elements in the first column of the *p* area (group IIIA) have an electron configuration ending in p^1. Elements in the second column of the *p* area (group IVA) have

FIGURE 6.14 Within the four periodic table areas are elements whose distinguishing electron is located, respectively, in *s, p, d,* and *f* subshells.

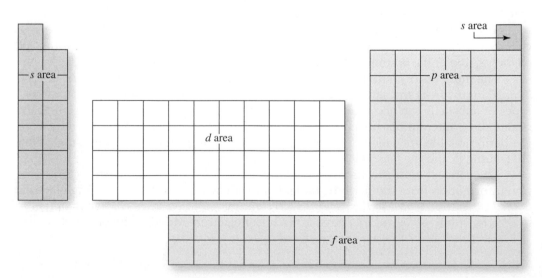

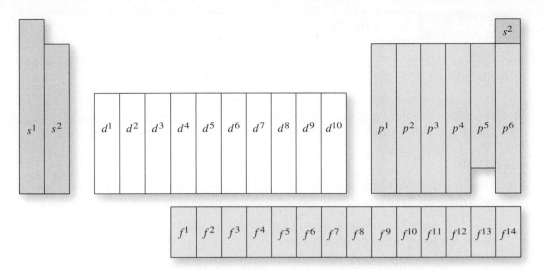

FIGURE 6.15
The extent of filling of the subshell containing the distinguishing electron can be determined from the element's position within a given area.

electron configurations ending in p^2, and so forth. Similar relationships hold in other areas of the table, as shown in Figure 6.15. A few exceptions to these generalizations do exist in the d and f areas (because of irregular electron configurations; see Sec. 6.7), but we will not be concerned with them in this text.

We can also use the periodic table to determine the shell in which the distinguishing electron is located. The relationship used involves the number of the period in which the element is found. In the s and p areas, the period number gives the shell number directly. In the d area, the period number minus one is equal to the shell number. (Remember that the $3d$ subshell is filled during the fourth period.) For similar reasons, in the f area the period number minus two equals the shell number. Thus, the subshell that contains the distinguishing electron for elements of period 6 may be the $6s$, $6p$, $5d$ (period number minus one) or $4f$ (period number minus two), depending on the location of the element in period 6. It must be remembered that even though the f area is located at the bottom of the table, it correctly belongs in periods 6 and 7. The complete matchup between period number and shell number for the distinguishing electron is given in the periodic table of Figure 6.16.

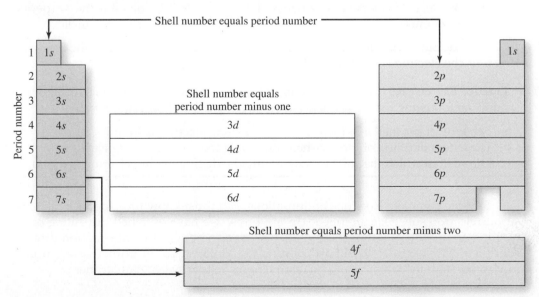

FIGURE 6.16
Relationship between the shell number for a distinguishing electron and the period number of the element's periodic table position.

EXAMPLE 6.6 | **Using the Periodic Table as a Guide in Obtaining Information about Distinguishing Electrons**

Using the periodic table and Figures 6.13 through 6.16, determine the following for the elements calcium ($Z = 20$), vanadium ($Z = 23$), and tellurium ($Z = 52$).

  a. The type of subshell in which the distinguishing electron is found
  b. The extent of filling of the subshell containing the distinguishing electron
  c. The shell in which the subshell containing the distinguishing electron is found

SOLUTION

  a. Knowing the area of the periodic table in which an element is found is sufficient to determine the type of subshell in which the distinguishing electron is found.

   Ca: Since this element is found in the s area of the periodic table, the distinguishing electron will be in an s subshell.

   V: Since this element is found in the d area of the periodic table, the distinguishing electron will be in a d subshell.

   Te: Since this element is found in the p area of the periodic table, the distinguishing electron will be in a p subshell.

  b. The extent of filling of the subshell containing the distinguishing electron is determined by noting the column in the area that the element occupies.

   Ca: Since this element is in the second column of the s area, the s subshell involved contains two electrons (s^2).

   V: Since this element is in the third column of the d area, the d subshell involved contains three electrons (d^3).

   Te: Since this element is in the fourth column of the p area, the p subshell involved contains four electrons (p^4).

  c. The shell number of the subshell containing the distinguishing electron is obtained from the period number, sometimes directly and sometimes with modifications.

   Ca: Since this element is in period 4, the s subshell involved is the $4s$; therefore, the electron configuration for Ca ends in $4s^2$.

   V: Since this element is in period 4, the d subshell involved is the $3d$ (period number minus one); therefore, the electron configuration for V ends in $3d^3$.

   Te: Since this element is in period 5, the p subshell involved is the $5p$; therefore, the electron configuration for Te ends in $5p^4$.

▶ **Practice Exercise 6.6**   Using the periodic table and Figures 6.13 through 6.16, determine the following for the elements krypton ($Z = 36$) and barium ($Z = 56$).

  a. The type of subshell in which the distinguishing electron is found
  b. The extent of filling of the subshell containing the distinguishing electron
  c. The shell in which the subshell containing the distinguishing electron is found

---

To write complete electron configurations, you must know the order in which the various electron subshells are filled. Up until now, we have obtained this filling order by using an aufbau diagram (Fig. 6.12). We can also obtain this information directly from the periodic table. To obtain it, we merely follow a path of increasing atomic number through the table, noting the various subshells as we encounter them.

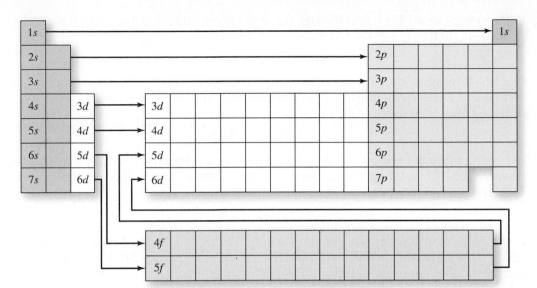

1. Begin with hydrogen and helium (the $1s$ elements, period 1).
2. Continue through the elements in the other periods, in order of increasing atomic number.
3. As you move across a period
   a. add electrons to the ns subshell as you pass through the s area.
   b. add electrons to the np subshell as you pass through the p area.
   c. add electrons to the $(n - 1)d$ subshell as you pass through the d area.
   d. add electrons to the $(n - 2)f$ subshell as you pass through the f area.
4. Continue until you reach the element whose electron configuration you are writing.

Figure 6.17 illustrates this way of using the periodic table. "Working our way" through successive periods of the periodic table (by following the arrows) gives the entire filling order for the electron subshells.

$$1s \rightarrow 2s \rightarrow 2p \rightarrow 3s \rightarrow 3p \rightarrow 4s \rightarrow 3d \rightarrow 4p \rightarrow 5s \rightarrow 4d \rightarrow$$

$$5p \rightarrow 6s \rightarrow 4f \rightarrow 5d \rightarrow 6p \rightarrow 7s \rightarrow 5f \rightarrow 6d \rightarrow 7p$$

Note particularly how the f area of the periodic table is worked into the filling scheme.

EXAMPLE 6.7 | **Using the Periodic Table as a Guide in Writing an Electron Configuration**

Write the complete electron configuration of $_{86}$Rn, using the periodic table as your guide.

SOLUTION

The needed information will be obtained by working our way through the periodic table. We will start at hydrogen and go from element to element in order of increasing atomic number until we arrive at position 86, radon. Every time we traverse the s area we will fill an s subshell; the p area, a p subshell; and so on. We will remember that the shell number for s and p subshells is the period number; for d subshells, the period number minus one; and for f subshells, the period number minus two.

Let us begin our journey through the periodic table. As we cross period 1, we encounter H and He, both $1s$ elements. We thus add the $1s$ electrons.

$$\text{Rn:} \quad 1s^2 \ldots$$

In traversing period 2, we pass through the s area (elements 3 and 4) and the p area (elements 5–10) in that order. We add $2s$ and $2p$ electrons.

$$\text{Rn:} \qquad 1s^2 2s^2 2p^6 \ldots$$

Our trip through period 3 is very similar to that of period 2. Only s-area elements (11 and 12) and p-area elements (13–18) are encountered. The message is to add s and p electrons—$3s$ and $3p$ because we are in period 3.

$$\text{Rn:} \qquad 1s^2 2s^2 2p^6 3s^2 3p^6 \ldots$$

In passing through period 4, we go through the s area (elements 19 and 20), the d area (elements 21–30), and the p area (elements 31–36). Electrons to be added are those in the $4s$, $3d$ (period number minus one), and $4p$ subshells, in that order.

$$\text{Rn:} \qquad 1s^2 2s^2 2p^6 3s^2 3p^6 4s^2 3d^{10} 4p^6 \ldots$$

In period 5 we encounter scenery similar to that in period 4—the s area (elements 37 and 38), the d area (elements 39–48), and the p area (elements 49–54). Hence, the $5s$, $4d$ (period number minus one), and $5p$ subshells are filled in order.

$$\text{Rn:} \qquad 1s^2 2s^2 2p^6 3s^2 3p^6 4s^2 3d^{10} 4p^6 5s^2 4d^{10} 5p^6 \ldots$$

Our journey ends in period 6, which we completely traverse. We go through the $6s$ area (elements 55 and 56), the $4f$ area (elements 58–71), the $5d$ area (elements 57 and 72–80), and finally the $6p$ area (elements 81–86). We will completely fill the $6p$ subshell since radon is in the last column of the p area.

$$\text{Rn:} \qquad 1s^2 2s^2 2p^6 3s^2 3p^6 4s^2 3d^{10} 4p^6 5s^2 4d^{10} 5p^6 6s^2 4f^{14} 5d^{10} 6p^6$$

Answer Double Check:

As a double-check that one or more subshells have not inadvertently been skipped, sum the superscripts to obtain the total number of electrons present. The total should match the atomic number for the element. Our answer is correct. The sum of the superscripts equals 86, the atomic number for radon.

▶ **Practice Exercise 6.7**   Write the complete electron configuration for $_{50}$Sn, using the periodic table as your guide.

Again, let us mention that a few slightly irregular electronic configurations are encountered in the d and f areas of the periodic table. The generalizations in this section do not address this problem. We will be working mostly with s- and p-area elements in future chapters. There are no irregularities in electron configurations for these elements.

6.11 CLASSIFICATION SYSTEMS FOR THE ELEMENTS

The elements can be classified in several ways. The two most common classification systems are the following:

1. A system based on selected physical properties of the elements, in which elements are described as *metals* or *nonmetals*
2. A system based on the electron configurations of the elements, in which elements are described as *noble gas, representative, transition,* or *inner transition* elements

On the basis of selected physical properties of the elements, the first of the two classification schemes divides the elements into the categories of metals and nonmetals. A **metal** *is an element that has the characteristic properties of luster, thermal conductivity, electrical conductivity, and malleability.* With the exception of mercury, all metals are solids at room

The electron configurations of the noble gases will be an important focal point when we consider chemical bonding in Chapter 7.

temperature (25°C). Metals are good conductors of heat and electricity. Most metals are ductile (they can be drawn into wires) and malleable (they can be rolled into sheets). Most metals have high luster (shine), high density, and high melting points. Among the more familiar metals are the elements iron, aluminum, copper, silver, and gold.

A **nonmetal** *is an element characterized by the absence of the properties of luster, thermal conductivity, electrical conductivity, and malleability.* Many of the nonmetals, such as hydrogen, oxygen, nitrogen, and the noble gases, are gases at room temperature (25°C). The only nonmetal that is a liquid at room temperature is bromine. Solid nonmetals include carbon, sulfur, and phosphorus. In general, the nonmetals have lower densities and melting points than metals.

The majority of the elements are metals: Only 23 elements are nonmetals, the rest (94) are metals. It is not necessary to memorize which elements are nonmetals and which are metals. As can be seen from Figure 6.18, the location of an element in the peri-odic table correlates directly with its classification as a metal or nonmetal. The steplike heavy line that runs through the *p* area of the periodic table separates the metals from the nonmetals; metals are on the left and nonmetals on the right. Note that the element hydrogen is a nonmetal even though it is located on the left side of the periodic table.

The fact that the vast majority of elements are metals in no way indicates that metals are more important than nonmetals. Most nonmetals are relatively abundant and are found in many important compounds. For example, water (H_2O) is a compound involving two nonmetals. An analysis of the previously given abundances of the elements in Earth's crust (Fig. 4.13) in terms of metals and nonmetals shows that three of the four most abundant elements, which account for 83% of all atoms, are nonmetals—oxygen, silicon, and hydrogen.

The classification scheme based on electron configurations of the elements is depicted in Figure 6.19. This type of classification system is used in numerous discus-sions in subsequent chapters.

A **noble gas element** *is an element located in the far right column (group VIIIA) of the periodic table.* Such elements are all gases at room temperature, and they have little tendency to form chemical compounds. With one exception, the distinguishing electron for a noble gas completes the *p* subshell. Therefore, they have electron configurations ending in p^6. The exception is helium, in which the distinguishing electron completes the first shell—a shell that has only two electrons. Helium's electron configuration is $1s^2$.

A **representative element** *is an element located in the s area or the first five columns of the p area of the periodic table.* The distinguishing electron for these elements partially or completely fills an *s* subshell or partially fills a *p* subshell. Some representative elements are nonmetals while others are metals. The four most abundant elements in the human body (Sec. 4.8)—hydrogen, oxygen, carbon, and nitrogen—are all nonmetallic represen-tative elements; they constitute more than 99% of all atoms in the body.

With two exceptions, all metals are silvery-white to silvery-gray to dull gray in color. The two exceptions are copper and gold.

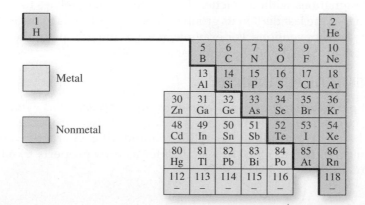

FIGURE 6.18 A portion of the periodic table, showing the dividing line between metals and nonmetals. All elements that are not shown are metals.

FIGURE 6.19
Elemental classification
scheme based on the
electron configurations
of the elements.

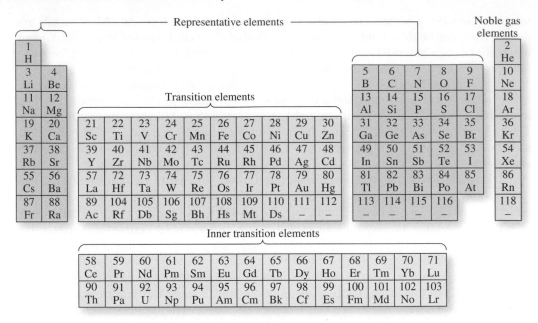

A **transition element** *is an element located in the* d *area of the periodic table.* The common feature in the electronic configurations of the transition elements is the presence of the distinguishing electron in a *d* subshell. All of the transition elements are metals. The most abundant transition element in the human body is iron.

An **inner-transition element** *is an element located in the* f *area of the periodic table.* The characteristic feature of the electronic configuration for such elements is the presence of the distinguishing electron in an *f* subshell. All of the inner-transition elements are metals. Many of them are laboratory-produced radioactive elements rather than naturally occurring elements (Sec. 5.10).

In Section 6.2 it was noted that three different conventions exist for designating the groups of the periodic table—an American convention, a European convention, and an IUPAC system. The basis for the American convention can now be addressed since it relates to the element classification systems just considered.

As shown in Figure 6.20a, the American group–numbering system assigns A group numbers to the representative elements and B group numbers to the transition elements.

As shown in Figure 6.20b, the European group–numbering system is not based on an element classification system. It has both representative and transition elements in both the A and B group designations. In this convention, the left side of the periodic table constitutes the A groups and the right side, the B groups.

The IUPAC convention, Figure 6.20c, avoids the A–B situation by using simple number designations without a letter. The IUPAC system relates to the other two conventions in that the last digit in its group number corresponds to the Roman numeral in these latter two conventions. For example, the 5 in the group number 15 matches with the American group VA and the European group VB.

6.12 CHEMICAL PERIODICITY

Chemical periodicity *is the variation in properties of elements as a function of their positions in the periodic table.* In this section we consider two properties of elements that exhibit chemical periodicity, metallic–nonmetallic character, and atomic size (atomic radius). In Section 7.10 we consider a third property exhibiting chemical periodicity—electronegativity.

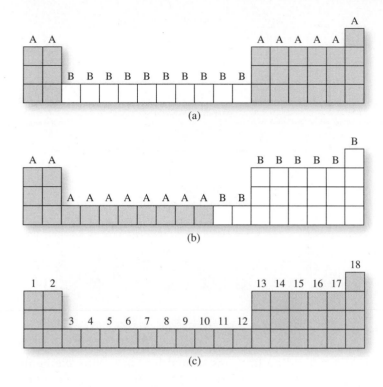

FIGURE 6.20
(a) American group–numbering convention. Representative elements are placed in A groups and transition elements in B groups. Roman numerals are used to distinguish among the A groups and among the B groups. **(b) European group–numbering convention.** Elements on the left side of the periodic table are placed in A groups, and elements on the right side are placed in B groups. Roman numerals are used to distinguish among the A groups and among the B groups. **(c) IUPAC group–numbering convention.** Groups of elements are numbered using a sequential numbering system that starts at the left side of the periodic table. The letters A and B are not part of the convention. Roman numerals are not part of this convention.

Metallic and Nonmetallic Character

The physical properties that distinguish metals and nonmetals were considered in Section 6.11. In general, these properties are opposites for the two classes of substances.

Not all metals possess metallic properties to the same extent; they have them, but to varying degrees. For example, some metals are better conductors of electricity than other metals. Similarly, not all nonmetals possess nonmetallic properties to the same extent.

Metallic and nonmetallic character for the various elements—the extent to which they possess metallic and nonmetallic properties—can be correlated with periodic table position for the elements:

1. Metallic character increases from right to left within a given period in the periodic table.
2. Metallic character increases from top to bottom within a group in the periodic table.

Thus, among the period 3 elements, Na is more metallic than Mg, which in turn is more metallic than Al. For the group IA elements, K is more metallic than Na, since it is farther down in group IA.

Similar, but opposite, trends exist for nonmetallic character:

1. Nonmetallic character increases from left to right within a given period in the periodic table.
2. Nonmetallic character increases from bottom to top within a group in the periodic table.

Figure 6.21 summarizes the chemical periodicity associated with the properties of metallic and nonmetallic character. The combined effect of these two trends is that the most nonmetallic elements are those in the upper right portion of the periodic table, a generalization consistent with Figure 6.18 in the preceding section of this chapter.

FIGURE 6.21
Chemical periodicity is associated with the properties of metallic character and nonmetallic character.

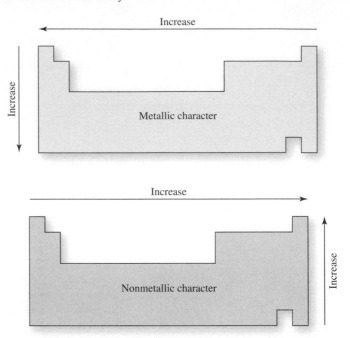

In Section 6.11 a dividing line was shown for classifying elements as metals or nonmetals (Fig. 6.18). Elements to the far left of this line have dominant metallic character. Elements to the far right of this line have dominant nonmetallic character. Most elements whose periodic table positions touch this metal–nonmetal dividing line actually show some properties that are characteristic of both metals and nonmetals. These elements are frequently given a classification of their own: *metalloids* (or semimetals). A **metalloid** *is an element with properties intermediate between those of metals and nonmetals.* Figure 6.22 identifies the metalloid elements. Several of the metalloids, including silicon, germanium, and antimony, are *semiconductors*. A **semiconductor** *is an element that does not conduct electrical current at room temperature but does so at higher temperatures.* Semiconductor elements are very important in the electronics industry.

Atomic Size

The quantum-mechanical model for an atom (Sec. 6.3) suggests that the shape of an atom is approximately spherical and that the size of the atom can be expressed in terms of the radius of a sphere. The most commonly used unit for expressing atomic sizes (atomic radii) is the picometer (Sec. 3.1). In this unit most atomic radii fall in the range of 50–200 pm.

FIGURE 6.22 A portion of the periodic table showing the metalloids (color), elements that show some properties characteristic of metals and some other properties characteristic of nonmetals.

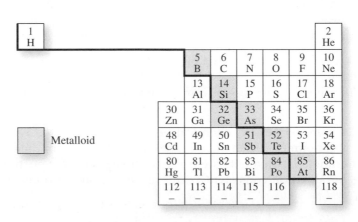

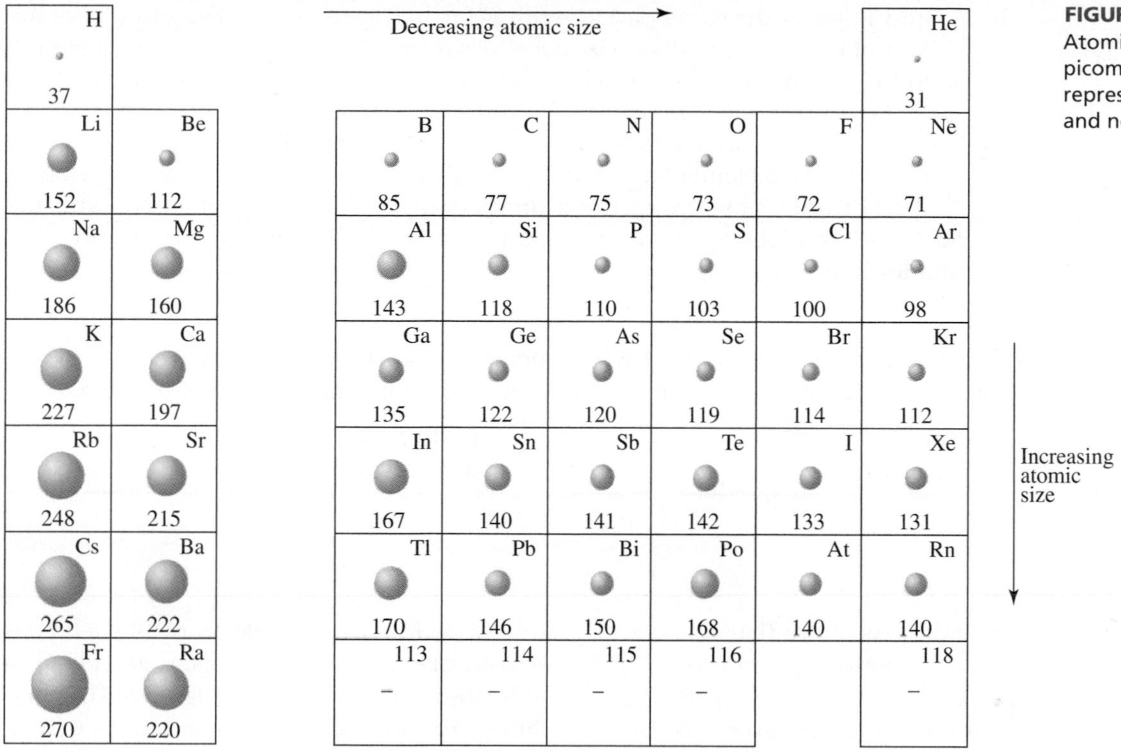

FIGURE 6.23
Atomic radii, in picometers, of the representative elements and noble gases.

Atomic sizes exhibit chemical periodicity, as is shown in Figure 6.23, which gives atomic radii for the representative elements and the noble gases. General periodic trends for atomic radii are the following:

1. Atomic radii tend to decrease from left to right within a period of the periodic table.
2. Atomic radii tend to increase from top to bottom within a periodic table group.

The trends in atomic radii values can be explained on the basis of the number of electrons present in the atom and their energies. In traversing a period in the periodic table (from left to right), we note that (1) nuclear charge (atomic number) increases and (2) the added electrons enter the same shell, the outermost shell. The increased nuclear charge draws the electrons in this outermost shell closer to the nucleus, resulting in smaller atomic radii. Down a group in the periodic table, atomic radii increase because the electrons are added to shells with higher numbers. Recall, from Section 6.4, that the higher the shell number, the greater the average distances between the electrons and the nucleus.

EXAMPLE 6.8   **Predicting Atomic Size Relationships, Using General Periodic Table Trends**

Predict which member of each of the following pairs of elements has the smaller atomic radius, based on periodic table atomic size trends.

a. Mg or Be   **c.** N or O   **b.** Al or P   **d.** Na or F

SOLUTION

In terms of periodic table trends, atomic size decreases in going from left to right across a period and increases in going down a group.

a. Mg and Be are in the same group, with Be above Mg. Therefore, Be is the smaller of the two elements as size increases going down a group.

    b. Al and P are in the same period, with P farther to the right. Therefore, P is the smaller of the two elements as size decreases from left to right going across a period.

    c. N and O are in the same period, with O farther to the right. Therefore, O is the smaller of the two elements as size decreases from left to right going across a period.

    d. Na is in period 3 and F is in period 2. The element Li can be used as a link between the two elements; it is in the same group as Na and the same period as F. The periodic table group trend predicts that Li is smaller than Na, and the periodic table period trend predicts that F is smaller than Li. Therefore, F is smaller than Na.

▶ **Practice Exercise 6.8**   Predict which member of each of the following pairs of elements has the smaller atomic radius, based on periodic table atomic size trends.

    a. S or Ar   **b.** Li or C

    c. Ne or Kr   **d.** Na or Be

Summary

1. **Periodic Law** The periodic law states that certain sets of physical and chemical properties occur at periodic (regularly occurring) intervals when the elements are arranged in order of increasing atomic number.

2. **Periodic Table** The periodic table is a tabular arrangement of the elements, in order of increasing atomic number, that places elements with similar properties in the same vertical columns of the table. It is a graphical portrayal of the periodic law. A group in the periodic table is a vertical column of elements. Roman numerals are used to distinguish the various periodic table groups. A period in the periodic table is a horizontal row of elements. Regular numbers are used to distinguish the various periodic table periods.

3. **Electron Shells** An electron shell is a region of space about a nucleus that contains electrons that have approximately the same energy and that spend most of their time approximately the same distance from the nucleus. The electron capacity of an electron shell is $2n^2$, where n is the shell number.

4. **Electron Subshells** An electron subshell is a region of space within an electron shell that contains electrons that have the same energy. The number of electron subshells in a particular electron shell is equal to the shell number. Each subshell can hold a specific maximum number of electrons. These values are 2, 6, 10, and 14 for $s, p, d,$ and f subshells, respectively.

5. **Electron Orbitals** An electron orbital is a region of space within an electron subshell where an electron with a specific energy is most likely to be found. Each electron subshell contains one or more orbitals. For $s, p, d,$ and f subshells there are one, three, five, and seven orbitals, respectively. No more than two electrons may occupy an orbital.

6. **Electron Configuration** An electron configuration is a statement of how many electrons an atom has in each of its electron subshells. The determination of an electron configuration is based on the aufbau principle, which is that electrons normally occupy the lowest-energy subshell available for occupancy.

7. **Orbital Diagram** An orbital diagram is a statement of how many electrons an atom has in each of its orbitals. Electrons occupy the orbitals of a subshell such that each orbital within the subshell acquires one electron before any orbital acquires a second electron. All electrons in such singly occupied orbitals have the same spin.

8. **Electron Configurations and the Periodic Law** Chemical properties repeat themselves in a regular manner among the elements because electron configurations repeat themselves in a regular manner among the elements.

9. **Electron Configurations and the Periodic Table** The groups of the periodic table contain elements with similar electron configurations. Thus, the location of an element in the periodic table can be used to obtain information about its electron configuration.

10. **Classification Systems for the Elements** On the basis of electron configuration, elements are classified into four categories: noble gases (far right

column of the periodic table), representative elements (*s* and *p* area of the periodic table, with the exception of the last *p*-area column), transition elements (*d* area of the periodic table), and inner transition elements (*f* area of the periodic table). On the basis of selected physical properties, elements are classified as metals or nonmetals. Metals exhibit luster, thermal conductivity, electrical conductivity, and malleability. Nonmetals are characterized by the absence of the properties associated with metals. The steplike heavy line that runs through the right third of the periodic table separates the metals on the left from the nonmetals on the right. The majority of elements are metals.

11. **Chemical Periodicity** Metallic and nonmetallic character of the elements can be correlated with periodic table position for the elements. Metallic character increases from right to left within a given period and increases from top to bottom within a group in the periodic table. Atom size (atomic radii) can also be correlated with periodic table position for the elements. Atomic radii tend to decrease from left to right within a period and increase from top to bottom within a periodic table group.

Key Terms

The new terms defined in this chapter are

alkali metal *Sec. 6.2*
alkaline earth metal *Sec. 6.2*
aufbau diagram *Sec. 6.7*
aufbau principle *Sec. 6.7*
chemical periodicity *Sec. 6.12*
condensed electron configuration *Sec. 6.7*
core electrons *Sec. 6.7*
diamagnetic atom *Sec. 6.8*

distinguishing electron *Sec. 6.10*
electron configuration *Sec. 6.7*
electron orbital *Sec. 6.6*
electron shell *Sec. 6.4*
electron spin *Sec. 6.6*
electron subshell *Sec. 6.5*
group *Sec. 6.2*
halogen *Sec. 6.2*
Hund's rule *Sec. 6.8*

inner-transition element *Sec. 6.11*
metal *Sec. 6.11*
metalloid *Sec. 6.12*
noble gas *Sec. 6.2*
noble gas element *Sec. 6.11*
nonmetal *Sec. 6.11*
orbital diagram *Sec. 6.8*
outer electrons *Sec. 6.7*
paired electrons *Sec. 6.8*
paramagnetic atom *Sec. 6.8*

period *Sec. 6.2*
periodic law *Sec. 6.1*
periodic table *Sec. 6.2*
quantized property *Sec. 6.3*
representative element *Sec. 6.11*
semiconductor *Sec. 6.12*
transition element *Sec. 6.11*
unpaired electron *Sec. 6.8*

Practice Problems

PERIODIC LAW AND PERIODIC TABLE (SECS. 6.1 AND 6.2)

6.1 Give the chemical symbol of the element that occupies each of the following positions in the periodic table.
  a. period 3 and group IIIA
  b. period 2 and group IIA
  c. group IVA and period 5
  d. group IA and period 4
6.2 Give the chemical symbol of the element that occupies each of the following positions in the periodic table.
  a. period 4 and group IIA
  b. period 5 and group IIIA
  c. group IVA and period 2
  d. group IA and period 3
6.3 Based on their periodic table positions, characterize each of the following pairs of elements as belonging to (1) the same group, (2) the same period, or (3) neither the same group nor the same period.
  a. $_{26}$Fe and $_{76}$Os   **b.** $_{20}$Ca and $_{33}$As
  c. $_{7}$N and $_{16}$S   **d.** $_{43}$Tc and $_{50}$Sn
6.4 Based on their periodic table positions, characterize each of the following pairs of elements as belonging to

(1) the same group, (2) the same period, or (3) neither the same group nor the same period.
  a. $_{23}$V and $_{42}$Mo   **b.** $_{19}$K and $_{55}$Cs
  c. $_{8}$O and $_{17}$Cl   **d.** $_{11}$Na and $_{18}$Ar
6.5 For each of the following sets of elements, choose the two that would be expected to have similar chemical properties.
  a. $_{19}$K, $_{29}$Cu, $_{37}$Rb, $_{41}$Nb
  b. $_{13}$Al, $_{14}$Si, $_{15}$P, $_{33}$As
  c. $_{9}$F, $_{40}$Zr, $_{50}$Sn, $_{53}$I
  d. $_{11}$Na, $_{12}$Mg, $_{54}$Xe, $_{55}$Cs
6.6 For each of the following sets of elements, choose the two that would be expected to have similar chemical properties.
  a. $_{11}$Na, $_{14}$Si, $_{23}$V, $_{55}$Cs   **b.** $_{13}$Al, $_{19}$K, $_{32}$Ge, $_{50}$Sn
  c. $_{37}$Rb, $_{38}$Sr, $_{54}$Xe, $_{56}$Ba   **d.** $_{2}$He, $_{6}$C, $_{8}$O, $_{10}$Ne
6.7 The following statements either define or are closely related to the terms *periodic law*, *period*, or *group*. Match the terms to the appropriate statements.
  a. This is a vertical arrangement of elements in the periodic table.
  b. The properties of the elements repeat in a regular way as the atomic numbers increase.

    c. The chemical properties of elements 12, 20, and 38 demonstrate this principle.

    d. Elements 24 and 33 belong to this arrangement.

6.8 The following statements either define or are closely related to the terms *periodic law, period*, and *group*. Match the terms to the appropriate statements.

    a. This is a horizontal arrangement of elements in the periodic table.

    b. Element 19 begins this arrangement in the periodic table.

    c. The element carbon is the first member of this arrangement.

    d. Elements 10, 18, 36, and 54 belong to this arrangement.

6.9 Identify each of the following elements by name.

    a. period 4 halogen

    b. period 2 alkali metal

    c. period 3 noble gas

    d. period 5 alkaline earth metal

6.10 Identify each of the following elements by name.

    a. period 3 alkali metal

    b. period 1 noble gas

    c. period 4 alkaline earth metal

    d. period 2 halogen

6.11 How many elements exist with an atomic number less than 40 that are

    a. halogens?   b. noble gases?

    c. alkali metals?   d. alkaline earth metals?

6.12 How many elements exist with an atomic number greater than 20 that are

    a. halogens?   b. noble gases?

    c. alkali metals?   d. alkaline earth metals?

6.13 Determine the following for the "highlighted" elements in the following periodic table. Specify your answer by giving the number of the element.

    a. Which highlighted element is a halogen?

    b. Which highlighted element is an alkali metal?

    c. Which highlighted element is in group IIA?

    d. Which highlighted element is in period 3?

6.14 Determine the following for the "highlighted" elements in the periodic table given in Problem 6.13. Specify your answer by giving the number of the element.

    a. Which highlighted element is an alkaline earth metal?

    b. Which highlighted element is a noble gas?

    c. Which highlighted element is in group VIIA?

    d. Which highlighted element is in period 2?

TERMINOLOGY ASSOCIATED WITH ELECTRON ARRANGEMENTS (SECS. 6.3–6.6)

6.15 What is the maximum number of electrons that can be found in each of the following electron subshells?

    a. $1s$ **b.** $3d$ **c.** $2p$ **d.** $5f$

6.16 What is the maximum number of electrons that can be found in each of the following electron subshells?

    a. $6p$ **b.** $5s$ **c.** $4f$ **d.** $4d$

6.17 What is the maximum number of electrons that can be found in an electron orbital of each of the following types?

    a. $2s$ **b.** $3d$ **c.** $5f$ **d.** $3p$

6.18 What is the maximum number of electrons that can be found in an electron orbital of each of the following types?

    a. $6s$ **b.** $6p$ **c.** $5d$ **d.** $5p$

6.19 How many electron orbitals are there of each of the following types?

    a. $1s$ **b.** $3s$ **c.** $5d$ **d.** $6p$

6.20 How many electron orbitals are there of each of the following types?

    a. $4d$ **b.** $3p$ **c.** $5p$ **d.** $7s$

6.21 In each of the following pairs of items, identify the item that can accommodate the most electrons.

    a. $3d$ subshell, second shell

    b. shell with $n = 1$, $2p$ subshell

    c. $3p$ orbital, $3p$ subshell

    d. $4f$ subshell, third shell

6.22 In each of the following pairs of items, identify the item that can accommodate the most electrons.

    a. first shell, third shell

    b. $4f$ subshell, $4d$ subshell

    c. second shell, $5f$ subshell

    d. $3d$ orbital, $3d$ subshell

6.23 Indicate whether each of the following statements is *true* or *false*.

    a. An orbital has a definite size and shape, which are related to the energy of the electrons it could contain.

    b. All shells accommodate the same number of electrons.

    c. All of the orbitals in a subshell have the same energy.

    d. A $2p$ and a $3p$ subshell would contain the same number of orbitals.

    e. The fourth shell is made up of six subshells.

6.24 Indicate whether each of the following statements is *true* or *false*.

    a. All the subshells in a shell have the same energy.

    b. A d subshell always contains five orbitals.

    c. An s orbital is shaped something like a four-leaf clover.

    d. The $n = 3$ shell can accommodate a maximum of 18 electrons.

    e. All subshells accommodate the same number of electrons.

6.25 Describe the general shape of each of the following orbitals.
  a. $4s$ **b.** $4p$ **c.** $4d$ **d.** $6s$

6.26 Describe the general shape of each of the following orbitals.
  a. $1s$ **b.** $2p$ **c.** $3d$ **d.** $5p$

6.27 Which of the following electron subshell and electron orbital designations is not allowed?
  a. $3s$ subshell   **b.** $3d$ orbital
  c. $1p$ subshell   **d.** $2f$ orbital

6.28 Which of the following electron subshell and electron orbital designations is not allowed?
  a. $2p$ orbital   **b.** $2d$ subshell
  c. $3f$ orbital   **d.** $6s$ subshell

ELECTRON CONFIGURATIONS (SEC. 6.7)

6.29 According to the aufbau principle, which electron subshell is filled immediately *after* each of the following electron subshells is filled?
  a. $3s$ **b.** $4p$ **c.** $5s$ **d.** $3d$

6.30 According to the aufbau principle, which electron subshell is filled immediately *after* each of the following electron subshells is filled?
  a. $3p$ **b.** $4s$ **c.** $5d$ **d.** $3p$

6.31 According to the aufbau principle, which electron subshell is filled immediately *before* each of the following electron subshells is filled?
  a. $3p$ **b.** $4s$ **c.** $5d$ **d.** $4p$

6.32 According to the aufbau principle, which electron subshell is filled immediately *before* each of the following electron subshells is filled?
  a. $3s$ **b.** $5p$ **c.** $5s$ **d.** $3d$

6.33 For a multi-electron atom, arrange the electron subshells in each of the following listings in order of increasing energy.
  a. $1s, 2s, 3s, 4s$ **b.** $4s, 4p, 4d, 4f$
  c. $6s, 4f, 2p, 5d$ **d.** $3p, 3d, 4s, 5p$

6.34 For a multi-electron atom, arrange the electron subshells in each of the following listings in order of increasing energy.
  a. $2p, 3p, 4p, 5p$ **b.** $5s, 5p, 5d, 5f$
  c. $4d, 5p, 4s, 3d$ **d.** $3p, 2s, 4f, 6d$

6.35 With the help of an aufbau diagram, write the complete electron configuration for atoms with each of the following atomic numbers.
  a. $Z = 13$ **b.** $Z = 7$ **c.** $Z = 10$ **d.** $Z = 15$

6.36 With the help of an aufbau diagram, write the complete electron configuration for atoms with each of the following atomic numbers.
  a. $Z = 12$ **b.** $Z = 18$ **c.** $Z = 36$ **d.** $Z = 6$

6.37 With the help of an aufbau diagram, write the complete electron configuration for each of the following atoms.
  a. $_{26}$Fe   **b.** $_{37}$Rb   **c.** $_{53}$I   **d.** $_{86}$Rn

6.38 With the help of an aufbau diagram, write the complete electron configuration for each of the following atoms.
  a. $_{31}$Ga   **b.** $_{38}$Sr   **c.** $_{48}$Cd   **d.** $_{88}$Ra

6.39 Based on total number of electrons present, identify the element represented by each of the following electron configurations.
  a. $1s^22s^22p^6$ **b.** $1s^22s^22p^63s^23p^64s^1$
  c. $1s^22s^22p^63s^23p^64s^23d^2$ **d.** $1s^22s^22p^63s^23p^64s^23d^{10}$

6.40 Based on total number of electrons present, identify the element represented by each of the following electron configurations.
  a. $1s^22s^22p^2$ **b.** $1s^22s^22p^63s^23p^3$
  c. $1s^22s^22p^63s^23p^64s^2$ **d.** $1s^22s^22p^63s^23p^64s^23d^6$

6.41 Convert each of the following electron configurations into condensed electron configurations.
  a. $_{13}$Al: $1s^22s^22p^63s^23p^1$
  b. $_{19}$K: $1s^22s^22p^63s^23p^64s^1$
  c. $_{30}$Zn: $1s^22s^22p^63s^23p^64s^23d^{10}$
  d. $_{50}$Sn: $1s^22s^22p^63s^23p^64s^23d^{10}4p^65s^24d^{10}5p^2$

6.42 Convert each of the following electron configurations into condensed electron configurations.
  a. $_{15}$P: $1s^22s^22p^63s^23p^3$
  b. $_{22}$Ti: $1s^22s^22p^63s^23p^64s^23d^2$
  c. $_{33}$As: $1s^22s^22p^63s^33p^64s^23d^{10}4p^3$
  d. $_{56}$Ba: $1s^22s^22p^63s^23p^64s^23d^{10}4p^65s^24d^{10}5p^66s^2$

6.43 Convert each of the following electron configurations into condensed electron configurations.
  a. $1s^22s^22p^4$
  b. $1s^22s^22p^63s^2$
  c. $1s^22s^22p^63s^23p^64s^2$
  d. $1s^22s^22p^63s^23p^64s^23d^{10}4p^5$

6.44 Convert each of the following electron configurations into condensed electron configurations.
  a. $1s^22s^22p^2$
  b. $1s^22s^22p^63s^23p^1$
  c. $1s^22s^22p^63s^23p^64s^1$
  d. $1s^22s^22p^63s^23p^24s^23d^{10}4p^65s^2$

6.45 Give the name of the element that has each of the following condensed electron configurations.
  a. [Ne]$3s^2$ **b.** [Ne]$3s^23p^5$
  c. [Kr]$5s^24d^{10}5p^4$ **d.** [Xe]$6s^2$

6.46 Give the name of the element that has each of the following condensed electron configurations.
  a. [Ne]$3s^23p^4$ **b.** [Ar]$4s^23d^3$
  c. [Ar]$4s^23d^{10}4p^5$ **d.** [Xe]$6s^1$

6.47 How many *core electrons* are present in each of the condensed electron configurations given in Problem 6.45?

6.48 How many *core electrons* are present in each of the condensed electron configurations given in Problem 6.46?

6.49 How many *outer electrons* are present in each of the condensed electron configurations given in Problem 6.45?

6.50 How many *outer electrons* are present in each of the condensed electron configurations given in Problem 6.46?

ORBITAL DIAGRAMS (SEC. 6.8)

6.51 Draw the electron orbital diagram associated with each of the following electron configurations. Use an up or down arrow to denote each electron.
  a. $1s^2 2s^1$ **b.** $1s^2 2s^2 2p^5$
  c. $1s^2 2s^2 2p^6 3s^2 3p^3$ **d.** $1s^2 2s^2 2p^6 3s^2 3p^6 4s^2 3d^8$

6.52 Draw the electron orbital diagram associated with each of the following electron configurations. Use an up or down arrow to denote each electron.
  a. $1s^2 2s^2 2p^3$ **b.** $1s^2 2s^2 2p^6 3s^2$
  c. $1s^2 2s^2 2p^6 3s^2 3p^5$ **d.** $1s^2 2s^2 2p^6 3s^2 3p^6 4s^2 3d^7$

6.53 Draw electron orbital diagrams for the following elements. Use an up or down arrow to denote each electron.
  a. $_6C$ **b.** $_{10}Ne$ **c.** $_{11}Na$ **d.** $_{15}P$

6.54 Draw electron orbital diagrams for the following elements. Use an up or down arrow to denote each electron.
  a. $_7N$ **b.** $_9F$ **c.** $_{12}Mg$ **d.** $_{21}Sc$

6.55 Draw the electron orbital diagram for atoms with the following atomic numbers. Use an up or down arrow to denote each electron, and use the abbreviation of the preceding noble gas to represent inner-shell (core) electrons.
  a. Z = 17   **b.** Z = 25   **c.** Z = 38   **d.** Z = 51

6.56 Draw the electron orbital diagram for atoms with the following atomic numbers. Use an up or down arrow to denote each electron, and use the abbreviation of the preceding noble gas to represent inner-shell (core) electrons.
  a. Z = 15   **b.** Z = 23   **c.** Z = 49   **d.** Z = 56

6.57 Draw electron orbital diagrams to show the distribution of electrons among orbitals in each of the following electron subshells.
  a. $3s$ subshell of Mg   **b.** $3p$ subshell of Cl
  c. $4s$ subshell of Ca   **d.** $3d$ subshell of Co

6.58 Draw electron orbital diagrams to show the distribution of electrons among orbitals in each of the following electron subshells.
  a. $2p$ subshell of N   **b.** $3p$ subshell of Ar
  c. $4s$ subshell of K   **d.** $4p$ subshell of Se

6.59 How many unpaired electrons are there in an atom of the following elements?
  a. carbon   **b.** sodium
  c. argon   **d.** titanium

6.60 How many unpaired electrons are there in an atom of the following elements?
  a. lithium   **b.** boron
  c. aluminum   **d.** iron

6.61 Indicate whether atoms of each of the elements in Problem 6.59 are paramagnetic or diamagnetic.

6.62 Indicate whether atoms of each of the elements in Problem 6.60 are paramagnetic or diamagnetic.

ELECTRON CONFIGURATIONS AND THE PERIODIC LAW (SEC. 6.9)

6.63 Indicate whether the elements represented by the given pairs of electron configurations have similar chemical properties.
  a. $1s^2 2s^1$ and $1s^2 2s^2$
  b. $1s^2 2s^2 2p^6$ and $1s^2 2s^2 2p^6 3s^2 3p^6$
  c. $1s^2 2s^2 2p^3$ and $1s^2 2s^2 2p^6 3s^2 3p^6 4s^2 3d^3$
  d. $1s^2 2s^2 2p^6 3s^2 3p^6$ and $1s^2 2s^2 2p^6 3s^2 3p^6 4s^2 3d^{10} 4p^6$

6.64 Indicate whether the elements represented by the given pairs of electron configurations have similar chemical properties.
  a. $1s^2 2s^2 2p^4$ and $1s^2 2s^2 2p^5$
  b. $1s^2 2s^2$ and $1s^2 2s^2 2p^2$
  c. $1s^2 2s^1$ and $1s^2 2s^2 2p^6 3s^2 3p^6 4s^1$
  d. $1s^2 2s^2 2p^6$ and $1s^2 2s^2 2p^6 3s^2 3p^6 4s^2 3d^6$

ELECTRON CONFIGURATIONS AND THE PERIODIC TABLE (SEC. 6.10)

6.65 Specify position in the periodic table, in terms of s, p, d, or f area, for each of the following elements.
  a. $_{16}S$ **b.** $_{24}Cr$ **c.** silver   **d.** barium

6.66 Specify position in the periodic table, in terms of s, p, d, or f area, for each of the following elements.
  a. $_{20}Ca$ **b.** $_{30}Zn$ **c.** bromine   **d.** gold

6.67 Identify the type of subshell $(s, p, d$ or $f)$ that contains the distinguishing electron for each of the elements in Problem 6.65.

6.68 Identify the type of subshell $(s, p, d$ or $f)$ that contains the distinguishing electron for each of the elements in Problem 6.66.

6.69 Identify the specific subshell $(3p, 4s, 5d,$ etc.) that contains the distinguishing electron for each of the elements in Problem 6.65.

6.70 Identify the specific subshell $(3p, 4s, 5d,$ etc.) that contains the distinguishing electron for each of the elements in Problem 6.66.

6.71 Determine the number of "highlighted" elements in the following periodic table that have each of the following electronic characteristics.

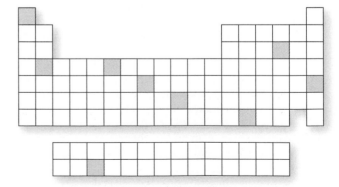

  a. located in the periodic table's d area
  b. distinguishing electron is in a p subshell

c. has two electrons in the subshell that contains the distinguishing electron
d. distinguishing electron is in shell 4

6.72 Determine the number of "highlighted" elements in the periodic table in Problem 6.71 that have each of the following electronic characteristics.
a. located in the periodic table's *f* area
b. distinguishing electron is in a *d* subshell
c. has six electrons in the subshell that contains the distinguishing electron
d. distinguishing electron is in shell 3

6.73 Indicate the position in the periodic table where each of the following occurs, by giving the symbol of the element.
a. The 3*p* subshell begins filling.
b. The 2*s* subshell begins filling.
c. The 5*d* subshell begins filling.
d. The 3*d* subshell begins filling.

6.74 Indicate the position in the periodic table where each of the following occurs, by giving the symbol of the element.
a. The 4*p* subshell begins filling.
b. The 5*f* subshell begins filling.
c. The 3*s* subshell begins filling.
d. The 7*s* subshell begins filling.

6.75 Indicate the position in the periodic table where each of the following occurs, by giving the symbol of the element.
a. The 4*p* subshell becomes completely filled.
b. The 2*s* subshell becomes half-filled.
c. The fourth shell begins filling.
d. The fourth shell becomes completely filled.

6.76 Indicate the position in the periodic table where each of the following occurs, by giving the symbol of the element.
a. The 3*d* subshell becomes completely filled.
b. The 4*p* subshell becomes half-filled.
c. The third shell becomes half-filled.
d. The third shell becomes completely filled.

6.77 Using only the periodic table, determine the complete electron configuration for each of the following elements.
a. $_{33}$As   **b.** $_{51}$Sb   **c.** $_{37}$Rb   **d.** $_{48}$Cd

6.78 Using only the periodic table, determine the complete electron configuration for each of the following elements.
a. $_{22}$Ti   **b.** $_{56}$Ba   **c.** $_{35}$Br   **d.** $_{19}$K

6.79 Using only the periodic table, determine the complete electron configuration for the elements with the following condensed electron configurations.
a. $[Ne]3s^2$ **b.** $[Ar]4s^2$
c. $[Ar]4s^23d^{10}$ **d.** $[Kr]5s^24d^{10}5p^2$

6.80 Using only the periodic table, determine the complete electron configuration for the elements with the following condensed electron configurations.
a. $[He]2s^2$ **b.** $[Ne]3s^23p^4$
c. $[Ar]4s^23d^5$ **d.** $[Kr]5s^24d^{10}$

6.81 How many 3*d* electrons are found in atoms of each of the following elements?
a. titanium   **b.** nickel
c. selenium   **d.** palladium

6.82 How many 4*d* electrons are found in atoms of each of the following elements?
a. zinc   **b.** yttrium   **c.** silver   **d.** iodine

6.83 Using the periodic table as a guide, indicate the number of
a. 3*p* electrons in a $_{16}$S atom.
b. 3*d* electrons in a $_{29}$Cu atom.
c. 4*s* electrons in a $_{37}$Rb atom.
d. 4*d* electrons in a $_{30}$Zn atom.

6.84 Using the periodic table as a guide, indicate the number of
a. 3*s* electrons in a $_{12}$Mg atom.
b. 4*p* electrons in a $_{32}$Ge atom.
c. 3*d* electrons in a $_{47}$Ag atom.
d. 4*p* electrons in a $_{15}$P atom.

CLASSIFICATION SYSTEMS FOR THE ELEMENTS (SEC. 6.11)

6.85 Classify each of the following general properties as characteristic of metallic elements or of nonmetallic elements.
a. ductile
b. low electrical conductivity
c. high thermal conductivity
d. good heat insulator

6.86 Classify each of the following general properties as characteristic of metallic elements or of nonmetallic elements.
a. nonmalleable
b. high luster
c. low thermal conductivity
d. brittle

6.87 Using the two-category "metal-nonmetal" classification system, in which of the following pairs of elements are both members of the pair metals?
a. $_{17}$Cl and $_{35}$Br   **b.** $_{13}$Al and $_{14}$Si
c. $_{29}$Cu and $_{42}$Mo   **d.** $_{30}$Zn and $_{32}$Ge

6.88 Using the two-category "metal-nonmetal" classification system, in which of the following pairs of elements are both members of the pair metals?
a. $_7$N and $_{34}$Se   **b.** $_{16}$S and $_{48}$Cd
c. $_3$Li and $_{26}$Fe   **d.** $_{51}$Sb and $_{53}$I

6.89 Using the two-category "metal-nonmetal" classification system, indicate whether each of the following groups in the periodic table contain (1) more metals than nonmetals or (2) more nonmetals than metals.
a. group IIIA   **b.** group IIIB
c. group VIA   **d.** group VIIIA

6.90 Using the two-category "metal-nonmetal" classification system, indicate whether each of the following groups in the periodic table contain (1) more metals than nonmetals or (2) more nonmetals than metals.
a. group IIA   **b.** group IIB
c. group IVA   **d.** group VIIA

6.91 Identify the metal in each of the following sets of elements.
  a. H, He, Li **b.** S, Cl, K **c.** N, Fe, O **d.** Hg, Ne, F

6.92 Identify the metal in each of the following sets of elements.
  a. C, Br, Pb   **b.** Ar, Kr, Na
  c. P, Ga, Se   **d.** Zn, I, Xe

6.93 Determine the following for the "highlighted" elements in the following periodic table.

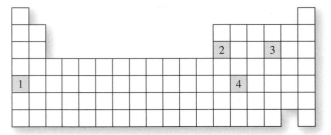

  a. Is ⬚1 a metal or nonmetal?

  b. Is ⬚3 a metal or nonmetal?

  c. Is ⬚2 a good or poor conductor of electricity?

  d. Is ⬚4 a good or poor conductor of heat?

6.94 Determine the following for the "highlighted" elements in the periodic table given in Problem 6.93.
  a. Is ⬚2 a metal or nonmetal?

  b. Is ⬚4 a metal or nonmetal?

  c. Is ⬚1 a good or poor conductor of electricity?

  d. Is ⬚3 a good or poor conductor of heat?

6.95 Classify each of the following elements as a noble gas, representative element, transition element, or inner-transition element.
  a. $_{29}$Cu   **b.** $_{32}$Ge   **c.** $_{54}$Xe   **d.** $_{72}$Hf

6.96 Classify each of the following elements as a noble gas, representative element, transition element, or inner-transition element.
  a. $_{27}$Co   **b.** $_{36}$Kr   **c.** $_{68}$Er   **d.** $_{79}$Au

6.97 Identify the lowest atomic-numbered element that is
  a. a nonmetal.
  b. a noble gas.
  c. a representative metal.
  d. an inner-transition element.

6.98 Identify the lowest atomic-numbered element that is
  a. a transition element.
  b. a representative nonmetal.
  c. an inner-transition metal.
  d. a metal.

6.99 Answer the following questions concerning groups and periods of the periodic table.
  a. How many periods are there that contain transition elements?
  b. How many groups are there that contain representative elements?
  c. How many periods are there that contain metals?
  d. How many periods are there that consist of all metals?

6.100 Answer the following questions concerning groups and periods of the periodic table.
  a. How many periods are there that contain inner-transition elements?
  b. How many groups are there that contain both metals and nonmetals?
  c. How many groups are there that consist of all nonmetals?
  d. How many periods are there that consist of all nonmetals?

6.101 Determine the number of "highlighted" elements in the following periodic table that belong to each of the following element classifications.

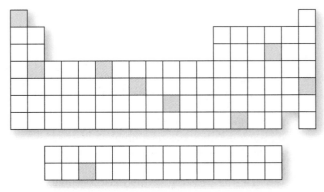

  a. noble gas
  b. inner-transition element
  c. nonmetal
  d. metallic representative element

6.102 Determine the number of "highlighted" elements in the periodic table in Problem 6.101 that belong to each of the following element classifications.
  a. transition element
  b. nonmetallic representative element
  c. metal
  d. metallic inner-transition element

CHEMICAL PERIODICITY (SEC. 6.12)

6.103 Based on general periodic table trends, indicate whether each of the following statements is *true* or *false*.
  a. The metallic character for a period of elements increases in proceeding from left to right in the periodic table.
  b. The nonmetallic character for a group of elements decreases in proceeding down a group in the periodic table.
  c. The atomic radius for a period of elements increases in proceeding from left to right in the periodic table.

6.104 Based on general periodic table trends, indicate whether each of the following statements is *true* or *false*.
  a. The nonmetallic character for a period of elements decreases in proceeding from left to right in the periodic table.

6.143 Refer
elem
confi
a. m
b. m

6.144 Refer
elem
confi
a. m
b. m

CUMULATI

6.145 Whic
boron
the f
a. ha
b. is
c. co
d. be

6.146 Whic
argon
of the
a. pe
b. rea
c. ha
d. ha

6.147 The
$1s^22s$
isoto

6.148 The
$1s^22s$
isoto

6.149 How
atom
whos

Multiple-

Use this bank
concepts pres
there may be
correct answe

6.157 Whic
an el
a. at
b. m
c. at
d. m
e. no

6.158 Whic
the p
a. "p
oc
b. "p
pe
c. "p

b. The atomic radius for a group of elements decreases in proceeding down a group in the periodic table.

c. The metallic character for a group of elements increases in proceeding down a group in the periodic table.

6.105 Identify the metalloid in each of the following sets of elements.
a. Ge, Ga, Zn **b.** B, Al, Ga
c. Pb, Po, P **d.** Te, I, Xe

6.106 Identify the metalloid in each of the following sets of elements.
a. Se, As, Br **b.** Br, I, At
c. Sn, Sb, Sm **d.** Al, Si, P

6.107 Based on periodic table trends, indicate which member of each pair of elements is the more metallic.
a. $_{11}$Na or $_{37}$Rb **b.** $_{22}$Ti or $_{32}$Ge
c. $_{34}$Se or $_{35}$Br **d.** $_{30}$Zn or $_{56}$Ba

6.108 Based on periodic table trends, indicate which member of each pair of elements is the more metallic.
a. $_{12}$Mg or $_{16}$S **b.** $_{47}$Ag or $_{79}$Au
c. $_{20}$Ca or $_{27}$Co **d.** $_{4}$Be or $_{37}$Rb

6.109 Using the periodic table, indicate which member of each pair is more nonmetallic.
a. $_{9}$F or $_{35}$Br **b.** $_{14}$Si or $_{15}$P
c. $_{25}$Mn or $_{30}$Zn **d.** $_{17}$Cl or $_{33}$As

6.110 Using the periodic table, indicate which member of each pair is more nonmetallic.
a. $_{6}$C or $_{8}$O **b.** $_{12}$Mg or $_{20}$Ca
c. $_{50}$Sn or $_{52}$Te **d.** $_{53}$I or $_{81}$Tl

6.111 Identify the element that fits each of the following descriptions.
a. most metallic element in group VA
b. most nonmetallic element in group IIA
c. most metallic element in period 4
d. most nonmetallic element in period 2

6.112 Identify the element that fits each of the following descriptions.
a. most nonmetallic element in group VIA
b. most metallic element in group IA
c. most nonmetallic element in period 3
d. most metallic element in period 2

6.113 Indicate which member of each of the following pairs of elements has the larger atomic radius.
a. $_{5}$B or $_{6}$C **b.** $_{17}$Cl or $_{31}$Ga
c. $_{19}$K or $_{35}$Br **d.** $_{19}$K or $_{37}$Rb

6.114 Indicate which member of each of the following pairs of elements has the larger atomic radius.
a. $_{15}$P or $_{16}$S **b.** $_{15}$P or $_{33}$As
c. $_{35}$Br or $_{52}$Te **d.** $_{37}$Rb or $_{53}$I

6.115 Identify the element that fits each of the following descriptions.
a. smallest atomic radius in group VA
b. largest atomic radius in group IIA
c. smallest atomic radius in period 4
d. largest atomic radius in period 2

6.116 Identify the element that fits each of the following descriptions.
a. largest atomic radius in group VIA
b. smallest atomic radius in group IA
c. largest atomic radius in period 3
d. smallest atomic radius in period 1

ADDITIONAL PROBLEMS

6.117 What is wrong with each of the following attempts to write an electron configuration?
a. $1s^22s^3$
b. $1s^22s^22p^23s^2$
c. $1s^22s^23s^2$
d. $1s^22s^22p^63s^23d^{10}$

6.118 What is wrong with each of the following attempts to write an electron configuration?
a. $1s^21p^6$ **b.** $1s^22s^4$
c. $1s^22s^22p^43s^2$ **d.** $1s^22s^22p^63s^23p^63d^{10}$

6.119 In what period and group in the periodic table is an element with each of the following electron configurations located?
a. $1s^22s^22p^63s^1$
b. $1s^22s^22p^63s^23p^1$
c. $1s^22s^22p^63s^23p^64s^23d^1$
d. $1s^22s^22p^63s^23p^64s^23d^{10}4p^5$

6.120 In what period and group in the periodic table is an element with each of the following electron configurations located?
a. $1s^22s^22p^2$
b. $1s^22s^22p^63s^2$
c. $1s^22s^22p^63s^23p^64s^23d^2$
d. $1s^22s^22p^63s^33p^64s^23d^{10}4p^65s^2$

6.121 Specify the number of unpaired electrons associated with each of the electron configurations given in Problem 6.119.

6.122 Specify the number of unpaired electrons associated with each of the electron configurations given in Problem 6.120.

6.123 Specify whether the atoms with the electron configurations given in Problem 6.119 are paramagnetic or diamagnetic.

6.124 Specify whether the atoms with the electron configurations given in Problem 6.120 are paramagnetic or diamagnetic.

6.125 Assign values to x and y in each of the following electron configurations.
a. Ca: $1s^22s^22p^63s^23p^x4s^y$
b. Al: $1s^22s^22p^63s^23p^y$
c. Zn: $1s^22s^22p^63s^23p^64s^x3d^y$
d. Kr: $1s^22s^22p^63s^23p^64s^x3d^{10}4p^y$

6.126 Assign values to x and y in each of the following electron configurations.
a. Ti: $1s^22s^22p^63s^23p^x4s^y3d^2$
b. Ar: $1s^22s^22p^63s^x3p^y$
c. Cl: $1s^22s^22p^x3s^23p^y$
d. Se: $1s^22s^22p^63s^23p^64s^23d^x4p^y$

6.127 Wri
mer
imn
the
a. [.
c. [
6.128 Wri
mer
imn
the
a. [.
c. [
6.129 Wri
mer
imn
the
a. [.
c. [
6.130 Wri
mer
imn
the
a. [.
c. [
6.131 Hov
ing
a. [
b. [
c. [
d. [
6.132 Hov
ing
a. [
b. [
c. [
d. [
6.133 Wri
ing
a. t
b. t
c. t
p
d. t
e
6.134 Wri
ing
a. t
b. t
c. t
d. t
e
6.135 Ref
eler
con
a. t
b. t

d. more than one correct response
e. no correct response

6.162 Both a number and a letter are used in designating an electron subshell. The letter
a. indicates the shell to which the subshell belongs.
b. may be s, p, d, or f.
c. gives information about the maximum number of electrons the subshell may hold.
d. more than one correct response
e. no correct response

6.163 Which of the following statements concerning electron shells are correct?
a. The shell number gives the number of electrons present.
b. All electrons present in a shell have the same energy.
c. The number of subshells present is the same as the shell number.
d. more than one correct response
e. no correct response

6.164 Which of the following statements about a d subshell is correct?
a. Seven orbitals are present within it.
b. The maximum number of electrons it can contain is 14.
c. All electrons present have the same energy.
d. more than one correct response
e. no correct response

6.165 Which of the following are incorrectly written electron configurations?
a. $1s^2 2s^2 2p^2 3s^2$
b. $1s^2 2s^2 2p^6 3s^2 3p^2$
c. $1s^2 2s^2 2p^6 3s^2 3p^6 3d^2$
d. more than one correct response
e. no correct response

6.166 Which of the following are incorrectly written condensed electron configurations?
a. $[Ne]3s^2 3p^2$
b. $[Ar]4s^2 3d^{10}$
c. $[Kr]5s^2$
d. more than one correct response
e. no correct response

6.167 How many *core* electrons are present in an atom whose condensed electron configuration is $[Ne]3s^2 3p^4$?
a. 6   **b.** 10   **c.** 16   **d.** 18
e. no correct response

6.168 Which electron subshell fills with electrons immediately before the $3d$ subshell fills with electrons?
a. $3s$ **b.** $3p$ **c.** $4s$ **d.** $4p$
e. no correct response

6.169 Which of the following statements concerning periodic table "position" and electron configurations is correct?
a. The $3p$ subshell begins filling at element 13.
b. The $4s$ subshell becomes completely filled at element 19.
c. The third shell becomes completely filled at element 18.

d. more than one correct response
e. no correct response

6.170 Which of the following statements concerning periodic table locations of elements is correct?
a. $_{27}Co$ and $_{30}Zn$ are in the d area of the periodic table.
b. $_{80}Hg$ and $_{82}Pb$ are in the p area of the periodic table.
c. $_{37}Rb$ and $_{56}Ba$ are in the s area of the periodic table.
d. more than one correct response
e. no correct response

6.171 Two unpaired electrons are present in the orbital diagram of which of the following elements?
a. $_{14}Si$ **b.** $_{20}Ca$ **c.** $_{34}Se$
d. more than one correct response
e. no correct response

6.172 Which of the following statements concerning atoms of the elements $_9F$ and $_{10}Ne$ is correct?
a. Both F and Ne atoms are diamagnetic.
b. Both F and Ne atoms are paramagnetic.
c. F atoms are diamagnetic and Ne atoms are paramagnetic.
d. more than one correct response
e. no correct response

6.173 In which of the following pairs of elements are both elements metals?
a. $_7N$ and $_{83}Bi$ **b.** $_{30}Zn$ and $_{34}Se$
c. $_{19}K$ and $_{58}Ce$
d. more than one correct response
e. no correct response

6.174 Which of the following statements about element classification systems is correct?
a. All of the elements in the p area of the periodic table are representative elements.
b. All of the elements in the d area of the periodic table are transition elements.
c. All of the elements in the s area of the periodic table are inner-transition elements.
d. more than one correct response
e. no correct response

6.175 Which of the following concept pairings relating to element classifications is incorrect?
a. inner-transition element—distinguishing electron is found in a f subshell
b. representative element—some are metals and some are nonmetals
c. noble gas element—electron configurations ends in p^5 or p^6
d. more than one correct response
e. no correct response

6.176 Which of the following statements concerning atomic radii is correct?
a. $_{17}Cl$ has a larger atomic radius than $_{13}Al$.
b. $_{17}Cl$ has a larger atomic radius than $_{35}Br$.
c. $_{17}Cl$ has a larger atomic radius than $_{18}Ar$.
d. more than one correct response
e. no correct response

Chemical Bonds

7.1 TYPES OF CHEMICAL BONDS

In Section 5.2 we considered the fact that chemical compounds are conveniently divided into two broad classes called *ionic compounds* and *molecular compounds*. Ionic and molecular compounds can be distinguished from each other on the basis of general physical properties. Ionic compounds tend to have high melting points (500–2000°C) and are good conductors of electricity when they are in a molten (liquid) state. Molecular compounds, on the other hand, generally have much lower melting points and tend to be gases, liquids, or low-melting solids. They do not conduct electricity in the molten state. Ionic compounds, unlike molecular compounds, do not have molecules as their basic structural unit. Instead, an extended array of positively and negatively charged particles called *ions* is present.

Some combinations of elements produce ionic compounds, whereas other combinations of elements form molecular compounds. What determines whether the interaction of two elements produces ions (an ionic compound) or molecules (a molecular compound)? An answer to this question requires information concerning the nature of *chemical bonds*. A **chemical bond** *is the attractive force that holds two atoms together in a more complex unit.* Chemical bonds form as the result of interactions between electrons found in the combining atoms. Thus, chemical bond considerations are closely linked to electron configurations (Sec. 6.7).

Corresponding to the two broad categories of chemical compounds are two types of chemical attractive forces (chemical bonds): *ionic bonds* and *covalent bonds*. An **ionic bond** *is a chemical bond formed through the transfer of one or more electrons from one atom or group of atoms to another*. As suggested by its name, the ionic bond model (electron transfer) is used in describing the attractive forces in ionic compounds. A **covalent bond** *is a chemical bond formed through the sharing of one or more pairs of electrons between two atoms*. The covalent bond model (electron sharing) is used in describing the attractions between atoms in molecular compounds.

It is important to emphasize, even before we consider the details of these two bond models, that the concepts of ionic and covalent bonds are actually "convenience" concepts. Most bonds are not 100% ionic or 100% covalent. Instead, most bonds have some degree of both ionic and covalent character, that is, some degree of both the transfer and sharing of electrons. However, it is easiest to understand these intermediate bonds (the real bonds) by relating them to the pure or ideal bond types called ionic and covalent.

There are two fundamental concepts that are common to and necessary for understanding both the ionic and covalent bonding models. These concepts are the following:

1. Not all electrons in an atom participate in bonding. Those that are available are called *valence electrons*.
2. Certain arrangements of electrons are more stable than other arrangements of electrons. The *octet rule* addresses this situation.

Section 7.2 deals with the concept of valence electrons and Section 7.3 discusses the octet rule.

7.2 VALENCE ELECTRONS AND LEWIS SYMBOLS

Certain electrons, called *valence electrons*, are particularly important in determining the bonding characteristics of a given atom. A **valence electron** *is an electron in the outermost electron shell of a representative element or noble gas element*. Valence electrons are always found in either *s* or *p* subshells or both. Note the restriction on the use of this definition; it applies only for representative and noble gas elements (Sec. 6.11). Many commonly encountered elements are representative elements; hence the definition finds much use. (We will not consider in this text the more complicated valence electron definitions for transition or inner transition elements (Sec. 6.11); the presence of incompletely filled *inner d* or *f* subshells is the complicating factor in definitions for these elements.)

The number of valence electrons in an atom of a representative element can be determined from the atom's electron configuration, as is illustrated in Example 7.1.

Another designation for a *molecular compound* is *covalent compound*. The modifier *molecular* draws attention to the basic structural unit present (the molecule) and the modifier *covalent* focuses on the mode of bond formation (electron sharing).

Purely ionic bonds involve complete transfer of electrons from one atom to another. *Purely* covalent bonds involve equal sharing of electrons. Experimentally, it is found that *most* actual bonds have some degree of both ionic and covalent character. The exception is bonds between identical atoms; here, the bonding is purely covalent.

The term *valence* is derived from the Latin *valentia*, meaning "capacity" (to form bonds).

EXAMPLE 7.1   **Determining the Number of Valence Electrons an Atom Possesses**

Determine the number of valence electrons present in atoms of each of the following elements.

   a. $_{12}Mg$ **b.** $_{17}Cl$ **c.** $_{34}Se$

SOLUTION

   a. The element magnesium has two valence electrons, as can be seen by examining its electron configuration.

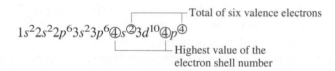

Number of valence electrons

$$1s^2 2s^2 2p^6 \textcircled{3}s^{\textcircled{2}}$$

Highest value of the electron shell number

     The highest value of the electron shell number is $n = 3$. Only two electrons are found in shell 3, two electrons in the $3s$ subshell.

   b. The element chlorine has seven valence electrons.

Total of seven valence electrons

$$1s^2 2s^2 2p^6 \textcircled{3}s^{\textcircled{2}} \textcircled{3}p^{\textcircled{5}}$$

Highest value of the electron shell number

     Electrons in two different subshells can simultaneously be valence electrons. The highest shell number is 3, and both the $3s$ and $3p$ subshells belong to shell number 3. Hence, all electrons in both subshells are valence electrons.

   c. The element selenium has six valence electrons.

Total of six valence electrons

$$1s^2 2s^2 2p^6 3s^2 3p^6 \textcircled{4}s^{\textcircled{2}} 3d^{10} \textcircled{4}p^{\textcircled{4}}$$

Highest value of the electron shell number

     The $3d$ electrons are not counted as valence electrons because the $3d$ subshell is in shell 3 and shell 3 is not the shell with maximum n value. Shell 4 is the outermost shell, the shell with maximum n value.

▶ **Practice Exercise 7.1**   Determine the number of valence electrons present in atoms of each of the following elements.

   a. $_{14}Si$ **b.** $_7N$ **c.** $_{19}K$

Answers to all practice exercises in this chapter are found in the back-of-the book answer section.

     It seems reasonable that the outermost electrons of atoms are those involved in bonding when you remember that when atoms collide—an event that is necessary before atoms can combine—it will be the outermost parts of the atoms that come into contact with each other. Also, since the outermost electrons are located the farthest from the nucleus, they are therefore the least tightly bound (attraction to the nucleus decreases with distance) and thus the most susceptible to change (transfer or sharing).

FIGURE 7.1 Lewis symbols of the first 20 elements.

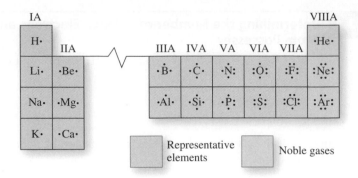

Chemists have developed a shorthand system for designating the number of valence electrons present in an atom. This system uses *Lewis symbols*. A **Lewis symbol** *is the chemical symbol of an element surrounded by dots equal in number to the number of valence electrons present in atoms of the element.* Lewis symbols for the first 20 elements, arranged as in the periodic table, are given in Figure 7.1. Lewis symbols, named in honor of the American chemist Gilbert Newton Lewis, an early contributor to chemical bonding theory and the person who first used them (see "The Human Side of Chemistry 7"), are also frequently called *electron-dot structures*.

The general practice when writing Lewis symbols is to place the first four dots separately on the four sides of the chemical symbol and then begin pairing the dots as further dots are added. It makes no difference on which side of the symbol the process of adding dots begins. The following notations for the Lewis symbol of the element magnesium are all equivalent.

> In a Lewis symbol, the chemical symbol represents the nucleus and all of the nonvalence electrons. The valence electrons are then shown as dots.

$$\text{M}\overset{..}{\text{g}}\cdot \qquad \text{M}\overset{.}{\underset{.}{\text{g}}}\cdot \qquad \cdot\text{M}\overset{.}{\underset{.}{\text{g}}} \qquad \cdot\overset{.}{\text{M}}\text{g} \qquad \overset{.}{\text{M}}\overset{.}{\text{g}} \qquad \cdot\overset{.}{\text{M}}\text{g}\cdot$$

Three important generalizations about valence electrons can be drawn from a study of the Lewis symbols in Figure 7.1.

1. *Representative elements in the same group of the periodic table have the same number of valence electrons.* This should not be surprising to you. Elements in the same group in the periodic table have similar chemical properties as a result of having similar outer shell electron configurations (Sec. 6.9). The electrons in the outermost shell are the valence electrons.
2. *The number of valence electrons for representative elements in a group is the same as the periodic table group number.* For example, the Lewis symbols for O and S, both members of group VIA, show six dots. Similarly, the Lewis symbols of H, Li, Na, and K, all members of group IA, show one dot.
3. *The maximum number of valence electrons for any element is eight.* Only the noble gases (Sec. 6.11), beginning with Ne, have the maximum number of eight electrons. Helium, with only two valence electrons, is the exception in the noble gas family; obviously an element with a grand total of two electrons cannot have eight valence electrons. Although shells with *n* greater than 2 are capable of holding more than eight electrons, they do so only when they are no longer the outermost shell and thus are not the valence shell. For example, selenium (Example 7.1) has 18 electrons in its third shell; however, shell 4 is the valence shell in selenium.

EXAMPLE 7.2 **Writing Lewis Symbols for Elements**

Write Lewis symbols for the following elements.

a. S, Se, and Te **b.** Be, B, and C

SOLUTION

a. These elements are all group VIA elements and thus possess the same number of valence electrons, which is six. (The number of valence electrons and the periodic table group number will always be the same for representative elements.) The Lewis symbols, which all have six dots, are

$$\cdot \ddot{\underset{..}{S}} \colon \qquad \cdot \ddot{\underset{..}{Se}} \colon \qquad \cdot \ddot{\underset{..}{Te}} \colon$$

b. These elements are sequential elements in period 2 of the periodic table; Be is in group IIA (two valence electrons), B is in group IIIA (three valence electrons), and C is in group IVA (four valence electrons). The Lewis symbols for these elements are

$$\cdot \dot{Be} \qquad \cdot \dot{B} \cdot \qquad \cdot \dot{\underset{.}{C}} \cdot$$

▶ **Practice Exercise 7.2**   Write Lewis symbols for the following elements.

a. Na, K, and Rb   **b.** N, O, and F

The Human Side of Chemistry 7

**Gilbert Newton Lewis
(1875–1946)**

Gilbert Newton Lewis, an American chemist, is recognized as one of the foremost chemists of the twentieth century. The son of a lawyer, he received his primary education at home. He read at age 3 and was intellectually precocious.

Lewis was born in Weymouth, Massachusetts (1875), but was raised in Nebraska where his family moved in 1884. In 1889, at age 14, he received his first formal education, attending the University of Nebraska preparatory school. He studied for two years at the University of Nebraska and then in 1893 transferred to Harvard University, where he obtained a Ph.D. in chemistry in 1899.

After two years of study in Germany and a stint as a government chemist in the Philippines, he became a professor of chemistry at the Massachusetts Institute of Technology (1905–1912). From 1912 to his death in 1946, the University of California at Berkeley was his home.

In 1916 he published a paper proposing that a chemical bond involved a pair of electrons shared or held jointly by two atoms. This proved to be one of the most fruitful ideas in the history of chemistry. Within a few years of his 1916 paper, with contributions from other scientists, a formal theory of chemical bonding based on sharing of electron pairs had been developed. Lewis is also the one who developed the symbolism now called Lewis symbols, or electron-dot structures. His memoirs indicate that he first had the idea of shared electrons while he was lecturing to an introductory chemistry class.

Lewis also made significant contributions in many other areas of chemistry besides bonding theory. Thermodynamic studies were a major interest during most of his career. In 1938 a generalized theory for acids and bases (now called Lewis acid–base theory) came from his mind and pen. Still later, he was the first to isolate heavy hydrogen ($^2_1$H).

Lewis was not only a profound researcher but also a teacher of renown. He strongly felt that students learn by doing. Among his teaching techniques was the use of large problem sets to reinforce concepts taught in lecture. This problem-set approach is still a part of most chemistry classes today.

7.3 THE OCTET RULE

A key concept in modern bonding theory is that certain arrangements of valence electrons are more stable than others. The term *stable* as used here refers to the idea that a system (in this case an arrangement of electrons) does not easily undergo spontaneous change.

The valence electron configurations possessed by the noble gases (He, Ne, Ar, Kr, Xe, and Rn; see Sec. 6.11) are considered to be the *most stable of all valence electron configurations*. For helium, this most stable electron configuration involves two outer shell electrons. The rest of the noble gases have eight outer shell electrons.

> *The outermost electron shell of an atom is also called the valence electron shell.*

He: $\left(\overline{1s^2}\right)$

Ne: $1s^2\left(\overline{2s^2 2p^6}\right)$

Ar: $1s^2 2s^2 2p^6 \left(\overline{3s^2 3p^6}\right)$

Kr: $1s^2 2s^2 2p^6 3s^2 3p^6 \left(\overline{4s^2}\right) 3d^{10} \left(\overline{4p^6}\right)$

Xe: $1s^2 2s^2 2p^6 3s^2 3p^6 4s^2 3d^{10} 4p^6 \left(\overline{5s^2}\right) 4d^{10} \left(\overline{5p^6}\right)$

Rn: $1s^2 2s^2 2p^6 3s^2 3p^6 4s^2 3d^{10} 4p^6 5s^2 4d^{10} 5p^6 \left(\overline{6s^2}\right) 4f^{14} 5d^{10} \left(\overline{6p^6}\right)$

> *The majority of noble gas compounds contain the element xenon. The first was synthesized in 1962. The three most studied xenon compounds are all xenon fluorides—XeF_2, XeF_4, and XeF_6. All three xenon fluorides are white solids at room temperature that react rapidly with water, even traces of moisture.*

The common feature among these noble gas electron configurations is *completely filled* outermost *s* and *p* subshells.

The conclusion that a noble gas configuration is the most stable of all outer shell electron configurations is based on the chemical properties of the noble gases. These elements are the *most unreactive* of all the elements. They are the only elemental gases found in nature in the form of individual uncombined atoms. There are no known compounds of He and Ne, and only a few compounds of Ar, Kr, Xe, and Rn. The noble gases appear to be "happy" the way they are. They have little or no "desire" to form bonds to other atoms.

Atoms of many elements that lack the very stable outer shell electron configuration of the noble gases tend to attain it in chemical reactions that result in compound formation. This observation has become known as the *octet rule* because of the eight outer shell electrons possessed by five of the six noble gases. A formal statement of the **octet rule** is: *In forming compounds, atoms of elements lose, gain, or share electrons in such a way as to produce a noble gas electron configuration for each of the atoms involved.*

Application of the octet rule to many different systems has shown that it has value in predicting correctly the observed combining ratios of atoms. For example, it explains why two hydrogen atoms rather than some other number are bonded to one oxygen atom in the molecular compound water. It explains why the formula of the ionic compound sodium chloride is NaCl rather than $NaCl_2$, $NaCl_3$, or Na_2Cl.

There are exceptions to the octet rule, but it is still used because of the large amount of information that it is able to correlate. It is particularly effective in explaining compound formation involving only representative elements. Often complications arise with transition and inner transition elements because of the involvement of *d* and *f* electrons in the bonding.

7.4 THE IONIC BOND MODEL

> *The pronunciation for the word ion is "eye-on."*

Electron transfer between two or more atoms is the basic premise for the ionic bond model. This electron transfer produces charged particles called ions. An **ion** *is an atom (or group of atoms) that is electrically charged as the result of loss or gain of electrons.* An atom is neutral only when the number of its protons (positive charges) is equal to the number of its electrons (negative charges). Loss or gain of electrons destroys this proton–electron balance and leaves a net charge on the atom.

If one or more electrons are gained by an atom, a negatively charged ion is produced; excess negative charge is present because electrons now outnumber protons. The loss of one or more electrons by an atom results in the formation of a positively charged ion; more protons than electrons are now present, resulting in excess positive charge. Note that the excess positive charge associated with a positive ion is never caused by proton gain but always by electron loss. If the number of protons remains constant and the number of electrons decreases, the result is net positive charge. The number of protons, which determines the identity of the element (Sec. 5.7), never changes during ion formation.

> An atom's nucleus *never* changes during the process of ion formation. The numbers of neutrons and protons remain constant.

The terms *anion* and *cation* describe ions of opposite charge. An **anion** *is a negatively charged ion.* A **cation** *is a positively charged ion.*

The charge on an ion is directly correlated with the number of electrons lost or gained. Loss of one, two, or three electrons gives cations with +1, +2, or +3 charges, respectively. Similarly, a gain of one, two, or three electrons gives anions with −1, −2, or −3 charges, respectively. (Atoms that have lost or gained more than three electrons are very seldom encountered.)

> The term *anion* is pronounced "an-eye-on," and the term *cation* is pronounced "cat-eye-on."

The notation for charges on ions is a superscript placed to the right of the elemental symbol. Some examples of ion symbols are

$$\text{positive ions (cations)} \quad Na^+, K^+, Ca^{2+}, Mg^{2+}, Al^{3+}$$
$$\text{negative ions (anions)} \quad Cl^-, Br^-, O^{2-}, S^{2-}, N^{3-}$$

Note that a single plus or minus sign is used to denote a charge of one, instead of using the notation 1+ or 1−. Also note that in multicharged ions the number precedes the charge sign; that is, the correct notation for a charge of plus two is 2+ rather than +2.

The chemical properties of a particle (atom or ion) depend on the particle's electron arrangement. Because an ion has a different electron configuration (fewer or more electrons) than the atom from which it was formed, it has different chemical properties as well. For example, water solutions containing Na^+ ions are very stable even though the element sodium (neutral Na) reacts vigorously with water.

EXAMPLE 7.3 **Writing Chemical Symbols for Ions**

Give the chemical symbol for each of the following ions:

  a. the ion formed when an aluminum atom loses three electrons
  b. the ion formed when a sulfur atom gains two electrons

SOLUTION

  a. A neutral aluminum atom contains 13 protons and 13 electrons, since the atomic number of aluminum is 13 (obtained from the periodic table). The aluminum ion formed by the loss of three electrons would still contain 13 protons but would have only 10 electrons because three electrons were lost.

$$13 \text{ protons} = 13 + \text{charges}$$
$$10 \text{ electrons} = \underline{10 - \text{charges}}$$
$$\text{net charge} = 3 +$$

The chemical symbol for the aluminum ion is Al^{3+}.

  b. The atomic number of sulfur is 16. Thus, 16 protons and 16 electrons are present in a neutral sulfur atom. A gain of two electrons raises the electron count to 18.

$$16 \text{ protons} = 16 + \text{charges}$$
$$18 \text{ electrons} = \underline{18 - \text{charges}}$$
$$\text{net charge} = 2 -$$

The chemical symbol for the sulfur ion is thus S^{2-}.

Answer Double Check:

Are the ion charges consistent with the concepts of loss and gain of electrons? Loss of electrons always produces a positively charged ion, and gain of electrons always produces a negatively charged ion. The answers are in accord with these generalizations.

▶ **Practice Exercise 7.3** Give the chemical symbol for each of the following ions.

    a. the ion formed when a magnesium atom loses two electrons
    b. the ion formed when a chlorine atom gains one electron

---

EXAMPLE 7.4 **Determining the Number of Protons and Electrons in Ions**

Determine the number of protons and electrons present in the following ions.

    a. P^{3-} **b.** Mg^{2+}

SOLUTION

    a. The number of protons present is the same as in a neutral atom and is therefore given by the atomic number, which is 15 for phosphorus.

$$\text{number of protons} = \text{atomic number} = 15$$

The number of electrons in a neutral phosphorus atom is 15. The charge of -3 on the ion indicates the gain of three electrons.

$$\text{number of electrons} = 15 + 3 = 18$$

    b. The atomic number of magnesium is 12.

$$\text{number of protons} = \text{atomic number} = 12$$

The number of electrons in a neutral Mg atom is 12. The charge of $+2$ on the ion indicates the loss of two electrons.

$$\text{number of electrons} = 12 - 2 = 10$$

Answer Double Check:

The number of protons never changes during ion formation. Is such the case here? Yes. The proton count remained constant in both cases. Electron count decreases when forming a positively charged ion and increases when forming a negatively charged ion. Is such the case here? Yes.

▶ **Practice Exercise 7.4** Determine the number of protons and electrons present in each of the following ions.

    a. Na^+ **b.** Se^{2-}

So far, our discussion about electron transfer and ion formation has focused on the loss or gain of electrons by isolated individual atoms. During ionic bond formation, ion formation occurs only when atoms of two elements are present—an element that can lose electrons and an element that can gain electrons. The total number of electrons lost by atoms of the one element is the same as the total number gained by atoms of the other element. Thus, positive and negative ions must always be formed at the same time (see Figure 7.2).

The mutual attraction between the positive and negative ions that results from electron transfer constitutes the force that holds the ions together as an ionic compound.

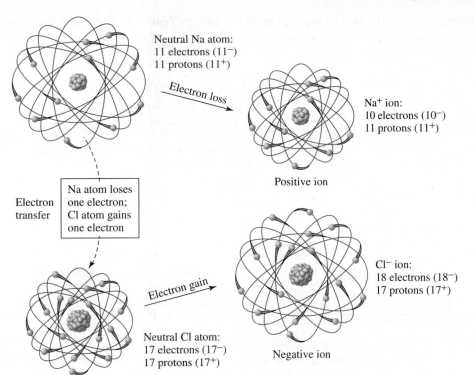

FIGURE 7.2 Electron transfer always produces positive and negative ions at the same time. You cannot form positive ions without also forming negative ions. Proton count does not change during ion formation. It is electrons that are transferred.

This force is referred to as an *ionic bond*. An **ionic bond** *is the chemical bond resulting from the attraction of positive and negative ions for each other*.

7.5 THE SIGN AND MAGNITUDE OF IONIC CHARGE

Figure 7.2 illustrates the ion formation process that occurs when the elements sodium and chlorine interact to form the compound NaCl. Sodium atoms lose (transfer) one electron to chlorine atoms, producing Na^+ and Cl^- ions. These ions combine in a 1:1 ratio to give the compound NaCl.

Why do sodium atoms form Na^+ and not Na^{2+} or Na^- ions? Why do chlorine atoms form Cl^- ions rather than Cl^{2-} or Cl^+ ions? In general, what determines the specific number of electrons lost or gained in electron transfer processes?

The octet rule (Sec. 7.3) provides a very simple and straightforward explanation for the charge magnitude associated with ions of the representative elements. *Atoms tend to gain or lose electrons until they have obtained an electron configuration that is the same as that of a noble gas.*

Consider the element sodium, which has the electron configuration

$$1s^2 2s^2 2p^6 3s^1$$

It can attain a noble gas configuration by losing one electron (to give it the electron configuration of neon) or by gaining seven electrons (to give it the electron configuration of argon).

Na $(1s^2 2s^2 2p^6 3s^1)$

Loss of 1 e$^-$ → Na^+ $(1s^2 2s^2 2p^6)$
Electron configuration of neon

Gain of 7 e$^-$ → Na^{7-} $(1s^2 2s^2 2p^6 3s^2 3p^6)$
Electron configuration of argon

The first process, the loss of one electron, is more energetically favorable than the gain of seven electrons and is the process that occurs. The process that involves the fewer number of electrons will always be the more energetically favorable process and will be the process that occurs.

Consider the element chlorine, which has the electron configuration

$$1s^2 2s^2 2p^6 3s^2 3p^5$$

It can attain a noble gas configuration by losing seven electrons (to give it the electron configuration of neon) or by gaining one electron (to give it the electron configuration of argon). The latter occurs for the reason cited previously.

$$Cl \quad (1s^2 2s^2 2p^6 3s^2 3p^5) \xrightarrow[\text{Gain of 1 } e^-]{\text{Loss of 7 } e^-} \begin{cases} Cl^{7+} \quad (1s^2 2s^2 2p^6) \\ \text{Electron configuration of neon} \\ Cl^- \quad (1s^2 2s^2 2p^6 3s^2 3p^6) \\ \text{Electron configuration of argon} \end{cases}$$

The type of consideration we have just used for the elements sodium and chlorine leads to the following generalizations:

1. Metal atoms containing one, two, or three valence electrons (the metals in groups IA, IIA, and IIIA of the periodic table) tend to lose electrons to acquire a noble gas electron configuration. The noble gas involved is the one preceding the metal in the periodic table.

> The positive charge on metal ions from groups IA, IIA, and IIIA has a magnitude equal to the metal's periodic table group number.

Group IA metals form +1 ions.

Group IIA metals form +2 ions.

Group IIIA metals form +3 ions.

Group IA metals are all located one periodic table position past a noble gas. Thus, they will each have one more electron than the preceding noble gas. This electron must be lost if a noble gas configuration is to be obtained. Group IIA and IIIA metals are two and three periodic table positions, respectively, beyond a noble gas. Consequently, two and three electrons, respectively, must be lost for these metals to attain a noble gas electron configuration.

2. Nonmetal atoms containing five, six, or seven valence electrons (the nonmetals in groups VA, VIA, and VIIA of the periodic table) tend to gain electrons to acquire a noble gas configuration. The noble gas involved is the one following the nonmetal in the periodic table.

> Nonmetals from groups VA, VIA, and VIIA form negative ions whose charge is equal to the group number minus 8. For example, S, in group VIA, forms S^{2-} ions (6 − 8 = −2).

Group VIIA nonmetals form −1 ions.

Group VIA nonmetals form −2 ions.

Group VA nonmetals form −3 ions.

The nonmetal ionic charge guidelines can be explained by reasoning similar to that for the metals; only this time, the periodic table positions are those immediately preceding the noble gases. Consequently, electrons must be gained to attain noble gas configurations.

3. Elements in group IVA occupy unique positions relative to the noble gases. They are located equidistant between two noble gases. For example, the element carbon is four positions beyond helium and four positions before neon. Theoretically, ions with charges of +4 or −4 could be formed by carbon, but in most cases the bonding that results is more adequately described by the covalent bond model to be discussed in Section 7.10.

EXAMPLE 7.5   Predicting Sign and Magnitude of Ionic Charge

How many electrons must each of the following atoms lose or gain in order to obtain a noble gas electron configuration and what is the chemical symbol for the ion produced from such loss or gain?

   a. N   **b.** S   **c.** K   **d.** Mg

SOLUTION

Elements with more than four valence electrons achieve an octet of electrons via electron gain. Conversely, electrons with less than four valence electrons achieve an octet of electrons through electron loss.

   a. Nitrogen, a group VA element, has five valence electrons and the electron configuration $1s^2 2s^2 2p^3$. It will need to gain three more electrons to achieve an octet of electrons. Gain of three electrons produces a triply-charged negative ion.

$$N^{3-} \quad 1s^2 2s^2 2p^6 \text{ (neon configuration)}$$

   b. Sulfur, a group VIA element, has six valence electrons and the electron configuration $1s^2 2s^2 2p^6 3s^2 3p^4$. An octet of electrons is achieved through gain of two electrons. Such electron gain produces a doubly-charged negative ion.

$$S^{2-} \quad 1s^2 2s^2 2p^6 3s^2 3p^6 \text{ (argon configuration)}$$

   c. Potassium, a member of group IA, possesses one valence electron. Loss of this one electron from its electron arrangement ($1s^2 2s^2 2p^6 3s^2 3p^6 4s^1$) produces an ion with a charge of +1.

$$K^{+} \quad 1s^2 2s^2 2p^6 3s^2 3p^6 \text{ (argon configuration)}$$

   d. Magnesium, a member of group IIA, possesses two valence electrons ($1s^2 2s^2 2p^6 3s^2$), both of which will be lost during ion formation. The resulting positive ion carries a +2 charge.

$$Mg^{2+} \quad 1s^2 2s^2 2p^6 \text{ (neon configuration)}$$

▶ **Practice Exercise 7.5**   How many electrons must each of the following atoms lose or gain in order to obtain a noble gas electron configuration and what is the chemical symbol for the ion produced from such loss or gain?

   a. O   **b.** F   **c.** Al   **d.** Ca

Isoelectronic Species

An ion formed in the preceding manner with an electronic configuration the same as that of a noble gas is said to be *isoelectronic* with the noble gas. **Isoelectronic species** *are ions, or an atom and ions, having the same number and configuration of electrons.* An atom and an ion or two ions may be isoelectronic. Numerous ions that are isoelectronic with a given noble gas exist, as can be seen from the entries in Table 7.1.

    It should be emphasized that an ion that is isoelectronic with a noble gas does not have the properties of the noble gas. It has not been converted into the noble gas. The number of protons in the nucleus of the isoelectronic ion is different from that in the noble gas. These points are emphasized by the comparison in Table 7.2 between Mg^{2+} and Ne, the noble gas with which Mg^{2+} is isoelectronic.

TABLE 7.1 Ions Isoelectronic with Selected Noble Gases

Electron Configuration	Anions			Noble Gas	Cations		
$1s^2$			H^-	He	Li^+	Be^{2+}	
$1s^2 2s^2 2p^6$	N^{3-}	O^{2-}	F^-	Ne	Na^+	Mg^{2+}	Al^{3+}
$1s^2 2s^2 2p^6 3s^2 3p^6$	P^{3-}	S^{2-}	Cl^-	Ar	K^+	Ca^{2+}	Sc^{3+}

TABLE 7.2 Comparison of the Structure of an Mg^{2+} Ion and an Ne Atom, the Noble Gas Atom Isoelectronic with the Ion

	Ne Atom	Mg^{2+} Ion
Protons (in the nucleus)	10	12
Electrons (around the nucleus)	10	10
Atomic number	10	12
Charge	0	2+

7.6 LEWIS STRUCTURES FOR IONIC COMPOUNDS

The use of *Lewis structures* is helpful in visualizing the formation of simple ionic compounds. A **Lewis structure** *is a grouping of Lewis symbols that shows either the transfer of electrons or the sharing of electrons in chemical bonds.* Lewis symbols involve atoms of individual elements. Lewis structures involve compounds. The reaction between sodium (with one valence electron) and chlorine (with seven valence electrons) is represented as follows with a Lewis structure.

$$Na\cdot + \cdot \ddot{\underset{\cdot\cdot}{Cl}}: \longrightarrow Na^+ \; [:\ddot{\underset{\cdot\cdot}{Cl}}:]^- \longrightarrow NaCl$$

The loss of an electron by sodium empties its valence shell. The next inner shell, which contains eight electrons (a noble gas configuration), then becomes the valence shell. After the valence shell of chlorine gains one electron, it then has the "desired" eight valence electrons.

When sodium, which has one valence electron, combines with oxygen, which has six valence electrons, two sodium atoms are needed to meet the "electron requirements" of one oxygen atom.

$$\begin{matrix} Na\cdot \searrow \\ & \ddot{\underset{\cdot\cdot}{O}}: \\ Na\cdot \nearrow \end{matrix} \longrightarrow \begin{matrix} Na^+ \\ Na^+ \end{matrix} [:\ddot{\underset{\cdot\cdot}{O}}:]^{2-} \longrightarrow Na_2O$$

Note how oxygen's need for two additional electrons dictates that each of the two sodium atoms supply one electron to the oxygen atom; hence the formula is Na_2O.

An opposite situation to that for Na_2O occurs in the reaction between calcium, which has two valence electrons, and chlorine, which has seven valence electrons. Here, two chlorine atoms are required to accommodate electrons transferred from one calcium atom because a chlorine atom can accept only one electron. (It has seven valence electrons and needs only eight.)

$$\text{Ca·} \quad \ddot{\text{Cl}}: \quad \ddot{\text{Cl}}: \longrightarrow \text{Ca}^{2+} \begin{matrix} [:\ddot{\text{Cl}}:]^- \\ [:\ddot{\text{Cl}}:]^- \end{matrix} \longrightarrow \text{CaCl}_2$$

We can usually tell whether a compound is ionic (consisting of ions) or covalent (consisting of molecules) from its composition. In general, metals tend to form cations (positive ions) and nonmetals tend to form anions (negative ions). Therefore, *ionic compounds are generally compounds in which both a metal and nonmetal are present*, as in NaCl, Na$_2$O, and CaCl$_2$. In contrast, *covalent compounds are generally compounds in which only nonmetals are present*, as in H$_2$O (see Sec. 7.10).

| EXAMPLE 7.6 | **Using Lewis Structures to Depict Ionic Compound Formation** |

Show the formation of the following ionic compounds using Lewis structures.

a. K$_3$P   **b.** NaF   **c.** Al$_2$O$_3$

SOLUTION

a. Potassium (a group IA element) has one valence electron, which it would "like" to lose. Phosphorus (a group VA element) has five valence electrons and would thus like to acquire three more. Three potassium atoms will be required to supply enough electrons for one phosphorus atom.

$$\begin{matrix} \text{K·} \\ \text{K·} \quad \cdot\ddot{\text{P}}: \\ \text{K·} \end{matrix} \longrightarrow \begin{matrix} \text{K}^+ \\ \text{K}^+ [:\ddot{\text{P}}:]^{3-} \\ \text{K}^+ \end{matrix} \longrightarrow \text{K}_3\text{P}$$

b. Sodium (a group IA element) has one valence electron, and fluorine (a group VIIA element) has seven valence electrons. The transfer of the one sodium valence electron to a fluorine atom will result in each atom's having a noble gas electron configuration. Thus, these two elements combine in a 1:1 ratio.

$$\text{Na} \quad \cdot\ddot{\text{F}}: \longrightarrow \text{Na}^+ [:\ddot{\text{F}}:]^- \longrightarrow \text{NaF}$$

c. Aluminum (a group IIIA element) has three valence electrons, all of which need to be lost through electron transfer. Oxygen (a group VIA element) has six valence electrons and thus needs to acquire two more. Three oxygen atoms are needed to accommodate the electrons given up by two aluminum atoms.

$$\begin{matrix} \text{Al·} \quad \cdot\ddot{\text{O}}: \\ \quad \cdot\ddot{\text{O}}: \longrightarrow \\ \text{Al·} \quad \cdot\ddot{\text{O}}: \end{matrix} \quad \begin{matrix} [:\ddot{\text{O}}:]^{2-} \\ \text{Al}^{3+} \\ [:\ddot{\text{O}}:]^{2-} \\ \text{Al}^{3+} \\ [:\ddot{\text{O}}:]^{2-} \end{matrix} \longrightarrow \text{Al}_2\text{O}_3$$

▶ **Practice Exercise 7.6** Show the formation of the following ionic compounds using Lewis structures.

a. $AlCl_3$ **b.** Na_2O **c.** Mg_3N_2

7.7 CHEMICAL FORMULAS FOR IONIC COMPOUNDS

Since total electron loss always equals total electron gain in an electron transfer process, ionic compounds are always neutral; no net charge is present. The total positive charge on the ions that have lost electrons is always exactly counterbalanced by the total negative charge on the ions that have gained electrons. Thus, *the ratio in which positive and negative ions combine is the ratio that achieves charge neutrality for the resulting compound.* This generalization can be used instead of Lewis structures to determine ionic compound formulas. Ions are combined in the ratio that causes the positive and negative charges to add to zero.

The correct combining ratio when K^+ and S^{2-} ions combine is two to one. Two K^+ ions (each of +1 charge) will be required to balance the charge on a single S^{2-} ion.

> In any ionic compound the total positive charge (from the cations) plus the total negative charge (from the anions) must add up to zero. Ionic compounds cannot have a net charge.

$$2(K^+): \quad (2\ \text{ions}) \times (\text{charge of} +1) = +2$$
$$S^{2-}: \quad (1\ \text{ion}) \times \underline{(\text{charge of} -2) = -2}$$
$$\text{net charge} = \quad 0$$

Hence, the formula is K_2S.

Example 7.7 gives further illustration of the procedures needed to determine correct combining ratios between ions and to write correct ionic formulas from the combining ratios. Written ionic formulas are consistent with the following guidelines:

1. The symbol for the positive ion is always written first.
2. The charges on the ions that are present are *not* shown in the chemical formula. Knowledge of charges is necessary to determine the chemical formula, but once it is determined, the charges are not explicitly written.
3. The numbers in the formula (the subscripts) give the combining ratio for the ions.

EXAMPLE 7.7 **Using Ionic Charges to Determine the Chemical Formula of an Ionic Compound**

Determine the chemical formula for the compound that is formed when each of the following types of ions interact.

a. Ba^{2+} and Cl^- **b.** Ba^{2+} and S^{2-} **c.** Ba^{2+} and N^{3-}

SOLUTION

a. Ba^{2+} and Cl^- ions will combine in a one-to-two ratio because this combination will cause the total charge to add up to zero. One Ba^{2+} ion gives a total positive charge of 2. Two Cl^- ions give a total negative charge of 2. Thus, the chemical formula of the compound is $BaCl_2$.

b. The chemical formula of this compound is simply BaS (a one-to-one ratio between ions). One Ba^{2+} ion contributes two units of positive charge, and that is counterbalanced by two units of negative charge from the S^{2-} ion.

c. The numbers in the charges for these ions are 2 and 3. The lowest common multiple of 2 and 3 is 6 ($2 \times 3 = 6$). Thus, we will need six units of positive charge and six units of negative charge. Three Ba^{2+} ions are needed to give the six units of positive charge, and two N^{3-} ions are needed to give the six units of negative charge. The combining ratio of ions is three to two, and the chemical formula is Ba_3N_2. The strategy used in determining this formula, finding the lowest common multiple in the charges of the ions, will always work.

Answer Double Check:

When writing the chemical formula for an ionic compound, the charges associated with the ions present are never explicitly shown in the finished chemical formula. Such is the case with these answers. Subscripts giving the combining ratio of the ions are used in the chemical formula, but the charges on the ions (even though they are needed to generate the chemical formula) are never shown.

▶ **Practice Exercise 7.7**   Determine the chemical formula for the compound that is formed when each of the following types of ions interact.

a. Na^+ and S^{2-} **b.** Mg^{2+} and F^- **c.** Ca^{2+} and N^{3-}

Before leaving the subject of ions, ionic bonds, and chemical formulas for ionic compounds, let us quickly review the key principles about ionic bonding that have been presented.

1. Ionic compounds usually contain both a metallic and a nonmetallic element.
2. The metallic element atoms lose electrons to produce positive ions, and the nonmetallic element atoms gain electrons to produce negative ions.
3. The electrons lost by the metal atoms are the same ones that are gained by the nonmetal atoms. Electron loss must always equal electron gain.
4. The ratio in which positive metal ions and negative nonmetal ions combine is the lowest whole-number ratio that achieves charge neutrality for the resulting compound.
5. Metals from groups IA, IIA, and IIIA of the periodic table form ions with charges of +1, +2, and +3, respectively. Nonmetals of groups VIIA, VIA, and VA of the periodic table form ions with charges of −1, −2, and −3, respectively. Table 7.3 lists, in general terms, all of the possible metal–nonmetal combinations from these periodic table groups that result in the formation of ionic compounds.

TABLE 7.3 General Formulas for Ionic Compounds as a Function of Periodic Table Position of the Metal and Nonmetal

Metals (M)	Nonmetals (X)		
	VIIA (−1 ions)	VIA (−2 ions)	VA (−3 ions)
IA (+1 ions)	MX	M_2X	M_3X
IIA (+2 ions)	MX_2	MX	M_3X_2
IIIA (+3 ions)	MX_3	M_2X_3	MX

FIGURE 7.3   The arrangement of ions in sodium chloride. Each of the Na$^+$ ions is surrounded on all sides by Cl$^-$ ions, and each of the Cl$^-$ ions is surrounded on all sides by Na$^+$ ions.

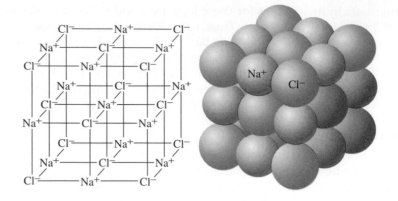

7.8 STRUCTURE OF IONIC COMPOUNDS

Ionic solids consist of positive and negative ions arranged in such a way that each ion is surrounded by nearest neighbors of the opposite charge. Any given ion is bonded by electrostatic (positive–negative) attractions to all of the other ions of opposite charge immediately surrounding it. Figure 7.3 gives two three-dimensional depictions of the arrangement of ions for the ionic compound NaCl (table salt).

The alternating array of positive and negative ions present in an ionic compound means that discrete molecules do not exist in such compounds (Sec. 5.4). Therefore, the chemical formulas of ionic compounds cannot represent the composition of molecules of these substances. Instead, such formulas represent the simplest combining ratio for the ions present. The chemical formula for sodium chloride, NaCl, indicates that sodium and chloride ions are present in a one-to-one ratio in this compound. Chemists use the term *formula unit*, rather than molecule, to refer to the smallest unit of an ionic compound. A **formula unit** *is the smallest whole-number repeating ratio of ions present in an ionic compound that results in charge neutrality.* A formula unit is hypothetic, because it does not exist as a separate entity; it is only a part of the extended array of ions that constitute an ionic solid (see Figure 7.4).

Although the chemical formulas for ionic compounds represent only ratios, they are used in equations and chemical calculations in the same way as the chemical formulas for molecular species. Remember, however, that they cannot be interpreted as indicating that molecules exist for these substances. They represent the simplest ratio of ions.

Looked at closely, salt grains reveal the cubic shape of the NaCl crystal. The intricate shapes of many gems such as rubies and emeralds reflect the arrangement of their microscopic ionic arrays.

The ions present in an ionic solid adopt an arrangement that maximizes attractions between ions of opposite charge and minimizes repulsions between ions of like charge. The specific arrangement that is adopted depends on ion sizes and on the ratio between positive and negative ions. Arrangements are usually very symmetrical and result in crystalline solids—that is, solids with highly regular shapes.

FIGURE 7.4   Two-dimensional cross section of an ionic solid (NaCl). No molecule can be distinguished in this structure. Instead, we can recognize a basic formula unit that is repeated indefinitely.

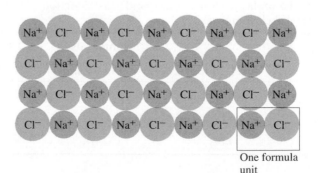

One formula unit

Crystalline solids usually have flat surfaces or faces that make definite angles with one another.

7.9 POLYATOMIC IONS

To this point in this chapter, all references to and comments about ions have involved *monoatomic ions*. A **monoatomic ion** *is an ion formed from a single atom through loss or gain of electrons.* Such ions are very common and very important. Another large and important category of ions, called *polyatomic ions*, exists. A **polyatomic ion** *is an ion formed from a group of atoms (held together by covalent bonds) through loss or gain of electrons.* Numerous ionic compounds exist in which the positive or negative ion (sometimes both) contains more than one atom. Polyatomic ions are very stable species, generally maintaining their identity during chemical reactions.

An example of a polyatomic ion is the sulfate ion, SO_4^{2-}. This ion contains four oxygen atoms and one sulfur atom, and the whole group of five atoms has acquired a -2 charge. The whole sulfate group is the ion rather than any one atom within the group. Covalent bonding, discussed in Section 7.10, holds the sulfur and oxygen atoms together.

Polyatomic ions are not molecules. They never occur alone as molecules do. Instead, they are always found associated with ions of opposite charge. Polyatomic ions are *pieces* of compounds, not compounds. Ionic compounds require the presence of both positive and negative ions and are neutral overall. Polyatomic ions are always charged species.

Chemical formulas for ionic compounds containing polyatomic ions are determined in the same way as those for ionic compounds containing monoatomic ions (Sec. 7.7). The basic rule is the same: The total positive and negative charge present must add up to zero.

Two conventions not encountered previously in chemical formula writing often arise when writing chemical formulas where polyatomic ions are present. They are the following:

1. When more than one polyatomic ion of a given kind is required in a chemical formula, the polyatomic ion is enclosed in parentheses, and a subscript is placed after the parentheses to indicate the number of polyatomic ions needed.
2. To preserve the identity of polyatomic ions, the same elemental symbol may be used more than once in a chemical formula.

Example 7.8 contains examples illustrating the use of both of these new conventions. Besides the sulfate ion, four other polyatomic ions are involved in this example: OH^- (hydroxide ion), NO_3^- (nitrate ion), NH_4^+ (ammonium ion), and CN^- (cyanide ion). The chemical formulas for numerous other polyatomic ions are considered in Section 8.4.

EXAMPLE 7.8 **Writing Chemical Formulas for Ionic Compounds Containing Polyatomic Ions**

Determine the chemical formulas for the ionic compounds containing the following pairs of ions, one or both of which are polyatomic.

a. K^+ and SO_4^{2-}
b. Na^+ and NO_3^-
c. Ca^{2+} and OH^-
d. NH_4^+ and CN^-

SOLUTION

a. To equalize the total positive and negative charges, we need two K^+ ions for each SO_4^{2-} ion. We indicate the presence of the two K^+ ions with the subscript 2 following the symbol of the ion. The chemical formula of the compound is K_2SO_4. The convention that the positive ion is always written first in the formula still holds when polyatomic ions are present.

b. Since both of these ions possess a charge of one, combining them in a one-to-one ratio will balance the charge. The chemical formula of the compound is $NaNO_3$.

c. Two OH^- ions are needed to balance the charge on one Ca^{2+} ion. Since more than one polyatomic ion is needed, the chemical formula will contain parentheses: $Ca(OH)_2$. The subscript 2 after the parentheses indicates two of what is inside the parentheses. If parentheses were not used, the chemical formula would appear to be $CaOH_2$, which is not intended and which actually conveys false information. The formula $Ca(OH)_2$ indicates a formula unit containing one Ca atom, two O atoms, and two H atoms (Sec. 5.4); the formula $CaOH_2$ would indicate a formula unit containing one Ca atom, one O atom, and two H atoms. Verbally the correct formula, $Ca(OH)_2$, would be read as "C-A" (pause) "O-H-taken-twice."

d. In this compound both ions are polyatomic, a perfectly legal situation. Since the ions have equal but opposite charges, they will combine in a one-to-one ratio. The chemical formula is thus NH_4CN. No parentheses are needed because we need only one polyatomic ion of each type in a formula unit. Parentheses are used only when there are two or more polyatomic ions of a given kind in a formula unit. What is different about this chemical formula is the appearance of the symbol for the element nitrogen (N) at two locations in the formula. This could be prevented by combining the two nitrogens, giving the chemical formula N_2H_4C. However, combining is not done in situations like this because the identity of the polyatomic ions present is lost in the resulting combined formula. The chemical formula N_2H_4C does not convey the message that NH_4^+ and CN^- ions are present; the formula NH_4CN does. Thus, in writing chemical formulas that contain polyatomic ions, we always maintain the identities of these ions even if it means having the same elemental symbol at more than one location in the chemical formula.

Answer Double Check:

Is the use of parentheses in the chemical formulas appropriate? Yes. Only one of the chemical formulas contains parentheses: the one in which more than one polyatomic ion of a given kind is present. In the other three chemical formulas parentheses are not needed, because only one polyatomic ion of a given kind is present.

▶ **Practice Exercise 7.8** Determine the chemical formulas for the ionic compounds containing the following pairs of ions, one or both of which are polyatomic.

a. Na^+ and SO_4^{2-}
b. K^+ and NO_3^-
c. Na^+ and OH^-
d. NH_4^+ and S^{2-}

Figure 7.5 gives pictorial representations of the ionic makeup of the four compounds whose formulas were determined in Example 7.8.

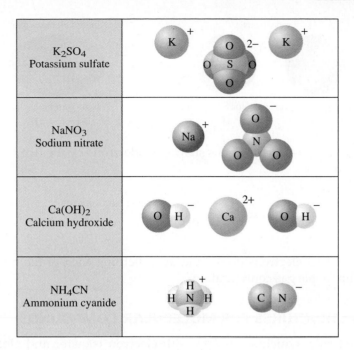

K₂SO₄ Potassium sulfate	
NaNO₃ Sodium nitrate	
Ca(OH)₂ Calcium hydroxide	
NH₄CN Ammonium cyanide	

FIGURE 7.5 Ionic components of selected compounds in which polyatomic ions are present.

7.10 THE COVALENT BOND MODEL

We begin our discussion of covalent bonding and the molecular compounds that result from such bonding by listing several key differences between ionic bonding (Sec. 7.4) and covalent bonding and the resulting ionic and molecular compounds.

1. Ionic bonds form between atoms of a metal and a nonmetal. Covalent bond formation occurs between two nonmetal atoms. The two nonmetal atoms can be identical but need not be so.
2. *Electron transfer* is the mechanism by which ionic bond formation occurs. Covalent bond formation involves *electron sharing*.
3. In an ionic compound, discrete molecules do not exist since such compounds involve an extended array of alternating positive and negative ions. In covalently bonded compounds, the basic structural unit is the molecule. Indeed, such compounds are called *molecular* compounds.
4. All ionic compounds are solids at room temperature. Molecular compounds may be solids (glucose), liquids (water), or gases (carbon dioxide).
5. An ionic solid, if soluble in water, forms an aqueous solution that conducts electricity. The electrical conductance is related to the presence of ions (charged particles) in the solution. A molecular compound, if soluble in water, usually produces a nonconducting aqueous solution.

A **covalent bond** *is a chemical bond resulting from two nuclei attracting the same shared electrons.* A consideration of the simple hydrogen (H_2) molecule provides initial insights into the nature of the covalent bond. When two hydrogen atoms, each with a single electron, are brought together, the orbitals containing these electrons *overlap* (as shown in Figure 7.6) to produce an orbital common to both atoms. The two electrons, one from each H atom, move throughout this new orbital and are said to be *shared* by the two nuclei.

Once two orbitals overlap, the most favorable location for the shared electrons is the area directly between the two nuclei. Here the two electrons can simultaneously interact with (be attracted to) both nuclei, a situation that produces increased stability.

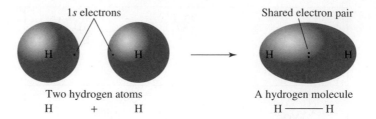

1s electrons

Shared electron pair

Two hydrogen atoms
H   +   H

A hydrogen molecule
H ———— H

In terms of Lewis structures, this sharing of electrons by the two hydrogen atoms is diagrammed as

Shared electron pair

H·H ⟶ H:H

The two shared electrons do double duty, helping each of the hydrogen atoms achieve a helium noble gas configuration.

7.11 LEWIS STRUCTURES FOR MOLECULAR COMPOUNDS

Using the octet rule, which applies to both electron transfer and electron sharing (Sec. 7.3) and Lewis symbols (Sec. 7.2), let us now consider the formation of selected simple covalently bonded molecules containing the element chlorine. Chlorine, located in group VIIA of the periodic table, has seven valence electrons. Its Lewis symbol is

·C̈l:

In covalent bonding, hydrogen atoms violate the "octet part" of the octet rule—they need only two electrons—but they do not violate the "noble gas part" of the octet rule—after sharing electrons, they are isoelectronic with helium.

Chlorine needs one additional electron to achieve the octet of electrons that makes it isoelectronic with the noble gas argon. In ionic compounds, where it bonds to metals, the Cl receives the needed electron via electron transfer. When Cl combines with another nonmetal, a common situation, the octet of electrons is completed via electron sharing. Representative of the situation where chlorine obtains its eighth valence electron through an electron-sharing process are the molecules HCl, Cl_2, and BrCl, whose Lewis structures are as follows.

H ⌒ ·C̈l: ⟶ H:C̈l:

:C̈l ⌒ ·C̈l: ⟶ :C̈l:C̈l:

:B̈r ⌒ ·C̈l: ⟶ :B̈r:C̈l:

The HCl and BrCl molecules illustrate the point that the two atoms involved in a covalent bond need not be identical (as in H_2 or Cl_2).

A common practice when writing Lewis structures for covalently bonded molecules is to represent the *shared* electron pairs with dashes. Using this notation, the previously discussed H_2, HCl, Cl_2, and BrCl molecules are written as

H—H   H—C̈l: :C̈l—C̈l: :B̈r—C̈l:

Nonbonding electron pairs are often also referred to as *unshared electron pairs* or *lone electron pairs* (or simply, *lone pairs*).

The atoms in covalently bonded molecules often possess both *bonding* and *nonbonding* electrons. **Bonding electrons** *are pairs of valence electrons that are shared between* *atoms in a covalent bond.* Each of the chlorine atoms in the molecules HCl, Cl_2, and BrCl possess one pair of bonding electrons. **Nonbonding electrons** *are pairs of valence electrons about an atom that are not involved in electron sharing.* Each of the chlorine atoms in HCl, Cl_2, and BrCl possesses three pairs of nonbonding electrons, as does the bromine atom in BrCl.

Bonding electrons (black)
Nonbonding electrons (blue)

The number of covalent bonds that an atom forms is equal to the number of electrons it needs to achieve a noble gas configuration. Note that the chlorine atoms in HCl, Cl_2, and BrCl all formed one covalent bond. For chlorine, seven valence electrons plus one electron acquired by electron sharing (one bond) give the eight valence electrons needed for a noble gas electronic configuration. The elements oxygen, nitrogen, and carbon have, respectively, six, five, and four valence electrons. Therefore, these elements form, respectively, two, three, and four covalent bonds. The number of covalent bonds these three elements form is reflected in the formulas of their simplest hydrogen compounds H_2O, NH_3, and CH_4. Lewis structures for these three molecules are as follows.

In Section 7.17 we will learn that nonbonding electron pairs play an important role in determining the shape (geometry) of molecules in which three or more atoms are present.

Oxygen has six valence electrons and gains two more through sharing

Nitrogen has five valence electrons and gains three more through sharing

Carbon has four valence electrons and gains four more through sharing

Thus, we see that just as the octet rule was useful in determining the ratio of ions in ionic compounds, we can use it to predict chemical formulas for molecular compounds. Example 7.9 gives additional illustrations of the use of the octet rule to determine chemical formulas for molecular compounds.

EXAMPLE 7.9   Writing Lewis Structures for Simple Molecular Compounds

Write Lewis structures for the simplest molecular compound formed from the following pairs of nonmetals.

  a. phosphorus and hydrogen
  b. sulfur and fluorine
  c. oxygen and chlorine

SOLUTION

a. Phosphorus is in group VA of the periodic table and thus has five valence electrons. It will form three covalent bonds, which through electron sharing will give it eight valence electrons (noble gas configuration). Hydrogen, in group IA of the periodic table, has one valence electron and will form only one covalent bond. For H an octet is two electrons; the noble gas that hydrogen "mimics" is helium, which has only two valence electrons. Using Lewis structures, we obtain a compound with the formula PH_3.

$$
\begin{array}{ccc}
H\!:\! & & H \\
H\!:\!P\!: \longrightarrow H\!:\!\ddot{P}\!: & \text{or} & H\!-\!P\!: \\
H\!:\! & \ddot{H} & \\
& & H
\end{array}
$$

b. Sulfur has six valence electrons, and fluorine has seven valence electrons. Thus sulfur will form two covalent bonds ($6 + 2 = 8$), and fluorine will form one covalent bond ($7 + 1 = 8$). The formula of the compound is SF_2.

$$
\begin{array}{ccc}
:\!\ddot{F}\! & :\!\ddot{F}\!: & :\!\ddot{F}\!: \\
:\!\ddot{S}\!: \longrightarrow :\!\ddot{F}\!:\!\ddot{S}\!: & \text{or} & :\!\ddot{F}\!-\!\ddot{S}\!: \\
:\!\ddot{F}\! & &
\end{array}
$$

c. Oxygen, with six valence electrons, will form two covalent bonds, and chlorine, with seven valence electrons, will form only one covalent bond. The formula of the compound is thus Cl_2O, which has the following Lewis structure.

$$
\begin{array}{ccc}
:\!\ddot{Cl}\! & :\!\ddot{Cl}\!:\!\ddot{O}\!: & :\!\ddot{Cl}\!-\!\ddot{O}\!: \\
:\!\ddot{O}\!: \longrightarrow & :\!\ddot{Cl}\!: & \text{or} \quad :\!\ddot{Cl}\!: \\
:\!\ddot{Cl}\! & &
\end{array}
$$

▶ **Practice Exercise 7.9** Write Lewis structures for the simplest molecular compound formed from the following pairs of nonmetals.

a. silicon and bromine
b. hydrogen and chlorine
c. arsenic and hydrogen

7.12 SINGLE, DOUBLE, AND TRIPLE COVALENT BONDS

A **single covalent bond** *is a covalent bond in which two atoms share one pair of valence electrons.* All bonds in all of the molecules discussed in the previous section were *single* covalent bonds.

Single covalent bonds are not adequate to explain covalent bonding in all molecules. Sometimes two atoms must share two or three pairs of electrons in order to provide a complete octet of electrons for each atom involved in the bonding. Such bonds are called *double* covalent bonds and *triple* covalent bonds. A **double covalent bond** *is a covalent bond in which two atoms share two pairs of valence electrons.* A double covalent bond between two atoms is stronger than a single covalent bond between the same two atoms; that is, it takes more energy to break the double bond than it takes to break the single bond. A **triple covalent bond** *is a covalent bond in which two atoms share three pairs of valence electrons.* For covalent bonds between the same two atoms, a triple covalent bond is stronger than a double covalent bond, which in turn is stronger than a single covalent bond. The term *multiple covalent bond* is a designation that applies collectively to both double and triple covalent bonds. This designation is often shortened to simply *multiple bond*.

One of the simplest molecules possessing a multiple covalent bond is the N_2 molecule; a triple covalent bond is present. A nitrogen atom has five valence electrons and needs three additional electrons to complete its octet.

$$\cdot\ddot{N}\cdot$$

In a N_2 molecule the only sharing that can take place is between the two nitrogen atoms. They are the only atoms present. Thus, to acquire a noble gas electron configuration, each nitrogen atom must share three of its electrons with the other nitrogen atom.

$$:\dot{N}\cdot \longrightarrow \longleftarrow \cdot\dot{N}: \longrightarrow :N:::N: \quad \text{or} \quad :N\equiv N:$$

Notice how all three shared electron pairs are placed in the area between the two nitrogen atoms in the above bonding diagrams. Note also that three lines are used to denote a triple covalent bond, paralleling the use of one line to denote a single covalent bond.

In "bookkeeping" electrons in a Lewis structure, to make sure that all atoms in the molecule have achieved their octet of electrons, *all* electrons in a multiple covalent bond are considered to "belong" to *both* of the atoms involved in that bond. The bookkeeping for the N_2 molecule would be

Each of the circles about a N atom contains eight valence electrons. Again, all the electrons in a multiple covalent bond are considered to belong to each of the atoms in the bond. Circles are never drawn to include just some of the electrons in a multiple covalent bond.

A slightly more complicated molecule containing a triple covalent bond is the molecule C_2H_2 (acetylene). A carbon–carbon triple covalent bond is present as well as two carbon–hydrogen single covalent bonds. The arrangement of valence electrons in C_2H_2 is as follows.

$$H:\dot{C}\cdot \longrightarrow \longleftarrow \cdot\dot{C}:H \longrightarrow H:C:::C:H \quad \text{or} \quad H-C\equiv C-H$$

The two atoms in a triple covalent bond are commonly the same element. However, they do not have to be. The molecule HCN (hydrogen cyanide) contains a heteroatomic triple covalent bond.

$$H:C:::N: \quad \text{or} \quad H-C\equiv N:$$

Double covalent bonds are found in numerous molecules. A very common molecule that contains bonding of this type is carbon dioxide (CO_2). In fact, there are two carbon–oxygen double covalent bonds present in CO_2.

$$:\ddot{O}::C::\ddot{O}: \longrightarrow :\ddot{O}::C::\ddot{O}: \quad \text{or} \quad :\ddot{O}=C=\ddot{O}:$$

The triple bond that holds the N_2 molecule together is very strong, and thus N_2 is a very unreactive substance. The nitrogen that makes up most of our atmosphere does not react with most living things. Many manufacturers package their products in nitrogen to keep them fresh.

A single line (dash) is used to denote a single covalent bond, two lines to denote a double covalent bond, and three lines to denote a triple covalent bond.

Note in the following diagram how the circles are drawn for the octet of electrons about each of the atoms in CO_2.

7.13 VALENCE ELECTRON COUNT AND NUMBER OF COVALENT BONDS FORMED

Not all elements can form multiple covalent bonds. There must be at least two vacancies in an atom's valence electron shell prior to bond formation if it is to participate in a multiple covalent bond. This requirement eliminates group VIIA elements (F, Cl, Br, I) and hydrogen from participating in such bonds. The group VIIA elements have seven valence electrons and one vacancy, and hydrogen has one valence electron and one vacancy. All bonds formed by these elements are single covalent bonds.

Double bonding becomes possible for elements needing two electrons to complete their octet, and triple bonding becomes possible when three or more electrons are needed to complete an octet. Note that the word *possible* was used twice in the previous sentence. Multiple bonding does not have to occur when an element has two or three or four vacancies in its octet; single covalent bonds can be formed instead. The bonding behavior of an element, when more than one behavior is possible, is determined by what other element or elements it is bonded to.

Let us consider the possible bonding behaviors for O (six valence electrons; two octet vacancies), N (five valence electrons; three octet vacancies), and C (four valence electrons; four octet vacancies).

To complete its octet by electron sharing, an oxygen atom can form either two single bonds or one double bond.

$$\overset{\displaystyle |}{\underset{\displaystyle\cdot\cdot}{\text{:O}}}\!-\qquad\qquad\underset{\displaystyle\cdot\cdot}{\text{:O}}\!=$$

Two single bonds   One double bond

Nitrogen is a very versatile element with respect to bonding. It can form single, double, or triple covalent bonds as dictated by the other atoms present in a molecule.

$$-\overset{\displaystyle\cdot\cdot}{\underset{\displaystyle |}{\text{N}}}\!-\qquad\qquad-\overset{\displaystyle\cdot\cdot}{\text{N}}\!=\qquad\qquad\text{:N}\!\equiv$$

Three single bonds   One single and   One triple bond
                      one double bond

Note that in each of these bonding situations a nitrogen atom forms three bonds. A double bond counts as two bonds, and a triple bond as three bonds. Since nitrogen has only five valence electrons, it must form three covalent bonds to complete its octet.

Carbon is an even more versatile element than nitrogen, with respect to variety of types of bonding as illustrated by the following possibilities for bonding.

$$-\overset{\displaystyle |}{\underset{\displaystyle |}{\text{C}}}\!-\qquad-\overset{\displaystyle |}{\text{C}}\!=\qquad=\text{C}\!=\qquad-\text{C}\!\equiv$$

Four single bonds   Two single bonds and   Two double bonds   One single bond and
                     one double bond   one triple bond

EXAMPLE 7.10   **Predicting Chemical Formulas for Simple Molecular Compounds**

For each of the following pairs of elements, predict the chemical formula of the compound formed when the two elements interact with each other.

a. carbon and bromine   **b.** sulfur and chlorine

SOLUTION

a. Carbon, with four valence electrons, will need to form four bonds. Bromine, with seven valence electrons, will need to form only one bond. Four bromine atoms are needed to meet carbon's need to form four bonds. Such interaction produces a compound with the chemical formula CBr_4.

b. Sulfur, with six valence electrons, will form two covalent bonds. Chlorine, with seven valence electrons, will form one covalent bond. Thus, two chlorine atoms are needed to react with one sulfur atom, giving a compound whose chemical formula is SCl_2.

▶ **Practice Exercise 7.10**   For each of the following pairs of elements, predict the chemical formula of the compound formed when the two elements interact with each other.

a. silicon and fluorine   **b.** oxygen and chlorine

7.14 COORDINATE COVALENT BONDS

In the covalent bonds considered so far (single, double, and triple), each of the participating atoms in the bond contributed an equal number of electrons to the bond. There is another *less common* way in which a covalent bond can form. It is possible for both electrons in a shared electron pair to come from the same atom; that is, one atom supplies two electrons and the other atom none. Such a covalent bond is called a *coordinate covalent bond*.

A **coordinate covalent bond** *is a covalent bond in which both electrons of a shared electron pair come from one of the two atoms involved in the bond.* Coordinate covalent bonding allows an atom that has two (or more) vacancies in its valence shell to share a pair of nonbonding electrons located on another atom.

The ammonium ion, NH_4^+, is an example of a species containing a coordinate covalent bond. The formation of an NH_4^+ ion can be viewed as resulting from the reaction of a hydrogen ion, H^+, with an ammonia molecule, NH_3. Doing the bookkeeping on all the valence electrons involved in this reaction, using x's for nitrogen electrons and dots for hydrogen electrons, we get

$$\text{H:}\overset{\times\times}{\underset{\times\times}{\text{N}}}\text{:} + \text{H}^+ \longrightarrow \left[\text{H:}\overset{\times\times}{\underset{}{\text{N}}}\text{:H} \right]^+$$

Coordinate covalent bond

An H$^+$ ion has no electrons, hydrogen having lost its only electron when it became an ion. The H$^+$ ion has two vacancies in its valence shell; that is, it needs two electrons to become isoelectronic with the noble gas helium. The nitrogen in NH_3 possesses a pair of nonbonding electrons. These electrons are used in forming the new nitrogen–hydrogen bond. The new species formed, the NH_4^+ ion, is charged, since the H$^+$ was charged. The +1 charge on the NH_4^+ ion is dispersed over the *entire* molecule; it is not localized on the "new" hydrogen atom.

An "ordinary" covalent bond can be thought of as a "Dutch treat" bond; each atom "pays" its part of the bill. A coordinate covalent bond can be thought of as a "you treat" bond; one atom pays the entire bill.

FIGURE 7.7 (a) A regular covalent single bond results from the overlap of two half-filled orbitals. (b) A coordinate covalent single bond results from the overlap of a filled and an empty orbital.

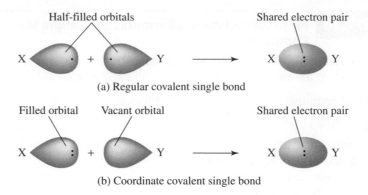

(a) Regular covalent single bond

(b) Coordinate covalent single bond

The element oxygen quite often forms coordinate covalent bonds. Consider the Lewis structures of the molecules HOCl (hypochlorous acid) and $HClO_2$ (chlorous acid).

$$H\!:\!\ddot{\underset{\cdot\cdot}{O}}\!:\!\ddot{\underset{\cdot\cdot}{Cl}}\!: \qquad H\!:\!\ddot{\underset{\cdot\cdot}{O}}\!:\!\ddot{Cl}\!:\!\ddot{\underset{\times\times}{O}}\!\times$$

Atoms participating in coordinate covalent bonds generally do not form their normal number of covalent bonds.

In the first structure, all the bonds are ordinary covalent bonds. In the second structure, which differs from the first in that a second oxygen atom is present, the new chlorine–oxygen bond is a coordinate covalent bond. The second oxygen atom with six valence electrons (denoted by x's) needs two more electrons for an octet. It shares one of the nonbonding electron pairs present on the chlorine atom. (The chlorine atom does not need any of the oxygen's electrons since it already has an octet.)

Atoms participating in coordinate covalent bonds generally deviate from the common bonding pattern (Sec. 7.13) expected for that type of atom. For example, oxygen normally forms two bonds; yet in the molecules N_2O (dinitrogen monoxide) and CO (carbon monoxide), which contain coordinate covalent bonds, oxygen forms one and three bonds, respectively.

$$:N\!:::\!N\!:\!\ddot{\underset{\cdot\cdot}{O}}\!: \quad \text{or} \quad :N\!\equiv\!N\!-\!\ddot{\underset{\cdot\cdot}{O}}\!:$$

$$:C\!:::\!O\!\times \quad \text{or} \quad :C\!\equiv\!O\!:$$

Once a coordinate covalent bond is formed, there is no way to distinguish it from any of the other covalent bonds in a molecule; all electrons are identical regardless of their source. The main use of the concept of coordinate covalency is to help rationalize the existence of certain molecules and polyatomic ions whose bonding electron arrangement would otherwise present problems. Figure 7.7 contrasts the formation of a regular covalent bond with that of a coordinate covalent bond.

7.15 RESONANCE STRUCTURES

In Section 7.12 it was noted that, in general, triple covalent bonds are stronger than double covalent bonds, which in turn are stronger than single covalent bonds. **Bond strength** *is a measure of the energy it takes to break a covalent bond; that is, to separate bonded atoms to give neutral particles.* It can be determined experimentally.

Another experimentally determinable parameter of bonds is bond length. **Bond length** *is the distance between the nuclei of covalently bonded atoms* (see Figure 7.8). A definite relationship exists between bond strength and bond length. It is found that as bond strength increases, bond length decreases; that is, the stronger the bond, the shorter the distance between the nuclei of the atoms of the bond. Thus, in general, triple covalent bonds are shorter than double covalent bonds, which are shorter than single covalent bonds.

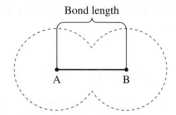

Bond length

A   B

Most Lewis structures for molecules give bonding pictures that are consistent with available experimental information on bond strength and bond length. However, there are some molecules for which no single Lewis structure that is consistent with such information can be written.

The molecule sulfur dioxide (SO_2) is an example of a situation in which a single Lewis structure does not adequately describe bonding. A plausible Lewis structure for SO_2, in which the octet rule is satisfied for all three atoms, is

$$\ddot{\underset{\cdot\cdot}{O}}:\ddot{S}::\ddot{O}: \quad \text{or} \quad :\ddot{\underset{\cdot\cdot}{O}}-\ddot{S}=\ddot{O}:$$

However, this Lewis structure suggests that one sulfur–oxygen bond, the double bond, should be stronger and shorter than the other sulfur–oxygen bond, the single bond. Experiment shows that this is not the case; both sulfur–oxygen bonds are equivalent, with both bond length and bond strength characteristics intermediate between those for sulfur–oxygen single and double bonds. A Lewis structure depicting this intermediate situation cannot be written.

The solution to the phenomenon in which no single Lewis structure adequately describes bonding involves the use of two or more Lewis structures, known as *resonance structures*, to represent the bonding in the molecule. **Resonance structures** *are two or more Lewis structures for a molecule or polyatomic ion that have the same arrangement of atoms, contain the same number of electrons, and differ only in the location of the electrons.*

Two resonance structures exist for an SO_2 molecule.

$$\ddot{\underset{\cdot\cdot}{O}}:\ddot{S}::\ddot{O}: \longleftrightarrow :\ddot{\underset{\cdot\cdot}{O}}::\ddot{S}:\ddot{\underset{\cdot\cdot}{O}}: \quad \text{or} \quad :\ddot{\underset{\cdot\cdot}{O}}-\ddot{S}=\ddot{O}: \longleftrightarrow :\ddot{O}=\ddot{S}-\ddot{\underset{\cdot\cdot}{O}}:$$

A double-headed arrow is used to connect resonance structures. The only difference between the two SO_2 resonance structures is in the location of two pairs of electrons. The positioning of these pairs of electrons determines whether the oxygen atom on the right or the left is the oxygen atom involved in the double bond.

The actual bonding in an SO_2 molecule is said to be a *resonance hybrid* of the two contributing resonance structures. Beginning chemistry students frequently misinterpret the concept of a resonance hybrid. They incorrectly envision that a molecule—SO_2 in this case—is constantly changing (resonating) between various resonance structure forms. This is not the case. For example, SO_2 is not a mixture of two kinds of molecules, nor does a single type of molecule flip-flop back and forth between the two resonance forms. There is only one kind of SO_2 molecule, and the bonding in it is an average of that depicted by the resonance structures. SO_2 molecules exist full time in this average state. A mule, the offspring of a donkey and horse, can be considered a hybrid of a donkey and a horse. However, it is not a horse at one instant and a donkey at another; it is always a mule. Likewise, a molecule has only one real structure, which is different from any of the resonance structures; it has characteristics of each one but does not match any one of them exactly.

Sometimes three, four, or even more resonance structures can be drawn for a molecule. Again, such resonance structures must all contain the same number of electrons and have the same arrangement of atoms; the structures may differ only in the location of the electrons about the atoms.

7.16 SYSTEMATIC PROCEDURES FOR DRAWING LEWIS STRUCTURES

The task of constructing Lewis structures for molecules and polyatomic ions containing many electrons or for which several resonance hybrids must be used to describe the bonding can become quite frustrating if it is approached in a nonsystematic trial-and-error manner. Use of a systematic approach to writing Lewis structures will enable a student to avoid most of this frustration. The following guidelines make the drawing of a Lewis structure for any molecule or polyatomic ion that obeys the octet rule, even a very complicated structure, a straightforward procedure.

Step 1 Determine the total number of valence electrons present in the molecule, that is, the total number of dots that must appear in the Lewis structure.

The total number of valence electrons is found by adding up the number of valence electrons that each atom in the molecule or ion possesses. If the species is a polyatomic ion, add one electron for each unit of negative charge present or subtract one electron for each positive charge. Do not worry about keeping track of which electrons come from which atoms. Only their total number is needed.

Step 2 Write the symbols of the atoms in the molecule, arranged in the way in which they are bonded to each other, and then place a single covalent bond (two electrons) between each pair of bonded atoms. Either a pair of dots or a dash can be used in denoting the single bond(s).

Determining which atom is the *central atom*, that is, which atom has the most other atoms bonded to it, is the key to determining the arrangement of atoms in a molecule or ion. Most other atoms present will be bonded to the central atom. For most molecular compounds containing just two elements, the molecular formula is of help in deciding the identity of the central atom. The central atom is the atom that appears only once in the formula; for example, S is the central atom in SO_3, O is the central atom in H_2O, and P is the central atom in PF_3. In a molecular compound containing hydrogen, oxygen, and an additional element, it is the additional element that is the central atom; for example, N is the central atom in HNO_3, and S is the central atom in H_2SO_4. In compounds of this type, the oxygen atoms are bonded to the central atom and the hydrogen atoms are bonded to the oxygens. Carbon is the central atom in almost all carbon-containing compounds. Hydrogen and fluorine are never the central atom.

Step 3 Add nonbonding electron pairs to the molecular structure such that each atom bonded to the central atom has an octet of electrons.

Remember that for hydrogen an octet of electrons is only two electrons. The noble gas electron configuration acquired by hydrogen is that of helium, and helium has only two electrons.

Step 4 Place any remaining electrons on the central atom in the structure.

The number of remaining electrons is obtained by subtracting from the total number of valence electrons (step 1) the number of electrons used in steps 2 and 3.

Step 5 If there are not enough electrons to give the central atom an octet, form multiple covalent bonds by shifting nonbonding electron pairs from surrounding atoms into bonding locations.

The elements C, N, and O are the elements most frequently involved in multiple bonding situations. The elements H, F, Cl, Br, and I in terminal atom positions do not participate in multiple bond formation.

Step 6 Count the total number of electrons in the completed Lewis structure to make sure it is equal to the total number of valence electrons available for bonding, as calculated in step 1.

This step serves as a double-check on the correctness of the Lewis structure.

The three examples that follow illustrate the use of the preceding procedures for drawing Lewis structures. In each part of each example, the outlined procedure is followed step by step so that you will become familiar with it.

EXAMPLE 7.11 **Drawing a Lewis Structure for a Molecular Compound, Using Systematic Procedures**

The acid present in the greatest amount in *acid rain* is sulfuric acid, a compound with the chemical formula H_2SO_4. Using systematic procedures, draw the Lewis structure for H_2SO_4.

SOLUTION

Step 1 Both sulfur and oxygen atoms have six valence electrons and hydrogen has one valence electron. The total number of valence electrons present in this molecule is 32.

$$1\,S: \quad 1 \times 6 = 6 \text{ valence electrons}$$
$$4\,O: \quad 4 \times 6 = 24 \text{ valence electrons}$$
$$2\,H: \quad 2 \times 1 = \underline{2 \text{ valence electrons}}$$
$$ 32 \text{ valence electrons}$$

Step 2 In compounds containing H, O, and an additional element, the additional element (S in this case) is the central atom. The oxygen atoms are attached to this central S atom, and the hydrogen atoms are attached to the oxygen atoms. Drawing this atomic arrangement with single covalent bonds (two electrons) placed between all bonded atoms gives

$$
\begin{array}{c}
\text{O} \\
| \\
\text{H}-\text{O}-\text{S}-\text{O}-\text{H} \\
| \\
\text{O}
\end{array}
$$

Step 3 Adding nonbonding electrons to the structure to complete the octets of all atoms bonded to the central atom gives

$$
\begin{array}{c}
\ddot{\text{O}}: \\
| \\
\text{H}-\ddot{\text{O}}-\text{S}-\ddot{\text{O}}-\text{H} \\
| \\
:\ddot{\text{O}}:
\end{array}
$$

Step 4 We started out with 32 valence electrons (step 1). Twelve were used in step 2 and 20 in step 3. This accounts for all of the available electrons. None is available to add to the central atom. This is fine because the central sulfur atom already has an octet of electrons; it is participating in four single bonds (8 electrons).

Step 5 No double or triple bonds are needed since all atoms have an octet of electrons with only single bonds present.

Step 6 There are 32 electrons in the Lewis structure, the same number of electrons as calculated in step 1.

Answer Double Check:

Does each atom, except for hydrogen atoms, have an octet of electrons? The answer is yes. Remembering that bonding electrons are counted as belonging to each atom in the bond, each oxygen has eight electrons as does the sulfur atom. Each hydrogen atom has two electrons (its octet), the two electrons of the single bond that the hydrogen participates in.

▶ **Practice Exercise 7.11** Using systematic procedures, draw the Lewis structure for the molecule H_3PO_4 (phosphoric acid).

EXAMPLE 7.12 **Drawing a Lewis Structure for a Polyatomic Ion, Using Systematic Procedures**

The compound sodium sulfite (Na_2SO_3), which contains the polyatomic sulfite ion ($SO_3{}^{2-}$), is used in many processed foods as an antioxidant. Using systematic procedures, draw the Lewis structure of the sulfite ion.

SOLUTION

Step 1 The sulfur atom has six valence electrons and each oxygen atom also has six valence electrons. (Sulfur and oxygen are in the same group in the periodic table.) There are two additional valence electrons present because of the −2 charge associated with this polyatomic ion.

$$1\,S:\ \ 1 \times 6 = \ \ 6 \text{ valence electrons}$$
$$3\,O:\ \ 3 \times 6 = 18 \text{ valence electrons}$$
$$\underline{\text{charge of } -2 = \ \ 2 \text{ valence electrons}}$$
$$26 \text{ valence electrons}$$

If the polyatomic ion had had a positive charge instead of a negative one, we would have had to subtract valence electrons from the total instead of adding. A positive charge would have denoted loss of electrons, and the electrons lost would have been valence electrons.

Step 2 The sulfur atom is the central atom with all three oxygen atoms individually attached to it. Drawing this atomic arrangement with single covalent bonds (two electrons) placed between all bonded atoms gives

Step 3 Adding nonbonding electrons to the structure to complete the octets of the oxygen atoms gives

$$\ddot{\text{:O}}-\text{S}-\ddot{\text{O}}\text{:} \\ | \\ \ddot{\text{:O:}}$$

Step 4 We started out with 26 electrons (step 1). Six electrons were used in step 2 (single bonds) and 18 electrons in step 3 (nonbonding electron pairs). This leaves 2 electrons not yet used. These 2 remaining electrons are available for placement on the central sulfur atom, which needs 2 more electrons to complete its octet.

$$\ddot{\text{:O}}-\ddot{\text{S}}-\ddot{\text{O}}\text{:} \\ | \\ \text{:O:}$$

Step 5 No double or triple bonds are needed since the action in step 4 causes the central sulfur atom to have an octet of electrons.

We must remember, however, that the Lewis structure we have just generated is that of a polyatomic ion. We therefore need to enclose it in large brackets and place the ionic charge for it outside the right bracket in a superscript position.

$$\left[\ddot{\text{:O}}-\ddot{\text{S}}-\ddot{\text{O}}\text{:} \\ | \\ \text{:O:}\right]^{2-}$$

Step 6 There are 26 electrons in the Lewis structure, the same number of electrons as calculated in step 1.

In this example we have shown the bonding within a polyatomic ion. This polyatomic ion, as well as all other polyatomic ions, is not a stable entity that exists alone. Polyatomic ions are *parts* of ionic compounds. The ion SO_3^{2-} would be found in ionic compounds such as Na_2SO_3, K_2SO_3, and $(NH_4)_2SO_3$. Ionic compounds containing polyatomic ions offer an interesting combination of both ionic and covalent bonds: covalent bonding *within* the polyatomic ion and ionic bonding *between* it and ions of opposite charge.

Answer Double Check:

Does each atom in the structure have an octet of electrons? Yes. Sulfur is involved in three single bonds (six electrons), and it has a nonbonding pair also (two electrons). Each oxygen atom participates in one single bond (two electrons) and has three nonbonding pairs (six electrons). Also, large square brackets are placed around the entire structure, as should be, and the charge on the polyatomic ion is placed outside the brackets as a superscript.

▶ **Practice Exercise 7.12** Using systematic procedures, draw the Lewis structure of the polyatomic NF_4^+ ion.

EXAMPLE 7.13 **Drawing a Lewis Structure for a Molecule That Has Resonance Forms**

Ozone (O_3), the triatomic form of oxygen, is the main irritant in photochemical smog, the type of smog that requires sunlight for its production. Draw the Lewis structures for O_3; there is more than one structure because of resonance.

SOLUTION

Step 1 There is a total of 18 valence electrons present, 6 from each oxygen atom.

Step 2 One of the oxygen atoms may be considered the central atom, and the other two oxygen atoms are attached to it. Drawing this atomic arrangement with single covalent bonds (two electrons) placed between all bonded atoms gives

$$O\!-\!O\!-\!O$$

Step 3 Adding six nonbonding electrons to each of the terminal oxygen atoms to complete their octets gives

$$:\!\ddot{O}\!-\!O\!-\!\ddot{O}\!:$$

Step 4 We started out with 18 electrons (step 1). Four electrons were used in step 2 (single bonds) and 12 in step 3 (nonbonding electron pairs). This leaves 2 electrons available for placement on the central oxygen atom.

$$:\!\ddot{O}\!-\!\ddot{O}\!-\!\ddot{O}\!:$$

Step 5 The addition of the two nonbonding electrons to the central oxygen atom is not sufficient to give this oxygen an octet of electrons. It still lacks two electrons. This problem is solved by moving a nonbonding pair of electrons from one of the terminal oxygen atoms into the oxygen–oxygen bonding region. This action, which produces a double bond, gives the central oxygen atom an octet of electrons.

$$:\!\ddot{O}\!-\!\ddot{O}\!\overset{\curvearrowleft}{\underset{\cdot\cdot}{O}}\!:$$

There are two choices for the terminal oxygen atom that is involved in double bond formation. The ramification of this situation is that resonance structures exist. The availability of choices for multiple bond formation (to give a central atom an octet of electrons) is always a signal for the existence of resonance structures. In this case, the resonance structures are

$$:\!\ddot{O}\!-\!\ddot{O}\!=\!\ddot{O}\!: \longleftrightarrow :\!\ddot{O}\!=\!\ddot{O}\!-\!\ddot{O}\!:$$

Step 6 There are 18 electrons in each of the resonance Lewis structures, the same number of electrons as calculated in step 1.

Answer Double Check:

Both structures have the same relative placement of atoms (a requirement for resonance structures). Both structures contain the same number of electrons, differing only in the placement of the electrons (another requirement for resonance structures).

▶ **Practice Exercise 7.13**   Draw the Lewis structures for the nitrite ion, NO_2^-; there is more than one structure because of resonance.

The systematic approach to drawing Lewis structures for molecules and poly-atomic ions illustrated in Examples 7.11 to 7.13 does not take into account the origin of the electrons in a chemical bond, that is, which atoms contribute which electrons to the bond. Thus, no distinction between normal covalent bonds and coordinate covalent bonds is made by the procedures of this system. This is acceptable. Each electron in a bond belongs to the bond as a whole. Electrons do not have labels of genealogy.

7.17 MOLECULAR GEOMETRY

Lewis structures give the number and types of bonds present in molecules. They do not, however, convey any information about molecular shape, that is, *molecular geometry*. **Molecular geometry** *is a description of the three-dimensional arrangement of atoms within a molecule*. Indeed, Lewis structures falsely imply that all molecules have flat, two-dimensional shapes.

Molecular geometry is an important factor in determining the physical and chem-ical properties of substances. Dramatic relationships between geometry and properties are often observed in research associated with the development of prescription drugs. A small change in overall molecular geometry, caused by addition or removal of atoms, can enhance drug effectiveness and/or decrease drug side effects. Studies also show that the human senses of taste and smell depend in part on the shapes of molecules.

For simple molecules (only a few atoms), molecular geometry can be predicted, using the information present in a molecule's Lewis structure and a procedure called *valence-shell-electron-pair-repulsion theory* (*VSEPR theory*). **VSEPR theory** *is a set of procedures for predicting the geometry of a molecule from the information contained in the molecule's Lewis structure.*

The central concept of VSEPR theory is that electron pairs in the valence shell of an atom adopt an arrangement in space about the nucleus that minimizes the repulsions between the like-charged (all negative) electron pairs. Minimization occurs when the electron pairs are as far away from each other as possible. Repulsion-minimizing arrangements about a nucleus for two, three, and four electron pairs are as follows:

1. Two electron pairs, to be as far apart as possible from each other, will be found on opposite sides of a nucleus, that is, at 180° angles to each other. Such an electron pair arrangement is said to be *linear*.

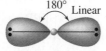

2. Three electron pairs are as far apart as possible when they are found at the corners of an equilateral triangle. In such an arrangement, they are separated by angles of 120°, giving a *trigonal planar* arrangement of the electron pairs.

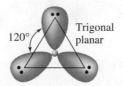

The acronym VSEPR is pronounced "vesper."

The preferred arrangement of a given number of valence electron pairs about a central atom is the one that maximizes the separation among them. Such an arrangement minimizes repulsions between electron pairs.

3. A *tetrahedral* arrangement of electron pairs minimizes repulsions between four sets of electron pairs. A tetrahedron is a four-sided geometric figure, all four sides being identical equilateral triangles. The angle between any two electron pairs is 109.5°.

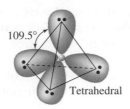

109.5°   Tetrahedral

Figure 7.9 shows these three electron-pair arrangements—linear, trigonal planar, and tetrahedral—using balloons that have their ends tied together. When two, three, and four like-sized balloons are tied together, they naturally assume the shapes we are talking about.

Most simple molecules have a central atom to which all other atoms are bonded. The first step in applying VSEPR theory to such molecules involves determining the number of electron pairs present on the *central atom*. This count is obtained from the molecule's Lewis structure, using an expanded concept of what constitutes an "electron pair."

Electron Pairs versus Electron Groups

Expansion of the concept of an electron pair to an *electron group* is needed prior to using VSEPR theory to predict molecular geometry. The electron group concept facilitates the extension of VSEPR theory to molecules in which double covalent bonds and triple covalent bonds are present. A **VSEPR electron group** *is a group of valence electrons present in a localized region about an atom in a molecule.* An electron group may contain two electrons (a single covalent bond or a nonbonding electron pair), four electrons (a double covalent bond) or six electrons (a triple covalent bond). Electron groups that contain four and six electrons repel other electron groups in the same way electron pairs do. This makes sense. The four electrons in a double covalent bond or the six electrons in a triple covalent bond are localized in the region between two bonded atoms in a manner similar to the two electrons in a single covalent bond.

Based on these conventions, the VSEPR electron group count about the *central atom* in molecules with the following Lewis structures is as indicated.

:N≡N—Ö: Central atom has two VSEPR electron groups (single bond and triple bond)

:Ö—S̈=Ö: Central atom has three VSEPR electron groups (single bond, double bond, and nonbonding electron pair)

H—Ö—H   Central atom has four VSEPR electron groups (two single bonds and two nonbonding electron pairs)

FIGURE 7.9   When balloons of the same size and shape are tied together, they will assume positions in space similar to those taken by pairs of valence electrons around a central atom.

Two balloons, linear   Three balloons, trigonal planar   Four balloons, tetrahedral

Let us now apply VSEPR theory to molecules in which two, three, and four VSEPR electron groups are present about a central atom. Our operational rules will be the following:

1. Draw a Lewis structure for the molecule and identify the specific atom for which geometric information is desired. (This atom will usually be the central atom in the molecule.)
2. Determine the number of VSEPR electron groups present about the central atom. The following conventions govern this determination.
   a. No distinction is made between bonding and nonbonding electron groups. Both are counted.
   b. Single, double, and triple covalent bonds are all counted equally as one electron group because each takes up only one region of space about a central atom.
3. Predict the VSEPR electron group arrangement about the atom by assuming that the electron groups orient themselves in a manner that minimizes repulsions (Figure 7.9).

Molecules with Two VSEPR Electron Groups

All molecules with two VSEPR electron groups are linear. Two common molecules in this category are carbon dioxide (CO_2) and hydrogen cyanide (HCN), whose Lewis structures are

$$:\ddot{O}=C=\ddot{O}: \qquad H-C\equiv N:$$

In CO_2 the central carbon atom's two VSEPR groups are the two double bonds. In HCN the central carbon atom's two VSEPR groups are a single bond and a triple bond. In both molecules the VSEPR electron groups arrange themselves on opposite sides of the carbon atom to produce a linear molecule.

Molecules with Three VSEPR Electron Groups

Two molecular geometries are associated with molecules that have three VSEPR electron groups: *trigonal planar* and *angular (or bent)* (see Figure 7.10). The former results when all three VSEPR groups are bonding, and the latter when one of the three VSEPR groups is nonbonding. Illustrative of these two situations are the molecules H_2CO (formaldehyde) and SO_2 (sulfur dioxide), whose Lewis structures are

Trigonal planar   Angular or bent

The shape of the SO_2 molecule is described as *angular* rather than *trigonal planar*, because molecular geometry describes only *atom positions*. The positions of nonbonding electron pairs are ignored when coining words to describe molecular geometry. Do not interpret this to mean that nonbonding electron pairs are unimportant in molecular geometry determinations; indeed, in the case of SO_2 it is the presence of the nonbonding electron pair that makes the molecule angular rather than linear.

VSEPR electron-group arrangement and molecular geometry are not the same when a central atom possesses nonbonding electron pairs. The word used to describe the geometry in such cases does not include the positions of the nonbonding electron pairs.

FIGURE 7.10

Molecular geometries associated with various combinations of bonding and nonbonding electrons about a central atom that obeys the octet rule.

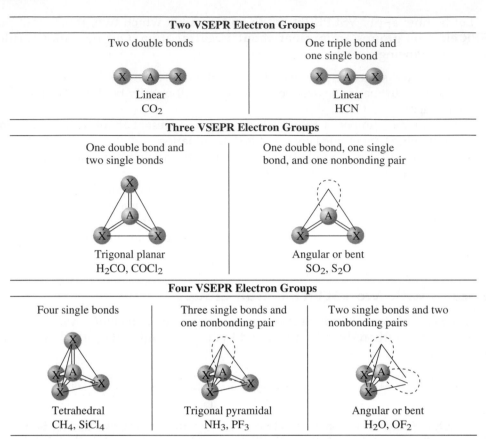

Two VSEPR Electron Groups

Two double bonds

Linear
CO_2

One triple bond and one single bond

Linear
HCN

Three VSEPR Electron Groups

One double bond and two single bonds

Trigonal planar
H_2CO, $COCl_2$

One double bond, one single bond, and one nonbonding pair

Angular or bent
SO_2, S_2O

Four VSEPR Electron Groups

Four single bonds

Tetrahedral
CH_4, $SiCl_4$

Three single bonds and one nonbonding pair

Trigonal pyramidal
NH_3, PF_3

Two single bonds and two nonbonding pairs

Angular or bent
H_2O, OF_2

Note: The figure does not consider cases where only one bond is present—for example, one single bond and three nonbonding pairs. If only one bond is present, the molecule is diatomic. All diatomic molecules have the same geometry: The two atoms lie along a straight line.

Molecules with Four VSEPR Electron Groups

Three molecular geometries are possible for molecules with four VSEPR electron groups: *tetrahedral* (no nonbonding electron pairs present), *trigonal pyramidal* (one nonbonding electron pair present), and *angular* (two nonbonding electron pairs present). The molecules CH_4 (methane), NH_3 (ammonia), and H_2O (water) illustrate this sequence of molecular geometries.

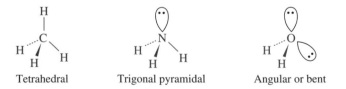

Tetrahedral   Trigonal pyramidal   Angular or bent

Again, note how the word that is used to describe the geometry of a molecule does not take into account the positions of nonbonding electron pairs.

Figure 7.10 summarizes the relationships among the number of VSEPR electron groups about a central atom, molecular geometry, and terminology used to describe molecular geometry.

VSEPR theory can also be used to predict the molecular geometry of molecules that contain more than one central atom. In such situations, each central atom is considered separately and then the results of the separate analyses are combined to obtain the

Molecular Formula	Lewis Structure and VSEPR Electron-Pair Analysis	Molecular Geometry
C_2H_2 (acetylene)	H—C≡C—H Two VSEPR electron groups; linear C center Two VSEPR electron groups; linear C center	H—C≡C—H Straight chain of four atoms—linear
HN_3 (hydrogen azide)	H—N̈=N=N̈: Three VSEPR electron groups; angular N center Two VSEPR electron groups; linear N center	N̈=N=N̈: H Chain of four atoms with one bend
H_2O_2 (hydrogen peroxide)	H—Ö—Ö—H Four VSEPR electron groups; angular O center Four VSEPR electron groups; angular O center	Ö—Ö H H Chain of four atoms with two bends

FIGURE 7.11 Using VSEPR theory to predict molecular geometry for molecules containing more than one central atom.

overall molecular geometry. Figure 7.11 shows the results of such procedures for selected molecules containing more than one central atom.

EXAMPLE 7.14 **Using VSEPR Theory to Predict Molecular Geometry**

Using VSEPR theory, predict the molecular geometry of each of the following molecules given their Lewis structures.

a.   :C̈l:
       |
   H—C—H
       |
     :C̈l:

b.   :F̈—P̈—F̈:
        :F̈:

c.   H H
       | |
   H—C=C—H

SOLUTION

a. The Lewis structure for this molecule shows that the central carbon atom has four VSEPR electron groups present (four single bonds). No nonbonding valence shell electrons are present on the carbon atom.

     The arrangement of the four electron groups about the central atom is tetrahedral.

Since each electron group is a bonding group, the molecular shape will also be tetrahedral.

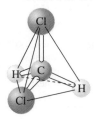

b. Electrons are present in four locations about the central atom—three electron groups involved in single bonds and one electron pair that is nonbonding.

The arrangement of four electron groups about a central atom is always tetrahedral. It does not matter whether the groups are bonding or nonbonding. Arrangement depends only on the number of groups and not on how the groups function.

However, the molecular geometry will be different from the electron group geometry since a nonbonding electron pair is present. The molecular geometry is trigonal pyramidal.

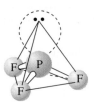

Note that the vacant corner of the tetrahedron (the corner that has a nonbonding electron pair instead of an atom) is not considered when coining a word to describe the molecular geometry. Molecular geometry describes the arrangement of *atoms only.*

c. This molecule has two central atoms, the two carbon atoms. A VSEPR theory analysis needs to be carried out for each of the carbon atoms, and then the results of the two analyses combined.

Each of the two carbon atoms has three VSEPR electron groups about it (two single bonds and a double bond). The geometry associated with three VSEPR electron groups is always trigonal planar.

Since all positions about the carbon atoms are bonding positions (no non-bonding electron pairs are present on the carbon atoms) there will also be a trigonal planar arrangement of atoms about each carbon atom. The geometry of the molecule is therefore the following.

A simpler representation of the geometry of this molecule is the following

Answer Double Check:

The answers are consistent with the information in Figure 7.10. Four VESPR electron groups, all of which are bonding, correlate with a tetrahedral geometry. Four VESPR electron groups, three of which are bonding, correlate with a trigonal pyramidal geometry. Three VESPR electron groups, all of which are bonding, correlate with a trigonal planar geometry.

▶ **Practice Exercise 7.14**   Using VSEPR theory, predict the molecular geometry of each of the following molecules or ions, given their Lewis structures.

a. $:\ddot{C}l—\ddot{N}—\ddot{C}l:$
   $\quad\quad |$
   $\quad\quad :\ddot{C}l:$

b. $\left[H—\ddot{O}—H \right]^{+}$
   $\quad\quad |$
   $\quad\quad H$

c.
   $\quad\quad H \quad H$
   $\quad\quad | \quad\ |$
   $H—C—C—H$
   $\quad\quad | \quad\ |$
   $\quad\quad H \quad H$

7.18 ELECTRONEGATIVITY

The ionic and covalent bonding models we have developed in this chapter seem to represent two very distinct forms of bonding. Actually, the two models are closely related to each other; they are the extremes of a broad continuum of bonding patterns. The close relationship between the two bonding models becomes apparent when the concepts of *electronegativity* (discussed in this section) and *bond polarity* (discussed in the next section) are considered.

The electronegativity concept has its origins in the fact that the nuclei of various elements have differing abilities to attract shared electrons (in a bond) to themselves. Different electron-attracting abilities result from differences in size, nuclear charge, and number of nonvalence electrons present for atoms of various elements. As a result of these factors, some elements are better electron attractors than other elements.

Electronegativity *is a measure of the relative attraction that an atom has for the shared electrons in a bond.* Electronegativity values are unitless numbers on a relative scale that are obtained from bond energies and other related experimental data. Electronegativities are calculated numbers that cannot be directly measured in the laboratory. A number of electronegativity scales exist. The most widely used one, a scale developed by the

American chemist Linus Carl Pauling (1901–1994; see "The Human Side of Chemistry 8"), is given in the format of a periodic table in Figure 7.12. On this scale, fluorine, the most electronegative of all elements, has arbitrarily been assigned a value of 4.0 (the maximum on the scale) and serves as the reference element. *The higher the electronegativity of an element, the greater the electron-attracting ability of atoms of that element.*

Patterns and trends in the electronegativity values given in Figure 7.12 include the following:

1. For representative elements (Sec. 6.11), electronegativity values generally increase from left to right within a period of the periodic table.
2. For the period 2 elements this left-to-right increase is regular, being 0.5 units.

Li	Be	B	C	N	O	F
1.0	1.5	2.0	2.5	3.0	3.5	4.0

The Human Side of Chemistry 8

Linus Carl Pauling (1901–1994)

In 1960, an undergraduate chemistry student asked of Linus Carl Pauling the question: "Professor Pauling, I'd like to know what goes through your mind when you decide that you understand something." Pauling's answer was: "When I say 'Aha! So that's how it works!' But it isn't long before I realize that what I thought I had gotten right begged a number of questions. I realize that I hadn't really understood; that there's more to the problem than I had thought. So I puzzle over the problem some more until I again say, 'Aha! So that's how it *really* works.' Of course, that isn't the end of it, because still more questions arise, still more 'Aha's!'"

Pauling, who died in 1994 at age 93, is considered one of the greatest chemists of all time. Born in Portland, Oregon, the son of a pharmacist, he graduated from Oregon Agricultural College (now Oregon State University) in 1922 (as a chemical engineer) and then three years later obtained a Ph.D. in chemistry from the California Institute of Technology (Caltech). After two years of further study in Europe, he returned to Caltech as a faculty member.

During the 1930s, Pauling made numerous key contributions to chemistry, including a scale of electronegativity that could be used to determine the ionic and covalent character of chemical bonds, the concept of bond orbital hybridization (the organization of orbitals of atoms in configurations that favor bonding), and the theory of resonance (distribution of electrons between two or more possible Lewis structures for a molecule). During this time period, the chairman of Caltech's chemistry department is said to have remarked, "Were all the rest of the chemistry department wiped away except Pauling, it would still be one of the most important departments of chemistry in the world."

In the 1940s, Pauling's interests included biological chemistry, where he is considered one of the founders of molecular biology. In this area he made important contributions concerning the three-dimensional structures of proteins and how enzymes and antibodies function.

In later years, he was concerned with what he called "orthomolecular medicine," in particular with the concept that large doses of ascorbic acid (vitamin C) could ward off common colds and prevent cancer.

Pauling also became an ardent advocate for peace and for a ban on the testing of nuclear weapons. He set in motion a petition that was signed by more than 11,000 scientists worldwide calling for such a ban. He also took vocal antiwar stands.

Pauling is the only person to receive two unshared Nobel Prizes. In 1954, for his contributions to chemistry, especially his work on chemical bonding theory, Pauling received the Nobel Prize in chemistry. In 1962, Pauling received the Nobel Peace Prize in recognition of his efforts to end nuclear weapons testing.

1	2	3	4	5	6	7	8	9	10	11	12	13	14	15	16	17	18
H 2.1																	He –
Li 1.0	Be 1.5											B 2.0	C 2.5	N 3.0	O 3.5	F 4.0	Ne –
Na 0.9	Mg 1.2											Al 1.5	Si 1.8	P 2.1	S 2.5	Cl 3.0	Ar –
K 0.8	Ca 1.0	Sc 1.3	Ti 1.5	V 1.6	Cr 1.6	Mn 1.5	Fe 1.8	Co 1.8	Ni 1.8	Cu 1.8	Zn 1.6	Ga 1.6	Ge 1.8	As 2.0	Se 2.4	Br 2.8	Kr –
Rb 0.8	Sr 1.0	Y 1.2	Zr 1.4	Nb 1.6	Mo 1.8	Tc 1.9	Ru 2.2	Rh 2.2	Pd 2.2	Ag 1.9	Cd 1.7	In 1.7	Sn 1.8	Sb 1.9	Te 2.1	I 2.5	Xe –
Cs 0.7	Ba 0.9	57–71 1.1–1.2	Hf 1.3	Ta 1.5	W 1.7	Re 1.9	Os 2.2	Ir 2.2	Pt 2.2	Au 2.4	Hg 1.9	Tl 1.8	Pb 1.8	Bi 1.9	Po 2.0	At 2.2	Rn –
Fr 0.7	Ra 0.9																

FIGURE 7.12
Electronegativities of the elements.

3. It is important to note how the electronegativity of hydrogen (period 1) compares with that of the period 2 elements. Hydrogen's value of electronegativity, 2.1, is between that of B and C.
4. For the period 3 elements the left-to-right increase is 0.3 units until the last two elements are reached, where it is 0.4 and 0.5.

Na	Mg	Al	Si	P	S	Cl
0.9	1.2	1.5	1.8	2.1	2.5	3.0

5. Electronegativity values generally increase from bottom to top within a periodic table group.

The group and period trends in electronegativity values result in nonmetals generally having higher electronegativities than metals. This fact is consistent with our previous generalization (Sec. 7.5) that metals tend to lose electrons and nonmetals tend to gain electrons when an ionic bond is formed. Metals (low electronegativities, poor electron attractors) will give up electrons to nonmetals (high electronegativities, good electron attractors).

Before VSEPR theory (Sec. 7.17) can be applied to a molecule, one must know which atom is the central atom in the molecule. In general, the *least* electronegative of the atoms present occupies the position of central atom.

EXAMPLE 7.15 **Predicting Electronegativity Relationships, Using General Periodic Table Trends**

Predict which member of each of the following pairs of elements has the greater electronegativity, based on periodic table electronegativity trends.

a. Be or C   b. O or S   c. Na or Cl   d. Mg or N

SOLUTION

In terms of periodic table trends, electronegativity increases in going from left to right across a period and decreases in going down a group.

a. Be and C are in the same period, with C being farther to the right. Therefore C has the greater electronegativity as electronegativity increases in going across a period.
b. O and S are in the same group, with O being above S. Therefore O is the more electronegative element as electronegativity decreases going down a group.
c. Na and Cl are in the same period, with Cl being farther to the right. Therefore Cl has the greater electronegativity as this property increases in going to the right in a period of the periodic table.
d. Mg is in period 3 and N is in period 2. The element Be can be used as a link between the two elements; it is in the same group as Mg and the same period as N. The periodic table group trend predicts that Be has a higher electronegativity than Mg and the periodic table period trend predicts that N is more electronegative than Be. Therefore, N is more electronegative than Mg.

▶ **Practice Exercise 7.15**   Predict which member of each of the following pairs of elements has the greater electronegativity, based on periodic table electronegativity trends.

a. F or Br   **b.** C or O   **c.** Li or K   **d.** Al or N

7.19 BOND POLARITY

Numerical quantification for the terms *nonpolar covalent bond* and *polar covalent bond* occurs later in this section. The electronegativity difference between bonded atoms will be used to distinguish nonpolar covalent bonds, polar covalent bonds, and ionic bonds from each other.

When two atoms of *equal* electronegativity share one or more pairs of electrons, each atom exerts the same attraction for the electrons, which results in the electrons' being *equally* shared. This type of bond is called a *nonpolar covalent bond*. A **nonpolar covalent bond** *is a covalent bond in which there is an equal sharing of electrons between two atoms.*

When the two atoms involved in a covalent bond have *different* electronegativities, the electron-sharing situation is not equal. The atom that has the higher electronegativity will attract the electrons more strongly than the other atom; this results in an unequal sharing of electrons. This type of covalent bond is called a *polar covalent bond*. A **polar covalent bond** *is a covalent bond in which there is an unequal sharing of electrons between two atoms.* Figure 7.13 pictorially contrasts a nonpolar covalent bond and a polar covalent bond using the molecules H_2 and HCl.

The significance of unequal sharing of electrons in a covalent bond is that it creates partial (fractional) positive and negative charges on the atoms involved in a bond. Although each atom involved in a polar covalent bond is initially uncharged, the unequal sharing of electrons means the electrons spend more time near the more electronegative atom of the bond (producing a partial negative charge). The presence of such partial charges on atoms within a molecule often significantly affects molecular properties (Sec. 7.20).

The unequal sharing of electrons in a covalent bond is often indicated by a notation that uses the lowercase Greek letter δ (delta). A $\delta-$ symbol, meaning a "partial negative charge," is placed above the relatively negative atom of the bond, and a $\delta+$ symbol, meaning a "partial positive charge," is placed above the relatively positive atom.

With delta notation, the bond in hydrogen chloride (HCl) would be depicted as

$$\overset{\delta+}{H}\text{---}\overset{\delta-}{Cl}$$

Different types of sharing are seen in everyday life. Partners in a business can own it equally, or one partner can have more than the other. It can be owned 60:40, 70:30, and so on.

Chlorine has an electronegativity of 3.0, and hydrogen's electronegativity is 2.1 (see Figure 7.12). Since Cl is the more electronegative of the two elements, it dominates the electron-sharing process and draws the electrons closer to it. Hence, the Cl end of the

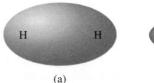
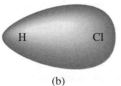

(a)   (b)

FIGURE 7.13   (a) In the nonpolar covalent bond present in H_2 (H—H), there is a symmetrical distribution of electron density between the two atoms, that is, equal sharing of electrons occurs. (b) In the polar covalent bond present in HCl (H—Cl), electron density is displaced toward the Cl atom because of its greater electronegativity, that is, unequal sharing of electrons occurs.

M⁺ X:⁻	δ+ δ− Y:X	X:X
Ionic (full charges)	Polar covalent (partial charges)	Nonpolar covalent (no charges)
Electron transfer	Unequal sharing of electrons	Equal sharing of electrons

FIGURE 7.14 Notation used to denote electron transfer, unequal sharing of electrons, and equal sharing of electrons in chemical bonds.

bond has the δ− designation. Again, the δ− over the Cl atom indicates a partial negative charge; that is, the chlorine end of the molecule is negative with respect to the hydrogen end. The meaning of the δ+ over the H is that the H end of the molecule is positive with respect to the Cl end. Partial charges are always charges less than +1 or −1. Charges of +1 and −1, full charges, would result when an electron is transferred from one atom to another. With partial charges we are talking about an intermediate charge state between 0 and 1. Figure 7.14 contrasts the use of formulas to denote electron transfer, unequal sharing of electrons, and equal sharing of electrons.

The direction of polarity of a polar covalent bond, that is, which atom of the bond bears the partial negative charge, can also be designated, using an arrow with a cross at its other end (⟵→). The crossed end of the arrow is placed near the atom bearing the partial positive charge, and the arrow's head is placed near the atom bearing the partial negative charge. Using this notation, the bond in the molecule HCl would be denoted as

$$\overset{\longleftrightarrow}{H-Cl}$$

Bond polarity *is a measure of the degree of inequality in the sharing of electrons in a chemical bond.* The absolute value of the difference in electronegativity of two bonded atoms gives a *rough* measure of the polarity to be expected in a bond. The greater the numerical difference in electronegativities, the greater the inequality of electron sharing and the greater the polarity of the bond. Note that it is the magnitude of the electronegativity *difference,* not the actual electronegativities of the bonded atoms, that determines the extent of polarity of a bond.

As the polarity of a bond increases, the bond is becoming increasingly ionic; that is, the *percent ionic character* of the bond is increasing. **Percent ionic character of a bond** *is a measure of how the actual charge separation (partial charge) in a bond because of electronegativity difference of the bonded atoms compares to the complete charge separation associated with ions.* Percent ionic character values are calculated from data obtained from experiments involving the behavior of molecules in electric fields.

A rough scale relating the difference in electronegativity to percent ionic character of a bond is as follows:

All bonds, except those between identical atoms, have some degree of polarity associated with them.

No known compound has a percent ionic character of 100%.

Electronegativity Difference	Percent Ionic Character
1.0	20%
1.5	40%
2.0	60%
2.5	80%

FIGURE 7.15 The relationships among the ionic and covalent characters of a chemical bond and electronegativity difference between the bonded atoms.

Electronegativity difference between the bonding atoms	None	Intermediate	Large
Bond type	Covalent	Polar covalent	Ionic
Covalent character		Decreases →	
Ionic character		Increases →	

The answer to the question "Is the bond ionic or covalent?" is usually "both." A better question is "How ionic or covalent is the bond?" The existence of partial charges on atoms because of unequal sharing of electrons means that all polar covalent bonds have partial ionic character.

A bond that has 50% ionic character correlates with an electronegativity difference of 1.7.

The message of the preceding discussion of bond polarity is that there is no natural boundary between ionic and covalent bonding. Most bonds are a mixture of pure ionic and pure covalent bonds; that is, unequal sharing of electrons occurs. Most bonds have both ionic and covalent character. Nevertheless, it is still convenient to use the terms *ionic* and *covalent* in describing chemical bonds, based on the following arbitrary, but useful (though not infallible), guidelines, which relate to electronegativity difference between bonded atoms.

1. Bonds that involve atoms with the same or very similar electronegativities are called *nonpolar covalent bonds*. "Similar" here means an electronegativity difference of 0.4 or less (less than 10% ionic character).

   Technically, the only *pure* nonpolar covalent bonds (totally equal sharing) are those between identical atoms. However, bonds with a small electronegativity difference behave very similarly to pure nonpolar covalent bonds.

2. Bonds with an electronegativity difference greater than 0.4 but less than 1.5 are called *polar covalent bonds*.

3. Bonds with an electronegativity difference of 2.0 or greater are called *ionic bonds*. Such bonds have 60% or greater ionic character.

4. Bonds with an electronegativity difference between 1.5 and 2.0 are considered to be *ionic* if the bond involves a metal and a nonmetal and *polar covalent* if the bond involves two nonmetals. In the 1.5–2.0 electronegativity-difference range some compounds exhibit characteristics associated with molecular compounds, while others exhibit characteristics associated with ionic compounds (see Section 7.1). This rule helps in dealing with this borderline area.

Figure 7.15 summarizes the interrelationships among the concepts of electronegativity difference between bonded atoms, bond type, a bond's covalent character, and a bond's ionic character.

EXAMPLE 7.16 **Predicting the Polarity Characteristics of Chemical Bonds**

Predict whether a chemical bond between each of the following pairs of atoms would be nonpolar covalent, polar covalent, or ionic. If polar covalent, specify which atom would carry the partial positive charge and which would carry the partial negative charge.

a. calcium and fluorine
b. carbon and oxygen
c. boron and hydrogen
d. phosphorus and bromine

SOLUTION

a. Calcium has an electronegativity of 1.0 and fluorine an electronegativity of 4.0. The electronegativity difference of 3.0 indicates that the bond is **ionic**. An electronegativity difference of 2.0 or greater produces an ionic bond.

b. The electronegativity difference between carbon (2.5) and oxygen (3.5) is 1.0. Bonds within the electronegativity difference range 0.4 to 1.5 are classified as **polar covalent**. In such bonds the more electronegative atom, oxygen in this case, bears the partial negative charge and the less electronegative atom, carbon in this case, bears the partial positive positive charge.

c. Boron (2.0) and hydrogren (2.1) have almost identical electronegatives. Such a situation, 0.4 difference or less, produces a bond classified as **nonpolar covalent**.

d. The electronegativity difference between phosphorus (2.1) and bromine (2.8) is sufficiently great to produce a **polar covalent** bond with phosphorus bearing the partial positive charge and bromine bearing the partial negative charge.

▶ **Practice Exercise 7.16** Predict whether a bond between each of the following pairs of atoms would be nonpolar covalent, polar covalent, or ionic. If polar covalent, specify which atom would carry the partial positive charge and which would carry the partial negative charge.

a. nitrogen and fluorine
b. silicon and chlorine
c. sodium and oxygen
d. aluminum and phosphorus

7.20 MOLECULAR POLARITY

Molecules, as well as bonds (Sec. 7.19), can have polarity. **Molecular polarity** *is a measure of the degree of inequality in the attraction of bonding electrons to various locations within a molecule.* In terms of electron attraction, if one part of a molecule is favored over other parts, then the molecule is *polar*. A **polar molecule** *is a molecule in which there is an unsymmetrical distribution of electronic charge.* In a polar molecule, bonding electrons are more attracted to one part of the molecule than to other parts. A **nonpolar molecule** *is a molecule in which there is a symmetrical distribution of electronic charge.* Attraction for bonding electrons is the same in all parts of the molecule.

A prerequisite for determining molecular polarity is a knowledge of molecular geometry.

Molecular polarity depends on two factors: (1) bond polarities and (2) molecular geometry (Sec. 7.17). The presence of polar bonds within a molecule does not automatically mean that the molecule as a whole is polar. In molecules that are symmetrical, the effects of polar bonds may cancel each other, resulting in a molecule as a whole having no polarity.

Determining the molecular polarity of a diatomic molecule is simple because only one bond is present. If that bond is nonpolar, the molecule is nonpolar; if the bond is polar, the molecule is polar.

Molecular polarity determination in triatomic molecules is more complicated. Two different molecular geometries are possible in triatomic molecules: linear and angular. In addition, the symmetrical or unsymmetrical nature of the molecule must be considered. Let us consider the polarities of three specific triatomic molecules: CO_2 (linear), H_2O (angular), and HCN (linear).

Both bonds in the symmetrical linear CO_2 are polar. For both bonds, the direction of polarity is toward oxygen since oxygen is more electronegative than carbon.

Molecules in which all bonds are polar can, as a whole, be nonpolar if the bonds are so oriented in direction that the polarity effects cancel each other.

$$\overset{\longleftarrow \;\; \longmapsto}{O=C=O}$$

Despite the presence of the two polar bonds, CO_2 molecules are *nonpolar*. The effects of the two polar bonds cancel. The shift of electronic charge toward one oxygen atom is exactly compensated for by the shift of electronic charge toward the other oxygen atom. Thus, one end of the molecule is not negatively charged relative to the other end (a requirement for polarity), and the molecule is nonpolar.

The nonlinear (angular) triatomic H_2O molecule is polar. The bond polarities associated with the two hydrogen–oxygen bonds do not cancel each other, because of the nonlinearity of the molecule.

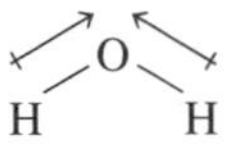

As a result of their orientation, both bonds contribute to an accumulation of negative charge on the oxygen atom. The two bond polarities are equal in magnitude but are not opposite in their direction.

The generalization that linear triatomic molecules are nonpolar and nonlinear triatomic molecules are polar, which is consistent with our discussion of CO_2 and H_2O molecular polarities, is not valid. The linear molecular HCN, which is polar, invalidates this generalization. Both bond polarities contribute to nitrogen's acquiring a partial negative charge relative to hydrogen in HCN.

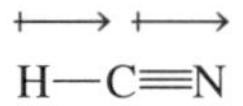

(Note that the two polarity arrows were drawn in the same direction because nitrogen is more electronegative than carbon and carbon is more electronegative than hydrogen.)

Commonly encountered geometries for molecules containing four and five atoms are, respectively, trigonal planar and tetrahedral. Molecules with those geometries in which all of the atoms attached to the central atom are identical, such as SO_3 (trigonal planar) and CH_4 (tetrahedral), are *nonpolar.* The individual bond polarities cancel as the result of the highly symmetrical arrangement of atoms around the central atom (see Figure 7.16).

If two or more kinds of atoms are attached to the central atom in a trigonal planar or tetrahedral molecule, the molecule will be polar. The high degree of symmetry required for cancellation of the individual bond polarities is no longer present. For example, if one of the hydrogen atoms in CH_4 (a nonpolar molecule) is replaced by a

> It is the polarity of water molecules that makes cooking in microwave ovens possible. The electromagnetic field of the microwave interacts with the polar water molecules causing them to shake. The food is cooked by the heat generated by the motion of the water molecules.

FIGURE 7.16
Examples of nonpolar molecules in which all bonds present are polar.

Type		Cancellation of Polar Bonds	Example
Linear molecules with two identical bonds	B—A—B		CO_2
Trigonal planar molecules with three identical bonds	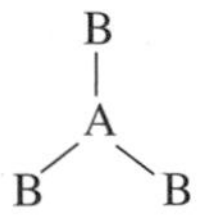		SO_3
Tetrahedral molecules with four identical bonds	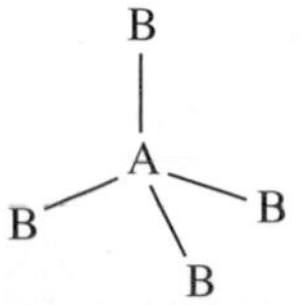		CH_4

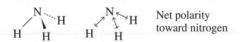

(a) CH₄, a nonpolar molecule   (b) CH₃Cl, a polar molecule

chlorine atom, a polar molecule results, even though the resulting CH_3Cl is still a tetrahedral molecule. A carbon–chlorine bond has a greater polarity than a carbon–hydrogen bond; chlorine has an electronegativity of 3.0 and hydrogen has an electronegativity of only 2.1. Figure 7.17 contrasts the polar CH_3Cl and nonpolar CH_4 molecules. Note that the direction of polarity of the carbon–chlorine bond is opposite that of the carbon–hydrogen bonds.

EXAMPLE 7.17 **Predicting the Polarity of Molecules Given Their Molecular Geometry**

Predict the polarity of each of the following molecules.

a. NH_3 (trigonal pyramidal)

b. H_2S (angular)

c. N_2O (linear)

:N≡N—Ö:

d. C_2H_2 (linear)

H—C≡C—H

SOLUTION

Knowledge of molecular geometry, which is given for each molecule in this example, is a prerequisite for predicting molecular polarity.

a. Noncancellation of the individual bond polarities in the trigonal pyramidal NH_3 molecule results in its being a polar molecule.

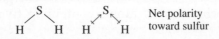

Net polarity toward nitrogen

The bond polarity arrows all point toward the nitrogen atom because nitrogen is more electronegative than hydrogen.

b. For the bent H_2S molecule, the shift in electron density in the polar sulfur–hydrogen bonds will be toward the sulfur atom because sulfur is more electronegative than hydrogen. The H_2S molecule as a whole is polar because of the noncancellation of the individual sulfur–hydrogen bond polarities.

Net polarity toward sulfur

c. The structure of the linear N_2O molecule is unsymmetrical; a nitrogen atom rather than the oxygen atom is the central atom. The nitrogen–nitrogen bond is nonpolar; the nitrogen–oxygen bond is polar. The molecule as a whole is polar because of the polarity of the nitrogen–oxygen bond.

$$N \equiv N - O \qquad N \quad \overset{\longrightarrow}{N} \quad O \qquad \begin{array}{l}\text{Net polarity}\\ \text{toward oxygen}\end{array}$$

d. The carbon–carbon bond is nonpolar. The two carbon–hydrogen bonds are polar and are "equal but opposite" in terms of effect; that is, they cancel. The molecule as a whole is thus nonpolar.

$$H - C \equiv C - H \qquad \overset{\longrightarrow}{H} \quad C \quad C \quad \overset{\longleftarrow}{H} \qquad \text{No net polarity}$$

▶ **Practice Exercise 7.17**   Predict the polarity of each of the following molecules.

a. PCl_3 (trigonal pyramidal)

$$:\!\overset{..}{\underset{}{Cl}}\!: \quad \overset{\overset{..}{P}}{\underset{:\overset{..}{Cl}:}{\blacktriangle}} \quad :\!\overset{..}{\underset{}{Cl}}\!:$$

b. SCl_2 (angular)

$$\overset{:\overset{..}{S}:}{:\!\overset{..}{\underset{..}{Cl}}\!: \quad :\!\overset{..}{\underset{..}{Cl}}\!:}$$

c. HOCl (angular)

$$\overset{:\overset{..}{O}:}{H \qquad :\!\overset{..}{\underset{..}{Cl}}\!:}$$

d. C_2H_4 (trigonal planar about C atoms)

$$\overset{H \qquad\qquad H}{\underset{H \qquad\qquad H}{C = C}}$$

Summary

1. **Chemical Bonds** Chemical bonds are the attractive forces that hold atoms together in more complex units. Chemical bonds result from the transfer of electrons between atoms (ionic bond) or from the sharing of electrons between atoms (covalent bond).

2. **Valence Electrons** Valence electrons, for representative elements, are the electrons in the outermost electron shell, which is the shell with the highest shell number. Valence electrons are particularly important in determining the bonding characteristics of a given atom.

3. **Lewis Structure** A Lewis structure for an atom is a notation for designating the number of valence electrons present in the atom. It consists of an element's symbol with one dot for each valence electron placed about the elemental symbol.

4. **Octet Rule** In compound formation, atoms of elements lose, gain, or share electrons in such a way as to produce a noble gas electron configuration for each of the atoms involved.

5. **Ions** An ion is an atom (or group of atoms) that is electrically charged as the result of loss or gain of electrons. The notation for charges on ions is a superscript placed to the right of the elemental symbol. An anion is a negatively charged ion. A cation is a positively charged ion.

6. **Ionic Compound** Ionic compounds usually contain a metallic element and a nonmetallic element. Metal atoms lose one or more electrons, producing cations. Nonmetal atoms acquire the electrons lost by the metal atoms, producing anions. The oppositely charged ions attract one another, creating ionic bonds.

7. **Charge Magnitude for Ions** Metal atoms containing one, two, or three valence electrons tend to lose such electrons, producing ions of +1, +2, and +3 charge, respectively. Nonmetal atoms containing five, six, or seven valence

electrons tend to gain electrons, producing ions of −3, −2, and −1 charge, respectively.

8. **Formulas for Ionic Compounds** The ratio in which positive and negative ions combine is the ratio that causes the total amount of positive and negative charges to add up to zero.

9. **Structure of Ionic Compounds** Ionic solids consist of positive and negative ions arranged in such a way that each ion is surrounded by ions of the opposite charge. Individual molecules are not present in an ionic compound.

10. **Polyatomic Ions** A polyatomic ion is a group of covalently bonded atoms that has acquired a charge through the loss or gain of electrons. Polyatomic ions are very stable entities that generally maintain their identity during chemical reactions. Polyatomic ions never occur alone as do molecules; they are a part of an ionic compound.

11. **Molecular Compound** Molecular compounds usually involve two or more nonmetals and covalent bonds. Covalent bond formation occurs through electron sharing between nonmetal atoms. The covalent bond results from the common attraction of two nuclei for the shared electrons.

12. **Bonding and Nonbonding Electron Pairs** Bonding electrons are pairs of valence electrons that are shared between atoms in a covalent bond. Nonbonding electrons are pairs of valence electrons about an atom that are not involved in electron sharing.

13. **Types of Covalent Bonds** One shared pair of electrons constitutes a single covalent bond. Two or three pairs of electrons may be shared between atoms to give double and triple covalent bonds. Most often, both atoms of the bond contribute an equal number of electrons to the bond. In a few cases, both electrons of a shared electron pair come from the same atom; this is a coordinate covalent bond.

14. **Number of Covalent Bonds Formed** There is a strong tendency for nonmetals to form a particular number of covalent bonds. The number of valence electrons the nonmetal has and the number of covalent bonds it forms gives a sum of eight.

15. **Resonance Structures** Resonance structures are two or more Lewis structures for a molecular compound that have the same arrangement of atoms, contain the same number of electrons, and differ only in the location of the electrons.

16. **Molecular Geometry and VSEPR Theory** Molecular geometry describes the way atoms in a molecule are arranged in space relative to one another. VSEPR theory is a set of procedures used to predict molecular geometry from a compound's Lewis structure. VSEPR theory is based on the concept that valence shell electron groups about an atom (bonding and nonbonding) orient themselves as far away from one another as possible (to minimize repulsions).

17. **Electronegativity** Electronegativity is a measure of the relative attraction that an atom has for the shared electrons in a bond. Electronegativity values are useful in predicting the polarity of a bond.

18. **Bond Polarity** Bond polarity is a function of the electronegativity difference between the bonded atoms and sometimes the type of atoms (metal or nonmetal) present in the bond. For small electronegativity differences (0.4 or less), the bond is classified as nonpolar covalent. In the electronegativity difference range greater than 0.4 but less than 1.5, the bond is considered polar covalent. In the range between 1.5 and 2.0, bond classification depends on metal–nonmetal makeup of the bond. The bond is considered polar covalent if two nonmetals are present, and ionic if a metal and nonmetal are present. For electronegativity differences of 2.0 or greater, the bond is always classified as ionic.

19. **Molecular Polarity** Molecules as a whole can have polarity. If individual bond polarities do not cancel because of the symmetrical nature of a molecule, then the molecule as a whole is polar. Molecular polarity is a measure of the degree of inequality in the attraction of bonding electrons to various locations within a molecule.

Key Terms

The new terms defined in this chapter are

anion *Sec. 7.4*
bond length *Sec. 7.15*
bond polarity *Sec. 7.19*
bond strength *Sec. 7.15*
bonding electrons *Sec. 7.11*
cation *Sec. 7.4*
chemical bond *Sec. 7.1*

coordinate covalent bond *Sec. 7.14*	isoelectronic species *Sec. 7.5*
covalent bond *Secs. 7.1 and 7.10*	Lewis structure *Sec. 7.6*
double covalent bond *Sec. 7.12*	Lewis symbol *Sec. 7.2*
electronegativity *Sec. 7.18*	molecular geometry *Sec. 7.17*
formula unit *Sec. 7.8*	molecular polarity *Sec. 7.20*
ion *Sec. 7.4*	monoatomic ion *Sec. 7.9*
ionic bond *Secs. 7.1 and 7.4*	nonbonding electrons *Sec. 7.11*
nonpolar covalent bond *Sec. 7.19*	resonance structures *Sec. 7.15*
nonpolar molecule *Sec. 7.20*	single covalent bond *Sec. 7.12*
octet rule *Sec. 7.3*	triple covalent bond *Sec. 7.12*
percent ionic character of a bond *Sec. 7.19*	valence electron *Sec. 7.2*
polar covalent bond *Sec. 7.19*	VSEPR electron group *Sec. 7.17*
polar molecule *Sec. 7.20*	VSEPR theory *Sec. 7.17*
polyatomic ion *Sec. 7.9*	

Practice Problems

VALENCE ELECTRONS (SEC. 7.2)

7.1 How many valence electrons do atoms with the following electron configurations have?
a. $1s^2 2s^2 2p^3$
b. $1s^2 2s^2 2p^6 3s^2$
c. $1s^2 2s^2 2p^6 3s^2 3p^6 4s^2 3d^5$
d. $1s^2 2s^2 2p^6 3s^2 3p^6 4s^2 3d^{10} 4p^4$

7.2 How many valence electrons do atoms with the following electron configurations have?
a. $1s^2 2s^2 2p^1$
b. $1s^2 2s^2 2p^5$
c. $1s^2 2s^2 2p^6 3s^2 3p^6 4s^2$
d. $1s^2 2s^2 2p^6 3s^2 3p^6 4s^2 3d^{10} 4p^3$

7.3 How many valence electrons do atoms with the following condensed electron configurations have?
a. $[He]2s^2 2p^2$
b. $[Ne]3s^2 3p^1$
c. $[Ar]4s^2 3d^{10} 4p^4$
d. $[Kr]5s^2 4d^7$

7.4 How many valence electrons do atoms with the following condensed electron configurations have?
a. $[He]2s^2 2p^5$
b. $[Ne]3s^2$
c. $[Ar]4s^2 3d^3$
d. $[Kr]5s^2 4d^{10} 5p^2$

7.5 For each of the following pairs of representative elements, indicate whether the first listed element has (1) more valence electrons, (2) fewer valence electrons, or (3) the same number of valence electrons compared to the second listed element.
a. Be, Li   **b.** N, P   **c.** F, O   **d.** Al, Cl

7.6 For each of the following pairs of representative elements, indicate whether the first listed element has (1) more valence electrons, (2) fewer valence electrons, or (3) the same number of valence electrons compared to the second listed element.
a. K, Ca   **b.** S, Se   **c.** N, B   **d.** Si, Na

7.7 How many of the "highlighted" elements in the following periodic table have each of the valence electron characteristics listed?

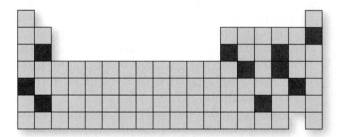

a. Five valence electrons are present.
b. Seven valence electrons are present.
c. Two valence electrons are present.
d. Eight valence electrons are present.

7.8 How many of the "highlighted" elements in the periodic table of Problem 7.7 have each of the valence electron characteristics listed?
a. Four valence electrons are present.
b. One valence electron is present.
c. Three valence electrons are present.
d. Six valence electrons are present.

7.9 Give the periodic table group number and the number of valence electrons present for each of the following representative elements.
a. lithium   **b.** neon
c. calcium   **d.** iodine

7.10 Give the periodic table group number and the number of valence electrons present for each of the following representative elements.
a. magnesium   **b.** potassium
c. phosphorus   **d.** bromine

7.11 Identify the element with each of the following characteristics.
a. period 2 element with four valence electrons
b. period 2 element with seven valence electrons

c. period 3 element with two valence electrons
d. period 3 element with five valence electrons

7.12 Identify the element with each of the following characteristics.
a. period 2 element with one valence electron
b. period 2 element with three valence electrons
c. period 3 element with three valence electrons
d. period 3 element with six valence electrons

LEWIS SYMBOLS FOR ATOMS (SEC. 7.2)

7.13 Draw Lewis symbols for the atoms of the following elements.
a. $_6$C   b. $_{14}$Si   c. $_{17}$Cl   d. $_{56}$Ba

7.14 Draw Lewis sybols for atoms of the following elements.
a. $_7$N   b. $_{15}$P   c. $_{35}$Br   d. $_{55}$Cs

7.15 Each of the following Lewis symbols represents a period 2 representative element. Determine the element's identity in each case.
a. ·X·   b. ·X·   c. ·X: d. X·

7.16 Each of the following Lewis symbols represents a period 3 representative element. Determine the element's identity in each case.
a. X·   b. ·X: c. ·X·   d. ·X·

7.17 Indicate whether each of the following Lewis symbols is correct or incorrect.
a. :Ö: b. ·Al·   c. ·Cl·   d. ·Se:

7.18 Indicate whether each of the following Lewis symbols is correct or incorrect.
a. :F: b. ·Si·   c. :Kr: d. Cs·

7.19 For each of the following pairs of elements, indicate whether the Lewis symbol for the first listed element contains (1) more dots, (2) fewer dots, or (3) the same number of dots, compared to the Lewis symbol for the second listed element.
a. Li, Be   b. Mg, Ca   c. Al, K   d. Ar, Ne

7.20 For each of the following pairs of elements, indicate whether the Lewis symbol for the first listed element contains (1) more dots, (2) fewer dots, or (3) the same number of dots compared to the Lewis symbol for the second listed element.
a. Si, P   b. O, S   c. F, N   d. Cs, Br

NOTATION FOR IONS (SEC. 7.4)

7.21 Give the symbol for each of the following ions.
a. a lithium atom that has lost one electron
b. a phosphorus atom that has gained three electrons
c. a bromine atom that has gained one electron
d. a barium atom that has lost two electrons

7.22 Give the symbol for each of the following ions.
a. an iodine atom that has gained one electron
b. a zinc atom that has lost two electrons

c. a gallium atom that has lost three electrons
d. a sodium atom that has lost one electron

7.23 Fill in the blanks in each line of the following table. The first line is already completed as an example.

	Chemical Symbol	Ion Formed	Number of Electrons in Ion	Number of Protons in Ion
	Ca	Ca^{2+}	18	20
a.		Be^{2+}		4
b.			54	53
c.	Al		10	
d.		S^{2-}		

7.24 Fill in the blanks in each line of the following table. The first line is already completed as an example.

	Chemical Symbol	Ion Formed	Number of Electrons in Ion	Number of Protons in Ion
	Mg	Mg^{2+}	10	12
a.		K^+		19
b.			36	34
c.	P		18	
d.		O^{2-}		

7.25 Indiciate whether each of the following atoms or ions is (1) a neutral species, (2) a negatively charged species, or (3) a positively charged species.
a. contains 5 electrons, 5 protons, and 5 neutrons
b. contains 18 electrons, 19 protons, and 20 neutrons
c. contains 18 electrons, 20 protons, and 20 neutrons
d. contains 36 electrons, 35 protons, and 46 neutrons

7.26 Indiciate whether each of the following atoms or ions is (1) a neutral species, (2) a negatively charged species, or (3) a positively charged species.
a. contains 6 electrons, 6 protons, and 6 neutrons
b. contains 18 electrons, 15 protons, and 16 neutrons
c. contains 18 electrons, 16 protons, and 16 neutrons
d. contains 36 electrons, 37 protons, and 48 neutrons

7.27 What would be the symbol for an ion with each of the following characteristics?
a. an aluminum ion with 10 electrons
b. an oxygen ion with 10 electrons
c. a magnesium ion with two more protons than electrons
d. a beryllium ion with two fewer electrons than protons

7.28 What would be the symbol for an ion with each of the following characteristics?
a. a sodium ion with 10 electrons
b. a fluorine ion with 10 electrons

c. a sulfur ion with two fewer protons than electrons
d. a calcium ion with two more protons than electrons

7.29   Classify each of the following species as (1) a cation, (2) an anion, or (3) not an ion.
a. Cl^- b. Na^+ c. F   d. N^{3-}

7.30   Classify each of the following species as (1) a cation, (2) an anion, or (3) not an ion.
a. O^{2-} b. Mg^{2+} c. K   d. F^-

THE SIGN AND MAGNITUDE OF IONIC CHARGE (SEC. 7.5)

7.31   Based on periodic table position, predict whether each of the following elements forms a positive ion or negative ion.
a. nitrogen   b. beryllium
c. magnesium   d. chlorine

7.32   Based on periodic table position, predict whether each of the following elements forms a positive ion or negative ion.
a. selenium   b. rubidium
c. iodine   d. strontium

7.33   Based on periodic table position, predict whether each of the elements in Problem 7.31 forms a cation or an anion.

7.34   Based on periodic table position, predict whether each of the elements in Problem 7.32 forms a cation or an anion.

7.35   Based on periodic table position, predict the actual number of electrons lost or gained when each of the elements in Problem 7.31 forms an ion.

7.36   Based on periodic table position, predict the actual number of electrons lost or gained when each of the elements in Problem 7.32 forms an ion.

7.37   Predict the general kind of behavior, that is, loss or gain of electrons, you would expect from atoms with the following electron configurations.
a. $1s^2 2s^2$ b. $1s^2 2s^2 2p^6 3s^2$
c. $1s^2 2s^2 2p^6 3s^2 3p^1$ d. $1s^2 2s^2 2p^6 3s^2 3p^5$

7.38   Predict the general kind of behavior, that is, loss or gain of electrons, you would expect from atoms with the following electron configurations.
a. $1s^2 2s^2 2p^4$ b. $1s^2 2s^2 2p^3$
c. $1s^2 2s^1$
d. $1s^2 2s^2 2p^6 3s^2 3p^6 4s^2 3d^{10} 4p^1$

7.39   Write the electron configuration for each of the following ions.
a. F^- b. Na^+ c. S^{2-} d. Ca^{2+}

7.40   Write the electron configuration for each of the following ions.
a. Cl^- b. Li^+ c. Al^{3+} d. N^{3-}

7.41   Write a condensed electron configuration for each of the ions in Problem 7.39.

7.42   Write a condensed electron configuration for each of the ions in Problem 7.40.

7.43   How many of the "highlighted" elements in the following periodic table will form ions with each of the following characteristics?

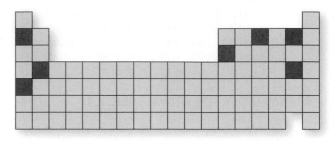

a. forms a positively charged ion
b. forms an anion
c. forms an ion that has a charge magnitude of 2
d. forms an ion that involves loss of two or more electrons

7.44   How many of the "highlighted" elements in the periodic table in Problem 7.43 will form ions with each of the following characteristics?
a. forms a negatively charged ion
b. forms a cation
c. forms an ion that has a charge magnitude of 1
d. forms an ion that involves gain of two or more electrons

7.45   In which group in the periodic table would representative elements that form ions with the following charges most likely be found?
a. +1   b. −2   c. −3   d. +4

7.46   In which group in the periodic table would representative elements that form ions with the following charges most likely be found?
a. +3   b. −4   c. +2   d. −1

7.47   Which noble gas has an electron configuration identical to that of each of the following ions?
a. Be^{2+} b. Ba^{2+} c. Br^- d. I^-

7.48   Which noble gas has an electron configuration identical to that of each of the following ions?
a. Na^+ b. Rb^+ c. S^{2-} d. Se^{2-}

7.49   Select from each of the following sets of ions the ion that is isoelectronic with the noble gas neon.
a. Li^+, Na^+, K^+ b. F^-, Cl^-, Br^-
c. Ca^{2+}, O^{2-}, Be^{2+} d. Mg^{2+}, P^{3-}, S^{2-}

7.50   Select from each of the following sets of ions the ion that is isoelectronic with the noble gas argon.
a. $Be^{2+}, Mg^{2+}, Ca^{2+}$ b. O^{2-}, S^{2-}, Se^{2-}
c. K^+, N^{3-}, Al^{3+} d. F^-, O^{2-}, P^{3-}

7.51   In which of the following are the two members of the pair isoelectronic with each other?
a. Ar and S^{2-} b. Ne and Be^{2+}
c. Ca^{2+} and P^{3-} d. Li^+ and O^{2-}

7.52 In which of the following are the two members of the pair isoelectronic with each other?
   a. He and Li$^+$ **b.** Kr and Ca^{2+}
   c. F$^-$ and Cl$^-$ **d.** Al^{3+} and N^{3-}

LEWIS STRUCTURES FOR IONIC COMPOUNDS (SEC. 7.6)

7.53 Show the formation of the following ionic compounds using Lewis structures.
   a. CaBr$_2$ **b.** MgS   **c.** Be$_3$N$_2$ **d.** Na$_3$P
7.54 Show the formation of the following ionic compounds using Lewis structures.
   a. K$_2$O   **b.** MgF$_2$ **c.** AlP   **d.** Na$_2$S
7.55 Using Lewis structures, show how ionic compounds are formed from the following pairs of elements.
   a. lithium and nitrogen
   b. magnesium and oxygen
   c. chlorine and barium
   d. fluorine and potassium
7.56 Using Lewis structures, show how ionic compounds are formed from the following pairs of elements.
   a. sodium and bromine
   b. aluminum and sulfur
   c. phosphorus and beryllium
   d. oxygen and calcium

CHEMICAL FORMULAS FOR IONIC COMPOUNDS (SEC. 7.7)

7.57 Write the chemical formula for the ionic compound formed from each of the following types of ions.
   a. Ca^{2+} and Cl$^-$
   b. Be^{2+} and O^{2-}
   c. Al^{3+} and N^{3-}
   d. K$^+$ and S^{2-}
7.58 Write the chemical formula for the ionic compound formed from each of the following types of ions.
   a. Mg^{2+} and S^{2-} **b.** Na$^+$ and Br$^-$
   c. Al^{3+} and O^{2-} **d.** Be^{2+} and N^{3-}
7.59 Fill in the blanks to complete the following table of chemical formulas for ionic compounds. For each compound the positive ion present is listed on the left side of the table and the negative ion present is listed at the top. The first line of the table is already completed as an example.

		F$^-$	O^{2-}	N^{3-}	C^{4-}
	Na$^+$	NaF	Na$_2$O	Na$_3$N	Na$_4$C
a.	Ca^{2+}				
b.	Al^{3+}				
c.	Ag$^+$				
d.	Zn^{2+}				

7.60 Fill in the blanks to complete the following table of chemical formulas for ionic compounds. For each compound the negative ion present is listed on the left side of the table and the positive ion present is listed at the top. The first line of the table is already completed as an example.

		Na$^+$	Mg^{2+}	Al^{3+}	Si^{4+}
	Cl$^-$	NaCl	MgCl$_2$	AlCl$_3$	SiCl$_4$
a.	Br$^-$				
b.	S^{2-}				
c.	P^{3-}				
d.	N^{3-}				

7.61 Write chemical formulas (symbol and charge) for both kinds of ions present in each of the following ionic compounds.
   a. MgS
   b. AlN
   c. Na$_2$O
   d. Ca$_3$N$_2$
7.62 Write chemical formulas (symbol and charge) for both kinds of ions present in each of the following ionic compounds.
   a. KCl
   b. CaS
   c. BeF$_2$
   d. Al$_2$S$_3$
7.63 Write the chemical formula of the ionic compound that could form from the elements X and Z if
   a. X has two valence electrons and Z has four valence electrons.
   b. X has three valence electrons and Z has seven valence electrons.
   c. X has one valence electron and Z has five valence electrons.
   d. X has seven valence electrons and Z has two valence electrons.
7.64 Write the chemical formula of the ionic compound that could form from the elements X and Z if
   a. X has two valence electrons and Z has seven valence electrons.
   b. X has one valence electron and Z has six valence electrons.
   c. X has three valence electrons and Z has five valence electrons.
   d. X has six valence electrons and Z has two valence electrons.

7.65 Determine the chemical formulas for the ionic compounds that form when each of the listed pairs of "highlighted" elements in the following periodic table interact with each other.

a. elements 1 and 5
b. elements 2 and 8
c. elements 3 and 6
d. elements 7 and 4

7.66 Determine the chemical formulas for the ionic compounds that form when each of the following pairs of "highlighted" elements in the periodic table of Problem 7.65 interact with each other.
a. elements 1 and 6
b. elements 2 and 7
c. elements 3 and 8
d. elements 5 and 4

POLYATOMIC-ION-CONTAINING IONIC COMPOUNDS (SEC. 7.9)

7.67 Determine the chemical formula for the ionic compound formed from each of the following types of ions.
a. Mg^{2+} and CN^- **b.** Ca^{2+} and SO_4^{2-}
c. Al^{3+} and OH^- **d.** NH_4^+ and NO_3^-

7.68 Determine the chemical formula for the ionic compound formed from each of the following types of ions.
a. Mg^{2+} and NO_3^- **b.** Ca^{2+} and CN^-
c. Be^{2+} and OH^- **d.** NH_4^+ and Cl^-

7.69 Determine the chemical formula for the ionic compound in which the Al^{3+} ion is combined with each of the following polyatomic ions.
a. PO_4^{3-} **b.** CO_3^{2-}
c. ClO_3^- **d.** $C_2H_3O_2^-$

7.70 Determine the chemical formula for the ionic compound in which the Mg^{2+} ion is combined with each of the following polyatomic ions.
a. PO_4^{3-} **b.** CO_3^{2-}
c. ClO_3^- **d.** $C_2H_3O_2^-$

7.71 Fill in the blanks to complete the following table of chemical formulas for polyatomic-ion-containing ionic compounds. For each compound the positive ion present is listed on the left side of the table and the negative ion present is listed at the top. The first line of the table is already completed as an example.

	OH^-	CN^-	NO_3^-	SO_4^{2-}
Na^+	NaOH	NaCN	$NaNO_3$	Na_2SO_4
a. K^+				
b. Mg^{2+}				
c. Al^{3+}				
d. NH_4^+				

7.72 Fill in the blanks to complete the following table of chemical formulas for polyatomic-ion-containing ionic compounds. For each compound the positive ion present is listed on the left side of the table and the negative ion present is listed at the top. The first line of the table already completed as an example.

	OH^-	CN^-	NO_3^-	SO_4^{2-}
Na^+	NaOH	NaCN	$NaNO_3$	Na_2SO_4
a. Ca^{2+}				
b. Li^+				
c. Ga^{3+}				
d. NH_4^+				

LEWIS STRUCTURES FOR COVALENT COMPOUNDS (SECS. 7.10–7.15)

7.73 Show the formation of the following covalent molecules, using Lewis structures.
a. I_2 **b.** ClF   **c.** H_2S **d.** PF_3

7.74 Show the formation of the following covalent molecules, using Lewis structures.
a. Br_2 **b.** IF   **c.** PCl_3 **d.** CBr_4

7.75 Write a Lewis structure for the simplest covalent compound most likely to be formed between each of these pairs of elements.
a. hydrogen and bromine
b. oxygen and fluorine
c. nitrogen and chlorine
d. silicon and iodine

7.76 Write a Lewis structure for the simplest covalent compound most likely to be formed between each of these pairs of elements.
a. hydrogen and iodine
b. nitrogen and bromine
c. phosphorus and chlorine
d. silicon and hydrogen

7.77 How many *bonding electron pairs* and *nonbonding electron pairs* are present in each of the following Lewis structures?

  a. :N:::N: **b.** H:C::C:H
                                    H H

  c. :Ö=C=Ö: **d.** :N≡C—C≡N:

7.78 How many *bonding electron pairs* and *nonbonding electron pairs* are present in each of the following Lewis structures?

  a. H:C:::C:H   **b.** H:P̈:C̈l:
                                    H

  c.   Ö: **d.**
              ‖
        H—C—H   :F̈—N=N̈—F̈:

7.79 How many *single covalent bonds, double covalent bonds*, and *triple covalent bonds* are present in each of the molecules in Problem 7.77?

7.80 How many *single covalent bonds, double covalent bonds*, and *triple covalent bonds* are present in each of the molecules in Problem 7.78?

7.81 Which of the following is a normally expected bonding pattern for the element shown?

  a. —Ö— **b.** :N≡ **c.** :C̈= **d.** :Ö=

7.82 Which of the following is a normally expected bonding pattern for the element shown?

  a. —N̈= **b.** :O≡ **c.** —C≡ **d.** =C=

7.83 What is a coordinate covalent bond?

7.84 Once formed, how (if at all) does a coordinate covalent bond differ from a regular covalent bond?

7.85 Identify the coordinate covalent bond(s) present, if any, in each of the following molecules by listing the two atoms involved in the bond. Name the atom on the left or below in the bond first.

  a. :N≡N—Ö: **b.** H—Ö—F̈:

  c. :Ö—C̈l—Ö—H   **d.** :Ö—B̈r—Ö—H
                                          |
                                         :O:

7.86 Identify the coordinate covalent bond(s) present, if any, in each of the following molecules by listing the two atoms involved in the bond. Name the atom on the left or below in the bond first.

  a. :S̈=S̈—Ö: **b.** H—Ö—B̈r:

  c. :Ö—Ï—Ö—H   **d.** :Ö—C̈l—Ö—H
                                          |
                                         :O:

7.87 What are resonance structures?

7.88 Explain why the Lewis structures H—C≡N: and H—N≡C: are not resonance structures.

7.89 The following is one of the three resonance structures that exist for the nitrate ion.

$$\left[:\ddot{O}—N—\ddot{O}: \right]^{-}$$
$$\qquad\qquad \overset{\|}{\underset{:O:}{}}$$

Draw the other two resonance structures.

7.90 The following is one of the three resonance structures that exist for the carbonate ion.

$$\left[:\ddot{O}—C=O: \right]^{2-}$$
$$\qquad\qquad \overset{|}{\underset{:O:}{}}$$

Draw the other two resonance structures.

SYSTEMATIC PROCEDURES FOR DRAWING LEWIS STRUCTURES (SEC. 7.16)

7.91 Without actually writing the Lewis structure, determine the total number of valence electrons available for use in drawing the Lewis structure of each of the following molecules or polyatomic ions.
  a. HNO_3 **b.** NCl_3 **c.** NO_3^- **d.** SO_4^{2-}

7.92 Without actually writing the Lewis structure, determine the total number of valence electrons available for use in drawing the Lewis structure of each of the following molecules or polyatomic ions.
  a. H_3PO_4 **b.** PH_3 **c.** PO_4^{3-} **d.** ClO_3^-

7.93 Without actually writing the Lewis structure, determine the total number of "dots" (electrons) that should be present in the Lewis structure of each of the molecules or polyatomic ions in Problem 7.91.

7.94 Without actually writing the Lewis structure, determine the total number of "dots" (electrons) that should be present in the Lewis structure of each of the molecules or polyatomic ions in Problem 7.92.

7.95 Without actually writing the Lewis structure, determine how many bonding and nonbonding electron pairs are present in each of the following diatomic molecules or ions.
  a. HF **b.** CO **c.** Cl_2 **d.** OH^-

7.96 Without actually writing the Lewis structure, determine how many bonding and nonbonding electron pairs are present in each of the following diatomic molecules or ions.
  a. ClF **b.** HS^- **c.** H_2 **d.** CN^-

7.97 Using systematic procedures, write the Lewis structure for each of the following molecules.
  a. H_3CCH_3 (C_2H_6) **b.** H_2NNH_2 (N_2H_4)
  c. F_2CH_2 (CH_2F_2) **d.** H_3CCCl_3 ($C_2H_3Cl_3$)
The above formulas are written in a form that specifies atomic arrangement. When an elemental symbol carries a subscript, these atoms are directly and separately bonded to the atom immediately following or immediately preceding the subscripted symbol.

7.98 Using systematic procedures, write the Lewis structure for each of the following molecules.
  a. H_2PPH_2 (P_2H_4)   **b.** H_3CCBr_3 ($C_2H_3Br_3$)
  c. F_2CCl_2 (CF_2Cl_2)   **d.** H_3SiSiH_3 (Si_2H_6)
  The formulas are written in a way that specifies atomic arrangement, as explained in Problem 7.97.

7.99 Using systematic procedures, write the Lewis structure for each of the following polyatomic ions.
  a. NH_4^+ **b.** BeH_4^{2-} **c.** ClO_3^- **d.** IO_4^-

7.100 Using systematic procedures, write the Lewis structure for each of the following polyatomic ions.
  a. BH_4^- **b.** $AlCl_4^-$ **c.** PF_4^+ **d.** ClO_2^-

7.101 Using systematic procedures, write the Lewis structure for each of the following molecules, each of which contains at least one multiple bond. The formulas of the molecules are written in a way that specifies atomic arrangement, as explained in Problem 7.97.
  a. Cl_2CCH_2 ($C_2H_2Cl_2$) **b.** H_3CCN (C_2H_3N)
  c. $HCCCH_3$ (C_3H_4)   **d.** $FNNF$ (N_2F_2)

7.102 Using systematic procedures, write the Lewis structure for each of the following molecules, each of which contains at least one multiple bond. The formulas of the molecules are written in a way that specifies atomic arrangement, as explained in Problem 7.97.
  a. H_2NCN (CH_2N_2)   **b.** Cl_2CO
  c. $ClCCCl$ (C_2Cl_2)   **d.** $NCCN$ (C_2N_2)

7.103 Using systematic procedures, write the Lewis structure for each of the following molecules or polyatomic ions. Resonance structures will be needed in each case. Some formulas are written in a way that specifies atomic arrangements, as explained in Problem 7.97.
  a. H_3CNO_2 **b.** NNO
  c. CO_3^{2-} **d.** SCN^-

7.104 Using systematic procedures, write the Lewis structure for each of the following molecules or polyatomic ions. Resonance structures will be needed in each case. Some formulas are written in a way that specifies atomic arrangements, as explained in Problem 7.97.
  a. $HONO$ **b.** H_2NNO_2
  c. NO_3^- **d.** OCN^-

7.105 The Lewis structure for the polyatomic ion ClO_4^{n-} is

$$\left[\begin{array}{c} \ddot{\text{O}} \\ | \\ \ddot{\text{O}} - \text{Cl} - \ddot{\text{O}} \\ | \\ \ddot{\text{O}} \end{array} \right]^{n-}$$

What is the value of n, the magnitude of the ionic charge, in the formula ClO_4^{n-}?

7.106 The Lewis structure for the polyatomic ion $BeCl_4^{n-}$ is

$$\left[\begin{array}{c} \ddot{\text{Cl}} \\ | \\ \ddot{\text{Cl}} - \text{Be} - \ddot{\text{Cl}} \\ | \\ \ddot{\text{Cl}} \end{array} \right]^{n-}$$

What is the value of n, the magnitude of the ionic charge, in the formula $BeCl_4^{n-}$?

MOLECULAR GEOMETRY (VSEPR THEORY) (SEC. 7.17)

7.107 What is the molecular geometry that VSEPR theory predicts for a molecule whose sole central atom has the environment of
  a. two single bonds and two nonbonding electron pairs?
  b. one single bond and one triple bond?
  c. one single bond, one double bond, and one nonbonding electron pair?
  d. one double bond and two nonbonding electron pairs?

7.108 What is the molecular geometry that VSEPR theory predicts for a molecule whose sole central atom has the environment of
  a. four single bonds?
  b. two double bond?
  c. three single bonds and one nonbonding electron pair?
  d. one triple bond and one nonbonding electron pair?

7.109 Using VSEPR theory, predict the geometry of the following triatomic molecules or ions.

  a. $H\!:\!\ddot{S}\!:\!H$ **b.** $:\!\ddot{\underset{..}{I}}\!:\!C\!:::\!N\!:$

  c. $[:\!\ddot{O}\!::\!N\!::\!\ddot{O}\!:]^+$ **d.** $:\!\ddot{O}\!:\!\ddot{O}\!::\!\ddot{O}\!:$

7.110 Using VSEPR theory, predict the geometry of the following triatomic molecules or ions.

  a. $:\!\ddot{N}\!::\!\ddot{S}\!:\!\ddot{F}\!:$ **b.** $:\!\ddot{N}\!::\!N\!::\!\ddot{O}\!:$

  c. $[H\!:\!\ddot{N}\!:\!H]^-$ **d.** $[:\!\ddot{\underset{..}{S}}\!:\!C\!:::\!N\!:]^-$

7.111 Using VSEPR theory, predict the geometry of the following sulfur-containing molecules or ions.
  a. $:\!\ddot{F}\!-\!\overset{\displaystyle}{\underset{\displaystyle :\!\ddot{F}\!:}{S}}\!-\!\ddot{O}\!:$ **b.** $[:\!\ddot{S}\!=\!N\!=\!\ddot{S}\!:]^+$

  c. $:\!\ddot{O}\!-\!\ddot{S}\!=\!\ddot{O}\!:$ **d.** $\left[\begin{array}{c} :\!\ddot{O}\!: \\ | \\ :\!\ddot{O}\!-\!S\!-\!\ddot{O}\!: \\ | \\ :\!\ddot{S}\!: \end{array} \right]^{2-}$

7.112 Using VSEPR theory, predict the geometry of the following nitrogen-containing molecules or ions.
  a. $\left[:\!\ddot{O}\!-\!\overset{\displaystyle}{\underset{\displaystyle :\!\ddot{O}\!:}{N}}\!=\!\ddot{O}\!: \right]^-$ **b.** $:\!\ddot{F}\!-\!\overset{\displaystyle}{\underset{\displaystyle :\!\ddot{F}\!:}{N}}\!-\!\ddot{F}\!:$

  c. $\left[\begin{array}{c} H \\ | \\ H\!-\!N\!-\!H \\ | \\ H \end{array} \right]^+$ **d.** $[:\!\ddot{O}\!-\!\ddot{N}\!=\!\ddot{O}\!:]^-$

7.113 Using VSEPR theory, predict the geometry of the following molecules or polyatomic ions. Some formulas are written in a way that specifies atomic arrangements as explained in Problem 7.97.
a. $AlCl_4^-$ b. CS_2 c. $OSCl_2$ d. PO_3^{3-}

7.114 Using VSEPR theory, predict the geometry of the following molecules or polyatomic ions. Some formulas are written in a way that specifies atomic arrangements as explained in Problem 7.97.
a. H_3CBr b. $BeCl_4^{2-}$ c. NO_2^- d. Cl_2O

7.115 Using VSEPR theory, predict the geometry of the following molecules, which have more than one central atom. The formulas are written in a way that specifies atomic arrangements as explained in Problem 7.97.
a. H_2NNH_2 b. H_3COH

7.116 Using VSEPR theory, predict the geometry of the following molecules, which have more than one central atom. The formulas are written in a way that specifies atomic arrangements as explained in Problem 7.97.
a. H_2CCCl_2 b. H_3COCH_3

ELECTRONEGATIVITY (SEC. 7.18)

7.117 What general trends in electronegativity values occur in the periodic table?

7.118 If an element has a low electronegativity, is it likely to be a metal or a nonmetal? Explain your answer.

7.119 Based on periodic table trends, select the more electronegative element in each of the following pairs of elements.
a. Be and O   b. Be and Ca
c. H and C   d. Cs and Ca

7.120 Based on periodic table trends, select the more electronegative element in each of the following pairs of elements.
a. Mg and S   b. Mg and Ca
c. H and B   d. Sb and Se

7.121 Using chemical periodicity trends, arrange each of the following sets of atoms in order of decreasing electronegativity.
a. F, B, O, Li   b. F, Cl, S, P
c. N, Sb, As, P   d. Fr, Mg, Si, F

7.122 Using chemical periodicity trends, arrange each of the following sets of atoms in order of decreasing electronegativity.
a. Cl, Br, I, F   b. B, Na, Al, Mg
c. S, P, O, As   d. O, N, Al, Na

7.123 Use the information in Figure 7.12 as a basis for answering the following questions.
a. Which elements have electronegativity values that exceed that of the element carbon?
b. Which elements have electronegativity values of less than 1.0?
c. What are the four most electronegative elements listed in Figure 7.12?
d. By what constant amount do the electronegativity values for sequential period 2 elements differ?

7.124 Use the information in Figure 7.12 as a basis for answering the following questions.
a. Which elements have electronegativity values that exceed that of the element sulfur?
b. What are the four least electronegative elements listed in Figure 7.12?
c. Which three elements in Figure 7.12 have an electronegativity value of 2.5?
d. Which period 2 element has an electronegativity value closest to that of hydrogen?

7.125 Indicate whether each of the statements relating to electronegativity values of the "highlighted" elements in the following periodic table are true or false. Use periodic table trends rather than actual electronegativity values in solving this problem.

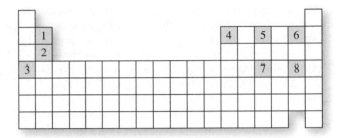

a. Element 1 has a higher electronegativity value than element 2.
b. Element 4 has a higher electronegativity value than element 5.
c. Element 3 has a higher electronegativity value than element 8.
d. Element 7 has a higher electronegativity value than element 6.

7.126 Indicate whether each of the statements relating to electronegativity values of the "highlighted" elements in the periodic table in Problem 7.125 are true or false. Use periodic table trends rather than actual electronegative values in solving this problem.
a. Element 2 has a higher electronegativity value than element 4.
b. Element 5 has a higher electronegativity value than element 6.
c. Element 7 has a higher electronegativity value than element 8.
d. Element 3 has a higher electronegativity value than element 1.

BOND POLARITY (SEC. 7.19)

7.127 Based on electronegativity values, determine which atom carries the partial positive charge in each of the following bonds.
a. N—S   b. N—Br   c. N—O   d. N—F

7.128 Based on electronegativity values, determine which atom carries the partial positive charge in each of the following bonds.
  a. $C-H$ **b.** $C-O$
  c. $C-N$ **d.** $C-P$

7.129 Identify the bond of *greatest* polarity in each of the following sets of bonds.
  a. $H-Cl, H-O, H-Br$
  b. $O-F, O-P, O-Al$
  c. $H-Cl, Br-Br, B-N$
  d. $Al-Cl, C-N, Cl-F$

7.130 Identify the bond of *greatest* polarity in each of the following sets of bonds.
  a. $H-C, H-N, H-B$
  b. $O-N, O-S, O-Br$
  c. $P-N, S-O, Br-F$
  d. $H-F, Cl-Cl, Si-O$

7.131 Identify the bond in each set of bonds in Problem 7.129 that has the *greatest* ionic character.

7.132 Identify the bond in each set of bonds in Problem 7.130 that has the *greatest* ionic character.

7.133 Identify the bond in each set of bonds in Problem 7.129 that has the *least* covalent character.

7.134 Identify the bond in each set of bonds in Problem 7.130 that has the *least* covalent character.

7.135 Identify the bond in each set of bonds in Problem 7.129 for which the positive partial charge (δ^+) on the less electronegative atom is the greatest.

7.136 Identify the bond in each set of bonds in Problem 7.130 for which the positive partial charge (δ^+) on the less electronegative atom is the greatest.

7.137 Based on the arbitrary generalizations given in Section 7.19, classify each of the following bonds as *nonpolar covalent, polar covalent, or ionic*.
  a. carbon–nitrogen
  b. beryllium–oxygen
  c. phosphorus–chlorine
  d. silicon–silicon

7.138 Based on the designations given in Section 7.19, classify each of the following bonds as *nonpolar covalent, polar covalent, or ionic*.
  a. cesium–fluorine
  b. potassium–chlorine
  c. hydrogen–carbon
  d. silicon–oxygen

7.139 Indicate whether bonds with the following electronegativity differences between atoms would be classified as *ionic, polar covalent, or nonpolar covalent*.
  a. 0.3   **b.** 1.2   **c.** 1.9   **d.** 2.1

7.140 Indicate whether bonds with the following electronegativity differences between atoms would be classified as *ionic, polar covalent, or nonpolar covalent*.
  a. 0.2   **b.** 0.9   **c.** 1.8   **d.** 2.2

7.141 Indicate whether bonds with the electronegativity differences given in Problem 7.139 would have more ionic character or more covalent character.

7.142 Indicate whether bonds with the electronegativity differences given in Problem 7.140 would have more ionic character or more covalent character.

7.143 Indicate whether each of the following bonds has more ionic character than covalent character, or vice versa.
  a. $P-I$ **b.** $H-F$ **c.** $B-O$ **d.** $Na-Cl$

7.144 Indicate whether each of the following bonds has more ionic character than covalent character or vice versa.
  a. $Cs-O$ **b.** $C-N$ **c.** $P-F$ **d.** $I-Cl$

MOLECULAR POLARITY (SEC. 7.20)

7.145 Indicate whether each of the following triatomic molecules is *polar* or *nonpolar*. The molecular geometry for each molecule is given in parentheses.
  a. CS_2 (linear with C in the center position)
  b. H_2Se (angular with Se in the center position)
  c. FNO (angular with N in the center position)
  d. N_2S (linear with N in the center position)

7.146 Indicate whether each of the following triatomic molecules is *polar* or *nonpolar*. The molecular geometry for each molecule is given in parentheses.
  a. SCl_2 (angular with S in the center position)
  b. OF_2 (angular with O in the center position)
  c. SO_2 (angular with S in the center position)
  d. ClCN (linear with C in the center position)

7.147 All of the following molecules have a tetrahedral geometry with carbon as the central atom. Classify each of the molecules as *polar* or *nonpolar*.
  a. CCl_4 **b.** CH_3Br **c.** CH_2F_2 **d.** CBr_4

7.148 All of the following molecules have a trigonal pyramidal geometry with nitrogen as the central atom. Classify each of the molecules as *polar* or *nonpolar*.
  a. NF_3 **b.** NF_2Cl **c.** NH_2Cl **d.** NCl_3

7.149 In which of the following pairs of molecules do both members of the pair have the same polarity, that is, both members are polar or both members are nonpolar?
  a. F_2 and BrF   **b.** HOCl and HCN
  c. CH_4 and CCl_4 **d.** SO_3 and NF_3

7.150 In which of the following pairs of molecules do both members of the pair have the same polarity, that is, both members are polar or both members are nonpolar?
  a. HBr and HCl   **b.** CO_2 and SO_2
  c. SO_2 and SO_3 **d.** CH_4 and CH_3Cl

7.151 What is the minimum number of different elements that must be present in molecules of compounds with the following molecular geometry characteristics?
  a. polar molecule with tetrahedral geometry
  b. nonpolar molecule with trigonal planar geometry
  c. nonpolar molecule with linear geometry
  d. polar molecule with angular geometry

7.152 What is the minimum number of different elements that must be present in molecules of compounds with the following molecular geometry characteristics?
a. nonpolar molecule with tetrahedral geometry
b. polar molecule with trigonal pyramidal geometry
c. polar molecule with linear geometry
_____ d. polar molecule with trigonal planar geometry

ADDITIONAL PROBLEMS

7.153 Write the electron configurations for the following atoms and ions.
a. Mg and Mg^{2+} b. F and F^-
c. N and N^{3-} d. Ca^{2+} and S^{2-}

7.154 Write the electron configurations for the following atoms and ions.
a. Na and Na^+ b. O and O^{2-}
_____ c. Al and Al^{3+} d. K^+ and Cl^-

7.155 Classify the members of each of the following pairs of ions as (1) isoelectronic cations, (2) isoelectronic anions, (3) nonisoelectronic cations, or (4) nonisoelectronic anions.
a. Li^+ and Na^+ b. F^- and Br^-
c. K^+ and Ca^{2+} d. S^{2-} and N^{3-}

7.156 Classify the members of each of the following pairs of ions as (1) isoelectronic cations, (2) isoelectronic anions, (3) nonisoelectronic cations, or (4) nonisoelectronic anions.
a. S^{2-} and Cl^- b. Mg^{2+} and Ca^{2+}
_____ c. N^{3-} and P^{3-} d. Na^+ and Be^{2+}

7.157 Indicate whether each of the following compounds is most likely to be an ionic compound or a molecular compound.
a. Al_2O_3 b. H_2O_2 c. K_2S d. N_2H_4

7.158 Indicate whether each of the following compounds is most likely to be an ionic compound or a molecular compound.
_____ a. Cu_2O b. CO c. NaBr d. Be_3P_2

7.159 Indicate whether the basic structural unit for each of the following compounds is a *formula unit* or a *molecule*.
a. CO_2 b. NaCl c. Li_2O d. HOCl

7.160 Indicate whether the basic structural unit for each of the following compounds is a *formula unit* or a *molecule*.
a. SO_2 b. KF
_____ c. $Mg(OH)_2$ d. BrCl

7.161 Indicate whether each of the following ionic compounds contains (1) only monoatomic ions, (2) only polyatomic ions, or (3) both monoatomic and polyatomic ions.
a. CaF_2 b. $NaNO_3$
c. $(NH_4)_2SO_4$ d. BaO

7.162 Indicate whether each of the following ionic compounds contains (1) only monoatomic ions, (2) only polyatomic ions, or (3) both monoatomic and polyatomic ions.
a. NH_4NO_3 b. $CaSO_4$
_____ c. K_2S d. LiCl

7.163 Indicate whether each of the following compounds contains ions with a charge of +2.
a. CaO   b. MgF_2
c. Na_2S d. NO

7.164 Indicate whether each of the following compounds contains ions with a charge of +2.
_____ a. AlN   b. MgS   c. KCl   d. $CaCl_2$

7.165 Identify the period 2 element that X represents in each of the following ionic compounds.
a. Na_2X b. Be_3X_2 c. CaX d. MgX_2

7.166 Identify the period 3 element that X represents in each of the following ionic compounds.
_____ a. Na_3X b. BeX c. CaX_2 d. Mg_3X_2

7.167 Specify the reason each of the following electron-dot structures is incorrect, using the choices (1) not enough electron dots, (2) too many electron dots, or (3) improper placement of a correct number of electron dots.
a. $:N\!\!=\!\!\ddot{N}:$ b. $F:\ddot{O}:F$
c. $:\ddot{Br}::\ddot{Cl}:$ d. $[H::\ddot{O}:]^-$

7.168 Specify the reason each of the following electron-dot structures is incorrect, using the choices (1) not enough electron-dots, (2) too many electron dots, or (3) improper placement of a correct number of electron dots.
a. $:O\!\!=\!\!O:$ b. $H:\ddot{O}:Cl$
c. $H:\ddot{O}:::\ddot{O}:H$ d. $\left[O:N::\ddot{O}: \atop \ddot{O} \right]^-$

7.169 Four hypothetical elements, A, B, C, and D, have electronegativities A = 3.8, B = 3.3, C = 2.8, and D = 1.3. These elements form the compounds BA, DA, DB, and CA. Arrange these compounds in order of increasing *ionic* bond character.

7.170 Four hypothetical elements, A, B, C, and D, have electronegativities A = 3.6, B = 3.0, C = 2.7, and D = 0.9. These elements form the compounds CA, CB, DA, and DC. Arrange these compounds in order of increasing *covalent* bond character.

7.171 In which of the following pairs of diatomic species do both members of the pair have bonds of the same multiplicity (single, double, triple)?
a. BrCl and ClF
b. F_2 and N_2
c. NO^+ and NO^-
d. CN^- and SN^-

7.172 In which of the following pairs of diatomic species do both members of the pair have bonds of the same multiplicity (single, double, triple)?
  a. HCl and HF   **b.** S_2 and Cl_2
  c. CO and NO^+ **d.** OH^- and HS^-

7.173 In each of the following Lewis structures, X represents a period 3 nonmetal. Identify the nonmetal in each case.

  a.
  $$H\!-\!\overset{..}{\underset{..}{O}}\!-\!X\!-\!\overset{..}{\underset{..}{O}}:$$
  with $:\overset{..}{\underset{..}{O}}:$ above X

  b. $\left[:\overset{..}{\underset{..}{O}}\!-\!\overset{..}{X}:\right]^-$ with $:\overset{..}{\underset{..}{O}}:$ below X

7.174 In each of the following Lewis structures, X represents a period 3 nonmetal. Identify the nonmetal in each case.

  a.
  $$:\overset{..}{\underset{..}{Cl}}\!-\!X\!-\!\overset{..}{\underset{..}{Cl}}:$$
  with $\overset{..}{O}:$ double bonded above X

  b. $H\!-\!\overset{..}{\underset{..}{O}}\!-\!\overset{..}{X}:$

7.175 Give a Lewis structure description of the bonding in each of the following compounds that takes into account the presence of both ionic and covalent bonding in each compound.
  a. $CaSO_4$ **b.** NH_4NO_3

7.176 Give a Lewis structure description of the bonding in each of the following compounds that takes into account the presence of both ionic and covalent bonding in each compound.
  a. NaOH   **b.** NH_4CN

7.177 Specify the electron-group geometry about the central atom and the molecular geometry for each of the following species.
  a. CF_4 **b.** NH_3 **c.** SCN^- **d.** PH_4^+

7.178 Specify the electron-group geometry about the central atom and the molecular geometry for each of the following species.
  a. SiH_4 **b.** NH_4^+ **c.** ClNO   **d.** NO_3^-

7.179 Give an approximate value for the indicated bond angle in each of the following molecules.

  a. $H\!-\!\overset{..}{\underset{..}{O}}\!-\!H$

  b. $:\overset{..}{\underset{..}{Cl}}\!-\!C\!=\!C\!-\!\overset{..}{\underset{..}{Cl}}:$ with H and H below the carbons

7.180 Give an approximate value for the indicated bond angle in each of the following molecules.

  a.
  $$H\!-\!N\!-\!H$$
  with H below N

  b.

  $$H\!-\!C\!-\!\overset{..}{\underset{..}{O}}\!-\!H$$
  with H above and H below C

CUMULATIVE PROBLEMS

7.181 How many unpaired electrons are present in the following atoms or ions?
  a. O atom
  b. O^{2-} ion
  c. Ca atom
  d. Ca^{2+} ion

7.182 How many unpaired electrons are present in the following atoms or ions?
  a. F atom   **b.** F^- ion
  c. Al atom   **d.** Al^{3+} ion

7.183 Identify the elements whose ions have the following characteristics.
  a. alkaline earth metal whose ion is isoelectronic with Kr
  b. element whose 3– ion is isoelectronic Ne
  c. element whose 2+ ion contains 18 electrons
  d. group IIIA element whose ion has the electron configuration $[Ar]3d^{10}$

7.184 Identify the elements whose ions have the following characteristics.
  a. halogen whose ion is isoelectronic with Kr
  b. element whose 2– ion is isoelectronic Ar
  c. element whose 3+ ion contains 10 electrons
  d. group IVA element whose ion has the electron configuration $[Ar]3d^{10}$

7.185 Determine the identities of the elements A and D in the ionic compound AD, given the following:
  1. The atomic number of element A is greater than that of element D.
  2. Three electrons are transferred from A to D during compound formation.
  3. The sum of the atomic numbers of A and D is 20.

7.186 Determine the identities of the elements A and D in the ionic compound AD, given the following:
  1. The atomic number of element A is less than that of element D.
  2. Two electrons are transferred from A to D during compound formation.
  3. The sum of the atomic numbers of A and D is 20.

7.187 Given the following information about two elements, A and D, determine their identities and the formula of the binary ionic compound that they form.
  1. A is located in period 3 of the periodic table.
  2. The occupied electron subshell of highest energy for element D contains two electrons.
  3. The complete electron configuration for element D contains four fewer electrons than that for element A.

7.188 Given the following information about two elements, A and D, determine their identities and the formula of the binary ionic compound that they form.
  1. Both A and D are in period 3 of the periodic table.
  2. The occupied electron subshell of highest energy for element A contains one electron, and the one for element D contains four electrons.
  3. The atom of the element that is a nonmetal contains three more total electrons than the atom of the element that is a metal.

7.189 Determine the Lewis structure for the covalent compound formed between the elements A and D given the following:
  1. Both elements A and D are located in the $2p$ area of the periodic table.

2. Element A has an electronegativity greater than 3.2.
3. Element D's periodic table position is next to that of a noble gas.

7.190 Determine the Lewis structure for the covalent compound formed between the elements A and D given the following:
1. Both elements A and D are located in the $3p$ area of the periodic table.
2. Element A has an electronegativity greater than 2.4, and element D has an electronegativity greater than 2.0.
3. Element A's periodic table position is two positions beyond that of element D.

7.191 A covalent compound containing the elements A and D has the formula A_xD_y. Determine the identities of the elements A and D and the values for x and y in the formula for the compound given the following information about A and D.
1. A and D are representative elements.
2. D atoms have electron configurations ending in p^4.
3. The isotope of D with a mass number of 16 has a percentage abundance greater than 95%.
4. The period number for element A is one less than that for element D.

7.192 A covalent compound containing the elements A and D has the formula A_xD_y. Determine the identities of the elements A and D and the values for x

and y in the formula for the compound, given the following information about A and D.
1. A and D are representative elements in the same period of the periodic table.
2. A atoms have electron configurations ending in p^2.
3. The isotope of A with a mass number of 12 has a percentage abundance greater than 95%.
4. The group number for element A is three less than the group number for element D.

7.193 A polyatomic ion containing the elements A and D has the formula AD_4^{2-}. Determine the identities of the elements A and D, and then draw the Lewis structure for the polyatomic ion given the following:
1. A, the central atom in the polyatomic ion, possesses only s electrons, and exactly one-half of these s electrons are valence electrons.
2. The D atoms present collectively possess 28 valence electrons.
3. The atomic number of D is less than 15.

7.194 A polyatomic ion containing the elements A and D has the formula AD_4^+. Determine the identities of the elements A and D, and then draw the Lewis structure for the polyatomic ion given the following:
1. A, the central atom, possesses nine p electrons.
2. For D atoms, the period number in the periodic table and number of valence electrons are the same.
3. For D atoms, the total number of electrons present is the same as the number of valence electrons present.

Multiple-Choice Practice Test

Use this bank of 20 multiple-choice questions as a review of key concepts presented in this chapter. For many of the questions, there may be more than one correct answer (choice d) or no correct answer (choice e).

7.195 Which of the following statements concerning valence electrons is correct?
a. Both $_8O$ and $_{16}S$ have the same number of valence electrons.
b. $_{12}Mg$ has fewer valence electrons than $_{11}Na$.
c. $_{14}Si$ has 14 valence electrons.
d. more than one correct response
e. no correct response

7.196 Which of the following elements would have a Lewis symbol that contains five dots?
a. $_5B$ b. $_7N$ c. $_{15}P$
d. more than one correct response
e. no correct response

7.197 Which of the following is an *incorrect* statement about the number of electrons lost or gained by a representative element during ion formation?
a. The number does not usually exceed three electrons.
b. The number is related to the position of the element in the periodic table.

c. The number is always the same as the number of valence electrons present.
d. more than one correct response
e. no correct response

7.198 Which of the following elements would be expected to form a monoatomic ion with a charge of -3?
a. $_{13}Al$ b. $_{15}P$ c. $_{20}Ca$
d. more than one correct response
e. no correct response

7.199 The atomic number of F is 9. How many protons and electrons are present in a F^- ion?
a. 9 protons and 9 electrons
b. 10 protons and 9 electrons
c. 9 protons and 10 electrons
d. 10 protons and 10 electrons
e. no correct response

7.200 In which of the following ion pairings is the chemical formula given *not* consistent with the ionic charges?
a. M^{2+} and X^{3-}, chemical formula M_3X_2
b. M^{2+} and X^-, chemical formula MX_2
c. M^+ and X^{3-}, chemical formula MX_3
d. more than one correct response
e. no correct response

7.201 In which of the following pairs of ionic compounds do both members of the pair contain positive ions with a +1 charge?
  a. CaS and MgO
  b. Na_3N and Li_2S
  c. $AlCl_3$ and $BaCl_2$
  d. more than one correct repsonse
  e. no correct response

7.202 The mechanism for ionic bond formation always involves the transferring of
  a. electrons from nonmetallic atoms to metallic atoms.
  b. protons from the nucleus of metallic atoms to the nucleus of nonmetallic atoms.
  c. sufficient electrons to produce ions of equal but opposite charge.
  d. more than one correct response
  e. no correct response

7.203 In which of the following pairs of ions are the two members of the pair isoelectronic?
  a. F^- and Cl^-
  b. Na^+ and Ca^{2+}
  c. Al^{3+} and O^{2-}
  d. more than one correct response
  e. no correct response

7.204 In which of the following ionic compounds are polyatomic ions present?
  a. NaOH
  b. Ba_3P_2
  c. $Ca(NO_3)_2$
  d. more than one correct repsonse
  e. no correct response

7.205 In which of the following pairs of compounds are both memebers of the pair *molecular* compounds?
  a. PCl_3 and LiBr
  b. CCl_4 and KOH
  c. CO_2 and NH_3
  d. more than one correct repsonse
  e. no correct response

7.206 Eighteen electrons are present in the Lewis structure of which of the following molecules?
  a. CO_2
  b. N_2O
  c. SO_2
  d. more than one correct repsonse
  e. no correct response

7.207 Which of the following statements concerning Lewis structures is correct?
  a. The Lewis structure for N_2 contains six nonbonding electrons.
  b. The Lewis structure for CH_4 contains eight bonding electrons.
  c. The Lewis structure for HCN contains a triple bond.

  d. more than one correct response
  e. no correct response

7.208 Which of the following is a possible bonding behavior for an element that has four valence electrons?
  a. formation of four single bonds
  b. formation of two double bonds
  c. formation of one single bond and one triple bond
  d. more than one correct response
  e. no correct response

7.209 The molecular geometry associated with three groups of bonding electrons and one pair of nonbonding electrons about a central atom in a molecule is
  a. tetrahedral.
  b. trigonal pyramidal.
  c. trigonal planar.
  d. angular.
  e. no correct response

7.210 In which of the following pairs of molecules do both members of the pair have the same molecular geometry?
  a. SO_2 and CO_2
  b. H_2S and HCN
  c. H_2O and OF_2
  d. more than one correct repsonse
  e. no correct response

7.211 Which of the following sets of elements is arranged in order of increasing electronegativity?
  a. Cl, S, P, Si
  b. B, O, H, F
  c. F, Cl, Br, I
  d. more than one correct response
  e. no correct response

7.212 When the electronegativity difference between two atoms in a bond is 0.9 the bond is classified as
  a. a nonpolar covalent bond.
  b. a polar covalent bond.
  c. a coordinate covalent bond.
  d. an ionic bond.
  e. no correct response

7.213 Which of the following molecules would be classified as a polar molecule?
  a. linear CO_2 molecule (O—C—O)
  b. linear HCN molecule (H—C—N)
  c. angular H_2O molecule (H—O—H)
  d. more than one correct response
  e. no correct response

7.214 Both ionic and covalent bonding are present in which of the following compounds?
  a. Mg_3N_2
  b. Na_2SO_4
  c. N_2H_4
  d. more than one correct repsonse
  e. no correct response

Chemical Nomenclature

8.1 CLASSIFICATION OF COMPOUNDS FOR NOMENCLATURE PURPOSES

Just as it is important to be able to write formulas for compounds, it is also important to be able to name compounds. When a chemist gives a description of a chemical compound, almost invariably he or she will use the compound's name and/or formula early in the description process.

    Chemical nomenclature *is the system of names used to distinguish compounds from each other and the rules needed to devise these names.* In the early history of chemistry, there was no system for naming compounds. Early names included quicksilver, blue vitriol, Glauber's salt, gypsum, sal ammoniac, and laughing gas. As chemistry grew, it became clear that the anything-goes system was not acceptable. Without a system for naming compounds, coping with the multitude of known substances would be a hopeless task.

    A systematic set of rules for naming compounds is available for use by scientists. These rules, known as *IUPAC rules,* have been formulated by the nomenclature committees of the International Union of Pure and Applied Chemistry (IUPAC). The committees of this international scientific organization meet periodically to revise and update the nomenclature rules to accommodate any newly discovered types of compounds.

    IUPAC rules are based on the premise that a compound's name should convey information about the composition of the compound. The majority of the nomenclature considered in this chapter will be based on IUPAC rules.

    Unsystematic names (common names) also exist for some compounds. A few such names will also be considered in this chapter since these common names are used more frequently than IUPAC names. The compound H_2O always goes by the name water (its common name) and not by hydrogen oxide (its IUPAC name).

Distinguishing between ionic compounds and molecular compounds (Secs. 4.11 and 7.1) is an important key to becoming successful at chemical nomenclature. The rules for naming ionic compounds differ from those for naming molecular compounds. Hence, the initial step in chemical nomenclature is classifying a compound as ionic or molecular.

Although there is no sharp dividing line between ionic and covalent bonds (Sec. 7.19) and thus between ionic and molecular compounds, we will create such a line for nomenclature purposes with the following simple generalizations:

1. Compounds resulting from the combination of a metal and one or more non-metals are considered *ionic.*
2. Compounds resulting from combinations of a nonmetal with other nonmetals are considered *molecular.*

An important exception to the generalization that compounds containing only nonmetals are molecular compounds is compounds in which a positively charged polyatomic ion is present. In such compounds, the positively charged polyatomic ion (a combination of nonmetals) is considered to have replaced the positively charged metallic ion ordinarily present in an ionic compound and to be functioning as a metal.

The most commonly encountered positive polyatomic ion is the ammonium ion (NH_4^+). Examples of compounds containing this ion are NH_4Cl, $(NH_4)_2S$, NH_4CN, and $(NH_4)_2SO_4$. Note that, in all cases, no metal atoms are present even though all of these compounds are ionic.

Metalloid elements (Sec. 6.12) are considered to be nonmetals for purposes of nomenclature. Thus, a compound resulting from the combination of a metalloid and a nonmetal is named as a molecular compound. A compound resulting from the combination of a metal and metalloid is named as an ionic compound.

EXAMPLE 8.1   **Classifying Compounds for Nomenclature Purposes**

Classify each of the following compounds as *ionic* or *molecular* for nomenclature purposes.

  a. Al_2S_3 **b.** NO_2 **c.** KF   **d.** NH_4Br

SOLUTION

  a. ionic: a metal (Al) and a nonmetal (S) are present
  b. molecular: two nonmetals are present
  c. ionic: a metal (K) and a nonmetal (F) are present
  d. ionic: the polyatomic NH_4^+ ion and the monoatomic Br^- ion are present

▶ **Practice Exercise 8.1**   Classify each of the following compounds as *ionic* or *molecular* for nomenclature purposes.

  a. CO_2 **b.** NaCl   **c.** K_2SO_4 **d.** N_2O_3

Answers to all practice exercises in this chapter are found in the back-of-the-book answer section.

8.2 TYPES OF BINARY IONIC COMPOUNDS

The term *binary* means "two." A **binary compound** *is a compound in which only two elements are present.* The compounds NaCl, CO_2, NH_3, and P_4O_{10} are all binary compounds. Any number of atoms of the two elements may be present in a molecule or formula unit of a binary compound, but only two elements may be present. A **binary ionic compound** *is an ionic compound in which one element present is a metal and the other element present is a nonmetal.* The metal is always present as the positive ion, and the nonmetal is always present as the negative ion. The joint presence of a metal and a nonmetal in a binary compound is the "recognition key" that the compound is a binary ionic compound. A consideration of nomenclature rules for binary ionic compounds will be our entry point into the realm of chemical nomenclature.

Binary ionic compounds, which are the simplest type of ionic compound, may be divided into two categories based on the identity of the metal ion that is present:

1. *Fixed-charge* binary ionic compounds contain a metal that can form only *one type* of positive ion.
2. *Variable-charge* binary ionic compounds contain a metal that can form *more than one type* of positive ion.

In general, all metals lose electrons when forming ions. Fixed-charge metals always exhibit the same behavior in ion formation; that is, they always lose the same number of electrons. A **fixed-charge metal** *is a metal that forms only one type of positive ion, which always has the same charge magnitude.*

Variable-charge metals do not always lose the same number of electrons upon ion formation. For example, the variable-charge metal iron sometimes forms an Fe^{2+} ion and at other times an Fe^{3+} ion. A **variable-charge metal** *is a metal that forms more than one type of positive ion, with the ion types differing in charge magnitude.*

Which metals are fixed-charge metals and which metals are variable-charge metals? This information is a prerequisite for naming binary ionic compounds. Fixed-charge binary ionic compounds are named in one way, and variable-charge binary ionic compounds are named in a slightly different way.

A limited number of fixed-charge metals exist. If these are learned, then all other metals will be variable-charge metals. The fixed-charge metals are the group IA and IIA metals plus Al, Ga, Zn, Cd, and Ag. Figure 8.1 shows the periodic table positions for these 15 fixed-charge metals. The charge magnitude for these metals can be related to their position in the periodic table. Looking at Figure 8.1, we see that all group IA elements form +1 ions. The charges for ions of elements in groups IIA and IIIA are +2 and +3, respectively. Group numbers and charge also directly correlate for Zn, Cd, and Ag, the other fixed-charge metallic ions. The reason for this charge–periodic table correlation, the octet rule, was considered in Section 7.3.

The vast majority of metals, all except the 15 fixed-charge metals, are variable-charge metals. Charge magnitude for ions of such metals cannot be easily related to periodic table position. (The presence of *d* or *f* electrons in most of these metals complicates octet rule considerations in a manner beyond what we consider in this book.) Table 8.1 gives the charges for selected commonly encountered variable-charge metal ions. Note that the charges of +2 and +3 are a common combination. However, there are also other combinations: Two +2 and +4 pairings (Pb and Sn), a +1 and +2 pairing (Cu), and a +1 and +3 pairing (Au) are listed in Table 8.1.

All the inner transition metals (*f* area of the periodic table), most of the transition metals (*d* area), and a few representative metals (*p* area) exhibit variable-charge behavior (see Fig. 8.1).

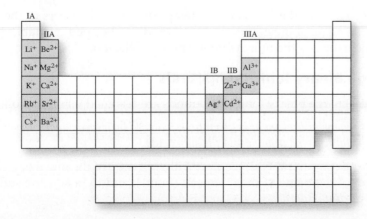

FIGURE 8.1 Periodic table showing the fixed-charge metallic ions. For these metals, ionic charge correlates with periodic table group number.

TABLE 8.1 Ionic Charges Associated with Ions of the More Common Variable-Charge Metals

Element	Ions Formed
Chromium	Cr^{2+} and Cr^{3+}
Cobalt	Co^{2+} and Co^{3+}
Copper	Cu^+ and Cu^{2+}
Gold	Au^+ and Au^{3+}
Iron	Fe^{2+} and Fe^{3+}
Lead	Pb^{2+} and Pb^{4+}
Manganese	Mn^{2+} and Mn^{3+}
Tin	Sn^{2+} and Sn^{4+}

8.3 NOMENCLATURE FOR BINARY IONIC COMPOUNDS

The names of binary ionic compounds are based on the names of the ions of which they are composed. Ion names are obtained as follows.

> A fixed-charge metal cation has the same name as its parent element.

1. Fixed-charge metal ions take the full name of the element plus the word "ion."

$$Na^+ \quad \text{sodium ion}$$
$$Ca^{2+} \quad \text{calcium ion}$$
$$Al^{3+} \quad \text{aluminum ion}$$

2. Variable-charge metal ions take the full name of the element followed by a Roman numeral (in parentheses) that gives the ionic charge plus the word "ion."

$$Fe^{2+} \quad \text{iron(II) ion}$$
$$Fe^{3+} \quad \text{iron(III) ion}$$
$$Cu^{2+} \quad \text{copper(II) ion}$$

3. Nonmetal ions take the *stem* of the name of the element followed by the suffix *-ide* plus the word "ion."

$$Cl^- \quad \text{chloride ion}$$
$$S^{2-} \quad \text{sulfide ion}$$
$$O^{2-} \quad \text{oxide ion}$$

The stem of the name of the nonmetal is always the first few letters of the nonmetal's name, that is, the name of the nonmetal with its ending chopped off. Table 8.2 gives the stem part of the name for the more common nonmetallic ions.

Fixed-Charge Binary Ionic Compounds

Names for fixed-charge binary ionic compounds are assigned using the following rule:

> Fixed-charge binary ionic compounds contain a metal that can form only one type of cation.

The full name of the metallic element is given first, followed by a separate word consisting of the stem of the nonmetallic element name and the suffix -ide.

Thus, to name the compound NaF, we start with the name of the metal (sodium), follow it with the stem of the name of the nonmetal (fluor-), and then add the suffix *-ide*. The name becomes *sodium fluoride*.

TABLE 8.2 Names for the More Common Nonmetal Ions

Element	Stem	Name of Ion	Formula
Bromine	brom-	bromide ion	Br^-
Carbon	carb-	carbide ion	C^{4-}
Chlorine	chlor-	chloride ion	Cl^-
Fluorine	fluor-	fluoride ion	F^-
Hydrogen	hydr-	hydride ion	H^-
Iodine	iod-	iodide ion	I^-
Nitrogen	nitr-	nitride ion	N^{3-}
Oxygen	ox-	oxide ion	O^{2-}
Phosphorus	phosph-	phosphide ion	P^{3-}
Sulfur	sulf-	sulfide ion	S^{2-}

The name of the metal ion present is always exactly the same as the name of the metal itself. The metal's name is never shortened as are nonmetal names. Example 8.2 illustrates further the use of the preceding rule in naming fixed-charge binary ionic compounds.

EXAMPLE 8.2 **Naming Fixed-Charge Binary Ionic Compounds**

Name the following fixed-charge binary ionic compounds.

  a. KCl **b.** MgF_2 **c.** Na_2O **d.** Be_3N_2

SOLUTION

The general pattern for naming compounds of this type is

> First word: name of metal
> Second word: stem of name of nonmetal + -*ide*

  a. The metal is potassium, and the nonmetal is chlorine. Thus, the compound's name is potassium <u>chlor</u>ide (the stem of the nonmetal name is underlined).
  b. The metal is magnesium, and the nonmetal is fluorine. Hence, the name is magnesium <u>fluor</u>ide. Note that no mention is made of the subscript 2 found after the symbol for fluorine in the formula. *The name of an ionic compound never contains any reference to formula subscript numbers.* Since magnesium is a fixed-charge metal, there is only one ratio in which magnesium and fluorine may combine. Thus, just telling the name of the elements present in the compound is adequate nomenclature.
  c. Sodium (Na) and oxygen (O) are present in the compound. Its name is sodium <u>ox</u>ide.
  d. This compound is named beryllium <u>nitr</u>ide.

Answer Double Check:

Is there consistency in the manner in which the compound names have been written? Yes. All of the names contain two separate words; the first word is the name of the metal, and the second word contains the stem of the name of the nonmetal followed by the suffix -*ide*.

▶ **Practice Exercise 8.2** Name the following fixed-charge binary ionic compounds.

  a. $MgCl_2$ **b.** K_3N **c.** Al_2O_3 **d.** BaS

A major consideration in formulating rules for naming compounds is that the name given a compound contain sufficient information that the chemical formula of the compound can be derived from the name. Example 8.3 considers this reverse process of going from compound name to chemical formula of the compound.

EXAMPLE 8.3 **Writing Chemical Formulas for Fixed-Charge Binary Ionic Compounds, Given Their Names**

Write the chemical formulas for the following fixed-charge binary ionic compounds, given their names.

a. magnesium oxide
b. sodium sulfide
c. aluminum chloride
d. lithium nitride

SOLUTION

a. The name magnesium oxide conveys the message that the metal magnesium is present as a positive ion, and the nonmetal oxygen in present as a negative ion. Magnesium is in group IIA of the periodic table and thus forms a +2 ion (Section 7.5). Oxygen is in group VIA of the periodic table and thus forms a −2 ion (Section 7.5). Combining the two ions, Mg^{2+} and O^{2-}, in a one-to-one ratio, the ratio that causes overall charge to add to zero (Section 7.7), gives the chemical formula MgO.

b. Sodium, a group IA element, forms an Na^+ ion. Sulfur, a group VIA element, forms an S^{2-} ion. The combining ratio for the ions is two-to-one (Section 7.7) giving the chemical formula Na_2S.

c. The metallic ion is Al^{3+} since aluminum is a group IIIA element. The nonmetallic ion is Cl^- since chlorine is a group VIIA element. Three Cl^- ions are needed to counterbalance the charge on one Al^{3+} ion. The chemical formula is, thus, $AlCl_3$.

d. The chemical formula resulting from the combination of Li^+ ions (group IA) and N^{3-} ions (group VA) is Li_3N.

Answer Double Check:

Is there consistency in the manner in which all of the chemical formulas have been written? Yes. In each case the positive ion has been written first in the formula as it should be.

▶ **Practice Exercise 8.3** Write the chemical formulas for the following fixed-charge binary ionic compounds, given their names.

a. calcium sulfide
b. potassium fluoride
c. magnesium phosphide
d. potassium oxide

Variable-Charge Binary Ionic Compounds

Variable-charge binary ionic compounds contain a metal that can form more than one type of cation.

Names for variable-charge binary ionic compounds are assigned using the following rule:
The full name of the metallic element with a Roman numeral appended to it is given first, followed by a separate word consisting of the stem of the nonmetallic element name and the suffix -ide.

Two different iron chlorides are known: One contains Fe^{2+} ions ($FeCl_2$), and the other contains Fe^{3+} ions ($FeCl_3$). The names of these two chlorides are, respectively, iron(II) chloride and iron(III) chloride. Without the use of Roman numerals (or some equivalent system), these two compounds would have the same name (iron chloride), an unacceptable situation.

In naming variable-charge binary ionic compounds, the Roman numeral to be used in the name is the metal ion charge, which can be calculated using the charge on the nonmetal ion and the principle of charge neutrality. The charge neutrality principle is that the total positive charge and the total negative charge must add to zero (Sec. 7.7). The calculation of metal ion charge (Roman numeral) using this technique, as well as the use of Roman numerals in the names of variable-charge binary ionic compounds, is illustrated in Example 8.4. As you work through Example 8.4, be sure to note that Roman numerals are *never* a part of the *formula* of a variable-charge binary ionic compound but *always* a part of the *name* of a variable-charge binary ionic compound.

> The cation name precedes the anion name in the names of *all* binary ionic compounds (both fixed-charge and variable-charge).

EXAMPLE 8.4 **Naming Variable-Charge Binary Ionic Compounds**

Name the following variable-charge binary ionic compounds.

  a. CuO   **b.** Cu_2O **c.** Mn_2S_3 **d.** $AuCl_3$

SOLUTION

The general pattern for naming compounds of this type is

First word: name of metal + Roman numeral

Second word: stem of name of nonmetal + -*ide*

a. The metal ion charge in this compound, a quantity needed to determine the Roman numeral to be used in the name, is easily calculated using the following procedure.

$$\text{copper charge} + \text{oxygen charge} = 0$$

The oxide ion has a –2 charge (Sec. 7.5). Therefore,

$$\text{copper charge} + (-2) = 0$$

Solving, we get

$$\text{copper charge} = +2$$

Therefore, the copper ions present are Cu^{2+}, and the name of the compound is copper(II) oxide.

b. For charge balance in this compound we have the equation

$$2(\text{copper charge}) + \text{oxygen charge} = 0$$

Note that we have to take into account the number of copper ions present, two in this case. The oxide ion carries a –2 charge (Sec. 7.5). Therefore,

$$2(\text{copper charge}) + (-2) = 0$$
$$2(\text{copper charge}) = +2$$
$$\text{copper charge} = +1$$

> The names copper(II) oxide and copper(I) oxide are pronounced "copper-two oxide" and "copper-one oxide."

Here we note that we are interested in the charge on a *single* copper ion (+1) and not in the total positive charge present (+2). Since Cu^+ ions are present, the compound is named copper(I) oxide. As is the case for all ionic compounds, the name does not contain any reference to the numerical subscripts in the compound's formula.

c. The charge balance equation is

$$2(\text{manganese charge}) + 3(\text{sulfur charge}) = 0$$

Substituting a charge of –2 into the equation for sulfur (Sec. 7.5), we get

$$2(\text{manganese}) + 3(-2) = 0$$
$$2(\text{manganese}) = +6$$
$$\text{manganese} = +3$$

The compound is thus named manganese(III) sulfide.

d. This compound is gold(III) chloride. Chloride ions carry a –1 charge (Sec. 7.5). Since the compound contains three chloride ions (–3 total charge), the single gold ion must bear a +3 charge to counterbalance the –3 charge of the chlorides. Hence, Au^{3+} ions are present.

Answer Double Check:

Is there consistency in the manner in which the compound names have been written? Yes. In each case the first word in the name, the name of the metal, has a Roman numeral appended to the metal name. The Roman numeral is needed, as all metals are variable-charge metals, and the Roman numeral specifies the charge on the metal ion.

▶ **Practice Exercise 8.4**   Name the following variable-charge binary ionic compounds.

a. PbS   b. PbS_2 c. $CoCl_2$ d. CuF

As was done when discussing nomenclature for fixed-charge binary ionic compounds, the reversal of the naming process, going from name to chemical formula is now considered. Here the reversed process is easier than in the previous consideration. The charge on the metal ion is obtained directly from the Roman numeral present in the name rather than from periodic table considerations. (The periodic table cannot be used to obtain the charge on the metal ion; that is why the Roman numeral is needed.) Example 8.5 illustrates this new situation.

EXAMPLE 8.5   **Writing Chemical Formulas for Variable-Charge Binary Ionic Compounds, Given Their Names**

Write the chemical formulas for the following variable-charge binary ionic compounds, given their names.

a. nickel(II) chloride
b. nickel(III) chloride
c. chromium(III) oxide
d. chromium(II) oxide

SOLUTION

a. The nickel ion present is Ni^{2+}; the Roman numeral (II) gives the message that the nickel ion has a +2 charge. Periodic table location, group VIIA, is used to derive the charge on the chloride ion (Cl^-). Combining the ions in the ratio that causes the charges to add to zero gives the chemical formula $NiCl_2$.
b. The only difference between this situation and part (a) is that nickel is present as the Ni^{3+} ion. The chemical formula is, thus, $NiCl_3$.

c. The ions present are Cr^{3+} (Roman numeral III) and O^{2-} (group VIA element). The chemical formula for the compound is Cr_2O_3 (six units of positive charge and six units of negative charge).

d. The ions present are Cr^{2+} (Roman numeral II) and O^{2-} (group VIA element). Combining the ions in a one-to-one ratio gives the chemical formula CrO.

▶ **Practice Exercise 8.5**   Write the chemical formulas for the following variable-charge binary ionic compounds, given their names.

a. iron(II) sulfide
b. iron(III) sulfide
c. lead(IV) chloride
d. lead(II) chloride

An older method exists, which still sees some use for indicating the charge on variable-charge metal ions. It uses suffixes rather than Roman numerals. This suffix system is often encountered on the labels of bottles of chemicals (see Fig. 8.2). When a metal has two common ionic charges, the suffix *-ous* is used for the ion of lower charge and the suffix *-ic* for the ion of higher charge. Table 8.3 compares the Roman numeral system and the suffix system for the metals, where the suffix system is most often encountered. Note, from this table, that in the suffix system the Latin names for metals are used (Sec. 4.9).

The *-ic, -ous* system for denoting metal ion charge has several limitations. First, the actual charge on the metal ions is not specified by the *-ic* and *-ous* suffixes; all that is specified is which is the higher charge and which is the lower charge. The *-ic, -ous* system is not useful when a metal can form ions with three or more different positive charges; such is the case for the metal manganese (Mn^{2+}, Mn^{3+}, and Mn^{4+}). In such a case the suffix system is ambiguous and is not used.

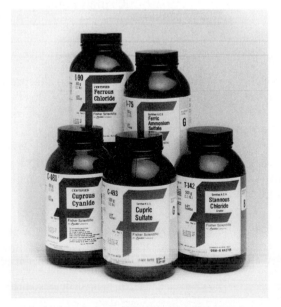

FIGURE 8.2   An older method of naming ionic compounds containing a variable-charge metal is still used on the labels of many laboratory chemicals. *(Fisher Scientific/ H. Stephen Stoker)*

The ruddy rust that forms on iron is ferric oxide (Fe_2O_3). This same compound gives red bricks and red clay sewer pipes their color. It also is the pigment of burnt sienna in artists' paints. Rouge, one of the most ancient of cosmetics, gets its color from ferric oxide as well. Barns were traditionally painted red because farmers used red ferric oxide paint as a way to preserve the wood. People still paint their barns red without knowing why.

TABLE 8.3 Comparison of Roman Numeral and Suffix System Names for Selected Metal Ions

Element	Ions	Preferred Name	Old System Name
Copper	Cu^+	copper(I) ion	cuprous ion
	Cu^{2+}	copper(II) ion	cupric ion
Iron	Fe^{2+}	iron(II) ion	ferrous ion
	Fe^{3+}	iron(III) ion	ferric ion
Tin	Sn^{2+}	tin(II) ion	stannous ion
	Sn^{4+}	tin(IV) ion	stannic ion
Lead	Pb^{2+}	lead(II) ion	plumbous ion
	Pb^{4+}	lead(IV) ion	plumbic ion
Gold	Au^+	gold(I) ion	aurous ion
	Au^{3+}	gold(III) ion	auric ion

EXAMPLE 8.6 Naming Variable-Charge Binary Ionic Compounds Using the *-ic,-ous* System

Two systems exist for indicating the charge on the metal ion present in variable-charge binary ionic compounds: (1) a system that uses Roman numerals and (2) a system that uses the suffixes *-ic* and *-ous*. Change the following compound names from one system to the other system.

a. iron(III) chloride
b. copper(I) fluoride
c. stannic bromide
d. plumbous fluoride

A mnemonic device for remembering the relationship between *-ic* and *-ous* and higher and lower ion charge is that there is an **o** in *-ous* and *lower*, and an **i** in *-ic* and *higher*.

SOLUTION

The information found in Table 8.3 is needed to bring about the desired name changes.

a. The given name indicates that Fe^{3+} ion is present. The alternative name for Fe^{3+} ion, from Table 8.3, is ferric ion. The name iron(III) chloride thus becomes ferric chloride.
b. The alternative name for the copper(I) ion, Cu^+, is cuprous ion. Therefore, copper(I) fluoride becomes cuprous fluoride.
c. The stannic ion, from Table 8.3, is Sn^{4+} ion; the alternative name for this ion is tin(IV). Therefore, the changed (and preferred) name for this compound is tin(IV) bromide.
d. Since the plumbous ion and Pb^{2+} ion are equivalent (Table 8.3), the names plumbous fluoride and lead(II) fluoride are equivalent.

▶ **Practice Exercise 8.6** Change each of the following variable-charge binary ionic compound names from the Roman numeral to the *-ic, -ous* system, or vice versa.

a. iron(III) oxide
b. gold(I) sulfide
c. cupric bromide
d. plumbic chloride

The overall strategy for naming binary ionic compounds, as discussed in this section, is summarized in the "nomenclature decision tree" of Figure 8.3.

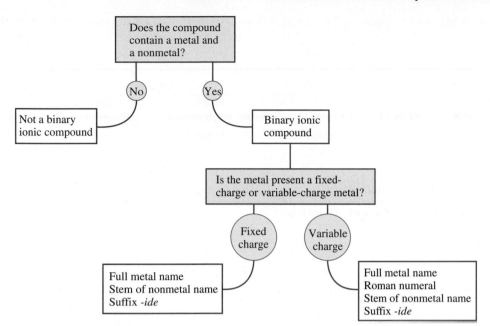

FIGURE 8.3 A nomenclature decision tree for naming binary ionic compounds.

8.4 NOMENCLATURE FOR IONIC COMPOUNDS CONTAINING POLYATOMIC IONS

In Section 7.9 the topic of polyatomic ions was considered, and at that time a limited number of such ions were introduced in the context of formula writing. We now consider some additional facts about polyatomic ions.

Hundreds of different polyatomic ions exist. Table 8.4 gives the names and formulas for the most common ones. Note that almost all the polyatomic ions listed in Table 8.4 contain oxygen atoms. The names, but not necessarily the formulas, of some of these common polyatomic ions should be familiar to you. Many of these ions are found in commercial products. Examples are fertilizers (phosphates, sulfates, nitrates), baking soda and baking powder (bicarbonates), and building materials (carbonates, sulfates).

There is no easy way to learn the formulas and names for all the common polyatomic ions. Memorization is required. The charges and formulas for the various polyatomic ions cannot be related easily to the periodic table as was the case for many of the monoatomic ions. In Table 8.4 the most frequently encountered polyatomic ions are in color. Their formulas and names should definitely be memorized. Your instructor may want you to memorize others, too. The inability to recognize the presence of polyatomic ions (both by name and by formula) in a compound is a major stumbling block for many chemistry students. It requires some effort to overcome this obstacle.

The following interrelationships among polyatomic ion names and formulas should be helpful as you become more familiar with the information in Table 8.4:

> The green corrosion on bronze statues is $Cu_2(OH)_2CO_3$, a substance that contains two different types of polyatomic ions. It forms from copper reacting with water, oxygen, and carbon dioxide in the air.

1. Most of the ions have a negative charge, which can vary from –1 to –3. Only two positive ions are listed in the table: NH_4^+ (ammonium) and H_3O^+ (hydronium).
2. Four of the polyatomic ions have names ending in *-ide*: OH^- (hydroxide), CN^- (cyanide), N_3^- (azide), and O_2^{2-} (peroxide). These names represent exceptions to the rule that the suffix *-ide* be reserved for use in naming monoatomic ions.
3. A number of *-ate, -ite* pairs of ions exist, for example, SO_4^{2-} (sulfate) and SO_3^{2-} (sulfite). The ion in the pair with the higher number of oxygens is always the *-ate* ion. The *-ite* ion always contains one less oxygen than the *-ate* ion.

The prefix *bi-* in polyatomic ion names means *hydrogen* rather than the number *two*.

4. A number of pairs of ions exist where one member of the pair differs from the other by having a hydrogen atom present, for example, CO_3^{2-} (carbonate) and HCO_3^- (hydrogen carbonate or bicarbonate). In such pairs, the charge on the hydrogen-containing ion is always one less than the charge on the other ion.

5. Two pairs of ions exist in which the difference between pair members is that a sulfur atom has replaced an oxygen atom in one member of the pair: $(SO_4^{2-}, S_2O_3^{2-})$ and (OCN^-, SCN^-). The prefix *thio-* is used to denote this replacement of oxygen by sulfur. The names of these pairs of ions, respectively, are sulfate–thiosulfate and cyanate–thiocyanate.

TABLE 8.4 Formulas and Names of Some Common Polyatomic Ions

Key Element Present	Formula	Name of Ion
Nitrogen	NO_3^-	nitrate ion
	NO_2^-	nitrite ion
	NH_4^+	ammonium ion
	N_3^-	azide ion
Sulfur	SO_4^{2-}	sulfate ion
	HSO_4^-	hydrogen sulfate (bisulfate ion)**
	SO_3^{2-}	sulfite ion
	HSO_3^-	hydrogen sulfite (bisulfite ion)**
	$S_2O_3^{2-}$	thiosulfate ion
Phosphorus	PO_4^{3-}	phosphate ion
	HPO_4^{2-}	hydrogen phosphate ion
	$H_2PO_4^-$	dihydrogen phosphate ion
	PO_3^{3-}	phosphite ion
Carbon	CO_3^{2-}	carbonate ion
	HCO_3^-	hydrogen carbonate (bicarbonate ion)**
	$C_2O_4^{2-}$	oxalate ion
	$C_2H_3O_2^-$	acetate ion
	CN^-	cyanide ion
	OCN^-	cyanate ion
	SCN^-	thiocyanate ion
Chlorine	ClO_4^-	perchlorate ion
	ClO_3^-	chlorate ion
	ClO_2^-	chlorite ion
	ClO^-	hypochlorite ion
Oxygen	O_2^{2-}	peroxide ion
Boron	BO_3^{3-}	borate ion
Hydrogen	H_3O^+	hydronium ion*
	OH^-	hydroxide ion
Metals	MnO_4^-	permanganate ion
	CrO_4^{2-}	chromate ion
	$Cr_2O_7^{2-}$	dichromate ion

*This ion is encountered only in aqueous solutions.
**These alternate names are often encountered.

Polyatomic-ion-containing compounds found in the home:

$NaHCO_3$ Baking soda
$MgSO_4$ Epsom salt
Na_3PO_4 Cleaning agent
$Al(OH)_3$ Antacid
$NaOH$ Drain cleaner
$MgCO_3$ Plaster of Paris
$NaClO$ Clorox
$CaCO_3$ Chalk

The names of ionic compounds containing polyatomic ions are derived in the same way as those of binary ionic compounds (Sec. 8.3). Recall that the rule for naming binary ionic compounds is to give the name of the metallic element first (including, when needed, a Roman numeral indicating ion charge), and then as a separate word give the stem of the nonmetallic element name to which the suffix -*ide* is appended.

For our present situation, *if the polyatomic ion is positive, its name is substituted for that of the metal. If the polyatomic ion is negative, its name is substituted for the nonmetal stem plus* -ide. In the case where both positive and negative ions are polyatomic, dual substitution occurs and the resulting name includes just the names of the polyatomic ions. Example 8.7 illustrates the use of these rules.

> Note that the word "ion" is never used in the name of an ionic compound. Ions are charged particles. When cations and anions combine to form an ionic compound, which is neutral, the word ion is no longer applicable.

EXAMPLE 8.7 Naming Polyatomic-Ion-Containing Compounds

Name the following polyatomic-ion-containing compounds.

a. Na_3PO_4 **b.** $Fe(NO_3)_3$ **c.** Cu_2SO_4 **d.** NH_4CN

SOLUTION

a. The positive ion present is the sodium ion (Na^+). The negative ion is the polyatomic phosphate ion (PO_4^{3-}). The name of the compound is sodium phosphate. No Roman numeral is needed in the name since sodium is a fixed-charge metal. As in naming binary ionic compounds (Sec. 8.3), subscripts in the formula are not incorporated into the name.

b. The positive ion present is iron, and the negative ion is the nitrate ion (NO_3^-). Since iron is a variable-charge metal, a Roman numeral must be used to indicate ionic charge. In this case, the Roman numeral is III. The fact that iron ions carry a +3 is deduced by noting that there are three nitrate ions present, each of which carries a –1 charge. The charge on the single iron ion present must be a +3 in order to counterbalance the total negative charge of –3. The name of the compound, therefore, is iron(III) nitrate.

c. The positive ion present is Cu(I) ion. The negative ion is the polyatomic sulfate ion (SO_4^{2-}). The name of the compound is copper(I) sulfate. The determination that copper (a variable-charge metal) is present as copper(I) involves the following calculation dealing with charge balance.

$$2(\text{copper charge}) + (\text{sulfate charge}) = 0$$
$$2(\text{copper charge}) + (-2) = 0$$
$$2(\text{copper charge}) = +2$$
$$\text{copper charge} = +1$$

d. Both the positive and negative ions in this compound are polyatomic—the ammonium ion (NH_4^+) and the cyanide ion (CN^-). The name of the compound is simply the combination of the names of the two polyatomic ions: ammonium cyanide.

Answer Double Check:

The presence of a negatively charged polyatomic ion in an ionic compound does not eliminate the need to classify the metal ion present as fixed-charge or variable-charge. For fixed-charged metals a Roman numeral must be included in the name. Parts (b) and (c) involve compounds with fixed-charge metals. In each case the name contains a Roman numeral, as it should. A Roman numeral is not needed in part (a) (fixed-charge metal) and in part (d) the compound contains a positively charged polyatomic ion in lieu of the metal ion.

▶ **Practice Exercise 8.7** Name the following polyatomic-ion-containing compounds.

a. $CaCO_3$ **b.** $Al_2(SO_4)_3$ **c.** $Mg(NO_3)_2$ **d.** NH_4Cl

---

EXAMPLE 8.8   **Writing Chemical Formulas for Polyatomic-Ion-Containing Compounds, Given Their Names**

Write the chemical formulas for the following polyatomic-ion-containing compounds, given their names.

     a. potassium cyanide   **b.** iron(II) hydroxide
     c. aluminum phosphate   **d.** copper(II) nitrate

SOLUTION

     a. Potassium and cyanide ions are present. Potassium is a fixed-charge metal; the charge is +1 (K^+) since potassium is located in group IA. The formula of the cyanide ion is CN^- (Table 8.4). Combining the like-charged ions in a one-to-one ratio produces the chemical formula KCN.

     b. The iron ion present is Fe^{2+} since the Roman numeral is (II). The hydroxide ion has the chemical formula OH^- (Table 8.4). Two hydroxide ions are needed per iron ion for charge balance. The chemical formula is $Fe(OH)_2$.

     c. The fixed-charge aluminum ion is present, Al^{3+} (group IIIA). The phosphate ion has the chemical formula $PO_4{}^{3-}$ (Table 8.4). The chemical formula of the compound is $AlPO_4$.

     d. The Roman numeral (II) indicates that the copper ion present is Cu^{2+}. The chemical formula of the nitrate ion is $NO_3{}^-$. Combining the ions in a one-to-two ratio produces the chemical formula $Cu(NO_3)_2$.

Answer Double Check:

Is the positive ion written first in each of the chemical formulas? Yes. Are parentheses present in each of the chemical formulas where more than one polyatomic ion is needed (Sec. 7.9)? Yes.

▶ **Practice Exercise 8.8** Write the chemical formulas for the following polyatomic-ion-containing compounds, given their names.

     a. ammonium bromide   **b.** silver phosphate
     c. copper(I) phosphate   **d.** magnesium perchlorate

A **ternary compound** *is a compound containing three different elements.* Many, but not all, polyatomic-ion-containing compounds are ternary compounds. Such is the case when one of the ions present is monoatomic and the other ion present is a polyatomic ion containing two elements. The compounds sodium sulfate [Na_2SO_4], potassium nitrate [KNO_3], and ammonium chloride [NH_4Cl] represent this situation. A **ternary ionic compound** *is an ionic compound in which three elements are present, one element in a monoatomic ion and two other elements in a polyatomic ion.*

Exceptions to the generalization that polyatomic-ion-containing compounds are *ternary ionic compounds* exist. Several common polyatomic-ion-containing compounds exist that have four elements present and there are a few polyatomic-ion containing compounds in which only two elements are present. The following are such exceptions:

     1. Compounds that contain four elements become possible when both ions present are polyatomic ions. Common examples of this situation are the compounds

ammonium chlorate [NH_4ClO_3] and ammonium sulfate [$(NH_4)_2SO_4$]. Note, however, that the compounds ammonium nitrate [NH_4NO_3] and ammonium cyanide [NH_4CN], where both polyatomic ions contain a common element (nitrogen in this case), contain only three elements, and are thus ternary ionic compounds.

2. Compounds that contain four elements are also possible when the polyatomic ion itself contains three elements as for the acetate ion ($C_2H_3O_2{}^-$) or bicarbonate ion ($HCO_3{}^-$). Examples of such compounds are $NaC_2H_3O_2$ and $KHCO_3$.

3. Interestingly, ionic compounds that contain only two elements, that is, are binary (Sec. 8.2), are possible with a polyatomic ion present. This occurs when the homoatomic polyatomic ions peroxide [$O_2{}^{2-}$] or azide [$N_3{}^-$] are present and the other ion present is monoatomic. Examples of such binary compounds include sodium peroxide [Na_2O_2] and potassium azide [KN_3]. Peroxide and azide are the only two polyatomic ions listed in Table 8.4 that are homoatomic; all others are heteroatomic.

Ionic peroxide compounds are not common. Only group IA and group IIA metals form such compounds. A number of metals besides those in groups IA and IIA form ionic azide compounds.

Nomenclature-wise, the set of rules presented in this section of the text for naming polyatomic-ion-containing compounds applies to all such compounds. The number of elements present—two, three, or four—does not affect the rules.

The overall strategy for naming ionic compounds—both binary (Sec. 8.3) and polyatomic-ion-containing (Sec. 8.4)—is summarized in the nomenclature decision tree of Figure 8.4.

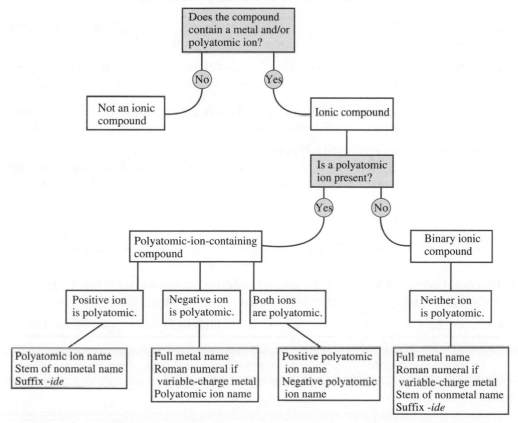

FIGURE 8.4 A nomenclature decision tree for naming binary and polyatomic-ion-containing ionic compounds.

8.5 NOMENCLATURE FOR BINARY MOLECULAR COMPOUNDS

A **binary molecular compound** *is a molecular compound in which only two nonmetallic elements are present.* The order in which the two nonmetals in a binary molecular compound are listed in its chemical formula is established by convention. This order convention, which approximately parallels increasing nonmetallic character of the elements (Sec. 6.12) is

Element	B	Si C	Sb As P	N H	Te Se S	I Br Cl	O F
Group	IIIA	IVA	VA		VIA	VIIA	

This order sequence is based, first, on periodic table group sequence (right to left, as is shown in the listing) and, second, on position within the group. Within a group, the elements are listed from bottom to top. Two important refinements to this order sequence are (1) the element hydrogen is placed in the order between groups VA and VIA, and (2) the element oxygen is moved from its group VIA position to a position immediately before fluorine (the last element in the order).

Using this order sequence, the first element listed in the chemical formula of a binary molecular compound is the more metallic of the two elements. This situation parallels, to a degree, the convention used for binary ionic compounds of listing the positive ion (the metal) first.

EXAMPLE 8.9 **Determining the Correct Order of Elemental Symbols in the Chemical Formula of a Binary Molecular Compound**

Select the correct ordering of elemental symbols, from the choices given, for the chemical formulas of the following binary molecular compounds.

a. HBr or BrH   **b.** CO_2 or O_2C **c.** ClF or FCl   **d.** SO_2 or O_2S

SOLUTION

The order determination for the chemical symbols is based on increasing nonmetallic character for the elements as is depicted in the order listing preceding this example.

a. HBr. Bromine is to the right of H in the order.
b. CO_2. Oxygen is next to the last in the order and will be last in all formulas except those that involve oxygen and fluorine. Fluorine is the last element in the order listing.
c. ClF. Chlorine and fluorine are in the same group. Since fluorine is higher up in the group, it is listed last.
d. SCl_2. Group VIIA (Cl) is farther to the right in the periodic table than group VIA (S).

▶ **Practice Exercise 8.9** Select the correct ordering of elemental symbols, from the choices given, for the chemical formulas of the following binary molecular compounds.

a. BrI or IBr   **b.** SO_3 or O_3S **c.** NH_3 or H_3N **d.** NO or ON

In binary molecular compound nomenclature, the two nonmetals present are named in the order in which they appear in the formula, in a manner consistent with the following rule.

The name of the first nonmetal is used in full, followed by a separate word containing the stem of the name of the second nonmetal and the suffix -ide. Numerical prefixes, giving numbers of atoms present, precede the names of both nonmetals.

TABLE 8.5 Common Numerical
Prefixes from 1 to 10

Prefix	Number
Mono-	1
Di-	2
Tri-	3
Tetra-	4
Penta-	5
Hexa-	6
Hepta-	7
Octa-	8
Nona-	9
Deca-	10

The use of numerical prefixes in binary molecular compound names is in direct contrast to the procedures used for naming ionic compounds. For ionic compounds, formula subscripts are not mentioned in the name.

Prefixes are needed in naming binary molecular compounds because numerous different compounds exist for many pairs of nonmetallic elements. For example, all of the following nitrogen–oxygen molecular compounds exist: NO, NO_2, N_2O, N_2O_3, N_2O_4, and N_2O_5. The prefixes used are standard numerical prefixes, which are given in Table 8.5 for the numbers 1 through 10, and which are used in words such as *mono*nucleosis, *di*chromatic, *tri*angle, *tetra*gram, *penta*gon, *hexa*pod, *hepta*thlon, *octa*ve, *nona*gon, and *deca*de. Example 8.10 shows how these prefixes are used in naming binary molecular compounds.

You cannot predict the formulas of most molecular compounds in the same way that you can predict the formulas of ionic compounds. That is why we name them using prefixes that explicitly indicate their composition.

EXAMPLE 8.10 Naming Binary Molecular Compounds

Name the following binary molecular compounds.

a. N_2O_5 **b.** PF_3 **c.** S_4N_4 **d.** $SiCl_4$

SOLUTION

The name of each of these compounds will consist of two words with the following general formats.

> First word: numerical prefix + full name of the first nonmetal
> Second word: numerical prefix + stem of name of second nonmetal + *-ide*

a. The elements present are nitrogen and oxygen. The two portions of the name before adding numerical prefixes are *nitrogen* and *oxide*. Adding the prefixes gives *dinitrogen* (two nitrogen atoms are present) and *pentoxide* (five oxygen atoms are present). (When an element name begins with an *a* or *o*, the *a* or *o* at the end of the numerical prefix is dropped for ease of pronunciation—pentoxide instead of pentaoxide.) The name of this compound is dinitrogen pentoxide.

b. When there is only one atom of the first nonmetal present, it is standard procedure to omit the prefix *mono-* for the element. Following this guideline, we have for the

An **acid** *is a hydrogen-containing molecular compound whose molecules yield hydrogen ions* (H^+) *when dissolved in water.* The production of the positive H^+ ion, which is the species that gives acids their characteristic properties, is common to all acids. The negative ion produced varies from acid to acid. For example,

HCl, in water, produces H^+ and Cl^- ions.

HNO_3, in water, produces H^+ and NO_3^- ions.

H_2SO_4, in water, produces H^+ and SO_4^{2-} ions.

Because of differences in the properties of acids and the anhydrous (without water) compounds from which they are produced, the acids and the anhydrous compounds are given different names. Acid nomenclature is derived from the names of the parent anhydrous compounds (which we considered in Sec. 8.5).

The names of acids are derived from the names of the *negative ions* produced from the acids' interaction with water. There are three nomenclature rules based on whether the name of the negative ion ends in *-ide, -ate,* or *-ite*.

> Not all acids are ternary molecular compounds. The acids HCl and HBr (hydrochloric acid and hydrobromic acid) are examples of binary molecular compounds.

1. *-ide* rule: When the name of the negative ion produced from the acid ends in *-ide* the acid name has four parts:
  1. the prefix *hydro-*
  2. the stem of the name of the negative ion
  3. the suffix *-ic*
  4. the word *acid*
Examples of the application of this rule are

HCl (chloride) is named *hydro*chlor*ic acid.*

HCN (cyanide) is named *hydro*cyan*ic acid.*

2. *-ate* rule: When the name of the negative ion produced from the acid ends in *-ate* the acid name has three parts:
  1. the name of the negative ion less the *-ate* ending
  2. the suffix *-ic*
  3. the word *acid*
Examples of the application of this rule are

HNO_3 (nitrate) is named nitr*ic acid.*

$HClO_4$ (perchlorate) is named perchlor*ic acid.*

3. *-ite* rule: When the name of the negative ion produced from the acid ends in *-ite* the acid name has three parts:
  1. the name of the negative ion less the *-ite* ending
  2. the suffix *-ous*
  3. the word *acid*
Examples of the application of this rule are

HNO_2 (nitrite) is named nitr*ous acid.*

HClO (hypochlorite) is named hypochlor*ous acid.*

Figure 8.6 summarizes the three ways in which anion names and acid names are related, and Examples 8.11 and 8.12 illustrate the use of these relationships in naming acids.

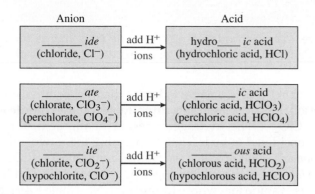

FIGURE 8.6 Summary of the ways in which anion names and acid names are related.

EXAMPLE 8.11   **Naming Hydrogen-Containing Compounds as Acids**

Name the following hydrogen-containing compounds as acids.

 a. H_2CO_3 **b.** HF   **c.** $HClO_2$ **d.** H_2SO_4

SOLUTION

 a. The negative ion present in solutions of this acid is the carbonate ion ($CO_3{}^{2-}$). Removing the -*ate* ending from the word carbonate and replacing it with the suffix -*ic* (rule 2) and then adding the word *acid* gives the name *carbonic acid*.

 b. The negative ion present in solutions of this acid is the fluoride ion (F^-). This is a rule 1 (-*ide*) situation. The name of the acid is formed from using the prefix *hydro-*, the suffix -*ic*, and the word *acid*. The acid name is *hydrofluoric acid*.

 c. The negative ion present in solutions of this acid is the chlorite ion ($ClO_2{}^-$). (If you are unsure of the identity of a particular negative ion, you can consult Table 8.4.) This is a rule 3 (-*ite*) situation. Removing the -*ite* ending from the word *chlorite* and replacing it with the suffix -*ous* and then adding the word *acid* produces the name *chlorous acid*.

 d. The negative ion present in solutions of this acid is the sulfate ion ($SO_4{}^{2-}$). Changing the -*ate* ending to -*ic* (rule 2) and adding the word *acid* gives the name *sulfuric acid*. (For acids involving sulfur, *ur* from sulfur is reinserted into the acid name for phonetic reasons.)

Answer Double Check:

Are the acid names consistent with the names of the negative ions that can be considered to be present in the acids? Yes. The two acids that can be considered to possess -*ate* anions have names that end, as they should, in -*ic acid*. The acid that can be considered to possess an -*ite* anion has a name that ends, as it should, in -*ous acid*. The acid that can be considered to possess an -*ide* anion, is named, as it should be, as a *hydro____ic acid*.

▶ **Practice Exercise 8.11** Name the following hydrogen-containing compounds as acids.

 a. HNO_2 **b.** H_3PO_3 **c.** HCN   **d.** $HClO_3$

EXAMPLE 8.12   **Writing Chemical Formulas for Acids, Given Their Names**

Give the chemical formulas of each of the following acids.

 a. chlorous acid   **b.** hydrocyanic acid

SOLUTION

In this problem we will go through a reasoning process that is the reverse of that in Example 8.11. Students often find this reverse process to be more challenging than the forward process.

a. The *-ous* ending in the acid's name indicates that the negative ion it forms in solution is an *-ite* ion. It is the *chlorite* ion. From Table 8.4, the chlorite ion has the chemical formula ClO_2^-. The hydrogen present in the acid, for formula-writing purposes, is considered to be H^+ ion. Combining these two ions, H^+ and ClO_2^-, in a one-to-one ratio will produce the neutral compound $HClO_2$ (chlorous acid).

b. Since this is a *hydro____ic* acid, the negative ion present in acid solution will be an *-ide* ion. It is the cyanide ion, whose formula is CN^- (Table 8.4). Combining the H^+ and CN^- ions in the ratio that produces neutrality, one to one, gives the chemical formula HCN for *hydrocyanic acid*.

▶ **Practice Exercise 8.12**   Give the chemical formulas for each of the following acids.

a. nitric acid   b. sulfurous acid

___

The overall strategy for identifying and naming acids is summarized in the nomenclature decision tree of Figure 8.7.

Acids are often subclassified into the categories *nonoxyacid* and *oxyacid*. A **nonoxyacid** *is a molecular compound composed of hydrogen and one or more nonmetals other than oxygen that produce H^+ ions in an aqueous solution.* All common nonoxyacids, with one

FIGURE 8.7 A decision tree for naming acids.

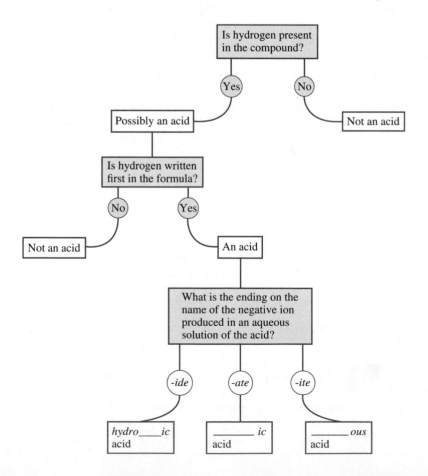

exception, HCN, are binary compounds. An **oxyacid** *is a molecular compound composed of hydrogen, oxygen, and one or more other elements that produces H^+ ions in an aqueous solution.*

Nonoxyacids, but not oxyacids, exist as both pure compounds and water solutions. Conventions for distinguishing between the two situations are as follows, using HCl (a gas in the pure state) as the example.

HCl in the pure state: HCl(g) is hydrogen chloride
HCl in an aqueous solution: HCl(aq) is hydrochloric acid

Table 8.7 gives other examples of this dual naming system for nonoxyacids. Since oxyacids are encountered only as aqueous solutions, a dual naming system is not needed for them.

For several nonmetals a series of oxyacids exist. The formulas of the members of the series differ from each other only in oxygen content. For example, there are four chlorine-containing oxyacids whose formulas are

$$HClO_4, HClO_3, HClO_2, \text{ and } HClO$$

In such a series of oxyacids, one of the acids is the *-ic* acid, the acid whose name has no prefix and ends in *-ic*. The names of the other acids in the series are related, in a constant manner, to the *-ic* acid by prefixes and suffixes as follows:

1. An oxyacid containing *one less oxygen atom* than the *-ic* acid is the *-ous* acid.
2. An oxyacid containing *two fewer oxygen atoms* than the *-ic* acid is the hypo-_____-ous acid.
3. An oxyacid containing *one more oxygen atom* than the *-ic* acid is the per-_____-ic acid.

These relationships between names for a series of oxyacids can be summarized as follows, where *n* represents the number of oxygen atoms in the *-ic* acid.

Number of Oxygen Atoms	Name
n + 1	per_____ic acid
n	_____ic acid
n − 1	_____ous acid
n − 2	hypo_____ous acid

Most nonmetals do not form a complete series of four simple oxyacids as does chlorine. Bromine is the only other nonmetal for which such a series of oxyacids exists. For both sulfur and nitrogen, only the *-ic* and *-ous* acids are known. For both phosphorus and iodine, there are three oxyacids; there is no *n* + 1 acid (*per_____ic acid*) for phosphorus and no *n* − 1 acid (_____*ous* acid) for iodine. Hypofluorous acid, HOF, is the only known fluorine-containing oxyacid.

TABLE 8.7 The Dual Naming System for Molecular Compounds Containing Hydrogen and a Nonmetal Other Than Oxygen

Formula	Name of Pure Compound	Name of Water Solution
HF	hydrogen fluoride	hydrofluoric acid
HBr	hydrogen bromide	hydrobromic acid
HI	hydrogen iodide	hydroiodic acid
H_2S	hydrogen sulfide	hydrosulfuric acid*

*For acids involving sulfur, ur from sulfur is reinserted in the acid name for pronunciation reasons.

FIGURE 8.8 The maximum oxygen content of and the charge on oxyanions that contain C, N, P, S, and Cl.

	Group IVA	Group VA	Group VIA	Group VIIA
Period 2	CO_3^{2-} Carbonate ion	NO_3^{-} Nitrate ion		
Period 3		PO_4^{3-} Phosphate ion	SO_4^{2-} Sulfate ion	ClO_4^{-} Perchlorate ion

To effectively navigate the interrelationships among the various members of a series of oxyanions (or their corresponding oxyacids), the oxygen content of one member of the series must be known. All the other relationships can then be derived from this starting point. Figure 8.8 can be used as an aid in remembering the oxygen content and charge of the one needed member of each of the common oxyanion series. Formatted as a small portion of the upper-right corner of the periodic table, Figure 8.8 lists the oxyanion that contains the *most oxygen atoms* for the five most common oxyanion series (C, N, P, S, and Cl).

The following periodic patterns exist among the five oxyanions shown in Figure 8.8:

1. For period 2 nonmetals (C and N), the maximum number of oxygen atoms present in an oxyanion is three. For period 3 nonmetals (P, S, and Cl), the maximum number of oxygen atoms present in an oxyanion is four.
2. For period 2 oxyanions, charge increases in going from right to left in the period (–1 to –2). For period 3 oxyanions, a similar charge increase pattern exists (–1 to –2 to –3).
3. Four of the five oxyanions in Figure 8.8 are -*ate* anions (carbonate, nitrate, phosphate, and sulfate). The fifth anion is a *per____ate* anion (perchlor*ate*).

Based on knowledge of the names and chemical formulas of these five anions, the names of all other members of these oxyanion series can easily be derived using the oxyanion interrelationship previously considered in this section.

A final item concerning oxyacids is to formally recognize that they are *ternary* molecular compounds, as contrasted to *binary* molecular compounds. A **ternary molecular compound** *is a molecular compound that contains three different nonmetallic elements.* Oxyacids are the only ternary molecular compounds that have been encountered is this chapter. All other molecular compounds have been binary, including nonoxyacids (except for HCN). It should also be noted that *ternary* molecular compounds that are *not* acids do exist. The compounds $POCl_3$ and SO_2Cl_2 (no H is present) are examples of such. Specialized rules, which we will not consider in this text, exist for naming such compounds.

8.7 NOMENCLATURE RULES—A SUMMARY

Within this chapter, various sets of nomenclature rules have been presented. Students sometimes have problems deciding which set of rules to use in a given situation. This dilemma can be avoided if, when confronted with the request to name a compound, the following reasoning pattern is used in this order:

1. Decide, first, whether the compound is ionic or molecular. If a metal or polyatomic ion is present, the compound is ionic. If not, the compound is molecular.
2. If the compound is ionic, then classify it as binary ionic or polyatomic-ion-containing ionic and use the rules appropriate for that classification.
3. If the compound is molecular, classify it as an acid or nonacid. An acid must have the element hydrogen present (written first in the chemical formula), and the compound must be in water solution. If the compound is a nonacid, name it according to the rules for binary molecular compounds.

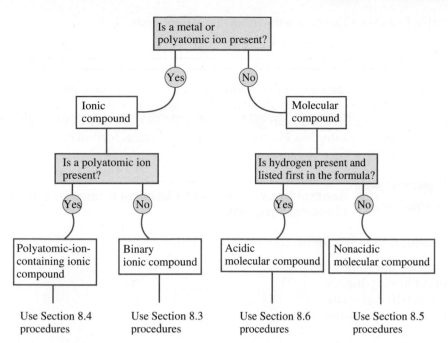

Figure 8.9 portrays the preceding information in the format of a nomenclature decision tree.

EXAMPLE 8.13 **General Exercise in Chemical Nomenclature**

Name the following compounds.

a. N_2O_4 **b.** $Cu(NO_3)_2$ **c.** $H_2S(aq)$ **d.** Na_3N

SOLUTION

a. Neither a metal nor a polyatomic ion is present. Thus, the compound is not ionic. If it is not ionic, it must be molecular. Since no hydrogen is present, the compound is a nonacidic molecular compound. Such compounds are named using numerical prefixes to denote the number of atoms present. The name of the compound is dinitrogen tetroxide.

b. This compound meets both tests for an ionic compound. Both a metal and a polyatomic ion are present. The metal present is a variable-charge metal, which means the name of the compound will have a Roman numeral in it. The metal ion present is copper(II), and the polyatomic ion is the nitrate ion. The name of the compound is copper(II) nitrate.

c. This compound does not meet either test for an ionic compound. It is thus a molecular compound. Hydrogen is present, and it is listed first in the formula. We have an acid. The (aq) means the compound is in solution rather than the pure state. The negative ion produced from the acid is sul*fide* ion. Using the -*ide* rule for naming acids, the name of the acid becomes hydrosulfuric acid.

d. A metal is present but no polyatomic ion is present. Thus, we have a binary ionic compound. The metal present is a fixed-charge metal, meaning that no Roman numeral is needed in the name. The name of the compound is sodium nitride. Numerical prefixes are never used in naming ionic compounds. (The name *trisodium nitride* would be wrong.)

▶ **Practice Exercise 8.13**   Name the following compounds.

   a. S_2O **b.** $Be_3(PO_4)_2$ **c.** H_2CO_3 **d.** CaS

Example 8.13 involved naming compounds, given their formulas. It is important to be able to do the reverse of this also—to write chemical formulas when given chemical names. Often students who can go forward without difficulty have problems with this reverse process. Example 8.14 involves writing chemical formulas given the names of the compounds.

EXAMPLE 8.14   **General Exercise in Using Chemical Names to Obtain Chemical Formulas**

Write chemical formulas for the following compounds.

   a. magnesium sulfide
   b. beryllium hydroxide
   c. copper(II) carbonate
   d. heptasulfur dioxide

SOLUTION

   a. Magnesium is a metal, and sulfur is a nonmetal. We have a binary ionic compound. The lack of a Roman numeral in the name indicates that the metal is a fixed-charge metal. Magnesium forms +2 ions (Fig. 8.1). The sulfide ion has a charge of –2 (Sec. 7.5). Combining the ions Mg^{2+} and S^{2-} in a one-to-one ratio, the ratio that will cause their charges to add to zero, gives the formula MgS.

   b. This is also an ionic compound. Both a metal (beryllium) and a polyatomic ion (hydroxide) are present. Beryllium is a fixed-charge metal, forming ions with a +2 charge (Sec. 7.5). The formula of the hydroxide ion is OH^- (Table 8.4). Combining these ions, Be^{2+} and OH^-, in the ratio that causes their charges to add to zero gives the formula $Be(OH)_2$. Remember that any time more than one polyatomic ion is needed in a formula, parentheses must be used. (Review Examples 7.5 and 7.6 if you are still having trouble determining the ratio in which ions combine.)

   c. This is another ionic compound containing a polyatomic ion. The Roman numeral associated with the word *copper* indicates that the copper ions present are Cu^{2+} ions. The Roman numeral gives the charge on the ions. The formula of the polyatomic carbonate ion is $CO_3{}^{2-}$ (Table 8.4). Combining Cu^{2+} and $CO_3{}^{2-}$ ions in a one-to-one ratio gives the formula $CuCO_3$. Note, again, that a Roman numeral in the name of a compound never becomes part of the formula of the compound. The Roman numeral's purpose is to give information about the ionic charge of a variable-charge metal.

   d. Both the presence of only nonmetals and the presence of numerical prefixes indicate that this is a molecular compound. The numerical prefixes give directly the number of each type of atom present. The chemical formula is S_7O_2.

▶ **Practice Exercise 8.14**   Write chemical formulas for the following compounds.

   a. silver nitrate
   b. tin(IV) chloride
   c. dinitrogen pentoxide
   d. potassium carbonate

Summary

1. **Chemical Nomenclature** Chemical nomenclature is the system of names used to distinguish compounds from each other and the rules needed to devise these names.
2. **IUPAC Nomenclature Rules** IUPAC nomenclature rules are a systematic set of rules for naming compounds based on the premise that a compound's name should convey information about the composition of the compound.
3. **Binary Ionic Compounds** Binary ionic compounds contain positively charged monoatomic metal ions and negatively charged monoatomic nonmetal ions. They are divided into two categories: those containing *fixed-charge* metal ions and those containing *variable-charge* metal ions. Binary ionic compounds are named by giving the full name of the metallic element first, followed by a separate word containing the stem of the nonmetallic element name and the suffix *-ide*. A Roman numeral (in parentheses) specifying ionic charge is appended to the name of the metallic element if it is a variable-charge metal.
4. **Polyatomic-Ion-Containing Ionic Compounds** Polyatomic-ion-containing ionic compounds are named in a way similar to binary ionic compounds, with the following modifications. If the polyatomic ion is positive, its name is substituted for that of the metal. If the polyatomic ion

is negative, its name is substituted for the non-metal stem plus *-ide*. When both positive and negative ions are polyatomic, dual substitution occurs and the resulting name includes just the names of the polyatomic ions.
5. **Binary Molecular Compounds** Binary molecular compounds are those in which just two nonmetallic elements are present. The two nonmetals are named in the order they occur in the formula of the compound. The name of the first nonmetal is used in full. The name of the second nonmetal is treated as was the nonmetal in binary ionic compounds; the stem of the name is given and the suffix *-ide* is added. Numerical prefixes, giving numbers of atoms, precede the names of both nonmetals. A few binary molecular compounds have common names—names not based on systematic rules.
6. **Acids** An acid is a hydrogen-containing molecular compound whose molecules yield H^+ ions when dissolved in water. Acids are subclassified into the categories nonoxyacids and oxyacids. Nonoxyacids have names that contain the prefix *hydro-* and the suffix *-ic acid*. Oxyacids that produce a negative ion in a solution that has a name ending in *-ate* are named *-ic acids*. Oxyacids that produce a negative ion in a solution that has a name ending in *-ite* are named *-ous acids*.

Key Terms

The new terms defined in this chapter are

acid *Sec. 8.6*
binary compound *Sec. 8.2*
binary ionic compound *Sec. 8.2*
binary molecular compound *Sec. 8.5*

chemical nomenclature *Sec. 8.1*
common name *Sec. 8.5*
fixed-charge metal *Sec. 8.2*
nonoxyacid *Sec. 8.6*

oxyacid *Sec. 8.6*
systematic name *Sec. 8.5*
ternary compound *Sec. 8.4*
ternary ionic compound *Sec 8.4*

ternary molecular compound *Sec 8.6*
variable-charge metal *Sec. 8.2*

Practice Problems

NOMENCLATURE CLASSIFICATIONS FOR COMPOUNDS (SEC. 8.1)

8.1 Classify each of the following compounds as *ionic* or *molecular* for nomenclature purposes.
a. SO_2 b. $BaCl_2$ c. Cu_2S d. NH_3
8.2 Classify each of the following compounds as *ionic* or *molecular* for nomenclature purposes.
a. CO_2 b. $AlBr_3$ c. Fe_2O_3 d. CH_4

8.3 Classify each of the following compounds as *ionic* or *molecular* for nomenclature purposes.
a. $NaCN$ b. $Al(NO_3)_3$
c. NH_4Cl d. $CuSO_4$
8.4 Classify each of the following compounds as *ionic* or *molecular* for nomenclature purposes.
a. KOH b. Na_2SO_4 c. P_4O_{10} d. NH_4CN

8.5 In which of the following pairs of compounds is one member of the pair an ionic compound and the other member of the pair a molecular compound?
a. AlN and HF **b.** $CuCO_3$ and H_2CO_3
c. SO_3 and H_2SO_4 **d.** NaCl and $Mg(OH)_2$

8.6 In which of the following pairs of compounds is one member of the pair an ionic compound and the other member of the pair a molecular compound?
a. CO and CO_2 **b.** Be_3N_2 and Cl_2O
c. HNO_3 and KNO_3 **d.** MgO and CaS

TYPES OF BINARY IONIC COMPOUNDS (SEC. 8.2)

8.7 Indicate whether or not each of the following compounds is a *binary* compound.
a. NO **b.** HCN **c.** Cu_2S **d.** CO_2

8.8 Indicate whether or not each of the following compounds is a *binary* compound.
a. N_2H_4 **b.** Na_2O **c.** KOH **d.** CO

8.9 Indicate whether or not each of the compounds in Problem 8.7 is an *ionic* compound.

8.10 Indicate whether or not each of the compounds in Problem 8.8 is an *ionic* compound.

8.11 Indicate whether or not each of the compounds in Problem 8.7 is a *binary ionic* compound.

8.12 Indicate whether or not each of the compounds in Problem 8.8 is a *binary ionic* compound.

8.13 Classify each of the following metals as a fixed-charge metal or a variable-charge metal.
a. Mg **b.** Ni **c.** Ca **d.** Au

8.14 Classify each of the following metals as a fixed-charge metal or a variable-charge metal.
a. Ba **b.** Ag **c.** Cr **d.** Li

8.15 In which of the following pairs of fixed-charge metals do both members of the pair form ions with the same charge?
a. Be and Mg **b.** Ag and Zn
c. Al and K **d.** Na and Li

8.16 In which of the following pairs of fixed-charge metals do both members of the pair form ions with the same charge?
a. Ag and Na **b.** Al and Ga
c. Be and Cd **d.** Ca and Zn

8.17 Classify each of the following compounds as a fixed-charge binary ionic compound or a variable-charge binary ionic compound.
a. FeO **b.** $NiCl_2$ **c.** K_2O **d.** Al_2S_3

8.18 Classify each of the following compounds as a fixed-charge binary ionic compound or a variable-charge binary ionic compound.
a. Cu_2S **b.** AgCl **c.** AuCl **d.** Na_3N

NOMENCLATURE FOR BINARY IONIC COMPOUNDS (SEC. 8.3)

8.19 Name each of the following fixed-charge metal ions.
a. Mg^{2+} ion **b.** K^+ ion
c. Zn^{2+} ion **d.** Ag^+ ion

8.20 Name each of the following fixed-charge metal ions.
a. Be^{2+} ion **b.** Li^+ ion **c.** Ga^{3+} ion **d.** Cd^{2+} ion

8.21 Name each of the following variable-charge metal ions.
a. Cu^{2+} ion **b.** Cu^+ ion
c. Co^{3+} ion **d.** Co^{2+} ion

8.22 Name each of the following variable-charge metal ions.
a. Pb^{4+} ion **b.** Pb^{2+} ion
c. Au^+ ion **d.** Au^{3+} ion

8.23 Name each of the following nonmetal ions.
a. Br^- ion **b.** N^{3-} ion
c. S^{2-} ion **d.** Cl^- ion

8.24 Name each of the following nonmetal ions.
a. F^- ion **b.** P^{3-} ion
c. O^{2-} ion **d.** I^- ion

8.25 What is the chemical formula of each of the following monoatomic ions?
a. zinc **b.** lead(II) **c.** calcium **d.** nitride

8.26 What is the chemical formula of each of the following monoatomic ions?
a. silver **b.** gold(I) **c.** selenide **d.** bromide

8.27 Name each of the following fixed-charge binary ionic compounds.
a. MgO **b.** Li_2S **c.** AgCl **d.** $ZnBr_2$

8.28 Name each of the following fixed-charge binary ionic compounds.
a. BeS **b.** $GaCl_3$ **c.** CaO **d.** Cd_3P_2

8.29 Indicate whether each of the following compounds is a fixed-charge or variable-charge binary ionic compound.
a. $AlCl_3$ **b.** $NiCl_3$ **c.** ZnO **d.** CoO

8.30 Indicate whether each of the following compounds is a fixed-charge or variable charge binary ionic compound.
a. Au_2O **b.** Ag_2O **c.** CuCl **d.** KCl

8.31 Indicate whether or not a Roman numeral is needed in the names of each of the compounds in Problem 8.29.

8.32 Indicate whether or not a Roman numeral is needed in the names of each of the compounds in Problem 8.30.

8.33 What is the charge on the metal ion in each of the following variable-charge binary ionic compounds?
a. FeO **b.** $NiCl_2$ **c.** Au_2O_3 **d.** Co_3N_2

8.34 What is the charge on the metal ion in each of the following variable-charge binary ionic compounds?
a. SnO **b.** PbS_2 **c.** Cu_2S **d.** Fe_2O_3

8.35 Name each of the variable-charge ionic compounds in Problem 8.33.

8.36 Name each of the variable-charge ionic compounds in Problem 8.34.

8.37 Name each compound in the following pairs of variable-charge binary ionic compounds.
a. $FeBr_2$ and $FeBr_3$ **b.** Cu_2S and CuS
c. SnS and SnS_2 **d.** NiO and Ni_2O_3

8.38 Name each compound in the following pairs of variable-charge binary ionic compounds.
a. $SnCl_4$ and $SnCl_2$ **b.** FeS and Fe_2S_3
c. Cu_3N and Cu_3N_2 **d.** NiI_2 and NiI_3

8.39 Name each of the following binary ionic compounds.
  a. Al_2O_3 **b.** CoF_3 **c.** Ag_3N **d.** BaS

8.40 Name each of the following binary ionic compounds.
  a. ZnO **b.** Ag_2S **c.** Cu_2O **d.** $FeBr_3$

8.41 Change each of the following compound names from the Roman numeral to the -ic, -ous system, or vice versa.
  a. lead(IV) oxide **b.** gold(III) chloride
  c. ferric iodide **d.** stannous bromide

8.42 Change each of the following compound names from the Roman numeral to the -ic, -ous system, or vice versa.
  a. cuprous chloride **b.** ferrous sulfide
  c. tin(IV) nitride **d.** lead(II) oxide

8.43 In which of the following pairs of compound names do the two names in the pair denote the same compound?
  a. cupric chloride and copper(I) chloride
  b. stannic bromide and tin(IV) bromide
  c. ferrous oxide and iron(II) oxide
  d. plumbous sulfide and lead(IV) sulfide

8.44 In which of the following pairs of compound names do the two names in the pair denote the same compound?
  a. auric chloride and gold(III) chloride
  b. ferric bromide and iron(III) bromide
  c. cuprous oxide and copper(I) oxide
  d. stannous sulfide and tin(II) sulfide

8.45 Write chemical formulas for the following fixed-charge binary ionic compounds, given their names.
  a. lithium sulfide **b.** zinc sulfide
  c. aluminum sulfide **d.** silver sulfide

8.46 Write chemical formulas for the following fixed-charge binary ionic compounds, given their names.
  a. sodium nitride **b.** calcium nitride
  c. aluminum nitride **d.** zinc nitride

8.47 Write chemical formulas for the following variable-charge binary ionic compounds, given their names.
  a. copper(II) sulfide **b.** copper(II) nitride
  c. tin(II) oxide **d.** tin(IV) oxide

8.48 Write chemical formulas for the following variable-charge binary ionic compounds, given their names.
  a. iron(II) chloride **b.** iron(III) chloride
  c. gold(I) oxide **d.** gold(I) iodide

8.49 Fill in the blanks in the following table of chemical formulas and compound names. The first line of the table is already completed as an example.

	Formula of Positive Ion	Formula of Negative Ion	Compound Formula	Compound Name
	Mg^{2+}	Cl^-	$MgCl_2$	magnesium chloride
a.	Al^{3+}	O^{2-}		
b.			$PbBr_2$	
c.		S^{2-}		iron(II) sulfide
d.				zinc bromide

8.50 Fill in the blanks in the following table of chemical formulas and compound names. The first line of the table is already completed as an example.

	Formula of Positive Ion	Formula of Negative Ion	Compound Formula	Compound Name
	Ca^{2+}	S^{2-}	CaS	calcium sulfide
a.	Fe^{3+}	O^{2-}		
b.			$SnBr_4$	
c.		N^{3-}		potassium nitride
d.				copper(I) chloride

NOMENCLATURE FOR IONIC COMPOUNDS CONTAINING POLYATOMIC IONS (SEC. 8.4)

8.51 Write the chemical formulas, including ionic charge, for each of the following polyatomic ions.
  a. hydroxide **b.** ammonium
  c. nitrate **d.** perchlorate

8.52 Write the chemical formulas, including ionic charge, for each of the following polyatomic ions.
  a. sulfate **b.** carbonate
  c. cyanide **d.** phosphate

8.53 With the help of Table 8.4, give the names for the following polyatomic ions.
  a. O_2^{2-} **b.** $S_2O_3^{2-}$ **c.** $C_2O_4^{2-}$ **d.** ClO_3^-

8.54 With the help of Table 8.4, give the names for the following polyatomic ions.
  a. N_3^- **b.** BO_3^{3-} **c.** SCN^- **d.** CrO_4^{2-}

8.55 Write chemical formulas for both ions in each of the following pairs of polyatomic ions.
  a. sulfate and sulfite
  b. phosphate and hydrogen phosphate
  c. hydroxide and peroxide
  d. chromate and dichromate

8.56 Write chemical formulas for both ions in each of the following pairs of polyatomic ions.
  a. nitrate and nitrite
  b. chlorate and perchlorate
  c. cyanide and azide
  d. cyanate and thiocyanate

8.57 What is the charge on the metal ion present in each of the following ionic compounds?
  a. $AgNO_3$ **b.** $ZnSO_4$ **c.** $Pb(CN)_2$ **d.** Cu_3PO_4

8.58 What is the charge on the metal ion present in each of the following ionic compounds?
  a. $Ba(OH)_2$ **b.** $FeCO_3$
  c. $Ni(NO_3)_2$ **d.** $KClO_4$

8.59 Classify each of the ionic compounds in Problem 8.57 as a fixed-charge ionic compound or a variable-charge ionic compound.

8.60 Classify each of the ionic compounds in Problem 8.58 as a fixed-charge ionic compound or a variable-charge ionic compound.

8.61 What is the name of each of the compounds in Problem 8.57?

8.62 What is the name of each of the compounds in Problem 8.58?

8.63 Name each compound in the following pairs of polyatomic-ion-containing compounds.
a. $Fe_2(CO_3)_3$ and $FeCO_3$
b. Au_2SO_4 and $Au_2(SO_4)_3$
c. $Sn(OH)_2$ and $Sn(OH)_4$
d. $Cr(C_2H_3O_2)_3$ and $Cr(C_2H_3O_2)_2$

8.64 Name each compound in the following pairs of polyatomic-ion-containing compounds.
a. $CuNO_3$ and $Cu(NO_3)_2$
b. $Pb_3(PO_4)_2$ and $Pb_3(PO_4)_4$
c. $Mn(CN)_3$ and $Mn(CN)_2$
d. $Co(ClO_3)_2$ and $Co(ClO_3)_3$

8.65 Write chemical formulas for the following polyatomic-ion-containing compounds.
a. silver carbonate   b. gold(I) nitrate
c. ferric sulfate   d. cuprous cyanide

8.66 Write chemical formulas for the following polyatomic-ion-containing compounds.
a. copper(II) sulfate   b. iron(III) hydroxide
c. cupric chlorate   d. ferrous nitrate

8.67 What is the charge on the negative ion present in each of the following polyatomic-ion-containing compounds?
a. ammonium sulfate   b. ammonium cyanide
c. sodium peroxide   d. potassium azide

8.68 What is the charge on the negative ion present in each of the following polyatomic-ion-containing compounds?
a. ammonium phosphate
b. ammonium nitrate
c. barium peroxide
d. calcium azide

8.69 What is the chemical formula for each of the compounds in Problem 8.67?

8.70 What is the chemical formula for each of the compounds in Problem 8.68?

8.71 Classify each of the compounds in Problem 8.67 as (1) a binary ionic compound, (2) a ternary ionic compound, or (3) an ionic compound with four different elements present.

8.72 Classify each of the compounds in Problem 8.68 as (1) a binary ionic compound, (2) a ternary ionic compound, or (3) an ionic compound with four different elements present.

8.73 Fill in the blanks in the following table of chemical formulas and compound names. The first line of the table is already completed as an example.

	Formula of Positive Ion	Formula of Negative Ion	Compound Formula	Compound Name
	Mg^{2+}	NO_3^-	$Mg(NO_3)_2$	magnesium nitrate
a.	Al^{3+}	SO_4^{2-}		
b.			$Cu(CN)_2$	
c.		OH^-		iron(II) hydroxide
d.				zinc azide

8.74 Fill in the blanks in the following table of chemical formulas and compound names. The first line of the table is already completed as an example.

	Formula of Positive Ion	Formula of Negative Ion	Compound Formula	Compound Name
	Ca^{2+}	NO_3^-	$Ca(NO_3)_2$	calcium nitrate
a.	Fe^{3+}	PO_4^{3-}		
b.			$SnCO_3$	
c.		ClO_4^-		aluminum perchlorate
d.				potassium peroxide

NOMENCLATURE FOR BINARY MOLECULAR COMPOUNDS (SEC. 8.5)

8.75 Select the correct ordering of elemental symbols, from the choices given, for the chemical formulas of the following binary molecular compounds.
a. CO or OC   b. Cl_2O or OCl_2
c. N_2O or ON_2 d. HI or IH

8.76 Select the correct ordering of elemental symbols, from the choices given, for the chemical formulas of the following binary molecular compounds.
a. SiH_4 or H_4Si b. Br_2O or OBr_2
c. NCl_3 or Cl_3N d. IBr or BrI

8.77 Write the number that corresponds to each of the following prefixes.
a. *hepta-*   b. *penta-*   c. *tri-*   d. *deca-*

8.78 Write the number that corresponds to each of the following prefixes.
a. *tetra-*   b. *octa-*   c. *hexa-*   d. *nona-*

8.79 Name the following binary molecular compounds.
a. P_4O_{10} b. SF_4 c. CBr_4 d. ClO_2

8.80 Name the following binary molecular compounds.
a. S_4N_2 b. SO_3 c. IF_7 d. N_2O_4

8.81 Write chemical formulas for the following binary molecular compounds, all of which are "sulfur oxides."
a. sulfur monoxide   b. disulfur monoxide
c. sulfur dioxide   d. heptasulfur dioxide

8.82 Write chemical formulas for the following binary molecular compounds, all of which are "nitrogen oxides."
a. dinitrogen monoxide
b. nitrogen dioxide
c. nitrogen monoxide
d. dinitrogen tetroxide

8.83 Name the following binary molecular compounds.
a. H_2S b. HF   c. NH_3 d. CH_4

8.84 Name the following binary molecular compounds.
a. HCl   b. H_2Se c. N_2H_4 d. H_2O_2

8.85 Write chemical formulas for the following binary molecular compounds.
a. phosphine   b. hydrogen bromide
c. ethane   d. hydrogen telluride

8.86 Write chemical formulas for the following binary molecular compounds.
  a. hydrogen iodide
  b. hydrazine
  c. hydrogen peroxide
  d. arsine

8.87 Explain why the name dicopper sulfur trioxide is not the correct name for the compound Cu_2SO_3.

8.88 Explain why the name tripotassium phosphorus trioxide is not the correct name for the compound K_3PO_3.

NOMENCLATURE FOR ACIDS (SEC. 8.6)

8.89 Indicate whether each of the following hydrogen-containing compounds forms an acid in an aqueous solution.
  a. CH_4 **b.** H_2S **c.** HCN   **d.** NH_3

8.90 Indicate whether each of the following hydrogen-containing compounds forms an acid in an aqueous solution.
  a. HCl   **b.** HClO   **c.** SiH_4 **d.** CH_4

8.91 What is the chemical formula for the negative ion produced when each of the following acids dissolve in water? (Assume all hydrogen atoms present become H^+ ions.)
  a. HNO_3 **b.** HI   **c.** HClO   **d.** H_3PO_3

8.92 What is the chemical formula for the negative ion produced when each of the following acids disolve in water? (Assume all hydrogen atoms present become H^+ ions.)
  a. $HClO_3$ **b.** $HClO_4$ **c.** H_2S **d.** HCl

8.93 What is the chemical formula for the acids that produce each of the following negative ions in an aqueous solution?
  a. CN^- **b.** $SO_4{}^{2-}$ **c.** $NO_2{}^-$ **d.** $BO_3{}^{3-}$

8.94 What is the chemical formula for the acids that produce each of the following negative ions in an aqueous solution?
  a. $NO_3{}^-$ **b.** I^- **c.** $PO_3{}^{3-}$ **d.** $C_2O_4{}^{2-}$

8.95 Name each of the compounds in Problem 8.91 as an acid.

8.96 Name each of the compounds in Problem 8.92 as an acid.

8.97 Name the acids that produce each of the negative ions in Problem 8.93.

8.98 Name the acids that produce each of the negative ions in Problem 8.94.

8.99 What is the chemical formula for each of the following acids?
  a. hydrochloric acid   **b.** carbonic acid
  c. chloric acid   **d.** sulfuric acid

8.100 What is the chemical formula for each of the following acids?
  a. hydrobromic acid
  b. nitrous acid
  c. hypochlorous acid
  d. phosphoric acid

8.101 Fill in the blanks in the following table of chemical formulas and names for acids. The first line of the table is already completed as an example.

	Positive Ion Present in Solution	Negative Ion Present in Solution	Chemical Formula of Acid	Name of Acid
	H^+	$SO_3{}^{2-}$	H_2SO_3	sulfurous acid
a.		$NO_2{}^-$		
b.			H_3PO_4	
c.	H^+			hydrocyanic acid
d.				hypochlorous acid

8.102 Fill in the blanks in the following table of chemical formulas and names for acids. The first line of the table is already completed as an example.

	Positive Ion Present in Solution	Negative Ion Present in Solution	Chemical Formula of Acid	Name of Acid
	H^+	$CO_3{}^{2-}$	H_2CO_3	carbonic acid
a.		$NO_3{}^-$		
b.			H_3PO_3	
c.	H^+			hydrobromic acid
d.				chlorous acid

8.103 Supply the missing name in each of the following pairs of name–formula combinations.
  a. H_3AsO_4 (arsenic acid); H_3AsO_3 (?)
  b. HIO_3 (iodic acid); HIO_4 (?)
  c. H_3PO_3 (phosphorous acid); H_3PO_2 (?)
  d. HBrO (hypobromous acid); $HBrO_2$ (?)

8.104 Supply the missing name in each of the following pairs of name–formula combinations.
  a. HIO_3 (iodic acid); HIO (?)
  b. H_2SeO_4 (selenic acid); H_2SeO_3 (?)
  c. HBrO (hypobromous acid); $HBrO_4$ (?)
  d. HNO_2 (nitrous acid); HNO_3 (?)

8.105 Name each of the following compounds.
  a. HBr(g)   **b.** HCN(aq)
  c. H_2S(g)   **d.** HI(aq)

8.106 Name each of the following compounds.
  a. HBr(aq)   **b.** HCN(g)
  c. H_2S(aq)   **d.** HI(g)

ADDITIONAL PROBLEMS

8.107 Element X forms the ionic compound CaX. Element Y forms the molecular compound NY_3. What would be the name for the simplest compound formed when elements X and Y interact if both X and Y are period 4 elements?

8.108 Element X forms the ionic compound K_2X. Element Y forms the molecular compound SY_2. What would be the name for the simplest compound formed when elements X and Y interact if both X and Y are period 3 elements?

8.109 Determine the name for the simplest compound that would be expected to form when each of the listed pairs of "highlighted" elements in the following periodic table interact with each other.

    a. elements 1 and 6   **b.** elements 3 and 7
    c. elements 2 and 5   **d.** elements 7 and 8

8.110 Determine the name for the simplest compound that would be expected to form when each of the listed pairs of "highlighted" elements in the periodic table in Problem 8.107 interact with each other.
    a. elements 1 and 5   **b.** elements 2 and 6
    c. elements 4 and 7   **d.** elements 6 and 8

8.111 In which of the following pairs of compounds would both members of the pair have names that contain Roman numerals?
    a. NaCl and CuCl   **b.** Al_2O_3 and Fe_2O_3
    c. $Cu(NO_3)_2$ and NiO   **d.** Ag_2SO_4 and $CuSO_4$

8.112 In which of the following pairs of compounds would both members of the pair have names that contain Roman numerals?
    a. $AuCl_3$ and $FeCl_3$ **b.** K_3N and AlN
    c. FeO and BaO   **d.** NiS and $Cu_3(PO_4)_2$

8.113 In which of the following pairs of compounds would both members of the pair have names that contain numerical prefixes?
    a. SO_3 and N_2O **b.** AlN and CO
    c. HCl and H_2S **d.** PCl_3 and NF_3

8.114 In which of the following pairs of compounds would both members of the pair have names that contain numerical prefixes?
    a. CO and CO_2 **b.** N_2O and K_2O
    c. OF_2 and BaF_2 **d.** HBr and H_2Se

8.115 In each of the following pairs of anions, select the one that contains the greatest number of oxygen atoms.
    a. sulfate and sulfite
    b. chlorate and perchlorate
    c. hydroxide and peroxide
    d. chromate and dichromate

8.116 In each of the following pairs of anions, select the one that contains the greatest number of oxygen atoms.
    a. nitride and nitrate
    b. phosphate and phosphite

    c. chlorite and hypochlorite
    d. acetate and cyanide

8.117 Write chemical formulas for the following compounds.
    a. calcium nitride   **b.** calcium nitrate
    c. calcium nitrite   **d.** calcium cyanide

8.118 Write chemical formulas for the following compounds.
    a. sodium sulfide   **b.** sodium sulfate
    c. sodium sulfite   **d.** sodium thiosulfate

8.119 Write chemical formulas for the following compounds.
    a. potassium phosphide
    b. potassium phosphate
    c. potassium hydrogen phosphate
    d. potassium dihydrogen phosphate

8.120 Write chemical formulas for the following compounds.
    a. magnesium carbide
    b. magnesium bicarbonate
    c. magnesium carbonate
    d. magnesium hydrogen carbonate

8.121 Indicate which compounds in each of the following groups have names that contain the prefix *di-*.
    a. N_2O, Li_2O, CO_2, K_2S
    b. K_2O, K_2CO_3, NO_2, SO_2
    c. BeF_2, $BeCl_2$, SF_2, SCl_2
    d. Au_2O_3, Fe_2O_3, Al_2O_3, N_2O_3

8.122 Indicate which compounds in each of the following groups have names that contain the suffix *-ide*.
    a. CaS, CO, SO_3, Be_3N_2
    b. K_3N, KNO_3, KNO_2, KClO
    c. MgO, AlN, KF, KOH
    d. NaCN, NaOH, Na_2CO_3, NaF

8.123 Indicate which compounds in each of the following groups have names that contain the suffix *-ous* or *-ate*.
    a. $CaCO_3$, H_2CO_3, $Ca(NO_2)_2$, HNO_2
    b. $NaClO_4$, $NaClO_3$, $NaClO_2$, NaClO
    c. $HClO_4$, $HClO_3$, $HClO_2$, HClO
    d. LiOH, LiCN, Li_2CO_3, Li_3PO_4

8.124 Indicate which compounds in each of the following groups have names that contain the suffix *-ic* or *-ite*.
    a. HBr(g), HBr(aq), HCN(g), HCN(aq)
    b. K_2SO_4, KCN, $KMnO_4$, K_3PO_3
    c. NH_4Cl, NH_4CN, $(NH_4)_2SO_4$, NH_4NO_2
    d. Li_3N, LiN_3, $LiNO_3$, $LiNO_2$

8.125 Indicate which compounds in each of the following groups possess molecules or formula units that are pentatomic.
    a. sodium cyanide, sodium thiocyanate, sodium hypochlorite, sodium nitrate
    b. aluminum sulfide, magnesium nitride, beryllium phosphide, potassium hydroxide
    c. beryllium oxide, iron(II) oxide, iron(III) oxide, sulfur dioxide
    d. gold(I) cyanide, gold(III) cyanide, gold(I) chlorate, gold(III) chlorate

8.126 Indicate which compounds in each of the following groups possess molecules or formula units that are heptatomic.
  a. magnesium cyanide, magnesium oxalate, magnesium chlorite, magnesium perchlorate
  b. aluminum oxide, calcium thiosulfate, beryllium thiocyanate, dinitrogen pentoxide
  c. iron(III) sulfide, iron(III) nitride, iron(III) sulfate, iron(III) hypochlorite
  d. dichlorine heptoxide, sulfur trioxide, sulfur hexafluoride, ammonium sulfide

8.127 The chemical formula of a hydroxide of nickel is $Ni(OH)_3$. What are the chemical formulas of the following nickel compounds in which nickel has the same ionic charge as in $Ni(OH)_3$?
  a. nickel sulfate   **b.** nickel oxide
  c. nickel oxalate   **d.** nickel nitrate

8.128 The chemical formula of a phosphate of vanadium is VPO_4. What are the chemical formulas of the following vanadium compounds in which vanadium has the same ionic charge as in VPO_4?
  a. vanadium sulfate   **b.** vanadium hydroxide
  c. vanadium oxide   **d.** vanadium nitrate

8.129 The chemical formula of the ionic compound potassium superoxide is KO_2. The chemical formula of the ionic compound nitronium perchlorate is NO_2ClO_4. What is the chemical formula for the ionic compound nitronium superoxide?

8.130 The chemical formula of the ionic compound calcium perrhenate is $Ca(ReO_4)_2$. The chemical formula of the ionic compound nitrosonium hydrogen sulfate is $NOHSO_4$. What is the chemical formula for the ionic compound nitrosonium perrhenate?

8.131 Classify each of the following compounds as (1) binary ionic, (2) ternary ionic, (3) binary molecular, or (4) ternary molecular.
  a. hydrocyanic acid   **b.** hydrobromic acid
  c. sodium peroxide   **d.** hydrogen peroxide

8.132 Classify each of the following compounds as (1) binary ionic, (2) ternary ionic, (3) binary molecular, or (4) ternary molecular.
  a. sodium azide
  b. ammonium azide
  c. beryllium nitride
  d. beryllium nitrate

CUMULATIVE PROBLEMS

8.133 Element X interacts with element Y to give a compound containing X^{2+} ions and Y^{3-} ions.
  a. Is element X a metal or nonmetal?
  b. Is element Y a metal or nonmetal?
  c. What is the chemical formula for the compound?
  d. What is the name of the compound if X is a period 2 element and Y is a period 3 element?

8.134 Element X interacts with element Y to give a compound containing X^{3+} ions and Y^{2-} ions.
  a. Is element X a metal or nonmetal?
  b. Is element Y a metal or nonmetal?
  c. What is the chemical formula for the compound?
  d. What is the name of the compound if X is a period 3 element and Y is a period 2 element?

8.135 Based on the information given, determine the name of each of the following compounds.
  a. Na_2XO_4, where X has an atomic number of 16
  b. Na_2Y, where Y has an electronegativity of 3.5
  c. Na_3Z, where the total number of protons present per formula unit is 40
  d. NaQ, where the total number of electrons present per formula unit is 46

8.136 Based on the information given, determine the name of each of the following compounds.
  a. Li_3XO_4, where X has an atomic number of 15
  b. LiY, where Y has an electronegativity of 2.8
  c. Li_2Z, where the total number of protons present per formula unit is 22
  d. LiQ, where the total number of electrons present per formula unit is 4

8.137 Give the name of the simplest binary compound that forms between elements with the following electron configurations.
  a. $1s^2 2s^2 2p^6 3s^2 3p^5$ and $1s^2 2s^2 2p^6 3s^2$
  b. $1s^2 2s^2 2p^4$ and $1s^2 2s^2 2p^5$

8.138 Give the name of the simplest binary compound that forms between elements with the following electron configurations.
  a. $1s^2 2s^2 2p^3$ and $1s^2 2s^2 2p^6 3s^2 3p^6 4s^1$
  b. $1s^2 2s^2 2p^6 3s^2 3p^4$ and $1s^2 2s^2 2p^6 3s^2 3p^5$

8.139 After determining the value of x in each of the following chemical formulas, name the compounds.
  a. $SiCl_x$ **b.** $MgCl_x$ **c.** K_xN **d.** NCl_x

8.140 After determining the value of x in each of the following chemical formulas, name the compounds.
  a. CCl_x **b.** BeO_x **c.** NF_x **d.** $AlBr_x$

8.141 A compound has the chemical formula $M(XO_3)_2$, where M is a metal and X is a nonmetal. Assign a name to this compound, given the following information.
  1. M is a metal with two valence electrons.
  2. X forms a monoatomic ion with a charge of –1.
  3. X has an electronegativity between 2.6 and 2.9.
  4. The sum of the periodic table period numbers for X and M is 6.

8.142 A compound has the chemical formula M_3XO_3, where M is a metal and X is a nonmetal. Assign a name to this compound, given the following information.
  1. M is a metal whose electron configuration ends in s^1.
  2. Both M and X are found in period 2 of the periodic table.
  3. There are 38 protons in one formula unit of M_3XO_3.

8.143 Determine the name for a binary ionic compound that has the following characteristics.
1. Positive and negative ions are present in a one-to-one ratio.
2. All ions present have the electron configuration $1s^2 2s^2 2p^6$.
3. One of the elements present has an atomic number that is six fewer than the atomic number of the other element.

8.144 Determine the name for a binary ionic compound that has the following characteristics.
1. Positive and negative ions are present in a one-to-two ratio.
2. All ions present have the electron configuration $1s^2 2s^2 2p^6 3s^2 3p^6$.
3. Neutral atoms of the more electronegative element present have 17 electrons.

8.145 Determine the name for a binary molecular compound that has the following characteristics.
1. Its molecules are triatomic.
2. There are twice as many atoms of the more electronegative element per molecule as there are of the less electronegative element.
3. Both elements are in period 2 of the periodic table.
4. Atoms of one element contain four valence electrons, and atoms of the other element contain six valence electrons.

8.146 Determine the name for a binary molecular compound that has the following characteristics.
1. Its molecules are hexatomic.
2. There are twice as many atoms of the more electronegative element per molecule as there are of the less electronegative element.

3. The two elements occupy adjacent positions in period 2 of the periodic table.
4. The sum of the valence electrons for an atom of each element is 11.

8.147 Determine the name for a compound that has the following characteristics.
1. Monoatomic and polyatomic ions are present.
2. The ratio between positive and negative ions is one to two.
3. A group IIA element that loses one half of its total electrons upon ion formation is present.
4. The polyatomic ion contains equal numbers of atoms of two nonmetallic elements.
5. The sum of the atomic numbers for the two elements involved in the polyatomic ion is 13.
6. One of the elements present in the compound forms a monoatomic ion with a charge of –3 that is isoelectronic with Ne.

8.148 Determine the name for a compound that has the following characteristics.
1. Monoatomic and polyatomic ions are present.
2. The ratio between positive and negative ions is one to three.
3. A group IIIA element whose ion electron configuration is isoelectronic with Ne is present.
4. The polyatomic ion contains equal numbers of atoms of two nonmetallic elements.
5. The sum of the atomic numbers for the two elements involved in the polyatomic ion is 9.

Multiple-Choice Practice Test

Use this bank of 20 multiple-choice questions as a review of key concepts presented in this chapter. For many of the questions, there may be more than one correct answer (choice d) or no correct answer (choice e).

8.149 The rules for naming binary ionic compounds would be used in naming which of the following compounds?
a. NaCl
b. N_2O_5
c. Al_2S_3
d. more than one correct response
e. no correct response

8.150 The correct name for the ionic compound $AlBr_3$ is
a. aluminum bromide.
b. aluminum bromate.
c. aluminum tribromide.
d. aluminum tribromate.
e. no correct response

8.151 In which of the following ionic compounds are three or fewer ions present per formula unit?
a. aluminum nitride
b. beryllium sulfide
c. potassium sulfide
d. more than one correct response
e. no correct response

8.152 Which of the following ionic compounds is named without using a Roman numeral?
a. $AuCl_3$
b. CuS
c. $MgCl_2$
d. more than one correct response
e. no correct response

8.153 A Roman numeral is required as part of the name for which of the following ionic compounds?
a. $CaCl_2$
b. $CuCl_2$
c. $ZnCl_2$

d. more than one correct response
e. no correct response

8.154 Which of the following pairings of compound formula and compound name is correct?
a. Cu$_2$O and copper(II) oxide
b. SnO$_2$ and tin(II) oxide
c. PbO and lead(II) oxide
d. more than one correct response
e. no correct response

8.155 In which of the following pairs of polyatomic ions do both members of the pair have the same charge?
a. nitrate and ammonium
b. sulfate and carbonate
c. hydroxide and cyanide
d. more than one correct response
e. no correct response

8.156 The correct name for the polyatomic-ion-containing compound CaCO$_3$ is
a. calcium carbon trioxide.
b. calcium monocarbon trioxide.
c. calcium carbonate.
d. calcium monocarbonate.
e. no correct response

8.157 Which of the following pairings of compound formula and compound name is correct?
a. lithium sulfate and Li$_2$SO$_4$
b. gold(I) cyanide and AuOH
c. ammonium nitrate and NH$_3$NO$_2$
d. more than one correct response
e. no correct response

8.158 Which of the following polyatomic-ion-containing compounds contain six or fewer atoms per formula unit?
a. sodium cyanide
b. aluminum phosphate
c. calcium sulfite
d. more than one correct response
e. no correct response

8.159 Which of the following is a binary molecular compound that contains five atoms per molecule?
a. silicon tetrachloride
b. dinitrogen pentoxide
c. dichlorine monoxide
d. more than one correct response
e. no correct response

8.160 Which of the following binary molecular compounds has a name in which no numerical prefixes are present?
a. CO
b. H$_2$S
c. SO$_2$
d. more than one correct response
e. no correct response

8.161 For which of the following pairs of compounds do both members of the pair have names that contain the prefix *di-*?
a. SCl$_2$ and BaCl$_2$

b. CO$_2$ and NO$_2$
c. N$_2$O$_3$ and Al$_2$O$_3$
d. more than one correct response
e. no correct response

8.162 For which of the following pairs of compounds do both members of the pair contain the same number of atoms per molecule?
a. hydrogen peroxide and hydrazine
b. methane and ethane
c. phosphine and ammonia
d. more than one correct response
e. no correct response

8.163 Which of the following compounds contains just two atoms per formula unit?
a. aluminum nitride
b. carbon monoxide
c. lead(II) iodide
d. more than one correct response
e. no correct response

8.164 Which of the following acids is named using the prefix *hydro-*?
a. HNO$_3$
b. H$_2$SO$_4$
c. H$_3$PO$_4$
d. more than one correct response
e. no correct response

8.165 Which of the following acids is named using the suffix *-ous*?
a. HClO$_3$
b. HClO$_2$
c. HClO
d. more than one correct response
e. no correct response

8.166 Which of the following is a nonoxy acid?
a. hydrocyanic acid
b. hypobromous acid
c. perchloric acid
d. more than one correct response
e. no correct response

8.167 An oxyacid containing two fewer oxygen atoms than the *-ic* acid would be the
a. per . . . ic acid.
b. hydro . . . ic acid.
c. hypo . . . ous acid.
d. . . . ous acid
e. no correct response

8.168 Which of the following compounds contains both ionic and covalent bonds?
a. sodium nitride
b. sodium nitrite
c. sodium nitrate
d. more than one correct response
e. no correct response

Chemical Calculations: The Mole Concept and Chemical Formulas

9.1 THE LAW OF DEFINITE PROPORTIONS

This chapter is the first of two chapters dealing with chemical arithmetic, that is, with the quantitative relationships between elements and compounds. The emphasis in this chapter will be on quantitative relationships that involve chemical formulas. In Chapter 10, the emphasis will be on quantitative relationships that involve chemical equations.

In Section 4.7 we saw that compounds are pure substances with the following characteristics:

1. They are chemical combinations of two or more elements.
2. They can be broken down into constituent elements by chemical, but not physical, means.
3. They have a definite, constant elemental composition.

In this section we consider further the fact that compounds have a definite composition.

The composition of a compound can be determined by decomposing a weighed amount of the compound into its elements and then determining the masses of the individual elements. Alternatively, determining the mass of a compound formed by the combination of known masses of elements will also allow the calculation of its composition.

Studies of composition data for many compounds have led to the conclusion that the percentage of each element present in a given compound does not vary. This conclusion has been formalized into a statement known as the **law of definite proportions**: *In a pure compound, the elements are always present in the same definite proportion by mass.*

The law of definite proportions is also called the *law of constant composition.*

The French chemist Joseph-Louis Proust (1754–1826; see "The Human Side of Chemistry 9") is responsible for the work that established this law as one of the fundamentals of chemistry.

Let us consider how composition data obtained from decomposing a compound are used to illustrate the law of definite proportions. Samples of the compound of *known masses* are decomposed. The masses of the constituent elements present in each sample are then obtained. This experimental information (elemental masses and the original sample mass) is then used to calculate the percent composition of the samples. Within the limits of experimental error, the calculated percentages for any given element present turn out to be the same, validating the law of definite proportions. Example 9.1 gives actual decomposition data for a compound and shows how they are used to verify the law of definite proportions.

The Human Side of Chemistry 9

Joseph-Louis Proust (1754–1826)

Joseph-Louis Proust (pronounced Proost), born in Angers, France, began his chemistry studies at home in his father's apothecary shop. Later he studied in Paris. It is Proust's research on the constancy of composition for various compounds that brought the law of definite proportions into focus.

Proust was a dedicated analyst, with little interest in theory. His most important work, during the period 1797–1807, involved copious experiments on the composition of various compounds. Other research involved studies on starch, camphor, preparation of mercury from cinnabar ore, and isolation of grape sugar (glucose) from grapes.

Almost all of Proust's work, including that related to the law of definite proportions, was carried out in Spain. At the invitation of the Spanish government, in 1791, he became director of the Royal Laboratory at Madrid. This laboratory, financed by the government, was elegantly equipped. Almost all the vessels, even those in common use, were made of platinum.

His research work in Madrid was stopped abruptly in 1808 when his laboratory was destroyed by French troops occupying Spain. Reduced to poverty, and plagued with ill health, he still refused an offer of 100,000 francs from Napoleon to come back to France and supervise the manufacture of glucose from grapes (a most abundant crop in both Spain and France).

Later, in 1817, after Louis XVIII had come to power, Proust returned to France, and in 1820, he took over the apothecary shop of his brother Joachim, who was in still poorer health. Proust remained in France until his death in 1826.

EXAMPLE 9.1 **Using Decomposition Data to Illustrate the Law of Definite Proportions**

Two samples of ammonia (NH_3) of differing mass and from different sources are individually decomposed to yield ammonia's constituent elements (nitrogen and hydrogen). The results of the decomposition experiments are as follows.

	Sample Mass Before Decomposition (g)	Mass of Nitrogen Produced (g)	Mass of Hydrogen Produced (g)
Sample 1	1.840	1.513	0.327
Sample 2	2.000	1.644	0.356

Show that these data are consistent with the law of definite proportions.

SOLUTION

Calculating the percent nitrogen in each sample will be sufficient to show whether the data are consistent with the law. Both nitrogen percentages should come out the same if the law is obeyed.

$$\text{percent nitrogen} = \frac{\text{mass of nitrogen obtained}}{\text{total sample mass}} \times 100$$

Sample 1

$$\%N = \frac{1.513 \text{ g}}{1.840 \text{ g}} \times 100 = 82.22826\% \quad \text{(calculator answer)}$$

$$= 82.23\% \quad \textbf{(correct answer)}$$

Sample 2

$$\% N = \frac{1.644 \text{ g}}{2.000 \text{ g}} \times 100 = 82.2\% \quad \text{(calculator answer)}$$

$$= 82.20\% \quad \textbf{(correct answer)}$$

Note that the percentages are close to being equal but are not identical. This is due to measuring uncertainties and rounding in the original experimental data. The two percentages are the same to three significant figures. The difference lies in the fourth significant digit. Recall from Section 2.4 that the last of the significant digits in a number (the fourth one here) has uncertainty in it. The two percentages are considered to be the same within experimental error.

An alternative way of treating the given data to illustrate the law of definite proportions involves calculating the mass ratio between nitrogen and hydrogen. This ratio should be the same for each sample; otherwise the two samples are not the same compound.

Sample 1

$$\frac{\text{mass of N}}{\text{mass of H}} = \frac{1.513 \text{ g}}{0.327 \text{ g}} = 4.6269113 \quad \text{(calculator answer)}$$

$$= 4.63 \quad \textbf{(correct answer)}$$

Sample 2

$$\frac{\text{mass of N}}{\text{mass of H}} = \frac{1.644 \text{ g}}{0.356 \text{ g}} = 4.6179775 \quad \text{(calculator answer)}$$

$$= 4.62 \quad \textbf{(correct answer)}$$

Again, slight differences in the ratios are caused by measuring errors and rounding in the original data.

▶ **Practice Exercise 9.1** Two samples of carbon dioxide (CO_2) of differing mass and from different sources are individually decomposed to yield the elements carbon and oxygen. The results of the decomposition experiments are as follows.

Answers to all practice exercises in this chapter are found in the back-of-the-book answer section.

	Sample Mass Before Decomposition (g)	Mass of Carbon Produced (g)	Mass of Oxygen Produced (g)
Sample 1	3.200	0.8733	2.327
Sample 2	4.424	1.207	3.217

Show that these data are consistent with the law of definite proportions.

The constancy of composition for compounds can also be examined by considering the mass ratios in which elements combine to form compounds. Let us consider the reaction between the elements calcium and sulfur to produce the compound calcium sulfide. Suppose an attempt is made to combine various masses of sulfur with a fixed mass of calcium. A set of possible experimental data for this attempt is given in the first four lines of Table 9.1. Note that, regardless of the mass of S present, only a certain amount, 44.4 g, reacts with the 55.6 g of Ca. The excess S is left over in an unreacted form. The data therefore illustrate that Ca and S will react in only one fixed mass ratio (55.6/44.4 = 1.25) to form CaS. This fact is consistent with the law of definite proportions. Note also that if the amount of Ca used is doubled (line 5 of Table 9.1), the amount of S with which it reacts also doubles (compare lines 1 and 5 of the table). Nevertheless, the ratio in which the substances react (111.2/88.8) still remains 1.25.

Figure 9.1 relates the law of definite proportions to Dalton's atomic theory (Sec. 4.10). This figure shows pictorially the formation of the compound tin(II) sulfide, SnS, from atoms of tin and sulfur. Note that a formula unit of SnS must always contain one atom of tin and one atom of sulfur.

TABLE 9.1 Data Illustrating the Law of Definite Proportions

Mass of Ca Used (g)	Mass of S Used (g)	Mass of CaS Formed (g)	Mass of Excess Unreacted Sulfur (g)	Ratio in Which Substances React
55.6	44.4	100.0	none	1.25
55.6	50.0	100.0	5.6	1.25
55.6	100.0	100.0	55.6	1.25
55.6	200.0	100.0	155.6	1.25
111.2	88.8	200.0	none	1.25

FIGURE 9.1
Illustration of the law of definite proportions at the level of atoms.

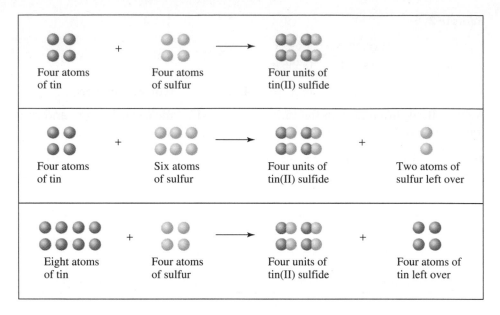

Four atoms of tin + Four atoms of sulfur → Four units of tin(II) sulfide

Four atoms of tin + Six atoms of sulfur → Four units of tin(II) sulfide + Two atoms of sulfur left over

Eight atoms of tin + Four atoms of sulfur → Four units of tin(II) sulfide + Four atoms of tin left over

9.2 CALCULATION OF FORMULA MASSES

Formula masses play a role in almost all chemical calculations and will be used extensively in later sections of this chapter and in succeeding chapters. A **formula mass** *is the sum of the atomic masses of all atoms present in one formula unit of a substance, expressed in atomic mass units.* Formula masses, like the atomic masses from which they are calculated, are relative masses based on the $^{12}_{6}C$ relative mass scale (Sec. 5.5). They can be calculated for compounds (both molecular and ionic) and for elements that exist in molecular form.

The term *molecular mass* is used interchangeably with *formula mass* by many chemists in referring to substances that contain discrete molecules. It is incorrect, however, to use the term *molecular mass* when talking about ionic substances, since molecules are not their basic structural unit (Sec. 7.8).

Some chemistry books use the terms *formula weight* and *molecular weight* rather than *formula mass* and *molecular mass*. This practice is not followed in this book because mass is technically more correct (Sec. 3.3).

Once the formula of a substance has been established, its formula mass is calculated by adding together the atomic masses of all the atoms in the formula. If more than one atom of any element is present, that element's atomic mass must be added as many times as there are atoms of the element present.

Example 9.2 contains sample formula mass calculations. Significant figures (Sec. 2.5) are a consideration because atomic masses are obtained from experimentally determined numbers (Sec. 5.4).

EXAMPLE 9.2 **Using Chemical Formulas and Atomic Masses to Calculate Formula Masses**

Calculate the formula masses for the following compounds.

   a. N_2H_4 (hydrazine, a rocket fuel)
   b. $ClNO_3$ (chlorine nitrate, a substance involved in the Antarctic ozone hole phenomenon)
   c. CHF_2Cl (chlorodifluoromethane, a Freon replacement)
   d. $C_9H_8O_4$ (aspirin, a mild pain reliever)

SOLUTION

Formula masses are obtained simply by adding the atomic masses of the constituent elements, counting each atomic mass as many times as the symbol for the element occurs in the formula.

a. A molecule of N_2H_4 contains six atoms: two atoms of N and four atoms of H. The formula mass, the collective mass of these six atoms, is calculated as follows:

$$2 \text{ atoms N} \times \frac{14.01 \text{ amu}}{1 \text{ atom N}} = 28.02 \text{ amu}$$

$$4 \text{ atoms H} \times \frac{1.01 \text{ amu}}{1 \text{ atom H}} = 4.04 \text{ amu}$$

$$\text{formula mass} = \overline{32.06 \text{ amu}} \quad \text{(calculator and \textbf{correct answer})}$$

The atomic mass values used in the conversion factors in this calculation were obtained from the periodic table found inside the front cover of the textbook. The atomic mass values found in this periodic table are values rounded to the hundredths place. Using atomic mass values rounded to hundredths will be standard procedure when calculating atomic masses in all calculations in the text. Additional comments concerning this operational rule are found in the next section of the text, which covers the topic of significant figures and atomic masses.

Often, conversion factors are not explicitly shown in a formula mass calculation as just shown; the calculation is simplified as follows:

N: 2 × 14.01 amu = 28.02 amu
H: 4 × 1.01 amu = 4.04 amu
      formula mass = 32.06 amu   (calculator and **correct answer**)

b. A molecule of $ClNO_3$ contains five atoms: one atom of Cl, one atom of N, and three atoms of O. Using the simplified calculation method, we calculate the atomic mass of $ClNO_3$ as

Cl: 1 × 35.45 amu = 35.45 amu
N: 1 × 14.01 amu = 14.01 amu
O: 3 × 16.00 amu = 48.00 amu
      formula mass = 97.46 amu   (calculator and **correct answer**)

c. A molecule of CHF_2Cl contains five atoms: one atom of C, one atom of H, two atoms of F, and one atom of Cl. Using the simplified calculation method, we calculate the atomic mass of CHF_2Cl as

C: 1 × 12.01 amu = 12.01 amu
H: 1 × 1.01 amu = 1.01 amu
F: 2 × 19.00 amu = 38.00 amu
Cl: 1 × 35.45 amu = 35.45 amu
      formula mass = 86.47 amu (calculator and **correct answer**)

d. There are 21 atoms in a molecule of this compound. Its formula mass is

C: 9 × 12.01 amu = 108.09 amu
H: 8 × 1.01 amu =   8.08 amu
O: 4 × 16.00 amu = 64.00 amu
      formula mass = 180.17 amu (calculator and **correct answer**)

An important point concerning significant figures in formula mass calculations is formally encountered in the carbon part of this calculation. Here, we have the multiplication (9×12.01). Nine is a counted (exact) number and 12.01 has four significant figures. The answer shown, 108.09, contains five significant figures rather than four allowed by the multiplication rule. Why? A multiplication involving a small whole number, 9 in this case, can be viewed as an addition, a situation previously considered in Section 2.6. We are repeatedly adding 12.01 to itself. As an addition, an uncertainty in the hundredth place is allowed in the answer (108.17).

Situations similar to that occurring here for carbon will be encountered repeatedly in future formula mass calculations. Our operational rule, for significant figure purposes, in all such situations will be to consider the calculation to be a pure addition problem rather than a combined multiplication–addition problem. Using this approach, we will often be able to justify one more significant figure in the calculated formula mass.

▶ **Practice Exercise 9.2** Calculate the formula masses for the following compounds.

   a. Al_2O_3 (aluminum oxide, the primary component of rubies)
   b. $(NH_4)_2C_2O_4$ (ammonium oxalate, used for stain and rust removal)

Formula masses, relative masses on the $^{12}_{6}C$ relative mass scale, can be used for mass comparisons. For example, using the formula masses calculated in Example 9.2, we can make the statement that one molecule of $C_9H_8O_4$ is 1.849 times as heavy as one molecule of $ClNO_3$ (180.17 amu/97.46 amu $= 1.849$).

9.3 SIGNIFICANT FIGURES AND ATOMIC MASS

The operational rules we will follow throughout the remainder of the text relative to calculating formula masses are as follows:

1. Atomic masses rounded to the hundredths place will be used as the input numbers for formula mass determinations. This rule allows us to use, without rounding, the atomic masses given in the periodic table inside the front cover. A benefit of this approach is that the same atomic mass value is always used for a given element and these atomic mass values, for the common elements, become very familiar entities.

2. Atomic masses, rounded to hundredths, almost always contain enough significant figures that they do not become the limiting factor in terms of significant figures for a calculation. If such should be the case, atomic masses with additional significant figures are available for use; they are found in the table inside the front cover located opposite the periodic table. The potential for significant-figure problems is greatest for the first four elements in the periodic table. These elements have atomic mass values of less than 10; hence their atomic masses rounded to hundredths contain only three significant figures. The atomic masses for these elements, rounded to thousandths to give four significant figures, are 1.008 amu (H), 4.003 amu (He), 6.941 amu (Li), and 9.012 amu (Be). In practice, these expanded values will very seldom be needed.

The atomic mass of lead (207.2 amu) is an "anomaly" in that, even though it contains four significant figures, its uncertainty lies in the tenths place. Lead is the only element whose atomic mass does not have an uncertainty of 0.01 or smaller. The larger uncertainty in lead's atomic mass value relates to a greater than normal variation in percentage abundances for lead's naturally occurring isotopes.

3. Formula mass calculations will always be considered pure addition problems rather than combination multiplication–addition problems for significant-figure purposes [see Example 9.2, part (d)]. Thus, the addition rule will always govern significant-figure determinations in formula mass calculations.

Example 9.3 illustrates the use of these operational rules in formula mass calculations.

EXAMPLE 9.3 **Treating Formula Mass Calculations as Addition Problems**

Calculate the formula masses for the following substances, using addition rather than multiplication significant figure rules.

a. $C_7H_6O_2$ (benzoic acid, a food preservative)
b. $PtCl_2(NH_3)_2$ (cisplatin, a chemotherapy agent)

SOLUTION

a. Atomic masses, rounded to hundredths, are the starting point for the calculation. From the periodic table (inside front cover), we obtain

C: 12.01 amu H: 1.01 amu O: 16.00 amu

We calculate the formula's mass, a sum of atomic masses, as

  C: 7×12.01 amu $= 84.07$ amu
  H: $6 \times \ \ 1.01$ amu $= \ \ 6.06$ amu
  O: 2×16.00 amu $= 32.00$ amu
    formula mass $= 122.13$ amu (calculator and **correct answer**)

Since all of the input atomic masses have an uncertainty of hundredths, the formula mass uncertainty will also be hundredths (addition rule for significant figures).

b. From the periodic table, we obtain the needed atomic masses rounded to hundredths.

Pt: 195.08 amu Cl: 35.45 amu N: 14.01 amu H: 1.01 amu

Summing the atomic masses, using each an appropriate number of times, we get

  Pt: 1×195.08 amu $= 195.08$ amu
  Cl: $2 \times \ \ 35.45$ amu $= \ \ 70.90$ amu
  N: $2 \times \ \ 14.01$ amu $= \ \ 28.02$ amu
  H: $6 \times \ \ \ \ 1.01$ amu $= \ \ \ \ 6.06$ amu
    formula mass $= 300.06$ amu (calculator and **correct answer**)

Based on the addition rule for significant figures, the uncertainty in the answer is hundredths since all of the input atomic masses had uncertainties of hundredths.

Answer Double Check:

Do the calculated atomic masses have the correct number of significant figures? Yes. Both answers are given to the hundredths place. Since all input atomic masses were known to hundredths and the calculation is treated as an addition problem (operational rule 3), the calculated atomic masses should also be values known to the hundredths place.

▶ **Practice Exercise 9.3** Calculate the formula masses for the following substances, using addition rather than multiplication significant figure rules.

a. $C_6H_8O_6$ (vitamin C)
b. $C_{17}H_{20}N_4O_6$ (riboflavin, a B-vitamin)

9.4 PERCENT COMPOSITION OF A COMPOUND

A useful piece of information about a compound is its percent composition. **Percent composition of a compound** *is the percent by mass of each element present in the compound.* For instance, the percent composition of water is 88.81% oxygen and 11.19% hydrogen.

Percent compositions are frequently used to compare compound compositions. The compounds gold(III) iodide (AuI_3), gold(III) nitrate [$Au(NO_3)_3$], and gold(I) cyanide (AuCN) contain, respectively, 34.10%, 51.43%, and 88.33% gold by mass. If you were given the choice of receiving a gift of 1 lb of one of these three gold compounds, which one would you choose?

The **percent by mass of an element in a compound** *is the number of grams of the element present in 100 grams of the compound.* The generalized equation for the percent by mass of an element in a compound is

$$\text{percent by mass of element} = \frac{\text{mass of element}}{\text{mass of compound sample}} \times 100$$

The percent composition of a compound can be calculated from experimental decomposition data as was done in Example 9.1. Such a calculation can be carried out even if the formula or the identity of the compound is unknown. A compound's chemical formula can also be used to obtain percentage composition information, as is illustrated in Example 9.4.

EXAMPLE 9.4 **Using a Compound's Formula to Calculate Percent Composition**

Calculate the percent composition (using atomic masses rounded to hundredths) of the pain reliever acetaminophen. Its chemical formula is $C_8H_9O_2N$.

SOLUTION

First, we calculate the formula mass of $C_8H_9O_2N$, using atomic masses rounded to the hundredths decimal place.

$$
\begin{aligned}
\text{C:} \quad & 8 \times 12.01 \text{ amu} = & 96.08 \text{ amu} \\
\text{H:} \quad & 9 \times 1.01 \text{ amu} = & 9.09 \text{ amu} \\
\text{O:} \quad & 2 \times 16.00 \text{ amu} = & 32.00 \text{ amu} \\
\text{N:} \quad & 1 \times 14.01 \text{ amu} = & \underline{14.01 \text{ amu}} \\
& \text{formula mass} = & 151.18 \text{ amu} \quad \text{(calculator and \textbf{correct answer})}
\end{aligned}
$$

The mass percent of each element in the compound is found by dividing the mass contribution of each element, in amu, by the total mass (formula mass), in amu, and multiplying by 100.

$$\text{mass \% element} = \frac{\text{mass of element in one formula unit}}{\text{formula mass}} \times 100$$

Finding percentages, we have

$$\text{mass \% C} = \frac{96.08 \text{ amu}}{151.18 \text{ amu}} \times 100 = 63.55338\% \quad \text{(calculator answer)}$$

$$= 63.55\% \quad \text{(\textbf{correct answer})}$$

$$\text{mass \% H} = \frac{9.09 \text{ amu}}{151.18 \text{ amu}} \times 100 = 6.0127\% \quad \text{(calculator answer)}$$

$$= 6.01\% \quad \text{(\textbf{correct answer})}$$

$$\text{mass \% O} = \frac{32.00 \text{ amu}}{151.18 \text{ amu}} \times 100 = 21.166821\% \quad \text{(calculator answer)}$$

$$= 21.17\% \qquad \text{(correct answer)}$$

$$\text{mass \% N} = \frac{14.01 \text{ amu}}{151.18 \text{ amu}} \times 100 = 9.2670988\% \quad \text{(calculator answer)}$$

$$= 9.267\% \qquad \text{(correct answer)}$$

Answer Double Check:

Does the sum of the calculated percentages add to 100%? Yes. The sum of the percentages is ($63.55\% + 6.01\% + 21.17\% + 9.267\% = 100.00\%$). On occasion, because of rounding errors, totals such as 99.99% or 100.01% may be obtained. Such was not the case here.

▶ **Practice Exercise 9.4**   What is the percent composition of lactic acid, $C_3H_6O_3$, a compound that builds up in muscles and causes them to hurt when they are worked hard? Use atomic masses rounded to hundredths in the calculation.

Percent compositions can also be calculated from mass data obtained from compound synthesis or compound decomposition experiments. Example 9.5 shows how synthesis data are treated to yield percent composition.

| **EXAMPLE 9.5** | **Using Synthesis Data to Calculate Percent Composition** |

The production of a 13.50 g sample of the hormone adrenaline requires 7.96 g of C, 0.96 g of H, 3.54 g of O, and 1.04 g of N. What is the percent composition of this compound?

SOLUTION

The total mass of the compound sample is given as 13.50 g. We divide the mass of each element present by this total mass (13.50 g) and multiply by 100 to give the percentage.

$$\text{mass \% element} = \frac{\text{mass of element}}{\text{total sample mass}} \times 100$$

Finding the percentages, we have

$$\text{mass \% C:} \quad \frac{7.96 \text{ g}}{13.50 \text{ g}} \times 100 = 58.962963\% \quad \text{(calculator answer)}$$

$$= 59.0\% \qquad \text{(correct answer)}$$

$$\text{mass \% H:} \quad \frac{0.96 \text{ g}}{13.50 \text{ g}} \times 100 = 7.1111111\% \quad \text{(calculator answer)}$$

$$= 7.1\% \qquad \text{(correct answer)}$$

$$\text{mass \% O:} \quad \frac{3.54 \text{ g}}{13.50 \text{ g}} \times 100 = 26.222222\% \quad \text{(calculator answer)}$$

$$= 26.2\% \qquad \text{(correct answer)}$$

$$\text{mass \% N:} \quad \frac{1.04 \text{ g}}{13.50 \text{ g}} \times 100 = 7.7037037\% \quad \text{(calculator answer)}$$

$$= 7.70\% \qquad \text{(correct answer)}$$

Answer Double Check:

Do the calculated percentages add up to 100%? Yes. Adding the percentages gives 59.0% + 7.1% + 26.2% + 7.70% = 100.0%

▶ **Practice Exercise 9.5**   The compound deoxyribose is an important component of DNA molecules, the molecules responsible for the transfer of genetic information from one generation to the next in living organisms. Given that a 2.341 g sample of deoxyribose contains 1.048 g C, 0.176 g H, and 1.117 g O, calculate the percent composition of the compound.

9.5 THE MOLE: THE CHEMIST'S COUNTING UNIT

Two common methods exist for specifying the quantity of material in a sample of a substance: (1) in terms of units of *mass* and (2) in terms of units of *amount*. We measure *mass* by using a balance (Sec. 3.3). Common mass units are gram, kilogram, and pound. For substances that consist of discrete units, we can specify the *amount* of substance present by indicating the number of units present—12, or 27, or 113, and so on.

We all use both units of mass and units of amount on a daily basis. We work well with this dual system. Sometimes it does not matter which type of unit is used; at other times one system is preferred over the other. When buying potatoes at the grocery store we can decide on quantity in either mass units (10 lb bag, 20 lb bag, etc.) or amount units (9 potatoes, 15 potatoes, etc.). When buying eggs, amount units are used almost exclusively—12 eggs (1 dozen), 24 eggs (2 dozen), and so on. On the other hand, peanuts and grapes are almost always purchased in weighed quantities. It is impractical to count the number of grapes in a bunch. Very few people go to the store with the idea of buying 117 grapes.

In chemistry, as in everyday life, both the mass and amount methods of specifying quantity find use. Again, the specific situation dictates the method used. In laboratory work, practicality dictates working with quantities of known mass (12.3 g, 0.1365 g, etc.). (Counting out a given number of atoms for a laboratory experiment is impossible, since we cannot see individual atoms.)

In performing chemical calculations, after the laboratory work has been done, it is often useful (even necessary) to think of quantities of substances present in terms of atoms or formula units. A problem exists when this is done—very very large numbers are always encountered. Any macroscopic sample of a chemical substance contains many trillions of atoms or formula units.

To cope with this large number problem, chemists have found it convenient to use a special counting unit. Employment of such a unit should not surprise you, as specialized counting units are used in many areas. The two most common counting units are *dozen* and *pair*. Other more specialized counting units exist. For example, at an office supply store, paper is sold by the *ream* (500 sheets), and pencils by the *gross* (144 pencils). (See Fig. 9.2.)

The chemist's counting unit is called a *mole*. What is unusual about the mole is its magnitude. A **mole** *is a counting unit based on the number* 6.022×10^{23}. This extremely large number is necessitated by the extremely small size of atoms, molecules, and ions. The use of a traditional counting unit, such as a dozen, would be, at best, only a slight improvement over counting atoms singly.

A more technical definition for a mole *will be given in Section 9.7.*

$$6.022 \times 10^{23} \text{ atoms} = 5.018 \times 10^{22} \text{ dozen atoms} = 1 \text{ mole atoms}$$

Note how the use of the mole counting unit decreases very significantly the magnitude of numbers encountered. The number 1 represents 6.022×10^{23} objects, while the number 2 represents double that number of objects. (Why the number 6.022×10^{23} was chosen as the counting unit rather than some other number will be discussed in Section 9.6.)

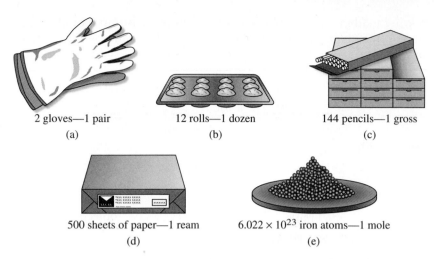

FIGURE 9.2 Some counting units used to denote quantities in terms of amount.

2 gloves—1 pair
(a)

12 rolls—1 dozen
(b)

144 pencils—1 gross
(c)

500 sheets of paper—1 ream
(d)

6.022×10^{23} iron atoms—1 mole
(e)

The number 6.022×10^{23} also has a special name. **Avogadro's number** *is the name given to the numerical value* 6.022×10^{23}, *the number of objects in a mole.* This designation honors the Italian physicist Lorenzo Romano Amedeo Carlo Avogadro (1776–1856; see "The Human Side of Chemistry 10"), whose pioneering work on gases later proved to be valuable in determining the number of particles present in a given volume of a substance.

The Human Side of Chemistry 10

Lorenzo Romano Amedeo Carlo Avogadro (1776–1856)

Amedeo Avogadro, born in Turin, Italy, followed in the footsteps of his father, obtaining a law degree at age 16 and for some years engaging in the practice of law. At age 24, he began to study, privately, mathematics and physics. Abandoning the law profession at age 30, he spent most of his later life as a professor of mathematical physics at the University of Turin.

Avogadro was the first scientist to distinguish between atoms and molecules. His most important paper, published in 1811, suggested that all gases (at a given temperature) contained the same number of particles per unit volume and that these particles need not be individual atoms but might be combinations of atoms. The thrust of this paper, known now as Avogadro's law, received little attention from his contemporaries. Turin, where Avogadro was a professor, was outside the mainstream of scientific activity in 1811.

It was not until after his death that his genius was recognized. In 1858, two years after his death, Stanislao Cannizzaro (1826–1910), a fellow countryman, pointed out the full significance of Avogadro's law at an international gathering of scientists. The "time was right," and acceptance was rapid. One chemist, after reading Cannizzaro's paper, wrote "The scales seemed to fall from my eyes. Doubts disappeared and a feeling of quiet certainty took their place." In 1911, in Turin, in commemoration of the 100th anniversary of the first publication of his law, a monument honoring Avogadro was unveiled by the king in one of the greatest posthumous tributes to a scientist in history.

Avogadro's name is attached to a number he never determined. During the early 1900s, chemical theory developed to the point that estimates could be made of the number of molecules in a given volume of gas. That equal volumes of all gases really contain the same number of molecules (Avogadro's law) was a provable reality. Jean-Baptiste Perrin (1870–1942), in 1909, was the first to use the phrase "Avogadro's number" to denote the calculated number of molecules in a specific volume of gas. The honor is appropriate since Avogadro's law was the springboard for Perrin's work, as well as that of many other scientists working in the same research area.

In solving mathematical problems dealing with the number of atoms, molecules, or ions present in a given amount of material, Avogadro's number becomes part of the conversion factor used to relate number of particles present to moles present.

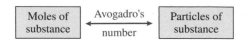

One mole of a chemical substance contains Avogadro's number of particles. The word *particles*, as used in the preceding sentence, can have a number of meanings, including

1. the number of *molecules* of a molecular compound
2. the number of *formula units* of an ionic compound
3. the number of *atoms* of an element

From the definition

$$1 \text{ mole} = 6.022 \times 10^{23} \text{ objects}$$

two conversion factors can be derived.

$$\frac{1 \text{ mole}}{6.022 \times 10^{23} \text{ objects}} \quad \text{and} \quad \frac{6.022 \times 10^{23} \text{ objects}}{1 \text{ mole}}$$

Example 9.6 illustrates the use of these particle-to-mole conversion factors.

EXAMPLE 9.6 | **Calculating the Number of Objects in a Molar Quantity**

How many objects are there in each of the following quantities?

a. 1.20 moles of carbon monoxide (CO) molecules
b. 2.53 moles of silver (Ag) atoms
c. 0.025 mole of magnesium sulfate ($MgSO_4$) formula units
d. 2.25 moles of watermelons

SOLUTION

We will use dimensional analysis (Sec. 3.7) in solving each part of this problem. All of the parts are similar in that we are given a certain number of moles of substance and want to find the number of particles contained in the given number of moles. All parts can be classified as moles-to-particles problems, and each solution will involve the use of Avogadro's number.

$$\boxed{\text{Moles of substance}} \xleftarrow[\text{number}]{\text{Avogadro's}} \boxed{\text{Particles of substance}}$$

a. The given quantity is 1.20 moles of CO molecules, and the desired quantity is the number of CO molecules.

$$1.20 \text{ moles CO} = ? \text{ CO molecules}$$

The setup, by dimensional analysis, involves only one conversion factor.

$$1.20 \text{ moles CO} \times \frac{6.022 \times 10^{23} \text{ CO molecules}}{1 \text{ mole CO}}$$

$$= 7.2264 \times 10^{23} \text{ CO molecules} \quad \text{(calculator answer)}$$

$$= 7.23 \times 10^{23} \text{ CO molecules} \quad \textbf{(correct answer)}$$

b. The given quantity is 2.53 moles of silver atoms, and the desired quantity is the actual number of silver atoms present.

$$2.53 \text{ moles Ag} = ? \text{ Ag atoms}$$

The setup, using the same conversion factor as in part (a), is

$$2.53 \; \text{moles Ag} \times \frac{6.022 \times 10^{23} \, \text{Ag atoms}}{1 \; \text{mole Ag}}$$

$$= 1.523566 \times 10^{24} \, \text{Ag atoms} \quad \text{(calculator answer)}$$

$$= 1.52 \times 10^{24} \, \text{Ag atoms} \qquad \textbf{(correct answer)}$$

c. The fact that we are dealing with formula units here (an ionic compound), rather than atoms or molecules, does not change the way the problem is solved.

$$0.025 \; \text{mole MgSO}_4 = ? \; \text{formula units MgSO}_4$$

The conversion factor setup is

$$0.025 \; \text{mole MgSO}_4 \times \frac{6.022 \times 10^{23} \, \text{MgSO}_4 \, \text{formula units}}{1 \; \text{mole MgSO}_4}$$

$$= 1.5055 \times 10^{22} \, \text{MgSO}_4 \, \text{formula units} \quad \text{(calculator answer)}$$

$$= 1.5 \times 10^{22} \, \text{MgSO}_4 \, \text{formula units} \qquad \textbf{(correct answer)}$$

d. Use of the mole as a counting unit is usually found only in a chemical context. Technically, however, any type of object can be counted in units of moles. One mole denotes 6.02×10^{23} objects; it does not matter what the objects are—even watermelons. Just as we can talk about dozens of watermelons, we can talk about moles of watermelons, although the latter involves a *very large* watermelon patch.

$$2.25 \; \text{moles watermelons} \times \frac{6.022 \times 10^{23} \, \text{watermelons}}{1 \; \text{mole watermelons}}$$

$$= 1.35495 \times 10^{24} \, \text{watermelons} \quad \text{(calculator answer)}$$

$$= 1.35 \times 10^{24} \, \text{watermelons} \qquad \textbf{(correct answer)}$$

Answer Double Check:

Are the answers reasonable in terms of their magnitude? The first answer involves the power 10^{23}, a power consistent with a 1.20 molar amount; the answer should be close in magnitude to Avogadro's number (10^{23}), the power associated with 1 mole. The second and fourth answers involve the power 10^{24}; they are consistent with molar amounts twice that of Avogadro's number (1 mole). The third answer involves the power 10^{22}; such an answer requires a molar amount of less than one, which is the case (0.025 mole).

▶ **Practice Exercise 9.6** How many objects (atoms or molecules) are there in each of the following molar quantities?

a. 3.21 moles of hydrogen peroxide (H_2O_2) molecules
b. 0.537 mole of copper (Cu) atoms

It is somewhat unfortunate, because of its similarity to the word *molecule*, that the name mole was selected as the name for the chemist's counting unit. Students often think that *mole* is an abbreviated form of the word *molecule*. That is not the case. The word mole comes from the Latin *moles*, which means "heap or pile." A mole is a macroscopic amount, a heap or pile of objects, that can easily be seen. A molecule is a particle too small to be seen with the naked eye.

In Example 9.6 we calculated the number of objects present in samples ranging in size from 0.025 mole to 2.53 moles. Our answers were numbers carrying the exponents

FIGURE 9.3 Word pictures of how big Avogadro's number is.

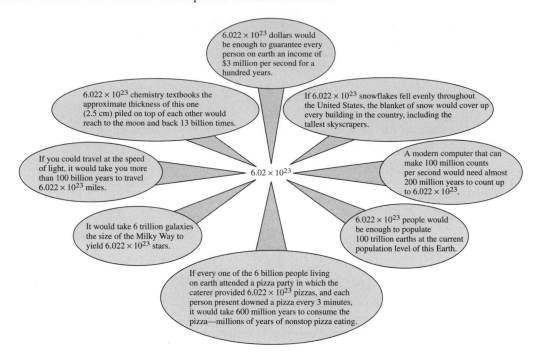

10^{22}, 10^{23}, or 10^{24}. Numbers with these exponents are inconceivably large. The magnitude of Avogadro's number itself is so large that it is almost incomprehensible. There is nothing in our experience to relate to it. (When chemists count, they count in really big jumps.) Many attempts have been made to create word pictures of the vast size of Avogadro's number. Such pictures, however, really only hint at its magnitude, since other large numbers must be used in the word pictures. Several such word pictures are given in Figure 9.3.

9.6 THE MASS OF A MOLE

How much does a mole weigh; that is, what is its mass? Consider a similar (but more familiar) question first: "How much does a dozen weigh?" Your response is now immediate. "A dozen what?" you reply. The mass of a dozen identical objects obviously depends on the identity of the object. For example, the mass of a dozen elephants will be "somewhat greater" than the mass of a dozen marshmallows. The mole, like the dozen, is a counting unit. Similarly, the mass of a mole of objects will depend on the identity of the object. Thus, the mass of a mole, *molar mass*, is not one set number; it varies, being different for each different chemical substance. This is in direct contrast to the *molar number*, Avogadro's number, which is the same for all chemical substances.

The **molar mass of an element** *is a mass in grams numerically equal to the atomic mass of the element when the element is present in atomic form.* Thus, if we know the atomic mass of an element, we also know the mass of one mole of atoms of the element. The two quantities are numerically the same, differing only in units. For the elements carbon, oxygen, and sodium, we can write the following numerical relationships.

The atomic mass of an element and the molar mass of an element in atomic form have the same numerical value. However, the units for the value differ. The unit for atomic mass is amu. The units for molar mass of an element are grams. The mathematical relationship between amu units and grams units is formally considered in Section 9.8.

$$\text{mass of 1 carbon atom} = 12.01 \text{ amu} \quad \text{(atomic mass)}$$
$$\text{mass of 1 mole of carbon atoms} = 12.01 \text{ g} \quad \textbf{(molar mass)}$$
$$\text{mass of 1 oxygen atom} = 16.00 \text{ amu} \quad \text{(atomic mass)}$$
$$\text{mass of 1 mole of oxygen atoms} = 16.00 \text{ g} \quad \textbf{(molar mass)}$$
$$\text{mass of 1 sodium atom} = 22.99 \text{ amu} \quad \text{(atomic mass)}$$
$$\text{mass of 1 mole of sodium atoms} = 22.99 \text{ g} \quad \textbf{(molar mass)}$$

It is not a coincidence that the mass in grams of a mole of atoms of an element and the element's atomic mass are numerically equal. The value selected for Avogadro's number is the one that validates this relationship; only for this unique value is the relationship valid. Experimentally it was determined that when 6.022×10^{23} atoms of an element are present, molar masses (in grams) and atomic masses (in amu) are numerically equal. Again, only when the chemist's counting unit has the value 6.022×10^{23} does this relationship hold. Avogadro's number is thus the "connecting link" between the chemist's microscopic mass scale (amu) and macroscopic mass scale (gram), a topic to be considered further in Section 9.8. The use of the mass unit grams in specifying the molar mass of an element is a must. The use of other mass units would require a different counting unit than 6.022×10^{23} for a numerical match between atomic mass and molar mass.

The molecular form of an element will have a different molar mass than its atomic form. Consider the element chlorine, which is found in nature in the form of diatomic molecules (Cl_2). The mass of one mole of chlorine atoms (Cl) is different from the mass of one mole of chlorine molecules (Cl_2). Since there are two atoms in each molecule of chlorine, the molar mass of molecular chlorine is twice the molar mass of atomic chlorine. The following relationships hold for chlorine, whose atomic mass is 35.45 amu.

$$6.022 \times 10^{23} \text{ Cl atoms} = 1 \text{ mole Cl atoms} = 35.45 \text{ g Cl}$$
$$6.022 \times 10^{23} \text{ Cl}_2 \text{ molecules} = 1 \text{ mole Cl}_2 \text{ molecules} = 70.90 \text{ g Cl}_2$$

Note that one mole of molecular chlorine contains twice as many atoms as one mole of atomic chlorine; however, the number of discrete particles present is the same (Avogadro's number) in both cases. For atomic chlorine, atoms are considered to be the object counted; for molecular chlorine, molecules are considered to be the discrete particle counted. There is the same number of atoms in the former case as there is of molecules in the latter case. This atomic–molecular chlorine situation is analogous to the difference between a dozen shoes and a dozen pairs of shoes. Both the mass and the actual number of shoes for the dozen pairs of shoes are double those for the dozen shoes.

The existence of some elements in molecular form becomes a source of error in chemical calculations if care is not taken to distinguish properly between atomic and molecular forms of the element. The phrase "one mole of chlorine" is an ambiguous term. Does it mean one mole of chlorine atoms (Cl), or does it mean one mole of chlorine molecules (Cl_2)? To avoid ambiguity, it is important to always state explicitly the chemical form of an element being discussed. Using the chemical formula Cl_2 or referring to the substance as molecular chlorine avoids ambiguity.

The **molar mass of a compound** *is a mass in grams that is numerically equal to the formula mass of the compound.* Thus, for compounds a numerical equivalence exists between molar mass and formula mass if the molar mass is specified in grams. When we add atomic masses to get the formula mass (in amu) of a compound, we are simultaneously finding the mass of one mole of compound (in grams). The molar mass–formula mass relationships for the compounds water (H_2O), ammonia (NH_3), and barium chloride ($BaCl_2$) are

mass of 1 H_2O molecule	= 18.02 amu	(formula mass)
mass of 1 mole of H_2O molecules	= 18.02 g	(molar mass)
mass of 1 NH_3 molecule	= 17.04 amu	(formula mass)
mass of 1 mole of NH_3 molecules	= 17.04 g	(molar mass)
mass of 1 $BaCl_2$ formula unit	= 208.23 amu	(formula mass)
mass of 1 mole of $BaCl_2$ formula units	= 208.23 g	(molar mass)

FIGURE 9.4 One mole each of seven different substances: the elements copper (Cu), mercury (Hg), and iodine (I_2) and the compounds water (H_2O), table salt (NaCl), table sugar ($C_{12}H_{22}O_{11}$), and aspirin ($C_9H_8O_4$).

Figure 9.4 pictures molar quantities of a number of common substances. Note again how the mass of a mole varies in numerical value. On the other hand, all of the pictured amounts of substances contain the same number of units—6.022×10^{23} atoms, molecules, or formula units.

A generalized definition for the *molar mass* of a substance, which is independent of whether the substance is an element or a compound, can be given. A **molar mass** *is the mass, in grams, of one mole of atoms, molecules, or formula units of a substance.*

It should now be very evident to you why the chemist's counting unit, the mole, has the value it does—6.022×10^{23}. *Avogadro's number represents the experimentally determined number of atoms, molecules, or formula units contained in a sample of a pure substance with a mass in grams numerically equal to the atomic mass or formula mass of the pure substance.*

The numerical match between molar mass and atomic or formula mass makes the calculation of the mass of any given number of moles of a substance a very simple procedure. In solving problems of this type, the numerical value of the molar mass becomes part of the conversion factor used to convert from moles to grams.

$$\boxed{\text{Moles of substance}} \xrightarrow[\text{mass}]{\text{Molar}} \boxed{\text{Grams of substance}}$$

For example, for the compound CO_2, which has a formula mass of 44.01 amu, we can write the equality

$$44.01 \text{ g } CO_2 = 1 \text{ mole } CO_2$$

From this statement two conversion factors can be written.

$$\frac{44.01 \text{ g } CO_2}{1 \text{ mole } CO_2} \quad \text{and} \quad \frac{1 \text{ mole } CO_2}{44.01 \text{ g } CO_2}$$

Example 9.7 illustrates the use of gram-mole conversion factors like these.

EXAMPLE 9.7 **Calculating the Mass, in Grams, of Molar Quantities**

Calculate the mass, in grams, of each of the following quantities of matter.

a. 1.50 moles of CH_4 molecules   **b.** 2.50 moles of NaCl formula units
c. 1.68 moles of N_2 molecules   **d.** 1.68 moles of N atoms

SOLUTION

We will use dimensional analysis to solve each of these problems. The relationship between molar mass and atomic or formula mass will serve as a conversion factor in the setup of each problem.

a. The given quantity is 1.50 moles of CH_4 molecules, and the desired quantity is grams of CH_4. Thus,

$$1.50 \text{ moles } CH_4 = ? \text{ g } CH_4$$

The formula mass of CH_4 is calculated to be 16.05 amu.

$$
\begin{array}{rl}
C: & 1 \times 12.01 \text{ amu} = 12.01 \text{ amu} \\
4 \text{ H}: & 4 \times 1.01 \text{ amu} = \underline{4.04 \text{ amu}} \\
& \text{formula mass} = 16.05 \text{ amu}
\end{array}
$$

Using the formula mass, we can write the equality

$$16.05 \text{ g } CH_4 = 1 \text{ mole } CH_4$$

The dimensional analysis setup for the problem with the gram-to-mole equation used as a conversion factor is

$$1.50 \text{ moles } CH_4 \times \frac{16.05 \text{ g } CH_4}{1 \text{ mole } CH_4} = 24.075 \text{ g } CH_4 \quad \text{(calculator answer)}$$

$$= 24.1 \text{ g } CH_4 \quad \textbf{(correct answer)}$$

b. The given quantity is 2.50 moles of NaCl formula units and the desired quantity is grams of NaCl.

$$2.50 \text{ moles NaCl} = ? \text{ g NaCl}$$

The calculated formula mass of NaCl is 58.44 amu. Thus,

$$58.44 \text{ g NaCl} = 1 \text{ mole NaCl}$$

With this relationship as a conversion factor, the setup for the problem becomes

$$2.50 \text{ moles NaCl} \times \frac{58.44 \text{ g NaCl}}{1 \text{ mole NaCl}} = 146.1 \text{ g NaCl} \quad \text{(calculator answer)}$$

$$= 146 \text{ g NaCl} \quad \textbf{(correct answer)}$$

c. The given quantity is 1.68 moles of N_2 molecules. The desired quantity is grams of N_2 molecules. Thus,

$$1.68 \text{ moles } N_2 = ? \text{ g } N_2$$

We are dealing here with diatomic nitrogen molecules (N_2) and not nitrogen atoms. Thus, 28.02 amu, twice the atomic mass of nitrogen, is the formula mass used in the mole-to-gram statement.

$$28.02 \text{ g } N_2 = 1 \text{ mole } N_2$$

With this relationship as a conversion factor, the setup becomes

$$1.68 \; \cancel{\text{moles } N_2} \times \frac{28.02 \text{ g } N_2}{1 \; \cancel{\text{mole } N_2}} = 47.0736 \text{ g } N_2 \quad \text{(calculator answer)}$$

$$= 47.1 \text{ g } N_2 \quad \text{(correct answer)}$$

d. The given quantity is 1.68 moles of N atoms, and the desired quantity is grams of N atoms. Thus,

$$1.68 \text{ moles } N = ? \text{ g } N$$

This problem differs from the previous one in that atoms, rather than molecules, of nitrogen are being counted. The atomic mass of nitrogen is 14.01 amu, and the mole-to-gram equality statement is

$$14.01 \text{ g } N = 1 \text{ mole } N$$

With this relationship as a conversion factor, the setup becomes

$$1.68 \; \cancel{\text{moles } N} \times \frac{14.01 \text{ g } N}{1 \; \cancel{\text{mole } N}} = 23.5368 \text{ g } N \quad \text{(calculator answer)}$$

$$= 23.5 \text{ g } N \quad \text{(correct answer)}$$

Notice that the mass of 1.68 moles of N_2 [part (c)] is twice the mass of 1.68 moles of N [part (d)]. This is as expected; there are twice as many atoms of N in 1.68 moles of N_2 as in 1.68 moles of N.

Answer Double Check:

Are the answers reasonable in terms of their magnitude? Yes. All four of the given molar amounts are roughly equal to 2 moles. The mass of 2 moles of a substance, in grams, should be double the molar mass of the substance. In all four parts of this problem this approximate two-to-one mass relationship holds.

▶ **Practice Exercise 9.7**   Calculate the mass, in grams, of each of the following molar quantities.

   a. 3.02 moles of carbon monoxide (CO) molecules.
   b. 3.02 moles of iron (Fe) atoms.

___

Summary of Mass Terminology

In a chemical context, *molar mass* is the fourth "mass term" that we have encountered. The previous three chemical contexts for mass were *atomic mass* (Sec. 5.4), *isotopic mass* (Sec. 5.4), and *formula mass* (Sec. 9.2). Contrasting the different meanings for these four terms is a useful summary for our discussion of molar mass.

   1. **Atomic mass:** The average mass of naturally occurring isotopes of an element. The unit for such masses is amu.
   2. **Isotopic mass:** The mass of a particular isotope of an element. The unit for such masses is also amu. Isotopic masses are used to calculate an atomic mass.
   3. **Formula mass:** The sum of the atomic masses of the atoms present in a formula unit of a chemical substance. The amu unit also is used in conjunction with this type of mass.
   4. **Molar mass:** The mass, in grams, of one mole of atoms, molecules, or formula units of a chemical substance. Molar mass and formula mass have the same numerical value. Units, are however, different for the two quantities. The unit for molar mass is grams and that for formula mass is amu. More specifically, the unit

for molar mass is actually grams/mole since by definition molar mass involves one mole of a chemical substance. Example 9.8 expands, in a problem-solving context, on the concept that the units for molar mass are actually grams/mole.

EXAMPLE 9.8 | **Calculating Molar Mass from Mass and Mole Data**

Calculate the molar mass of ammonia, a gaseous substance with a sharp odor, given that 0.2310 mole of ammonia weighs 3.936 grams.

SOLUTION

The chemical formula of ammonia was not given in the problem statement. It is not needed to do the molar mass calculation. The given data is sufficient to calculate the molar mass.

The units for molar mass are grams per mole. Both of these quantities are known. Thus, the given gram amount is divided by the given mole amount to give grams per mole.

$$\text{molar mass} \left(\frac{\text{grams}}{\text{mole}} \right) = \frac{3.936 \text{ g}}{0.2310 \text{ mole}}$$

$$= 17.038961 \frac{\text{g}}{\text{mole}} \text{ (calculator answer)}$$

$$= 17.04 \frac{\text{g}}{\text{mole}} \quad \textbf{(correct answer)}$$

Calculating the molar mass of a compound in this manner is similar to the manner in which density is calculated (Sec. 3.8). Density is calculated by dividing mass by volume. Molar mass is calculated by dividing mass by moles.

The chemical formula for ammonia is known to be NH_3. This chemical formula can be used as a double check on the preceding calculation. The formula mass of NH_3 (Sec. 9.2) is

$$\text{formula mass} = (\text{atomic mass N}) + 3(\text{atomic mass H})$$

$$= 14.01 \text{ amu} + 3(1.01 \text{ amu})$$

$$= 17.04 \text{ amu (calculator and correct answer)}$$

The molar mass of NH_3 will be, therefore (Sec 9.2), 17.04 g/mole, which is the same value that was calculated using mass/mole data rather than the chemical formula.

▶ **Practice Exercise 9.8** Calculate the molar mass of acetaminophen, a mild non-prescription pain reliever, given that 0.200 mole of acetaminophen weighs 30.24 grams.

9.7 SIGNIFICANT FIGURES AND AVOGADRO'S NUMBER

Now that the mass of a molar amount of a substance has been considered (Sec. 9.6), let us revisit the concept of Avogadro's number.

In Section 9.5 we defined the mole simply as

$$1 \text{ mole} = 6.022 \times 10^{23} \text{ objects}$$

Although this statement conveys correct information (the value of Avogadro's number to four significant figures is 6.022×10^{23}), it is not the officially accepted definition for Avogadro's number. The official definition, which is mass-based, is that the **mole** *is the amount of a substance that contains as many particles (atoms, molecules, or formula units) as there are* $^{12}_{6}C$ *atoms in 12.000000 grams of* $^{12}_{6}C$. The value of Avogadro's number is an experimentally determined quantity (the number of atoms in 12.000000 g of $^{12}_{6}C$ atoms)

One mole of a substance represents a *fixed number* of chemical entities and has a *fixed mass*. It is more than just a counting unit (Sec. 9.5) that specifies the number of objects. The official definition of the mole specifies the number of objects in a *fixed mass* of a substance.

rather than an exactly defined quantity. Its value is not even mentioned in the definition. The most up-to-date experimental value for Avogadro's number is 6.0221415×10^{23}, which is consistent with our previous definition. In calculations we will never need a value with as many significant figures as the experimentally determined one.

In problem solving, conversion factors that involve Avogadro's number should never be the "limiting factor" in significant figure considerations. To ensure that this is the case, our approach will be to carry more digits than required by the data in Avogadro's number conversion factors. Our operational rule will be to use the value 6.022×10^{23} for Avogadro's number in problem-solving situations. This will suffice in most situations. However, remember that more significant figures are available for use if the data ever require such.

9.8 RELATIONSHIP BETWEEN ATOMIC MASS UNITS AND GRAM UNITS

The atomic mass unit (amu) and the grams unit (g) are related to one another through Avogadro's number.

$$6.022 \times 10^{23} \text{ amu} = 1.000 \text{ g}$$

That this is the case can be deduced from the following equalities:

$$\text{atomic mass of N} = \text{mass of 1 N atom} = 14.01 \text{ amu}$$

$$\text{molar mass of N} = \text{mass of } 6.022 \times 10^{23} \text{ N atoms} = 14.01 \text{ g}$$

Because the second equality involves 6.022×10^{23} times as many atoms as the first equality and the masses come out numerically equal, the gram unit must be 6.022×10^{23} times larger than the amu unit.

Using a dimensional analysis setup, the relationship

$$6.022 \times 10^{23} \text{ amu} = 1.000 \text{ g}$$

can be derived from the preceding data on nitrogen in the following manner:

$$\frac{6.022 \times 10^{23} \text{ atoms N}}{1 \text{ mole N}} \times \frac{1 \text{ mole N}}{14.01 \text{ g N}} \times \frac{14.01 \text{ amu}}{1 \text{ atom N}}$$

$$= 6.022 \times 10^{23} \text{ amu/g} \quad \text{(calculator and correct answer)}$$

The first conversion factor is based on the number of atoms in 1 mole of N, the second on the molar mass of N, and the third on the atomic mass of N.

EXAMPLE 9.9 **Using the Relationship between Grams and Atomic Mass Units as a Conversion Factor**

What is the mass, in grams, of a molecule whose mass on the amu scale is 104.0 amu?

SOLUTION

This is a one-step problem based on the conversion factor

$$6.022 \times 10^{23} \text{ amu} = 1.000 \text{ g}$$

The dimensional analysis setup is

$$104.0 \text{ amu} \times \frac{1.000 \text{ g}}{6.022 \times 10^{23} \text{ amu}} = 1.7270009 \times 10^{-22} \text{ g} \quad \text{(calculator answer)}$$

$$= 1.727 \times 10^{-22} \text{ g} \quad \text{(correct answer)}$$

▶ **Practice Exercise 9.9** What is the mass, in grams, of an atom whose mass on the amu scale is 22.99 amu?

9.9 THE MOLE AND CHEMICAL FORMULAS

A chemical formula has two meanings or interpretations: (1) a microscopic-level interpretation and (2) a macroscopic-level interpretation.

The first of these two interpretations was discussed in Section 4.13. At the *microscopic level* a chemical formula indicates the number of atoms of each element present in one molecule or formula unit of a substance. The subscripts in the formula are interpreted to mean the numbers of atoms of the various elements present in one unit of the substance. The formula C_2H_6, interpreted at the microscopic level, conveys the information that two atoms of C and six atoms of H are present in one molecule of C_2H_6.

Now that the mole concept has been introduced, a macroscopic interpretation of formulas is possible. At the *macroscopic level* a chemical formula indicates the number of moles of atoms of each element present in one mole of a substance. The subscripts in the formula are interpreted to mean the numbers of moles of atoms of the various elements present in one mole of the substance. The designation "macroscopic" is given to this molar interpretation, since moles are laboratory-sized quantities of atoms. The formula C_2H_6, interpreted at the macroscopic level, conveys the information that two moles of C atoms and six moles of H atoms are present in one mole of C_2H_6.

It is now evident, then, that the subscripts in a formula always carry a dual meaning: "atom" at the microscopic level and "moles of atoms" at the macroscopic level.

The validity of the molar interpretation for subscripts in a formula derives from the following line of reasoning. In x molecules of C_2H_6, where x is any number, there are $2x$ atoms of C and $6x$ atoms of H. Regardless of the value of x, there must always be two times as many C atoms as molecules and six times as many H atoms as molecules; that is,

$$\text{number of } C_2H_6 \text{ molecules} = x$$
$$\text{number of C atoms} = 2x$$
$$\text{number of H atoms} = 6x$$

Now let x equal 6.022×10^{23}, the value of Avogadro's number. With this x value, the following statements are true.

$$\text{number of } C_2H_6 \text{ molecules} = 6.022 \times 10^{23}$$
$$\text{number of C atoms} = 2 \times 6.022 \times 10^{23} = 1.2044 \times 10^{24} \text{ (calculator answer)}$$
$$= 1.204 \times 10^{24} \text{ (correct answer)}$$
$$\text{number of H atoms} = 6 \times 6.022 \times 10^{23} = 3.6132 \times 10^{24} \text{ (calculator answer)}$$
$$= 3.613 \times 10^{24} \text{ (correct answer)}$$

Since 6.022×10^{23} is equal to 1 mole, 1.204×10^{24} to 2 moles, and 3.613×10^{24} to 6 moles, these statements may be changed to read

$$\text{number of } C_2H_6 \text{ molecules} = 1 \text{ mole}$$
$$\text{number of C atoms} = 2 \text{ moles}$$
$$\text{number of H atoms} = 6 \text{ moles}$$

Thus, the mole ratio is the same as the subscript ratio: 2:6.

In calculations where the moles of a particular element within a compound are asked for, the subscript of that particular element in the chemical formula of the compound becomes part of the conversion factor used to convert from moles of compound to moles of element within the compound.

$$\boxed{\text{Moles of compound}} \xrightarrow[\text{subscript}]{\text{Formula}} \boxed{\text{Moles of element within compound}}$$

For example, again using C_2H_6 as our chemical formula, we can write the following conversion factors.

For C: $\dfrac{2 \text{ moles C atoms}}{1 \text{ mole } C_2H_6 \text{ molecules}}$ or $\dfrac{1 \text{ mole } C_2H_6 \text{ molecules}}{2 \text{ moles C atoms}}$

For H: $\dfrac{6 \text{ moles H atoms}}{1 \text{ mole } C_2H_6 \text{ molecules}}$ or $\dfrac{1 \text{ mole } C_2H_6 \text{ molecules}}{6 \text{ moles H atoms}}$

Example 9.10 illustrates the use of conversion factors of this type in a problem-solving context.

EXAMPLE 9.10   Calculating Molar Quantities of Compound Components

The characteristic odor of pineapple is due to ethyl butyrate, a compound with the formula $C_6H_{12}O_2$. How many moles of each type of atom present in $C_6H_{12}O_2$ are contained in a 2.65 mole sample of this compound?

SOLUTION

The formula $C_6H_{12}O_2$ specifies that 1 mole of this substance will contain 6 moles of carbon atoms, 12 moles of hydrogen atoms, and 2 moles of oxygen atoms. This information leads to the following conversion factors:

$$\dfrac{6 \text{ moles C atoms}}{1 \text{ mole } C_6H_{12}O_2} \qquad \dfrac{12 \text{ moles H atoms}}{1 \text{ mole } C_6H_{12}O_2} \qquad \dfrac{2 \text{ moles O atoms}}{1 \text{ mole } C_6H_{12}O_2}$$

Using the first of these conversion factors, the moles of C atoms present are calculated as follows:

$$2.65 \text{ moles } C_6H_{12}O_2 \times \dfrac{6 \text{ moles C atoms}}{1 \text{ mole } C_6H_{12}O_2} = 15.9 \text{ moles C atoms}$$

(calculator and **correct answer**)

Similarly, using the second conversion factor, the moles of H atoms present are calculated.

$$2.65 \text{ moles } C_6H_{12}O_2 \times \dfrac{12 \text{ moles H atoms}}{1 \text{ mole } C_6H_{12}O_2} = 31.8 \text{ moles H atoms}$$

(calculator and **correct answer**)

Finally, using the third conversion factor, we obtain the moles of O atoms present.

$$2.65 \text{ moles } C_6H_{12}O_2 \times \dfrac{2 \text{ moles O atoms}}{1 \text{ mole } C_6H_{12}O_2} = 5.3 \text{ moles O atoms} \quad \text{(calculator answer)}$$

$$= 5.30 \text{ moles O atoms} \quad \textbf{(correct answer)}$$

If the question "How many total moles of atoms are present?" were asked, we could obtain the answer by adding the moles of C, H, and O atoms just calculated.

$$(15.9 + 31.8 + 5.30) \text{ moles atoms} = 53 \text{ moles atoms} \quad \text{(calculator answer)}$$
$$= 53.0 \text{ moles atoms} \quad \textbf{(correct answer)}$$

Alternatively, by noting that there are a total of 20 moles of atoms present in 1 mole of $C_6H_{12}O_2$ (the sum of the subscripts in the formula), we could calculate the total moles of atoms present, using the following setup:

$$2.65 \text{ moles } \cancel{C_6H_{12}O_2} \times \frac{20 \text{ moles of atoms}}{1 \text{ mole } \cancel{C_6H_{12}O_2}} = 53 \text{ moles of atoms}$$
$$= 53.0 \text{ moles of atoms} \quad \textbf{(correct answer)}$$

Answer Double Check:

The relationships among the three answers should be the same as the relationships among the subscripts in the given chemical formula. Is such the case? Yes. The second answer should be double the first one (subscripts 12 and 6), which it is (31.8 g and 15.9 g), and the third answer should be one-third the first answer (subscripts 2 and 6), which it is (5.30 g and 15.9 g).

▶ **Practice Exercise 9.10** How many moles of each type of atom are present in each of the following molar quantities?

    a. 2.043 moles of H_2O molecules
    b. 2.043 moles of H_2O_2 molecules

      Table 9.2 serves as a summary of the mole relationships we have considered up to this point in the chapter. Note, through comparison of the third and fourth entries in Table 9.2 that the same-size molar quantity of a given type of atom and its monoatomic ion are considered to have the same mass. This is because the mass of electrons is so very small (Sec. 5.1).

TABLE 9.2 Mole Relationships

Name	Formula	Formula Mass (amu)	Mass of 1 Mole Formula Units (g)	Number and Kind of Particles in 1 Mole
Atomic nitrogen	N	14.01	14.01	6.022×10^{23} N atoms
Molecular nitrogen	N_2	28.02	28.02	$\begin{cases} 6.022 \times 10^{23} \ N_2 \text{ molecules} \\ 2(6.022 \times 10^{23}) \text{ N atoms} \end{cases}$
Zinc	Zn	65.38	65.38	6.022×10^{23} Zn atoms
Zinc ions	Zn^{2+}	65.38*	65.38	$6.022 \times 10^{23} \ Zn^{2+}$ ions
Calcium chloride	$CaCl_2$	110.98	110.98	$\begin{cases} 6.022 \times 10^{23} \ CaCl_2 \text{ units} \\ 6.022 \times 10^{23} \ Ca^{2+} \text{ ions} \\ 2(6.022 \times 10^{23}) \ Cl^- \text{ ions} \end{cases}$

Recall that the electron has negligible mass; thus ions and atoms have essentially the same mass.

9.10 THE MOLE AND CHEMICAL CALCULATIONS

In this section we combine the major points we have learned about moles in previous sections to produce a general approach to problem solving that is applicable to a variety of types of chemical calculations.

The three quantities most often calculated in chemical problems are

1. The number of *particles* of a substance, that is, the number of atoms, molecules, or formula units.
2. The number of *moles* of a substance.
3. The number of *grams* (mass) of a substance.

These quantities are interrelated. The conversion factors dealing with these relationships, as previously noted, involve the concepts of (1) Avogadro's number, (2) molar mass, and (3) molar interpretation of chemical formula subscripts.

1. Avogadro's number (Sec. 9.5) provides a relationship between the number of particles of a substance and the number of moles of the same substance.

| Particles of substance | ←Avogadro's number→ | Moles of substance |

2. Molar mass (Sec. 9.6) provides a relationship between the number of grams of a substance and the number of moles of the same substance.

| Grams of substance | ←Molar mass→ | Moles of substance |

3. Molar interpretation of chemical formula subscripts (Sec. 9.9) provides a relationship between the number of moles of a substance and the number of moles of its component parts.

| Moles of compound | ←Formula subscript→ | Moles of element within compound |

The preceding three concepts can be combined into a single diagram that is very useful in problem solving. This diagram, Figure 9.5, can be viewed as a road map from which conversion factor sequences (pathways) can be obtained. It gives all the needed relationships for solving two general types of problems.

1. Calculations for which information (moles, grams, particles) is given about a particular substance and additional information (moles, grams, particles) is needed concerning the *same* substance.
2. Calculations for which information (moles, grams, particles) is given about a particular substance and information (moles, grams, particles) is needed concerning a *component* of that same substance.

FIGURE 9.5 Useful relationships for solving chemical formula–based problems.

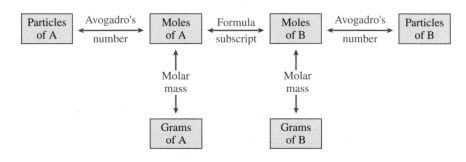

For the first type of problem, only the left side of Figure 9.5 (the A boxes) is needed. For problems of the second type, both sides of the diagram (both A and B boxes) are used.

The thinking pattern needed to use Figure 9.5 is very simple.

1. Determine which box in the diagram represents the *given* quantity in the problem.
2. Next, locate the box that represents the *desired* quantity.
3. Finally, follow the indicated pathway that takes you from the *given* quantity to the *desired* quantity. This involves simply following the arrows. There will always be only one pathway possible for the needed transition.

Examples 9.11 to 9.15 illustrate a few of the types of problems that can be solved using the relationships shown in Figure 9.5. In the first three examples we will need only the A side of the diagram; the following two examples make use of both the A and B sides of the diagram.

EXAMPLE 9.11 | **Calculating the Number of Particles in a Given Mass of Substance**

Nicotine, the second most widely used central nervous system stimulant in our society (caffeine is first), occurs naturally in tobacco leaves. Its chemical formula is $C_{10}H_{14}N_2$. How many nicotine molecules are present in a 0.0015 g sample of nicotine (a typical amount in a cigarette)?

SOLUTION

We will solve this problem by using the three steps of dimensional analysis (Sec. 3.7) and Figure 9.5.

Example 9.11 illustrates the important concept that Avogadro's number is a means for relating macroscopic scale (laboratory scale) measurements to microscopic scale (atomic scale) measurements or vice versa.

Step 1 The given quantity is 0.0015 g of $C_{10}H_{14}N_2$, and the desired quantity is molecules of $C_{10}H_{14}N_2$.

$$0.0015 \text{ g } C_{10}H_{14}N_2 = ? \text{ molecules } C_{10}H_{14}N_2$$

In terms of Figure 9.5, this is a grams-of-A to particles-of-A problem. We are given grams of substance A and desire to find particles (molecules) of that same substance.

Step 2 Figure 9.5 gives us the pathway (sequence of conversion factors) needed to work the problem. We want to start in the "grams of A" box and end up in the "particles of A" box. The pathway is

$$\boxed{\begin{array}{c} \text{Grams} \\ \text{of A} \end{array}} \xrightarrow[\text{mass}]{\text{Molar}} \boxed{\begin{array}{c} \text{Moles} \\ \text{of A} \end{array}} \xrightarrow[\text{number}]{\text{Avogadro's}} \boxed{\begin{array}{c} \text{Particles} \\ \text{of A} \end{array}}$$

Using dimensional analysis, the setup for this sequence of conversion factors is

$$0.0015 \text{ g } C_{10}H_{14}N_2 \times \frac{1 \text{ mole } C_{10}H_{14}N_2}{162.26 \text{ g } C_{10}H_{14}N_2} \times \frac{6.022 \times 10^{23} \text{ molecules } C_{10}H_{14}N_2}{1 \text{ mole } C_{10}H_{14}N_2}$$

$$\text{grams A} \longrightarrow \text{moles A} \longrightarrow \text{particles A}$$

The number 162.26 that was used in the first conversion factor is the formula mass of $C_{10}H_{14}N_2$. It is not given in the problem, but had to be calculated using atomic masses and the operational rules given in Section 9.3.

Step 3 The solution to the problem, obtained by doing the arithmetic, is

$$\frac{0.0015 \times 1 \times 6.022 \times 10^{23}}{162.26 \times 1} \text{ molecules } C_{10}H_{14}N_2$$

$$= 5.5669912 \times 10^{18} \text{ molecules } C_{10}H_{14}N_2 \quad \text{(calculator answer)}$$

$$= 5.6 \times 10^{18} \text{ molecules } C_{10}H_{14}N_2 \quad \textbf{(correct answer)}$$

Answer Double Check:

Does the answer have the correct magnitude? Yes. Since the given mass (approximately 10^{-3} g) is less than the molar mass (approximately 10^2 g) by a factor of 10^5, the number of molecules present should be less than Avogadro's number by a factor of 10^5. The power of ten in the answer, 10^{18}, is consistent with this answer $(10^{23}/10^5 = 10^{18})$.

▶ **Practice Exercise 9.11** The compound 1-propanethiol, which is the eye irritant released when fresh onions are chopped up, has the chemical formula C_3H_8S. How many molecules of this irritant compound are present in a 5.000 g sample of the compound?

EXAMPLE 9.12 **Calculating the Mass of a Given Number of Particles of a Substance**

The starting material for the production of photochemical smog is the air pollutant nitrogen dioxide (NO_2). Its interaction with ultraviolet light from the sun produces the chemical species needed for smog production. What would be the mass, in grams, of an NO_2 sample in which one hundred billion (1.000×10^{11}) molecules are present?

SOLUTION

Step 1 The given quantity is 1.000×10^{11} NO_2 molecules. The desired quantity is grams of NO_2

$$1.000 \times 10^{11} \text{ } NO_2 \text{ molecules} = ? \text{ g } NO_2$$

In terms of Figure 9.5, this is a particles-of-A to grams-of-A problem.

Step 2 The pathway for this problem is the exact reverse of the one used in the previous example. We are given particles and asked to find grams of the same substance.

$$\boxed{\begin{array}{c}\text{Particles}\\\text{of A}\end{array}} \xrightarrow[\text{number}]{\text{Avogadro's}} \boxed{\begin{array}{c}\text{Moles}\\\text{of A}\end{array}} \xrightarrow[\text{mass}]{\text{Molar}} \boxed{\begin{array}{c}\text{Grams}\\\text{of A}\end{array}}$$

Using dimensional analysis, the setup is

$$1.000 \times 10^{11} \text{ } \cancel{NO_2 \text{ molecules}} \times \frac{1 \text{ } \cancel{\text{mole } NO_2}}{6.022 \times 10^{23} \text{ } \cancel{NO_2 \text{ molecules}}} \times \frac{46.01 \text{ g } NO_2}{1 \text{ } \cancel{\text{mole } NO_2}}$$

$$\text{particles A} \longrightarrow \text{moles A} \longrightarrow \text{grams A}$$

The molar mass of NO_2, 46.01 g, which was used in the second conversion factor, was calculated from the atomic masses of nitrogen and oxygen. Avogadro's number in the first conversion factor, specified to four significant figures, is consistent with the given number 1.000×10^{11} having four significant figures.

Step 3 The final answer is obtained by doing the arithmetic.

$$\frac{1.000 \times 10^{11} \times 1 \times 46.01}{6.022 \times 10^{23} \times 1} \text{ g NO}_2$$

$$= 7.6403118 \times 10^{-12} \text{ g NO}_2 \quad \text{(calculator answer)}$$

$$= 7.640 \times 10^{-12} \text{ g NO}_2 \quad \text{(correct answer)}$$

Answer Double Check:

Is the magnitude of the answer reasonable? Yes. The given number of molecules differs from Avogadro's number by a factor of approximately 10^{-12} ($10^{11}/10^{23} = 10^{-12}$). The actual mass and the molar mass should differ by the same factor. The calculated mass of 10^{-12} g is consistent with a molar mass of 46.0 g.

▶ **Practice Exercise 9.12** How many grams of methane (CH_4), a greenhouse gas that contributes to global warming, are present in a methane sample that contains six billion (6.00×10^9) molecules? (Six billion is the approximate human population of Earth.)

EXAMPLE 9.13 | **Calculating the Mass of an Atom or a Molecule**

Calculate the mass, in grams, of each of the following chemical entities.

   a. a single atom of Cu   **b.** a single molecule of CO_2

SOLUTION

  a.

Step 1 The given quantity is 1 atom of Cu, and the desired quantity is grams of Cu.

$$1 \text{ atom Cu} = ? \text{ g Cu}$$

In terms of Figure 9.5, this is a particles-of-A to grams-of-A problem.

Step 2 The pathway for this problem is identical to that used in the previous example. We are given particles (one particle in this case), and asked to find grams of the same substance.

$$\boxed{\text{Particles of A}} \xrightarrow[\text{number}]{\text{Avogadro's}} \boxed{\text{Moles of A}} \xrightarrow[\text{mass}]{\text{Molar}} \boxed{\text{Grams of A}}$$

Using dimensional analysis, the setup is

$$1 \text{ atom Cu} \quad \times \quad \frac{1 \text{ mole Cu}}{6.022 \times 10^{23} \text{ atoms Cu}} \quad \times \quad \frac{63.55 \text{ g Cu}}{1 \text{ mole Cu}}$$

$$\text{particles A} \quad \longrightarrow \quad \text{moles A} \quad \longrightarrow \quad \text{grams A}$$

The second conversion factor contains the atomic mass of Cu, which was obtained from the periodic table.

Step 3 The solution to the problem, obtained by doing the arithmetic, is

$$\frac{1 \times 1 \times 63.55}{6.022 \times 10^{23} \times 1} \text{ g Cu} = 1.0552972 \times 10^{-22} \text{ g Cu} \quad \text{(calculator answer)}$$

$$= 1.055 \times 10^{-22} \text{ g Cu} \quad \text{(correct answer)}$$

The mass of a single atom is, indeed, a very small mass in the unit of grams—0.0000000000000000000001055 g. Commonly used analytical balances are capable of weighing to ±0.0001 g. The mass of a single atom can be calculated but, obviously, it cannot be determined using an analytical balance.

b.

Step 1 The given quantity is 1 molecule of CO_2 and the desired quantity is grams of CO_2.

$$1 \text{ molecule } CO_2 = ? \text{ g } CO_2$$

In terms of Figure 9.5, this problem, like part (a), is a particles-of-A to grams-of-A problem.

Step 2 We are dealing with an identical setup to that in part (a), except that a molecule rather than an atom is the particle of concern.

Particles of A	Avogadro's number	Moles of A	Molar mass	Grams of A

Using dimensional analysis, the conversion factor setup becomes

$$1 \text{ molecule } CO_2 \quad \times \quad \frac{1 \text{ mole } CO_2}{6.022 \times 10^{23} \text{ molecules } CO_2} \quad \times \quad \frac{44.01 \text{ g } CO_2}{1 \text{ mole } CO_2}$$

$$\text{particles A} \quad \longrightarrow \quad \text{moles A} \quad \longrightarrow \quad \text{grams A}$$

The second conversion factor involves the molar mass of CO_2 which is calculated using the atomic masses of carbon and oxygen obtained from the periodic table.

Step 3 Collecting the conversion factors and doing the arithmetic gives

$$\frac{1 \times 1 \times 44.01}{6.022 \times 10^{23} \times 1} \text{ g } CO_2 = 7.3082032 \times 10^{-23} \text{ g } CO_2 \quad \text{(calculator answer)}$$

$$= 7.308 \times 10^{-23} \text{ g } CO_2 \quad \textbf{(correct answer)}$$

Note that a Cu atom [part (a)] weighs more than a CO_2 molecule [part (b)]. The atoms present in a CO_2 molecule are small atoms when compared to a Cu atom.

Answer Double Check:

Do the answers have a magnitude that is reasonable? Yes. One amu is equivalent to approximately 10^{-24} g (the reciprocal of Avogadro's number). Since most atoms and molecules have masses of 100 amu or less, their masses in grams should be in the 10^{-22} to 10^{-24} g range. The answers, 10^{-22} and 10^{-23} g, fall within this approximate range.

▶ **Practice Exercise 9.13** Calculate the mass, in grams, of the following chemical entities.

a. a single atom of Au (gold)
b. a single molecule of N_2H_4 (hydrazine)

EXAMPLE 9.14 **Calculating the Mass of an Element Present in a Given Mass of Compound**

Caffeine, the stimulant in coffee and tea, has the formula $C_8H_{10}N_4O_2$. How many grams of nitrogen are present in a 50.0 g sample of caffeine?

SOLUTION

Step 1 There is an important difference between this problem and the preceding three; here we are dealing with not one but two substances, caffeine and nitrogen. The given quantity is grams of caffeine (substance A), and we are asked to find the grams of nitrogen (substance B). This is a grams-of-A to grams-of-B problem.

$$50.0 \text{ g } C_8H_{10}N_4O_2 = ? \text{ g } N$$

Step 2 The appropriate set of conversions for a grams-of-A to grams-of-B problem, from Figure 9.5, is

$$\boxed{\begin{array}{c}\text{Grams}\\\text{of A}\end{array}}\xrightarrow[\text{mass}]{\text{Molar}}\boxed{\begin{array}{c}\text{Moles}\\\text{of A}\end{array}}\xrightarrow[\text{subscript}]{\text{Formula}}\boxed{\begin{array}{c}\text{Moles}\\\text{of B}\end{array}}\xrightarrow[\text{mass}]{\text{Molar}}\boxed{\begin{array}{c}\text{Grams}\\\text{of B}\end{array}}$$

The mathematical setup involving conversion factors is

$$50.0 \text{ g } C_8H_{10}N_4O_2 \times \underbrace{\frac{1 \text{ mole } C_8H_{10}N_4O_2}{194.22 \text{ g } C_8H_{10}N_4O_2}}_{\text{moles A}} \times \underbrace{\frac{4 \text{ moles N}}{1 \text{ mole } C_8H_{10}N_4O_2}}_{\text{moles B}} \times \underbrace{\frac{14.01 \text{ g N}}{1 \text{ mole N}}}_{\text{grams B}}$$

$$\underbrace{}_{\text{grams A}} \longrightarrow \longrightarrow \longrightarrow$$

The number 194.22 that is used in the first conversion factor is the formula mass for caffeine. The conversion from moles-of-A to moles-of-B (the second conversion factor) is derived using the information contained in the formula for caffeine. One mole of caffeine contains 4 moles of N because the subscript for N in the formula is 4. The number 14.01 in the final conversion factor is the atomic mass of nitrogen.

Step 3 Collecting the numbers from the various conversion factors together and doing the arithmetic gives us our answer.

$$\frac{50.0 \times 1 \times 4 \times 14.01}{194.22 \times 1 \times 1} \text{ g N} = 14.426938 \text{ g N} \quad \text{(calculator answer)}$$

$$= 14.4 \text{ g N} \quad \quad \textbf{(correct answer)}$$

Answer Double Check:

Is the magnitude of the answer reasonable? Yes. The mass of the entire sample is 50.0 g. The nitrogen content of the sample should be much less than the total mass. Such is the case; the N content is 14.4 g.

▶ **Practice Exercise 9.14**   Citric acid, the substance responsible for the sour taste of citrus fruits, has the chemical formula $C_6H_8O_7$. How many grams of carbon are present in a 25.0 g sample of citric acid?

EXAMPLE 9.15   **Calculating the Number of Particles of an Element Present in a Given Mass of a Compound**

The compound cholesterol has the chemical formula $C_{27}H_{46}O$. How many hydrogen atoms would be present in a 2.000 g sample of cholesterol?

SOLUTION

Step 1 The given quantity is 2.000 g of $C_{27}H_{46}O$ (cholesterol) and the desired quantity is atoms of hydrogen.

$$2.000 \text{ g } C_{27}H_{46}O = ? \text{ atoms H}$$

In the jargon of Figure 9.5, this is a grams-of-A to particles-of-B problem.

Step 2 From Figure 9.5, the appropriate pathway for solving this problem is

$$\boxed{\begin{array}{c}\text{Grams}\\\text{of A}\end{array}}\xrightarrow[\text{mass}]{\text{Molar}}\boxed{\begin{array}{c}\text{Moles}\\\text{of A}\end{array}}\xrightarrow[\text{subscript}]{\text{Formula}}\boxed{\begin{array}{c}\text{Moles}\\\text{of B}\end{array}}\xrightarrow[\text{number}]{\text{Avogadro's}}\boxed{\begin{array}{c}\text{Particles}\\\text{of B}\end{array}}$$

Using dimensional analysis, the conversion factor setup becomes

$$2.000 \text{ g } C_{27}H_{46}O \times \frac{1 \text{ mole } C_{27}H_{46}O}{386.73 \text{ g } C_{27}H_{46}O} \times \frac{46 \text{ moles } H}{1 \text{ mole } C_{27}H_{46}O} \times \frac{6.022 \times 10^{23} \text{ atoms H}}{1 \text{ mole H}}$$

grams A $\longrightarrow$ moles A $\longrightarrow$ moles B $\longrightarrow$ particles B

The numbers in the second conversion factor (46 and 1) were obtained from the chemical formula for cholesterol; they are, thus, exact numbers. One mole of $C_{27}H_{46}O$ contains 46 moles of hydrogen; the hydrogen subscript is 46.

Step 3 The solution to the problem, obtained by doing the arithmetic, is

$$\frac{2.000 \times 1 \times 46 \times 6.022 \times 10^{23}}{386.73 \times 1 \times 1} \text{ atoms H} = 1.432586 \times 10^{23} \text{ atoms H}$$

(calculator answer)

$$= 1.433 \times 10^{23} \text{ atoms H}$$

(**correct answer**)

▶ **Practice Exercise 9.15** Octane, a component of gasoline, has the chemical formula C_8H_{18}. How many hydrogen atoms are present in a 1.00 g sample of octane?

9.11 PURITY OF SAMPLES

Most substances used in chemical laboratories and most substances encountered in everyday situations contain impurities; that is, they are not 100% pure. Sometimes the impurities are naturally present substances, and other times impurities have been added on purpose to give a desired effect.

Drinking water contains naturally present minerals (impurities) that give drinking water its taste; distilled water, which lacks such minerals, has a very flat taste. Table salt contains potassium iodide as an additive (an impurity). This additive serves as an iodine source for the human body; iodine is necessary for the proper functioning of the thyroid gland.

The amount of impurities present in a sample of a substance is specified using *percent purity*. **Percent purity** *is the percent by mass of a specified substance in an impure sample of the substance.* For example, the purity of laboratory-grade sodium hydroxide (NaOH) is 98.2% by mass.

A general discussion of the use of percent as a conversion factor has been previously given. It is found in Section 3.10.

Percent purity values can be used as conversion factors in several types of problem-solving situations. The information that the purity of a sodium chloride (NaCl) sample is 99.23% by mass is a source for six conversion factors. Three of the conversion factors are

$$\frac{99.23 \text{ g NaCl}}{100 \text{ g impure NaCl}} \qquad \frac{0.77 \text{ g impurities}}{100 \text{ g impure NaCl}} \qquad \frac{0.77 \text{ g impurities}}{99.23 \text{ g NaCl}}$$

The reciprocals of these conversion factors are the other three of the six relationships.

EXAMPLE 9.16 **Calculating Masses of Substances in an Impure Sample**

A 32.00 g sample of nitric acid (HNO_3) has a purity of 96.20% by mass. Calculate the following for this sample of nitric acid.

a. the mass, in grams, of HNO_3 present
b. the mass, in grams, of impurities present

SOLUTION

a. From the given percent purity, the following needed conversion factor is obtained.

$$\frac{96.20 \text{ g HNO}_3}{100 \text{ g impure HNO}_3}$$

The dimensional analysis setup, using this conversion factor, is

$$32.00 \text{ g impure HNO}_3 \times \frac{96.20 \text{ g HNO}_3}{100 \text{ g impure HNO}_3}$$

Cancelling units and doing the arithmetic gives

$$\frac{32.00 \times 96.20}{100} \text{ g HNO}_3 = 30.784 \text{ g HNO}_3 \qquad \text{(calculator answer)}$$

$$= 30.78 \text{ g HNO}_3 \qquad \text{(\textbf{correct answer})}$$

b. Similarly for the amount of impurities present, we have the dimensional analysis setup

$$32.00 \text{ g impure HNO}_3 \times \frac{3.80 \text{ g impurities}}{100 \text{ g impure HNO}_3}$$

$$= 1.216 \text{ g impurities} \qquad \text{(calculator answer)}$$

$$= 1.22 \text{ g impurities} \qquad \text{(\textbf{correct answer})}$$

Alternatively, the mass, in grams, of impurities present can be determined using subtraction.

$$\text{total sample mass} - \text{HNO}_3 \text{ mass} = \text{impurity mass}$$

$$32.00 \text{ g} - 30.78 \text{ g} = 1.22 \text{ g (calculator and \textbf{correct answer})}$$

Answer Double Check:

Are the magnitudes of the answers reasonable? Yes. With a purity of 96.20%, the amount of nitric acid present should be almost the same as the total mass, which it is (30.78 g and 32.00 g). Also the sum of the nitric acid mass and the impurity mass should give the total mass, which is the case (30.78 g + 1.22 g = 32.00 g).

▶ **Practice Exercise 9.16**   A 40.00 g sample of phosphoric acid (H_3PO_4) has a purity of 97.32% by mass. Calculate the following for this sample of phosphoric acid:

a. the mass, in grams, of H_3PO_4 present
b. the mass, in grams, of impurities present

Percent purity, when appropriate, can be incorporated into grams-to-moles-to-particles problems like those previously considered in Section 9.10. Example 9.17 illustrates how this is done.

EXAMPLE 9.17   **Calculating Atoms of a Substance in an Impure Sample**

How many iron (Fe) atoms are present in a 25.00 g impure iron sample (iron ore) that has a purity of 87.70% by mass iron?

SOLUTION

If the iron sample were 100% pure, this calculation would be a standard "grams-of-A" to particles of A problem, in terms of Figure 9.5. The pathway would be

$$\boxed{\text{Grams of A}} \xrightarrow[\text{mass}]{\text{Molar}} \boxed{\text{Moles of A}} \xrightarrow[\text{number}]{\text{Avogadro's}} \boxed{\text{Particles of A}}$$

To take into account the fact that the sample is impure, we add a new step at the start of the pathway that involves grams-of-impure-A to grams-of-A. The enhanced pathway is

$$\boxed{\text{Grams of Impure A}} \xrightarrow[\text{purity}]{\text{Percent}} \boxed{\text{Grams of A}} \xrightarrow[\text{mass}]{\text{Molar}} \boxed{\text{Moles of A}} \xrightarrow[\text{number}]{\text{Avogadro's}} \boxed{\text{Particles of A}}$$

The dimensional analysis setup consistent with the new pathway is

$$25.00 \text{ g impure Fe} \times \frac{87.70 \text{ g Fe}}{100 \text{ g impure Fe}} \times \frac{1 \text{ mole Fe}}{55.85 \text{ g Fe}} \times \frac{6.022 \times 10^{23} \text{ atoms Fe}}{1 \text{ mole Fe}}$$

Collecting the numbers from the various conversion factors together and doing the arithmetic gives us our answer.

$$\frac{25.00 \times 87.70 \times 1 \times 6.022 \times 10^{23}}{100 \times 55.85 \times 1} \text{ atoms Fe} = 2.36405282 \times 10^{23} \text{ atoms Fe}$$

$$\text{(calculator answer)}$$
$$= 2.364 \times 10^{23} \text{ atoms}$$
$$\text{(\textbf{correct answer})}$$

Answer Double Check:

Is the magnitude of the answer reasonable? Yes. Rounding numbers for ease of analysis, we note that if the sample were pure iron, the sample mass of 25.00 g would be equal to a little less than one-half mole of iron (25.00 g/55.85 g). One-half mole of iron would contain one-half mole of iron atoms (3.01×10^{23} atoms, one-half of Avogadro's number). The answer, 2.36×10^{23} atoms, is smaller than this one-half number, as it should be.

▶ **Practice Exercise 9.17** Ethylene glycol, the major component of automotive antifreeze, has the chemical formula $C_2H_6O_2$. How many ethylene glycol molecules are present in a 22.45 g sample of ethylene glycol that has a purity of 87.6%, by mass?

9.12 EMPIRICAL AND MOLECULAR FORMULAS

Chemical formulas provide a great deal of useful information about the substances they represent and, as seen in previous sections of this chapter, are key entities in many types of calculations. How are chemical formulas themselves determined? They are determined by calculation using experimentally obtained information.

Depending on the amount of experimental information available, two types of chemical formulas may be obtained: an *empirical formula* or a *molecular formula*. The **empirical formula** (or *simplest formula*) *is a chemical formula that gives the* smallest *whole-number ratio of atoms present in a formula unit of a compound*. In empirical formulas the subscripts in the formula cannot be reduced to a simpler set of numbers by division with a small integer. The **molecular formula** (or *true formula*) *is a chemical formula that gives the* actual *number of atoms present in a formula unit of a compound*.

For ionic compounds the empirical and molecular formulas are almost always the same; that is, the actual ratio of ions present in a formula unit and the smallest ratio of

TABLE 9.3   A Comparison of Empirical and Molecular Formulas for Selected Compounds

Compound	Empirical Formula	Molecular Formula	Whole-Number Multiplier
Dinitrogen tetrafluoride	NF_2	N_2F_4	2
Hydrogen peroxide	HO	H_2O_2	2
Sodium chloride	$NaCl$	$NaCl$	1
Benzene	CH	C_6H_6	6

ions present are the same. For molecular compounds the two types of formulas may be the same, but frequently they are not. When they are not the same, the molecular formula is a multiple of the empirical formula.

$$\text{molecular formula} = \text{whole-number multiplier} \times \text{empirical formula}$$

Table 9.3 contrasts the differences between the two types of formulas for selected compounds.

EXAMPLE 9.18   Obtaining Empirical Formulas from Molecular Formulas

Write the empirical formula for each of the following compounds.

  a. C_2H_4 (ethene, used to make polyethylene)
  b. C_8H_{18} (octane, a component of gasoline)
  c. CH_4O (methanol, an industrial solvent)
  d. $C_6H_{12}O_6$ (glucose, often called blood sugar)

SOLUTION

  a. Each of the formula subscripts is divided by 2 to give the empirical formula CH_2.
  b. Each of the formula subscripts is divided by 2 to give the empirical formula C_4H_9.
  c. There is no small number that will divide evenly into each of the formula subscripts other than 1. Thus, the empirical formula is the same as the given formula: CH_4O.
  d. Each of the formula subscripts is divided by 6 to give the empirical formula CH_2O.

Answer Double Check:

Do the empirical formulas have the correct characteristics for empirical formulas? Yes. Each of the empirical formulas has a set of subscripts (1:2, 4:9, 1:4:1, and 1:2:1) that cannot be reduced further by division by a small whole number. If such division was possible, the formulas would not be correct empirical formulas.

▶ **Practice Exercise 9.18**   What is the empirical formula of each of the following compounds?

  a. N_2O_4 **b.** C_6H_{14} **c.** C_6H_{12} **d.** NH_3

9.13 DETERMINATION OF EMPIRICAL FORMULAS

The determination of a compound's chemical formula from experimental data is usually carried out in two calculational steps: (1) Elemental composition data are used to determine the compound's empirical formula (simplest ratio of atoms present—Sec. 9.12), and (2) this empirical formula and molar mass data are used to determine the compound's molecular formula (actual ratio of atoms present—Sec. 9.12).

In this section we consider the first of these two steps, and then in Section 9.14 we learn the additional procedures needed to go from an empirical formula to a molecular formula.

Empirical Formulas from Direct Analysis Data

Elemental composition data, obtained by direct analysis, are sufficient for an empirical formula determination. Such data may be in the form of percent composition or simply the mass of each element present in a known mass of compound. Example 9.19 shows the steps used in obtaining an empirical formula from percent composition data. Example 9.20 uses the mass of the elements present in a compound sample of known mass to obtain an empirical formula.

EXAMPLE 9.19 **Determining an Empirical Formula from Percent Composition Data**

Percent composition by mass data for the chlorofluorocarbon Freon-12 are 9.933% carbon, 58.63% chlorine, and 31.44% fluorine. Based on these data, what is the empirical formula of Freon-12?

SOLUTION

The problem of calculating the empirical formula of a compound from percent composition data can be broken down into three steps.

1. Determine the number of grams of each element in a sample of the compound.
2. Convert the grams of each element to moles of element.
3. Express the mole ratio between the elements in terms of *small whole numbers*.

Step 1 Mass percentage values are independent of sample size; that is, they apply to samples of all sizes. In working with mass percentages in empirical formula calculations, it is convenient to assume a 100.0 g sample size. When this is done, mass percentages translate directly into gram amounts. The mass of each element in the 100.0 g sample is numerically equal to the percentage value.

$$\text{C:} \quad 9.933\% \text{ of } 100.0 \text{ g} = 9.933 \text{ g}$$

$$\text{Cl:} \quad 58.63\% \text{ of } 100.0 \text{ g} = 58.63 \text{ g}$$

$$\text{F:} \quad 31.44\% \text{ of } 100.0 \text{ g} = 31.44 \text{ g}$$

Step 2 We next convert the grams (from step 1) to moles. We need moles information to determine the subscripts in the formula of the compound. Formula subscripts give the ratio of the number of moles of each element present in a compound (Sec. 9.9).

$$9.933 \text{ g C} \times \frac{1 \text{ mole C}}{12.01 \text{ g C}} = 0.82706078 \text{ mole C} \quad \text{(calculator answer)}$$

$$= 0.8271 \text{ mole C} \quad \text{(\textbf{correct answer})}$$

$$58.63 \text{ g Cl} \times \frac{1 \text{ mole Cl}}{35.45 \text{ g Cl}} = 1.6538787 \text{ moles Cl} \quad \text{(calculator answer)}$$

$$= 1.654 \text{ moles Cl} \quad \text{(\textbf{correct answer})}$$

$$31.44 \text{ g F} \times \frac{1 \text{ mole F}}{19.00 \text{ g F}} = 1.6547368 \text{ moles F} \quad \text{(calculator answer)}$$

$$= 1.655 \text{ moles F} \quad \text{(\textbf{correct answer})}$$

Thus, in 100.0 g of compound there are 0.8271 mole of C, 1.654 moles of Cl, and 1.655 moles of F.

Step 3 The subscripts in a formula are expressed as whole numbers, not as decimals. To obtain whole numbers from the decimals of step 2, each of the numbers is divided by the smallest of the numbers.

$$C: \frac{0.8271 \text{ mole}}{0.8271 \text{ mole}} = 1 \qquad \text{(calculator answer)}$$

$$= 1.000 \qquad \text{(\textbf{correct answer})}$$

$$Cl: \frac{1.654 \text{ mole}}{0.8271 \text{ mole}} = 1.9997581 \qquad \text{(calculator answer)}$$

$$= 2.000 \qquad \text{(\textbf{correct answer})}$$

$$F: \frac{1.655 \text{ mole}}{0.8271 \text{ mole}} = 2.0009672 \qquad \text{(calculator answer)}$$

$$= 2.001 \qquad \text{(\textbf{correct answer})}$$

Thus the ratio of carbon to chlorine to fluorine in this compound is 1:2:2, and the empirical formula is CCl_2F_2.

The calculated value for one of the three formula subscripts (2.001) has a nonzero digit to the right of the decimal place. This is because of experimental error in the original mass percentages. This calculated value is, however, sufficiently close to a whole number that we easily recognize what the whole number is.

▶ **Practice Exercise 9.19**   Percent composition data by mass for the prescription drug Bupropion, a compound used as an aid in quitting smoking, is 65.13% C, 7.57% H, 14.79% Cl, 5.84% N, and 6.67% O. Use these data to calculate the empirical formula for Bupropion.

| **EXAMPLE 9.20** | **Determining an Empirical Formula Using Mass Data** |

Analysis of a sample of ibuprofen, the active ingredient in Advil, shows that the sample contains 7.568 g of carbon, 0.881 g of hydrogen, and 1.551 g of oxygen. Use these data to calculate the empirical formula of ibuprofen.

SOLUTION

Having data given in grams instead of as percent composition (Example 9.18) simplifies the calculation of an empirical formula. The grams are changed to moles, and we are ready to find the smallest whole-number ratio between elements.

Step 1 Each of the given gram amounts is converted to moles, using dimensional analysis.

$$7.568 \text{ g C} \times \frac{1 \text{ mole C}}{12.01 \text{ g C}} = 0.63014154 \text{ mole C} \quad \text{(calculator answer)}$$

$$= 0.6301 \text{ mole C} \qquad \text{(\textbf{correct answer})}$$

$$0.881 \text{ g H} \times \frac{1 \text{ mole H}}{1.01 \text{ g H}} = 0.87227722 \text{ mole H} \quad \text{(calculator answer)}$$

$$= 0.872 \text{ mole H} \qquad \text{(\textbf{correct answer})}$$

$$1.551 \text{ g O} \times \frac{1 \text{ mole O}}{16.00 \text{ g O}} = 0.0969375 \text{ mole O} \quad \text{(calculator answer)}$$

$$= 0.09694 \text{ mole O} \qquad \text{(\textbf{correct answer})}$$

Step 2 Dividing the mole quantities from step 1 by the *smallest* of the three numbers gives

$$C: \quad \frac{0.6301 \text{ mole}}{0.09694 \text{ mole}} = 6.4998968 \quad \text{(calculator answer)}$$

$$= 6.500 \qquad \textbf{(correct answer)}$$

$$H: \quad \frac{0.872 \text{ mole}}{0.09694 \text{ mole}} = 8.9952547 \quad \text{(calculator answer)}$$

$$= 9.00 \qquad \textbf{(correct answer)}$$

$$O: \quad \frac{0.09694 \text{ mole}}{0.09694 \text{ mole}} = 1 \qquad \text{(calculator answer)}$$

$$= 1.000 \qquad \textbf{(correct answer)}$$

Frequently all whole numbers or near-whole numbers are obtained at this point (as was the case in Example 9.19), and the calculation is finished. Sometimes, however, as is the case in this example, we obtain one or more numbers that are not even close to being whole numbers. The number we obtained for carbon, 6.50, is such a number. What do we do?

The number 6.50 is recognizable as being close to a simple fraction; it is $6\frac{1}{2}$. When we obtain simple fractions, we clear them by multiplying *all of the numbers* in our set of numbers by a common factor. We multiply all numbers by 2 if we are dealing with halves, 3 if we are dealing with thirds, 4 if we are dealing with fourths, and so on. Following this procedure, we obtain for the situation in this example the following:

$$C: \quad 6\frac{1}{2} \times 2 = 13$$
$$H: \quad 9 \times 2 = 18$$
$$O: \quad 1 \times 2 = 2$$

We now have a whole-number ratio. The empirical formula of ibuprofen is $C_{13}H_{18}O_2$.

The fractions that most commonly occur in empirical formula calculations are fourths, $0.25 \left(\frac{1}{4}\right)$ and $0.75 \left(\frac{3}{4}\right)$, which are multiplied by 4 to clear; thirds, $0.33 \left(\frac{1}{3}\right)$ and $0.67 \left(\frac{2}{3}\right)$, which are multiplied by 3 to clear; and half, $0.50 \left(\frac{1}{2}\right)$, which is multiplied by 2 to clear.

Answer Double Check:

Are the subscripts obtained for the empirical formula appropriate for an empirical formula? Yes. The set of three subscripts (13, 18, and 2) are not divisible by any small whole number.

▶ **Practice Exercise 9.20** Iron and structural steel, when exposed to a moist atmosphere, rust. The reddish-brown rust that forms is an iron–oxygen compound. Analysis of a 1.386 g sample of dry rust indicates that 0.9694 g of Fe and 0.4166 g of O are present. What is the empirical formula of rust?

FIGURE 9.6 Outline of the procedure used to calculate the empirical formula of a compound from (a) percentage composition data (start in box 1) and (b) mass data (start in box 2).

Figure 9.6 summarizes the procedures used in Examples 9.19 and 9.20 for calculating empirical formulas. If mass percents are the starting point for the calculation (Example 9.19), you start in box 1. If mass data is the starting point (Example 9.20), the procedure is shortened by one step as you start in box 2.

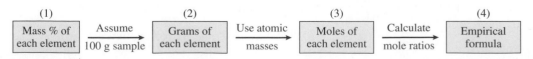

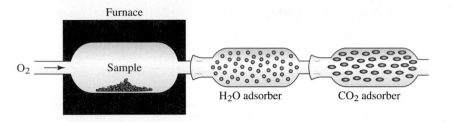

Furnace

O_2 → | Sample | H$_2$O adsorber | CO$_2$ adsorber

FIGURE 9.7
Combustion analysis method for determining the percentages of carbon and hydrogen in a compound. A weighed sample of the compound is burned in a stream of oxygen, producing gaseous CO_2 and H_2O. These gases then pass through a series of tubes. One tube contains a substance that adsorbs H_2O; the substance in the other tube adsorbs CO_2. By comparing the masses of these tubes before and after the reaction, the analyst can determine the masses of hydrogen and carbon present in the compound that was burned.

Empirical Formulas From Indirect Analysis Data

Examples 9.19 and 9.20 suggest that compounds are always broken down completely into their elements when they are analyzed. This is not always the case. Sometimes a compound whose empirical formula is to be determined is changed into other compounds, with each element in the original compound becoming part of a *separate* product compound whose chemical formula is known. Such indirect analysis data is sufficient to determine an empirical formula.

A common indirect analysis procedure is *combustion analysis*. This procedure is used extensively for compounds containing carbon, hydrogen, and oxygen (or just carbon and hydrogen). **Combustion analysis** *is a method used to measure the amounts of carbon and hydrogen present in a combustible compound that contains these two elements (and perhaps other elements) when that compound is burned in pure O_2.* The basic idea in combustion analysis is as follows. A sample of the compound is burned completely in pure oxygen. The products of the reaction are CO_2 and H_2O. All carbon atoms originally present end up in CO_2 molecules, and all hydrogen atoms originally present end up in H_2O molecules (see Fig. 9.7). From the mass of CO_2 and H_2O produced, it is possible to calculate the empirical formula of the original compound. Example 9.21 shows how this is done for a carbon–hydrogen compound, and Example 9.22 deals with the same situation for a carbon–hydrogen–oxygen compound.

EXAMPLE 9.21 | **Determining an Empirical Formula Using Combustion Analysis Data**

Ethylene, a compound that contains only carbon and hydrogen, is commercially used as a fruit-ripening agent. A sample of this compound is burned in a combustion analysis apparatus and 3.14 g of CO_2 and 1.29 g of H_2O are produced. What is the empirical formula of ethylene?

SOLUTION

Every carbon atom in the CO_2 and every hydrogen atom in the H_2O came from ethylene. We will need to calculate the number of moles of carbon present in the CO_2 and then the number of moles of hydrogen present in the H_2O. Both these calculations are grams-of-A to moles-of-B problems in terms of the jargon of Figure 9.5.

The dimensional analysis setup from which we obtain the moles of carbon is

$$3.14 \text{ g } CO_2 \times \frac{1 \text{ mole } CO_2}{44.01 \text{ g } CO_2} \times \frac{1 \text{ mole C}}{1 \text{ mole } CO_2}$$

$$= 0.07134742 \text{ mole C} \quad \text{(calculator answer)}$$
$$= 0.0713 \text{ mole C} \quad \text{(\textbf{correct answer})}$$

The last conversion factor used in this calculation comes from the formula CO_2. One molecule of CO_2 contains one atom of carbon. Thus, one mole of CO_2 will contain one mole of carbon.

The dimensional analysis setup from which we obtain the moles of hydrogen is

$$1.29 \text{ g } H_2O \times \frac{1 \text{ mole } H_2O}{18.02 \text{ g } H_2O} \times \frac{2 \text{ moles } H}{1 \text{ mole } H_2O}$$

$$= 0.14317425 \text{ mole H (calculator answer)}$$
$$= 0.143 \text{ mole H} \qquad \textbf{(correct answer)}$$

The last conversion factor in this setup is obtained from the information in the formula H_2O. There are 2 moles of H in 1 mole of H_2O.

With the moles of C and of H both known, we can now calculate the whole-number mole ratio by dividing each mole amount by the smallest of these amounts.

$$C: \quad \frac{0.0713 \text{ mole}}{0.0713 \text{ mole}} = 1 \qquad \text{(calculator answer)}$$
$$= 1.00 \qquad \textbf{(correct answer)}$$

$$H: \quad \frac{0.143 \text{ mole}}{0.0713 \text{ mole}} = 2.00561 \quad \text{(calculator answer)}$$
$$= 2.01 \qquad \textbf{(correct answer)}$$

Carbon and hydrogen are present in the compound in a one-to-two molar ratio. The empirical formula of the compound is CH_2.

▶ **Practice Exercise 9.21** Naphthalene is a carbon–hydrogen compound that finds use as mothballs. A sample of naphthalene is subjected to combustion analysis, producing 1.100 g of CO_2 and 0.1802 g of H_2O. Based on these data, calculate the empirical formula of naphthalene.

When the compound to be analyzed using combustion analysis contains oxygen in addition to carbon and hydrogen, the calculation of the amount (or percent) of oxygen in the compound proceeds in a different manner than that shown in Example 9.21. Only part of the oxygen involved in the production of CO_2 and H_2O comes from the compound itself, and part comes from the O_2 used to support combustion of the compound. In this situation, the amount of hydrogen and carbon present is calculated as before. Then the sum of the hydrogen and carbon masses is subtracted from the original compound mass to obtain the mass of oxygen present in the original sample of the compound.

EXAMPLE 9.22 **Determining an Empirical Formula Using Combustion Analysis Data**

The compound ascorbic acid (vitamin C) is a carbon–hydrogen–oxygen compound. This water-soluble vitamin cannot be stored in the body and thus must continuously be supplied by the diet. Combustion analysis of a 3.08 g ascorbic acid sample yields 6.17 g of CO_2 and 2.52 g of H_2O. What is the empirical formula for ascorbic acid?

SOLUTION

First, we will calculate the grams of C in the original sample from the grams of CO_2 and the grams of H in the original sample from the grams of H_2O. The grams of O present in the original sample can then be calculated by difference: the difference between the original sample mass and the masses of C and H. We cannot calculate the grams of oxygen present in the original sample from the CO_2 and H_2O masses because the oxygen atoms in these compounds came from two sources: (1) ascorbic acid and (2) the oxygen in air.

With masses of the individual elements known, we then obtain moles of the individual elements. Dividing these molar amounts by the smallest number of moles of an element present leads to the subscripts for the empirical formula. The following flowchart outlines the overall calculational process.

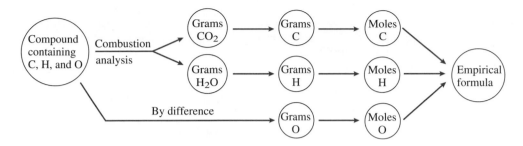

Converting the grams of CO_2 and grams of H_2O to grams of C and grams of H, respectively, involves grams-of-A to grams-of-B calculations.

$$6.17 \text{ g } CO_2 \times \frac{1 \text{ mole } CO_2}{44.01 \text{ g } CO_2} \times \frac{1 \text{ mole C}}{1 \text{ mole } CO_2} \times \frac{12.01 \text{ g C}}{1 \text{ mole C}}$$

$$= 1.6837468 \text{ g C} \qquad \text{(calculator answer)}$$
$$= 1.68 \text{ g C} \qquad \textbf{(correct answer)}$$

$$2.52 \text{ g } H_2O \times \frac{1 \text{ mole } H_2O}{18.02 \text{ g } H_2O} \times \frac{2 \text{ mole H}}{1 \text{ mole } H_2O} \times \frac{1.01 \text{ g H}}{1 \text{ mole H}}$$

$$= 0.28248612 \text{ g H} \quad \text{(calculator answer)}$$
$$= 0.282 \text{ g H} \qquad \textbf{(correct answer)}$$

The next-to-last conversion factor in the CO_2 calculation comes from the formula CO_2. One CO_2 molecule contains one C atom. Likewise, the formula H_2O is the key to obtaining information about hydrogen; one H_2O molecule contains two H atoms.

The difference between the mass of the sample and the masses of C and H in the sample is the mass of O in the sample.

$$\text{grams O} = \text{grams of sample} - \text{grams C} - \text{grams H}$$
$$= 3.08 \text{ g} - 1.68 \text{ g} - 0.282 \text{ g}$$
$$= 1.118 \text{ g} \qquad \text{(calculator answer)}$$
$$= 1.12 \text{ g} \qquad \textbf{(correct answer)}$$

The next step is to go from grams of element to moles of element. These are simple one-step grams-of-A to moles-of-A calculations.

$$1.68 \text{ g C} \times \frac{1 \text{ mole C}}{12.01 \text{ g C}} = 0.13988343 \text{ mole C} \quad \text{(calculator answer)}$$

$$= 0.140 \text{ mole C} \qquad \textbf{(correct answer)}$$

$$0.282 \text{ g H} \times \frac{1 \text{ mole H}}{1.01 \text{ g H}} = 0.27920792 \text{ mole H} \quad \text{(calculator answer)}$$

$$= 0.279 \text{ mole H} \qquad \textbf{(correct answer)}$$

$$1.12 \text{ g O} \times \frac{1 \text{ mole O}}{16.00 \text{ g O}} = 0.07 \text{ mole O} \qquad \text{(calculator answer)}$$

$$= 0.0700 \text{ mole O} \qquad \textbf{(correct answer)}$$

Dividing each of these molar values by the smallest among them, 0.0700 mole O, gives us the small whole numbers that lead to the empirical formula of the compound.

$$C: \quad \frac{0.140 \ \text{mole}}{0.0700 \ \text{mole}} = 2 \qquad \text{(calculator answer)}$$

$$= 2.00 \qquad \text{(correct answer)}$$

$$H: \quad \frac{0.279 \ \text{mole}}{0.0700 \ \text{mole}} = 3.9857142 \qquad \text{(calculator answer)}$$

$$= 3.99 \qquad \text{(correct answer)}$$

$$O: \quad \frac{0.0700 \ \text{mole}}{0.0700 \ \text{mole}} = 1 \qquad \text{(calculator answer)}$$

$$= 1.00 \qquad \text{(correct answer)}$$

The empirical formula of ascorbic acid is C_2H_4O.

Answer Double Check:

Are the calculated mass amounts needed to generate the empirical formula reasonable? Yes. The mass of C (1.68 g) must be less than the mass of CO_2 (6.17 g). It is. The mass of H (0.282 g) must be less than the mass of H_2O (2.52 g). It is. The mass of O (1.12 g) must be less than the total sample mass (3.08 g). It is.

▶ **Practice Exercise 9.22** Ethyl alcohol, the alcohol present in alcoholic beverages, is a carbon–hydrogen–oxygen compound. Combustion analysis of a 1.000 g sample of ethyl alcohol produces 1.913 g of CO_2 and 1.174 g of H_2O. Based on these data, calculate the empirical formula of ethyl alcohol.

9.14 DETERMINATION OF MOLECULAR FORMULAS

To determine the molecular formula of a compound we need additional information besides its empirical formula. That additional information is the compound's molar mass. A number of experimental methods exist for obtaining molar mass information. One such method, applicable to compounds in the gaseous state, will be considered in Section 12.13.

A compound's molecular formula will always be a *whole-number multiple* of its empirical formula (Sec. 9.12).

molecular formula = (empirical formula)$_x$ where x = whole number

The value of x is calculated by dividing the compound's molar mass by the compound's empirical formula mass.

$$x = \frac{\text{molar mass (experimentally determined quantity)}}{\text{empirical formula mass (calculated from atomic masses)}}$$

If the value of x was found to be 5.00 and the compound's empirical formula was CH_2, then the compound's molecular formula would be

$$(CH_2)_5 = C_5H_{10}$$

Example 9.23 illustrates further the mechanics involved in obtaining a molecular formula from empirical formula and molar mass data.

EXAMPLE 9.23   **Calculating a Molecular Formula from an Empirical Formula and Molar Mass Data**

Determine the molecular formula of each of the following compounds from the given empirical formula and molar mass information.

a. empirical formula = CH; molar mass = 78.12 amu
b. empirical formula = NH_2; molar mass = 32.06 amu
c. empirical formula = CO; molar mass = 28.01 amu
d. empirical formula = C_4H_9; molar mass = 114.26 amu

SOLUTION

In each case we will determine the whole-number multiplier (x) that relates empirical and molecular formulas to each other.

$$x = \frac{\text{molar mass (MM)}}{\text{empirical formula mass (EFM)}}$$

a. First, we calculate the empirical formula mass of CH from atomic masses. The atomic mass of C is 12.01 amu and that of H is 1.01 amu.

$$\text{EFM of CH} = 12.01 \text{ amu} + 1.01 \text{ amu} = 13.02 \text{ amu}$$

The whole-number multiplier is

$$x = \frac{\text{MM}}{\text{EFM}} = \frac{78.12 \text{ amu}}{13.02 \text{ amu}} = 6.000$$

Therefore, the molecular formula is

$$(CH)_6 = C_6H_6$$

b. The empirical formula mass of NH_2 is 16.03 amu. Nitrogen has an atomic mass of 14.01 amu, and each hydrogen has an atomic mass of 1.01 amu. The whole-number multiplier, x, is

$$x = \frac{\text{MM}}{\text{EFM}} = \frac{32.06 \text{ amu}}{16.03 \text{ amu}} = 2.000$$

The molecular formula is $(NH_2)_2 = N_2H_4$

c. The empirical formula mass is 28.01 amu. The whole-number multiplier, x, is

$$x = \frac{\text{MM}}{\text{EFM}} = \frac{28.01 \text{ amu}}{28.01 \text{ amu}} = 1.000$$

An x value of 1.000 means that the empirical formula and the molecular formula are one and the same. Both are CO.

d. The empirical formula mass is 57.13 amu, the value of x is 2.000, and the molecular formula is $(C_4H_9)_2 = C_8H_{18}$.

Answer Double Check:

Are the molecular formula–empirical formula relationships appropriate? Yes. Dividing the molar mass by the empirical formula mass must always yield a number that is close to a whole number. Such is the case here for all four compounds. Also, the subscripts in the molecular formula, when divided by a small whole number, must produce the subscripts in the empirical formula. Such is the case here for all four compounds.

▶ **Practice Exercise 9.23** Determine the molecular formula of each of the following compounds from the given empirical formula and molar mass information.

a. empirical formula = NO_2; molar mass = 92.02 amu
b. empirical formula = SO_2; molar mass = 64.06 amu

Example 9.23 showed how a molecular formula is determined when the empirical formula is already known. Example 9.24 illustrates a complete molecular formula determination, one in which we begin at the beginning—with percent composition data. There are two different approaches to such a calculation, and both approaches are illustrated in Example 9.24. The two approaches differ in the selection of the basis (amount of compound) for the calculation. The two approaches are the following:

1. *Select 100.0 g of compound as the basis for the calculation*, determine the empirical formula, and then use the empirical formula and the molar mass to determine the molecular formula.
2. *Select 1.000 mole of compound as the basis for the calculation*, and directly determine the molecular formula.

In the first approach, the molar mass data are used at the end of the problem. In the second approach, the molar mass data are used at the beginning of the calculation. Either way, the answer is the same.

| **EXAMPLE 9.24** | **Calculating a Molecular Formula from Percent Composition and Molar Mass Data** |

Butane, a compound containing only carbon and hydrogen, is the fuel used in many camp stoves. Butane has a percent composition by mass of 82.63% C and 17.37% H and a molar mass of 58.14 amu. Determine the molecular formula of butane using

a. 100.0 g of butane as the basis for the calculation.
b. 1.000 mole of butane as the basis for the calculation.

SOLUTION

a. Taking 100.0 g of compound as our basis, we will have the following amounts of carbon and hydrogen present.

$$\text{C:} \quad 82.63\% \text{ of } 100.0 \text{ g} = 82.63 \text{ g}$$

$$\text{H:} \quad 17.37\% \text{ of } 100.0 \text{ g} = 17.37 \text{ g}$$

We next convert these gram amounts of C and H to moles of the same.

$$82.63 \text{ g C} \times \frac{1 \text{ mole C}}{12.01 \text{ g C}} = 6.8800999 \text{ moles C} \quad \text{(calculator answer)}$$

$$= 6.880 \text{ moles C} \quad \textbf{(correct answer)}$$

$$17.37 \text{ g H} \times \frac{1 \text{ mole H}}{1.008 \text{ g H}} = 17.232142 \text{ moles H} \quad \text{(calculator answer)}$$

$$= 17.23 \text{ moles H} \quad \textbf{(correct answer)}$$

Note that the atomic mass of hydrogen is specified to thousandths (1.008) rather than the usual hundredths (1.01). The given amount for hydrogen (17.37 g) requires an atomic mass with at least four significant figures.

Dividing each of these molar values by the smallest one (6.880) gives the following results.

$$\text{C:} \quad \frac{6.880 \ \cancel{\text{mole}}}{6.880 \ \cancel{\text{mole}}} = 1 \qquad \text{(calculator answer)}$$

$$= 1.000 \qquad \textbf{(correct answer)}$$

$$\text{H:} \quad \frac{17.23 \ \cancel{\text{mole}}}{6.880 \ \cancel{\text{mole}}} = 2.5043604 \quad \text{(calculator answer)}$$

$$= 2.504 \qquad \textbf{(correct answer)}$$

Our result for hydrogen is not a whole number or a near-whole number that can be rounded to a whole number. For hydrogen, we should recognize that the number we are dealing with is $2\frac{1}{2}$. Multiplication of both molar ratios by 2 will clear the fraction. This gives us our needed whole-number ratio.

$$\text{C:} \quad 1 \times 2 = 2$$
$$\text{H:} \quad 2\tfrac{1}{2} \times 2 = 5$$

The empirical formula of butane is C_2H_5.

To make the transition from empirical formula to molecular formula, we first determine the formula mass for the empirical formula.

$$\text{C:} \quad 2 \times 12.01 \ \text{amu} = 24.02 \ \text{amu}$$
$$\text{H:} \quad 5 \times 1.01 \ \text{amu} = \underline{5.05 \ \text{amu}}$$
$$29.07 \ \text{amu}$$

The molar mass of butane, given in the problem statement, is 58.14 amu. We next determine how many times larger the molar mass is than the empirical formula mass. This is done by dividing the molar mass by the empirical formula mass.

$$\frac{58.14 \ \cancel{\text{amu}}}{29.07 \ \cancel{\text{amu}}} = 2.000$$

We have just determined the multiplication factor that converts the empirical formula into a molecular formula. Each of the subscripts in the empirical formula is multiplied by 2.

$$(C_2H_5)_2 = C_4H_{10}$$

The molecular formula of butane is C_4H_{10}.

b. We will go through this same calculation again, this time using 1.000 mole of butane as our basis. The mass of 1.000 mole of butane is 58.14 g. We know this because it was given in the problem statement; the molar mass of butane is 58.14 amu.

We first determine the number of grams of each element present in our basis amount of 58.14 g of butane.

$$\text{C:} \quad 82.63\% \text{ of } 58.14 \ \text{g} = 48.041082 \ \text{g} \quad \text{(calculator answer)}$$
$$= 48.04 \ \text{g} \qquad \textbf{(correct answer)}$$
$$\text{H:} \quad 17.37\% \text{ of } 58.14 \ \text{g} = 10.098918 \ \text{g} \quad \text{(calculator answer)}$$
$$= 10.10 \ \text{g} \qquad \textbf{(correct answer)}$$

We next change grams of element to moles of element. Formula subscripts are always determined from molar information.

$$48.04 \ \cancel{\text{g C}} \times \frac{1 \ \text{mole C}}{12.01 \ \cancel{\text{g C}}} = 4 \ \text{moles C} \qquad \text{(calculator answer)}$$

$$= 4.000 \ \text{moles C} \qquad \textbf{(correct answer)}$$

$$10.10 \ \cancel{g \ H} \times \frac{1 \ \text{mole H}}{1.008 \ \cancel{g \ H}} = 10.019841 \ \text{moles H} \quad \text{(calculator answer)}$$

$$= 10.02 \ \text{moles H} \quad \textbf{(correct answer)}$$

At this stage in the calculation, because the basis for the calculation is one mole of compound, we will always obtain whole-number molar amounts or near-whole-number molar amounts that can be rounded to whole numbers. (If we do not, we have a mistake somewhere in our calculation.) These whole-number molar amounts are the subscripts in the molecular formula. Thus, the molecular formula of butane is C_4H_{10}.

What caused this formula, C_4H_{10}, to be a molecular formula rather than an empirical formula? Again, it is the fact that the basis for the calculation was 1 mole of compound.

This second method for calculating a molecular formula is usually shorter than the first method.

▶ **Practice Exercise 9.24** The mass percent composition for propane, a fuel used for home heating in rural areas, is 81.7% carbon and 18.3% hydrogen, and its molar mass is 44.11 amu. Determine the molecular formula of propane using

a. 100.0 g of propane as the basis for the calculation.
b. 1.000 mole of propane as the basis for the calculation.

Figure 9.8 summarizes the methods used in this section and the preceding one to obtain empirical and molecular formulas. It contrasts each of the example calculations we carried out in these sections.

FIGURE 9.8 A summary of how empirical formulas and molecular formulas are calculated from various types of data.

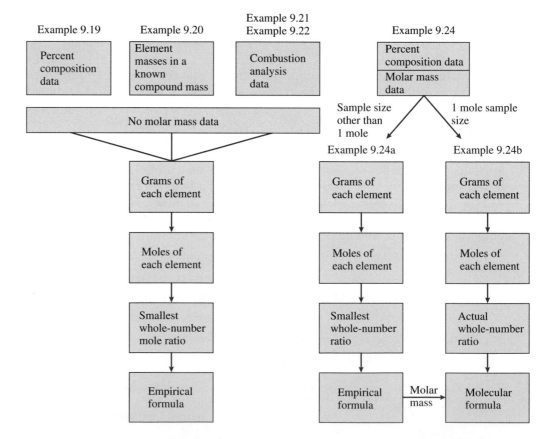

Summary

1. **Law of Definite Proportions** Study of composition data for many compounds led to the law of definite proportions: In a pure compound, the elements are always present in the same definite proportion by mass.

2. **Formula Mass** The formula mass of a substance is the sum of the atomic masses of the atoms present in one formula unit of the substance. An operational rule, in this text, for calculation of formula masses is that all atomic masses are rounded to the hundredths place before they are used as input numbers in a formula mass calculation.

3. **Percent Composition** The percent composition of a compound specifies the percent by mass of each element present in the compound. The formula mass and chemical formula of a compound can be used to calculate percent composition.

4. **The Mole** The mole is the chemist's counting unit. One mole of any substance—element or compound—consists of 6.022×10^{23} formula units of the substance.

5. **Avogadro's Number** Avogadro's number is the name given to the numerical value 6.022×10^{23}.

6. **Molar Mass** The molar mass of a substance is the mass in grams of the substance that is numerically equal to the substance's formula mass. Molar mass is not a set number; it is different for each chemical substance.

7. **The Mole and Chemical Formulas** The numerical subscripts in a chemical formula give the number of moles of atoms of the various elements present in one mole of the substance.

8. **The Mole and Chemical Calculations** The three quantities most often calculated in chemical problems are the number of particles present (atoms or molecules), the number of moles present, and the number of grams present. These quantities are interrelated. The necessary conversion factors relate to the concepts of (1) Avogadro's number, (2) molar mass, and (3) molar interpretation of chemical formula subscripts.

9. **Percent Purity** Most substances are not 100% pure. In accurate work, account must be taken of the impurities present. Percent purity is the mass percentage of a specified substance in an impure sample of the substance.

10. **Empirical and Molecular Formulas** An empirical formula gives the *smallest* whole-number ratio of atoms present in a formula unit of a compound. A molecular formula gives the *actual* ratio of atoms present in a formula unit of a compound. The empirical formula of a compound can be determined from percent composition data or combustion analysis data. The molecular formula can be determined from the empirical formula if the molar mass of the substance is known.

Key Terms

The new terms defined in this chapter are

Avogadro's number *Sec. 9.5*
combustion analysis *Sec. 9.13*
empirical formula *Sec. 9.12*

formula mass *Sec. 9.2*
law of definite proportions *Sec. 9.1*
molar mass *Sec. 9.6*
molar mass of a compound *Sec. 9.6*

molar mass of an element *Sec. 9.6*
mole *Secs. 9.5 and 9.7*
molecular formula *Sec. 9.12*

percent by mass of an element in a compound *Sec. 9.4*
percent composition of a compound *Sec. 9.4*
percent purity *Sec. 9.11*

Practice Problems

LAW OF DEFINITE PROPORTIONS (SEC. 9.1)

9.1 The air pollutant carbon monoxide, CO, is produced when gasoline is burned. A 3.00 g sample of CO is found to contain 42.9% C and 57.1% O by mass. What would be the percent composition of an 8.00 g sample of CO?

9.2 The compound carbon dioxide, CO_2, a substance that contributes to the global warming process, is produced when coal is burned. A 2.00 g sample of CO_2 is found to contain 27.3% C and 72.7% O by mass. What would be the percent composition of a 13.00 g sample of CO_2?

9.3 Two different samples of a pure compound (containing elements A and D) were analyzed with the following results.

Sample I: 17.35 g of compound yielded 10.03 g of A and 7.32 g of D.
Sample II: 22.78 g of compound yielded 13.17 g of A and 9.61 g of D.

Show that these data are consistent with the law of definite proportions.

9.4 Two different samples of a pure compound (containing elements E and D) were analyzed with the following results.

Sample I: 11.21 g of compound yielded 5.965 g of E and 5.245 g of D.
Sample II: 13.65 g of compound yielded 7.263 g of E and 6.387 g of D.

Show that these data are consistent with the law of definite proportions.

9.5 It has been found experimentally that the elements X and Q react to produce two different compounds, depending on conditions. Some sample experimental data are as follows.

Experiment	Grams of X	Grams of Q	Grams of Compound
1	3.37	8.90	12.27
2	0.561	1.711	2.272
3	26.9	71.0	97.9

Which two of the three experiments produced the same compound?

9.6 It has been found experimentally that the elements R and T react to produce two different compounds, depending upon conditions. Some sample experimental data are as follows.

Experiment	Grams of R	Grams of T	Grams of Compound
1	1.926	1.074	3.000
2	3.440	1.560	5.000
3	4.494	2.506	7.000

Which two of the three experiments produced the same compound?

9.7 The elemental composition of the gaseous compound SO_2 is 50.1% S and 49.1% O by mass. What is the maximum amount of SO_2 that could be formed from 50.1 g of S and 75.0 g of O?

9.8 The elemental composition of the gaseous compound NO is 46.7% N and 53.3% O by mass. What is the maximum amount of NO that could be formed from 65.3 g of N and 53.3 g of O?

FORMULA MASSES (SECS. 9.2 AND 9.3)

9.9 Calculate the formula mass of each of the following compounds, using the operational rules given in Section 9.3.
a. SnF_2 (tin(II) fluoride, a toothpaste additive)
b. $FeSO_4$ (iron(II) sulfate, used in treatment of iron deficiency)
c. $C_{12}H_{22}O_{11}$ (sucrose, table sugar)
d. $C_{14}H_9Cl_5$ (DDT, a banned insecticide)

9.10 Calculate the formula mass of each of the following compounds, using the operational rules given in Section 9.3.
a. $C_{10}H_8$ (naphthalene, ingredient in some mothballs)
b. NaCN (sodium cyanide, used to extract gold from ores)
c. Na_4SiO_4 (sodium silicate, a flame retardant)
d. $C_{20}H_{24}N_2O_2$ (quinine, an antimalarial drug)

9.11 Calculate the formula mass of each of the following substances, using the operational rules given in Section 9.3.
a. $C_2H_4(OH)_2$ (ethylene glycol, an automotive antifreeze component)
b. $(H_2N)_2CO$ (urea, a component of urine)
c. $Mg_3(Si_2O_5)_2(OH)_2$ (talc, a mineral from which talcum powder comes)
d. $Al_2Si_2O_5(OH)_2$ (kaolinite, a form of clay)

9.12 Calculate the formula mass of each of the following substances, using the operational rules given in Section 9.3.
a. $Al(OH)_3$ (aluminum hydroxide, a water purification chemical)
b. $C_3H_5(OH)_3$ (glycerin, a substance derived from fats)
c. $KLi_2Al(Si_2O_5)_2(OH)_2$ (a form of the mineral mica)
d. $Ca_3Be(OH)_2(Si_3O_{10})$ (a form of the mineral aminoffite)

9.13 Calculate the molecular mass of each of the following molecules.
a. ethylene glycol (present in antifreeze)

b. lactic acid (present in sour milk)

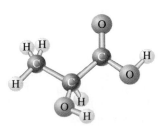

9.14 Calculate the molecular mass of each of the following molecules.

a. acetic acid (present in vinegar)

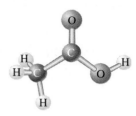

b. isopropyl alcohol (present in rubbing alcohol)

9.15 The compound 2-butene-1-thiol, which is responsible in part for the characteristic odor of skunks, has a formula mass of 88.19 amu and the formula C_yH_8S. What number does y stand for in this formula?

9.16 The compound 1-propanethiol, which is the eye irritant that is released when fresh onions are chopped up, has a formula mass of 76.18 amu and the formula C_3H_yS. What number does y stand for in this formula?

9.17 A pentaatomic molecule with one carbon atom present is 74.8% C by mass. What is the formula mass, in amu, for the molecule?

9.18 A hexaatomic molecule with one carbon atom present is 85.6% C by mass. What is the formula mass, in amu, for the molecule?

9.19 The mass percent nitrogen in a molecule in which two nitrogen atoms are present is 63.6%. What is the formula mass, in amu, of the compound?

9.20 The mass percent oxygen in a molecule in which two oxygen atoms are present is 69.6%. What is the formula mass, in amu, of the compound?

9.21 What is the chemical formula of the compound in Problem 9.19 given that all other atoms present are oxygen atoms?

9.22 What is the chemical formula of the compound in Problem 9.20 given that all other atoms present are nitrogen atoms?

PERCENT COMPOSITION (SEC. 9.4)

9.23 Calculate the percent composition for each of the following compounds. (Use atomic masses rounded to hundredths in doing the calculations.)
a. H_3BO_3 (boric acid, a mild antiseptic)
b. C_7H_{16} (heptane, a component of gasoline)

c. $C_{12}H_{11}NO_2$ (Sevin, an insecticide)
d. $C_{22}H_{30}ClNO_2$ (Darvon, a prescription pain killer)

9.24 Calculate the percent composition for each of the following compounds. (Use atomic masses rounded to hundredths in doing the calculations.)
a. $NaHCO_3$ (sodium bicarbonate, baking soda)
b. Tl_2SO_4 (thallium(I) sulfate, an ant poison)
c. $C_5H_8NO_4Na$ (MSG, a flavor enhancer used in Chinese cooking)
d. $C_{55}H_{72}MgN_4O_5$ (chlorophyll, present in all green plants)

9.25 Calculate the percent composition of each of the following compounds from the given information.
a. 1.271 g of Cu and 0.320 g of O completely react to produce a sample of the compound.
b. Decomposition of a 49.31 g sample of the compound yields 15.96 g of Na, 11.13 g of S, and 22.22 g of O.
c. A 6.48 g sample of N is reacted with oxygen to give 25.00 g of compound.
d. The reaction of 3.34 g of S with 10.00 g of O produces 10.00 g of compound and 3.34 g of unreacted (leftover) O.

9.26 Calculate the percent composition of each of the following compounds from the given information.
a. 1.12 g of Fe and 0.48 g of O completely react to produce a sample of the compound.
b. Decomposition of an 18.03 g sample of the compound yields 4.79 g of K, 6.38 g of Cr, and 6.86 g of O.
c. A 5.76 g sample of P is reacted with oxygen to give 13.20 g of compound.
d. The reaction of 4.67 g of N with 10.00 g of O produces 10.00 g of compound and 4.67 g of unreacted (leftover) O.

9.27 Indicate whether the mass percent of H is less than 10% in both members of each of the following pairs of compounds.
a. NH_3 and CH_4 **b.** LiH and BeH_2
c. H_2SO_4 and HNO_3 **d.** $Mg(OH)_2$ and NaOH

9.28 Indicate whether the mass percent of H is less than 10% in both members of each of the following pairs of compounds.
a. H_2O and H_2O_2 **b.** NH_3 and N_2H_4
c. H_3PO_4 and H_3PO_3 **d.** $Ca(OH)_2$ and $Al(OH)_3$

9.29 Do the mass percentages 15.77% C and 84.23% S match the chemical formula CS_2?

9.30 Do the mass percentages 30.45% N and 69.55% O match the chemical formula N_2O_3?

9.31 Calculate the percent compositions of acetylene (C_2H_2) and benzene (C_6H_6). Explain your results.

9.32 Calculate the percent compositions of propene (C_3H_6) and cyclobutane (C_4H_8). Explain your results.

9.33 The hydrogen content of a hydrogen-nitrogen compound is 12.6% by mass. How many grams of N are present in a sample of the compound that contains 52.34 g of H?

9.34 The sodium content of a sodium-nitrogen compound is 35.36% by mass. How many grams of N are present in a sample of the compound that contains 57.08 g of Na?

THE MOLE: THE CHEMIST'S COUNTING UNIT (SEC. 9.5)

9.35 How many particles (atoms, molecules, formula units, or ions) are present in 1.00 mole of each of the following?
a. silver (Ag) atoms
b. water (H_2O) molecules
c. sodium nitrate ($NaNO_3$) formula units
d. sulfate (SO_4^{2-}) ions

9.36 How many particles (atoms, molecules, formula units, or ions) are present in 1.00 mole of each of the following?
a. copper (Cu) atoms
b. ammonia (NH_3) molecules
c. potassium carbonate (K_2CO_3) formula units
d. phosphate (PO_4^{3-}) ions

9.37 How many silicon atoms are present in each of the following molar quantities of silicon?
a. 3.20 moles   **b.** 0.36 mole
c. 1.21 moles   **d.** 16.7 moles

9.38 How many sulfur atoms are present in each of the following molar quantities of sulfur?
a. 3.02 moles   **b.** 0.45 mole
c. 1.31 moles   **d.** 18.2 moles

9.39 Calculate the number of molecules present in each of the following samples of molecular compounds.
a. 1.50 moles CO_2 **b.** 0.500 mole NH_3
c. 2.33 moles PF_3 **d.** 1.115 moles N_2H_4

9.40 Calculate the number of molecules present in each of the following samples of molecular compounds.
a. 4.69 moles CO   **b.** 0.433 mole SO_3
c. 1.44 moles P_2H_4 **d.** 2.307 moles H_2O_2

9.41 The population of Europe is estimated to be 719 million people. How many moles of people is this equivalent to?

9.42 The memory capacity of a new computer is stated as 1.50 terabytes. How many moles of bytes is this equivalent to?

MOLAR MASS (SEC. 9.6)

9.43 What is the mass, in grams, of 1.000 mole of each of the following elements?
a. Ca   **b.** Si   **c.** Co   **d.** Ag

9.44 What is the mass, in grams, of 1.000 mole of each of the following elements?
a. Mg   **b.** Cl   **c.** Fe   **d.** Au

9.45 What is the molar mass, to hundredths, of each of the following compounds?
a. NaCl   **b.** Na_2S **c.** $NaNO_3$ **d.** Na_3PO_4

9.46 What is the molar mass, to hundredths, of each of the following compounds?
a. SO_2 **b.** SO_2Cl_2 **c.** S_4N_4 **d.** $Li_2S_2O_3$

9.47 What is the mass, in grams, of 1.000 mole of each of the compounds in Problem 9.45?

9.48 What is the mass, in grams, of 1.000 mole of each of the compounds in Problem 9.46?

9.49 What is the mass, in grams, of 2.314 moles of each of the following compounds?
a. $Al(OH)_3$ **b.** Mg_3N_2
c. $Cu(NO_3)_2$ **d.** La_2O_3

9.50 What is the mass, in grams, of 1.782 moles of each of the following compounds?
a. NH_4Cl **b.** $HClO_4$
c. $Ca(ClO_2)_2$ **d.** HfO_2

9.51 Molar masses of compounds are not unique. Different compounds can have almost identical molar masses. Calculate the atomic masses, to hundredths, of the following compounds.
a. CO_2 **b.** N_2O **c.** C_3H_8

9.52 Molar masses of compounds are not unique. Different compounds can have almost identical molar masses. Calculate the atomic masses, to hundredths, of the following compounds.
a. CO   **b.** N_2 **c.** C_2H_4

9.53 In each of the following pairs of molar-sized quantities, select the quantity that has the greater mass, in grams.
a. 2.00 moles of Cu and 2.00 moles of O
b. 1.00 mole of Br and 5.00 moles of Be
c. 2.00 moles of CO and 1.50 moles of N_2O
d. 4.87 moles of B_2H_6 and 0.35 mole of U

9.54 In each of the following pairs of molar-sized quantities, select the quantity that has the greater mass, in grams.
a. 3.00 moles of Na and 3.00 moles of Al
b. 1.00 mole of I and 9.00 moles of Li
c. 2.00 moles of CO_2 and 1.00 mole of SO_3
d. 9.00 moles of Be and 1.00 mole of HCl

9.55 Calculate the molar mass of compounds for which each of the following experimentally determined mole–mass relationships apply.
a. 0.6232 mole = 11.23 g
b. 0.5111 mole = 20.14 g
c. 2.357 moles = 352.6 g
d. 1.253 moles = 100.0 g

9.56 Calculate the molar mass of compounds for which each of the following experimentally determined mole–mass relationships apply.
a. 0.6232 mole = 15.07 g
b. 0.5111 mole = 40.73 g
c. 2.357 moles = 123.7 g
d. 1.253 moles = 300.0 g

9.57 A 0.0521 mole sample of compound A has a mass of 3.62 g. A 1.23 mole sample of compound B has a mass of 83.64 g. Does compound A or compound B have the greater molar mass?

9.58 A 0.0721 mole sample of compound A has a mass of 4.62 g. A 1.75 mole sample of compound B has a mass of 114.03 g. Does compound A or compound B have the greater molar mass?

RELATIONSHIP BETWEEN ATOMIC MASS UNITS AND GRAM UNITS (SEC. 9.8)

9.59 What is the mass, in grams, of atoms whose masses on the amu scale have each of the following values?
a. 19.00 amu   b. 52.00 amu
c. 118.71 amu   d. 196.97 amu

9.60 What is the mass, in grams, of atoms whose masses on the amu scale have each of the following values?
a. 9.012 amu   b. 40.08 amu
c. 107.87 amu   d. 207.2 amu

9.61 Identify the element that contains atoms with an average mass of 5.143×10^{-23} g.

9.62 Identify the element that contains atoms with an average mass of 2.326×10^{-23} g.

9.63 How many Al atoms of average mass 26.98 amu are present in 26.98 g of Al?

9.64 How many P atoms of average mass 30.97 amu are present in 30.97 g of P?

THE MOLE AND CHEMICAL FORMULAS (SEC. 9.9)

9.65 Write the six mole-to-mole conversion factors that can be derived from the formula Na_3PO_4.

9.66 Write the six mole-to-mole conversion factors that can be derived from the formula K_2SO_4.

9.67 Based on the chemical formula $S_4N_4Cl_2$, write the conversion factor that would be needed to do each of the following one-step conversions.
a. moles of $S_4N_4Cl_2$ to moles of N atoms
b. moles of $S_4N_4Cl_2$ to moles of Cl atoms
c. moles of $S_4N_4Cl_2$ to total moles of atoms
d. moles of S atoms to moles of Cl atoms

9.68 Based on the chemical formula SO_2Cl_2, write the conversion factor that would be needed to do each of the following one-step conversions.
a. moles of SO_2Cl_2 to moles of O atoms
b. moles of SO_2Cl_2 to moles of Cl atoms
c. moles of SO_2Cl_2 to total moles of atoms
d. moles of S atoms to moles of Cl atoms

9.69 In which of the following pairs of compound amounts do both members of the pair contain the same number of moles of sulfur atoms?
a. 1.0 mole Na_2SO_4 and 0.50 mole $Na_2S_2O_3$
b. 2.00 moles S_3Cl_2 and 1.50 moles S_2O
c. 3.00 moles $H_2S_2O_5$ and 6.00 moles H_2SO_4
d. 1.00 mole $Na_3Ag(S_2O_3)_2$ and 2.00 moles S_2F_{10}

9.70 In which of the following pairs of compound amounts do both members of the pair contain the same number of moles of nitrogen atoms?
a. 0.50 mole N_2O_5 and 1.0 mole of N_2O_4
b. 2.00 moles HNO_3 and 2.00 moles HNO_2

c. 3.00 moles NH_3 and 1.00 mole HN_3
d. 1.50 moles $(NH_4)_2SO_4$ and 1.00 mole $(NH_4)_3PO_4$

9.71 Which amount in each of the following pairs of amounts contains the greater number of total moles of atoms?
a. 2.00 moles $NaAuBr_4$ and 2.00 moles Au_2Te_3
b. 1.00 mole $C_2H_2Cl_4$ and 1.00 mole CCl_4
c. 3.00 moles $Ba(NO_3)_2$ and 3.00 moles $BaSO_4$
d. 1.20 moles NH_4CN and 1.30 moles NH_4Cl

9.72 Which amount in each of the following pairs of amounts contains the greater number of total moles of atoms?
a. 3.00 moles Cl_2O and 3.00 moles Cl_2O_3
b. 1.00 mole C_2H_6 and 1.00 mole $SOCl_2$
c. 2.00 moles $CaSO_4$ and 2.00 moles NH_4Br
d. 1.00 mole $(NH_4)_2CO_3$ and 3.00 moles $Ba(OH)_2$

THE MOLE AND CHEMICAL CALCULATIONS (SEC. 9.10)

9.73 Calculate the number of atoms present in a 23.0 g sample of each of the following group IIA elements.
a. Be   b. Mg   c. Ca   d. Sr

9.74 Calculate the number of atoms present in a 17.0 g sample of each of the following group IA elements.
a. Li   b. Na   c. K   d. Rb

9.75 Calculate the number of molecules present in a 25.0 g sample of each of the following nitrogen oxides.
a. NO   b. N_2O c. N_2O_3 d. N_2O_5

9.76 Calculate the number of molecules present in a 50.0 g sample of each of the following sulfur oxides.
a. SO_2 b. SO_3 c. S_2O d. S_8O

9.77 Without actually doing a calculation, indicate whether each of the molecular counts in Problem 9.75 is greater than or less than a mole of molecules.

9.78 Without actually doing a calculation, indicate whether each of the molecular counts in Problem 9.76 is greater than or less than a mole of molecules.

9.79 What is the mass, in grams, of each of the following quantities of chemical substance?
a. 3.333×10^{23} atoms of P
b. 3.333×10^{23} molecules of PH_3
c. 2431 atoms of P
d. 2431 molecules of PH_3

9.80 What is the mass, in grams, of each of the following quantities of chemical substance?
a. 4.444×10^{23} atoms of Si
b. 4.444×10^{23} molecules of SiH_4
c. 1764 atoms of Si
d. 1764 molecules of SiH_4

9.81 What is the mass, in grams (to four significant figures), of a single atom (for elements) or single molecule (for compounds) of the following substances?
a. Na   b. Mg   c. C_4H_{10} d. C_6H_6

9.82 What is the mass, in grams (to four significant figures), of a single atom (for elements) or single molecule (for compounds) of the following substances?
a. Cu   b. B   c. C_2H_6 d. $C_{10}H_{20}$

9.83 Determine the number of grams of Cl present in each of the following amounts of chlorine-containing compounds.
a. 100.0 g of NaCl   b. 980.0 g of CCl_4
c. 10.0 g of HCl   d. 50.0 g of $BaCl_2$

9.84 Determine the number of grams of N present in each of the following amounts of nitrogen-containing compounds.
a. 100.0 g of $NaNO_3$ b. 980.0 g of N_2H_4
c. 10.0 g of Na_3N d. 50.0 g of KCN

9.85 Determine the number of phosphorus atoms present in 25.0 g of each of the following substances.
a. PF_3 b. Be_3P_2 c. $POCl_3$ d. $Na_5P_3O_{10}$

9.86 Determine the number of sulfur atoms present in 35.0 g of each of the following substances.
a. CS_2 b. S_4N_4 c. SF_6 d. $Al_2(SO_4)_3$

9.87 Determine the number of grams of O present in each of the following samples of oxygen-containing compounds.
a. 4.7×10^{24} molecules of XeO_3
b. 55.00 g of SO_2Cl_2
c. 0.30 mole of $C_6H_{12}O_6$
d. 475 g of Na_2CO_3

9.88 Determine the number of grams of S present in each of the following samples of sulfur-containing compounds.
a. 4.5×10^{25} molecules of H_2S
b. 65.00 g of H_2SO_3
c. 0.75 mole of $S_3N_3O_3Cl_3$
d. 675.0 g of SF_4

9.89 What amount or mass of each of the following substances is required to obtain 1.000 g of B?
a. moles of $B(OH)_3$
b. molecules of B_4H_{10}
c. grams of $C_2B_4H_6$
d. atoms of B

9.90 What amount or mass of each of the following substances is required to obtain 1.000 g of C?
a. moles of C_2H_6
b. molecules of $COCl_2$
c. grams of $(NH_4)_2CO_3$
d. atoms of C

9.91 What would be the mass, in grams, of an NH_3 sample in which each of the following were present?
a. 2.70×10^{23} atoms of nitrogen
b. 2.70×10^{23} atoms of hydrogen
c. 2.70×10^{23} total atoms
d. 2.70×10^{23} molecules

9.92 What would be the mass, in grams, of an N_2H_4 sample in which each of the following were present?
a. 6.77×10^{23} atoms of nitrogen
b. 6.77×10^{23} atoms of hydrogen
c. 6.77×10^{23} total atoms
d. 6.77×10^{23} molecules

9.93 What is the mass, in grams, of an SiH_4 sample with each of the following characteristics?
a. 20.0 g of silicon are present.
b. 5.75 moles of molecules are present.
c. 5.75 total moles of atoms are present.
d. A combined total of 25.0 g of silicon and hydrogen are present.

9.94 What is the mass, in grams, of an Si_2H_6 sample with each of the following characteristics?
a. 20.0 g of silicon are present.
b. 5.75 moles of molecules are present.
c. 5.75 total moles of atoms are present.
d. A combined total of 25.0 g of silicon and hydrogen are present.

9.95 Progesterone, a female hormone, has the formula $C_{21}H_{30}O_2$. In a 25.00 g sample of this compound,
a. how many moles of atoms are present?
b. how many atoms of C are present?
c. how many grams of O are present?
d. how many progesterone molecules are present?

9.96 Testosterone, a male hormone, has the formula $C_{19}H_{28}O_2$. In a 30.00 g sample of this compound,
a. how many moles of atoms are present?
b. how many atoms of H are present?
c. how many grams of C are present?
d. how many testosterone molecules are present?

9.97 A compound has a molar mass of 120.0 g and has molecules that are triatomic.
a. How many total atoms are present in a 60.0 g sample of the compound?
b. How many total molecules are present in a 90.0 g sample of the compound?

9.98 A compound has a molar mass of 80.0 g and has molecules that are tetratomic.
a. How many total atoms are present in a 120.0 g sample of the compound?
b. How many total molecules are present in a 40.0 g sample of the compound?

PURITY OF SAMPLES (SEC. 9.11)

9.99 Calculate the following for a 325 g sample of 95.4% pure Fe_2S_3.
a. mass, in grams, of Fe_2S_3 present
b. mass, in grams, of impurities present

9.100 Calculate the following for a 25.4 g sample of 88.7% pure Cu_2S.
a. mass, in grams, of Cu_2S present
b. mass, in grams, of impurities present

9.101 What is the percent purity by mass of a $CaCO_3$ sample in which 0.23 g of impurities are present per 32.21 g of $CaCO_3$?

9.102 What is the percent purity by mass of an $Mg(OH)_2$ sample in which 2.45 g of impurities are present per 63.41 g of $Mg(OH)_2$?

9.103 The mass percent purity of a Cu_2S sample is 85.0%. What is the mass, in grams, of the sample if 3.00 g of impurities are present?

9.104 The mass percent purity of a CuS sample is 95.0%. What is the mass, in grams, of the sample if 0.520 g of impurities are present?

9.105 How many copper atoms are present in a 35.00 g sample of impure copper that has a purity of 92.35% copper by mass? Assume that there are no copper-containing impurities present in the sample.

9.106 How many silver atoms are present in a 50.00 g sample of impure silver that has a purity of 89.98% silver by mass? Assume that there are no silver-containing impurities present in the sample.

9.107 What mass, in grams, of chromium is present in a 25.00 g sample of chromium ore that is 64.0% Cr_2O_3? Assume that impurities present do not contain any chromium.

9.108 What mass, in grams, of nickel is present in a 25.00 g sample of nickel ore that is 77.25% NiS? Assume that impurities present do not contain any nickel.

DETERMINATION OF EMPIRICAL FORMULAS (SECS. 9.13 AND 9.14)

9.109 What is the empirical formula for each of the following correctly written molecular formulas?
  a. C_2H_2 **b.** N_4S_4 **c.** C_2H_6O **d.** $B_3N_3H_6$

9.110 What is the empirical formula for each of the following correctly written empirical formulas?
  a. C_9H_{20} **b.** P_4O_{10} **c.** H_2O_2 **d.** SO_2Cl_2

9.111 Indicate whether both members of each of the following pairs of molecular formulas have the same empirical formula.
  a. C_6H_{14} and C_6H_{12} **b.** NH_3 and N_2H_4
  c. C_8H_8 and C_6H_6 **d.** NO_2 and N_2O_4

9.112 Indicate whether both members of each of the following pairs of molecular formulas have the same empirical formula.
  a. C_2H_4 and CH_4 **b.** N_2H_4 and C_2H_4
  c. C_3H_6 and C_4H_8 **d.** N_2O and NO_2

9.113 Determine both the molecular formula and the empirical formula for each of the following molecules.
  a. citric acid

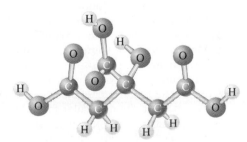

b. glucose

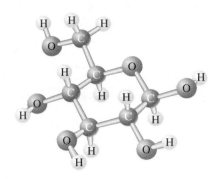

9.114 Determine both the molecular formula and the empirical formula for each of the following molecules.
  a. ribose

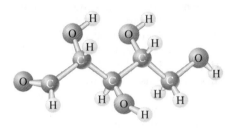

  b. vitamin C (ascorbic acid)

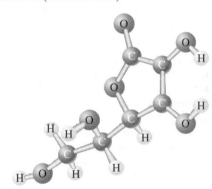

9.115 Given the following percent compositions, determine the empirical formula.
  a. 58.91% Na and 41.09% S
  b. 24.74% K, 34.76% Mn, and 40.50% O
  c. 2.06% H, 32.69% S, and 65.25% O
  d. 19.84% C, 2.50% H, 66.08% O, and 11.57% N

9.116 Given the following percent compositions, determine the empirical formula.
  a. 47.26% Cu and 52.74% Cl
  b. 40.27% K, 26.78% Cr, and 32.96% O

    c. 40.04% Ca, 12.00% C, and 47.96% O

    d. 28.03% Na, 29.28% C, 3.69% H, and 39.01% O

9.117 Convert each of the following fractional molar ratios to a whole-number molar ratio.

    a. 1.00 to 1.67   b. 1.00 to 1.50

    c. 2.00 to 2.33   d. 1.33 to 2.33 to 2.00

9.118 Convert each of the following fractional molar ratios to a whole-number molar ratio.

    a. 1.00 to 1.25   b. 1.00 to 2.33

    c. 3.00 to 4.50   d. 1.50 to 1.00 to 1.67

9.119 Determine the empirical formula for substances with each of the following percent compositions.

    a. 43.64% P and 56.36% O

    b. 72.24% Mg and 27.76% N

    c. 29.08% Na, 40.56% S, and 30.36% O

    d. 21.85% Mg, 27.83% P, and 50.32% O

9.120 Determine the empirical formula for substances with each of the following percent compositions.

    a. 54.88% Cr and 45.12% S

    b. 38.76% Cl and 61.24% O

    c. 59.99% C, 4.485% H, and 35.52% O

    d. 26.58% K, 35.35% Cr, and 38.06% O

9.121 Copper forms two sulfides. One has 33.54% S by mass, and the other 20.15% S by mass. What are the empirical formulas of these two sulfides?

9.122 Iron forms two sulfides. One has 36.48% S by mass, and the other 46.27% S by mass. What are the empirical formulas of these two sulfides?

9.123 Ethyl mercaptan is an odorous substance added to natural gas to make leaks easily detectable. Analysis of a sample of ethyl mercaptan indicates that 5.798 g of C, 1.46 g of H, and 7.740 g of S are present. What is the empirical formula of this compound?

9.124 Hydroquinone is a compound used in developing photographic film. Analysis of a sample of hydroquinone indicates that 16.36 g of C, 1.38 g of H, and 7.265 g of O are present. What is the empirical formula of this compound?

9.125 A 2.00 g sample of beryllium metal is burned in an oxygen atmosphere to produce 5.55 g of a beryllium–oxygen compound. Determine the compound's empirical formula.

9.126 A 2.00 g sample of lithium metal is burned in an oxygen atmosphere to produce 4.31 g of a lithium–oxygen compound. Determine the compound's empirical formula.

FORMULA DETERMINATION USING COMBUSTION ANALYSIS (SEC. 9.13)

9.127 Determine the empirical formula of the carbon–hydrogen compound that, upon combustion in a combustion analysis apparatus, generates each of the following sets of CO_2–H_2O data.

    a. 0.338 g of CO_2 and 0.277 g of H_2O

    b. 0.303 g of CO_2 and 0.0621 g of H_2O

    c. 0.225 g of CO_2 and 0.115 g of H_2O

    d. 0.314 g of CO_2 and 0.192 g of H_2O

9.128 Determine the empirical formula of the carbon–hydrogen compound that, upon combustion in a combustion analysis apparatus, generates each of the following sets of CO_2–H_2O data.

    a. 0.269 g of CO_2 and 0.221 g of H_2O

    b. 0.294 g of CO_2 and 0.120 g of H_2O

    c. 0.600 g of CO_2 and 0.184 g of H_2O

    d. 0.471 g of CO_2 and 0.0963 g of H_2O

9.129 Determine the empirical formula of the carbon–hydrogen compound that, upon combustion of a 0.7420 mg sample in a combustion analysis apparatus, generates 2.328 mg of CO_2.

9.130 Determine the empirical formula of the carbon–hydrogen compound that, upon combustion of a 0.4244 mg sample in a combustion analysis apparatus, generates 1.164 mg of CO_2.

9.131 A 3.750 g sample of the compound responsible for the odor of cloves (containing only C, H, and O) is burned in a combustion analysis apparatus. The mass of CO_2 produced is 10.05 g, and the mass of H_2O produced is 2.47 g. What is the empirical formula of the compound?

9.132 A 0.8640 g sample of the compound responsible for the pungent odor of rancid butter (containing only C, H, and O) is burned in a combustion analysis apparatus. The mass of CO_2 produced is 1.727 g, and the mass of H_2O produced is 0.7068 g. What is the empirical formula of the compound?

DETERMINATION OF MOLECULAR FORMULAS (SEC. 9.14)

9.133 Determine the molecular formulas of compounds with the following empirical formulas and molar masses.

    a. CH_2, 42.08 amu   b. NaS_2O_3, 270.26 amu

    c. $C_3H_6O_2$, 74.09 amu   d. CHN, 135.15 amu

9.134 Determine the molecular formulas of compounds with the following empirical formulas and molar masses.

    a. C_2HCl, 181.44 amu   b. NO_2, 92.02 amu

    c. CB_2H_3, 73.3 amu   d. $SNCl_2$, 350.94 amu

9.135 A compound has the generalized empirical formula XY_3. Calculate the molecular formula of the compound for each of the following cases:

    a. Molecules of the compound contain three X atoms.

    b. Molecules of the compound contain three Y atoms.

    c. Molecules of the compound contain a total of eight atoms.

    d. Molecules of the compound are tetratomic.

9.136 A compound has the generalized empirical formula X_2Y_3. Calculate the molecular formula of the compound for each of the following cases:

    a. Molecules of the compound contain four X atoms.

    b. Molecules of the compound contain six Y atoms.

    c. Molecules of the compound contain a total of fifteen atoms.

    d. Molecules of the compound are pentatomic.

9.137 A compound has an empirical formula of C_2H_3O. Calculate the molecular formula of the compound for each of the following cases:
  a. The compound's molar mass is twice the compound's empirical formula mass.
  b. Molecules of the compound contain 18 atoms.
  c. The sum of the carbon and oxygen atoms in a molecule of the compound is 18.
  d. The mass of 0.010 mole of the compound is 0.86 g.

9.138 A compound has an empirical formula of C_3H_5O. Calculate the molecular formula of the compound for each of the following cases:
  a. The compound's molar mass and empirical formula mass differ by a factor of 2.
  b. Molecules of the compound contain 18 atoms.
  c. Molecules of the compound contain more than 20 atoms but fewer than 30 atoms.
  d. The mass of 0.010 mole of the compound is 0.57 g.

9.139 Methyl benzoate, a compound used in the manufacture of perfumes, has a molar mass of 136 amu, and its percent composition by mass is 70.57% C, 5.93% H, and 23.49% O. Determine the molecular formula of methyl benzoate using
  a. 100.0 g of methyl benzoate as the basis for the calculation.
  b. 1.00 mole of methyl benzoate as the basis for the calculation.

9.140 Adipic acid, a compound used as a raw material for the manufacture of nylon, has a molar mass of 146 amu, and its percent composition by mass is 49.30% C, 6.91% H, and 43.79% O. Determine the molecular formula of adipic acid, using
  a. 100.0 g of adipic acid as the basis for the calculation.
  b. 1.00 mole of adipic acid as the basis for the calculation.

9.141 Lactic acid, the substance that builds up in muscles and causes them to hurt when they are worked hard, has a molar mass of 90.0 amu and a percent composition by mass of 40.0% C, 6.71% H, and 53.3% O. Determine the molecular formula of lactic acid, using
  a. 100.0 g of lactic acid as the basis for the calculation.
  b. 1.00 mole of lactic acid as the basis for the calculation.

9.142 Citric acid, a flavoring agent in many carbonated beverages, has a molar mass of 192 amu and a percent composition by mass of 37.50% C, 4.21% H, and 58.29% O. Determine the molecular formula of citric acid, using
  a. 100.0 g of citric acid as the basis for the calculation.
  b. 1.00 mole of citric acid as the basis for the calculation.

ADDITIONAL PROBLEMS

9.143 Select the quantity that has the greater mass in each of the following pairs of quantities. Make your selection using the periodic table but without performing an actual calculation.
  a. 1.00 mole of silver or 1.00 mole of gold
  b. 1.00 mole of sulfur or 6.022×10^{23} atoms of carbon
  c. 1.00 mole of Cl atoms or 1.00 mole of Cl_2 molecules
  d. 8.00 g of He or 6.022×10^{23} atoms of Ne

9.144 Select the quantity that has the greater mass in each of the following pairs of quantities. Make your selection using the periodic table but without performing an actual calculation.
  a. 1.00 mole of tin or 1.00 mole of lead
  b. 1.00 mole of phosphorus or 6.022×10^{23} atoms of boron
  c. 1.00 mole of O atoms or 1.00 mole of O_2 molecules
  d. 7.5 g of Be or 6.022×10^{23} atoms of Li

9.145 Select the quantity that has the greater number of atoms in each of the following pairs of quantities. Make your selection using the periodic table but without performing an actual calculation.
  a. 1.00 mole of P or 1.00 mole of P_4
  b. 21.0 g of Na or 1.00 mole of Na
  c. 63.5 g of Cu or 8.0 g of B
  d. 1.00 g of K or 6.022×10^{23} atoms of Be

9.146 Select the quantity that has the greater number of atoms in each of the following pairs of quantities. Make your selection using the periodic table but without performing an actual calculation.
  a. 1.00 mole of S or 1.00 mole of S_8
  b. 28.0 g of Al or 1.00 mole of Al
  c. 28.1 g of Si or 30.0 g of Mg
  d. 2.00 g of Na or 6.022×10^{23} atoms of He

9.147 Indicate whether each of the following statements concerning O_2 and O_3 molecules is *true* or *false*.
  a. The mass of one mole of each of these molecules is the same.
  b. The number of molecules in one mole of each of these molecules is the same.
  c. The mass of oxygen in one mole of each of these molecules is the same.
  d. The number of atoms of oxygen in one mole of each of these molecules is the same.

9.148 Indicate whether each of the following statements concerning S_2 and S_8 molecules is *true* or *false*.
  a. The mass of one mole of each of these molecules is the same.
  b. The number of molecules in one mole of each of these molecules is the same.
  c. The mass of sulfur in one mole of each of these molecules is the same.
  d. The number of atoms of sulfur in one mole of each of these molecules is the same.

9.149 How many grams of potassium and sulfur are theoretically needed to make 4.000 g of K_2S?

9.150 How many grams of beryllium and nitrogen are theoretically needed to make 3.00 g of Be_3N_2?

9.151 How many grams of B would contain the same number of atoms as there are in 3.50 moles of Xe?

9.152 How many grams of Si would contain the same number of atoms as there are in 2.10 moles of Ar?

9.153 How many grams of glucose, $C_6H_{12}O_6$, would contain the same mass of carbon as there is in 3.44 g of ethanol, C_2H_6O?

9.154 How many grams of ethanol, C_2H_6O, would contain the same mass of oxygen as there is in 7.08 g of glucose, $C_6H_{12}O_6$?

9.155 How many grams of O are combined with 7.23×10^{24} atoms of Al in the compound aluminum oxide (Al_2O_3)?

9.156 How many grams of O are combined with 6.67×10^{25} atoms of K in the compound potassium oxide (K_2O)?

9.157 For the compound $(CH_3)_3SiCl$, calculate, to two significant figures, the
   a. mass percent of H present.
   b. atom percent of H present.
   c. mole percent of H present.

9.158 For the compound $(CH_3)_2SiCl_2$, calculate, to two significant figures, the
   a. mass percent of H present.
   b. atom percent of H present.
   c. mole percent of H present.

9.159 What are the empirical formulas for the compounds that contain each of the following?
   a. 9.0×10^{23} atoms of Na, 3.0×10^{23} atoms of Al, and 1.8×10^{24} atoms of F
   b. 3.2 g of S and 1.20×10^{23} atoms of O
   c. 0.36 mole of Ba, 0.36 mole of C, and 17.2 g of O
   d. 1.81×10^{23} atoms of H, 10.65 g of Cl, and 0.30 mole of O atoms

9.160 What are the empirical formulas for the compounds that contain each of the following?
   a. 3.0×10^{30} atoms of Fe, 3.0×10^{30} atoms of Cr, and 1.2×10^{31} atoms of O
   b. 0.0023 g of N and 2.0×10^{20} atoms of O
   c. 0.40 mole of Li, 6.4 g of S, and 0.80 mole of O
   d. 0.15 mole of S, 1.8×10^{23} atoms of O, and 5.7 g of F

9.161 A sample of a hydrogen–oxygen compound with a mass of 0.331 g contains 0.311 g of O. The compound's molar mass is found to be 34.02 g/mole. Determine the empirical and molecular formulas of the compound.

9.162 A sample of a hydrogen–nitrogen compound with a mass of 0.778 g contains 0.680 g of N. The compound's molar mass is found to be 32.06 g/mole. Determine the empirical and molecular formulas of the compound.

9.163 Calculate the number of carbon atoms in 5.25 g of a compound that contains 92.26% C and 7.74% H by mass.

9.164 Calculate the number of nitrogen atoms in 5.25 g of a compound that contains 87.39% N and 12.61% H by mass.

9.165 A sample of a compound containing only C and H is burned in oxygen, and 13.75 g of CO_2 and 11.25 g of H_2O are obtained. What was the mass of the sample, in grams, that was burned?

9.166 A sample of a compound containing only C and H is burned in oxygen, and 14.66 g of CO_2 and 9.00 g of H_2O are obtained. What was the mass of the sample, in grams, that was burned?

9.167 A sample of a compound containing only C, H, and S was burned in oxygen, and 6.60 g of CO_2, 5.41 g of H_2O, and 9.61 g of SO_2 were obtained.
   a. What is the empirical formula of the compound?
   b. What was the mass, in grams, of the sample that was burned?

9.168 A sample of a compound containing only C, H, and N was burned in oxygen, and 6.60 g of CO_2, 6.76 g of H_2O, and 4.50 g of NO were obtained.
   a. What is the empirical formula of the compound?
   b. What was the mass, in grams, of the sample that was burned?

9.169 A sample containing NaF, Na_2SO_4, and $NaNO_3$ gives the following elemental analysis by mass: 18.1% F and 6.60% N. Calculate the mass percent of each compound in the mixture.

9.170 A sample containing NaF, Na_2SO_4, and $NaNO_3$ gives the following elemental analysis by mass: 11.3% F and 5.65% S. Calculate the mass percent of each compound in the mixture.

9.171 A gaseous mixture is 5.000% by mass CO_2, 10.00% by mass N_2O, and the remainder is H_2O. What is the mass percent of each element present in the mixture?

9.172 A gaseous mixture is 5.000% by mass CO, 10.00% by mass NO_2, and the remainder is H_2O. What is the mass percent of each element present in the mixture?

9.173 By analysis, a compound with the formula $KClO_x$ is found to contain 28.9% chlorine by mass. What is the value of the integer x in the compound's formula?

9.174 By analysis, a compound with the formula H_3AsO_x is found to contain 52.78% arsenic by mass. What is the value of the integer x in the compound's formula?

9.175 A 7.503 g sample of metal is reacted with excess oxygen to yield 10.498 g of the oxide MO. Calculate the molar mass of the element M.

9.176 An 11.17 g sample of metal is reacted with excess oxygen to yield 15.97 g of the oxide M_2O_3. Calculate the molar mass of the element M.

9.177 A certain compound contains only carbon, hydrogen, and oxygen. If it contains 47.4% carbon by mass and if there is one oxygen atom present for every four hydrogen atoms, what is its empirical formula?

9.178 A certain compound contains only lead, carbon, and hydrogen. If it contains 64.07% lead by mass and if there are two carbon atoms present for every five hydrogen atoms, what is its empirical formula?

CUMULATIVE PROBLEMS

9.179 Calculate the molar mass, in amu, of the compound H_3AsO_4 to the following number of significant figures.
a. three **b.** four **c.** five **d.** six

9.180 Calculate the molar mass, in amu, of the compound H_3AsO_3 to the following number of significant figures.
a. three **b.** four **c.** five **d.** six

9.181 Calculate the molar mass, to four significant figures, of each of the following compounds.
a. magnesium sulfate
b. potassium phosphide
c. copper(II) bromide
d. ammonium dihydrogen phosphate

9.182 Calculate the molar mass, to four significant figures, of each of the following compounds.
a. calcium thiosulfate
b. nickel(II) hydroxide
c. ammonium nitrate
d. aluminum hydrogen phosphate

9.183 Calculate the percent by mass of sulfur in each of the following compounds.
a. disulfur monoxide **b.** sulfurous acid
c. ammonium sulfate **d.** hydrosulfuric acid

9.184 Calculate the percent by mass of nitrogen in each of the following compounds.
a. sodium nitride **b.** potassium azide
c. ammonium cyanide **d.** nitrous acid

9.185 Calculate the density of the metal lead, in g/cm^3, given that 0.422 mole of lead occupies a volume of 7.74 cm^3.

9.186 Calculate the density of the metal potassium, in g/cm^3, given that 0.674 mole of potassium occupies a volume of 30.8 cm^3.

9.187 What is the volume, in liters, of 3.752 moles of carbon dioxide gas (CO_2) at 25°C if the density of carbon dioxide is 1.96 g/L at that temperature?

9.188 What is the volume, in liters, of 2.573 moles of methane gas (CH_4) at 25°C if the density of methane is 0.66 g/L at that temperature?

9.189 At a certain temperature, the density of water is 1.00 g/mL and the density of ethyl alcohol (C_2H_6O) is 0.789 g/mL. At this temperature, what volume of water contains the same number of molecules as are present in 225 mL of ethyl alcohol?

9.190 At a certain temperature, the density of water is 1.00 g/mL and the density of ethyl alcohol (C_2H_6O) is 0.789 g/mL. At this temperature, what volume of ethyl alcohol contains the same number of molecules as are present in 122 mL of water?

9.191 The percent natural abundance of $^{40}_{19}K$ is 0.012%. How many $^{40}_{19}K$ atoms does a person ingest by drinking one cup of whole milk containing 371 mg of K per cup?

9.192 The percent natural abundance of $^{41}_{19}K$ is 6.88%. How many $^{41}_{19}K$ atoms does a person ingest by drinking one cup of whole milk containing 392 mg of K?

9.193 A 2.33 mole sample of the ionic compound $Al_2(SO_4)_3$ contains how many of the following?
a. $Al_2(SO_4)_3$ formula units
b. Al^{3+} ions
c. SO_4^{2-} ions
d. total ions

9.194 A 3.50 mole sample of the ionic compound $(NH_4)_3PO_4$ contains how many of the following?
a. $(NH_4)_3PO_4$ formula units
b. NH_4^+ ions
c. PO_4^{3-} ions
d. total ions

9.195 What mass of KCl would contain the same *total* number of *ions* as 32.5 g of $CaCl_2$?

9.196 What mass of LiF would contain the same *total* number of *ions* as 32.5 g of $MgCl_2$?

9.197 A mixture consists of 42.0% NaCl and 58.0% $CaCl_2$, by mass. What is the total number of chloride ions (Cl^-) present in 425 g of mixture?

9.198 A mixture consists of 22.0% $Cu(NO_3)_2$ and 78.0% $Fe(NO_3)_3$, by mass. What is the total number of nitrate ions (NO_3^-) present in 25.00 g of mixture?

9.199 The mass percent composition for an alloy that has a density of 8.31 g/cm^3 is 56.0% copper, 43.0% nickel, and 1.0% manganese. How many nickel atoms are in a block of this alloy measuring 10.0 cm × 30.0 cm × 62.0 cm?

9.200 The mass percent composition for an alloy that has a density of 8.28 g/cm^3 is 64.0% iron, 12.0% cobalt, and 24.0% molybdenum. How many molybdenum atoms are in a block of this alloy measuring 2.0 cm × 1.5 cm × 6.7 cm?

9.201 What volume, in milliliters, of an NaOH solution that is 12.0% NaOH, by mass, contains 0.275 mole of NaOH? The density of the solution is 1.131 g/mL.

9.202 What volume, in milliliters, of an H_3PO_4 solution that is 85.5% H_3PO_4, by mass, contains 0.100 mole of H_3PO_4? The density of the solution is 1.70 g/mL.

Multiple-Choice Practice Test

Use this bank of 20 multiple-choice questions as a review of key concepts presented in this chapter. For many of the questions, there may be more than one correct answer (choice d) or no correct answer (choice e).

9.203 Which of the following are the values of the formula masses, respectively, of the compounds NO and N_2O?
   a. 15.00 amu and 22.00 amu
   b. 30.01 amu and 60.02 amu
   c. 22.01 amu and 44.02 amu
   d. more than one correct response
   e. no correct response

9.204 A compound with the chemical formula SO_n has a formula mass of 80.07 amu. What is the value for n in the formula SO_n?
   a. one   **b.** two   **c.** three   **d.** four
   e. no correct response

9.205 The mass percentage of O present in the compound CO_2 is
   a. 27.29%.   **b.** 36.36%.   **c.** 57.12%.   **d.** 72.71%.
   e. no correct response

9.206 Which of the following statements concerning *Avogadro's number* is correct?
   a. It has the numerical value of 6.022×10^{23}.
   b. It denotes the number of molecules present in 1 mole of any molecular compound.
   c. It is the mass, in grams, of 1 mole of any substance.
   d. more than one correct response
   e. no correct response

9.207 One mole of a chemical compound is the mass of the compound, in grams, that
   a. will combine with 12.00 g of C.
   b. will combine with 100.00 g of O.
   c. is numerically equal to its molecular mass.
   d. more than one correct response
   e. no correct response

9.208 Avogadro's number of Al atoms has a mass equal to
   a. 6.022×10^{23} g Al.
   b. 6.022×10^{-23} g Al.
   c. 26.98 g Al.
   d. more than one correct response
   e. no correct response

9.209 Which of the following is a correct statement?
   a. A nitrogen atom has a mass of 2.57×10^{-23} g.
   b. An aluminum atom has a mass of 4.48×10^{-23} g.
   c. A copper atom has a mass of 9.97×10^{-23} g.
   d. more than one correct response
   e. no correct response

9.210 Which of the following sets of information about a sample of a compound is sufficient to calculate the compound's molar mass?
   a. contains nitrogen and oxygen; 1.23 moles are present
   b. 0.300 mole is present; 24.36 g are present
   c. 5.6×10^{23} atoms are present; 37.23 g are present

   d. more than one correct response
   e. no correct response

9.211 For which of the following compounds does 1.9 g represent 4.3×10^{-2} moles?
   a. CO_2
   b. C_3H_8
   c. H_2O_2
   d. more than one correct response
   e. no correct response

9.212 Which of the following samples contains 6 moles of atoms?
   a. 2 moles of CO_2
   b. 3 moles of NaCl
   c. 6 moles of O_2
   d. more than one correct response
   e. no correct response

9.213 One mole of Cl_2O molecules contains
   a. 16.00 g of O.
   b. 35.45 g of Cl.
   c. 70.90 g of Cl.
   d. more than one correct response
   e. no correct response

9.214 Which of the following is the correct dimensional analysis setup for the problem "How many atoms are present in 10.00 g of S?"
   a. $10.00 \text{ g S} \times \left(\dfrac{6.022 \times 10^{23} \text{ atoms S}}{1.000 \text{ g S}} \right)$

   b.
$$10.00 \text{ g S} \times \left(\frac{1 \text{ mole S}}{32.07 \text{ g S}} \right) \times \left(\frac{1 \text{ atom S}}{6.022 \times 10^{23} \text{ moles S}} \right)$$

   c.
$$10.00 \text{ g S} \times \left(\frac{1 \text{ mole S}}{32.07 \text{ g S}} \right) \times \left(\frac{6.022 \times 10^{23} \text{ atoms S}}{1 \text{ mole S}} \right)$$
   d. more than one correct response
   e. no correct response

9.215 The given dimensional analysis set for the problem "What is the mass, in grams, of 3.002×10^6 atoms of F?" is correct, except numbers in the middle conversion factor have been replaced by the letters A and B. What are the numerical values of A and B, respectively?

$$3.002 \times 10^6 \text{ atoms F} \times \left(\frac{A \text{ moles F}}{B \text{ atoms F}} \right) \times \left(\frac{19.00 \text{ g F}}{1 \text{ mole F}} \right)$$
   a. 1 and 6.022×10^{23}
   b. 6.022×10^{23} and 1
   c. 19.00 and 6.022×10^{23}
   d. more than one correct response
   e. no correct response

9.216 How many molecules of H_2O are present in 28.01 g of H_2O?
   a. 1.193×10^{21} molecules
   b. 3.874×10^{23} molecules

c. 9.361×10^{23} molecules
d. 3.040×10^{26} molecules
e. no correct response

9.217 How many moles of Na atom are present in 100.0 g of Na_2S?
a. 0.7804 mole Na   b. 1.281 moles Na
c. 1.561 moles Na   d. 2.563 moles Na
e. no correct response

9.218 What is the empirical formula for a compound whose molecular formula is $C_2H_2O_4$?
a. CHO   b. $C_2H_2O_2$ c. $C_4H_4O_8$ d. $C_4H_2O_4$
e. no correct response

9.219 Analysis of a sample of a compound shows that 5.00 moles of C, 6.50 moles of H, and 2.00 moles of O are present. What is the empirical formula of the compound?
a. C_2H_3O b. C_3H_3O c. $C_5H_7O_2$ d. $C_{10}H_{13}O_4$
e. no correct response

9.220 What is the empirical formula of a compound that contains 63.65% N and 36.35% O?
a. NO   b. NO_2 c. N_2O_3 d. N_2O_5
e. no correct response

9.221 If a 1.00 g sample of a carbon-containing compound is burned in air, 0.75 g of CO_2 (the only carbon-containing product) is produced. What is the mass percent carbon in the compound?
a. 20.5% by mass   b. 41.0% by mass
c. 57.7% by mass   d. 75.0% by mass
e. no correct response

9.222 A 2.00 mole sample of H_2O contains
a. 2.02 g of H atoms.
b. 6.022×10^{23} atoms of H.
c. 32.00 g of O atoms.
d. more than one correct response
e. no correct response

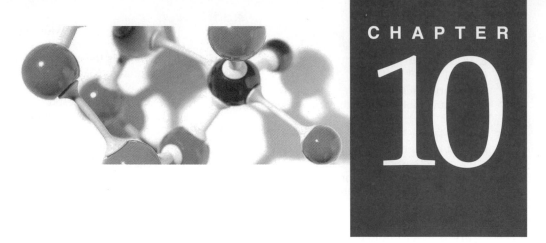

Chemical Calculations Involving Chemical Equations

10.1 THE LAW OF CONSERVATION OF MASS

In an earlier consideration of chemical change (Sec. 4.4), it was noted that chemical changes are referred to as chemical reactions. As we learned there, a *chemical reaction* is a process in which at least one new substance is produced as a result of chemical change. It is usually easy to see that a chemical reaction has occurred. Color change, emission of heat and/or light, gas evolution, and solid formation are an indication that a chemical reaction has taken place.

The starting materials for a chemical reaction are known as *reactants*. **Reactants** *are the starting substances that undergo change in a chemical reaction*. As a chemical reaction proceeds, reactants are consumed (used up) and new materials with new chemical properties are produced. **Products** *are the substances produced as a result of a chemical reaction*.

From a molecular viewpoint, a chemical reaction (chemical change) involves the union, separation, or rearrangement of atoms to produce new substances. Figure 10.1 shows the rearrangement of atoms that occurs when methane (CH_4) reacts with oxygen (O_2) to produce carbon dioxide (CO_2) and water (H_2O). Hydrogen atoms originally associated with carbon atoms (CH_4) became associated with oxygen

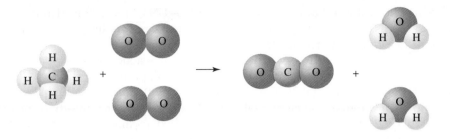

FIGURE 10.1
Rearrangement of atoms that occurs when methane (CH_4) reacts with oxygen (O_2). The products are carbon dioxide (CO_2) and water (H_2O).

atoms (H_2O) as the result of the chemical reaction. This chemical reaction is an example of a combustion reaction (Sec. 10.5); energy is also a product of the chemical reaction.

Studies of countless chemical reactions over a period of more than 200 years have shown that there is no detectable change in the quantity of matter present during an ordinary chemical reaction. This generalization concerning chemical reactions has been formalized into a statement known as the **law of conservation of mass**: *mass is neither created nor destroyed in any ordinary chemical reaction*. To demonstrate the validity of this law, the masses of all reactants (substances that react together) and all products (substances formed) in a chemical reaction are carefully determined. It is found that the sum of the masses of the products is always the same as the sum of the masses of the reactants. The

The Human Side of Chemistry 11

Antoine-Laurent Lavoisier (1743–1794)

Antoine-Laurent Lavoisier (pronounced Lav-wazy-ay), the son of a wealthy Parisian lawyer, is often called the "father of modern chemistry." It is he who first appreciated the importance of carrying out very accurate (quantitative) measurements of chemical change. His work was performed on balances he had specially designed that were more accurate than any then known. From his balance work came the discovery of the law of conservation of mass. He was also the first to make quantitative measurements on the heat produced during chemical reactions.

Originally trained as a lawyer, Lavoisier entered the field of science through geology and from there became interested in chemistry. This change in focus took place during his early twenties.

Lavoisier's studies on combustion are considered his major work. He was the first to realize that combustion involves the reaction of oxygen from air with the substance that is burned. He gave the name oxygen to the gas then known as "dephlogisticated air." In later years, he became interested in physiological chemistry.

Lavoisier's *Elementary Treatise on Chemistry*, published in 1789, was the first textbook based on quantitative experiments. He was also one of the first to use systematic nomenclature for elements and a few compounds.

In 1790 he was appointed secretary and treasurer of a commission established to standardize weights and measures used throughout France. The outgrowth of this commission's work was the metric system.

Early in life he invested money in a private tax-collecting firm and married the daughter of one of the company executives. Firms of this type were licensed by the state to collect taxes and keep a portion of the proceeds. Lavoisier used his earnings from this endeavor (about 100,000 francs a year) to support his scientific work. This connection with tax collecting proved fatal.

During the French Revolution all associated with this type of activity, known as "tax-farming," were denounced, arrested, and guillotined. Lavoisier's death, at age 51, came just two months before the end of the Revolution. On the day of his death, one of Lavoisier's scientific colleagues gave this tribute: "It took but a moment to cut off that head; perhaps a hundred years will be required to produce another like it."

French chemist Antoine-Laurent Lavoisier (1743–1794; see "The Human Side of Chemistry 11") is given credit for being the first to state this important relationship between the reactants and products of a chemical reaction.

Consider, as an illustrative example of this law, the reaction of known masses of the elements beryllium (Be) and oxygen (O) to form the compound beryllium oxide (BeO). Experimentally, it is found that 36.03 g of Be will react with *exactly* 63.97 g of O. After the reaction, no Be or O remains in elemental form; the only substance present is the product BeO, combined Be and O. When this product is weighed, its mass is found to be 100.00 g, which is the sum of the masses of the reactants (36.03 g + 63.97 g = 100.00 g). It is also found that when the 100.00 g of product BeO is heated to a high temperature in the absence of air, the BeO decomposes into Be and O, producing 36.03 g of Be and 63.97 g of O. Once again, no detectable mass change is observed; the mass of the reactants is equal to the mass of the products.

The law of conservation of mass is consistent with the statements of atomic theory (Sec. 4.10). Since all reacting chemical substances are made up of atoms (statement 1), each with its unique identity (statement 2), and these atoms can be neither created nor destroyed in a chemical reaction but merely rearranged (statement 4), it follows that the total mass after the reaction must equal the total mass before the reaction. We have the same number of atoms of each kind after the reaction as we started out with. An alternative way of stating the law of conservation of mass is *The total mass of reactants and the total mass of products in a chemical reaction are always equal.*

The law of conservation of mass applies to all ordinary chemical reactions, and there is no known case of a *measurable* change in total mass during an *ordinary* chemical reaction.* This law will be a guiding principle for the discussion that follows about chemical equations and their use.

10.2 WRITING CHEMICAL EQUATIONS

A **chemical equation** *is a representation for a chemical reaction that uses chemical symbols and chemical formulas instead of words to describe the changes that occur in a chemical reaction.* The following example shows the contrast between a word description of a chemical reaction and a chemical equation for the same reaction.

Word description: Magnesium oxide reacts with carbon to produce carbon monoxide and magnesium.

Chemical equation: $MgO + C \longrightarrow CO + Mg$

In the same way that chemical symbols are considered the *letters* of chemical language and chemical formulas the *words* of the language, chemical equations can be considered the *sentences* of chemical language.

The conventions used in writing chemical equations are

1. The correct formulas of the *reactants* are always written on the *left* side of the equation.

$$MgO + C \longrightarrow CO + Mg$$

---

*In the last half-century the law of conservation of mass has had to be qualified. Certain types of reactions that involve radioactive processes have been found to deviate from this law. In these processes, there is a conversion of a small amount of matter into energy rather than into another form of matter. A more general law incorporates this apparent discrepancy—the law of conservation of mass and energy. This law takes into account the fact that matter and energy are interconvertible. Note that the statement of the law of conservation of mass, as given at the start of this discussion, contains the phrase "ordinary chemical reaction." Radioactive processes are not considered to be ordinary chemical reactions.

2. The correct formulas of the *products* are always written on the *right* side of the equation.

$$MgO + C \longrightarrow CO + Mg$$

3. The reactants and products are separated by an arrow pointing toward the products.

$$MgO + C \longrightarrow CO + Mg$$

4. Plus signs are used to separate different reactants or different products from each other.

$$MgO + C \longrightarrow CO + Mg$$

In reading chemical equations, plus signs on the reactant side of the equation are taken to mean "reacts with"; the arrow, "to produce"; and plus signs on the product side, "and."

A catchy, informal way of defining a chemical equation is to say that it gives the before-and-after picture of a chemical reaction. *Before* the reaction starts, only reactants are present—the left side of the equation. *After* the reaction is completed, products are present—the right side of the equation.

For a chemical equation to be *valid*, it must satisfy two conditions.

1. *It must be consistent with experimental facts.* Only the reactants and products actually involved in a reaction are shown in an equation. An accurate chemical formula must be used for each of these substances. For compounds, molecular rather than empirical formulas (Sec. 9.11) are always used. Elements in the solid and liquid states are represented in equations by the chemical symbol for the element. Elements that are gases at room temperature are represented by the molecular form in which they actually occur in nature. Monoatomic, diatomic, and tetratomic elemental gases are known.

Monoatomic: He, Ne, Ar, Kr, Xe, Rn
Diatomic: H_2, O_2, N_2, F_2, Cl_2, Br_2 (vapor)*, I_2 (vapor)*
Tetratomic: P_4(vapor)*, As_4 (vapor)*

2. *It must be consistent with the law of conservation of mass* (Sec. 10.1). There must be the same number of product atoms of each kind as there are reactant atoms of each kind, because atoms are neither created nor destroyed in an ordinary chemical reaction. Equations that satisfy the conditions of this law are said to be *balanced*. Using the four conventions previously listed for writing equations does not guarantee a balanced equation. Sections 10.3 and 10.4 consider the steps that must be taken to ensure that an equation is balanced.

> In a chemical equation, the *reactants* (starting materials in a chemical reaction) are always written on the left side of the equation, and the *products* (substances produced in a chemical reaction) are always written on the right side of the equation.

> The diatomic elemental gases are the elements whose names end in *-gen* (hydrogen, oxygen, and nitrogen) or *-ine* (fluorine, chlorine, bromine, and iodine).

10.3 CHEMICAL EQUATION COEFFICIENTS

A **balanced chemical equation** *is a chemical equation that has the same number of atoms of each element involved in the reaction on each side of the equation.* It is therefore an equation consistent with the law of conservation of mass (Sec. 10.1).

An unbalanced equation is brought into balance by adding coefficients to the equation; such coefficients adjust the number of reactant and/or product molecules (or formula units) present. An **equation coefficient** *is a number placed to the left of a chemical formula in a chemical equation that changes the amount but not the identity of a substance.* In the notation 2 H_2O, the 2 on the left is a coefficient; 2 H_2O means two molecules of H_2O, and

---

*The four elements listed as vapors are not gases at room temperature but vaporize at slightly higher temperatures. The resultant vapors contain molecules with the formulas indicated. Even if these elements do not vaporize, they are still represented with these formulas.

3 H_2O means three molecules of H_2O. Equation coefficients tell how many formula units of a given substance are present.

The following is a balanced chemical equation with the equation coefficients shown in color.

$$3\,Cu + 8\,HNO_3 \longrightarrow 3\,Cu(NO_3)_2 + 2\,NO + 4\,H_2O$$

The message of this balanced equation is "three Cu atoms react with eight HNO_3 molecules to produce three $Cu(NO_3)_2$ formula units, two NO molecules, and four H_2O molecules." A coefficient of 1 in a balanced equation is not explicitly written; it is considered to be understood. Both PCl_3 and H_3PO_3 have understood coefficients of 1 in the following balanced equation.

$$PCl_3 + 3\,H_2O \longrightarrow H_3PO_3 + 3\,HCl$$

The distinction between an *equation coefficient* placed in front of a chemical formula and a *subscript* in a chemical formula is a very important difference. An equation coefficient placed in front of a chemical formula applies to the whole formula. In contrast, subscripts, also present in formulas, affect only parts of a chemical formula.

The above notation denotes two molecules of H_2O; it also denotes a total of four H atoms and two O atoms.

Changing a chemical formula subscript affects the *identity* of the substance. Changing the subscript 2 in the formula SO_2 to a 3 to give SO_3 produces an identity change. The substances SO_2 (sulfur dioxide) and SO_3 (sulfur trioxide) are different compounds with distinctly different properties. In contrast, placing a coefficient in front of a chemical formula does not change *identity* but rather changes *amount*. The notation 2 SO_2 means two molecules of SO_2 and the notation 3 SO_2 means three molecules of SO_2.

Example 10.1 illustrates further the mathematical significance of equation coefficients.

EXAMPLE 10.1 | Using Both Equation Coefficients and Chemical Formulas to Determine Number of Atoms Present

How many oxygen atoms are part of each of the following chemical expressions?

a. 3 SO_2 **b.** 4 SO_3 **c.** 7 $H_2S_2O_3$ **d.** 6 $Al_2(SO_4)_3$

SOLUTION

In each case the numerical coefficient that precedes the chemical formula gives the number of molecules or formula units of compound present. Equation coefficients affect all parts of the chemical formulas with which they are associated.

a. The expression 3 SO_2 denotes three SO_2 molecules. The total number of oxygen atoms present is

$$3\;\text{SO}_2\;\text{molecules} \times \frac{2\,\text{O atoms}}{1\;\text{SO}_2\;\text{molecule}} = 6\,\text{O atoms} \quad \left(\begin{array}{l}\text{calculator and}\\\text{correct answer}\end{array}\right)$$

b. Similarly, we have, for four SO_3 molecules

$$4\;\text{SO}_3\;\text{molecules} \times \frac{3\,\text{O atoms}}{1\;\text{SO}_3\;\text{molecule}} = 12\,\text{O atoms} \quad \left(\begin{array}{l}\text{calculator and}\\\text{correct answer}\end{array}\right)$$

c. Similarly, we have

$$7 \text{ H}_2\text{S}_2\text{O}_3 \text{ molecules} \times \frac{3 \text{ O atoms}}{1 \text{ H}_2\text{S}_2\text{O}_3 \text{ molecule}} = 21 \text{ O atoms} \begin{pmatrix} \text{calculator and} \\ \text{correct answer} \end{pmatrix}$$

d. $6 \text{ Al}_2(\text{SO}_4)_3 \text{ formula units} \times \dfrac{12 \text{ O atoms}}{1 \text{ Al}_2(\text{SO}_4)_3 \text{ formula unit}}$

$$= 72 \text{ O atoms (calculator and \textbf{correct answer})}$$

Here the number of oxygen atoms present is affected by three numbers: the coefficient preceding the chemical formula, the subscript 4 immediately following the chemical symbol of O, and the subscript 3 outside the parenthesis in the chemical formula.

▶ **Practice Exercise 10.1** How many nitrogen atoms are part of each of the follwing chemical expressions?

a. 3 NO_2 b. 4 HNO_3 c. $7 \text{ N}_2\text{F}_3\text{Cl}$ d. $6 \text{ Al}(\text{NO}_3)_3$

Answers to all practice exercises in this chapter are found in the back-of-the-book answer section.

10.4 BALANCING PROCEDURES FOR CHEMICAL EQUATIONS

We now proceed to the procedures needed for determining the equation coefficients needed to bring a given chemical equation into balance. They are introduced in the context of actually balancing two chemical equations (Examples 10.2 and 10.3). Both of these examples should be studied carefully, because each includes detailed commentary concerning the "ins and outs" of balancing chemical equations.

EXAMPLE 10.2 | **Balancing a Chemical Equation**

Balance the following chemical equation:

$$\text{Fe}_3\text{O}_4 + \text{H}_2 \longrightarrow \text{Fe} + \text{H}_2\text{O}$$

SOLUTION

Step 1 *Examine the chemical equation, and pick one element to balance first.* It is often convenient to identify the most complex substance first, that is, the substance with the greatest number of atoms per formula unit. For this most complex substance, whether a reactant or product, "key in" on the element within it that is present in the greatest amount (greatest number of atoms). Using this guideline, we select Fe_3O_4 and the element oxygen.

We note that there are four oxygen atoms on the left side of the equation (in Fe_3O_4) and only one oxygen atom on the right side (in H_2O). For the oxygen atoms to balance, we will need four on each side. To obtain four atoms of oxygen on each side of the equation, we place the coefficient 1 in front of Fe_3O_4 and the coefficient 4 in front of H_2O.

$$1 \text{ Fe}_3\text{O}_4 + \text{H}_4 \longrightarrow \text{Fe} + 4 \text{ H}_2\text{O}$$

The coefficient 1 (in front of Fe_3O_4) has been explicitly shown in the preceding equation to remind us that the Fe_3O_4 coefficient has been determined. We now have four oxygen atoms on each side of the equation.

$$1 \text{ Fe}_3\text{O}_4: \quad 1 \times 4 = 4$$
$$4 \text{ H}_2\text{O}: \quad 4 \times 1 = 4$$

In balancing a chemical equation, chemical formula subscripts are *never changed*. You must leave the chemical formulas just as they are given. The only thing you can do is place coefficients in front of the chemical formulas.

Step 2 *Now pick a second element to balance.* We will balance the element Fe next. (In this particular equation it does not matter whether we balance Fe or H second.) The number of Fe atoms on the left side of the equation is three; the coefficient 1 in front of Fe_3O_4 sets the Fe atom number at three. We will need three Fe atoms on the product side. This is accomplished by placing the coefficient 3 in front of Fe.

$$1\ Fe_3O_4 + H_2 \longrightarrow 3\ Fe + 4\ H_2O$$

Now there are three Fe atoms on each side of the equation.

Step 3 *Now pick a third element to balance.* The only element left to balance is H. There are two H atoms on the left and eight H atoms on the right ($4\ H_2O$ involves 8 H atoms). Placing the coefficient 4 in front of H_2 on the left side gives eight H atoms on that side.

$$1\ Fe_3O_4 + 4\ H_2 \longrightarrow 3\ Fe + 4\ H_2O$$

Step 4 *As a final check on the correctness of the balancing procedure, count atoms on each side of the chemical equation.* The following table can be constructed from our balanced equation.

$$Fe_3O_4 + 4\ H_2 \longrightarrow 3\ Fe + 4\ H_2O$$

Atom	Left Side	Right Side
Fe	$1 \times 3 = 3$	$3 \times 1 = 3$
O	$1 \times 4 = 4$	$4 \times 1 = 4$
H	$4 \times 2 = 8$	$4 \times 2 = 8$

Note that in the preceding final form of the balanced equation, the subscript 1 in front of Fe_3O_4 has been dropped. We carried this subscript 1 in the individual steps of the balancing procedure to remind us of which elements had been balanced and which had not been balanced. Once the balancing procedure has been completed, a 1 coefficient need not be shown since a 1 is implied just by the presence of the element's symbol in the chemical formula. This convention parallels that used in writing chemical formulas themselves: The subscript 1 is implied rather than explicitly written, so the formula for water is written as H_2O, not as H_2O_1.

Answer Double Check:

Is the chemical equation balanced? Yes. There are the same number of atoms of each type on each side of the chemical equation; there are three Fe atoms on each side, four O atoms on each side, and eight H atoms on each side.

▶ **Practice Exercise 10.2**   Balance the following chemical equation.

$$SiO_2 + C \longrightarrow SiC + CO$$

EXAMPLE 10.3   **Balancing a Chemical Equation**

Balance the following chemical equation

$$C_4H_{10} + O_2 \longrightarrow CO_2 + H_2O$$

SOLUTION

Step 1 *Examine the chemical equation, and pick one element to balance first.* The formula containing the most atoms is C_4H_{10}. We will balance the element H first. We have 10 H atoms on the left and 2 H atoms on the right. The two sides are brought into balance by placing the coefficient 5 in front of H_2O on the right side. We now have 10 H atoms on each side.

$$1 \, C_4H_{10}: \qquad 1 \times 10 = 10$$

$$5 \, H_2O: \qquad 5 \times 2 = 10$$

Our chemical equation now has the following appearance:

$$1 \, C_4H_{10} + O_2 \longrightarrow CO_2 + 5 \, H_2O$$

In setting the H balance at 10 atoms, we are setting the coefficient in front of C_4H_{10} at 1. The 1 has been explicitly shown in the above equation to remind us that the C_4H_{10} coefficient has been determined. (In the final balanced equation the 1 should not be shown.)

Step 2 *Now pick a second element to balance.* We will balance C next. You always balance the elements that appear in only one reactant and one product before trying to balance any elements appearing in several formulas on one side of the equation. Oxygen, our other choice for an element to balance at this stage, appears in two places on the product side of the equation. The number of carbon atoms is already set at four on the left side of the equation.

$$1 \, C_4H_{10}: \qquad 1 \times 4 = 4$$

We obtain a balance of four carbon atoms on each side of the equation by placing the coefficient 4 in front of CO_2.

$$1 \, C_4H_{10} + O_2 \longrightarrow 4 \, CO_2 + 5 \, H_2O$$

Step 3 *Now pick a third element to balance.* Only one element is left to balance—oxygen. The number of oxygen atoms on the right side of the equation is already set at 13: eight O atoms from the CO_2 and five O atoms from the H_2O.

$$4 \, CO_2: \qquad 4 \times 2 = 8$$

$$5 \, H_2O: \qquad 5 \times 1 = 5$$

To obtain 13 O atoms on the left side of the equation, we need a fractional coefficient, $6\frac{1}{2}$.

$$6\tfrac{1}{2} \, O_2: \qquad 6\tfrac{1}{2} \times 2 = 13$$

The coefficient 6 gives 12 atoms and the coefficient 7 gives 14 atoms. The only way we can get 13 atoms is by using $6\frac{1}{2}$.

All the coefficients in the equation have now been determined.

$$1 \, C_4H_{10} + 6\tfrac{1}{2} \, O_2 \longrightarrow 4 \, CO_2 + 5 \, H_2O$$

Chemical equations containing fractional coefficients are not considered to be written in their most conventional form. Although such equations are mathematically correct, they have some problems chemically. The above equation indicates the need for $6\frac{1}{2}$ O_2 molecules among the reactants, but half an O_2 molecule does not exist as such. Step 4 shows how to take care of this "problem."

Step 4 *After all coefficients have been determined, clear any fractional coefficients that are present.* We can clear the fractional coefficient present in this equation, $6\frac{1}{2}$, by multiplying *each* of the coefficients in the equation by the factor 2.

$$2 \, C_4H_{10} + 13 \, O_2 \longrightarrow 8 \, CO_2 + 10 \, H_2O$$

Now we have the equation in its conventional form. Note that *all the coefficients* had to be multiplied by 2, not just the fractional one. It will always be the case that whatever is done to a fractional coefficient to make it a whole number must also be carried out on all of the other coefficients.

If a coefficient involving $\frac{1}{3}$ had been present in the equation, we would have multiplied by 3 instead of by 2.

Step 5 *As a final check on the correctness of the balancing procedure, count atoms on each side of the chemical equation.* The following table can be constructed from our balanced equation.

$$2\,C_4H_{10} + 13\,O_2 \longrightarrow 8\,CO_2 + 10\,H_2O$$

Atom	Left Side	Right Side
C	$2 \times 4 = 8$	$8 \times 1 = 8$
H	$2 \times 10 = 20$	$10 \times 2 = 20$
O	$13 \times 2 = 26$	$(8 \times 2) + (10 \times 1) = 26$

Answer Double Check:

Is the chemical equation balanced? Yes. There are the same number of atoms of each type on each side of the chemical equation; there are 8 C atoms on each side, 20 H atoms on each side, and 26 O atoms on each side.

▶ **Practice Exercise 10.3** Balance the following chemical equation:

$$HNO_3 \longrightarrow NO_2 + O_2 + H_2O$$

Some additional comments and guidelines concerning chemical equations and the process of balancing them are in order.

It is important to remember that the ultimate source of the information conveyed by a chemical equation is experimental data. Before you can write a chemical equation for a reaction, the identity and formulas of the reactants and products must be determined *by experiment*. Once you know these things, then you can write the balanced equation.

There are four additional guidelines concerning chemical equations and the balancing process:

1. *The coefficients in a balanced equation are always the smallest set of whole numbers that will balance the equation.* This is significant because more than one set of coefficients will balance an equation. Consider the following three equations.

$$2\,H_2 + O_2 \longrightarrow 2\,H_2O$$

$$4\,H_2 + 2\,O_2 \longrightarrow 4\,H_2O$$

$$8\,H_2 + 4\,O_2 \longrightarrow 8\,H_2O$$

The only way to learn to balance chemical equations is through practice. The problems at the end of this chapter contain numerous chemical equations for your practice.

All three of these equations are mathematically correct; there are equal numbers of H and O atoms on each side of the equation. The first equation, however, is considered the conventional form because the coefficients used there are the smallest set of whole numbers that will balance the equation. The coefficients in the second equation are double those in the first, and the third equation has coefficients 4 times those of the first equation.

2. *It is useful to consider polyatomic ions as single entities in balancing a chemical equation, provided they maintain their identities in the chemical reaction.* Polyatomic ions have maintained their identity in a chemical reaction if they appear on both sides of the equation. For example, in an equation where sulfate units are present (SO_4^{2-}), both as reactants and products, balance them as a unit rather than trying to balance S and O separately. The reasoning would be "We have two sulfates on this side, so we need two sulfates on that side," and so on.

A chemical equation where this guideline applies is

$$K_2SO_4 + BaCl_2 \longrightarrow 2\,KCl + BaSO_4$$

On each side of the equation we have 2 K, 1 SO_4, 1 Ba, and 2 Cl.

3. *Subscripts in the chemical formula of a reactant or product should never be altered (changed) during the balancing process.* A student might try to balance the atoms of C in CO_2 at two by using the notation C_2O_4 instead of $2\,CO_2$. This is incorrect. The notation C_2O_4 denotes a molecule containing six atoms, whereas the notation $2\,CO_2$ denotes two molecules, each containing three atoms. The experimental fact is that a molecule of CO_2 contains three rather than six atoms. *A coefficient deals with the number of formula units of a substance, and a subscript deals with the composition of the substance.* Subscripts illustrate the law of definite proportions (Sec. 9.1); coefficients relate to the law of conservation of mass (Sec. 10.1).

4. *Knowing the procedures for balancing chemical equations does not enable you to predict what the products of a chemical reaction will be.* You are not expected, at this point, to be able to write down the products for a chemical reaction, given what the reactants are. After learning how to balance equations, students sometimes get the mistaken idea that they ought to be able to write down equations from scratch. This is not so. At this stage you should be able to balance simple equations given *all* of the reactants and *all* of the products.

> Some chemical equations are much more difficult to balance than those you encounter in this chapter's examples and end-of-chapter problems. The procedures discussed here simply are not adequate for those more difficult chemical equations. In Chapter 15 a more systematic method for balancing chemical equations, specifically designed for those more difficult situations, will be presented.

A final insightful aspect of our consideration of balanced chemical equations is a look at the concept of a balanced chemical equation at the molecular level. For the balanced chemical equation

$$CH_4 + 2\,O_2 \longrightarrow CO_2 + 2\,H_2O$$

such a molecular view is given in Figure 10.2. Note visually the rearrangement of atoms that occurs as reactants are changed to products. Also important is the atom balance between the two "boxes." There are the same number of atoms of each kind in the reactant and product boxes.

Example 10.4 considers the opposite situation to that depicated in Figure 10.2. In this example a balanced chemical equation is derived using a "molecular view" of the reactants and products.

$$CH_4 + 2\,O_2 \longrightarrow CO_2 + 2\,H_2O$$

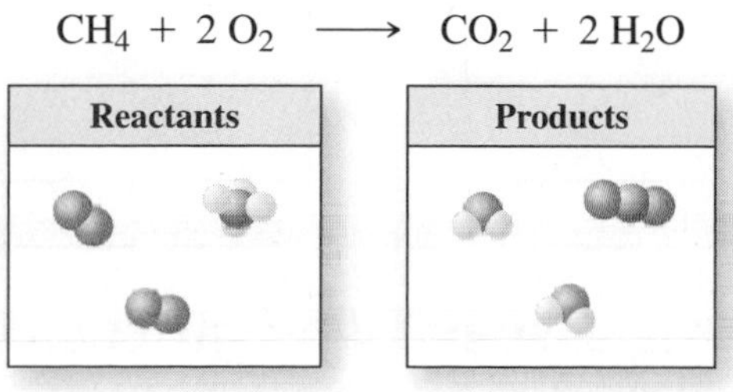

FIGURE 10.2 Molecular view of a balanced chemical equation.

EXAMPLE 10.4 | **Deriving a Balanced Chemical Equation from Molecular Models of Reactants and Products**

Using the following visual information about a chemical reaction, write a balanced chemical equation for the reaction given that the chemical species involved in the reaction are CO, O_2, and CO_2.

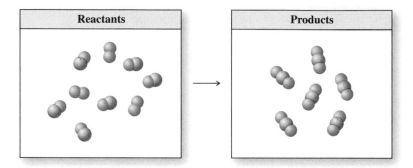

SOLUTION

Present in the reactant "box" are 3 O_2 molecules and 6 CO molecules. Within the product "box" are 6 CO_2 molecules. Atom balance is present—6 C atoms and 12 O atoms in each box. All of the reactant molecules react since none of them appear in the product box. The balanced chemical equation can, thus, be written as

$$6\,CO + 3\,O_2 \longrightarrow 6\,CO_2$$

The equation coefficients in this chemical equation are all divisible by 3. Carrying out this operation produces the chemical equation

$$2\,CO + O_2 \longrightarrow 2\,CO_2$$

The preferred form for a balanced chemical equation is the one that uses the smallest set of whole number equation coefficients that balance the equation.

▶ **Practice Exercise 10.4** Using the following visual information about a chemical reaction, write a balanced chemical equation for the reaction given that the chemical species involved in the reaction are N_2, H_2, and NH_3.

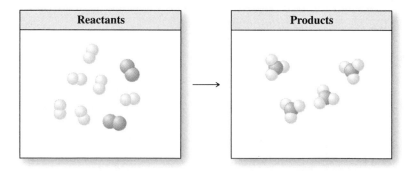

10.5 SPECIAL SYMBOLS USED IN CHEMICAL EQUATIONS

In addition to the essential plus sign and arrow notation used in chemical equations, a number of optional symbols convey more information about a chemical reaction than just the chemical species involved. In particular, it is often useful to

TABLE 10.1 Symbols Used in Equations

Symbol	Meaning
Essential	
$\rightarrow$	"to produce"
+	"reacts with" or "and"
Optional	
(s)	solid
(l)	liquid
(g)	gas
(aq)	aqueous solution (a substance dissolved in water)

know the physical state of the substances involved in a chemical reaction. The optional symbols listed in Table 10.1 are used to specify the physical state of reactants and products.

The chemical equations we balanced in Section 10.4 (Examples 10.2 and 10.3) are written as follows when the optional symbols are included.

$$Fe_3O_4(s) + 4\,H_2(g) \longrightarrow 3\,Fe(s) + 4\,H_2O(l)$$

$$2\,C_4H_{10}(g) + 13\,O_2(g) \longrightarrow 8\,CO_2(g) + 10\,H_2O(g)$$

Two more examples of the use of optional symbols are

$$NaCl(aq) + AgNO_3(aq) \longrightarrow AgCl(s) + NaNO_3(aq)$$

$$NaOH(aq) + HCl(aq) \longrightarrow NaCl(aq) + H_2O(l)$$

The optional symbols in these latter two chemical equations indicate that both reactions take place in an aqueous solution. In the first reaction, one of the products, AgCl, is insoluble, being present in the mixture as a solid. In the second reaction, the product NaCl is soluble and thus remains in the solution.

Also, note that in the last equation the reactant HCl is functioning as an acid and must be designated as such (Sec. 8.6). The notation HCl(g) indicates hydrogen chloride in its gaseous state; the notation HCl(aq), hydrogen chloride dissolved in water, denotes hydrochloric acid.

10.6 CLASSES OF CHEMICAL REACTIONS

An almost inconceivable number of chemical reactions are possible. The problems associated with organizing our knowledge about them are diminished considerably by grouping the reactions into classes based on common characteristics. In this section we examine five classes of chemical reactions. These classes are

1. Synthesis reactions
2. Decomposition reactions
3. Single-replacement reactions
4. Double-replacement reactions
5. Combustion reactions

Although there are other categories of reactions besides these five, most of the reactions discussed in this text fall into one of these five categories.

FIGURE 10.3 Sulfuric acid (H_2SO_4) is the ingredient in acid rain most responsible for the eating away (deterioration) of marble statues. It is produced in air when fuel containing sulfur as an impurity is burned. Sulfuric acid production involves a series of three synthesis reactions. First, sulfur combines with oxygen to form sulfur dioxide ($S + O_2 \longrightarrow SO_2$). Then the sulfur dioxide reacts with another molecule of oxygen to form sulfur trioxide ($2 SO_2 + O_2 \longrightarrow 2 SO_3$). Finally, sulfur trioxide reacts with moisture in the air to form sulfuric acid ($SO_3 + H_2O \longrightarrow H_2SO_4$). The sulfuric acid so formed reacts with marble ($CaCO_3$) to produce $CaSO_4$ (a soft, crumbly substance) and CO_2. The chemical equation is $CaCO_3 + H_2SO_4 \longrightarrow CaSO_4 + CO_2 + H_2O$. *(Spencer Platt/Getty Images)*

Synthesis Reactions

The first of the five categories of reactions is the *synthesis reaction*. A **synthesis reaction** *is a chemical reaction in which a single product is produced from two (or more) reactants.*

$$X + Y \longrightarrow XY$$

Synthesis reactions always involve simpler substances being combined into a more complex substance. The reactants X and Y may be elements or compounds or an element and a compound. The product of the reaction, XY, is always a compound.

Some representative synthesis reactions with elements as the reactants are

$$H_2 + Cl_2 \longrightarrow 2 HCl$$

$$S + O_2 \longrightarrow SO_2$$

$$Ni + S \longrightarrow NiS$$

Some examples of synthesis reactions in which compounds are involved as reactants are

$$SO_3 + H_2O \longrightarrow H_2SO_4$$

$$K_2O + H_2O \longrightarrow 2 KOH$$

$$2 NO + O_2 \longrightarrow 2 NO_2$$

$$2 NO_2 + H_2O_2 \longrightarrow 2 HNO_3$$

The atmospheric production of sulfuric acid (H_2SO_4), the "active ingredient" in acid rain, involves a sequence of three synthesis reactions (see Fig. 10.3).

Decomposition Reactions

A **decomposition reaction** *is a chemical reaction in which a single reactant is converted into two or more simpler substances.* It is thus the exact opposite of a synthesis reaction. The general equation for a decomposition reaction is

$$XY \longrightarrow X + Y$$

Heat or light is often the stimulus used to effect a decomposition reaction.

At sufficiently high temperatures all compounds can be broken down (decomposed) into their constituent elements. Such reactions include

$$2 CuO \longrightarrow 2 Cu + O_2$$

$$2 H_2O \longrightarrow 2 H_2 + O_2$$

Examples of decomposition reactions that result in at least one compound as a product are

$$NH_4NO_3 \longrightarrow N_2O + 2\,H_2O$$

$$2\,KClO_3 \longrightarrow 2\,KCl + 3\,O_2$$

A most common type of decomposition reaction is that which involves metal carbonates. Metal carbonates, when heated to a high temperature (the temperature needed varies with the metal), break down, releasing carbon dioxide gas and producing the metal oxide. Typical carbonate decomposition equations include the following.

$$Na_2CO_3 \longrightarrow Na_2O + CO_2$$

$$MgCO_3 \longrightarrow MgO + CO_2$$

$$Al_2(CO_3)_3 \longrightarrow Al_2O_3 + 3\,CO_2$$

When hydrogen peroxide is poured on a wound, it decomposes into water and oxygen:
$2\,H_2O_2 \longrightarrow 2\,H_2O + O_2$
The bubbles that form are oxygen gas.

The decomposition reaction $CaCO_3 \longrightarrow CaO + CO_2$ occurs when limestone ($CaCO_3$) is heated in cement factories to produce calcium oxide, one of the components of cement.

Single-Replacement Reactions

A **single-replacement reaction** *is a chemical reaction in which one element within a compound is replaced by another element.* In this type of reaction there are always two reactants, an element and a compound, and two products, also an element and a compound. The general equation for a single-replacement reaction is

$$X + YZ \longrightarrow Y + XZ$$

Both metals and nonmetals can be replaced in this manner. Such reactions usually involve aqueous solutions. Examples include

$$Zn + H_2SO_4 \longrightarrow H_2 + ZnSO_4$$

$$Ni + 2\,HCl \longrightarrow H_2 + NiCl_2$$

$$Fe + CuSO_4 \longrightarrow Cu + FeSO_4$$

$$Mg + Ni(NO_3)_2 \longrightarrow Ni + Mg(NO_3)_2$$

Double-Replacement Reactions

A **double-replacement reaction** *is a chemical reaction in which two compounds exchange parts with each other and form two new compounds.* The general equation for such a reaction is

$$AX + BY \longrightarrow AY + BX$$

Double-replacement reactions generally involve ionic compounds in an aqueous solution. Most often the positive ion from one compound exchanges with the positive ion of the other. The process may be thought of as "partner swapping," since each negative ion ends up paired with a new partner (positive ion). Such reactions include

$$AgNO_3 + NaCl \longrightarrow AgCl + NaNO_3$$

$$NaF + HCl \longrightarrow NaCl + HF$$

$$AgNO_3 + HCl \longrightarrow AgCl + HNO_3$$

Figure 10.4 summarizes in pictorial form the four classes of chemical reactions we have so far considered. Example 10.5 is an exercise in classifying reactions into these four general types.

FIGURE 10.4 Schematic diagrams for synthesis, decomposition, single-replacement, and double-replacement types of chemical reactions.

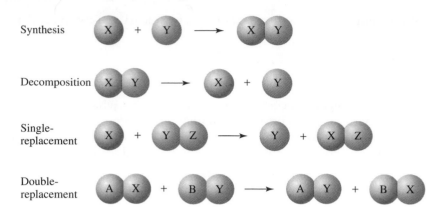

EXAMPLE 10.5 **Classification of Chemical Reactions into the Categories Synthesis, Decomposition, Single-Replacement, and Double-Replacement**

Classify each of the following chemical reactions as synthesis, decomposition, single-replacement, or double-replacement.

    a. $CuCO_3 \longrightarrow CuO + CO_2$
    b. $Fe + Cu(NO_3)_2 \longrightarrow Cu + Fe(NO_3)_2$
    c. $3\,Mg + N_2 \longrightarrow Mg_3N_2$
    d. $NH_4Cl + AgNO_3 \longrightarrow NH_4NO_3 + AgCl$

SOLUTION

    a. Since two substances are produced from a single substance, this reaction is a decomposition reaction.
    b. Having an element and a compound as reactants and an element and a compound as products is a characteristic of a single-replacement reaction.
    c. Two substances combine to form a single substance; hence, this reaction is classified as a synthesis reaction.
    d. This reaction is a double-replacement reaction. Ammonium ion and silver ion are changing places, that is, swapping partners.

▶ **Practice Exercise 10.5** Classify each of the following chemical reactions as synthesis, decomposition, single-replacement, or double-replacement.

    a. $H_2SO_4 \longrightarrow SO_3 + H_2O$
    b. $2\,PCl_3 + O_2 \longrightarrow 2\,POCl_3$
    c. $ZnO + 2\,HCl \longrightarrow ZnCl_2 + H_2O$
    d. $2\,Al + 6\,HCl \longrightarrow 2\,AlCl_3 + 3\,H_2$

Combustion Reactions

Combustion reactions are a most common type of reaction. A **combustion reaction** *is a chemical reaction in which a substance reacts with oxygen (usually from air) that proceeds with evolution of heat and usually also a flame.* Hydrocarbons, binary compounds of carbon and hydrogen (of which many exist), are the most common type of compound that undergoes combustion. In hydrocarbon combustion, the carbon of the hydrocarbon combines with oxygen to produce carbon dioxide (CO_2). The hydrogen of the hydrocarbon also interacts with oxygen of air to give water (H_2O) as a product. The relative amounts of CO_2 and H_2O produced depends on the composition of the hydrocarbon.

$$2\,C_2H_2 + 5\,O_2 \longrightarrow 4\,CO_2 + 2\,H_2O$$

$$C_3H_8 + 5\,O_2 \longrightarrow 3\,CO_2 + 4\,H_2O$$

$$C_4H_8 + 6\,O_2 \longrightarrow 4\,CO_2 + 4\,H_2O$$

Note that the equation that was balanced in Example 10.3 was a hydrocarbon combustion reaction.

Combustion of compounds containing oxygen as well as carbon and hydrogen (for example, CH_4O or C_3H_8O) also produce CO_2 and H_2O as products.

$$2\,CH_4O + 3\,O_2 \longrightarrow 2\,CO_2 + 4\,H_2O$$

$$2\,C_3H_8O + 9\,O_2 \longrightarrow 6\,CO_2 + 8\,H_2O$$

Examples of combustion reactions in which products other than CO_2 and H_2O are produced include

$$CS_2 + 3\,O_2 \longrightarrow CO_2 + 2\,SO_2$$

$$2\,H_2S + 3\,O_2 \longrightarrow 2\,SO_2 + 2\,H_2O$$

$$4\,NH_3 + 5\,O_2 \longrightarrow 4\,NO + 6\,H_2O$$

Example 10.6 is an equation-balancing exercise involving a combustion reaction.

Hydrocarbon combustion reactions are the basis of an industrial society, making possible the burning of gasoline in cars, of natural gas in homes, and of coal in factories. Gasoline, natural gas, and coal all contain hydrocarbons.

Unlike most other chemical reactions, hydrocarbon combustion reactions are carried out for the energy they produce rather than for the material products.

EXAMPLE 10.6 Writing and Balancing an Equation for a Combustion Reaction

Diethyl ether, $C_4H_{10}O$, a very flammable compound, was one of the first general anesthetics. Its use for this purpose was first demonstrated by the Boston dentist William Morton, in 1846. Write the balanced chemical equation for the combustion reaction that occurs when diethyl ether burns, that is, reacts with the oxygen (O_2) in air.

SOLUTION

The C atoms in the diethyl ether will end up in product CO_2 and the H atoms of the diethyl ether in product H_2O. There are two sources for the oxygen present in the product CO_2 and H_2O. Most of it comes from the O_2 of air; however, the reactant $C_4H_{10}O$ is also an oxygen source. The unbalanced equation for this reaction is

$$C_4H_{10}O + O_2 \longrightarrow CO_2 + H_2O$$

The equation is balanced using the procedures of Section 10.4.

Step 1 *Balancing of H atoms:* There are 10 H atoms on the left and only 2 H atoms on the right. Placing the coefficient 5 in front of H_2O balances the H atoms at 10 on each side.

$$1\,C_4H_{10}O + O_2 \longrightarrow CO_2 + 5\,H_2O$$

Step 2 *Balancing of C atoms:* An effect of balancing the H atoms at 10 (step 1) is the setting of the C atoms on the left side of the equation at 4; the coefficient in front of $C_4H_{10}O$ is 1. Placing the coefficient 4 in front of CO_2 causes the carbon atoms to balance at 4 on each side of the equation.

$$1\,C_4H_{10}O + O_2 \longrightarrow 4\,CO_2 + 5\,H_2O$$

Step 3 *Balancing of O atoms:* The oxygen content of the right side of the equation is set at 13 atoms: 8 oxygen atoms from 4 CO_2 and 5 oxygen atoms from 5 H_2O. To obtain

13 oxygen atoms on the left side of the equation, the coefficient 6 is placed in front of O_2; 6 O_2 gives 12 oxygen atoms, and there is an additional O in 1 C_2H_6O. Note that the element oxygen is present in all four formulas in the equation.

$$1\,C_4H_{10}O + 6\,O_2 \longrightarrow 4\,CO_2 + 5\,H_2O$$

Step 4 *Final check*: The equation is balanced. There are 4 carbon atoms, 10 hydrogen atoms, and 13 oxygen atoms on each side of the equation.

$$C_4H_{10}O + 6\,O_2 \longrightarrow 4\,CO_2 + 5\,H_2O$$

Answer Double Check:

Are the products listed for the combustion reaction, CO_2 and H_2O, reasonable products? Yes. The compound undergoing combustion, $C_4H_{10}O$, contains C and H. The C will combine with oxygen to give an oxide of carbon, CO_2, and the H will combine with oxygen to give an oxide of hydrogen, H_2O.

▶ **Practice Exercise 10.6**   Toluene, C_7H_8, is one of many types of hydrocarbon molecules present in gasoline. Write the balanced chemical equation for the combustion reaction that occurs when toluene burns, that is, reacts with the O_2 in air.

Many, but not all, chemical reactions fall into one of the five categories we have discussed in this section. Even though this classification system is not all-inclusive, it is still very useful because of the many reactions it does help correlate.

In later chapters, two additional categories of chemical reactions are considered. *Acid–base reactions* are the topic of Chapter 14, and *oxidation–reduction reactions* are considered in Chapter 15.

10.7 CHEMICAL EQUATIONS AND THE MOLE CONCEPT

The coefficients in a balanced chemical equation, like the subscripts in a chemical formula (Sec. 9.9), have two levels of interpretation—a microscopic level of meaning and a macroscopic level of meaning.

The first of these two interpretations, the microscopic level, has been used in the previous sections of this chapter. At the *microscopic* level of interpretation the coefficients in a balanced chemical equation give directly the numerical relationships among formula units consumed (used up) and/or produced in the chemical reaction. Interpreted at the microscopic level, the chemical equation

$$4\,NH_3 + 5\,O_2 \longrightarrow 4\,NO + 6\,H_2O$$

conveys the information that four molecules of NH_3 react with five molecules of O_2 to produce four molecules of NO and six molecules of H_2O.

At the *macroscopic* level of interpretation, the coefficients in a balanced chemical equation give the fixed molar ratios between substances consumed and/or produced in the chemical reaction. Interpreted at the macroscopic level, the chemical equation

$$4\,NH_3 + 5\,O_2 \longrightarrow 4\,NO + 6\,H_2O$$

conveys the information that four moles of NH_3 react with five moles of O_2 to produce four moles of NO and six moles of H_2O.

The validity of the molar (macroscopic) interpretation of coefficients in an equation can be derived very straightforwardly from the microscopic level of interpretation. A balanced chemical equation remains valid (mathematically correct) when all of its coefficients are multiplied by the same number. (If molecules react in a three-to-one ratio, they will also react in a six-to-two or nine-to-three ratio.) Multiplying the previous equation by y, where y is any number, we have

$$4y\,NH_3 + 5y\,O_2 \longrightarrow 4y\,NO + 6y\,H_2O$$

The situation for $y = 6.022 \times 10^{23}$ is of particular interest because $6.022 \times 10^{23} = 1$ mole. Using $y = 1$ mole, we have by substitution

$$4 \text{ moles } NH_3 + 5 \text{ moles } O_2 \longrightarrow 4 \text{ moles } NO + 6 \text{ moles } H_2O$$

Thus, as with the subscripts in chemical formulas (Sec. 9.9), the coefficients in chemical equations carry a dual meaning. "Number of formula units" is the microscopic-level interpretation for equation coefficients and "moles of formula units" is the macroscopic-level interpretation.

In Section 10.4 it was noted that fractional equation coefficients are often obtained in the equation-balancing process. We can now further note that such fractional coefficients do have valid meaning for the macroscopic-level interpretation of a chemical equation ($3\frac{1}{2}$ moles, and so on), whereas they are totally unacceptable for the microscopic level of interpretation ($3\frac{1}{2}$ molecules, and so on).

The coefficients in a balanced chemical equation may be used to generate conversion factors used in problem solving. Numerous conversion factors are obtainable from a single balanced equation. Consider the balanced chemical equation

$$P_4O_{10} + 6\,H_2O \longrightarrow 4\,H_3PO_4$$

Three mole-to-mole relationships can be obtained from this chemical equation.

> One mole of P_4O_{10} produces four moles of H_3PO_4.

> Six moles of H_2O produce four moles of H_3PO_4.

> One mole of P_4O_{10} reacts with six moles H_2O.

From these three macroscopic-level relationships, six conversion factors can be written. From the first relationship,

$$\frac{1 \text{ mole } P_4O_{10}}{4 \text{ moles } H_3PO_4} \quad \text{and} \quad \frac{4 \text{ moles } H_3PO_4}{1 \text{ mole } P_4O_{10}}$$

From the second relationship,

$$\frac{6 \text{ moles } H_2O}{4 \text{ moles } H_3PO_4} \quad \text{and} \quad \frac{4 \text{ moles } H_3PO_4}{6 \text{ moles } H_2O}$$

From the third relationship,

$$\frac{1 \text{ mole } P_4O_{10}}{6 \text{ moles } H_2O} \quad \text{and} \quad \frac{6 \text{ moles } H_2O}{1 \text{ mole } P_4O_{10}}$$

Any chemical equation can be the source of numerous conversion factors. The more reactants and products there are in the equation, the greater the number of conversion factors.

Conversion factors obtained from chemical equations are used in several different types of calculations. Example 10.7 illustrates some very simple applications of their use. In Section 10.10 we explore their use in more complicated problem-solving situations.

To a chemist, a chemical equation is a recipe. Just as the (very simple) recipe 3 cups flour + 1 cup milk = 2 cakes says that if 3 cups of flour are mixed with 1 cup of milk, 2 cakes can be made. The chemical equation $3\,H_2 + N_2 \longrightarrow 2\,NH_3$ tells a chemist that when 3 moles of H_2 are reacted with 1 mole of nitrogen, 2 moles of NH_3 are formed.

EXAMPLE 10.7 **Calculating Molar Quantities Using a Balanced Chemical Equation**

Two air pollutants present in automobile exhaust are carbon monoxide (CO) and nitrogen monoxide (NO). Within an automobile's catalytic converter, these two pollutants react with each other to produce carbon dioxide (CO_2) and nitrogen (N_2). The chemical equation for the reaction is

$$2\,CO + 2\,NO \longrightarrow 2\,CO_2 + N_2$$

a. How many moles of N_2 are produced when 3.50 moles of CO reacts?
b. How many moles of NO are needed to react with 2.31 moles of CO?

SOLUTION

Both parts of this problem are one-step mole-to-mole calculations. In each case the needed conversion factor is derived from the coefficients of the chemical equation.

a.

Step 1 The given quantity is 3.50 moles of CO, and the desired quantity is moles of N_2.

$$3.50 \text{ moles CO} = ? \text{ moles } N_2$$

Step 2 The conversion factor needed to convert from moles of CO to moles of N_2 is derived from the coefficients of CO and N_2 in the balanced equation. This chemical equation tells us that two moles of CO produce one mole of N_2. From this relationship two conversion factors are obtainable:

$$\frac{2 \text{ moles CO}}{1 \text{ mole } N_2} \quad \text{and} \quad \frac{1 \text{ mole } N_2}{2 \text{ moles CO}}$$

We will use the second of these conversion factors in solving the problem. The setup is

$$3.50 \text{ moles CO} \times \frac{1 \text{ mole } N_2}{2 \text{ moles CO}}$$

We used the second of the two conversion factors because it had moles of CO in the denominator, a requirement for the unit moles of CO to cancel.

Step 3 Collecting numerical terms, after cancellation of units, gives

$$\frac{3.50 \times 1}{2} \text{ moles } N_2 = 1.75 \text{ moles } N_2 \text{ (calculator and } \textbf{correct answer)}$$

Note that the coefficients in the chemical equation enter directly into the numerical calculation. Having a correctly balanced equation is therefore of vital importance. Using an unbalanced or misbalanced chemical equation as a source of a conversion factor will lead to a wrong numerical answer.

b.

Step 1 Both the given species, CO, and the desired species, NO, are reactants.

$$2.31 \text{ moles CO} = ? \text{ moles NO}$$

Step 2 Molar relationships obtained from a chemical equation are not required to always involve one reactant and one product, as was the case in part (a). Molar relationships involving only reactants or only products are often needed and used. In this problem we will need the molar relationship between the two reactants, CO and NO, which is two to two. From this ratio, the conversion factor

$$\frac{2 \text{ moles NO}}{2 \text{ moles CO}}$$

can be constructed, which is used in the setup of the problem as follows:

$$2.31 \text{ ~~moles CO~~} \times \frac{2 \text{ moles NO}}{2 \text{ ~~moles CO~~}}$$

Step 3 Collecting numerical terms, after cancellation of units, gives

$$\frac{2.31 \times 2}{2} \text{ moles NO} = 2.31 \text{ moles NO} \quad \text{(calculator and \textbf{correct answer})}$$

In terms of significant figures, the numbers in conversion factors obtained from equation coefficients are considered exact numbers. Thus, since 2.31 contains three significant figures, the answer to this problem should also contain three significant figures.

Answer Double Check:

Are the magnitudes of the numerical results reasonable? Yes. In part (a), using rounded numbers, 4 moles of CO should produce half as many moles of N_2 (2 moles). The calculated answer 1.75 moles N_2 is close to this rounded answer. In part (b), using rounded numbers, 2 moles of CO should react with 2 moles of NO (the equation coefficients are the same). The calculated answer of 2.31 is close to this rounded answer.

▶ **Practice Exercise 10.7**   Ammonia, NH_3, a colorless, flammable gas with a *very sharp* and penetrating odor, reacts with oxygen in the following manner:

$$4 \, NH_3 + 5 \, O_2 \longrightarrow 4 \, NO + 6 \, H_2O$$

   a. How many moles of H_2O are produced when 2.50 moles of NH_3 reacts?
   b. How many moles of O_2 are needed to react with 7.50 moles of NH_3?

10.8 BALANCED CHEMICAL EQUATIONS AND THE LAW OF CONSERVATION OF MASS

We began this chapter with a discussion of the law of conservation of mass (Sec. 10.1), which became the basis for the concepts involved in balancing chemical equations (Sec. 10.4). Now that the relationship between chemical equation coefficients and moles has been addressed (Sec. 10.7), it is time to revisit the law of conservation of mass.

   A balanced chemical equation can be used to verify the law of conservation of mass. Let us consider this verification process using the balanced equation

$$4 \, NH_3 + 5 \, O_2 \longrightarrow 4 \, NO + 6 \, H_2O$$

According to the coefficients in this balanced equation, the molar ratio among the reactants and products is 4:5 to 4:6, that is,

$$4 \text{ moles } NH_3 + 5 \text{ moles } O_2 \longrightarrow 4 \text{ moles } NO + 6 \text{ moles } H_2O$$

Substituting molar masses (Sec. 9.6) into this equation, which are, respectively, 17.04 g/mole for NH_3, 32.00 g/mole for O_2, 30.01 g/mole for NO, and 18.02 g/mole for H_2O, we obtain

$$4\,(17.04\text{ g}) + 5\,(32.00\text{ g}) = 4\,(30.01\text{ g}) + 6\,(18.02\text{ g})$$

Simplifying, we get

$$68.16\text{ g} + 160.00\text{ g} = 120.04\text{ g} + 108.12\text{ g}$$
$$228.16\text{ g} = 228.16\text{ g}$$

Thus, based on the molar interpretation of equation coefficients, the sum of the reactant masses is equal to the sum of the product masses, as it should be according to the law of conservation of mass.

Example 10.8 further illustrates the relationship between the coefficients in a balanced equation and the law of conservation of mass.

EXAMPLE 10.8 | **Verifying the Law of Conservation of Mass Using a Balanced Chemical Equation**

Verify the law of conservation of mass using the balanced chemical equation

$$3\,FeO + 2\,Al \longrightarrow 3\,Fe + Al_2O_3$$

SOLUTION

The given balanced chemical equation, interpreted in terms of moles, is

$$3 \text{ moles } FeO + 2 \text{ moles } Al \longrightarrow 3 \text{ moles } Fe + 1 \text{ mole } Al_2O_3$$

The molar masses of the substances involved in this reaction are

$$\begin{aligned} FeO\!: \quad 1 \text{ mole} &= \ \ 71.85 \text{ g} \\ Al\!: \quad 1 \text{ mole} &= \ \ 26.98 \text{ g} \\ Fe\!: \quad 1 \text{ mole} &= \ \ 55.85 \text{ g} \\ Al_2O_3\!: \quad 1 \text{ mole} &= 101.96 \text{ g} \end{aligned}$$

Substituting these molar masses into the given equation in place of moles gives

$$3\,(71.85\text{ g}) + 2\,(26.98\text{ g}) = 3\,(55.85\text{ g}) + 1\,(101.96\text{ g})$$

Simplifying, we obtain

$$215.55\text{ g} + 53.96\text{ g} = 167.55\text{ g} + 101.96\text{ g}$$
$$269.51\text{ g} = 269.51\text{ g}$$

The sum of the reactant masses is equal to the sum of the product masses, a verification of the law of conservation of mass.

▶ **Practice Exercise 10.8**   Verify the law of conservation of mass, using the balanced chemical equation

$$2\,KOH + CO_2 \longrightarrow K_2CO_3 + H_2O$$

10.9 CALCULATIONS BASED ON CHEMICAL EQUATIONS—STOICHIOMETRY

A major area of concern for chemists is the quantities of materials consumed and produced in chemical reactions. This area of study is called *chemical stoichiometry*. **Chemical stoichiometry** *is the study of the quantitative relationships among reactants and*

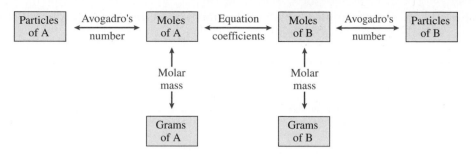

FIGURE 10.5
Conversion-factor relationships needed for solving chemical equation–based problems.

products in a chemical reaction. The word *stoichiometry*, pronounced stoy-kee-OM-eh-tree, is derived from the Greek *stoicheion* (element) and *metron* (measure). The stoichiometry of a chemical reaction always involves the *molar relationships* between reactants and products (Sec. 10.7) and thus is given by the coefficients in the balanced equation for the chemical reaction.

In a typical stoichiometric calculation, information is given about one reactant or product of a reaction (number of grams, moles, or particles), and similar information is requested concerning another reactant or product of the same reaction. The substances involved in such a calculation may both be reactants, may both be products, or may be one of each.

The conversion-factor relationships needed to solve problems of the above general type are given in Figure 10.5. This diagram should seem very familiar to you, for it is almost identical to Figure 9.5, with which you have worked repeatedly. There is only one difference between the two. In the Chapter 9 diagram, the subscripts in a chemical formula were listed as the basis for relating moles of given and desired substances to each other. In this new diagram, these same two quantities are related using the coefficients of a balanced chemical equation.

The most common type of stoichiometric calculation is a mass-to-mass (gram-to-gram) problem. In such problems the mass of one substance involved in a chemical reaction (either reactant or product) is given, and information is requested about the mass of another of the substances involved in the reaction (either a reactant or product). Situations requiring the solution of problems of this type are frequently encountered in laboratory settings. For example, a chemist has available so many grams of a certain chemical and wants to know how many grams of another substance can be produced from it, or how many grams of a third substance are needed to react with it. Examples 10.9 and 10.10 are both problem-solving situations of the gram-to-gram type.

EXAMPLE 10.9   **Calculating the Needed Mass of a Reactant in a Chemical Reaction**

A mixture of hydrazine (N_2H_4) and hydrogen peroxide (H_2O_2) is used as a fuel for rocket engines. These two substances react as shown by the chemical equation

$$N_2H_4(l) + 2\,H_2O_2(l) \longrightarrow N_2(g) + 4\,H_2O(g)$$

How many grams of H_2O_2 are needed to completely react with 50.0 g of N_2H_4?

SOLUTION

Step 1   Here we are given information about one reactant (50.0 g of N_2H_4) and asked to calculate information about the other reactant (H_2O_2).

$$50.0 \text{ g } N_2H_4 = ? \text{ g } H_2O_2$$

Step 2 This problem is of the grams-of-A to grams-of-B type. The pathway to be used in solving this type of problem, in terms of Figure 10.5, is

$$\boxed{\text{Grams of A}} \xrightarrow[\text{mass}]{\text{Molar}} \boxed{\text{Moles of A}} \xrightarrow[\text{coefficients}]{\text{Equation}} \boxed{\text{Moles of B}} \xrightarrow[\text{mass}]{\text{Molar}} \boxed{\text{Grams of B}}$$

Step 3 The dimensional analysis setup for this pathway is

$$50.0 \ \text{g N}_2\text{H}_4 \times \frac{1 \ \text{mole N}_2\text{H}_4}{32.06 \ \text{g N}_2\text{H}_4} \times \frac{2 \ \text{moles H}_2\text{O}_2}{1 \ \text{mole N}_2\text{H}_4} \times \frac{34.02 \ \text{g H}_2\text{O}_2}{1 \ \text{mole H}_2\text{O}_2}$$

$$\text{grams A} \longrightarrow \text{moles A} \longrightarrow \text{moles B} \longrightarrow \text{grams B}$$

The number 32.06 in the first conversion factor is the molar mass of N_2H_4; the 2 and 1 in the second conversion factor are the coefficients of H_2O_2 and N_2H_4, respectively, in the balanced chemical equation; and the number 34.02 in the last conversion factor is the molar mass of H_2O_2.

Step 4 The solution, obtained from combining all of the numerical factors, is

$$\frac{50.0 \times 1 \times 2 \times 34.02}{32.06 \times 1 \times 1} \ \text{g H}_2\text{O}_2 = 106.11353 \ \text{g H}_2\text{O}_2 \quad \text{(calculator answer)}$$

$$= 106 \ \text{g H}_2\text{O}_2 \quad \textbf{(correct answer)}$$

Answer Double Check:

Is the magnitude of the answer reasonable? Yes. Considering the numbers in the conversion factors, in rounded form, the math is 2 times 34/32, which is roughly 2. Multiplying 50 by 2 gives 100, which is consistent with the calculated answer of 106.

▶ **Practice Exercise 10.9** Sulfur dioxide, SO_2, a gaseous air pollutant that is a precursor of acid rain is produced when sulfur-containing fossil fuels are burned and when sulfide ores are roasted. The chemical equation for the roasting of ZnS-containing ores is

$$2 \ \text{ZnS(s)} + 3 \ \text{O}_2\text{(g)} \longrightarrow 2 \ \text{ZnO(s)} + 2 \ \text{SO}_2\text{(g)}$$

How many grams of ZnS are needed to produce 1.00 g of SO_2 gas?

EXAMPLE 10.10 | **Calculating the Mass of a Product in a Chemical Reaction**

When baking soda ($NaHCO_3$) is heated, it decomposes, producing carbon dioxide gas (CO_2). This carbon dioxide is responsible for the rising of bread, doughnuts, and cookies. The equation for baking soda decomposition reaction is

$$2 \ \text{NaHCO}_3\text{(s)} \longrightarrow \text{Na}_2\text{CO}_3\text{(s)} + \text{CO}_2\text{(g)} + \text{H}_2\text{O(l)}$$

How many grams of CO_2 are produced when 1.00 g of $NaHCO_3$ decomposes?

SOLUTION

Step 1 Here we are given information about the reactant (1.00 g of $NaHCO_3$) and asked to calculate information about one of the products (CO_2).

$$1.00 \ \text{g NaHCO}_3 = ? \ \text{g CO}_2$$

Step 2 This problem, like Example 10.9, is a grams-of-A to grams-of-B problem. The pathway used in solving it will be the same, which, in terms of Figure 10.5, is

$$\boxed{\text{Grams of A}} \xrightarrow[\text{mass}]{\text{Molar}} \boxed{\text{Moles of A}} \xrightarrow[\text{coefficients}]{\text{Equation}} \boxed{\text{Moles of B}} \xrightarrow[\text{mass}]{\text{Molar}} \boxed{\text{Grams of B}}$$

Step 3 The dimensional analysis setup is

$$1.00 \text{ g NaHCO}_3 \times \frac{1 \text{ mole NaHCO}_3}{84.01 \text{ g NaHCO}_3} \times \frac{1 \text{ mole CO}_2}{2 \text{ moles NaHCO}_3} \times \frac{44.01 \text{ g CO}_2}{1 \text{ mole CO}_2}$$

$$\text{grams A} \longrightarrow \text{moles A} \longrightarrow \text{moles B} \longrightarrow \text{grams B}$$

The chemical equation is the bridge that enables us to go from $NaHCO_3$ to CO_2. The numbers in the second conversion factor, the bridge factor, are coefficients from this equation.

Step 4 The solution, obtained from combining all of the numerical factors in the setup, is

$$\frac{1.00 \times 1 \times 1 \times 44.01}{84.01 \times 2 \times 1} \text{ g CO}_2 = 0.2619331 \text{ g CO}_2 \quad \text{(calculator answer)}$$

$$= 0.262 \text{ g CO}_2 \quad \textbf{(correct answer)}$$

Answer Double Check:

Is the magnitude of the answer reasonable? Yes. Considering the numbers in the conversion factors, in rounded form, the math is one-half (44/88) divided by 2, which is roughly one-fourth. One-fourth times 1 is one-fourth, which is consistent with the calculated answer of 0.262.

▶ **Practice Exercise 10.10** The element iodine, I_2, occurs in seawater and seaweed, from which it can be commercially extracted. The chemical equation for the reaction of elemental I_2 with hydrogen sulfide gas, H_2S, is

$$I_2(g) + H_2S(g) \longrightarrow S(s) + 2 HI(g)$$

For this reaction, how many grams of I_2 are needed to produce 25.00 g of S?

Grams-of-A to grams-of-B problems (Examples 10.9 and 10.10) are not the only type of problem for which the coefficients in a balanced equation can be used to relate quantities of two substances. As further examples of the use of equation coefficients in problem solving, consider Example 10.11 (a grams-of-A to moles-of-B problem) and Example 10.12 (a particles-of-A to grams-of-B problem).

EXAMPLE 10.11 **Calculating Moles of Product Produced in a Chemical Reaction**

Automotive air bags inflate when sodium azide, NaN_3, rapidly decomposes to its constituent elements. The equation for the chemical reaction is

$$2 NaN_3(s) \longrightarrow 2 Na(s) + 3 N_2(g)$$

How many moles of N_2 are produced when 1.000 g of NaN_3 decomposes?

SOLUTION

Step 1 The given quantity is 1.000 g of NaN_3, and the desired quantity is moles of N_2.

$$1.000 \text{ g NaN}_3 = ? \text{ moles N}_2$$

Step 2 This is a grams-of-A to moles-of-B problem. The pathway used to solve such a problem is, according to Figure 10.5,

$$\boxed{\begin{array}{c}\text{Grams} \\ \text{of A}\end{array}} \xrightarrow[\text{mass}]{\text{Molar}} \boxed{\begin{array}{c}\text{Moles} \\ \text{of A}\end{array}} \xrightarrow[\text{coefficients}]{\text{Equation}} \boxed{\begin{array}{c}\text{Moles} \\ \text{of B}\end{array}}$$

Step 3 The dimensional analysis setup is

$$1.000 \; \cancel{\text{g NaN}_3} \times \frac{1 \; \cancel{\text{mole NaN}_3}}{65.02 \; \cancel{\text{g NaN}_3}} \times \frac{3 \text{ moles N}_2}{2 \; \cancel{\text{moles NaN}_3}}$$

$$\text{grams A} \longrightarrow \text{moles A} \longrightarrow \text{moles B}$$

The number 65.02 in the first conversion factor is the molar mass of NaN_3.

Step 4 The solution, obtained from combining all of the numbers in the manner indicated in the setup, is

$$\frac{1.000 \times 1 \times 3}{65.02 \times 2} \text{ moles N}_2 = 0.023069824 \text{ mole N}_2 \quad \text{(calculator answer)}$$

$$= 0.02307 \text{ mole N}_2 \quad \textbf{(correct answer)}$$

Answer Double Check:

Is the numerical answer correct in terms of the number of significant figures present? Yes. The given data, 1.000 g, as well as the molar mass needed contain four significant figures. The other numbers in the calculation, coefficients from the chemical equation, are exact. The answer should contain four significant figures, which it does (0.02307).

▶ **Practice Exercise 10.11** Nitrous oxide, N_2O, is a gas used as an anesthetic in minor surgical procedures. It is known as "laughing gas" because a person inhaling it becomes somewhat giddy. The decomposition reaction for nitrous oxide is

$$2 \text{ N}_2\text{O(g)} \longrightarrow 2 \text{ N}_2\text{(s)} + \text{O}_2\text{(g)}$$

How many moles of N_2O must be decomposed to produce 10.00 g of N_2?

EXAMPLE 10.12 **Calculating the Amount of a Substance Produced in a Chemical Reaction**

The reaction of sulfuric acid (H_2SO_4) with elemental copper (Cu) produces three products—sulfur dioxide (SO_2), water (H_2O), and copper(II) sulfate ($CuSO_4$). How many grams of water will be produced at the same time that five billion (5.00×10^9) sulfur dioxide molecules are produced?

SOLUTION

Although a calculation of this type will not have a lot of practical significance, it will test your understanding of the problem-solving relationships under discussion in this section of the text.

The specifics of the chemical reaction of concern to us in this problem were given in word rather than equation form in the problem statement. These words must be translated into an equation before we can proceed with the problem solving. The equation is

$$\text{H}_2\text{SO}_4 + \text{Cu} \longrightarrow \text{SO}_2 + \text{H}_2\text{O} + \text{CuSO}_4$$

Having a chemical equation is not enough. It must be a *balanced* chemical equation. Using the balancing procedures of Section 10.4, the preceding equation in balanced form becomes

$$2\,H_2SO_4 + Cu \longrightarrow SO_2 + 2\,H_2O + CuSO_4$$

Now we are ready to proceed with the solving of our problem.

Step 1 We are given a certain number of particles (molecules) and asked to find the number of grams of a related substance.

$$5.00 \times 10^9 \text{ molecules } SO_2 = ?\text{ g } H_2O$$

Step 2 This is a particles-of-A to grams-of-B problem. Even though SO_2 and H_2O are both products, we can still work this problem in a manner similar to previous problems. The coefficients in a balanced equation relate reactants to products, reactants to reactants, and *products to products*. The pathway for this problem (see Fig. 10.5) is

$$\boxed{\text{Particles of A}} \xrightarrow[\text{number}]{\text{Avogadro's}} \boxed{\text{Moles of A}} \xrightarrow[\text{coefficients}]{\text{Equation}} \boxed{\text{Moles of B}} \xrightarrow[\text{mass}]{\text{Molar}} \boxed{\text{Grams of B}}$$

Step 3 The dimensional analysis setup is

$$5.00 \times 10^9 \text{ molecules } SO_2 \times \frac{1 \text{ mole } SO_2}{6.022 \times 10^{23} \text{ molecules } SO_2} \times \frac{2 \text{ moles } H_2O}{1 \text{ mole } SO_2} \times \frac{18.02 \text{ g } H_2O}{1 \text{ mole } H_2O}$$

$$\text{particles A} \longrightarrow \text{moles A} \longrightarrow \text{moles B} \longrightarrow \text{grams B}$$

Step 4 The solution, obtained by combining all of the numerical factors in the setup, is

$$\frac{5.00 \times 10^9 \times 1 \times 2 \times 18.02}{6.022 \times 10^{23} \times 1 \times 1} \text{ g } H_2O = 2.9923613 \times 10^{-13} \text{ g } H_2O$$

(calculator answer)

$$= 2.99 \times 10^{-13} \text{ g } H_2O$$

(**correct answer**)

Answer Double Check:

Is the magnitude of the numerical answer reasonable? Yes. The major factors contributing to answer magnitude are the two powers of 10. The division $10^9/10^{23}$ gives 10^{-14}. The calculated answer 2.99×10^{-13} g, is consistent with this magnitude estimate.

▶ **Practice Exercise 10.12**   Aluminum, the most abundant metal on Earth, has a silvery appearance when freshly cut. However, its surface quickly changes to a dull gray as a thin film of aluminum oxide, Al_2O_3, forms.

$$Al + O_2 \longrightarrow Al_2O_3 \quad \text{(unbalanced equation)}$$

How many O_2 molecules are needed to produce 1.000 g of Al_2O_3?

10.10 THE LIMITING REACTANT CONCEPT

When a chemical reaction is carried out in a laboratory or industrial setting, the reactants are not usually present in the exact molar ratios specified in the balanced chemical equation for the reaction. Most often, on purpose, excess quantities of one or more of the reactants are present.

Numerous reasons exist for having some reactants present in excess. Sometimes such a procedure will cause a reaction to occur more rapidly. For example, large amounts of oxygen make combustible materials burn faster. Sometimes an excess of one reactant will ensure that another reactant, perhaps a very expensive one, is completely consumed.

When reactants are not present in the exact molar ratios specified by the balanced chemical equation for the reaction, the reaction proceeds only until one of the reactants is depleted, that is, completely used up. At this point, the reaction stops. This reaction-stopping reactant, which has been entirely consumed, is called the *limiting reactant*. A **limiting reactant** *is the reactant in a chemical reaction that is entirely consumed when the reaction goes to completion (stops)*. The limiting reactant limits the amount of product(s) formed. Other reactants present are often called *excess reactants*; they are not entirely consumed. For excess reactants, the excess remains unreacted because there is not enough of the limiting reactant present to react with the excess.

The concept of a limiting reactant plays a major role in chemical calculations of certain types. It must be thoroughly understood. Let us consider some simple but analogous non-chemical examples of a limiting reactant before we go on to limiting reactant calculations.

Suppose we have a vending machine that contains forty 50-cent candy bars and we have 30 quarters. In this case we can purchase only 15 candy bars. The quarters are the limiting reactant. The candy bars are present in excess. Suppose we have 10 slices of cheese and 18 slices of bread and we want to make as many cheese sandwiches as possible using 1 slice of cheese and 2 slices of bread per sandwich. The 18 slices of bread limit us to 9 sandwiches; 1 slice of cheese is left over. The bread is the limiting reactant in this case even though initially there was more bread (18 slices) than cheese (10 slices) present. The bread is still limiting because it is used up first.

In combustion reactions, oxygen is not usually a limiting reactant, because it is so plentiful in the air. But if a jar is put over a burning candle, oxygen becomes a limiting reactant. As soon as all the oxygen in the jar is used up, the candle sputters out.

FIGURE 10.6 Starting with 10 nuts and 10 bolts, we can make (a) 10 one-nut, one-bolt combinations, (b) 5 two-nut, one-bolt combinations with 5 bolts left over (the nuts are the limiting reactant); and (c) 5 one-nut, two-bolt combinations with 5 nuts left over (the bolts are the limiting reactant).

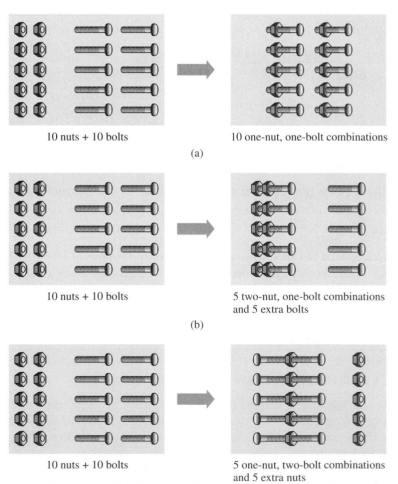

10 nuts + 10 bolts   10 one-nut, one-bolt combinations

(a)

10 nuts + 10 bolts   5 two-nut, one-bolt combinations and 5 extra bolts

(b)

10 nuts + 10 bolts   5 one-nut, two-bolt combinations and 5 extra nuts

(c)

An additional limiting reactant analogy that comes closer to the realm of molecules and atoms involves nuts and bolts. Assume we have 10 identical nuts and 10 identical bolts. From this collection we can make 10 one-nut, one-bolt entities by screwing a nut on each bolt. This situation is depicted in Figure 10.6a.

Next, let us make "two-nut, one-bolt" entities from our same collection of nuts and bolts. This time we can make only five combinations, and we will have five bolts left over, as is shown in Figure 10.6b. We run out of nuts before all the bolts are used up. In chemical jargon, we say that the nuts are the limiting reactant.

Finally, let us consider making one-nut, two-bolt combinations. As is shown in Figure 10.6c, this time we do not have enough bolts; we can make five combinations, and we will have five nuts left over. The bolts are the limiting reactant.

Now let us consider, as is presented in Example 10.13, an extension of our nut–bolt discussion to a situation that cannot easily be reasoned out in one's head. Instead, a calculation must be performed.

EXAMPLE 10.13 **Determining a Limiting Reactant in a Nonchemical Context**

What will be the limiting reactant in the production of two-nut, three-bolt combinations from a collection of 284 nuts and 414 bolts?

SOLUTION

There are three possible answers to this problem.

1. We will run out of bolts first; bolts are the limiting reactant.
2. We will run out of nuts first; nuts are the limiting reactant.
3. The ratio of nuts to bolts is such that we run out of both at the same time; both are limiting reactants.

We determine which of these answers is the correct one by calculating how many nut–bolt combinations can be made from each of the "ingredients," assuming an excess of the other.

$$284 \text{ nuts} \times \frac{1 \text{ combination}}{2 \text{ nuts}} = 142 \text{ combinatons} \quad \text{(calculator and \textbf{correct answer})}$$

$$414 \text{ bolts} \times \frac{1 \text{ combination}}{3 \text{ bolts}} = 138 \text{ combinatons} \quad \text{(calculator and \textbf{correct answer})}$$

Because fewer combinations can be made from the bolts, the bolts are the limiting reactant.

▶ **Practice Exercise 10.13** What is the limiting reactant in the production of four-nut, three-bolt combinations from a collection of 230 nuts and 175 bolts?

Now let us proceed to chemical calculations that involve a limiting reactant. *Whenever the quantities of two or more reactants in a chemical reaction are given, it is necessary to determine which of the given quantities is the limiting reactant.*

Determining the limiting reactant, that is, the reactant in shortest supply, can be accomplished by the following procedure.

1. Determine the number of *moles* of each of the reactants present.
2. Calculate the number of moles of product *each* of the molar amounts of reactant would produce if it were the only reactant amount given. If more than one product is formed in the reaction, you need to do this mole calculation for only one of the products.
3. The reactant that produces the *lower number* of moles of product is the limiting reactant.

EXAMPLE 10.14 **Determining the Limiting Reactant from Given Reactant Amounts**

Silver and silver-plated objects tarnish in the presence of hydrogen sulfide (H_2S), a gas that originates from the decay of food, because of the reaction

$$4\,Ag + 2\,H_2S + O_2 \longrightarrow 2\,Ag_2S + 2\,H_2O$$

The black product Ag_2S is the tarnish. If 25.00 g of Ag, 5.00 g of H_2S, and 4.00 g of O_2 are present in a reaction mixture, which is the limiting reactant for tarnish formation?

SOLUTION

To determine the limiting reactant, we determine how many moles of product each of the reactants can form. In this particular problem there are two products: Ag_2S and H_2O. It is sufficient to calculate how many moles of either Ag_2S or H_2O are formed. The decision as to which product to use is arbitrary; we will choose H_2O.

The calculation type will be grams-of-A to moles-of-B. We start with a given number of grams of reactant, and we want to calculate moles of product. The pathway for the calculation, in terms of Figure 10.5, is

$$\boxed{\text{Grams of A}} \xrightarrow[\text{mass}]{\text{Molar}} \boxed{\text{Moles of A}} \xrightarrow[\text{coefficients}]{\text{Equation}} \boxed{\text{Moles of B}}$$

Note that we will have to go through this type of calculation three times because we have three reactants: once for Ag, once for H_2S, and once for O_2.
For Ag,

$$25.00\ \cancel{\text{g Ag}} \times \frac{1\ \cancel{\text{mole Ag}}}{107.87\ \cancel{\text{g Ag}}} \times \frac{2\ \text{moles } H_2O}{4\ \cancel{\text{moles Ag}}} = 0.11588022\ \text{mole } H_2O$$

(calculator answer)

$$= 0.1159\ \text{mole } H_2O$$

(**correct answer**)

For H_2S,

$$5.00\ \cancel{\text{g } H_2S} \times \frac{1\ \cancel{\text{mole } H_2S}}{34.09\ \cancel{\text{g } H_2S}} \times \frac{2\ \text{moles } H_2O}{2\ \cancel{\text{moles } H_2S}} = 0.14667058\ \text{mole } H_2O$$

(calculator answer)

$$= 0.147\ \text{mole } H_2O$$

(**correct answer**)

For O_2,

$$4.00\ \cancel{\text{g } O_2} \times \frac{1\ \cancel{\text{mole } O_2}}{32.00\ \cancel{\text{g } O_2}} \times \frac{2\ \text{moles } H_2O}{1\ \cancel{\text{mole } O_2}} = 0.25\ \text{mole } H_2O$$

(calculator answer)

$$= 0.250\ \text{mole } H_2O$$

(**correct answer**)

The limiting reactant is the reactant that will produce the fewest number of moles of H_2O. Looking at the numbers just calculated, we see that Ag will be the limiting reactant. Once the limiting reactant has been determined, the amount of that reactant present becomes the starting point for any further calculations about the chemical reaction under consideration. Example 10.15 illustrates this point.

Answer Double Check:

Limiting reactant calculations are always grams-of-A to moles-of-B calculations (where A is a reactant and B is a product). The same product must be used for each of the parallel calculations. Such is the case here; moles of H_2O was the end result of each calculation. Was the limiting reactant properly identified? Yes. The reactant that produces the least number of moles of product, Ag in this case, is the limiting reactant.

▶ **Practice Exercise 10.14** Adipic acid, $C_6H_{10}O_4$, a substance needed for the production of nylon, is prepared commercially by the reaction of cyclohexane (C_6H_{12}) with oxygen (O_2).

$$2\,C_6H_{12} + 5\,O_2 \longrightarrow 2\,C_6H_{10}O_4 + 2\,H_2O$$

If 15.00 g of cyclohexane and 25.00 g of oxygen are present in a reaction mixture, which is the limiting reactant for the production of adipic acid?

EXAMPLE 10.15 **Calculating the Mass of Product from Masses of Reactants**

Aluminum metal, in the form of a powder, reacts vigorously with iron(III) oxide when the two substances are heated. The equation for the reaction is

$$2\,Al(s) \longrightarrow Fe_2O_3(s) \longrightarrow Al_2O_3(s) + 2\,Fe(l)$$

How many grams of molten Fe can be formed from a reaction mixture containing 30.0 g Al and 90.0 g of Fe_2O_3?

SOLUTION

First, we must determine the limiting reactant since specific amounts of both reactants are given in the problem. This determination involves two grams-of-A to moles-of-B calculations—one for the reactant Al and one for the reactant Fe_2O_3. The product we "key in" on is Fe because our final goal is the mass of Fe produced.

$$\boxed{\begin{array}{c}\text{Grams}\\\text{of A}\end{array}} \xrightarrow{\substack{\text{Molar}\\\text{mass}}} \boxed{\begin{array}{c}\text{Moles}\\\text{of A}\end{array}} \xrightarrow{\substack{\text{Equation}\\\text{coefficients}}} \boxed{\begin{array}{c}\text{Moles}\\\text{of B}\end{array}}$$

For Al,

$$30.0\ \cancel{\text{g Al}} \times \frac{1\ \cancel{\text{mole Al}}}{26.98\ \cancel{\text{g Al}}} \times \frac{2\ \text{moles Fe}}{2\ \cancel{\text{moles Al}}} = 1.1119347\ \text{moles Fe} \qquad \text{(calculator answer)}$$

$$= 1.11\ \text{moles Fe} \qquad \textbf{(correct answer)}$$

For Fe_2O_3,

$$90.0\ \cancel{\text{g Fe}_2\text{O}_3} \times \frac{1\ \cancel{\text{mole Fe}_2\text{O}_3}}{159.70\ \cancel{\text{g Fe}_2\text{O}_3}} \times \frac{2\ \text{moles Fe}}{1\ \cancel{\text{mole Fe}_2\text{O}_3}} = 1.1271133\ \text{moles Fe}$$

$$\text{(calculator answer)}$$

$$= 1.13\ \text{moles Fe}$$

$$\textbf{(correct answer)}$$

Thus, Al is the limiting reactant since fewer moles of Fe can be produced from it (1.11 moles) than from the Fe_2O_3 (1.13 moles).

We can now calculate the grams of Fe formed in the reaction, using the 1.11 moles of Fe (formed from our limiting reactant) as our starting factor. The calculation will be a simple one-step moles-of-A to grams-of-A conversion.

$$\boxed{\text{Moles of Fe}} \xrightarrow[\text{mass}]{\text{Molar}} \boxed{\text{Grams of Fe}}$$

$$1.11 \ \cancel{\text{moles Fe}} \times \frac{55.85 \text{ g Fe}}{1 \ \cancel{\text{mole Fe}}} = 61.9935 \text{ g Fe} \quad \text{(calculator answer)}$$

$$= 62.0 \text{ g Fe} \quad \textbf{(correct answer)}$$

Answer Double Check:

Was the correct moles-of-Fe value selected for the final calculation step? Yes. The smaller of the two moles-of-Fe values, the limiting reactant, is the correct choice.

▶ **Practice Exercise 10.15** Chloroform ($CHCl_3$), an important industrial solvent, can be made by the reaction of chlorine and methane (CH_4), according to the following chemical equation:

$$3 \text{ Cl}_2 + \text{CH}_4 \longrightarrow \text{CHCl}_3 + 3 \text{ HCl}$$

How many grams of $CHCl_3$ can be made from a reaction mixture containing 150.0 g Cl_2 and 50.0 g CH_4?

10.11 YIELDS: THEORETICAL, ACTUAL, AND PERCENT

When stoichiometric relationships (Secs. 10.9 and 10.10) are used to calculate the amount of a product that will be produced in a chemical reaction from given amounts of reactants, the answer obtained represents a *theoretical yield*. A **theoretical yield** *is the maximum amount of a product that can be obtained from given amounts of reactants in a chemical reaction if no losses or inefficiencies of any kind occur.* Examples 10.10 and 10.15 are theoretical yield calculations, although this fact was not noted at the time they were presented.

In most chemical reactions the amount of a given product experimentally isolated from the reaction mixture is less than that predicted by calculation. Why is this so? Two major factors contribute to this situation.

1. Some product is almost always lost in the process of its isolation and purification and in such mechanical operations as transferring materials from one container to another.
2. Often a particular set of reactants undergoes two or more reactions simultaneously, forming undesired products (in small amounts) as well as the desired products. Reactants consumed in these side reactions obviously will not end up in the form of the desired products.

The net effect of these factors is that the actual quantity of product experimentally isolated—that is, the *actual yield*—is almost always less, sometimes far less, than the theoretically calculated amount. An **actual yield** *is the amount of a product actually obtained from a chemical reaction.* Actual yield is always an experimentally determined number; it cannot be calculated.

Product loss is specified in terms of *percent yield*. The **percent yield** *is the ratio of the actual (experimental) yield of a product in a chemical reaction to its theoretical (calculated) yield multiplied by 100 (to give percent).* The mathematical equation for percent yield is

$$\text{percent yield} = \frac{\text{actual yield}}{\text{theoretical yield}} \times 100$$

If the theoretical yield of a product for a reaction is calculated to be 17.9 g and the amount of product actually obtained (the actual yield) is 15.8 g, the percent yield is 88.3%.

$$\text{percent yield} = \frac{15.8\ \text{g}}{17.9\ \text{g}} \times 100 = 88.268156\% \quad \text{(calculator answer)}$$

$$= 88.3\% \quad \textbf{(correct answer)}$$

Note that percent yield cannot exceed 100%; you cannot obtain a yield greater than that which is theoretically possible.

EXAMPLE 10.16 **Calculating the Theoretical Yield and Percent Yield for a Chemical Reaction**

Mixing aluminum oxide with sulfuric acid produces aluminum sulfate, a compound that is used in the paper industry to strengthen paper and make it water-resistant. The chemical equation for this reaction is

$$Al_2O_3(s) + 3\ H_2SO_4(aq) \longrightarrow Al_2(SO_4)_3(aq) + 3\ H_2O(l)$$

a. What is the theoretical yield of $Al_2(SO_4)_3$ that can be obtained from a reaction mixture containing 75.0 g of Al_2O_3 and 150.0 g of H_2SO_4?

b. If the actual yield of $Al_2(SO_4)_3$ for the reaction mixture in part (a) is 136 g, what is the percent yield of $Al_2(SO_4)_3$ for the reaction?

SOLUTION

a. The limiting reactant must be determined before the theoretical yield can be calculated. Recalling the procedures of Examples 10.14 and 10.15 for determining the limiting reactant, we calculate the number of moles of $Al_2(SO_4)_3$ that can be produced from each individual reactant amount using a grams-of-A to moles-of-B type of calculation.

$$\boxed{\begin{array}{c}\text{Grams}\\\text{of A}\end{array}} \xrightarrow{\begin{array}{c}\text{Molar}\\\text{mass}\end{array}} \boxed{\begin{array}{c}\text{Moles}\\\text{of A}\end{array}} \xrightarrow{\begin{array}{c}\text{Equation}\\\text{coefficients}\end{array}} \boxed{\begin{array}{c}\text{Moles}\\\text{of B}\end{array}}$$

For Al_2O_3,

$$75.0\ \text{g}\ Al_2O_3 \times \frac{1\ \text{mole}\ Al_2O_3}{101.96\ \text{g}\ Al_2O_3} \times \frac{1\ \text{mole}\ Al_2(SO_4)_3}{1\ \text{mole}\ Al_2O_3} = 0.73558258\ \text{mole}\ Al_2(SO_4)_3$$

$$\text{(calculator answer)}$$

$$= 0.736\ \text{mole}\ Al_2(SO_4)_3$$

$$\textbf{(correct answer)}$$

For H_2SO_4,

$$150.0\ \text{g}\ H_2SO_4 \times \frac{1\ \text{mole}\ H_2SO_4}{98.09\ \text{g}\ H_2SO_4} \times \frac{1\ \text{mole}\ Al_2(SO_4)_3}{3\ \text{moles}\ H_2SO_4}$$

$$= 0.50973596\ \text{mole}\ Al_2(SO_4)_3$$

$$\text{(calculator answer)}$$

$$= 0.5097\ \text{mole}\ Al_2(SO_4)_3$$

$$\textbf{(correct answer)}$$

The calculations show that H_2SO_4 is the limiting reactant.

The maximum number of grams of $Al_2(SO_4)_3$ obtainable from the limiting reactant, that is, the theoretical yield, can now be calculated. It is done using a one-step moles-of-A to grams-of-A setup.

$$\boxed{\begin{array}{c} \text{Moles of} \\ Al_2(SO_4)_3 \end{array}} \xrightarrow[\text{mass}]{\text{Molar}} \boxed{\begin{array}{c} \text{Grams of} \\ Al_2(SO_4)_3 \end{array}}$$

$$0.5097 \text{ mole } Al_2(SO_4)_3 \times \frac{342.17 \text{ g } Al_2(SO_4)_3}{1 \text{ mole } Al_2(SO_4)_3} = 174.40405 \text{ g } Al_2(SO_4)_3$$

$$\text{(calculator answer)}$$

$$= 174.4 \text{ g } Al_2(SO_4)_3$$

(correct answer)

b. The percent yield is obtained by dividing the actual yield by the theoretical yield and multiplying by 100.

$$\text{percent yield} = \frac{\text{actual yield}}{\text{theoretical yield}} \times 100 = \frac{136 \text{ g}}{174.4 \text{ g}} \times 100 = 77.981651\%$$

$$\text{(calculator answer)}$$

$$= 78.0\%$$

(correct answer)

Answer Double Check:

Is the magnitude of the percent yield value reasonable? Yes. With an actual yield of 150 g (rounded) and a theoretical yield of 200 g (rounded), a yield of 78.0% is realistic because 150/200 times 100 is equal to 75.

▶ **Practice Exercise 10.16** Solutions of sodium hypochlorite, NaClO, are sold as laundry bleach. This bleaching agent can be produced by the reaction

$$2 \text{ NaOH} + Cl_2 \longrightarrow \text{NaCl} + \text{NaClO} + H_2O$$

a. What is the theoretical yield of NaClO that can be obtained from a reaction mixture containing 75.0 g of NaOH and 50.0 g of Cl_2?
b. The actual yield is 43.2 g of NaClO. What is the percent yield of NaClO from this reaction mixture?

When synthesizing a compound with a complex structure in a laboratory or industrial setting, a multistep procedure (a series of reactions) is often required.

$$A \longrightarrow B \longrightarrow C \longrightarrow D \longrightarrow E \longrightarrow F$$

This "line" equation denotes a five-step process needed to produce a product F from a reactant A. The intermediate compounds (B, C, D, and E) are each produced in one step and consumed in the next step.

For such multistep situations, there is a percent yield factor associated with each step in the process. The percent yield for the overall synthesis is the product of the percent yields (expressed as decimals) for the individual steps. For example, suppose a five-step reaction sequence is required to produce a particular compound and the percent yield for each step is 85% (0.85). The overall percent yield is much lower than the individual 85% yields, as shown by the following calculation.

$$\text{Overall percent yield} = (0.85 \times 0.85 \times 0.85 \times 0.85 \times 0.85) \times 100 = 44\%$$

Consideration of overall percent yields in a multistep synthesis is an important factor in determining the economic feasibility (commercial potential) for producing a complex product. Section 10.12 considers further the topic of sequential (multistep) chemical reactions.

10.12 SIMULTANEOUS AND SEQUENTIAL CHEMICAL REACTIONS

The concepts presented so far in this chapter can easily be adapted to problem-solving situations that involve two or more chemical reactions. In some cases the two or more chemical reactions occur simultaneously, and in other cases they occur sequentially (one right after the other). Example 10.17 deals with a pair of simultaneous reactions, and Example 10.18 deals with three sequential reactions.

EXAMPLE 10.17 | **A Calculation Based on Simultaneous Chemical Reactions**

A mixture contains 47.3% by mass magnesium carbonate ($MgCO_3$) and 52.7% by mass calcium carbonate ($CaCO_3$). The mixture is heated until both carbonates completely decompose as shown by the following equations.

$$MgCO_3 \longrightarrow MgO + CO_2$$

$$CaCO_3 \longrightarrow CaO + CO_2$$

How many grams of CO_2 are produced from decomposition of 78.3 g of the mixture?

SOLUTION

In solving this problem we will need to carry out two parallel calculations. In the one calculation we will determine the grams of CO_2 produced from the $MgCO_3$ component of the mixture and in the other, the grams of CO_2 produced from the $CaCO_3$ component. Then we will add together the answers from the two parallel calculations to get our final answer, the total grams of CO_2 produced. The sequence of conversion factors for each setup is derived from the following pathway.

| Grams of mixture | $\xrightarrow{\text{Percent composition}}$ | Grams of carbonate | $\xrightarrow{\text{Molar mass}}$ | Moles of carbonate | $\xrightarrow{\text{Equation coefficients}}$ | Moles of CO_2 | $\xrightarrow{\text{Molar mass}}$ | Grams of CO_2 |

The number of grams of CO_2 produced from the $MgCO_3$ is given by the following setup:

$$78.3 \text{ g mixture} \times \frac{47.3 \text{ g } MgCO_3}{100.0 \text{ g mixture}} \times \frac{1 \text{ mole } MgCO_3}{84.32 \text{ g } MgCO_3} \times \frac{1 \text{ mole } CO_2}{1 \text{ mole } MgCO_3} \times \frac{44.01 \text{ g } CO_2}{1 \text{ mole } CO_2}$$

$$= 19.330526 \text{ g } CO_2 \quad \text{(calculator answer)}$$

$$= 19.3 \text{ g } CO_2 \quad \text{(correct answer)}$$

The number of grams of CO_2 produced from the $CaCO_3$ is given by the following setup.

$$78.3 \text{ g mixture} \times \frac{52.7 \text{ g } CaCO_3}{100.0 \text{ g mixture}} \times \frac{1 \text{ mole } CaCO_3}{100.09 \text{ g } CaCO_3} \times \frac{1 \text{ mole } CO_2}{1 \text{ mole } CaCO_3} \times \frac{44.01 \text{ g } CO_2}{1 \text{ mole } CO_2}$$

$$= 18.144 \text{ g } CO_2 \quad \text{(calculator answer)}$$

$$= 18.1 \text{ g } CO_2 \quad \text{(correct answer)}$$

The first conversion factor in each setup is derived from the given percentage of that compound in the mixture. The use of percentages as conversion factors was covered in Section 3.9.

The total number of grams of CO_2 produced is the sum of the grams of CO_2 in the individual reactions.

$$(19.3 + 18.1) \text{ g } CO_2 = 37.4 \text{ g } CO_2 \qquad \text{(calculator and \textbf{correct answer})}$$

Answer Double Check:

Were the given percentages for the components of the mixture used properly in setting up the calculations? Yes. The percentages, as conversion factors, were applied to the same number (78.3 g), the total amount of mixture present.

▶ **Practice Exercise 10.17**   A mixture contains 75.0%, by mass, carbon disulfide (CS_2) and 25.0%, by mass, hydrogen sulfide (H_2S). The mixture is burned in oxygen to produce SO_2, CO_2, and H_2O. The combustion reactions are

$$CS_2 + 3\,O_2 \longrightarrow CO_2 + 2\,SO_2$$

$$2\,H_2S + 3\,O_2 \longrightarrow 2\,SO_2 + 2\,H_2O$$

How many moles of SO_2 can be produced from the combustion of 75.0 g of mixture?

Often, particularly in industrial processes, more than one chemical reaction is needed to change starting materials into desired products; a series of sequential chemical reactions is required. The product amount from each of the chemical reactions in the sequence becomes the starting material for the next chemical reaction in the sequence.

EXAMPLE 10.18   **A Calculation Based on Sequential Chemical Reactions**

In steelmaking, a series of three chemical reactions is needed to convert Fe_2O_3 (the iron-containing component of iron ore) to molten iron.

$$\text{reaction (1):} \quad 3\,Fe_2O_3 + CO \longrightarrow 2\,Fe_3O_4 + CO_2$$

$$\text{reaction (2):} \quad Fe_3O_4 + CO \longrightarrow 3\,FeO + CO_2$$

$$\text{reaction (3):} \quad FeO + CO \longrightarrow Fe + CO_2$$

Assuming that the reactant CO is present in excess, how many grams of Fe can be produced from 125 g of Fe_2O_3?

SOLUTION

The key substances in this set of reactions, from a calculational point of view, are the iron-containing species: Fe_2O_3, Fe_3O_4, FeO, and Fe. Note that the iron-containing species produced in the first and second reactions (Fe_3O_4 and FeO, respectively) are the reactants for the second and third reactions, respectively.

$$Fe_2O_3 \xrightarrow{\text{reaction(1)}} Fe_3O_4 \xrightarrow{\text{reaction(2)}} FeO \xrightarrow{\text{reaction(3)}} Fe$$

We can solve this problem using a single multiple-step setup. The sequence of conversion factors needed is that for a gram-to-gram problem with two additional intermediate mole-to-mole steps added.

The dimensional analysis setup is

$$125 \text{ g Fe}_2O_3 \times \frac{1 \text{ mole Fe}_2O_3}{159.70 \text{ g Fe}_2O_3} \times \frac{2 \text{ mole Fe}_3O_4}{3 \text{ mole Fe}_2O_3} \times \frac{3 \text{ moles FeO}}{1 \text{ mole Fe}_3O_4}$$

$$\times \frac{1 \text{ mole Fe}}{1 \text{ mole FeO}} \times \frac{55.85 \text{ g Fe}}{1 \text{ mole Fe}}$$

$$= 87.429555 \text{ g Fe} \quad \text{(calculator answer)}$$

$$= 87.4 \text{ g Fe} \qquad \text{(correct answer)}$$

An alternative approach to solving this problem would involve setting up a separate calculation for each equation. As a first step, the number of moles of Fe_3O_4 produced in the first reaction would be calculated. In the second step, one would determine the moles of FeO obtained if all the Fe_3O_4 produced in the first reaction entered into the second reaction. In the final step, one would determine the grams of Fe derivable from the FeO produced in the second reaction. The answer obtained from this three-setup method is the same as that obtained from the one-setup method.

▶ **Practice Exercise 10.18**   A three-step process for producing sulfuric acid, H_2SO_4, the industrial chemical produced in the greatest amount in the United States, is

$$2 SO_2 + O_2 \longrightarrow 2 SO_3$$
$$SO_3 + H_2SO_4 \longrightarrow H_2S_2O_7$$
$$H_2S_2O_7 + H_2O \longrightarrow 2 H_2SO_4$$

Assuming that the reactant O_2 in the first step is present in excess, how many grams of H_2SO_4 can be produced from 25.0 g of SO_2?

Combining Sequential Chemical Reaction Equations into a Single Overall Chemical Reaction Equation

A common characteristic of all sets of sequential chemical reaction equations is that a product from one reaction becomes the reactant for the next reaction. This characteristic facilitates the combination of a set of sequential chemical reaction equations into a *single overall* chemical reaction equation. In this combining process intermediate chemical substances in the sequence, those that are produced first as a product and then used secondly as a reactant, are eliminated from the chemical equations.

The combining procedure used to produce the single overall chemical equation involves the following steps.

1. Write the chemical equations for the sequential steps of the reaction.
2. Identify and then equalize chemical equation coefficients for *intermediate substances* so that these substance can be canceled from the equations.
3. Carry out the "cancellations" by adding the step-equations together to obtain the single overall equation.

Example 10.19 illustrates this just described "cancellation-addition" procedure.

EXAMPLE 10.19   **Combining Sequential Chemical Equations to Produce a Single Overall Chemical Equation**

Sulfur trioxide, SO_3, can be produced from sulfur, S, in the following two-step process.

$$S + O_2 \longrightarrow SO_2$$
$$2 SO_2 + O_2 \longrightarrow 2 SO_3$$

Combine these sequential chemical equations into a single overall chemical equation for the process.

SOLUTION

Step 1 Write the sequential chemical equations that are to be combined.

$$S + O_2 \longrightarrow SO_2$$
$$2\,SO_2 + O_2 \longrightarrow 2\,SO_3$$

Step 2 Identify the substance that "links" the chemical equations together. In our case, it is SO_2, the product in the first equation and a reactant in the second equation. For this SO_2 "link" to cancel out, which needs to occur during the combining process, the coefficient for SO_2 must be the same in both equations.

The equalizing of SO_2 coefficients is obtained by doubling *all* of the coefficients in the first equation while leaving all of the coefficients in the second equation as is.

$$2\,S + 2\,O_2 \longrightarrow 2\,SO_2 \text{ (coefficients have been doubled)}$$
$$2\,SO_2 + O_2 \longrightarrow 2\,SO_3$$

Step 3 Having the coefficients for SO_2 the same means that the amount of SO_2 formed in the first reaction is the same as that which reacts in the second reaction. With the amounts the same, the SO_2 can be canceled from the equation as the two equations are added together

$$2\,S + 2\,O_2 \longrightarrow 2\,\cancel{SO_2}$$
$$\underline{2\,\cancel{SO_2} + O_2 \longrightarrow 2\,SO_3}$$
$$2\,S + 3\,O_2 \longrightarrow 2\,SO_3$$

In adding the two equations, the reactants from *both* equations appear on the left side of the equation and the products from *both* equations appear on the right side of the equation. The first equation has $2\,O_2$ on the reactant side and the second has O_2 on the reactant side which gives a total of $3\,O_2$.

The single equation

$$2\,S + 3\,O_2 \longrightarrow 2\,SO_3$$

is equivalent to the two chemical equations we started with and can be used in lieu of the two-step formulation in many types of calculations. This single equation, however, does not convey the important fact that SO_2 is produced as an "intermediate" in this two-step reaction process.

▶ **Practice Exercise 10.19** Disulfuric acid, $H_2S_2O_7$, can be produced from sulfur dioxide, SO_2, in the following two-step process.

$$2\,SO_2 + O_2 \longrightarrow 2\,SO_3$$
$$SO_3 + H_2SO_4 \longrightarrow H_2S_2O_7$$

Combine these sequential chemical equations into a single overall chemical equation for the process.

Summary

1. **Reactants and Products** The starting materials in a chemical reaction are called reactants and the substances produced as a result of a chemical reaction are called products.

2. **Law of Conservation of Mass** Mass is neither created nor destroyed in any ordinary chemical reaction. In a chemical reaction the sum of the

masses of the products is always equal to the sum of the masses of the reactants.

3. **Chemical Equation** A chemical equation is a representation for a chemical reaction that uses chemical symbols and chemical formulas for the reactants and products involved in the chemical reaction. The formulas of the reactants are always written on the left side of the chemical equation and the formulas of the products are always found on the right side of the equation.

4. **Balanced Chemical Equation** A balanced chemical equation has the same number of atoms of each element involved in the reaction on each side of the equation. An unbalanced equation is brought into balance through the use of coefficients. A coefficient is a number that is placed to the left of the formula of a substance and that changes the amount, but not the identity, of the substance.

5. **Classes of Chemical Reactions** Chemical reactions are grouped into classes based on common characteristics. A synthesis reaction is a reaction in which a single product is produced from two (or more) reactants. A decomposition reaction is one in which a single reactant is converted into two or more simpler products. A single-replacement reaction is a reaction in which one element within a compound is replaced by another element. A double-replacement reaction is one in which two compounds exchange parts with each other and

form two different compounds. A combustion reaction involves the reaction of a substance with oxygen (usually from air) that proceeds with evolution of heat and usually also a flame.

6. **The Mole and Chemical Equations** The coefficients in a balanced chemical equation give the molar ratios between substances consumed or produced in the chemical reaction described by the chemical equation.

7. **Chemical Stoichiometry** Chemical stoichiometry is the study of the quantitative relationships among reactants and products in a chemical reaction. The stoichiometry of a chemical reaction always involves the molar relationships (coefficients) among reactants and products.

8. **Limiting Reactant** A limiting reactant is completely consumed in a reaction. When it is used up, the reaction stops, thus limiting the quantities of products formed. Other reactants present are often called excess reactants; they are not entirely consumed.

9. **Theoretical Yield, Actual Yield, and Percent Yield** The theoretical yield is the maximum amount of a product that can be obtained from given amounts of reactants in a chemical reaction if no losses or inefficiencies of any kind occur. In most chemical reactions the amount of a given product isolated from the reaction mixture, the actual yield, is less than that which is theoretically possible. The percent yield compares the actual and theoretical yields.

Key Terms

The new terms defined in this chapter are

actual yield *Sec. 10.11*
balanced chemical equation *Sec. 10.3*
chemical equation *Sec. 10.2*
chemical stoichiometry *Sec. 10.9*

combustion reaction *Sec. 10.6*
decomposition reaction *Sec. 10.6*
double-replacement reaction *Sec. 10.6*

equation coefficient *Sec. 10.3*
law of conservation of mass *Sec. 10.1*
limiting reactant *Sec. 10.10*
percent yield *Sec. 10.11*

products *Sec. 10.1*
reactants *Sec. 10.1*
single-replacement reaction *Sec. 10.6*
synthesis reaction *Sec. 10.6*
theoretical yield *Sec. 10.11*

Practice Problems

THE LAW OF CONSERVATION OF MASS (SEC. 10.1)

10.1 Determine whether each of the following statements is consistent with or inconsistent with the law of conservation of mass.
a. 127.10 g of Cu reacts with 34.00 g of O_2 to produce 159.10 g of CuO.
b. 76.15 g CS_2 reacts with 96.00 g O_2 to produce 44.01 g of CO_2 and 128.14 g of SO_2.

10.2 Determine whether each of the following statements is consistent with or inconsistent with the law of conservation of mass.
a. 80.07 g of SO_3 reacts with 18.02 g of H_2O to produce 98.09 g of H_2SO_4.
b. 81.55 g of CuO reacts with 4.04 g of H_2 to produce 63.55 g of Cu and 18.02 g of H_2O.

10.3 Determine the value of x in the following otherwise correct description of a chemical reaction.

$$(16.05 \text{ g CH}_4) + (64.00 \text{ g O}_2) \longrightarrow (x \text{ g CO}_2) + (36.04 \text{ g H}_2\text{O})$$

10.4 Determine the value of x in the following otherwise correct description of a chemical reaction.

$$(137.32 \text{ g PCl}_3) + (x \text{ g H}_2) \longrightarrow (34.00 \text{ g PH}_3) + (109.38 \text{ g HCl})$$

10.5 A 4.2 g sample of sodium hydrogen carbonate is added to a solution of acetic acid weighing 10.0 g. The two substances react, releasing carbon dioxide gas to the atmosphere. After the reaction, the contents of the reaction vessel weigh 12.0 g. What is the mass of carbon dioxide given off during the reaction?

10.6 A 1.00 g sample of solid calcium carbonate is added to a reaction flask containing 10.00 g of hydrochloric acid solution. The calcium carbonate slowly dissolves in the acid solution as evidenced by the generation of carbon dioxide gas. After 5 minutes of reaction, 0.21 g of carbon dioxide gas has been given off. At that time, what is the mass, in grams, of the reaction flask contents?

10.7 Diagram I represents the reactant mixture for a chemical reaction. Select from diagrams II through IV the product mixture that is consistent with both diagram I and the concepts associated with the law of conservation of mass.

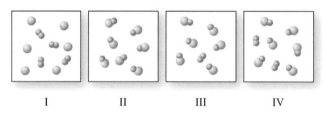

I   II   III   IV

10.8 Diagram I represents the reactant mixture for a chemical reaction. Select from diagrams II through IV the product mixture that is consistent with both diagram I and the concepts associated with the law of conservation of mass.

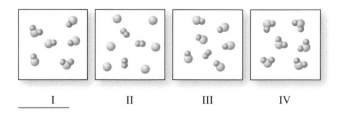

I   II   III   IV

CHEMICAL EQUATION NOTATION (SECS. 10.2 AND 10.5)

10.9 In which of the following pairs of symbols and/or formulas for gaseous elements are both members of the pair written appropriately for use in chemical equations?
a. O_2 and H_2 b. Xe and Cl_2
c. Kr and Rn   d. N_2 and Br

10.10 In which of the following pairs of symbols and/or formulas for gaseous elements are both members of the pair written appropriately for use in chemical equations?
a. Xe_2 and Br_2 b. Kr and N
c. H_2 and F_2 d. I_2 and Rn

10.11 What do the symbols in parentheses stand for in the following chemical equations?
a. $Li_3N(s) + 3 H_2O(l) \longrightarrow 3 LiOH(aq) + NH_3(g)$
b. $CaCO_3(s) + H_2SO_4(aq) \longrightarrow CaSO_4(s) + CO_2(g) + H_2O(l)$

10.12 What do the symbols in parentheses stand for in the following chemical equations?
a. $2 Na_2O_2(s) + 2 H_2O(l) \longrightarrow 4 NaOH(aq) + O_2(g)$
b. $Na_2CO_3(s) + 2 HCl(aq) \longrightarrow 2 NaCl(aq) + CO_2(g) + H_2O(l)$

BALANCING CHEMICAL EQUATIONS (SEC. 10.4)

10.13 Indicate whether each of the following chemical equations is *balanced* or *unbalanced*.
a. $FeO + CO \longrightarrow Fe + CO_2$
b. $CH_4 + O_2 \longrightarrow CO_2 + H_2O$
c. $NH_3 + HNO_3 \longrightarrow NH_4NO_3$
d. $KCl + O_2 \longrightarrow KClO_3$

10.14 Indicate whether each of the following chemical equations is *balanced* or *unbalanced*.
a. $Fe_2O_3 + CO \longrightarrow Fe + CO_2$
b. $NaBr + AgNO_3 \longrightarrow AgBr + NaNO_3$
c. $SO_3 + H_2SO_4 \longrightarrow H_2S_2O_7$
d. $PCl_3 + H_2 \longrightarrow PH_3 + HCl$

10.15 For each of the following balanced equations, indicate how many atoms of each element are present on the reactant and product sides of the chemical equation.
a. $2 Cu + O_2 \longrightarrow 2 CuO$
b. $2 H_2O \longrightarrow 2 H_2 + O_2$
c. $BaCl_2 + Na_2S \longrightarrow BaS + 2 NaCl$
d. $2 C_3H_8O + 9 O_2 \longrightarrow 6 CO_2 + 8 H_2O$

10.16 For each of the following balanced equations, indicate how many atoms of each element are present on the reactant and product sides of the chemical equation.
a. $N_2 + 3 H_2 \longrightarrow 2 NH_3$
b. $4 Fe + 3 O_2 \longrightarrow 2 Fe_2O_3$
c. $Au_2S_3 + 3 H_2 \longrightarrow 3 H_2S + 2 Au$
d. $C_4H_{10}O + 6 O_2 \longrightarrow 4 CO_2 + 5 H_2O$

10.17 Balance each of the following chemical equations.
a. $N_2 + O_2 \longrightarrow N_2O_3$
b. $NH_3 + NO \longrightarrow N_2 + H_2O$
c. $CS_2 + O_2 \longrightarrow CO_2 + SO_2$
d. $Mg + HBr \longrightarrow MgBr_2 + H_2$

10.18 Balance each of the following chemical equations.
  a. $Al + O_2 \longrightarrow Al_2O_3$
  b. $Na + H_2O \longrightarrow NaOH + H_2$
  c. $Co + HgCl_2 \longrightarrow CoCl_3 + Hg$
  d. $NH_3 + O_2 \longrightarrow N_2O + H_2O$

10.19 Balance the following chemical equations.
  a. $PbO + NH_3 \rightarrow Pb + N_2 + H_2O$
  b. $NaHCO_3 + H_2SO_4 \rightarrow Na_2SO_4 + H_2O + CO_2$
  c. $TiO_2 + C + Cl_2 \rightarrow TiCl_4 + CO_2$
  d. $NBr_3 + NaOH \rightarrow N_2 + NaBr + HBrO$

10.20 Balance the following chemical equations.
  a. $NH_3 + O_2 + CH_4 \rightarrow HCN + H_2O$
  b. $KClO_3 + HCl \rightarrow KCl + Cl_2 + H_2O$
  c. $SO_2Cl_2 + HI \rightarrow H_2S + H_2O + HCl + I_2$
  d. $NO + CH_4 \rightarrow HCN + H_2O + H_2$

10.21 Balance the following chemical equations.
  a. $C_5H_{12} + O_2 \longrightarrow CO_2 + H_2O$
  b. $C_5H_{10} + O_2 \longrightarrow CO_2 + H_2O$
  c. $C_5H_8 + O_2 \longrightarrow CO_2 + H_2O$
  d. $C_5H_{10}O + O_2 \longrightarrow CO_2 + H_2O$

10.22 Balance the following chemical equations.
  a. $C_6H_{14} + O_2 \longrightarrow CO_2 + H_2O$
  b. $C_6H_{12} + O_2 \longrightarrow CO_2 + H_2O$
  c. $C_6H_{10} + O_2 \longrightarrow CO_2 + H_2O$
  d. $C_6H_{10}O_2 + O_2 \longrightarrow CO_2 + H_2O$

10.23 Balance the following chemical equations.
  a. $Ca(OH)_2 + HNO_3 \rightarrow Ca(NO_3)_2 + H_2O$
  b. $BaCl_2 + (NH_4)_2SO_4 \rightarrow BaSO_4 + NH_4Cl$
  c. $Fe(OH)_3 + H_2SO_4 \rightarrow Fe_2(SO_4)_3 + H_2O$
  d. $Na_3PO_4 + AgNO_3 \rightarrow NaNO_3 + Ag_3PO_4$

10.24 Balance the following chemical equations.
  a. $Al + Sn(NO_3)_2 \rightarrow Al(NO_3)_3 + Sn$
  b. $Na_2CO_3 + Mg(NO_3)_2 \rightarrow MgCO_3 + NaNO_3$
  c. $Al(NO_3)_3 + H_2SO_4 \rightarrow Al_2(SO_4)_3 + HNO_3$
  d. $Ba(C_2H_3O_2)_2 + (NH_4)_3PO_4 \rightarrow Ba_3(PO_4)_2 + NH_4C_2H_3O_2$

10.25 Each of the following *mathematically balanced* chemical equations is in a *nonconventional* form. Through coefficient adjustment, change each of these equations to conventional form without unbalancing them.
  a. $3\,AgNO_3 + 3\,KCl \rightarrow 3\,KNO_3 + 3\,AgCl$
  b. $2\,CS_2 + 6\,O_2 \rightarrow 2\,CO_2 + 4\,SO_2$
  c. $H_2 + \frac{1}{2}O_2 \rightarrow H_2O$
  d. $Ag_2CO_3 \rightarrow 2\,Ag + CO_2 + \frac{1}{2}O_2$

10.26 Each of the following *mathematically balanced* chemical equations is in a *nonconventional* form. Through coefficient adjustment, change each of these equations to conventional form without unbalancing them.
  a. $2\,Cu(NO_3)_2 + 2\,Fe \rightarrow 2\,Cu + 2\,Fe(NO_3)_2$
  b. $2\,N_2H_4 + 4\,H_2O_2 \rightarrow 2\,N_2 + 8\,H_2O$
  c. $Li_3N \rightarrow 3\,Li + \frac{1}{2}N_2$
  d. $2\,HNO_3 \rightarrow 2\,NO_2 + H_2O + \frac{1}{2}O_2$

10.27 The following diagrams represent the reaction of A_2 (light spheres) with B_2 (dark spheres) to give specific products. Write a balanced chemical equation, in terms of A's and B's for each reaction based on the information in the diagram. Allow for the possibility that not all of the reactant molecules react.

a.

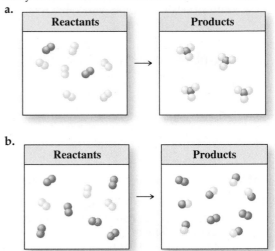

b.

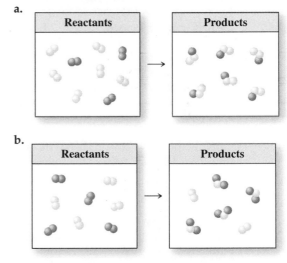

10.28 The following diagrams represent the reaction of A_2 (light spheres) with B_2 (dark spheres) to give specific products. Write a balanced chemical equation, in terms of A's and B's for each reaction based on the information in the diagram. Allow for the possibility that not all of the reactant molecules react.

a.

b.

10.29 Diagram I represents a reactant mixture for a chemical reaction. Select from diagrams II through IV the product mixture that is consistent with both diagram I and the concepts associated with a balanced chemical equation.

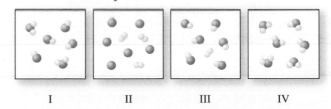

I   II   III   IV

10.30 Diagram I represents a reactant mixture for a chemical reaction. Select from diagrams II through IV the product mixture that is consistent with both diagram I and the concepts associated with a balanced chemical equation.

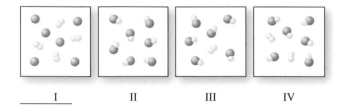

I	II	III	IV

CLASSES OF CHEMICAL REACTIONS (SEC. 10.6)

10.31 Classify each of the following chemical reactions as synthesis, decomposition, single-replacement, or double-replacement.
   a. $SO_3 + H_2O \rightarrow H_2SO_4$
   b. $2 H_2 + O_2 \rightarrow 2 H_2O$
   c. $Na_2CO_3 + Ca(OH)_2 \rightarrow CaCO_3 + 2 NaOH$
   d. $Cu(NO_3)_2 + Fe \rightarrow Cu + Fe(NO_3)_2$

10.32 Classify each of the following chemical reactions as synthesis, decomposition, single-replacement, or double-replacement.
   a. $3 CuSO_4 + 2 Al \rightarrow Al_2(SO_4)_3 + 3 Cu$
   b. $K_2CO_3 \rightarrow K_2O + CO_2$
   c. $2 AgNO_3 + K_2SO_4 \rightarrow Ag_2SO_4 + 2 KNO_3$
   d. $2 SO_2 + O_2 \rightarrow 2 SO_3$

10.33 Classify each of the chemical reactions depicted in Problem 10.27 as a synthesis, decomposition, single-replacement, or double-replacement reaction.

10.34 Classify each of the chemical reactions depicted in Problem 10.28 as a synthesis, decomposition, single-replacement, or double-replacement reaction.

10.35 Identify the products of, and then write a balanced chemical equation for, each of the following chemical reactions.
   a. $Zn + Cu(NO_3)_2 \rightarrow ? + ?$ (single-replacement reaction)
   b. $Ca + O_2 \rightarrow ?$ (synthesis reaction)
   c. $K_2SO_4 + Ba(NO_3)_2 \rightarrow ? + ?$ (double-replacement reaction)
   d. $Ag_2O \rightarrow ? + ?$ (decomposition reaction)

10.36 Identify the products of, and then write a balanced chemical equation for, each of the following chemical reactions.
   a. $AlCl_3 \rightarrow ? + ?$ (decomposition reaction)
   b. $Cu(NO_3)_2 + Na_2CO_3 \rightarrow ? + ?$ (double-replacement reaction)
   c. $Al + Ni(NO_3)_2 \rightarrow ? + ?$ (single-replacement reaction)
   d. $Be + N_2 \rightarrow ?$ (synthesis reaction)

10.37 Write a balanced chemical equation for the thermal decomposition of each of the following metal carbonates to its metal oxide and carbon dioxide.
   a. Rb_2CO_3 **b.** $SrCO_3$

   c. $Al_2(CO_3)_3$ **d.** Cu_2CO_3

10.38 Write a balanced chemical equation for the thermal decomposition of each of the following metal carbonates to its metal oxide and carbon dioxide.
   a. Ag_2CO_3 **b.** $BaCO_3$
   c. $Ga_2(CO_3)_3$ **d.** $ZnCO_3$

10.39 Write a balanced chemical equation for the combustion of each of the following hydrogen-oxygen compounds in air (O_2).
   a. C_3H_6 **b.** C_2H_4 **c.** C_7H_{16} **d.** C_8H_{16}

10.40 Write a balanced chemical equation for the combustion of each of the following hydrogen-oxygen compounds in air (O_2).
   a. C_2H_6 **b.** C_4H_6 **c.** C_7H_{14} **d.** C_8H_{14}

10.41 Write a balanced chemical equation for the combustion of each of the following carbon-hydrogen-oxygen compounds in air (O_2).
   a. CH_2O **b.** $C_5H_{12}O$ **c.** $C_5H_{10}O$ **d.** $C_5H_{10}O_2$

10.42 Write a balanced chemical equation for the combustion of each of the following carbon-hydrogen-oxygen compounds in air (O_2).
   a. CH_2O_2 **b.** $C_6H_{14}O$ **c.** $C_6H_{12}O$ **d.** $C_6H_{12}O_2$

10.43 Write a balanced chemical equation for the combustion in air of each of the following compounds.
   a. C_2H_7N, where NO_2 is one of the products
   b. CH_4S, where SO_2 is one of the products

10.44 Write a balanced chemical equation for the combustion in air of each of the following compounds.
   a. C_2H_6S, where SO_2 is one of the products
   b. CH_5N, where NO_2 is one of the products

10.45 Indicate to which of the following types of chemical reactions each of the statements listed applies: *synthesis, decomposition, single-replacement, double-replacement,* or *combustion.*
   a. An element may be a reactant.
   b. An element may be a product.
   c. A compound may be a reactant.
   d. A compound may be a product.

10.46 Indicate to which of the following types of chemical reactions each of the statements listed applies: *synthesis, decomposition, single-replacement, double-replacement,* or *combustion.*
   a. Two reactants are required.
   b. Only one reactant is present.
   c. Two products are required.
   d. Only one product is present.

CHEMICAL EQUATIONS AND THE MOLE CONCEPT (SEC. 10.7)

10.47 Consider the general chemical equation

$$4 A + 2 B \longrightarrow 3 C + D$$

How many moles of B will react with the following molar amounts of A?
   a. 1.0 mole   **b.** 2.0 moles
   c. 5.0 moles   **d.** 7.0 moles

10.48 Consider the general chemical equation

$$2\,A + B \longrightarrow C + 3\,D$$

How many moles of A will react with the following molar amounts of B?
a. 1.0 mole   **b.** 2.0 moles
c. 4.0 moles   **d.** 5.0 moles

10.49 Write the 12 mole-to-mole conversion factors that can be derived from the balanced equation.

$$4\,NH_3 + 3\,O_2 \rightarrow 2\,N_2 + 6\,H_2O$$

10.50 Write the 12 mole-to-mole conversion factors that can be derived from the balanced equation.

$$CS_2 + 3\,O_2 \rightarrow CO_2 + 2\,SO_2$$

10.51 For the chemical reaction

$$6\,ClO_2 + 3\,H_2O \longrightarrow 5\,HClO_3 + HCl$$

write the conversion factor that would be needed to affect the conversion from
a. moles of ClO_2 to moles H_2O.
b. moles of $HClO_3$ to moles HCl.
c. moles of H_2O to moles HCl.
d. moles of $HClO_3$ to moles ClO_2.

10.52 For the chemical reaction

$$2\,MnI_2 + 13\,F_2 \longrightarrow 2\,MnF_3 + 4\,IF_5$$

write the conversion factor that would be needed to affect the conversion from
a. moles of MnI_2 to moles of F_2.
b. moles of IF_5 to moles of MnF_3.
c. moles of MnF_3 to moles of MnI_2.
d. moles of F_2 to moles of IF_5.

10.53 Using each of the following chemical equations, calculate the number of moles of the first-listed reactant that are needed to produce 3.00 moles of N_2.
a. $2\,NaN_3 \rightarrow 2\,Na + 3\,N_2$
b. $3\,CO + 2\,NaCN \rightarrow Na_2CO_3 + 4\,C + N_2$
c. $2\,NH_2Cl + N_2H_4 \rightarrow 2\,NH_4Cl + N_2$
d. $4\,C_3H_5O_9N_3 \rightarrow 12\,CO_2 + 6\,N_2 + O_2 + 10\,H_2O$

10.54 Using each of the following chemical equations, calculate the number of moles of the first-listed reactant that are needed to produce 4.00 moles of N_2.
a. $4\,NH_3 + 3\,O_2 \rightarrow 2\,N_2 + 6\,H_2O$
b. $(NH_4)_2Cr_2O_7 \rightarrow Cr_2O_3 + N_2 + 4\,H_2O$
c. $N_2H_4 + 2\,H_2O_2 \rightarrow N_2 + 4\,H_2O$
d. $2\,Li_3N \rightarrow 6\,Li + N_2$

10.55 How many moles of the first-listed reactant in each of the following chemical equations will completely react with 1.42 moles of the second-listed reactant?
a. $H_2O_2 + H_2S \longrightarrow 2\,H_2O + S$
b. $CS_2 + 3\,O_2 \longrightarrow 2\,SO_2 + CO_2$
c. $Mg + 2\,HCl \longrightarrow MgCl_2 + H_2$
d. $6\,HCl + 2\,Al \longrightarrow 3\,H_2 + 2\,AlCl_3$

10.56 How many moles of the first-listed reactant in each of the following chemical equations will completely react with 2.03 moles of the second-listed reactant?
a. $SiO_2 + 3\,C \longrightarrow 2\,CO + SiC$
b. $5\,O_2 + C_3H_8 \longrightarrow 3\,CO_2 + 4\,H_2O$
c. $CH_4 + 4\,Cl_2 \longrightarrow 4\,HCl + CCl_4$
d. $3\,NO_2 + H_2O \longrightarrow 2\,HNO_3 + NO$

10.57 Using each of the following equations, calculate the total number of moles of products that can be obtained from the decomposition of 1.75 moles of the reactant.
a. $2\,NH_4NO_3 \rightarrow 2\,N_2 + O_2 + 4\,H_2O$
b. $2\,NaClO_3 \rightarrow 2\,NaCl + 3\,O_2$
c. $2\,KNO_3 \rightarrow 2\,KNO_2 + O_2$
d. $4\,I_4O_9 \rightarrow 6\,I_2O_5 + 2\,I_2 + 3\,O_2$

10.58 Using each of the following equations, calculate the total number of moles of products that can be obtained from the decomposition of 2.25 moles of the reactant.
a. $2\,Ag_2CO_3 \rightarrow 4\,Ag + 2\,CO_2 + O_2$
b. $2\,KClO_3 \rightarrow 2\,KCl + 3\,O_2$
c. $4\,HNO_3 \rightarrow 4\,NO_2 + 2\,H_2O + O_2$
d. $2\,H_2O_2 \rightarrow 2\,H_2O + O_2$

10.59 For the decomposition reaction

$$2\,Ag_2CO_3 \longrightarrow 4\,Ag + 2\,CO_2 + O_2$$

how many moles of Ag_2CO_3 undergo decomposition in order to produce
a. 6.0 moles of Ag?
b. 8.0 moles of CO_2?
c. 2.4 moles of O_2?
d. 15.0 total moles of products?

10.60 For the decomposition reaction

$$3\,HNO_2 \longrightarrow 2\,NO + HNO_3 + H_2O$$

how many moles of HNO_2 undergo decomposition in order to produce
a. 6.0 moles of NO?
b. 7.0 moles of HNO_3?
c. 3.5 moles of H_2O?
d. 17.0 total moles of products?

BALANCED CHEMICAL EQUATIONS AND THE LAW OF CONSERVATION OF MASS (SEC. 10.8)

10.61 Consider the general chemical reaction

$$4\,A + 3\,B \longrightarrow 2\,C$$

a. If 4.00 g of A reacts with 1.67 g of B, what is the mass, in grams, of C that is produced?
b. If the reaction of 3.76 g of A with B produces 7.02 g of C, what is the mass, in grams, of B that reacted?

10.62 Consider the general chemical reaction

$$2\,A + 3\,B \longrightarrow C$$

a. If 5.20 g of A reacts with 6.23 g of B, what is the mass, in grams, of C that is produced?

b. If the reaction of 4.50 g of B with A produces 7.23 g of C, what is the mass, in grams, of A that reacted? _____

10.63 Verify the law of conservation of mass, using the molar masses of reactants and products for each substance in the following balanced equations.
   a. $C_7H_{16} + 11\,O_2 \longrightarrow 7\,CO_2 + 8\,H_2O$
   b. $2\,HCl + CaCO_3 \longrightarrow CaCl_2 + CO_2 + H_2O$
   c. $Na_2SO_4 + 2\,C \longrightarrow Na_2S + 2\,CO_2$
   d. $4\,Na_2CO_3 + Fe_3Br_8 \longrightarrow 8\,NaBr + 4\,CO_2 + Fe_3O_4$

10.64 Verify the law of conservation of mass, using the molar masses of reactants and products for each substance in the following balanced equations.
   a. $3\,O_2 + CS_2 \longrightarrow CO_2 + 2\,SO_2$
   b. $2\,FeI_2 + 3\,Cl_2 \longrightarrow 2\,FeCl_3 + 2\,I_2$
   c. $2\,C_8H_{18} + 25\,O_2 \longrightarrow 16\,CO_2 + 18\,H_2O$
   d. $2\,NH_3 + 3\,O_2 + 2\,CH_4 \longrightarrow 2\,HCN + 6\,H_2O$

CALCULATIONS BASED ON CHEMICAL EQUATIONS (SEC. 10.9)

10.65 How many grams of oxygen can be obtained by the decomposition of 7.00 moles of reactant in each of the following chemical reactions?
   a. $2\,KClO_3 \rightarrow 2\,KCl + 3\,O_2$
   b. $2\,CuO \rightarrow 2\,Cu + O_2$
   c. $2\,NaNO_3 \rightarrow 2\,NaNO_2 + O_2$
   d. $4\,HNO_3 \rightarrow 4\,NO_2 + 2\,H_2O + O_2$

10.66 How many grams of oxygen can be obtained by the decomposition of 2.50 moles of reactant in each of the following chemical reactions?
   a. $2\,KClO_4 \rightarrow 2\,KCl + 4\,O_2$
   b. $2\,HgO \rightarrow 2\,Hg + O_2$
   c. $2\,H_2O \rightarrow 2\,H_2 + O_2$
   d. $2\,Ag_2CO_3 \rightarrow 4\,Ag + 2\,CO_2 + O_2$

10.67 How much nitric acid (HNO_3), in grams, is needed to produce 1.00 mole of water in each of the following chemical reactions?
   a. $Cu + 4\,HNO_3 \rightarrow Cu(NO_3)_2 + 2\,NO_2 + 2\,H_2O$
   b. $Al_2O_3 + 6\,HNO_3 \rightarrow 2\,Al(NO_3)_3 + 3\,H_2O$
   c. $Au + HNO_3 + 4\,HCl \rightarrow HAuCl_4 + NO + 2\,H_2O$
   d. $4\,Zn + 10\,HNO_3 \rightarrow 4\,Zn(NO_3)_2 + NH_4NO_3 + 3\,H_2O$

10.68 How much nitric acid (HNO_3), in grams, is needed to produce 2.00 moles of water in each of the following chemical reactions?
   a. $Fe_2O_3 + 6\,HNO_3 \rightarrow 2\,Fe(NO_3)_3 + 3\,H_2O$
   b. $4\,HNO_3 \rightarrow 4\,NO_2 + O_2 + 2\,H_2O$
   c. $3\,Cu + 8\,HNO_3 \rightarrow 3\,Cu(NO_3)_2 + 2\,NO + 4\,H_2O$
   d. $8\,Al + 30\,HNO_3 \rightarrow 8\,Al(NO_3)_3 + 3\,NH_4NO_3 + 9\,H_2O$

10.69 How many grams of the second-listed reactant in each of the following reactions are needed to react completely with 1.772 g of the first-listed reactant?
   a. $CaO + 2\,HNO_3 \longrightarrow Ca(NO_3)_2 + H_2O$
   b. $PCl_5 + 4\,H_2O \longrightarrow H_3PO_4 + 5\,HCl$
   c. $CuCl_2 + 2\,NaOH \longrightarrow Cu(OH)_2 + 2\,NaCl$
   d. $Na_2S + 2\,AgC_2H_3O_2 \longrightarrow 2\,NaC_2H_3O_2 + Ag_2S$

10.70 How many grams of the second-listed reactant in each of the following reactions are needed to react completely with 12.56 g of the first-listed reactant?
   a. $CrCl_3 + 3\,NaOH \longrightarrow Cr(OH)_3 + 3\,NaCl$
   b. $16\,H_2S + 8\,SO_2 \longrightarrow 3\,S_8 + 16\,H_2O$
   c. $2\,Cr_2O_3 + 3\,Si \longrightarrow 4\,Cr + 3\,SiO_2$
   d. $3\,CCl_4 + 2\,SbF_3 \longrightarrow 3\,CCl_2F_2 + 2\,SbCl_3$

10.71 Silicon carbide, SiC, used as an abrasive on sandpaper, is prepared using the following chemical reaction

$$SiO_2(s) + 3\,C(s) \rightarrow SiC(s) + 2\,CO(g)$$

   a. How many grams of SiO_2 are needed to react with 1.50 moles of C?
   b. How many grams of CO are produced when 1.37 moles of SiO_2 react?
   c. How many grams of SiC are produced at the same time that 3.33 moles of CO are produced?
   d. How many grams of C must react in order to produce 0.575 mole of SiC?

10.72 In the atmosphere, the air pollutant nitrogen dioxide (NO_2) reacts with water to produce nitric acid (HNO_3). The reaction for the formation of nitric acid is

$$3\,NO_2(g) + H_2O(l) \rightarrow 2\,HNO_3(aq) + NO(g)$$

   a. How many grams of NO_2 are needed to react with 2.30 moles of H_2O?
   b. How many grams of NO are produced when 2.04 moles of H_2O react?
   c. How many grams of HNO_3 are produced at the same time that 0.500 mole of NO is produced?
   d. How many grams of NO_2 must react in order to produce 1.23 moles of HNO_3?

10.73 One way to remove gaseous carbon dioxide (CO_2) from the air in a spacecraft is to let canisters of solid lithium hydroxide (LiOH) absorb it according to the reaction

$$2\,LiOH(s) + CO_2(g) \rightarrow Li_2CO_3(s) + H_2O(l)$$

Based on this equation, how many grams of LiOH must be used to achieve the following?
   a. absorb 4.50 moles of CO_2
   b. absorb 3.00×10^{24} molecules of CO_2
   c. produce 10.0 g of H_2O
   d. produce 10.0 g of Li_2CO_3

10.74 Tungsten (W) metal, used to make incandescent light-bulb filaments, is produced by the reaction

$$WO_3(s) + 3\,H_2(g) \rightarrow W(s) + 3\,H_2O(l)$$

Based on this equation, how many grams of WO_3 are needed to produce each of the following?
a. 10.00 g of W
b. 1 billion (1.00×10^9) molecules of H_2O
c. 2.53 moles of H_2O
d. 250,000 atoms of W

10.75 Hydrofluoric acid, HF, cannot be stored in glass bottles because it attacks silicate compounds present in the glass. For example, sodium silicate, Na_2SiO_3, reacts with HF in the following way:

$$Na_2SiO_3 + 8\,HF \rightarrow H_2SiF_6 + 2\,NaF + 3\,H_2O$$

a. How many moles of Na_2SiO_3 must react to produce 25.00 g of NaF?
b. How many grams of HF must react to produce 27.00 g of H_2O?
c. How many molecules of H_2SiF_6 are produced from the reaction of 2.000 g of Na_2SiO_3?
d. How many grams of HF are needed to react with 50.00 g of Na_2SiO_3?

10.76 Potassium thiosulfate, $K_2S_2O_3$, is used to remove any excess chlorine from fibers and fabrics that have been bleached with that gas.

$$K_2S_2O_3 + 4\,Cl_2 + 5\,H_2O \rightarrow 2\,KHSO_4 + 8\,HCl$$

a. How many moles of $K_2S_2O_3$ must react to produce 2.500 g of HCl?
b. How many grams of Cl_2 must react to produce 20.00 g of $KHSO_4$?
c. How many molecules of HCl are produced at the same time that 2.000 g of $KHSO_4$ are produced?
d. How many grams of H_2O are consumed as 12.50 g of Cl_2 reacts?

10.77 How many grams of aluminum (Al) are needed to react completely with 23.7 grams of oxygen (O_2) in the synthesis of aluminum oxide (Al_2O_3)?

10.78 How many grams of potassium (K) are needed to react completely with 17.8 g of nitrogen (N_2) in the synthesis of potassium nitride (K_3N)?

10.79 When chromium metal reacts with chlorine gas, a violet solid with the formula $CrCl_3$ is formed.

$$2\,Cr + 3\,Cl_2 \rightarrow 2\,CrCl_3$$

How many grams of Cr and how many grams of Cl_2 are needed to produce 200.0 g of $CrCl_3$?

10.80 Black silver sulfide can be produced from the reaction of silver metal with sulfur.

$$2\,Ag + S \rightarrow Ag_2S$$

How many grams of Ag and how many grams of S are needed to produce 150.0 g of Ag_2S?

LIMITING REACTANT CALCULATIONS (SEC. 10.10)

10.81 What will be the limiting reactant in the production of three-nut, four-bolt combinations from a collection of 216 nuts and 284 bolts?

10.82 What will be the limiting reactant in the production of five-nut, four-bolt combinations from a collection of 785 nuts and 660 bolts?

10.83 A model airplane kit is designed to contain 2 wings, 1 fuselage, 4 engines, and 6 wheels. How many model airplane kits can a manufacturer produce from a parts inventory of 426 wings, 224 fuselages, 860 engines, and 1578 wheels?

10.84 A model car kit is designed to contain 1 body, 4 wheels, 2 bumpers, and 1 steering wheel. How many model car kits can a manufacturer produce from a parts inventory of 137 bodies, 532 wheels, 246 bumpers, and 139 steering wheels?

10.85 Beryllium nitride can be prepared by the reaction of the metallic beryllium with ammonia at elevated temperatures as shown by the chemical equation

$$3\,Be + 2\,NH_3 \longrightarrow Be_3N_2 + 3\,H_2$$

For each of the following combinations of reactants, decide which reactant is the limiting reactant.
a. 2.00 moles of Be and 0.500 mole of NH_3
b. 3.00 moles of Be and 3.00 moles of NH_3
c. 3.00 g of Be and 0.100 mole of NH_3
d. 20.00 g of Be and 60.00 g of NH_3

10.86 Under appropriate conditions water can be produced from the reaction of the elements hydrogen and oxygen as shown by the chemical equation

$$2\,H_2 + O_2 \longrightarrow 2\,H_2O$$

For each of the following combinations of reactants, decide which reactant is the limiting reactant.
a. 1.750 moles of H_2 and 1.00 mole of O_2
b. 2.50 moles of H_2 and 2.00 moles of O_2
c. 6.00 g of H_2 and 1.25 moles of O_2
d. 1.00 g of H_2 and 7.00 g of O_2

10.87 At high temperatures and pressures nitrogen will react with hydrogen to produce ammonia as shown by the chemical equation

$$N_2 + 3\,H_2 \longrightarrow 2\,NH_3$$

How many grams of ammonia can be produced from the following amounts of reactants?
a. 3.0 g of N_2 and 5.0 g of H_2
b. 30.0 g of N_2 and 10.0 g of H_2
c. 50.0 g of N_2 and 8.00 g of H_2
d. 56 g of N_2 and 12 g of H_2

10.88 Aluminum oxide can be prepared by the direct reaction of the elements as shown by the chemical equation

$$4\,Al + 3\,O_2 \longrightarrow 2\,Al_2O_3$$

How many grams of aluminum oxide can be produced from the following amounts of reactants?
a. 11.0 g of Al and 9.40 g of O_2
b. 10.0 g of Al and 10.0 g of O_2

c. 50.0 g of Al and 100.0 g of O_2
d. 6.20 g of Al and 5.51 g of O_2

10.89 Calculate, in grams, the amount of excess reactant present in each of the reaction mixtures specified in Problem 10.87.

10.90 Calculate, in grams, the amount of excess reactant present in each of the reaction mixtures specified in Problem 10.88.

10.91 The following diagrams represent the reaction of A_2 (light spheres) with B_2 (dark spheres) to give specific products. Identify the limiting reactant in each of the reactions based on the information in the diagram.

a.

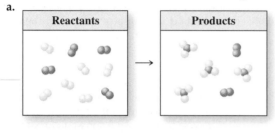

b.

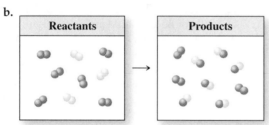

10.92 The following diagrams represent the reaction of A_2 (light spheres) with B_2 (dark spheres) to give specific products. Identify the limiting reactant in each of the reactions based on the information in the diagram.

a.

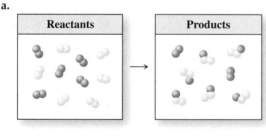

b.

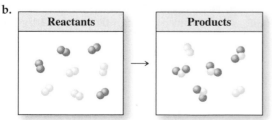

10.93 Determine how many $CoCl_3$ formula units can be produced from a reaction mixture containing 525 cobalt atoms and 525 HCl molecules according to the following reaction.

$$2\,Co + 6\,HCl \rightarrow 2\,CoCl_3 + 3\,H_2$$

10.94 Determine how many $NiCl_2$ formula units can be produced from a reaction mixture containing 782 nickel atoms and 782 HCl molecules according to the following reaction.

$$Ni + 2\,HCl \rightarrow NiCl_2 + H_2$$

10.95 If 70.0 g of Fe_3O_4 and 12.0 g of O_2 are present in a reaction mixture, determine how many grams of each reactant will be left unreacted upon completion of the following reaction.

$$4\,Fe_3O_4 + O_2 \rightarrow 6\,Fe_2O_3$$

10.96 If 70.0 g of $TiCl_4$ and 16.0 g of Ti are present in a reaction mixture, determine how many grams of each reactant will be left unreacted upon completion of the following reaction.

$$3\,TiCl_4 + Ti \rightarrow 4\,TiCl_3$$

10.97 Determine the number of grams of each of the products that can be made from 8.00 g of SCl_2 and 4.00 g of NaF by the following reaction.

$$3\,SCl_2 + 4\,NaF \rightarrow SF_4 + S_2Cl_2 + 4\,NaCl$$

10.98 Determine the number of grams of each of the products that can be made from 100.0 g of Na_2CO_3 and 300.0 g of Fe_3Br_8 by the following reaction.

$$4\,Na_2CO_3 + Fe_3Br_8 \rightarrow 8\,NaBr + 4\,CO_2 + Fe_3O_4$$

THEORETICAL YIELD AND PERCENT YIELD (SEC. 10.11)

10.99 Because of sloppiness in his procedures, a student was able to isolate only 16.0 g of a desired product from a chemical reaction rather than the 52.0 g that were theoretically possible. What was the percent yield of product that the student obtained?

10.100 The theoretical yield of product for a particular reaction is 25.31 g. A very meticulous student isolates 24.79 g of product when the reaction is run. What is the percent yield that this student obtained?

10.101 In an experiment designed to produce magnesium oxide by the chemical reaction

$$2\,Mg + O_2 \longrightarrow 2\,MgO$$

125.6 g of MgO is isolated out of a possible 172.2 g MgO.

a. What is the theoretical yield of MgO?

b. What is the actual yield of MgO?

c. What is the percent yield of MgO?

10.102 In an experiment designed to produce calcium oxide by the chemical reaction

$$2\,Ca + O_2 \longrightarrow 2\,CaO$$

115.6 g of CaO is isolated out of a possible 162.2 g CaO.

a. What is the theoretical yield of CaO?

b. What is the actual yield of CaO?

c. What is the percent yield of CaO?

10.103 Diagram I represents the reactant mixture for the chemical reaction

$$2AB + B_2 \longrightarrow 2\,AB_2$$

Select from diagrams II through IV the product mixture that is consistent with the experimental observation that the percent yield for the reaction is 67%.

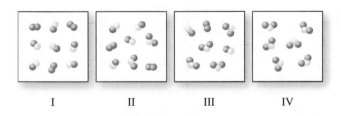

| I | II | III | IV |

10.104 Diagram I represents the reactant mixture for the chemical reaction

$$A_2 + 2\,B_2 \longrightarrow 2\,AB_2$$

Select from diagrams II through IV the product mixture that is consistent with the experimental observation that the percent yield for the reaction is 67%.

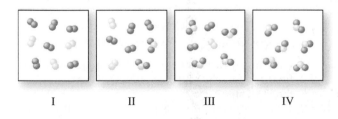

| I | II | III | IV |

10.105 If 74.30 g of HCl were produced from 2.13 g of H_2 and an excess of Cl_2 according to the reaction

$$H_2 + Cl_2 \rightarrow 2\,HCl$$

what was the percent yield of HCl?

10.106 If 115.7 g of Ca_3N_2 were produced from 28.2 g of N_2 and an excess of Ca according to the reaction

$$3\,Ca + N_2 \rightarrow Ca_3N_2$$

what was the percent yield of Ca_3N_2?

10.107 Under appropriate reaction conditions, Al and S produce Al_2S_3 according to the equation

$$2\,Al + 3\,S \rightarrow Al_2S_3$$

In a certain experiment with 55.0 g of Al and an excess of S, a percent yield of 85.6% was obtained. What was the actual yield of Al_2S_3, in grams, for this experiment?

10.108 Under appropriate reaction conditions, Ag and S produce Ag_2S according to the equation

$$2\,Ag + S \rightarrow Ag_2S$$

In a certain experiment with 75.0 g of Ag and an excess of S, a percent yield of 72.9% was obtained. What was the actual yield of Ag_2S, in grams, for this experiment?

10.109 If the percent yield for the reaction

$$2\,CO + O_2 \rightarrow 2\,CO_2$$

was 57.8%, what mass of product, in grams, could be produced from a reactant mixture containing 35.0 g of each reactant?

10.110 If the percent yield for the reaction

$$2\,C + O_2 \rightarrow 2\,CO$$

was 34.3%, what mass of product, in grams, could be produced from a reactant mixture containing 2.25 g of each reactant?

10.111 Three sequential reactions, A $\longrightarrow$ B, B $\longrightarrow$ C, and C $\longrightarrow$ D, have yields of 92%, 87%, and 43%, respectively. What is the overall percent yield for the conversion of A to D?

10.112 Three sequential reactions, E $\longrightarrow$ F, F $\longrightarrow$ G, and G $\longrightarrow$ H, have yields of 95%, 93%, and 89%, respectively. What is the overall percent yield for the conversion of E to H?

SIMULTANEOUS CHEMICAL REACTIONS (SEC. 10.12)

10.113 A mixture of composition 60.0% ZnS and 40.0% CuS is heated in air until the sulfides are completely converted to oxides as shown by the following equations.

$$2\,ZnS + 3\,O_2 \longrightarrow 2\,ZnO + 2\,SO_2$$
$$2\,CuS + 3\,O_2 \rightarrow 2\,CuO + 2\,SO_2$$

How many grams of SO_2 are produced from the reaction of 82.5 g of the sulfide mixture?

10.114 A mixture of composition 50.0% H_2S and 50.0% CH_4 is reacted with oxygen, producing SO_2, CO_2, and H_2O. The equations for the reactions are

$$2 H_2S + 3 O_2 \rightarrow 2 SO_2 + 2 H_2O$$
$$CH_4 + 2 O_2 \rightarrow CO_2 + 2 H_2O$$

How many grams of H_2O are produced from the reaction of 65.0 g of mixture?

10.115 A mixture of composition 70.0% methane (CH_4) and 30.0% ethane (C_2H_6), by mass, is burned in oxygen to produce CO_2 and H_2O. The reactions that occur are

$$CH_4 + 2 O_2 \rightarrow CO_2 + 2 H_2O$$
$$2 C_2H_6 + 7 O_2 \rightarrow 4 CO_2 + 6 H_2O$$

How many grams of O_2 are needed to react completely with 75.0 g of mixture?

10.116 A mixture of composition 64.0% Zn and 36.0% Sn, by mass, is dissolved in hydrochloric acid (HCl) to produce the metal chlorides and H_2. The reactions that occur are

$$Zn + 2 HCl \rightarrow ZnCl_2 + H_2$$
$$Sn + 2 HCl \rightarrow SnCl_2 + H_2$$

How many grams of HCl are needed to react completely with 50.0 g of mixture?

SEQUENTIAL CHEMICAL REACTIONS (SEC. 10.12)

10.117 Consider the following two-step reaction sequence.

$$2 NaClO_3 \rightarrow 2 NaCl + 3 O_2$$
$$S + O_2 \rightarrow SO_2$$

Assuming that all of the oxygen generated in the first step is consumed in the second step,
   a. how many moles of sodium chlorate ($NaClO_3$) are needed in the first step to produce 6.00 moles of sulfur dioxide (SO_2) in the second step?
   b. how many grams of sodium chlorate ($NaClO_3$) are needed in the first step to produce 20.0 grams of sulfur dioxide (SO_2) in the second step?

10.118 Consider the following two-step reaction sequence.

$$2 NaCl + 2 H_2O \rightarrow 2 NaOH + H_2 + Cl_2$$
$$Ni + Cl_2 \rightarrow NiCl_2$$

Assuming that all of the chlorine generated in the first step is consumed in the second step,
   a. how many moles of sodium chloride (NaCl) are needed in the first step to produce 8.00 moles of nickel(II) chloride ($NiCl_2$) in the second step?
   b. how many grams of sodium chloride (NaCl) are needed in the first step to produce 50.0 grams of nickel(II) chloride ($NiCl_2$) in the second step?

10.119 Acid rain contains both nitric acid and sulfuric acid. The nitric acid is formed from atmospheric N_2 in a three-step process.

$$N_2 + O_2 \rightarrow 2 NO$$
$$2 NO + O_2 \rightarrow 2 NO_2$$
$$4 NO_2 + 2 H_2O + O_2 \rightarrow 4 HNO_3$$

   a. How many moles of HNO_3 result from the reaction of 2.00 moles of N_2 in the first step?
   b. How many grams of HNO_3 result from the reaction of 2.00 grams of N_2 in the first step?

10.120 Acid rain contains both nitric acid and sulfuric acid. The sulfuric acid is formed from S (primarily in coal) in a three-step process.

$$S + O_2 \rightarrow SO_2$$
$$2 SO_2 + O_2 \rightarrow 2 SO_3$$
$$SO_3 + H_2O \rightarrow H_2SO_4$$

   a. How many moles of H_2SO_4 result from the reaction of 5.00 moles of S in the first step?
   b. How many grams of H_2SO_4 result from the reaction of 5.00 grams of S in the first step?

10.121 The following process has been used for obtaining iodine from oil-field brines:

$$NaI + AgNO_3 \rightarrow AgI + NaNO_3$$
$$2 AgI + Fe \rightarrow FeI_2 + 2 Ag$$
$$2 FeI_2 + 3 Cl_2 \rightarrow 2 FeCl_3 + 2 I_2$$

How much $AgNO_3$, in grams, is required in the first step for every 5.00 g of I_2 produced in the third step?

10.122 Sodium bicarbonate, $NaHCO_3$, can be prepared from sodium sulfate, Na_2SO_4, using the following three-step process.

$$Na_2SO_4 + 4 C \rightarrow Na_2S + 4 CO$$
$$Na_2S + CaCO_3 \rightarrow CaS + Na_2CO_3$$
$$Na_2CO_3 + H_2O + CO_2 \rightarrow 2 NaHCO_3$$

How much carbon, C, in grams, is required in the first step for every 10.00 g of $NaHCO_3$ produced in the third step?

10.123 Combine, through the process of addition, the two sequential chemical equations given in Problem 10.117 into a single chemical equation that represents the *overall* chemical process.

10.124 Combine, through the process of addition, the two sequential chemical equations given in Problem 10.118 into a single chemical equation that represents the *overall* chemical process.

10.125 Combine, through the process of addition, the three sequential chemical equations given in Problem 10.119 into a single chemical equation that represents the *overall* chemical process.

10.126 Combine, through the process of addition, the three sequential chemical equations given in Problem 10.120 into a single chemical equation that represents the *overall* chemical process.

ADDITIONAL PROBLEMS

10.127 For each of the following *unbalanced* chemical equations, determine how many moles of the second-listed reactant would be required to react exactly with 3.00 moles of the first-listed reactant.
   a. $CH_4 + O_2 \longrightarrow CO_2 + H_2O$
   b. $PCl_3 + H_2O \longrightarrow H_3PO_3 + HCl$
   c. $NaOH + H_3PO_4 \longrightarrow Na_3PO_4 + H_2O$
   d. $NaClO_2 + Cl_2 \longrightarrow ClO_2 + NaCl$

10.128 For each of the following *unbalanced* chemical equations, determine how many moles of the second-listed reactant would be required to react exactly with 3.00 moles of the first-listed reactant.
   a. $CH_4 + H_2O \longrightarrow H_2 + CO$
   b. $C_3H_8 + O_2 \longrightarrow CO_2 + H_2O$
   c. $LiOH + H_2SO_4 \longrightarrow Li_2SO_4 + H_2O$
   d. $SCl_4 + H_2O \longrightarrow SO_2 + HCl$

10.129 Ammonium dichromate decomposes according to the following reaction:

$$(NH_4)_2 Cr_2O_7 \rightarrow N_2 + 4 H_2O + Cr_2O_3$$

How many grams of each of the products can be formed from the decomposition of 75.0 g of $(NH_4)_2Cr_2O_7$?

10.130 Nitrous acid decomposes according to the following reaction:

$$3 HNO_2 \rightarrow 2 NO + HNO_3 + H_2O$$

How many grams of each of the products can be formed from the decomposition of 63.5 g of HNO_2?

10.131 Hydrogen sulfide burns in oxygen to form sulfur dioxide and water.

$$2 H_2S + 3 O_2 \rightarrow 2 SO_2 + 2 H_2O$$

How many grams of hydrogen sulfide must react in order to produce a total of 100.0 g of products?

10.132 Carbon disulfide burns in oxygen to form carbon dioxide and sulfur dioxide.

$$CS_2 + 3 O_2 \rightarrow CO_2 + 2 SO_2$$

How many grams of carbon disulfide must react in order to produce a total of 50.0 g of products?

10.133 Pure, dry NO gas can be made by the following reaction.

$$3 KNO_2 + KNO_3 + Cr_2O_3 \rightarrow 4 NO + 2 K_2CrO_4$$

How many grams of NO can be produced from a reaction mixture containing 2.00 moles each of KNO_2, KNO_3, and Cr_2O_3?

10.134 Sodium cyanide, NaCN, can be made by the following reaction.

$$Na_2CO_3 + 4 C + N_2 \rightarrow 2 NaCN + 3 CO$$

How many grams of CO can be produced from a reaction mixture containing 3.00 moles each of Na_2CO_3, C, and N_2?

10.135 An *impure* sample of $CuSO_4$ weighing 7.53 g was dissolved in water. The dissolved $CuSO_4$, but not the impurities, then reacted with excess zinc.

$$CuSO_4 + Zn \rightarrow ZnSO_4 + Cu$$

What was the mass percent $CuSO_4$ in the sample if 1.33 g of Cu were produced?

10.136 An *impure* sample of $Hg(NO_3)_2$ weighing 64.5 g was dissolved in water. The dissolved $Hg(NO_3)_2$, but not the impurities, then reacted with excess Mg metal.

$$Hg(NO_3)_2 + Mg \rightarrow Mg(NO_3)_2 + Hg$$

What was the mass percent $Hg(NO_3)_2$ in the sample if 23.6 g of Hg were produced?

10.137 The reaction between 113.4 g of I_2O_5 and 132.2 g of BrF_3 was found to produce 97.0 g of IF_5. The equation for the reaction is

$$6 I_2O_5 + 20 BrF_3 \rightarrow 12 IF_5 + 15 O_2 + 10 Br_2$$

What is the percent yield of IF_5?

10.138 The reaction between 20.0 g of NH_3 and 20.0 g of CH_4 with an excess of oxygen was found to produce 15.0 g of HCN. The equation for the reaction is

$$2 NH_3 + 3 O_2 + 2 CH_4 \rightarrow 2 HCN + 6 H_2O$$

What is the percent yield of HCN?

10.139 A 13.20 g sample of a mixture of $CaCO_3$ and $NaHCO_3$ was heated, and the compounds decomposed as follows.

$$CaCO_3 \rightarrow CaO + CO_2$$
$$2 NaHCO_3 \rightarrow Na_2CO_3 + CO_2 + H_2O$$

The decomposition of the sample yields 4.35 g of CO_2 and 0.873 g of H_2O. What percentage, by mass, of the original sample was $CaCO_3$?

10.140 A 4.00 g sample of a mixture of H_2S and CS_2 was burned in oxygen. The equations for the reactions are

$$2\,H_2S + 3\,O_2 \rightarrow 2\,H_2O + 2\,SO_2$$
$$CS_2 + 3\,O_2 \rightarrow CO_2 + 2\,SO_2$$

If 7.32 g of SO_2 and 0.577 g of CO_2 were produced along with some H_2O, what percentage, by mass, of the original sample was H_2S?

CUMULATIVE PROBLEMS

10.141 Write a balanced equation for each of the following chemical reactions.
   a. zinc + silver nitrate $\rightarrow$ zinc nitrate + silver
   b. hydrochloric acid + sodium hydroxide $\rightarrow$ sodium chloride + water
   c. phosphorus trichloride + chlorine $\rightarrow$ phosphorus pentachloride
   d. copper + oxygen $\rightarrow$ copper(II) oxide

10.142 Write a balanced equation for each of the following chemical reactions.
   a. sodium oxide + sulfur trioxide $\rightarrow$ sodium sulfate
   b. barium carbonate $\rightarrow$ barium oxide + carbon dioxide
   c. aluminum + iron(II) oxide $\rightarrow$ iron + aluminum oxide
   d. ammonia + phosphoric acid $\rightarrow$ ammonium phosphate

10.143 After the following equation was balanced, the name of one of the reactants was substituted for its formula.

$$2\text{ cyclopropane} + 9\,O_2 \rightarrow 6\,CO_2 + 6\,H_2O$$

Using only the information found within this equation, determine the molecular formula of cyclopropane.

10.144 After the following equation was balanced, the name of one of the reactants was substituted for its formula.

$$2\text{ butyne} + 11\,O_2 \rightarrow 8\,CO_2 + 6\,H_2O$$

Using only the information found within this equation, determine the molecular formula of butyne.

10.145 What is the percent yield for a chemical reaction in which 30.0 g of tin (Sn) reacts with excess hydrogen fluoride (HF) to produce tin(II) fluoride (SnF_2) and 2.50 L of hydrogen gas (H_2) of density 0.090 g/L?

10.146 What is the percent yield for a chemical reaction in which 10.0 g of magnesium (Mg) reacts with excess hydrochloric acid (HCl) to produce magnesium chloride ($MgCl_2$) and 6.98 L of hydrogen gas (H_2) of density 0.090 g/L?

10.147 Write a balanced chemical equation for the reaction in which Cu_2S and O_2 are reactants and SO_2 and a copper oxide containing 88.82% copper, by mass, are products. The molecular and empirical formulas of the copper oxide are the same.

10.148 Write a balanced chemical equation for the reaction in which NH_3 and O_2 are reactants and H_2O and a nitrogen oxide containing 46.68% nitrogen by mass are products. The molecular and empirical formulas of the nitrogen oxide are the same.

10.149 Write a balanced chemical equation for the reaction in which CO_2 and H_2O are the products, O_2 is one of the reactants, and a compound with an empirical formula of CH and a molar mass of 78.12 amu is the other reactant.

10.150 Write a balanced chemical equation for the reaction in which CO_2 and H_2O are the products, O_2 is one of the reactants, and a compound with an empirical formula of C_3H_5 and a molar mass of 82.16 amu is the other reactant.

10.151 Fifty (50.000) grams of Be are reacted with an excess of F_2 to produce BeF_2. The equation for the reaction is

$$Be + F_2 \rightarrow BeF_2$$

Calculate the mass of BeF_2 produced, using each of the following specifications:
   a. mass, in grams, to three significant figures
   b. mass, in kilograms, to five significant figures
   c. mass, in micrograms, to four significant figures
   d. mass, in pounds, to four significant figures

10.152 Eighty (80.000) grams of Na are reacted with an excess of P to produce Na_3P. The equation for the reaction is

$$3\,Na + P \rightarrow Na_3P$$

Calculate the mass of Na_3P produced, using each of the following specifications:
   a. mass, in grams, to four significant figures
   b. mass, in milligrams, to three significant figures
   c. mass, in pounds, to five significant figures
   d. mass, in ounces, to four significant figures

10.153 If 100.0 g of $KClO_3$ and 200.0 g of HCl are allowed to react according to the equation

$$2\,KClO_3 + 4\,HCl \rightarrow 2\,KCl + 2\,ClO_2 + Cl_2 + 2\,H_2O$$

what is the combined total number of moles of chlorine-containing products produced?

10.154 If 50.0 g of SO_2Cl_2 and 200.0 g of HI are allowed to react according to the equation

$$SO_2Cl_2 + 8\,HI \rightarrow H_2S + 2\,H_2O + 2\,HCl + 4\,I_2$$

what is the combined total number of moles of hydrogen-containing products produced?

10.155 The reusable booster rockets of the U.S. space shuttle employ a mixture of aluminum and ammonium perchlorate for fuel. The chemical reaction that occurs is

$$3\,Al + 3\,NH_4ClO_4 \rightarrow Al_2O_3 + AlCl_3 + 3\,NO + 6\,H_2O$$

How many moles of electrons are present in the $AlCl_3$ produced when 10.0 g of Al react?

10.156 The fluoride in many toothpastes is tin(II) fluoride produced by the reaction of Sn metal with gaseous HF.

$$Sn + 2\,HF \rightarrow SnF_2 + H_2$$

How many moles of electrons are present in the HF consumed when 25.0 g of H_2 are produced?

10.157 If the products produced by the reaction of 500.0 g of $CaCl_2$ with excess Na_2CO_3, according to the equation

$$CaCl_2 + Na_2CO_3 \rightarrow 2\,NaCl + CaCO_3$$

were broken up into ions, how many positive ions would result?

10.158 If the products produced by the reaction of 220.0 g of $AgNO_3$ with excess K_3PO_4, according to the equation

$$K_3PO_4 + 3\,AgNO_3 \rightarrow Ag_3PO_4 + 3\,KNO_3$$

were broken up into ions, how many negative ions would result?

10.159 The concentration of an aqueous NaBr solution, whose density is 1.046 g/mL, is 6.00% by mass. Determine the volume, in milliliters, of NaBr solution needed to prepare 10.0 g of AgBr by the following reaction.

$$NaBr + AgNO_3 \rightarrow AgBr + NaNO_3$$

10.160 The concentration of an aqueous NH_3 solution, whose density is 0.979 g/mL, is 4.50% by mass. Determine the volume, in milliliters, of NH_3 solution needed to prepare 10.0 g of NH_4NO_3 by the following reaction.

$$NH_3 + HNO_3 \rightarrow NH_4NO_3$$

10.161 A three-step process for producing nitric acid, HNO_3, from gaseous ammonia, NH_3, is

$$4\,NH_3 + 5\,O_2 \rightarrow 4\,NO + 6\,H_2O$$
$$2\,NO + O_2 \rightarrow 2\,NO_2$$
$$3\,NO_2 + H_2O \rightarrow 2\,HNO_3 + NO$$

Assuming yields, respectively, of 85.2%, 82.7%, and 87.0% for the nitrogen-containing products in the three steps, how many grams of nitric acid can be produced from 75.0 mL of ammonia with a density of 0.695 g/L?

10.162 A three-step process for producing sulfuric acid, H_2SO_4, from gaseous sulfur dioxide, SO_2, is

$$2\,SO_2 + O_2 \rightarrow 2\,SO_3$$
$$SO_3 + H_2SO_4 \rightarrow H_2S_2O_7$$
$$H_2S_2O_7 + H_2O \rightarrow 2\,H_2SO_4$$

Assuming yields, respectively, of 63.1%, 87.5%, and 73.8% for the three steps, how many grams of sulfuric acid can be produced from 125 mL of sulfur dioxide with a density of 0.773 g/L?

10.163 A particular coal contains 4.3% sulfur by mass as an impurity. When the coal is burned, the S is converted to gaseous SO_2. The SO_2 enters the exhaust gases, where it is removed by reaction with powdered CaO to produce solid $CaSO_3$. How much by-product $CaSO_3$, in tons, is produced by the burning of 1.0 ton of coal?

10.164 A particular coal contains 3.5% sulfur by mass as an impurity. When the coal is burned, the S is converted to gaseous SO_2. The SO_2 enters the exhaust gases, where it is removed by reaction with powdered CaO to produce solid $CaSO_3$. How much coal would have to be burned, in tons, to produce 2.0 tons of by-product $CaSO_3$?

Multiple-Choice Practice Test

Use this bank of 20 multiple-choice questions as a review of key concepts presented in this chapter. For many of the questions, there may be more than one correct answer (choice d) or no correct answer (choice e).

10.165 Which of the following are balanced chemical equations?
 a. $Cu + S \longrightarrow Cu_2S$
 b. $N_2 + H_2 \longrightarrow 2\,NH_3$
 c. $2\,SO_2 + O_2 \longrightarrow 2\,SO_3$
 d. more than one correct response
 e. no correct response

10.166 When the equation $N_2 + H_2 \longrightarrow NH_3$ is correctly balanced, which of the following statements about equation coefficients is correct?
 a. The coefficient for N_2 is 2.
 b. The coefficient for H_2 is 2.
 c. The coefficient for NH_3 is 2.
 d. more than one correct response
 e. no correct response

10.167 When the equation $C_2H_6 + O_2 \longrightarrow CO_2 + H_2O$ is correctly balanced, which of the following statements about equation coefficients is correct?
  a. Two of the coefficients have the same numerical vlaue.
  b. The sum of all four coefficients is 16.
  c. The coefficient for H_2O is 3.
  d. more than one correct response
  e. no correct response

10.168 Which of the following statements is correct for all balanced chemical equations?
  a. The number of products must equal the number of reactants.
  b. The sum of the subscripts on each side of the equation must be equal.
  c. The sum of the coefficients on each side of the equation must be equal.
  d. more than one correct response
  e. no correct response

10.169 What is the chemical formula of the substance that reacts with O_2 in the reaction described by the following balanced chemical equation?

$$2 \text{ (reactant)} + 11 O_2 \longrightarrow 8 CO_2 + 6 H_2O$$

  a. C_2H_4 b. C_2H_6 c. C_4H_8 d. C_4H_6
  e. no correct response

10.170 The balanced equation $2 CO + O_2 \longrightarrow 2 CO_2$ conveys the information that
  a. two grams of CO will produce two grams of CO_2.
  b. two molecules of CO will produce two molecules of CO_2.
  c. two moles of CO will produce two moles of CO_2.
  d. more than one correct response
  e. no correct response

10.171 Which of the following is a double-replacement chemical reaction?
  a. $CuO + H_2 \longrightarrow Cu + H_2O$
  b. $NaCl + AgNO_3 \longrightarrow NaNO_3 + AgCl$
  c. $Mg + 2 HCl \longrightarrow MgCl_2 + H_2$
  d. more than one correct response
  e. no correct response

10.172 Which of the following is a synthesis chemical reaction?
  a. $Al(OH)_3 + 3 HCl \longrightarrow AlCl_3 + 3 H_2O$
  b. $C_3H_8 + 5 O_2 \longrightarrow 3 CO_2 + 4 H_2O$
  c. $2 NaHCO_3 \longrightarrow Na_2CO_3 + CO_2 + H_2O$
  d. more than one correct response
  e. no correct response

10.173 An element may be a reactant in which of the following types of equations?
  a. single-replacement reaction
  b. double-replacement reaction
  c. decomposition reaction
  d. more than one correct response
  e. no correct response

10.174 How many mole-to-mole conversion factors can be derived from the information in the following balanced chemical equation?

$$CS_2 + 3 O_2 \longrightarrow CO_2 + 2 SO_2$$

  a. 4   b. 6   c. 8   d. 12
  e. no correct response

10.175 How many moles of Fe are needed to react exactly with 4.00 moles of HCl according to the following chemical reaction?

$$Fe + 2 HCl \longrightarrow FeCl_2 + H_2$$

  a. 2.00 moles Fe   b. 4.00 moles Fe
  c. 6.00 moles Fe   d. 8.00 moles Fe
  e. no correct response

10.176 Which of the following is the correct setup for the problem "How many grams of H_2O can be produced from 3.2 moles of O_2 and an excess of H_2S?" according to the reaction

$$2 H_2S + 3 O_2 \longrightarrow 2 H_2O + 2 SO_2$$

  a. $3.2 \text{ moles } O_2 \times \dfrac{32.00 \text{ g } O_2}{1 \text{ mole } O_2} \times \dfrac{18.02 \text{ g } H_2O}{32.00 \text{ g } O_2}$

  b. $3.2 \text{ moles } O_2 \times \dfrac{2 \text{ moles } H_2O}{3 \text{ moles } O_2} \times \dfrac{18.02 \text{ g } H_2O}{1 \text{ mole } H_2O}$

  c. $3.2 \text{ moles } O_2 \times \dfrac{32.00 \text{ g } O_2}{1 \text{ mole } O_2} \times \dfrac{2 \text{ moles } H_2O}{3 \text{ moles } O_2}$

  d. more than one correct response
  e. no correct response

10.177 In the following chemical reaction, how many grams of H_2O are produced if 64.00 g of O_2 react?

$$2 H_2 + O_2 \longrightarrow 2 H_2O$$

  a. 18.02 g H_2O b. 36.04 g H_2O
  c. 64.00 g H_2O d. 72.08 g H_2O
  e. no correct response

10.178 For the chemical reaction

$$2 NH_4NO_3 \longrightarrow 2 N_2 + O_2 + 4 H_2O$$

which of the following statements concerning a product is correct if 2.00 moles of reactant reacts?
  a. 3.01×10^{23} molecules of N_2 are produced.
  b. 6.02×10^{23} molecules of O_2 are produced.
  c. 1.20×10^{24} molecules of H_2O are produced.
  d. more than one correct response
  e. no correct response

10.179 If a mixture containing 2.0 moles of A, 2.0 moles of B, and 2.0 moles of C is allowed to react according to the following balanced equation?

$$2 A + 2 B + C \longrightarrow 2 D + 4 E$$

  a. C is the limiting reactant.
  b. Both B and C are limiting reactants.

c. Both A and B are limiting reactants.

d. more than one correct response

e. no correct response

10.180 How many grams of CO_2 will be produced from 33.0 g of CO and 28.0 g of O_2, according to the following chemical reaction?

$$2\,CO + O_2 \longrightarrow 2\,CO_2$$

a. 28.0 g CO_2

b. 33.0 g CO_2

c. 51.9 g CO_2

d. 77.0 g CO_2

e. no correct response

10.181 Which of the following statements concerning *yields* is correct?

a. Percent yield is the ratio of actual yield to the theoretical yield times 100.

b. Theoretical yield is the difference between the calculated yield and the actual yield.

c. Actual yield is a quantity that cannot be calculated.

d. more than one correct response

e. no correct response

10.182 If a reaction mixture contains equal *moles* of Ca and S, the mass of CaS produced, with a percent yield of 80.0%, would be equal to

a. the sum of the masses of Ca and S.

b. 80.0% of the sum of the masses of Ca and S.

c. 20.0% of the sum of the masses of Ca and S.

d. more than one correct response

e. no correct response

10.183 If a reaction mixture contains equal *masses* of Ca and S, the mass of CaS produced, with a percent yield of 80.0%, would be equal to

a. the sum of the masses of Ca and S.

b. 80.0% of the sum of the masses of Ca and S.

c. 20.0% of the sum of the masses of Ca and S.

d. more than one correct response

e. no correct response

10.184 How many moles of $NaClO_3$ are needed to produce 6.00 moles of SO_2 in the following two-step reaction?

$$2\,NaClO_3 \longrightarrow 2\,NaCl + 3\,O_2$$
$$S + O_2 \longrightarrow SO_2$$

a. 2.00 moles $NaClO_3$

b. 3.00 moles $NaClO_3$

c. 4.00 moles $NaClO_3$

d. 18.00 moles $NaClO_3$

e. no correct response

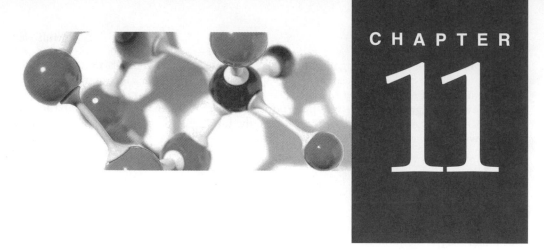

States of Matter

11.1 FACTORS THAT DETERMINE PHYSICAL STATE

The physical state of a substance is determined by (1) what it is, that is, its chemical identity; (2) its temperature; and (3) the pressure to which it is subjected.

At room temperature and pressure, some substances are solids (gold, sodium chloride, etc.), others are liquids (water, mercury, etc.), and still others are gases (oxygen, carbon dioxide, etc.). Thus chemical identity must be a determining factor for physical state, since all three states are observed at room temperature and pressure. On the other hand, when the physical state of a single substance is considered, temperature and pressure are determining variables. Liquid water can be changed to a solid by lowering the temperature or to a gas by raising the temperature.

We tend to characterize a substance almost exclusively in terms of its most common physical state, that is, the state in which it is found at room temperature and pressure. Oxygen is almost always thought of as a gas, its most common state; gold is almost always thought of as a solid, its most common state. A major reason for such single-state characterization is the narrow range of temperatures encountered on this planet. Most substances are never encountered in more than one physical state under natural conditions. We must be careful not to fall into the error of assuming that the commonly observed state of a substance is the *only* state in which it can exist. Under laboratory conditions, states other than the "natural" one can be obtained for almost all substances. Figure 11.1 shows the temperature ranges for the solid, liquid, and gaseous states of a few elements and compounds. As can be seen in this figure, the size and location of the physical-state temperature ranges vary widely among chemical substances. Extremely high temperatures are required to obtain some substances in the gaseous state; other substances are gases at temperatures below room temperature. The size of a given physical-state range also varies dramatically. For example, the elements H_2 and O_2 are liquids over a very narrow temperature range, whereas the elements Ga and Au remain liquids over ranges of hundreds of degrees.

Water is the only substance that is commonly encountered in all three physical states at temperatures normally found on Earth.

A large majority of the naturally occurring elements (75 out of 88) are solids at room temperature and pressure. Of the remaining 13 naturally occurring elements, 11 are gases and 2 (bromine and mercury) are liquids. The abbreviated periodic table in

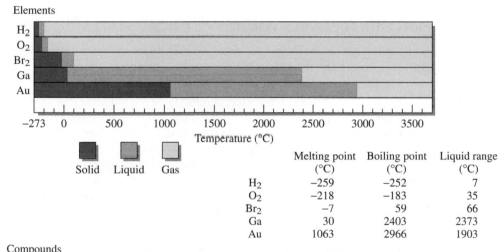

FIGURE 11.1
Solid, liquid, and gaseous state temperature ranges for selected elements and compounds.

	Melting point (°C)	Boiling point (°C)	Liquid range (°C)
H_2	−259	−252	7
O_2	−218	−183	35
Br_2	−7	59	66
Ga	30	2403	2373
Au	1063	2966	1903

	Melting point (°C)	Boiling point (°C)	Liquid range (°C)
CO	−199	−192	7
C_2H_5OH	−115	78	193
H_2O	0	100	100
NaCl	801	1413	612
MgO	2800	3600	800

FIGURE 11.2 Physical states of the elements at room temperature and pressure.

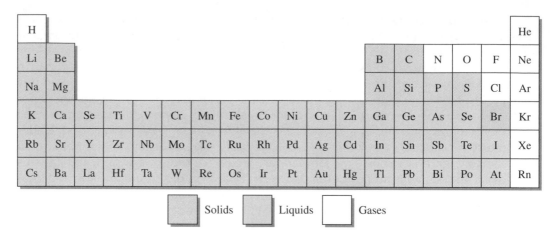

Solids Liquids Gases

Figure 11.2 identifies the nonsolid elements. There are two naturally occurring elements, cesium of group IA and gallium of group IIIA, that have melting points between 25°C and 30°C. Temperatures in this range are reached on hot summer days. Thus, at such times, there could be four elements rather than two that are liquids.

Explanations for the experimentally observed physical-state variations among substances can be derived from the concepts of atomic structure and bonding we have considered in previous chapters. Such explanations are found in later sections of this chapter.

11.2 PROPERTY DIFFERENCES AMONG PHYSICAL STATES

The differences among solids, liquids, and gases are so great that only a few gross distinguishing features need to be mentioned to differentiate them clearly. Certain obvious differences among the three states of matter are apparent to even the most casual observer—differences related to (1) volume and shape, (2) density, (3) compressibility, and (4) thermal expansion. These distinguishing properties are compared in Table 11.1 for the three states of matter. The properties of volume and density have been discussed in detail previously (Secs. 3.4 and 3.8, respectively). **Compressibility** *is a measure of the volume change resulting from a pressure change.* **Thermal expansion** *is a measure of the volume change resulting from a temperature change.*

The contents of Table 11.1 should be studied in detail; the data given will serve as the starting point for further discussions about the states of matter.

TABLE 11.1 Distinguishing Properties of Solids, Liquids, and Gases

Property	Solid State	Liquid State	Gaseous State
Volume and shape	definite volume and definite shape	definite volume and indefinite shape; takes the shape of container to the extent it is filled	indefinite volume and indefinite shape; takes the volume and shape of container that it fills
Density	high	high, but usually lower than corresponding solid	low
Compressibility	small	small, but usually greater than corresponding solid	large
Thermal expansion	very small: about 0.01% per °C	small: about 0.1% per °C	moderate: about 0.3% per °C

11.3 THE KINETIC MOLECULAR THEORY OF MATTER

The **kinetic molecular theory of matter** *is a set of five statements used to explain the physical behavior of the three states of matter (solids, liquids, and gases).* The central idea of this theory is that the particles (atoms, molecules, or ions) present in a substance, independent of the physical state of the substance, are always in motion.

The five statements of the kinetic molecular theory of matter are as follows:

Statement 1 *Matter is ultimately composed of tiny particles (atoms, molecules, or ions) that have definite and characteristic sizes that do not change.*

Statement 2 *The particles are in constant random motion and therefore possess kinetic energy.*

Kinetic energy *is energy that matter possesses because of particle motion.* An object that is in motion has the ability to transfer its kinetic energy to another object upon collision with that object (see Fig. 11.3).

Statement 3 *The particles interact with each other through attractions and repulsions and therefore possess potential energy.*

Potential energy *is stored energy that matter possesses as a result of its position, condition, and/or composition.* The potential energy of greatest importance when considering the three states of matter is that which originates from *electrostatic forces* among charged particles. An **electrostatic force** *is an attractive force or repulsive force that occurs among charged particles.* Particles of opposite charge (one positive and one negative) attract one another, and particles of like charge (both positive or both negative) repel one another.

Statement 4 *The kinetic energy (velocity) of the particles increases as the temperature is increased.*

The *average* kinetic energy (velocity) of all particles in a system depends on temperature; the higher the temperature, the higher the average kinetic energy of the system.

Statement 5 *The particles in a system transfer energy to each other through elastic collisions.*

In an *elastic* collision, the total kinetic energy remains constant; no kinetic energy is lost. The difference between an *elastic* and an *inelastic* collision is illustrated by comparing the collision of two hard steel spheres with the collision of two masses of putty. The collision of spheres approximates an elastic collision (the spheres bounce off one another and continue moving); the putty collision has none of these characteristics (the masses glob together with no resulting movement).

The word *kinetic* comes from the Greek *kinesis*, which means "movement." The kinetic molecular theory deals with the movement of particles.

Water behind a dam represents potential energy because of *position*. Uncontrolled water release can cause massive destruction (a flood) as the potential energy is converted to other forms of energy. A compressed spring can spontaneously expand and do work as the result of potential energy associated with *condition*. Potential energy associated with *composition* is released when gasoline is burned. Electrostatic interactions are also potential energy associated with *composition* of a substance.

FIGURE 11.3 The kinetic energy of water falling through the penstocks of a large dam is used to turn huge turbines and generate electricity. *(George Stringham—US Army Corp of Engineers)*

Two consequences of the elasticity of particle collisions are that (1) the energy of any given particle is continually changing and (2) particle energies in a system are not all the same; a range of particle energies is always encountered.

The relative influence of kinetic energy and potential energy in a chemical system is the major consideration in using kinetic molecular theory to explain the general properties of the solid, liquid, and gaseous states of matter. The important question is whether the kinetic energy or the potential energy dominates the energetics of the chemical system under study.

Kinetic energy may be considered a *disruptive force* within the chemical system, tending to make the particles of the system increasingly independent of each other. As the result of energy of motion, the particles will tend to move away from each other. Potential energy may be considered a *cohesive force* tending to cause order and stability among the particles of the system.

The role that temperature plays in determining the state of a system is related to kinetic energy magnitude. Kinetic energy increases as temperature increases (statement 4 of the kinetic molecular theory). Thus, the higher the temperature, the greater the magnitude of disruptive influences within a chemical system. The magnitude of potential energy is essentially independent of temperature change. Neither charge nor separation distance, the two factors on which the magnitude of potential energy depends, is affected significantly by temperature change.

Sections 11.4, 11.5, and 11.6 deal with kinetic molecular theory explanations for the general properties of the solid, liquid, and gaseous states, respectively.

11.4 THE SOLID STATE

A **solid** *is the physical state characterized by a dominance of potential energy (cohesive forces) over kinetic energy (disruptive forces).* The particles in a solid are drawn close together in a regular pattern by the strong cohesive forces present. Each particle occupies a fixed position about which it vibrates. An explanation of the characteristic properties of solids is obtained from this model.

1. *Definite volume and definite shape.*   The strong cohesive forces hold the particles in essentially fixed positions, resulting in definite volume and definite shape.
2. *High density.*   The constituent particles of solids are located as close together as possible. Therefore, large numbers of particles are contained in a unit volume, resulting in a high density.
3. *Small compressibility.*   Since there is very little space between particles, increased pressure cannot push them any closer together and therefore has little effect on the solid's volume.
4. *Very small thermal expansion.*   An increase in temperature increases the kinetic energy (disruptive forces), thereby causing more vibrational motion of the particles. Each particle occupies a slightly larger volume. The result is a slight expansion of the solid. The strong cohesive forces prevent this effect from becoming very large.

Some solids, such as foam rubber, are compressible because they are full of gas (air).

11.5 THE LIQUID STATE

The liquid state consists of particles randomly packed relatively close to each other. The molecules are in constant random motion, freely sliding over one another but without sufficient energy to separate from each other. A **liquid** *is the physical state characterized by potential energy (cohesive forces) and kinetic energy (disruptive forces) of about the same magnitude.* The fact that the particles freely slide over each other indicates the influence of disruptive forces, but the fact that the particles do not separate indicates a fairly strong influence from cohesive forces. The characteristic properties of liquids are explained by this model.

1. *Definite volume and indefinite shape.*   Attractive forces are strong enough to restrict particles to movement within a definite volume. They are not strong enough, however, to prevent the particles from moving over each other in a random manner,

limited only by the container walls. Thus liquids have no definite shape, with the exception that they maintain a horizontal upper surface in containers that are not completely filled.

2. *High density.*   The particles in a liquid are not widely separated; they essentially touch each other. Therefore, there will be a large number of particles per unit volume and a resultant high density.

3. *Small compressibility.*   Since the particles in a liquid essentially touch each other, there is very little empty space. Therefore, a pressure increase cannot squeeze the particles much closer together.

4. *Small thermal expansion.*   Most of the particle movement in a liquid involves particles sliding over each other. The increased particle velocity that accompanies a temperature increase results in a small increase in such motion. The net effect is an increase in the effective volume a particle occupies, which causes a slight volume increase in the liquid.

The measurement of temperature using a thermometer is based on the thermal expansion of liquid mercury in a narrow tube. As the temperature increases, the increased mercury volume raises the height of the mercury column in the tube.

11.6 THE GASEOUS STATE

A **gas** *is the physical state characterized by a complete dominance of kinetic energy (disruptive forces) over potential energy (cohesive forces).* As a result, the particles of a gas are essentially independent of one another and move in a totally random manner. Under ordinary pressure, the particles are relatively far apart except, of course, when they collide with each other. In between collisions with each other or with the container walls, gas particles travel in straight lines. The particle velocities and resultant collision frequencies are extremely high; at room temperature and pressure, the collisions experienced by one molecule are of the order of 10^{10} collisions per second.

The kinetic theory explanation of gaseous state properties follows the same pattern we saw earlier for solids and liquids.

1. *Indefinite volume and indefinite shape.*   The attractive (cohesive) forces between particles have been overcome by kinetic energy, and the particles are free to travel in all directions. Therefore, the particles completely fill the container the gas is in, and they assume its shape.

2. *Low density.*   The particles of a gas are widely separated. There are relatively few of them in a given volume, which means little mass per unit volume.

3. *Large compressibility.*   Particles in a gas are widely separated; a gas is mostly empty space. When pressure is applied, the particles are easily pushed closer together, decreasing the amount of empty space and the volume of the gas (see Fig. 11.4).

4. *Moderate thermal expansion.*   An increase in temperature means an increase in particle velocity. The increased kinetic energy of the particles enables them to push

FIGURE 11.4 The compression of a gas involves decreasing the amount of empty space in the container. Particles present do not change in size.

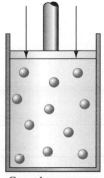
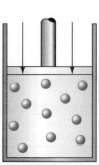

Gas at low pressure   Gas at higher pressure

back whatever barrier is confining them into a given volume. Hence, the volume increases.

It must be understood that the size of the particles is not changed during expansion or compression of gases, solids, or liquids. The particles merely move farther apart or closer together; the space between them is what changes.

11.7 A COMPARISON OF SOLIDS, LIQUIDS, AND GASES

Two obvious conclusions about the similarities and differences between the various states of matter may be drawn from a comparison of the descriptive materials in Sections 11.4 through 11.6.

1. One of the states of matter, the gaseous state, is markedly different from the other two states.
2. Two of the states of matter, the solid and the liquid states, have many similar characteristics.

> The difference in distance between molecules in gases and liquids explains why it is easier to walk through air than through water.

These two conclusions are illustrated diagrammatically in Figure 11.5.

The average distance between particles is only slightly different in the solid and liquid states but markedly different in the gaseous state. Roughly speaking, at ordinary temperatures and pressures, particles in a liquid are about 10% and particles in a gas about 1000% farther apart than those in the solid state. The distance ratio between particles in the three states (solid to liquid to gas) is thus 1 to 1.1 to 10.

11.8 ENDOTHERMIC AND EXOTHERMIC CHANGES OF STATE

> Although the processes of sublimation and deposition are not common in everyday life, they are still encountered. Dry ice sublimes, as do mothballs placed in a clothing-storage area. It is because of sublimation that ice cubes left in a freezer get smaller as time passes. Ice or snow forming in clouds (from water vapor) during the winter season is an example of deposition.

A **change of state** *is a process in which a substance is transformed from one physical state to another physical state.* Changes of state were previously considered in Section 4.4. The terminology associated with various changes of state—evaporation, condensation, sublimation, and so on—was introduced at that time (Fig. 4.4).

Changes of state are usually accomplished through heating or cooling a substance. (Pressure change is also a factor in some systems.) Changes of state may be classified according to whether heat (thermal energy) is absorbed or released. An **endothermic change of state** *is a change of state that requires the input (absorption) of heat energy.* The endothermic changes of state are melting, sublimation, and evaporation. An **exothermic change of state** *is a change of state that requires heat energy to be given up (released).* Exothermic changes of state are the reverse of endothermic changes of state and include deposition, condensation, and freezing. Figure 11.6 summarizes the classification of changes of state as exothermic or endothermic.

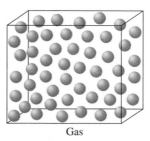

Gas
Molecules far apart and disordered
Negligible interactions among molecules

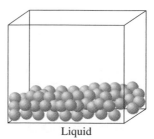

Liquid
Molecules close together and disordered
Moderately strong interactions among molecules

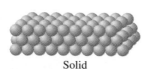

Solid
Molecules close together and ordered
Strong interactions among molecules

FIGURE 11.5 Similarities and differences, at the molecular level, among the three states of matter.

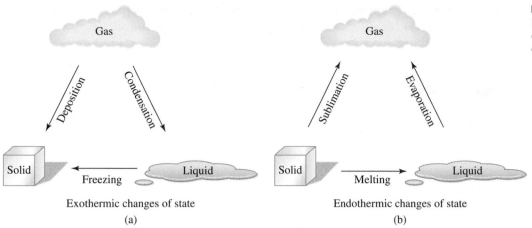

FIGURE 11.6
Exothermic (a) and endothermic (b) changes of state.

Exothermic changes of state
(a)

Endothermic changes of state
(b)

11.9 HEAT ENERGY AND SPECIFIC HEAT

Energy can exist in any of several forms. Common forms include radiant (light) energy, chemical energy, heat (thermal) energy, electrical energy, and mechanical energy. These forms of energy are interconvertible. The heating of a home is a process that illustrates energy interconversion. As the result of burning natural gas or some other fuel, chemical energy is converted into heat energy. In large conventional power plants that are used to produce electricity, the heat energy obtained from burning coal is used to change water into steam, which can then turn a turbine (mechanical energy) to produce electricity (electrical energy).

> In addition to the various *forms* of energy, there are two *types* of energy: potential and kinetic (Sec. 11.3). The basis for determining energy types depends on whether the energy is available but not being used (potential) or is actually in use (kinetic).

The form of energy most often encountered when considering physical and chemical changes is *heat energy*. For this reason, we will consider further particulars about this form of energy.

Heat Energy Units

The two most commonly used units for expressing the amount of heat energy released or absorbed in a process are the *joule* and the *calorie*. Today, most heat energy calculations are carried out using the joule unit because this unit is more compatible with other metric system units than is the calorie.

A **joule (J)** *is the base unit for energy in the metric system.* The joule unit, abbreviated as J, is suitable for measuring all types of energy, not just heat energy. The name *joule* (pronounced *jool*, rhymes with pool) comes from the name of the English physicist James Prescott Joule (1818–1889), an early researcher in the area of energy studies.

It is convenient to think of the joule in terms of the amount of heat energy required to raise the temperature of one gram of water from 14.5°C to 15.5°C (one degree Celsius), which is 4.184 J. In this context, the joule unit and the calorie unit are easily related. The calorie was originally defined as the amount of heat energy necessary to raise the temperature of one gram of water from 14.5°C to 15.5°C. Thus,

$$4.184 \text{ joules} = 1 \text{ calorie}$$

Both the joule and the calorie involve relatively small amounts of energy, so the kilojoule (kJ) and the kilocalorie (kcal) are often used instead. It follows from the relationship between joules and calories that

$$4.184 \text{ kJ} = 1 \text{ kcal}$$

In discussions involving nutrition, the energy content of foods, or dietary tables, the unit *Calories* (spelled with a capital C) is used. The dietetic Calorie is actually 1 kilocalorie

The Btu unit that is used to rate air conditioners and heaters is an English system unit for heat. Btu stands for British thermal unit and is the amount of heat energy it takes to raise the temperature of one pound of water one degree Fahrenheit. One Btu is the equivalent of 3.42 kJ.

(1000 cal). The statement that an oatmeal raisin cookie contains 60 Calories means that 60 kcal (60,000 cal) of energy is released when the cookie is metabolized (undergoes chemical change) within the body.

EXAMPLE 11.1 Interrelationships among Heat Energy Units

One gram of a combustible substance burns to produce 55.2 kJ of heat energy. Express this amount of heat energy, 55.2 kJ, in the following units:

    a. joules   **b.** kilocalories   **c.** calories

SOLUTION

    a. The relationship between joules and kilojoules is

$$10^3 \, J = 1 \, kJ$$

Therefore, using dimensional analysis, we have

$$55.2 \, \cancel{kJ} \times \frac{10^3 \, J}{1 \, \cancel{kJ}} = 55{,}200 \, J \quad \text{(calculator and \textbf{correct answer})}$$

    b. The relationship between kilocalories and kilojoules is

$$1 \, kcal = 4.184 \, kJ$$

The one-step dimensional analysis setup for this problem is

$$55.2 \, \cancel{kJ} \times \frac{1 \, kcal}{4.184 \, \cancel{kJ}} = 13.193116 \, kcal \quad \text{(calculator answer)}$$

$$= 13.2 \, kcal \quad \text{(\textbf{correct answer})}$$

    c. There will be two conversion factors in this part, derived from the relationship 1 kcal = 4.184 kJ and 10^3 cal = 1 kcal, respectively. The pathway will be

$$kJ \rightarrow kcal \rightarrow cal$$

$$55.2 \, \cancel{kJ} \times \frac{1 \, \cancel{kcal}}{4.184 \, \cancel{kJ}} \times \frac{10^3 \, cal}{1 \, \cancel{kcal}} = 13193.116 \, cal \quad \text{(calculator answer)}$$

$$= 13{,}200 \, cal \quad \text{(\textbf{correct answer})}$$

There can be only three significant figures in the correct answer.

Answer Double Check:

Answers to all practice exercises in this chapter are found in the back-of-the-book answer section.

Are the relationships among answers reasonable? Yes. An answer in joules should be approximately four times larger than the same answer expressed in calories. Such is the relationship between the part (a) and part (c) answers (55,200 and 13,200). The part (c) answer is 1000 times larger than the part (b) answer; such should be the case for a comparison of calorie and kilocalorie values for the same quantity.

▶ **Practice Exercise 11.1** Benzene, in the liquid state, readily undergoes combustion. Combustion of 1.00 g of benzene produces 83.8 kJ of heat energy. Express this heat energy amount in the units of

    a. joules   **b.** calories   **c.** kilocalories

Specific Heat

For every pure substance in a given state (solid, liquid, or gas), we can measure a physical property called the *specific heat* of that substance. **Specific heat** *is the amount of heat energy needed to raise the temperature of one gram of a substance in a specific physical state by 1°C.* The units most commonly used for specific heat, in scientific work, are joules per gram per degree Celsius [J/(g · °C)]. Specific heats for a number of substances in various states are given in Table 11.2. Note the three different entries for water in this table—one for each of the physical states. As these entries point out, the magnitude of the specific heat for a substance changes when the physical state of the substance changes. (Water is one of the few substances that we routinely encounter in all three physical states, hence the three specific-heat values in the table for this substance.)

The lower the specific heat of a substance, the more its temperature will change when it absorbs a given amount of heat. Metals generally have low specific heats. This means they heat up quickly and also cool quickly.

Liquid water has one of the highest specific heats known (see Table 11.2). This high specific heat makes water a very effective coolant. The moderate climates of geographical areas where large amounts of water are present—for example, the Hawaiian Islands—are related to water's ability to absorb large amounts of heat without undergoing drastic temperature changes. Desert areas, areas that lack water, are the areas where the extremes of high temperature are encountered on Earth. The temperature of a living organism remains relatively constant because of the large amounts of water present in it.

The amount of heat energy needed to cause a fixed amount of a substance to undergo a specific temperature change (within a range that causes no change of state) can easily be calculated if the substance's specific heat is known. The specific heat [in

A walk on the beach on a sunny day will demonstrate differences in specific heat. The sand, which has a relatively low specific heat, heats up rapidly in the sun. The water, with its higher specific heat, stays cool.

TABLE 11.2 Specific Heats of Selected Pure Substances

Substance	Physical State	Specific Heat [J/(g · °C)]
Aluminum	solid	0.908
Copper	solid	0.382
Ethyl alcohol	liquid	2.42
Gold	solid	0.13
Iron	solid	0.444
Nitrogen	gas	1.0
Oxygen	gas	0.92
Silver	solid	0.24
Sodium chloride	solid	0.88
Water (ice)	solid	2.09
Water	liquid	4.18
Water (steam)	gas	2.03
Zinc	solid	0.388

J/(g·°C)] is multiplied by the mass (in grams) and by the temperature change (in degrees Celsius) to eliminate the units of g and °C and obtain the unit joules.

$$\text{heat energy change} = \text{specific heat} \times \text{mass} \times \text{temperature change}$$

$$= \frac{J}{\cancel{g} \cdot \cancel{°C}} \times \cancel{g} \times \cancel{°C}$$

$$= J$$

The temperature change, denoted as ΔT, is defined as $T_f - T_i$, where T_f is the final temperature of the system and T_i is the initial temperature of the system.

$$\Delta T = T_f - T_i$$

This definition for ΔT allows for both positive and negative ΔT values, depending on whether the final temperature is higher or lower than the initial temperature.

Depending on the sign of ΔT (positive or negative), the heat energy change for the system will be positive or negative. A positive heat energy change value is indicative of heat energy absorption, a characteristic of an *endothermic* change (Sec. 11.8). A negative heat energy change value is indicative of heat energy release, a characteristic of an *exothermic* change (Sec. 11.8).

EXAMPLE 11.2 **Calculating the Amount of Heat Energy Absorbed by a Substance Undergoing a Specific Temperature Increase**

The element beryllium is a relatively unreactive steel gray metal with a specific heat of 1.8 J/g·°C. How many joules of heat energy must 24.0 g of beryllium metal absorb for its temperature to increase from 25°C to 55°C?

SOLUTION

We will use the equation

$$\text{heat energy change} = \text{specific heat} \times \text{mass} \times \text{temperature change}$$

in solving the problem. All the quantities on the right-hand side of the equation are known. Both specific heat and mass are directly given in the problem statement. The temperature change can be calculated using the two temperatures given in the problem statement.

Substituting the known quantities into the equation gives

$$\text{heat energy change} = \frac{1.8\ J}{\cancel{g} \cdot \cancel{°C}} \times 24.0\ \cancel{g} \times (55 - 25)\cancel{°C}$$

$$= 1296\ J \quad \text{(calculator answer)}$$

$$= 1300\ J \quad \text{(\textbf{correct answer})}$$

Since the heat energy change value has a positive sign, an endothermic change has occurred.

Answer Double Check:

Is the positive sign for the heat energy change reasonable? Yes. Raising the temperature of a metal requires the addition of heat energy (energy absorption). A positive heat energy value indicates heat absorption, that is, an endothermic process.

▶ **Practice Exercise 11.2** Calculate the number of joules of heat energy needed to increase the temperature of a 25.0 g piece of aluminum metal from 35.0°C to 57.0°C. The specific heat of aluminum is 0.908 J/g°C.

EXAMPLE 11.3 **Calculating the Temperature Change Caused by Addition of a Specific Amount of Heat Energy**

One cup of dry-roasted, salted, shelled pistachio nuts has a caloric value of 78.7 Cal (78,700 cal). What would be the temperature change in a quart of water (944 g) at 10°C if the same amount of energy were added to it?

SOLUTION

The equation

$$\text{heat energy change} = \text{specific heat} \times \text{mass} \times \text{temperature change}$$

is rearranged to isolate temperature change on the left side of the equation.

$$\text{temperature change (°C)} = \frac{\text{heat absorbed (J)}}{\text{mass (g)} \times \text{specific heat [J/(g} \cdot \text{°C)]}}$$

In this equation the heat absorbed will be the heat energy (caloric value) supplied by the pistachio nuts. Since this is given in calories, we will need to change it to joule units.

$$78,700 \text{ cal} \times \frac{4.184 \text{ J}}{1 \text{ cal}} = 329,280.8 \text{ J} \quad \text{(calculator answer)}$$

$$= 329,000 \text{ J} \quad \textbf{(correct answer)}$$

Substituting the values 329,000 J (heat absorbed), 944 g (mass of water), and 4.18 J/g·°C (specific heat of water; see Table 11.2) in the previously derived equation for change in temperature (°C) gives

$$\text{temperature change (°C)} = \frac{329,000 \text{ J}}{944 \text{ g} \times 4.184 \text{ J/(g} \cdot \text{°C)}}$$

$$= 83.297549 \text{°C} \quad \text{(calculator answer)}$$

$$= 83.3 \text{°C} \quad \textbf{(correct answer)}$$

The temperature of the water will increase from 10°C to 93°C, an increase of 83°.

Answer Double Check:

Is the magnitude of the temperature change reasonable? Yes. Doing a rough calculation, using rounded numbers, gives 300,000/1000, which is equal to 300, and 300 divided by 4 is 75. The calculated temperature change of 83° is consistent with this rough estimate of 75° for the answer.

▶ **Practice Exercise 11.3** A 1 oz piece of milk chocolate has a caloric value of approximately 145 Cal (145,000 calories). What would be the temperature change in a 10.0 lb (4540 g) piece of aluminum metal at 20°C if the same amount of energy were added to it? The specific heat of Al is 0.908 J/g°C.

A quantity closely related to specific heat is that of *heat capacity*. **Heat capacity** *is the amount of heat energy needed to raise the temperature of a given quantity of a substance in a specific physical state by 1°C.* The relationship between heat capacity and specific heat is

$$\text{heat capacity} = \text{grams} \times \text{specific heat}$$

Common units for heat capacity are J/°C. Heat capacity refers to a property of a whole object (its entire mass), while specific heat refers to the heat capacity per unit mass (1 g). If either heat capacity or specific heat is known, the other quantity can always be calculated from the known quantity.

EXAMPLE 11.4 | **Using Heat Capacity Data to Calculate Specific Heat**

An 80.0 g sample of the metal gold has a heat capacity of 10.5 J/°C. What is the specific heat, in J/g°C, of gold?

SOLUTION

The relationship between heat capacity and specific heat is

$$\text{heat capacity} = \text{mass} \times \text{specific heat}$$

Rearranging this equation to isolate specific heat on one side gives

$$
\begin{aligned}
\text{specific heat} &= \frac{\text{heat capacity}}{\text{mass}} \\
&= \frac{10.5 \text{ J/°C}}{80.0 \text{ g}} \\
&= 0.13125 \text{ J/(g} \cdot \text{°C)} \quad \text{(calculator answer)} \\
&= 0.131 \text{ J/(g} \cdot \text{°C)} \quad \textbf{(correct answer)}
\end{aligned}
$$

Answer Double Check:

Is the numerical value for the specific heat reasonable, based on the numbers used to calculate it? Yes. Using rounded input numbers, a rough estimate for the numerical value is 10/100, which gives 0.1. The calculated answer of 0.131 is of the same order as 0.1.

▶ **Practice Exercise 11.4**   A 100.0 g sample of ethyl alcohol, also called grain alcohol because it can be produced from the fermentation of plant residues, has a heat capacity of 242 J/°C. What is the specific heat of ethyl alcohol, in J/g°C?

Another common type of heat energy calculation involves the transfer of heat from one substance to another. The following two generalizations always apply to such situations.

1. Heat always flows from the warmer body to the colder body.
2. The heat lost by the warmer body is equal to the heat gained by the colder body.

EXAMPLE 11.5 | **Calculation Involving "Heat Lost Is Equal to Heat Gained."**

A 125 g piece of a rock of unknown specific heat is heated to 93°C and then the rock is dropped into 100.0 g of water at 19°C. The temperature of the water rises to 31°C. What is the specific heat, in J/g · °C, of the rock?

SOLUTION

The heat lost by the rock (the hotter body) is equal in magnitude but opposite in sign to the heat gained by the water (the colder body). Thus,

$$-[\text{heat lost (rock)}] = \text{heat gained (water)}$$

Using the general equation for heat lost or heat gained, which is

$$\text{heat lost or gained} = \text{specific heat} \times \text{mass} \times \text{temperature change}$$

we have for our present situation

$$-\left[\text{specific heat} \times 125\text{ g} \times (-62°C)\right] = 4.18\frac{J}{g\cdot°C} \times 100.0\text{ g} \times (12°C)$$

$$\underbrace{\qquad\qquad\qquad\qquad\qquad}_{\text{heat lost by rock}} \qquad \underbrace{\qquad\qquad\qquad\qquad}_{\text{heat gained by water}}$$

Solving this equation for the specific heat of the rock gives

$$\text{specific heat of rock} = \frac{\left[4.18\dfrac{J}{g\cdot°C} \times 100.0\cancel{g} \times (12°\cancel{C})\right]}{-[125\cancel{g} \times (-62°\cancel{C})]}$$

$$= 0.6472258\text{ J/g}\cdot°C \qquad \text{(calculator answer)}$$

$$= 0.65\text{ J/g}\cdot°C \qquad \text{(correct answer)}$$

Answer Double Check:

Is the magnitude of the specific heat value reasonable? Yes. The water and rock have roughly the same mass, and the rock temperature drops more than the water temperature increases. This indicates that the specific heat of the rock should be less than the specific heat of the water. The calculated answer is consistent with this line of reasoning.

▶ **Practice Exercise 11.5**   The specific heat of a sample of wood (1.76 J/g°C) is found to be twice that of a sample of concrete (0.88 J/g°C). How many grams of the wood sample can be heated from 25°C to 35°C by the heat energy released when 10.0 g of the concrete sample is cooled from 50°C to 40°C?

11.10 TEMPERATURE CHANGES AS A SUBSTANCE IS HEATED

For a given pure substance, molecules in the gaseous state possess more energy than molecules in the liquid state, which in turn possess more energy than molecules in the solid state. This fact is obvious: We know that it takes energy (heat) to melt a solid and still more energy (heat) to change the resulting liquid to a gas. Additional information concerning the relationship of energy to the states of matter can be obtained by a closer examination of what happens, step-by-step, to a solid (which is below its melting point) as heat is continuously supplied, causing it to melt and ultimately to change to a gas.

The heating curve shown in Figure 11.7 gives the steps involved in changing a solid to a gas, using water as the example. As the ice (solid) is heated, its temperature rises until the melting point for ice is reached. The temperature increase indicates that the added heat causes an increase in the kinetic energy of the water molecules (recall Sec. 11.3, kinetic molecular theory). Once the melting point is reached, the temperature remains constant while the ice melts. The constant temperature during the melting process indicates that the added heat has decreased the potential energy of the water molecules without increasing their kinetic energy—the intermolecular attractions are being weakened by the decrease in potential energy. The net result is that the ice becomes liquid water. The addition of more heat to the liquid water increases its temperature until the boiling point is reached. During this stage the water molecules are again gaining kinetic energy. At the boiling point another state change occurs (liquid to gas) as heat is added with the temperature remaining constant. The constant temperature again indicates a decrease in potential energy. Once the system is completely changed to steam (water vapor), the temperature again increases with further heating.

FIGURE 11.7 A heating curve depicting the addition of heat (at a constant rate) to a solid (ice) at a temperature below its melting point (−20°C) until it becomes a gas (steam) at a temperature above its boiling point (120°C). Note that when heat is added during a transition between physical states, the temperature does not change.

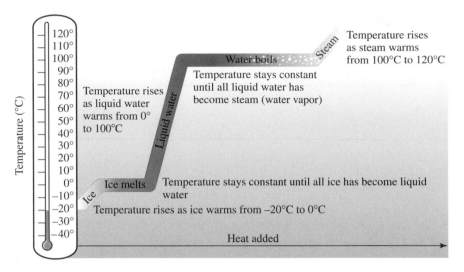

11.11 ENERGY AND CHANGES OF STATE

When heat energy is added to a solid, its temperature rises until the melting point is reached, with the amount of heat added governed by the specific heat (Sec. 11.9) of the solid. Once the melting point is reached, the temperature then remains constant while the solid changes to a liquid (Sec. 11.10). The energetics of the system during this transition from the solid state to the liquid state depend on the value of the substance's *heat of fusion*. (The term *fusion* means melting.)

The **heat of fusion** *is the amount of heat energy absorbed in the conversion of one gram of a solid to a liquid at the solid's melting point.* Units for heat of fusion are joules per gram (J/g). Note that these units do not involve temperature (degrees), as was the case for specific heat [J/(g·°C)]. No temperature units are needed because the temperature remains constant during a change of state.

The reverse of the fusion process is solidification (or freezing). The **heat of solidification** *is the amount of heat energy evolved in the conversion of one gram of a liquid to a solid at the liquid's freezing point.* The heat of solidification always has the *same numerical value* as the heat of fusion. The only difference between these two entities is in the *direction* of heat flow (in or out), which is indicated by a sign (positive or negative). Heat of solidification has a negative sign because it is associated with an exothermic process, and heat of fusion has a positive sign because it is associated with an endothermic process. The amount of heat required to melt 50.0 g of ice at its melting point is the same as the amount of heat that must be removed to freeze 50.0 g of water at its freezing point. The heats of fusion for selected substances are given in Table 11.3 in the units joules per gram.

Table 11.3 also gives a second heat of fusion value for each substance, the *molar heat of fusion*, which has the units kJ/mole. In some types of calculations, kJ/mole are more convenient units to use.

The magnitude of the heat of fusion for a given solid depends on the intermolecular forces of attraction in the solid state. The strength of such forces is the subject of Section 11.16.

Ice cubes cool drinks not only by being cold, but also by absorbing heat as they melt. Each gram of ice absorbs 334 joules from the liquid as it melts. The amount of heat it takes to melt 1 gram of ice can cool 1 gram of water from 80°C to 0°C.

TABLE 11.3 Heats of Fusion for Various Substances at Their Melting Points

Solid	Melting Point (°C)	Heat of Fusion J/g	Heat of Fusion kJ/mole
Methane	−182	59	0.94
Ethyl alcohol	−117	109	5.01
Carbon tetrachloride	−23	16.3	2.51
Water	0	334	6.01
Benzene	6	126	9.87
Aluminum	658	393	10.6
Copper	1083	205	13.0

11.12

In doin
followi

1. Tl
g
s
s

2. Tl
g
of
sa
of

Fi
and ene
the key
sented

The general equation for calculating the amount of heat absorbed as a substance changes from a solid to a liquid is

$$\text{heat absorbed (J)} = \text{heat of fusion (J/g)} \times \text{mass(g)}$$

Similarly, for the amount of heat released as a liquid freezes to a solid, we have

$$\text{heat released (J)} = \text{heat of solidification (J/g)} \times \text{mass(g)}$$

Example 11.6 illustrates the use of the first of these two equations.

EXAMPLE 11.6 **Calculating the Heat Energy Absorbed as a Substance Melts**

How much heat energy, in joules, is required to melt 25.1 g of aluminum at its melting point of 658°C?

SOLUTION

The heat of fusion for aluminum, from Table 11.3, is 393 J/g. Therefore, the total amount of heat energy required is

$$\text{Heat energy change} = 25.1 \, g \times \frac{393 \, \text{J}}{g} = 9864.3 \, \text{J} \quad \text{(calculator answer)}$$

$$= 9860 \, \text{J} \quad \text{(correct answer)}$$

In
ature ch
equatic
trate pr

▶ **Practice Exercise 11.6**   How much heat energy, in joules, is required to melt 35.0 g of copper at its melting point of 1083°C? The heat of fusion of copper is 205 J/g.

EXA

Calcula
the resu

SOLUI

We ma

1. C
2. C

W
amount

Principles similar to those just considered for solid–liquid or liquid–solid changes apply to changes between the liquid and gaseous states. Here the specific energy quantities involved are *heats of vaporization* and *heats of condensation*. The **heat of vaporization** *is the amount of heat energy absorbed in the conversion of one gram of a liquid to a gas at the liquid's boiling point*. Heats of vaporization are usually measured at the normal boiling point (Sec. 11.15) of the liquid.

The reverse of the vaporization (evaporation) process is condensation (liquefaction). The **heat of condensation** *is the amount of heat energy evolved in the conversion of one gram of a gas to a liquid at the liquid's boiling point*. The heat of vaporization and the heat of condensation will always have the same numerical value because these two quantities characterize processes that are the reverse of each other; they will also always have opposite signs—the former is positive and the latter is negative. The only difference is direction of

Steam at 100°C is much more dangerous than water at 100°C. If steam contacts the skin, it is not only hot, it also releases its heat of condensation as it condenses on the skin. This exothermic release of energy can cause severe burns.

FIGURE 11.10   Kinetic energy distributions of molecules of a liquid at two different temperatures. The dashed line represents the minimum kinetic energy required for molecules of the liquid to overcome attractive forces and escape into the gas phase. Molecules in the shaded area have the necessary energy to overcome attractions.

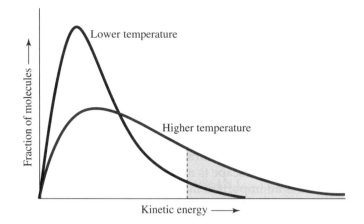

Evaporative cooling is important in many processes. Our own bodies use evaporation to maintain a constant temperature. In hot summer weather and also during strenuous exercise, the human body produces perspiration. Perspiration production is an effective method for body temperature regulation because evaporation of the perspiration cools our skin. The cooling effect of evaporation is quite noticeable when someone first comes out of a swimming pool on a hot day (especially if a breeze is blowing). A canvas water bag keeps water cool because some of the water seeps through the canvas and evaporates, a process that removes heat from the remaining water. In medicine, the local skin anesthetic ethyl chloride (C_2H_5Cl) exerts its effect through evaporative cooling (freezing). The minimum kinetic energy that molecules of this substance need to acquire to escape (evaporate) is very low, resulting in a very rapid evaporation rate. Evaporation is so fast that the cooling "freezes" tissue near the surface of the skin, with temporary loss of feeling in the region of application. Certain alcohols also evaporate quite rapidly. An alcohol rub is sometimes used to reduce body temperature when a high fever is present.

For a liquid in a container, the decrease in temperature that occurs as a result of evaporation can be actually measured only if the container is an *insulated* one. When a liquid evaporates from a noninsulated container, there is sufficient heat flow from the surroundings into the container to counterbalance the loss of energy in the escaping molecules and thus prevent any cooling effect. The thermos bottle is an insulated container that minimizes heat transfer, making it useful for maintaining liquids at a cool temperature for a short period of time. Figure 11.11 contrasts the evaporation process occurring in an uninsulated container with that which occurs within an insulated one. The temperature of liquid and surroundings is the same in an uninsulated container; in an insulated one, liquid temperature drops below that of the surroundings because of evaporative cooling.

Refrigera
letting a f
in coils in
refrigerat
endotherr
absorbs h
the refrig
compress
the gas ba
outside th
(This is w
electricity
The heat
condensa
out of the
That is wl
warm at t
refrigerat

FIGURE 11.11   For an evaporative cooling effect to be measured, a container that minimizes heat flow into the container from its surroundings must be used.

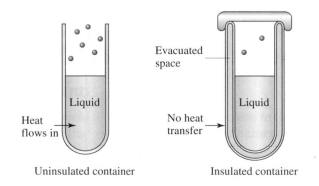

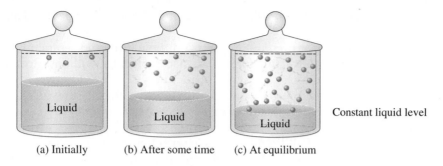

FIGURE 11.12
Evaporation of a liquid in a closed container.

(a) Initially   (b) After some time   (c) At equilibrium

Collectively, the molecules that escape from an evaporating liquid are often referred to as *vapor* rather than gas. A **vapor** *is the gaseous state of a substance at a temperature and pressure at which the substance is normally a liquid or solid*. For example, at room temperature and atmospheric pressure, the normal state for water is the liquid state. Molecules that escape (evaporate) from liquid water at these conditions are called water vapor.

Evaporation and Equilibrium

The evaporative behavior of a liquid in a *closed* container is quite different from that in an *open* container. In a closed container we observe that some liquid evaporation occurs, as indicated by a drop in liquid level. However, unlike the open container system, the liquid level, with time, ceases to drop (becomes constant), an indication that not all of the liquid will evaporate.

Kinetic molecular theory explains these observations in the following way. The molecules that do evaporate are unable to move completely away from the liquid as they did in the open container. They find themselves confined in a fixed space immediately above the liquid (see Fig. 11.12a). These trapped vapor molecules undergo many random collisions with the container walls, other vapor molecules, and the liquid surface. Molecules colliding with the liquid surface may be recaptured by the liquid. Thus, two processes—evaporation (escape) and condensation (recapture)—take place in the closed container.

In a closed container, for a short time, the rate of evaporation exceeds the rate of condensation and the liquid level drops. However, as more and more of the liquid evaporates, the number of vapor molecules increases and the chance of their recapture through striking the liquid surface also increases. Eventually the rate of condensation becomes equal to the rate of evaporation and the liquid level stops dropping (see Fig. 11.12c). At this point, the number of molecules that escape in a given time is the same as the number recaptured; a steady-state condition has been reached. The amounts of liquid and vapor in the container are not changing, even though both evaporation and condensation are still occurring.

This steady-state situation, which will continue as long as the temperature of the system remains constant, is an example of an *equilibrium state*. An **equilibrium state** *is a situation in which two opposite processes take place at equal rates*. For systems in a state of equilibrium, no net macroscopic changes can be detected. However, the system is dynamic; both forward and reverse processes are still occurring but in a manner such that they balance each other.

Keeping a lid on a pot of water allows the pot of water to boil faster. Without a lid, the heat of vaporization absorbed when water turns to steam escapes into the air. With a lid on, the water recondenses inside the lid and releases its heat of condensation back into the pot.

11.14 VAPOR PRESSURE OF LIQUIDS

For a liquid–vapor system in equilibrium in a closed container, the vapor in the fixed space immediately above the liquid exerts a constant pressure on the liquid surface and the walls of the container. This pressure is called the liquid's *vapor pressure*.

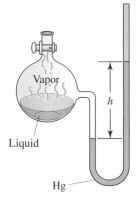

FIGURE 11.13
Apparatus for determining the vapor pressure of a liquid.

Vapor pressure *is the pressure exerted by a vapor above a liquid when the liquid and vapor are in equilibrium.*

The magnitude of vapor pressure depends on the nature and temperature of the liquid. Liquids with strong attractive forces between molecules will have lower vapor pressures than liquids in which only weak attractive forces exist between particles. Substances with high vapor pressures evaporate readily; that is, they are *volatile*. A **volatile substance** *is a substance that readily evaporates at room temperature because of a high vapor pressure.*

The vapor pressures of all liquids increase with temperature. Why? An increase in temperature results in more molecules having the minimum energy required for evaporation. Hence, at equilibrium the pressure of the vapor is greater. Table 11.5 shows the variation in vapor pressure with increasing temperature for water.

When the temperature of a liquid–vapor system in equilibrium is changed, the equilibrium is upset. The system immediately begins the process of establishing a new equilibrium. Let us consider a case where the temperature is increased. The higher temperature signifies that energy has been added to the system. More molecules will have the minimum energy needed to escape. Thus, immediately after the temperature is increased, molecules begin to escape at a rate greater than that at which they are recaptured. With time, however, the rates of escape and recapture will again become equal. The new rates will, however, be different from those for the previous equilibrium. Since energy has been added to the system, the rates will be higher, resulting in higher vapor pressure.

The size (volume) of the space that the vapor occupies does not affect the magnitude of the vapor pressure. A larger fixed space will enable more molecules to be present in the vapor at equilibrium. However, the larger number of molecules spread over a larger volume results in the same pressure as a small number of molecules in small volume.

Vapor pressures for liquids are commonly measured in an apparatus similar to that pictured in Figure 11.13. At the moment liquid is added to the vessel, the space above the liquid is filled only with air and the levels of mercury in the U-tube are equal. With time, liquid evaporates and equilibrium is established. The vapor pressure of the liquid is proportional to the difference between the heights of the mercury columns. The larger the vapor pressure, the greater the extent to which the mercury column is pushed up.

11.15 BOILING AND BOILING POINTS

Usually, for a molecule to escape from the liquid state, it must be on the surface of the liquid. **Boiling** *is a special form of evaporation in which conversion from the liquid to the vapor state occurs within the body of a liquid through bubble formation.* This phenomenon begins to

TABLE 11.5 Vapor Pressure of Water at Various Temperatures			
Temperature (°C)	Vapor Pressure (mm Hg)*	Temperature (°C)	Vapor Pressure (mm Hg)*
0	4.6	60	149.4
10	9.2	70	233.7
20	17.5	80	355.1
30	31.8	90	525.8
40	55.3	100	760.0
50	92.5		

The units used to specify vapor pressure in this table will be discussed in detail in Section 12.2.

FIGURE 11.14
Bubble formation associated with a liquid that is boiling.

occur when the vapor pressure of a liquid, which is steadily increasing as a liquid is heated, reaches a value equal to that of the prevailing external pressure on the liquid; for liquids in open containers, this value is atmospheric pressure. When these two pressures become equal, bubbles of vapor form around any speck of dust or any irregularity associated with the surface of the container. Being less dense than the liquid itself, these vapor bubbles quickly rise to the surface and escape. The quick ascent of the bubbles causes the agitation associated with a boiling liquid.

Let us consider this bubble phenomenon in more detail. As the heating of a liquid begins, the first small bubbles that form on the bottom and sides of the container are bubbles of dissolved air (oxygen and nitrogen), which have been driven out of the solution by the rising liquid temperature. (The solubilities of oxygen and nitrogen in liquids decrease with increasing temperature.)

As the liquid is heated further, larger bubbles form and begin to rise. These bubbles, usually also formed on the bottom of the container (where the heat is being applied and the liquid is hottest), are vapor bubbles rather than air bubbles. Initially, these vapor bubbles disappear as they rise, never reaching the liquid surface. Their disappearance is related to vapor pressure. In the hotter lower portions of the liquid, the liquid's vapor pressure is high enough to sustain bubble formation. [For a bubble to exist, the pressure within it (vapor pressure) must equal external pressure (atmospheric pressure).] In the cooler, higher portions of the liquid, where the vapor pressure is lower, the bubbles are collapsed by external pressure. Finally, with further heating, the temperature throughout the liquid becomes high enough to sustain bubble formation. Then bubbles rise all the way to the surface and escape (see Figure 11.14). At this point we say the liquid is boiling.

Like evaporation, boiling is actually a cooling process. When heat is taken away from a boiling liquid, boiling ceases almost immediately. It is the highest-energy molecules that are escaping. Quickly the temperature of the remaining molecules drops below the *boiling point* of the liquid.

Factors That Affect Boiling Point

A **boiling point** *is the temperature of a liquid at which the vapor pressure of the liquid becomes equal to the external (atmospheric) pressure exerted on the liquid.* Since atmospheric pressure fluctuates from day to day, so does the boiling point of a liquid. To compare the boiling points of different liquids, the external pressure must be the same. The boiling point of a liquid most often used for comparison and tabulation purposes (in reference books, for example) is the *normal boiling point.* A **normal boiling point** *is the temperature of a liquid at which the liquid boils under a pressure of 760 mm Hg.*

At any given location, the changes in the boiling point of liquids due to *natural variation* in atmospheric pressure seldom exceed a few degrees; in the case of water the

Anyone who has let a pan of water boil dry knows that boiling is an endothermic process. As long as the water is boiling, the pan stays at a "cool" 100°C. Once all the water boils away, the pan turns red-hot.

TABLE 11.6 Variation of the Boiling Point of Water with Elevation

Location	Elevation (ft above sea level)	Boiling Point of Water (°C)
San Francisco, CA	0	100.0
Salt Lake City, UT	4,390	95.6
Denver, CO	5,280	95.0
La Paz, Bolivia	12,795	91.4
Mount Everest	28,028	76.5

The converse of the pressure cooker phenomenon is that food cooks more slowly at reduced pressures. The pressure reduction associated with higher altitudes means that food cooked over a campfire in the mountains requires longer cooking times.

maximum is about 2°C. However, variations in boiling points *between* locations at different elevations can be quite striking, as shown by the data in Table 11.6.

The boiling point of a liquid can be increased by increasing the external pressure. Use is made of this principle in the operation of a pressure cooker. Foods cook faster in pressure cookers because the elevated pressure causes water to boil above 100°C. An increase in temperature of only 10°C will cause food to cook in approximately half the normal time. (Cooking involves chemical reactions, and the rate of a chemical reaction generally doubles with every 10°C increase in temperature.) Table 11.7 gives the boiling temperatures reached by water in normal household pressure cookers. Hospitals use the same principle in sterilizing instruments and laundry in autoclaves; sufficiently high temperatures are reached to destroy bacteria.

Liquids that have high normal boiling points or that undergo undesirable chemical reactions at boiling temperatures can be made to boil at low temperatures by reducing the external pressure. This principle is used in the preparation of numerous food products including frozen fruit juice concentrates. At a reduced pressure some of the water in a fruit juice is boiled away, concentrating the juice without having to heat it to a high temperature. Heating to a high temperature would cause changes that would spoil the taste of the juice and/or reduce its nutritional value.

11.16 INTERMOLECULAR FORCES IN LIQUIDS

For a liquid in an open container to boil, its vapor pressure must reach atmospheric pressure. For some substances this occurs at temperatures well below zero; for example, oxygen has a boiling point of –183°C. Other substances do not boil until the temperature is much higher. Mercury, for example, has a boiling point of 357°C, which is 540°C higher than that of oxygen. An explanation for this variation involves a consideration of the nature of the *intermolecular forces* that must be overcome in order for molecules (or atoms) to escape from the liquid state to the vapor state. An **intermolecular force** *is an attractive force that acts between a molecule and another molecule.* [Strictly speaking, the term *intermolecular force* refers

TABLE 11.7 Boiling Points of Water in a Pressure Cooker

Pressure above Atmospheric		Boiling Point of Water (°C)
lb/in.^{2}	mm Hg	
5	259	108
10	517	116
15	776	121

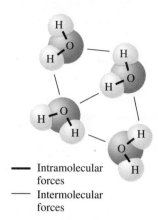

- Intramolecular forces
- Intermolecular forces

FIGURE 11.15
Intermolecular forces exist *between* molecules. *Intramolecular* forces exist *within* molecules and hold the molecules together; they are the chemical bonds within the molecule. Intermolecular forces are much weaker than intramolecular forces.

only to the attraction between *molecules*, but in practice it is used to refer to interactions in liquids that involve all types of particles (molecules, ions, and atoms).]

Intermolecular forces are similar in one way to the previously discussed *intra*molecular forces (*within* molecules) involved in covalent bonding (Sec. 7.10). They are electrostatic in origin; that is, they involve positive–negative interactions (Sec. 11.3). A major difference between inter- and intramolecular forces is their magnitude; the former are much weaker. However, intermolecular forces, despite their relative weakness, are sufficiently strong to influence the behavior of liquids, often in a very dramatic way. Figure 11.15 pictorially contrasts *intermolecular* forces and *intramolecular* forces for water molecules.

Five types of intermolecular forces that can be present in a liquid are considered in this section. They are (1) dipole–dipole interactions, (2) hydrogen bonds, (3) London forces, (4) ion–dipole interactions, and (5) ion–ion interactions.

Dipole–Dipole Interactions

A **dipole–dipole interaction** *is an intermolecular attractive force that occurs between polar molecules.* Polar molecules, it should be recalled, are electrically unsymmetrical (Sec. 7.20); that is, they have a positive end and a negative end. This uneven charge distribution in a polar molecule is called a *dipole;* the molecule has two poles or ends, one pole being more negative than the other, hence the terminology *dipole–dipole interactions* for interactions between polar molecules.

When polar molecules approach each other, they tend to line up so that the relatively positive end of one molecule is directed toward the relatively negative end of the other molecule. As a result, there is an electrostatic attraction between the molecules. The greater the polarity of the molecules, the greater the strength of the dipole–dipole interaction. Figure 11.16 shows the many dipole–dipole interactions possible for a random arrangement of polar ClF molecules.

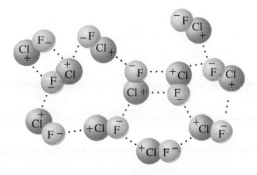

FIGURE 11.16
Dipole–dipole interactions between randomly arranged ClF molecules.

Hydrogen Bonds

Unusually strong dipole–dipole interactions are found among hydrogen-containing molecules in which hydrogen is covalently bonded to a highly electronegative element of small atomic size (fluorine, oxygen, and nitrogen). Two factors account for the extra strength of these dipole–dipole interactions.

1. The highly electronegative element to which hydrogen is covalently bonded dominates the electron-sharing process to such a degree that the hydrogen atom is left with significant partial positive charge (Sec. 7.19).

$$\overset{\delta^+ \quad \delta^-}{H—F} \qquad \overset{\delta^+ \quad \delta^-}{H—O} \qquad \overset{\delta^+ \quad \delta^-}{H—N}$$

The hydrogen atom is essentially a "bare" nucleus because it has no electrons besides the one attracted to the electronegative element—a unique property of hydrogen.

2. The small size of the hydrogen atom allows the bare nucleus to approach closely and be strongly attracted to an unshared pair of electrons on the electronegative atom of another molecule.

Dipole–dipole interactions of the type we are now considering are given a special name, *hydrogen bonds*. A **hydrogen bond** *is an extra strong dipole–dipole interaction involving a hydrogen atom covalently bonded to a small, very electronegative atom (F, O, or N) and an unshared pair of electrons on another small, very electronegative atom (F, O, or N).*

Water is the most commonly encountered substance wherein hydrogen bonding is significant. Figure 11.17 depicts the process of hydrogen bonding among water molecules. Note that the oxygen atom in water can participate in two hydrogen bonds—one involving each of its nonbonding electron pairs and that a given water molecule can participate in four hydrogen bonds (see the "center" water molecule in Figure 11.17).

The effects of hydrogen bonding on the properties of water are significant. It is not an overstatement to say that hydrogen bonding makes life possible. Were it not for hydrogen bonding, water, the most abundant compound in the human body, would be a gas at room temperature. Life as we know it could not exist under such a condition.

The three elements that have significant hydrogen bonding ability are fluorine, oxygen, and nitrogen. They are all very electronegative elements of small atomic size. Chlorine has the same electronegativity as nitrogen, but its larger atomic size causes it to have little hydrogen bonding ability.

FIGURE 11.17 Depiction of hydrogen bonding among water molecules. The dotted lines are the hydrogen bonds.

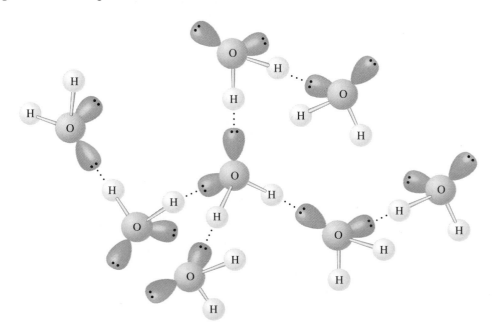

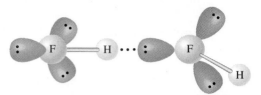

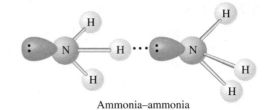

Hydrogen fluoride–hydrogen fluoride   Ammonia–ammonia

FIGURE 11.18
Depiction of hydrogen
bonding between
various simple
molecules. The
dotted lines are the
hydrogen bonds.

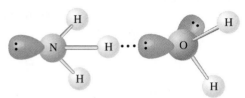

Hydrogen fluoride–water   Ammonia–water

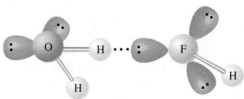

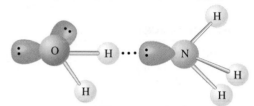

Water–hydrogen fluoride   Water–ammonia

Section 11.17 discusses in detail how water's properties are significantly changed as a result of hydrogen bonding.

  The two molecules that participate in a hydrogen bond need not be identical as was the case with the water molecules whose hydrogen bonding is depicted in Figure 11.17. Hydrogen bond formation is possible whenever two molecules, the same or different, have the following characteristics:

1. One molecule has a hydrogen atom attached by a covalent bond to an atom of nitrogen, oxygen, or fluorine.
2. The other molecule has a nitrogen, oxygen, or fluorine atom present that possesses one or more nonbonding electron pairs.

  Figure 11.18 shows additional examples of hydrogen bonding between simple molecules.

Hydrogen bonding also plays a major role in the behavior of DNA and proteins. Both DNA and proteins contain O—H and N—H bonds, which participate in hydrogen bonding. Such hydrogen bonds have just the right strength that they are capable of being broken and then re-formed with relative ease in a biochemical setting.

EXAMPLE 11.10   **Predicting Whether Molecules Can Participate in Hydrogen Bonding**

Indicate whether hydrogen bonding is possible between two molecules of each of the following substances:

a. Nitrogen trifluoride   **b.** Ethyl alcohol   **c.** Formaldehyde

SOLUTION

a. Hydrogen bonding cannot occur. No hydrogen atoms are present.
b. Hydrogen bonding should occur, because we have an O—H bond and an oxygen atom with unshared electron pairs.
c. Hydrogen bonding cannot occur. We have an oxygen atom with unshared electron pairs, but no N—H, O—H, or F—H bond is present.

▶ **Practice Exercise 11.10**   Indicate whether hydrogen bonding is possible between two molecules of each of the following substances:

a. hydrogen sulfide   b. methyl ethyl ether   c. propyl amine

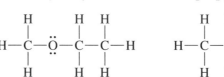

London Forces

Named after the German physicist Fritz London (1900–1954), who first postulated their existence, *London forces* are the weakest type of intermolecular force. A **London force** *is a weak temporary dipole–dipole interaction that occurs between an atom or molecule (polar or nonpolar) and another atom or molecule (polar or nonpolar).* The origin of London forces is more difficult to visualize than that of dipole–dipole interactions.

London forces result from momentary (temporary) uneven electron distributions in molecules. Most of the time electrons can be visualized as being distributed in a molecule in a definite pattern determined by their energies and the electronegativities of the atoms. However, there is a small statistical chance (probability) that the electrons will deviate from their normal pattern. For example, in the case of a nonpolar diatomic molecule, more electron density may temporarily be located on one side of the molecule than on the other. This condition causes the molecule to become polar for an instant. The negative side of this instantaneously polar molecule will tend to repel electrons of adjoining molecules and cause these molecules to also become polar (an *induced polarity*). The original (statistical) polar molecule and all the molecules with induced polarity are then attracted to each other. This happens many, many times per second throughout the liquid, resulting in a net attractive force. Figure 11.19 depicts the situation present when London forces exist.

The strength of London forces depends on the ease with which an electron distribution in a molecule can be distorted (polarized) by the polarity present in another molecule. In large-diameter molecules the outermost electrons are necessarily located farther from the nucleus than are the outermost electrons in small molecules. The farther electrons are located from the nucleus, the weaker the attractive forces that act on them, the more freedom they have, and the more susceptible they are to polarization. This leads to the observation that for *related* molecules, boiling points increase with

As an analogy for London forces, consider what happens when a bucket filled with water is moved. The water will slosh from side to side. This is similar to the movement of electrons. The sloshing from side to side is instantaneous; a given slosh quickly disappears. Uneven electron distribution is likewise an instantaneous situation.

FIGURE 11.19 A London force. The instantaneous (temporary) polarity present in atom A results in induced polarity in atom B.

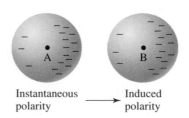

TABLE 11.8 Boiling Point Trends for Related Series of Nonpolar Molecules

Substance	Molecular Mass (amu)	Boiling Point (°C)
Noble Gases		
He	4.0	−269
Ne	20.2	−246
Ar	39.9	−186
Kr	83.8	−153
Xe	131.3	−107
Rn	222	−62
Group VIIA Elements		
F_2	39.0	−187
Cl_2	70.9	−35
Br_2	159.8	+59
I_2	253.8	+184

molecular mass, which usually parallels size. This trend is reflected in the boiling points given in Table 11.8 for two series of related substances: the noble gases and the elements of group VIIA.

If two molecules have approximately the same molecular mass and polarity but are of different diameter, the smaller one will have the lower boiling point because its electrons are less susceptible to polarization. The nonpolar SF_6 and $C_{10}H_{22}$ molecules (molecular masses 146 and 142 amu, respectively) reflect this trend with boiling points of −64 and 174°C, respectively.

Of the three types of intermolecular forces discussed so far (dipole–dipole interactions, hydrogen bonds, and London forces), the London forces are the most common and, in the majority of cases, the most prevalent. They are the only attractive forces present between *nonpolar* molecules, and it is estimated that London forces contribute 85% of the total intermolecular force in the *polar* HCl molecule. Only where hydrogen bonding is involved do London forces play a minor role. In water, for example, about 80% of the intermolecular attraction is attributed to hydrogen bonding and only 20% to London forces.

Ion–Dipole Interactions

An **ion–dipole interaction** *is an intermolecular attractive force between an ion and a polar molecule.* Polar molecules are dipoles; they have a positive end and a negative end. A positive ion (cation) can attract the negative end of dipoles and a negative ion (anion) can attract the positive end of dipoles, as shown in Figure 11.20. Ion–dipole interactions are an important consideration in aqueous solutions in which dissolved ionic compounds are present, a topic to be considered in Chapter 13.

Ion–Ion Interactions

Our discussion of intermolecular forces to this point has assumed that the particles making up a liquid are primarily molecules or atoms. This is a valid assumption for liquids encountered at normal temperatures. However, at extremely high temperatures,

FIGURE 11.20
Ion–dipole interactions. (a) The negative ends of polar molecules are attracted by a positive ion, and (b) the positive ends of polar molecules are attracted by a negative ion. Such attractions are important when ionic compounds dissolve in water, because water is a polar molecule.

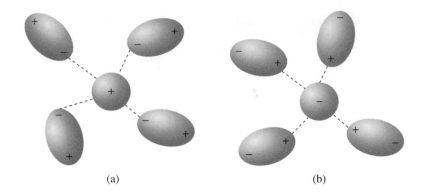

(a)　　　　　　(b)

liquids exist in which no molecules are present. Liquids obtained by melting ionic compounds, which always requires an extremely high temperature, contain only ions. The attractive forces in such liquids are those that result from *ion–ion interactions* among positive and negative ions. An **ion–ion interaction** *is an intermolecular attractive force between oppositely charged ions present in liquid state (molten) ionic compounds.*

11.17 HYDROGEN BONDING AND THE PROPERTIES OF WATER

Water is the most abundant and most essential compound known to human beings. This substance covers approximately 75% of Earth's surface and constitutes 60%–70% of the mass of a human body. In this section we consider four "unusual" hydrogen-bonding-caused properties of water:

1. A lower-than-expected vapor pressure
2. Higher-than-expected thermal properties
3. A seldom-encountered temperature dependence for density
4. A higher-than-expected surface tension

Vapor Pressure

Liquids in which significant hydrogen bonding occurs have vapor pressures that are significantly lower than those of similar liquids where little or no hydrogen bonding occurs. The presence of hydrogen bonds makes it more difficult for molecules to escape from the condensed state; additional energy is needed to overcome the hydrogen bonds. The greater the hydrogen bond strength, the lower the vapor pressure at any given temperature.

Since the boiling point of a liquid depends upon vapor pressure (Sec. 11.15), liquids with low vapor pressures will have to be heated to higher temperatures to bring their vapor pressures up to the point where boiling occurs (atmospheric pressure). Hence, boiling points are much higher for liquids in which hydrogen bonding occurs.

The effect that hydrogen bonding has on water's boiling point can be seen by comparing it with the boiling points of other hydrogen compounds of group VIA elements—H_2S, H_2Se, and H_2Te. In this series of compounds—H_2O, H_2S, H_2Se, and H_2Te—water is the only one in which significant hydrogen bonding occurs (Sec 11.16). Normally, the boiling points of a series of compounds containing elements in the same periodic table group increase with increasing molar mass. Thus, in the hydrogen-group VIA element series, we would expect that H_2Te, the heaviest member of the series, would have the highest boiling point and that water, the compound of lowest molar mass, would have the lowest boiling point. Contrary to expectation, H_2O has the highest boiling point, as can be seen from the boiling-point data shown in Figure 11.21. The data in Figure 11.21 indicate that water "should have" a boiling point of approximately –80°C, a value obtained by extrapolation of the line connecting the three heavier compounds. The

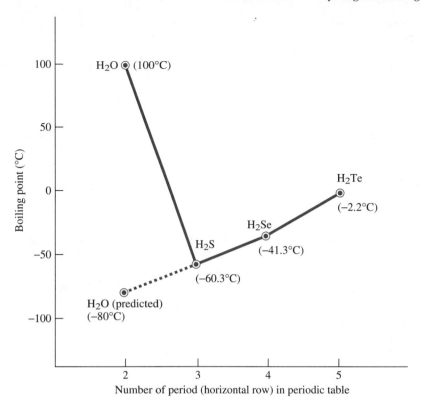

FIGURE 11.21 Boiling points of the hydrogen compounds of group VIA elements. Water is the only one of the four compounds in which significant hydrogen bonding occurs.

actual boiling point of water, 100°C, is nearly 200°C higher than predicted. Indeed, in the absence of hydrogen bonding, water would be a gas at room temperature, as mentioned in Section 11.16.

A higher-than-expected freezing point is also characteristic of water, and a plot of freezing points for similar compounds would have the same general shape as that in Figure 11.21.

Thermal Properties

The extensive hydrogen bonding between water molecules increases significantly the ability of water to absorb heat energy as it warms and then evaporates (endothermic processes) and to release energy when freezing (an exothermic process). Thus, water has higher-than-normal specific heat, heat of vaporization, and heat of solidification values, as can be seen by comparing numerical values for water with those of other substances [see Table 11.2 (specific heat), Table 11.3 (heat of fusion/solidification), and Table 11.4 (heat of vaporization/condensation)]. These higher-than-normal thermal properties for water, together with its abundance, largely account for water's widespread use as a coolant.

Large bodies of water exert a temperature-moderating effect on their surroundings. In the heat of a summer day, extensive water evaporation occurs, and in the process energy is absorbed from the surroundings. The net effect of the evaporation is a lowering of the temperature of the surroundings. In the cool of evening, some of this water vapor condenses back to the liquid state, releasing heat that raises the temperature of the surroundings. In this manner the temperature variation between night and day is reduced. In the winter a similar process occurs; water freezes on cold days and releases heat energy to the surroundings. The hottest and coldest regions on Earth are all inland regions, those distant from the moderating effects of large bodies of water.

The large release of heat from freezing water is the basis for the practice of spraying orange trees with water when freezing temperatures are expected. The freezing water liberates enough heat energy to keep the temperature of the air higher than the freezing point of the fruit.

Water's thermal properties are a factor in the cooling of the human body. The coolant in this case is perspiration, which evaporates, absorbs heat in the process, and lowers skin temperature. For similar reasons, a swimmer upon emerging from the water on a warm but windy day will feel cold and will shiver as excess water rapidly evaporates from his or her body. Additionally, water's ability to absorb large amounts of heat helps keep water loss to a minimum, thus making it easier for humans, animals, and plants to exist in environments where water is scarce.

Density

A very striking and unusual behavior pattern occurs in the variation of the density of water with temperature. For most liquids, density increases with decreasing temperature, reaching a maximum for the liquid at its freezing point. The density pattern for water is different. Maximum density is reached not at its freezing point, but at a temperature a few degrees higher than the freezing point. As shown in Figure 11.22 the maximum density for liquid water occurs at 4°C. This abnormality, that water at its freezing point is less dense than water at slightly higher temperatures, has tremendous ecological significance. Furthermore, at 0°C, solid water (ice), is significantly less dense than liquid water—0.9170 g/mL versus 0.9998 g/mL. All of this "strange" density behavior of water is directly related to hydrogen bonding.

Of primary importance is the fact that hydrogen bonding between water molecules is directional. Hydrogen bonds can form only at certain angles between molecules because of the angular geometry of the water molecule. The net result is that water molecules that are hydrogen bonded are farther apart than those that are not.

From a molecular viewpoint, let us now consider what happens to water molecules as the temperature of water is lowered. At high temperatures, such as 80 °C, the kinetic energy of the water molecules is sufficiently great to prevent hydrogen bonding from having much of an orientation effect on the molecules. Hydrogen bonds are rapidly and continually being formed and broken. As the temperature is lowered, the accompanying decrease in kinetic energy decreases molecular motion and the molecules move closer together. This results in an increase in density. The kinetic energy is still sufficient to negate most of the orientation effects from hydrogen bonding. When the temperature is lowered still further, the kinetic energy finally becomes insufficient to prevent hydrogen bonding from orienting molecules into definite patterns that require "open spaces"

FIGURE 11.22
A plot of density versus temperature for liquid water.

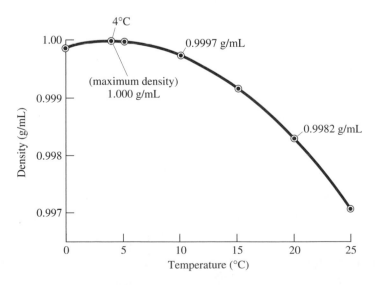

between molecules. At 4°C, the temperature at which water has its maximum density, the trade-off between random motion from kinetic energy and orientation from hydrogen bonding is such that the molecules are as close together as they will ever be. Cooling below 4°C causes a decrease in density as hydrogen bonding causes more and more "open spaces" to be present in the liquid. Density decrease continues down to the freezing point, at which temperature the hydrogen bonding causes molecular orientation to the maximum degree, and the solid crystal lattice of ice is formed. This solid crystal lattice of ice (Fig. 11.23) is an open structure, with the result that solid ice has a lower density than liquid water. Water is one of only a few substances known whose solid phase is less dense than the liquid phase at the freezing point.

Two important consequences of ice being less dense than water are that ice floats in liquid water and liquid water expands upon freezing.

When ice floats in liquid water, approximately 8% of its volume is above water. In the case of icebergs found in far northern locations, 92% of the volume of an iceberg is located below the liquid surface. The common expression "it is only the tip of the iceberg" is literally true; what is seen is only a very small part of what is actually there.

If a container is filled with water and sealed, the force generated from the expansion of the water upon freezing will break the container. Antifreeze is added to car radiators in the winter to prevent the water present from freezing and cracking the engine block; sufficient force is generated from water expansion upon freezing to burst even iron or copper parts. During the winter season, also, the weathering of rocks and concrete and the formation of "potholes" in the streets are hastened by the expansion of freezing water in cracks.

Water's density pattern also explains why lakes freeze from top to bottom and not vice versa, and why aquatic life can continue to exist for extended periods of time in bodies of water that are "frozen over." In the fall of the year, surface water is cooled through contact with cold air. This water becomes denser than the warmer water underneath and sinks. In this way, cool water is circulated from the top of a lake to the bottom until the entire lake has reached the temperature of water's maximum density, 4°C. During the circulation process, oxygen and nutrients are distributed throughout the water. Upon further cooling, below 4°C, a new behavior pattern emerges. Surface water, upon cooling, no longer sinks, since it is less dense than the water underneath. Eventually a thin layer of surface water is cooled to the freezing point and changed to ice, which floats because of its still lower density. Even in the coldest winters, lakes will usually not freeze to a depth of more than a few feet because the ice forms an insulating layer over the water. Thus, aquatic life can live throughout the winter, under the ice, in water that is "thermally insulated" and contains nutrients. If water behaved as "normal" substances do, freezing would occur from the bottom up, and most, if not all, aquatic life would be destroyed.

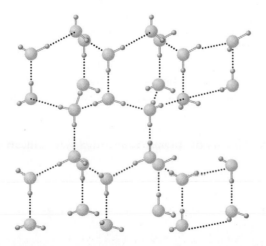

FIGURE 11.23 The crystalline structure of ice. Normal covalent bonds between oxygen and hydrogen, which hold water molecules together, are shown by solid short lines. The weaker hydrogen bonds are shown by dotted lines.

FIGURE 11.24
Molecules on the
surface of a liquid
experience an
imbalance in
intermolecular forces.
Molecules within the
liquid are uniformly
attracted in
all directions.

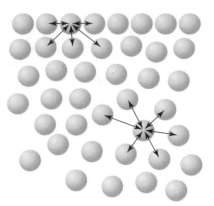

For a few weeks in the early spring, circulation again occurs in a body of water during the melting and warming of the surface water. This process stops when the surface water is warmed above 4°C. Above this temperature surface water is less dense and forms a layer over the colder water of higher density. This causes a thermal stratification that persists through the hot summer months. In the fall, circulation again begins, and the cycle is repeated.

Surface Tension

Surface tension is a property of liquids that is directly related to intermolecular forces. Molecules *within* a liquid, because they are completely surrounded by other molecules, experience equal intermolecular attractions in all directions. Molecules *on the surface* of a liquid, because there are no molecules above them, experience an imbalance in intermolecular attractions that pulls them toward the interior of the liquid. Figure 11.24 illustrates this unbalance of force. **Surface tension** *is a measure of the inward force on the surface of a liquid caused by unbalanced intermolecular forces.*

The strong attraction of water molecules for each other as a result of hydrogen bonding gives water a very high surface tension. This high surface tension causes water surfaces to seem to have a "membrane" or "skin" covering them. A steel needle or a razor blade will float on the surface of water because of surface tension even though each has a density greater than that of water. Water bugs are able to skitter across water because of surface tension.

Because of surface tension it is possible to fill a container with water slightly above the rim (see Fig. 11.25). Surface tension prevents the water from running over. It is also surface tension that causes water to form droplets in many situations. Almost everyone has observed water droplets that have formed on a car windshield or on greasy or waxed surfaces. Surface tension causes water to resist increasing its surface area. For a given volume of any liquid, the geometrical shape having the minimum surface area is a sphere; hence, water tends to form spherical droplets.

The property of surface tension helps water to rise in the narrow vessels in plant stems and roots. High surface tension also helps to hold water in the small spaces between soil particles.

The surface tension of a liquid can be "broken" by dissolved substances. Soaps and detergents are used for this purpose. Water bugs will sink in soapy water. Soapy water spreads out over a glass surface instead of forming droplets as pure water would.

FIGURE 11.25 Surface tension enables one to fill a glass with water above the rim. (*Eric Schrader–Pearson Science*)

Summary

1. **Physical State Differences** The physical state of a substance is determined by (1) its chemical identity, (2) its temperature, and (3) the pressure to which it is subjected. Distinguishing properties of the physical states include (1) definiteness of volume and shape, (2) relative density, (3) relative compressibility, and (4) relative thermal expansion.

2. **Kinetic Molecular Theory of Matter** The kinetic molecular theory of matter is a set of five statements that explains the physical behavior of the three states of matter. Two premises of this theory are that the particles (atoms, molecules, or ions) present in a substance (1) are in constant motion and therefore possess kinetic energy and (2) interact with each other through attractions and repulsions and therefore possess potential energy.

3. **Kinetic and Potential Energy** Kinetic energy is energy that matter possesses because of the movement (motion) of its particles. Potential energy is energy that matter possesses as a result of its position, condition, and/or composition. The solid state is characterized by a dominance of potential energy over kinetic energy. The liquid state is characterized by potential energy and kinetic energy of about the same magnitude. The gaseous state is characterized by a complete dominance of kinetic energy over potential energy.

4. **Endothermic and Exothermic Changes of State** Endothermic changes of state, which require the input of heat energy, are melting, sublimation, and evaporation. Exothermic changes of state, which require the release of heat energy, are deposition, condensation, and freezing.

5. **Specific Heat and Heat Capacity** The specific heat of a substance is the amount of heat energy that is necessary to raise the temperature of 1 gram of the substance in a specific physical state by 1°C. Heat capacity is the amount of heat energy needed to raise the temperature of a given quantity of a substance in a specific physical state by 1°C.

6. **Temperature Changes as a Substance Is Heated** As a solid is heated, its temperature rises until the melting point is reached, and then it remains constant during the change of state from solid to liquid. Heating the resultant liquid causes the temperature to rise again until the boiling point is reached. The temperature then remains constant during the change of state from liquid to gas. Heating the resultant gas again causes the temperature to rise.

7. **Heat Energy Calculations** Important physical quantities used when calculating the heat energy added or removed from a chemical system are (1) heat of fusion/solidification, (2) heat of vaporization/condensation, and (3) specific heat for each of the three physical states.

8. **Evaporation** The evaporation rate for a liquid depends on the liquid's temperature and on its surface area. The evaporative behavior of a liquid in a closed container is dependent on both the rate of evaporation and rate of condensation of the liquid. With time, a steady-state situation (an equilibrium state) is reached in which the rates of these two opposite processes become equal.

9. **Vapor Pressure** The pressure exerted by a vapor in equilibrium with its liquid is the vapor pressure of the liquid. Vapor pressure increases as liquid temperature increases.

10. **Boiling and Boiling Points** Boiling is a special form of evaporation in which bubbles of vapor form within the liquid and rise to the surface. The boiling point of a liquid is the temperature at which the vapor pressure of the liquid becomes equal to the external (atmospheric) pressure exerted on the liquid. The boiling point of a liquid increases or decreases as the prevailing atmospheric pressure increases or decreases.

11. **Intermolecular Forces** Intermolecular forces are the attractive interactions between the particles (molecules, ions, atoms) present in a chemical system. In the liquid state, five different types of intermolecular forces can be present, depending on the composition of the liquid: (1) dipole–dipole interactions, (2) hydrogen bonds, (3) London forces, (4) ion–dipole interactions, and (5) ion–ion interactions.

12. **Properties of Water** Because of hydrogen bonding the substance water has (1) a lower-than-expected vapor pressure and a higher-than-expected boiling point; (2) higher-than-expected specific heat, heat of vaporization, and heat of solidification; (3) a seldom-encountered temperature dependence for density resulting in the solid state being less dense than the liquid state; and (4) a higher-than-expected surface tension.

Key Terms

The new terms defined in this chapter are

boiling *Sec. 11.15*
boiling point *Sec. 11.15*
change of state *Sec. 11.8*
compressibility *Sec. 11.2*
dipole–dipole interaction
  Sec. 11.16
electrostatic force *Sec. 11.3*
endothermic change of
  state *Sec. 11.8*
equilibrium state *Sec. 11.13*
evaporation *Sec. 11.13*

exothermic change of
  state *Sec. 11.8*
gas *Sec. 11.6*
heat capacity *Sec. 11.9*
heat of condensation
  Sec. 11.11
heat of fusion *Sec. 11.11*
heat of solidification
  Sec. 11.11
heat of vaporization
  Sec. 11.11
hydrogen bond *Sec. 11.16*

intermolecular force
  Sec. 11.16
ion–dipole interaction
  Sec. 11.16
ion–ion interaction
  Sec. 11.16
joule *Sec. 11.9*
kinetic energy *Sec. 11.3*
kinetic molecular theory
  of matter *Sec. 11.3*
liquid *Sec. 11.5*
London force *Sec. 11.16*

normal boiling point
  Sec. 11.15
potential energy *Sec. 11.3*
solid *Sec. 11.4*
specific heat *Sec. 11.9*
surface tension *Sec. 11.17*
thermal expansion
  Sec. 11.2
vapor *Sec. 11.13*
vapor pressure *Sec. 11.14*
volatile substance
  Sec. 11.14

Practice Problems

PHYSICAL STATES OF MATTER (SECS. 11.1 AND 11.2)

11.1   The following statements relate to the terms *solid state, liquid state,* and *gaseous state.* Match the terms to the appropriate statements.
  a. This state is characterized by the lowest density of the three.
  b. This state is characterized by an indefinite shape and a definite volume.
  c. Temperature changes influence the volume of this state significantly.
  d. In this state, constituent particles are more free to move about than in other states.

11.2   The following statements relate to the terms *solid state, liquid state,* and *gaseous state.* Match the terms to the appropriate statements.
  a. This state is characterized by an indefinite shape and high density.
  b. Pressure changes influence the volume of this state more than that of the other two.
  c. In this state, constituent particles are less free to move about than in other states.
  d. This state is characterized by a definite shape and a definite volume.

11.3   For which of the following pairs of elements do both members of the pair have the same physical state (solid, liquid, gas) at room temperature and pressure?
  a. chlorine and oxygen
  b. mercury and silver
  c. krypton and neon
  d. phosphorus and aluminum

11.4   For which of the following pairs of elements do both members of the pair have the same physical state (solid, liquid, gas) at room temperature and pressure?
  a. fluorine and iodine
  b. nitrogen and hydrogen
  c. sodium and sulfur
  d. bromine and mercury

11.5   How many of the "highlighted" elements in the following periodic table have the listed physical state characterizations?
  a. gases at room temperature and pressure
  b. liquids at room temperature and pressure

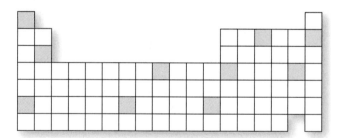

11.6   How many of the "highlighted" elements in the periodic table in Problem 11.5 have the listed physical state characterizations?
  a. solids at room temperature and pressure
  b. liquids at temperatures reached on a hot summer day

KINETIC MOLECULAR THEORY OF MATTER (SECS. 11.3–11.6)

11.7   Identify the principal *type* of energy (kinetic or potential) that is exhibited by each of the following.
  a. a car parked on a hill
  b. a car traveling 65 mi/hr on a level road
  c. an elevator stopped at the 35th floor
  d. water behind a dam

11.8 Identify the principal *type* of energy (kinetic or potential) that is exhibited by each of the following.
a. a piece of coal
b. a falling rock
c. a compressed metal spring
d. a rolling soccer ball on a level field

11.9 Using kinetic molecular theory concepts, answer the following questions.
a. What is the relationship between temperature and the average velocity at which particles move?
b. What type of energy is related to cohesive forces?
c. What effect does temperature have on the magnitude of disruptive forces?
d. In which of the three states of matter are disruptive forces present?

11.10 Using kinetic molecular theory concepts, answer the following questions.
a. How do molecules transfer energy from one to another?
b. What type of energy is related to disruptive forces?
c. What effect does temperature have on the magnitude of cohesive forces?
d. In which of the three states of matter are cohesive forces present?

11.11 Classify each of the following as a description of the solid, liquid, or gaseous state.
a. Cohesive forces dominate over disruptive forces.
b. Neither potential energy nor kinetic energy dominates.
c. Potential energy dominates over kinetic energy.
d. There is very small thermal expansion and small compressibility.

11.12 Classify each of the following as a description of the solid, liquid, or gaseous state.
a. Disruptive forces dominate over cohesive forces.
b. Kinetic energy dominates over potential energy.
c. Neither disruptive forces nor cohesive forces dominate.
d. There is large compressibility and moderate thermal expansion.

11.13 Explain each of the following observations, using the kinetic molecular theory.
a. Particles in a solid occupy essentially fixed positions.
b. The compressibility of gases is much greater than that of liquids and solids.
c. Liquids show little change in volume with changes in temperature.
d. In gases, particles are free to move about in all directions.

11.14 Explain each of the following observations, using the kinetic molecular theory.
a. A gas always exerts a pressure on the object or container with which it is in contact.
b. The particles in both liquids and solids essentially touch each other.
c. An aerosol can heated in an open fire may explode.
d. Solids maintain characteristic shapes.

11.15 Identify each of the following as a characteristic of the *solid*, *liquid*, or *gaseous* state. More than one choice may be correct for some characterizations.
a. definite volume and indefinite shape
b. indefinite shape
c. definite shape

11.16 Identify each of the following as a characteristic of the *solid*, *liquid*, or *gaseous* state. More than one choice may be correct for some characterizations.
a. indefinite volume and indefinite shape
b. definite volume
c. indefinite volume

PHYSICAL CHANGES OF STATE (SEC. 11.8)

11.17 In which of the following pairs of physical state changes are both members of the pair of the same thermicity (both exothermic or both endothermic)?
a. evaporation and freezing
b. melting and deposition
c. freezing and condensation
d. sublimation and evaporation

11.18 In which of the following pairs of physical state changes are both members of the pair of the same thermicity (both exothermic or both endothermic)?
a. sublimation and condensation
b. freezing and deposition
c. melting and evaporation
d. deposition and condensation

11.19 In which of the pairs of state changes in Problem 11.17 is the final state (solid, liquid, gas) the same for both members of the pair?

11.20 In which of the pairs of state changes in Problem 11.18 is the final state (solid, liquid, gas) the same for both members of the pair?

11.21 In which of the pairs of state changes in Problem 11.17 are the two members of the pair opposite changes?

11.22 In which of the pairs of state changes in Problem 11.18 are the two members of the pair opposite changes?

HEAT ENERGY AND SPECIFIC HEAT (SEC. 11.9)

11.23 Which quantity of heat energy in each of the following pairs of heat energy values is the larger?
a. 2.0 joules or 2.0 calories
b. 1.0 kilocalorie or 92 calories
c. 100 Calories or 100 calories
d. 2.3 Calories or 1000 kilocalories

11.24 Which quantity of heat energy in each of the following pairs of heat energy values is the larger?
a. 6.0 Joules or 20.0 calories
b. 1.0 kilocalorie or 785 calories
c. 10 Calories or 1000 calories
d. 4.4 Calories or 2.5 kilocalories

11.25 The energy content of a chicken sandwich is found to be 2290 kJ. Specify this amount of energy in
a. joules.   b. kilocalories.
c. calories.   d. Calories.

11.26 The energy content of a quarter-pound hamburger is found to be 211,000 J. Specify this amount of energy in
a. kilojoules.   b. kilocalories.
c. calories.   d. Calories.

11.27 Calculate the number of joules of heat energy required to change the temperature of 40.0 g of each of the following substances from 25.0°C to 35.0°C.
a. copper   b. iron
c. nitrogen   d. ethyl alcohol

11.28 Calculate the number of joules of heat energy required to change the temperature of 30.0 g of each of the following substances from 32.0°C to 44.0°C.
a. silver   b. gold
c. oxygen   d. liquid water

11.29 Calculate the final temperature of a 7.73 g sample of aluminum metal, originally at a temperature of 43.2°C, after each of the following heat energy changes occur.
a. 145 J of heat energy is added.
b. 145 J of heat energy is removed.
c. 325 J of heat energy is added.
d. 525 J of heat energy is removed.

11.30 Calculate the final temperature of a 6.44 g sample of zinc metal, originally at a temperature of 33.4°C, after each of the following heat energy changes occur.
a. 163 J of heat energy is added.
b. 163 J of heat energy is removed.
c. 425 J of heat energy is added.
d. 273 J of heat energy is removed.

11.31 What is the mass, in grams, of a sample of each of the following substances if its temperature changes from 20.0°C to 95°C when it absorbs 371 J of heat energy?
a. iron   b. zinc
c. liquid water   d. ethyl alcohol

11.32 What is the mass, in grams, of a sample of each of the following substances if its temperature changes from 30.0°C to 85°C when it absorbs 402 J of heat energy?
a. copper   b. gold
c. liquid water   d. sodium chloride

11.33 Calculate the specific heat of a substance, if each of the following amounts of heat energy is required to raise the temperature of 40.0 g of the substance by 3.0°C.
a. 39.8 J   b. 46.9 J   c. 57.8 J   d. 75.0 J

11.34 Calculate the specific heat of a substance, if each of the following amounts of heat energy is required to raise the temperature of 60.0 g of the substance by 5.0°C.
a. 33.8 J   b. 48.7 J   c. 61.3 g   d. 83.0 J

11.35 Based on the data in Table 11.2, calculate the heat capacity of 40.0 g of each of the following substances.
a. zinc   b. aluminum
c. water (liquid) d. sodium chloride

11.36 Based on the data in Table 11.2, calculate the heat capacity of 50.0 g of each of the following substances.
a. iron   b. silver
c. water (ice)   d. ethyl alcohol

11.37 Which has the higher heat capacity, 40.0 g of gold or 80.0 g of copper?

11.38 Which has the higher heat capacity, 20.0 g of liquid water or 30.0 g of ethyl alcohol?

11.39 The heat capacity of 75.1 g of a metal is 28.7 J/°C. What is the specific heat, in J/g · °C, for this metal?

11.40 The heat capacity of 4.71 g of a metal is 0.60 J/°C. What is the specific heat, in J/g · °C, for this metal?

11.41 How many grams of copper can be heated from 30°C to 50°C when 1.0 g of liquid water cools from 100°C to 15°C?

11.42 How many grams of aluminum can be heated from 20°C to 45°C when 20.0 g of oxygen gas cools from 285°C to 15°C?

ENERGY AND CHANGES OF STATE (SEC. 11.11)

11.43 Which one of the quantities *heat of fusion, heat of solidification, heat of vaporization*, or *heat of condensation* is needed to calculate how much energy is absorbed or released during each of the following changes? (Do not actually carry out the energy calculation.)
a. changing of 50.0 g of molten aluminum at 658°C (its melting point) to solid aluminum at the same temperature
b. changing of 50.0 g of steam at 100°C to liquid water at 100°C
c. changing of 50.0 g of solid copper at its melting point (1083°C) to molten copper at the same temperature
d. changing of 50.0 g of liquid water at 100°C to steam at 100°C

11.44 Which one of the quantities *heat of fusion, heat of solidification, heat of vaporization*, or *heat of condensation* is needed to calculate how much energy is absorbed or released during each of the following changes? (Do not actually carry out the energy calculation.)
a. changing of 35.0 g of liquid water at its boiling point to steam at the same temperature
b. changing of 35.0 g of molten copper at 1083°C (its melting point) to solid copper at 1083°C
c. changing of 35.0 g of steam at 100°C to liquid water at the same temperature
d. changing of 35.0 g of solid aluminum at its melting point (658°C) to molten aluminum at 658°C

11.45 Calculate how much heat energy, in joules, would be absorbed or evolved in each of the changes listed in Problem 11.43.

11.46 Calculate how much heat energy, in joules, would be absorbed or evolved in each of the changes listed in Problem 11.44.

11.47 If 3075 J of energy are absorbed in changing 15.0 g of solid Cu at its melting point into molten (liquid) copper at the same temperature, how much energy, in joules, is released when 15.0 g of molten copper at its freezing point are changed to solid copper at the same temperature?

11.48 If 34,500 J of energy are evolved in changing 25.0 g of gaseous ammonia at its boiling point temperature to the liquid state at the same temperature, how much energy, in joules, is needed to change 25.0 g of liquid ammonia at its boiling point to gaseous ammonia at the same temperature?

11.49 The heats of fusion of sodium hydroxide and sodium sulfate, respectively, are 15.79 kJ/mole and 80.93 kJ/mole. How many times greater is the energy input needed to melt 1.00 mole of sodium sulfate than 1.00 mole of sodium hydroxide at their melting points?

11.50 The heats of fusion of sodium thiosulfate and sodium carbonate, respectively, are 49.75 kJ/mole and 71.88 kJ/mole. How many times greater is the energy input needed to melt 1.00 mole of sodium carbonate than 1.00 mole of sodium thiosulfate at their melting points?

11.51 Based on the data in Table 11.4, calculate how many moles of gaseous methane at its boiling point must be condensed to the liquid state at the same temperature in order to release 432 kJ of heat energy.

11.52 Based on the data in Table 11.4, calculate how many moles of gaseous diethyl ether at its boiling point must be condensed to the liquid state at the same temperature in order to release 327 kJ of heat energy.

11.53 Experiments involving two samples of the same substance, each at its melting point temperature, are carried out with the following results.
  1. The amount of heat energy required to melt 6.21 g of the substance is 1273 J.
  2. The amount of heat energy required to melt 4.46 g of the substance is 914 J.
  With the help of Table 11.3, determine the chemical identity of the substance.

11.54 Experiments involving two samples of the same substance, each at its melting point temperature, are carried out with the following results.
  1. The amount of heat energy required to melt 5.33 g of the substance is 2095 J.
  2. The amount of heat energy required to melt 2.22 g of the substance is 872 J.
  With the help of Table 11.3, determine the chemical identity of the substance.

11.55 The heat of fusion of Na at its melting point is 2.40 kJ/mole. How much heat, in joules, must be absorbed by 7.00 g of solid Na at its melting point to convert it to molten Na?

11.56 The heat of vaporization of Hg at its boiling point is 58.6 kJ/mole. How much heat, in joules, must be absorbed by 13.0 g of liquid Hg at its boiling point to convert it to gaseous Hg?

HEAT ENERGY CALCULATIONS (SEC. 11.12)

11.57 Draw the general shape of the temperature–energy graph (heating curve; Figure 11.7) for carbon tetrachloride from –100°C to 100°C. The melting and boiling points of carbon tetrachloride are, respectively, –23°C and 77°C.

11.58 Draw the general shape of the temperature–energy graph (heating curve; Figure 11.7) for benzene from –50°C to 150°C. The melting and boiling points of benzene are, respectively, 6°C and 80°C.

11.59 Calculate the amount of heat energy needed, in joules, to convert 52.0 g of ice at –18°C to each of the following conditions of state and temperature.
  a. ice at –3°C   **b.** water at 37°C
  c. water at 100°C   **d.** steam at 115°C

11.60 Calculate the amount of heat energy needed, in joules, to convert 87.0 g of ice at –28°C to each of the following conditions of state and temperature.
  a. ice at 0°C   **b.** water at 0°C
  c. steam at 100°C   **d.** steam at 120°C

11.61 The melting point of iron is 1530°C, its solid-state specific heat is 0.444 J/(g · °C), and its heat of fusion is 247 J/g. How much heat energy, in joules, is required to melt 35.2 g of iron that is at each of the following initial temperatures?
  a. its melting point
  b. 25C° below its melting point
  c. 1120°C
  d. 27°C

11.62 The melting point of calcium is 851°C, its solid-state specific heat is 0.632 J/(g · °C), and its heat of fusion is 233 J/g. How much heat energy, in joules, is required to melt 52.0 g of calcium that is at each of the following initial temperatures?
  a. its melting point
  b. 42C° below its melting point
  c. 657°C
  d. 27°C

11.63 Calculate the heat required to convert 15.0 g of ethyl alcohol, C_2H_6O, from a solid at –135°C into the gaseous state at 95°C. The normal melting and boiling points of this substance are –117°C and 78°C, respectively. The heat of fusion is 109 J/g, and the heat of vaporization is 837 J/g. The specific heats of the solid, liquid, and gaseous states are 0.97, 2.3, and 0.95 J/(g · °C).

11.64 Calculate the heat required to convert 25.0 g of propyl alcohol, C_3H_8O, from a solid at –150°C into the gaseous state at 115°C. The normal melting and boiling points of this substance are –127°C and 97°C, respectively. The heat of fusion is 86.2 J/g, and the heat of vaporization is 694 J/g. The specific heats of the solid, liquid, and gaseous states are 2.36, 2.83, and 1.76 J/(g · °C).

PROPERTIES OF LIQUIDS (SECS. 11.13–11.15)

11.65 Match the following statements to the appropriate term: *vapor, vapor pressure, volatile, boiling, boiling point.*
  a. This is a temperature at which the liquid vapor pressure is equal to the external pressure on a liquid.
  b. This property can be measured by allowing a liquid to evaporate in a closed container.

c. In this process, bubbles of vapor form within a liquid.

d. This temperature changes with changes in atmospheric pressure.

11.66 Match the following statements to the appropriate term: *vapor, vapor pressure, volatile, boiling, boiling point.*

a. This state involves gaseous molecules of a substance at a temperature and pressure where we would ordinarily expect the substance to be a liquid.

b. A substance that readily evaporates at room temperature because of a high vapor pressure has this property.

c. This process is a special form of evaporation.

d. This property always increases in magnitude with increasing temperature.

11.67 Offer a concise, clear explanation for each of the following observations.

a. Increasing the temperature of a liquid increases its vapor pressure.

b. It takes more time to cook an egg in boiling water on a mountain top than at sea level.

c. Food cooks faster in a pressure cooker than in an open pan.

d. Evaporation is a cooling process.

11.68 Offer a concise, clear explanation for each of the following observations.

a. Liquids do not all have the same vapor pressure at a given temperature.

b. The boiling point of a liquid varies with atmospheric pressure.

c. A person emerging from an outdoor swimming pool on a breezy warm day gets the shivers.

d. Food will cook just as fast in boiling water with the stove set at low heat as in boiling water at high heat.

11.69 What effect (increase, decrease, or no change) will each of the following changes have on the *rate of evaporation* of a liquid in an open cylindrical container?

a. increasing the temperature of the liquid by 10°C at constant pressure

b. moving the container to a higher elevation (altitude) at constant temperature

c. transferring the liquid, at constant temperature and pressure, to a new container whose dimensions are double those of the old container

d. doubling the amount of liquid in the container at constant temperature and pressure

11.70 What effect (increase, decrease, or no change) will each of the following changes have on the *rate of evaporation* of a liquid in an open cylindrical container?

a. decreasing the external pressure on the surface of the liquid at constant temperature

b. transferring the liquid, at constant temperature and pressure, to a new container that doubles the surface area for the liquid

c. decreasing, at constant temperature and pressure, the amount of liquid in the container

d. decreasing the temperature of the liquid by 20°C at constant pressure

11.71 What effect (increase, decrease, or no change) will each of the changes in Problem 11.69 have on the *magnitude of the boiling point* of a liquid in an open container?

11.72 What effect (increase, decrease, or no change) will each of the changes in Problem 11.70 have on the *magnitude of the boiling point* of a liquid in an open container?

11.73 What effect (increase, decrease, or no change) will each of the changes in Problem 11.69 have on the *magnitude of the vapor pressure* of a liquid in a closed, rigid container?

11.74 What effect (increase, decrease, or no change) will each of the changes in Problem 11.70 have on the *magnitude of the vapor pressure* of a liquid in a closed, rigid container?

11.75 Identical amounts of liquids A and B are placed in *identical* open containers on a tabletop. Liquid B evaporates at a faster rate than liquid A even though both liquids are at the same temperature. Explain why this could be so.

11.76 Two liquid state samples of the same substance, of identical volume, are placed in open cylindrical containers of different diameters on a tabletop. The liquid in one container evaporates faster than the liquid in the other container even though both liquids are at the same temperature. Explain why this could be so.

11.77 Given that the vapor pressures, at 25°C, of CS_2 and CCl_4 are, respectively, 309 and 107 mm Hg, which substance is more volatile? Explain your answer.

11.78 Given that the vapor pressures, at 25°C, of CS_2 and CCl_4 are, respectively, 309 and 107 mm Hg, which substance would you predict to have the lower boiling point? Explain your answer.

INTERMOLECULAR FORCES IN LIQUIDS (SEC. 11.16)

11.79 Describe the molecular conditions necessary for the existence of a dipole–dipole interaction.

11.80 Describe the molecular conditions necessary for the existence of a London force.

11.81 In liquids, what is the relationship between boiling point and the strength of intermolecular forces?

11.82 In liquids, what is the relationship between vapor pressure magnitude and the strength of intermolecular forces?

11.83 For liquid state samples of the following substances, classify the dominant intermolecular forces present as London forces, dipole–dipole interactions, or hydrogen bonds.

a. N_2 b. He   c. HCl   d. Cl_2

11.84 For liquid state samples of the following substances, classify the dominant intermolecular forces present as London forces, dipole–dipole interactions, or hydrogen bonds.

a. H_2 b. HBr   c. Ne   d. ClF

11.85 In which of the following substances, in the pure liquid state, would hydrogen bonding occur?

a.

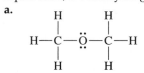

b.

H—C—N̈—H with H below C, H below and to right of N

c.

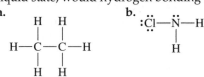

d. H—Ï̈:

11.86 In which of the following substances, in the pure liquid state, would hydrogen bonding occur?

a.

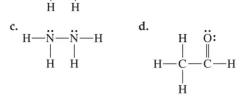

b. :C̈l—N̈—H with H below N

c. H—N̈—N̈—H with H below each N

d. H—C—C—H with Ö: double bonded above right C, and H below C

11.87 In each of the following pairs of substances, predict which member of the pair would be expected to have the higher boiling point, and indicate why.
a. F_2 and H_2 (both are nonpolar substances)
b. HF and HBr (both are polar substances)
c. CO and N_2 (CO is polar and N_2 is nonpolar)
d. C_2H_6 and O_2 (both are nonpolar substances)

11.88 In each of the following pairs of substances, predict which member of the pair would be expected to have the higher boiling point, and indicate why.
a. Cl_2 and Br_2 (both are nonpolar substances)
b. O_2 and CO (O_2 is nonpolar and CO is polar)
c. H_2O and H_2S (both are polar substances)
d. C_4H_{10} and CO_2 (both are nonpolar substances)

11.89 In each of the following pairs of molecules, predict which member of the pair would be expected to have the greatest polarizability.
a. CH_4 and SiH_4 **b.** SiH_4 and $SiCl_4$
c. $SiCl_4$ and $GeBr_4$ **d.** N_2 and C_2H_4

11.90 In each of the following pairs of molecules, predict which member of the pair would be expected to have the greatest polarizability.
a. SiH_4 and GeH_4 **b.** $GeCl_4$ and $GeBr_4$
c. F_2 and Cl_2 **d.** CO_2 and C_3H_8

HYDROGEN BONDING AND THE PROPERTIES OF WATER (SEC 11.17)

11.91 Explain, in terms of hydrogen bonding, why water has a higher-than-expected boiling point.

11.92 What would be the approximate boiling point of water if hydrogen bonding effects were not present?

11.93 Explain why the maximum density of water occurs at a temperature that is higher than its freezing point.

11.94 What is the "density behavior" pattern of a sample of water as its temperature is lowered from 10°C to 0°C?

11.95 What structural feature of ice causes ice to be less dense than liquid water at water's freezing point?

11.96 What would be the consequences, to humans, of solid ice being more dense than liquid water?

11.97 Explain why large bodies of water have a moderating effect on the climate of the surrounding area.

11.98 Explain why the formation of "potholes" in streets is hastened during the winter season.

11.99 Define the property of liquids called surface tension.

11.100 What is the relationship between intermolecular force strength and the property of surface tension?

ADDITIONAL PROBLEMS

11.101 The vapor pressure of $SnCl_4$ reaches 400 mm Hg at 92°C. The vapor pressure of SnI_4 reaches 400 mm Hg at 315°C.
a. At 100°C, which substance should evaporate at the faster rate?
b. Which substance should have the higher normal boiling point?
c. Which substance should have the stronger intermolecular forces?
d. At 80°C, which substance should have the lower vapor pressure?

11.102 The vapor pressure of PBr_3 reaches 400 mm Hg at 150°C. The vapor pressure of PCl_3 reaches 400 mm Hg at 57°C.
a. At 100°C, which substance should evaporate at the faster rate?
b. Which substance should have the lower normal boiling point?
c. Which substance should have the weaker intermolecular forces?
d. At 50°C, which substance should have the higher vapor pressure?

11.103 If heat is supplied at an identical, constant rate in each case, which would take longer, heating 50.0 g of water from 0.0 to 80.0°C or vaporizing 50.0 g of water at 100.0°C?

11.104 If heat is supplied at an identical, constant rate in each case, which would take longer, heating 80.0 g of water from 20.0 to 100.0°C or melting 80.0 g of ice at 0.0°C?

11.105 A quantity of ice at 0.0°C was added to 40.0 g of water at 19.0°C in an insulated container. All of the ice melted, and the water temperature decreased to 0.0°C. How many grams of ice were added?

11.106 A quantity of ice at 0.0°C was added to 55.0 g of water at 25.3°C in an insulated container. All of the ice melted, and the water temperature decreased to 0.0°C. How many grams of ice were added?

11.107 If 10.0 g of ice at –10.0°C and 30.0 g of liquid water at 80.0°C are mixed in an insulated container, what will the final temperature of the liquid be?

11.108 If 10.0 g of ice at –20.0°C and 60.0 g of liquid water at 70.0°C are mixed in an insulated container, what will the final temperature of the liquid be?

11.109 A 500.0 g piece of metal at 50.0°C is placed in 100.0 g of water at 10.0°C in an insulated container. The metal and water come to the same temperature of 23.4°C. What is the specific heat, in joules per gram per degree Celsius, of the metal?

11.110 A 100.0 g piece of metal at 80.0°C is placed in 200.0 g of water at 20.0°C in an insulated container. The metal and water come to the same temperature of 33.6°C. What is the specific heat, in joules per gram per degree Celsius, of the metal?

11.111 Heat is added to a 25.3 g sample of a metal at the constant rate of 136 J/sec. After the sample reaches the metal's melting point temperature, the temperature remains constant for 4.3 min. Calculate the heat of fusion for this metal in joules per gram.

11.112 Heat is added to a 35.2 g sample of a metal at the constant rate of 189 J/sec. After the sample reaches the metal's melting point temperature, the temperature remains constant for 5.2 min. Calculate the heat of fusion for this metal in joules per gram.

CUMULATIVE PROBLEMS

11.113 The specific heat of the metal silver is 0.24 J/(g · °C). How many joules of energy are needed to raise the temperature of a block of silver containing 2.50 moles of silver atoms by 10.0°C?

11.114 The specific heat of the metal gold is 0.13 J/(g · °C). How many joules of energy are needed to raise the temperature of a block of gold containing 1.40 moles of gold atoms by 15.0°C?

11.115 The human body contains approximately 5.7 L of blood. Assuming that the density of blood is 1.06 g/mL and that the specific heats of blood and water are the same, how many kilojoules of energy are required to raise the temperature of this amount of blood by 1.0°C?

11.116 Assuming that the density of blood is 1.06 g/mL and that the specific heats of blood and water are the same, how many liters of blood are present in a human body if 36.0 kJ of energy raises the temperature of the blood present by 1.5°C?

11.117 The specific heat of substance A is 0.88 J/(g · °C) and that of substance B is 2.1 J/(g · °C). You are given an unknown that could be pure substance A, pure substance B, or a homogeneous mixture of A and B. In the laboratory you determine that it requires 59.5 J of heat energy to raise the temperature of a 35.0 g sample of the unknown by 1.0°C. What conclusions can you make about the identity of your unknown from these data?

11.118 The specific heat of substance C is 0.93 J/(g · °C) and that of substance D is 1.8 J/(g · °C). You are given an unknown that could be pure substance C, pure substance D, or a homogeneous mixture of C and D. In the laboratory you determine that it requires 23.3 J of heat energy to raise the temperature of a 25.0 g sample of the unknown by 1.0°C. What conclusions can you make about the identity of your unknown from these data?

11.119 To solidify a 10.0 g sample of a liquid at its freezing point, 1485 J of heat must be removed. To solidify a 10.0 mL sample of this same liquid at its freezing point, 1151 J of heat must be removed. What is the density, in grams per milliliter, of this liquid?

11.120 To solidify a 10.0 mL sample of a liquid at its freezing point, 1063 J of heat must be removed. To solidify a 15.0 g sample of this same liquid at its freezing point, 1276 J of heat must be removed. What is the density, in grams per milliliter, of this liquid?

11.121 The heat of vaporization for a substance is 18.1 kJ/mole. Determine the molar mass of this substance, given that 1.00 g at its boiling point releases 602 J of heat as it condenses.

11.122 The heat of vaporization for a substance is 36.6 kJ/mole. Determine the molar mass of this substance, given that 1.00 g at its boiling point releases 314 J of heat as it condenses.

11.123 How many kilojoules of heat energy are needed to melt completely at its melting point a pure sample of copper that was obtained from the complete decomposition of 52.0 g of Cu_2O?

11.124 How many kilojoules of heat energy are needed to melt completely at its melting point a pure sample of aluminum that was obtained from the complete decomposition of 75.0 g of Al_2O_3?

11.125 How many joules of heat energy are needed to heat a sample of liquid benzene (C_6H_6) from a temperature 10C° below its boiling point to its boiling point, given that the sample contains 6.32×10^{24} hydrogen atoms and that the specific heat of liquid benzene is 1.74 J/(g · °C)?

11.126 How many joules of heat energy must be removed from a sample of liquid benzene (C_6H_6) to lower its temperature from 10C° above its melting point to its melting point, given that the sample contains 4.03×10^{23} carbon atoms and the specific heat of liquid benzene is 1.74 J/(g · °C)?

Multiple-Choice Practice Test

Use this bank of 20 multiple-choice questions as a review of key concepts presented in this chapter. For many of the questions, there may be more than one correct answer (choice d) or no correct answer (choice e).

11.127 Which of the following statements concerning the physical states of the elements at room temperature and pressure is correct?
  a. The vast majority of the elements are solids.
  b. Elements in the liquid state are more numerous than those in the gaseous state.
  c. Only four of the elements are gases.
  d. more than one correct response
  e. no correct response

11.128 Which of the following statements is *not* part of the kinetic molecular theory of matter?
  a. Matter is composed of particles that are in constant motion.
  b. Particle velocity increases as the temperature increases.
  c. Particles in a system cannot transfer energy to each other.
  d. more than one correct response
  e. no correct response

11.129 Which of the following is a property of both solids and liquids?
  a. small degree of thermal expansion
  b. large degree of compressibility
  c. a relatively low density
  d. more than one correct response
  e. no correct response

11.130 In which of the following groupings of terms are the three terms closely related?
  a. kinetic energy, energy of motion, cohesive forces
  b. potential energy, energy of attraction, disruptive forces
  c. kinetic energy, electrostatic attractions, cohesive forces
  d. more than one correct response
  e. no correct response

11.131 In the liquid state, disruptive forces are
  a. roughly of the same magnitude as cohesive forces.
  b. very weak compared to cohesive forces.
  c. more important than electrostatic attractions.
  d. more than one correct response
  e. no correct response

11.132 The phrases "particles close together and held in fixed positions" and "particles occupy the entire volume of the container" apply, respectively, to
  a. liquids and solids.   **b.** solids and liquids.

  c. gases and liquids.   **d.** liquids and gases.
  e. no correct response

11.133 In which of the following pairs of physical changes are both members of the pair exothermic changes?
  a. deposition and sublimation
  b. melting and evaporation
  c. freezing and condensation
  d. more than one correct response
  e. no correct response

11.134 Which of the following statements concerning specific heat are correct?
  a. Specific heat magnitude depends on the physical state of a substance.
  b. J/g is a correct set of units for specific heat.
  c. The lower its specific heat the greater the temperature change when heat is absorbed by a substance.
  d. more than one correct response
  e. no correct response

11.135 Which of the following statements concerning temperature change as a substance is heated is *incorrect*?
  a. As a solid is heated, its temperature rises until its melting point is reached.
  b. During the time a solid changes to a liquid its temperature remains constant.
  c. During the time a liquid changes to a solid its temperature decreases.
  d. more than one correct response
  e. no correct response

11.136 Which of the following would have the same numerical magnitude?
  a. heats of fusion and solidification
  b. heats of vaporization and condensation
  c. heat of fusion and condensation
  d. more than one correct response
  e. no correct response

11.137 Which of the following quantities is needed in calculating the amount of heat energy released as liquid water is cooled from 100°C to 15°C?
  a. heat of condensation of water
  b. heat of fusion for water
  c. specific heat for liquid water
  d. more than one correct response
  e. no correct response

11.138 What is the final temperature of 30.1 g of water at 25.1°C after 455 J of heat energy is added? The specific heat of liquid water is 4.184 $J/(g \cdot °C)$.
  a. 28.7°C   **b.** 30.1°C   **c.** 42.5°C   **d.** 52.3°C
  e. no correct response

11.139 Which of the following statements concerning the process of evaporation is correct?
   a. Evaporation causes the liquid temperature to increase.
   b. Increasing the liquid temperature increases the rate of evaporation.
   c. Decreasing the liquid surface area decreases the rate of evaporation.
   d. more than one correct response
   e. no correct response

11.140 A volatile liquid would
   a. have weak attractive forces between molecules.
   b. evaporate rapidly at room temperature.
   c. have a low vapor pressure at room temperature.
   d. more than one correct response
   e. no correct response

11.141 The boiling point of a liquid is the temperature at which
   a. the rates of sublimation and evaporation become equal.
   b. the vapor pressure of the liquid becomes equal to the external pressure on the liquid.
   c. a state of liquid–vapor equilibrium is reached.
   d. more than one correct response
   e. no correct response

11.142 *Inter*molecular forces differ from *intra*molecular forces in that the former
   a. occur only in liquids.
   b. are much stronger.
   c. occur only between molecules that contain hydrogen atoms.

   d. more than one correct response
   e. no correct response

11.143 Which of the following statements about intermolecular forces is correct?
   a. They must be overcome in order for molecules to escape from the liquid to the vapor state.
   b. They are electrostatic in origin.
   c. They occur only among polar molecules.
   d. more than one correct response
   e. no correct response

11.144 Which of the following statements about various types of intermolecular forces are correct?
   a. Dipole–dipole interactions occur only between nonpolar molecules.
   b. Hydrogen bonds are very strong dipole–dipole interactions.
   c. London forces are instantaneous dipole–dipole interactions.
   d. more than one correct response
   e. no correct response

11.145 In which of the following liquids would London forces be the predominant intermolecular force?
   a. HF   b. Br_2 c. H_2O d. ClF

11.146 Hydrogen bonding among water molecules produces which of the following property effects?
   a. causes water's boiling point to be lower than predicted
   b. causes ice to be less dense than liquid water
   c. causes water's specific heat to be higher than predicted
   d. more than one correct response
   e. no correct response

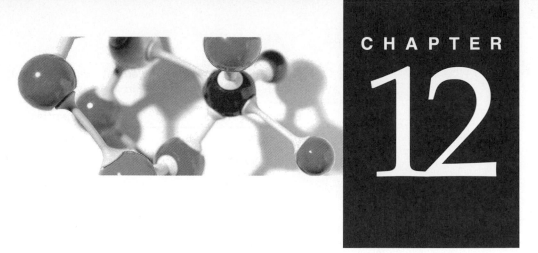

Gas Laws

12.1 PROPERTIES OF SOME COMMON GASES

The word *gas* is used to refer to a substance that is normally in the gaseous state at ordinary temperatures and pressures. The word *vapor* (Sec. 11.13) describes the gaseous form of any substance that is a liquid or solid at normal temperatures and pressures. Thus, we speak of oxygen gas and water vapor.

The normal state for 11 of the elements is that of a gas (Sec. 11.1). A listing of these elements along with some of their properties is given in Table 12.1. This table also lists some common compounds that are gases at room temperature and pressure.

The three most commonly encountered elemental gases—hydrogen, oxygen, and nitrogen—are colorless and odorless. So also are all of the noble gases (Sec. 6.11). From these observations one should not conclude that all gases, or even all elemental gases, are colorless and odorless. Note from Table 12.1 the pale yellow and greenish-yellow colors associated with elemental fluorine and chlorine and the irritating odors of both of them. Many gaseous compounds have pungent odors.

Table 12.1 also brings to your attention again (recall Sec. 10.2) that all of the elemental gases, except for the noble gases, exist in the form of diatomic molecules (H_2, O_2, N_2, F_2, Cl_2).

TABLE 12.1 Color, Odor, and Toxicity of Elements and Common Compounds That Are Gases at Ordinary Temperatures and Pressures

Element		Properties
H_2	hydrogen	colorless, odorless
O_2	oxygen	colorless, odorless
N_2	nitrogen	colorless, odorless
Cl_2	chlorine	greenish-yellow, choking odor, toxic
F_2	fluorine	pale yellow, pungent-odor, toxic
He	helium	colorless, odorless
Ne	neon	colorless, odorless
Ar	argon	colorless, odorless
Kr	krypton	colorless, odorless
Xe	xenon	colorless, odorless
Rn	radon	colorless, odorless

Compound		Properties
CO_2	carbon dioxide	colorless, faintly pungent odor
CO	carbon monoxide	colorless, odorless, toxic
NH_3	ammonia	colorless, pungent odor, toxic
CH_4	methane	colorless, odorless
SO_2	sulfur dioxide	colorless, pungent choking odor, toxic
H_2S	hydrogen sulfide	colorless, rotten egg odor, toxic
HCl	hydrogen chloride	colorless, choking odor, toxic
NO_2	nitrogen dioxide	reddish-brown, irritating odor, toxic

Many of the gaseous compounds listed in Table 12.1 are colorless, have odors, and are toxic. Note that odor and toxicity do not have to go together. Carbon monoxide, a deadly poison, is odorless but toxic.

12.2 GAS LAW VARIABLES

The behavior of a gas can be described reasonably well by *simple* quantitative relationships called *gas laws*. **Gas laws** *are generalizations that summarize in mathematical terms experimental observations about the relationships among the amount, pressure, temperature, and volume of a gas.*

It is only the gaseous state that is describable by *simple* mathematical laws. Laws describing liquid and solid-state behavior are mathematically more complex. Consequently, quantitative treatments of these latter behaviors will not be given in this text.

Before we discuss the mathematical form of the various gas laws, some comments concerning the major variables involved in gas law calculations—amount, volume, temperature, and pressure—are in order. Three of these four variables, amount, volume, and temperature, have been discussed previously (Secs. 9.5, 3.4, and 3.11, respectively). Amount is usually specified in terms of *moles* of gas present. The units of *liter* or *milliliter* are usually used in specifying gas volume. Only one of the three temperature scales discussed in Section 3.11, the *Kelvin* scale, can be used in gas law calculations if the results are to be valid. Therefore, you should be thoroughly familiar with the conversion of Celsius and Fahrenheit scale readings to kelvins (Sec. 3.11). We have not yet discussed *pressure*, the fourth variable. Comments concerning pressure occupy the remainder of this section.

Pressure *is the force applied per unit area, that is, the total force on a surface divided by the area of that surface.*

$$P \text{ (pressure)} = \frac{F \text{ (force)}}{A \text{ (area)}}$$

Note that pressure and force are not the same. Identical forces give rise to different pressures if they are acting on areas of different size. For areas of the same size, the larger the force, the greater the pressure.

Barometers, manometers, and gauges are the instruments commonly used by chemists to measure gas pressure. Barometers and manometers measure pressure in terms of the height of a column of mercury, whereas gauges are usually calibrated in terms of force per area, for example, in pounds per square inch (psi).

The air that surrounds Earth exerts a pressure on all objects with which it has contact. A **barometer** *is a device used for measuring atmospheric pressure.* It was invented by the Italian physicist Evangelista Torricelli (1608–1647) in 1643. The essential components of a barometer are shown in Figure 12.1. A barometer can be constructed by filling a long glass tube, sealed at one end, all the way to the top with mercury and then inverting the tube (without letting any air in) into a dish of mercury. The mercury in the tube falls until the pressure from the mass of the mercury in the tube is just balanced by the pressure of the atmosphere on the mercury in the dish. The pressure of the atmosphere is then expressed in terms of the height of the supported column of mercury.

Mercury is the liquid of choice in a barometer for two reasons: (1) it is a very dense liquid, and therefore only a short glass tube is needed; and (2) it has a very low vapor pressure, so the pressure reading does not have to be corrected for vapor pressure.

The height of the mercury column is most often expressed in millimeters or inches. Millimeters of mercury (mm Hg) are used in laboratory work. The most common use of inches of mercury (in. Hg) is in weather reporting.

A sharp knife cuts better than a dull knife. Because the pushing force acts on a smaller area in a sharp knife, the pressure is greater and the knife cuts better with less effort.

Blood pressure is measured with the aid of an apparatus known as a sphygmomanometer, which is essentially a barometer tube connected to an inflatable cuff by a hollow tube. A normal blood pressure is 120:80; this ratio means a systolic pressure of 120 mm Hg above atmospheric pressure and a diastolic pressure of 80 mm Hg above atmospheric pressure.

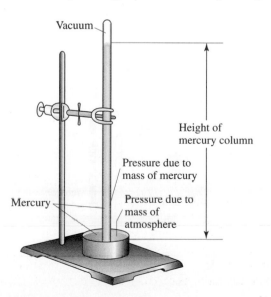

FIGURE 12.1 The essential components of a mercury barometer.

Vacuum

Height of mercury column

Pressure due to mass of mercury

Mercury

Pressure due to mass of atmosphere

The pressure of the atmosphere varies with altitude, decreasing at the rate of approximately 25 mm Hg per 1000 ft increase in altitude. It also fluctuates with weather conditions. Recall the terminology used in a weather report: high-pressure front, low-pressure front, and so on. At sea level the height of a column of mercury in a barometer fluctuates with weather conditions between about 740 and 770 mm Hg and averages about 760 mm Hg. Another pressure unit, the atmosphere (atm), is defined in terms of this average sea-level pressure.

$$1 \text{ atm} = 760 \text{ mm Hg}$$

Because of its size, 760 times larger than 1 mm Hg, the atmosphere unit is frequently used to express the high pressures encountered in many industrial processes and in some experimental work.

It is impractical to measure the pressure of gases other than air with a barometer. One cannot usually introduce a mercury barometer directly into a container of a gas. A **manometer** *is a device used to measure gas pressures in a laboratory.* It is a U-tube filled with mercury. One side of the U-tube is connected to the container in which the pressure is to be measured, and the other side is connected to a region of known pressure. Such an arrangement is depicted in Figure 12.2. The gas in the container exerts a pressure on the Hg on that side of the tube, while atmospheric pressure pushes on the open end. A difference in the heights of the Hg columns in the two arms of the manometer indicates a pressure difference between the gas and the atmosphere. In Figure 12.2 the gas pressure in the container exceeds atmospheric pressure by an amount, ΔP, that is equal to the difference in the heights of the Hg columns in the two arms.

$$P_{\text{container}} = P_{\text{atm}} + \Delta P$$

Pressures of contained gas samples can also be measured by special gas pressure gauges attached to their containers. Such gauges are commonly found on tanks (cylinders) of gas purchased commercially. These gauges are most often calibrated in terms of pounds per square inch (lb/in.^{2} or psi). The relationship between psi and atmospheres is

$$14.68 \text{ psi} = 1 \text{ atm}$$

Table 12.2 summarizes the relationships among the various pressure units we have discussed and also lists one additional pressure unit, the *pascal*. The pascal is the SI unit (Sec. 3.1) of pressure.

The popping feeling in the ears encountered when driving up a steep mountainous road is caused by the decrease in pressure with altitude.

Automobile tire pressure is measured in psi units. The pressure inside the tire is the pressure shown by the gauge plus atmospheric pressure.

FIGURE 12.2 Pressure measurement by means of a manometer.

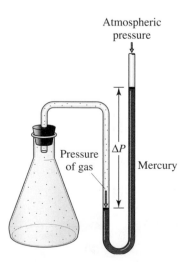

TABLE 12.2 Units of Pressure and Their Relationship to the Unit Atmosphere

Unit	Relationship to Atmosphere	Area of Use
Atmosphere	—	gas law calculations
Millimeters of mercury	760 mm Hg = 1 atm	gas law calculations
Inches of mercury	29.92 in. Hg = 1 atm	weather reports
Pounds per square inch	14.68 psi = 1 atm	stored or bottled gases
Pascal	1.013×10^5 Pa = 1 atm	calculations requiring SI units

The values 1 atm and 760 mm Hg are exact numbers since they arise from a definition. Consequently, these two values have an infinite number of significant figures. All other values in Table 12.2 are not exact numbers; they are given to four significant figures in the table.

The equalities of Table 12.2 can be used to construct conversion factors for problem solving in the usual way (Sec. 3.6). Example 12.1 illustrates the use of such conversion factors in the context of change in pressure unit problems.

EXAMPLE 12.1   **Interconverting between Various Pressure Units**

At a certain altitude, a weather balloon recorded a barometric pressure of 367 mm Hg. Express this barometric pressure in the following units:

a. atmospheres   **b.** inches Hg   **c.** pounds per square inch

SOLUTION

a. The given quantity is 367 mm Hg, and the unit of the desired quantity is atmospheres.

$$367 \text{ mm Hg} = ? \text{ atm}$$

The conversion factor relating these two units is

$$\frac{1 \text{ atm}}{760 \text{ mm Hg}}$$

The result of using this conversion factor is

$$367 \text{ mm Hg} \times \frac{1 \text{ atm}}{760 \text{ mm Hg}} = 0.48289473 \text{ atm} \qquad \text{(calculator answer)}$$

$$= 0.483 \text{ atm} \qquad \textbf{(correct answer)}$$

b. This time, the unit for the desired quantity is in. Hg.

$$367 \text{ mm Hg} = ? \text{ in. Hg}$$

A direct relationship between mm Hg and in. Hg is not given in Table 12.1. However, the table does give the relationships of both inches and millimeters of mercury to atmospheres. Hence, we can use atmospheres as an intermediate step in the unit conversion sequence.

$$\text{mm Hg} \longrightarrow \text{atm} \longrightarrow \text{in. Hg}$$

The dimensional analysis setup is

$$367 \text{ mm Hg} \times \frac{1 \text{ atm}}{760 \text{ mm Hg}} \times \frac{29.92 \text{ in. Hg}}{1 \text{ atm}}$$

$$= 14.44821 \text{ in. Hg} \quad \text{(calculator answer)}$$

$$= 14.4 \text{ in. Hg} \quad \text{(\textbf{correct answer})}$$

c. The problem to be solved here is

$$367 \text{ mm Hg} = ? \text{ psi}$$

The unit conversion sequence will be

$$\text{mm Hg} \longrightarrow \text{atm} \longrightarrow \text{psi}$$

The dimensional analysis sequence of conversion factors, for this pathway, is

$$367 \text{ mm Hg} \times \frac{1 \text{ atm}}{760 \text{ mm Hg}} \times \frac{14.68 \text{ psi}}{1 \text{ atm}} = 7.0888947 \text{ psi} \quad \text{(calculator answer)}$$

$$= 7.09 \text{ psi} \quad \text{(\textbf{correct answer})}$$

Answer Double Check:

Are the magnitudes of the answers reasonable? Yes. Since the given pressure is approximately one-half atmosphere ($760/2 = 380$), all of the answers should be numbers that are approximately one-half the numbers in the following chain of equalities.

$$760 \text{ mm Hg} = 1 \text{ atm} = 29.92 \text{ in. Hg} = 14.68 \text{ psi}$$

Such is the case, with answers of 0.483 atm, 14.4 in. Hg, and 7.09 psi.

Answers to all practice exercises in this chapter are found in the back-of-the-book answer section.

▶ **Practice Exercise 12.1** The barometric pressure in a university chemistry laboratory on Wednesday of the third week of fall semester is found to be 674.3 mm Hg. Express this pressure in the following units:

a. atmospheres   b. inches of mercury   c. pounds per square inch

Pressure Readings and Significant Figures

Standard procedure in obtaining pressures that are based on the height of a column of mercury (barometric and manometric readings) is to estimate the column height to the closest millimeter. Thus, such pressure readings have an uncertainty in the ones place, that is, to the closest millimeter of mercury.

A similar common sense approach to significant figures for temperature readings was considered in Section 3.11.

When considering significant figures for pressure readings, the preceding operational procedure must be taken into account. Millimeter of mercury pressure readings of 750, 730, 720, 650, etc., are considered to have three significant figures even though no decimal point is explicitly shown after the zero (Sec. 2.5). A pressure reading of 700 mm Hg or 600 mm Hg is considered to possess three significant figures.

Note that the previous paragraph applies to ordinary pressures obtained using mercury column heights. A very high pressure, such as 19,000 mm Hg (25 atmospheres), would be assumed to be a two-significant-figure pressure reading rather than a five-significant-figure reading, without further information being given. Obviously, this pressure was not determined by reading the height of a column of mercury.

12.3 BOYLE'S LAW: A PRESSURE–VOLUME RELATIONSHIP

Of the several relationships that exist between gas law variables, the first to be discovered was the one that relates gas pressure to gas volume. It was formulated over 300 years ago, in 1662, by the British chemist and physicist Robert Boyle (1627–1691; see "The Human Side of Chemistry 12") and is known as *Boyle's law*. **Boyle's law** *states that the volume of a fixed mass of gas is* inversely proportional *to the pressure applied to the gas if the temperature is kept constant.* This means that if the pressure on the gas increases, the volume decreases proportionally; and conversely, if the pressure is decreased, the volume will increase. Doubling the pressure cuts the volume in half, tripling the pressure cuts the volume to one-third its original value, quadrupling the pressure cuts the volume to one-fourth, and so on. Any time two quantities are *inversely proportional,* as pressure and volume are (Boyle's law), one increases as the other decreases. Data illustrating Boyle's law are given in Figure 12.3.

The phrase "fixed mass of gas" in the statement of Boyle's law means that the number of moles of gas present is a constant.

Boyle's law can be illustrated physically quite simply with the J-tube apparatus shown in Figure 12.4. The pressure on the trapped gas is increased by adding mercury to the J-tube. The volume of the trapped gas decreases as the pressure is raised.

Boyle's law can be stated mathematically as

$$P \times V = \text{constant}$$

In this expression, V is the volume of the gas at a given temperature and P is the pressure. This expression thus indicates that at constant temperature the product of the pressure times the volume is always the same (or constant). (Note that Boyle's law is valid only if the temperature of the gas does not change.)

The Human Side of Chemistry 12

Robert Boyle (1627–1691)

Robert Boyle, one of the most prominent of seventeenth-century scientists, spent most of his life in England although he was born in Ireland. Born to wealth and nobility, the 14th of 15 children of the earl of Cork, he was an infant prodigy. At the age of 14 he was in Italy studying the works of the recently deceased Galileo.

Like most men of the seventeenth century who devoted themselves to science, Boyle was self-taught. Primarily known as a chemist and natural philosopher, he was best known for the gas law that bears his name. However, he made many other significant contributions in chemistry and physics.

During his lifetime his name was known throughout the learned world as a leading advocate of the "experimental philosophy." Through his efforts, the true value of experimental investigation was first realized.

Historically, before chemistry could become a flourishing science, a period of "housecleaning" was needed to get rid of accumulated false concepts of previous centuries. Perhaps Boyle's greatest work in chemistry was in this area. He was very instrumental in demolishing the false dogma of only four elements—earth, air, water, and fire.

His book *The Sceptical Chymist* (1661), which attacked many of the false concepts of previous years, influenced generations of chemists. In this book he developed the concept of "primary particles," a forerunner of our present concept of atoms and elements. He attempted to explain different kinds of matter in terms of the organization and motion of these primary particles.

Boyle was a prolific writer on not only scientific but also philosophical and religious topics. In later years he became increasingly interested in religion. Upon his death, Boyle left a sum of money to fund the Boyle lectures, which were to be not on science, but on the defense of Christianity against unbelievers.

FIGURE 12.3 Data
illustrating the inverse
proportionality
associated with
Boyle's law.

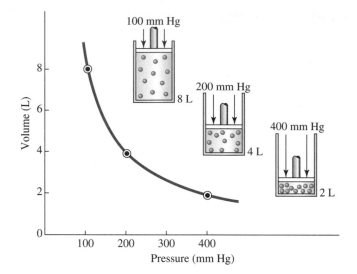

FIGURE 12.3 Data illustrating the inverse proportionality associated with Boyle's law.

An alternative, and more useful, mathematical form of Boyle's law can be derived by considering the following situation. Suppose a gas is at an initial pressure P_1 and has a volume V_1. (We will use the subscript 1 to indicate the initial conditions.) Now imagine that the pressure is changed to some final pressure P_2. The volume will also change, and we will call the final volume V_2. (We will use the subscript 2 to indicate the final conditions.) According to Boyle's law,

$$P_1 \times V_1 = \text{constant}$$

After the change in pressure and volume, we have

$$P_2 \times V_2 = \text{constant}$$

The constant is the same in both cases; we are dealing with the same sample of gas. Thus, we can combine the two PV products to give the equation

$$P_1 \times V_1 = P_2 \times V_2$$

This PV product equation is more commonly written without multiplication signs as

$$P_1V_1 = P_2V_2$$

When we know any three of the four quantities in this equation, we can calculate the fourth, which will usually be the final pressure, P_2, or the final volume, V_2, as illustrated in Examples 12.2 and 12.3.

FIGURE 12.4 An apparatus for demonstrating Boyle's law. The volume of the trapped gas in the closed end of the tube (V_1, V_2, and V_3) decreases as mercury is added through the open end of the tube.

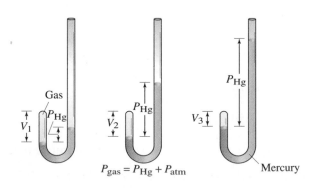

EXAMPLE 12.2	**Calculating Final Volume Using Boyle's Law**

An air bubble forms at the bottom of a lake, where the total pressure is 2.45 atm. At this pressure, the bubble has a volume of 3.1 mL. What volume, in milliliters, will it have when it rises to the surface, where the pressure is 0.93 atm? Assume that the temperature remains constant, as does the amount of gas in the bubble.

SOLUTION

A suggested first step in working all gas law problems involving two sets of conditions is to analyze the given data in terms of initial and final conditions. Doing this, we find that

$$P_1 = 2.45 \text{ atm} \qquad P_2 = 0.93 \text{ atm}$$
$$V_1 = 3.1 \text{ mL} \qquad V_2 = ? \text{ mL}$$

Next, we rearrange Boyle's law to isolate V_2 (the quantity to be calculated) on one side of the equation. This is accomplished by dividing both sides of the Boyle's law equation by P_2.

$$P_1 V_1 = P_2 V_2 \qquad \textbf{(Boyle's law)}$$

$$\frac{P_1 V_1}{P_2} = \frac{P_2 V_2}{P_2} \qquad \text{(division of each side of the equation by } P_2)$$

$$V_2 = V_1 \times \frac{P_1}{P_2}$$

Substituting the given data in the rearranged equation and doing the arithmetic gives

$$V_2 = 3.1 \text{ mL} \times \frac{2.45 \text{ atm}}{0.93 \text{ atm}} = 8.1666666 \text{ mL} \qquad \text{(calculator answer)}$$

$$= 8.2 \text{ mL} \qquad \text{(correct answer)}$$

Answer Double Check:

Is the answer, in terms of magnitude, reasonable? Yes. Boyle's law involves an inverse proportion; as one variable increases, the other variable must decrease. Decreasing the pressure on a fixed amount of gas, which is the case in this problem, should result in a volume increase for the gas. Our answer is consistent with this conclusion.

▶ **Practice Exercise 12.2** A sample of pure Cl_2 gas, which has a pale yellow-green color, has a volume of 2.30 L at a pressure of 1.20 atm. What will be the sample volume, in liters, if the pressure is increased to 2.11 atm, assuming that the temperature remains constant?

EXAMPLE 12.3	**Calculating Final Volume Using Boyle's Law with Pressure Values Given in Different Units**

On an airliner, an inflated toy has a volume of 0.310 L at 25°C and 745 mm Hg. During flight, the cabin pressure unexpectedly drops to 0.810 atm, with the temperature remaining the same. What is the volume, in liters, of the toy at the new conditions?

SOLUTION

Analyzing the given data in terms of initial and final conditions gives

$$P_1 = 745 \text{ mm Hg} \qquad P_2 = 0.810 \text{ atm}$$
$$V_1 = 0.310 \text{ L} \qquad V_2 = ? \text{ L}$$

The temperature is constant; it will not enter into the calculation.

A slight complication exists with the pressure values; they are not given in the same units. We must make the units the same before proceeding with the calculation. It does not matter whether they are both millimeters of mercury or both atmospheres. The same answer is obtained with either unit. Let us arbitrarily decide to change atmospheres to mm Hg.

$$0.810 \ \text{atm} \times \frac{760 \ \text{mm Hg}}{1 \ \text{atm}} = 615.6 \ \text{mm Hg} \qquad \text{(calculator answer)}$$

$$= 616 \ \text{mm Hg} \qquad \textbf{(correct answer)}$$

Our given conditions are now

$$P_1 = 745 \ \text{mm Hg} \qquad P_2 = 616 \ \text{mm Hg}$$
$$V_1 = 0.310 \ \text{L} \qquad V_2 = ? \ \text{L}$$

Boyle's law with V_2 isolated on the left side has the form

$$V_2 = V_1 \times \frac{P_1}{P_2}$$

Plugging the given quantities into this equation and doing the arithmetic gives

$$V_2 = 0.310 \ \text{L} \times \frac{745 \ \text{mm Hg}}{616 \ \text{mm Hg}} = 0.37491883 \ \text{L} \qquad \text{(calculator answer)}$$

$$= 0.375 \ \text{L} \qquad \textbf{(correct answer)}$$

The answer is reasonable. Decreased pressure, at constant temperature, should produce a volume increase.

Rearrangement of Boyle's law from the form $P_1V_1 = P_2V_2$ to the form $V_1/V_2 = P_2/P_1$ shows that the law involves a ratio of volumes and a ratio of pressures. What units should be used for the volumes and the pressures? Any volume unit may be used as long as it is used for both volumes, because the units will cancel. Similarly, because of unit cancellation, any pressure unit may be used as long as it is used for both pressures.

In solving this problem, we arbitrarily chose to use mm Hg pressure units. If we had used atmosphere pressure units instead, the answer obtained would still be the same, as we now illustrate.

$$745 \ \text{mm Hg} \times \frac{1 \ \text{atm}}{760 \ \text{mm Hg}} = 0.98026315 \ \text{atm} \qquad \text{(calculator answer)}$$

$$= 0.980 \ \text{atm} \qquad \textbf{(correct answer)}$$

Using $P_1 = 0.980$ atm and $P_2 = 0.810$ atm, we get

$$V_2 = 0.310 \ \text{L} \times \frac{0.980 \ \text{atm}}{0.810 \ \text{atm}} = 0.37506172 \ \text{L} \qquad \text{(calculator answer)}$$

$$= 0.375 \ \text{L} \qquad \textbf{(correct answer)}$$

Answer Double Check:

Is the direction of change for the volume of the gas reasonable? Yes. A pressure increase, under Boyle's law conditions (fixed quantity of gas, constant temperature), should result in a volume increase. Such is the case.

▶ **Practice Exercise 12.3**   Ethylene, C_2H_4, is a gas that occurs in small amounts in plants, where it functions as a plant hormone that stimulates the ripening of fruit. If a sample of ethylene gas occupies a volume of 25.0 mL at a pressure of 735 mm Hg, what volume, in milliliters, will the sample occupy when the pressure is decreased to 0.526 atm?

Boyle's law is consistent with kinetic molecular theory (Sec. 11.3). The theory states that the pressure a gas exerts results from collisions of the gas molecules with the sides of the container. The pressure of the gas at a given temperature is proportional to the number of collisions within a unit area on the container wall at a given instant. If the volume of a container holding a specific number of gas molecules is increased, the total wall area of the container will also increase and the number of collisions in a given area (the pressure) will decrease due to the greater wall area. Conversely, if the volume of the container is decreased, the wall area will be smaller and there will be more collisions in a given wall area. This means an increase in pressure. Figure 12.5 illustrates this idea.

The phenomenon described by Boyle's law has practical importance. Helium-filled research balloons, used to study the upper atmosphere, are only half-filled with helium when launched. As the balloon ascends, it encounters lower and lower pressures. As the pressure decreases, the balloon expands until it reaches full inflation. A balloon launched at full inflation would burst in the upper atmosphere because of the reduced external pressure.

Breathing is an example of Boyle's law in action, as is the operation of a respirator, a machine designed to help patients with respiration difficulties to breathe. A respirator contains a movable diaphragm that works in opposition to the patient's lungs. When the diaphragm is moved out so that the volume inside the respirator increases, the lower pressure in the respirator allows air to expand out of the patient's lungs. When the diaphragm is moved in the opposite direction, the higher pressure inside the respirator compresses the air into the lungs and causes them to increase in volume.

Filling a medical syringe with a liquid demonstrates Boyle's law. As the plunger is drawn out of the syringe, the increase in volume inside the syringe chamber results in decreased pressure there. The liquid, which is at atmospheric pressure, flows into this reduced-pressure area. This liquid is then expelled from the chamber by pushing the plunger back in. This ejection of liquid does not involve Boyle's law; a liquid is incompressible and mechanical force pushes it out.

12.4 CHARLES'S LAW: A TEMPERATURE–VOLUME RELATIONSHIP

The relationship between the temperature and the volume of a fixed amount of gas at constant pressure is called *Charles's law*, after the French scientist Jacques Charles (1746–1823; see "The Human Side of Chemistry 13"). In 1787, over 100 years after the discovery of Boyle's law, Charles showed that a simple mathematical relationship exists between the volume and temperature of a fixed amount of gas, at constant pressure, *provided the temperature is expressed in kelvins*. **Charles's law** *states that the volume of a fixed mass of gas is* directly proportional *to its Kelvin temperature if the pressure is kept constant*. Contained within the wording of this law is the phrase *directly proportional*; this contrasts with Boyle's law, which contains the phrase *inversely proportional*. Any time a *direct* proportion exists between two quantities, one increases when the other increases, and one decreases when the other decreases. Thus a direct proportion and an inverse proportion portray "opposite" behaviors. The direct proportion relationship of Charles's law means that if the temperature increases the volume will also increase and if the temperature decreases the volume will also decrease. Data illustrating Charles's law are given in Figure 12.6.

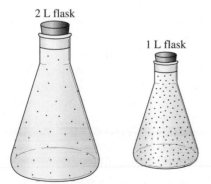

2 L flask

1 L flask

The volume is decreased by one-half.

(a)

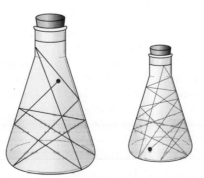

A given molecule hits container walls twice as often.

(b)

FIGURE 12.5 The pressure exerted by a gas is doubled when the volume of the gas, at constant temperature, is cut by one-half.

FIGURE 12.6 Data illustrating the direct proportionality associated with Charles's law.

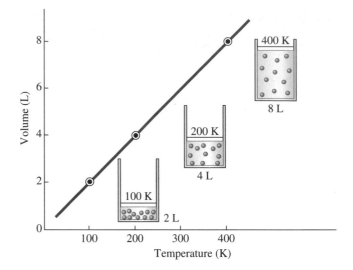

Charles's law can be qualitatively illustrated by a balloon filled with air. If the balloon is placed near a heat source, such as a lightbulb that has been on for some time, the heat will cause the balloon to increase in size (volume). The change in volume is usually apparent. Cooling the air in the balloon will cause the balloon to shrink (see Fig. 12.7).

Charles's law can be stated mathematically as

$$\frac{V}{T} = \text{constant}$$

The Human Side of Chemistry 13

Jacques Alexandre César Charles (1746–1823)

Jacques Alexandre César Charles, a French physicist, had an early career completely removed from science. He worked at a routine job in the Bureau of Finances in Paris. All was well until, during a period of government austerity, he was discharged from his job. Unemployed, he decided to study physics.

During this time of transition in Charles's life, hot-air ballooning, both unmanned and manned, was in its first stages of development. Charles was best known to his colleagues for his contributions to the developing science of ballooning. He was the first to design a hydrogen-filled, rather than air-filled, balloon for manned flight. In constructing the balloon Charles showed much innovation in the design of scientific apparatus.

The first hot-air balloon flight with passengers was made November 17, 1783. Burning straw was the source of heat for such hot-air balloons. A few days later Charles made an ascent in his hydrogen-filled balloon. His landing put him in a rural area. The peasants in the area, terrified at their first exposure to manned flight, attacked and destroyed the balloon, using pitchforks.

In the process of working with balloons, about 1787, Charles observed that several different gases expanded the same way when heated. From further observations he was able to estimate the degree of thermal expansion for gases as a function of temperature. This led to what is now known as Charles's law.

Charles never published his work. He did, however, tell another scientist, Joseph Louis Gay-Lussac ("The Human Side of Chemistry 14"), about it, and it was Gay-Lussac who made it public. In an article on the expansion of gases by heat, published in 1802, Gay-Lussac described, criticized, and considerably improved upon Charles's experimental procedures.

Charles's contributions to science were limited to those associated with his ballooning.

FIGURE 12.7 As liquid nitrogen (–196°C) is poured over a balloon, the gas in the balloon is cooled and the volume decreases. *(Richard Megna/Fundamental Photographs, NYC)*

In this expression V is the volume of a gas at a given pressure and T is the temperature, expressed on the Kelvin temperature scale. A consideration of two sets of temperature–volume conditions for a gas, in a manner similar to that done for Boyle's law, leads to the following useful form of the law.

When you use the mathematical form of Charles's law, the temperatures used *must always be* Kelvin scale temperatures.

$$\frac{V_1}{T_1} = \frac{V_2}{T_2}$$

Again, note that in any of the mathematical expressions of Charles's laws, or any other gas law, the symbol T is understood to mean the Kelvin temperature.

EXAMPLE 12.4 **Calculating Final Volume Using Charles's Law**

A helium-filled birthday balloon has a volume of 223 mL at a temperature of 24°C (room temperature). The balloon is placed in a car overnight during the winter where the temperature drops to –13°C. What is the balloon's volume, in milliliters, when it is removed from the car in the morning (assuming no helium has escaped)?

SOLUTION

Writing all the given data in the form of initial and final conditions, we have

$$V_1 = 223 \text{ mL} \qquad V_2 = ? \text{ mL}$$
$$T_1 = 24°C = 297 \text{ K} \qquad T_2 = -13°C = 260 \text{ K}$$

Note that both of the given temperatures have been converted to Kelvin scale readings by adding 273 to them.

Rearranging Charles's law to isolate V_2, the desired quantity, on one side of the equation is accomplished by multiplying each side of the equation by T_2.

$$\frac{V_1}{T_1} = \frac{V_2}{T_2} \qquad \textbf{(Charles's law)}$$

$$\frac{V_1 T_2}{T_1} = \frac{V_2 \cancel{T_2}}{\cancel{T_2}} \qquad \text{(multiplication of each side by } T_2\text{)}$$

$$V_2 = V_1 \times \frac{T_2}{T_1}$$

Substituting the given data into the equation and doing the arithmetic gives

$$V_2 = 223 \text{ mL} \times \frac{260 \cancel{K}}{297 \cancel{K}}$$

$$= 195.21885 \text{ mL} \qquad \text{(calculator answer)}$$

$$= 195 \text{ mL} \qquad \textbf{(correct answer)}$$

Answer Double Check:

Is the direction of change for the temperature reasonable? Yes. Since Charles's law involves a direct proportion, as the temperature decreases, at constant pressure, the volume should also decrease. Such is the case.

▶ **Practice Exercise 12.4** Hydrogen sulfide, H_2S, is a colorless gas with a strong offensive odor akin to that of rotten eggs. At constant pressure, a 725 mL sample of H_2S gas is warmed from 125°C to 225°C. What is the new volume, in milliliters, of the H_2S?

Figure 12.8 is a graph showing four sets of volume–temperature data for a gas, each data set being at a different constant pressure. The plot of each data set gives a straight line. These four straight lines can be used to show the basis for the use of the Kelvin temperature scale in gas law calculations. If each of these lines is extrapolated (extended) to lower temperatures (the dashed portions of the lines in Fig. 12.8), we find that they all intersect at a common point on the temperature axis. This point of intersection, corresponding to a temperature value of –273°C, is the point at which the volume of the gas "would become" zero. Notice that the words "would become" in the last sentence are in quotation marks. In reality, the volume of a gas never reaches zero; all gases would liquefy before they reach a temperature of –273°C, and Charles's law would no longer apply. Thus, the extrapolated portions of the four straight lines portray a hypothetical situation. It is not, however, a situation without significance.

The Scottish mathematician and physicist William Thomson (1824–1907), better known as Lord Kelvin, was the first to recognize the importance of the "zero-volume" temperature value of –273°C. It is the lowest temperature that is theoretically attainable, a temperature now referred to as *absolute zero*. It is this temperature that is now used as the starting point for the temperature scale called the Kelvin temperature scale. When

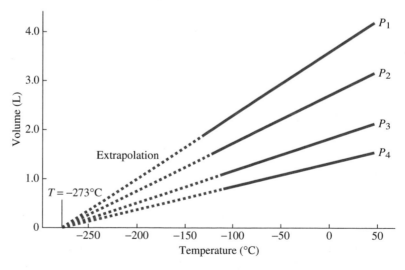

FIGURE 12.8 A graph showing the variation of the volume of a gas with temperature at a constant pressure. Each line in the graph represents a different fixed pressure for the gas. The pressures increase from P_1 to P_4. The solid portion of each line represents the temperature region before the gas condenses to a liquid. The extrapolated (dashed) portion of each line represents the hypothetical volume of the gas if it had not condensed. All extrapolated lines intersect at the same point, a point corresponding to zero volume and a temperature of –273°C.

temperatures are specified in kelvins, Charles's law assumes the simple mathematical form in which it was previously given in this section.

Charles's law is readily understood in terms of kinetic molecular theory. The theory states that when the temperature of a gas increases, the velocity of the gas particles increases. The speedier particles hit the container walls harder. For the pressure of the gas to remain constant, it is necessary for the container volume to increase. This will result in fewer particles hitting a unit area of wall at a given instant. A similar argument applies if the temperature of the gas is lowered, only this time the velocity of the molecules decreases and the wall area (volume) must decrease to increase the number of collisions in a given area in a given time.

Charles's law is the principle used in the operation of a convection heater. When air comes in contact with the heating element, it expands (its density becomes less). The hot, less dense air rises, causing continuous circulation of warm air. This same principle has ramifications in closed rooms in which there is not effective air circulation. The warmer and less dense air stays near the top of the room. This is desirable in the summer but not in the winter.

Thunder can be related to Charles's law. A typical lightning bolt heats the air around it to 30,000°C in a few microseconds. The heated air expands rapidly and pushes against the cooler surrounding air. This creates the shock wave of sound known as thunder.

12.5 GAY-LUSSAC'S LAW: A TEMPERATURE–PRESSURE RELATIONSHIP

If a gas is placed in a rigid container, one that cannot expand, the volume of the gas must remain constant. What is the relationship between the pressure and temperature of a fixed amount of gas in such a situation? This question was answered in 1802 by the French scientist Joseph Louis Gay-Lussac (1778–1850; see "The Human Side of Chemistry 14"). Performing a number of experiments similar to those performed by Charles, he discovered the relationship between pressure and the temperature of a fixed

The Human Side of Chemistry 14

Joseph Louis Gay-Lussac (1778–1850)

Joseph Louis Gay-Lussac, the eldest of five children, is considered one of the greatest scholars France has produced. Because of the French Revolution and the uncertainties created by it, Gay-Lussac's formal education had to be delayed until he was 16. He quickly distinguished himself as a student.

His grandfather was a physician and his father a lawyer. The family surname was actually Gay, the surname by which he was baptized. His father added the appendage Lussac (which has geographical connotations) to the family name to distinguish his family from many others in the area with the same surname.

Gay-Lussac devoted nearly all of his life to pure and applied science. Much of it, from 1809 on, was spent as a professor at the Ecole Polytechnique in Paris. He did have a brief political career (1831–1838) as an elected French legislator.

Gay-Lussac's first major research (1801–1802) involved the thermal expansion of gases. Here he followed up on work performed earlier by Charles. His greatest achievement, which came in 1808, was work involving the ratio, by volume, in which gases combine during chemical reactions. From this work came the law of combining volumes that now bears his name (Sec. 12.9).

He was the first to isolate the elements potassium and boron. With a coworker, he made the first thorough studies of iodine and cyanogen. He was an early pioneer in the area of volumetric analysis. The Gay-Lussac tower, an essential part of sulfuric acid plants for many years, is evidence of his interest in chemical technology.

amount of gas when it is held at constant volume. **Gay-Lussac's law** *states that the pressure of a fixed mass of gas is* directly proportional *to its Kelvin temperature if the volume is kept constant.* Thus, as the temperature of the gas increases, the pressure increases; conversely, as temperature is decreased, pressure decreases.

The kinetic molecular theory explanation for Gay-Lussac's law is very simple. As the temperature increases, the velocity of the gas molecules increases. This increases the number of times the molecules hit the container walls in a given time, which translates into an increase in pressure.

Gay-Lussac's law explains the observed pressure increase inside an automobile tire after a car has been driven for a period of time. It also explains why aerosol cans should not be disposed of in a fire. The aerosol can keeps the gas at a constant volume. The elevated temperatures encountered in the fire increase the pressure of the confined gas to the point that the can will probably explode.

Mathematical forms of Gay-Lussac's law are

$$\frac{P}{T} = \text{constant} \quad \text{and} \quad \frac{P_1}{T_1} = \frac{P_2}{T_2}$$

Again, note that the symbol T is understood to mean the Kelvin temperature. Any pressure unit is acceptable; P_1 and P_2 must, however, be in the same units.

EXAMPLE 12.5 | Calculating Final Temperature Using Gay-Lussac's Law

A pressurized can of shaving cream contains a gas under an internal pressure of 1.065 atm at 24°C (room temperature). The can itself is able to withstand an internal pressure of 3.00 atm before it explodes. What temperature, in degrees Celsius, is needed to elevate the internal pressure within the can to its maximum value?

SOLUTION

Writing all the given data in the form of initial and final conditions, we have

$$P_1 = 1.065 \text{ atm} \qquad P_2 = 3.00 \text{ atm}$$
$$T_1 = 24°C = 297 \text{ K} \qquad T_2 = ? \text{ K}$$

Rearrangement of Gay-Lussac's law to isolate T_2, the quantity we desire, on the left side of the equation gives

$$T_2 = T_1 \times \frac{P_2}{P_1}$$

Substituting the given data into the equation and doing the arithmetic gives

$$T_2 = 297 \text{ K} \times \frac{3.00 \text{ atm}}{1.065 \text{ atm}} = 836.61971 \text{ K} \qquad \text{(calculator answer)}$$
$$= 837 \text{ K} \qquad \text{(\textbf{correct answer})}$$

Our answer is in kelvins. We were asked for the temperature in degrees Celsius. Subtracting 273 from the Kelvin temperature will give us the Celsius temperature.

$$T(°C) = 837 \text{ K} - 273 = 564°C$$

Answer Double Check:

Is the direction of change for the temperature reasonable? Yes. Because pressure and temperature are directly proportional under Gay-Lussac law conditions, an increased pressure should produce a temperature increase. Such is the case.

▶ **Practice Exercise 12.5**   The gas SO_2 finds use as a food additive because it is toxic to molds and certain bacteria. To what temperature, in °C, will a 70°C sample of SO_2 gas under a pressure of 6.00 atm in a constant-volume container need to be heated in order to increase its pressure to 8.00 atm?

12.6 THE COMBINED GAS LAW

The **combined gas law** *states that the product of the pressure and volume of a fixed amount of gas is* directly proportional *to its Kelvin temperature.* It is obtained by mathematically combining Boyle's, Charles's, and Gay-Lussac's laws. The combined gas law's mathematical form is

$$\frac{P_1V_1}{T_1} = \frac{P_2V_2}{T_2}$$

This combined gas law is a much more versatile equation than the individual gas laws. With it, for a fixed amount of gas, a change in any one of the other three gas law variables brought about by changes in *both* of the remaining two variables can be calculated. Each of the individual gas laws requires that an additional variable besides amount of gas be held constant.

The combined gas law reduces (simplifies) to each of the equations for the individual gas laws when the appropriate variable is held constant. These reduction relationships are given in Table 12.3. Because of the ease with which they can be derived from the combined gas law, you need not memorize the mathematical forms for the individual gas laws if you know the mathematical form for the combined gas law.

TABLE 12.3 Relationship of the Individual Gas Laws to the Combined Gas Law

Law	Constancy Requirement (for a fixed mass of gas)	Mathematical Form of the Law
Combined gas law	none	$\frac{P_1V_1}{T_1} = \frac{P_2V_2}{T_2}$
Boyle's law	$T_1 = T_2$	Since T_1 and T_2 are equal, substitute T_1 for T_2 in the combined gas law and cancel. $\frac{P_1V_1}{\cancel{T_1}} = \frac{P_2V_2}{\cancel{T_1}}$ or $P_1V_1 = P_2V_2$
Charles's law	$P_1 = P_2$	Since P_1 and P_2 are equal, substitute P_1 for P_2 in the combined gas law and cancel. $\frac{P_1V_1}{T_1} = \frac{P_1V_2}{T_2}$ or $\frac{V_1}{T_1} = \frac{V_2}{T_2}$
Gay-Lussac's law	$V_1 = V_2$	Since V_1 and V_2 are equal, substitute V_1 for V_2 in the combined gas law and cancel. $\frac{P_1\cancel{V_1}}{T_1} = \frac{P_2\cancel{V_1}}{T_2}$ or $\frac{P_1}{T_1} = \frac{P_2}{T_2}$

The three most used forms of the combined gas law are those that isolate V_2, P_2, and T_2 on the left side of the equation.

$$V_2 = V_1 \times \frac{P_1}{P_2} \times \frac{T_2}{T_1}$$

$$P_2 = P_1 \times \frac{V_1}{V_2} \times \frac{T_2}{T_1}$$

$$T_2 = T_1 \times \frac{P_2}{P_1} \times \frac{V_2}{V_1}$$

Students frequently have questions about the algebra involved in accomplishing these rearrangements. Example 12.6 should clear up any such questions. Examples 12.7 and 12.8 illustrate the use of the combined gas law equation.

EXAMPLE 12.6 Mathematical Manipulation of the Combined Gas Law

Rearrange the standard form of the combined gas law equation such that the variable P_2 is by itself on the left side of the equation.

SOLUTION

In rearranging the standard form of the combined gas law into various formats, the following rule from algebra is useful.

In rearranging the standard form of the combined gas law into various formats, the following rule from algebra is useful.
    If two fractions are equal,

$$\frac{a}{b} = \frac{c}{d}$$

then the numerator of the first fraction (a) times the denominator of the second fraction (d) is equal to the numerator of the second fraction (c) times the denominator of the first fraction (b).

$$\text{If } \frac{a}{b} = \frac{c}{d} \quad \text{then } a \times d = c \times b$$

Applying this rule to the standard form of the combined gas law gives

$$\frac{P_1V_1}{T_1} = \frac{P_2V_2}{T_2} \longrightarrow P_1V_1T_2 = P_2V_2T_1$$

With the combined gas law in the form $P_1V_1T_2 = P_2V_2T_1$, any of the six variables can be isolated by a simple division. To isolate P_2, we divide both sides of the equation by V_2T_1 (the other quantities on the same side of the equation as P_2).

$$\frac{P_1V_1T_2}{V_2T_1} = \frac{P_2\cancel{V_2}\cancel{T_1}}{\cancel{V_2}\cancel{T_1}}$$

$$P_2 = \frac{P_1V_1T_2}{V_2T_1} \quad \text{or} \quad P_2 = P_1 \times \frac{V_1}{V_2} \times \frac{T_2}{T_1}$$

▶ **Practice Exercise 12.6**   Rearrange the standard form of the combined gas law equation such that the variable V_1 is by itself of the left side of the equation.

EXAMPLE 12.7 **Calculating Final Volume Using the Combined Gas Law**

Hydrogen gas (H_2) is the least dense of all gases. A sample of H_2 gas is found to occupy a volume of 1.23 L at 755 mm Hg and 0°C. What volume, in liters, will this same gas sample occupy at 735 mm Hg pressure and a temperature of 50°C?

SOLUTION

Writing all the given data in the form of initial and final conditions, we have

$$P_1 = 755 \text{ mm Hg} \qquad P_2 = 735 \text{ mm Hg}$$
$$V_1 = 1.23 \text{ L} \qquad V_2 = ? \text{ L}$$
$$T_1 = 0°C = 273 \text{ K} \qquad T_2 = 50°C = 323 \text{ K}$$

Rearrangement of the combined gas law expression to isolate V_2 on the left side gives

$$V_2 = V_1 \times \frac{P_1}{P_2} \times \frac{T_2}{T_1}$$

Substituting numerical values into this equation and doing the arithmetic gives

$$V_2 = 1.23 \text{ L} \times \frac{755 \text{ mm Hg}}{735 \text{ mm Hg}} \times \frac{323 \text{ K}}{273 \text{ K}}$$
$$= 1.494874 \text{ L} \qquad \text{(calculator answer)}$$
$$= 1.49 \text{ L} \qquad \textbf{(correct answer)}$$

Answer Double Check:

Is the setup for solving the problem reasonable? Yes. The original volume, V_1, is multiplied by a pressure factor and a temperature factor. The pressure factor has a value greater than one because volume increases with a decrease in pressure. The temperature factor also has a value greater than one because volume increases as temperature increases.

▶ **Practice Exercise 12.7**   With a density of 5.89 g/mL, five times that of air, xenon is the densest gaseous-state element. A sample of Xe gas is found to occupy a volume of 3.52 L at 775 mm Hg and 20°C. What volume, in liters, will this same gas sample occupy at 785 mm Hg and a temperature of 22°C?

EXAMPLE 12.8 **Calculating Final Temperature Using the Combined Gas Law**

A sample of helium gas [the gas used in lighter-than-air airships (blimps)] occupies a volume of 180.0 mL at a pressure of 0.800 atm and a temperature of 29°C. What will be the temperature of the gas, in degrees Celsius, if the volume of the container is decreased to 90.0 mL and the pressure is increased to 1.60 atm?

SOLUTION

Writing all the given data in the form of initial and final conditions, we have

$$P_1 = 0.800 \text{ atm} \qquad P_2 = 1.60 \text{ atm}$$
$$V_1 = 180.0 \text{ mL} \qquad V_2 = 90.0 \text{ mL}$$
$$T_1 = 29°C = 302 \text{ K} \qquad T_2 = ?°C$$

Rearrangement of the combined gas law to isolate T_2 on the left side of the equation gives

$$T_2 = T_1 \times \frac{P_2}{P_1} \times \frac{V_2}{V_1}$$

Substituting the given data into this equation and doing the arithmetic gives

$$T_2 = 302 \text{ K} \times \frac{1.60 \text{ atm}}{0.800 \text{ atm}} \times \frac{90.0 \text{ mL}}{180.0 \text{ mL}}$$

$$= 302 \text{ K} \quad \text{(calculator and \textbf{correct answer} in kelvins)}$$

Converting the temperature to the Celsius scale by subtracting 273 gives 29°C as the final answer.

$$\text{T(°C)} = 302 \text{ K} - 273 = 29°\text{C}$$

Answer Double Check:

The temperature did not change! Is this a reasonable answer? Yes. The pressure correction factor, considered by itself, would cause the temperature to increase by a factor of 2; the pressure was doubled. The volume correction factor, considered by itself, would cause the temperature to decrease by a factor of 2; the volume was halved. Considered together, the effects of the two factors cancel each other and result in the temperature's not changing.

▶ **Practice Exercise 12.8** A sample of O_2 gas occupies a volume of 225 mL at a pressure of 1.20 atm and a temperature of 33°C. What will be the temperature, in °C, of the gas if the volume is decreased to 205 mL and the pressure in decreased to 1.00 atm?

12.7 AVOGADRO'S LAW

In the year 1811, Amedeo Avogadro, the same Avogadro who is the namesake for Avogadro's number (Sec. 9.5), published observations dealing with the relationship between the volume and amount of a gas. His observations are now embodied in what is called *Avogadro's law*. Several different ways of stating this law exist, two of which are considered in this section.

Avogadro's earliest formulation of the law, which is in terms of the number of molecules present in equal volumes of *different* gases, is considered first. **Avogadro's law** *states that equal volumes of different gases, measured at the same temperature and pressure, contain equal numbers of molecules.* Although Avogadro's law seems very simple in terms of today's scientific knowledge, it was a very astute conclusion at the time it was first proposed. At that time scientists were still struggling with the differences between atoms and molecules, and Avogadro's reasoning eventually led to the realization that many of the common gaseous elements, such as hydrogen, oxygen, nitrogen, and chlorine, occur naturally as diatomic molecules (H_2, O_2, N_2, Cl_2) rather than single atoms.

Equal volumes of gases at the same temperature and pressure, because they contain equal numbers of molecules, must also contain equal numbers of moles of molecules. Thus, Avogadro's law can also be stated in terms of volumes and moles rather than volumes and molecules.

Avogadro's law can also be stated in another alternative form, which uses terminology similar to that used in Boyle's, Charles's, and Gay-Lussac's laws. This alternative

Boyle's law, Charles's law, Gay-Lussac's law, and the combined gas law all share a common feature—in each case a fixed amount of gas is present. What happens if the amount of gas present in a sample is changed? Avogadro's law deals with this situation.

form deals with two samples of the same gas rather than two samples of different gases. **Avogadro's law** *states that the volume of a gas is* directly proportional *to the number of moles of gas present if the temperature and pressure are kept constant.* When the original number of moles of gas present is doubled, the volume of the gas increases twofold. Halving the number of moles present halves the volume.

Experimentally, it is easy to show the direct relationship between volume and amount of gas present at constant temperature and pressure. All that is needed is an inflated balloon. Adding more gas to the balloon increases the volume of the balloon. The more gas added to the balloon, the greater the volume increase. Conversely, letting gas out of the balloon decreases its volume. It should be noted that this balloon demonstration is an approximate rather than exact demonstration for the law, since the tension of the rubber material from which the balloon is made exerts an effect on the proportionality between volume and moles.

Mathematical statements of Avogadro's law, where n represents the number of moles, are

$$\frac{V}{n} = \text{constant} \quad \text{and} \quad \frac{V_1}{n_1} = \frac{V_2}{n_2}$$

Note the similarity in mathematical form between this law and the laws of Charles and Gay-Lussac. The similarity results from all three laws' being direct proportionality relationships between two variables.

Avogadro's law in the form

$$\frac{V_1}{n_1} = \frac{V_2}{n_2}$$

can be used to calculate the change in the volume of a gas when the molar amount of the gas changes. Example 12.9 illustrates the use of Avogadro's law in this context, which involves the comparison of two different samples of the *same* gas.

Avogadro's law can also be used in comparisons involving samples of two *different* gases as long as both samples are at the same temperature and pressure. The original formulation of Avogadro's law, as discussed at the start of this section, came from comparing samples of different gases. The message of Avogadro's law, in mathematical terms, is that the volume-to-mole ratio for all samples of gases, at a given temperature and pressure, is the same; that is, this ratio is a constant. Volume depends only on the number of moles present and not on the chemical identity of the gas.

This constancy of the volume-to-mole ratio at a given set of conditions means that if 1.00 mole of carbon dioxide gas (CO_2) has a volume of 4.00 L, then 2.00 moles of ammonia gas (NH_3) will have a volume of 8.00 L at the same temperature and pressure. The volume-to-mole ratio for both gas samples is 4.00

$$\frac{V_{CO_2}}{n_{CO_2}} = \frac{4.00 \text{ L}}{1.00 \text{ mole}} = 4.00 \, \frac{\text{L}}{\text{mole}} \quad \text{and} \quad \frac{V_{NH_3}}{n_{NH_3}} = \frac{8.00 \text{ L}}{2.00 \text{ mole}} = 4.00 \, \frac{\text{L}}{\text{mole}}$$

This ratio constancy means that Avogadro's law can be viewed as having the following form when samples of different gases are compared:

$$\frac{V_{\text{gas A}}}{n_{\text{gas A}}} = \frac{V_{\text{gas B}}}{n_{\text{gas B}}}$$

This interpretation of Avogadro's law can be used in applicable problem-solving situations that involve different gaseous systems.

EXAMPLE 12.9 Calculating Final Volume Using Avogadro's Law

A balloon containing 2.00 moles of He has a volume of 0.500 L at a given temperature and pressure. If 0.48 mole of He gas is removed from the balloon without changing temperature and pressure conditions, what will be the new volume, in liters, of the balloon?

SOLUTION

We will use the equation

$$\frac{V_1}{n_1} = \frac{V_2}{n_2}$$

in solving the problem.

Writing the given data in terms of initial and final conditions, we have

$$V_1 = 0.500 \text{ L} \qquad V_2 = ? \text{ L}$$
$$n_1 = 2.00 \text{ moles} \qquad n_2 = (2.00 - 0.48) \text{ moles} = 1.52 \text{ moles}$$

Rearrangement of the Avogadro's law expression to isolate V_2 on the left side gives

$$V_2 = V_1 \times \frac{n_2}{n_1}$$

Substituting numerical values into the equation gives

$$V_2 = 0.500 \text{ L} \times \frac{1.52 \text{ moles}}{2.00 \text{ moles}} = 0.38 \text{ L} \qquad \text{(calculator answer)}$$
$$= 0.380 \text{ L} \qquad \textbf{(correct answer)}$$

Answer Double Check:

Is our answer consistent with reasoning? Yes. Decreasing the number of moles of gas present in the balloon at constant temperature and pressure should decrease the volume of the balloon.

▶ **Practice Exercise 12.9**   A balloon containing 3.50 moles of N_2 gas has a volume of 25.0 L at a given temperature and pressure. If 1.75 moles of N_2 gas is added to the balloon without changing temperature and pressure conditions, what will be the new volume, in liters, of the balloon?

Avogadro's law and the combined gas laws can be combined to give the expression

$$\frac{P_1 V_1}{n_1 T_1} = \frac{P_2 V_2}{n_2 T_2}$$

This equation covers the situation where none of the four variables, P, T, V, and n, is constant. With it, a change in any one of the four variables brought about by changes in the other three variables can be calculated.

EXAMPLE 12.10 Calculating Final Volume When None of the Other Gas Law Variables Are Constant

A balloon containing 4.00 moles of N_2 gas has a volume of 0.750 L at a pressure of 1.00 atm and a temperature of 27°C. What will the new volume of the balloon be, in liters, if the pressure is doubled to 2.00 atm, the temperature is increased to 127°C, and 1.50 moles of N_2 is removed from the balloon?

SOLUTION

We will use the equation

$$\frac{P_1V_1}{n_1T_1} = \frac{P_2V_2}{n_2T_2}$$

in the rearranged form

$$V_2 = V_1 \times \frac{P_1}{P_2} \times \frac{n_2}{n_1} \times \frac{T_2}{T_1}$$

in solving the problem.

Writing the given data in terms of initial and final conditions, we have

$V_1 = 0.750$ L	$V_2 = ?$ L
$n_1 = 4.00$ moles	$n_2 = (4.00 - 1.50)$ moles $= 2.50$ moles
$P_1 = 1.00$ atm	$P_2 = 2.00$ atm
$T_1 = 27°C + 273 = 300$ K	$T_2 = 127°C + 273 = 400$ K

Substituting numerical values into the rearranged equation where V_1 is alone on the left side of the equation gives

$$V_2 = 0.750 \text{ L} \times \frac{1.00 \text{ atm}}{2.00 \text{ atm}} \times \frac{2.50 \text{ moles}}{4.00 \text{ moles}} \times \frac{400 \text{ K}}{300 \text{ K}}$$

$$= 0.3125 \text{ L} \quad \text{(calculator answer)}$$

$$= 0.312 \text{ L} \quad \text{(correct answer)}$$

Answer Double Check:

Is the answer correct in terms of the number of significant figures present? Yes. Every one of the numbers used in the calculation contains three significant figures. Therefore, having three significant figures in the answer is correct.

▶ **Practice Exercise 12.10**   A balloon containing 2.50 moles of He gas has a volume of 30.0 L at a pressure of 5.00 atm and a temperature of 27°C. What will the new volume be, in liters, if the pressure is decreased by a factor of 2, the temperature is decreased by 100 degrees Celsius, and an additional 2.50 moles of He gas is added to the balloon?

12.8 AN IDEAL GAS

Most simple gases such as N_2, O_2, H_2, CO, CO_2, and the noble gases obey the gas laws we have so far considered as well as those we have yet to consider very closely at the temperatures and pressures at which these gases are normally encountered. However, when very precise measurements are made on a gaseous system, it is found that slight deviations from the gas laws do occur. These deviations increase in magnitude when gas pressures are very high or when temperatures are near the point where the gas is about to change to a liquid.

A gas that would *exactly* obey all gas laws, a hypothetical situation, is called an *ideal gas*. An **ideal gas** *is a gas that would obey gas laws exactly over all temperatures and all pressures.* Real gases exhibit behavior more and more like an ideal gas as pressure decreases (from high values) and temperature increases (away from the condensation point). Real gases, under the conditions at which they are normally encountered can be considered to be ideal gases.

Remarkably, the individual gas laws are singular expressions that apply to all gases. Different expressions are not needed for different gases. The behavior of N_2, O_2, H_2, He, CO, and CO_2 as well as other gases can be predicted using a single set of gas laws as long as temperature and pressure conditions are those normally encountered.

12.9 THE IDEAL GAS LAW

All of the gas laws so far considered in this chapter are used to describe gaseous systems where change occurs. Two sets of conditions, with one unknown variable, is the common feature of the systems that they describe. It is useful to also have a gas law that describes a chemical system where no changes in conditions occurs. Such a law exists and is known as the *ideal gas law*. The name given to this law, *ideal* gas law, relates to the previous discussion in Section 12.8 of the concept of an *ideal* gas. The **ideal gas law** is *a gas law that describes the relationships among the four variables pressure, volume, molar amount, and temperature for a gaseous substance at a given set of conditions.*

The ideal gas law is easily derived from the gas laws we have previously considered. Specifically, Boyle's law (Sec. 12.3), Charles's law (Sec 12.4), and Avogadro's law (Sec. 12.7), which constitute three independent relationships that deal with the volume of a gas, can be used to derive the ideal gas law.

$$\text{Boyle's law:} \quad V = k\left(\frac{1}{P}\right) \quad (n \text{ and } T \text{ constant})$$

$$\text{Charles's law:} \quad V = kT \quad (n \text{ and } P \text{ constant})$$

$$\text{Avogadro's law:} \quad V = kn \quad (P \text{ and } T \text{ constant})$$

We can combine these three equations into a single expression

$$V = \frac{kTn}{P}$$

since we know (from mathematics) that if a quantity is independently proportional to two or more quantities, it is also proportional to their product. This combined equation is a mathematical statement of the ideal gas law. Any gas that obeys the individual laws of Boyle, Charles, and Avogadro will also obey the ideal gas law.

An alternate statement of the ideal gas law is

$$PV = nRT$$

The ideal gas law applies to systems that do not undergo changes in pressure, volume, temperature, and amount (moles) of gas.

This form of the ideal gas law, besides being rearranged, differs from the previous form in that the proportionality constant k has been given the symbol R. The constant R is called the *ideal gas constant*. To use the ideal gas law in chemical calculations, the numerical value of R must be known. Its value is dependent on the units chosen to express P and V. (The variables n and T will always have the units moles and K, respectively.) Volume, for gases, is most often expressed in liters. There are two commonly encountered units for gas pressure—atm and mm Hg (Sec 12.2). The two most commonly used numerical values for R are

$$R = 0.082057 \, \frac{\text{atm} \cdot \text{L}}{\text{mole} \cdot \text{K}}$$

and

$$R = 62.364 \, \frac{\text{mm Hg} \cdot \text{L}}{\text{mole} \cdot \text{K}}$$

The difference in the two R values relates to the choice of pressure unit—atm or mm Hg. The two values are related to each other by the number 760, which is the number of millimeters of mercury in an atmosphere.

Note the complex units associated with the ideal gas law constant R; all four of the variables pressure, volume, moles, and K are involved since no cancellation of units occurs. Remembering the arrangement of these units is facilitated by recalling the mathematical expression for the ideal gas law when it is solved for R.

$$R = \frac{PV}{nT}$$

Pressure and volume units are in the numerator and mole and K units are in the denominator.

The preceding two R values have an experimental basis. Laboratory measurements indicate that 1 mole of a gas at a temperature of 0°C (273.15 K) and a pressure of 1 atm (or 760 mm Hg) occupies a volume of 22.414 L. Substituting these values into the ideal gas equation, solved for R, generates the R values that were previously given, as is shown by the following equations.

$$R = \frac{PV}{nT} = \frac{(1 \text{ atm}) (22.414 \text{ L})}{(1 \text{ mole}) (273.15 \text{ K})} = 0.082057477 \frac{\text{atm} \cdot \text{L}}{\text{mole} \cdot \text{K}} \qquad \text{(calculator answer)}$$

$$= 0.082057 \frac{\text{atm} \cdot \text{L}}{\text{mole} \cdot \text{K}} \qquad \textbf{(correct answer)}$$

$$R = \frac{PV}{nT} = \frac{(760 \text{ mm Hg}) (22.414 \text{ L})}{(1 \text{ mole}) (273.15 \text{ K})} = 62.363682 \frac{\text{mm Hg} \cdot \text{L}}{\text{mole} \cdot \text{K}} \qquad \text{(calculator answer)}$$

$$= 62.364 \frac{\text{mm Hg} \cdot \text{L}}{\text{mole} \cdot \text{K}} \qquad \textbf{(correct answer)}$$

In these two expressions, from a significant figure standpoint, the value 22.414 L is treated as a measured quantity and the other three variables are treated as defined quantities; thus the volume measurement determines the number of significant figures in the value of R.

These two values of R, to four significant figures, should be memorized, since it is these unit combinations that are usually used in problem solving.

$$0.08206 \frac{\text{atm} \cdot \text{L}}{\text{mole} \cdot \text{K}} \qquad 62.36 \frac{\text{mm Hg} \cdot \text{L}}{\text{mole} \cdot \text{K}}$$

When pressure units other than millimeters of mercury or atmosphere and volume units other than liters are encountered, convert them to one of these units and then use the appropriate known R value.

If three of the four variables in the ideal gas law are known, the fourth can be calculated. Importantly, this fourth variable can have only one value. This means that the state of a gas can be determined by specifying any three of the four variables in the ideal gas law.

The ideal gas law is used in chemical calculations, when *one* set of conditions is given (P, V, n, T) with the value of one variable not known. The combined gas law (Sec 12.6) is used when *two* sets of conditions are given with the value of one variable not known. Examples 12.11 and 12.12 illustrate the use of the ideal gas law in the manner just described.

EXAMPLE 12.11 Calculating the Volume of a Gas Using the Ideal Gas Law

The colorless, odorless, tasteless, toxic gas carbon monoxide, CO, is a by-product of incomplete combustion of any material that contains the element carbon. Calculate the volume, in liters, occupied by 1.52 moles of this gas at 0.992 atm pressure and a temperature of 65°C.

SOLUTION

This problem deals with only one set of conditions, a situation where the ideal gas equation is applicable. Three of the four variables in the ideal gas equation (P, n, and T) are given, and the fourth (V) is to be calculated.

$$P = 0.992 \text{ atm} \qquad n = 1.52 \text{ moles}$$
$$V = ? \text{ L} \qquad T = 65°C = 338 \text{ K}$$

Rearranging the ideal gas equation to isolate V on the left side of the equation gives

$$V = \frac{nRT}{P}$$

Since the pressure is given in atmospheres and the volume unit is in liters, the appropriate R value is

$$R = 0.08206 \frac{\text{atm} \cdot \text{L}}{\text{mole} \cdot \text{K}}$$

Substituting the given numerical values into the equation and cancelling units gives

$$V = \frac{(1.52 \text{ moles})\left(0.08206 \dfrac{\text{atm} \cdot \text{L}}{\text{mole} \cdot \text{K}}\right)(338 \text{ K})}{0.992 \text{ atm}}$$

Note how all of the parts of the ideal gas constant unit system cancel except for one, the volume part.

Doing the arithmetic, we get as an answer 42.5 L CO.

$$V = \frac{1.52 \times 0.08206 \times 338}{0.992} \text{ L CO}$$
$$= 42.499138 \text{ L CO} \qquad \text{(calculator answer)}$$
$$= 42.5 \text{ L CO} \qquad \text{(correct answer)}$$

Answer Double Check:

Is the ideal gas law, in its rearranged form, in a form such that all of the units cancel except for that of volume (liters)? The answer is yes.

▶ **Practice Exercise 12.11**   Hydrogen chloride (HCl) is a colorless gas that dissolves in water to produce hydrochloric acid. Calculate the volume, in liters, occupied by 0.345 mole of HCl gas at 1.03 atm pressure and a temperature of 43°C.

EXAMPLE 12.12 **Calculating the Temperature of a Gas Using the Ideal Gas Law**

Dinitrogen monoxide (nitrous oxide), N_2O, is naturally present, in trace amounts, in the atmosphere. Its source is decomposition reactions occurring in soils. What would be the temperature, in degrees Celsius, of a 67.4 g sample of N_2O gas under a pressure of 5.00 atm in a 7.00 L container?

SOLUTION

The amount of N_2O present is given in grams rather than moles. The grams need to be changed to moles prior to using the ideal gas law.

$$67.4 \text{ g } N_2O \times \frac{1 \text{ mole } N_2O}{44.02 \text{ g } N_2O} = 1.5311222 \text{ moles } N_2O \qquad \text{(calculator answer)}$$
$$= 1.53 \text{ moles } N_2O \qquad \text{(correct answer)}$$

Three of the four variables in the ideal gas equation (P, V, and n) are now known, and the fourth (T) is to be calculated.

$$P = 5.00 \text{ atm} \qquad n = 1.53 \text{ moles}$$
$$V = 7.00 \text{ L} \qquad T = ? \text{ K}$$

Rearranging the ideal gas equation to isolate T on the left side gives

$$T = \frac{PV}{nR}$$

Since the pressure is given in atmospheres and the volume in liters, the value of R to be used is

$$R = 0.08206 \frac{\text{atm} \cdot \text{L}}{\text{mole} \cdot \text{K}}$$

Substituting numerical values into the equation gives

$$T = \frac{(5.00 \text{ atm})(7.00 \text{ L})}{(1.53 \text{ moles})\left(0.08206 \dfrac{\text{atm} \cdot \text{L}}{\text{mole} \cdot \text{K}}\right)}$$

Notice again how the gas constant units, except for K, cancel. After cancellation, the expression $1/(1/\text{K})$ remains. This expression is equivalent to K. That this is the case can be easily shown. All we need to do is multiply both the numerator and denominator of the fraction by K.

$$\frac{1 \times \text{K}}{\dfrac{1}{\text{K}} \times \text{K}} = \text{K}$$

Doing the arithmetic, we get as an answer 279 K for the temperature of the O_2 gas.

$$T = \frac{(5.00)(7.00)}{(1.53)(0.08206)} \text{ K} = 278.7694 \text{ K} \qquad \text{(calculator answer)}$$

$$= 279 \text{ K} \qquad \text{(correct answer)}$$

The calculated temperature is in kelvins. To convert to degrees Celsius, the unit specified in the problem statement, we subtract 273 from the Kelvin temperature.

$$T(^\circ\text{C}) = 279 \text{ K} - 273 = 6 ^\circ\text{C}$$

Answer Double Check:

Is the magnitude of the numerical answer reasonable? Yes. Rounding off the input numbers for the calculation to numbers easier to work with — 5, 10, 2, and 0.1 — the ballpark answer for the problem is $(5 \times 10)/(2 \times 0.1) = 250$. The value of the calculated answer, in kelvins, is of the same order as the rounded answer.

▶ **Practice Exercise 12.12** The toxic air pollutant carbon monoxide (CO) gives humans no warning of its presence because it is colorless, odorless, and tasteless; it is these properties that make CO so dangerous. What would be the temperature, in degrees Celsius, of a 32.0 g sample of CO gas under a pressure of 6.00 atm in a 6.00 L container?

12.10 MODIFIED FORMS OF THE IDEAL GAS LAW EQUATION

In its standard form, $PV = nRT$, the ideal gas law equation is a very useful equation, as was considered in the previous section. Recasting the ideal gas law equation into slightly modified forms expands the uses that can be made of this equation. In this section the calculation of molar masses for gases and gas densities using modified forms of the ideal gas law equation are considered.

The Molar Mass of a Gas

For compounds that are gases at convenient temperatures and pressures, the ideal gas law provides a basis for determining molar mass. Required information is the mass of the gas and the conditions it is under (P, T, and V). The quantity molar mass (M) is incorporated directly into the ideal gas equation before using it, and then the equation is solved directly for the molar mass.

Recall, from Section 9.8 (Example 9.8), that the units for molar mass are grams/mole and that the molar mass of a substance is given by the expression

$$\text{molar mass } (M) = \frac{\text{mass } (m)}{\text{moles } (n)}$$

Rearranging this expression to isolate moles on the left side of the equation gives

$$n = \frac{m}{M}$$

Replacing n in the ideal gas equation with this equivalent expression gives

$$PV = nRT$$
$$= \left(\frac{m}{M}\right)RT$$

This equation, in the rearranged form where molar mass is isolated on the left side of the equation, can be used to calculate the molar mass of a gas.

$$M = \frac{mRT}{PV}$$

Example 12.13 illustrates the use of this equation in calculating a molar mass for a gaseous substance.

EXAMPLE 12.13 **Calculating the Molar Mass of a Gas Using the Ideal Gas Law in Modified Form**

A 3.30 g sample of chlorine gas (Cl_2) occupies a volume of 1.20 L at 741 mm Hg and 33°C. Based on this data, calculate the molar mass of chlorine gas.

SOLUTION

Rearranging the modified ideal gas equation

$$PV = \left(\frac{m}{M}\right)RT$$

to isolate M on the left side of the equation gives

$$M = \frac{mRT}{PV}$$

This equation can be used to directly calculate molar mass since all quantities on the right side of the equation are known.

$$m = 3.30 \text{ g}, \quad R = 62.36 \, \frac{\text{mm Hg} \cdot \text{L}}{\text{mole} \cdot \text{K}}, \quad T = 33°\text{C} + 273 = 306 \text{ K}$$

$$P = 741 \text{ mm Hg}, \quad V = 1.20 \text{ L}$$

Substitution of these known values in the equation gives

$$M = \frac{(3.30 \text{ g})\left(62.36 \, \frac{\text{mm Hg} \cdot \text{L}}{\text{mole} \cdot \text{K}}\right)(306 \text{ K})}{(741 \text{ mm Hg})(1.20 \text{ L})}$$

All units cancel except for grams per mole, the units of molar mass.

Doing the arithmetic, we obtain a value of 70.8 for the molar mass of Cl_2.

$$M = \frac{3.30 \times 62.36 \times 306}{741 \times 1.20} \, \frac{\text{g}}{\text{mole}} = 70.817732 \, \frac{\text{g}}{\text{mole}} \qquad \text{(calculator answer)}$$

$$= 70.8 \, \frac{\text{g}}{\text{mole}} \qquad \text{(correct answer)}$$

Answer Double Check:

Is the answer reasonable? Yes. The calculated molar mass can be compared to that obtained using atomic masses from the periodic table. The periodic table molar mass of chlorine (Cl_2) is 70.90 g/mole. The answer obtained, using the ideal gas law and allowing for experimental error, is consistent with the periodic table value.

▶ **Practice Exercise 12.13**   A 0.276 g sample of oxygen gas (O_2) occupies a volume of 0.270 L at 739 mm Hg and 98°C. Calculate from these data the molar mass of gaseous O_2.

The Density of a Gas

The density of a substance is equal to its mass divided by its volume (Sec. 3.8).

$$d = \frac{m \, (\text{gram})}{V \, (\text{liter})}$$

The ideal gas law in the form

$$PV = \left(\frac{m}{M}\right)RT$$

can be rearranged to calculate density by isolating m/V on the left side of the equation.

$$\frac{m}{V} = \frac{PM}{RT}$$

Thus, the density form of the ideal gas equation is

$$d = \frac{PM}{RT}$$

Before doing a calculation using this density equation, let us consider messages about the density of a gas conveyed by this equation.

1. The density of a gas is directly proportional (Sec. 12.4) to pressure. The higher the pressure, the more dense the gas.
2. The density of a gas is directly proportional to its molar mass. The higher the molar mass, the more dense the gas. At a given temperature and pressure, CO_2 gas (with a molar mass of 44.01 g/mole) will have a higher density than O_2 gas (with a molar mass of 32.00 g/mole).
3. The density of a gas is inversely proportional (Sec. 12.3) to temperature. The higher the temperature, the less dense the gas.

EXAMPLE 12.14 **Calculating the Density of a Gas Using the Ideal Gas Law in Modified Form**

In gas law calculations, air is often considered to be a single gas with a molar mass of 29 g/mole. On this basis, calculate the density of air, in grams per liter, on a hot summer day (41°C) when the atmospheric pressure is 0.91 atm.

SOLUTION

The ideal gas equation in the modified form

$$d = \frac{PM}{RT}$$

is used to calculate the density of a gas.

All the quantities on the right side of this equation are known.

$$P = 0.91 \text{ atm,} \qquad M = 29 \text{ g/mole}$$

$$R = 0.08206 \frac{\text{atm} \cdot \text{L}}{\text{mole} \cdot \text{K}} \qquad T = 41°C = 314 \text{ K}$$

Substitution of these values into the equation gives

$$d = \frac{(0.91 \text{ atm})\left(29 \frac{\text{g}}{\text{mole}}\right)}{\left(0.08206 \frac{\text{atm} \cdot \text{L}}{\text{mole} \cdot \text{K}}\right)(314 \text{ K})}$$

All units cancel except for the desired ones, grams per liter.

Doing the arithmetic, we obtain a value of 1.0 g/L for the density of air at the specified temperature and pressure.

$$d = \frac{0.91 \times 29}{0.08206 \times 314} \frac{\text{g}}{\text{L}} = 1.0241845 \frac{\text{g}}{\text{L}} \qquad \text{(calculator answer)}$$

$$= 1.0 \frac{\text{g}}{\text{L}} \qquad \text{(correct answer)}$$

▶ **Practice Exercise 12.14** Calculate the density of carbon tetrachloride (CCl_4) vapor, in grams per liter, at 1.20 atm and 230°C.

Using Density to Calculate Molar Mass

If the modified form of the ideal gas law in which both molar mass (*M*) and density (*d*) are present, which is

$$d = \frac{PM}{RT}$$

is rearranged to isolate molar mass on the left side of the equation, an expression is generated by which the molar mass of a gas can be calculated given its density at a given temperature and pressure.

$$M = \frac{dRT}{P}$$

Example 12.15 illustrates a molar mass calculation using this form of the modified ideal gas law.

| EXAMPLE 12.15 | Calculating the Molar Mass of a Gas Given its Density |

At 25°C and a pressure of 1.14 atm, a gas is found to have a density of 1.40 g/L. Calculate the molar mass of the gas.

SOLUTION

The ideal gas law in the modified form

$$M = \frac{dRT}{P}$$

is used to calculate the molar mass of a gas given its density at a given temperature and pressure.

All of the quantities on the right side of this equation are known.

$$d = 1.40 \text{ g/L} \qquad P = 1.14 \text{ atm}$$

$$R = 0.08206 \; \frac{\text{atm} \cdot \text{L}}{\text{mole} \cdot \text{K}} \qquad T = 25°C = 398 \text{ K}$$

Substitution of these values into the equation gives

$$M = \frac{\left(1.40 \; \dfrac{\text{g}}{\text{L}}\right)\left(0.08206 \; \dfrac{\text{atm} \cdot \text{L}}{\text{mole} \cdot \text{K}}\right)(298 \text{ K})}{1.14 \text{ atm}}$$

All units cancel except for the desired ones, which are grams per mole.

Doing the arithmetic, we obtain a value of

$$M = \frac{1.40 \times 0.08206 \times 298}{1.14} \; \frac{\text{g}}{\text{mole}} = 30.031081 \; \frac{\text{g}}{\text{mole}} \qquad \text{(calculator answer)}$$

$$= 30.0 \; \frac{\text{g}}{\text{mole}} \qquad \textbf{(correct answer)}$$

Answer Double Check:

Is the magnitude of the answer reasonable? Yes. There are a number of common gases that have molar masses of about 30 g/mole. Included among them are O_2 (32.00 g/mole), N_2 (28.02 g/mole), CO (28.01 g/mole) and NO (30.01 g/mole).

▶ **Practice Exercise 12.15** At 30°C and a pressure of 1.12 atm, a gas is found to have a density of 1.64 g/L. Calculate the molar mass of the gas.

12.11 VOLUMES OF GASES IN CHEMICAL REACTIONS

Gases are involved as reactants or products in many chemical reactions. In such reactions it is often easier to determine the volumes rather than the masses of the gases involved.

A very simple relationship exists between the volumes of different gases consumed or produced in chemical reactions, provided the volumes are all determined at the same temperature and pressure. This relationship, known as *Gay-Lussac's law of combining volumes*, was first formulated in 1808 by Joseph Louis Gay-Lussac (the same Gay-Lussac discussed in Sec. 12.5). **Gay-Lussac's law of combining volumes** *states that the volumes of different gases that participate in a chemical reaction, measured at the same temperature and pressure, are in the same ratio as the coefficients for these gases in the balanced* *equation for the reaction.* For example, 1 volume of nitrogen and 3 volumes of hydrogen react to give 2 volumes of ammonia (NH_3).

$$N_2(g) + 3\,H_2(g) \longrightarrow 2\,NH_3(g)$$
1 volume 3 volumes   2 volumes

"Volume" is used here in the general sense of relative volume in any units. It could, for example, be 1 L of N_2, 3 L of H_2, and 2 L of NH_3 or 0.1 mL of N_2, 0.3 mL of H_2, and 0.2 mL of NH_3. Note that in making volume comparisons, the units must always be the same; if the volume of N_2 is measured in milliliters, the volumes of H_2 and NH_3 must also be in milliliters.

It is important to remember that these volume relationships apply only to gases, and then only when all gaseous volumes are measured at the same temperature and pressure. The volumes of solids and liquids involved in reactions cannot be treated this way.

The previously given statement of the law of combining volumes is a modern version of the law. At the time the law was first formulated by Gay-Lussac, chemists were still struggling with the difference between atoms and molecules. Equations as we now write them were unknown; formulas for substances were still to be determined. The original statement of the law simply noted that the ratio of volumes was always a ratio of small whole numbers. Gay-Lussac's work was based solely on volume measurements and had nothing to do with chemical equations. The explanation for the small whole numbers and the linkup with equations came later.

This volume interpretation for the coefficients in a balanced chemical equation involving gaseous substances is the third interpretation we have encountered for equation coefficients. In Section 10.4 we learned that equation coefficients can be interpreted in terms of molecules, and in Section 10.7 a molar interpretation for equation coefficients was considered. Thus, for the equation

$$4\,NH_3(g) + 3\,O_2(g) \longrightarrow 2\,N_2(g) + 6\,H_2O(g)$$

it is correct to say

  4 *molecules* $NH_3(g)$ + 3 *molecules* $O_2(g) \longrightarrow$ 2 *molecules* $N_2(g)$ + 6 *molecules* $H_2O(g)$

and

  4 *moles* $NH_3(g)$ + 3 *moles* $O_2(g) \longrightarrow$ 2 *moles* $N_2(g)$ + 6 *moles* $H_2O(g)$

and, as a result of Gay-Lussac's law of combining volumes,

  4 *volumes* $NH_3(g)$ + 3 *volumes* $O_2(g) \longrightarrow$ 2 *volumes* $N_2(g)$ + 6 *volumes* $H_2O(g)$

Again, all the volumes must be measured at the same temperature and pressure for these volume relationships to be valid.

Table 12.4 summarizes the three ways of interpreting equation coefficients in chemical equations where gaseous substances are involved.

TABLE 12.4 Ways in Which Equation Coefficients May Be Interpreted

For the general equation	2A (g)	+	3B (g)	$\longrightarrow$	C (g)	+	2D (g)
The ratio of molecules is	2	:	3	:	1	:	2
The ratio of moles is	2	:	3	:	1	:	2
The ratio of volumes of gas (at the same temperature and pressure) is	2	:	3	:	1	:	2

In calculations involving chemical reactions, where two or more gases are participants, this volume interpretation of coefficients can be used to generate conversion factors useful in problem solving. Consider the balanced equation

$$2\,CO(g)\,+\,O_2(g) \rightarrow 2\,CO_2(g)$$

At constant temperature and pressure, three volume–volume relationships are obtainable from this equation.

2 volumes CO	produce	2 volumes CO_2
1 volume O_2	produces	2 volumes CO_2
2 volumes CO	react with	1 volume O_2

From these three relationships, six conversion factors can be written. From the first relationship,

$$\frac{2 \text{ volumes CO}}{2 \text{ volumes } CO_2} \quad \text{and} \quad \frac{2 \text{ volumes } CO_2}{2 \text{ volumes CO}}$$

From the second relationship,

$$\frac{1 \text{ volume } O_2}{2 \text{ volumes } CO_2} \quad \text{and} \quad \frac{2 \text{ volumes } CO_2}{1 \text{ volume } O_2}$$

From the third relationship,

$$\frac{2 \text{ volumes CO}}{1 \text{ volume } O_2} \quad \text{and} \quad \frac{1 \text{ volume } O_2}{2 \text{ volumes CO}}$$

The more gaseous reactants and products there are in a chemical reaction, the greater the number of volume–volume conversion factors obtainable from the equation for the chemical reaction. The use of volume–volume conversion factors is illustrated in Example 12.16.

EXAMPLE 12.16   **Using Gay-Lussac's Law of Combining Volumes to Determine Reactant–Product Volume Relationships**

Nitrogen (N_2) reacts with hydrogen (H_2) to produce ammonia (NH_3) as shown by the equation

$$N_2(g)\,+\,3\,H_2(g) \longrightarrow 2\,NH_3(g)$$

What volume of H_2, in liters, at 750 mm Hg and 25°C is required to produce 1.75 L of NH_3 at the same temperature and pressure?

SOLUTION

Step 1 The given quantity is 1.75 L of NH_3, and the desired quantity is liters of H_2.

$$1.75 \text{ L } NH_3 = ? \text{ L } H_2$$

Step 2 This is a volume–volume problem. The conversion factor needed for this one-step problem is derived from the coefficients of H_2 and NH_3 in the equation for the chemical reaction. The equation tells us that at constant temperature and pressure, three volumes of H_2 are needed to prepare two volumes of NH_3. From this relationship, the conversion factor

$$\frac{3 \text{ L } H_2}{2 \text{ L } NH_3}$$

is obtained.

Note that when *Gay-Lussac's law of combining volumes* does apply, you do not need to know the exact temperature and pressure values for the gases since they are not used in the calculation. All you need to know is that the temperature and pressure, whatever they may be, are the same for all of the gases.

If gases are not at identical temperatures and pressures, *Gay-Lussac's law of combining volumes* does not apply. In such cases, the best approach is to convert volume information to a mole basis, using other gas laws. Then use the mole interpretation for the coefficients in the chemical equation. Section 12.14 considers this approach in detail.

Step 3 The dimensional analysis setup for this problem is

$$1.75 \text{ L NH}_3 \times \frac{3 \text{ L H}_2}{2 \text{ L NH}_3}$$

Note that it makes no difference what the temperature and pressure are, as long as they are the same for the two gases involved in the calculation.

Step 4 Doing the arithmetic, after cancellation of units, gives

$$\frac{1.75 \times 3}{2} \text{ L H}_2 = 2.625 \text{ L H}_2 \qquad \text{(calculator answer)}$$

$$= 2.62 \text{ L H}_2 \qquad \textbf{(correct answer)}$$

Answer Double Check:

Is the calculated answer consistent with the fact that the two gases react in a 3:2 volume ratio? Yes. The calculated volume of 2.62 L is 1.5 times as large as the given volume of 1.75 L.

▶ **Practice Exercise 12.16** At high temperatures, sulfur dioxide gas (SO_2) reacts with oxygen gas (O_2) to produce gaseous sulfur trioxide (SO_3) according to the equation

$$2 \, SO_2(g) + O_2(g) \longrightarrow 2 \, SO_3(g)$$

What volume of SO_2, in liters is required to react with 2.77 L of O_2? Assume that both gases are at the same conditions of temperature and pressure.

12.12 VOLUMES OF GASES AND THE LIMITING REACTANT CONCEPT

In Section 10.10, we considered the concept of a limiting reactant and did calculations that determined limiting reactant identity. These calculations were based on moles of product formed from given masses of various reactants. Limiting reactant calculations can also be made using volumes of gases and the volume interpretation of equation coefficients. Example 12.17 illustrates how volume-based limiting reactant calculations are carried out.

| **EXAMPLE 12.17** | Calculating a Limiting Reactant Using Volumes of Gases |

Many high-temperature welding torches use an acetylene–oxygen mixture as a fuel. The equation for the reaction between these two substances is

$$2 \, C_2H_2(g) + 5 \, O_2(g) \longrightarrow 4 \, CO_2(g) + 2 \, H_2O(g)$$

If a fuel mixture contains 30.0 L of C_2H_2 and 80.0 L of O_2, which mixture component is the limiting reactant?

SOLUTION

To determine a limiting reactant by using volumes, we calculate how many volumes of gaseous product can be formed from given volumes of reactants, assuming constant conditions of temperature and pressure. In this particular problem there are two gaseous products: CO_2 and H_2O. It is sufficient to calculate the volume of just one of the products. The decision as to which product is used for the calculation is arbitrary; we will choose CO_2. We go through the calculation twice, once for each reactant. For C_2H_2,

$$30.0 \text{ L C}_2H_2 \times \frac{4 \text{ L CO}_2}{2 \text{ L C}_2H_2} = 60 \text{ L CO}_2 \qquad \text{(calculator answer)}$$

$$= 60.0 \text{ L CO}_2 \qquad \textbf{(correct answer)}$$

The conversion factor used was obtained from the coefficients of the balanced equation. For O_2,

$$80.0 \text{ L } O_2 \times \frac{4 \text{ L } CO_2}{5 \text{ L } O_2} = 64 \text{ L } CO_2 \qquad \text{(calculator answer)}$$

$$= 64.0 \text{ L } CO_2 \qquad \textbf{(correct answer)}$$

The limiting reactant is the reactant that will produce the fewest number of liters of CO_2, which in this case is C_2H_2.

Answer Double Check:

Limiting reactant calculations are always calculations in which amounts (volumes in this case) of the same product are compared from each of several parallel calculations. Such is the case here; volumes of the product CO_2 are compared.

▶ **Practice Exercise 12.17**   Chlorine and iodine, both diatomic molecules in the gaseous state, react according to the equation

$$3 \text{ Cl}_2 + \text{I}_2(g) \longrightarrow 2 \text{ ICl}_3(g)$$

If 55 mL of Cl_2 and 17 mL of I_2 at the same temperature and pressure are allowed to react, which gas is the limiting reactant?

12.13 MOLAR VOLUME OF A GAS

Chemical-equation-based calculations that involve volumes of gases is the topic of the next section in the text. A prerequisite for doing such calculations is an understanding of the concept of *molar volume*, the topic of this section. The **molar volume of a gas** *is the volume occupied by one mole of a gas at a specified temperature and pressure.* There are many different molar volumes for a gas since there are many different temperature–pressure combinations under which a sample of the gas can exist. Since gaseous volume is usually specified in liters, the most common units for molar volume are liters per mole.

Molar volume is a quite different quantity from the previously discussed concept of molar mass (Secs. 9.6 and 9.8). *Molar mass* has the units of grams per mole. Every compound has a specific molar mass that does not vary with external conditions such as temperature and pressure. There is only one molar mass for a given substance. Molar mass is easily calculated from a substance's chemical formula (Sec. 9.6). *Molar volume* has the units liters per mole. Molar volumes are not unique for substances as are molar masses since they depend on temperature and pressure, which can vary. Molar volumes for a substance are calculated using the ideal gas law. Examples 12.18 and 12.19 illustrate how such molar volume calculations proceed. Example 12.18 emphasizes the point that the molar volume of a gas changes as temperature and volume change. Example 12.19 emphasizes the point that all gases have the same molar volume at the same temperature and pressure.

EXAMPLE 12.18   **Calculating the Molar Volume of a Gas**

What is the molar volume of O_2 gas at each of the following sets of pressure–temperature conditions?

    a. 1.20 atm and 25°C
    b. 1.20 atm and 125°C
    c. 1.70 atm and 25°C
    d. 1.70 atm and 125°C

SOLUTION

The molar volume of a sample of any gas is calculated using the ideal gas law with volume isolated on the left side of the equation.

$$V = \frac{nRT}{P}$$

By definition, molar volume involves 1 mole of a gas. Molar volume is calculated by substituting this molar amount and the given vlaues for temperature and pressure, into the preceding equation.

a. molar volume $(V) = \dfrac{nRT}{P} = \dfrac{(1 \text{ mole})\left(0.08206 \dfrac{\text{atm} \cdot \text{L}}{\text{mole} \cdot \text{K}}\right)(298 \text{ K})}{(1.20 \text{ atm})}$

$\qquad\qquad\qquad = 20.378233 \text{ L}$ (calculator answer)

$\qquad\qquad\qquad = 20.4 \text{ L}$ **(correct answer)**

b. molar volume $(V) = \dfrac{nRT}{P} = \dfrac{(1 \text{ mole})\left(0.08206 \dfrac{\text{atm} \cdot \text{L}}{\text{mole} \cdot \text{K}}\right)(398 \text{ K})}{(1.20 \text{ atm})}$

$\qquad\qquad\qquad = 27.216567 \text{ L}$ (calculator answer)

$\qquad\qquad\qquad = 27.2 \text{ L}$ **(correct answer)**

c. molar volume $(V) = \dfrac{nRT}{P} = \dfrac{(1 \text{ mole})\left(0.08206 \dfrac{\text{atm} \cdot \text{L}}{\text{mole} \cdot \text{K}}\right)(298 \text{ K})}{(1.70 \text{ atm})}$

$\qquad\qquad\qquad = 14.384635 \text{ L}$ (calculator answer)

$\qquad\qquad\qquad = 14.4 \text{ L}$ **(correct answer)**

d. molar volume $(V) = \dfrac{nRT}{P} = \dfrac{(1 \text{ mole})\left(0.08206 \dfrac{\text{atm} \cdot \text{L}}{\text{mole} \cdot \text{K}}\right)(398 \text{ K})}{(1.70 \text{ atm})}$

$\qquad\qquad\qquad = 19.211694 \text{ L}$ (calculator answer)

$\qquad\qquad\qquad = 19.2 \text{ L}$ **(correct answer)**

Answer Double Check:

As the pressure on a gas is increased, its volume should decrease with other variables being held constant. Comparing the answers to parts (a) and (c) or to parts (b) and (d) shows that this is the case. As the temperature of a gas increases, the volume of the gas should increase with other variables being held constant. Comparing the answers to parts (a) and (b) or to parts (c) and (d) shows that this is the case.

▶ **Practice Exercise 12.18**   What is the molar volume of N_2 gas at each of the following sets of pressure–temperature conditions?

a. 0.988 atm and 37°C
b. 1.98 atm and 137°C

EXAMPLE 12.19 **Calculating the Molar Volume of a Gas**

What is the molar volume of each of the following gases at a temperature of 25°C and a pressure of 0.673 atm?

   a. H_2
   b. He
   c. CO
   d. HCN

SOLUTION

As in the previous example, the ideal gas law, rearranged in the form in which volume is isolated on the left side of the equation, is used to do each of the calculations. The temperature, pressure, and number of moles (1 mole by definition) are known. Note that the chemical formula of the gas does not enter into the calculation as there is nothing in the ideal gas law equation that relates specifically to a gas's chemical identity. This means that the calculation in each part will be identical. All gases at a given temperature and pressure will have the same molar volume as is shown in this set of calculations.

a. molar volume $(V) = \dfrac{nRT}{P} = \dfrac{(1 \text{ mole})\left(0.08206\ \dfrac{\text{atm} \cdot \text{L}}{\text{mole} \cdot \text{K}}\right)(298 \text{ K})}{(0.673 \text{ atm})}$

$\quad\quad\quad\quad\quad\quad\quad\quad = 36.335631 \text{ L}\quad$ (calculator answer)
$\quad\quad\quad\quad\quad\quad\quad\quad = 36.3 \text{ L}\quad\quad\quad$ **(correct answer)**

b. molar volume $(V) = \dfrac{nRT}{P} = \dfrac{(1 \text{ mole})\left(0.08206\ \dfrac{\text{atm} \cdot \text{L}}{\text{mole} \cdot \text{K}}\right)(298 \text{ K})}{(0.673 \text{ atm})}$

$\quad\quad\quad\quad\quad\quad\quad\quad = 36.335631 \text{ L}\quad$ (calculator answer)
$\quad\quad\quad\quad\quad\quad\quad\quad = 36.3 \text{ L}\quad\quad\quad$ **(correct answer)**

c. molar volume $(V) = \dfrac{nRT}{P} = \dfrac{(1 \text{ mole})\left(0.08206\ \dfrac{\text{atm} \cdot \text{L}}{\text{mole} \cdot \text{K}}\right)(298 \text{ K})}{(0.673 \text{ atm})}$

$\quad\quad\quad\quad\quad\quad\quad\quad = 36.335631 \text{ L}\quad$ (calculator answer)
$\quad\quad\quad\quad\quad\quad\quad\quad = 36.3 \text{ L}\quad\quad\quad$ **(correct answer)**

d. molar volume $(V) = \dfrac{nRT}{P} = \dfrac{(1 \text{ mole})\left(0.08206\ \dfrac{\text{atm} \cdot \text{L}}{\text{mole} \cdot \text{K}}\right)(298 \text{ K})}{(0.673 \text{ atm})}$

$\quad\quad\quad\quad\quad\quad\quad\quad = 36.335631 \text{ L}\quad$ (calculator answer)
$\quad\quad\quad\quad\quad\quad\quad\quad = 36.3 \text{ L}\quad\quad\quad$ **(correct answer)**

Answer Double Check:

Different gaseous substances at the same temperature and pressure should have the same molar volume. Thus, answers to all parts of this Example should be the same. Such is the case.

▶ **Practice Exercise 12.19** What is the molar volume of each of the following gases at a temperature of 175°C and a pressure of 1.21 atm?

a. F_2
b. Ne
c. CO_2
d. NH_3

Standard Temperature and Standard Pressure Conditions

The volumes of liquids and solids change only slightly with temperature and pressure changes (Sec. 11.2). This is not the case for volumes of gases. As was illustrated in Example 12.18, the molar volume of a gas is always dependent of its temperature and pressure.

Volume of different gases can be compared only if the gases are at the same temperatures and pressures. It is convenient to specify a particular temperature and pressure as standards for comparison purposes. **Standard temperature for gases** *is defined as 0°C (273.15 K).* **Standard pressure for gases** *is defined as 1 atm (760 mm Hg).* Gases that are at standard temperature and standard pressure conditions are said to be at *STP conditions.* **STP conditions for gases** *are those of standard temperature (0°C) and standard pressure (1 atm).*

In terms of significant figures, we note that in the definitions for standard temperature and standard pressure, the values 0°C, 1 atm, and 760 mm Hg are defined (exact) numbers and the value 273.15 K is an experimental value. In problem solving, standard temperature on the Kelvin scale is usually specified to three significant figures (273 K) rather than five significant figures (273.15 K). This operational rule is based on the general practice of specifying laboratory-measured temperatures to the closest number of degrees. The value 273.15 K is available, however, for use when temperatures involve tenths or hundredths of a degree.

The molar volume of any gas at STP conditions, to five significant figures, is 22.414 L.

$$V = \frac{nRT}{P} = \frac{(1 \text{ mole})\left(0.082057 \frac{\text{atm} \cdot \text{L}}{\text{mole} \cdot \text{K}}\right)(273.15 \text{ K})}{1 \text{ atm}}$$

$$= 22.413870 \text{ L (calculator answer)}$$

$$= 22.414 \text{ L} \quad \textbf{(correct answer)}$$

The selection of 0°C and 1 atm pressure as the values for standard temperature and standard pressure for gases is an arbitrary decision in the same way that the selection of ^{12}C as the reference point for the amu scale (Sec. 5.9) was an arbitrary decision.

Because STP conditions are so frequently encountered, it is useful to remember (memorize) this molar volume value. The value 22.414 is known as the *standard molar volume of a gas.* The **standard molar volume of a gas** *is the volume occupied by one mole of gas at STP conditions, which is 22.414 L.* To visualize a volume equal to 22.414 L, think of standard-size basketballs; the volume occupied by three standard-size basketballs is very close to 22.414 L. When used in chemical calculations, standard molar volume is usually specified to four significant figures (22.41 L) rather than five significant figures (24.414 L).

Using Molar Volume to Calculate Density

The molar volume of a gas, in conjunction with its molar mass, can be used to calculate the gas's density. Dividing the molar mass (grams/mole) of the substance by its molar volume (liters/mole) gives density (grams/liter). The mole unit cancels from the equation, leaving grams per liter.

$$\text{density} = \frac{\text{molar mass}}{\text{molar volume}} = \frac{\frac{\text{grams}}{\text{mole}}}{\frac{\text{liters}}{\text{mole}}} = \frac{\text{grams}}{\text{liter}}$$

Since molar volume is the same for all gases at a given temperature and pressure, the density of a gas is solely a function of (depends only on) the molar mass. Density for a gas does change, however, if the temperature and pressure are changed. Example 12.20 shows a molar-volume-based density calculation for the gas hydrogen cyanide (HCN).

EXAMPLE 12.20 **Calculating Density Using Molar Volume and Molar Mass**

Calculate the density of HCN, a highly poisonous gas, at each of the following sets of conditions.

a. 250°C and 2.00 atm pressure
b. 0°C and 1.00 atm pressure (STP conditions)

SOLUTION

a. The molar mass and molar volume relationships needed for this HCN density calculation are

$$\text{molar mass} = 27.03 \text{ g/mole} \qquad \text{molar volume} = 21.5 \text{ L/mole}$$

The molar volume value was obtained in the same manner as that shown in Examples 12.18 and 12.19.

$$V = \frac{nRT}{P} = \frac{(1 \text{ mole})\left(0.08206 \dfrac{\text{atm} \cdot \text{L}}{\text{mole} \cdot \text{K}}\right)(523 \text{ K})}{2.00 \text{ atm}}$$

$$= 21.45869 \text{ L} \quad \text{(calculator answer)}$$

$$= 21.5 \text{ L} \qquad \text{(correct answer)}$$

Substituting molar volume and molar mass into the density equation gives

$$\text{density} = \frac{\text{molar mass}}{\text{molar volume}} = \frac{\dfrac{27.03 \text{ g}}{\text{mole}}}{\dfrac{21.5 \text{ L}}{\text{mole}}}$$

$$= 1.2572093 \dfrac{\text{g}}{\text{L}} \quad \text{(calculator answer)}$$

$$= 1.26 \dfrac{\text{g}}{\text{L}} \qquad \text{(correct answer)}$$

b. The molar mass of a gas is independent of temperature and pressure. Thus, molar mass for HCN remains at 27.03 grams/mole. The molar volume of HCN will, however, be different than it was in part (a) since the temperature and pressure have changed. Molar volume now has the value 22.41 liters/mole, the molar volume value associated with STP conditions, which are the conditions for this part of the example. Substituting into the density equation gives

$$\text{density} = \frac{\text{molar mass}}{\text{molar volume}} = \frac{\dfrac{27.03 \text{ g}}{\text{mole}}}{\dfrac{22.41 \text{ L}}{\text{mole}}}$$

$$= 1.2061580 \dfrac{\text{g}}{\text{L}} \quad \text{(calculator answer)}$$

$$= 1.206 \dfrac{\text{g}}{\text{L}} \qquad \text{(correct answer)}$$

▶ **Practice Exercise 12.20** Calculate the density of SO_2 gas, an air pollutant produced when coal is burned, at each of the following sets of conditions.

a. 125°C and 1.25 atm pressure
b. 0°C and 1 atm pressure (STP conditions)

12.14 CHEMICAL CALCULATIONS USING MOLAR VOLUME

Molar volume, in the form of a conversion factor, is a useful entity in a variety of chemical calculations that involve volumes of gases. An example of a reciprocal pair of molar volume conversion factors is

$$\frac{19.82 \text{ L gas}}{1 \text{ mole gas}} \quad \text{or} \quad \frac{1 \text{ mole gas}}{19.82 \text{ L gas}}$$

A common type of problem requiring molar volume conversion factors is one where the volume of gas is known at a given set of conditions and you are asked to calculate from it moles, grams, or particles of gas present, or vice versa. Figure 12.9 summarizes the relationships needed in performing calculations of this general type.

Perhaps you recognize Figure 12.9 as being very similar to a diagram you have already encountered many times in problem solving, Figure 9.5. This new diagram differs from Figure 9.5 in only one way: Volume boxes have been added at the top of the diagram. The diagram in Figure 12.9 is used in the same way as the earlier one. The given and desired quantities are determined, and the arrows of the diagram are used to map out the pathway to be used in going from the given quantity to the desired quantity.

Examples 12.21 through 12.25, which follow, illustrate the usefulness of Figure 12.9 in several different contexts. We begin with a simple volume-of-gas-A to grams-of-gas-A problem, Example 12.21, which has the simplifying feature that STP conditions are present. At STP, the molar volume of a gas is known to be 22.41 L (Sec. 12.13).

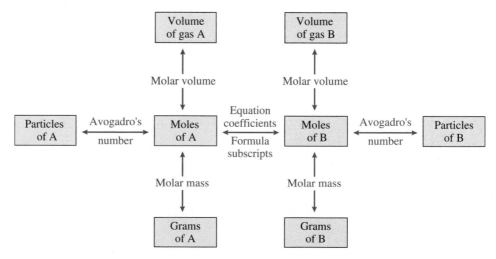

FIGURE 12.9 Quantitative relationships necessary for solving mass-to-volume and volume-to-mass chemical-formula-based problems.

EXAMPLE 12.21 **Calculating the Volume of a Gaseous Sample at STP Given Its Mass**

Fluorine gas, F_2, is one of the most reactive of all gases. What would be the volume, in liters, of a 15.0 g sample of this gas at STP conditions?

SOLUTION

Step 1 The given quantity is 15.0 g of F_2, and the unit of the desired quantity is liters of F_2.

$$15.0 \text{ g } F_2 = ? \text{ L } F_2$$

Step 2 In terms of Figure 12.9, this problem is a grams-of-A to volume-of-gas-A problem. The pathway is

$$\boxed{\begin{array}{c}\text{Grams}\\\text{of A}\end{array}} \xrightarrow[\text{mass}]{\text{Molar}} \boxed{\begin{array}{c}\text{Moles}\\\text{of A}\end{array}} \xrightarrow[\text{volume}]{\text{Molar}} \boxed{\begin{array}{c}\text{Volume}\\\text{of gas A}\end{array}}$$

Using dimensional analysis, we set up this sequence of conversion factors as

$$15.0 \text{ g } F_2 \times \frac{1 \text{ mole } F_2}{38.00 \text{ g } F_2} \times \frac{22.41 \text{ L } F_2}{1 \text{ mole } F_2}$$

$$\text{grams A} \longrightarrow \text{moles A} \longrightarrow \text{volume A}$$

Note how the standard molar volume relationship that 1 mole of gas is equal to 22.41 L is used as the last conversion factor.

Step 3 The solution, obtained by doing the arithmetic, is

$$\frac{15.0 \times 1 \times 22.41}{38.00 \times 1} \text{ L } F_2 = 8.8460526 \text{ L } F_2 \quad \text{(calculator answer)}$$

$$= 8.85 \text{ L } F_2 \quad \text{(correct answer)}$$

Answer Double Check:

Is the magnitude of the answer reasonable? Yes. The given amount of fluorine gas, 15.0 g, is less than one-half mole of this gas. Thus, the volume should be less than one-half of the standard molar volume of 22.4 liters. Such is the case; one-half of 22.4 L is 11.2 L, and 8.85 L (the answer) is less than that.

▶ Practice Exercise 12.21 Phosgene, a substance used in poisonous gas warfare in World War I, has the chemical formula $COCl_2$. What would be the volume, in liters, of a 40.0 g sample of this gas at STP conditions?

Example 12.22, which follows, is a problem in which the molar volume conversion factor needed must be determined in a "side calculation" because the conditions are other than STP conditions. This problem also illustrates the important point that equal volumes of different gases at the same set of conditions do not have equal masses since the molecules present have different masses.

The use of a side calculation to determine a quantity needed for a conversion factor, in this case molar volume, is not a new concept. In Chapters 9 and 10, any time a conversion from moles-of-A to grams-of-A (or vice versa) was carried out the molar mass of A had to be determined in a side calculation. In a similar manner, a side calculation is needed to determine molar volume any time a change from volume-of-gas-A to moles-of-A, at non STP conditions, is carried out. The molar volume side calculation involves use of the ideal gas law (Sec. 12.13).

EXAMPLE 12.22 **Calculating the Mass of a Gaseous Sample Given Its Volume**

What is the mass, in grams, of 2.50 L of each of the following gases at 28°C and 1.05 atm pressure?

    a. carbon monoxide (CO)   **b.** carbon dioxide (CO_2)

SOLUTION

    a. Calculation of the mass of CO proceeds as follows:

Step 1 The given quantity is 2.50 L of CO and the desired quantity is grams of CO.

$$2.50 \text{ L CO} = \text{g CO}$$

Step 2 This is a volume-of-gas-A to grams-of-A problem. The pathway appropriate for solving this problem, as indicated in Figure 12.9, is

$$\boxed{\begin{array}{c}\text{Volume}\\\text{of gas A}\end{array}} \xrightarrow[\text{volume}]{\text{Molar}} \boxed{\begin{array}{c}\text{Moles}\\\text{of A}\end{array}} \xrightarrow[\text{mass}]{\text{Molar}} \boxed{\begin{array}{c}\text{Grams}\\\text{of A}\end{array}}$$

This pathway shows that two conversion factors are needed in the dimensional analysis setup for the problem: one involving molar volume and the other involving molar mass. The numerical values for both of these conversion factors are obtained using side calculations. Using atomic masses, the molar mass of CO is determined to be 28.01 g. Using the ideal gas law, the molar volume of CO is determined to be 23.5 L. The setup for this latter determination is

$$V = \frac{nRT}{P} = \frac{(1 \text{ mole})\left(0.08206 \; \dfrac{\text{atm} \cdot \text{L}}{\text{mole} \cdot \text{K}}\right)(301 \text{ K})}{1.05 \text{ atm}}$$

$$= 23.523867 \text{ L} \quad \text{(calculator answer)}$$

$$= 23.5 \text{ L} \qquad \textbf{(correct answer)}$$

Using these conversion factors, the setup for the problem from dimensional analysis becomes

$$2.50 \text{ L CO} \times \frac{1 \text{ mole CO}}{23.5 \text{ L CO}} \times \frac{28.01 \text{ g CO}}{1 \text{ mole CO}}$$

Step 3 Collecting the numerical terms after cancellation of units and doing the arithmetic gives

$$\frac{2.50 \times 1 \times 28.01}{23.5 \times 1} \text{ g CO} = 2.9797872 \text{ g CO} \quad \text{(calculator answer)}$$

$$= 2.98 \text{ g CO} \qquad \textbf{(correct answer)}$$

    b. The analysis and setup for this part are identical to those in (a), except that CO_2 and its molecular mass replace CO and its molecular mass. The molar volume factor remains the same because the temperature–pressure conditions in both parts are the same. Thus, the setup is

$$2.50 \text{ L } \text{CO}_2 \times \frac{1 \text{ mole } \text{CO}_2}{23.5 \text{ L } \text{CO}_2} \times \frac{44.01 \text{ g } \text{CO}_2}{1 \text{ mole } \text{CO}_2} = 4.6819149 \text{ g } \text{CO}_2 \quad \text{(calculator answer)}$$

$$= 4.68 \text{ g } \text{CO}_2 \qquad \textbf{(correct answer)}$$

Equal volumes of different gases at STP do not have equal masses; 2.50 L of CO has a mass of 2.98 g, and 2.50 L of CO_2 has a mass of 4.68 g.

Answer Double Check:

Are the magnitudes of the two answers reasonable when compared to each other? Yes. For equal volumes of the two gases, the mass of the CO_2 should be greater than the mass of CO because a CO_2 molecule is heavier than a CO molecule.

▶ **Practice Exercise 12.22** What is the mass, in grams, of 3.75 L of each of the following gases at 35°C and 0.987 atm pressure?

  a. ammonia (NH_3)
  b. phosphine (PH_3)

Example 12.23 illustrates the use of the molar volume concept in the context of a chemical reaction calculation where information is given about a solid-state reactant and information is desired about a gaseous product.

EXAMPLE 12.23 **Calculating the STP Volume of a Gas that Reacts in a Chemical Reaction**

Lithium metal is one of the few substances with which nitrogen gas (N_2) will react.

$$6 \text{ Li}(s) + \text{N}_2(g) \longrightarrow 2 \text{ Li}_3\text{N}(s)$$

What volume, in liters, of N_2 at STP will completely react with 75.0 g of Li metal?

SOLUTION

Step 1 The given quantity is 75.0 g of Li, and the desired quantity is liters of N_2 at STP conditions.

$$75.0 \text{ g Li} = ? \text{ L } \text{N}_2$$

Step 2 This is a grams-of-A to volume-of-gas-B problem. The pathway used in solving it, in terms of Figure 12.9, is

Grams of A	→ Molar mass →	Moles of A	→ Equation coefficients →	Moles of B	→ Molar volume →	Volume of gas B

The dimensional analysis setup for the calculation is

$$75.0 \text{ g } \text{Li} \times \frac{1 \text{ mole } \text{Li}}{6.94 \text{ g } \text{Li}} \times \frac{1 \text{ mole } \text{N}_2}{6 \text{ moles } \text{Li}} \times \frac{22.41 \text{ L } \text{N}_2}{1 \text{ mole } \text{N}_2}$$

Step 3 The solution, obtained from combining all the numerical factors, is

$$\frac{75.0 \times 1 \times 1 \times 22.41}{6.94 \times 6 \times 1} \text{ L N}_2 = 40.363832 \text{ L N}_2 \quad \text{(calculator answer)}$$

$$= 40.4 \text{ L N}_2 \qquad \text{(\textbf{correct answer})}$$

Answer Double Check:

Is the magnitude of the numerical answer of the right order? Yes. Rounding numbers in the numerical setup, the numerator (other than 22.41) is about 80 and the denominator is about 40. This division gives a factor of 2; therefore, the answer should be approximately twice the standard molar volume value (2 × 22). The calculated answer of 40.4 is consistent with this analysis.

▶ **Practice Exercise 12.23**   Chlorine gas reacts with elemental phosphorus as follows:

$$3 \text{ Cl}_2(g) + 2 \text{ P}(s) \longrightarrow 2 \text{ PCl}_3(g)$$

What volume, in liters, of Cl_2 gas at STP conditions will completely react with 85.0 g of P?

Examples 12.24 and 12.25 are both chemical-equation-based calculations. Example 12.24 involves a situation where information is given about a gaseous product and information is desired about a solid reactant. Example 12.25 brings limiting reactant considerations into the calculational picture.

EXAMPLE 12.24   **Calculating the Mass of a Reactant from the Volume of a Gaseous Product**

Oxygen gas can be generated by heating KClO_3 to a high temperature.

$$2 \text{ KClO}_3(s) \longrightarrow 2 \text{ KCl}(s) + 3 \text{ O}_2(g)$$

How much KClO_3, in grams, is needed to generate 7.50 L of O_2 at a pressure of 1.00 atm and a temperature of 37°C?

SOLUTION

Step 1   The given quantity is 7.50 L of O_2, and the desired quantity is grams of KClO_3.

$$7.50 \text{ L O}_2 = ? \text{ g KClO}_3$$

Step 2   This is a volume-of-gas-A to grams-of-B problem. The pathway, in terms of Figure 12.9, is

For the first unit change (volume-of-gas-A to moles-of-A) the molar volume conversion factor relationship needed, calculated using the ideal gas law, is

$$\frac{1 \text{ mole O}_2}{25.4 \text{ L O}_2}$$

The complete dimensional analysis setup for the problem becomes

$$7.50 \ \cancel{L\text{-}O_2} \times \frac{1 \ \cancel{\text{mole } O_2}}{25.4 \ \cancel{L\text{-}O_2}} \times \frac{2 \ \cancel{\text{moles } KClO_3}}{3 \ \cancel{\text{moles } O_2}} \times \frac{122.55 \ g \ KClO_3}{1 \ \cancel{\text{mole } KClO_3}}$$

$$\text{volume A} \longrightarrow \text{moles A} \longrightarrow \text{moles B} \longrightarrow \text{grams B}$$

Step 3 The solution, obtained from combining all the numerical factors, is

$$\frac{7.50 \times 1 \times 2 \times 122.55}{25.4 \times 3 \times 1} \ g \ KClO_3 = 24.124016 \ g \ KClO_3 \quad \text{(calculator answer)}$$

$$= 24.1 \ g \ KClO_3 \quad \text{(correct answer)}$$

▶ **Practice Exercise 12.24** Carbon dioxide gas can be generated by heating copper(II) carbonate to a high temperature.

$$CuCO_3(s) \longrightarrow CuO(s) + CO_2(g)$$

How much $CuCO_3$, in grams, is needed to generate 2.00 L of CO_2 at a pressure of 2.00 atm and a temperature of 27°C?

| EXAMPLE 12.25 | **Calculating Gaseous Product Volume from Gaseous Reactant Volumes** |

How many liters of NH_3 gas at 3.00 atm pressure and 55°C can be produced from 10.0 L of N_2 gas at 1.00 atm pressure and 35°C and 30.0 L of H_2 gas at 2.00 atm pressure and 45°C, according to the following chemical reaction?

$$N_2(g) + 3 H_2(g) \longrightarrow 2 NH_3(g)$$

SOLUTION

First, we must determine the limiting reactant because specific volumes of both reactants are given in the problem. This determination involves two separate volume-of-gas-A to moles-of-B calculations—one for each reactant. The pathway for these parallel calculations is

$$\boxed{\begin{array}{c} \text{Volume} \\ \text{of gas A} \end{array}} \xleftrightarrow[\text{volume}]{\text{Molar}} \boxed{\begin{array}{c} \text{Moles} \\ \text{of A} \end{array}} \xleftrightarrow[\text{coefficients}]{\text{Equation}} \boxed{\begin{array}{c} \text{Moles} \\ \text{of B} \end{array}}$$

The molar volume conversion factors for N_2 and H_2 will be different because the two gases are under different temperature–pressure conditions. Using the ideal gas law, we obtain the needed molar volume conversion factors as follows:

$$V_{N_2} = \frac{nRT}{P} = \frac{(1 \ \cancel{\text{mole}})\left(0.08206 \ \dfrac{\cancel{\text{atm}} \cdot L}{\cancel{\text{mole}} \cdot \cancel{K}}\right)(308 \ \cancel{K})}{(1.00 \ \cancel{\text{atm}})}$$

$$= 25.27448 \ L \quad \text{(calculator answer)}$$

$$= 25.3 \ L \quad \text{(correct answer)}$$

$$V_{H_2} = \frac{nRT}{P} = \frac{(1 \ \cancel{\text{mole}})\left(0.08206 \ \dfrac{\cancel{\text{atm}} \cdot L}{\cancel{\text{mole}} \cdot \cancel{K}}\right)(318 \ \cancel{K})}{(2.00 \ \cancel{\text{atm}})}$$

$$= 13.04754 \ L \quad \text{(calculator answer)}$$

$$= 13.0 \ L \quad \text{(correct answer)}$$

With these two conversion factors known, the limiting reactant calculation proceeds in the normal manner.

For N_2,

$$10.0 \text{ L } N_2 \times \frac{1 \text{ mole } N_2}{25.3 \text{ L } N_2} \times \frac{2 \text{ moles } NH_3}{1 \text{ mole } N_2}$$

$$= 0.79051383 \text{ mole } NH_3 \quad \text{(calculator answer)}$$

$$= 0.791 \text{ mole } NH_3 \qquad \textbf{(correct answer)}$$

For H_2,

$$30.0 \text{ L } H_2 \times \frac{1 \text{ mole } H_2}{13.0 \text{ L } H_2} \times \frac{2 \text{ moles } NH_3}{3 \text{ moles } H_2}$$

$$= 1.5384615 \text{ moles } NH_3 \quad \text{(calculator answer)}$$

$$= 1.54 \text{ moles } NH_3 \qquad \textbf{(correct answer)}$$

Nitrogen (N_2) is the limiting reactant since it produces fewer moles of NH_3.

The overall calculation is a volume-of-gas-A to volume-of-gas-B calculation.

Volume of gas A	$\xrightarrow{\text{Molar volume}}$	Moles of A	$\xrightarrow{\text{Equation coefficients}}$	Moles of B	$\xrightarrow{\text{Molar volume}}$	Volume of gas B

In doing the limiting reactant calculation we obtained moles of B (0.791 mole of NH_3). Thus, all that is left to do is to go from moles-of-B to volume-of-gas-B. A new molar volume conversion factor is needed since NH_3 is present at different temperature–pressure conditions than was either N_2 or H_2.

This molar volume conversion factor is

$$V_{NH_3} = \frac{nRT}{P} = \frac{(1 \text{ mole})\left(0.08206 \frac{\text{atm} \cdot \text{L}}{\text{mole} \cdot \text{K}}\right)(328 \text{ K})}{(3.00 \text{ atm})}$$

$$= 8.9718933 \text{ L } NH_3 \quad \text{(calculator answer)}$$

$$= 8.97 \text{ L } NH_3 \qquad \textbf{(correct answer)}$$

The moles-of-B to grams-of-B calculation, a one-step calculation, becomes

$$0.791 \text{ mole } NH_3 \times \frac{8.97 \text{ L } NH_3}{1 \text{ mole } NH_3}$$

$$= 7.09527 \text{ N L } NH_3 \text{ (calculator answer)}$$

$$= 7.10 \text{ L } NH_3 \qquad \textbf{(correct answer)}$$

▶ **Practice Exercise 12.25** How many liters of CO_2 gas at 1.50 atm pressure and 50°C can be produced from 20.0 L of CO gas at 2.00 atm pressure and 40°C and 40.0 L of O_2 gas at 5.00 atm pressure and 250°C according to the following chemical equation?

$$2\,CO(g) + O_2(g) \longrightarrow 2\,CO_2(g)$$

12.15 MIXTURES OF GASES

Many of the original experiments from which the gas laws we have considered in this chapter were "discovered" were based on the behavior of samples of air. Air is a mixture of gases. The simple gas laws and the ideal gas law apply not only to individual gases but also to *mixtures* of gases that do not react with each other.

Often the gas laws can be applied to gaseous mixtures by using, for the value of n (moles of gas), the sum of the moles of the various components present in a gaseous mixture. This approach to using gas laws for gaseous mixtures is illustrated in Example 12.26.

EXAMPLE 12.26 | **Applying the Ideal Gas Law to a Mixture of Gases**

A gaseous mixture consists of 3.00 g of N_2 and 7.00 g of Ne. What is the volume, in liters, occupied by this mixture if it is under a pressure of 2.00 atm and at a temperature of 27°C?

SOLUTION

We first calculate the number of moles of each of the gases present and then add these quantities together to obtain the total moles of gas present.

$$n_{N_2}: \quad 3.00 \text{ g } N_2 \times \frac{1 \text{ mole } N_2}{28.02 \text{ g } N_2} = 0.10706638 \text{ mole } N_2 \quad \text{(calculator answer)}$$

$$= 0.107 \text{ mole } N_2 \quad \text{(correct answer)}$$

$$n_{Ne}: \quad 7.00 \text{ g Ne} \times \frac{1 \text{ mole Ne}}{20.18 \text{ g Ne}} = 0.34687809 \text{ mole Ne} \quad \text{(calculator answer)}$$

$$= 0.347 \text{ mole Ne} \quad \text{(correct answer)}$$

$$n_{total} = n_{N_2} + n_{Ne} = (0.107 + 0.347) \text{ mole} = 0.454 \text{ mole}$$

$$\text{(calculator and correct answer)}$$

Rearranging the ideal gas law to isolate V on the left side of the equation gives

$$V = \frac{nRT}{p}$$

All of the quantities on the right side of the equation are known.

$$n_{total} = 0.454 \text{ mole}, \quad R = 0.08206 \frac{\text{atm} \cdot \text{L}}{\text{mole} \cdot \text{K}} \quad T = 300 \text{ K}, \quad P = 2.00 \text{ atm}$$

Substituting in our equation, cancelling units, and doing the arithmetic gives

$$V = \frac{(0.454 \text{ mole})\left(0.08206 \frac{\text{atm} \cdot \text{L}}{\text{mole} \cdot \text{K}}\right)(300 \text{ K})}{2.00 \text{ atm}}$$

$$= 5.588286 \text{ L} \quad \text{(calculator answer)}$$

$$= 5.59 \text{ L} \quad \text{(correct answer)}$$

Answer Double Check:

Is the magnitude of the numerical answer reasonable? Yes. Approximately 0.5 total moles of gas are present, which would give 11.2 L of gas at STP conditions. The conditions differ from STP mainly in pressure, which is double that at STP. This would decrease the volume by a factor of 2 to 5.6 L. The calculated answer is consistent with this rough analysis.

▶ **Practice Exercise 12.26** A gaseous mixture consists of 5.00 g of Xe and 5.00 g of Ne. What is the volume, in liters, occupied by this mixture if it is under a pressure of 1.00 atm and at a temperature of 127°C?

12.16 DALTON'S LAW OF PARTIAL PRESSURES

In a mixture of gases that do not react with each other, each type of molecule moves about in the container as if the other kinds were not there. This type of behavior is possible because attractions between molecules in the gaseous state are negligible at most temperatures and pressures and because a gas is mostly empty space (Sec. 11.6). Each gas in the mixture occupies the entire volume of the container; that is, it distributes itself uniformly throughout the container. The molecules of each type strike the walls of the container as frequently and with the same energy as though they were the only gas in the mixture. Consequently, the pressure exerted by a gas in a mixture is the same as it would be if the gas were alone in the same container under the same conditions.

John Dalton—the same John Dalton discussed in Section 4.10—was the first to notice this independent behavior of gases in mixtures. In 1803 he published a summary statement concerning such behavior, which is now known as *Dalton's law of partial pressures*. **Dalton's law of partial pressures** *states that the total pressure exerted by a mixture of gases is the sum of the partial pressures of the individual gases.* A new term, *partial pressure*, is used in stating Dalton's law. A **partial pressure** *is the pressure that a gas in a gaseous mixture would exert if it were the only gas present under the same conditions.*

Expressed mathematically, Dalton's law states that

$$P_{total} = P_1 + P_2 + P_3 + \cdots$$

where P_{total} is the total pressure of a gaseous mixture and P_1, P_2, P_3, and so on are the partial pressures of the individual gaseous components of the mixture. (When the identity of a gas is known, its molecular formula is used as a subscript in the partial-pressure notation; for example, P_{CO_2} is the partial pressure of carbon dioxide in a mixture.)

To illustrate Dalton's law, consider the four identical gas containers shown in Figure 12.10. Suppose we place amounts of three different gases (represented by A, B, and C) into three of the containers and measure the pressure exerted by each sample. We then place all three samples in the fourth container and measure the pressure exerted by this mixture of gases. It is found that

$$P_{total} = P_A + P_B + P_C$$

Using the pressures given in Figure 12.10, we see that

$$P_{total} = 1 + 3 + 2 = 6$$

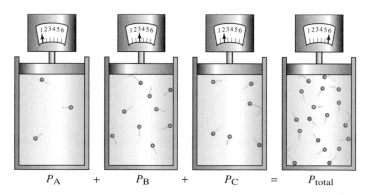

$$P_A \quad + \quad P_B \quad + \quad P_C \quad = \quad P_{total}$$

FIGURE 12.10 An illustration of Dalton's law of partial pressures.

EXAMPLE 12.27   Using Dalton's Law to Calculate a Partial Pressure

An unknown quantity of the noble gas xenon (Xe) is added to a cylinder already containing a mixture of the noble gases helium (He) and argon (Ar) at partial pressures, respectively, of 3.00 atm and 1.00 atm. After the Xe addition, the total pressure in the cylinder is 5.80 atm. What is the partial pressure, in atmospheres, of the Xe gas?

SOLUTION

The partial pressures of the He and Ar in the mixture will not be affected by the addition of the Xe (Dalton's law). Thus, the partial pressures of the He and Ar remain at 3.00 atm and 1.00 atm, respectively.

The sum of the partial pressures of He and Ar is 4.00 atm.

$$3.00 \text{ atm} + 1.00 \text{ atm} = 4.00 \text{ atm}$$

The difference between this pressure sum and the total pressure in the cylinder is caused by the Xe present. Thus, the partial pressure of the Xe is

$$P_{Xe} = P_{total} - (P_{He} + P_{Ar})$$
$$= 5.80 \text{ atm} - 4.00 \text{ atm} = 1.8 \text{ atm} \quad \text{(calculator answer)}$$
$$= 1.80 \text{ atm} \quad \textbf{(correct answer)}$$

▶ **Practice Exercise 12.27**   The total pressure for a mixture of three noble gases is 3.00 atm, with each noble gas having the same partial pressure. If one of the noble gases is completely removed from the gaseous mixture, what are the partial pressures, in atm, for each of the remaining noble gases?

The validity of Dalton's law of partial pressures is easily demonstrated by using the ideal gas law (Sec. 12.9). For a mixture of three gases (A, B, and C), the total pressure is given by the expression

$$P_{total} = n_{total} \frac{RT}{V}$$

The total number of moles present is the sum of the moles of A, B, and C; that is,

$$n_{total} = n_A + n_B + n_C$$

Substituting this equation into the previous one gives

$$P_{total} = (n_A + n_B + n_C) \frac{RT}{V}$$

Expanding the right side of this equation results in the expression

$$P_{total} = n_A \frac{RT}{V} + n_B \frac{RT}{V} + n_C \frac{RT}{V}$$

The three individual terms on the right side of this equation are, respectively, the partial pressure, of A, B, and C.

$$P_A = n_A \frac{RT}{V}, \; P_B = n_B \frac{RT}{V}, \; P_C = n_C \frac{RT}{V}$$

Substitution of this information into the previous equation gives

$$P_{total} = P_A + P_B + P_C$$

which is a statement of Dalton's law of partial pressures for a mixture of three gases.

In the preceding derivation, two of the expressions that are encountered are

$$P_A = n_A \frac{RT}{V} \quad \text{and} \quad P_{total} = n_{total} \frac{RT}{V}$$

If we divide the first of these expressions by the second, we obtain

$$\frac{P_A}{P_{total}} = \frac{n_A \dfrac{RT}{V}}{n_{total} \dfrac{RT}{V}} = \frac{n_A}{n_{total}} \quad \text{or} \quad \frac{P_A}{P_{total}} = \frac{n_A}{n_{total}}$$

Rearrangement of this equation gives

$$P_A = \frac{n_A}{n_{total}} \times P_{total}$$

The fraction n_A/n_{total} in this equation is called the *mole fraction* of A in the mixture. It is the fraction of the total moles that is accounted for by gas A. A **mole fraction** *is a dimensionless quantity that gives the ratio of the number of moles of a component in a mixture to the number of moles of all components present.* The symbol X is used to denote a mole fraction.

$$X_A = \frac{n_A}{n_{total}}$$

Substituting this equation into the preceding equation gives

$$P_A = X_A \times P_{total}$$

The partial pressure of a gas in a mixture is equal to its mole fraction multiplied by the total pressure. Example 12.28 shows how mole fractions are calculated, and Example 12.29 shows how mole fractions are used in obtaining partial pressures.

EXAMPLE 12.28 **Calculating the Mole Fraction for a Gas in a Gaseous Mixture**

A gaseous mixture contains 10.0 g each of the gases N_2, O_2, and Ar. What is the mole fraction of each gas in the mixture?

SOLUTION

We first calculate the number of moles of each gas present.

N_2: $10.0 \text{ g } N_2 \times \dfrac{1 \text{ mole } N_2}{28.02 \text{ g } N_2} = 0.35688793 \text{ mole } N_2$ (calculator answer)

$= 0.357 \text{ mole } N_2$ (**correct answer**)

O_2: $10.0 \text{ g } O_2 \times \dfrac{1 \text{ mole } O_2}{32.00 \text{ g } O_2} = 0.3125 \text{ mole } O_2$ (calculator answer)

$= 0.312 \text{ mole } O_2$ (**correct answer**)

Ar: $10.0 \text{ g Ar} \times \dfrac{1 \text{ mole Ar}}{39.95 \text{ g Ar}} = 0.25031289 \text{ mole Ar}$ (calculator answer)

$= 0.250 \text{ mole Ar}$ (**correct answer**)

The total number of moles of gas present is

$$n_{\text{total}} = n_{N_2} + n_{O_2} + n_{Ar}$$
$$= (0.357 + 0.312 + 0.250) \text{ mole}$$
$$= 0.919 \text{ mole} \quad \text{(calculator and \textbf{correct answer})}$$

Mole fractions are calculated as ratios of individual component moles to total moles.

$$X_{N_2} = \frac{0.357 \text{ mole}}{0.919 \text{ mole}} = 0.38846572 \quad \text{(calculator answer)}$$
$$= 0.388 \qquad \text{(\textbf{correct answer})}$$

$$X_{O_2} = \frac{0.312 \text{ mole}}{0.919 \text{ mole}} = 0.33949945 \quad \text{(calculator answer)}$$
$$= 0.339 \qquad \text{(\textbf{correct answer})}$$

$$X_{Ar} = \frac{0.250 \text{ mole}}{0.919 \text{ mole}} = 0.27203482 \quad \text{(calculator answer)}$$
$$= 0.272 \qquad \text{(\textbf{correct answer})}$$

Answer Double Check:

The sum of all mole fractions should always add to one. Such is the case here.

$$0.388 + 0.339 + 0.272 = 0.999$$

Rounding errors cause the sum to be 0.999 rather than 1.000.

▶ **Practice Exercise 12.28** A gaseous mixture contains 2.00 g of N_2, 1.00 g of O_2, and 10.0 g of He. What is the mole fraction of each gas in the mixture?

EXAMPLE 12.29 **Obtaining Partial Pressures Using Mole Fractions**

A mixture of gases contains 4.23 moles of neon (Ne), 0.93 mole of argon (Ar), and 7.65 moles of hydrogen (H_2). Calculate the partial pressures of the gases if the total pressure is 5.00 atm at a certain temperature.

SOLUTION

We first calculate the mole fraction of each gas.

$$X_{Ne} = \frac{4.23 \text{ moles}}{(4.23 + 0.93 + 7.65) \text{ moles}} = 0.33021077 \quad \text{(calculator answer)}$$
$$= 0.330 \qquad \text{(\textbf{correct answer})}$$

$$X_{Ar} = \frac{0.93 \text{ mole}}{(4.23 + 0.93 + 7.65) \text{ moles}} = 0.072599531 \quad \text{(calculator answer)}$$
$$= 0.073 \qquad \text{(\textbf{correct answer})}$$

$$X_{H_2} = \frac{7.65 \text{ moles}}{(4.23 + 0.93 + 7.65) \text{ moles}} = 0.59718969 \quad \text{(calculator answer)}$$
$$= 0.597 \qquad \text{(\textbf{correct answer})}$$

To calculate partial pressures, we rearrange the equation

$$\frac{P_A}{P_{total}} = X_A$$

to isolate the partial pressure on a side by itself.

$$P_A = X_A \times P_{total}$$

Substituting known quantities into this equation gives the partial pressures.

$P_{Ne} = 0.330 \times 5.00$ atm $= 1.65$ atm   (calculator and **correct answer**)

$P_{Ar} = 0.073 \times 5.00$ atm $= 0.365$ atm   (calculator answer)

                       $= 0.36$ atm   **(correct answer)**

$P_{H_2} = 0.597 \times 5.00$ atm $= 2.985$ atm   (calculator answer)

                       $= 2.98$ atm   **(correct answer)**

Answer Double Check:

The sum of the partial pressures of the gases in the mixture should equal the total pressure of the mixture. Such is the case here.

$$(1.65 + 0.36 + 2.98) \text{ atm} = 4.99 \text{ atm}$$

The sum 4.99 atm is consistent with the given total pressure of 5.00 atm; rounding errors are the basis for the two pressures differing in the hundredths digit.

▶ **Practice Exercise 12.29**   A mixture of gases contains 5.00 moles of CO, 2.00 moles of Ne, and 1.00 mole of Ar. Calculate the partial pressures of the gases if the total pressure is 10.0 atm at a certain temperature.

EXAMPLE 12.30   **Calculating the Partial Pressure of a Gas**

Calculate the partial pressure of O_2, in atm, in a gaseous mixture with a volume of 2.50 L at 20°C, given that the mixture composition is 0.50 mole O_2 and 0.75 mole N_2.

SOLUTION

This problem differs from Example 12.29 in that both the temperature and volume of the gaseous mixture are given. This added information will enable us to calculate the partial pressure of O_2 using two different methods: (1) mole fractions, and (2) the ideal gas law.

Using the mole fraction method, we calculate first the total pressure of the gaseous mixture based on the presence of 1.25 total moles of gas (0.50 mole + 0.75 mole):

$$P_{total} = \frac{n_{total}RT}{V} = \frac{(1.25 \text{ moles})\left(0.08206 \dfrac{\text{atm} \cdot \text{L}}{\text{mole} \cdot \text{K}}\right)(293 \text{ K})}{2.50 \text{ L}}$$

$$= 12.02179 \text{ atm} \quad \text{(calculator answer)}$$

$$= 12.0 \text{ atm} \qquad \text{(correct answer)}$$

Next we calculate the mole fraction of O_2:

$$X_{O_2} = \frac{n_{O_2}}{n_{total}} = \frac{0.50 \text{ mole}}{(0.50 + 0.75) \text{ mole}} = 0.4 \quad \text{(calculator answer)}$$

$$= 0.40 \text{ } \textbf{(correct answer)}$$

The partial pressure of the O_2 is obtained using the mole fraction O_2 and the total pressure:

$$P_{O_2} = X_{O_2} P_{total} = 0.40 \times 12.0 \text{ atm} = 4.8 \text{ atm} \quad \text{(calculator and \textbf{correct answer})}$$

This problem can also be solved without the use of mole fractions by using the concept that the partial pressure exerted by the O_2 is determined by the number of moles of O_2 present. With this approach, we substitute directly into the ideal gas law the quantities given in the problem statement.

$$P_{O_2} = \frac{n_{O_2} RT}{V} = \frac{(0.50 \text{ mole})\left(0.08206 \frac{\text{atm} \cdot \text{L}}{\text{mole} \cdot \text{K}}\right)(293 \text{ K})}{2.50 \text{ L}}$$

$$= 4.808716 \text{ atm} \quad \text{(calculator answer)}$$

$$= 4.8 \text{ atm} \quad \quad \text{(\textbf{correct answer})}$$

Answer Double Check:

The partial pressure of the O_2 gas should be 40% of the total pressure, based on its mole fraction of 0.40. Such is the case; 0.40×12.0 atm (total pressure) = 4.8 atm.

▶ **Practice Exercise 12.30** Calculate the partial pressure of N_2, in atm, in a gaseous mixture with a volume of 2.20 L at 50°C, given that the mixture composition is 0.20 mole N_2 and 2.20 mole NH_3.

The air we breathe is a most important mixture of gases. The composition of clean air from which all water vapor has been removed (dry air) is found to be virtually constant over the entire earth. Table 12.5 gives the composition of clean, dry air in terms of mole fractions. All components that have a mole fraction of at least 1×10^{-5} (0.001%) are listed.

Atmospheric pressure is the sum of the partial pressures of the gaseous components present in air. Table 12.5 also gives the partial pressure of each component of air in a situation where total atmospheric pressure is 760 mm Hg.

The composition of air is not absolutely constant. The variability in composition is caused predominantly by the presence of water vapor, a substance not listed in Table 12.5 because those statistics are for *dry* air. The amount of water vapor in air varies between almost zero and a mole fraction of 0.05–0.06, depending on weather and temperature.

TABLE 12.5 The Major Components of Clean, Dry Air

Gaseous Component	Formula	Mole Fraction	Partial Pressure (mm Hg) When Total Pressure Is 760.0 mm Hg
Nitrogen	N_2	0.78084	593.4
Oxygen	O_2	0.20948	159.2
Argon	Ar	9.34×10^{-3}	7.1
Carbon dioxide	CO_2	3.1×10^{-4}	0.24
Neon	Ne	2×10^{-5}	0.02
Helium	He	1×10^{-5}	0.01

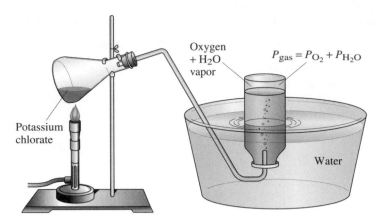

FIGURE 12.11 Collection of oxygen gas by water displacement. Potassium chlorate (KClO₃) decomposes to form oxygen (O₂), which is collected over water.

A common application of Dalton's law of partial pressures is encountered in the laboratory preparation of gases. Such gases are often collected by displacement of water. Figure 12.11 shows O₂, prepared from the decomposition of KClO₃, being collected by water displacement. A gas collected by water displacement is never pure. It always contains some water vapor. The total pressure exerted by the gaseous mixture is the sum of the partial pressures of the gas being collected and the water vapor.

$$P_{total} = P_{gas} + P_{H_2O}$$

The pressure exerted by the water vapor in the mixture will be constant at any given temperature if sufficient time has been allowed to establish equilibrium conditions.

For gases collected by water displacement, the partial pressure of the water vapor can be obtained from a table showing the variation of water vapor pressure with temperature (see Table 12.6). Thus, the partial pressure of the collected gas is easily determined.

$$P_{gas} = P_{atm} - P_{H_2O}$$

TABLE 12.6 Vapor Pressure of Water at Various Temperatures

$T(°C)$	Vapor Pressure (mm Hg)	$T(°C)$	Vapor Pressure (mm Hg)	$T(°C)$	Vapor Pressure (mm Hg)
15	12.8	22	19.8	29	30.0
16	13.6	23	21.1	30	31.8
17	14.5	24	22.4	31	33.7
18	15.5	25	23.8	32	35.7
19	16.5	26	25.2	33	37.7
20	17.5	27	26.7	34	39.9
21	18.7	28	28.3	35	42.2

EXAMPLE 12.31 **Calculating the Partial Pressure of a Gas Collected over Water**

What is the partial pressure of oxygen gas collected over water at 17°C on a day when the barometric pressure is 743 mm Hg?

SOLUTION

From Table 12.6, we determine that water has a vapor pressure of 14.5 mm Hg at 17°C.
Using the equation

$$P_{O_2} = P_{atm} - P_{H_2O}$$

we find the partial pressure of the oxygen to be 728 mm Hg.

$$P_{O_2} = (743 - 14.5) \text{ mm Hg} = 728.5 \text{ mm Hg} \quad \text{(calculator answer)}$$
$$= 728 \text{ mm Hg} \quad \textbf{(correct answer)}$$

▶ **Practice Exercise 12.31**   Calculate the partial pressure of hydrogen gas collected over water at 27°C on a day when the barometric pressure is 689 mm Hg.

When considering the components of a gaseous mixture, a useful set of equalities for any given component (A) of the gaseous mixture, at a given temperature and pressure, is

mole percent A = pressure percent A = volume percent A

where

$$\text{mole \% A} = \frac{n_A}{n_{total}} \times 100$$

$$\text{pressure \% A} = \frac{P_A}{P_{total}} \times 100$$

$$\text{volume \% A} = \frac{V_A}{V_{total}} \times 100$$

Example 12.32 is a calculation showing the equality of these three quantities.

EXAMPLE 12.32 **Expressing Gaseous Mixture Composition in Mole Percent, Pressure Percent, and Volume Percent**

A 17.92 L flask, at STP, contains 0.200 mole of O_2, 0.300 mole of N_2, and 0.300 mole of Ar. For this gaseous mixture, calculate the

a. mole percent of O_2 present.
b. pressure percent of O_2 present.
c. volume percent of O_2 present.

SOLUTION

a. The total number of moles of gas present is 0.800 mole (0.200 mole + 0.300 mole + 0.300 mole). The mole percent O_2 is

$$\text{mole \% } O_2 = \frac{n_{O_2}}{n_{total}} \times 100 = \frac{0.200 \text{ mole}}{0.800 \text{ mole}} \times 100 = 25\% \quad \text{(calculator answer)}$$

$$= 25.0\% \textbf{ (correct answer)}$$

b. Since conditions are specified as STP, the total pressure in the flask is 1.00 atm. The partial pressure of the O_2 is

$$P_{O_2} = X_{O_2} \times P_{total}$$

$$= \frac{0.200 \text{ mole}}{0.800 \text{ mole}} \times 1.00 \text{ atm}$$

$$= 0.25 \text{ atm} \qquad \text{(calculator answer)}$$

$$= 0.250 \text{ atm} \qquad \textbf{(correct answer)}$$

The pressure percent O_2 is

$$\text{pressure \% } O_2 = \frac{P_{O_2}}{P_{total}} \times 100 = \frac{0.250 \text{ atm}}{1.00 \text{ atm}} \times 100$$

$$= 25\% \qquad \text{(calculator answer)}$$

$$= 25.0\% \qquad \textbf{(correct answer)}$$

c. The total volume of the flask is given as 17.92 L. The volume of the oxygen present, if it were alone at STP conditions in a different container, is

$$0.200 \text{ mole } O_2 \times \frac{22.41 \text{ L } O_2}{1 \text{ mole } O_2} = 4.482 \text{ L } O_2 \quad \text{(calculator answer)}$$

$$= 4.48 \text{ L } O_2 \quad \textbf{(correct answer)}$$

The volume percent O_2 is

$$\text{volume \% } O_2 = \frac{V_{O_2}}{V_{total}} \times 100 = \frac{4.48 \text{ L}}{17.92 \text{ L}} \times 100$$

$$= 25\% \qquad \text{(calculator answer)}$$

$$= 25.0\% \qquad \textbf{(correct answer)}$$

Note, from the answers to parts (a), (b), and (c), that for O_2 the

$$\text{mole percent} = \text{pressure percent} = \text{volume percent}$$

▶ **Practice Exercise 12.32** A 10.00 L flask, at STP, contains 0.200 mole of H_2, 1.25 mole of Ne, and 2.50 mole of He. For this gaseous mixture, calculate the

a. mole percent Ne.
b. pressure percent Ne.
c. volume percent Ne.

Summary

1. Gas Laws Gas laws are generalizations that describe in mathematical terms the relationships among the amount, pressure, temperature, and volume of a gas. When gas laws are used, it is necessary to express the temperature of the gas on the Kelvin scale. Pressure is usually expressed in atmospheres or millimeters of mercury.

2. Boyle's Law Boyle's law, the pressure–volume law, states that the volume of a fixed quantity of a gas is *inversely proportional* to the pressure applied to the gas if the temperature is kept constant. This means that when the pressure on the gas increases, the volume decreases proportionally; conversely, when the volume decreases, the pressure increases.

3. **Charles's Law** Charles's law, the temperature–volume law, states that the volume of a fixed quantity of gas is *directly proportional* to its Kelvin temperature if the pressure is kept constant. This means that when the temperature increases, the volume also increases and that when the temperature decreases, the volume also decreases.

4. **Gay-Lussac's Law** Gay-Lussac's law, the temperature–pressure law, states that the pressure of a fixed quantity of gas is *directly proportional* to its Kelvin temperature if the volume is kept constant. As the temperature of the gas increases, the pressure also increases; conversely, as the temperature is decreased, pressure decreases.

5. **Combined Gas Law** The combined gas law is an expression obtained by mathematically combining Boyle's, Charles's, and Gay-Lussac's laws. A change in pressure, temperature, or volume of a fixed quantity of gas that is brought about by changes in the other two variables can be calculated by using this law.

6. **Avogadro's Law** Avogadro's law, the volume–quantity law, states that equal volumes of different gases, measured at the same temperature and pressure, contain equal numbers of molecules. An alternative statement of the law is that the volume of a gas, at constant temperature and pressure, is *directly proportional* to the number of moles of gas present.

7. **Ideal Gas** An ideal gas is a gas that obeys exactly all of the gas laws. Real gases approximate the behavior of an ideal gas at normally encountered temperatures and pressures.

8. **Ideal Gas Law** The ideal gas law describes the relationships among the four gas law variables temperature, pressure, volume, and moles for a gas under one set of conditions. This law is used to calculate any one of the gas law variables (*P, V, T, n*), given the other three. Equations derived from the ideal gas law by substitution of variables can be used to calculate the density and molar mass of a gas.

9. **Gay-Lussac's Law of Combining Volumes** The law of combining volumes states that the volumes of different gases involved in a reaction, measured at the same temperature and pressure, are in the same ratio as the coefficients for these gases in the balanced equation for the reaction.

10. **Molar Volume of a Gas** The molar volume of a gas is the volume occupied by one mole of a gas at a specific temperature and pressure. The ideal gas law is used to calculate a molar volume. The *standard molar volume of a gas*, which has the value 22.414 L, is the volume of one mole of a gas at STP conditions (0°C and 1 atm).

11. **Chemical Calculations using Molar Volume** A simple extension of the procedures used in mass-to-mass stoichiometric calculations enables mass-to-volume and volume-to-mass calculations to be carried out for reactions where at least one gas is involved. The extension is based on the concept of molar volume for a gas.

12. **Dalton's Law of Partial Pressures** Dalton's law of partial pressures states that the total pressure exerted by a mixture of gases that do not react with each other is the sum of the partial pressures of the individual gases present. A partial pressure is the pressure that a gas in a mixture would exert if it were present alone under the same conditions. Dalton's law of partial pressures, in modified form, may also be used to specify gas mixture concentrations in terms of mole percent, volume percent, and pressure percent.

Key Terms

The new terms defined in this chapter are

Practice Problems

MEASUREMENT OF PRESSURE (SEC. 12.2)

12.1   If the helium gas in a steel cylinder is at a pressure of 6.20 atm, what is the pressure in each of the following pressure units?
   a. millimeters of mercury
   b. inches of mercury
   c. pounds per square inch
   d. centimeters of mercury

12.2   If the oxygen gas in a steel cylinder is at a pressure of 9570 millimeters of mercury, what is the pressure in each of the following pressure units?
   a. inches of mercury
   b. pounds per square inch
   c. atmospheres
   d. centimeters of mercury

12.3   For each of the following pairs of pressure measurements, indicate whether the first listed measurement is larger than, equal to, or smaller than the second listed measurement.
   a. 578 mm Hg and 1.01 atm
   b. 14.21 lb/in.^{2} and 775 mm Hg
   c. 29.92 in. Hg and 760 mm Hg
   d. 3.57 atm and 48.7 psi

12.4   For each of the following pairs of pressure measurements, indicate whether the first listed measurement is larger than, equal to, or smaller than the second listed measurement.
   a. 1.01 atm and 759 mm Hg
   b. 27.00 psi and 29.92 in. Hg
   c. 14.68 psi and 1 atm
   d. 760 mm Hg and 35.0 in. Hg

12.5   The mercury level in the arm of a manometer (see Fig. 12.2) that is open to the atmosphere is found to be 237 mm higher than the mercury level in the arm of the manometer connected to the container of gas. Measured barometric pressure is 762 mm Hg. What is the pressure, in millimeters of mercury, of the gas in the container?

12.6   The mercury level in the arm of a manometer (see Fig. 12.2) that is open to the atmosphere is found to be 35 mm lower than the mercury level in the arm of the manometer connected to the container of gas. Measured barometric pressure is 743 mm Hg. What is the pressure, in millimeters of mercury, of the gas in the container?

BOYLE'S LAW (SEC. 12.3)

12.7   A sample of gas occupies a volume of 3.00 L at 27°C and a pressure of 1.00 atm. Without actually doing a calculation, predict whether the volume will increase or decrease when the pressure is changed, at constant temperature, to the following values.
   a. 0.75 atm   **b.** 0.333 atm
   c. 1.25 atm   **d.** 200 mm Hg

12.8   A sample of gas occupies a volume of 2.00 L at 33°C and a pressure of 1.50 atm. Without actually doing a calculation, predict whether the volume will increase or decrease when the pressure is changed, at constant temperature, to the following values.
   a. 2.00 atm   **b.** 0.952 atm
   c. 1.20 atm   **d.** 775 mm Hg

12.9   A sample of O_2 gas occupies a volume of 2.00 L at 27°C and 2.00 atm pressure. What volume, in liters, will this O_2 sample occupy at the same temperature but at each of the following pressures?
   a. 2.98 atm   **b.** 10.5 atm
   c. 453 mm Hg   **d.** 54.2 mm Hg

12.10   A sample of N_2 gas occupies a volume of 3.00 L at 37°C and 3.00 atm pressure. What volume, in liters, will this N_2 sample occupy at the same temperature but at each of the following pressures?
   a. 3.13 atm   **b.** 0.723 atm
   c. 762 mm Hg   **d.** 37.2 mm Hg

12.11   At constant temperature, the pressure on a sample of H_2 gas is decreased from 4.0 atm to 2.5 atm. What was the original volume, in milliliters, of the gas sample if this action increases the sample volume to each of the following amounts?
   a. 322 mL   **b.** 15.0 mL
   c. 2.24 L   **d.** 0.88 L

12.12   At constant temperature, the pressure on a sample of H_2 gas is increased from 2.5 atm to 4.0 atm. What was the original volume, in milliliters, of the gas sample if this action decreases the sample volume to each of the following amounts?
   a. 425 mL   **b.** 25.4 mL
   c. 1.08 L   **d.** 4.68 L

12.13   Diagram I depicts a gas, in a cylinder of variable volume, at the conditions specified in the diagram. Which of the diagrams II through IV depicts the result of doubling the pressure on the gas at constant temperature and constant number of moles of gas?

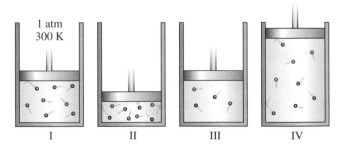

I   II   III   IV

12.14   Using the diagrams given in Problem 12.13, which of the diagrams II through IV correctly depicts the result of halving the pressure on the gas at constant temperature and constant number of moles of gas?

12.15 A sample of Cl_2 gas at a pressure of 645 mm Hg is transferred to a new container having a volume one-third that of the original container. What pressure, in millimeters of mercury, does the Cl_2 exert in the new container? Assume that the temperature does not change.

12.16 A sample of F_2 gas at a pressure of 1.03 atm is transferred to a new container having a volume 2.50 times that of the original container. What pressure, in atmospheres, does the F_2 exert in the new container? Assume that the temperature does not change.

CHARLES'S LAW (SEC. 12.4)

12.17 A sample of gas occupies a volume of 3.00 L at 27°C and a pressure of 1.00 atm. Without actually doing a calculation, predict whether the volume will increase or decrease when the temperature is changed, at constant pressure, to the following values.
a. 23°C   **b.** 63°C   **c.** −23°C   **d.** 303 K

12.18 A sample of gas occupies a volume of 2.00 L at 33°C and a pressure of 0.50 atm. Without actually doing a calculation, predict whether the volume will increase or decrease when the temperature is changed, at constant pressure, to the following values.
a. 37°C   **b.** 17°C   **c.** 375°C   **d.** 295 K

12.19 A sample of carbon dioxide gas, CO_2, has a volume of 5.00 L at 35°C. What volume, in liters, will this CO_2 gas occupy at each of the following temperatures if the pressure is held constant?
a. 123°C   **b.** 223°C   **c.** −25°C   **d.** 883°C

12.20 A sample of nitrogen dioxide gas, NO_2, has a volume of 1.50 L at 23°C. What volume, in liters, will this NO_2 gas occupy at each of the following temperatures if the pressure is held constant?
a. 125°C   **b.** 5°C
c. −5°C   **d.** 985°C

12.21 At constant pressure, the temperature of a sample of He gas is decreased from 73°C to 0°C. What was the original volume of the gas sample, in milliliters, if this action decreases the sample volume to each of the following amounts?
a. 17.5 mL   **b.** 742 mL
c. 3.42 L   **d.** 0.90 L

12.22 At constant pressure, the temperature of a sample of Ne gas is increased from 0°C to 73°C. What was the original volume of the gas sample, in milliliters, if this action increases the sample volume to each of the following amounts?
a. 15.2 mL   **b.** 879 mL
c. 1.20 L   **d.** 10.7 L

12.23 Diagram I depicts a gas, in a cylinder of variable volume, at the conditions specified in the diagram. Which of the diagrams II through IV depicts the results of decreasing the Kelvin temperature of the gas by a factor of two at constant pressure and constant number of moles of gas?

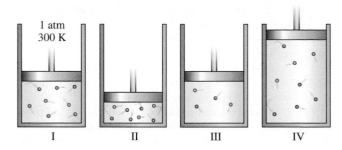

12.24 Using the diagrams in Problem 12.23, which of the diagrams II through IV correctly depicts the result of increasing the Kelvin temperature of the gas by a factor of two at constant pressure and constant number of moles of gas?

12.25 At constant pressure, at what temperature, in degrees Celsius, will a gas have exactly one-half the volume that it has at room temperature (24°C)?

12.26 At constant pressure, at what temperature, in degrees Celsius, will a gas have exactly double the volume that it has at room temperature (24°C)?

GAY-LUSSAC'S LAW (SEC. 12.5)

12.27 A sample of gas occupies a volume of 3.00 L at 27°C and a pressure of 1.00 atm. Without actually doing a calculation, predict whether the pressure will increase or decrease when the temperature is changed, at constant volume, to the following values.
a. 23°C   **b.** 63°C   **c.** −23°C   **d.** 303 K

12.28 A sample of gas occupies a volume of 2.00 L at 33°C and a pressure of 0.50 atm. Without actually doing a calculation, predict whether the pressure will increase or decrease when the temperature is changed, at constant value, to the following values.
a. 37°C   **b.** 17°C   **c.** 375°C   **d.** 295 K

12.29 A sample of air exerts a pressure of 1.00 atm at 22°C. If the sample volume remains constant, what is the new pressure, in atmospheres, exerted by the air when it is heated or cooled to each of the following temperatures?
a. 31°C   **b.** 6°C   **c.** −37°C   **d.** −137°C

12.30 A sample of air exerts a pressure of 2.00 atm at 37°C. If the sample volume remains constant, what is the new pressure, in atmospheres, exerted by the air when it is heated to each of the following temperatures?
a. 122°C   **b.** 222°C   **c.** 422°C   **d.** 722°C

12.31 At constant volume, the temperature of a sample of sulfur dioxide gas, SO_2, is decreased from 97°C to 27°C. What was the original pressure of the gas, in atmospheres, if this action decreases the sample pressure to the following?
a. 762 mm Hg   **b.** 662 mm Hg
c. 1.05 atm   **d.** 25.0 atm

12.32 At constant volume, the temperature of a sample of carbon monoxide gas, CO, is decreased from 122°C

to 22°. What was the original pressure of the gas, in millimeters of mercury, if this action decreases the sample pressure to the following?
a. 3.00 atm   b. 1.00 atm
c. 1394 mm Hg   d. 375 mm Hg

12.33 A spray can is empty except for the propellant gas, which exerts a pressure of 1.2 atm at 24°C. If the can is thrown into a fire (485°C), what will be the pressure, in atmospheres, inside the hot can?

12.34 A spray can is empty except for the propellant gas, which exerts a pressure of 1.2 atm at 24°C. If the can is placed in a refrigerator (3°C), what will be the pressure, in atmospheres, inside the cold can?

THE COMBINED GAS LAW (SEC. 12.6)

12.35 Rearrange the standard form of the combined gas law to result in the following.
a. The variable T_2 is isolated on the left side of the equation.
b. The quantity V_2/P_1 is isolated on the left side of the equation.

12.36 Rearrange the standard form of the combined gas law to result in the following.
a. The variable T_1 is isolated on the left side of the equation.
b. The quantity P_2/V_1 is isolated on the left side of the equation.

12.37 What is the new volume, in milliliters, of a 3.00 mL sample of air at 0.980 atm and 230°C that is compressed and cooled to each of the following sets of conditions?
a. 25°C and 2.00 atm   b. –25°C and 3.00 atm
c. –75°C and 5.00 atm   d. –125°C and 7.75 atm

12.38 What is the new volume, in liters, of a 25.0 L sample of air at 1.11 atm and 152°C that is compressed and cooled to each of the following sets of conditions?
a. 142°C and 1.50 atm   b. 35°C and 2.00 atm
c. –35°C and 4.00 atm   d. –125°C and 5.67 atm

12.39 A sample of CO_2 gas has a volume of 15.2 L at a pressure of 1.35 atm and a temperature of 33°C. Determine the following for this gas sample.
a. volume, in liters, at $T = 35°C$ and $P = 3.50$ atm
b. volume, in milliliters, at $T = 97°C$ and $P = 6.70$ atm
c. pressure, in atmospheres, at $T = 42°C$ and $V = 10.0$ L
d. temperature, in degrees Celsius, at $P = 7.00$ atm and $V = 0.973$ L

12.40 A sample of NO_2 gas has a volume of 37.3 mL at a pressure of 621 mm Hg and a temperature of 52°C. Determine the following for this gas sample.
a. volume, in milliliters, at $T = 35°C$ and $P = 650$ mm Hg
b. volume, in liters, at $T = 43°C$ and $P = 1.11$ atm

c. pressure, in millimeters of mercury, at $T = 125°C$ and $V = 52.4$ mL
d. temperature, in degrees Celsius, at $P = 775$ mm Hg and $V = 23.0$ mL

12.41 A sample of ammonia gas, NH_3, in a 375 mL container at a pressure of 1.03 atm and a temperature of 27°C is transferred to a container with a volume of 1.25 L.
a. What is the new pressure, in millimeters of mercury, if no change in temperature occurs?
b. What is the new temperature, in degrees Celsius, if no change in pressure occurs?

12.42 A sample of nitrous oxide gas, N_2O, in a 475 mL container at a pressure of 676 mm Hg and a temperature of 22°C is transferred to a container with a volume of 5.00 L.
a. What is the new pressure, in atmospheres, if no change in temperature occurs?
b. What is the new temperature, in degrees Celsius, if no change in pressure occurs?

12.43 A sample of nitrous oxide gas, N_2O, in a nonrigid container at a temperature of 33°C occupies a certain volume at a certain pressure. What will be its temperature, in degrees Celsius, in each of the following situations?
a. Both pressure and volume are tripled.
b. Both pressure and volume are cut in half.
c. The pressure is tripled and the volume is cut in half.
d. The pressure is cut in half and the volume is doubled.

12.44 A sample of nitric oxide gas, NO, in a nonrigid container, occupies a volume of 2.50 L at a certain temperature and pressure. What will be its volume, in liters, in each of the following situations?
a. Both pressure and Kelvin temperature are doubled.
b. Both pressure and Kelvin temperature are cut by one-third.
c. The pressure is doubled, and the Kelvin temperature is cut by one-third.
d. The pressure is cut in half, and the Kelvin temperature is tripled.

AVOGADRO'S LAW (SEC. 12.7)

12.45 If 1.00 mole of Cl_2 gas occupies a volume of 24.0 L at a certain temperature and pressure, what would be the volume, in liters, of the following molar-sized samples of Cl_2 gas at the same temperature and pressure?
a. 2.00 moles   b. 2.32 moles
c. 5.40 moles   d. 0.500 mole

12.46 If 3.00 moles of F_2 gas occupies a volume of 18.0 L at a certain temperature and pressure, what would be the volume, in liters, of the following molar-sized samples of F_2 gas at the same temperature and pressure?
a. 1.00 mole   b. 2.50 moles
c. 3.50 moles   d. 5.00 moles

12.47 A 2.00 mole sample of Xe gas in a balloon at a specific temperature and pressure has a volume of 2.33 L. How many moles of Xe gas would have to be present in the balloon in order to increase its volume to the following values at the same temperature and pressure?
  a. 3.00 L   **b.** 4.00 L   **c.** 5.00 L   **d.** 6.00 L

12.48 A 3.00 mole sample of Ar gas in a balloon at a specific temperature and pressure has a volume of 0.500 L. How many moles of Ar gas would have to be present in the balloon in order to increase its volume to the following values at the same temperature and pressure?
  a. 1.50 L   **b.** 2.00 L   **c.** 2.50 L   **d.** 3.00 L

12.49 Diagram I depicts a gas, in a cylinder of variable volume, at the conditions specified in the diagram. Which of the diagrams II through IV depicts the results of adding more gas to the container until the volume has increased to twice its original value, at constant temperature and pressure?

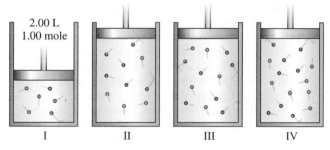

| I | II | III | IV |

2.00 L
1.00 mole

12.50 Diagram I depicts a gas, in a cylinder of variable volume, at the conditions specified in the diagram. Which of the diagrams II through IV depicts the results of decreasing the amount of gas in the container until the volume has decreased to one-half its original value, at constant temperature and pressure?

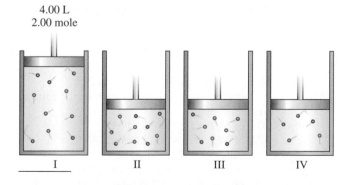

| I | II | III | IV |

4.00 L
2.00 mole

12.51 A balloon containing 7.83 moles of He has a volume of 27.5 L at a certain temperature and pressure. What would be the volume of the balloon, in liters, if the following adjustments are made to the contents of the balloon at constant temperature and pressure?
  a. 1.00 mole of He is added to the balloon.
  b. 3.00 moles of He are added to the balloon.
  c. 1.00 mole of He is removed from the balloon.
  d. 3.00 moles of He are removed from the balloon.

12.52 A balloon containing 5.25 moles of H_2 has a volume of 22.3 L at a certain temperature and pressure. What would be the volume of the balloon, in liters, if the following adjustments are made to the contents of the balloon at constant temperature and pressure?
  a. 2.00 moles of H_2 are added to the balloon.
  b. 4.00 moles of H_2 are added to the balloon.
  c. 2.00 moles of H_2 are removed from the balloon.
  d. 4.00 moles of H_2 are removed from the balloon.

12.53 If 1.00 mole of Cl_2 gas occupies a volume of 24.0 L at a certain temperature and pressure, what would be the volume, in liters, of the following *gram*-sized samples of Cl_2 gas at the same temperature and pressure?
  a. 2.00 grams   **b.** 2.32 grams
  c. 5.40 grams   **d.** 0.500 gram

12.54 If 3.00 moles of F_2 gas occupies a volume of 18.0 L at a certain temperature and pressure, what would be the volume, in liters, of the following *gram*-sized samples of F_2 gas at the same temperature and pressure?
  a. 1.00 gram   **b.** 2.50 grams
  c. 3.50 grams   **d.** 5.00 grams

12.55 A 0.625 mole sample of N_2 gas at 1.50 atm and 45°C occupies a volume of 10.9 L. What volume, in liters, would a 0.625 mole sample of each of the following gases occupy at the same temperature and pressure?
  a. NO   **b.** N_2O **c.** NO_2 **d.** N_2O_3

12.56 A 1.25 mole sample of O_2 gas at 2.50 atm and 75°C occupies a volume of 14.3 L. What volume, in liters, would a 1.25 mole sample of each of the following gases occupy at the same temperature and pressure?
  a. CO_2 **b.** CO   **c.** SO_2 **d.** SO_3

12.57 A 0.625 mole sample of N_2 gas at 1.50 atm and 45°C occupies a volume of 10.9 L. What volume, in liters, would each of the following molar-sized samples of gas occupy at the same temperature and pressure?
  a. 2.00 moles CO_2 **b.** 3.00 moles of N_2O
  c. 2.50 moles O_2 **d.** 6.00 moles NH_3

12.58 A 1.25 mole sample of O_2 gas at 2.50 atm and 75°C occupies a volume of 14.3 L. What volume, in liters, would each of the following molar-sized samples of gas occupy at the same temperature and pressure?
  a. 1.75 moles SO_2 **b.** 2.00 moles CO
  c. 2.33 moles NO_2 **d.** 3.75 moles PH_3

12.59 A balloon containing 2.00 moles of N_2 and 1.50 moles of O_2 has a volume of 0.500 L at a given temperature and pressure. If 0.48 mole of N_2 is removed from the balloon and 0.76 mole of O_2 is added to the balloon without changing the temperature and pressure conditions, what will be the new volume, in liters, in the balloon?

12.60 A balloon containing 2.00 moles of He and 3.00 moles of Ne has a volume of 20.0 L at a given temperature and pressure. If 1.00 mole of He is removed from the balloon and 2.00 moles of Ne are added to the balloon without changing the temperature and pressure conditions, what is the new volume, in liters, in the balloon?

12.61 In each of the following pairs of gas samples, select the pair member that would have the larger volume at 27°C and 1.00 atm.
  a. 0.400 mole N_2 and 0.450 mole O_2
  b. 2.32 moles CH_4 and 2.00 moles C_2H_6
  c. 100.0 g NO_2 and 100.0 g N_2O
  d. 100.0 g CO and 100.0 g CO_2

12.62 In each of the following pairs of gas samples, select the pair member that would have the smaller volume at 33°C and 2.00 atm.
  a. 0.300 mole He and 0.200 mole H_2
  b. 1.05 moles SO_2 and 2.10 moles SO_3
  c. 50.0 g F_2 and 50.0 g OF_2
  d. 50.0 g NH_3 and 50.0 g PH_3

12.63 A 0.500 mole sample of a gas has a volume of 2.0 L at a pressure of 1.5 atm and a temperature of 27°C. What will be the volume of the gas, in liters, if the following changes are made to the system: The amount of gas is decreased by 0.100 mole, the pressure is decreased by 0.50 atm, and the temperature is increased by 20°C?

12.64 A 2.00 mole sample of a gas has a volume of 4.0 L at a pressure of 0.50 atm and a temperature of 35°C. What will be the volume of the gas, in liters, if the following changes are made to the system: The amount of gas is increased by 0.100 mole, the pressure is increased by 0.50 atm, and the temperature is decreased by 20°C?

THE IDEAL GAS LAW (SEC. 12.9)

12.65 Using the ideal gas law, calculate the volume, in liters, of 1.20 moles of Cl_2 gas at each of the following sets of conditions.
  a. 0°C and 1.00 atm   **b.** 73°C and 1.54 atm
  c. 525°C and 15.0 atm   **d.** −23°C and 765 mm Hg

12.66 Using the ideal gas law, calculate the volume, in liters, of 1.15 moles of F_2 gas at each of the following sets of conditions.
  a. 0°C and 1.25 atm   **b.** 315°C and 456 mm Hg
  c. −45°C and 4.50 atm   **d.** 23°C and 762 mm Hg

12.67 How many moles of H_2 gas are present in a 6.00 L container at a pressure of 3.67 atm at 25°C?

12.68 How many moles of He gas are present in a 5.00 L container at a pressure of 2.50 atm at 35°C?

12.69 If 0.332 mole of He gas has a volume of 1275 mL and a pressure of 5.78 atm, what is its temperature in degrees Celsius?

12.70 If 0.504 mole of Ar gas has a volume of 1975 mL and a pressure of 4.60 atm, what is its temperature in degrees Celsius?

12.71 What is the pressure, in atmospheres, inside a 6.00 L container that contains the following amounts of N_2 gas at 25°C?
  a. 0.30 mole   **b.** 1.20 moles
  c. 0.30 g   **d.** 1.20 g

12.72 What is the pressure, in millimeters of mercury, inside a 4.00 L container that contains the following amounts of O_2 gas at 40°C?
  a. 0.72 mole   **b.** 4.5 moles
  c. 0.72 g   **d.** 4.5 g

12.73 At a temperature of 152°C and a pressure of 13.3 atm 1.14 moles of the noble gas argon occupy a volume of 3.00 L. Use this information to calculate the value of the ideal gas constant R in the units of atm · L/mole · K.

12.74 At a temperature of 25°C and a pressure of 679 mm Hg 0.317 mole of the noble gas helium occupies a volume of 8.67 L. Use this information to calculate the value of the ideal gas constant R in the units of mm Hg · L/mole · K.

12.75 A 1.00 mole sample of liquid water is placed in a flexible sealed container and allowed to evaporate. After complete evaporation, what will be the container volume, in liters, at 127°C and 0.908 atm pressure?

12.76 A 1.00 mole sample of dry ice (solid CO_2) is placed in a flexible sealed container and allowed to sublime. After all the CO_2 has changed from solid to gas, what will be the container volume, in liters, at 23°C and 0.983 atm pressure?

12.77 Calculate the mass, in grams, of each of the following quantities of gas.
  a. 25.0 L of C_2H_6 at 0.972 atm and 29°C
  b. 2.22 L of HCl at 854 mm Hg and 75°C
  c. 5.50 L of SO_2 at 0°C and 2.00 atm
  d. 783 mL of N_2O at 359 mm Hg and 273°C

12.78 Calculate the mass, in grams, of each of the following quantities of gas.
  a. 3.00 L of NO at 1.23 atm and 35°C
  b. 1.37 L of CO_2 at 498 mm Hg and 285°C
  c. 3.50 L of HF at 0°C and 1.00 atm
  d. 1780 mL of C_2H_2 at 3.00 atm and 585°C

12.79 A 30.0 L cylinder contains 100.0 g of Cl_2 at 23°C. How many grams of Cl_2 must be added to the container to increase the pressure in the cylinder to 1.65 atm? Assume that the temperature remains constant.

12.80 A 50.0 L cylinder contains 100.0 g of F_2 at 30°C. How many grams of F_2 must be removed from the container to decrease the pressure in the cylinder to 1.00 atm? Assume that the temperature remains constant.

MODIFIED FORMS OF THE IDEAL GAS LAW EQUATION (SEC. 12.10)

12.81 A 1.305 g sample of carbon monoxide gas (CO) occupies a volume of 1.20 L at 741 mm Hg pressure and 33°C. Based on these data, calculate the molar mass of carbon monoxide.

12.82 A 1.305 g sample of nitrogen gas (N_2) occupies a volume of 1.20 L at 0.975 atm pressure and 33°C. Based on these data, calculate the molar mass of nitrogen gas.

12.83 A 125 mL flask contains 0.450 g of a gaseous compound at 75°C and 1.00 atm pressure. What is the molar mass of the compound?

12.84 A 250 mL flask contains 0.350 g of a gaseous compound at 85°C and 1.00 atm pressure. What is the molar mass of the compound?

12.85 Determine the molar mass of a liquid whose vapor filled a 125 mL flask at 90°C and 1.20 atm pressure. The mass of the vapor is 0.4537 g.

12.86 Determine the molar mass of a liquid whose vapor filled a 125 mL flask at 75°C and 1.00 atm pressure. The mass of the vapor is 0.5545 g.

12.87 A gas is either CO or CO_2. Based on the following information about a sample of the gas, what is its identity? A 0.902 L sample of the gas exerts a pressure of 1.65 atm, has a mass of 1.733 g, and is at a temperature of 20°C.

12.88 A gas is either CH_4 or HF. Based on the following information about a sample of the gas, what is its identity? A 1.98 L sample of the gas exerts a pressure of 2.00 atm, has a mass of 3.20 g, and is at a temperature of 30°C.

12.89 Calculate the density of H_2S gas, in grams per liter, at each of the following temperature–pressure conditions.
  a. 24°C and 675 mm Hg **b.** 24°C and 1.20 atm
  c. 370°C and 1.30 atm **d.** –25°C and 452 mm Hg

12.90 Calculate the density of SCl_2 gas, in grams per liter, at each of the following temperature–pressure conditions.
  a. 27°C and 1.20 atm **b.** 37°C and 5.00 atm
  c. 227°C and 4.00 atm **d.** 550°C and 1.00 atm

12.91 What pressure, in atmospheres, is required to cause each of the following gases to have a density of 1.00 g/L at 47°C?
  a. N_2 **b.** Xe **c.** ClF **d.** N_2O

12.92 What pressure, in atmospheres, is required to cause each of the following gases to have a density of 1.00 g/L at 53°C?
  a. O_2 **b.** O_3 **c.** SO_2 **d.** CH_4

12.93 A gas is either NO or CO. Based on the following information about a sample of the gas, what is its identity? A sample of the gas at a temperature

of 27°C and a pressure of 2.00 atm has a density of 2.27 g/L.

12.94 A gas is either H_2S or HF. Based on the following information about a sample of the gas, what is its identity? A sample of the gas at a temperature of 127°C and a pressure of 3.00 atm has a density of 3.12 g/L.

12.95 Calculate the molar mass of a gas whose density is 1.35 g/L at 25°C and a pressure of 1.32 atm.

12.96 Calculate the molar mass of a gas whose density is 1.21 g/L at 35°C and a pressure of 0.955 atm.

12.97 A gas with a molar mass of 30.01 g/mole has a density of 1.25 g/L at a pressure of 1.20 atm. What is the temperature, in °C, of this gas?

12.98 A gas with a molar mass of 28.01 g/mole has a density of 1.15 g/L at a pressure of 1.15 atm. What is the temperature, in °C, of this gas?

VOLUMES OF GASES IN CHEMICAL REACTIONS (SEC. 12.11)

12.99 Ammonia reacts with oxygen to form nitrogen and water:

$$4 NH_3(g) + 3 O_2(g) \longrightarrow 2 N_2(g) + 6 H_2O(g)$$

It can also react with oxygen to form nitric oxide (NO) and water:

$$4 NH_3(g) + 5 O_2(g) \longrightarrow 4 NO(g) + 6 H_2O(g)$$

Which of these two reactions occurred if 1.60 liters of NH_3 are found by experiment to react with 2.00 liters of O_2? Assume that both gas volumes are measured at the same temperature and pressure.

12.100 Methane, CH_4, reacts with steam to form hydrogen and carbon monoxide:

$$CH_4(g) + H_2O(g) \longrightarrow 3 H_2(g) + CO(g)$$

It can also react with steam to form hydrogen and carbon dioxide:

$$CH_4(g) + 2 H_2O(g) \longrightarrow 4 H_2(g) + CO_2(g)$$

Which of these two reactions occurred if 2.33 L of methane is found by experiment to react with 2.33 L of steam? Assume that both gas volumes are measured at the same temperature and pressure.

12.101 The equation for the combustion of the fuel propane, C_3H_8, is

$$C_3H_8(g) + 5 O_2(g) \longrightarrow 3 CO_2(g) + 4 H_2O(g)$$

  a. How many liters of C_3H_8 must be burned to produce 1.30 L of CO_2 if both volumes are measured at 25°C and 1.00 atm?
  b. How many liters of C_3H_8 must be burned to produce 1.30 L of H_2O if both volumes are measured at 2.00 atm and 56°C?

12.102 The equation for the combustion of the fuel butane, C_4H_{10}, is

$$2\,C_4H_{10}(g) + 13\,O_2(g) \longrightarrow 8\,CO_2(g) + 10\,H_2O(g)$$

a. How many liters of C_4H_{10} must be burned to produce 2.60 L of CO_2 if both volumes are measured at 15°C and 1.00 atm?
b. How many liters of C_4H_{10} must be burned to produce 2.60 L of H_2O if both volumes are measured at 1.75 atm and 43°C?

12.103 For the reaction in Problem 12.101, how many liters of C_3H_8 must be burned to produce a combined total of 0.75 liter of gaseous products if all volumes are measured at the same temperature and pressure?

12.104 For the reaction in Problem 12.102, how many liters of C_4H_{10} must be burned to produce a combined total of 1.75 liters of gaseous products if all volumes are measured at the same temperature and pressure?

VOLUMES OF GASES AND THE LIMITING REACTANT CONCEPT (SEC. 12.12)

12.105 At high temperatures and pressures, nitrogen will react with hydrogen to produce ammonia, as shown by the equation

$$N_2(g) + 3\,H_2(g) \rightarrow 2\,NH_3(g)$$

For each of the following combinations of volumes of gases, decide which is the limiting reactant. (Assume all gases are at the same conditions.)
a. 1.00 L of N_2 and 1.50 L of H_2
b. 2.00 L of N_2 and 5.50 L of H_2
c. 1.00 L of N_2 and 4.00 L of H_2
d. 3.00 L of N_2 and 1.00 L of H_2

12.106 At normal atmospheric conditions, nitric oxide will react with oxygen (from the air) to produce nitrogen dioxide, as shown by the equation

$$2\,NO(g) + O_2(g) \rightarrow 2\,NO_2(g)$$

For each of the following combinations of volumes of gases, decide which is the limiting reactant. (Assume all gases are at the same conditions.)
a. 2.00 L of NO and 1.50 L of O_2
b. 2.00 L of NO and 5.50 L of O_2
c. 5.00 L of NO and 2.00 L of O_2
d. 3.00 L of NO and 1.00 L of O_2

12.107 Under appropriate conditions, the reaction between methane gas and steam proceeds as shown by the equation

$$CH_4(g) + 2\,H_2O(g) \rightarrow CO_2(g) + 4\,H_2(g)$$

How many liters of each of the products can be produced from the following volumes of reactants? Assume all gases are at the same conditions.
a. 45.0 L of CH_4 and 45.0 L of H_2O
b. 45.0 L of CH_4 and 66.0 L of H_2O

c. 64.0 L of CH_4 and 134 L of H_2O
d. 16.0 L of CH_4 and 32.0 L of H_2O

12.108 Under appropriate conditions, the reaction between methane gas and oxygen gas proceeds as shown by the equation

$$CH_4(g) + 2\,O_2(g) \rightarrow CO_2(g) + 2\,H_2O(g)$$

How many liters of each of the products can be produced from the following volumes of reactants? Assume all gases are at the same conditions.
a. 45.0 L of CH_4 and 45.0 L of O_2
b. 45.0 L of CH_4 and 65.0 L of O_2
c. 64.0 L of CH_4 and 142 L of O_2
d. 56.0 L of CH_4 and 112 L of O_2

MOLAR VOLUME OF A GAS (SEC. 12.13)

12.109 What is the molar volume of a sample of NH_3 gas at each of the following sets of pressure–temperature conditions?
a. 1.20 atm and 33°C　**b.** 1.20 atm and 45°C
c. 123°C and 1.00 atm　**d.** 123°C and 2.00 atm

12.110 What is the molar volume of a sample of CO_2 gas at each of the following sets of temperature–pressure conditions?
a. 0.975 atm and 50°C　**b.** 0.975 atm and 75°C
c. 45°C and 2.33 atm　**d.** 45°C and 2.75 atm

12.111 What is the molar volume of each of the following gases at a temperature of 22°C and a pressure of 2.00 atm?
a. N_2　**b.** Ar　**c.** SO_2　**d.** SO_3

12.112 What is the molar volume of each of the following gases at a temperature of 27°C and a pressure of 1.50 atm?
a. O_2　**b.** Kr　**c.** Cl_2　**d.** NO

12.113 The molar volume of a gas is 44.01 L. What pressure does 1.00 L of the gas exert, in atm, at each of the following temperatures?
a. 0°C　**b.** 30°C　**c.** 275°C　**d.** 525°C

12.114 The molar volume of a gas is 28.01 L. What pressure does 1.00 L of the gas exert, in atm, at each of the following temperatures?
a. –25°C　**b.** 25°C　**c.** 100°C　**d.** 372°C

12.115 What is the molar volume of each of the gases in Problem 12.111 at STP conditions?

12.116 What is the molar volume of each of the gases in Problem 12.112 at STP conditions?

12.117 What is the density of a gas, in g/L, given that it has a molar mass of 17.04 g/mole and that its molar volume has each of the following values?
a. 17.21 L/mole　**b.** 22.41 L/mole
c. 23.34 L/mole　**d.** 35.00 L/mole

12.118 What is the density of a gas, in g/L, given that it has a molar mass of 18.02 g/mole and that its molar volume has each of the following values?
a. 18.07 L/mole　**b.** 22.76 L/mole
c. 28.75 L/mole　**d.** 32.00 L/mole

12.119 Based on the ratio between molar mass and molar volume, what is the density, in g/L, of each of the following gases at the specified conditions?
   a. CO_2 at 25°C and 1.20 atm pressure
   b. CO at 35°C and 1.33 atm pressure
   c. CH_4 gas at STP conditions
   d. C_2H_6 gas at STP conditions

12.120 Based on the ratio between molar mass and molar volume, what is the density, in g/L, of each of the following gases at the specified conditions?
   a. NO_2 gas at 27°C and 2.00 atm pressure
   b. N_2O gas at 37°C and 0.750 atm pressure
   c. NH_3 gas at STP conditions
   d. PH_3 gas at STP conditions

12.121 Which gas in each of the following pairs of gases will have the greater density at STP conditions?
   a. O_2 and O_3 **b.** NH_3 and PH_3
   c. CO and CO_2 **d.** F_2 and CH_4

12.122 Which gas in each of the following pairs of gases will have the greater density at STP conditions?
   a. SO_2 and S_2O **b.** F_2 and OF_2
   c. NO and CO   **d.** SF_6 and UF_6

12.123 Calculate molar masses for gases with the following densities at STP conditions.
   a. 1.97 g/L   **b.** 1.25 g/L
   c. 0.714 g/L   **d.** 3.17 g/L

12.124 Calculate molar masses for gases with the following densities at STP conditions.
   a. 1.70 g/L   **b.** 0.897 g/L
   c. 1.16 g/L   **d.** 0.759 g/L

CHEMICAL CALCULATIONS USING MOLAR VOLUME (SEC. 12.14)

12.125 What is the mass, in grams, of 23.7 L samples of each of the following gases at STP conditions?
   a. Ar   **b.** N_2O **c.** SO_3 **d.** PH_3

12.126 What is the mass, in grams, of 35.2 L samples of each of the following gases at STP conditions?
   a. Kr   **b.** S_2O **c.** O_3 **d.** NH_3

12.127 What is the mass, in grams, of 23.7 L samples of each of the gases in Problem 12.125 at 44°C and 1.15 atm?

12.128 What is the mass, in grams, of 35.2 L samples of each of the gases in Problem 12.126 at 54°C and 1.25 atm?

12.129 What is the volume, in liters, occupied by 1.25 mole samples of each of the following gases at STP conditions?
   a. N_2 **b.** NH_3 **c.** CO_2 **d.** Cl_2

12.130 What is the volume, in liters, occupied by 1.75 mole samples of each of the following gases at STP conditions?
   a. O_3 **b.** O_2 **c.** NO_2 **d.** NO

12.131 What is the volume, in liters, occupied by 1.25 mole samples of each of the gases in Problem 12.129 at 46°C and 1.73 atm pressure?

12.132 What is the volume, in liters, occupied by 1.75 mole samples of each of the gases in Problem 12.130 at 53°C and 1.42 atm pressure?

12.133 In each of the following pairs of gas samples, select the pair member that occupies the larger volume at STP conditions.
   a. 24.5 g N_2 and 24.5 g NH_3
   b. 30.0 g O_2 and 30.0 g O_3
   c. 10.0 g SO_2 and 20.0 g NO_2
   d. 15.0 g N_2O and 20.0 g NO

12.134 In each of the following pairs of gas samples, select the pair member that occupies the larger volume at STP conditions.
   a. 10.0 g H_2 and 10.0 g He
   b. 25.0 g F_2 and 25.0 g Cl_2
   c. 15.0 g CH_4 and 5.00 g PH_3
   d. 100.0 g HCN and 2.00 g UF_6

12.135 A mixture of 25.0 g of NO and an excess of O_2 reacts according to the balanced equation

$$2\,NO(g) + O_2(g) \longrightarrow 2\,NO_2(g)$$

How many liters of NO_2, at STP, are produced?

12.136 A mixture of 25.0 g of H_2 and an excess of N_2 reacts according to the balanced equation

$$3\,H_2(g) + N_2(g) \longrightarrow 2\,NH_3(g)$$

How many liters of NH_3, at STP, are produced?

12.137 A sample of O_2 with a volume of 25.0 L at 27°C and 1.00 atm is reacted with excess N_2 to produce NO. The equation for the reaction is

$$O_2(g) + N_2(g) \longrightarrow 2\,NO(g)$$

How many grams of NO are produced?

12.138 A sample of H_2 with a volume of 35.0 L at 35°C and 1.35 atm is reacted with excess O_2 to produce H_2O. The equation for the reaction is

$$2\,H_2(g) + O_2(g) \longrightarrow 2\,H_2O(g)$$

How many grams of H_2O are produced?

12.139 Hydrogen gas can be produced in the laboratory through reaction of magnesium metal with hydrochloric acid:

$$Mg(s) + 2\,HCl(aq) \longrightarrow MgCl_2(aq) + H_2(g)$$

What volume, in liters, of H_2 at 23°C and 0.980 atm pressure can be produced from the reaction of 12.0 g of Mg with an excess of HCl?

12.140 A common laboratory preparation for O_2 gas involves the thermal decomposition of potassium nitrate:

$$2\,KNO_3(s) \longrightarrow 2\,KNO_2(s) + O_2(g)$$

What volume, in liters, of O_2 at 35°C and 1.31 atm pressure can be produced from the decomposition of 35.0 g of KNO_3?

12.141 Ammonium nitrate, NH_4NO_3, can decompose explosively when heated to a high temperature according to the equation

$$2\,NH_4NO_3(s) \longrightarrow 2\,N_2(g) + 4\,H_2O(g) + O_2(g)$$

560 Chapter 12 • Gas Laws

If a 100.0 g sample of ammonium nitrate decomposes at 450°C, how many liters of gaseous products would be formed? Assume that atmospheric pressure is 1.00 atm.

12.142 The industrial explosive nitroglycerin, $C_3H_5N_3O_9$, detonates according to the equation

$$4 C_3H_5N_3O_9(s) \longrightarrow 6 N_2(g) + O_2(g) + 12CO_2(g) + 10H_2O(g)$$

At a detonation temperature of 1950°C, how many liters of gaseous products would be formed from 100.0 g of nitroglycerin? Assume that atmospheric pressure is 1.00 atm.

12.143 How many liters of NO_2 gas at 21°C and 2.31 atm must be consumed in producing 75.0 L of NO gas at 38°C and 645 mm Hg according to the following reaction?

$$3 NO_2(g) + H_2O(l) \longrightarrow 2 HNO_3(aq) + NO(g)$$

12.144 How many liters of Cl_2 gas at 25°C and 1.50 atm are needed to react completely with 3.42 L of NH_3 gas at 50°C and 2.50 atm according to the following reaction?

$$2 NH_3(g) + 3 Cl_2(g) \rightarrow N_2(g) + 6 HCl(g)$$

12.145 How many liters of CO_2 gas at 2.00 atm pressure and 75°C can be produced from 10.0 L of C_2H_2 gas at 3.00 atm pressure and 125°C and 2.50 L of O_2 gas at a pressure of 0.750 atm and 20°C, according to the following reaction?

$$2 C_2H_2(g) + 5 O_2(g) \longrightarrow 4 CO_2(g) + 2 H_2O(g)$$

12.146 How many liters of CO_2 gas at 2.00 atm pressure and 75°C can be produced from 10.0 L of C_2H_6 gas at 3.00 atm pressure and 125°C and 12.50 L of O_2 gas at a pressure of 0.750 atm and 20°C, according to the following reaction?

$$2 C_2H_6(g) + 7 O_2(g) \longrightarrow 4 CO_2(g) + 6 H_2O(g)$$

MIXTURES OF GASES (SEC. 12.15)

12.147 A gaseous mixture consists of 3.00 moles of N_2 and 3.00 moles of O_2. What is the volume, in liters, occupied by this mixture if it is under a pressure of 20.00 atm and at a temperature of 27°C?

12.148 A gaseous mixture consists of 4.00 moles of N_2 and 5.00 moles of O_2. What is the volume, in liters, occupied by this mixture if it is under a pressure of 20.00 atm and at a temperature of 27°C?

12.149 What would be the volume, in liters, of a gaseous mixture at 1.00 atm pressure and 27°C if its composition is each of the following?
   a. 3.00 g Ne and 3.00 g Ar
   b. 4.00 g Ne and 2.00 g Ar

   c. 3.00 moles Ne and 5.00 g Ar
   d. 4.00 moles Ne and 4.00 moles Ar

12.150 What would be the volume, in liters, of a gaseous mixture at 1.00 atm pressure and 27°C if its composition is each of the following?
   a. 4.00 g He and 8.00 g H_2
   b. 4.00 g He and 8.00 g N_2
   c. 3.00 moles He and 3.00 moles N_2
   d. 3.00 moles He and 3.00 moles H_2

12.151 What would be the pressure, in atm, of a gaseous mixture that occupies a volume of 27.0 L at 20°C if 3.00 moles each of Ar, Ne, and He are present?

12.152 What would be the pressure, in atm, of a gaseous mixture that occupies a volume of 2.00 L at 40°C if 3.00 grams each of Ar, Ne, and He are present?

DALTON'S LAW OF PARTIAL PRESSURES (SEC. 12.16)

12.153 Helium gas is added to an empty gas cylinder until the pressure reaches 9.0 atm. Neon gas is then added to the cylinder until the total pressure is 14.0 atm. Argon gas is then added to the cylinder until the total cylinder pressure reaches 29.0 atm. What is the partial pressure of each gas in the cylinder?

12.154 Nitrogen gas (N_2) is added to an empty gas cylinder until the pressure reaches 8.0 atm. Helium gas is then added to the cylinder until the total pressure is 10.0 atm. Carbon monoxide gas (CO) is then added to the cylinder until the total cylinder pressure reaches 20.0 atm. What is the partial pressure of each gas in the cylinder?

12.155 A mixture of H_2, N_2, and Ar gases is present in a steel cylinder. The total pressure within the cylinder is 675 mm Hg and the partial pressures of N_2 and Ar are, respectively, 354 mm Hg and 235 mm Hg. If CO_2 gas is added to the mixture at constant temperature until the total pressure reaches 842 mm Hg, what is the partial pressure, in millimeters of Hg, of the following?
   a. CO_2
   b. N_2
   c. Ar
   d. H_2

12.156 A mixture of O_2, He, and Ne gases is present in a steel cylinder. The total pressure within the cylinder is 652 mm Hg and the partial pressures of He and Ne are, respectively, 251 mm Hg and 152 mm Hg. If CO_2 gas is added to the mixture at constant temperature until the total pressure reaches 704 mm Hg, what is the partial pressure, in millimeters of Hg, of the following?
   a. CO_2
   b. He
   c. Ne
   d. O_2

12.157 The following diagram depicts a gaseous mixture of neon (blue spheres), argon (gray spheres), and krypton (white spheres).

If the total pressure in the container is 0.060 atm, what is the partial pressure, in atm, of the following gases?
a. neon **b.** argon **c.** krypton

12.158 The following diagram depicts a gaseous mixture of O_2 (blue spheres), N_2 (gray spheres), and H_2 (white spheres).

Assuming that the gases do not react with each other, what is the total pressure, in atm, in the container if the partial pressure of the O_2 gas is 0.020 atm?

12.159 A gaseous mixture contains 25.0 g of each of the gases CO, CO_2, and H_2S. Assuming that the gases do not react with each other, calculate the following:
a. the mole fraction of each gas present in the mixture
b. the partial pressure of each gas present given that the total pressure exerted by the gaseous mixture is 1.72 atm

12.160 A gaseous mixture contains 15.0 g of each of the gases HCl, H_2S, and Xe. Assuming that the gases do not react with each other, calculate the following:
a. the mole fraction of each gas present in the mixture
b. the partial pressure of each gas present, given that the total pressure exerted by the gaseous mixture is 2.24 atm

12.161 Calculate the partial pressure of O_2, in atmospheres, in a gaseous mixture with a volume of 2.50 L at 20°C, given that the mixture composition is
a. 0.50 mole O_2 and 0.50 mole N_2.
b. 0.50 mole O_2 and 0.75 mole N_2.
c. 0.50 mole O_2, 0.75 mole N_2, and 0.75 mole Ar.
d. 0.50 g O_2 and 0.75 g N_2.

12.162 Calculate the partial pressure of Xe, in atmospheres, in a gaseous mixture with a volume of 1.20 L at 32°C, given that the mixture composition is
a. 0.40 mole Xe and 0.40 mole Ne.
b. 0.40 mole Xe and 0.60 mole Ne.
c. 0.40 mole Xe, 0.60 mole Ne, and 1.25 moles O_2.
d. 0.40 g Xe and 0.60 g Ne.

12.163 What is the partial pressure of O_2 in a gaseous mixture whose total pressure is 1.20 atm, given the following mixture compositions?
a. 0.40 mole O_2 and 0.40 mole Ne
b. 0.40 mole O_2 and 0.80 mole Ne
c. an equal number of moles of O_2, N_2, and H_2
d. an equal number of molecules of O_2, N_2, and H_2

12.164 What is the partial pressure of Xe in a gaseous mixture whose total pressure is 1.55 atm, given the following mixture compositions?
a. 0.50 mole Xe and 0.50 mole Ne
b. 0.50 mole Xe and 1.00 mole Ne
c. an equal number of moles of Xe, Ne, and He
d. an equal number of atoms of Xe, Ne, and He

12.165 What is the total pressure in a flask that contains 4.0 moles He, 2.0 moles Ne, and 0.50 mole Ar, and in which the partial pressure of Ar is 0.40 atm?

12.166 What is the total pressure in a flask that contains 2.0 moles H_2, 6.0 moles O_2, and 0.50 mole N_2, and in which the partial pressure of H_2 is 0.80 atm?

12.167 Calculate the partial pressure of O_2, in atm, in a gaseous mixture of O_2 and N_2, given that
a. the total pressure is 6.00 atm and the mole fraction of N_2 is 0.150.
b. the partial pressure of N_2 is 2.26 atm and the mole fraction N_2 is 0.180.

12.168 Calculate the partial pressure of O_2, in atm, in a gaseous mixture of O_2 and N_2, given that
a. the total pressure is 5.00 atm and the mole fraction of N_2 is 0.800.
b. the partial pressure of N_2 is 3.00 atm and the mole fraction N_2 is 0.220.

12.169 What is the mole fraction of each gas in a mixture having the partial pressures of 0.500 atm of He, 0.250 atm of Ar, and 0.350 atm of Xe?

12.170 What is the mole fraction of each gas in a mixture having the particle pressures of 0.350 atm of Ne, 0.550 atm of Kr, and 0.750 atm of He?

12.171 A sample of N_2 gas of mass 20.0 g is present in a vessel at 0°C and 1.00 atm. What will be the final pressure, in atm, in the vessel after 8.00 g of Ar is pumped into the vessel at constant temperature?

12.172 A sample of O_2 gas of mass 25.0 g is present in a vessel at 35°C and 2.00 atm. What will be the final pressure, in atm, in the vessel after 20.00 g of He is pumped into the vessel at constant temperature?

12.173 A sample of ammonia, NH_3, is *completely* decomposed to its constituent elements.

$$2 NH_3(g) \longrightarrow N_2(g) + 3 H_2(g)$$

If the total pressure of the N_2 and H_2 produced is 852 mm Hg, calculate the partial pressures, in millimeters of mercury, of N_2 and H_2.

12.174 A sample of steam, H_2O, is *completely* decomposed to its constituent elements.

$$2\,H_2O(g) \longrightarrow 2\,H_2(g) + O_2(g)$$

If the total pressure of the H_2 and O_2 produced is 1.35 atm, calculate the partial pressures, in atmospheres, of H_2 and O_2.

12.175 What would be the partial pressure, in millimeters of mercury, of O_2 collected over water at the following conditions of temperature and atmospheric pressure?
   a. 19°C and 743 mm Hg
   b. 28°C and 645 mm Hg
   c. 34°C and 762 mm Hg
   d. 21°C and 0.933 atm

12.176 What would be the partial pressure, in millimeters of mercury, of O_2 collected over water at the following conditions of temperature and atmospheric pressure?
   a. 15°C and 632 mm Hg
   b. 35°C and 749 mm Hg
   c. 31°C and 682 mm Hg
   d. 26°C and 0.975 atm

12.177 A 24.64 L flask, at STP, contains 0.100 mole of He, 0.200 mole of Ne, and 0.800 mole of Ar. For this gaseous mixture, calculate the
   a. mole fraction He.   **b.** mole percent Ar.
   c. pressure percent Ne. **d.** volume percent He.

12.178 A 15.68 L flask, at STP, contains 0.200 mole of Ar, 0.400 mole of Kr, and 0.100 mole of Xe. For this gaseous mixture, calculate the
   a. mole fraction Ar.   **b.** mole percent Kr.
   c. pressure percent Xe. **d.** volume percent Ar.

12.179 Three containers of gases are combined into a single large container; that is, 2.0 L of O_2 at STP, 3.0 L of Ar at STP, and 3.0 L of Ne at STP are put into an 8.0 L container at STP. Calculate the following items pertaining to the gaseous mixture.
   a. volume percent O_2 **b.** mole percent Ar
   c. pressure percent Ne **d.** partial pressure O_2

12.180 Three containers of gases are combined into a single large container; that is, 1.0 L of N_2 at STP, 4.0 L of He at STP, and 1.0 L of Xe at STP are put into a 6.0 L container at STP. Calculate the following items pertaining to the gaseous mixture.
   a. volume percent He   **b.** mole percent Xe
   c. pressure percent He **d.** partial pressure N_2

ADDITIONAL PROBLEMS

12.181 How many molecules of carbon dioxide (CO_2) gas are contained in 1.00 L of CO_2 at STP?

12.182 How many molecules of hydrogen sulfide (H_2S) gas are contained in 2.00 L of H_2S at STP?

12.183 At a particular temperature and pressure, 8.00 g of N_2 gas occupy 6.00 L. What would be the volume, in liters, occupied by 1.00×10^{23} molecules of NH_3 at the same temperature and pressure?

12.184 At a particular temperature and pressure, 2.00×10^{23} molecules of N_2 gas occupy 5.00 L. What would be the volume, in liters, occupied by 25.7 g of SO_2 at the same temperature and pressure?

12.185 A piece of Al is placed in a 1.00 L container with pure O_2. The O_2 is at a pressure of 1.00 atm and a temperature of 25°C. One hour later the pressure has dropped to 0.880 atm and the temperature has dropped to 22°C. Calculate the number of grams of O_2 that reacted with the Al.

12.186 A piece of Ca is placed in a 1.00 L container with pure N_2. The N_2 is at a pressure of 1.12 atm and a temperature of 26°C. One hour later the pressure has dropped to 0.924 atm and the temperature has dropped to 24°C. Calculate the number of grams of N_2 that reacted with the Ca.

12.187 At constant temperature, the pressure on a sample of H_2 gas is increased from 1.50 atm to 3.50 atm. This action decreases the sample volume by 25.0 mL. What was the original volume, in milliliters, of the gas sample?

12.188 At constant temperature, the pressure on a sample of HCl gas is decreased from 3.50 atm to 1.50 atm. This action increases the sample volume by 75.0 mL. What was the original volume, in milliliters, of the gas sample?

12.189 The volume of a fixed quantity of gas, at constant temperature, is decreased by 20.0%. What is the resulting percentage increase in the pressure of the gas?

12.190 The volume of a fixed quantity of gas, at constant temperature, is increased by 30.0%. What is the resulting percentage decrease in the pressure of the gas?

12.191 The volume of a fixed quantity of gas, at constant pressure, is increased by 50.0%. What is the resulting percentage increase in the Kelvin temperature of the gas?

12.192 The volume of a fixed quantity of gas, at constant pressure, is decreased by 10.0%. What is the resulting percentage decrease in the Kelvin temperature of the gas?

12.193 Calculate the ratio of the densities of O_2 and N_2 at
   a. STP conditions.
   b. 1.25 atm and 25°C.

12.194 Calculate the ratio of the densities of He and Ne at
   a. STP conditions.
   b. 2.30 atm and 57°C.

12.195 At constant temperature, a 2.000 L mixture of gases is produced by combining 1.000 L of N_2 at 350.0 mm Hg, 6.000 L of O_2 at 300.0 mm Hg, and 1.000 L of H_2 at 250.0 mm Hg. What is the pressure, in millimeters of mercury, of the mixture? Assume that no chemical reactions occur.

12.196 At constant temperature, a 2.000 L mixture of gases is produced by combining 2.000 L of N_2 at 250.0 mm Hg, 4.000 L of O_2 at 250.0 mm Hg, and 1.000 L of H_2 at 600.0 mm Hg. What is the pressure, in millimeters of mercury, of the mixture? Assume that no chemical reactions occur.

12.197 Suppose 532 mL of Ne gas at 20°C and 1.04 atm and 376 mL of SF_6 gas at 20°C and 0.97 atm are put into a 275 mL flask. Assuming that the temperature remains constant, calculate the partial pressure, in atmospheres, of the SF_6 gas in the mixture.

12.198 Suppose 734 mL of Ar gas at 18°C and 1.25 atm and 252 mL of HF gas at 18°C and 2.25 atm are put into a 545 mL flask. Assuming that the temperature remains constant, calculate the partial pressure, in atmospheres, of the HF gas in the mixture.

12.199 A mixture of 15.0 g of Ar and 15.0 g of CH_4 occupies a 4.0 L container at 8.80 atm and 54°C. What is the partial pressure, in atmospheres, of Ar in the mixture?

12.200 A mixture of 30.0 g of Ar and 15.0 g of CH_4 occupies a 4.0 L container at 11.9 atm and 27°C. What is the partial pressure, in atmospheres, of CH_4 in the mixture?

12.201 Suppose 30.0 mL of N_2 gas at 27°C and 645 mm Hg pressure is added to a 40.0 mL container that already contains He at 37°C and 765 mm Hg. If the resulting mixture is brought to 32°C, what is the total pressure, in millimeters of mercury, of the mixture?

12.202 Suppose 50.0 mL of Xe gas at 45°C and 0.998 atm pressure is added to a 100.0 mL container that already contains He at 37°C and 765 mm Hg. If the resulting mixture is warmed to 75°C, what is the total pressure, in millimeters of mercury, of the mixture?

CUMULATIVE PROBLEMS

12.203 A 2.24 L sample of a gaseous compound has a mass of 7.5 g at STP. What is the molecular mass of a molecule of this compound in atomic mass units?

12.204 A 4.48 L sample of a gaseous compound has a mass of 20.0 g at STP. What is the molecular mass of a molecule of this compound in atomic mass units?

12.205 What is the value of x in the formula PH_x if the density of the PH_x gas is 1.517 g/L at 0°C and 1.00 atm?

12.206 What is the value of x in the formula P_2H_x if the density of P_2H_x gas is 2.944 g/L at 0°C and 1.00 atm?

12.207 The composition of a gaseous mixture in terms of mass percent is 75.0% HCl, 5.00% H_2, and 20.0% He. For this mixture, calculate the following:
  a. the mole fraction of each component
  b. the partial pressure of each component, given that the total pressure is 1.20 atm
  c. a weighted average molar mass for the mixture as a whole
  d. the density of the mixture, at STP, based on the weighted average molar mass

12.208 The composition of a gaseous mixture in terms of mass percent is 43.0% Ar, 15.0% He, and 42.0% H_2S. For this mixture, calculate the following:
  a. the mole fraction of each component
  b. the partial pressure of each component, given that the total pressure is 3.50 atm
  c. a weighted average molar mass for the mixture as a whole
  d. the density of the mixture, at STP, based on the weighted average molar mass

12.209 A 0.581 g sample of a gaseous compound containing only carbon and hydrogen contains 0.480 g of carbon and 0.101 g of hydrogen. At STP, 33.6 mL of the gas have a mass of 0.0869 g. What is the molecular formula for the compound?

12.210 A 6.01 g sample of a gaseous compound containing only carbon and hydrogen contains 4.80 g of carbon and 1.21 g of hydrogen. At STP, 762 mL of the gas have a mass of 1.02 g. What is the molecular formula for the compound?

12.211 Ammonia, NH_3, burns in oxygen to form nitric oxide, NO, and water.

$$4 NH_3(g) + 5 O_2(g) \longrightarrow 4 NO(g) + 6 H_2O(g)$$

Calculate the total volume, in liters, of products formed when 60.0 L of NH_3 burn in the presence of 60.0 L of O_2. Assume that all volume measurements are made at the same temperature and pressure.

12.212 Methane, CH_4, burns in oxygen to form carbon dioxide, CO_2, and water.

$$CH_4(g) + 2 O_2(g) \longrightarrow CO_2(g) + 2 H_2O(g)$$

Calculate the total volume, in liters, of products formed when 30.0 L of CH_4 burn in the presence of 50.0 L of O_2. Assume that all volume measurements are made at the same temperature and pressure.

12.213 A 22.0 g sample of NH_3 reacts with an excess of Cl_2 gas, according to the equation

$$2 NH_3(g) + 3 Cl_2(g) \longrightarrow N_2(g) + 6 HCl(g)$$

What volume, in cubic meters at STP, of HCl gas is produced?

12.214 A 43.0 g sample of NO reacts with an excess of O_2 gas, according to the equation

$$2 NO(g) + O_2(g) \longrightarrow 2 NO_2(g)$$

What volume, in cubic meters at STP, of NO_2 gas is produced?

12.215 Ammonia gas reacts with hydrogen chloride gas according to the equation

$$NH_3(g) + HCl(g) \longrightarrow NH_4Cl(s)$$

If 7.00 g of NH_3 is reacted with 12.0 g of HCl in a 1.00 L container at 25°C, what will be the final pressure, in atmospheres, in the reaction container?

12.216 At elevated temperatures, phosphorus reacts with oxygen according to the equation

$$4\,P(g) + 5\,O_2(g) \longrightarrow P_4O_{10}(s)$$

If 50.0 g of P is reacted with 25.0 g of O_2 in an 8.00 L container at 175°C, what will be the final pressure, in atmospheres, in the reaction container?

Multiple-Choice Practice Test

Use this bank of 20 multiple-choice questions as a review of key concepts presented in this chapter. For many of the questions, there may be more than one correct answer (choice d) or no correct answer (choice e).

12.217 In terms of size, the mm Hg pressure unit is
  a. 100 times larger than the torr unit.
  b. 760 times larger than the atmosphere unit.
  c. 14.68 times larger than the psi unit.
  d. more than one correct response
  e. no correct response

12.218 Which of the following gas laws involves a direct proportion and a constant pressure?
  a. Boyle's law
  b. Charles's law
  c. Gay-Lussac's law
  d. more than one correct response
  e. no correct response

12.219 A sample of O_2 gas in a 6.00 L container is under a pressure of 4.00 atm. If the volume of the container is decreased to 2.00 L at constant temperature, what is the new pressure, in atm?
  a. 3.00 atm   **b.** 6.00 atm
  c. 9.00 atm   **d.** 12.0 atm
  e. no correct response

12.220 A mathematical statement of Charles's law is
  a. $V_1T_1 = V_2T_2$.
  b. $V_1T_2 = V_2T_1$.
  c. $V_1 + T_1 = V_2 + T_2$.
  d. more than one correct response
  e. no correct response

12.221 If the volume of a sample of F_2 gas is reduced by one-fourth at constant pressure, what happens to its Kelvin temperature?
  a. It remains constant.
  b. It decreases by one-fourth.
  c. It increases by a factor of 4.
  d. more than one correct response
  e. no correct response

12.222 The Gay-Lussac's law observation that an increased temperature causes an increased pressure is explained using kinetic molecular theory in the following manner: The molecules of the gas
  a. strike the container walls less often.
  b. strike the container walls more often.
  c. move at a slower speed.
  d. more than one correct response
  e. no correct response

12.223 Based on the combined gas law, if both the volume and Kelvin temperature of a gas sample are doubled, the pressure will
  a. double.   **b.** quadruple.
  c. be halved.   **d.** remain the same.
  e. no correct response

12.224 Which of the following sets of variables are present in mathematical statements of Avogadro's law?
  a. n and V **b.** n and T **c.** n, T, and V
  d. more than one correct response
  e. no correct response

12.225 Which of the following statements concerning molar volume is correct?
  a. It has the value 22.414 L at all temperature–pressure combinations.
  b. It cannot be determined without knowing the chemical identity of the gas.
  c. It has the value 13.3 L at 25°C and 2.00 atm pressure for all gases.
  d. more than one correct response
  e. no correct response

12.226 What is the density of a gas for which molar mass is 30.01 g/mole and molar volume is 12.23 L/mole?
  a. (molar mass) x (molar volume)
  b. (molar mass) ÷ (molar volume)
  c. (molar mass) + (molar volume)
  d. more than one correct response
  e. no correct response

12.227 The conditions known as STP are
  a. 760 mm Hg and 273 K.
  b. 1 atm and 0°C.
  c. 760 atm and 273°C.
  d. more than one correct response
  e. no correct response

12.228 At STP conditions for gases, what is the volume, in liters, of H_2O vapor (steam) that is produced from the reaction of 8.00 L of H_2 and 4.00 L of O_2 according to the reaction

$$2\,H_2(g) + O_2(g) \longrightarrow 2\,H_2O(g)$$

  a. 4.00 L   **b.** 6.00 L
  c. 8.00 L   **d.** 12.0 L
  e. no correct response

12.229 At STP conditions, one mole of any gas will occupy a volume of
  a. 1.000 milliliter.
  b. 1.000 liter.

    c. 22.41 milliliters.
    d. 22.41 liters.
    e. no correct response

12.230 A correct form of the ideal gas law equation is
    a. $PV = nRT$.
    b. $P/V = nRT$.
    c. $P = nRT/V$.
    d. more than one correct response
    e. no correct response

12.231 What is the volume, in liters, occupied by 1.73 moles of N_2 gas at 0.992 atm pressure and a temperature of 75°C?
    a. 10.7 L
    b. 33.8 L
    c. 49.8 L
    d. 52.2 L
    e. no correct response

12.232 What is the density of F_2 gas at 24°C and a pressure of 0.986 atm?
    a. 1.54 g/L
    b. cannot be determined without knowing the volume of the gas
    c. cannot be determined without knowing the amount of gas present
    d. more than one correct response
    e. no correct response

12.233 In the reaction

$$Fe_2O_3(s) + 3 H_2(g) \longrightarrow 2 Fe(s) + 3 H_2O(l)$$

how many grams of Fe can be produced from 17.4 L of H_2 at STP?
    a. 17.4 g
    b. 28.9 g

    c. 44.4 g
    d. 65.3 g
    e. no correct response

12.234 In the reaction

$$Fe_2O_3(s) + 3 H_2(g) \longrightarrow 2 Fe(s) + 3 H_2O(g)$$

how many moles of Fe_2O_3 must react to produce 10.8 L of H_2O at 125°C and 1.00 atm pressure?
    a. 0.110 mole
    b. 0.331 mole
    c. 0.917 mole
    d. 1.01 moles
    e. no correct response

12.235 The partial pressure of He gas in a gaseous mixture of He and H_2 gas is
    a. the pressure that the He would exert in the absence of H_2 under the same conditions.
    b. equal to the total pressure divided by helium's atomic mass.
    c. equal to the total pressure multiplied by the mole fraction of He present.
    d. more than one correct response
    e. no correct response

12.236 What is the partial pressure of O_2 gas, in atm, in a mixture of 0.40 mole He, 0.20 mole Ne, and 0.60 mole O_2 if the total pressure exerted by the mixture is 2.00 atm?
    a. 0.80 atm
    b. 1.0 atm
    c. 1.2 atm
    d. 1.4 atm
    e. no correct response

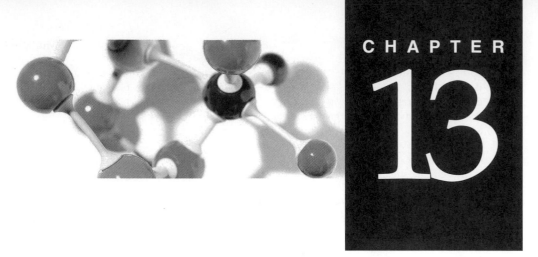

13

Solutions

13.1 CHARACTERISTICS OF SOLUTIONS

A **solution** *is a homogeneous mixture of two or more substances, with each substance retaining its chemical identity.* To form a homogeneous mixture, the intermingling of components must be on the molecular level; that is, the particles present must be of atomic or molecular size.

In discussing solutions it is often convenient to categorize the components of a solution as *solvent* and *solute(s)*. A **solvent** *is the component of the solution present in the greatest amount.* The solvent may be thought of as the medium in which the other substances present are *dissolved.* A **solute** *is a solution component present in a small amount relative to that of a solvent.* More than one solute may be present in the same solution. For example, both sugar and salt (two solutes) may be dissolved in water (a solvent).

In most situations we will encounter, the solutes present in a solution will be of more interest to us than the solvent. The solutes are the "active ingredients" in the solution. They are the substances that may react when solutions are mixed.

Solutions used in the laboratory are often liquids, and the solvent is almost always water. However, as we shall see shortly, gaseous solutions and solid solutions of several types do exist.

A solution, since it is homogeneous, will have the same properties throughout. No matter from where we take a sample in a solution, it will have the same composition as that of any other sample from the solution. The composition of a solution can be varied, usually within certain limits, by changing the

TABLE 13.1 **Examples of Various Types of Solutions**

Solution Type (solute listed first)	Example
Gaseous Solutions	
Gas dissolved in gas	Dry air (oxygen and other gases dissolved in nitrogen)
Liquid dissolved in gas*	Wet air (water vapor in air)
Solid dissolved in gas*	Moth repellent (or mothballs) sublimed into air
Liquid Solutions	
Gas dissolved in liquid	Carbonated beverage (carbon dioxide in water)
Liquid dissolved in liquid	Vinegar (acetic acid dissolved in water)
Solid dissolved in liquid	Salt water
Solid Solutions	
Gas dissolved in solid	Hydrogen in platinum
Liquid dissolved in solid	Dental filling (mercury dissolved in silver)
Solid dissolved in solid	Sterling silver (copper dissolved in silver)

An alternative viewpoint is that liquid-in-gas and solid-in-gas solutions do not actually exist as true solutions. From this viewpoint, water vapor or moth repellent in air is considered to be a gas-in-gas solution since the water or moth repellent must evaporate or sublime first in order to enter the air.

relative amounts of solvent and solute present. (If the composition limits are violated, a heterogeneous mixture is formed.)

Two-component solutions can be classified into nine types according to the physical states of the solvent and solute before mixing. These types, along with an example of each, are listed in Table 13.1. Solutions in which the final state of the solution components is liquid are the most common and are the type that are emphasized in this text.

The physical state of the solute becomes that of the solvent when a solution is formed. For example, solid naphthalene (moth repellent) must be sublimed (Sec. 4.4) for it to dissolve in air. Pulverizing a solid to a fine powder and dispersing it in air does not produce a solution. (Dust particles in air would be an example of this.) The particles of the solid must be subdivided to the molecular level; the solid must sublime. Similarly, fog is a suspension of water droplets in air; the droplets are large enough to reflect light, a fact that becomes evident when we drive an automobile on a foggy night. Thus, fog is not a solution. Water vapor, however, is present in solution form in air. When hydrogen gas dissolves in platinum metal (a gas-in-solid solution), the gas molecules take up fixed positions in the metal lattice. The gas is "solidified" as a result.

"All solutions are mixtures" is a valid statement. However, the reverse statement, "All mixtures are solutions," is not valid. Only those mixtures that are *homogeneous* are solutions.

13.2 SOLUBILITY

In addition to *solvent* and *solute*, several other terms are useful in describing characteristics of solutions. **Solubility** *is the maximum amount of solute that will dissolve in a given amount of solvent.* Numerous factors affect the numerical value of a solute's solubility in a given solvent, including the nature of the solvent itself, the temperature, and in some cases the pressure and the presence of other solutes. Common units for expressing solubility are grams of solute per 100 g of solvent.

Effect of Temperature on Solubility

Most solids become more soluble in water with increasing temperature. The data in Table 13.2 illustrate this temperature–solubility pattern. Here, the solubilities of selected ionic solids in water are given at three different temperatures. Note from the values in this table that a temperature-caused solubility increase can be dramatic ($AgNO_3$,

TABLE 13.2 Solubilities of Various Compounds in Water at 0°C, 50°C, and 100°C

Solute	Solubility (g solute/100 g H$_2$O)		
	0°C	50°C	100°C
Lead(II) bromide (PbBr$_2$)	0.455	1.94	4.75
Silver sulfate (Ag$_2$SO$_4$)	0.573	1.08	1.41
Copper(II) sulfate (CuSO$_4$)	14.3	33.3	75.4
Sodium chloride (NaCl)	35.7	37.0	39.8
Silver nitrate (AgNO$_3$)	122	455	952
Cesium chloride (CsCl)	161.4	218.5	270.5

680% increase by going from 0°C to 100°C) to slight (NaCl, 11% increase in going from 0°C to 100°C). Figure 13.1 shows the solubility change with temperature, at 10°C intervals, for the molecular substance sucrose (C$_{12}$H$_{22}$O$_{11}$, table sugar) in water.

A few ionic compounds, Na$_2$SO$_4$ for example, have solubilities that decrease with increasing temperature.

The use of specific units for solubility, as in Table 13.2, allows us to compare solubilities quite accurately. Such precision is often unnecessary, and instead *qualitative* statements about solubilities are made by using terms such as *very soluble, slightly soluble*, and so forth. The guidelines for the use of such terms are given in Table 13.3.

In contrast to the solubilities of solids, gas solubilities in water decrease with increasing temperature. For example, both N$_2$ and O$_2$, the major components of air, are less soluble in hot water than in cold water.

Effect of Pressure on Solubility

Pressure has little effect on the solubility of solids and liquids in water. However, it has a major effect on the solubility of gases in water. The amount of gas that will dissolve in a liquid at a given temperature is directly proportional to the pressure of the gas above the liquid. In other words, as the pressure of a gas above a liquid increases, the solubility of the gas increases; conversely, as the pressure of the gas decreases, its solubility decreases.

Respiratory therapy takes advantage of the fact that increased pressure increases the solubility of a gas. Patients with lung problems who are unable to get sufficient oxygen from air are given an oxygen-enriched mixture of gases to breathe. The larger oxygen partial pressure in the enriched mixture translates into increased oxygen uptake in the patient's lungs.

FIGURE 13.1 Solubility in water of the molecular compound sucrose (C$_{12}$H$_{22}$O$_{11}$, table sugar) as a function of temperature in the range 0°C to 100°C.

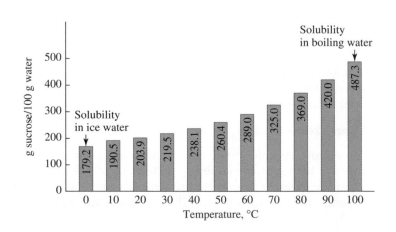

TABLE 13.3 Qualitative Solubility Terms

Solute Solubility (g solute/100 g solvent)	Qualitative Solubility Description
Less than 0.1	insoluble
0.1–1	slightly soluble
1–10	soluble
Greater than 10	very soluble

Terminology for Relative Amount of Solute in a Solution

A **saturated solution** *is a solution that contains the maximum amount of solute that can be dissolved under the conditions at which the solution exists.* A saturated solution together with excess undissolved solute is an equilibrium situation where the rate of dissolution of undissolved solute is equal to the rate of crystallization of dissolved solute. Consider the process of adding table sugar (sucrose) to a container of water. Initially the sugar dissolves as the solution is stirred. Finally, as we add more sugar, a point is reached where no amount of stirring will cause the added sugar to dissolve. Sugar remains as a solid on the bottom of the container; the solution is *saturated*. Although it appears to the eye that nothing is happening once the saturation point is reached, on the molecular level this is not the case. Solid sugar from the bottom of the container is continuously dissolving in the water, and an equal amount of sugar is coming out of the solution. Accordingly, the net number of sugar molecules in the liquid remains the same, and outwardly it appears that the dissolution process has stopped. This equilibrium situation in the saturated solution is somewhat similar to the previously discussed evaporation of a liquid in a closed container (Sec. 11.12). Figure 13.2 illustrates the dynamic equilibrium process occurring in a saturated solution in the presence of undissolved excess solute.

> When the amount of dissolved solute in a solution corresponds to the solute's solubility in the solvent, the solution is a saturated solution.

An **unsaturated solution** *is a solution that contains less solute than the maximum amount that could dissolve.* The majority of solutions encountered are *unsaturated* solutions.

Sometimes it is possible to exceed the maximum solubility of a compound, producing a *supersaturated solution.* A **supersaturated solution** *is a solution that contains more dissolved solute than that needed for a saturated solution.* An indirect rather than a direct procedure is needed for preparing a supersaturated solution; it involves the slow cooling, without agitation of any kind, of a high-temperature saturated solution in which no excess solid solute is present. Even though solute solubility decreases as the temperature is reduced, the excess solute remains in solution. A supersaturated solution is an unstable

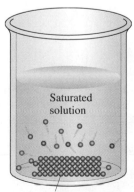

Saturated solution

Undissolved solute

FIGURE 13.2 The dynamic equilibrium process occurring in a saturated solution that contains undissolved solute.

situation; with time, excess solute will crystallize out, and the solution will revert to a saturated solution. A supersaturated solution will produce crystals rapidly, often in a dramatic manner, if it is slightly disturbed or if it is seeded with a tiny crystal of solute.

The terms *dilute* and *concentrated* are also used to convey qualitative information about the degree of saturation of a solution. A **dilute solution** *is a solution that contains a small amount of solute in solution relative to the amount that could dissolve.* On the other hand, a **concentrated solution** *is a solution that contains a large amount of solute relative to the amount that could dissolve.* A concentrated solution need not be a saturated solution.

Aqueous and Nonaqueous Solutions

A set of solution terms that relates to the identity of the solvent present is *aqueous* and *nonaqueous.* An **aqueous solution** *is a solution in which water is the solvent.* The presence of water is not a prerequisite for a solution, however. A **nonaqueous solution** *is a solution in which a substance other than water is the solvent.* Alcohol-based solutions are often encountered in a medical setting.

> When the term *solution* is used, it is generally assumed that "aqueous solution" is meant, unless the context makes it clear that the solvent is not water.

13.3 SOLUTION FORMATION

In a solution, solute particles are uniformly dispersed throughout the solvent. Considering what happens at the molecular level during the solution process will help us to understand how this is achieved. Let us consider in detail the process of dissolving sodium chloride, a typical ionic solid, in water.

Figure 13.3 shows what is thought to happen when sodium chloride is placed in water. The polar water molecules become oriented so that the negative oxygen portion points toward positive sodium ions and the positive hydrogen portions point toward negative chloride ions. As the polar water molecules begin to surround ions on the crystal surface, they exert sufficient attraction to cause these ions to break away from the crystal surface. After leaving the crystal, an ion retains its surrounding group of water molecules; it has become a *hydrated ion.* As each hydrated ion leaves the surface, other ions are exposed to the water, and the crystal is picked apart ion by ion. Once in solution,

FIGURE 13.3   The solution process for an ionic solid in water.

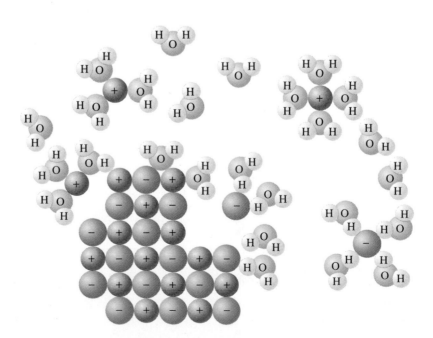

the hydrated ions are uniformly distributed by stirring or by random collisions with other molecules or ions.

The random motion of solute ions in solution causes them to collide with each other, with solvent molecules, and occasionally with the surface of the undissolved solute. Ions undergoing this last type of collision occasionally stick to the solid surface and thus leave the solution. When the number of ions in solution is low, the chances for collision with the undissolved solute are low. However, as the number of ions in solution increases, so do the chances for such collisions, and more ions are recaptured by the undissolved solute. Eventually, the number of ions in solution reaches such a level that ions return to the undissolved solute at the same rate as other ions leave. At this point the solution is saturated, and the equilibrium process discussed in the last section is in operation.

Factors Affecting the Rate of Solution Formation

The rate at which a solution forms is governed by how rapidly the solute particles are distributed throughout the solvent. Three factors that affect the rate of solution formation are

1. *The state of subdivision of the solute.* A crushed aspirin tablet will dissolve in water more rapidly than a whole aspirin tablet. The more compact whole aspirin tablet has less surface area, and thus fewer solvent molecules can interact with it at a given time.
2. *The degree of agitation during solution preparation.* Stirring solution components disperses the solute particles more rapidly, increasing the possibilities for solute–solvent interactions. Hence, the rate of solution formation is increased.
3. *The temperature of the solution components.* Solution formation occurs more rapidly as the temperature is increased. At a higher temperature, both solute and solvent molecules move more rapidly (Sec. 11.3) so more interactions between them occur within a given time period.

13.4 SOLUBILITY RULES

In this section rules are presented for qualitatively predicting solubilities. They summarize in a concise form the results of thousands of experimental solute–solvent solubility determinations. The basis for the rules is *polarity* considerations—specifically, the magnitude of the difference between the polarity of the solute and solvent. In general, it is found that the greater the difference in solute–solvent polarity, the less soluble the solute.

A simple summary statement of the polarity–solubility relationship is *substances of like polarity tend to be more soluble in each other than substances that differ in polarity.* This conclusion is often expressed as the even simpler phrase *"like dissolves like."* Polar substances, in general, are good solvents for other polar substances but not for nonpolar substances. Similarly, nonpolar substances exhibit greater solubility in nonpolar solvents than they do in polar solvents.

A consideration of the intermolecular forces present in a solution reveals the basis for the like-dissolves-like rule. The same types of intermolecular forces present in pure liquids (Sec. 11.16) also operate in solutions. However, the situation is more complex for solutions because there are three types of interactions present: solute–solute interactions, solute–solvent interactions, and solvent–solvent interactions.

Solutions form when these three types of interactions are similar in nature and in magnitude (see Fig. 13.4). Cholesterol ($C_{27}H_{46}O$), a nonpolar substance, is soluble in fat (also a nonpolar material) because of the similar magnitude of the London forces (Sec. 11.16) present. Cholesterol has limited solubility in water because of the dissimilarity of the intermolecular forces, which are London forces for cholesterol and hydrogen bonding for water (Sec. 11.16). Sodium chloride is soluble in water because the

FIGURE 13.4 Solution formation occurs when these three kinds of intermolecular forces are similar in nature and in magnitude.

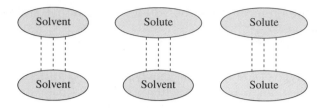

strong Na^+ ion-water and Cl^- ion-water attractions (ion–dipole attractions; Sec. 11.16) are similar in nature to the strong attractions between Na^+ and Cl^- ions (ion–ion attractions; Sec. 11.16) and the strong hydrogen bonds between water molecules.

The generalization "like dissolves like" is a useful tool for predicting solubility behavior in many, but not all, solute–solvent situations. Results that agree with the generalization are almost always obtained in the cases of gas-in-liquid and liquid-in-liquid solutions and for solid-in-liquid solutions in which the solute is not an ionic compound. For example, NH_3 gas (a polar gas) is much more soluble in H_2O (a polar liquid) than is O_2 gas (a nonpolar gas). (The actual solubilities of NH_3 and O_2 in water at 20°C are, respectively, 51.8 g/100 g H_2O and 0.0043 g/100 g H_2O.)

For those solid-in-liquid solutions in which the solute is an ionic compound—a very common situation—the rule like dissolves like is not adequate. One would predict that since all ionic compounds are polar, they all would dissolve in polar solvents such as water. This is not the case. The failure of the generalization here is related to the complexity of the factors involved in determining the magnitude of the solute–solute (ion–ion) and solute–solvent (ion–polar solvent molecule) interactions. Among other things, both the charge on and the size of the ions in the solute must be considered. Changes in these factors affect both types of interactions, but not to the same extent.

Some guidelines concerning the solubility of ionic compounds in water, which should be used in place of like dissolves like, are given in Table 13.4.

There is no such thing as an absolutely insoluble ionic compound; all dissolve to a slight extent. Thus, the insoluble classification in Table 13.4 really means ionic compounds that have a very limited solubility in water.

TABLE 13.4 Solubility Guidelines for Ionic Compounds in Water

Soluble Compounds	**Important Exceptions**
Compounds containing the following ions are soluble, with exceptions as noted.	
Group IA (Li^+, Na^+, K^+, etc.)	none
Ammonium (NH_4^+)	none
Acetate ($C_2H_3O_2^-$)	none
Nitrate (NO_3^-)	none
Chloride (Cl^-), bromide (Br^-), and iodide (I^-)	Ag^+, Pb^{2+}, Hg_2^{2+}
Sulfate (SO_4^{2-})	Ca^{2+}, Sr^{2+}, Ba^{2+}, Pb^{2+}
Insoluble Compounds	**Important Exceptions**
Compounds containing the following ions are insoluble, with exceptions as noted.	
Carbonate (CO_3^{2-})	group IA and NH_4^+
Phosphate (PO_4^{3-})	group IA and NH_4^+
Sulfide (S^{2-})	groups IA and IIA and NH_4^+
Hydroxide (OH^-)	group IA, Ca^{2+}, Sr^{2+}, Ba^{2+}

You should become thoroughly familiar with the rules in Table 13.4. They find extensive use in chemical discussions. We will next encounter them in Section 15.3 when the topic of net ionic equations is presented.

EXAMPLE 13.1   **Using Solubility Guidelines to Predict Compound Solubilities**

Predict the solubility of each of the following solutes in the solvent indicated.

   a. acetone (a polar liquid) in water
   b. ammonia (a polar gas) in benzene (a nonpolar liquid)
   c. NaBr (an ionic solid) in water
   d. $CaCO_3$ (an ionic solid) in water
   e. AgCl (an ionic solid) in water

SOLUTION

   a. Soluble. Acetone is polar, as is water. Like dissolves like.
   b. Insoluble. Since the two substances are of unlike polarity, they should be relatively insoluble in each other.
   c. Soluble. Table 13.4 indicates that all compounds containing Na^+ ion are soluble.
   d. Insoluble. Table 13.4 indicates that all carbonates are insoluble except those of group IA ions and NH_4^+ ions.
   e. Insoluble. Table 13.4 indicates that silver is an exception to the rule that all chlorides are soluble.

▶ **Practice Exercise 13.1**   Predict the solubility of each of the following solutes in the solvent indicated.

   a. ethyl alcohol (a polar liquid) in water
   b. ethane (a nonpolar gas) in benzene (a nonpolar liquid)
   c. $NaNO_3$ (an ionic solid) in water
   d. $(NH_4)_3PO_4$ (an ionic solid) in water
   e. Al_2S_3 (an ionic solid) in water

Answers to all practice exercises in this chapter are found in the back-of-the-book answer section.

13.5 SOLUTION CONCENTRATIONS

In Section 13.2 we learned that, in general, there is a limit to the amount of solute that can be dissolved in a specified amount of solvent and also that a solution is said to be saturated when this maximum amount of solute has been dissolved. The amount of dissolved solute in a saturated solution is given by the solute's solubility.

Most solutions chemists deal with are *unsaturated* rather than saturated solutions. The amount of solute present in an unsaturated solution is specified by stating the *concentration* of the solution, an entity that describes the composition of the solution. **Concentration** *is the amount of solute present in a specified amount of solvent or a specified amount of solution.* Thus, concentration is a ratio of two quantities, being either the ratio

$$\frac{\text{amount of solute}}{\text{amount of solvent}} \quad \text{or} \quad \frac{\text{amount of solute}}{\text{amount of solution}}$$

In specifying a concentration, what are the units used to indicate the amounts of solute and solvent or solution present? In practice, a number of different unit combinations are used, with the choice of units depending on the use to be made of the concentration units. In each of the next four sections we shall discuss a commonly encountered set of units used to express solution concentration. The concentration expressions to be discussed are (1) percentage of solute (Sec. 13.6), (2) parts per million and parts per billion (Sec. 13.7), (3) molarity (Sec. 13.8), and (4) molality (Sec. 13.9).

13.6 CONCENTRATION: PERCENTAGE OF SOLUTE

The concentration of a solution is often specified in terms of the percentage of solute in the total amount of solution. Since the amounts of solute and solution present can be stated in terms of either mass or volume, different types of percent units exist. The three most common are

1. Percent by mass concentration (or mass–mass percent concentration)
2. Percent by volume concentration (or volume–volume percent concentration)
3. Mass–volume percent concentration

The concentration percent unit most frequently used by chemists is *percent by mass* (or mass–mass percent). **Percent by mass concentration** *is the mass of solute divided by the total mass of solution multiplied by 100 (to put the value in terms of percentage).* (Percentage is always part of the whole divided by the whole times 100; see Sec. 3.10.)

$$\text{percent by mass} = \frac{\text{mass of solute}}{\text{mass of solution}} \times 100$$

The solute and solution masses must be in the same units but any units are allowed. The mass of solution is equal to the mass of *solute* plus the mass of *solvent*.

$$\text{percent by mass} = \frac{\text{mass of solute}}{\text{mass of solute} + \text{mass of solvent}} \times 100$$

The concentration of butterfat in milk is expressed in terms of mass percent. When you buy 2% milk, you are buying milk that contains 2 grams of butterfat per 100 grams of milk.

A solution of 5.0% by mass concentration would contain 5.0 g of solute in 100.0 g of solution (5.0 g of solute and 95.0 g of solvent). Thus, percent by mass gives directly the number of grams of solute in 100 g of solution. The abbreviation for percent by mass is % (m/m).

EXAMPLE 13.2 **Calculating Mass Percent Concentration from Mass of Solute and Mass of Solvent**

What is the percent by mass, % (m/m), concentration of sucrose (table sugar) in a solution made by dissolving 5.4 g of sucrose in 75.0 g of water?

SOLUTION

To calculate percent by mass, we need both mass of solute and mass of solution.

$$\text{percent by mass} = \frac{\text{mass of solute}}{\text{mass of solution}} \times 100$$

The mass of solute is given (5.4 g) and the mass of solution is calculated by adding together mass of solute and mass of solvent.

mass of solution = 5.4 g + 75.0 g = 80.4 g   (calculator and **correct answer**)

Substituting these values into the defining equation for percent by mass gives

$$\text{percent by mass} = \frac{5.4 \text{ g}}{80.4 \text{ g}} \times 100 = 6.7164179\% \quad \text{(calculator answer)}$$

$$= 6.7\% \qquad \textbf{(correct answer)}$$

Answer Double Check:

Is the magnitude of the answer reasonable? Yes. A concentration of 6.7 mass percent means 6.7 g of solute per 100 g of solution. The ratio 6.7/100 is consistent with the given ratio of 5.4/80. The larger amount of solution should contain the larger amount of solute.

▶ **Practice Exercise 13.2** What is the percent by mass, % (m/m), concentration of sodium chloride (table salt) in a solution made by dissolving 13.0 g of sodium chloride in 275.0 g of water?

Percent by volume (or volume–volume percent) finds use as a concentration unit when both the solute and solvent are liquids or gases. In such cases it is often more convenient to measure volumes than masses. **Percent by volume concentration** *is the volume of solute divided by the total volume of solution multiplied by 100.*

$$\text{percent by volume} = \frac{\text{volume of solute}}{\text{volume of solution}} \times 100$$

The proof system used for alcoholic beverages is twice the volume–volume percent. Forty proof is 20% (v/v) alcohol; 100 proof is 50% (v/v) alcohol.

Solute and solution volumes must always be expressed in the same units when this expression is used. The abbreviation for percent by volume is % (v/v).

The numerical value of a concentration expressed as a percent by volume gives directly the number of milliliters of solute in 100 mL of solution. Thus, a 100 mL sample of a 5.0% alcohol-in-water solution contains 5.0 mL of alcohol dissolved in enough water to give 100 mL of solution. Note that such a 5.0%-by-volume solution could not be made by adding 5 mL of alcohol to 95 mL of water, since volumes of liquids are not usually additive. Differences in the way molecules are packed as well as in the distances between molecules almost always result in the volume of a solution being less than the sum of the volumes of solute and solvent. For example, the final volume resulting from the addition of 50.0 mL of ethyl alcohol to 50.0 mL of water is 96.5 mL of solution.

When volumes of two different liquids are combined, the volumes are not additive. This process is somewhat analogous to pouring marbles and golf balls together. The marbles can fill in the spaces between the golf balls. This results in the "mixed" volume being less than the sum of the "premixed" volumes.

EXAMPLE 13.3 **Calculating the Percent by Volume Concentration of a Solution**

A windshield washer solution is made by mixing 37.8 mL of methyl alcohol (CH_4O) with 46.2 mL of water to produce 80.0 mL of solution. What is the concentration of methyl alcohol in the solution expressed as percent by volume methyl alcohol?

SOLUTION

To calculate this percent by volume, the volumes of methyl alcohol and of solution are needed. Both are given in this problem.

$$\text{methyl alcohol volume} = 37.8 \text{ mL}$$
$$\text{solution volume} = 80.0 \text{ mL}$$

Note that the solution volume is not the sum of the solute and solvent volumes. As previously mentioned, liquid volumes of different substances are generally not additive.

Substituting the given values into the equation

$$\text{percent by volume} = \frac{\text{volume of methyl alcohol}}{\text{volume of solution}} \times 100$$

gives

$$\text{percent by volume} = \frac{37.8 \text{ mL}}{80.0 \text{ mL}} \times 100 = 47.25\% \quad \text{(calculator answer)}$$

$$= 47.2\% \quad \textbf{(correct answer)}$$

Answer Double Check:

Since the solute and solvent volumes are roughly equal, the volume percent should be a percentage close to the 50% mark; such is the case, 47.0%.

▶ **Practice Exercise 13.3**   The final volume of a solution resulting from the addition of 90.0 mL of methyl alcohol (CH_4O) to 167.0 mL of water is 250.0 mL. What is the volume percent of methyl alcohol in the solution?

When a percent concentration is given without specifying which of the three types of percent concentration it is (not a desirable situation), it is assumed to mean percent by mass. Thus, a 5% NaCl solution is assumed to be a 5% (m/m) NaCl solution.

The third type of percentage unit in common use is *mass–volume* percent. This unit, which is often encountered in hospital and industrial settings, is particularly convenient to use when working with a solid solute (which is easily weighed) and a liquid solvent. Concentrations are specified using this unit when dealing with physiological fluids such as blood and urine. **Mass–volume percent concentration** *is the mass of solute (in grams) divided by the total volume of solution (in milliliters) multiplied by 100.*

$$\text{mass–volume percent} = \frac{\text{mass of solute (g)}}{\text{volume of solution (mL)}} \times 100$$

Note that specific mass and volume units are given in the definition of mass–volume percent. This is necessary because the units do not cancel as was the case with mass percent and volume percent. The abbreviation for mass–volume percent is % (m/v).

For dilute aqueous solutions, % (m/m) and % (m/v) concentrations are almost the same because mass in grams of the solution equals the volume in milliliters when the density is close to 1.00 g/mL, as it is for pure water and for dilute aqueous solutions.

Using Percent Concentrations as Conversion Factors

Preparation of a solution with a specific percent concentration requires knowledge about the amount of solute needed and/or the final volume of solution needed. Such information can be calculated using conversion factors obtained from the desired percent concentration value. Table 13.5 shows the relationship between the definitions for the three types of percent concentrations and the conversion factors that can be derived from them. Examples 13.4 through 13.6 illustrate how these definition-derived conversion factors are used in a problem-solving context.

TABLE 13.5   Conversion Factors Obtained from Percent Concentration Units

Percent Concentration	Meaning in Words	Conversion Factors	
12%(m/m) NaCl solution	There are 12 g of NaCl in 100 g of solution.	$\dfrac{12 \text{ g NaCl}}{100 \text{ g solution}}$ and	$\dfrac{100 \text{ g solution}}{12 \text{ g NaCl}}$
8%(v/v) ethanol solution	There are 8 mL of ethanol in 100 mL of solution.	$\dfrac{8 \text{ mL ethanol}}{100 \text{ mL solution}}$ and	$\dfrac{100 \text{ mL solution}}{8 \text{ mL ethanol}}$
22%(m/v) sucrose solution	There are 22 g of sucrose in 100 mL of solution.	$\dfrac{22 \text{ g sucrose}}{100 \text{ mL solution}}$ and	$\dfrac{100 \text{ mL solution}}{22 \text{ g sucrose}}$

EXAMPLE 13.4 **Calculating the Mass of Solute Necessary to Produce a Solution of a Given Mass Percent Concentration**

How many grams of iodine must be added to 25.0 g of ethyl alcohol to prepare a 5.00% (m/m) ethyl alcohol solution of iodine?

SOLUTION

Often, when a solution concentration is given as part of a problem statement, the concentration information is used in the form of a conversion factor in solving the problem. That will be the case in this problem.

The given quantity is 25.0 g of ethyl alcohol (grams of solvent), and the desired quantity is grams of iodine (grams of solute).

$$25.0 \text{ g ethyl alcohol} = ? \text{ g iodine}$$

The conversion factor relating these two quantities (solvent and solute) is obtained from the given concentration. In this 5.00% (m/m) iodine solution, there are 5.00 g of iodine for every 95.00 g of ethyl alcohol.

$$100.00 \text{ g solution} - 5.00 \text{ g iodine} = 95.00 \text{ g ethyl alcohol}$$

This relationship between grams of solute and grams of solvent (5.00 to 95.00) gives us the needed conversion factor

$$\frac{5.00 \text{ g iodine}}{95.00 \text{ g ethyl alcohol}}$$

Dimensional analysis gives the problem setup, which is solved in the following manner.

$$25.0 \text{ g ethyl alcohol} \times \frac{5.00 \text{ g iodine}}{95.00 \text{ g ethyl alcohol}} = 1.3157894 \text{ g iodine} \quad \text{(calculator answer)}$$

$$= 1.32 \text{ g iodine} \qquad \textbf{(correct answer)}$$

Answer Double Check:

Is the magnitude of the answer reasonable? Yes. A 5.00% (m/m) solution contains 5 g of solute per 95 g of solvent. For 25 g of solvent, approximately one-fourth of 95 g, the amount of solute present should be about one-fourth of 5 g, which it is (1.32 g).

▶ Practice Exercise 13.4   How many grams of ammonium chloride (NH_4Cl) must be added to 2255 g of water to prepare a 10.00% (m/m) aqueous solution of ammonium chloride?

EXAMPLE 13.5 **Calculating the Mass of Solute Present in a Solution of a Given Mass–Volume Percent Concentration**

Vinegar is a 5.0% (m/v) aqueous solution of acetic acid ($HC_2H_3O_2$). How much acetic acid, in grams, is present in one teaspoon (5.0 mL) of vinegar?

SOLUTION

The given quantity is 5.0 mL of vinegar, and the desired quantity is grams of acetic acid.

$$5.0 \text{ mL vinegar} = ? \text{ g acetic acid}$$

The given concentration of 5.0% (m/v), which means 5.0 g acetic acid per 100 mL vinegar, can be used as a conversion factor to go from milliliters of vinegar to grams of acetic acid. The setup for the conversion is

$$5.0 \ \text{mL vinegar} \times \frac{5.0 \ \text{g acetic acid}}{100 \ \text{mL vinegar}}$$

Doing the arithmetic, after cancellation of units, gives

$$\frac{5.0 \times 5.0}{100} \ \text{g acetic acid} = 0.25 \ \text{g acetic acid} \quad \text{(calculator and \textbf{correct answer})}$$

Answer Double Check:

Using whole numbers, a 5% mass–volume percent denotes 5 g of solvent per 100 mL of solution. For 10 mL of solution, the grams should be one-tenth of 5, which is 0.5. For 5 mL of solution, which is the given amount, the grams should be 0.25. This is the answer that was obtained.

▶ **Practice Exercise 13.5**   How many grams of nickel(II) sulfate, $NiSO_4$, are required to prepare 455 mL of a 6.00 % (m/v) nickel sulfate solution?

As was illustrated in Example 13.5, the conversion between mass of solute and volume of solution is easily accomplished when the solution concentration unit is mass–volume percent. To do mass–volume conversions, when the concentration is given in mass percent or volume percent, requires density information in addition to the concentration. Example 13.6 illustrates how density is involved in such a calculation.

EXAMPLE 13.6   **Calculating the Volume of Solution Necessary to Supply a Given Mass of Solute Using Density and Mass Percent Concentration**

An 18.0% by mass solution of ammonium sulfate $[(NH_4)_2SO_4]$ has a density of 1.10 g/mL. What volume, in mL, of this solution contains 10.0 g of $(NH_4)_2SO_4$?

SOLUTION

The given quantity is 10.0 g of $(NH_4)_2SO_4$ and the desired quantity is milliliters of $(NH_4)_2SO_4$ solution.

$$10.0 \ \text{g} \ (NH_4)_2SO_4 = ? \ \text{mL} \ (NH_4)_2SO_4 \ \text{solution}$$

The pathway needed to solve the problem using dimensional analysis is

$$\text{g} \ (NH_4)_2SO_4 \longrightarrow \text{g} \ (NH_4)_2SO_4 \ \text{solution} \longrightarrow \text{mL} \ (NH_4)_2SO_4 \ \text{solution}$$

The given concentration is the conversion factor needed for the first unit change and the density is the conversion factor needed for the second unit change.

$$10.0 \ \text{g} \ (NH_4)_2SO_4 \times \frac{100 \ \text{g solution}}{18.0 \ \text{g} \ (NH_4)_2SO_4} \times \frac{1 \ \text{mL solution}}{1.10 \ \text{g solution}}$$

Doing the arithmetic, after cancellation of units, gives

$$\frac{10.0 \times 100 \times 1}{18.0 \times 1.10} \text{ mL solution} = 50.505050 \text{ mL solution (calculator answer)}$$

$$= 50.5 \text{ mL solution} \qquad \textbf{(correct answer)}$$

Answer Double Check:

Using rounded numbers, if the concentration was 20% by mass (20 g per 100 g), and the density was 1.00 g/mL, 50 mL of solution would supply 10 g of solute. Since the actual numbers (18.0% by mass, and 1.10 g/mL) are close to these rounded values, the answer should still be about 50 mL. The actual answer of 50.5 mL is consistent with this analysis.

▶ **Practice Exercise 13.6** A 30.0% by mass nitric acid (HNO_3) solution has a density of 1.18 g/mL. What volume, in mL, of HNO_3 solution contains 20.0 g of HNO_3?

13.7 CONCENTRATION: PARTS PER MILLION AND PARTS PER BILLION

The concentration units parts per million (ppm) and parts per billion (ppb) find use when dealing with extremely dilute solutions. Environmental chemists frequently use such units in specifying the concentrations of the minute amounts of trace pollutants or toxic chemicals in air and water samples.

Parts per million and *parts per billion* units are closely related to percentage concentration units. Not only are the defining equations very similar, but also various forms of the units exist. Because amounts of solute and solution present may be stated in terms of either mass or volume, there are three different forms for each unit: mass–mass (m/m), volume–volume (v/v), and mass–volume (m/v).

A **part per million concentration** *(ppm) is one part of solute per million parts of solution.* In terms of defining equations, we can write

> A ppm is the equivalent of 1 second in 11 days and 14 hours.

$$\text{ppm (m/m)} = \frac{\text{mass of solute}}{\text{mass of solution}} \times 10^6$$

$$\text{ppm (v/v)} = \frac{\text{volume of solute}}{\text{volume of solution}} \times 10^6$$

$$\text{ppm (m/v)} = \frac{\text{mass of solute (g)}}{\text{volume of solution (mL)}} \times 10^6$$

Note that the units of grams and milliliters are specified in the last of the three defining equations, but that no units are given in the first two equations. For the first two equations, the only unit restriction is that the units be the same for both numerator and denominator.

A **part per billion concentration** *(ppb) is one part of solute per billion parts of solution.* The mathematical defining equations for the three types of part-per-billion units are identical to those just shown for parts per million except that a multiplicative factor of 10^9 instead of 10^6 is used.

> A ppb is the equivalent of 1 second in 31 years and 8 months.

The use of parts per million and parts per billion in specifying concentration often avoids the very small numbers that result when other concentration units are used. For example, a pollutant in water might be present at a level of 0.0013 g per 100 mL of solution. In terms of mass–volume percent, this concentration is 0.0013%. In parts per million, however, the concentration is 13.

$$\text{ppm (m/v)} = \frac{0.0013 \text{ g}}{100 \text{ mL}} \times 10^6 = 13$$

The only difference in the ways in which percent concentrations and parts per million or billion are calculated is in the multiplicative factor used. For percentages it is 10^2; for parts per million, 10^6; and for parts per billion, 10^9. An alternative name for percentage concentration units would be *parts per hundred*.

For very dilute aqueous solutions, because water has a density of 1.00 g/mL, the following relationships are valid.

$$1 \text{ ppm (m/v)} = 1 \text{ milligram/liter (mg/L)}$$
$$1 \text{ ppb (m/v)} = 1 \text{ microgram/liter } (\mu\text{g/L})$$

EXAMPLE 13.7 **Expressing Concentrations in Parts per Million and Parts per Billion**

The concentration of sodium fluoride, NaF, in a town's fluoridated tap water is found to be 32.3 mg of NaF per 20.0 kg of tap water. Express this NaF concentration in

a. ppm (m/m).   **b.** ppb (m/m).

SOLUTION

a. The defining equation for ppm (m/m) is

$$\text{ppm (m/m)} = \frac{\text{mass of solute}}{\text{mass of solution}} \times 10^6$$

The two masses in this equation must be in the same units. Let us use grams as our mass unit. Expressing the given quantities in terms of grams, we have

$$32.3 \text{ mg NaF} = 3.23 \times 10^{-2} \text{ g NaF}$$
$$20.0 \text{ kg tap water} = 2.00 \times 10^4 \text{ g tap water}$$

Substituting these gram quantities into the defining equation for ppm (m/m) gives

$$\text{ppm (m/m)} = \frac{3.23 \times 10^{-2} \text{ g}}{2.00 \times 10^4 \text{ g}} \times 10^6$$
$$= 1.615 \quad \text{(calculator answer)}$$
$$= 1.62 \quad \textbf{(correct answer)}$$

b. For parts per billion, we have

$$\text{ppb (m/m)} = \frac{3.23 \times 10^{-2} \text{ g}}{2.00 \times 10^4 \text{ g}} \times 10^9$$
$$= 1615 \quad \text{(calculator answer)}$$
$$= 1620 \quad \textbf{(correct answer)}$$

Answer Double Check:

Is the relationship between the two answers correct? Yes. Any time a concentration is expressed in both parts per million and parts per billion, the parts per billion value will be 1000 times larger than the parts per million value.

▶ **Practice Exercise 13.7**   A 4.00 g water sample contains a small amount (50.0 mg) of Mg. What is the Mg concentration in the sample in (a) ppm (m/m) and (b) ppb (m/m)?

EXAMPLE 13.8   **Calculating the Volume of Solute Present Given a Parts per Million Concentration**

The carbon monoxide, CO, content of the tobacco smoke that reaches a smoker's lungs can be as high as 375 ppm (v/v). At this concentration, how much CO, in milliliters, would be present in a sample of air the size of a standard-size basketball (7.5 L)?

SOLUTION

The defining equation for ppm (v/v) is

$$\text{ppm (v/v)} = \frac{\text{volume solute}}{\text{volume solution}} \times 10^6$$

The solute is CO. Rearranging the equation to isolate volume of solute on the left gives

$$\text{volume solute (mL)} = \frac{\text{ppm (v/v)} \times \text{volume solution (mL)}}{10^6}$$

Substituting the known values into the equation, remembering to change 7.5 L to 7500 mL so that the answer will have the unit of milliliters, gives

$$\text{volume CO} = \frac{375 \times 7500}{10^6} \text{ mL}$$
$$= 2.8125 \text{ mL} \quad \text{(calculator answer)}$$
$$= 2.8 \text{ mL} \quad \textbf{(correct answer)}$$

▶ **Practice Exercise 13.8**   Waste water from an industrial complex is found to contain 7.5 ppm (m/v) of dissolved copper. At this concentration, how much copper, in grams, would be present in a 20.0 mL sample of the waste water?

13.8 CONCENTRATION: MOLARITY

Molarity concentration *is the number of moles of solute per liter of solution.* The defining equation for molarity is

$$\text{molarity (M)} = \frac{\text{moles of solute}}{\text{liters of solution}}$$

A solution containing 1 mole of KBr in 1 L of solution has a molarity of 1 and is said to be a 1 M (1 *molar*) solution. Note that the abbreviation for molarity is a capital M.

When a solution is to be used for a chemical reaction, concentration is almost always expressed in units of molarity. A major reason for this is the fact that the amount of solute is expressed in moles, a most convenient unit for dealing with stoichiometry in chemical reactions. Because chemical reactions occur between molecules and atoms, a unit that counts particles, as the mole does, is desirable.

To find the molarity of a solution, we need to know the number of moles of solute present and the solution volume in liters and then take the ratio of the two quantities. An alternative to knowing the number of moles of solute is knowledge about the grams of solute present and the solute's molar mass.

In preparing 100 mL of a solution of a specific molarity, enough solvent is added to a weighed amount of solute to give a *final* volume of 100 mL. The weighed solute is not added to a *starting* volume of 100 mL; this would produce a final volume greater than 100 mL, because the solute increases the total volume.

EXAMPLE 13.9 **Calculating the Molarity of a Solution from Mass/Amount and Volume Data**

Determine the molarities of the following solutions.

    a. 1.45 moles of KCl dissolved in enough water to give 875 mL of solution
    b. 57.2 g of NH_4Br dissolved in enough water to give 2.15 L of solution

SOLUTION

    a. The number of moles of solute is given in the problem statement.

$$\text{moles of solute} = 1.45 \text{ moles KCl}$$

The volume of the solution is also given in the problem statement, but not in the right units. Molarity requires liters for volume units. Making the unit change gives

$$875 \text{ mL} \times \frac{10^{-3} \text{ L}}{1 \text{ mL}} = 0.875 \text{ L} \quad \text{(calculator and \textbf{correct answer})}$$

The molarity of the solution is obtained by substituting the known quantities into the equation

$$\text{molarity (M)} = \frac{\text{moles of solute}}{\text{L of solution}}$$

which gives

$$M = \frac{1.45 \text{ moles of KCl}}{0.875 \text{ L solution}} = 1.6571428 \frac{\text{moles KCl}}{\text{L solution}} \quad \text{(calculator answer)}$$

$$= 1.66 \frac{\text{moles KCl}}{\text{L solution}} \quad \text{(\textbf{correct answer})}$$

Note that the units of molarity are always moles per liter.

    b. This time the volume of solution is given in the right units, liters.

$$\text{volume of solution} = 2.15 \text{ L}$$

The moles of solute must be calculated from the grams of solute (given) and the solute's molar mass, which is 97.95 grams (calculated from a table of atomic masses).

$$57.2 \text{ g } NH_4Br \times \frac{1 \text{ mole } NH_4Br}{97.95 \text{ g } NH_4Br} = 0.58397141 \text{ mole } NH_4Br \text{ (calculator answer)}$$

$$= 0.584 \text{ mole } NH_4Br \quad \text{(\textbf{correct answer})}$$

Substituting the known quantities into the defining equation for molarity gives

$$M = \frac{0.584 \text{ mole } NH_4Br}{2.15 \text{ L solution}} = 0.2716279 \frac{\text{mole } NH_4Br}{\text{L solution}} \quad \text{(calculator answer)}$$

$$= 0.272 \frac{\text{mole } NH_4Br}{\text{L solution}} \quad \text{(\textbf{correct answer})}$$

Answer Double Check:

Is the magnitude of the answer in part (a) reasonable? Yes. If the volume was 1000 mL instead of 875 mL, the molarity would be 1.45, because 1.45 moles of solute are present. With 1.45 moles of solute present in a smaller volume (875 mL), the molarity will be slightly greater than 1.45. The answer, 1.66 molar, is consistent with this analysis.

▶ **Practice Exercise 13.9** Calculate the molarity of a solution prepared by dissolving 1.00 g of ethyl alcohol (C_2H_6O) in enough water to give 25.0 mL of solution.

As the previous example indicates, when you perform a molarity calculation that involves grams of solute, the identity (chemical formula) of the solute is always needed. You cannot calculate moles of solute from grams of solute without knowing the chemical formula of the solute. In contrast, when you perform percent concentration calculations (or parts per million or billion)—Sections 13.6 and 13.7—the chemical formula of the solute is not used in the calculation. All that is needed is the mass (grams) of the solute; solute identity (chemical formula) is not required for the calculation.

Using Molarity as a Conversion Factor

The mass of solute present in a known volume of solution is an easily calculable quantity if the molarity of the solution is known. When such is the case, molarity serves as a conversion factor that relates liters of solution to moles of solute. In a similar manner, the volume of solution needed to supply a given amount of solute can be calculated using the solution's molarity as a conversion factor.

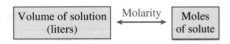

Examples 13.10 and 13.11 show, respectively, the two previously mentioned uses for molarity as a conversion factor.

EXAMPLE 13.10 **Calculating the Amount of Solute Present in a Given Amount of Solution Using Molarity as a Conversion Factor**

Citric acid, $H_3C_6H_5O_7$, is the substance that gives lemon juice and other citrus fruit juices a sour taste. How many grams of citric acid are present in 125 mL of a 0.400 M citric acid solution?

SOLUTION

The given quantity is 125 mL of solution, and the desired quantity is grams of $H_3C_6H_5O_7$.

$$125 \text{ mL solution} = ? \text{ g } H_3C_6H_5O_7$$

The pathway to be used in solving this problem is

$$\text{mL solution} \longrightarrow \text{L solution} \longrightarrow \text{moles } H_3C_6H_5O_7 \longrightarrow \text{g } H_3C_6H_5O_7$$

The given molarity (0.400 M) will serve as the conversion factor for the second unit change; the molar mass of citric acid (which must be calculated because it is not given) is used in accomplishing the third unit change.

The dimensional analysis setup for this pathway is

$$125 \text{ mL solution} \times \frac{10^{-3} \text{ L solution}}{1 \text{ mL solution}} \times \frac{0.400 \text{ mole } H_3C_6H_5O_7}{1 \text{ L solution}} \times \frac{192.14 \text{ g } H_3C_6H_5O_7}{1 \text{ mole } H_3C_6H_5O_7}$$

Cancelling units and doing the arithmetic gives

$$\frac{125 \times 10^{-3} \times 0.400 \times 192.14}{1 \times 1 \times 1} \text{ g } H_3C_6H_5O_7$$

$$= 9.607 \text{ g } H_3C_6H_5O_7 \text{ (calculator answer)}$$
$$= 9.61 \text{ g } H_3C_6H_5O_7 \quad \textbf{(correct answer)}$$

Answer Double Check:

Is the magnitude of the numerical answer reasonable? Yes, 125 mL of solution is one-eighth of a liter. If a liter of solution contains 0.40 of a mole of solute (the given molarity), then one-eighth of a liter will contain 0.050 mole (0.40/8). The molar mass is approximately 200 g, and 0.050 times 200 g is 10 g. The calculated answer of 9.61 g is consistent with this estimated answer of 10 g.

▶ **Practice Exercise 13.10** How many grams of table sugar (sucrose), $C_{12}H_{22}O_{11}$, are present in 25.0 mL of a 2.00 M sucrose solution?

EXAMPLE 13.11 **Calculating the Amount of Solution Necessary to Supply a Given Amount of Solute**

A typical dose of iron(II) sulfate ($FeSO_4$) used in the treatment of iron-deficiency anemia is 0.35 g. How many milliliters of a 0.10 M iron(II) sulfate solution would be needed to supply this dose?

SOLUTION

The given quantity is 0.35 g of $FeSO_4$, and the desired quantity is milliliters of $FeSO_4$ solution.

$$0.35 \text{ g } FeSO_4 = ? \text{ mL } FeSO_4 \text{ solution}$$

The pathway to be used to solve this problem is

$$\text{g } FeSO_4 \longrightarrow \text{ moles } FeSO_4 \longrightarrow \text{ L } FeSO_4 \text{ solution} \longrightarrow \text{ mL } FeSO_4 \text{ solution}$$

We accomplish the first unit conversion by using the molar mass of $FeSO_4$ (which must be calculated) as a conversion factor. The second unit conversion involves the use of the given molarity as a conversion factor.

$$0.35 \text{ g } FeSO_4 \times \left(\frac{1 \text{ mole } FeSO_4}{151.92 \text{ g } FeSO_4}\right) \times \left(\frac{1 \text{ L solution}}{0.10 \text{ mole } FeSO_4}\right) \times \left(\frac{1 \text{ mL solution}}{10^{-3} \text{ L solution}}\right)$$

Cancelling units and doing the arithmetic, we find that

$$\left(\frac{0.35 \times 1 \times 1 \times 1}{151.92 \times 0.10 \times 10^{-3}}\right) \text{ mL solution}$$

$$= 23.038441 \text{ mL solution} \quad \text{(calculator answer)}$$
$$= 23 \text{ mL solution} \qquad \textbf{(correct answer)}$$

▶ **Practice Exercise 13.11** How many milliliters of 1.00 M hydrogen peroxide solution, H_2O_2, can be prepared from 10.0 g of pure hydrogen peroxide?

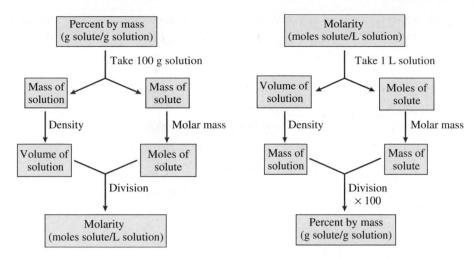

FIGURE 13.5 "Road-map" diagrams showing the steps involved in converting from percent by mass to molarity, or vice versa.

Molarity and mass percent are probably the two most commonly used concentration units. The need to convert from one to the other often arises. Such a conversion can easily be done, provided the density of the solution is known. Figure 13.5 shows schematically the steps involved in converting one of these concentration units to the other.

EXAMPLE 13.12   **Calculating Molarity from Density and Percent by Mass**

The Great Salt Lake in Utah is second only to the Dead Sea in salt content. A water sample from the Great Salt Lake is found to contain 21.2 g of NaCl per 100.0 g of lake water, which is a 21.2 percent by mass NaCl concentration. Calculate the molarity of NaCl in the water given that the density of the lake water is 1.160 g/mL.

SOLUTION

Calculate the moles of solute and liters of solution present in a sample of this solution. Since solution concentration is independent of sample size, any size sample can be the basis for the calculation. To simplify the math, take a 100.0 g sample of solution.

Step 1 *Moles of solute.* The given quantity is 100.0 g of solution, and the desired quantity is moles of NaCl.

$$100.0 \text{ g solution} = ? \text{ moles NaCl}$$

The known mass percent concentration will be the basis for the conversion factor that takes us from grams of solution to grams of solute. The pathway to be used in solving this problem is

$$\text{g solution} \longrightarrow \text{g solute} \longrightarrow \text{moles solute}$$

The setup is

$$100.0 \text{ g solution} \times \frac{21.2 \text{ g NaCl}}{100 \text{ g solution}} \times \frac{1 \text{ mole NaCl}}{58.44 \text{ g NaCl}}$$

$$= 0.36276523 \text{ mole NaCl} \quad \text{(calculator answer)}$$

$$= 0.363 \text{ mole NaCl} \qquad \textbf{(correct answer)}$$

Step 2 *Liters of solution.* The density of the solution is used as a conversion factor in obtaining the volume of solution. The pathway for the calculation is

$$\text{g solution} \longrightarrow \text{mL solution} \longrightarrow \text{L solution}$$

The setup is

$$100.0 \; \cancel{\text{g solution}} \times \frac{1 \; \cancel{\text{mL solution}}}{1.160 \; \cancel{\text{g solution}}} \times \frac{10^{-3} \; \text{L solution}}{1 \; \cancel{\text{mL solution}}}$$

$$= 0.086206897 \; \text{L solution} \quad \text{(calculator answer)}$$

$$= 0.08621 \; \text{L solution} \qquad \textbf{(correct answer)}$$

Step 3 *Molarity.* With both moles of solute and liters of solution known, the molarity is obtained by substitution into the defining equation for molarity:

$$M = \frac{\text{moles NaCl}}{\text{L solution}} = \frac{0.363 \; \text{mole NaCl}}{0.08621 \; \text{L solution}}$$

$$= 4.2106484 \; \frac{\text{moles NaCl}}{\text{L solution}} \quad \text{(calculator answer)}$$

$$= 4.21 \; \frac{\text{moles NaCl}}{\text{L solution}} \qquad \textbf{(correct answer)}$$

An alternate method for solving this problem, which involves one "continuous chain" of conversion factors, exists. In this method, the mass percent as a conversion factor (g solute/g solution) is the starting point for the calculation, and the unit conversions used produce the units for molarity (moles solute/liter solution). The conversion factor sequence necessary to produce this change is

$$\frac{21.2 \; \cancel{\text{g NaCl}}}{100 \; \cancel{\text{g solution}}} \times \frac{1.160 \; \cancel{\text{g solution}}}{1 \; \cancel{\text{mL solution}}} \times \frac{1 \; \cancel{\text{mL solution}}}{10^{-3} \; \text{L solution}} \times \frac{1 \; \text{mole NaCl}}{58.44 \; \cancel{\text{g NaCl}}}$$

$$= 4.2080767 \; \frac{\text{moles NaCl}}{\text{L solution}} \quad \text{(calculator answer)}$$

$$= 4.21 \; \frac{\text{moles NaCl}}{\text{L solution}} \qquad \textbf{(correct answer)}$$

▶ **Practice Exercise 13.12**   A common strength for industrially produced sulfuric acid, H_2SO_4, is 96.0% by mass. What is the molarity of such an acid, given that its density is 1.84 g/mL?

Molar concentrations do not give information about the amount of *solvent* present. All that is known is that enough solvent is present to give a specific volume of *solution*. The amount of solvent present in a solution of a known molarity can be calculated if the density of the solution is known. Without the density, it cannot be calculated. Example 13.13 illustrates how this is done.

EXAMPLE 13.13   **Using Molarity and Density to Calculate the Amount of Solvent Present in a Solution**

Large amounts of sulfuric acid (H_2SO_4) are used in the production of phosphate fertilizers. A 2.324 M H_2SO_4 solution has a density of 1.142 g/mL. How many grams of solvent (water) are present in 25.0 mL of this solution?

SOLUTION

To find the grams of solvent present, we must first find the grams of solute (H_2SO_4) and the grams of solution. The grams of solvent present is then obtained by subtraction.

$$\text{g solvent} = \text{g solution} - \text{g solute}$$

Step 1 *Grams of solution.* The volume of solution is given. Density, used as a conversion factor, will enable us to convert this volume to grams of solution.

$$25.0 \text{ mL solution} \times \frac{1.142 \text{ g solution}}{1 \text{ mL solution}} = 28.55 \text{ g solution} \quad \text{(calculator answer)}$$

$$= 28.6 \text{ g solution} \quad \textbf{(correct answer)}$$

Step 2 *Grams of solute.* We will use the molarity of the solution as a conversion factor in obtaining the grams of solute. The setup for this calculation is similar to that in Example 13.10.

$$25.0 \text{ mL solution} \times \frac{10^{-3} \text{ L solution}}{1 \text{ mL solution}} \times \frac{2.324 \text{ moles } H_2SO_4}{1 \text{ L solution}} \times \frac{98.09 \text{ g } H_2SO_4}{1 \text{ mole } H_2SO_4}$$

$$= 5.699029 \text{ g } H_2SO_4 \quad \text{(calculator answer)}$$

$$= 5.70 \text{ g } H_2SO_4 \quad \textbf{(correct answer)}$$

Step 3 *Grams of solvent.* The grams of solvent will be the difference in mass between the grams of solution and the grams of solute.

$$28.6 \text{ g solution} - 5.70 \text{ g solute} = 22.9 \text{ g solvent} \quad \text{(calculator and \textbf{correct answer})}$$

Answer Double Check:

Is the magnitude of the calculated answer reasonable? Yes, 25.0 mL of solution with a density of 1.142 g/mL would have a mass of 28.6 grams (25.0 × 1.142). The major part of this mass is due to water since the solution is relatively dilute (2 M). Thus, the mass of water present should be about 20–25 grams. Such is the case; the calculated answer is 22.9 grams.

▶ **Practice Exercise 13.13** The density of a 2.00 M hydrochloric acid solution (HCl) is 1.03 g/mL. How many grams of solvent (water) are present in 50.0 mL of this solution?

13.9 CONCENTRATION: MOLALITY

Molality is a concentration unit based on a fixed amount of *solvent* and is used in areas where this is a concern. Despite this unit having a name very similar to molarity, molality differs distinctly from molarity. Molarity is a unit based on a fixed amount of *solution* rather than a fixed amount of *solvent*. **Molality concentration** *is the number of moles of solute per kilogram of solvent.* The defining equation for molality is

$$\text{molality } (m) = \frac{\text{moles of solute}}{\text{kilograms of solvent}}$$

Note that the abbreviation for molality is an italic lowercase m.

It is important not to confuse *molarity* with *molality*.

$$\text{Molality} = \frac{\text{moles of solute}}{\text{kilograms of solvent}} \qquad \text{Molarity} = \frac{\text{moles of solute}}{\text{liters of solution}}$$

The numerators of the two defining equations are the same. There are, however, major differences in the denominators of the defining equations. One involves *solvent* and the other involves *solution*. One involves mass (*kilograms*) and the other involves volume (*liters*).

Molality also finds use, in preference to molarity, in experimental situations where changes in temperature are of concern. Molality is a temperature-independent concentration unit; molarity is a temperature-dependent concentration unit. To be temperature independent, a concentration unit cannot involve a volume measurement. Volumes of solutions change (expand or contract) with changes in temperature. A change in temperature thus means a change in concentration, even though the amount of solute remains constant, if a concentration unit has a volume dependency. Volume changes caused by temperature change are usually very very small; consequently, temperature independence or dependence is a factor in only the most accurate experimental measurements.

Careful note should be taken of the fact that the same letter of the alphabet is used as an abbreviation for both molality and molarity—a lowercase, italic m for molality (m) and a capitalized M for molarity (M).

In dilute aqueous solutions molarity and molality are practically identical in numerical value. This results because dilute aqueous solutions have a density of 1.0 g/mL. Molarity and molality have significantly different values when the solvent has a density that is not equal to unity or when the solution is concentrated.

EXAMPLE 13.14 Calculating the Molality of a Solution

What would be the molality of a solution made by dissolving 3.50 g of sodium chloride (NaCl) in 225 g of H_2O?

SOLUTION

To calculate molality, the number of moles of solute and the solvent mass in kilograms must be known.

In this problem the solvent mass is given, but in grams rather than kilograms. We thus need to change the grams unit to kilograms.

$$225 \text{ g } H_2O \times \frac{1 \text{ kg } H_2O}{10^3 \text{ g } H_2O} = 0.255 \text{ kg } H_2O \quad \text{(calculator and \textbf{correct answer})}$$

Information about the solute is given in terms of grams. We can calculate moles of solute from the given information by using molar mass (which is not given and must be calculated) as a conversion factor.

$$3.50 \text{ g NaCl} \times \frac{1 \text{ mole NaCl}}{58.44 \text{ g NaCl}} = 0.059890485 \text{ mole NaCl} \quad \text{(calculator answer)}$$

$$= 0.0599 \text{ mole NaCl} \quad \text{(\textbf{correct answer})}$$

Substituting moles of solute and kilograms of solvent into the defining equation for molality gives

$$m = \frac{\text{moles solute}}{\text{kg solvent}} = \frac{0.0599 \text{ mole NaCl}}{0.225 \text{ kg } H_2O}$$

$$= 0.26622222 \frac{\text{mole NaCl}}{\text{kg } H_2O} \quad \text{(calculator answer)}$$

$$= 0.266 \frac{\text{mole NaCl}}{\text{kg } H_2O}$$

$$= 0.266 \, m \quad \text{(\textbf{correct answer})}$$

Answer Double Check:

Is the magnitude for the numerical value of the molality reasonable? Yes. Approximately 0.06 mole of NaCl (3.50/60) is present in approximately one-fourth of a kilogram (250 g) of solvent. One kilogram of solvent would contain 0.24 mole of solute (0.06 × 4), and the resulting molality would be 0.24 m. The calculated answer, 0.266 m, is consistent with this rough calculation for the molality.

▶ **Practice Exercise 13.14**   Calculate the molality of an antifreeze solution made by dissolving 50.0 g of ethylene glycol ($C_2H_6O_2$) in 50.0 g of H_2O.

Example 13.15 illustrates the use of molality as a conversion factor in obtaining the amount of solute needed to prepare a solution of specified molality.

EXAMPLE 13.15   **Calculating the Amount of Solute Needed to Prepare a Solution of Specified Molality**

Calculate the number of grams of isopropyl alcohol, C_3H_8O, that must be added to 275 g of water to prepare a 2.00 m solution of isopropyl alcohol.

SOLUTION

The given quantity is 275 g of H_2O, and the desired quantity is grams of C_3H_8O.

$$275 \text{ g } H_2O = ? \text{ g } C_3H_8O$$

The pathway to be used in solving this problem is

$$\text{g solvent} \longrightarrow \text{kg solvent} \longrightarrow \text{moles solute} \longrightarrow \text{g solute}$$

The molality of the solution, which is given, will serve as a conversion factor to effect the change from kilograms of solvent to moles of solute.

The dimensional analysis setup for the problem is

$$275 \text{ g } H_2O \times \frac{1 \text{ kg } H_2O}{10^3 \text{ g } H_2O} \times \frac{2.00 \text{ moles } C_3H_8O}{1 \text{ kg } H_2O} \times \frac{60.11 \text{ g } C_3H_8O}{1 \text{ mole } C_3H_8O}$$

The second conversion factor involves the numerical value of the molality, and the third conversion factor is based on the molar mass of C_3H_8O.

Cancelling units and then doing the arithmetic gives

$$\frac{275 \times 1 \times 2.00 \times 60.11}{10^3 \times 1 \times 1} \text{ g } C_3H_8O = 33.0605 \text{ g } C_3H_8O \quad \text{(calculator answer)}$$

$$= 33.1 \text{ g } C_3H_8O \quad \textbf{(correct answer)}$$

Answer Double Check:

Is the magnitude of the numerical answer reasonable? Yes. The mass of solute present in 1000 g (1 kg) of solvent would be about 120 g, since the molar mass is 60 g and 2 moles are present. The amount of solvent actually present is about one-fourth kilogram (275 g). One-fourth of 120 g is 30 g. The calculated answer of 33.1 g is consistent with this analysis.

▶ **Practice Exercise 13.15**   Calculate the number of grams of glucose, $C_6H_{12}O_6$, that must be added to 1000.0 g of water to prepare a 1.500 m solution.

Interconversion between molarity and molality concentration units requires knowledge of solution density. Example 13.16 is a sample molality-to-molarity interconversion, and Example 13.17 is the opposite process, a sample molarity-to-molality interconversion.

EXAMPLE 13.16 Calculating Molarity from Molality and Density

Calculate the molarity of an 8.92 m (molal) ethyl alcohol (C_2H_6O) solution whose density is 0.927 g/mL.

SOLUTION

The defining equation for molarity involves moles of solute (numerator) and liters of solution (denominator). The defining equation for molality involves moles of solute (numerator) and kilograms of solvent (denominator). The numerators of the two defining equations are the same and the denominators are different. The essence of this problem is, thus, the conversion of kilograms of solvent to liters of solution.

$$\text{kg solvent} \longrightarrow \text{L solution}$$

To convert kilograms of solvent to liters of solution requires three steps, with each step involving a different type of calculation.

Step 1 kilograms solvent $\longrightarrow$ mass of solute
Step 2 mass of solute $\longrightarrow$ mass of solution
Step 3 mass of solution $\longrightarrow$ liters of solution

The given molality (8.92 m), which has 1 kilogram of solvent as its basis, is the starting point for the calculation. Information associated with this molality is

$$1 \text{ kilogram } H_2O = 1000 \text{ grams } H_2O = 8.92 \text{ moles } C_2H_6O$$

Step 1 *Calculation of mass of solute*

The mass of solute needed is that present in 1 kilogram of solvent. It is calculated using the given molality and the molar mass of the solute.

$$1 \text{ kg } H_2O \times \frac{8.92 \text{ moles } C_2H_6O}{1 \text{ kg } H_2O} \times \frac{46.08 \text{ g } C_2H_6O}{1 \text{ mole } C_2H_6O}$$

$$= 411.0336 \text{ g } C_2H_6O \quad \text{(calculator answer)}$$
$$= 411 \text{ g } C_2H_6O \qquad \textbf{(correct answer)}$$

Step 2 *Calculation of the mass of solution*

The mass of solution is obtained from the mass of solute (calculated in step 1) and the mass of solvent (our basis amount). This is a simple addition problem.

$$\underset{\text{solvent}}{1000 \text{ g}} + \underset{\text{solute}}{411 \text{ g}} = \underset{\text{solution}}{1411 \text{ g}} \quad \text{(calculator and \textbf{correct answer})}$$

Step 3 *Calculation of liters of solution*

This calculation requires use of the mass of solution (calculated in step 2) and the density of the solution (given in the problem statement).

$$1411 \text{ g solution} \times \frac{1 \text{ mL solution}}{0.927 \text{ g solution}} \times \frac{10^{-3} \text{ L solution}}{1 \text{ mL solution}}$$

$$= 1.5221143 \text{ L solution} \quad \text{(calculator answer)}$$
$$= 1.52 \text{ L solution} \qquad \textbf{(correct answer)}$$

We are now ready to calculate the molarity of the solution.

$$\text{molarity} = \frac{\text{moles of solute (given in the basis statement)}}{\text{liters of solution (calculated in step 3)}}$$

$$= \frac{8.92 \text{ moles}}{1.52 \text{ L}} = 5.868421 \frac{\text{moles solute}}{\text{L solution}} \quad \text{(calculator answer)}$$

$$= 5.87 \frac{\text{moles solute}}{\text{L solution}} \quad \textbf{(correct answer)}$$

Answer Double Check:

Is the magnitude of the numerical answer reasonable? Yes. Numerically, the molarity of a solution will always be less that the molality of the same solution. Such is the case here: 8.92 *m* and 5.87 M. Comparing the defining equations of molarity and molality shows that both equations have the same numerator (moles of solute) and that they differ in their denominators. The molarity denominator involves liters of *solution* (solute plus solvent), and the molality denominator involves kilograms of *solvent*. The former will always be numerically larger than the latter because it involves both solvent and solute. Division by the larger molarity denominator produces the smaller number for the concentration.

▶ **Practice Exercise 13.16** Calculate the molarity of a 12.6 *m* hydrogen peroxide solution (H_2O_2) that has a density of 1.11 g/mL.

EXAMPLE 13.17 Calculating Molality from Molarity and Density

Calculate the molality of an 1.08 M lactose ($C_{12}H_{22}O_{11}$) solution that has a density of 1.102 g/mL.

SOLUTION

The defining equation for molarity involves moles of solute (numerator) and liters of solution (denominator). The defining equation for molality involves moles of solute (numerator) and kilograms of solvent (denominator). The numerators of the two defining equations are the same, and the denominators are different. The essence of this problem is, thus, the conversion of liters of solution to kilograms of solvent.

$$\text{L solution} \longrightarrow \text{kg solvent}$$

To convert liters of solution to kilograms of solvent requires three steps, with each step involving a different type of calculation.

Step 1 liters of solution ⟶ mass of solute
Step 2 liters of solution ⟶ mass of solution
Step 3 mass of solution ⟶ kilograms of solvent

The given molarity (1.08 M), which has 1 liter of solution as its basis, is the starting point for the calculation. Information associated with this molarity is

$$1 \text{ L solution} = 1000 \text{ mL solution} = 1.08 \text{ moles } C_{12}H_{22}O_{11}$$

Step 1 *Calculation of mass of solute*

The mass of solute needed is that present in 1 L of solution. It is calculated using the given molarity and the molar mass of the solute.

$$1 \text{ L solution} \times \frac{1.08 \text{ moles } C_{12}H_{22}O_{11}}{1 \text{ L solution}} \times \frac{342.34 \text{ g } C_{12}H_{22}O_{11}}{1 \text{ mole } C_{12}H_{22}O_{11}}$$

$$= 369.7272 \text{ g } C_{12}H_{22}O_{11} \qquad \text{(calculator answer)}$$

$$= 37\overline{0} \text{ g } C_{12}H_{22}O_{11} \qquad \textbf{(correct answer)}$$

Step 2 *Calculation of the mass of solution*

The mass of solution is obtained using the given density for the solution.

$$1 \text{ L solution} \times \frac{1 \text{ mL solution}}{10^{-3} \text{ L solution}} \times \frac{1.102 \text{ g solution}}{1 \text{ mL solution}}$$

$$= 1102 \text{ g solution} \quad \text{(calculator and } \textbf{correct answer)}$$

Step 3 *Calculation of kilograms of solvent*

This calculation requires the mass of solute (step 1) and the mass of solution (step 2). The former is subtracted from the latter to give mass of solvent.

$$1102 \text{ g} - 37\overline{0} \text{ g} = 732 \text{ g} \quad \text{(calculator and } \textbf{correct answer)}$$

$$\quad\text{solution} \quad \text{solute} \quad \text{solvent}$$

$$732 \text{ g solvent} \times \frac{1 \text{ kg solvent}}{10^3 \text{ g solvent}} = 0.732 \text{ kg solvent}$$

$$\text{(calculator and } \textbf{correct answer)}$$

We are now ready to calculate the molality of the solution.

$$\text{molality} = \frac{\text{moles solute (given in the basis statement)}}{\text{kilograms solvent (calculated in step 3)}}$$

$$= \frac{1.08 \text{ moles}}{0.732 \text{ kg}} = 14754098 \frac{\text{moles solute}}{\text{kg solvent}} \qquad \text{(calculator answer)}$$

$$= 1.48 \frac{\text{moles solute}}{\text{kg solvent}} \qquad \textbf{(correct answer)}$$

Answer Double Check:

Is the magnitude of the numerical answer reasonable? Yes. The molarity concentration, 1.08 M, is numerically less than the molality concentration, 1.48 *m*, as it should be. Molarity and molality values are close to being equal when a solution is dilute and the solution's density is about 1.0 g/mL. These conditions are approached in this problem, so the difference in magnitude between the two concentrations is small (1.48 *m* and 1.08 M) when compared with values at other conditions, such as those in Example 13.16 (8.92 *m* and 5.87 M).

▶ **Practice Exercise 13.17** Calculate the molality of a 0.744 M sodium bicarbonate solution ($NaHCO_3$) that has a density of 1.04 g/mL.

Examples 13.16 and 13.17 illustrate opposite processes: molality to molarity versus molarity to molality. The similarities and differences in these opposite processes are contrasted in Figure 13.6. In both cases, the second and third steps involve calculation of mass of solute and mass of solution, respectively. Note,

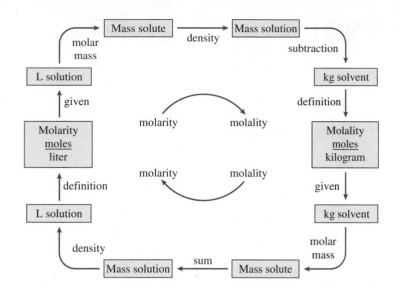

FIGURE 13.6 The steps involved in converting from molality to molarity concentration units and vice versa are similar but not identical. Density is needed in both cases, as are the mass of solute and mass of solvent. How the latter two quantities are calculated and used differs in the two cases.

however, that calculation of these two quantities is carried out in a different manner for the two processes.

13.10 DILUTION

A common procedure encountered when working with solutions in the laboratory is that of *diluting* a solution of known concentration (usually called a stock solution) to a lower concentration. **Dilution** *is the process in which more solvent is added to a specific volume of solution to lower its concentration.* Dilution *always* lowers the concentration of a solution. The same amount of solute is present, but it is now distributed in a larger amount of solvent (the original solvent plus the added solvent).

Since laboratory solutions are almost always liquids, dilution is normally a volumetric procedure. Most often, a solution of a specific molarity is prepared by adding solvent to a solution until a specific total volume is reached.

With molar concentration units, a very simple mathematical relationship exists between the volumes and molarities of the diluted and stock solutions. This relationship is derived from the fact that the same amount of solute is present in both solutions; only solvent is added in a dilution procedure.

$$\text{moles solute}_{\text{stock solution}} = \text{moles solute}_{\text{diluted solution}}$$

The number of moles of solute in both solutions is given by the expression

$$\text{moles solute} = \text{molarity } (M) \times \text{liters of solution } (V)$$

(This equation is just a rearrangement of the defining equation for molarity to isolate moles of solute on the left side.) Substitution of this second expression into the first one gives the equation

$$M_s \times V_s = M_d \times V_d$$

In this equation M_s and V_s are the molarity and volume of the stock solution (the solution to be diluted) and M_d and V_d the molarity and volume of the solution resulting from the dilution. Because volume appears on both sides of the equation, any volume unit, not just liters, may be used as long as it is the same on both sides of the equation. Again, the validity of this equation is based on the fact that there is no change in the amount of solute present.

EXAMPLE 13.18 **Calculating the Molarity of a Solution After It Has Been Diluted**

What is the molarity of the solution prepared by diluting 65 mL of 0.95 M nitric acid (HNO_3) solution to a final volume of 135 mL through addition of solvent?

SOLUTION

Three of the four variables in the equation

$$M_s \times V_s = M_d \times V_d$$

are known.

$$M_s = 0.95 \text{ M} \quad M_d = ? \text{ M}$$
$$V_s = 65 \text{ mL} \quad V_d = 135 \text{ mL}$$

Rearranging the equation to isolate M_d on the left side and substituting the known variables into it gives

$$M_d = M_s \times \frac{V_s}{V_d}$$

$$= 0.95 \text{ M} \times \frac{65 \text{ mL}}{135 \text{ mL}} = 0.4574074 \text{ M} \quad \text{(calculator answer)}$$

$$= 0.46 \text{ M} \quad \textbf{(correct answer)}$$

Thus, the diluted solution's concentration is 0.46 M.

Answer Double Check:

Is the magnitude of the numerical answer reasonable? Yes. A volume increase by a factor of approximately 2 (65 mL to 135 mL) should produce a concentration decrease by a factor of approximately 2. Such is the case. The concentration decreases from 0.95 M to 0.46 M.

▶ **Practice Exercise 13.18** What is the molarity of a solution prepared by diluting 225 mL of 2.00 M nitric acid solution (HNO_3) to a final volume of 275 mL?

EXAMPLE 13.19 **Calculating the Amount of Solvent That Must Be Added to a Solution to Dilute It to a Specified Concentration**

How much solvent, in milliliters, must be added to 200.0 mL of a 1.25 M sodium chloride (NaCl) solution to decrease its concentration to 0.770 M?

SOLUTION

The volume of solvent added is equal to the difference between the final and initial volumes. The initial volume is known. The final volume can be calculated using the equation

$$M_s \times V_s = M_d \times V_d$$

Once the final volume is known, the difference between the two volumes can be obtained. Substituting the known quantities into the dilution equation, rearranged to isolate V_d on the left side, gives

$$V_d = V_s \times \frac{M_s}{M_d}$$

$$= 200.0 \text{ mL} \times \frac{1.25 \text{ M}}{0.770 \text{ M}} = 324.67532 \text{ mL} \quad \text{(calculator answer)}$$

$$= 325 \text{ mL} \quad \textbf{(correct answer)}$$

The solvent added is

$$V_d - V_s = (325 - 200.0) \text{ mL} = 125 \text{ mL} \quad \text{(calculator and \textbf{correct answer})}$$

Answer Double Check:

Is the magnitude of the numerical answer reasonable? Yes. The concentrations are in an approximate three-to-two ratio (1.20 M and 0.80 M). Thus, the initial and final volumes should be in an approximately two-to-three ratio, which they are (200 mL and 300 mL). The change from 200 mL to 300 mL requires adding 100 mL of solvent. The calculated answer, 125 mL, is consistent with this analysis.

▶ **Practice Exercise 13.19**   How much solvent, in milliliters, must be added to 25.0 mL of 1.50 M citric acid solution ($C_6H_8O_7$) to decrease its concentration to 0.0750 M?

When two "like" solutions—that is, solutions that contain the same solute and the same solvent—of differing known molarities and volumes are mixed together, the molarity of the newly formed solution can be calculated by using the same principles that apply in a simple dilution problem.

Again, the key concept involves the amount of solute present; it is constant. The sum of the amounts of solute present in the individual solutions prior to mixing is the same as the total amount of solute present in the solution after mixing. No solute is lost or gained in the mixing process. Thus, we can write

moles of solute$_{\text{first solution}}$ + moles of solute$_{\text{second solution}}$ = moles of solute$_{\text{combined solution}}$

Substituting the expression $(M \times V)$ for moles of solute in this equation gives

$$(M_1 \times V_1) + (M_2 \times V_2) = M_3 \times V_3$$

where the subscripts 1 and 2 denote the solutions to be mixed and the subscript 3 is the solution resulting from the mixing. Again, this expression is valid only when the solutions that are mixed are like solutions.

EXAMPLE 13.20   **Calculating Molarity when Two Like Solutions of Differing Concentration Are Mixed**

What is the molarity of the solution obtained by mixing 50.0 mL of 2.25 M hydrochloric acid (HCl) solution with 160.0 mL of 1.25 M hydrochloric acid solution?

SOLUTION

Five of the six variables in the equation

$$(M_1 \times V_1) + (M_2 \times V_2) = M_3 \times V_3$$

are known:

$$M_1 = 2.25 \text{ M} \quad V_1 = 50.0 \text{ mL}$$
$$M_2 = 1.25 \text{ M} \quad V_2 = 160.0 \text{ mL}$$
$$M_3 = ? \text{ M} \qquad V_3 = 210.0 \text{ mL}$$

Note that in the mixing process we consider the volumes of the solution to be additive; that is,

$$V_3 = V_1 + V_2$$

This is a valid assumption for like solutions.

Solving our equation for M_3 and then substituting the known quantities into it gives

$$M_3 = \frac{(M_1 \times V_1) + (M_2 \times V_2)}{V_3} = \frac{(2.25\ \text{M} \times 50.0\ \text{mL}) + (1.25\ \text{M} \times 160.0\ \text{mL})}{(210.0\ \text{mL})}$$

$$= 1.4880952\ \text{M} \quad \text{(calculator answer)}$$

$$= 1.49\ \text{M} \quad\quad \text{(\textbf{correct answer})}$$

Answer Double Check:

Is the magnitude of the numerical answer reasonable? Yes. It is reasonable in two aspects. First, the final concentration must be in the range between the two starting concentrations, which it is. The value 1.49 M is between 1.25 M and 2.25 M. Second, the final concentration should be closer to the concentration 1.25 M than to the concentration 2.25 M because a larger volume of the weaker solution is used. Such is the case; 1.49 M is closer to 1.25 M than to 2.25 M.

▶ **Practice Exercise 13.20**   What is the molarity of the solution obtained by mixing 25.0 mL of a 3.00 M methyl alcohol solution with 225.0 mL of a 0.100 M methyl alcohol solution?

In the solution of Example 13.20 the given liquid volumes were considered additive. In Section 13.6, when discussing volume percent, it was stressed that volumes were not additive. Why the difference? Volumes of different liquids (Sec. 13.6) are not additive; volumes of the same liquid (Example 13.20) are additive.

13.11 MOLARITY AND CHEMICAL EQUATIONS

Section 10.9 introduced a general problem-solving procedure for setting up problems that involve chemical equations. With this procedure, if information is given about one reactant or product in a chemical reaction (number of grams, moles, or particles), similar information can easily be obtained for any other reactant or product.

In Section 12.14 this procedure was refined to allow us to do mass-to-volume or volume-to-mass calculations for reactions when at least one reactant or product is a gas.

This section further refines our problem-solving procedure to deal efficiently with reactions that occur in aqueous solution. Of primary importance in this new area of problem solving will be *solution volume.* In most situations, solution volume is more conveniently determined than solution mass.

When solution concentrations are expressed in terms of molarity, a direct relationship exists between solution volume (in liters) and moles of solute present. The definition of molarity itself gives the relationship; molarity is the ratio of moles of solute to volume (in liters) of solution. Thus, molarity is the connection that links volume of solution to the other common problem-solving parameters, such as moles and grams. Figure 13.7 shows diagrammatically the place that volume of solution occupies relative to other parameters in the overall scheme of chemical equation–based problem solving. This diagram is a simple extension of Figure 12.9; "volume of solution" boxes have been added. It is used in the same way as Figure 12.9 was.

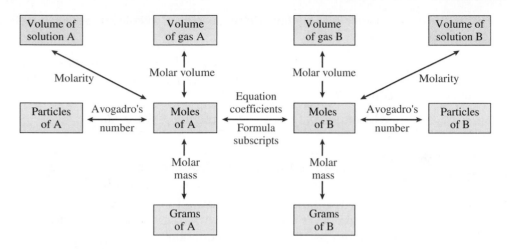

FIGURE 13.7
Conversion factor relationships necessary to solve problems involving chemical reactions that occur in aqueous solution.

EXAMPLE 13.21   **Calculating the Volume of a Reactant Given the Volume and Concentration of Another Reactant**

The fizz produced when an Alka-Seltzer tablet is dissolved in water is due to the reaction between sodium bicarbonate, $NaHCO_3$, and citric acid, $C_6H_8O_7$.

$$3\,NaHCO_3(aq) + C_6H_8O_7(aq) \longrightarrow 3\,CO_2(g) + 3\,H_2O(l) + Na_3C_6H_5O_7(aq)$$

If this reaction were run in a laboratory, what volume, in liters, of 2.50 M $NaHCO_3$ solution is needed to react completely with 0.025 L of 3.50 M $C_6H_8O_7$ solution?

SOLUTION

Step 1 The given quantity is 0.025 L of $C_6H_8O_7$ solution, and the desired quantity is liters of $NaHCO_3$ solution.

$$0.025\;L\;C_6H_8O_7 = ?\;L\;NaHCO_3$$

Step 2 This problem is a volume-of-solution-A to volume-of-solution-B problem. The pathway used in solving it, in terms of Figure 13.7, is

$$\boxed{\begin{array}{c}\text{Volume of}\\\text{solution A}\end{array}} \xrightarrow{\text{Molarity}} \boxed{\begin{array}{c}\text{Moles}\\\text{of A}\end{array}} \xrightarrow[\text{coefficients}]{\text{Equation}} \boxed{\begin{array}{c}\text{Moles}\\\text{of B}\end{array}} \xrightarrow{\text{Molarity}} \boxed{\begin{array}{c}\text{Volume of}\\\text{solution B}\end{array}}$$

Step 3 The dimensional analysis setup for the calculation is

$$0.025\;\text{L}\,C_6H_8O_7 \times \frac{3.50\;\text{moles}\,C_6H_8O_7}{1\;\text{L}\,C_6H_8O_7} \times \frac{3\;\text{moles NaHCO}_3}{1\;\text{mole}\,C_6H_8O_7} \times \frac{1\;\text{L NaHCO}_3}{2.50\;\text{moles NaHCO}_3}$$

Step 4 Combining all the numerical factors gives

$$\frac{0.025 \times 3.50 \times 3 \times 1}{1 \times 1 \times 2.50}\;\text{L NaHCO}_3 = 0.105\;\text{L NaHCO}_3 \;\;(\text{calculator answer})$$

$$= 0.10\;\text{L NaHCO}_3 \;\;(\textbf{correct answer})$$

Answer Double Check:

Is the magnitude of the numerical answer reasonable? Yes. If the two solutions were of equal strength (same molarities), three time as much $NaHCO_3$ would be needed as $C_6H_8O_7$, based on the chemical equation coefficients. Three times 0.025 L is 0.075 L. Since the $C_6H_8O_7$ is the stronger solution, even more than 0.075 L of the weaker

NaHCO$_3$ solution is needed for the chemical reaction. The calculated answer, 0.10 L, is consistent with this analysis.

▶ **Practice Exercise 13.21**   What volume, in liters, of a 1.50 M Na$_2$S solution is needed to react completely with 0.750 L of a 0.400 M Cu(NO$_3$)$_2$ solution, according to the following chemical equation?

$$Cu(NO_3)_2(aq) + Na_2S(aq) \longrightarrow CuS(s) + 2\,NaNO_3(aq)$$

EXAMPLE 13.22   **Calculating the Mass of Product Produced from a Reactant of Known Volume and Concentration**

How many grams of lead(II) chloride can be produced from the reaction of 1.05 L of 0.470 M potassium chloride (KCl) solution with an excess of 4.00 M lead(II) nitrate $\left[(Pb(NO_3)_2)\right]$ solution according to the following equation?

$$2\,KCl(aq) + Pb(NO_3)_2(aq) \longrightarrow PbCl_2(s) + 2\,KNO_3(aq)$$

SOLUTION

Step 1 The given quantity is 1.05 L of KCl solution, and the desired quantity is grams of PbCl$_2$.

$$1.05\ L\ KCl = ?\ g\ PbCl_2$$

Step 2 This is a volume-of-solution-A to grams-of-B problem. The pathway, in terms of Figure 13.7, is

| Volume of solution A | $\xrightarrow{\text{Molarity}}$ | Moles of A | $\xrightarrow[\text{coefficients}]{\text{Equation}}$ | Moles of B | $\xrightarrow{\text{Molar mass}}$ | Grams of B |

Step 3 The dimensional analysis setup for the calculation is

$$1.05\ \text{L KCl} \times \frac{0.470\ \text{mole KCl}}{1\ \text{L KCl}} \times \frac{1\ \text{mole PbCl}_2}{2\ \text{mole KCl}} \times \frac{278.1\ \text{g PbCl}_2}{1\ \text{mole PbCl}_2}$$

Step 4 The answer, obtained from combining all of the numerical factors, is

$$\frac{1.05 \times 0.470 \times 1 \times 278.1}{1 \times 2 \times 1}\ \text{g PbCl}_2 = 68.621175\ \text{g PbCl}_2 \quad \text{(calculator answer)}$$

$$= 68.6\ \text{g PbCl}_2 \quad \textbf{(correct answer)}$$

Note that the concentration of Pb(NO$_3$)$_2$ solution, given as 4.00 M in the problem statement, did not enter into the calculation. This is because the Pb(NO$_3$)$_2$ solution is present in excess; we know that we have enough of it. If a specific volume of Pb(NO$_3$)$_2$ solution had been given in the problem statement, we would have had to determine the limiting reactant [Pb(NO$_3$)$_2$ or KCl] as the first step in working the problem. The concept of a limiting reactant was discussed in Section 10.10 and is considered further in Example 13.23, which involves a limiting reactant calculation involving two solutions.

▶ **Practice Exercise 13.22**   How many grams of silver chloride (AgCl) can be produced from the reaction of 3.25 L of 2.00 M silver nitrate solution (AgNO$_3$) with an excess of 0.200 M aluminum chloride solution (AlCl$_3$), according to the following chemical equation?

$$AlCl_3(aq) + 3\,AgNO_3(aq) \longrightarrow 3\,AgCl(s) + Al(NO_3)_3(aq)$$

EXAMPLE 13.23 **Calculating a Product Amount Given Two Reactant Solution Volumes**

How many grams of solid copper(II) sulfide (CuS) product can be produced when 1.20 L of 0.20 M copper nitrate [$Cu(NO_3)_2$] solution and 1.50 L of 0.15 M soldium sulfide (Na_2S) solution are mixed? The chemical equation for the reaction that occurs is

$$Cu(NO_3)_2(aq) + Na_2S(aq) \longrightarrow CuS(s) + 2\,NaNO_3(aq)$$

SOLUTION

First, we must determine the limiting reactant since specific volumes of both reactant solutions are given in the problem statement. This determination involves two volume-of-solution-A to moles-of-B calculations—one for each reactant. The pathway in terms of Figure 13.7, for these parallel calculations is

$$\boxed{\text{Volume of solution A}} \xrightarrow{\text{Molarity}} \boxed{\text{Moles of A}} \xrightarrow[\text{coefficients}]{\text{Equation}} \boxed{\text{Moles of B}}$$

The dimensional analysis setup for these parallel calculations follows.

For $Cu(NO_3)_2$

$$1.20\;\text{L Cu(NO}_3)_2\;\text{solution} \times \frac{0.20\;\text{mole Cu(NO}_3)_2}{1\;\text{L Cu(NO}_3)_2\;\text{solution}} \times \frac{1\;\text{mole CuS}}{1\;\text{mole Cu(NO}_3)_2}$$
$$= 0.24\;\text{mole CuS (calculator and \textbf{correct answer})}$$

For Na_2S

$$1.50\;\text{L Na}_2\text{S solution} \times \frac{0.15\;\text{mole Na}_2\text{S}}{1\;\text{L Na}_2\text{S solution}} \times \frac{1\;\text{mole CuS}}{1\;\text{mole Na}_2\text{S}}$$
$$= 0.225\;\text{mole CuS}\quad \text{(calculator answer)}$$
$$= 0.22\;\text{mole CuS}\quad \text{(\textbf{correct answer})}$$

Based on these two calculations, Na_2S is the limiting reactant since it produces fewer moles of CuS.

The overall calculation is a volume-of-solution-A to grams-of-B calculation (see Figure 13.7).

$$\boxed{\text{Volume of solution A}} \xrightarrow{\text{Molarity}} \boxed{\text{Moles of A}} \xrightarrow[\text{coefficients}]{\text{Equation}} \boxed{\text{Moles of B}} \xrightarrow[\text{mass}]{\text{Molar}} \boxed{\text{Grams of B}}$$

In doing the limiting reactant calculation, we obtained moles of B (moles of CuS). Thus, all that is left to do is to go from moles of B to grams of B, using the limiting reactant molar amount of CuS and a conversion factor based on molar mass.

$$0.22\;\text{mole CuS} \times \frac{95.62\;\text{g CuS}}{1\;\text{mole CuS}} = 21.0364\;\text{g CuS}\quad \text{(calculator answer)}$$
$$= 21\;\text{g Cu}\quad\quad \text{(\textbf{correct answer})}$$

Answer Double Check:

In each of the three dimensional analysis setups used in solving this problem, do all of the units cancel except for the needed one? Yes, all of the units cancel correctly. Does the final numerical answer possess the correct number of significant figures? Yes. This answer is limited to two significant figures because each of the molar concentrations given in the problem statement contained only two significant figures.

▶ **Practice Exercise 13.23** How many grams of solid silver phosphate (Ag_3PO_4) product can be produced when 1.50 L of 0.26 M silver nitrate ($AgNO_3$) solution and 1.50 L of 0.50 M sodium phosphate (Na_3PO_4) solution are mixed? The chemical equation for the reaction that occurs is

$$3\ AgNO_3(aq) + Na_3PO_4(aq) \longrightarrow Ag_3PO_4(s) + 3\ NaNO_3(aq)$$

Example 13.24, the final example problem in this section, involves both volume of solution and volume of gas. Based on the relationships shown in Figure 13.7, this problem is no more complicated than the previous examples in this section since both of the volume quantities are "box locations" in Figure 13.7.

EXAMPLE 13.24 **Calculating the Gaseous Volume of a Product from the Solution Volume of a Reactant**

What volume, in liters of nitrogen monoxide gas, NO, measured at STP, can be produced from 1.75 L of 0.550 M nitric acid, HNO_3, and an excess of 0.650 M hydrosulfuric acid, H_2S, according to the following reaction?

$$2\ HNO_3(aq) + 3\ H_2S(aq) \longrightarrow 2\ NO(g) + 2\ S(s) + 4\ H_2O(l)$$

SOLUTION

Step 1 The given quantity is 1.75 L of HNO_3 solution, and the desired quantity is liters of NO gas at STP.

$$1.75\ L\ HNO_3 = ?\ L\ NO\ (at\ STP)$$

Step 2 This is a volume-of-solution-A to volume-of-gas-B problem. From Figure 13.7 the pathway for solving the problem is

Volume of solution A	→Molarity→	Moles of A	→Equation coefficients→	Moles of B	→Molar volume→	Volume of gas B

Step 3 The dimensional analysis setup for the calculation is

$$1.75\ L\ HNO_3 \times \frac{0.550\ mole\ HNO_3}{1\ L\ HNO_3} \times \frac{2\ moles\ NO}{2\ moles\ HNO_3} \times \frac{22.41\ L\ NO}{1\ mole\ NO}$$

The last conversion factor is derived from the fact that one mole of any gas occupies 22.41 L at STP conditions (Sec. 12.13).

Step 4 The result, obtained by combining all the numerical factors, is

$$\frac{1.75 \times 0.550 \times 2 \times 22.41}{1 \times 2 \times 1}\ L\ NO = 21.569625\ L\ NO \quad \text{(calculator answer)}$$

$$= 21.6\ L\ NO \quad \textbf{(correct answer)}$$

▶ **Practice Exercise 13.24** What volume, in liters, of $CO_2(g)$ measured at STP can be produced from 0.500 L of 0.500 M HCl solution and an excess of $CaCO_3$, according to the following chemical equation?

$$CaCO_3(s) + 2\ HCl(aq) \longrightarrow CaCl_2(aq) + CO_2(g) + H_2O(l)$$

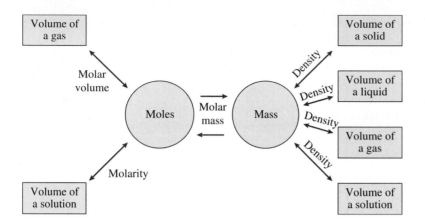

FIGURE 13.8 A summary of the various ways in which the quantity *volume* enters into chemical calculations.

13.12 CALCULATIONS INVOLVING VOLUME: A SUMMARY

Many calculations in this chapter have involved the quantity called *volume*. Such was the case also in the previous chapter, where calculations involving the gas laws were considered. Even earlier, in Chapter 3, volume was a central concept when calculations involving density were first considered. Figure 13.8 summarizes, in diagrammatic form, the various ways we have encountered volume to this point in a calculational setting.

Summary

1. **Solution Characteristics and Components** A solution is a homogeneous (uniform) mixture. The component of a solution that is present in the greatest amount is the *solvent*. A *solute* is a solution component that is present in a small amount relative to the solvent. The composition and properties of a solution are dependent on the ratio of solutes(s) to solvent. Solutes are present as individual particles (molecules, atoms, or ions).

2. **Solubility** The solubility of a solute is the maximum amount of solute that will dissolve in a given amount of solvent. The extent to which a solute dissolves in a solvent depends on the polarities of solute(s) and solvent, the temperature, and the pressure. A saturated solution contains the maximum amount of solute that can be dissolved under the conditions at which the solution exists. A supersaturated solution, an unstable situation, contains more dissolved solute than that needed for a saturated solution.

3. **Solubility Rules** Substances of like polarity tend to be more soluble in each other than substances that differ in polarity. This conclusion is often expressed as the simple phrase "like dissolves like." This generalization is not adequate for predicting the solubility of ionic compounds in water. More detailed guidelines are needed for this situation.

4. **Solution Concentration** The concentration of a solution is the amount of solute present in a specified amount of solvent or a specified amount of solution. A number of different concentration units are used, with the choice of units depending on the use to be made of the concentration units.

5. **Percent Concentration Units** The concentration of a solution in terms of percentage solute can be expressed in three different ways: (1) percent by mass (mass–mass percent), (2) percent by volume (volume–volume percent), and (3) mass–volume percent. All three types of percentages are in common use.

6. **Parts per Million and Parts per Billion** The concentration units parts per million and parts per billion find use when dealing with extremely dilute solutions. These units are closely related to percentage concentration units, differing only in the multiplicative factors used.

7. **Molarity** The molarity concentration unit, abbreviated M, is a ratio giving the number of moles of solute per liter of solution. It is the most used concentration unit in a chemical laboratory. Molarity finds use as a conversion factor in mass-to-volume calculations that involve solution volume.

8. **Molality** The molality concentration unit, abbreviated *m*, is a ratio giving the number of moles of solute per kilogram of solvent. Molality finds use, in preference to molarity, in experimental situations where changes in temperature are of concern since solution volume is dependent on temperature. In dilute solution, molality and molarity are practically identical in numerical value.

9. **Dilution** Dilution is the process in which more solvent is added to a solution to lower its concentration. Most dilutions are carried out by adding a predetermined volume of solvent to a specific volume of stock solution of known concentration.

Key Terms

The new terms defined in this chapter are

aqueous solution *Sec. 13.2*
concentrated solution
 Sec. 13.2
concentration *Sec. 13.5*
dilute solution *Sec. 13.2*
dilution *Sec. 13.10*
mass–volume percent
 concentration *Sec. 13.6*

molality concentration
 Sec. 13.9
molarity concentration
 Sec. 13.8
nonaqueous solution
 Sec. 13.2
parts per billion concen-
 tration *Sec. 13.7*

parts per million concen-
 tration *Sec. 13.7*
percent by mass concen-
 tration *Sec. 13.6*
percent by volume con-
 centration *Sec. 13.6*
saturated solution
 Sec. 13.2

solubility *Sec. 13.2*
solute *Sec. 13.1*
solution *Sec. 13.1*
solvent *Sec. 13.1*
supersaturated solution
 Sec. 13.2
unsaturated solution
 Sec. 13.2

Practice Problems

CHARACTERISTICS OF SOLUTIONS (SEC. 13.1)

13.1 Indicate whether each of the following statements about the general properties of solutions is *true* or *false*.
 a. A solution may contain more than one solute.
 b. All solutions are homogeneous mixtures.
 c. Every part of a solution has exactly the same properties as every other part.
 d. The solutes present in a solution will settle out with time if the solution is left undisturbed.

13.2 Indicate whether each of the following statements about the general properties of solutions is *true* or *false*.
 a. All solutions have a variable composition.
 b. For a solution to form, the solute and solvent must chemically react with each other.
 c. Solutes are present as individual particles (mole-cules, atoms, ions) in a solution.
 d. A general characteristic of all solutions is the liquid state.

13.3 Identify the *solute* and the *solvent* in solutions with the following compositions.
 a. 5.00 g of sodium chloride and 50.0 g of water
 b. 4.00 g of sucrose and 1000 g of water
 c. 2.00 mL of water and 20.0 mL of ethyl alcohol
 d. 60.0 mL of methyl alcohol and 20.0 mL of ethyl alcohol

13.4 Identify the *solute* and the *solvent* in solutions with the following compositions.
 a. 5.00 g of sodium bromide and 200.0 g of water
 b. 50.0 g of silver nitrate and 1000 g of water

 c. 50.0 mL of water and 100.0 mL of methyl alcohol
 d. 50.0 mL of isopropyl alcohol and 20.0 mL of ethyl alcohol

SOLUBILITY (SEC. 13.2)

13.5 Using the solubilities given in Table 13.2, character-ize each of the following solids as *insoluble, slightly soluble, soluble, or very soluble* in water at the indi-cated temperature.
 a. lead(II) bromide at 0°C
 b. silver sulfate at 50°C
 c. sodium chloride at 100°C
 d. silver nitrate at 0°C

13.6 Using the solubilities given in Table 13.2, character-ize each of the following solids as *insoluble, slightly soluble, soluble, or very soluble* in water at the indi-cated temperature.
 a. lead(II) bromide at 50°C
 b. silver sulfate at 0°C
 c. cesium chloride at 50°C
 d. copper(II) sulfate at 0°C

13.7 Classify the original solution as *unsaturated, saturated, or supersaturated* based on the following observations and/or changes.
 a. Agitation of the solution produces a large amount of solid crystals.
 b. Excess undissolved solute is present at the bottom of the solution container.

c. Amount of solute dissolved is less than the maximum amount that could dissolve under the conditions at which the solution exists.

d. Undissolved solute present in the solution dissolves when the solution is heated.

13.8 Classify the original solution as *unsaturated, saturated,* or *supersaturated* based on the following observations and/or changes.

a. Additional solute added rapidly dissolves.

b. Additional solute added falls to the bottom of the container where it remains without any decrease in amount.

c. Additional solute falls to the bottom of the container where it decreases in size for several hours and thereafter its size remains constant.

d. A small amount of added solute causes the production of a large amount of solid white crystals.

13.9 Use Table 13.2 to determine whether each of the following CsCl solutions is *unsaturated, saturated,* or *supersaturated.*

a. 161.4 g CsCl in 100 g H_2O at 0°C

b. 161.4 g CsCl in 100 g H_2O at 50°C

c. 161.4 g CsCl in 200 g H_2O at 0°C

d. 161.4 g CsCl in 50 g H_2O at 100°C

13.10 Use Table 13.2 to determine whether each of the following NaCl solutions is *unsaturated, saturated,* or *supersaturated.*

a. 37.0 g NaCl in 100 g H_2O at 0°C

b. 37.0 g NaCl in 100 g H_2O at 50°C

c. 37.0 g NaCl in 100 g H_2O at 100°C

d. 37.0 g NaCl in 200 g H_2O at 50°C

13.11 Based on the solubilities in Table 13.2, characterize each of the following silver sulfate (Ag_2SO_4) solutions as *dilute* or *concentrated.*

a. 0.50 g Ag_2SO_4 in 100 g H_2O at 100°C

b. 0.50 g Ag_2SO_4 in 100 g H_2O at 0°C

c. 0.50 g Ag_2SO_4 in 50 g H_2O at 50°C

d. 0.050 g Ag_2SO_4 in 10 g H_2O at 0°C

13.12 Based on the solubilities in Table 13.2, characterize each of the following copper(II) sulfate solutions as *dilute* or *concentrated.*

a. 14.3 g of $CuSO_4$ in 100 g of H_2O at 100°C

b. 14.3 g of $CuSO_4$ in 100 g of H_2O at 0°C

c. 14.3 g of $CuSO_4$ in 50 g of H_2O at 50°C

d. 1.43 g of $CuSO_4$ in 10 g of H_2O at 0°C

13.13 For each of the following pairs of solutions, select the solution for which solute solubility in water is the greatest at the given temperatures and pressures.

a. NH_3 gas at 90°C and 1 atm
   NH_3 gas at 50°C and 1 atm

b. CO_2 gas at 75°C and 2 atm
   CO_2 gas at 75°C and 1 atm

c. table salt at 40°C and 1 atm
   table salt at 60°C and 1 atm

d. table sugar at 70°C and 2 atm
   table sugar at 95°C and 1 atm

13.14 For each of the following pairs of solutions, select the solution for which solute solubility in water is the greatest at the given temperatures and pressures.

a. O_2 gas at 45°C and 1 atm
   O_2 gas at 35°C and 1 atm

b. N_2 gas at 35°C and 2 atm
   N_2 gas at 45°C and 1 atm

c. table sugar at 25°C and 1 atm
   table sugar at 65°C and 1 atm

d. table salt at 40°C and 2 atm
   table salt at 65°C and 1 atm

13.15 The solubility of $Pb(NO_3)_2$ at 70°C is 110 g per 100 g of water. At 40°C the solubility drops to 78 g per 100 g of water. A 200 g quantity of $Pb(NO_3)_2$ is stirred into 200 mL of water.

a. At 70°C how many grams, if any, of crystals of $Pb(NO_3)_2$ will settle out of solution?

b. After cooling the solution to 40°C, how many grams, if any, of crystals of $Pb(NO_3)_2$ will settle out of solution?

13.16 The solubility of KNO_3 at 60°C is 94 g per 100 g of water. At 20°C the solubility drops to 56 g per 100 g of water. A 150 g quantity of KNO_3 is stirred into 200 mL of water.

a. At 60°C how many grams, if any, of crystals of KNO_3 will settle out of solution?

b. After cooling the solution to 20°C, how many grams, if any, of crystals of KNO_3 will settle out of solution?

SOLUTION FORMATION (SEC. 13.3)

13.17 Match each of the following statements about the dissolving of the ionic solid NaCl in water with the term *hydrated ion, hydrogen atom, or oxygen atom.*

a. a Na^+ ion surrounded with water molecules

b. a Cl^- ion surrounded with water molecules

c. the portion of a water molecule that is attracted to an Na^+ ion

d. the portion of a water molecule that is attracted to a Cl^- ion

13.18 Match each of the following statements about the dissolving of the ionic solid KBr in water with the term *hydrated ion, hydrogen atom, or oxygen atom.*

a. a K^+ ion surrounded with water molecules

b. a Br^- ion surrounded with water molecules

c. the portion of a water molecule that is attracted to a K^+ ion

d. the portion of a water molecule that is attracted to a Br^- ion

13.19 Indicate whether each of the following actions will *increase* or *decrease* the rate of dissolving of a sugar cube in water.

a. cooling the sugar cube–water mixture

b. stirring the sugar cube–water mixture

c. breaking the sugar cube up into smaller chunks

d. crushing the sugar cube to give a granulated form of sugar

13.20 Indicate whether each of the following actions will *increase* or *decrease* the rate of dissolving of table salt in water.
  a. heating the table salt–water mixture
  b. agitating the table salt–water mixture
  c. heating the table salt prior to adding it to the water
  ___ d. heating the water prior to adding the table salt to it

SOLUBILITY RULES (SEC. 13.4)

13.21 Ethanol is a polar solvent and carbon tetrachloride is a nonpolar solvent. In which of these two solvents are each of the following solutes more likely to be soluble?
  a. NaCl, ionic
  b. cooking oil, nonpolar
  c. sucrose, polar
  d. $LiNO_3$, ionic
13.22 Methanol is a polar solvent and benzene is a nonpolar solvent. In which of these two solvents are each of the following solutes more likely to be soluble?
  a. KCl, ionic
  b. rubbing alcohol, polar
  c. gasoline, nonpolar
  ___ d. $NaNO_3$, ionic
13.23 Indicate whether each of the following anions forms compounds that are *generally soluble* or *generally insoluble* in water.
  a. NO_3^- b. Cl^- c. S^{2-} d. SO_4^{2-}
13.24 Indicate whether each of the following anions forms compounds that are *generally soluble* or *generally insoluble* in water.
  a. $C_2H_3O_2^-$ b. CO_3^{2-}
  c. PO_4^{3-} d. Br^-
13.25 On the basis of the general solubility rules for ionic compounds given in Table 13.4, indicate which of the rules covers each of the following substances' solubility situation.
  a. silver carbonate
  b. magnesium sulfide
  c. ammonium cyanide
  d. calcium sulfate
13.26 On the basis of the general solubility rules for ionic compounds given in Table 13.4, indicate which of the rules covers each of the following substances' solubility situation.
  a. copper(II) carbonate
  b. potassium cyanide
  c. aluminum nitrate
  ___ d. silver phosphate
13.27 Classify each of the following pairs of types of ionic compounds into the solubility categories *soluble, soluble with exceptions, insoluble,* and *insoluble with exceptions.*
  a. acetates and nitrates
  b. sulfates and chlorides
  c. sodium-ion-containing and ammonium-ion-containing
  d. phosphate and hydroxides

13.28 Classify each of the following pairs of types of ionic compounds into the solubility categories *soluble, soluble with exceptions, insoluble,* and *insoluble with exceptions.*
  a. chlorides and iodides
  b. sodium-ion-containing and potassium-ion-containing
  c. carbonates and sulfides
  ___ d. bromides and sulfates
13.29 In which of the following pairs of compounds do both members of the pair have like solubility (both soluble or both insoluble)?
  a. NH_4Cl and NH_4Br b. KNO_3 and Na_2SO_4
  c. $CaCO_3$ and CaS d. $Ni(OH)_2$ and $Ni_3(PO_4)_2$
13.30 In which of the following pairs of compounds do both members of the pair have like solubility (both soluble or both insoluble)?
  a. $NaNO_3$ and K_3PO_4 b. NH_4Br and $CuBr_2$
  ___ c. $Pb(NO_3)_2$ and $PbCl_2$ d. $FeCO_3$ and FeS
13.31 Which of the following ions would react with both Mg^{2+} and Ca^{2+} ions to form water-insoluble compounds?
  a. sulfate b. phosphate
  c. sulfide d. nitrate
13.32 Which of the following ions would react with both Cu^{2+} and Ba^{2+} ions to form water-insoluble compounds?
  a. chloride b. carbonate
  ___ c. hydroxide d. acetate

MASS PERCENT (SEC. 13.6)

13.33 Calculate the mass percent of solute in each of the following solutions.
  a. 7.37 g NaCl dissolved in 95.0 g H_2O
  b. 3.73 g KBr dissolved in 131 g H_2O
  c. 10.3 g NH_4NO_3 dissolved in 53.0 g of solution
  d. 25.0 g $MgSO_4$ dissolved in 275.0 g of solution
13.34 Calculate the mass percent of solute in each of the following solutions.
  a. 1.13 g $AgNO_3$ dissolved in 20.0 g H_2O
  b. 218 g CsCl dissolved in 102 g H_2O
  c. 10.3 g K_2SO_4 dissolved in 95.2 g of solution
  ___ d. 27.0 g of $(NH_4)_2S$ dissolved in 975 g of solution
13.35 Using Table 13.2, calculate the percent by mass of solute in each of the following solutions.
  a. NaCl at 0°C b. NaCl at 100°C
  c. $AgNO_3$ at 0°C d. $AgNO_3$ at 100°C
13.36 Using Table 13.2, calculate the percent by mass of solute in each of the following solutions.
  a. $CuSO_4$ at 0°C b. $CuSO_4$ at 100°C
  ___ c. CsCl at 0°C d. CsCl at 100°C
13.37 How many grams of solute are dissolved in 125.0 g of the following solutions?
  a. 2.00% (m/m) NaCl b. 3.50% (m/m) $AgNO_3$
  c. 10.00% (m/m) K_2SO_4 d. 8.25% (m/m)HCl

13.38 How many grams of solute are dissolved in 337.2 g of the following solutions?
a. 3.00% (m/m) KNO_3 **b.** 9.735% (m/m) NaOH
c. 0.800% (m/m) HI **d.** 12.0% (m/m) NH_4Cl

13.39 What mass of water, in grams, is needed to prepare each of the following calcium chloride ($CaCl_2$) solutions?
a. 5.75 g of 10.00% (m/m) $CaCl_2$ solution
b. 57.5 g of 10.00% (m/m) $CaCl_2$ solution
c. 57.5 g of 1.00% (m/m) $CaCl_2$ solution
d. 2.3 g of 0.80% (m/m) $CaCl_2$ solution

13.40 What mass of water, in grams, is needed to prepare each of the following lithium nitrate ($LiNO_3$) solutions?
a. 34.7 g of 5.00% (m/m) $LiNO_3$ solution
b. 3.47 g of 5.00% (m/m) $LiNO_3$ solution
c. 235 g of 12.75% (m/m) $LiNO_3$ solution
d. 1352 g of 0.0032% (m/m) $LiNO_3$ solution

13.41 How many grams of water must be added to 50.0 g of each of the following solutes to prepare a 5.00% (m/m) solution?
a. NaCl **b.** KCl **c.** Na_2SO_4 **d.** $LiNO_3$

13.42 How many grams of water must be added to 20.0 g of each of the following solutes to prepare a 2.00% (m/m) solution?
a. NaOH **b.** LiBr **c.** Li_2SO_4 **d.** $Ca(NO_3)_2$

VOLUME PERCENT (SEC. 13.6)

13.43 What is the volume percent ethyl alcohol in a solution containing 257 mL of ethyl alcohol and enough water to give the following amounts of solution?
a. 325 mL **b.** 675 mL **c.** 1.23 L **d.** 5.000 L

13.44 What is the volume percent water in a solution containing 35.0 mL of water and enough ethyl alcohol to give the following amounts of solution?
a. 45.0 mL **b.** 675 mL **c.** 1.08 L **d.** 4.500 L

13.45 The final volume of a solution made by adding 360.6 mL of methyl alcohol to 667.2 mL of water is 1000.0 mL. Determine the volume percent of the following:
a. methyl alcohol in the solution
b. water in the solution

13.46 The final volume of a solution made by adding 678.2 mL of methyl alcohol to 358.4 mL of water is 1000.0 mL. Determine the volume percent of the following:
a. methyl alcohol in the solution
b. water in the solution

13.47 How much isopropyl alcohol (C_3H_8O), in milliliters, is needed to prepare 225 mL of a 1.25% (v/v) solution of isopropyl alcohol in water?

13.48 How much ethyl alcohol (C_3H_8O), in milliliters, is needed to prepare 125 mL of a 2.25% (v/v) solution of ethyl alcohol in water?

13.49 What volume of water, in gallons, is contained in 4.00 gal of a 35.0% (v/v) solution of water in acetone?

13.50 What volume of water, in quarts, is contained in 3.50 qt of a 2.00% (v/v) solution of water in acetone?

MASS–VOLUME PERCENT (SEC. 13.6)

13.51 Calculate the mass–volume percent concentration for sodium nitrate ($NaNO_3$) solutions in which each of the following amounts of solute are present in 375 mL of solution.
a. 0.325 g **b.** 1.75 g **c.** 8.43 g **d.** 23.6 g

13.52 Calculate the mass–volume percent concentration for ammonium chloride (NH_4Cl) solutions in which each of the following amounts of solute are present in 525 mL of solution.
a. 0.475 g **b.** 2.02 g **c.** 7.50 g **d.** 21.3 g

13.53 Calculate the mass–volume percent concentration for potassium bromide (KBr) solutions with the following characteristics.
a. 4.00 g solute, 55.0 mL solution
b. 15.00 g solute, 1.75 L solution
c. 15.00 mg solute, 1.75 mL solution
d. 0.0300 mole solute, 52.0 mL solution

13.54 Calculate the mass–volume percent concentration for magnesium sulfate ($MgSO_4$) solutions with the following characteristics.
a. 3.00 g solute, 75.0 mL solution
b. 25.00 g solute, 2.25 L solution
c. 25.00 mg solute, 22.25 mL solution
d. 0.0500 mole solute, 67.2 mL solution

13.55 How many milliliters of a 6.00% (m/v) sodium chloride (NaCl) solution is required to supply the following amounts of solute?
a. 5.00 g **b.** 7.00 g **c.** 225 g **d.** 225 mg

13.56 How many milliliters of a 9.50% (m/v) potassium nitrate (KNO_3) solution is required to supply the following amounts of solute?
a. 7.25 g **b.** 9.25 g **c.** 457 g **d.** 457 mg

13.57 Determine how many grams of sodium phosphate (Na_3PO_4) would be
a. needed to prepare 455 mL of a 2.50% (m/v) Na_3PO_4 solution.
b. present in 50.0 L of a 7.50% (m/v) Na_3PO_4 solution.

13.58 Determine how many grams of potassium carbonate (K_2CO_3) would be
a. needed to prepare 4.55 mL of a 15.00% (m/v) K_2CO_3 solution.
b. present in 1.06 L of a 0.800% (m/v) K_2CO_3 solution.

13.59 Calculate the concentration, as mass–volume percent cesium chloride (CsCl), for a solution prepared by adding 5.0 g of CsCl to 20.0 g of H_2O to give a solution with a density of 1.18 g/mL.

13.60 Calculate the concentration, as mass–volume percent ammonium sulfate [$(NH_4)_2SO_4$], for a solution prepared by adding 3.0 g of $(NH_4)_2SO_4$ to 17.0 g of H_2O to give a solution with a density of 1.09 g/mL.

PARTS PER MILLION AND PARTS PER BILLION (SEC. 13.7)

13.61 What is the concentration of sodium chloride (NaCl), in ppm (m/m), in each of the following NaCl solutions?
 a. 37.5 mg of NaCl in 21.0 kg of water
 b. 2.12 cg of NaCl in 125 g of water
 c. 1.00 μg of NaCl in 32.0 dg of water
 d. 35.7 mg of NaCl in 15.7 g of water

13.62 What is the concentration of sodium bromide (NaBr), in ppm (m/m), in each of the following NaBr solutions?
 a. 37.5 cg of NaBr in 33.0 kg of water
 b. 2.12 mg of NaBr in 375 dg of water
 c. 3.00 μg of NaBr in 45.0 g of water
 d. 125 dg of NaBr in 255 kg of water

13.63 What is the concentration of each of the solutions in Problem 13.61 in parts per billion (m/m)?

13.64 What is the concentration of each of the solutions in Problem 13.62 in parts per billion (m/m)?

13.65 Fish generally need an oxygen concentration in water of at least 5 ppm (m/v) for survival. Will river water that contains 7 mg of O_2 per liter contain sufficient O_2 to sustain fish life?

13.66 A carbon dioxide concentration in water of 200 ppm (m/v) or higher is lethal to fish. Will river water that contains 0.62 g of dissolved CO_2 per 2.0 L be toxic to fish?

13.67 A typical concentration of the air pollutant sulfur dioxide (SO_2) in urban atmospheres is 0.087 ppm(v/v). At this concentration, how many liters of air would be needed to obtain 1.00 mL of SO_2?

13.68 A typical concentration of the air pollutant carbon monoxide (CO) in urban atmospheres is 2.8 ppm(v/v). At this concentration, how many liters of air would be needed to obtain 1.00 mL of CO?

13.69 Determine how much carbon dioxide (CO_2), in grams, must be present in a 523 mL sample of air to give the following mass/volume CO_2 concentrations.
 a. 3.0 ppb   **b.** 6.0 ppm   **c.** 2.5 pph
 d. 5.2 mass/volume percent

13.70 Determine how much hydrogen sulfide (H_2S), in grams, must be present in a 625 mL sample of air to give the following mass/volume H_2S concentrations.
 a. 3.5 ppb   **b.** 7.2 ppm   **c.** 4.0 pph
 d. 6.0 mass/volume percent

MOLARITY (SEC. 13.8)

13.71 Calculate the molarity of each of the following aqueous sodium hydroxide (NaOH) solutions.
 a. 2.0 moles NaOH in 0.50 L of solution
 b. 13.7 g NaOH in 90.0 mL of solution
 c. 53.0 g NaOH in 1.255 L of solution
 d. 0.0020 mole NaOH in 5.00 mL of solution

13.72 Calculate the molarity of each of the following aqueous potassium chloride (KCl) solutions.
 a. 1.45 moles KCl in 2.50 L of solution
 b. 12.5 g KCl in 85.0 mL of solution
 c. 27.0 g KCl in 1.055 L of solution
 d. 0.0500 mole KCl in 12.0 mL of solution

13.73 Calculate the number of grams of solute in each of the following sodium sulfate (Na_2SO_4) solutions.
 a. 35.0 mL of a 6.00 M solution
 b. 10.0 mL of a 0.600 M solution
 c. 375 L of a 1.00 M solution
 d. 375 g of a 7.91 M solution with a density of 1.25 g/mL

13.74 Calculate the number of grams of solute in each of the following sodium thiosulfate ($Na_2S_2O_3$) solutions.
 a. 27.0 mL of a 3.00 M solution
 b. 20.0 mL of a 6.00 M solution
 c. 125 L of a 0.100 M solution
 d. 125 g of a 7.50 M solution with a density of 1.42 g/mL

13.75 Calculate the volume, in milliliters, of the following nitric acid (HNO_3) solutions needed to provide the indicated amounts of solute.
 a. 2.50 g of HNO_3 from a 0.468 M solution
 b. 125 g of HNO_3 from a 3.50 M solution
 c. 4.50 moles of HNO_3 from a 2.50 M solution
 d. 0.0015 mole of HNO_3 from a 0.990 M solution

13.76 Calculate the volume, in milliliters, of the following sulfuric acid (H_2SO_4) solutions needed to provide the indicated amounts of solute.
 a. 2.50 g of H_2SO_4 from a 0.468 M solution
 b. 125 g of H_2SO_4 from a 3.50 M solution
 c. 4.50 moles of H_2SO_4 from a 2.50 M solution
 d. 0.0015 mole of H_2SO_4 from a 0.990 M solution

13.77 The following diagrams show varying amounts of the same solute (the blue spheres) in varying amounts of solution.

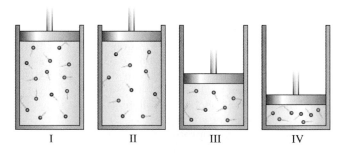

I   II   III   IV

 a. In which of the diagrams is the molarity concentration the greatest?
 b. In which two of the diagrams are the molarity concentrations the same?

13.78 The following diagrams show varying amounts of the same solute (the blue spheres) in varying amounts of solution.

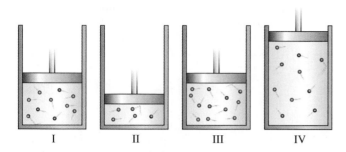

I II III IV

a. In which of the diagrams is the molarity concentration the greatest?
b. In which two of the diagrams are the molarity concentrations the same?

13.79 How many liters of 1.25 M aqueous solution can be prepared from 67.0 g each of the following solutes?
a. $NaNO_3$ **b.** KNO_3 **c.** NH_4NO_3 **d.** $Ca(NO_3)_2$

13.80 How many liters of 2.30 M aqueous solution can be prepared from 23.4 g of each of the following solutes?
a. NH_4Cl
b. NH_4Br
c. $(NH_4)_2SO_4$
d. NH_4CN

13.81 The density of an 88.00% (m/m) methyl alcohol (CH_4O) solution is 0.8274 g/mL. What is the molarity of the solution?

13.82 The density of a 60.00% (m/m) ethyl alcohol (C_2H_6O) solution is 0.8937 g/mL. What is the molarity of the solution?

13.83 The density of a 2.019 M sodium bromide (NaBr) solution is 1.157 g/mL. What is the concentration of this solution expressed as % (m/m) NaBr?

13.84 The density of a 2.687 M sodium acetate ($NaC_2H_3O_2$) solution is 1.104 g/mL. What is the concentration of this solution expressed as % (m/m) $NaC_2H_3O_2$?

13.85 What is the molarity of a 20.0% (m/v) hydrochloric acid (HCl) solution?

13.86 What is the molarity of a 25.0% (m/v) potassium hydroxide (KOH) solution?

MOLALITY (SEC. 13.9)

13.87 Calculate the molality of each of the following sucrose ($C_{12}H_{22}O_{11}$) solutions.
a. 16.5 g of sucrose in 1.35 kg of water
b. 3.15 moles of sucrose in 455 g of water
c. 0.0356 g of sucrose in 13.0 g of water
d. 45.0 g of sucrose in enough water to give 318 mL of solution with a density of 1.06 g/mL

13.88 Calculate the molality of each of the following glucose ($C_6H_{12}O_6$) solutions.
a. 23.0 g of glucose in 2.40 kg of water
b. 2.00 moles of glucose in 975 g of water
c. 0.230 g of glucose in 22.0 g of water
d. 30.0 g of glucose in enough water to give 312 mL of solution with a density of 1.04 g/mL

13.89 Calculate the number of grams of each solute that must be added to 234 g of water to prepare a 0.600 m solution of
a. NH_4NO_3. **b.** NaOH.
c. K_3PO_4. **d.** $Al_2(SO_4)_3$.

13.90 Calculate the number of grams of each solute that must be added to 153 g of water to prepare a 0.750 m solution of
a. $Be(OH)_2$. **b.** Na_2CO_3.
c. $CaCl_2$. **d.** $(NH_4)_3PO_4$.

13.91 How many grams of water must be added to 30.0 g of sodium bromide (NaBr) to prepare the following molal solutions?
a. 0.150 m **b.** 0.43 m **c.** 2.435 m **d.** 4.0 m

13.92 How many grams of water must be added to 40.0 g of potassium iodide (KI) to prepare the following molal solutions?
a. 0.0330 m **b.** 0.97 m **c.** 1.337 m **d.** 3.0 m

13.93 What is the molality for a solution consisting of 75.0 mL of cyclohexane (C_6H_{12}; density = 0.779 g/mL) dissolved in 175.0 mL of hexane (C_6H_{14}; density = 0.659 g/mL)?

13.94 What is the molality for a solution consisting of 60.0 mL of toluene (C_7H_8; density = 0.867 g/mL) dissolved in 120.0 mL of benzene (C_6H_6; density = 0.877 g/mL)?

13.95 An aqueous solution of oxalic acid ($H_2C_2O_4$) is 0.568 M and has a density of 1.022 g/mL. What is the molality of the solution?

13.96 An aqueous solution of citric acid ($H_3C_6H_5O_7$) is 0.655 M and has a density of 1.049 g/mL. What is the molality of the solution?

13.97 An aqueous solution of acetic acid ($HC_2H_3O_2$) is 0.796 m and has a density of 1.004 g/mL. What is the molality of the solution?

13.98 An aqueous solution of tartaric acid ($H_2C_4H_4O_6$) is 0.278 m and has a density of 1.006 g/mL. What is the molality of the solution?

13.99 What is the density, in grams per milliliter, for a hydrochloric acid (HCl) solution whose concentration is 15.5 molal and 11.8 molar?

13.100 What is the density, in grams per milliliter, for a hydrogen peroxide (H_2O_2) solution whose concentration is 12.6 molal and 9.7 molar?

13.101 Calculate the molality of a 22.0% by mass citric acid ($H_3C_6H_5O_7$) solution.

13.102 Calculate the molality of a 27.0% by mass oxalic acid ($H_2C_2O_4$) solution.

DILUTION (SEC. 13.10)

13.103 What is the molarity of a solution prepared by diluting 25.0 mL of 0.400 M potassium hydroxide (KOH) to each of the following volumes?
a. 40.0 mL **b.** 87.0 mL **c.** 225 mL **d.** 1.45 L

13.104 What is the molarity of a solution prepared by diluting 50.0 mL of 0.300 M sodium nitrate ($NaNO_3$) to each of the following volumes?
a. 85.0 mL **b.** 123 mL **c.** 128 mL **d.** 3.24 L

13.105 What is the molarity of the solution prepared by concentrating, by evaporation of solvent, 1353 mL of 0.500 M ammonium chloride (NH_4Cl) solution to each of the following final volumes?
a. 1223 mL **b.** 1.12 L **c.** 853 mL **d.** 302.5 mL

13.106 What is the molarity of the solution prepared by concentrating, by evaporation of solvent, 2212 mL of 0.400 M potassium sulfate (K_2SO_4) solution to each of the following final volumes?
a. 2145 mL **b.** 1.45 L **c.** 977 mL **d.** 453 mL

13.107 How many milliliters of 4.05 M hydrochloric acid (HCl) solution are required to produce, using dilution, the following hydrochloric acid solutions?
a. 45.0 mL of 3.90 M solution
b. 7.2 L of 2.00 M solution
c. 345 mL of 1.00 M solution
d. 3.0 mL of 0.20 M solution

13.108 How many milliliters of 6.02 M nitric acid (HNO_3) solution are required to produce, using dilution, the following nitric acid solutions?
a. 35.0 mL of 5.87 M solution
b. 63.0 mL of 3.01 M solution
c. 3.2 L of 1.00 M solution
d. 7.5 mL of 0.11 M solution

13.109 The following diagrams show various amounts of the same solute (blue spheres) in varying amounts of solution. If one-half of the solution in diagram I is withdrawn and then diluted by a factor of 4, which of the other diagrams (II–IV) represents the newly formed solution?

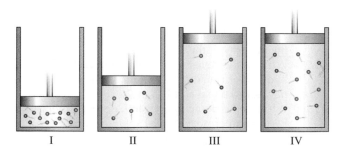

I　　　　II　　　　III　　　　IV

13.110 The following diagrams show various amounts of the same solute (blue spheres) in varying amounts of solution. If one-half of the solution in diagram I is withdrawn and then diluted by a factor of 2, which of the other diagrams (II–IV) represents the newly formed solution?

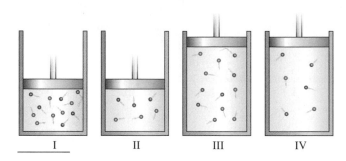

I　　　　II　　　　III　　　　IV

13.111 In each of the following silver nitrate ($AgNO_3$) solutions, how many milliliters of water should be added to obtain a solution that has a concentration of 0.100 M?
a. 20.0 mL of a 2.00 M solution
b. 20.0 mL of a 0.250 M solution
c. 358 mL of a 0.950 M solution
d. 2.3 L of a 6.00 M solution

13.112 In each of the following sodium nitrate ($NaNO_3$) solutions, how many milliliters of water should be added to obtain a solution that has a concentration of 0.200 M?
a. 30.0 mL of a 4.00 M solution
b. 30.0 mL of a 0.400 M solution
c. 785 mL of a 0.230 M solution
d. 1.25 L of a 1.50 M solution

13.113 What will be the final concentration of each of the following solutions if the volume of the solution is increased by 20.0 mL by adding water?
a. 25.0 mL of 6.0 M Na_2SO_4
b. 100.0 mL of 3.0 M K_2SO_4
c. 0.155 L of 10.0 M CsCl
d. 2.00 mL of 0.100 M $MgCl_2$

13.114 What will be the final concentration of each of the following solutions if the volume of the solution is increased by 20.0 mL by adding water?
a. 50.0 mL of 2.0 M KNO_3
b. 50.0 mL of 3.0 M $AgNO_3$
c. 1.0000 L of 1.2131 M $NaNO_3$
d. 1.0000 mL of 1.000 M $LiNO_3$

13.115 What would be the molarity of a solution obtained when 275 mL of 6.00 M sodium hydroxide (NaOH) solution is mixed with each of the following?
  a. 3.254 L of H_2O
  b. 125 mL of 6.00 M NaOH solution
  c. 125 mL of 2.00 M NaOH solution
  d. 27 mL of 5.80 M NaOH solution

13.116 What would be the molarity of a solution obtained when 352 mL of 4.00 M sodium bromide (NaBr) solution is mixed with each of the following?
  a. 425 mL of water
  b. 225 mL of 4.00 M NaBr solution
  c. 225 mL of 2.00 M NaBr solution
  d. 15 mL of 4.20 M NaBr solution

13.117 How many liters of a 3.00 M NaCl solution must be added to 25.0 L of a 5.00 M NaCl solution to reduce the concentration of the solution to 4.50 M?

13.118 How many liters of a 2.00 M KCl solution must be added to 15.0 L of a 4.50 M KCl solution to reduce the concentration of the solution to 2.50 M?

MOLARITY AND CHEMICAL EQUATIONS (SEC. 13.11)

13.119 What volume, in liters, of 1.00 M $Pb(NO_3)_2$ is needed to react completely with 0.500 L of 4.00 M NaCl, according to the following equation?

$$Pb(NO_3)_2(aq) + 2\,NaCl(aq) \rightarrow PbCl_2(s) + 2\,NaNO_3(aq)$$

13.120 What volume, in milliliters, of 0.300 M $CaCl_2$ is needed to react completely with 40.0 mL of 0.200 M H_3PO_4 according to the following equation?

$$3\,CaCl_2(aq) + 2\,H_3PO_4(aq) \rightarrow Ca_3(PO_4)_2(s) + 6\,HCl(aq)$$

13.121 How many grams of S can be produced from the reaction of 30.0 mL of 12.0 M HNO_3 with an excess of 0.035 M H_2S solution according to the following equation?

$$2\,HNO_3(aq) + 3\,H_2S(aq) \rightarrow 2\,NO(g) + 3\,S(s) + 4\,H_2O(l)$$

13.122 How many grams of Ag_3PO_4 can be produced from the reaction of 2.50 L of 0.200 M $AgNO_3$ with an excess of 0.750 M K_3PO_4 solution according to the following equation?

$$3\,AgNO_3(aq) + K_3PO_4(aq) \rightarrow Ag_3PO_4(s) + 3\,KNO_3(aq)$$

13.123 What volume, in milliliters, of 0.50 M H_2SO_4 is required to react with 18.0 g of nickel, according to the following equation?

$$Ni(s) + H_2SO_4(aq) \longrightarrow NiSO_4(aq) + H_2(g)$$

13.124 What volume, in milliliters, of is required to react with 100.0 g of tin, according to the following equation?

$$\underline{\hspace{1cm}}\ Sn(s) + 2\,HNO_3(aq) \longrightarrow Sn(NO_3)_2(aq) + H_2(g)$$

13.125 What is the molarity of a 37.5 mL sample of HNO_3 solution that will completely react with 23.7 mL of 0.100 M NaOH, according to the following equation?

$$HNO_3(aq) + NaOH(aq) \longrightarrow NaNO_3(aq) + H_2O(l)$$

13.126 What is the molarity of a 50.0 mL sample of H_2SO_4 solution that will completely react with 40.0 mL of 0.200 M $Mg(OH)_2$ according to the following equation?

$$\underline{\hspace{0.5cm}}H_2SO_4(aq) + Mg(OH)_2(aq) \longrightarrow MgSO_4(aq) + 2\,H_2O(l)$$

13.127 What volume, in liters, of NO gas measured at STP can be produced from 50.0 mL of 6.0 M HNO_3 solution and an excess of Cu metal, according to the following reaction?

$$8\,HNO_3(aq) + 3\,Cu(s) \longrightarrow 3\,Cu(NO_3)_2(aq) + 2\,NO(g) + 4\,H_2O(l)$$

13.128 What volume, in liters, of H_2 gas measured at STP can be produced from 50.0 mL of 3.0 M HBr solution and an excess of Zn metal, according to the following reaction?

$$\underline{\hspace{1cm}}\ 2\,HBr(aq) + Zn(s) \longrightarrow ZnBr_2(aq) + H_2(g)$$

13.129 What is the molarity of a 1.75 L $Ca(OH)_2$ solution that would completely react with 2.00 L of CO_2 gas measured at STP according to the following reaction?

$$CO_2(g) + Ca(OH)_2(aq) \longrightarrow CaCO_3(s) + H_2O(l)$$

13.130 What is the molarity of a 5.00 L NaOH solution that would completely react with 4.00 L of CO_2 gas measured at STP, according to the following reaction?

$$\underline{\hspace{1cm}}\ CO_2(g) + 2\,NaOH(aq) \longrightarrow Na_2CO_3(aq) + H_2O(l)$$

ADDITIONAL PROBLEMS

13.131 In each of the following sets of ionic compounds, identify the members of the set that are soluble in water.
  a. $Be_3(PO_4)_2$, $AlPO_4$, $FePO_4$, $(NH_4)_3PO_4$
  b. $Cu(OH)_2$, $Be(OH)_2$, $Ca(OH)_2$, $Zn(OH)_2$
  c. Ag_3PO_4, $AgNO_3$, $AgCl$, $AgBr$
  d. CaS, $Ca(NO_3)_2$, $CaSO_4$, $Ca(C_2H_3O_2)_2$

13.132 In each of the following sets of ionic compounds, identify the members of the set that are soluble in water.
a. K_2CO_3, $MgCO_3$, $NiCO_3$, $Al_2(CO_3)_3$
b. MgS, Rb_2S, Al_2S_3, CaS
c. $Pb(OH)_2$, $PbCl_2$, $Pb(NO_3)_2$, $PbSO_4$
d. BaS, $BaCl_2$, $BaSO_4$, $Ba(OH)_2$

13.133 The solubility of $CuSO_4$ in water at 50°C is 33.3 g/100 g H_2O. If 400.0 g of a 75% saturated $CuSO_4$ solution at 50°C is heated to evaporate the water completely, how much solid $CuSO_4$ should be recovered?

13.134 The solubility of $NaCl$ in water at 50°C is 37.0 g/100 g H_2O. If 300.0 g of an 85% saturated $NaCl$ solution at 50°C is heated to evaporate the water completely, how much solid $NaCl$ should be recovered?

13.135 After all the water is evaporated from 254 mL of a $AgNO_3$ solution, 45.2 g of $AgNO_3$ remain. Express the original concentration of the $AgNO_3$ solution in each of the following units.
a. mass–volume percent
b. molarity

13.136 After all the water is evaporated from 10.0 mL of a $CsCl$ solution, 3.75 g of $CsCl$ remains. Express the original concentration of $CsCl$ solution in each of the following units.
a. mass–volume percent
b. molarity

13.137 What mass, in grams, of Na_2SO_4 would be required to prepare 425 mL of a 1.55% (m/m) Na_2SO_4 solution whose density is 1.02 g/mL?

13.138 What mass, in grams, of $NaCl$ would be required to prepare 275 mL of a 30.0% (m/m) $NaCl$ solution whose density is 1.18 g/mL?

13.139 A 3.000 M $NaNO_3$ solution has a density of 1.161 g/mL at 20°C. How many grams of solvent are present in 1.375 L of this solution?

13.140 A 0.157 M $NaCl$ solution has a density of 1.09 g/mL at 20°C. How many grams of solvent are present in 80.0 mL of this solution?

13.141 A solute concentration is 3.74 ppm (m/m). What would this concentration be in the units of milligram of solute per kilogram of solution?

13.142 A solute concentration is 5.14 ppm (m/m). What would this concentration be in the units of microgram of solute per milligram of solution?

13.143 A solution is prepared by dissolving 1.00 g of $NaCl$ in enough water to make 10.00 mL of solution. A 1.00 mL portion of this solution is then diluted to a final volume of 10.00 mL. What is the molarity of the final $NaCl$ solution?

13.144 A solution is prepared by dissolving 30.0 g of Na_2SO_4 in enough water to make 750.0 mL of solution. A 10.00 mL portion of this solution is then diluted to a final volume of 100.0 mL. What is the molarity of the final Na_2SO_4 solution?

13.145 How many milliliters of 38.0% (m/m) HCl solution (density of 1.19 g/mL) are needed to make, using a dilution procedure, 1.00 L of 0.100 M HCl?

13.146 How many milliliters of 20.0% (m/m) $NaCl$ solution (density of 1.15 g/mL) are needed to make, using a dilution procedure, 3.50 L of 0.150 M $NaCl$?

13.147 How many grams of water should you add to a 1.23 m $NaCl$ solution containing 1.50 kg H_2O to reduce the molality to 1.00 m?

13.148 How many grams of water should you add to a 0.0883 m $NaCl$ solution containing 0.650 kg H_2O to reduce the molality to 0.0100 m?

13.149 Calculate the total mass, in grams, and the total volume, in milliliters, of a 2.16 m H_3PO_4 solution containing 52.0 g of solute. The density of the solution is 1.12 g/mL.

13.150 Calculate the total mass, in grams, and the total volume, in milliliters, of a 0.710 m $H_3C_6H_5O_7$ (citric acid) solution containing 23.0 g of solute. The density of the solution is 1.05 g/mL.

13.151 An aqueous solution having a density of 0.980 g/mL is prepared by dissolving 11.3 mL of CH_3OH (density of 0.793 g/mL) in enough water to produce 75.0 mL of solution. Express the percent CH_3OH in this solution as
a. % (m/v).
b. % (m/m).
c. % (v/v).

13.152 An aqueous solution having a density of 0.993 g/mL is prepared by dissolving 20.0 mL of C_2H_6O (density of 0.789 g/mL) in enough water to produce 85.0 mL of solution. Express the percent C_2H_6O in this solution as
a. % (m/v).
b. % (m/m).
c. % (v/v).

13.153 The concentration of a KCl solution is 0.273 molal and 0.271 molar. What is the density of the solution in grams per milliliter?

13.154 The concentration of a $Pb(NO_3)_2$ solution is 0.953 molal and 0.907 molar. What is the density of the solution in grams per milliliter?

CUMULATIVE PROBLEMS

13.155 Identify the insoluble substance(s) formed when each of the following pairs of soluble substances react in aqueous solution through a double-replacement reaction.
a. $NaCl$ and $AgNO_3$
b. $Ba(C_2H_3O_2)_2$ and K_3PO_4
c. $Pb(NO_3)_2$ and Ag_2SO_4
d. $CuSO_4$ and BaS

13.156 Identify the insoluble substance(s) formed when each of the following pairs of soluble substances react in aqueous solution through a double-replacement reaction.
a. $MgCl_2$ and $Ba(OH)_2$
b. NH_4Cl and $Pb(NO_3)_2$
c. MgS and Na_2CO_3
d. $SrCl_2$ and Ag_2SO_4

13.157 How many liters of gas at 25°C and 1.46 atm pressure are required to prepare 2.00 L of a 3.50 M solution of NH_3?

13.158 How many liters of HCl gas at 35°C and 1.05 atm pressure are required to prepare 4.00 L of a 0.500 M solution of HCl?

13.159 Calculate the theoretical yield, in grams, of AgCl formed from the reaction of 6.41 g of $ZnCl_2$ with 40.0 mL of a 0.404 M $AgNO_3$ solution according to the reaction

$$ZnCl_2(s) + 2\,AgNO_3(aq) \longrightarrow Zn(NO_3)_2(aq) + 2\,AgCl(s)$$

13.160 Calculate the theoretical yield, in grams, of AgCl formed from the reaction of 1.00 g of KCl with 100.0 mL of a 0.0250 M $AgC_2H_3O_2$ solution, according to the reaction

$$KCl(s) + AgC_2H_3O_2(aq) \longrightarrow KC_2H_3O_2(aq) + AgCl(s)$$

13.161 What mass, in grams, of $BaCrO_4$ would be produced by mixing 0.350 L of a 3.25 M $BaCl_2$ solution with 0.450 L of a 4.50 M K_2CrO_4 solution? The two solutions react according to the equation

$$BaCl_2(aq) + K_2CrO_4(aq) \longrightarrow BaCrO_4(s) + 2\,KCl(aq)$$

13.162 What mass, in grams, of $BaSO_4$ would be produced by mixing 1.53 L of a 4.50 M Na_2SO_4 solution with 3.20 L of a 2.50 M $Ba(NO_3)_2$ solution? The two solutions react according to the equation

$$Na_2SO_4(aq) + Ba(NO_3)_2(aq) \longrightarrow 2\,NaNO_3(aq) + BaSO_4(s)$$

13.163 A 1.25 g sample of *impure* Na_2CO_3 is found to react completely with 70.0 mL of 0.125 M HCl. The equation for the reaction is

$$Na_2CO_3(s) + 2\,HCl(aq) \longrightarrow 2\,NaCl(aq) + CO_2(g) + H_2O(l)$$

What is the mass percent Na_2CO_3 in the impure sample?

13.164 A 5.00 g sample of *impure* $CaCO_3$ is found to react completely with 100.0 mL of 0.100 M H_2SO_4. The equation for the reaction is

$$CaCO_3(s) + H_2SO_4(aq) \longrightarrow CaSO_4(s) + CO_2(g) + H_2O(l)$$

What is the mass percent $CaCO_3$ in the impure sample?

13.165 Magnesium, calcium, and zinc all react with hydrochloric acid as follows (where M represents any of these metals).

$$M(s) + 2\,HCl(aq) \longrightarrow MCl_2(aq) + H_2(g)$$

A sample of one of these metals reacts completely with the acid in 27.9 mL of 2.48 M HCl, and the resulting solution is evaporated to dryness. The residue MCl_2 has a mass of 4.72 g. What is the identity of the metal used?

13.166 Iron, nickel, and tin all react with hydrochloric acid as follows (where M represents any of these metals).

$$M(s) + 2\,HCl(aq) \longrightarrow MCl_2(aq) + H_2(g)$$

A sample of one of these metals reacts completely with the acid in 34.2 mL of 4.00 M HCl, and the resulting solution is evaporated to dryness. The residue MCl_2 has a mass of 8.87 g. What is the identity of the metal used?

13.167 A quantity of sodium peroxide (Na_2O_2) is added to water, and the following reaction occurs.

$$2\,Na_2O_2(s) + 2\,H_2O(l) \longrightarrow 4\,NaOH(aq) + O_2(g)$$

If 70.0 mL of O_2 gas (at STP) and 150.0 mL of NaOH solution are produced, what is the molarity of the NaOH solution?

13.168 A quantity of lithium nitride (Li_3N) is added to water, and the following reaction occurs.

$$Li_3N(s) + 3\,H_2O(l) \longrightarrow 3\,LiOH(aq) + NH_3(g)$$

If 100.0 mL of NH_3 gas (at STP) and 255 mL of LiOH solution are produced, what is the molarity of the LiOH solution?

Multiple-Choice Practice Test

Use this bank of 20 multiple-choice questions as a review of key concepts presented in this chapter. For many of the questions, there may be more than one correct answer (choice d) or no correct answer (choice e).

13.169 Which of the following statements about solutions is correct?
  a. The solvent must always be a liquid.
  b. They can alternatively be called homogeneous mixtures.
  c. The components present readily separate into solute and solvent if left undisturbed for 24 hours.

  d. more than one correct response
  e. no correct response

13.170 What type of solution is characterized by the rate of crystallization of dissolved solute being equal to the rate of dissolution of undissolved solute?
  a. unsaturated solution
  b. saturated solution
  c. dilute solution
  d. more than one correct response
  e. no correct response

13.171 Which of the following is a characteristic of a super-saturated solution?
  a. Dissolved solute is in equilibrium with undissolved solute.
  b. The solubility limit for the solute has been exceeded.
  c. Solute will rapidly precipitate if a seed crystal is added.
  d. more than one correct response
  e. no correct response

13.172 Which of the following generalizations concerning factors that affect solute solubility is correct?
  a. The solubility of most solid-state solutes decreases as temperature increases.
  b. The solubility of gaseous-state solutes increases with increasing temperature.
  c. The solubility of gaseous-state solutes decreases with increasing pressure.
  d. more than one correct response
  e. no correct response

13.173 The solubility rule "like dissolves like" is not adequate for predicting solubilities when the solute is a(n)
  a. nonpolar gas.
  b. ionic solid.
  c. polar liquid.
  d. more than one correct response
  e. no correct response

13.174 In which of the following pairs of solid-state solutes are both members of the pair *soluble* in water?
  a. $AgNO_3$ and $AgCl$
  b. Na_2CO_3 and Na_3PO_4
  c. $CuSO_4$ and $Cu(OH)_2$
  d. more than one correct response
  e. no correct response

13.175 What is the mass percent concentration of a solution containing 20.0 g $NaNO_3$ and 250.0 g H_2O?
  a. 6.76 mass percent
  b. 7.41 mass percent
  c. 8.00 mass percent
  d. 8.25 mass percent
  e. no correct response

13.176 A 2.0% (m/v) NaCl solution contains 2.0 g of NaCl per
  a. 100.0 g of solution.
  b. 100.0 mL of solution.
  c. 1.0 L of solution.
  d. more than one correct response
  e. no correct response

13.177 What volume, in mL, of an 8.50% (m/v) glucose solution contains 50.0 grams of glucose?
  a. 135 mL   **b.** 279 mL   **c.** 356 mL   **d.** 588 mL
  e. no correct response

13.178 In which of the following pairs of concentrations are the two concentrations equivalent to each other?
  a. 7 ppm (v/v) and 7000 ppb (v/v)
  b. 7 pph (v/v) and 7000 ppm (v/v)

  c. 7% (v/v) and 70,000 ppm (v/v)
  d. more than one correct response
  e. no correct response

13.179 Which of the following solutions has a molarity of 2.00 M?
  a. 2.00 L of solution containing 2.00 moles of solute
  b. 0.500 L of solution containing 2.00 moles of solute
  c. 1.50 L of solution containing 3.00 moles of solute
  d. more than one correct response
  e. no correct response

13.180 An aqueous solution contains 17.04 g NH_3 per 500.0 mL of solution. The molarity of this solution is
  a. 0.06800 M.   **b.** 1.000 M.
  c. 2.000 M.   **d.** 8.520 M.
  e. no correct response

13.181 For which of the following unit changes can molarity be directly used as the conversion factor?
  a. grams of solute to liters of solvent
  b. moles of solute to liters of solution
  c. liters of solution to moles of solute
  d. more than one correct response
  e. no correct response

13.182 How many grams of LiF are needed to make 325 mL of a 0.100 M LiF solution?
  a. 0.390 g
  b. 0.843 g
  c. 2.67 g
  d. 8.43 g
  e. no correct response

13.183 Which of the following is the correct defining equation for the molality concentration unit?
  a. moles of solute per liter of solution
  b. moles of solute per kilogram of solution
  c. moles of solute per liter of solvent
  d. moles of solute per kilogram of solvent
  e. no correct answer

13.184 What is the molality of a solution made by dissolving 16.84 g of NaF in enough water to give 325 g of solution?
  a. 0.555 m   **b.** 0.713 m   **c.** 1.23 m   **d.** 1.30 m
  e. no correct response

13.185 If 400.0 mL of 1.00 M NaOH is diluted to 2.00 L, the concentration of the resulting solution is
  a. 0.0400 M.
  b. 0.200 M.
  c. 0.400 M.
  d. 0.800 M.
  e. no correct response

13.186 What is the volume, in mL, of a 2.75 M solution NaCl that must be used to make 1.25 L of a 0.150 M NaCl solution by dilution?
  a. 0.0682 mL
  b. 0.682 mL
  c. 68.2 mL
  d. 682 mL
  e. no correct response

13.187 What is the volume ratio in which the two reactants in the chemical reaction

$$H_2CO_3(aq) + 2\ KOH(aq) \longrightarrow K_2CO_3(aq) + 2\ H_2O(l)$$

react if the molarities of the two reactants are the same?

a. 2 to 1
b. 1 to 2
c. cannot be determined without knowing the actual molarities.
d. more than one correct response
e. no correct response

13.188 What volume, in liters, of 0.600 M KOH solution is needed to react completely with 0.100 L of 2.50 M H_2SO_4 solution, according to the equation

$$2\ KOH(aq) + H_2SO_4(aq) \longrightarrow K_2SO_4(aq) + 2\ H_2O(l)$$

a. 0.208 L
b. 0.417 L
c. 0.537 L
d. 0.833 L
e. no correct response

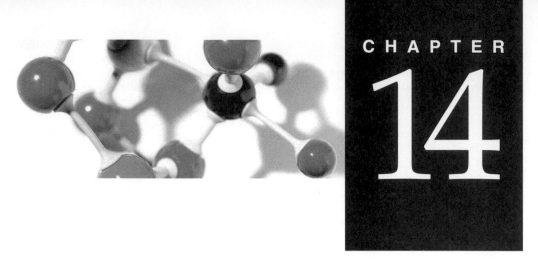

Acids, Bases, and Salts

14.1 ARRHENIUS ACID–BASE THEORY

Acids and bases are among the most common and important compounds known. Aqueous solutions of acids and bases are key materials in both biological systems and chemical industrial processes.

Historically, as early as the seventeenth century, acids and bases were recognized as important groups of compounds. Such early recognition was based on what the substances did, rather than on their chemical composition.

Early known facts about acids include the following.

1. Acids, when dissolved in water, have a sour taste. (The name acid comes from the Latin word *acidus*, which means "sour.") Note that although taste was once an acceptable criterion for identifying a chemical, it is not anymore. It is not a wise idea to taste chemicals when in a chemical laboratory.
2. Acids cause the dye litmus to change from blue to red. (Litmus is a naturally occurring vegetable dye obtained from lichens.)
3. When certain metals, such as zinc and iron, are placed in acids, they dissolve, liberating hydrogen gas.

Early known characteristics of bases include the following.

1. Bases, when dissolved in water, have a bitter taste.
2. Bases cause the dye litmus to change from red to blue.

3. When fats are placed in base solutions, they dissolve.
4. Base solutions feel slippery or soapy to the touch. (The bases themselves are not slippery, but react with fats in the skin to form new slippery or soapy compounds.)

It was not until 1884 that acids and bases were defined in terms of chemical composition. In that year, the Swedish chemist Svante August Arrhenius (1859–1927; see "The Human Side of Chemistry 15") proposed that acids and bases be defined in terms of the species they form upon dissolution in water. An **Arrhenius acid** *is a hydrogen-containing compound that, in water, produces hydrogen ions* (H^+). The acidic species in Arrhenius theory is, thus, the hydrogen ion. An **Arrhenius base** *is a hydroxide-ion-containing compound that, in water, produces hydroxide ions* (OH^-). The basic species in Arrhenius theory is, thus, the hydroxide ion.

Two common examples of acids, according to the Arrhenius definition, are the substances HNO_3 and HCl.

$$HNO_3(l) \xrightarrow{H_2O} H^+(aq) + NO_3{}^-(aq)$$

$$HCl(g) \xrightarrow{H_2O} H^+(aq) + Cl^-(aq)$$

Arrhenius acids in the pure state (not in solution) are covalent compounds; that is, they do not contain H^+ ions. The H^+ ions are produced when the Arrhenius acid interacts with the water, a process called *ionization*. **Ionization** *is the process whereby ions are produced from a molecular compound when it is dissolved in a solvent.*

Two common examples of Arrhenius bases are NaOH and KOH.

$$NaOH(s) \xrightarrow{H_2O} Na^+(aq) + OH^-(aq)$$

$$KOH(s) \xrightarrow{H_2O} K^+(aq) + OH^-(aq)$$

Arrhenius bases are ionic compounds in the pure state, in direct contrast to acids. When such compounds dissolve in water, the already existent OH^- ions are released, a process called *dissociation*. **Dissociation** *is the process whereby the already existent ions in an ionic compound separate when the ionic compound is dissolved in a solvent.* Figure 14.1 contrasts the processes of ionization (Arrhenius acids) and dissociation (Arrhenius bases).

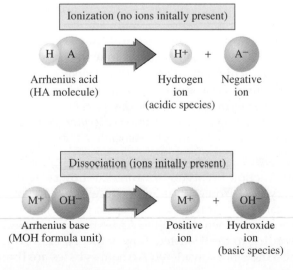

FIGURE 14.1 The difference between the aqueous solution processes of ionization (Arrhenius acids) and dissociation (Arrhenius bases). Ionization is the production of ions from a *molecular* compound that has been dissolved in a solvent. Dissociation is the production of ions from an *ionic* compound that has been dissolved in a solvent.

Ionization (no ions initally present)

H A
Arrhenius acid
(HA molecule)

H+ + A⁻
Hydrogen   Negative
ion   ion
(acidic species)

Dissociation (ions initally present)

M+ OH⁻
Arrhenius base
(MOH formula unit)

M+ + OH⁻
Positive   Hydroxide
ion   ion
(basic species)

The Human Side of Chemistry 15

**Svante August Arrhenius
(1859–1927)**

Svante August Arrhenius, born in 1859 near Uppsala, Sweden, is considered one of the founders of modern physical chemistry.

His roots are those of a Swedish farming family. An infant prodigy, on his own accord and against his parents' wishes, he taught himself to read at age 3.

In 1884, he proposed his definitions for acids and bases. Simultaneously he shook the world of chemistry by presenting his theory of ionic dissociation, which was that ionic substances, when dissolved in water, dissociate into ions. This theory, which came directly from his university doctoral work in Stockholm, was given a hostile reception by many other chemists, including his mentors at the University of Stockholm. He was awarded his doctoral degree with reluctance and given the lowest possible passing grade. It was the opinion of his teachers that his theory was "too farfetched."

Completely rebuffed by Swedish scientists, he decided to approach the scientific world elsewhere. He did find acceptance in some places and, building on this acceptance, he further developed and refined his theory.

Acceptance in his homeland eventually came. In 1903, Arrhenius was awarded the Nobel Prize in chemistry for his "farfetched" ideas concerning ions in solution. In 1905, the king of Sweden founded the Nobel Institute for Physical Research at Stockholm and installed Arrhenius as director. This appointment came as a counteroffer to that of a major professorship in Berlin. Arrhenius remained at the Nobel Institute until shortly before his death.

Arrhenius made other major contributions to chemistry besides those dealing with ions, acids, and bases. In 1889, while studying how rates of reactions increased with temperature, he worked out the concept of energy of activation (to be discussed in Chapter 16). In later years he became interested in such diverse things as serum chemistry and astronomy. He spent considerable time speculating on the origin of life on Earth.

14.2 BRØNSTED–LOWRY ACID–BASE THEORY

Although widely used, Arrhenius acid–base theory has some shortcomings. Two disadvantages are that it is restricted to aqueous solutions and it does not explain why compounds like ammonia (NH_3), which do not contain hydroxide ion, produce a basic water solution.

In 1923, Johannes Nicolaus Brønsted (1879–1947), a Danish chemist, and Thomas Martin Lowry (1874–1936), a British chemist, independently and almost simultaneously proposed broadened definitions for acids and bases—definitions that applied in both aqueous and nonaqueous solutions and that also explained how some non-hydroxide-containing substances, when added to water, produce basic solutions.

A **Brønsted–Lowry acid** *is any substance that can donate a proton (H^+) to some other substance.* A **Brønsted–Lowry base** *is any substance that can accept a proton (H^+) from some other substance.* In simpler terms, a Brønsted–Lowry *acid* is a *proton donor* (or hydrogen ion donor) and a Brønsted–Lowry *base* is a *proton acceptor* (or hydrogen ion acceptor).

Three important additional concepts associated with Brønsted–Lowry acid–base theory are as follows:

The terms *hydrogen ion* and *proton* are used synonymously in acid–base discussions. Why? The predominant (99.98%) hydrogen isotope, ^1_1H, is unique in that no neutrons are present; it consists of a proton and an electron. Thus, the ion $^1_1H^+$, a hydrogen atom that has lost its only electron, is simply a proton.

1. Any chemical reaction involving a Brønsted–Lowry acid must also involve a Brønsted–Lowry base. You cannot have one without the other. Proton donation (from an acid) cannot occur unless an acceptor (a base) is present.
2. All the acids and bases included in the Arrhenius theory (Sec. 14.1) are also acids and bases according to the Brønsted–Lowry theory. However, the converse is not true; some substances not considered Arrhenius bases are Brønsted–Lowry bases.

3. The identity of the acidic species in *aqueous solution* is not the Arrhenius H^+ ion, but the H_3O^+ ion. Hydrogen ions in solution react with water. The attraction between a hydrogen ion and a water molecule is sufficiently strong to bond the hydrogen ion to the water molecule to form a *hydronium ion* (H_3O^+). The bond between them is a coordinate covalent bond (Sec. 7.14) because both electrons are furnished by the oxygen atom.

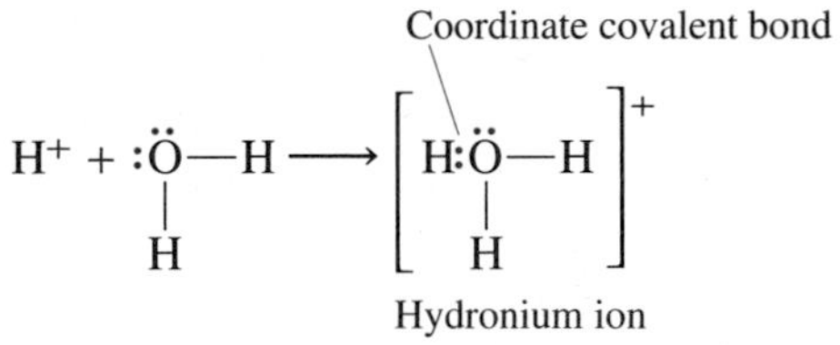

Coordinate covalent bond

$$H^+ + \;:\!\ddot{O}\!-\!H \longrightarrow \left[H\!:\!\ddot{O}\!-\!H \right]^+$$

Hydronium ion

The Brønsted–Lowry acid–base definitions can best be illustrated by example. Consider the formation reaction for hydrochloric acid, which involves the dissolving of hydrogen chloride gas in water.

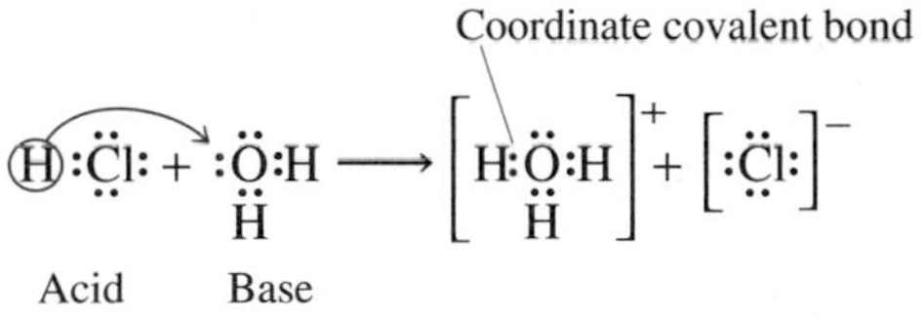

Coordinate covalent bond

$$(\!H\!)\!:\!\ddot{C}\!l\!: + \;:\!\ddot{O}\!:\!H \longrightarrow \left[H\!:\!\ddot{O}\!:\!H \right]^+ + \left[:\!\ddot{C}\!l\!: \right]^-$$

Acid   Base

The HCl behaves as a Brønsted–Lowry acid by donating a proton to a water molecule. Note that a hydronium ion is formed as a result. The base in this reaction is water because it has accepted a proton; no hydroxide ions are involved. The Brønsted–Lowry definition of a base includes all species that accept a proton; hydroxide ions can do this, but so can many other substances.

It is not necessary that a water molecule be one of the reactants in a Brønsted–Lowry acid–base reaction or that the reaction take place in the liquid state. An important application of Brønsted–Lowry acid–base theory is to gas-phase reactions. The white solid haze that often covers glassware in a chemistry laboratory results from the gas-phase reaction between HCl and NH_3.

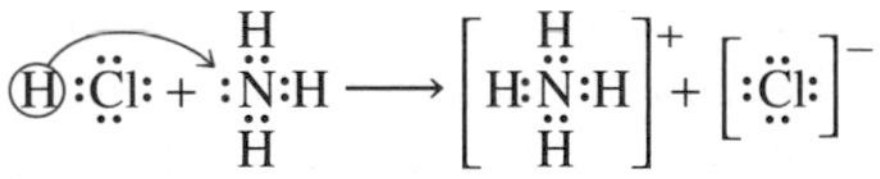

$$(\!H\!)\!:\!\ddot{C}\!l\!: + \;:\!\underset{H}{\overset{H}{N}}\!:\!H \longrightarrow \left[H\!:\!\underset{H}{\overset{H}{N}}\!:\!H \right]^+ + \left[:\!\ddot{C}\!l\!: \right]^-$$

This is a Brønsted–Lowry acid–base reaction, because the HCl molecules donate protons to the NH_3, forming NH_4^+ and Cl^- ions. These ions instantaneously combine to form the white solid NH_4Cl.

Another example of a Brønsted–Lowry acid–base reaction involves the dissolving of ammonia (a nonhydroxide base) in water. In the following equation, note how a hydroxide ion is produced as the result of the transfer of a proton from water (written as HOH) to the ammonia.

$$NH_3(g) + (\!H\!)OH(l) \longrightarrow NH_4^+(aq) + OH^-(aq)$$

Base   Acid

A Brønsted–Lowry base—a proton acceptor—must contain an atom that possesses a pair of unshared electrons that can be used in forming a coordinate covalent bond to an incoming proton (from a Brønsted–Lowry acid).

As an ionic solid dissolves in water to produce an aqueous solution, the solid ionic lattice breaks up, producing individual ions that are free to move about in the solution (Sec. 13.3). Ions so formed can function as Brønsted–Lowry acids or bases, as is illustrated in the following two equations.

$$\overset{\frown}{H}CO_3^-(aq) + H_2O(l) \longrightarrow CO_3^{2-}(aq) + H_3O^+(aq)$$
$$\quad\ \text{Acid} \qquad\quad \text{Base}$$

$$\overset{\frown}{H_2}PO_4^-(aq) + \overset{\frown}{O}H^-(aq) \longrightarrow HPO_4^{2-}(aq) + H_2O(l)$$
$$\quad\ \text{Acid} \qquad\qquad \text{Base}$$

14.3 CONJUGATE ACIDS AND BASES

For most Brønsted–Lowry acid–base reactions, 100% proton transfer does not occur. Instead, a state of equilibrium (Sec. 11.13) is reached in which a forward reaction and a reverse reaction are occurring at an equal rate.

The equilibrium mixture for a Brønsted–Lowry acid–base reaction always has *two* acids and *two* bases present. To illustrate this, consider the acid–base reaction involving hydrogen fluoride and water.

$$HF(aq) + H_2O(l) \rightleftharpoons H_3O^+(aq) + F^-(aq)$$

(The double arrows in this equation indicate a state of equilibrium—both a forward and a reverse reaction are occurring.) For the forward reaction, the HF molecules donate protons to water molecules. Thus, the HF is functioning as an acid and the H_2O is functioning as a base.

$$HF(aq) + H_2O(l) \longrightarrow H_3O^+(aq) + F^-(aq)$$
$$\quad\ \text{Acid} \qquad \text{Base}$$

For the reverse reaction, the one going from right to left, a different picture emerges. Here, H_3O^+ is functioning as an acid (by donating a proton), and F^- behaves as a base (by accepting the proton).

$$H_3O^+(aq) + F^-(aq) \longrightarrow HF(aq) + H_2O(l)$$
$$\quad\ \text{Acid} \qquad \text{Base}$$

Conjugate means "coupled" or "joined together" (as in a pair).

The two acids and two bases involved in a Brønsted–Lowry equilibrium situation can be grouped into two *conjugate acid–base pairs*. A **conjugate acid–base pair** *is two species, one an acid and one a base, that differ from each other through the loss or gain of a proton (H^+ ion).* The two conjugate acid–base pairs in our example are (HF and F^-) and (H_3O^+ and H_2O).

$$\overbrace{\qquad\qquad\qquad\qquad\text{Conjugate pair}\qquad\qquad\qquad\qquad}$$
$$HF(aq) + H_2O(l) \rightleftharpoons H_3O^+(aq) + F^-(aq)$$
$$\text{Acid} \qquad \text{Base} \qquad\quad \text{Acid} \qquad \text{Base}$$
$$\underbrace{\qquad\qquad\qquad\text{Conjugate pair}\qquad\qquad\qquad}$$

Abbreviated notation for specifying a conjugate acid–base pair is "acid/base." Using this notation, the two conjugate acid–base pairs in the preceding example are HF/F^- and H_3O^+/H_2O. The acid is always written first in such notation.

For any given conjugate acid–base pair,

1. the acid in the acid–base pair always has one *more* H atom and one *fewer* negative charge than the base. Note this relationship for the HF/F^- conjugate acid–base pair.
2. the base in the acid–base pair always has one *fewer* H atom and one *more* negative charge than the acid. Note this relationship for the HF/F^- conjugate acid–base pair.

The acid in a conjugate acid–base pair is called the *conjugate acid* of the base, and the base in the conjugate acid–base pair is called the *conjugate base* of the acid. A **conjugate acid** *is the species formed when a proton (H^+ ion) is added to a Brønsted–Lowry base.* The H_3O^+ ion is the conjugate acid of an H_2O molecule. A **conjugate base** *is the species that remains when a proton (H^+ ion) is removed from a Brønsted–Lowry acid.* The H_2O molecule is the conjugate base of the H_3O^+ ion. Every acid has a conjugate base, and every base has a conjugate acid. In general terms, these relationships can be diagrammed as follows.

$$HA \quad + \quad B \quad \rightleftharpoons \quad HB^+ \quad + \quad A^-$$

| Acid | Base | Conjugate acid | Conjugate base |

EXAMPLE 14.1 Determining the Members of a Conjugate Acid–Base Pair

Identify the conjugate acid–base pairs in the following reaction.

$$HBr(aq) + H_2O(l) \longrightarrow H_3O^+(aq) + Br^-(aq)$$

SOLUTION

To determine the conjugate acid–base pairs, we look for formulas that differ by only one H^+ ion. For this reaction, one pair must be HBr and Br^- and the other pair must be H_3O^+ and H_2O. In each pair, the acid is the substance with one more hydrogen atom, so the two *acids* are HBr and H_3O^+, and the two *bases* are Br^- and H_2O.

Conjugate pair

$$HBr(aq) + H_2O(l) \rightleftharpoons H_3O^+(aq) + Br^-(aq)$$

Conjugate pair

Answer Double Check:

In a conjugate acid–base pair, the acid always contains one more hydrogen atom than the base; such is the case here. There is also a definite charge relationship between the two members of a conjugate acid–base pair. The acid, with its one more hydrogen, will always have a charge that is one unit greater than the charge associated with the base; such is the case here. For the HBr/Br^- pair, the charges are 0 and –1; for the H_3O^+/H_2O pair, the charges are +1 and 0.

▶ Practice Exercise 14.1   Identify the conjugate acid–base pairs in the following reaction.

$$HClO_2(aq) + H_2O(l) \longrightarrow H_3O^+(aq) + ClO_2^-(aq)$$

Answers to all practice exercises in this chapter are found in the back-of-the-book answer section.

EXAMPLE 14.2 Determining the Formula of One Member of a Conjugate Acid–Base Pair when Given the Other Member

Write chemical formulas for the following:

a. the conjugate base of HCO_3^- **b.** the conjugate acid of PO_4^{3-}

SOLUTION

a. A conjugate base is formed by removing one H^+ ion from a given acid. Removing one H^+ (both the atom and the charge) from HCO_3^- leaves CO_3^{2-}. Thus, CO_3^{2-} is the conjugate base of HCO_3^-.

b. A conjugate acid is formed by adding one H^+ ion to a given base. Adding one H^+ (both the atom and the charge) to PO_4^{3-} produces HPO_4^{2-}. Thus, HPO_4^{2-} is the conjugate acid of PO_4^{3-}.

Answer Double Check:

Are the acid/base interrelationships correct? Yes. The conjugate *base* (CO_3^{2-}) contains one fewer hydrogen and has a charge one unit more negative than its conjugate *acid* (HCO_3^-). The conjugate *acid* (HPO_4^{2-}) contains one more hydrogen and has a charge one unit more positive than its conjugate *base* (PO_4^{3-}).

▶ **Practice Exercise 14.2**   Write chemical formulas for the following:

a. the conjugate acid of CO_3^{2-} **b.** the conjugate base of H_2CO_3

Amphiprotic Substances

The term *amphiprotic* comes from the Greek *amphoteres*, which means "partly one and partly the other." Just as an amphibian is an animal that lives partly on land and partly in the water, an amphiprotic substance is sometimes an acid and sometimes a base.

Some molecules or ions are able to function as either an acid or a base, depending on the kind of substance with which they react. Such molecules are said to be *amphiprotic*. An **amphiprotic substance** *is a substance that can either lose or accept a proton (H^+ ion) and thus can function as either an acid or a base.*

Water is the most common example of an amphiprotic substance. In the first of the following two reactions, water functions as a base, and in the second, it functions as an acid.

$$HNO_3(l) + H_2O(l) \rightleftharpoons H_3O^+(aq) + NO_3^-(aq)$$
$$\text{Acid} \qquad \text{Base}$$

$$NH_3(g) + H_2O(l) \rightleftharpoons NH_4^+(aq) + OH^-(aq)$$
$$\text{Base} \qquad \text{Acid}$$

Another example of an amphiprotic substance is the hydrogen carbonate ion.

$$HCO_3^-(aq) + OH^-(aq) \rightleftharpoons CO_3^{2-}(aq) + H_2O(l)$$
$$\text{Acid} \qquad\quad \text{Base}$$

$$HCO_3^-(aq) + H_3O^+(aq) \rightleftharpoons H_2CO_3(aq) + H_2O(l)$$
$$\text{Base} \qquad\quad \text{Acid}$$

14.4 MONO-, DI-, AND TRIPROTIC ACIDS

Acids can be classified according to the number of hydrogen ions (protons) they can transfer per molecule during an acid–base reaction. A **monoprotic acid** *is an acid that can transfer only one H^+ ion (proton) per molecule during an acid–base reaction.* Hydrochloric acid (HCl) and nitric acid (HNO_3) are both monoprotic acids.

A **diprotic acid** *is an acid that can transfer two H^+ ions (two protons) per molecule during an acid–base reaction.* Sulfuric acid (H_2SO_4) and carbonic acid (H_2CO_3) are examples of diprotic acids. The transfer of protons for a diprotic acid always occurs in steps. For H_2SO_4, the two steps are as follows:

$$H_2SO_4(aq) + H_2O(l) \longrightarrow H_3O^+(aq) + HSO_4^-(aq)$$

$$HSO_4^-(aq) + H_2O(l) \longrightarrow H_3O^+(aq) + SO_4^{2-}(aq)$$

A few *triprotic acids* exist. A **triprotic acid** *is an acid that can transfer three* H^+ *ions (three protons) per molecule during an acid–base reaction.* Phosphoric acid, H_3PO_4, is the most common triprotic acid. The three proton-transfer steps for this acid are as follows:

$$H_3PO_4(aq) + H_2O(l) \longrightarrow H_3O^+(aq) + H_2PO_4^-(aq)$$

$$H_2PO_4^-(aq) + H_2O(l) \longrightarrow H_3O^+(aq) + HPO_4^{2-}(aq)$$

$$HPO_4^{2-}(aq) + H_2O(l) \longrightarrow H_3O^+(aq) + PO_4^{3-}(aq)$$

A **polyprotic acid** *is an acid that can transfer two or more* H^+ *ions (protons) per molecule during an acid–base reaction.* Both diprotic acids and triprotic acids are examples of polyprotic acids.

The number of hydrogen atoms present in one molecule of an acid cannot always be used to classify the acid as mono-, di-, or triprotic. For example, a molecule of acetic acid contains four hydrogen atoms, and yet it is a monoprotic acid. Only one of the hydrogen atoms in acetic acid is acidic. An **acidic hydrogen atom** *is a hydrogen atom in an acid molecule that can be transferred to a base during an acid–base reaction.*

Whether or not a hydrogen atom is acidic is related to its location in a molecule, that is, to which other atom it is bonded. Let us consider our previously mentioned acetic acid example in more detail by looking at the structure of this acid. A *structural equation* for the acidic behavior of acetic acid is

Note the structure of the acetic acid molecule (reactant side of the equation): One hydrogen atom is bonded to an oxygen atom, and the other three hydrogen atoms are each bonded to a carbon atom. It is only the hydrogen atom bonded to the oxygen atom that is acidic. The hydrogen atoms bonded to the carbon atom are too tightly held to be removed by reaction with water molecules. Water has very little effect on a carbon–hydrogen bond because it is essentially nonpolar (Sec. 7.19). On the other hand, the hydrogen bonded to oxygen is involved in a very polar bond because of oxygen's large electronegativity (Sec. 7.18). Water, which is a polar molecule, readily attacks polar bonds but has very little effect on nonpolar bonds.

We now see why the formula for acetic acid is usually written as $HC_2H_3O_2$ rather than as $C_2H_4O_2$. In the situation where some hydrogens are easily removed (acidic) and others are not (nonacidic), it is accepted procedure to write the acidic hydrogens first, separated from the other hydrogens in the formula. Citric acid, the principal acid in citrus fruits, is another example of an acid that contains both acidic and nonacidic hydrogens. Its formula, $H_3C_6H_5O_7$, indicates that three of the eight hydrogen atoms present in a molecule are acidic. Table 14.1 gives the formulas, classifications, and common occurrences of selected mono-, di-, and triprotic acids, many of which contain nonacidic hydrogen atoms.

We have focused our attention on acids in the preceding discussion. It should be noted that similar concepts can be applied to bases. From an Arrhenius standpoint, bases can release more than one hydroxide ion; for example, $Ca(OH)_2$ is a base that produces two OH^- ions per molecule. From a Brønsted–Lowry viewpoint, bases exist that can accept more than one proton, in a stepwise manner; for example, the PO_4^{3-} ion is a Brønsted–Lowry base that can ultimately accept three protons through reaction with three H_3O^+ ions:

$$PO_4^{3-} \xrightarrow{H_3O^+} HPO_4^{2-} \xrightarrow{H_3O^+} H_2PO_4^- \xrightarrow{H_3O^+} H_3PO_4$$

TABLE 14.1 Selected Common Mono-, Di-, and Triprotic Acids

Name	Formula	Classification	Number of Nonacidic Hydrogen Atoms	Common Occurrence
Acetic acid	$HC_2H_3O_2$	monoprotic	three	vinegar
Lactic acid	$HC_3H_5O_3$	monoprotic	five	sour milk, cheese; produced during muscle contraction
Salicylic acid	$HC_7H_5O_3$	monoprotic	five	present in chemically combined form in aspirin
Hydrochloric acid	HCl	monoprotic	zero	constituent of gastric juice; industrial cleaning agent
Nitric acid	HNO_3	monoprotic	zero	used in urinalysis test for protein; used in manufacture of dyes and explosives
Tartaric acid	$H_2C_4H_4O_6$	diprotic	four	grapes
Carbonic acid	H_2CO_3	diprotic	zero	carbonated beverages; produced in the body from carbon dioxide
Sulfuric acid	H_2SO_4	diprotic	zero	storage batteries; manufacture of fertilizer
Citric acid	$H_3C_6H_5O_7$	triprotic	five	citrus fruits
Phosphoric acid	H_3PO_4	triprotic	zero	found in dissociated form (HPO_4^{2-}, $H_2PO_4^{-}$) in intracellular fluid; component of DNA

14.5 STRENGTHS OF ACIDS AND BASES

Brønsted–Lowry acids vary in their ability to transfer protons and produce hydronium ions in aqueous solution. Such acids are classified as strong or weak on the basis of the extent that proton transfer occurs in an aqueous solution. A **strong acid** *is an acid that, in an aqueous solution, transfers 100%, or very nearly 100%, of its acidic hydrogen atoms to water.* Thus, if an acid is strong, almost all of the acid molecules present give up protons to water. This extensive transfer of protons produces many hydronium ions (the acidic species) within the solution. A **weak acid** *is an acid that, in an aqueous solution, transfers only a small percentage of its acidic hydrogen atoms to water.* The extent of proton transfer for weak acids is usually less than 5%. The actual percentage of molecules involved in proton transfer to water depends on the molecular structure of the acid; molecular polarity and the strength and polarity of individual bonds are important factors in determining whether an acid is strong or weak.

A graphical representation of the differences between strong and weak acids, in terms of species present in solution, is given in Figure 14.2. The formula HA represents the acid, and H_3O^+ and A^- are the products from the proton transfer to H_2O.

A 0.1 M solution of nitric acid (HNO_3) or sulfuric acid (H_2SO_4), when spilled on your clothes and not immediately washed off, will eat holes in your clothing. If 0.1 M solutions of either acetic acid ($HC_2H_3O_2$) or carbonic acid (H_2CO_3) were spilled on your clothes, the previously noted corrosive effects would not be observed. Why? All four acid solutions are of equal concentration; all are 0.1 M solutions. The difference in behavior relates to the *strength* of the acids; nitric and sulfuric acids are *strong* acids, whereas acetic and carbonic acids are *weak* acids. The number of H_3O^+ ions (the active species) present in the strong acid solutions is many times greater than for the weak acid solutions even though all the solutions had the same number of acid molecules present (before reaction with water).

It is important not to confuse the terms *strong* and *weak* with the terms *concentrated* and *dilute*. *Strong* and *weak* apply to the *extent of proton transfer*, not to the concentration of acid or base. *Concentrated* and *dilute* are relative concentration terms. Stomach acid (gastric juice) is a dilute (not weak) solution of a strong acid (HCl); it is 5% by mass hydrochloric acid.

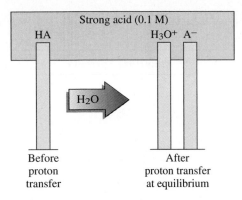

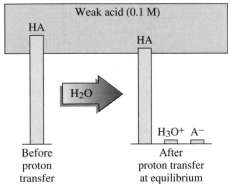

FIGURE 14.2 A comparison of the number of H_3O^+ ions (the acidic species) present in strong acid and weak acid solutions of the same concentration.

There are very few strong acids; the formulas and structures of the seven most commonly encountered strong acids are given in Table 14.2. You should know the identity of these seven strong acids; you will need such knowledge to write net ionic equations, the topic of Section 15.3.

TABLE 14.2 Commonly Encountered Strong Acids

Name*	Molecular Formula	Molecular Structure
Nitric acid	HNO_3	H—O—N—O ‖ O
Sulfuric acid	H_2SO_4	O ‖ H—O—S—O—H ‖ O
Perchloric acid	$HClO_4$	O ‖ H—O—Cl—O ‖ O
Chloric acid	$HClO_3$	H—O—Cl—O ‖ O
Hydrochloric acid	HCl	H—Cl
Hydrobromic acid	HBr	H—Br
Hydroiodic acid	HI	H—I

Nomenclature for acids was discussed in Section 8.6.

TABLE 14.3 Percent Proton-Transfer Values for 1.0 M Solutions (at 25°C) of Selected Weak Acids

Name of Acid	Molecular Formula	Percent Proton Transfer
Phosphoric acid	H_3PO_4	8.3
Nitrous acid	HNO_2	2.7
Hydrofluoric acid	HF	2.5
Acetic acid	$HC_2H_3O_2$	0.42
Carbonic acid	H_2CO_3	0.065
Dihydrogen phosphate ion	$H_2PO_4^-$	0.025
Hydrocyanic acid	HCN	0.0020
Hydrogen carbonate ion	HCO_3^-	0.00075
Hydrogen phosphate ion	HPO_4^{2-}	0.000047

The oxyacids HNO_3, $HClO_4$, $HClO_3$, and H_2SO_4 are strong acids. The oxyacids HNO_2, $HClO_2$, $HClO$, H_2SO_3, H_2CO_3, and H_3PO_4 are weak acids. A generalization exists relative to this situation. For simple oxyacids (H, O, and one other nonmetal), if the number of oxygen atoms present exceeds the number of acid hydrogen atoms present by two or more, the acid strength is strong. For oxyacids where the oxygen-hydrogen difference is less than two, the acid strength is weak.

The vast majority of acids that exist are weak acids. Familiar weak acids include acetic acid ($HC_2H_3O_2$), the acidic component of vinegar, and carbonic acid (H_2CO_3), found in carbonated beverages. Weak acids are not all equally weak; proton transfer occurs to a greater extent for some weak acids than for others. Table 14.3 gives percent proton-transfer values for selected weak acids. The calculational techniques needed to determine percent proton-transfer values, such as those in Table 14.3, will not be considered in this text.

For polyprotic acids, the stepwise proton-transfer sequence that occurs (Sec. 14.4) can be used to determine relative acid strengths for the related acidic species. Consider the two-step proton-transfer process for carbonic acid.

$$H_2CO_3(aq) + H_2O(l) \longrightarrow H_3O^+(aq) + HCO_3^-(aq)$$

$$HCO_3^-(aq) + H_2O(l) \longrightarrow H_3O^+(aq) + CO_3^{2-}(aq)$$

The second proton is not as easily removed as the first because it must be pulled away from a negatively charged particle, HCO_3^-. Accordingly, HCO_3^- is a weaker acid than H_2CO_3, In general, each successive step in a stepwise proton-transfer process occurs to a lesser extent than the previous step. Thus, for triprotic H_3PO_4, the parent H_3PO_4 species (first-step reactant) is a stronger acid than $H_2PO_4^-$ (second-step reactant), which in turn is a stronger acid than HPO_4^{2-} (third-step reactant). This ordering of the phosphoric acid–derived species is reflected in the values given in Table 14.3.

Just as there are strong acids and weak acids, there are also strong bases and weak bases. As with acids, there are only a few strong bases. Strong bases are limited to the hydroxides of groups IA and IIA of the periodic table and are listed in Table 14.4. Of the strong bases, only NaOH and KOH are commonly used in the chemical laboratory. The low solubility of the group IIA hydroxides in water limits their use. However, despite this low solubility, these hydroxides are still considered to be strong bases because whatever dissolves dissociates into ions 100%.

Only one of the many weak bases that exist is fairly common—aqueous ammonia. In this solution of ammonia gas (NH_3) in water, small amounts of OH^- ions are produced through the reaction of NH_3 molecules with water.

$$NH_3(g) + H_2O(l) \longrightarrow NH_4^+(aq) + OH^-(aq)$$

TABLE 14.4 Common Strong Bases

Group IA Hydroxides	Group IIA Hydroxides
LiOH	
NaOH	
KOH	$Ca(OH)_2$
RbOH	$Sr(OH)_2$
CsOH	$Ba(OH)_2$

A solution of ammonia in water is most properly called *aqueous ammonia,* although it is also occasionally called ammonium hydroxide. Aqueous ammonia is the preferred designation, since most of the NH_3 present is in molecular form. Only a very few NH_3 molecules have reacted with the water to give ammonium (NH_4^+) and hydroxide (OH^-) ions.

14.6 SALTS

The title of this chapter is "Acids, Bases, and Salts." In preceding sections, we have discussed acids and bases, but not salts. What is a salt? To a nonscientist, the word *salt* connotes a white granular substance used as a seasoning for food. To the chemist, it has a much broader meaning. Sodium chloride, or table salt, is only one of thousands of salts known to a chemist. "Pass the salt" is a very ambiguous request to a chemist.

From a chemical viewpoint, a **salt** *is an ionic compound containing a metal ion or polyatomic ion as the positive ion and a nonmetal ion or polyatomic ion (except hydroxide) as the negative ion.* (Ionic compounds containing hydroxide ion are bases rather than salts.)

Many salts occur in nature, and numerous others have been prepared in the laboratory. The wide variety of uses found for salts can be seen from Table 14.5, a listing of selected salts and their uses.

TABLE 14.5 Some Common Salts and Their Uses

Name	Formula	Uses
Ammonium nitrate	NH_4NO_3	fertilizer; explosives
Barium sulfate	$BaSO_4$	enhancer for X-rays of gastrointestinal tract
Calcium carbonate	$CaCO_3$	chalk; limestone
Calcium chloride	$CaCl_2$	drying agent for removal of small amounts of water
Iron(II) sulfate	$FeSO_4$	treatment for anemia
Potassium chloride	KCl	salt substitute for low-sodium diets
Sodium chloride	NaCl	table salt; used as a deicer (to melt ice)
Sodium bicarbonate	$NaHCO_3$	ingredient in baking powder
Sodium hypochlorite	NaClO	bleaching agent
Silver bromide	AgBr	light-sensitive material in photographic film
Tin(II) fluoride	SnF_2	toothpaste additive

Much information concerning salts has been presented in previous chapters, although the term *salt* was not explicitly used in those discussions. Formula writing and nomenclature for binary ionic compounds (salts) were covered in Sections 7.7 and 8.3. Many salts, as shown in Table 14.5, contain polyatomic ions such as nitrate and sulfate. Such ions were discussed in Sections 7.9 and 8.4. The solubility of ionic compounds (salts) in water was the topic of Section 13.4.

In solution all common salts are dissociated into ions (Sec. 13.3). Even if a salt is only slightly soluble, the small amount that does dissolve completely dissociates. Thus, the terms *weak* and *strong*, used to denote qualitatively the percent dissociation of acids and bases, are not applicable to common salts. We do not use the terms *strong salt* and *weak salt*.

Acids, bases, and salts are related in that a salt is one of the products resulting from the reaction of an acid with a hydroxide base. This particular type of reaction, called neutralization, is discussed in Section 14.7.

14.7 REACTIONS OF ACIDS

All acids have some unique properties that adapt them for use in specific situations. In addition, all acids have certain chemical properties in common, properties related to the presence of H_3O^+ ions in aqueous solution. In this section we consider three types of chemical reactions that acids characteristically undergo.

1. Acids react with active metals to produce hydrogen gas and a salt.
2. Acids react with hydroxide bases to produce a salt and water.
3. Acids react with carbonates and bicarbonates to produce carbon dioxide, a salt, and water.

Reaction with Metals

Acids react with many, but not all, metals. When they do react, the metal dissolves and hydrogen gas (H_2) is liberated. In the reaction, the metal atoms lose electrons and become, metal ions. The lost electrons are taken up by the hydrogen ions (protons) of the acid; the hydrogen ions become electrically neutral, combine into molecules, and emerge from the reaction mixture as hydrogen gas. Illustrative of the reaction of an acid and a metal is the reaction between zinc and sulfuric acid.

$$Zn + H_2SO_4 \longrightarrow ZnSO_4 + H_2$$

In terms of the reaction types discussed in Section 10.6, the reaction of an acid with a metal to produce hydrogen gas is a *single-replacement reaction;* the metal replaces the hydrogen from the acid. Recall, from Section 10.6, that a single-replacement reaction has the general form

$$X + YZ \longrightarrow Y + XZ$$

Metals can be arranged in a reactivity order based on their ability to react with acids. Such an ordering for the more common metals is given in Table 14.6. Any metal above hydrogen in the activity series will dissolve in an acid solution and form H_2. The closer a metal is to the top of the series, the more rapid the reaction. Those metals below hydrogen in the series do not dissolve in an acid to form H_2.

As noted in Table 14.6, the most active metals (those nearest the top in the activity series) also react with water. Again, hydrogen gas is produced. In the cases of potassium and sodium, the reaction is sometimes violent enough to cause explosions, as the result of H_2 ignition. The equation for the reaction of potassium with water, which is also a single-replacement reaction, is

$$2\,K + 2\,H_2O \longrightarrow 2\,KOH + H_2$$

TABLE 14.6   Activity Series for Common Metals

		Metal	Symbol	Remarks
↑ Increasing tendency to react	React with H⁺ ions to liberate hydrogen gas	Potassium	K }	react violently with cold water
		Sodium	Na }	
		Calcium	Ca	reacts slowly with cold water
		Magnesium	Mg ⎫	
		Aluminum	Al ⎪	react slowly with hot water (steam)
		Zinc	Zn ⎬	
		Chromium	Cr ⎭	
		Iron	Fe	
		Nickel	Ni	
		Tin	Sn	
		Lead	Pb	
		Hydrogen	H	
	Do not react with H⁺ ions	Copper	Cu	
		Mercury	Hg	
		Silver	Ag	
		Platinum	Pt	
		Gold	Au	

Because they do not react with the acidic components of skin secretions (sweat, etc.), gold, platinum, and silver are good metals for jewelry. Jewelry made with these metals will not tarnish like jewelry made from other metals.

Note that the resulting solution is basic when a metal reacts with water; hydroxide ions are produced.

Reaction with Bases

When Arrhenius acids and bases are mixed, they react with each other; their acidic and basic properties disappear, and we say that they have *neutralized* each other. **Neutralization** *is the reaction between an acid and a hydroxide base to form a salt and water.* The hydrogen ions from the acid combine with the hydroxide ions from the base to form water. The salt formed contains the negative ion from the acid and the positive ion from the base. Neutralization is a *double-replacement* reaction (Sec. 10.6).

$$AX + BY \longrightarrow AY + BX$$

$$HCl + KOH \longrightarrow HOH + KCl$$

$$\text{acid} \quad \text{base} \qquad \text{water} \quad \text{salt}$$

Any time an acid is completely reacted with a base, neutralization occurs. It does not matter whether the acid and base are strong or weak. Sodium hydroxide (a strong base) and nitric acid (a strong acid) react as follows.

$$HNO_3 + NaOH \longrightarrow NaNO_3 + H_2O$$

The equation for the reaction of potassium hydroxide (a strong base) with hydrocyanic acid (a weak acid) is

$$HCN + KOH \longrightarrow KCN + H_2O$$

Note that in each case the products are a salt ($NaNO_3$ in the first reaction, KCN in the second) and water.

In any acid–base neutralization reaction, the amounts of H^+ ion and OH^- ion that react are equal. These two ions always react in a one-to-one ratio to form water.

$$H^+ + OH^- \longrightarrow H_2O \text{ (HOH)}$$

Hydrochloric acid (HCl), which is necessary for proper digestion of food, is present in the gastric juices of the human stomach. Overeating and emotional factors can cause the stomach to produce too much hydrochloric acid, a condition called acid indigestion or heartburn. Substances known as antacids provide symptomatic relief from this condition. Over-the-counter antacids such as Maalox, Tums, and Alka-Seltzer contain one or more basic substances [often $Mg(OH)_2$] that are capable of neutralizing the excess hydrochloric acid present. The neutralization reaction that occurs when stomach acid reacts with a typical antacid is $2\,HCl + Mg(OH)_2 \longrightarrow MgCl_2 + 2\,H_2O$

This constant reaction ratio between the two ions enables us to balance chemical equations for neutralization reactions quickly.

Let us consider the neutralization reaction between H_3PO_4 and KOH to give a salt and water.

$$H_3PO_4 + KOH \longrightarrow salt + H_2O$$

Because H_3PO_4 is triprotic and the base KOH contains only one OH^- ion, we will need three times as many base molecules as acid molecules. Thus, we place the coefficient 3 in front of the formula for KOH in the chemical equations; this gives three H^+ reacting with three OH^- to produce three H_2O molecules.

$$H_3PO_4 + 3\,KOH \longrightarrow salt + 3\,H_2O$$

The salt formed is K_3PO_4; there are three K^+ ions and one PO_4^{3-} ion on the left side of the equation, which combine to give the salt. The balanced equation for the neutralization is thus

$$H_3PO_4 + 3\,KOH \longrightarrow K_3PO_4 + 3\,H_2O$$

Reaction with Carbonates and Bicarbonates

The reaction equation for the "volcanoes" that children enjoy making by mixing vinegar [5% acetic acid (v/v)] and baking soda is $HC_2H_3O_2 + NaHCO_3 \longrightarrow CO_2 + H_2O + NaC_2H_3O_2$ The carbon dioxide gas makes the foamy volcano.

Carbon dioxide gas (CO_2), water, and a salt are always the products of the reaction of acids with carbonates or bicarbonates, as illustrated by the following equations.

$$2\,HCl + Na_2CO_3 \longrightarrow 2\,NaCl + CO_2 + H_2O$$
$$HCl + NaHCO_3 \longrightarrow NaCl + CO_2 + H_2O$$

Baking powder is a mixture of a bicarbonate and an acid-forming solid. The addition of water to this mixture generates the acid that then reacts with the bicarbonate to release carbon dioxide into the batter. It is the generated carbon dioxide that causes the batter to rise. Baking soda is pure $NaHCO_3$. To cause it to release carbon dioxide, an acid-containing substance, such as buttermilk, sour milk, or fruit juice, must be added to it.

14.8 REACTIONS OF BASES

The most important characteristic reaction of bases is their reaction with acids (neutralization), discussed in the preceding section. Another characteristic reaction, that of bases with certain salts, is discussed in Section 14.9.

Bases react with fats and oils and convert them into smaller, soluble molecules. For this reason most household cleaning products contain basic substances. Lye (impure NaOH) is an active ingredient in numerous drain cleaners. Also, many advertisements for liquid household cleaners emphasize the fact that aqueous ammonia (a weak base) is present in the product.

In Section 14.1 we noted that one of the general properties of bases is a slippery or soapy feeling to the touch. The bases themselves are not slippery; the slipperiness results as the bases react with fats and oils in the skin to form slippery or soapy compounds.

14.9 REACTIONS OF SALTS

Dissolved salts will react with metals, acids, bases, and other salts under specific conditions.

1. Salts react with some metals to convert the metallic ion of the salt to free metal and the free metal to its salt.
2. Salts react with some acid solutions to form other acids and salts.

3. Salts react with some base solutions to form other bases and salts.

4. Salts react with some solutions of other salts to form new salts.

The tendency for salts to react with metals is related to the relative positions of the two involved metals in the activity series (Table 14.6). For salts to react with acids, bases, or other salts, one of the reaction products must be (1) an insoluble salt, (2) a gas that is evolved from the solution, or (3) an undissociated soluble species, such as a weak acid or a weak base. The formation of any of these products serves as the driving force to cause the reaction to occur.

Reaction with Metals

If an iron nail is placed in a solution of copper sulfate ($CuSO_4$), metallic copper will be deposited on the nail and some of the iron will dissolve.

$$Fe(s) + CuSO_4(aq) \longrightarrow Cu(s) + FeSO_4(aq)$$

One metal has replaced the other: A single-replacement reaction (Sec. 10.6) has occurred. This type of reaction will occur only if the metal going into solution is above the replaced metal in the activity series. Iron is above copper in the activity series and can replace it. If a strip of copper were placed in a solution of $FeSO_4$—just the opposite situation to what we have been discussing—no reaction would occur because copper is below iron in the activity series.

Reaction with Acids

For a salt to react with an acid, a new weaker acid, a new insoluble salt, or a gaseous compound must be one of the products.

An example of a reaction in which the formation of an *insoluble salt* is the driving force for the reaction to occur is

$$AgNO_3(aq) + HCl(aq) \longrightarrow AgCl(s) + HNO_3(aq)$$

This is a double-replacement reaction (Sec. 10.6); the silver and hydrogen have traded partners.

The conclusion that this reaction will occur comes from a consideration of the possible recombinations of the reacting species. In a solution made by mixing silver nitrate ($AgNO_3$) and hydrochloric acid (HCl), four kinds of ions are present initially (before any reaction occurs): Ag^+ and NO_3^- (since $AgNO_3$ is a soluble salt) and H^+ and Cl^- (since HCl is a strong acid). The question is whether these ions can get together in new appropriate combinations. The possible new combinations of oppositely charged ions are $H^+NO_3^-$ and Ag^+Cl^-. The first of these combinations would result in the formation of the strong acid HNO_3. Strong acids in solution exist in dissociated form; therefore, these ions will not combine. The second combination does occur because AgCl is an insoluble salt. Thus, the overall reaction takes place as a result of the formation of this insoluble salt. The net result of the reaction is that the original ions exchange partners.

The double-replacement reaction of sodium fluoride (a soluble salt) with hydrochloric acid (a strong acid) illustrates the case where formation of a *new weaker acid* is the driving force for the reaction.

$$HCl(aq) + NaF(aq) \longrightarrow NaCl(aq) + HF(aq)$$

Using an analysis pattern similar to that in the previous example, we find that four types of ions are present initially: Na^+ and F^- (from the soluble salt) and H^+ and Cl^- (from the strong acid). Possible new combinations are Na^+Cl^- and H^+F^-. Sodium chloride, the result of the first combination, will not form because this salt is soluble. The combination of H^+ ion with F^- ion does occur because it yields the weak acid HF. In solution, weak acids exist predominantly in molecular form. In all reactions of this general type, the acid formed in the reaction must be weaker than the reactant acid. If the reactant acid is strong, as in this example, such a determination is obvious. If both the reactant and product acids are weak, information such as that given in Table 14.3 would be needed to predict which of the two acids is the weaker.

Note that a reaction does not always occur when acid and salt solutions are mixed. Consider the possible reaction of NaCl and HNO_3 solutions. Initially, four types of ions are present: Na^+ and Cl^- (from the soluble salt) and H^+ and NO_3^- (from the strong acid). The new combinations, if a reaction did occur, would be $Na^+NO_3^-$ (a soluble salt) and H^+Cl^- (a strong acid). Since both the products would exist in dissociated form in solution, no recombination of ions occurs; hence, no reaction occurs.

The most common type of reaction in which the driving force is the *evolution of a gas* involves a carbonate or bicarbonate salt. This type of reaction was discussed in Section 14.7.

Reaction with Bases

The criteria for the reaction of bases with salts are similar to those for acid–salt reactions, except that weaker base formation replaces weaker acid formation as one of the three driving forces. An example of a base–salt reaction involving the formation of an *insoluble salt* is

$$Ba(OH)_2(aq) + Na_2SO_4(aq) \longrightarrow BaSO_4(s) + 2\,NaOH(aq)$$

The most common situation in which *gas evolution* is the driving force for base–salt reactions is where ammonium salts are involved. In such cases, ammonia gas is given off, as illustrated by the reaction of NH_4Cl and KOH.

$$NH_4Cl(aq) + KOH(aq) \longrightarrow KCl(aq) + NH_3(g) + H_2O(l)$$

Reaction of Salts with Each Other

Two different salt solutions will react when mixed, in a double-replacement reaction, only if an *insoluble salt* can be formed. Consider the following possible reactions:

$$AgNO_3(aq) + NaCl(aq) \longrightarrow AgCl(s) + NaNO_3(aq)$$
$$KNO_3(aq) + NaCl(aq) \longrightarrow KCl(aq) + NaNO_3(aq)$$

The first reaction occurs because AgCl is an insoluble salt. The second reaction does not occur since both of the possible products are soluble salts, which means there is no driving force for the reaction.

EXAMPLE 14.3 | **Predicting Whether a Reaction Will Occur and Writing an Equation for the Reaction If It Does Occur**

Write a chemical equation for the chemical reaction that occurs, if any, when 0.1 M solutions of the following substances are mixed.

a. $Fe(NO_3)_2$ and K_2S
b. $CaCl_2$ and H_2SO_4
c. HNO_3 and $NaC_2H_3O_2$

SOLUTION

a. Both the reactants are soluble salts. Two different salt solutions will react when mixed, only if an insoluble salt can be formed.

In a solution made by mixing $Fe(NO_3)_2$ and K_2S, four kinds of ions are present initially (before any reaction occurs): Fe^{2+} and NO_3^- [from the $Fe(NO_3)_2$] and K^+ and S^{2-} (from the K_2S). The possible new combinations of oppositely charged ions are Fe^{2+} with S^{2-} and K^+ with NO_3^-.

Original ion combinations   *Possible new combinations*

   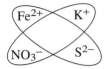

The first one of these new combinations, the formation of FeS, is the one that will be the driving force for the reaction to occur. FeS is an insoluble salt. The second new combination, the formation of KNO_3 does not occur because KNO_3 is a soluble salt and soluble salts exist in dissociated form in solution. The equation for the reaction is

$$Fe(NO_3)_2 + K_2S \longrightarrow FeS + 2 KNO_3$$

b. One of the reactants, $CaCl_2$, is a soluble salt, and the other reactant, H_2SO_4, is a strong acid. Both reactants exist in solution in dissociated form; thus, four types of ions are present in the mixed solution (before any reaction occurs): Ca^{2+}, Cl^-, H^+, and SO_4^{2-}. The conclusion that a reaction will occur comes from a consideration of the possible new combinations of the reacting species.

Original ion combinations   *Possible new combinations*

   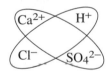

The first one of these new combinations, the formation of $CaSO_4$, is the driving force for the reaction to occur; $CaSO_4$ is an insoluble salt. The other new combination, the formation of HCl, does not occur because HCl is a strong acid and will exist in solution in dissociated form. The equation for the reaction is

$$CaCl_2 + H_2SO_4 \longrightarrow CaSO_4 + 2 HCl$$

c. The reactants are a strong acid (HNO_3) and a soluble salt ($NaC_2H_3O_2$). Both are dissociated in solution; hence, H^+, NO_3^-, Na^+, and $C_2H_3O_2^-$ ions are present in the reaction mixture (before any reaction occurs). Possible new combinations of the reacting species are

Original ion combinations   *Possible new combinations*

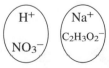   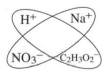

Weak acid formation is the driving force for the reaction; acetic acid ($HC_2H_3O_2$) forms from the combination of H^+ and $C_2H_3O_2^-$ ions. The Na^+ and NO_3^- will not combine because a soluble salt would be the product. The equation for the reaction is

$$HNO_3 + NaC_2H_3O_2 \longrightarrow HC_2H_3O_2 + NaNO_3$$

Answer Double Check:

Are the answers reasonable in terms of at least one product being a weak acid, weak base, insoluble salt, or a gas? Yes. In part (a) the product is FeS; all sulfides are insoluble except for groups IA and IIA and NH_4^+ (Table 13.4). The product in part (b) is $CaSO_4$; the insoluble sulfates are those of calcium, strontium, barium, and lead (Table 13.4). The part (c) product is $HC_2H_3O_2$ (acetic acid), which is a weak acid.

▶ **Practice Exercise 14.3** Write a chemical equation for the reaction that occurs, if any, when 0.1 M solutions of the following substances are mixed.

  a. $FeSO_4$ and NaCl   **b.** $Al(NO_3)_3$ and K_2S **c.** H_3PO_4 and Na_2CO_3

14.10 SELF-IONIZATION OF WATER

Although we usually think of water as a molecular (covalent) substance, experiments show that a *small* percentage of water molecules in pure water interact with one another to form ions, a process that is called *self-ionization.* This interaction can be thought of as a Brønsted–Lowry acid–base reaction (Sec. 14.2) involving the transfer of a proton from one water molecule to another (see Figure 14.3):

$$H_2O + H_2O \longrightarrow H_3O^+ + OH^-$$

or simply as the formation of ions from a single water molecule (Arrhenius theory; Sec. 14.1):

$$H_2O\ (HOH) \longrightarrow H^+ + OH^-$$

The amount of H^+ ion present in water $(1.00 \times 10^{-7}\,M)$ is a very small amount. If one molecule (or ion) was removed from a liter of water, examined, and then returned every second for an extended period of time, how often would an H^+ ion be encountered? The answer is once every 17.4 years.

From either viewpoint, the net result is the formation of *equal amounts* of hydronium (hydrogen) ion and hydroxide ion.

    The dissociation of water molecules is part of an equilibrium situation. Individual water molecules are continually dissociating. This process is balanced by hydroxide and hydronium ions recombining to form water at the same rate. At equilibrium, at 24°C, the H_3O^+ and OH^- ion concentrations are each $1.00 \times 10^{-7}\,M$ (0.000000100 M). This very very small concentration is equivalent to there being one H_3O^+ and one OH^- ion present for every 550,000,000 undissociated water molecules. Even though the H_3O^+ and OH^- ion concentrations are very minute, they are important, as we shall shortly see.

FIGURE 14.3
Brønsted–Lowry acid–base reaction between two water molecules to produce hydronium ion and hydroxide ion.

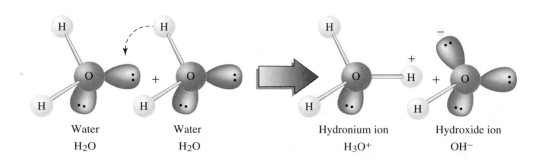

| Water | Water | Hydronium ion | Hydroxide ion |
| H_2O | H_2O | H_3O^+ | OH^- |

Ion Product Constant for Water

Experimentally, it is found that at any given temperature, the product of the concentrations of H_3O^+ ion and OH^- ion in water is a constant. We can calculate the value of this constant at 24°C because we know that the concentration of each ion is 1.00×10^{-7} M at this temperature.

$$[H_3O^+] \times [OH^-] = \text{constant}$$
$$(1.00 \times 10^{-7}) \times (1.00 \times 10^{-7}) = 1.00 \times 10^{-14}$$

Ion product constant for water *is the numerical value* (1.00×10^{-14}) *associated with the product of the* H_3O^+ *ion and* OH^- *ion molar concentrations in water at 24°C.* Note that the ion concentrations must be expressed in moles per liter (M) in order to obtain the value 1.00×10^{-14} for the ion product for water. The general expression for the ion product constant for water is

$$[H_3O^+] \times [OH^-] = \text{ion product for water} = 1.00 \times 10^{-14}$$

The brackets [ ] specifically denote ion concentration in moles per liter. The symbol K_w, where K stands for a constant and w stands for water, is used as a designation for the value 1.00×10^{-14}.

Effect of Solutes on Water Self-Ionization

The water self-ionization process occurs not only in pure water but also in *any* aqueous solution. The solute(s) present in an aqueous solution can cause the $[H_3O^+]$ and $[OH^-]$ values to not be equal. However, whatever they are, their product will still be equal to K_w. Solutes do not alter the value of K_w.

If the $[H_3O^+]$ is increased by addition of an acidic solute, the $[OH^-]$ must decrease until the expression

$$[H_3O^+] \times [OH^-] = 1.00 \times 10^{-14}$$

is satisfied. Similarly, if OH^- ions are added to the water, the $[H_3O^+]$ must correspondingly decrease. The extent of the decrease in $[H_3O^+]$ or $[OH^-]$ as the result of the addition of a quantity of the other ion, is easily calculated using the ion product expression.

If we know the concentration of the other ion, we can calculate the concentration of either $[H_3O^+]$ or $[OH^-]$ present in an aqueous solution by simply rearranging the ion product expression.

$$[H_3O^+] = \frac{1.00 \times 10^{-14}}{[OH^-]} \quad \text{or} \quad [OH^-] = \frac{1.00 \times 10^{-14}}{[H_3O^+]}$$

EXAMPLE 14.4   **Calculating the Hydroxide Ion Concentration of a Solution with a Known Hydronium Ion Concentration**

Sufficient acidic solute is added to a quantity of water to produce $[H_3O^+] = 7.50 \times 10^{-5}$. What is the $[OH^-]$ in this solution?

SOLUTION

The $[OH^-]$ can be calculated using the ion product constant expression for water. Solving this expression for $[OH^-]$ gives

$$[OH^-] = \frac{1.00 \times 10^{-14}}{[H_3O^+]}$$

Substituting into this expression the known $[H_3O^+]$ and doing the arithmetic gives

$$[OH^-] = \frac{1.00 \times 10^{-14}}{7.50 \times 10^{-5}} = 1.3333333 \times 10^{-10} \quad \text{(calculator answer)}$$

$$= 1.33 \times 10^{-10} \quad \text{(correct answer)}$$

Answer Double Check:

The product of the $[OH^-]$ and $[H_3O^+]$ for any aqueous solution should be 1.00×10^{-14} M. Is the calculated answer consistent with this generalization? Yes. The calculated $[OH^-]$, 1.33×10^{-10} M, times the given $[H_3O^+]$, 7.50×10^{-15} M, is equal to 1.00×10^{-14} M, when rounding errors are taken into account.

▶ **Practice Exercise 14.4**   Sufficient acidic solute is added to a quantity of water to produce $[H_3O^+] = 3.33 \times 10^{-3}$. What is the $[OH^-]$ in this solution?

Neither $[H_3O^+]$ nor $[OH^-]$ is ever zero in an aqueous solution.

The relationship between $[H_3O^+]$ and $[OH^-]$ is that of an inverse proportion; when one increases, the other decreases. If $[H_3O^+]$ increases by a factor of 10^2, the $[OH^-]$ decreases by the same factor, 10^2. A graphical portrayal of this increase–decrease relationship for $[H_3O^+]$ and $[OH^-]$ is given in Figure 14.4.

Effect to Temperature on Water Self-Ionization

Although the value of K_w does not change when a solute is added to water, its value is temperature dependent. The K_w value 1.0×10^{-14} used in the preceding discussion is that for a temperature of 24°C. As temperature increases, K_w values increase slightly, and conversely K_w values decrease slightly with temperature decrease. At 37°C, average

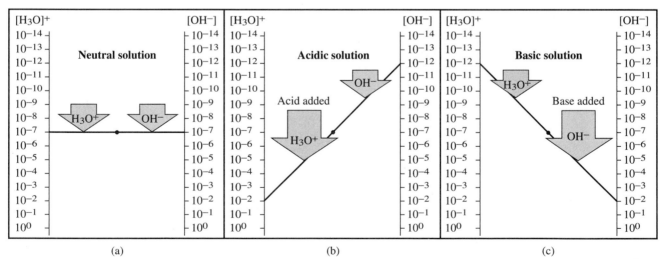

(a)
In pure water the concentration of hydronium ions, $[H_3O^+]$, and that of hydroxide ions, $[OH^-]$, are equal. Both are 1.00×10^{-7} M at 24°C.

(b)
If $[H_3O^+]$ is increased by a factor of 10^5 (from 10^{-7} M to 10^{-2} M), then $[OH^-]$ is decreased by a factor of 10^5 (from 10^{-7} M to 10^{-12} M).

(c)
If $[OH^-]$ is increased by a factor of 10^5 (from 10^{-7} M to 10^{-2} M), then $[H_3O^+]$ is decreased by a factor of 10^5 (from 10^{-7} M to 10^{-12} M).

FIGURE 14.4 The relationship between $[H_3O^+]$ and $[OH^-]$ in an aqueous solution is an inverse proportion; when $[H_3O^+]$ is increased, $[OH^-]$ decreases, and vice versa.

human body temperature, the value of K_w is of 2.6×10^{-14}. The $[H_3O^+]$ and $[OH^-]$ in pure water at this temperature are 1.6×10^{-7} M rather than 1.0×10^{-7} M (24°C).

$$\underset{[H_3O^+]}{[1.6 \times 10^{-7}]} \times \underset{[OH^-]}{[1.6 \times 10^{-7}]} = \underset{K_w}{2.6 \times 10^{-14}}$$

Unless otherwise indicated, we will always assume a temperature of 24°C in problem-solving situations.

Acidic, Basic, and Neutral Solutions

Small amounts of both H_3O^+ ion and OH^- ion are present in all aqueous solutions. What, then, determines whether a given solution is acidic or basic? It is the relative amounts of these two ions present. An **acidic solution** *is a solution in which the concentration of H_3O^+ ion is greater than that of OH^- ion.* Acids are substances that, when added to water, increase the concentration of H_3O^+ ion. A **basic solution** *is a solution in which the concentration of OH^- ion is greater than that of H_3O^+ ion.* Bases are substances that, when added to water, increase the concentration of OH^- ion. A **neutral solution** *is a solution in which the concentrations of H_3O^+ and OH^- ions are equal.* Figure 14.5 summarizes the relationships between $[H_3O^+]$ and $[OH^-]$ that we have just considered.

Acidic solutions contain both H_3O^+ and OH^- ions. Basic solutions contain both H_3O^+ and OH^- ions. The acidic or basic classification is determined by which ion is present in the greater amount. In acidic solutions, the concentration of H_3O^+ ion is greater than the concentration of OH^- ion. In a basic solution the opposite is true.

14.11 THE PH SCALE

Hydronium ion concentrations in aqueous solutions range from relatively high values (10 M) to extremely small ones (10^{-14} M). It is inconvenient to work with numbers that extend over such a wide range; a hydronium ion concentration of 10 M is 1000 trillion times larger than a hydronium ion concentration of 10^{-14} M. The *pH scale*, proposed by the Danish chemist Sören Peter Lauritz Sörensen (1868–1939) in 1909, is a more practical way to handle such a wide range of numbers. The **pH scale** *is a scale of small numbers that is used to specify molar hydronium ion concentration in an aqueous solution.*

The calculation of pH scale values involves the use of logarithms. The **pH** *of a solution is the negative logarithm of the solution's molar hydronium ion concentration.* Expressed mathematically, the definition of pH is

$$pH = -\log [H_3O^+]$$

Logarithms are simply exponents. The *common logarithm,* abbreviated *log,* which is the type of logarithm used in the definition of pH, is based on powers of 10. *For a number expressed in scientific notation that has a coefficient of 1, the log of that number is the value of the exponent.* For instance, the log of 1×10^{-8} is –8.0, and the log of 1×10^6 is 6.0. Table 14.7

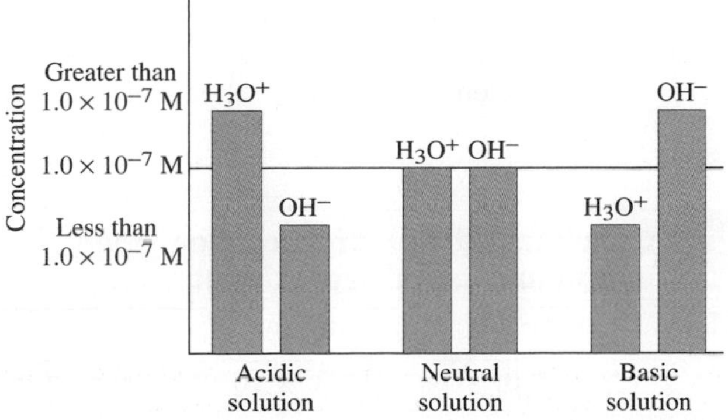

FIGURE 14.5 Relative molar concentrations of H_3O^+ and OH^- ions at 24°C in acidic, neutral, and basic solutions.

TABLE 14.7 Logarithm Values for Selected Numbers

Number	Number Expressed as a Power of 10	Common Logarithm
10,000	1×10^4	4.0
1,000	1×10^3	3.0
100	1×10^2	2.0
10	1×10^1	1.0
1	1×10^0	0.0
0.1	1×10^{-1}	−1.0
0.01	1×10^{-2}	−2.0
0.001	1×10^{-3}	−3.0
0.0001	1×10^{-4}	−4.0

gives more examples of the relationship between powers of 10 and logarithmic values for numbers in scientific notation whose coefficients are 1. Note from this table that log values may be either positive or negative depending on the sign of the exponent. (How we determine significant figures in logarithmic calculations is discussed later in this section.)

Integral pH Values

It is easy to calculate the pH value for a solution when the molar hydronium ion concentration is simply a power of 10, for example, 1×10^{-4}. In this situation the pH is given directly by the negative of the exponent value on the power of 10.

$$[H_3O^+] = 1 \times 10^{-x}$$
$$pH = x$$

Thus, if the hydronium ion concentration is 1×10^{-9} the pH will be 9.0.

This simple relationship between pH and power of 10 is obtained from the formal definition of pH as follows.

$$pH = -\log[H_3O^+]$$
$$= -\log[1 \times 10^{-x}]$$
$$= -(-x)$$
$$= x$$

The *p* in pH comes from the German word *potenz*, which means "power," as in "power of 10."

Again, it should be noted that this simple relationship is valid only when the coefficient in the exponential expression for the hydronium ion concentration is 1. How the pH is calculated when the coefficient is a number other than one will be covered later in this section.

EXAMPLE 14.5 Calculating the pH of a Solution when Given Its Hydronium Ion or Hydroxide Ion Concentration

Calculate the pH for each of the following solutions.

a. $[H_3O^+] = 1 \times 10^{-3}$　　　**b.** $[H_3O^+] = 1 \times 10^{-9}$　　　**c.** $[OH^-] = 1 \times 10^{-4}$

SOLUTION

a. Let us use the formal definition of pH in obtaining this first pH value.

$$pH = -\log[H_3O^+]$$

This expression indicates that to obtain a pH, we must first take the logarithm of the molar hydronium ion concentration and then change the sign of that logarithm.

The logarithm of 1×10^{-3} is –3.0. Thus, we have

$$pH = -\log(1 \times 10^{-3})$$
$$= -(-3.0)$$
$$= 3.0$$

b. Let us use the shorter, more direct way for obtaining pH this time, the method based on the relationship

$$[H_3O^+] = 1 \times 10^{-x}$$
$$pH = x$$

Because the power of 10 is –9 in this case, the pH will be 9.0.

c. The given quantity involves hydroxide ion rather than hydronium ion. Thus, we must first calculate the hydronium ion concentration, and then the pH.

$$[H_3O^+] = \frac{1.00 \times 10^{-14}}{[OH^-]} = \frac{1.00 \times 10^{-14}}{1 \times 10^{-4}}$$
$$= 1 \times 10^{-10} \quad \text{(calculator and **correct answer**)}$$

A solution with a hydronium ion concentration of 1×10^{-10} M will have a pH of 10.0.

> **Practice Exercise 14.5**   Calculate the pH for each of the following solutions.
>
> a. $[H_3O^+] = 1 \times 10^{-5}$ b. $[H_3O^+] = 1 \times 10^{-10}$ c. $[OH^-] = 1 \times 10^{-8}$

A formal discussion of significant figure rules as they apply to logarithms is presented later in this section.

Because pH is simply another way of expressing hydronium ion concentration, acidic, neutral, and basic solutions can be identified by their pH values. At 24°C, a neutral solution ($[H_3O^+ = 1.0 \times 10^{-7})$ has a pH of 7.00. At 24°C, values of pH less than 7.00 correspond to acidic solutions. The lower the pH value, the greater the acidity. At 24°C, values of pH greater than 7.00 represent basic solutions. The higher the pH value, the greater the basicity. The relationships between $[H_3O^+]$, $[OH^-]$, and pH are summarized in Table 14.8. Note that a change of *one* unit in pH corresponds to a *tenfold* increase or decrease in $[H_3O^+]$. Also note that *lowering* the pH always corresponds to *increasing* the H_3O^+ ion concentration.

Table 14.9 lists the pH values of a number of common substances. Except for gastric juice, most human body fluids have pH values within a couple of units of neutrality. Almost all foods are acidic. Tart taste is associated with foods having a low pH.

The pH of a neutral solution at 37°C is 6.80 rather than the 7.00 observed at 24°C. This is the result of K_w values differing at the two temperatures (Sec. 14.11). Both solutions are, however, neutral solutions because in both solutions the hydronium and hydroxide ion concentrations are equal.

Nonintegral pH Values

Note that some of the pH values in Table 14.9 are nonintegral, that is, not whole numbers. Nonintegral pH values result from molar hydronium ion concentrations where the coefficient in the exponential expression for concentration has a value

TABLE 14.8 The pH Scale

pH	$[H_3O^+]$	$[OH^-]$	
0.0	10^0	10^{-14}	
1.0	10^{-1}	10^{-13}	
2.0	10^{-2}	10^{-12}	
3.0	10^{-3}	10^{-11}	Acidic
4.0	10^{-4}	10^{-10}	
5.0	10^{-5}	10^{-9}	
6.0	10^{-6}	10^{-8}	
7.0	10^{-7}	10^{-7}	Neutral
8.0	10^{-8}	10^{-6}	
9.0	10^{-9}	10^{-5}	
10.0	10^{-10}	10^{-4}	
11.0	10^{-11}	10^{-3}	Basic
12.0	10^{-12}	10^{-2}	
13.0	10^{-13}	10^{-1}	
14.0	10^{-14}	10^0	

TABLE 14.9 Approximate pH Values of Some Common Substances

	pH	
	0	1 M HCl (0.0)
	1	0.1 M HCl (1.0), gastric juice (1.6–1.8), lime juice (1.8–2.0)
	2	soft drinks (2.0–4.0), vinegar (2.4–3.4)
Acidic	3	grapefruit (3.0–3.3), peaches (3.4–3.6)
	4	tomatoes (4.0–4.4), human urine (4.8–8.4)
	5	carrots (4.9–5.3), peas (5.8–6.4)
	6	human saliva (6.2–7.4), cow's milk (6.3–6.6), drinking water (6.5–8.0)
Neutral	7	pure water (7.0), human blood (7.35–7.45), fresh eggs (7.6–8.0)
	8	seawater (8.3), soaps, shampoos (8.0–9.0)
	9	detergents (9.0–10.0)
	10	milk of magnesia (9.9–10.1)
Basic	11	household ammonia (11.5–12.0)
	12	liquid bleach (12.0)
	13	0.1 M NaOH (13.0)
	14	1 M NaOH (14.0)

other than 1. For example, consider the following matchups between hydronium ion concentration and pH:

$$[H_3O^+] = 6.3 \times 10^{-5} \qquad pH = 4.20$$

$$[H_3O^+] = 4.0 \times 10^{-5} \qquad pH = 4.40$$

$$[H_3O^+] = 2.0 \times 10^{-5} \qquad pH = 4.70$$

Obtaining nonintegral pH values like these from hydronium ion concentrations requires an electronic calculator that allows for the input of exponential numbers and has a base 10 logarithm key (log).

In using an electronic calculator, depending on the model you have, you can obtain logarithm values simply by pressing the log key after having entered the number whose log value is desired, or vice versa. For pH, you must remember that after obtaining the log value, you must change its sign because of the negative sign in the defining equation for pH.

Significant figure considerations for log values involve a concept not previously encountered. It can best be illustrated by considering some actual log values. Consider the following five related numbers and their log values.

Number	Logarithm
2.43×10^0	0.38560627
2.43×10^2	2.38560627
2.43×10^4	4.38560627
2.43×10^7	7.38560627
2.43×10^{11}	11.38560627

This tabulation shows that

1. The number to the left of the decimal point in each logarithm (called the *characteristic*) is related only to the exponent of 10 in the number whose logarithm was taken.
2. The number to the right of the decimal point in each logarithm (called the *mantissa*) is related only to the coefficient in the exponential notation form of the number. Because the coefficient is 2.43 in each case, the mantissas are all the same (0.38560627).

Combining generalizations (1) and (2) gives us the significant figure rule for logarithms. *In a logarithm, the digits to the left of the decimal point are not counted as significant figures.* These digits relate to the placement of the decimal point in the number. From our previous significant figure work (Sec. 2.5), they are somewhat analogous to the leading zeros (which are not significant) in a number such as 0.0000243.

Thus, the *coefficient* of the number whose logarithm has been taken and the *mantissa* of the logarithm should have the same number of digits. Rewriting the previous tabulation of logarithms to the correct number of significant figures gives

$$\begin{array}{ll} 2.43 \times 10^0 & 0.386 \\ 2.43 \times 10^2 & 2.386 \\ 2.43 \times 10^4 & 4.386 \\ 2.43 \times 10^7 & 7.386 \\ \underbrace{2.43} \times 10^{11} & 11.\underbrace{386} \\ \text{3 digits} \qquad \text{3 digits} \end{array}$$

Now we can understand the following number–logarithm relationships that were given, without explanation, earlier in this section.

$$\log (1 \times 10^{-4}) = -4.0$$

$$\log (1.0 \times 10^{-9}) = -9.00$$

The first exponential number has one significant figure, and the second exponential number has two significant figures.

EXAMPLE 14.6 **Calculating the pH of a Solution Given Its Hydronium Ion Concentration**

Calculate the pH of a solution with $[H_3O^+] = 3.9 \times 10^{-5}$.

SOLUTION

The pH of concentrated acid solutions can be negative. A solution with a hydronium ion concentration of 1.0 M $(1.0 \times 10^0$ M) has a pH of 0.00, a 2.0 M solution has a pH of –0.30, and a 10.0 M $(1.00 \times 10^1$ M) acidic solution has a pH of –1.00. Most pH calculations deal with hydronium ion concentrations that are smaller than 1.0 M; for such solutions, the pH is always positive.

With an electronic calculator, we first enter the number 3.9×10^{-5}. We then use the log key to obtain the logarithm value, –4.4089353. (With some calculators, the log key is pressed before entering the number.)

$$\begin{aligned} pH &= -\log (3.9 \times 10^{-5}) \\ &= -(-4.4089353) \\ &= 4.4089353 \qquad \text{(calculator answer)} \\ &= 4.41 \qquad \text{(correct answer)} \end{aligned}$$

The given hydronium ion concentration has two significant figures. Therefore, the logarithm should have two significant figures, as 4.41 does. In a logarithm, only the digits to the right of the decimal place are considered significant.

Answer Double Check:

Is the magnitude of the numerical answer reasonable? Yes. A solution with a hydronium ion concentration of 1.0×10^{-5} M would have a pH of 5.00. The given hydronium ion concentration of 3.9×10^{-5} is slightly larger than this value, and thus the pH will have a value that is less than 5.00; as hydronium ion concentration increases, the pH value decreases. The answer 4.41 is reasonable. All hydronium ion concentrations where the power of 10 is 10^{-5} will have pH values of in the range of 4.0 to 5.0.

▶ **Practice Exercise 14.6** Calculate the pH of a solution with $[H_3O^+] = 9.3 \times 10^{-8}$.

It is frequently necessary to calculate the hydronium ion concentration for a solution from its pH value. This type of calculation, which is the reverse of that just illustrated, is shown in Example 14.7.

EXAMPLE 14.7 **Calculating the Molar Hydronium Ion Concentration of a Solution from the Solution's pH**

The pH of a solution is 5.70. What is the molar hydronium ion concentration for this solution?

SOLUTION

Because the pH is between 5 and 6, we know immediately that $[H_3O^+]$ will be between 10^{-5} and 10^{-6} M. From the defining equation for pH, we have

$$[pH] = -\log [H_3O^+] = 5.70$$

$$\log [H_3O^+] = -5.70$$

To find $[H_3O^+]$ we need to determine the *antilog* of –5.70.

How an antilog is obtained using a calculator depends on the type of calculator you have. Many calculators have an antilog function (sometimes labed INV log) that performs this operation. If this key is present, then

1. Enter the number –5.70. Note that it is the *negative* of the pH that is entered into the calculator.
2. Press the INV log key (or an inverse key and then a log key). The result is the desired hydronium ion concentration.

$$\log [H_3O^+] = -5.70$$

$$\text{antilog} [H_3O^+] = 1.9952623 \times 10^{-6} \quad \text{(calculator answer)}$$

$$[H_3O^+] = 2.0 \times 10^{-6} \quad \text{(\textbf{correct answer})}$$

(With some calculators, step 2 and step 1 are reversed.) Remember that the original pH value was a two-significant-figure pH.

Some calculators use a 10^x key to perform the antilog operation. Use of this key is based on the mathematical identity

$$\text{antilog } X = 10^x$$

For our case, this means

$$\text{antilog} - 5.70 = 10^{-5.70}$$

If the 10^x key is present, then

1. Enter the number –5.70 (the negative of the pH).
2. Press the function key 10^x. The result is the desired hydronium ion concentration.

$$[H_3O^+] = 10^{-5.70} = 1.9952623 \times 10^{-6} \quad \text{(calculator answer)}$$

$$= 2.0 \times 10^{-6} \quad \text{(\textbf{correct answer})}$$

Answer Double Check:

Is the magnitude of the numerical answer reasonable? Yes. For a pH value between 5.0 and 6.0, the hydronim ion concentration will always be between 1×10^{-5} M and 1×10^{-6} M. The answer, 2.0×10^{-6} M, falls within this range.

▶ **Practice Exercise 14.7** The pH of a solution is 7.61. What is the molar hydronium ion concentration for this solution?

14.12 HYDROLYSIS OF SALTS

The addition of an acid to water produces an acidic solution. The addition of a base to water produces a basic solution. What type of solution is produced when a salt is added to water? Since salts are the products of acid–base neutralizations, a logical supposition would be that salts dissolve in water to produce neutral (pH = 7) solutions. Such is the case for a *few* salts. Aqueous solutions of *most* salts, however, are either acidic or basic, rather than neutral. Let us consider why this is so.

When a salt is dissolved in water, it completely dissociates; that is, it completely breaks up into the ions of which it is composed (Sec. 13.3). For many salts, one or more of the ions so produced is reactive toward water. The ensuing reaction, which is called *hydrolysis*, causes the solution to have a nonneutral pH. **Hydrolysis** *is the reaction of a substance with water to produce hydronium ions or hydroxide ions or both.*

Types of Salt Hydrolysis

Not all salts hydrolyze. Which ones do, and which ones do not? Of those salts that do hydrolyze, which ones produce acidic solutions and which ones produce basic solutions? The following guidelines, based on the neutralization "parentage" of a salt, that is, on the acid and base that will produce the salt through neutralization, can be used to answer these questions.

1. The salt of a *strong acid* and a *strong base* does not hydrolyze, and therefore its aqueous solution is neutral.
2. The salt of a *strong acid* and a *weak base* hydrolyzes to produce an acidic solution.
3. The salt of a *weak acid* and a *strong base* hydrolyzes to produce a basic solution.
4. The salt of a *weak acid* and a *weak base* hydrolyzes to produce a slightly acidic, neutral, or slightly basic solution, depending on the relative weaknesses of the acid and base.

The first prerequisite for using these guidelines is the ability to classify a salt into one of the four categories mentioned in the guidelines. This classification is accomplished by writing the neutralization equation (Sec. 14.7) that produces the salt and then specifying the strength (strong or weak) of the involved acid and base. The "parent" acid and base for the salt are identified by pairing the negative ion of the salt with H^+ (to form the acid) and pairing the positive ion of the salt with OH^- (to form the base). The following two equations illustrate the overall procedure.

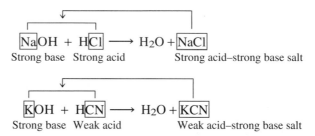

Note that knowledge of which acids and bases are strong and which are weak (Sec. 14.5) is a necessary part of the classification process. Once the salt has been classified, the guideline that is appropriate for the situation is easily selected.

Table 14.10 summarizes the concepts of this section to this point in our discussion.

TABLE 14.10 Neutralization Parentage of Salts and the Nature of the Aqueous Solutions They Form

Type of Salt	Nature of Aqueous Solution	Examples
Strong acid–strong base	neutral	NaCl, KBr
Strong acid–weak base	acidic	NH_4Cl, NH_4NO_3
Weak acid–strong base	basic	$NaC_2H_3O_2$, K_2CO_3
Weak acid–weak base	depends on the salt	$NH_4C_2H_3O_2$, NH_4NO_2

EXAMPLE 14.8 Predicting Whether a Salt's Aqueous Solution Will Be Acidic, Basic, or Neutral

Determine the acid–base parentage of each of the following salts, and then use this information to predict whether each salt's aqueous solution is acidic, basic, or neutral.

a. sodium acetate, $NaC_2H_3O_2$

b. ammonium chloride, NH_4Cl

c. potassium chloride, KCl

d. ammonium fluoride, NH_4F

SOLUTION

a. The ions present are Na^+ and $C_2H_3O_2^-$. The "parent" base of Na^+ is NaOH, a strong base. The parent acid of $C_2H_3O_2^-$ is $HC_2H_3O_2$, a weak acid. Thus, the acid–base neutralization that produces this salt is

$$NaOH + HC_2H_3O_2 \longrightarrow H_2O + NaC_2H_3O$$
$$\text{Strong base} \quad \text{Weak acid} \qquad\qquad \text{Weak acid–strong base salt}$$

The solution of a weak acid–strong base salt (guideline 3) produces a basic solution.

b. The ions present are NH_4^+ and Cl^-. The parent base of NH_4^+ is NH_3, a weak base. The parent acid of Cl^- is HCl, a strong acid. This parentage will produce a strong acid–weak base salt through neutralization. Such a salt gives an acidic solution upon hydrolysis (guideline 2).

c. The ions present are K^+ and Cl^-. The parent base is KOH (a strong base) and the parent acid is HCl (a strong acid). The salt produced from neutralization involving this acid–base pair will be a strong acid–strong base salt. Such salts do not hydrolyze. The aqueous solution is neutral (guideline 1).

d. The ions present are NH_4^+ and F^-. Both ions are of weak parentage; NH_3 is a weak base and HF is a weak acid. Thus, NH_4F is a weak acid–weak base salt. This is a guideline 4 situation. In this situation you cannot predict the effect of hydrolysis unless you know the relative strengths of the weak acid and weak base, that is, which is the weaker of the two. Guidelines to determine such information are not given in this text and thus we cannot predict the final acidity of the solution.

▶ **Practice Exercise 14.8** Determine the acid–base parentage of each of the following salts, and then use this information to predict whether each salt's aqueous solution is acidic, basic, or neutral.

a. ammonium nitrate, NH_4NO_3
b. sodium iodide, NaI
c. potassium fluoride, KF
d. lithium carbonate, Li_2CO_3

Chemical Equations for Salt Hydrolysis Reactions

Salt hydrolysis reactions are Brønsted–Lowry acid–base (proton transfer) reactions (Sec. 14.2). Such reactions are of the following two general types:

1. *Basic hydrolysis*: The reaction of the *negative ion* from a salt with water to produce the ion's conjugate acid and hydroxide ion. Examples of such reactions are

$$\text{Conjugate acid–base pair}$$
$$CN^- + H_2O \longrightarrow HCN + OH^-$$
$$\text{Proton} \quad \text{Proton} \qquad \text{Weak} \quad \text{Makes}$$
$$\text{acceptor} \quad \text{donor} \qquad \text{acid} \quad \text{solution basic}$$

$$\text{Conjugate acid–base pair}$$
$$F^- + H_2O \longrightarrow HF + OH^-$$
$$\text{Proton} \quad \text{Proton} \qquad \text{Weak} \quad \text{Makes}$$
$$\text{acceptor} \quad \text{donor} \qquad \text{acid} \quad \text{solution basic}$$

The only negative ions that undergo hydrolysis are those of "weak acid parentage." The driving force for the reaction is the formation of the weak acid.

2. *Acidic hydrolysis*: The reaction of the *positive ion* from a salt with water to produce the ion's conjugate base and hydronium ion. The most common ion to undergo this type of reaction is the NH_4^+ ion.

Conjugate acid–base pair

$$NH_4^+ + H_2O \longrightarrow NH_3 + H_3O^+$$

| Proton donor | Proton acceptor | Weak base | Makes solution acidic |

The only positive ions that undergo hydrolysis are those of "weak base parentage." The driving force for the reaction is the formation of the weak base.

EXAMPLE 14.9   **Writing Chemical Equations for Hydrolysis Reactions**

For each of the following salts, identify the ion or ions present that will hydrolyze and then write Brønsted–Lowry acid-base equations for hydrolysis reactions that occur.

   a. sodium fluoride, NaF
   b. potassium bromide, KBr
   c. ammonium nitrate, NH_4NO_3
   d. ammonium cyanide, NH_4CN

SOLUTION

   a. The ions produced when NaF dissolves are Na^+ and F^-. The Na^+ ion will not hydrolyze since its parent base, NaOH, is *strong*. The F^- ion will hydrolyze since its parent acid, HF, is *weak*. The equation for the hydrolysis reaction is

$$F^- + H_2O \longrightarrow HF + OH^-$$

The product OH^- causes the solution to be basic.

Water functions as Brønsted–Lowry acid when a negative ion hydrolyzes. Conversely, water functions as a Brønsted–Lowry base when a positive ion hydrolyzes.

   b. Dissolution of KBr in water produces K^+ and Br^- ions. Neither of these ions will hydrolyze. The parent base of K^+ is KOH (strong) and the parent acid of Br^- is HBr (strong).

   c. This salt ionizes to produce NH_4^+ and NO_3^- ions. The NH_4^+ ion is associated with the *weak* base NH_3, and the NO_3^- ion is associated with the *strong* acid HNO_3. The NH_4^+ ion will hydrolyze; the NO_3^- ions will not. The hydrolysis reaction, which produces an acidic solution (H_3O^+), is

$$NH_4^+ + H_2O \longrightarrow NH_3 + H_3O^+$$

   d. Both the NH_4^+ ion (from the weak base NH_3) and the CN^- ion (from the weak acid HCN) will hydrolyze.

$$NH_4^+ + H_2O \longrightarrow NH_3 + H_3O^+$$
$$CN^- + H_2O \longrightarrow HCN + OH^-$$

The pH of the solution will be determined by the reaction that occurs to the greater extent. If the first reaction occurs to the greater extent, the solution will be acidic; conversely, if the second reaction is dominant, a basic solution results. (In this course you are not expected to be able to make such a determination that involves comparison of the relative acid and base strengths of HCN and NH_3. For the record, the CN^- hydrolysis dominates and the solution is basic.)

▶ **Practice Exercise 14.9**   For each of the following salts, identify the ion or ions present that will hydrolyze, and then write Brønsted-Lowry acid-base equations for the hydrolysis reactions that occur.

a. sodium cyanide, NaCN

b. ammonium sulfate, $(NH_4)_2SO_4$

The pH change that accompanies salt hydrolysis can be significant—differing from neutrality by 2 to 4 units. Table 14.11 shows the range of pH values encountered for selected 0.1 M aqueous salt solutions after hydrolysis has occurred.

14.13 BUFFERS

A **buffer** *is a solution that resists major changes in pH when small amounts of acid or base are added to it.* Buffers are used in a laboratory setting to maintain optimum pH conditions for chemical reactions. Many commercial products contain buffers. Examples include buffered aspirin (Bufferin) and pH-controlled hair shampoos. Naturally occurring biochemical buffers are a necessary aspect of life processes. Most human body fluids are buffer solutions. For example, a buffer system maintains blood's pH at a value close to 7.4, an optimum pH for oxygen transport through the blood.

Buffers contain two chemical species: (1) a substance to react with and remove added base, and (2) a substance to react with and remove added acid. Typically, a buffer system is composed of a weak acid *and* its conjugate base—that is, a conjugate acid–base pair (Section 14.3). Conjugate acid–base pairs commonly used as buffers include $HC_2H_3O_2/C_2H_3O_2^-$, $H_2PO_4^-/HPO_4^{2-}$, and H_2CO_3/HCO_3^-.

A less common type of buffer involves a weak base and its conjugate acid. We will not consider this type of buffer here.

EXAMPLE 14.10   **Recognizing Pairs of Chemical Substances that Can Function as a Buffer in an Aqueous Solution**

Predict whether each of the following pairs of substances could function as a buffer system in an aqueous solution.

a. HCl and NaCl

b. HCN and KCN

c. HCl and HCN

d. NaCN and KCN

TABLE 14.11   pH Values of Selected 0.1 M Aqueous Salt Solutions at 24°C

Name of Salt	Formula of Salt	pH	Category of Salt
Ammonium nitrate	NH_4NO_3	5.1	strong acid–weak base
Ammonium nitrite	NH_4NO_2	6.3	weak acid–weak base
Ammonium acetate	$NH_4C_2H_3O_2$	7.0	weak acid–weak base
Sodium chloride	NaCl	7.0	strong acid–strong base
Sodium fluoride	NaF	8.1	weak acid–strong base
Sodium acetate	$NaC_2H_3O_2$	8.9	weak acid–strong base
Ammonium cyanide	NH_4CN	9.3	weak acid–weak base
Sodium cyanide	NaCN	11.1	weak acid–strong base

SOLUTION

Buffer solutions contain either a weak acid and a salt of that weak acid or a weak base and a salt of that weak base. The salt supplies the conjugate base of the acid or the conjugate acid of the base.

a. No. We have an acid and a salt of that acid. However, the acid is a strong acid rather than a weak acid.
b. Yes. HCN is a weak acid, and KCN is a salt of that weak acid. The conjugate acid–base pair HCN/CN^- is present.
c. No. Both HCl and HCN are acids. No salt is present.
d. No. Both NaCN and KCN are salts. No weak acid is present.

Answer Double Check:

Are the answers reasonable? Yes. One of the substances present in a buffer must be a weak acid or a weak base. None of the four pairs contains a weak base. Only one of the four pairs contains a weak acid; in part (b) HCN is the weak acid. Thus, part (b) is the only combination where a buffer is produced. Since the "partner species" for HCN is KCN, a salt that can be produced from HCN, this combination can function as a buffer.

▶ **Practice Exercise 14.10**   Predict whether each of the following pairs of substances could function as a buffer system in an aqueous solution.

a. $NaNO_3$ and KNO_3 b. HCl and HF
c. H_2SO_4 and Na_2SO_4 d. H_2CO_3 and $NaHCO_3$

Chemical Equations for Buffer Action

As an illustration of buffer action, consider a buffer solution containing approximately equal concentrations of acetic acid (a weak acid) and sodium acetate (a salt of this weak acid). This buffer solution resists pH change by the following mechanisms.

1. When a small amount of a strong acid such as HCl is added to this buffer solution, the newly added H_3O^+ ions react with the acetate ions from the sodium acetate to give acetic acid.

$$H_3O^+ + C_2H_3O_2{}^- \longrightarrow HC_2H_3O_2 + H_2O$$

Most of the added H_3O^+ ions are incorporated into acetic acid molecules, and the pH changes very little.

2. When a small amount of a strong base such as NaOH is added to this buffer solution, the newly added OH^- ions react with the acetic acid (neutralization) to give acetate ions and water.

$$OH^- + HC_2H_3O_2 \longrightarrow C_2H_3O_2{}^- + H_2O$$

Most of the added OH^- ions are converted to water, and the pH changes only slightly.

The reactions that are responsible for the buffering action in the acetic acid/acetate ion system can be summarized as follows:

$$C_2H_3O_2{}^- \underset{OH^-}{\overset{H_3O^+}{\rightleftharpoons}} HC_2H_3O_2$$

Note that one member of the buffer pair (acetate ion) removes excess H_3O^+ ion and that the other (acetic acid) removes excess OH^- ion. The buffering action always results in the active species being converted into its partner species.

| EXAMPLE 14.11 | **Writing Equations for Reactions that Occur in a Buffered Solution** |

Write a chemical equation for each of the following buffering actions.

    a. The response of $H_2PO_4^-/HPO_4^{2-}$ buffer to the addition of H_3O^+ ions

    b. The response of HCN/CN^- buffer to the addition of OH^- ions

SOLUTION

    a. The base in a conjugate acid–base pair is the species that responds to the addition of acid. (Recall, from Section 14.3, that the base in a conjugate acid–base pair always has one fewer hydrogens than the acid.) The base for this reaction is HPO_4^{2-}. The equation for the buffering action is

$$H_3O^+ + HPO_4^{2-} \longrightarrow H_2PO_4^- + H_2O$$

    In the buffering response, the base is always converted into its conjugate acid.

    b. The acid in a conjugate acid–base pair is the species that responds to the addition of base. The acid for this reaction is HCN. The equation for the buffering action is

$$OH^- + HCN \longrightarrow CN^- + H_2O$$

    Water will always be one of the products of the buffering action in an aqueous solution.

Answer Double Check:

Are the two equations correct in terms of appropriate reactants? Yes. In part (a), it is the base of the conjugate acid–base pair (HPO_4^{2-}) that should respond to the addition of acid. In part (b), it is the acid of the acid–base pair (HCN) that should respond to the addition of base.

▶ **Practice Exercise 14.11** Write a chemical equation for each of the following buffering actions.

    a. The response of HF/F^- buffer to the addition of H_3O^+ ions

    b. The response of H_2CO_3/HCO_3^- buffer to the addition of OH^- ions

---

    A false notion about buffers is that they will hold the pH of a solution *absolutely* constant. The addition of even small amounts of a strong acid or a strong base to any solution, buffered or not, will lead to a change in pH. The important concept is that the change in pH will be much less when a buffer is present than in an unbuffered solution (see Table 14.12).

    Buffer systems have their limits. If large amounts of H_3O^+ or OH^- ions are added to a buffer, the buffer capacity can be exceeded; then the buffer system is overwhelmed and the pH changes. For example, if large amounts of H_3O^+ were added to the acetate/acetic acid buffer previously discussed, the H_3O^+ ion would react with the acetate ion until the acetate ion was depleted. Then the pH would begin to decrease rapidly as free H_3O^+ ions accumulated in the solution.

    A common misconception about buffers is that the starting pH of a buffered solution is always 7.0 (a neutral solution). This is false. Buffer solutions can be made for any desired pH. A pH 7.4 buffer will hold the pH of the solution near pH 7.4, whereas a pH 9.3 buffer will tend to hold the pH of a solution near pH 9.3. The starting pH of a buffer is determined by the degree of weakness of the weak acid used and by the concentration of the acid and its conjugate base.

TABLE 14.12 A Comparison of pH Changes in Buffered and Unbuffered Solutions

Unbuffered solution	
1 liter water	pH = 7.0
1 liter water + 0.01 mole strong base (NaOH)	pH = 12.0
1 liter water + 0.01 mole strong acid (HCl)	pH = 2.0
Buffered solution	
1 liter buffer*	pH = 7.2
1 liter buffer* + 0.01 mole strong base (NaOH)	pH = 7.3
1 liter buffer* + 0.01 mole strong acid (HCl)	pH = 7.1

*Buffer = 0.1 M HPO_4^{2-} + 0.1 M $H_2PO_4^-$

14.14 ACID–BASE TITRATIONS

Determining the concentration of acid or base in a solution is a regular activity in many laboratories. The concentration of acid or base in a solution and the solution's pH are two different entities. The pH of a solution gives information about the concentration of hydrogen (hydronium) ions in solution. Only dissociated molecules influence the pH value. The concentration of an acidic or a basic solution gives information about the *total number* of acid or base molecules present; both dissociated and undissociated molecules are counted.

The procedure most frequently used to determine the concentration of an acidic or basic solution is an *acid–base titration*. An **acid–base titration** *is a procedure in which a measured volume of acid or base of known concentration is exactly reacted with a measured volume of a base or an acid of unknown concentration.*

Suppose we want to determine the concentration of an acid solution by titration. We would first measure out a *known volume* of the acid solution into a flask. We would then slowly add a solution of base of *known concentration* to the flask by means of a buret (see Fig. 14.6). Base addition continues until all the acid has completely reacted with

FIGURE 14.6 Use of a buret in a titration procedure.

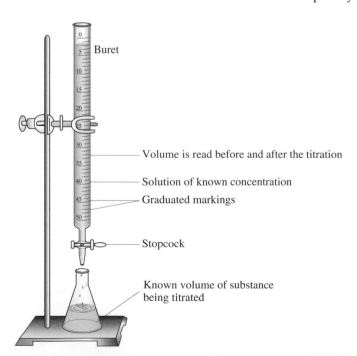

Buret

Volume is read before and after the titration

Solution of known concentration

Graduated markings

Stopcock

Known volume of substance being titrated

added base. The *volume of base* needed to reach this point is obtained from the buret readings. Knowing the original volume of acid, the concentration of the base, and the volume of added base, we can calculate the concentration of the acid.

To complete a titration successfully, we must be able to detect when the reaction between acid and base is complete. One way to do this is to add an *indicator* to the solution being titrated. An **indicator** *is a compound that exhibits different colors, depending on the pH of its surroundings.* Typically, an indicator is a weak acid or weak base whose conjugate base or acid exhibits a different color in a solution. An indicator is selected that will change color at a pH corresponding as nearly as possible to the pH of the solution when the titration is complete. This pH can be calculated ahead of time based on the identities of the acid and base involved in the titration.

Example 14.12 shows how titration data are used to calculate the molarity of an acid solution of unknown concentration.

EXAMPLE 14.12   Calculating an Unknown Molarity, Using Acid–Base Titration Data

In an acid–base titration, 32.7 mL of 0.100 M KOH is required to neutralize completely 50.0 mL of H_3PO_4. Calculate the molarity of the H_3PO_4 solution.

SOLUTION

The first thing we must do is write the balanced equation for the neutralization reaction. Since the acid is triprotic and the base is monobasic, it will take three moles of base to neutralize one mole of acid. The neutralization equation is

$$H_3PO_4 + 3\,KOH \longrightarrow K_3PO_4 + 3\,H_2O$$

Next, we calculate the number of moles of H_3PO_4 that reacted with the KOH. This is a volume-of-solution-A to moles-of-B problem (Fig. 13.7).

$$32.7 \text{ mL KOH} = ?\ \text{moles } H_3PO_4$$

The pathway for the calculation, using dimensional analysis, is

$$\text{mL KOH} \longrightarrow \text{L KOH} \longrightarrow \text{moles KOH} \longrightarrow \text{moles } H_3PO_4$$

The sequence of conversion factors that effect this series of unit changes is

$$32.7 \text{ mL KOH} \times \frac{10^{-3} \text{ L KOH}}{1 \text{ mL KOH}} \times \frac{0.100 \text{ mole KOH}}{1 \text{ L KOH}} \times \frac{1 \text{ mole } H_3PO_4}{3 \text{ moles KOH}}$$

The first conversion factor comes from the definition of a milliliter, the second conversion factor derives from the definition of molarity, and the third conversion factor uses the coefficients in the balanced chemical equation for the titration.

The number of moles of H_3PO_4 that react is obtained by combining the numbers in the dimensional analysis setup as indicated.

$$\frac{32.7 \times 10^{-3} \times 0.100 \times 1}{1 \times 1 \times 3} \text{ mole } H_3PO_4 = 0.00109 \text{ mole } H_3PO_4$$

(calculator and **correct answer**)

Now that we know how many moles of H_3PO_4 reacted, we calculate the molarity of the H_3PO_4 solution, using the definition for molarity and the volume of acid given in the problem statement.

$$\text{Molarity } H_3PO_4 = \frac{\text{moles } H_3PO_4}{\text{L solution}} = \frac{0.00109 \text{ mole } H_3PO_4}{0.0500 \text{ L solution}}$$
$$= 0.0218 \text{ M } H_3PO_4$$

(calculator and **correct answer**)

Note that the units in the denominator of the molarity equation must be liters (0.0500) rather than milliliters (50.0).

Answer Double Check:

If the acid and base were of equal molarity, it would take three times as much base as acid because the acid is triprotic and the base is monobasic. For equal volumes of acid and base, the acid concentration would be one-third that of the base (0.033 M). Since the base amount actually required is less than the acid volume, the acid concentration is even less than 0.033 M. The answer, 0.0218 M acid, is consistent with this analysis.

▶ **Practice Exercise 14.12**   In an acid–base titration, 13.6 mL of 0.200 M HNO_3 is required to neutralize completely 75.0 mL of $Ba(OH)_2$. Calculate the molarity of the $Ba(OH)_2$ solution.

Summary

1. **Arrhenius Acid–Base Theory** An Arrhenius acid is a hydrogen-containing compound that, in water, ionizes to produce hydrogen (H^+) ions. An Arrhenius base is a hydroxide-ion-containing compound that, in water, dissociates to release hydroxide (OH^-) ions.

2. **Brønsted–Lowry Acid–Base Theory** A Brønsted–Lowry acid is any substance that can donate a proton (H^+) to another substance. A Brønsted–Lowry base is any substance that can accept a proton (H^+) from another substance. Proton donation (from an acid) does not occur unless an acceptor (a base) is present.

3. **Conjugate Acids and Conjugate Bases** The conjugate base of an acid is the species that remains when the acid loses a proton. The conjugate acid of a base is the species formed when the base accepts a proton. A conjugate acid–base pair is two species that differ by one proton.

4. **Acids Classification** Acids can be classified, according to the number of hydrogen ions (protons) they can transfer per molecule during an acid–base reaction, as *monoprotic, diprotic,* and *triprotic. Polyprotic* acids are acids that can transfer two or more hydrogen ions during an acid–base reaction. Acids can be classified as *strong* or *weak,* based on the extent to which proton transfer occurs in an aqueous solution. A strong acid completely transfers its protons to water. A weak acid transfers only a small percentage of its protons to water.

5. **Salts** A salt is an ionic compound containing a metal ion or a polyatomic ion as the positive ion and a nonmetal ion or a polyatomic ion (except hydroxide) as the negative ion. Ionic

compounds containing hydroxide ion are bases rather than salts.

6. **Reactions of Acids** Acids react with active metals to produce hydrogen gas and a salt. They react with hydroxide bases to produce a salt and water (neutralization). They react with carbonates and bicarbonates to produce carbon dioxide, a salt, and water.

7. **Reactions of Bases** The most characteristic reaction of bases is their reaction with acids (neutralization). They also react with fats and oils and convert them into smaller, soluble molecules.

8. **Reactions of Salts** Salts react with some metals to convert the metallic ion of the salt to free metal and the free metal to its salt. They react with some acid solutions to form other acids and salts, and with some base solutions to form other bases and salts. Salts react with some solutions of other salts to form new salts.

9. **Self-Ionization of Water** In pure water, a small number of water molecules [1.0×10^{-7} M] donate protons to other water molecules to produce small concentrations [1.0×10^{-7} M] of hydronium and hydroxide ions. Ion product constant for water is the name given to the numerical value 1.00×10^{-14} associated with the product of the hydronium and hydroxide ion molar concentrations in pure water.

10. **The pH Scale** The pH scale is a scale of small numbers that is used to specify molar hydronium ion concentration in an aqueous solution. The calculation of pH scale values involves the use of logarithms. The pH of a solution is the negative logarithm of the solution's molar hydronium ion concentration.

11. **Acidic and Basic Solutions** An acidic solution has a higher hydronium ion concentration than hydroxide ion concentration. Conversely, a basic solution has a higher hydroxide ion concentration than hydronium ion concentration. An acidic solution has a pH of less than 7.00. A basic solution has a pH greater than 7.00. A neutral solution has a pH of 7.00.

12. **Hydrolysis of Salts** Salt hydrolysis is a reaction in which a salt interacts with water to produce an acidic or a basic solution. Only salts that contain the conjugate base of a weak acid and/or the conjugate acid of a weak base hydrolyze.

13. **Buffers** A buffer is a solution that resists major changes in pH when small amounts of acid or base are added to the solution. The resistance to pH change in most buffers is caused by the presence of a weak acid and a salt of its conjugate base.

14. **Acid–Base Titrations** An acid–base titration is a procedure in which an acid–base neutralization reaction is used to determine the unknown concentration of an acid or base. A measured volume of an acid or a base of known concentration is exactly reacted with a measured volume of a base or an acid of unknown concentration. An indicator is used to detect when the neutralization process is complete.

Key Terms

The new terms defined in this chapter are

acid–base titration *Sec 14.14*
acidic hydrogen atom *Sec. 14.4*
acidic solution *Sec. 14.10*
amphiprotic substance *Sec. 14.3*
Arrhenius acid *Sec. 14.1*
Arrhenius base *Sec. 14.1*

basic solution *Sec. 14.10*
Brønsted–Lowry acid *Sec. 14.2*
Brønsted–Lowry base *Sec. 14.2*
buffer *Sec 14.13*
conjugate acid *Sec. 14.3*
conjugate acid–base pair *Sec. 14.3*

conjugate base *Sec. 14.3*
diprotic acid *Sec. 14.4*
dissociation *Sec. 14.1*
hydrolysis *Sec. 14.12*
indicator *Sec. 14.14*
ion product constant for water *Sec. 14.10*
ionization *Sec. 14.1*
monoprotic acid *Sec. 14.4*

neutralization *Sec. 14.7*
neutral solution *Sec. 14.10*
pH *Sec. 14.11*
pH scale *Sec. 14.11*
polyprotic acid *Sec. 14.4*
salt *Sec. 14.6*
strong acid *Sec. 14.5*
triprotic acid *Sec. 14.4*
weak acid *Sec. 14.5*

Practice Problems

ACID–BASE DEFINITIONS (SECS. 14.1 AND 14.2)

14.1 In Arrhenius acid–base theory:
   a. What ion is responsible for the properties of acidic solutions?
   b. What term is used to describe the formation of ions, in an aqueous solution, from an ionic compound?

14.2 In Arrhenius acid–base theory:
   a. What ion is responsible for the properties of basic solutions?
   b. What term is used to describe the formation of ions, in an aqueous solution, from a molecular compound?

14.3 Classify each of the following properties as that of an Arrhenius acid or that of an Arrhenius base.
   a. has a sour taste
   b. changes the color of blue litmus paper to red

14.4 Classify each of the following properties as that of an Arrhenius acid or that of an Arrhenius base.
   a. has a bitter taste
   b. changes the color of red litmus paper to blue

14.5 Based on Arrhenius acid–base theory, indicate whether each of the following chemical reactions is an example of ionization or dissociation.
   a. $HNO_3(aq) \longrightarrow H^+(aq) + NO_3{}^-(aq)$
   b. $NaOH(aq) \longrightarrow Na^+(aq) + OH^-(aq)$
   c. $RbOH(aq) \longrightarrow Rb^+(aq) + OH^-(aq)$
   d. $HCN(aq) \longrightarrow H^+(aq) + CN^-(aq)$

14.6 Based on Arrhenius acid–base theory, indicate whether each of the following chemical reactions is an example of ionization or dissociation.
   a. $HI(aq) \longrightarrow H^+(aq) + I^-(aq)$
   b. $KOH(aq) \longrightarrow K^+(aq) + OH^-(aq)$
   c. $HC_2H_3O_2(aq) \longrightarrow H^+(aq) + C_2H_3O_2{}^-(aq)$
   d. $Ca(OH)_2(aq) \longrightarrow Ca^{2+}(aq) + 2\,OH^-(aq)$

14.7 Write equations for the ionization/dissociation of the following Arrhenius acids and bases in water.
   a. HBr (hydrobromic acid)
   b. $HClO_2$ (chlorous acid)
   c. LiOH (lithium hydroxide)
   d. $Ba(OH)_2$ (barium hydroxide)

14.8 Write equations for the ionization/dissociation of the following Arrhenius acids and bases in water.
  a. $HClO_3$ (chloric acid)
  b. HI (hydroiodic acid)
  c. H_2SO_4 (sulfuric acid)
  d. CsOH (cesium hydroxide)

14.9 Identify the Brønsted–Lowry acid and the Brønsted–Lowry base in each of the following chemical reactions.
  a. $HBr + H_2O \longrightarrow H_3O^+ + Br^-$
  b. $H_2O + N_3^- \longrightarrow HN_3 + OH^-$
  c. $H_2O + H_2S \longrightarrow H_3O^+ + HS^-$
  d. $HS^- + H_2O \longrightarrow H_3O + S^{2-}$

14.10 Identify the Brønsted–Lowry acid and the Brønsted–Lowry base in each of the following chemical reactions.
  a. $NH_3 + H_3PO_4 \longrightarrow NH_4^+ + H_2PO_4^-$
  b. $H_3O^+ + OH^- \longrightarrow H_2O + H_2O$
  c. $HSO_4^- + H_2O \longrightarrow SO_4^{2-} + H_3O^+$
  d. $H_2O + S^{2-} \longrightarrow HS^- + OH^-$

14.11 Write equations to illustrate the acid–base reactions that can take place between the following Brønsted–Lowry acids and bases.
  a. acid, HOCl; base, NH_3
  b. acid, H_2CO_3; base, H_2O
  c. acid, H_2O; base, H_2O
  d. acid, $HC_2O_4^-$; base, H_2O

14.12 Write equations to illustrate the acid–base reactions that can take place between the following Brønsted–Lowry acids and bases.
  a. acid, H_3O^+; base, NH_3
  b. acid, H_2CO_3; base, H_2O
  c. acid, H_2O; base, F^-
  d. acid, $H_2PO_4^-$; base, H_2O

CONJUGATE ACIDS AND BASES (SEC. 14.3)

14.13 Give the chemical formula of the conjugate base of each of the following Brønsted–Lowry acids.
  a. HNO_2 **b.** H_3PO_4 **c.** HPO_4^{2-} **d.** NH_4^+

14.14 Give the chemical formula of the conjugate base of each of the following Brønsted–Lowry acids.
  a. H_2S **b.** $HClO_2$ **c.** $H_2PO_4^-$ **d.** PH_4^+

14.15 Give the chemical formula of the conjugate acid of each of the following Brønsted–Lowry bases.
  a. ClO_2^- **b.** HSO_4^- **c.** SO_3^{2-} **d.** PH_3

14.16 Give the chemical formula of the conjugate acid of each of the following Brønsted–Lowry bases.
  a. ClO^- **b.** HS^- **c.** NH_2^- **d.** NH_3

14.17 Identify the two conjugate acid–base pairs involved in each of the following reactions.
  a. $H_2C_2O_4 + ClO^- \rightleftharpoons HC_2O_4^- + HClO$
  b. $HSO_4^- + H_2O \rightleftharpoons H_3O^+ + SO_4^{2-}$
  c. $HPO_4^{2-} + NH_4^+ \rightleftharpoons NH_3 + H_2PO_4^-$
  d. $HCO_3^- + H_2O \rightleftharpoons OH^- + H_2CO_3$

14.18 Identify the two conjugate acid–base pairs involved in each of the following reactions.
  a. $SO_4^{2-} + H_2O \rightleftharpoons HSO_4^- + OH^-$
  b. $CN^- + H_2O \rightleftharpoons HCN + OH^-$

  c. $HSO_4^- + HCO_3^- \rightleftharpoons SO_4^{2-} + H_2CO_3$
  d. $H_3PO_4 + PO_4^{3-} \rightleftharpoons H_2PO_4^- + HPO_4^{2-}$

14.19 In which of the following pairs of substances do the two members of the pair constitute a conjugate acid–base pair?
  a. HCN and CN^-
  b. H_3PO_4 and PO_4^{3-}
  c. HCO_3^- and HSO_4^-
  d. NH_4^+ and NH_3

14.20 In which of the following pairs of substances do the two members of the pair constitute a conjugate acid–base pair?
  a. HN_3 and N_3^- **b.** H_2SO_4 and SO_4^{2-}
  c. H_2CO_3 and HSO_4^- **d.** NH_3 and NH_2^-

14.21 For each of the following amphiprotic substances, write the two equations needed to describe its behavior in an aqueous solution.
  a. HS^- **b.** HPO_4^{2-} **c.** HCO_3^- **d.** $H_2PO_3^-$

14.22 For each of the following amphiprotic substances, write the two equations needed to describe its behavior in an aqueous solution.
  a. $H_2PO_4^-$ **b.** HSO_3^- **c.** $HC_2O_4^-$ **d.** PH_3

MONO-, DI-, AND TRIPROTIC ACIDS (SEC. 14.4)

14.23 Classify each of the following acids as monoprotic, diprotic, or triprotic.
  a. $HClO_3$ (perchloric acid)
  b. $HC_3H_5O_4$ (glyceric acid)
  c. $HC_2H_3O_2$ (acetic acid)
  d. $H_2C_4H_2O_4$ (fumaric acid)

14.24 Classify each of the following acids as monoprotic, diprotic, or triprotic.
  a. H_2SO_4 (sulfuric acid)
  b. $H_2C_5H_6O_4$ (glutaric acid)
  c. $H_3C_6H_5O_7$ (citric acid)
  d. $HC_7H_5O_3$ (salicylic acid)

14.25 How many acidic hydrogen atoms are present in each of the acids in Problem 14.23?

14.26 How many acidic hydrogen atoms are present in each of the acids in Problem 14.24?

14.27 How many nonacidic hydrogen atoms are present in each of the acids in Problem 14.23?

14.28 How many nonacidic hydrogen atoms are present in each of the acids in Problem 14.24?

14.29 Write chemical equations for the stepwise proton transfer process that occurs in an aqueous solution for each of the following acids.
  a. $H_2C_2O_4$ (oxalic acid)
  b. $H_2C_3H_2O_4$ (malonic acid)

14.30 Write chemical equations for the stepwise proton transfer process that occurs in an aqueous solution for each of the following acids.
  a. H_2CO_3 (carbonic acid)
  b. $H_2C_4H_4O_6$ (tartaric acid)

14.31 The chemical formula for lactic acid is preferably written as $HC_3H_5O_3$ rather than as $H_6C_3O_3$. Explain why this is so.

14.32 The chemical formula for malonic acid is preferably written as $H_2C_3H_2O_4$ rather than as $H_4C_3O_4$. Explain why this is so.

14.33 Pyruvic acid, which is produced in metabolic reactions within the human body, has the following structure:

$$
\begin{array}{ccccc}
\text{H} & \text{O} & \text{O} \\
| & || & || \\
\text{H}-\text{C}-\text{C}-\text{C}-\text{O}-\text{H} \\
| \\
\text{H}
\end{array}
$$

How many acidic hydrogen atoms are present in this structure? Give the reasoning for your answer.

14.34 Succinic acid, a biologically important substance, has the following structure:

$$
\begin{array}{ccccc}
\text{O} & \text{H} & \text{H} & \text{O} \\
|| & | & | & || \\
\text{H}-\text{O}-\text{C}-\text{C}-\text{C}-\text{C}-\text{O}-\text{H} \\
& | & | \\
& \text{H} & \text{H}
\end{array}
$$

How many acidic hydrogen atoms are present in the structure? Give the reasoning for your answer.

STRENGTH OF ACIDS AND BASES (SEC. 14.5)

14.35 Classify each of the acids in Problem 14.23 as a *strong acid* or a *weak acid*.

14.36 Classify each of the acids in Problem 14.24 as a *strong acid* or a *weak acid*.

14.37 For each of the following pairs of acids, indicate whether the first member of the pair is stronger or weaker than the second member of the pair.
a. H_2SO_4 or H_2SO_3
b. $HClO_4$ and HCN
c. HF and HBr
d. HNO_2 and $HClO_3$

14.38 For each of the following pairs of acids, indicate whether the first member of the pair is stronger or weaker than the second member of the pair.
a. HNO_3 and HNO_2
b. HCl and HClO
c. H_3PO_4 and $HClO_3$
d. H_2CO_3 and HI

14.39 In which of the following acid and base combinations are both a strong acid and a strong base present?
a. H_2SO_4 and KOH
b. HNO_3 and NaOH
c. H_3PO_4 and LiOH
d. HF and $Ba(OH)_2$

14.40 In which of the following acid and base combinations are both a strong acid and a strong base present?
a. HCl and NaOH
b. H_2CO_3 and $Ca(OH)_2$
c. $HC_2H_3O_2$ and KOH
d. $HClO_4$ and $Sr(OH)_2$

14.41 With the help of Table 14.3 when necessary, indicate which acid in each of the following pairs of acids is the stronger.
a. HNO_3 and HNO_2
b. $HClO_4$ and $HC_2H_3O_2$
c. H_3PO_4 and HCN
d. H_2CO_3 and HF

14.42 With the help of Table 14.3 when necessary, indicate which acid in each of the following pairs of acids is the stronger.
a. H_2SO_4 and H_2SO_3
b. HCl and HF
c. HCN and HNO_2
d. $HC_2H_3O_2$ and H_3PO_4

SALTS (SEC. 14.6)

14.43 What is the distinction between the terms *weak acid* and *dilute acid*?

14.44 What is the distinction between the terms *strong acid* and *concentrated acid*?

14.45 The following four diagrams represent aqueous solutions of four different acids with the general formula HA. Which of the four acids is the strongest acid?

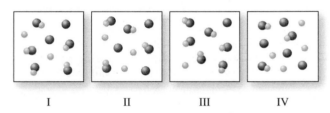

I   II   III   IV

14.46 Using the diagrams shown in Problem 14.45, which of the four acids is the weakest acid?

14.47 Identify each of the following substances as an acid, a base, or a salt.
a. NH_4Cl b. HCl   c. KCl   d. $NaNO_3$

14.48 Identify each of the following substances as an acid, a base, of a salt.
a. NaOH   b. Na_2SO_4 c. K_3PO_4 d. NH_4CN

14.49 What is the name for each of the following salts?
a. $CaSO_4$ b. Li_2CO_3 c. NaBr   d. Al_2S_3

14.50 What is the name for each of the following salts?
a. NH_4Cl b. $Be(C_2H_3O_2)_2$
c. AgI   d. KNO_3

14.51 With the help of Table 13.4, indicate whether each of the salts in Problem 14.49 is soluble or insoluble.

14.52 With the help of Table 13.4, indicate whether each of the salts in Problem 14.50 is soluble or insoluble.

14.53 How many ions are produced, per formula unit, when each of the following soluble salts dissolve in water?
a. $NaNO_3$ b. $CuCO_3$ c. $BaCl_2$ d. $Al_2(SO_4)_3$

14.54 How many ions are produced, per formula unit, when each of the following soluble salts dissolve in water?
a. $Ba(NO_3)_2$ b. K_3PO_4
c. NH_4CN d. KF

14.55 Write a balanced chemical equation for the dissociation in water of each of the salts in Problem 14.53.

14.56 Write a balanced chemical equation for the dissociation in water of each of the salts in Problem 14.54.

REACTIONS OF ACIDS AND BASES (SECS. 14.7 AND 14.8)

14.57 On the basis of the activity series (Table 14.6), predict whether a reaction takes place when
  a. iron metal is added to hydrochloric acid.
  b. potassium metal is added to cold water.
  c. gold metal is added to hydrochloric acid.
  d. aluminum metal is added to hot water (steam).

14.58 On the basis of the activity series (Table 14.6), predict whether a reaction takes place when
  a. chromium metal is added to hydrochloric acid.
  b. silver metal is added to hot water (steam).
  c. sodium metal is added to cold water.
  d. copper metal is added to hydrochloric acid.

14.59 Write the balanced chemical equation for each of the following hydrogen-gas-producing reactions.
  a. Nickel metal is added to hydrochloric acid, giving $NiCl_2$.
  b. Calcium metal is added to cold water.
  c. Magnesium metal is added to hydrochloric acid.
  d. Zinc metal is added to hot water (steam).

14.60 Write the balanced chemical equation for each of the following hydrogen-gas-producing reactions.
  a. Magnesium metal is added to hot water (steam).
  b. Calcium metal is added to hydrochloric acid.
  c. Tin metal is added to hydrochloric acid, giving $SnCl_2$.
  d. Lead metal is added to hydrochloric acid, giving $PbCl_2$.

14.61 Indicate whether each of the following reactions is an acid–base neutralization reaction.
  a. $NaCl + AgNO_3 \longrightarrow AgCl + NaNO_3$
  b. $HNO_3 + KOH \longrightarrow NaNO_3 + H_2O$
  c. $HBr + KOH \longrightarrow KBr + H_2O$
  d. $H_2SO_4 + Pb(NO_3)_2 \longrightarrow PbSO_4 + 2\ HNO_3$

14.62 Indicate whether each of the following reactions is an acid–base neutralization reaction.
  a. $H_2S + CuSO_4 \longrightarrow H_2SO_4 + CuS$
  b. $HCN + LiOH \longrightarrow LiCN + H_2O$
  c. $H_2SO_4 + Ba(OH)_2 \longrightarrow BaSO_4 + 2\ H_2O$
  d. $Ni + 2\ HCl \longrightarrow NiCl_2 + H_2$

14.63 Without writing a chemical equation, specify the molar ratio in which each of the following acids and bases will react in a neutralization reaction.
  a. HBr and $Sr(OH)_2$
  b. HCN and LiOH
  c. H_2SO_4 and $Mg(OH)_2$
  d. H_3PO_4 and KOH

14.64 Without writing a chemical equation, specify the molar ratio in which each of the following acids and bases will react in a neutralization reaction.
  a. HCl and NaOH
  b. HNO_3 and KOH

  c. H_2S and $Ba(OH)_2$
  d. H_2CO_3 and LiOH

14.65 Write a balanced chemical equation to represent each of the acid–base reactions in Problem 14.63.

14.66 Write a balanced chemical equation to represent each of the acid–base reactions in Problem 14.64.

14.67 Write a balanced chemical equation for the neutralization of each of the following acids with the base sodium hydroxide (NaOH).
  a. $HC_3H_5O_3$ (lactic acid)
  b. $H_2C_4H_4O_5$ (malic acid)
  c. $H_2C_4H_4O_4$ (succinic acid)
  d. $H_3C_6H_5O_7$ (citric acid)

14.68 Write a balanced chemical equation for the neutralization of each of the following acids with the base potassium hydroxide (KOH).
  a. $HC_3H_3O_3$ (pyruvic acid)
  b. $H_2C_4H_3O_5$ (oxaloacetic acid)
  c. $H_2C_4H_2O_4$ (fumaric acid)
  d. $H_2C_5H_6O_4$ (glutaric acid)

14.69 What is the chemical formula of the salt produced from each of the following acid–base neutralization reactions?
  a. HNO_3 and NaOH
  b. HCN and KOH
  c. H_2SO_4 and LiOH
  d. H_2CO_3 and $Ba(OH)_2$

14.70 What is the chemical formula of the salt produced from each of the following acid–base neutralization reactions?
  a. HCl and $Mg(OH)_2$
  b. $HClO_4$ and NaOH
  c. HClO and CsOH
  d. H_2CO_3 and $Al(OH)_3$

14.71 Give the chemical formulas of the acid and base needed to prepare each of the following salts using an acid–base neutralization reaction.
  a. Na_2S **b.** KNO_3
  c. $Al(ClO_3)_3$ **d.** $CaBr_2$

14.72 Give the chemical formulas of the acid and base needed to prepare each of the following salts using an acid–base neutralization reaction.
  a. Na_3PO_4 **b.** KCN **c.** $AlCl_3$ **d.** $CaSO_4$

14.73 Write a chemical equation for the action of HCl on each of the following. (Zn metal forms a +2 ion in a solution.)
  a. Zn   **b.** NaOH   **c.** Na_2CO_3 **d.** $NaHCO_3$

14.74 Write a chemical equation for the action of HNO_3 on each of the following. (Ni metal forms a +2 ion in a solution.)
  a. Ni   **b.** KOH   **c.** Li_2CO_3 **d.** $LiHCO_3$

REACTIONS OF SALTS (SEC. 14.9)

14.75 On the basis of the activity series (Table 14.6), predict whether a reaction occurs when each of the following metals is added to a nickel(II) nitrate solution.
  a. Cu   **b.** Zn   **c.** Au   **d.** Fe

14.76 On the basis of the activity series (Table 14.6), predict whether a reaction occurs when each of the following metals is added to an iron(II) nitrate solution.
  a. Ag **b.** Ni **c.** Cr **d.** Pb

14.77 Write a chemical equation for each of the following metal-replacement reactions. Assume that metals going into solution form +2 ions.
  a. Iron is added to $CuSO_4$ solution.
  b. Tin is added to $AgNO_3$ solution.
  c. Zinc is added to $NiCl_2$ solution.
  d. Chromium is added to $Pb(C_2H_3O_2)_2$ solution.

14.78 Write a chemical equation for each of the following metal-replacement reactions. Assume that metals going into solution form +2 ions.
  a. Lead is added to Cu_2SO_4 solution.
  b. Mercury is added to $Au(NO_3)_3$ solution.
  c. Chromium is added to $FeCl_2$ solution.
  d. Iron is added to $Ni(C_2H_3O_3)_2$ solution.

14.79 Indicate the driving force (condition) that causes each of the following acid–salt reactions to occur.
  a. $Ba(NO_3)_2 + H_2SO_4 \longrightarrow BaSO_4 + 2\,HNO_3$
  b. $3\,CaCl_2 + 2\,H_3PO_4 \longrightarrow Ca_3(PO_4)_2 + 6\,HCl$
  c. $AgC_2H_3O_2 + HCl \longrightarrow AgCl + HC_2H_3O_2$
  d. $K_2CO_3 + 2\,HNO_3 \longrightarrow 2\,KNO_3 + CO_2 + H_2O$

14.80 Indicate the driving force (condition) that causes each of the following acid–salt reactions to occur.
  a. $NaCN + HCl \longrightarrow NaCl + HCN$
  b. $3\,MgSO_4 + 2\,H_3PO_4 \longrightarrow Mg_3(PO_4)_2 + 3\,H_2SO_4$
  c. $Li_2CO_3 + 2\,HBr \longrightarrow 2\,LiBr + CO_2 + H_2O$
  d. $Na_3PO_4 + 3\,HCl \longrightarrow 3\,NaCl + H_3PO_4$

14.81 Write a chemical equation for the reaction, if any, between each of the following pairs of aqueous solutions. If no reaction occurs, write "no reaction."
  a. $Al(NO_3)_3$ and $(NH_4)_2S$
  b. HCl and $Ba(OH)_2$
  c. Na_2SO_4 and HNO_3
  d. KNO_3 and $HC_2H_3O_2$

14.82 Write a chemical equation for the reaction, if any, between each of the following pairs of aqueous solutions. If no reaction occurs, write "no reaction."
  a. H_2CO_3 and KOH
  b. K_3PO_4 and HCl
  c. $CaCl_2$ and HNO_3
  d. $Fe(NO_3)_2$ and $MgSO_4$

HYDRONIUM ION AND HYDROXIDE ION CONCENTRATIONS (SEC. 14.10)

14.83 What is the molar H_3O^+ ion concentration in aqueous solutions with the following OH^- ion concentrations?
  a. 3.5×10^{-3} M **b.** 4.7×10^{-6} M
  c. 1.1×10^{-8} M **d.** 8.7×10^{-10} M

14.84 What is the molar H_3O^+ ion concentration in aqueous solution with the following OH^- ion concentrations?
  a. 4.2×10^{-4} M **b.** 6.0×10^{-5} M
  c. 3.4×10^{-9} M **d.** 7.3×10^{-11} M

14.85 Indicate whether each of the solutions in Problem 14.83 is acidic, basic, or neutral.

14.86 Indicate whether each of the solutions in Problem 14.84 is acidic, basic, or neutral.

14.87 What is the molar OH^- ion concentration in aqueous solutions with the following H_3O^+ ion concentrations?
  a. 5.5×10^{-2} M **b.** 9.4×10^{-5} M
  c. 2.3×10^{-7} M **d.** 6.6×10^{-12} M

14.88 What is the molar OH^- ion concentration in aqueous solutions with the following H_3O^+ ion concentrations?
  a. 2.4×10^{-3} M **b.** 7.5×10^{-6} M
  c. 6.7×10^{-8} M **d.** 5.0×10^{-10} M

14.89 Indicate whether each of the solutions in Problem 14.87 is acidic, basic, or neutral.

14.90 Indicate whether each of the solutions in Problem 14.88 is acidic, basic, or neutral.

14.91 Selected information about five aqueous solutions, each at 24°C, is given in the following table. Fill in the blanks in each line in the table. The first line is already completed as an example.

	$[H_3O^+]$	$[OH^-]$	Acidic or Basic
	2.2×10^{-2}	4.5×10^{-13}	acidic
a.		3.3×10^{-3}	
b.	6.8×10^{-8}		
c.		7.2×10^{-8}	
d.	4.5×10^{-6}		

14.92 Selected information about five aqueous solutions, each at 24°C, is given in the following table. Fill in the blanks in each line in the table. The first line is already completed as an example.

	$[H_3O^+]$	$[OH^-]$	Acidic or Basic
	1.3×10^{-13}	7.7×10^{-2}	basic
a.	6.3×10^{-8}		
b.	4.2×10^{-6}		
c.		3.3×10^{-10}	
d.		6.6×10^{-5}	

THE PH SCALE (SEC. 14.11)

14.93 Calculate the pH of solutions with the following hydronium ion concentrations.
  a. 1×10^{-4} M **b.** 1×10^{-11} M
  c. 0.00001 M **d.** 0.000000001 M

14.94 Calculate the pH of solutions with the following hydronium ion concentrations.
  a. 1×10^{-2} M **b.** 1×10^{-11} M
  c. 0.0001 M **d.** 0.0000001 M

14.95 Calculate the pH of solutions with the following hydronium ion concentrations.
  a. 6×10^{-3} M   **b.** 6×10^{-4} M
  c. 3×10^{-8} M   **d.** 7×10^{-10} M

14.96 Calculate the pH of solutions with the following hydronium ion concentrations.
  a. 5×10^{-4} M   **b.** 5×10^{-5} M
  c. 2×10^{-9} M   **d.** 8×10^{-12} M

14.97 Calculate the pH of solutions with the following hydronium ion concentrations.
  a. 3×10^{-3} M   **b.** 3.0×10^{-3} M
  c. 3.00×10^{-3} M   **d.** 3.000×10^{-3} M

14.98 Calculate the pH of solutions with the following hydronium ion concentrations.
  a. 7×10^{-6} M   **b.** 7.0×10^{-6} M
  c. 7.00×10^{-3} M   **d.** 7.000×10^{-3} M

14.99 In which of the following pairs of pH values do both values represent acidic solution conditions?
  a. 5.31 and 6.31   **b.** 6.31 and 7.31
  c. 7.31 and 8.31   **d.** 6.90 and 7.00

14.100 In which of the following pairs of pH values do both values represent basic solution conditions?
  a. 5.92 and 6.92   **b.** 6.92 and 7.92
  c. 7.92 and 8.92   **d.** 7.01 and 7.10

14.101 What is the molar hydronium ion concentration associated with each of the following pH values?
  a. 4.0   **b.** 4.2   **c.** 4.5   **d.** 4.8

14.102 What is the molar hydronium ion concentration associated with each of the following pH values?
  a. 6.0   **b.** 6.3   **c.** 6.7   **d.** 6.9

14.103 What is the molar hydronium ion concentration associated with each of the following pH values?
  a. 2.43   **b.** 5.72   **c.** 7.73   **d.** 8.750

14.104 What is the molar hydronium ion concentration associated with each of the following pH values?
  a. 3.57   **b.** 6.03   **c.** 8.51   **d.** 9.555

14.105 What is the molar hydroxide ion concentration associated with each of the pH values in Problem 14.103?

14.106 What is the molar hydroxide ion concentration associated with each of the pH values in Problem 14.104?

14.107 Selected information about five solutions, each at 24°C, is given in the following table. Fill in the blanks in each line in the table. The first line is already completed as an example.

	$[H_3O^+]$	$[OH^-]$	pH	Acidic or Basic
	6.2×10^{-8}	1.6×10^{-7}	7.21	basic
a.	7.2×10^{-10}			
b.			5.30	
c.		7.2×10^{-10}		
d.			8.23	

14.108 Selected information about five solutions, each at 24°C, is given in the following table. Fill in the blanks in each line in the table. The first line is already completed as an example.

	$[H_3O^+]$	$[OH^-]$	pH	Acidic or Basic
	1.9×10^{-9}	5.4×10^{-6}	8.73	basic
a.		7.2×10^{-5}		
b.	6.3×10^{-3}			
c.			9.03	
d.			5.77	

14.109 Solution A has $[OH^-] = 4.3 \times 10^{-4}$. Solution B has $[H_3O^+] = 7.3 \times 10^{-10}$.
  a. Which solution is more basic?
  b. Which solution has the lower pH?

14.110 Solution A has $[H_3O^+] = 2.7 \times 10^{-6}$. Solution B has $[OH^-] = 4.5 \times 10^{-8}$.
  a. Which solution is more acidic?
  b. Which solution has the higher pH?

14.111 A solution has a pH of 4.500. What will be the pH of this solution if the hydronium ion concentration is
  a. doubled?
  b. quadrupled?
  c. increased by a factor of 10?
  d. increased by a factor of 10^3?

14.112 A solution has a pH of 3.699. What will be the pH of this solution if the hydronium ion concentration is
  a. tripled?
  b. cut in half?
  c. increased by a factor of 100?
  d. decreased by a factor of 10^4?

14.113 Calculate the pH of each of the following solutions of strong acids or strong bases.
  a. 6.3×10^{-3} M HNO_3 **b.** 0.20 M HCl
  c. 0.000021 M H_2SO_4 **d.** 2.3×10^{-4} M NaOH

14.114 Calculate the pH of each of the following solutions of strong acids or strong bases.
  a. 4.02×10^{-5} M HCl
  b. 4.02×10^{-5} M H_2SO_4
  c. 0.00035 M HNO_3
  d. 5.7×10^{-3} M KOH

HYDROLYSIS OF SALTS (SEC. 14.12)

14.115 Identify the ion (or ions) present, if any, that will undergo hydrolysis in an aqueous solution in each of the following salts.
  a. Na_3PO_4 **b.** NaCN   **c.** NH_4Cl **d.** LiCl

14.116 Identify the ion (or ions) present, if any, that will undergo hydrolysis in an aqueous solution in each of the following salts.
  a. $KC_2H_3O_2$ **b.** NH_4F
  c. $Ca(CN)_2$ **d.** NaBr

14.117 Predict whether each of the following aqueous salt solutions will be acidic, basic, or neutral.
  a. Na_2SO_4 **b.** LiCN   **c.** NH_4Br **d.** KI

14.118 Predict whether each of the following aqueous salt solutions will be acidic, basic, or neutral.
  a. KNO_3 **b.** NaCN
  c. $LiC_2H_3O_2$ **d.** $(NH_4)_2SO_4$

14.119 Write a chemical equation for the hydrolysis of each of the following salts in an aqueous solution.
   a. NH_4Br **b.** $NaC_2H_3O_2$
   c. KF   **d.** LiCN

14.120 Write a chemical equation for the hydrolysis of each of the following salts in an aqueous solution.
   a. MgF_2 **b.** KCN   **c.** $LiNO_2$ **d.** NH_4I

BUFFERS (SEC. 14.13)

14.121 Predict whether each of the following pairs of substances could function as a buffer system in an aqueous solution.
   a. HNO_3 and $NaNO_3$
   b. HF and NaF
   c. KCl and KCN
   d. H_2CO_3 and $NaHCO_3$

14.122 Predict whether each of the following pairs of substances could function as a buffer system in an aqueous solution.
   a. HNO_3 and HCl
   b. HNO_2 and KNO_2
   c. $NaC_2H_3O_2$ and $KC_2H_3O_2$
   d. $HC_2H_3O_2$ and $NaNO_3$

14.123 Identify the two "active species" in each of the following buffer systems.
   a. HCN and KCN   **b.** H_3PO_4 and NaH_2PO_4
   c. H_2CO_3 and $KHCO_3$ **d.** $NaHCO_3$ and K_2CO_3

14.124 Identify the two "active species" in each of the following buffer systems.
   a. HF and LiF
   b. Na_2HPO_4 and KH_2PO_4
   c. K_2CO_3 and $KHCO_3$
   d. $NaNO_2$ and HNO_2

14.125 The following four diagrams represent aqueous solutions containing a weak acid (HA) and/or its conjugate base (A^-). Which of the four solutions is a buffer solution? There may be more than one correct answer.

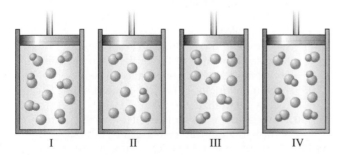

   I   II   III   IV

14.126 Using the diagrams shown in Problem 14.125, which of the solutions would have the greatest buffer capacity, that is, the greatest protection against pH change, when the following occurs?
   a. A small amount of strong acid is added to the solution.
   b. A small amount of strong base is added to the solution.

14.127 Write an equation for each of the following buffering actions.
   a. the response of an HF/F^- buffer to the addition of OH^- ions
   b. the response of an $HPO_4{}^{2-}$/$PO_4{}^{3-}$ buffer to the addition of H_3O^+ ions
   c. the response of an $HCO_3{}^-$/$CO_3{}^{2-}$ buffer to the addition of acid
   d. the response of an H_3PO_4/$H_2PO_4{}^-$ buffer to the addition of base

14.128 Write an equation for each of the following buffering actions.
   a. the response of an $HPO_4{}^{2-}$/$PO_4{}^{3-}$ buffer to the addition of OH^- ions
   b. the response of an H_3PO_4/$H_2PO_4{}^-$ buffer to the addition of H_3O^+ ions
   c. the response of an HCN/CN^- buffer to the addition of acid
   d. the response of an $HCO_3{}^-$/$CO_3{}^{2-}$ buffer to the addition of base

ACID–BASE TITRATIONS (SEC. 14.14)

14.129 What volume, in milliliters, of a 0.100 M HCl solution would be needed to neutralize 20.00 mL samples of the following bases?
   a. 0.100 M NaOH
   b. 0.200 M NaOH
   c. 0.100 M $Ba(OH)_2$
   d. 0.200 M $Ba(OH)_2$

14.130 What volume, in milliliters, of a 0.200 M HNO_3 solution would be needed to neutralize 10.00 mL samples of the following bases?
   a. 0.100 M KOH
   b. 0.200 M KOH
   c. 0.100 M $Sr(OH)_2$
   d. 0.200 M $Sr(OH)_2$

14.131 What volume, in milliliters, of a 0.200 M KOH solution would be needed to neutralize each of the following acid samples?
   a. 20.00 mL of 0.200 M HCl
   b. 20.00 mL of 0.200 M H_2SO_4
   c. 50 00 mL of 0.400 M HNO_3
   d. 50.00 mL of 0.300 M H_2CO_3

14.132 What volume, in milliliters, of a 0.300 M NaOH solution would be needed to neutralize each of the following acid samples?
   a. 10.00 mL of 0.100 M HCl
   b. 10.00 mL of 0.200 M H_2SO_4
   c. 30.00 mL of 0.500 M HNO_3
   d. 30.00 mL of 0.600 M H_2CO_3

14.133 It requires 34.5 mL of 0.102 M NaOH to neutralize each of the following acid solution samples. What is the molarity of each of the acid samples?
   a. 25.0 mL of H_2SO_4
   b. 20.0 mL of HClO
   c. 20.0 mL of H_3PO_4
   d. 10.0 mL of HNO_3

14.134 It requires 21.4 mL of 0.198 M NaOH to neutralize each of the following acid solution samples. What is the molarity of each of the acid samples?
 a. 30.0 mL of H_2CO_3
 b. 25.0 mL of $H_2C_2O_4$
 c. 25.0 mL of $HC_2H_3O_2$
 d. 20.0 mL of HCl

ADDITIONAL PROBLEMS

14.135 Identify each of the following species as the conjugate base of a strong acid or the conjugate base of a weak acid.
 a. Br^- **b.** CN^-
 c. $H_2PO_4^-$ **d.** NO_3^-

14.136 Identify each of the following species as the conjugate base of a strong acid or the conjugate base of a weak acid.
 a. $C_2H_3O_2^-$ **b.** Cl^-
 c. F^- **d.** HPO_4^{2-}

14.137 A solution has a pH of 2.2. Another solution has a pH of 4.5. How many times greater is the $[H_3O^+]$ in the first solution than in the second one?

14.138 A solution has a pH of 3.4. Another solution has a pH of 6.7. How many times greater is the $[H_3O^+]$ in the first solution than in the second one?

14.139 What is the pH of an aqueous solution in which there are three times as many hydronium ions as hydroxide ions?

14.140 What is the pH of an aqueous solution in which there are three times as many hydroxide ions as hydronium ions?

14.141 Solution A has a pH of 3.20, solution B a pH of 4.20, solution C a pH of 11.20, and solution D a pH of 7.20. Arrange the four solutions in order of
 a. decreasing acidity. **b.** increasing $[H_3O^+]$.
 c. decreasing $[OH^-]$. **d.** increasing basicity.

14.142 Solution A has a pH of 1.25, solution B a pH of 12.50, solution C a pH of 7.00, and solution D a pH of 4.44. Arrange the four solutions in order of
 a. increasing acidity.
 b. decreasing $[H_3O^+]$.
 c. increasing $[OH^-]$.
 d. decreasing basicity.

14.143 At a temperature of 37°C, the value of K_w for water is 2.6×10^{-14}. For a sample of pure water, at 37°C, determine the following.
 a. molar hydronium ion concentration
 b. molar hydroxide ion concentration
 c. pH
 d. whether the solution is neutral, acidic, or basis

14.144 At a temperature of 50°C, the value of K_w for water is 5.5×10^{-14}. For a sample of pure water, at 50°C, determine the following.
 a. molar hydronium ion concentration
 b. molar hydroxide ion cencentration
 c. pH
 d. whether the solution is neutral, acidic, or basis

14.145 In which of the following pairs of solutions does the first-listed solution have a lower pH than the second-listed solution?
 a. 0.1 M HCl and 0.2 M HCl
 b. 0.1 M HCl and 0.1 M H_2SO_4
 c. 0.20 M H_2SO_4 and 0.25 M HNO_3
 d. 0.2 M H_2SO_4 and 0.2 M H_2CO_3

14.146 In which of the following pairs of solutions does the first-listed solution have a lower pH than the second-listed solution?
 a. 0.2 M HNO_3 and 0.3 M HNO_3
 b. 0.1 M HNO_3 and 0.1 M H_2SO_4
 c. 0.1 M H_2SO_4 and 0.12 M HCl
 d. 0.2 M H_2SO_4 and 0.2 M $H_2C_2O_4$

14.147 What would be the pH of a solution that contains 0.1 mole of each of the solutes NaCl, HNO_3, HCl, and NaOH in enough water to give 3.00 L of solution?

14.148 What would be the pH of a solution that contains 0.1 mole of each of the solutes NaBr, HBr, KOH, and NaOH in enough water to give 5.00 L of solution?

14.149 Arrange the following 0.1 M aqueous solutions in order of decreasing pH: NH_4Br, $Ba(OH)_2$, $HClO_4$, K_2SO_4, and LiCN.

14.150 Arrange the following 0.1 M aqueous solutions in order of increasing pH: HBr, $HC_2H_3O_2$, $NaC_2H_3O_2$, KOH, and $Ca(NO_3)_2$.

14.151 Identify the buffer system(s) [conjugate acid–base pair(s)] present in solutions that contain equal molar amounts of the following.
 a. HCN, KCN, NaCN, and NaCl
 b. HF, HCl, $NaC_2H_3O_2$, and NaF

14.152 Identify the buffer system(s) [conjugate acid–base pair(s)] present in solutions that contain equal molar amounts of the following.
 a. HF, HCN, NaCN, and KF
 b. H_2CO_3, Na_2CO_3, KCN, and HCN

14.153 It is possible to make two completely different buffers that involve the dihydrogen phosphate ion, $H_2PO_4^-$. Characterize each of the buffers by specifying the conjugate acid–base pair that is present.

14.154 It is possible to make two completely different buffers that involve the hydrogen carbonate ion, HCO_3^-. Characterize each of the buffers by specifying the conjugate acid–base pair that is present.

14.155 How many moles of $Ca(OH)_2$ would it take to completely neutralize 0.40 gram of H_3PO_4?

14.156 How many moles of $Sr(OH)_2$ would it take to completely neutralize 5.00 grams of HNO_3?

14.157 If 125 mL of 5.00 M HNO_3 is mixed with 125 mL of 6.00 M NaOH, what will be the
 a. $[H_3O^+]$ for the solution?
 b. $[OH^-]$ for the solution?
 c. pH of the solution?

14.158 If 125 mL of 6.00 M HNO_3 is mixed with 125 mL of 5.00 M NaOH, what will be the
 a. $[H_3O^+]$ for the solution?
 b. $[OH^-]$ for the solution?
 c. pH of the solution?

CUMULATIVE PROBLEMS

14.159 Name and give the formula for each of the following species.
  a. conjugate base of hydroiodic acid
  b. conjugate base of the dihydrogen phosphate ion
  c. conjugate acid of the oxide ion
  d. conjugate acid of water

14.160 Name and give the formula for each of the following species.
  a. conjugate base of perchloric acid
  b. conjugate base of the bicarbonate ion
  c. conjugate acid of the hydroxide ion
  d. conjugate acid of ammonia

14.161 What is the pH of a solution obtained by each of the following operations?
  a. dissolving 4.8 g of HCl in enough water to obtain 0.40 L of solution
  b. dissolving 12.5 g of LiOH in enough water to obtain 255 mL of solution
  c. diluting 75 mL of 0.10 M HCl to a volume of 125 mL
  d. mixing equal volumes of 0.20 M HCl and 0.50 M HNO_3

14.162 What is the pH of a solution obtained by each of the following operations?
  a. dissolving 4.8 g of HBr in enough water to obtain 0.30 L of solution
  b. dissolving 3.50 g of NaOH in enough water to obtain 45 mL of solution
  c. diluting 25 mL of 0.10 HNO_3 to a volume of 375 mL
  d. mixing equal volumes of 0.20 M HCl and 0.20 M HNO_3

14.163 If 1.00 mL of 0.10 M HNO_3 is diluted to 100.0 mL, what is
  a. $[H_3O^+]$ before dilution?
  b. $[H_3O^+]$ after dilution?
  c. the pH before dilution?
  d. the pH after dilution?

14.164 If 100.0 mL of 0.10 M HCl is diluted to 500.0 mL, what is
  a. $[H_3O^+]$ before dilution?
  b. $[H_3O^+]$ after dilution?
  c. the pH before dilution?
  d. the pH after dilution?

14.165 How many hydronium ions are present in a 10.0 mL sample of hydrochloric acid that has a pH of 5.42?

14.166 How many hydronium ions are present in a 50.0 mL sample of nitric acid that has a pH of 1.32?

14.167 How many total ions are present in a 236 mL sample of a HNO_3 solution with a pH of 2.37 to which 0.100 mole of Na_2SO_4 has been added and dissolved?

14.168 How many total ions are present in a 435 mL sample of an HCl solution with a pH of 1.54 to which 0.050 mole of $Mg(NO_3)_2$ has been added and dissolved?

14.169 How many liters of HCl gas, at STP, are dissolved in 7.50 L of aqueous HCl solution if the solution has a pH of 2.40?

14.170 How many liters of NH_3 gas, at STP, are dissolved in 7.50 L of aqueous NH_3 solution if the solution has a pH of 8.20?

Multiple-Choice Practice Test

Use this bank of 20 multiple-choice questions as a review of key concepts presented in this chapter. For many of the questions, there may be more than one correct answer (choice d) or no correct answer (choice e).

14.171 Which of the following statements concerning Arrhenius acid–base theory is correct?
  a. Arrhenius acid–base definitions are based on water as the solvent.
  b. In the pure state, Arrhenius acids are ionic compounds.
  c. In the pure state, Arrhenius bases are molecular compounds.
  d. more than one correct response
  e. no correct response

14.172 According to Brønsted–Lowry acid–base theory, a base is a
  a. proton donor.
  b. proton acceptor.
  c. hydroxide ion donor.

  d. more than one correct response
  e. no correct response

14.173 The Brønsted–Lowry acid and Brønsted–Lowry base for the reaction

$$N_3^- + H_2O \longrightarrow HN_3 + OH^-$$

are, respectively,
  a. N_3^- and H_2O.   **b.** H_2O and N_3^-.
  c. N_3^- and HN_3.   **d.** H_2O and OH^-.
  e. no correct response

14.174 Which of the following is *not* a Brønsted–Lowry conjugate acid–base pair?
  a. PH_4^+/PH_3
  b. H_2O/OH^-
  c. H_2S/S^{2-}
  d. more than one correct response
  e. no correct response

14.175 In which of the following pairs of acids are both members of the pair *polyprotic* acids?
  a. H_3PO_4 and $HC_2H_3O_2$
  b. H_2SO_4 and $H_2C_4H_4O_6$

c. HNO_3 and HNO_2
d. more than one correct response
e. no correct response

14.176 Which of the following is a species formed in the second step of the dissociation of H_3PO_4?
   a. $H_2PO_4{}^-$
   b. $HPO_4{}^{2-}$
   c. $PO_4{}^{3-}$
   d. more than one correct response
   e. no correct response

14.177 In which of the following pairs of acids are both members of the pair *strong* acids?
   a. H_2CO_3 and H_2SO_4
   b. HCN and HCl
   c. $HClO_3$ and HNO_3
   d. more than one correct response
   e. no correct response

14.178 Which of the following statement about *weak* acids is correct?
   a. Weak acids always contain carbon atoms.
   b. The percent dissociation for a weak acid is in the range 40%–60%.
   c. Weak acid molecules have a strong affinity for acidic hydrogen atoms.
   d. more than one correct response
   e. no correct response

14.179 In which of the following pairs of substance are both species in the pair salts?
   a. LiOH and LiCN
   b. NH_4F and KF
   c. HNO_3 and KNO_3
   d. more than one correct response
   e. no correct response

14.180 Based on the way the chemical formula is written, which of the following acids contain nonacidic hydrogen atoms?
   a. HCN
   b. H_2SO_4
   c. $H_2C_2O_4$
   d. more than one correct response
   e. no correct response

14.181 Which of the following statements concerning reactions of acids is correct?
   a. Acids react with hydroxide ion bases to produce a salt and water.
   b. Acids react with all metals to produce hydrogen gas as a product.
   c. Acids do not react with salts at ordinary conditions.
   d. more than one correct response
   e. no correct response

14.182 A solution with $[H_3O^+] = 1.00 \times 10^{-9}$ would have $[OH^-]$ equal to
   a. 1.00×10^{-5}.
   b. 2.00×10^{-5}.
   c. 1.00×10^{-9}.
   d. 2.00×10^{-14}.
   e. no correct response

14.183 Which of the following solutions is acidic?
   a. $[H_3O^+] = 1.00 \times 10^{-5}$
   b. $[H_3O^+] = 1.00 \times 10^{-7}$
   c. $[H_3O^+] = 1.00 \times 10^{-9}$
   d. more than one correct response
   e. no correct response

14.184 Which of the following does *not* describe an acidic solution?
   a. The pH is less than 7.0
   b. The $[OH^-]$ is less than the $[H_3O^+]$.
   c. The hydronium ion concentration is twice the hydroxide ion concentration.
   d. more than one correct response
   e. no correct response

14.185 If the pH of a solution is increased from 6.0 to 8.0 the $[H_3O^+]$
   a. increases by a factor of 2.
   b. increases by a factor of 20.
   c. decreases by a factor of 20.
   d. decreases by a factor of 100.
   e. no correct response

14.186 If the pH of a solution is 3.26, the molar hydronium ion concentration for the solution is
   a. 2.91×10^{-4}.
   b. 3.26×10^{-4}.
   c. 5.50×10^{-4}.
   d. 8.71×10^{-4}.
   e. no correct response

14.187 Which of the following salts, upon hydrolysis, produces a basic solution?
   a. $LiNO_3$
   b. NH_4Cl
   c. K_2SO_4
   d. more than one correct response
   e. no correct response

14.188 Which of the following combinations of substances would produce a buffer?
   a. a strong acid and a salt of the strong acid
   b. a salt of a strong acid and a salt of a weak acid
   c. a strong acid and a weak acid
   d. more than one correct response
   e. no correct response

14.189 Which of the following pairs of compounds could be used to prepare a buffer solution?
   a. NaCl and HCl
   b. NaCN and HCN
   c. NaOH and HCl
   d. more than one correct response
   e. no correct response

14.190 In an acid–base titration, 25.8 mL of 0.1000 M KOH is required to neutralize completely 50.0 mL of H_2SO_4. What is the molarity of the H_2SO_4 solution?
   a. 1.29×10^{-3} M
   b. 2.58×10^{-3} M
   c. 2.58×10^{-2} M
   d. 5.16×10^{-2} M
   e. no correct response

Chemical Equations: Net Ionic and Oxidation-Reduction

15.1 TYPES OF CHEMICAL EQUATIONS

Two major chemical-equation-related topics constitute the bulk of the subject matter for this chapter. These topics are *net ionic chemical equations* and *oxidation-reduction chemical equations*.

Net ionic chemical equations represent a method for writing chemical equations for reactions that emphasizes that ions are often the reactants in chemical reactions that occur in an *aqueous solution*. The chemical equations encountered to this point in the text have been *molecular* equations in which the *complete* chemical formulas for reactants and products have been used. With net ionic equations, the focus is on ionic species present rather than on the substances from which the ions were produced via dissociation or ionization (Sec. 10.1).

Oxidation-reduction chemical equations describe chemical reactions in which electron transfer from one reactant to another reactant occurs. The equations for such oxidation-reduction processes are often more difficult to balance than those for non-oxidation-reduction reactions. The equation balancing procedures previously considered (Sec. 10.3) are often not adequate for balancing many of these equations. Two systematic methods for balancing oxidation-reduction equations, both of which focus on the electron-transfer that has occurred, are available for use with such equations.

Net ionic equations are the subject matter for the early parts of the chapter and discussion involving oxidation-reduction equations then follows.

15.2 ELECTROLYTES

Numerous water-soluble compounds form aqueous solutions in which ions, generated by ionization or dissociation, are present. This situation was encountered several times in the previous chapter in the discussions about acids, bases, and salts (Secs. 14.5 and 14.6).

Aqueous solutions in which ions are present conduct electricity. The greater the number of ions present, the greater the conducting ability of the solution. Since acids, bases, and water-soluble salts all produce ions in solution, solutions of these substances conduct electricity. Acids, bases, and water-soluble salts are thus *electrolytes*. An **electrolyte** *is a substance whose aqueous solution conducts electricity*. It is the presence of ions (charged particles) that causes a solution to be electrically conductive.

Some water-soluble substances, such as sucrose (table sugar), glucose, and isopropyl alcohol, do not conduct electricity. These substances are called *nonelectrolytes*. A **nonelectrolyte** *is a substance whose aqueous solution does not conduct electricity*.

Electroloytes can be divided into two groups—*strong* electrolytes and *weak* electrolytes. A **strong electrolyte** *is a substance that ionizes/dissociates to a large extent (70%–100%) into ions in an aqueous solution.* Strong electrolytes produce strongly conducting solutions. Strong acids (Sec. 14.5), strong bases (Sec. 14.5), and soluble salts (Sec 13.4) are strong electrolytes. A **weak electrolyte** *is a substance that ionizes/dissociates to a small extent into ions in an aqueous solution.* Weak electrolytes produce solutions that are intermediate between those containing strong electrolytes and those containing nonelectrolytes in their ability to conduct an electric current. Weak acids and weak bases (Sec. 14.5) constitute the weak electrolytes.

Whether a substance is an electrolyte in a solution can be determined by using a conductivity apparatus that involves a battery, a light bulb, and two metal electrodes connected as shown in Figure 15.1. If the medium between the electrodes (the solution) is a conductor of electricity, the light bulb glows. A strong glow indicates a strong electrolyte, a faint glow indicates a weak electrolyte, and no glow indicates a nonelectrolyte.

The existence of solutes that in a solution produce many ions (strong electrolytes), a few ions (weak electrolytes), or no ions (nonelectrolytes) becomes the basis for the writing of net ionic equations. The focus for such equations is the ions produced by a strong electrolyte.

FIGURE 15.1 This simple conductivity apparatus can be used to distinguish among strong electrolyes, weak electrolytes, and nonelectrolytes. The light bulb glows strongly for strong electrolytes (left), weakly for weak electrolytes (center), and not at all for nonelectrolytes (right).

Richard Megna - Fundamental Photographs

15.3 IONIC AND NET IONIC EQUATIONS

Up to this point in the text, most equations we have used have been *molecular equations,* equations where the complete formulas of all reactants and products are shown. Molecular equations are the starting point for deriving *ionic equations,* which in turn lead to *net ionic equations.* An **ionic equation** *is an equation in which the formulas of the predominant form of each compound in an aqueous solution are used; strong electrolytes are written in ionic form, and weak electrolytes and nonelectrolytes are written in molecular form.* Net ionic equations are derived from ionic equations. A **net ionic equation** *is an ionic equation from which nonparticipating (spectator) species have been eliminated.*

The differences between molecular, ionic, and net ionic equations can best be illustrated by examples. Let us consider the chemical reaction that results when a solution of potassium chloride (KCl) is mixed with a solution of silver nitrate ($AgNO_3$). An insoluble salt, silver chloride (AgCl), is produced as a result of the mixing. The *molecular equation* for this reaction is

$$AgNO_3(aq) + KCl(aq) \longrightarrow KNO_3(aq) + AgCl(s)$$

Three of the four substances involved in this reaction—$AgNO_3$, KCl, and KNO_3— are soluble salts (strong electrolytes) and thus exist in solution in ionic form. This is shown by writing the *ionic equation* for the reaction.

$$\underbrace{Ag^+(aq) + NO_3^-(aq)}_{AgNO_3 \text{ in ionic form}} + \underbrace{K^+(aq) + Cl^-(aq)}_{KCl \text{ in ionic form}} \longrightarrow \underbrace{K^+(aq) + NO_3^-(aq)}_{KNO_3 \text{ in ionic form}} + AgCl(s)$$

A chemical reaction such as the one discussed here (the reaction of aqueous solutions of $AgNO_3$ and KCl) in which a solid product is produced is called a *precipitation reaction.* The *precipitate* is the solid produced in the chemical reaction. The solid is said to have *precipitated* out of the solution.

In this equation each of the three soluble salts is shown in dissociated (ionic) form rather than in undissociated form. A close look at this ionic equation shows that the potassium ions (K^+) and nitrate ions (NO_3^-) appear on both sides of the equation, indicating that they did not undergo any chemical change. In other words, they are *spectator ions;* they did not participate in the reaction. The *net ionic equation* for this reaction is written by dropping (cancelling) all spectator ions from the ionic equation. (The spectator ions dropped must occur in equal numbers on both sides of the equation.) In our case, the net ionic equation becomes

$$Ag^+(aq) + Cl^-(aq) \longrightarrow AgCl(s)$$

This net ionic equation indicates that the product AgCl was formed by the reaction of silver ions (Ag^+) with chloride ions (Cl^-). It totally ignores the presence of those ions that are not taking part in the reaction. Thus, a net ionic equation focuses on only those species in a solution actually involved in a chemical reaction. It does not give all species present in the solution.

If you can write equations in molecular form, you will find it a straightforward process to convert such equations to net ionic form. Follow these three steps:

1. Check the given molecular equation to make sure that it is balanced.
2. Expand the molecular equation into an ionic equation.
3. Convert the ionic equation into a net ionic equation by eliminating spectator ions.

In expanding a molecular equation into an ionic equation (step 2), you must decide whether to write each reactant and product in dissociated (ionic) form or undissociated (molecular) form. The following rules serve as guidelines in making such decisions.

1. Strong electrolytes (Sec 15.2), substances that dissociate/ionize completely or to a large extent in an aqueous solution, are written in ionic form. They include
   a. all soluble salts (see Sec. 13.4 for solubility rules).
   b. all strong acids (see Table 14.2).
   c. all strong bases (see Table 14.4).

2. Weak electrolytes (Sec 15.2), substances that dissociate/ionize to a small extent in an aqueous solution are written in molecular form as they exist primarily in a solution in molecular form: only a few ions are produced. Soluble weak acids and soluble weak bases are weak electrolytes. The following acids and bases fall into this "weak" category:

a. All acids not listed in Table 14.2 as strong acids. Common examples are HNO_2, HF, H_2S, $HC_2H_3O_2$, H_2CO_3, and H_3PO_4.

b. All bases not listed in Table 14.4 as strong bases. Aqueous ammonia (NH_3) is the most common weak base.

3. All insoluble substances (solids, liquids, and gases), whether ionic or covalent, exist as molecules or neutral ionic units and are written as such.

4. All soluble covalent substances, for example, carbon dioxide (CO_2) or sucrose ($C_{12}H_{22}O_{11}$), are written in molecular form as they are nonelectrolytes.

5. If water, the solvent, appears in the equation, it is written in molecular form.

Now, we apply these guidelines by writing some net ionic equations.

EXAMPLE 15.1 **Converting a Molecular Equation into a Net Ionic Equation**

Write the net ionic equation for the following aqueous solution reaction.

$$MgCl_2 + AgNO_3 \longrightarrow Mg(NO_3)_2 + AgCl \qquad \text{(unbalanced equation)}$$

SOLUTION

Step 1 To balance the given molecular equation, the coefficient 2 must be placed in front of both $AgNO_3$ and AgCl.

$$MgCl_2 + 2\,AgNO_3 \longrightarrow Mg(NO_3)_2 + 2\,AgCl$$

Step 2 A decision must be made whether to write each reactant and product in ionic or molecular form. Let us consider them one by one.

$MgCl_2$: This is a soluble salt. All chloride salts are soluble except AgCl, $PbCl_2$, and Hg_2Cl_2. Thus, $MgCl_2$ will be written in ionic form: $Mg^{2+} + 2\,Cl^-$. Note that three ions (one Mg^{2+} ion and two Cl^- ions) are produced from the dissociation of one $MgCl_2$ formula unit.

$AgNO_3$: This is a soluble salt; all nitrates are soluble. Thus, it is written in ionic form. Each $AgNO_3$ formula unit (there are two) produces one Ag^+ ion and one NO_3^- ion.

$Mg(NO_3)_2$: All nitrate salts are soluble. Thus, $Mg(NO_3)_2$ will be written in ionic form. Three ions are produced upon dissociation of one $Mg(NO_3)_2$ formula unit: one Mg^{2+} ion and two NO_3^- ions.

AgCl: The solubility rules indicate that this compound is an insoluble salt. Thus, it will be written in molecular form in the ionic equation.

The ionic equation will have the form

$$Mg^{2+} + 2\,Cl^- + 2\,Ag^+ + 2\,NO_3^- \longrightarrow Mg^{2+} + 2\,NO_3^- + 2\,AgCl$$

Note how the coefficient 2 in front of $AgNO_3$ and AgCl in the molecular equation affects the ionic equation. The dissociation of two $AgNO_3$ formula units produces two Ag^+ ions and two NO_3^- ions. Similarly, two AgCl formula units are present.

Step 3 Inspection of the ionic equation shows that the Mg^{2+} ion and the two NO_3^- ions are spectator ions. Cancellation of these ions from the equation will give the net ionic equation.

$$\cancel{Mg^{2+}} + 2\,Cl^- + 2\,Ag^+ + \cancel{2\,NO_3^-} \longrightarrow \cancel{Mg^{2+}} + \cancel{2\,NO_3^-} + 2\,AgCl$$

$$2\,Ag^+ + 2\,Cl^- \longrightarrow 2\,AgCl$$

The coefficients in the net ionic equation should be the smallest set of numbers that correctly balance the equation. In this case, all the coefficients are divisible by two. Dividing by two, we get

$$Ag^+ + Cl^- \longrightarrow AgCl$$

Answer Double Check:

Is the net ionic equation a balanced equation? Yes, it is. To be balanced, it must pass two checks: (1) an atom balance and (2) a charge balance. Atom-wise, the equation is balanced; there is one atom of Ag and one atom of Cl on each side of the equation. Charge-wise, the equation is also balanced. On the left side of the equation, the $+1$ and -1 charges associated with the ions add up to zero. The charge on the other side of the equation is zero because no ions are present.

▶ **Practice Exercise 15.1**   Write the net ionic equation for the following aqueous solution reaction.

$$Pb(NO_3)_2 + Na_2SO_4 \longrightarrow NaNO_3 + PbSO_4 \text{ (unbalanced equation)}$$

Answers to all practice exercises in this chapter are found in the back-of-the-book answer section.

EXAMPLE 15.2   **Converting a Molecular Equation into a Net Ionic Equation**

Write the net ionic equation for the following aqueous solution reaction.

$$H_2S + AlI_3 \longrightarrow Al_2S_3 + HI \qquad \text{(unbalanced equation)}$$

SOLUTION

Step 1 Balancing the molecular equation, we get

$$3\,H_2S + 2\,AlI_3 \longrightarrow Al_2S_3 + 6\,HI$$

Step 2 The expansion of the molecular equation into an ionic equation is accomplished by the following analysis.

H_2S: This is a weak acid. All weak acids are written in molecular form in ionic equations.

AlI_3: This is a soluble salt. All iodide salts are soluble, with three exceptions; this is not one of the exceptions. Soluble salts are written in ionic form in ionic equations.

Al_2S_3: This is an insoluble salt. All sulfides are insoluble except for groups IA and IIA and NH_4^+. Thus, Al_2S_3 will remain in molecular form in the ionic equation.

HI: This compound is an acid. It is one of the seven strong acids listed in Table 14.2. Strong acids are written in ionic form.

The ionic equation for the reaction is

$$3\,H_2S + 2\,Al^{3+} + 6\,I^- \longrightarrow Al_2S_3 + 6\,H^+ + 6\,I^-$$

Note again that the coefficients present in the balanced molecular equation must be taken into consideration when determining the total number of ions produced from dissociation. On dissociation, an AlI_3 formula unit produces four ions: one Al^{3+} ion and three I^- ions. This number must be doubled for the ionic equation because AlI_3 carries the coefficient 2 in the balanced molecular equation. Similar considerations apply to the HI in this equation.

Step 3 Inspection of the ionic equation shows that only I^- ions (six of them) are spectator ions. Cancellation of these ions from the equation gives the net ionic equation.

$$3\,H_2S + 2\,Al^{3+} + \cancel{6\,I^-} \longrightarrow Al_2S_3 + 6\,H^+ + \cancel{6\,I^-}$$

$$3\,H_2S + 2\,Al^{3+} \longrightarrow Al_2S_3 + 6\,H^+$$

Answer Double Check:

Is the net ionic equation balanced relative to atoms and also relative to charge? Yes. The atom balance is six hydrogen atoms, three sulfur atoms, and two aluminum atoms on each side of the equation. The charge balance is a +6 charge on each side of the equation. The charge balance on each side of the equation can be at any number; there is no requirement that the charge be zero on each side. Two aluminum ions (3+) generate a 6+ charge and six hydrogen ions (1+) generate a 6+ charge.

▶ **Practice Exercise 15.2**   Write the net ionic equation for the following aqueous solution reaction.

$$H_3PO_4 + CuCl_2 \longrightarrow Cu_3(PO_4)_2 + HCl \text{ (unbalanced equation)}$$

EXAMPLE 15.3 **Converting a Molecular Equation into a Net Ionic Equation**

Write the net ionic equation for the following aqueous solution reaction.

$$HNO_3 + LiOH \longrightarrow LiNO_3 + H_2O$$

SOLUTION

Step 1 All coefficients in this equation are 1; the equation is balanced as written.

$$HNO_3 + LiOH \longrightarrow LiNO_3 + H_2O$$

Step 2 The expansion of the molecular equation into an ionic equation is based on the following analysis.

HNO_3: This compound is an acid. It is one of the seven strong acids listed in Table 14.2. Strong acids are written in ionic form in ionic equations.

$LiOH$: This compound is a base. It is one of the strong bases listed in Table 14.4. Strong bases are written in ionic form in ionic equations.

$LiNO_3$: This compound is a soluble salt. All nitrate salts are soluble. Thus, $LiNO_3$ is written in ionic form.

H_2O: This compound is a covalent compound; two nonmetals are present. Covalent compounds are always written in molecular form.

The ionic equation for the reaction, using the above information, is

$$H^+ + NO_3^- + Li^+ + OH^- \longrightarrow Li^+ + NO_3^- + H_2O$$

Step 3 Inspection of the ionic equation shows that NO_3^- ions and Li^+ ions are spectator ions. Cancellation of these ions from the equation gives the net ionic equation.

$$H^+ + \cancel{NO_3^-} + \cancel{Li^+} + OH^- \longrightarrow \cancel{Li^+} + \cancel{NO_3^-} + H_2O$$

$$H^+ + OH^- \longrightarrow H_2O$$

Answer Double Check:

Is the net ionic equation balanced relative to atoms and charge? Yes. Atom-wise, there are two hydrogens and one oxygen on each side of the equation. Charge-wise, the net charge on each side of the equation is zero; that 1+ and 1– ionic charges on the left side add up to zero.

▶ **Practice Exercise 15.3**   Write the net ionic equation for the following aqueous solution reaction.

$$H_2SO_4 + KOH \longrightarrow K_2SO_4 + H_2O \text{ (unbalanced equation)}$$

The net ionic equation for the reaction of a strong acid with a strong base, independent of the identity of the strong acid and the strong base, will always be $H^+ + OH^- \longrightarrow H_2O$.

15.4 OXIDATION–REDUCTION TERMINOLOGY

Oxidation–reduction reactions are a very important class of chemical reactions. They occur all around us and even within us. The bulk of the energy needed for the functioning of all living organisms, including humans, is obtained from food via oxidation–reduction processes. Such diverse phenomena as the electricity obtained from a battery to start a car, the use of natural gas to heat a home, iron rusting, and the functioning of antiseptic agents to kill or prevent the growth of bacteria all involve oxidation–reduction reactions. In short, knowledge of this type of reaction is fundamental to understanding many biological and technological processes.

The terms *oxidation* and *reduction*, like the terms *acid* and *base* (Sec. 14.1), have several definitions. Historically, the word *oxidation* was first used to describe the reaction of a substance with oxygen. According to this historical definition, each of the following reactions involves oxidation.

$$4\,Fe + 3\,O_2 \longrightarrow 2\,Fe_2O_3$$
$$S + O_2 \longrightarrow SO_2$$
$$CH_4 + 2\,O_2 \longrightarrow CO_2 + 2\,H_2O$$

The substance on the far left in each of these equations is said to have been *oxidized*.

Originally, the term *reduction* referred to processes where oxygen was removed from a compound. A particularly common type of reduction reaction, according to this original definition, is the removal of oxygen from a metal oxide to produce the free metal.

$$CuO + H_2 \longrightarrow Cu + H_2O$$
$$2\,Fe_2O_3 + 3\,C \longrightarrow 4\,Fe + 3\,CO_2$$

The word *reduction* comes from the reduction in mass of the metal-containing species; the metal has a mass less than that of the metal oxide.

Today the words *oxidation* and *reduction* are used in a much broader sense. Current definitions include the previous examples but also much more. It is now recognized that the same changes brought about in a substance from reaction with oxygen can be

Oxidation involves the *loss* of electrons, and reduction involves the *gain* of electrons. Students often have trouble remembering which is which. Two helpful mnemonic devices follow.

LEO the lion says *GER*. Loss of Electrons: Oxidation. Gain of Electrons: Reduction. *OIL RIG* Oxidation *Is* Loss (of electrons). Reduction *Is* Gain (of electrons).

caused by reaction with numerous non-oxygen-containing substances. For example, consider the following reactions:

$$2\,Mg + O_2 \longrightarrow 2\,MgO$$

$$Mg + S \longrightarrow MgS$$

$$Mg + F_2 \longrightarrow MgF_2$$

$$3\,Mg + N_2 \longrightarrow Mg_3N_2$$

In each of these reactions, magnesium metal is converted to a magnesium compound that contains Mg^{2+} ions. The process is the same—the changing of magnesium atoms, through the loss of two electrons, to magnesium ions; the only difference is the identity of the substance that causes magnesium to undergo the change. All these reactions are considered to involve *oxidation* by the current definition. **Oxidation** *is the process whereby a substance in a chemical reaction loses one or more electrons*. The current definition for *reduction* involves the use of similar terminology. **Reduction** *is the process whereby a substance in a chemical reaction gains one or more electrons*.

Oxidation and reduction are complementary processes rather than isolated phenomena. They *always* occur together; you cannot have one without the other. If electrons are lost by one species, they cannot just disappear; they must be gained by another species. Electron transfer, then, is the basis for oxidation and reduction. An **oxidation–reduction reaction** *is a chemical reaction in which transfer of electrons between reactants occurs*. The term *oxidation–reduction reaction* is often shortened to the term *redox reaction*. **Redox reaction** *is abbreviated terminology for an oxidation-reduction reaction*.

The terms *oxidizing agent* and *reducing agent* sometimes cause confusion because the oxidizing agent is not oxidized (it is reduced) and the reducing agent is not reduced (it is oxidized). By simple analogy, a travel agent is not the one who takes a trip—he or she is the one who causes the trip to be taken.

There are two different ways of looking at the reactants in a redox reaction. First, the reactants can be viewed as being acted upon. From this perspective one reactant is *oxidized* (the one that loses electrons) and one is *reduced* (the one that gains electrons). Second, the reactants can be looked at as bringing about the reaction. In this approach the terms *oxidizing agent* and *reducing agent* are used. An **oxidizing agent** *is the reactant in a redox reaction that causes oxidation by accepting electrons from the other reactant*. Such acceptance, the gain of electrons, means that the oxidizing agent itself is reduced. Similarly, the **reducing agent** *is the reactant in a redox reaction that causes reduction by providing electrons for the other reactant to accept*. As a result of providing electrons, the reducing agent itself becomes oxidized. Note, then, that the reducing agent and substance oxidized are one and the same, as are the oxidizing agent and substance reduced.

$$\text{Substance oxidized} = \text{reducing agent}$$

$$\text{Substance reduced} = \text{oxiding agent}$$

Table 15.1 summarizes the terms presented in this section.

TABLE 15.1 Oxidation–Reduction Terminology in Terms of Loss and Gain of Electrons

Terms Associated with the Loss of Electrons	Terms Associated with the Gain of Electrons
Process of oxidation	Process of reduction
Substance oxidized	Substance reduced
Reducing agent	Oxidizing agent

15.5 OXIDATION NUMBERS

Oxidation numbers are used to help determine whether oxidation or reduction has occurred in a reaction and, if such is the case, the identity of the oxidizing or reducing agents. An **oxidation number** *is the charge that an atom appears to have when the electrons in each bond it is participating in are assigned to the more electronegative of the two atoms involved in the bond.**

Consider an HCl molecule, a molecule in which there is one single bond involving two shared electrons.

$$\text{H:}\overset{\cdot\cdot}{\underset{\cdot\cdot}{\text{Cl}}}\text{:}$$

According to the definition for oxidation number, the electrons in this bond are assigned to the chlorine atom (the more electronegative atom; Sec. 7.18). This results in the chlorine atom having one more electron than a neutral Cl atom; hence, the oxidation number of chlorine is –1 (one extra electron). At the same time, the H atom in the HCl molecule has one fewer electron than a neutral H atom; its electron was given to the chlorine. This electron deficiency of one results in an oxidation number of +1 for hydrogen.

As a second example, consider the molecule CF_4.

$$\overset{\cdot\cdot}{\underset{\cdot\cdot}{\text{F}}}$$
$$\overset{\cdot\cdot}{\underset{\cdot\cdot}{\text{F}}}\text{:C:}\overset{\cdot\cdot}{\underset{\cdot\cdot}{\text{F}}}$$
$$\overset{\cdot\cdot}{\underset{\cdot\cdot}{\text{F}}}$$

Fluorine is more electronegative than carbon. Hence, the two shared electrons in each of the four carbon–fluorine bonds are assigned to the fluorine atom. Each F atom thus gains an extra electron, resulting in each F atom's having a –1 oxidation number. The carbon atom loses a total of four electrons, one to each F atom, as a result of the electron "assignments." Hence, its oxidation number is +4, indicating the loss of the four electrons.

As a third example, consider the N_2 molecule where like atoms are involved in a triple bond.

$$\text{:N:::N:}$$

Since the identical atoms are of equal electronegativity, the shared electrons are "divided" equally between the two atoms; each N receives three of the bonding electrons to count as its own. This results in each N atom having five valence electrons (three from the triple bond and two nonbonding electrons), the same number of valence electrons as in a neutral N atom. Hence, the oxidation number of N in N_2 is zero.

Before going any further in our discussion of oxidation numbers, it should be noted that *calculated* oxidation numbers are *not* actual charges on atoms. This is why the phrase "appears to have" is found in the definition of oxidation number given at the start of this section. In assigning oxidation numbers, we assume when we give the bonding electrons to the more electronegative element that each bond is ionic (complete transfer of electrons). We know that this is not always the case. Sometimes it is a good approximation, sometimes it is not. Why, then, do we do this when we know that

1. If the oxidation state of an atom in a molecule is +*n*, then the atom "owns" *n* fewer electrons in the molecule than it would as a free atom.
2. If the oxidation state of an atom in a molecule is –*n*, then the atom "owns" *n* more electrons in the molecule than it would as a free atom.

---

*In some textbooks the term *oxidation state* is used in place of oxidation number. In other textbooks the two terms are used interchangeably. We will use oxidation number.

it does not always correspond to reality? Oxidation numbers, as we shall see shortly, serve as a very convenient device for keeping track of electron transfer in redox reactions. Even though they do not always correspond to physical reality, they are very useful entities.

In principle, the procedures used to determine oxidation numbers for the atoms in the molecules HCl, CF_4, and N_2 can be used to determine oxidation numbers in all compounds. However, the procedures become very laborious for substances that have complicated Lewis structures. In practice, an alternative, much simpler procedure that does not require the drawing of Lewis structures is used to obtain oxidation numbers. This alternative procedure is based on a set of operational rules that are consistent with and derivable from the general definition for oxidation numbers. The operational rules are as follows:

Rule 1 The oxidation number of any *free element* (an element not combined chemically with another element) is zero.

For example, O in O_2, P in P_4, and S in S_8 all have an oxidation number of zero. This rule is independent of the molecular complexity of the element.

Rule 2 The oxidation number of any monoatomic ion is equal to the charge on the ion.

For example, the Na^+ ion has an oxidation number of +1, and the S^{2-} ion has an oxidation number of –2.

Rule 3 The oxidation numbers of groups IA and IIA elements in compounds are always +1 and +2, respectively.

Rule 4 The oxidation number of fluorine in compounds is always –1 and that of the other group VIIA elements (Cl, Br, and I) is usually –1.

The exception for these latter elements is when they are bonded to more electronegative elements. In this case they are assigned positive oxidation numbers.

Rule 5 The usual oxidation number for oxygen in compounds is –2.

The exceptions occur when oxygen is bonded to the more electronegative fluorine (O then is assigned a positive oxidation number) or found in compounds containing oxygen–oxygen bonds (peroxides). In peroxides the oxidation number –1 is assigned to oxygen. Peroxides exist for hydrogen (H_2O_2), group IA elements (Na_2O_2, etc.), and group IIA elements (BaO_2, etc.).

Rule 6 The usual oxidation number for hydrogen in compounds is +1.

The exception occurs in hydrides, compounds where hydrogen is bonded to a metal of lower electronegativity. In such compounds hydrogen is assigned an oxidation number of –1. Examples of hydrides are NaH, CaH_2, and LiH.

Rule 7 In binary compounds, the element with the greater electronegativity is assigned a negative oxidation number equal to its charge as an anion in its ionic compounds.

For example, in the compound AlN, N (the more electronegative element) is assigned an oxidation number of –3, the charge on a nitride ion (N^{3-}).

Rule 8 The algebraic sum of the oxidation numbers of all atoms in a neutral molecule must be zero.

Rule 9 The algebraic sum of the oxidation numbers of all atoms in a polyatomic ion is equal to the charge on the ion.

The use of these rules is illustrated in Example 15.4.

EXAMPLE 15.4 **Assigning Oxidation Numbers to Elements in a Compound or Polyatomic Ion**

Assign oxidation numbers to each element in the following chemical species.

    a. SO_3 **b.** N_2H_4 **c.** $KMnO_4$ **d.** ClO_4^-

SOLUTION

    a. Oxygen has an oxidation number of –2 (rule 5 or rule 7). The oxidation number of S can be calculated by rule 8. Letting x equal the oxidation number of S, we have

$$\text{S: 1 atom} \times (x) \quad = x$$
$$\text{O: 3 atoms} \times (-2) = \underline{-6}$$
$$\text{sum} = 0 \quad \text{(rule 8)}$$

Solving for x algebraically, we get

$$x + (-6) = 0$$
$$x = +6$$

Consequently, the oxidation number of sulfur is +6 in the compound SO_3.

    b. Hydrogen has an oxidation number of +1 (rule 6). Rule 8 will allow us to calculate the oxidation number of N; the sum of the oxidation numbers must be zero. Letting x equal the oxidation number of N, we have

$$\text{H: 4 atoms} \times (+1) = +4$$
$$\text{N: 2 atoms} \times (x) \quad = \underline{2x}$$
$$\text{sum} = 0 \quad \text{(rule 8)}$$

Solving for x algebraically, we get

$$2x + (+4) = 0$$
$$x = -2$$

Thus, the oxidation number of nitrogen in N_2H_4 is –2. Note that the oxidation number of N is not –4 (the calculated charge associated with two N atoms). Oxidation number is always specified on a *per atom* basis.

    c. Potassium has an oxidation number of +1 (rule 3), and oxygen has an oxidation number of –2 (rule 5 or rule 7). Letting x equal the oxidation number of manganese and using rule 8, we get

$$\text{K: 1 atom} \times (+1) \quad = +1$$
$$\text{Mn: 1 atom} \times (x) \quad = x$$
$$\text{O: 4 atoms} \times (-2) = \underline{-8}$$
$$\text{sum} = 0 \quad \text{(rule 8)}$$

Solving for x algebraically, we get

$$(+1) + x + (-8) = 0$$
$$x = +7$$

Thus, the oxidation number of manganese in $KMnO_4$ is +7.

d. According to rule 9, the sum of the oxidation numbers must equal −1, the charge on this polyatomic ion. The oxidation number of oxygen is −2 (rule 5). Chlorine will have a positive oxidation number because it is bonded to a more electronegative element (rule 4). Letting x equal the oxidation number of chlorine, we have

$$\text{Cl: 1 atom} \times (x) = x$$
$$\text{O: 4 atoms} \times (-2) = \underline{-8}$$
$$\text{sum} = -1 \quad \text{(rule 8)}$$

Solving for x algebraically, we get

$$x + (-8) = -1$$
$$x = +7$$

Thus, chlorine has an oxidation number of +7 in this ion.

Answer Double Check:

For each of the chemical species, does the sum of the oxidation numbers of all atoms equal the net charge on the species? Yes. In the first three parts, the oxidation number sum is zero, consistent with these chemical species having zero net charge. In the fourth part, the oxidation number sum adds to −1, the charge on the ClO_4^- ion [$4(-2) + 7 = -1$].

▶ **Practice Exercise 15.4** Assign oxidation numbers to each element in the following chemical species.

a. H_2SO_4 b. CS_2 c. $KClO_3$ d. NH_2^-

Application of oxidation number rules in the manner illustrated in Example 15.4 enables one to quickly assign oxidation numbers to the elements in a wide variety of compounds.

Many elements display a range of oxidation numbers in their various compounds. For example, nitrogen exhibits oxidation numbers ranging from −3 to +5 in various compounds. Selected examples are

NH_3	N_2H_4	N_2O	NO	N_2O_3	NO_2	HNO_3
−3	−2	+1	+2	+3	+4	+5

As shown in this listing of nitrogen-containing compounds, the oxidation number of an atom is written *underneath* the atom in the formula. This convention is used to avoid confusion with the charge on an ion.

Although not common, nonintegral oxidation numbers are possible. For example, the oxidation number of iron in the compound Fe_3O_4 is +2.67. The oxidation numbers of the oxygens in the compound add up to −8. Therefore, the iron atoms must have an oxidation number sum of +8. Dividing +8 by 3 (the number of iron atoms) gives +2.67.

Oxidizing and reducing agents were previously defined in terms of loss and gain of electrons (Table 15.1). They can also be defined in terms of changes in oxidation numbers.

TABLE 15.2 Oxidation–Reduction Terminology in Terms of Oxidation Number Change

Terms Associated with an Increase in Oxidation Number	Terms Associated with a Decrease in Oxidation Number
Process of oxidation	Process of reduction
Substance oxidized	Substance reduced
Reducing agent	Oxidizing agent

An **oxidizing agent** *is the reactant in a redox reaction that contains the element that shows a decrease in oxidation number.* Since the oxidizing agent is the substance reduced in a reaction, *reduction involves a decrease in oxidation number.* A **reducing agent** *is the reactant in a redox reaction that contains the element that shows an increase in oxidation number.* Since the reducing agent is the substance oxidized in a reaction, *oxidation involves an increase in oxidation number.*

Table 15.2 summarizes the relationships between oxidation–reduction terminology and oxidation number changes. A comparison of Table 15.2 with Table 15.1 shows that the loss of electrons and oxidation number increases are synonymous as are the gain of electrons and oxidation number decreases. The fact that the oxidation number becomes more positive (increases) as electrons are lost is consistent with our understanding that electrons are negatively charged.

EXAMPLE 15.5	**Determining Oxidation Numbers and Identifying Oxidizing Agents and Reducing Agents**

Determine oxidation numbers for each atom in the following reactions, and identify the oxidizing and reducing agents.

a. $2\,NO + O_2 \longrightarrow 2\,NO_2$
b. $Zn + 2\,HCl \longrightarrow ZnCl_2 + H_2$
c. $Cl_2 + 2\,I^- \longrightarrow I_2 + 2\,Cl^-$

SOLUTION

The oxidation numbers are calculated by the methods illustrated in Example 15.4.

a. $2\,NO\ +\ O_2\ \longrightarrow\ 2\,NO_2$
  $+2\ -20+4\ -2$
  rules 5, 8   rule 1   rules 5, 8

The oxidation number of N has increased from +2 to +4. Therefore, the substance that contains N, NO, has been oxidized and is the reducing agent.

  The oxidation number of the O in O_2 has decreased from 0 to –2. Therefore, the O_2 has been reduced and is the oxidizing agent.

b. $Zn\ +\ 2\,HCl\ \longrightarrow\ ZnCl_2\ +\ H_2$
  $0+1\ -1+2\ -10$
  rule 1   rules 6, 8   rules 4, 8   rule 1

The oxidation number of Zn has increased from 0 to +2. An increase in oxidation number is associated with oxidation. Therefore, the element Zn has been oxidized and is the reducing agent.

The oxidation number of H has decreased from +1 to 0. A decrease in oxidation number is associated with reduction. Therefore, the HCl, the hydrogen-containing compound, is the oxidizing agent.

c. $Cl_2 + 2I^- \longrightarrow I_2 + Cl^-$

$0 -1 0 -1$

rule 1 rule 2   rule 1 rule 2

The oxidation number of I has increased from –1 to 0. Thus, I^-, the iodine-containing reactant, has been oxidized and is the reducing agent.

The oxidation number of Cl has decreased from 0 to –1. Thus, Cl_2, the chlorine-containing reactant, has been reduced and is the oxidizing agent.

Answer Double Check:

Are the answers consistent with the concept that oxidizing agents contain the element that shows a decrease in oxidation number and reducing agents contain the element that shows an increase in oxidation number? Yes. In all cases this is the situation: oxidizing agent—oxidation number decreases, and reducing agent—oxidation number increases.

▶ **Practice Exercise 15.5** Determine the oxidation numbers for each atom in the following reactions, and identify the oxidizing and reducing agents.

   a. $Fe + Cu^{2+} \longrightarrow Cu + Fe^{2+}$
   b. $2SO_2 + O_2 \longrightarrow 2SO_3$
   c. $PH_3 + 2NO_2 \longrightarrow H_3PO_4 + N_2$

15.6 REDOX AND NONREDOX CHEMICAL REACTIONS

Two classification systems for chemical reactions are in common use. We have now encountered both of them.

The first system, presented initially in Section 10.6, recognized five types of reactions:

   1. Synthesis $(X + Y \longrightarrow XY)$
   2. Decomposition $(XY \longrightarrow X + Y)$
   3. Single-replacement $(X + YZ \longrightarrow Y + XZ)$
   4. Double-replacement $(AX + BY \longrightarrow AY + BX)$
   5. Combustion (reaction with O_2)

The second system involves two reaction types:

   1. Oxidation–reduction (or redox)
   2. Non-oxidation–reduction (or nonredox)

As we have just learned (Sec. 15.5), reactions in which oxidation numbers change are called oxidation–reduction reactions. A **non-oxidation–reduction reaction** *is a chemical reaction in which oxidation numbers do not change.*

These two classification systems are not mutually exclusive and are commonly used together. For example, a particular reaction may be characterized as a single-replacement redox reaction.

Synthesis reactions with only elements as reactants are always oxidation–reduction reactions. Oxidation-number changes must occur because all elements (the reactants) have an oxidation number of zero and all of the constituent elements of a compound *cannot* have oxidation numbers of zero. Synthesis reactions in which compounds are the reactants may or may not be redox reactions.

$$S + O_2 \longrightarrow SO_2 \qquad \text{(redox synthesis)}$$

$$K_2O + H_2O \longrightarrow 2\,KOH \qquad \text{(nonredox synthesis)}$$

$$2\,NO + O_2 \longrightarrow 2\,NO_2 \qquad \text{(redox synthesis)}$$

Both redox and nonredox decomposition reactions are common. At sufficiently high temperatures, all compounds can be broken down (decomposed) into their constituent elements. Such reactions, where only elements are the products, are always redox reactions. Decomposition reactions where compounds are the products can be redox or nonredox reactions.

$$2\,CuO \longrightarrow 2\,Cu + O_2 \qquad \text{(redox decomposition)}$$

$$2\,KClO_3 \longrightarrow 2\,KCl + 3\,O_2 \qquad \text{(redox decomposition)}$$

$$CaCO_3 \longrightarrow CaO + CO_2 \qquad \text{(nonredox decomposition)}$$

Single-replacement reactions are always redox reactions. By definition, an element and a compound are reactants and an element and a compound are products. The elements always undergo oxidation number change (see Fig. 15.2). Two of the reaction types studied in Chapter 14 are redox single-replacement reactions—the reaction between an acid and an active metal (Sec. 14.7) and the reaction between a metal and an aqueous salt solution (Sec. 14.9).

Double-replacement reactions generally involve acids, bases, and salts in an aqueous solution. In such reactions ions, which maintain their identity, are generally trading places. Such reactions will always be nonredox reactions. All acid–base neutralization reactions (Sec. 14.7) are nonredox double-replacement reactions.

Combustion reactions (Sec. 10.6) are always redox reactions. However, as mentioned in Section 10.6, they do not fit any of the four general reaction patterns of synthesis, decomposition, single-replacement, and double-replacement. Features common to all combustion reactions are the necessity of oxygen (O_2) as a reactant and the presence of one or more oxides among the products.

FIGURE 15.2 Spears like this one were made by prehistoric people by the reduction of iron ore with charcoal. Such a reaction is a single-replacement redox reaction:

$$Fe_2O_3 + 3\,C \longrightarrow 2\,Fe + 3\,CO$$

(Lance head from Zur Bachar, near Jerusalem. Iron, 22 cm, 9th century BC. Reuben and Edith Hecht Collection, Haifa University, Haifa, Israel. Photo by Erich Lessing/Art Resource).

EXAMPLE 15.6 **Classifying Chemical Reactions as Redox or Nonredox**

Classify the following chemical reactions as redox or nonredox. Further classify them as synthesis, decomposition, single-replacement, double-replacement, or combustion.

a. $Ni + F_2 \longrightarrow NiF_2$

b. $Fe_2O_3 + 3\,C \longrightarrow 2\,Fe + 3\,CO$

c. $C_4H_8 + 6\,O_2 \longrightarrow 4\,CO_2 + 4\,H_2O$

d. $H_2SO_4 + 2\,NaOH \longrightarrow Na_2SO_4 + 2\,H_2O$

SOLUTION

The oxidation numbers are calculated by the method illustrated in Example 15.4.

a. $\quad Ni \quad + \quad F_2 \quad \longrightarrow \quad NiF_2$
$\qquad 0 \qquad\quad 0 \qquad\qquad +2\ -1$
$\quad$ rule 1 $\quad$ rule 1 $\qquad$ rules 4, 8

This is a *redox* reaction; the oxidation numbers of both Ni and F change. Since one substance is produced from two substances, it is also a *synthesis* reaction. We thus have a redox synthesis reaction.

b. $\quad Fe_2O_3 \quad + \quad 3\,C \quad \longrightarrow \quad 2\,Fe \quad + \quad 3\,CO$
$\quad\, +3\ -2 \qquad\quad 0 \qquad\qquad 0 \qquad\quad +2\ -2$
$\quad$ rules 5, 8 $\quad$ rule 1 $\qquad$ rule 1 $\quad$ rules 5, 8

This is a *redox* reaction; carbon is oxidized, iron is reduced. Having an element and a compound as reactants and an element and compound as products is a characteristic of a *single-replacement* reaction. That is the type of reaction we have here: Iron and carbon are exchanging places. We thus have a redox single-replacement reaction.

c. $\quad C_4H_8 \quad + \quad 6\,O_2 \quad \longrightarrow \quad 4\,CO_2 \quad + \quad 4\,H_2O$
$\quad\, -2\ +1 \qquad\quad 0 \qquad\qquad +4\ -2 \qquad\quad +1\ -2$
$\quad$ rules 6, 8 $\quad$ rule 1 $\qquad$ rules 5, 7 $\quad$ rules 5, 8

This is a *redox* reaction; the oxidation numbers of both carbon and oxygen change. This reaction is also a *combustion* reaction. We thus have a redox combustion reaction.

d. $\quad H_2SO_4 \quad + \quad 2\,NaOH \quad \longrightarrow \quad Na_2SO_4 + 2\,H_2O$
$\quad +1\ +6\ -2 \quad +1\ -2\ +1 \qquad\qquad +1\ +6\ -2 \quad +1\ -2$
$\quad$ rules 5, 6, 8 $\quad$ rules 3, 5, 6 $\qquad\quad$ rules 3, 5, 8 $\quad$ rules 5, 6

This is a *nonredox* reaction; there are no oxidation number changes. The reaction is also a *double-replacement* reaction; hydrogen and sodium are changing places, that is, "swapping partners." Thus we have a nonredox double-replacement reaction.

▶ **Practice Exercise 15.6** Classify the following reactions as redox or nonredox. Further classify them as synthesis, decomposition, single-replacement, double-replacement, or combustion.

a. $CaCO_3 \longrightarrow CaO + CO_2$

b. $2\,NH_3 \longrightarrow 3\,H_2 + N_2$

c. $Zn + 2\,HCl \longrightarrow ZnCl_2 + H_2$

d. $C_3H_8 + 5\,O_2 \longrightarrow 3\,CO_2 + 4\,H_2O$

15.7 BALANCING OXIDATION–REDUCTION EQUATIONS

Balancing an equation is not a new topic to us. In Section 10.4 we learned how to balance equations by the *inspection method*. With that method, we start with the most complicated compound within the equation and balance one of the elements in it. Then we balance the atoms of a second element, then a third, and so on until all elements are balanced. This inspection procedure is a useful method for balancing simple equations with small coefficients. However, it breaks down when applied to complicated equations.

Equations for redox reactions are often quite complicated and contain numerous reactants and products and large coefficients. Trying to balance redox equations such as

$$PH_3 + CrO_4{}^{2-} + H_2O \longrightarrow P_4 + Cr(OH)_4{}^- + OH^-$$

or

$$As_4O_6 + MnO_4{}^- + H_2O \longrightarrow AsO_4{}^{3-} + H^+ + Mn^{2+}$$

by inspection is a tedious, time-consuming, frustrating experience. Balancing such equations is, however, easily accomplished by systematic equation-balancing procedures that use oxidation numbers and focus on the fact that the numbers of electrons lost and gained in a redox reaction must be equal.

Two distinctly different approaches for systematically balancing redox equations are in common use; the *oxidation number method* and the *half-reaction method*. Each method has advantages and disadvantages. We will consider both methods.

15.8 OXIDATION NUMBER METHOD FOR BALANCING REDOX EQUATIONS

A useful feature of oxidation numbers is that they provide a rather easy method for recognizing and balancing redox equations. The steps involved in their use in this balancing process are as follows:

Step 1 Assign oxidation numbers to all atoms in the equation and determine which atoms are undergoing a change in oxidation number.

Step 2 Determine the magnitude of the change in oxidation number *per atom* for the elements undergoing a change in oxidation number.

Draw two brackets—one connecting the substance oxidized to its product and the other connecting the substance reduced to its product. Then place the oxidation number change is the middle of the bracket.

Step 3 When more than one atom of an element that changes oxidation number is present in a formula unit (of either reactant or product), determine the change in oxidation number per *formula unit*.

Indicate this change per formula unit by multiplying the oxidation number change per atom, already written on the brackets, by an appropriate factor.

Step 4 Determine multiplying factors that make the total increase in oxidation number equal to the total decrease in oxidation number.

Place them on the bracket also.

Step 5 Place in front of the oxidizing and reducing agents and their products in the equation coefficients that are consistent with the total number of atoms of the elements undergoing oxidation number change.

Step 6 Balance all other atoms in the equation except those of hydrogen and oxygen.

In doing this, do not alter the coefficients determined previously.

Step 7 Balance the charge (the sum of all the ionic charges) so that it is the same on both sides of the equation by adding H^+ or OH^- ions.

This step is necessary only when dealing with net ionic equations describing aqueous solution reactions. If the reaction takes place in an acidic solution, add H^+ ion to the side deficient in positive charge. If the reaction takes place in a basic solution, add OH^- ion to the side deficient in negative charge.

Step 8 Balance the hydrogen atoms.

For net ionic equations, H_2O must usually be added to an appropriate side of the equation to achieve hydrogen balance. Water is, of course, present in all aqueous solutions and can be either a reactant or a product.

Step 9 Balance the oxygen atoms.

The oxygens should automatically be balanced. If oxygens do not balance, there is a mistake in a previous step. Check your work.

Now let us consider some examples where these rules are applied. The first example involves a molecular equation. The second and third examples involve net ionic equations. In balancing net ionic equations, any H_2O, H^+, or OH^- present is usually left out of the unbalanced equation that we start with and then added as needed during the balancing process.

> For net ionic redox reactions, two "balances" must be made: atoms and charge. That is, there must be the same number of atoms of each element in the reactants and products, and the total charge on the reactants must equal the total charge on the products.

EXAMPLE 15.7 **Balancing a Molecular Redox Equation Using the Oxidation Number Method**

Balance the following molecular redox equation using the oxidation number method of balancing.

$$Cr + O_2 + HBr \longrightarrow CrBr_3 + H_2O$$

SOLUTION

Step 1 We identify the elements being oxidized and reduced by assigning oxidation numbers.

$$Cr + O_2 + HBr \longrightarrow CrBr_3 + H_2O$$
$$0 \quad\ \ 0 \quad +1-1 \quad\ +3-1 \quad +1-2$$

Chromium (Cr) and oxygen (O) are the elements that undergo oxidation number change.

Step 2 The change in oxidation number *per atom* is shown by drawing brackets connecting the oxidizing and reducing agents to their products and indicating the change at the middle of the bracket.

$$
\overset{0}{\mathrm{Cr}} + \overset{0}{\mathrm{O_2}} + \mathrm{HBr} \xrightarrow{\quad} \overset{+3}{\mathrm{CrBr_3}} + \overset{-2}{\mathrm{H_2O}}
$$

Change in oxidation number per atom

Step 3 For Cr the change in oxidation number per formula unit is the same as the change per atom because both Cr and $CrBr_3$, the two Cr-containing species, contain only one Cr atom. For O, the change in oxidation number per formula unit will be double the change per atom since O_2 contains two atoms. The change per formula unit is indicated by multiplying the per-atom change by an appropriate numerical factor, which is 2 in this case.

$$
\mathrm{Cr} + \mathrm{O_2} + \mathrm{HBr} \xrightarrow{\quad} \mathrm{CrBr_3} + \mathrm{H_2O}
$$

Change in oxidation number per formula unit

Step 4 For Cr, the total increase in oxidation number per formula unit is +3. For oxygen, the total decrease in oxidation number per formula unit is –4. To make the increase equal to the decrease, we must multiply the oxidation number change for the element oxidized (Cr) by 4 and the oxidation number change for the element reduced (O) by 3. This will make the increase and decrease both numerically equal to 12.

$$
\mathrm{Cr} + \mathrm{O_2} + \mathrm{HBr} \xrightarrow{\quad} \mathrm{CrBr_3} + \mathrm{H_2O}
$$

Oxidation number increase equals oxidation number decrease

Step 5 We are now ready to place coefficients in the equation in front of the oxidizing and reducing agents and their products. The bracket notation indicates that four Cr atoms undergo an oxidation number change. Place the coefficient 4 in front of both Cr and $CrBr_3$. The bracket notation also indicates that six O atoms (3×2) undergo an oxidation number decrease of two units. Thus, we need six oxygen atoms on each side. Place the coefficient 3 in front of O_2 (six atoms of O), and the coefficient 6 in front of H_2O (six atoms of O).

$$4\,\mathrm{Cr} + 3\,\mathrm{O_2} + \mathrm{HBr} \xrightarrow{\quad} 4\,\mathrm{CrBr_3} + 6\,\mathrm{H_2O}$$

The equation is only partially balanced at this point; only Cr and O atoms are balanced.

Step 6 We next balance the element Br (by inspection). There are 12 Br atoms on the right side. Thus, to obtain 12 Br atoms on the left side, we place the coefficient 12 in front of HBr.

$$4\,\mathrm{Cr} + 3\,\mathrm{O_2} + 12\,\mathrm{HBr} \xrightarrow{\quad} 4\,\mathrm{CrBr_3} + 6\,\mathrm{H_2O}$$

Step 7 This step is not needed when the equation is a molecular equation.

Step 8 In this particular equation the H atoms are already balanced. There are 12 hydrogen atoms on each side of the equation.

Step 9 If all of the previous procedures (steps) have been carried out correctly, the O atoms should automatically balance. They do. There are six O atoms on each side of the equation. The balanced equation is thus

$$4\,Cr + 3\,O_2 + 12\,HBr \longrightarrow 4\,CrBr_3 + 6\,H_2O$$

Answer Double Check:

Are there the same number of atoms of each kind on each side of the equation? Yes. The atom balance is 4 Cr, 6 O, 12 H, and 12 Br on each side of the equation.

▶ **Practice Exercise 15.7** Balance the following molecular redox equation using the oxidation number method of balancing.

$$PbO + NH_3 \longrightarrow N_2 + H_2O + Pb$$

EXAMPLE 15.8 | **Balancing a Net Ionic Redox Equation Using the Oxidation Number Method**

Balance the following net ionic redox equation using the oxidation number method of balancing.

$$Cu + NO_3{}^- \longrightarrow Cu^{2+} + NO_2$$

This reaction occurs in an acidic solution.

SOLUTION

Step 1 The elements undergoing oxidation and reduction are identified by assigning oxidation numbers

$$Cu + NO_3{}^- \longrightarrow Cu^{2+} + NO_2$$
$$0 \quad +5-2 \qquad +2 \qquad +4-2$$

The two elements undergoing oxidation number change are Cu and N.

Step 2 The change in oxidation number *per atom* is determined.

$$
\begin{array}{cccc}
0 & (+2) & +2 \\
\overbrace{\phantom{Cu + NO_3{}^- \longrightarrow Cu^{2+}}} \\
Cu + NO_3{}^- \longrightarrow Cu^{2+} + NO_2 \\
\underbrace{\phantom{Cu + NO_3{}^- \longrightarrow Cu^{2+} + NO_2}} \\
+5 \quad (-1) \qquad +4
\end{array}
$$

Change in oxidation number per atom

Step 3 For both Cu and N the oxidation number change per formula unit is the same as per atom. Both Cu and Cu^{2+} contain only one Cu atom; similarly, both $NO_3{}^-$ and NO_2 contain only one N atom.

Step 4 By multiplying the N oxidation number decrease by 2, we make the oxidation number increase and decrease per formula unit the same—two units.

$$
\begin{array}{ccc}
& (+2) & \\
\overbrace{\phantom{Cu + NO_3{}^- \longrightarrow Cu^{2+}}} \\
Cu + NO_3{}^- \longrightarrow Cu^{2+} + NO_2 \\
\underbrace{\phantom{Cu + NO_3{}^-}} \\
2(-1)
\end{array}
$$

Oxidation number increase equals oxidation number decrease

Step 5 The bracket notation indicates that two N atoms and one Cu atom undergo an oxidation number change. Translating this information into coefficients, we get

$$1\,Cu + 2\,NO_3^- \longrightarrow 1\,Cu^{2+} + 2\,NO_2$$

Step 6 The only atoms left to balance are those of O.

Step 7 Since this is a net ionic equation, the charges must balance; that is, the sum of the ionic charges of all species on each side of the equation must be equal. (They do not have to add up to zero; they just have to be equal.) In an acidic solution, which is the case in this example, charge balance is accomplished by adding H^+ ion.

As the equation now stands, we have a charge of –2 on the left side (two nitrate ions each with a –1 charge) and a charge of +2 on the right side (one copper ion). By adding four H^+ ions to the left side, we balance the charge at +2.

$$-2 + (+4) = +2$$

The equation at this point becomes

$$1\,Cu + 2\,NO_3^- + 4\,H^+ \longrightarrow 1\,Cu^{2+} + 2\,NO_2$$

Step 8 The hydrogen atoms are balanced through the addition of H_2O molecules. There are four H atoms on the left side (four H^+ ions) and none on the right side. Addition of two H_2O molecules to the right side will balance the H atoms at four per side.

$$1\,Cu + 2\,NO_3^- + 4\,H^+ \longrightarrow 1\,Cu^{2+} + 2\,NO_2 + 2\,H_2O$$

Step 9 The O atoms automatically balance at six atoms on each side. This is our double-check that previous steps have been correctly carried out. The balanced net ionic equation is thus

$$Cu + 2\,NO_3^- + 4\,H^+ \longrightarrow Cu^{2+} + 2\,NO_2 + 2\,H_2O$$

Answer Double Check:

Is the final equation balanced relative to the charge associated with the ions present? Yes. The charge balance is at +2. On the reactant side of the equation, two –1 ions (NO_3^-) and four +1 ions (H^+) give a net charge of +2. On the product side of the equation, the one Cu^{2+} ion gives a charge of +2.

▶ **Practice Exercise 15.8**   Balance the following net ionic redox equation using the oxidation number method of balancing.

$$Sb + NO_3^- \longrightarrow Sb_4O_6 + NO$$

The reaction occurs in acidic solution.

EXAMPLE 15.9 **Balancing a Net Ionic Redox Equation Using the Oxidation Number Method**

Balance the following net ionic redox equation using the oxidation number method of balancing.

$$MnO_4^- + C_2O_4^{2-} \longrightarrow MnO_2 + CO_3^{2-}$$

This reaction occurs in basic solution.

SOLUTION

Step 1 The elements being oxidized and reduced are identified by assigning oxidation numbers.

$$MnO_4^- + C_2O_4^{2-} \longrightarrow MnO_2 + CO_3^{2-}$$
$$+7\ +2 \qquad +3\ -2 \qquad\quad +4\ -2 \quad +4\ -2$$

The two elements undergoing oxidation number change are Mn and C.

Step 2 The change in oxidation number *per atom* is determined.

$$\overset{+7}{MnO_4^-} + \overset{(-3)}{C_2O_4^{2-}} \longrightarrow \overset{+4}{MnO_2} + CO_3^{2-}$$
$$\ +3 \qquad (+1) \qquad +4$$

Change in oxidation number per atom

Step 3 The carbon oxidation number change is multiplied by 2, since there are two carbon atoms in the $C_2O_4^{2-}$, to obtain the oxidation number change *per formula unit*.

$$MnO_4^- + \overset{(-3)}{C_2O_4^{2-}} \longrightarrow MnO_2 + CO_3^{2-}$$
$$2(+1)$$

Change in oxidation number per formula unit

Step 4 By multiplying the Mn per formula unit oxidation number decrease of –3 by 2 and multiplying the C per formula unit oxidation number increase of +2 by 3, the oxidation number increase and the oxidation number decrease become equal; both are at six units.

$$MnO_4^- + \overset{2(-3)}{C_2O_4^{2-}} \longrightarrow MnO_2 + CO_3^{2-}$$
$$3[2(+1)]$$

Oxidation number increase equals oxidation number decrease

Step 5 The bracket notation indicates that two Mn atoms and six C atoms undergo an oxidation number change. Translating this information into coefficients produces the equation

$$2\,MnO_4^- + 3\,C_2O_4^{2-} \longrightarrow 2\,MnO_2 + 6\,CO_3^{2-}$$

Step 6 The only atoms left to balance are those of O.

Step 7 Since this is a net ionic equation, the ionic charges must balance. In a basic solution, which is the situation in this example, charge balance is accomplished by adding OH^- ions.

The equation, at present, has a net charge on the reactant side of –8 (two –1 ions and three –2 ions). The charge on the product side of the equation is –12 (six –2 ions). By adding four OH^- ions to the reactant side, the charge balances at –12 on each side of the equation.

The equation, with charge balance included, becomes

$$\underbrace{2\,MnO_4^- + 3\,C_2O_4^{2-} + 4\,OH^-}_{-12\ charge} \longrightarrow \underbrace{2\,MnO_2 + 6\,CO_3^{2-}}_{-12\ charge}$$

Step 8 H atom balance is achieved by addition of H_2O molecules. Two H_2O molecules are added to the product side of the equation to counterbalance the four H atoms already present on the reactant side of the equation (four OH^- ions).

$$2\,MnO_4^- + 3\,C_2O_4^{2-} + 4\,OH^- \longrightarrow 2\,MnO_2 + 6\,CO_3^{2-} + 2\,H_2O$$

Step 9 The oxygen atoms should automatically balance if all of the previous steps have been carried out correctly. Such is the case. There are 24 O atoms on each side of the equation.

The final balanced equation is

$$2\,MnO_4^- + 3\,C_2O_4^{2-} + 4\,OH^- \longrightarrow 2\,MnO_2 + 6\,CO_3^{2-} + 2\,H_2O$$

Answer Double Check:

Is the equation balanced relative to both atoms and charge? Yes.

$$(2\,Mn,\, 6\,C,\, 4\,H,\, 24\,O,\, -12\,\text{charge}) = (2\,Mn,\, 6\,C,\, 4\,H,\, 24\,O,\, -12\,\text{charge})$$

▶ **Practice Exercise 15.9** Balance the following net ionic redox equation using the oxidation number method of balancing.

$$SeO_3^{2-} + Cl_2 \longrightarrow SeO_4^{2-} + Cl^-$$

This reaction occurs in basic solution.

15.9 HALF-REACTION METHOD FOR BALANCING REDOX EQUATIONS

The basis for the half-reaction method for balancing redox equations is the separation of the unbalanced redox equation into two *half-reactions*, one for oxidation and one for reduction. A **half-reaction** *is a chemical equation that describes one of the two parts of an overall oxidation–reduction reaction, either the oxidation part or the reduction part*. There are always two half-reactions associated with a given redox equation. These half-reactions, once obtained, are then balanced separately. The two balanced half-reactions are then added together to generate the overall balanced equation.

The division of the original unbalanced redox equation into two parts (two half-reactions) is artificial. One half-reaction does not really take place independently of the other; we cannot have oxidation without reduction. Nevertheless, this method of balancing redox equations is preferred in certain areas of redox chemistry. In particular, it leads to an increased understanding of the reactions that take place in electrochemical cells such as batteries.

As we did with the oxidation number method for balancing redox equations, we will break the half-reaction balancing process into a series of steps.

Step 1 Using oxidation numbers, determine which atoms are oxidized and which are reduced. Based on this information, split the redox equation into two skeletal half-reaction equations:
  a. An *oxidation* half-reaction equation, which involves the formula of the substance containing the element oxidized, along with other species associated with it
  b. A *reduction* half-reaction equation, which involves the formula of the substance containing the element reduced, along with other species associated with it

Step 2 Balance each of the half-reactions.

  a. First, balance the element oxidized or reduced, and then balance any other elements present in the skeletal equation other than oxygen or hydrogen.

  b. Next, show the number of electrons lost or gained in the oxidation or reduction. Use the change in oxidation number and the number of atoms oxidized or reduced to determine the number of electrons lost or gained. Electrons lost (oxidation) are shown on the product side of the equation, and electrons gained (reduction) on the reactant side of the equation.

  c. Balance the ionic charge by adding H^+ ions (acidic solution) or OH^- ions (basic solution) as reactant or product. (Remember that the electrons previously added must be considered in balancing charge.)

  d. Balance the hydrogen atoms by adding H_2O molecules as reactant or product.

  e. Verify that the oxygen atoms are balanced. (If they do not balance, a mistake has been made in a previous step.)

Step 3 Multiply each balanced half-reaction by appropriate integers to make the total number of electrons lost equal the total number of electrons gained.

Step 4 Add the two half-reactions together and cancel identical species, including electrons, on each side of the equation. See if the coefficients obtained can be simplified.

> In a half-reaction equation the electron placement is
> 1. product side for an oxidation process (loss of electrons).
> 2. reactant side for a reduction process (gain of electrons).

Examples 15.10 through 15.12 illustrate how the preceding guidelines for balancing redox equations are applied.

EXAMPLE 15.10 **Balancing a Net Ionic Redox Equation Using the Half-Reaction Method**

Balance the following net ionic redox equation that occurs in an acidic solution using the half-reaction method of balancing.

$$S^{2-} + NO_3^- \longrightarrow S + NO \quad \text{(acidic solution)}$$

SOLUTION

Step 1 *Determine the oxidation and reduction skeletal half-reactions.* Assigning oxidation numbers, we get

$$\begin{array}{ccccc} S^{2-} & + & NO_3^- & \longrightarrow & S + NO \\ -2 & & +5-2 & & 0 \ +2-2 \end{array}$$

Sulfur is oxidized, increasing in oxidation number from –2 to 0. Nitrogen is reduced, decreasing in oxidation number from +5 to +2.
The skeletal half-reactions for oxidation and reduction are

$$\text{Oxidation:} \quad S^{2-} \longrightarrow S$$

$$\text{Reduction:} \quad NO_3^- \longrightarrow NO$$

Step 2 *Balance the individual half-reactions.*

  a. In both half-reactions, the element being oxidized or reduced is already balanced—one atom of S on both sides in the first half-reaction and one atom of N on both sides in the second half-reaction. There are no other elements present except oxygen.

$$\text{Oxidation:} \quad S^{2-} \longrightarrow S$$

$$\text{Reduction:} \quad NO_3^- \longrightarrow NO$$

b. The oxidation number increase for S is +2. This is caused by the loss of two electrons, which are shown on the product side of the oxidation half-reaction.

$$\text{Oxidation:}\quad S^{2-} \longrightarrow S + 2\,e^-$$

The oxidation number decrease for N is –3. This results from the gain of three electrons, which are shown on the reactant side of the reduction half-reaction.

$$\text{Reduction:}\quad NO_3^- + 3\,e^- \longrightarrow NO$$

c. Since this is an acidic solution reaction, charge balance is achieved by adding H^+ ions. In the oxidation half-reaction there is a charge of –2 on each side of the equation. No H^+ ions are needed since the charge is already in balance.

$$\text{Oxidation:}\quad S^{2-} \longrightarrow S + 2\,e^-$$

In the reduction half-reaction there is a charge of –4 on the left side of the equation (–1 from the NO_3^- ion and –3 from the three electrons). There is no charge on the right side of the equation. Charge balance is achieved by adding four H^+ ions to the left side of the equation. Each side of the equation will now have zero charge.

$$\text{Reduction:}\quad NO_3^- + 3\,e^- + 4\,H^+ \longrightarrow NO$$

d. Water molecules are used to achieve hydrogen balance. Since no hydrogen is present in the oxidation half-reaction, no water molecules are needed.

$$\text{Oxidation:}\quad S^{2-} \longrightarrow S + 2\,e^-$$

In the reduction half-reaction, two water molecules are added to the right side of the equation. We now have four hydrogen atoms on each side of the equation.

$$\text{Reduction:}\quad NO_3^- + 3\,e^- + 4\,H^+ \longrightarrow NO + 2\,H_2O$$

e. There are no oxygen atoms present in the oxidation half-reaction. In the reduction half-reaction, the oxygen balances at three atoms on each side of the equation. The two balanced half-reactions are

$$\text{Oxidation:}\quad S^{2-} \longrightarrow S + 2e^-$$
$$\text{Reduction:}\quad NO_3^- + 3\,e^- + 4\,H^+ \longrightarrow NO + 2\,H_2O$$

Step 3 *Equalize electron loss and electron gain.*
Two electrons are produced in the oxidation half-reaction and three electrons are gained in the reduction half-reaction. To equalize electron loss and electron gain, we multiply the oxidation half-reaction by three and the reduction half-reaction by two. We have then an electron loss of 6 and an electron gain of 6.

$$\text{Oxidation:}\quad 3(S^{2-} \longrightarrow S + 2\,e^-)$$
$$\text{Reduction:}\quad 2(NO_3^- + 3\,e^- + 4\,H^+ \longrightarrow NO + 2\,H_2O)$$

Step 4 *Add the half-reactions and cancel identical species.*
Adding the two half-reactions together, we get

$$\text{Oxidation:}\quad 3\,S^{2-} \longrightarrow 3\,S + \cancel{6\,e^-}$$
$$\text{Reduction:}\quad 2\,NO_3^- + \cancel{6\,e^-} + 8\,H^+ \longrightarrow 2\,NO + 4\,H_2O$$
$$\overline{\quad 3\,S^{2-} + 2\,NO_3^- + 8\,H^+ \longrightarrow 3\,S + 2\,NO + 4\,H_2O \quad}$$

There are no species to cancel other than the electrons. The electrons must always cancel. If they do not, we have made a mistake in step 3.

Two different aspects of a half-reaction equation must be balanced.

1. *Atoms.* The same number of atoms of each kind must be present on both sides of the equation.

2. *Charge.* The net electrical charge must be the same on each side of the equation. The net charge does not have to be zero. It can be any value as long as it is the same on both sides of the equation (+2 and +2, or –4 and –4, etc.).

Answer Double Check:

Do the reactant and product sides of the equation have the same net ionic charge? Yes. There is no charge associated with the product side of the equation since none of the products is an ion. Thus, the sum of ionic charges on the reactant side of the equation must add to zero. Such is the case; $3(-2) + 2(-1) + 8(+1) = 0$.

▶ **Practice Exercise 15.10** Balance the following net ionic redox equation that occurs in an acidic solution, using the half-reaction method of balancing.

$$CrO_4{}^{2-} + N_2O \longrightarrow Cr^{3+} + NO$$

EXAMPLE 15.11 **Balancing a Net Ionic Redox Equation Using the Half-Reaction Method**

Balance the following net ionic redox equation that occurs in a basic solution using the half-reaction method of balancing.

$$S^{2-} + Cl_2 \longrightarrow SO_4{}^{2-} + Cl^- \qquad \text{(basic solution)}$$

SOLUTION

Step 1 *Determine the oxidation and reduction skeletal half-reactions.*
Assigning oxidation numbers, we get

$$\begin{array}{cccc} S^{2-} + & Cl_2 \longrightarrow & SO_4{}^{2-} + & Cl^- \\ -2 & 0 & +6-2 & -1 \end{array}$$

Sulfur is oxidized, increasing in oxidation number from –2 to +6. Chlorine is reduced, decreasing in oxidation number from 0 to –1. The skeletal half-reactions for oxidation and reduction are

$$\text{Oxidation:} \quad S^{2-} \longrightarrow SO_4{}^{2-}$$

$$\text{Reduction:} \quad Cl_2 \longrightarrow Cl^-$$

Step 2 *Balance the individual half-reactions.*
  a. In the oxidation half-reaction the S is already balanced.

$$\text{Oxidation:} \ S^{2-} \longrightarrow SO_4{}^{2-}$$

To balance the Cl in the reduction half-reaction, the coefficient 2 must be added on the right side.

$$\text{Reduction:} \quad Cl_2 \longrightarrow 2\,Cl^-$$

  b. The oxidation number increase for S is +8, which corresponds to the loss of eight electrons.

$$\text{Oxidation:} \ S^{2-} \longrightarrow SO_4{}^{2-} + 8\,e^-$$

The oxidation number decrease for Cl is –1, which corresponds to a gain of one electron. Since there are two Cl atoms changing, the total electron gain is two electrons.

$$\text{Reduction:} \quad Cl_2 + 2\,e^- \longrightarrow 2\,Cl^-$$

  c. Since this reaction occurs in a basic solution, charge balance is achieved by adding OH^- ions. In the oxidation half-reaction there is a charge of –2 on the left side and a charge of −10 (one SO_4^{2-} ion and eight electrons) on the right side. The charge is brought into balance, at a −10, by adding eight OH^- ions to the left side of the equation.

$$\text{Oxidation: } S^{2-} + 8\,OH^- \longrightarrow SO_4^{2-} + 8\,e^-$$

In the reduction half-reaction, the charge is already balanced at a −2 on each side. No OH^- ions are needed.

$$\text{Reduction: } Cl_2 + 2\,e^- \longrightarrow 2\,Cl^-$$

d. Hydrogen balance is achieved in the oxidation half-reaction by adding four H_2O molecules to the right side of the equation.

$$\text{Oxidation: } S^{2-} + 8\,OH^- \longrightarrow SO_4^{2-} + 8\,e^- + 4\,H_2O$$

Hydrogen balance is not needed in the reduction half-reaction since no hydrogen is present.

$$\text{Reduction: } Cl_2 + 2\,e^- \longrightarrow 2\,Cl^-$$

e. Oxygen balances at eight atoms on each side of the equation in the oxidation half-reaction. Oxygen is not present in the reduction half-reaction. The two balanced half-reactions are

$$\text{Oxidation:} \quad S^{2-} + 8\,OH^- \longrightarrow SO_4^{2-} + 8\,e^- + 4\,H_2O$$

$$\text{Reduction:} \quad Cl_2 + 2\,e^- \longrightarrow 2\,Cl^-$$

Step 3 *Equalize electron loss and electron gain.*
Eight electrons are produced in the oxidation half-reaction and two electrons are gained in the reduction half-reaction. Multiplying the reduction half-reaction by 4 will cause electron loss and electron gain to be equal at eight electrons.

$$\text{Oxidation:} \quad S^{2-} + 8\,OH^- \longrightarrow SO_4^{2-} + 8\,e^- + 4\,H_2O$$

$$\text{Reduction:} \quad 4(Cl_2 + 2e^- \longrightarrow 2\,Cl^-)$$

Step 4 *Add the half-reactions and cancel identical species.*
Adding the two half-reactions together, we get

$$\text{Oxidation:} \quad S^{2-} + 8\,OH^- \longrightarrow SO_4^{2-} + \cancel{8\,e^-} + 4\,H_2O$$

$$\text{Reduction:} \quad 4\,Cl_2 + \cancel{8\,e^-} \longrightarrow 8\,Cl^-$$

$$S^{2-} + 8\,OH^- + 4\,Cl_2 \longrightarrow SO_4^{2-} + 4\,H_2O + 8\,Cl^-$$

There are no species to cancel other than the electrons.

Answer Double Check:

Do the product and reactant sides of the equation have the same net charge? Yes. Both sides of the equation have a net charge of −10.

▶ **Practice Exercise 15.11** Balance the following net ionic redox equation that occurs in a basic solution using the half-reaction method of balancing.

$$MnO_4^- + SO_3^{2-} \longrightarrow MnO_2 + SO_4^{2-}$$

EXAMPLE 15.12 **Balancing a Net Ionic Redox Equation Using the Half-Reaction Method**

Balance the following net ionic redox equation that occurs in an acidic solution using the half-reaction method of balancing.

$$H_3AsO_3 + MnO_4^- \longrightarrow H_3AsO_4 + Mn^{2+} \qquad \text{(acidic solution)}$$

SOLUTION

Step 1 *Determine the oxidation and reduction skeletal half-reactions.*
Assigning oxidation numbers, we get

$$\begin{array}{ccccc} H_3AsO_3 & + MnO_4^- & \longrightarrow & H_3AsO_4 & + Mn^{2+} \\ {}_{+1\ +3\ -2} & {}_{+7\ -2} & & {}_{+1\ +5\ -2} & {}_{+2} \end{array}$$

Arsenic is oxidized, increasing in oxidation number from +3 to +5. Manganese is reduced, decreasing in oxidation number from +7 to +2. The skeletal half-reactions for oxidation and reduction are

$$\text{Oxidation:} \qquad H_3AsO_3 \longrightarrow H_3AsO_4$$

$$\text{Reduction:} \qquad MnO_4^- \longrightarrow Mn^{2+}$$

Step 2 *Balance the individual half-reactions.*

a. In both half-reactions, the element being oxidized or reduced is already balanced—one atom of As on both sides in the oxidation half-reaction and one atom of Mn on both sides in the reduction half-reaction.

$$\text{Oxidation:} \qquad H_3AsO_3 \longrightarrow H_3AsO_4$$

$$\text{Reduction:} \qquad MnO_4^- \longrightarrow Mn^{2+}$$

b. The oxidation number increase for As is +2, which corresponds to the loss of two electrons.

$$\text{Oxidation:} \qquad H_3AsO_3 \longrightarrow H_3AsO_4 + 2\,e^-$$

The oxidation number decrease for Mn is –5, which corresponds to the gain of five electrons.

$$\text{Reduction:} \qquad MnO_4^- + 5\,e^- \longrightarrow Mn^{2+}$$

c. Since this reaction occurs in an acidic solution, charge balance is achieved by adding H^+ ions. In the oxidation half-reaction there is a charge of zero on the left side and a charge of –2 (two electrons) on the right side. Charge balance, at zero, is achieved by adding two H^+ ions to the right side of the equation.

$$\text{Oxidation:} \qquad H_3AsO_3 \longrightarrow H_3AsO_4 + 2\,e^- + 2\,H^+$$

In the reduction half-reaction there is a charge of –6 on the left side (one MnO_4^- ion and five electrons) and a charge of +2 on the right side. Charge balance, at a +2, is achieved by adding eight H^+ ions to the left side of the equation.

$$\text{Reduction:} \qquad MnO_4^- + 5\,e^- + 8\,H^+ \longrightarrow Mn^{2+}$$

d. Hydrogen balance is obtained in the oxidation half-reaction by adding one H_2O molecule to the left side of the equation.

$$\text{Oxidation:} \qquad H_3AsO_3 + H_2O \longrightarrow H_3As\,O_4 + 2\,e^- + 2\,H^+$$

Hydrogen balance is obtained in the reduction half-reaction by adding four H_2O molecules to the right side of the equation.

Reduction: $MnO_4^- + 5\,e^- + 8\,H^+ \longrightarrow Mn^{2+} + 4\,H_2O$

e. Oxygen balances at four atoms on each side in both the oxidation and reduction half-reactions. The two balanced half-reactions are

Oxidation: $H_3AsO_3 + H_2O \longrightarrow H_3AsO_4 + 2\,e^- + 2\,H^+$

Reduction: $MnO_4^- + 5\,e^- + 8\,H^+ \longrightarrow Mn^{2+} + 4\,H_2O$

Step 3 *Equalize the electron loss and electron gain.*
The lowest common multiple for an electron loss of 2 and an electron gain of 5 is 10 electrons. Thus, we multiply the oxidation half-reaction by 5 and the reduction half-reaction by 2.

Oxidation: $5(H_3AsO_3 + H_2O \longrightarrow H_3AsO_4 + 2\,e^- + 2\,H^+)$

Reduction: $2(MnO_4^- + 5\,e^- + 8\,H^+ \longrightarrow Mn^{2+} + 4\,H_2O)$

Step 4 *Add the half-reactions, and cancel identical species.*
Adding the two half-reactions together, we get

Oxidation: $5\,H_3AsO_3 + 5\,H_2O \longrightarrow 5\,H_3AsO_4 + \cancel{10\,e^-} + 10\,H^+$

Reduction: $2\,MnO_4^- + \cancel{10\,e^-} + 16\,H^+ \longrightarrow 2\,Mn^{2+} + 8\,H_2O$

$5\,H_3AsO_3 + 5\,H_2O + 2\,MnO_4^- + 16\,H^+ \longrightarrow 5\,H_3AsO_4 + 10\,H^+ + 2\,Mn^{2+} + 8\,H_2O$

Both H^+ ion and H_2O are on both sides of the equation. We can cancel $5\,H_2O$ molecules from each side and $10\,H^+$ ions from each side. The final balanced equation becomes

$5\,H_3AsO_3 + 2\,MnO_4^- + 6\,H^+ \longrightarrow 5\,H_3AsO_4 + 2\,Mn^{2+} + 3\,H_2O$

Answer Double Check:

Were H_2O and H^+ added to the equation in the appropriate manner to achieve H and O balance? Yes. There are 21 H atoms on each side of the equation and 23 O atoms on each side of the equation.

▶ **Practice Exercise 15.12**   Balance the following net ionic redox equation that occurs in an acidic solution using the half-reaction method of balancing.

$MnO_2 + NO_2^- \longrightarrow Mn^{2+} + NO_3^-$

In each of the three examples we have just considered, the oxidation and reduction half-reactions were simultaneously balanced. This approach was used in the examples to enable us to make comparisons. In practice, particularly when you are thoroughly familiar with the balancing procedure, one half-reaction is usually completely balanced before work begins on balancing the other half-reaction. Usually, it is better to work on just one reaction at a time.

A comparison of the two methods for balancing redox equations is in order. Basic to each method is being able to recognize the elements involved in the actual oxidation–reduction process. The oxidation number method works on the principle that the increase in oxidation number must equal the decrease in oxidation number. The half-reaction method involves equalizing the number of electrons lost by the substance oxidized with the number of electrons gained by the substance reduced.

The oxidation number method is usually faster, particularly for simple equations. This potential speed is considered the major advantage of the oxidation number method. The half-reaction method's focus on electron transfer is its major advantage. The feature becomes particularly important in electrochemistry. In this field it is most useful to discuss chemical reactions in terms of half-reactions occurring at different locations (electrodes) in an electrochemical cell.

15.10 DISPROPORTIONATION REACTIONS

A *disproportionation reaction* is a special type of oxidation–reduction reaction. A **disproportionation reaction** *is a redox reaction in which some atoms of a single element in a reactant are oxidized and others are reduced.* For such reactant behavior to be possible, the reactant must contain an element that is capable of having at least three oxidation numbers: its original number plus one higher and one lower oxidation number. Note that any given atom is not both oxidized and reduced. Some atoms are oxidized, and other atoms of the same element are reduced.

An example of a disproportionation reaction is

$$3\,Br_2 + 3\,H_2O \longrightarrow HBrO_3 + 5\,HBr$$

Note that two bromine-containing products have been produced from one bromine-containing reactant. The reactant bromine atoms have an oxidation number of zero. Bromine in $HBrO_3$ has a +5 oxidation number (it has been oxidized), and bromine in HBr has a –1 oxidation number (it has been reduced).

$$3\,Br_2 + 3\,H_2O \longrightarrow HBrO_3 + 5\,HBr$$
$$0 \qquad\qquad +5 \qquad -1$$

Thus, some of the reactant bromine atoms have been oxidized, while others have been reduced. A disproportionation reaction has taken place.

Example 15.13 shows how the procedures for balancing redox equations (Sec. 15.8 and 15.9) are slightly modified to balance disproportionation reaction equations.

EXAMPLE 15.13 **Balancing a Disproportionation Redox Equation**

Balance the following disproportionation redox reaction using

- **a.** the oxidation number method of balancing.
- **b.** the half-reaction method of balancing.

$$NO_2 \longrightarrow NO_3^- + NO \qquad \text{(acidic solution)}$$

SOLUTION

a. *Oxidation Number Method*

Step 1 In assigning oxidation numbers, we immediately become aware that this is a disproportionation reaction. Nitrogen is the only element for which an oxidation number change occurs.

$$NO_2 \longrightarrow NO_3^- + NO$$
$$+4-2 \qquad +5-2 \qquad +2-2$$

Step 2 Since the species NO_2 is undergoing both oxidation and reduction, for balancing purposes we will write it twice on the reactant side of the equation. With the in two places, brackets can then be drawn in the normal manner to connect the substances involved in oxidation and reduction.

$$\overset{+4}{NO_2} + \overset{(+1)}{NO_2} \longrightarrow \overset{+5}{NO_3^-} + NO$$

Change in oxidation number per atom

$+4 \quad (-2) \quad +2$

(The NO_2 molecules will be recombined later into one location.)

Step 3 The change in oxidation number per formula unit in both cases is the same as per atom.

Step 4 By multiplying the oxidation number increase by 2, we equalize the oxidation number increase and decrease per formula unit.

$$\overset{2(+1)}{NO_2 + NO_2} \longrightarrow NO_3^- + NO$$

Oxidation number increase equals oxidation number decrease

(-2)

Step 5 The bracket notation indicates that two N atoms undergo an increase in oxidation number for every one that undergoes a decrease in oxidation number. Translating this information into equation coefficients gives

$$2\,NO_2 + 1\,NO_2 \longrightarrow 2\,NO_3^- + 1\,NO$$

Now that the equation coefficients for the substance involved in oxidation and reduction, NO_2, have been determined, we can combine the NO_2 into one location, reversing the process carried out in step 2.

$$3\,NO_2 \longrightarrow 2\,NO_3^- + 1\,NO$$

Step 6 The only atoms left to balance are oxygen atoms.

Step 7 Since this is a net ionic equation, charge must be balanced. In an acidic solution, which is the case here, we balance the charge by adding H^+ ion. As the equation now stands, we have a charge of –2 on the right side (2 NO_3^- ions). By adding two H^+ ions to the right side of the equation we balance the charge at zero

$$3\,NO_2 \longrightarrow 2\,NO_3^- + 1\,NO + 2\,H^+$$

Step 8 Hydrogen atom balance is achieved through the addition of H_2O molecules. There are no H atoms on the left side and two H atoms on the right side. Addition of one H_2O molecule to the left side will balance the H atoms at two on each side.

$$1\,H_2O + 3\,NO_2 \longrightarrow 2\,NO_3^- + 1\,NO + 2\,H^+$$

Step 9 The oxygen atoms should automatically balance. They do, at seven atoms on each side.

$$H_2O + 3\,NO_2 \longrightarrow 2\,NO_3^- + NO + 2\,H^+$$

b. *Half-Reaction Method*

Step 1 *Determine the oxidation and reduction skeletal half-reactions.*
Assignment of oxidation numbers is the same as in part (a).

$$NO_2 \longrightarrow NO_3^- + NO$$
$$+4-2 \quad +5-2 \quad +2-2$$

Nitrogen is undergoing both oxidation and reduction. The skeleton half-reactions for oxidation and reduction are

$$\text{Oxidation:} \quad NO_2 \longrightarrow NO_3^-$$

$$\text{Reduction:} \quad NO_2 \longrightarrow NO$$

Note how disproportionation is handled at this point. The substance undergoing disproportionation appears as a reactant in both the oxidation and reduction half-reactions.

Step 2 *Balance the individual half-reactions.*

a. In both half-reactions, the element being oxidized or reduced is already balanced—one atom of N in both cases.

$$\text{Oxidation:} \quad NO_2 \longrightarrow NO_3^-$$

$$\text{Reduction:} \quad NO_2 \longrightarrow NO$$

b. The oxidation number increase for the oxidized N is +1, which corresponds to the loss of one electron.

$$\text{Oxidation:} \quad NO_2 \longrightarrow NO_3^- + 1\,e^-$$

The oxidation number decrease for the reduced N is –2, which corresponds to the gain of two electrons.

$$\text{Reduction:} \quad NO_2 + 2\,e^- \longrightarrow NO$$

c. Since this reaction occurs in an acidic solution, charge balance is achieved by adding H^+ ions. In the oxidation half-reaction there is no charge on the left side of the equation and a charge of –2 on the right side. Adding two H^+ ions to the right side of the equation will balance the charge at zero on both sides.

$$\text{Oxidation:} \quad NO_2 \longrightarrow NO_3^- + 1\,e^- + 2\,H^+$$

In the reduction half-reaction there is a charge of –2 on the left side of the equation and no charge on the right side. Charge balance is achieved by adding two H^+ ions to the left side of the equation.

$$\text{Reduction:} \quad NO_2 + 2\,e^- + 2\,H^+ \longrightarrow NO$$

d. Hydrogen balance is obtained in the oxidation half-reaction by adding one H_2O to the left side of the equation.

$$\text{Oxidation:} \quad NO_2 + H_2O \longrightarrow NO_3^- + 1\,e^- + 2\,H^+$$

Hydrogen balance is obtained in the reduction half-reaction by adding one H_2O to the right side of the equation.

$$\text{Reduction:} \quad NO_2 + 2\,e^- + 2\,H^+ \longrightarrow NO + H_2O$$

e. Oxygen balances at two atoms on each side in both the oxidation and reduction half-reactions. The two balanced half-reactions are

$$\text{Oxidation:} \quad NO_2 + H_2O \longrightarrow NO_3^- + 1\,e^- + 2\,H^+$$

$$\text{Reduction:} \quad NO_2 + 2\,e^- + 2\,H^+ \longrightarrow NO + H_2O$$

Step 3 *Equalize the electron loss and electron gain.*

The oxidation half-reaction involves the loss of one electron. The reduction half-reaction involves the gain of two electrons. Multiplying the oxidation half-reaction by a factor of 2 will cause electron loss and gain to be equal at two electrons.

Oxidation: $2(NO_2 + H_2O \longrightarrow NO_3^- + 1\,e^- + 2\,H^+)$

Reduction: $NO_2 + 2\,e^- + 2\,H^+ \longrightarrow NO + H_2O$

Step 4 *Add the half-reactions, and cancel identical species.*
Adding the two half-reactions together, we get

Oxidation: $2\,NO_2 + 2\,H_2O \longrightarrow 2\,NO_3^- + \cancel{2\,e^-} + 4\,H^+$

Reduction: $NO_2 + \cancel{2\,e^-} + 2\,H^+ \longrightarrow NO + H_2O$

$$2\,NO_2 + 2\,H_2O + NO_2 + 2\,H^+ \longrightarrow 2\,NO_3^- + 4\,H^+ + NO + H_2O$$

Both H_2O and H^+ are on both sides of the equation, and some of each can be cancelled. Also, NO_2 appears in two places on the left side of the equation and needs to be combined. The final balanced equation is

$$3\,NO_2 + H_2O \longrightarrow 2\,NO_3^- + NO + 2\,H^+$$

Answer Double Check:

Have both atom balance and charge balance been achieved for this chemical equation? Yes. Atom balance is at 3 N, 7 O, and 2 H. Charge balance is at zero. On the product side of the equation there are two –1 ions and two +1 ions, producing a net charge of zero.

▶ **Practice Exercise 15.13**   Balance the following disproportionation redox reaction using

a. the oxidation number method of balancing.
b. the half-reaction method of balancing.

$$NO_2 \longrightarrow NO_3^- + NO_2^- \qquad \text{(basic solution)}$$

15.11 STOICHIOMETRIC CALCULATIONS INVOLVING IONS

Grams-to-moles-to-particles stoichiometric calculations were the subject of the latter portion of Chapter 10 (Secs. 10.8 through 10.11). The chemical equation basis for these calculations was always a balanced *molecular* equation. How does the presence of ions in a chemical equation, either a net ionic redox equation or a net ionic non-redox equation, affect the stoichiometric calculation process? What modifications need to be made? The answer to these questions is simple. The calculational process previously used for molecular-equation-based problems does not need to be modified.

The lack of need for modification results from the fact that ions have essentially the same mass as the parent neutral species from which they can be considered derived. For example, let us formally consider the mass difference between a sodium atom (Na) and a sodium ion (Na^+). Subatomic-particlewise, these two species differ by one electon. The mass of an electron, to six decimal places on the amu scale (Sec. 5.8), is 0.000549 amu. Using this value and the mass of a sodium atom to six decimal places, which is 22.989769 amu (see inside front cover listing), we have for the mass of an Na atom and an Na^+ ion, respectively,

Na atom = 22.989769 amu

Na^+ ion = (22.989769 − 0.000549) amu = 22.989220 amu

FIGURE 15.3
Conversion-factor relationships needed for solving chemical-equation-based problems that involve ions.

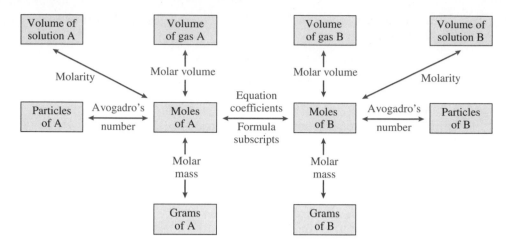

The mass difference between the two species lies in the ten-thousandths place, a difference too small to affect the way standard stoichiometric calculations are carried out.

The operational rule for use of atomic masses in chemical calculations is to use values rounded to the hundredths place (Sec. 9.3). Based on this operational rule, we have

$$\text{mass of a sodium atom (Na)} = 22.99 \text{ amu}$$

$$\text{mass of a sodium ion (Na}^+) = 22.99 \text{ amu}$$

Thus, in general, in chemical-equation-based calculations that involve ions an ion is always considered to have the same mass as the neutral "parent" species from which it can be considered to have been derived.

With this concept in place, ion-based stoichiometric calculations proceed in the same manner as atom-based stoichiometric calculations. The relationships needed are those between grams, moles, particles, solution volume, and gaseous volume, the same relationships previously used in Chapter 13 in solving solution-based stoichiometric problems. These relationships are summarized in Figure 15.3.

EXAMPLE 15.14   **Calculating the Mass of an Ion Present in a Solution**

How many grams of Ca^{2+} ions are present in an aqueous $CaCl_2$ solution that contains 9.7×10^{22} Cl^- ions?

SOLUTION

The equation that governs the ions present in an aqueous $CaCl_2$ solution is

$$CaCl_2(aq) \longrightarrow Ca^{2+}(aq) + 2\,Cl^-(aq)$$

The extent of this dissociation is 100% as $CaCl_2$ is a strong electrolyte (Sec. 15.2).

Step 1 The given quantity is 9.7×10^{22} Cl^- ions and the desired quantity is grams of Ca^{2+} ions present.

$$9.7 \times 10^{22} \text{ Cl}^- \text{ ions} = ? \text{ g Ca}^{2+} \text{ ions}$$

Step 2 This is a particles-of-A to grams-of-B problem. The pathway, in terms of Figure 15.3, is

| Particles of A | Avogadro's number | Moles of A | Equation coefficients | Moles of B | Molar mass | Grams of B |

Step 3 The dimensional analysis setup for the calculation is

$$9.7 \times 10^{22} \text{ Cl}^- \text{ ions} \times \frac{1 \text{ mole Cl}^-}{6.022 \times 10^{23} \text{ Cl}^- \text{ ions}} \times \frac{1 \text{ mole Ca}^{2+}}{2 \text{ mole Cl}^-} \times \frac{40.08 \text{ g Ca}^{2+}}{1 \text{ mole Ca}^{2+}}$$

particles A $\longrightarrow$ moles A $\longrightarrow$ moles B $\longrightarrow$ grams B

Note the number 40.08 in the last conversion factor. The molar mass of Ca^{2+} ions is the same as the molar mass of Ca atoms, as was explained in the discussion preceding this worked-out example.

Step 4 The result, obtained by combining the numerical factors is

$$\frac{9.7 \times 10^{22} \times 1 \times 1 \times 40.08}{6.022 \times 10^{23} \times 2 \times 1} \text{ g Ca}^{2+} = 3.2279641 \text{ g Ca}^{2+}$$

(calculator answer)

$$= 3.2 \text{ g Ca}^{2+}$$

(correct answer)

▶ **Practice Exercise 15.14** How many grams of K^+ ions are present in an aqueous K_2S solution that contains $7.6 \times 10^{23} S^{2-}$ ions?

EXAMPLE 15.15 **Calculating the Volume of Solution Needed to Supply a Given Amount of an Ion**

What volume, in milliliters, of a 0.300 M $Al(NO_3)_3$ solution is needed to supply 0.782 mole of NO_3^- ion?

SOLUTION

The equation that governs the ions present in an aqueous $Al(NO_3)_3$ solution is

$$Al(NO_3)_3(aq) \longrightarrow Al^{3+}(aq) + 3 NO_3^-(aq)$$

The extent of this dissociation is 100% as $Al(NO_3)_3$ is a strong electrolyte (Sec. 15.2).

Step 1 The given quantity is 0.782 mole of NO_3^- ion and the desired quantity is volume of $Al(NO_3)_3$ solution.

$$0.782 \text{ mole NO}_3^- \text{ ion} = ? \text{ mL Al(NO}_3)_3 \text{ solution}$$

Step 2 This is moles-of-A to volume-of-solution B problem. The pathway, in terms of Figure 15.3, is

| Moles of A | Equation coefficients | Moles of B | Molarity | Volume of solution B |

Step 3 The dimensional analysis setup for the calculation is

$$0.782 \text{ mole NO}_3^- \text{ ion} \times \frac{1 \text{ mole Al(NO}_3)_3}{3 \text{ moles NO}_3^- \text{ ion}} \times \frac{1000 \text{ mL Al(NO}_3)_3}{0.300 \text{ mole Al(NO}_3)_3}$$

moles A $\longrightarrow$ moles B $\longrightarrow$ solution volume B

The numbers in the first conversion factor come from the dissociation equation and the numbers in the second conversion factor come from the given molarity of the solution.

Step 4 The result, obtained by combining the numerical factors, is

$$\frac{0.782 \times 1 \times 1000}{3 \times 0.300} \text{ mL Al(NO}_3)_3 = 868.8889 \text{ mL Al(NO}_3)_3$$

(calculator answer)

$$= 869 \text{ mL Al(NO}_3)_3$$

(**correct answer**)

Answer Double Check:

Is the magnitude of the answer reasonable? Yes. The 0.300 M $Al(NO_3)_3$ is 0.900 molar in NO_3^- ion. Thus, 1 liter (1000 mL) of solution would supply 0.900 M of NO_3^- ion. A volume less than 1000 mL would be needed to supply the needed 0.782 mole of NO_3^- ion, which is the case here.

▶ **Practice Exercise 15.15** What volume, in milliliters, of a 1.50 M $(NH_4)_3PO_4$ solution is needed to supply 0.750 mole of NH_4^+ ion?

EXAMPLE 15.16 **Calculating the Volume of Solution Needed for a Reaction Involving Ions**

A solution is 1.20 M in NO_3^- ion. What volume of this solution, in milliliters, is needed to react with 10.0 g of Cu in the following redox reaction?

$$Cu(s) + 2\,NO_3^-(aq) + 4\,H^+(aq) \longrightarrow Cu^{2+}(aq) + 2\,NO_2(g) + 2\,H_2O(l)$$

SOLUTION

Step 1 The given quantity is 10.0 g of Cu and the desired quantity is milliliters of NO_3^- ion solution.

$$10.0 \text{ g Cu} = ? \text{ mL NO}_3^- \text{ solution}$$

Step 2 In terms of Figure 15.3, this is a grams-of-A to volume-of-solution B

Grams of A	Molar mass →	Moles of A	Equation coefficients →	Moles of B	Molarity →	Volume of solution B

Step 3 The dimensional analysis setup for the calculation is

$$10.0 \text{ g Cu} \times \frac{1 \text{ mole Cu}}{63.55 \text{ g Cu}} \times \frac{2 \text{ moles NO}_3^-}{1 \text{ mole Cu}} \times \frac{1000 \text{ mL NO}_3^-}{1.20 \text{ moles NO}_3^-}$$

grams A $\longrightarrow$ moles A $\longrightarrow$ moles B $\longrightarrow$ volume solution B

Step 4 The result, obtained by combining the numerical factors is

$$\frac{10.0 \times 1 \times 2 \times 1000}{63.55 \times 1 \times 1.20} \text{ mL NO}_3^- = 262.26069 \text{ mL NO}_3^-$$

(calculator answer)

$$= 262 \text{ mL NO}_3^-$$

(**correct answer**)

Answer Double Check:

Is the answer reasonable? Yes. If the given amount of Cu is approximated as 0.20 of a mole the needed amount of NO_3^- would be 0.40 of a mole. This NO_3^- amount would require one-third (333 mL) of 1.20 M NO_3 solution. The actual molar amount of Cu is a little less than 0.20 mole. Thus, the volume of NO_3^- solution needed should be less than 333 mL, which it is.

▶ **Practice Exercise 15.16**   A solution is 1.50 M in SeO_3^{2-} ion (selenite ion). What volume of this solution, in milliliters, is needed to react with 70.0 g of Cl_2 in the following redox reaction?

$$SeO_3^{2-}(aq) + Cl_2(g) + 2\,OH^-(aq) \longrightarrow SeO_4^{2-}(aq) + 2\,Cl^-(aq) + H_2O(l)$$

Summary

1. **Electrolytes** An electrolyte is a substance whose solution conducts electricity. Strong electrolytes, which include strong acids, strong bases, and soluble salts, produce strongly conducting solutions as a result of complete or almost complete dissociation/ionization in a solution. Weak electrolytes, which include weak acids and weak bases, produce weakly conducting solutions as a result of dissociation/ionization occurring to only a small extent.

2. **Ionic and Net Ionic Chemical Equations** An ionic chemical equation is an equation in which the chemical formulas of the predominant form of each compound in a solution are used; dissociated and ionized compounds are written as ions, and undissociated and un-ionized compounds are written in molecular form. A net ionic chemical equation is an ionic equation from which nonparticipating (spectator) ions have been eliminated.

3. **Oxidation–Reduction Reaction Terminology** Oxidation is the loss of electrons by a reactant; reduction is the gain of electrons by a reactant. An oxidizing agent causes oxidation by accepting electrons from another reactant. A reducing agent causes reduction by providing electrons for another reactant to accept. An oxidation–reduction reaction is any reaction involving the transfer of electrons between reactants. Shortened terminology for an oxidation–reduction reaction is redox reaction.

4. **Oxidation Numbers** An oxidation number for an atom is a number that represents the charge that the atom appears to have when

the electrons in each bond it is participating in are assigned to the more electronegative of the two atoms involved in the bond. Oxidation numbers are used to identify the electron transfer that occurs in a redox reaction. In terms of oxidation numbers, oxidation is associated with an increase in oxidation number and reduction involves a decrease in oxidation number.

5. **Redox and Nonredox Chemical Reactions** A redox reaction is a chemical reaction in which oxidation numbers change. A nonredox reaction is a chemical reaction in which there is no change in oxidation numbers.

6. **Methods for Balancing Redox Reactions** Two distinctly different approaches for systematically balancing redox equations are in common use: the oxidation number method and the half-reaction method. Each method has advantages and disadvantages. The basis for the oxidation number method is that the oxidation number increase must equal the oxidation number decrease. In the half-reaction method for balancing redox equations the unbalanced redox equation is separated into two half-reactions, one for oxidation and one for reduction.

7. **Disproportionation Reactions** A disproportionation reaction is a redox reaction in which some atoms of a single element in a reactant are oxidized and others are reduced. For such reactant behavior to be possible, the reactant must contain an element that is capable of having at least three oxidation numbers: its original number plus one higher and one lower oxidation number.

Key Terms

The new terms defined in this chapter are

disproportionation reaction *Sec. 15.10*
electrolyte *Sec. 15.2*
half-reaction *Sec. 15.9*
ionic equation *Sec. 15.3*
net ionic equation *Sec. 15.3*

nonelectrolyte *Sec. 15.2*
non-oxidation–reduction reaction *Sec. 15.6*
oxidation *Sec 15.4*
oxidation number *Sec. 15.5*

oxidation–reduction reaction *Sec. 15.4*
oxidizing agent *Secs. 15.4 and 15.5*
redox reaction *Sec. 15.4*

reducing agent *Secs. 15.4 and 15.5*
reduction *Sec. 15.4*
strong electrolyte *Sec. 15.2*
weak electrolyte *Sec. 15.2*

Practice Problems

ELECTROLYTES (SEC. 15.2)

15.1 Classify each of the following acids or bases as a strong electrolyte or a weak electrolyte.
  a. H_2CO_3 **b.** KOH   **c.** H_2SO_4 **d.** HCN

15.2 Classify each of the following acids or bases as a strong electrolyte or a weak electrolyte.
  a. H_3PO_4 **b.** HNO_3 **c.** KOH   **d.** $H_2C_2O_4$

15.3 Indicate whether each of the following substances is present in a solution in ionic form, molecular form, or both.
  a. acetic acid, a weak acid
  b. sucrose, a nonelectrolyte
  c. sodium sulfate, a soluble salt
  d. hydrofluoric acid, a weak electrolyte

15.4 Indicate whether each of the following substances is present in solution in ionic form, molecular form, or both.
  a. hydrochloric acid, a strong acid
  b. sodium nitrate, a soluble salt
  c. potassium chloride, a soluble salt
  d. ethanol, a nonelectrolyte

15.5 How many ions are produced, per formula unit, when each of the following strong electrolytes are dissolved in water?
  a. NaCl   **b.** $Mg(NO_3)_2$
  c. NH_4CN **d.** $HClO_4$

15.6 How many ions are produced, per formula unit, when each of the following strong electrolytes are dissolved in water?
  a. KNO_3 **b.** Na_2CO_3
  c. $Al(OH)_3$ **d.** K_3N

15.7 Write a balanced chemical equation for the dissolution in water of each of the substances in Problem 15.5.

15.8 Write a balanced chemical equation for the dissolution in water of each of the substances in Problem 15.6.

15.9 Four different substances of the generalized formula HA were dissolved in water with the results shown in the following diagrams. Which of the diagrams represents the substance that is the strongest electrolyte of the four substances?

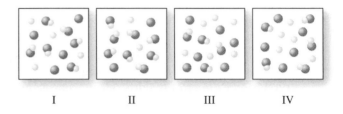

I   II   III   IV

15.10 Which of the diagrams in Problem 15.9 represents the substance that is the weakest electrolyte of the four substances?

IONIC AND NET IONIC EQUATIONS (SEC. 15.3)

15.11 Indicate whether or not each of the following types of substances would be written in ionic form in an ionic equation.
  a. strong electrolyte   **b.** weak acid
  c. soluble salt   **d.** nonelectrolyte

15.12 Indicate whether or not each of the following types of substances would be written in ionic form in an ionic equation.
  a. weak electrolyte   **b.** strong acid
  c. insoluble salt   **d.** an element

15.13 Indicate whether or not each of the following compounds would be written in ionic form in an ionic equation.
  a. NaBr   **b.** KOH   **c.** $HClO_3$ **d.** $HClO_2$

15.14 Indicate whether or not each of the following compounds would be written in ionic form in an ionic equation.
  a. $NaNO_3$ **b.** CO_2 **c.** HClO   **d.** $HClO_4$

15.15 Classify the following equations for reactions occurring in an aqueous solution as *molecular, ionic,* or *net ionic.*
  a. $MgCO_3 + 2\,HBr \longrightarrow MgBr_2 + H_2O + CO_2$
  b. $2\,OH^- + H_2CO_3 \longrightarrow CO_3^{2-} + 2\,H_2O$
  c. $K^+ + OH^- + H^+ + I^- \longrightarrow K^+ + I^- + H_2O$
  d. $Ag^+ + Cl^- \longrightarrow AgCl$

15.16 Classify the following equations for reactions occurring in an aqueous solution as *molecular, ionic,* or *net ionic.*
  a. $CaCO_3 + 2\,H^+ + 2\,NO_3^- \longrightarrow Ca^{2+} + 2\,NO_3^- + H_2O + CO_2$
  b. $Ni + Cu^{2+} \longrightarrow Cu + Ni^{2+}$
  c. $NaCl + AgNO_3 \longrightarrow NaNO_3 + AgCl$
  d. $Ca^{2+} + 2\,OH^- \longrightarrow Ca(OH)_2$

15.17 Write a balanced net ionic equation for each of the following reactions, each of which occurs in an aqueous solution.
  a. $2\,NaBr + Pb(NO_3)_2 \longrightarrow 2\,NaNO_3 + PbBr_2$
  b. $FeCl_3 + 3\,NaOH \longrightarrow Fe(OH)_3 + 3\,NaCl$
  c. $Zn + 2\,HCl \longrightarrow ZnCl_2 + H_2$
  d. $H_2S + 2\,KOH \longrightarrow K_2S + 2\,H_2O$

15.18 Write a balanced net ionic equation for each of the following reactions, each of which occurs in an aqueous solution.
  a. $CaCl_2 + CuSO_4 \longrightarrow CaSO_4 + CuCl_2$
  b. $Ca(NO_3)_2 + K_2CO_3 \longrightarrow CaCO_3 + 2\,KNO_3$
  c. $Mg + 2\,HBr \longrightarrow MgBr_2 + H_2$
  d. $H_3PO_4 + 3\,NaOH \longrightarrow Na_3PO_4 + 3\,H_2O$

15.19 Write a balanced net ionic equation for each of the following reactions, each of which occurs in an aqueous solution.
  a. $Pb + 2\,AgNO_3 \longrightarrow 2\,Ag + Pb(NO_3)_2$
  b. $Cl_2 + 2\,NaBr \longrightarrow 2\,NaCl + Br_2$
  c. $2\,Al(NO_3)_3 + 3\,Na_2S \longrightarrow Al_2S_3 + 6\,NaNO_3$
  d. $NaC_2H_3O_2 + NH_4Cl \longrightarrow NH_4C_2H_3O_2 + NaCl$

15.20 Write a balanced net ionic equation for each of the following reactions, each of which occurs in an aqueous solution.
  a. $Ni + Cu(NO_3)_2 \longrightarrow Ni(NO_3)_2 + Cu$
  b. $Br_2 + 2\,NaI \longrightarrow 2\,NaBr + I_2$
  c. $Hg(NO_3)_2 + K_2S \longrightarrow 2\,KNO_3 + HgS$
  d. $(NH_4)_2SO_4 + 2\,NaBr \longrightarrow 2\,NH_4Br + Na_2SO_4$

OXIDATION–REDUCTION TERMINOLOGY (SECS. 15.4 AND 15.5)

15.21 Give definitions of *oxidation* in terms of
  a. loss or gain of electrons.
  b. increase or decrease in oxidation number.

15.22 Give definitions of *reduction* in terms of
  a. loss or gain of electrons.
  b. increase or decrease in oxidation number.

15.23 Give definitions of *oxidizing agent* in terms of
  a. loss or gain of electrons.
  b. increase or decrease in oxidation number.
  c. substance oxidized or substance reduced.

15.24 Give definitions of *reducing agent* in terms of
  a. loss or gain of electrons.
  b. increase or decrease in oxidation number.
  c. substance oxidized or substance reduced.

15.25 In each of the following statements, choose the word in parentheses that best completes the statement.
  a. An element that has lost electrons in a redox reaction is said to have been (oxidized, reduced).
  b. Reduction always results in an (increase, decrease) in the oxidation number.
  c. The substance oxidized in a redox reaction is the (oxidizing, reducing) agent.
  d. The reducing agent (gains, loses) electrons during a redox reaction.

15.26 In each of the following statements, choose the word in parentheses that best completes the statement.
  a. The reducing agent causes an (increase, decrease) in the oxidation number of the oxidizing agent in a redox reaction.
  b. The oxidizing agent (gains, loses) electrons during a redox reaction.
  c. Oxidation always results in an (increase, decrease) in the oxidation number.
  d. An element that has gained electrons in a redox reaction is said to have been (oxidized, reduced).

OXIDATION NUMBERS (SEC. 15.5)

15.27 Give the oxidation number for the element present in the following species.
  a. O_2 **b.** P_4 **c.** S_8 **d.** Cu

15.28 Give the oxidation number for the element present in the following species.
  a. O_3 **b.** U **c.** As_4 **d.** B_{12}

15.29 What is the oxidation number of the indicated element in each of the following compounds?
  a. P in PCl_3 **b.** Si in SiH_4
  c. C in CO_2 **d.** N in N_2H_4

15.30 What is the oxidation number of the indicated element in each of the following compounds?
  a. N in NO **b.** C in CF_4
  c. Si in SiO_2 **d.** P in P_2H_4

15.31 What is the oxidation number of the indicated element in each of the following ions?
  a. Ca in Ca^{2+} **b.** Br in Br^-
  c. P in PO_4^{3-} **d.** C in CO_3^{2-}

15.32 What is the oxidation number of the indicated element in each of the following ions?
  a. Al in Al^{3+} **b.** N in N^{3-}
  c. Cl in ClO_3^- **d.** N in NH_4^+

15.33 Determine the oxidation number of Cl in each of the following chlorine-containing species.
  a. $BeCl_2$ **b.** ClF
  c. $AlCl_4^-$ **d.** ClO^-

15.34 Determine the oxidation number of Cr in each of the following chromium-containing species.
  a. Cr_2O_3 **b.** $BaCrO_4$
  c. CrF_5 **d.** $Cr_2O_7^{2-}$

15.35 What is the oxidation number of each element in each of the following compounds?

    a. H_3PO_4 **b.** $BaCr_2O_7$

    c. NH_4ClO_4 **d.** $H_4P_2O_7$

15.36 What is the oxidation number of each element in each of the following compounds?

    a. $H_2S_2O_7$ **b.** $KMnO_4$

    c. Na_2CrO_4 **d.** $Ba(ClO)_2$

15.37 Arrange the following sulfur-containing species in order of increasing oxidation number.

    a. H_2S **b.** H_2SO_3 **c.** SO_3 **d.** S_8

15.38 Arrange the following sulfur-containing species in order of increasing oxidation number.

    a. S_2 **b.** SO_2 **c.** S^{2-} **d.** H_2SO_4

15.39 Determine the oxidation number of the metal in each of the following polyatomic-ion containing compounds.

    a. $Rh_2(CO_3)_2$ **b.** $Cr_2(SO_4)_3$

    c. $Cu(ClO_2)_2$ **d.** $Co_3(PO_4)_2$

15.40 Determine the oxidation number of the metal in each of the following polyatomic-ion containing compounds.

    a. $Ni_2(SO_4)_3$ **b.** $Ru_3(PO_4)_2$

    c. $Fe_2(CO_3)_3$ **d.** $Al(NO_3)_3$

15.41 Indicate whether oxygen has a –2, –1, or positive oxidation number in each of the following oxygen-containing species.

    a. Na_2O **b.** OF_2 **c.** Na_2O_2 **d.** BaO

15.42 Indicate whether oxygen has a –2, –1, or positive oxidation number in each of the following oxygen-containing species.

    a. O_2F_2 **b.** SO_3 **c.** BaO_2 **d.** BeO

15.43 Indicate whether hydrogen has a +1 or –1 oxidation number in each of the following hydrogen-containing species.

    a. NaH **b.** CH_4 **c.** HCl **d.** CaH_2

15.44 Indicate whether hydrogen has a +1 or –1 oxidation number in each of the following hydrogen-containing species.

    a. H_2Se **b.** N_2H_2 **c.** KH **d.** MgH_2

CHARACTERISTICS OF OXIDATION–REDUCTION REACTIONS (SEC. 15.5)

15.45 Identify which substance is oxidized and which substance is reduced in each of the following redox reactions.

    a. $N_2 + 3\,H_2 \longrightarrow 2\,NH_3$

    b. $Cl_2 + 2\,KI \longrightarrow 2\,KCl + I_2$

    c. $Sb_2O_3 + 3\,Fe \longrightarrow 2\,Sb + 3\,FeO$

    d. $3\,H_2SO_3 + 2\,HNO_3 \longrightarrow 2\,NO + H_2O +$
                                             $3\,H_2SO_4$

15.46 Identify which substance is oxidized and which substance is reduced in each of the following redox reactions.

    a. $2\,Al + 3\,Cl_2 \longrightarrow 2\,AlCl_3$

    b. $Zn + CuCl_2 \longrightarrow ZnCl_2 + Cu$

    c. $2\,NiS + 3\,O_2 \longrightarrow 2\,NiO + 2\,SO_2$

    d. $3\,H_2S + 2\,HNO_3 \longrightarrow 3\,S + 2\,NO + 4\,H_2O$

15.47 Identify which substance is the oxidizing agent and which substance is the reducing agent in each of the redox reactions in Problem 15.45.

15.48 Identify which substance is the oxidizing agent and which substance is the reducing agent in each of the redox reactions in Problem 15.46.

15.49 Identify the following species for the redox reaction

$$2\,HNO_3 + SO_2 \longrightarrow H_2SO_4 + 2\,NO_2$$

    a. substance that is oxidized

    b. oxidizing agent

    c. substance that contains the element that decreases in oxidation number

    d. substance that contains the element that loses electrons during the oxidation–reduction reaction

15.50 Identify the following species for the redox reaction

$$PH_3 + 2\,NO_2 \longrightarrow H_3PO_4 + N_2$$

    a. substance that is reduced

    b. reducing agent

    c. substance that contains the element that increases in oxidation number

    d. substance that contains the element that loses electrons during the oxidation–reduction reaction

15.51 For each of the following redox reactions that involve the reaction of a metallic element with a nonmetallic element, indicate whether the metal is oxidized or reduced and whether the nonmetal is oxidized or reduced.

    a. $3\,Zn + N_2 \longrightarrow Zn_3N_2$

    b. $2\,Ca + O_2 \longrightarrow 2\,CaO$

    c. $2\,Na + S \longrightarrow Na_2S$

    d. $Mg + Cl_2 \longrightarrow MgCl_2$

15.52 For each of the following redox reactions that involve the reaction of a metallic element with a nonmetallic element, indicate whether the metal is oxidized or reduced and whether the nonmetal is oxidized or reduced.

    a. $Cu + S \longrightarrow CuS$

    b. $4\,Ag + O_2 \longrightarrow 2\,Ag_2O$

    c. $Ni + F_2 \longrightarrow NiF_2$

    d. $2\,Al + N_2 \longrightarrow 2\,AlN$

REDOX AND NONREDOX CHEMICAL REACTIONS (SEC. 15.6)

15.53 Characterize each of the following chemical reactions as a redox or a nonredox reaction.

    a. $2\,FeBr_3 \longrightarrow 2\,FeBr_2 + Br_2$

    b. $K_2O + H_2O \longrightarrow 2\,KOH$

    c. $2\,KClO_3 \longrightarrow 2\,KCl + 3\,O_2$

    d. $CH_4 + 2\,O_2 \longrightarrow CO_2 + 2\,H_2O$

15.54 Characterize each of the following chemical reactions as a redox reaction or a nonredox reaction.

    a. $2\,NO + O_2 \longrightarrow 2\,NO_2$

    b. $CO_2 + H_2O \longrightarrow H_2CO_3$

    c. $Zn + 2\,AgNO_3 \longrightarrow Zn(NO_3)_2 + 2\,Ag$
    d. $HNO_3 + NaOH \longrightarrow NaNO_3 + H_2O$

15.55 Characterize each of the following reactions, using one selection from the choices *redox* and *nonredox* combined with one selection from the choices *synthesis, decomposition, single-replacement,* and *double-replacement*.
    a. $H_2 + Cl_2 \longrightarrow 2\,HCl$
    b. $2\,HBr + Mg \longrightarrow MgBr_2 + H_2$
    c. $MgCO_3 \longrightarrow MgO + CO_2$
    d. $2\,KOH + H_2SO_4 \longrightarrow K_2SO_4 + 2\,H_2O$

15.56 Characterize each of the following reactions, using one selection from the choices *redox* and *nonredox* combined with one selection from the choices *synthesis, decomposition, single-replacement,* and *double-replacement*.
    a. $Zn + Cu(NO_3)_2 \longrightarrow Zn(NO_3)_2 + Cu$
    b. $2\,SO_2 + O_2 \longrightarrow 2\,SO_3$
    c. $2\,CuO \longrightarrow 2\,Cu + O_2$
    d. $NaCl + AgNO_3 \longrightarrow AgCl + NaNO_3$

15.57 Characterize each of the following chemical reactions as (1) a redox reaction, (2) a nonredox reaction, or (3) "can't classify because of insufficient information."
    a. a synthesis reaction in which both reactants are elements
    b. a combustion reaction
    c. a decomposition reaction in which the products are all compounds
    d. a decomposition reaction in which an element and a compound are products

15.58 Characterize each of the following chemical reactions as (1) a redox reaction, (2) a nonredox reaction, or (3) "can't classify because of insufficient information."
    a. a synthesis reaction in which one reactant is an element and the other is a compound
    b. an acid–base neutralization reaction
    c. a decomposition reaction in which the products are all elements
    d. a single-replacement reaction involving an active metal and an acid

BALANCING REDOX EQUATIONS: OXIDATION NUMBER METHOD (SEC. 15.8)

15.59 Balance the following equations by the oxidation number method.
    a. $Cr + HCl \longrightarrow CrCl_3 + H_2$
    b. $Cr_2O_3 + C \longrightarrow Cr + CO_2$
    c. $SO_2 + NO_2 \longrightarrow SO_3 + NO$
    d. $BaSO_4 + C \longrightarrow BaS + CO$

15.60 Balance the following equations by the oxidation number method.
    a. $Fe_2O_3 + CO \longrightarrow Fe + CO_2$
    b. $Al + MnO_2 \longrightarrow Al_2O_3 + Mn$
    c. $I_2O_5 + CO \longrightarrow I_2 + CO_2$
    d. $N_2H_4 + O_2 \longrightarrow N_2 + H_2O$

15.61 Balance the following equations by the oxidation number method.
    a. $Br_2 + H_2O + SO_2 \longrightarrow HBr + H_2SO_4$
    b. $H_2S + HNO_3 \longrightarrow S + NO + H_2O$
    c. $SnSO_4 + FeSO_4 \longrightarrow Sn + Fe_2(SO_4)_3$
    d. $Na_2TeO_3 + NaI + HCl \longrightarrow NaCl + Te + H_2O + I_2$

15.62 Balance the following equations by the oxidation number method.
    a. $HNO_3 + I_2 \longrightarrow NO_2 + H_2O + HIO_3$
    b. $As_4O_6 + Cl_2 + H_2O \longrightarrow H_3AsO_4 + HCl$
    c. $HI + HNO_3 \longrightarrow I_2 + NO + H_2O$
    d. $PbO_2 + Sb + NaOH \longrightarrow PbO + NaSbO_2 + H_2O$

15.63 Balance the following equations by the oxidation number method. All reactions occur in an acidic solution.
    a. $I_2 + Cl_2 \longrightarrow HIO_3 + Cl^-$
    b. $MnO_4^- + AsH_3 \longrightarrow H_3AsO_4 + Mn^{2+}$
    c. $Br^- + SO_4^{2-} \longrightarrow Br_2 + SO_2$
    d. $Au + Cl^- + NO_3^- \longrightarrow AuCl_4^- + NO_2$

15.64 Balance the following equations by the oxidation number method. All reactions occur in an acidic solution.
    a. $I^- + SO_4^{2-} \longrightarrow H_2S + I_2$
    b. $Mn^{2+} + BiO_3^- \longrightarrow MnO_4^- + Bi^{3+}$
    c. $Fe^{2+} + ClO_3^- \longrightarrow Fe^{3+} + Cl^-$
    d. $Pt + Cl^- + NO_3^- \longrightarrow PtCl_6^{2-} + NO_2$

15.65 Balance the following equations by the oxidation number method. All reactions occur in a basic solution.
    a. $S^{2-} + Cl_2 \longrightarrow SO_4^{2-} + Cl^-$
    b. $SO_3^{2-} + CrO_4^{2-} \longrightarrow Cr(OH)_4^- + SO_4^{2-}$
    c. $MnO_4^- + IO_3^- + MnO_2 + IO_4^-$
    d. $I_2 + Cl_2 \longrightarrow H_3IO_6^{2-} + Cl^-$

15.66 Balance the following equations by the oxidation number method. All reactions occur in a basic solution.
    a. $Zn + MnO_4^- \longrightarrow Zn(OH)_2 + MnO_2$
    b. $NO_2^- + Al \longrightarrow NH_3 + AlO_2^-$
    c. $NO_2^- + MnO_4^- \longrightarrow NO_3^- + MnO_2$
    d. $Al + NO_3^- \longrightarrow Al(OH)_4^- + NH_3$

BALANCING REDOX EQUATIONS: HALF-REACTION METHOD (SEC. 15.9)

15.67 Classify each of the following unbalanced half-reactions as either oxidation or reduction.
    a. $NO_3^- \longrightarrow NO$
    b. $Zn \longrightarrow Zn^{2+}$
    c. $Ti^{3+} \longrightarrow TiO_2$
    d. $Cr_2O_7^{2-} \longrightarrow Cr^{3+}$

15.68 Classify each of the following unbalanced half-reactions as either oxidation or reduction.
    a. $O_2 \longrightarrow OH^-$ **b.** $I^- \longrightarrow I_2$
    c. $Sn^{2+} \longrightarrow Sn^{4+}$ **d.** $H_2S \longrightarrow S$

15.69 What is the unbalanced oxidation half-reaction associated with each of the following processes?
   a. $Te + NO_3^- \longrightarrow TeO_2 + NO$
   b. $H_2O_2 + Fe^{2+} \longrightarrow Fe^{3+} + H_2O$
   c. $CN^- + ClO_2^- \longrightarrow CNO^- + Cl^-$
   d. $ClO^- + Cl^- \longrightarrow Cl_2$

15.70 What is the unbalanced oxidation half-reaction associated with each of the following processes?
   a. $Zn + NO_3^- \longrightarrow Zn^{2+} + NH_4^+$
   b. $I^- + MnO_4^- \longrightarrow I_2 + Mn^{2+}$
   c. $ClO^- + Br^- \longrightarrow Cl^- + BrO_3^-$
   d. $H_5IO_6 + I^- \longrightarrow I_2$

15.71 Balance the following half-reactions occurring in an acidic solution.
   a. $MnO_2 \longrightarrow Mn^{3+}$
   b. $H_3MnO_4 \longrightarrow Mn$
   c. $MnO_4^- \longrightarrow Mn^{2+}$
   d. $MnO_4^- \longrightarrow MnO_2$

15.72 Balance the following half-reactions occurring in an acidic solution.
   a. $V^{2+} \longrightarrow VO_2^+$ **b.** $V^{3+} \longrightarrow VO^{2+}$
   c. $VO^{2+} \longrightarrow VO_2^+$ **d.** $V \longrightarrow VO_2^+$

15.73 Balance the following half-reactions occurring in a basic solution.
   a. $SeO_4^{2-} \longrightarrow Se$
   b. $Se^{2-} \longrightarrow SeO_3^{2-}$
   c. $SeO_4^{2-} \longrightarrow SeO_3^{2-}$
   d. $Se \longrightarrow SeO_3^{2-}$

15.74 Balance the following half-reactions occurring in a basic solution.
   a. $H_3IO_6^{2-} \longrightarrow I_2$ **b.** $IO_3^- \longrightarrow IO^-$
   c. $I^- \longrightarrow IO^-$ **d.** $IO^- \longrightarrow H_3IO_6^{2-}$

15.75 Balance each of the following redox reactions by the half-reaction method. Each reaction occurs in an acidic solution.
   a. $Zn + Cu^{2+} \longrightarrow Cu + Zn^{2+}$
   b. $Br_2 + I^- \longrightarrow Br^- + I_2$
   c. $S_2O_3^{2-} + Cl_2 \longrightarrow HSO_4^- + Cl^-$
   d. $Zn + As_2O_3 \longrightarrow AsH_3 + Zn^{2+}$

15.76 Balance each of the following redox reactions by the half-reaction method. Each reaction occurs in an acidic solution.
   a. $Fe + Ag^+ \longrightarrow Fe^{3+} + Ag$
   b. $Cl_2 + Br^- \longrightarrow Cl^- + Br_2$
   c. $S_2O_3^{2-} + Cu^{2+} \longrightarrow S_4O_6^{2-} + Cu$
   d. $C_2O_4^{2-} + MnO_4^- \longrightarrow CO_2 + Mn^{2+}$

15.77 Balance each of the equations in Problem 15.63, using the half-reaction method for balancing.

15.78 Balance each of the equations in Problem 15.64, using the half-reaction method for balancing.

15.79 Balance each of the following redox reactions by the half-reaction method. Each reaction occurs in a basic solution.
   a. $NH_3 + ClO^- \longrightarrow N_2H_4 + Cl^-$
   b. $Cr(OH)_2 + BrO^- \longrightarrow CrO_4^{2-} + Br^-$
   c. $CrO_2^- + H_2O_2 \longrightarrow CrO_4^{2-} + OH^-$
   d. $Bi(OH)_3 + Sn(OH)_3^- \longrightarrow Sn(OH)_6^{2-} + Bi$

15.80 Balance each of the following redox reactions by the half-reaction method. Each reaction occurs in a basic solution.
   a. $Cr_2O_3 + ClO^- \longrightarrow CrO_4^{2-} + Cl^-$
   b. $NO + MnO_4^- \longrightarrow NO_3^- + MnO_2$
   c. $Al + PO_3^{3-} \longrightarrow PH_3 + AlO_2^-$
   d. $Bi(OH)_3 + SnO_2^{2-} \longrightarrow SnO_3^{2-} + Bi$

15.81 Balance each of the equations in Problem 15.65, using the half-reaction method for balancing.

15.82 Balance each of the equations in Problem 15.66, using the half-reaction method for balancing.

BALANCING REDOX EQUATIONS: DISPROPORTIONATION REACTIONS (SEC. 15.10)

15.83 Balance each of the following redox reactions by the oxidation number method.
   a. $HNO_2 \longrightarrow NO + NO_3^-$ (acidic solution)
   b. $ClO^- + Cl^- \longrightarrow Cl_2$ (acidic solution)
   c. $S \longrightarrow S^{2-} + SO_3^{2-}$ (basic solution)
   d. $Br_2 \longrightarrow BrO_3^- + Br^-$ (basic solution)

15.84 Balance each of the following redox reactions by the oxidation number method.
   a. $NO + NO_3^- \longrightarrow N_2O_4$ (acidic solution)
   b. $H_5IO_6 + I^- \longrightarrow I_2$ (acidic solution)
   c. $P_4 \longrightarrow HPO_3^{2-} + PH_3$ (basic solution)
   d. $HClO_2 \longrightarrow ClO_2 + Cl^-$ (basic solution)

15.85 Balance each of the redox reactions in Problem 15.83 using the half-reaction method.

15.86 Balance each of the redox reactions in Problem 15.84 using the half-reaction method.

STOICHIOMETRIC CALCULATIONS INVOLVING IONS (SEC 15.11)

15.87 What is the molar mass for each of the following ions?
   a. Cl^- **b.** NH_4^+ **c.** $Cr_2O_7^{2-}$ **d.** MnO_4^-

15.88 What is the molar mass for each of the following ions?
   a. Mg^{2+} **b.** PO_4^{3-} **c.** $S_2O_3^{2-}$ **d.** HCO_3^-

15.89 How many grams of Mg^{2+} ions are present in the following solutions?
   a. $MgCO_3$ solution that contains 1.00×10^{24} CO_3^{2-} ions
   b. $MgSO_4$ solution that contains 1.00×10^{24} SO_4^{2-} ions
   c. $MgCl_2$ solution that contains 1.00×10^{24} Cl^- ions
   d. $Mg(OH)_2$ solution that contains 1.00×10^{24} total ions

15.90 How many grams of K^+ ions are present in the following solutions?
   a. K_2CO_3 solution that contains 1.00×10^{24} CO_3^{2-} ions
   b. K_3PO_4 solution that contains 1.00×10^{24} PO_4^{3-} ions
   c. KCl solution that contains 1.00×10^{24} Cl^- ions
   d. KNO_3 solution that contains 1.00×10^{24} total ions

15.91 What volume, in milliliters, of a 0.500 M $Ca(CN)_2$ solution is needed to supply
  a. 0.634 mole of Ca^{2+} ion?
  b. 0.634 mole of CN^- ion?
  c. 0.553 mole of Ca^{2+} ion?
  d. 1.20 total moles of ions?

15.92 What volume, in milliliters, of a 0.750 M $(NH_4)_2SO_4$ solution is needed to supply
  a. 0.766 mole of NH_4^+ ion?
  b. 0.766 mole of SO_4^{2-} ion?
  c. 0.544 mole of NH_4^+ ion?
  d. 0.321 total mole of ions?

15.93 A solution is 1.35 M in Cu^{2+} ion. What volume of this solution, in milliliters, is needed to react completely with 20.0 g of Fe in the following redox reaction?

$$Fe(s) + Cu^{2+}(aq) \longrightarrow Fe^{2+}(aq) + Cu(s)$$

15.94 A solution is 1.35 M in Sn^{2+} ion. What volume of this solution, in milliliters, is needed to react completely with 25.0 g of Cr in the following redox reaction?

$$Cr(s) + Sn^{2+}(aq) \longrightarrow Cr^2(aq) + Sn(s)$$

15.95 What is the molar concentration of each ion in the following solutions?
  a. 0.20 M NaCl   **b.** 0.20 M K_2SO_4
  c. 0.20 M $Al(NO_3)_3$ **d.** 0.20 $MgCl_2$

15.96 What is the molar concentration of each ion in the following solutions?
  a. 0.36 M $CaBr_2$
  b. 0.36 M $NaNO_3$
  c. 0.36 M K_3PO_4
  d. 0.36 M $(NH_4)_2SO_4$

15.97 Calculate the total number of ions present in solutions in which each of the following solutes are present in the given amounts.
  a. 8.45 g NaBr   **b.** 3.20 g K_2SO_4
  c. 30.0 g HCl   **d.** 40.0 g KOH

15.98 Calculate the total number of ions present in solutions in which each of the following solutes are present in the given amounts.
  a. 1.25 g NaOH   **b.** 4.30 g Na_2SO_4
  c. 25.0 g HNO_3 **d.** 32.7 g $AlBr_3$

15.99 Calculate the volume, in milliliters, of 0.125 M Na_2SO_4 solution needed to provide each of the following.
  a. 10.0 g of Na_2SO_4
  b. 2.5 g of Na^+ ion
  c. 0.567 mole of Na_2SO_4
  d. 0.112 mole of SO_4^{2-}

15.100 Calculate the volume, in milliliters, of 0.125 M $Mg(NO_3)_2$ solution needed to provide each of the following.
  a. 15.7 g of $Mg(NO_3)_2$
  b. 3.57 g of Mg^{2+} ion
  c. 1.2 moles of $Mg(NO_3)_2$
  d. 0.57 mole of NO_3^- ion

15.101 A solution is made by mixing 175 mL of 0.100 M K_3PO_4 with 27 mL of 0.200 M KCl. Assuming that the volumes are additive, what are the molar concentrations of the following ions in the new solution?
  a. K^+ ion   **b.** Cl^- ion   **c.** PO_4^{3-} ion

15.102 A solution is made by mixing 50.0 mL of 0.300 M Na_2SO_4 with 30.0 mL of 0.900 M K_2SO_4. Assuming that the volumes are additive, what are the molar concentrations of the following ions in the new solution?
  a. Na^+ ion **b.** K^+ ion   **c.** SO_4^{2-} ion

ADDITIONAL PROBLEMS

15.103 Nitrogen forms a number of oxides including NO_2, N_2O_3, NO, N_2O, and N_2O_5. Arrange these oxides in order of increasing oxidation number of nitrogen.

15.104 Sulfur forms a number of oxides including S_2O, S_7O_2, SO_2, SO_3, and S_6O. Arrange these oxides in order of increasing oxidation number of sulfur.

15.105 Possible oxidation numbers for the element S range from +6 to –2. Based on this information, explain each of the following observations.
  a. The S^{2-} ion functions only as a reducing agent.
  b. The SO_4^{2-} ion functions only as an oxidizing agent.
  c. The SO_2 molecule can function as either a reducing agent or an oxidizing agent.
  d. The SO_3 molecule functions only as an oxidizing agent.

15.106 Possible oxidation numbers for the element N range from +5 to –3. Based on this information, explain each of the following observations.
  a. The N^{3-} ion functions only as a reducing agent.
  b. The NO_3^- ion functions only as an oxidizing agent.
  c. The NO_2^- molecule can function as either a reducing agent or an oxidizing agent.
  d. The NH_3 molecule can function as either a reducing agent or an oxidizing agent.

15.107 In which of the following pairs of ionic compounds is the oxidation number of the metal the same in both members of the pair?
  a. $CuSO_4$ and Cu_2SO_4 **b.** $Fe(NO_3)_3$ and $FePO_4$
  c. AuCl and $AgNO_3$ **d.** $AlPO_4$ and GaN

15.108 In which of the following pairs of ionic compounds is the oxidation number of the metal the same in both members of the pair?
  a. CuO and $CuCl_2$ **b.** Ni_2O_3 and NiN
  c. PbO_2 and $SnCl_4$ **d.** Be_3N_2 and MgO

15.109 Classify each of the following pairs of balanced half-reactions as (1) two reduction half-reactions, (2) two oxidation half-reactions, or (3) one reduction and one oxidation half-reaction.
  a. $Fe^{3+} + e^- \longrightarrow Fe^{2+}$ and $Fe^{2+} + 2e^- \longrightarrow Fe$
  b. $Ni^{3+} + e^- \longrightarrow Ni^{2+}$ and $Ni \longrightarrow Ni^{2+} + 2e^-$
  c. $Cu \longrightarrow Cu^+ + e^-$ and $Cu \longrightarrow Cu^2 + 2e^-$
  d. $Au \longrightarrow Au^{3+} + 3e^-$ and $Au^{3+} + 3e^- \longrightarrow Au$

15.110 Classify each of the following pairs of balanced half-reactions as (1) two reduction half-reactions, (2) two oxidation half-reactions, or (3) one reduction and one oxidation half-reaction.

a. $Sn^{2+} \longrightarrow Sn^{4+} + 2\,e^-$ and $Sn \longrightarrow Sn^{2+} + 2\,e^-$

b. $Pb^{2+} \longrightarrow Pb^{4+} + 2\,e^-$ and $Pb^{2+} + 2\,e^- \longrightarrow Pb$

c. $Co^{2+} + 2\,e^- \longrightarrow Co$ and $Co^{3+} + 3\,e^- \longrightarrow Co$

d. $Ag \longrightarrow Ag^+ + e^-$ and $Ag^+ + e^- \longrightarrow Ag$

15.111 Write balanced equations for all possible redox reactions obtainable by combining the following balanced half-reactions in sets of two. The half-reactions must be used as written; they cannot be reversed in direction.

1. $2\,H_2O + PH_3 \longrightarrow H_3PO_2 + 4\,H^+ + 4\,e^-$
2. $3\,H_2O + As \longrightarrow H_3AsO_3 + 3\,H^+ + 3\,e^-$
3. $MnO_4^- + 8\,H^+ + 5\,e^- \longrightarrow Mn^{2+} + 4\,H_2O$
4. $SO_4^{2-} + 4\,H^+ + 2\,e^- \longrightarrow SO_2 + 2\,H_2O$

15.112 Write balanced equations for all possible redox reactions obtainable by combining the following balanced half-reactions in sets of two. The half-reactions must be used as written; they cannot be reversed in direction.

1. $4\,OH^- + ClO_2^- \longrightarrow ClO_4^- + 2\,H_2O + 4\,e^-$
2. $2\,H_2O + MnO_4^- + 3\,e^- \longrightarrow MnO_2 + 4\,OH^-$
3. $6\,H_2O + NO_3^- + 8\,e^- \longrightarrow NH_3 + 9\,OH^-$
4. $4\,OH^- + Al \longrightarrow AlO_2^- + 2\,H_2O + 3\,e^-$

15.113 Write the two balanced half-reactions associated with the following redox reaction.

$$4\,Zn + 10\,H^+ + NO_3^- \longrightarrow 4\,Zn^{2+} + NH_4^+ + 3\,H_2O$$

15.114 Write the two balanced half-reactions associated with the following redox reaction.

$$2\,NO_3^- + 3\,Cu_2O + 14\,H^+ \longrightarrow 6\,Cu^{2+} + 2\,NO + 7\,H_2O$$

15.115 Using the half-reaction method, balance the following redox equation, which occurs in an acidic solution. Then, using the reactant identities shown in brackets convert the balanced net ionic redox equation into a balanced molecular redox equation.

$$MnO_4^- + C_2O_4^{2-} \longrightarrow Mn^2 + CO_2 \ [KMnO_4, HCl, K_2C_2O_4]$$

15.116 Using the half-reaction method, balance the following redox equation, which occurs in an acidic solution. Then, using the reactant identities shown in brackets convert the balanced net ionic redox equation into a balanced molecular redox equation.

$$Zn + NO_3^- \longrightarrow Zn^{2+} + NH_4^+ \ [Zn, HNO_3]$$

CUMULATIVE PROBLEMS

15.117 Classify each of the following water-producing reactions as an acid–base reaction or as an oxidation–reduction reaction. If it is an acid–base reaction, identify the acid; if it is an oxidation–reduction reaction, identify the oxidizing agent.

a. $2\,HNO_3 + 3\,H_2S \longrightarrow 3\,S + 2\,NO + 4\,H_2O$

b. $2\,KOH + H_2S \longrightarrow K_2S + 2\,H_2O$

c. $2\,HI + H_2O_2 \longrightarrow I_2 + 2\,H_2O$

d. $H_2SO_4 + 2\,NaOH \longrightarrow Na_2SO_4 + 2\,H_2O$

15.118 Classify each of the following water-producing reactions as an acid–base reaction or as an oxidation–reduction reaction. If it is an acid–base reaction, identify the acid; if it is an oxidation–reduction reaction, identify the oxidizing agent.

a. $7\,HI + H_5IO_6 \longrightarrow 4\,I_2 + 6\,H_2O$

b. $H_2CO_3 + 2\,KOH \longrightarrow K_2CO_3 + 2\,H_2O$

c. $5\,HClO_2 + NaOH \longrightarrow 4\,ClO_2 + 3\,H_2O + NaCl$

d. $14\,HNO_3 + 3\,Cu_2O \longrightarrow 6\,Cu(NO_3)_2 + 2\,NO + 7\,H_2O$

15.119 Convert each of the following balanced molecular redox equations to balanced net ionic redox equations.

a. $SnSO_4(aq) + 2\,FeSO_4(aq) \longrightarrow Sn(s) + Fe_2(SO_4)_3(aq)$

b. $PH_3(g) + 2\,NO_2(g) \longrightarrow H_3PO_4(aq) + N_2(g)$

c. $S(s) + 3\,H_2O(l) + 2\,Pb(NO_3)_2 \longrightarrow 2\,Pb(s) + H_2SO_3(aq) + 4\,HNO_3(aq)$

d. $4\,Zn\,(s) + 10\,HNO_3(aq) \longrightarrow 4\,Zn(NO_3)_2(aq) + NH_4NO_3(aq) + 3\,H_2O(l)$

15.120 Convert each of the following balanced molecular redox equations to balanced net ionic redox equations.

a. $NaNO_3(aq) + Pb(s) \longrightarrow NaNO_2(aq) + PbO(s)$

b. $3\,H_2S(aq) + 2\,HNO_3(aq) \longrightarrow 3\,S(s) + 2\,NO(g) + 4\,H_2O(l)$

c. $2\,HNO_3(aq) + SO_2(g) \longrightarrow H_2SO_4(aq) + 2\,NO_2(g)$

d. $10\,FeSO_4(aq) + 2\,KMnO_4(aq) + 8\,H_2SO_4(aq) \longrightarrow 5\,Fe_2(SO_4)_3(aq) + 2\,MnO_4(aq) + K_2SO_4(aq) + 8\,H_2O(l)$

15.121 Balance each of the following net ionic redox reactions.

a. Hydrosulfuric acid plus dichromate ion produces chromium(III) ion plus sulfur (acidic solution).

b. Chlorate ion plus iodine produces iodate ion plus chloride ion (acidic solution).

c. Sulfide ion plus bromine produces sulfate ion plus bromide ion (basic solution).

d. Nitrogen dioxide disproportionates to produce nitrate ion plus nitrite ion (basic solution).

15.122 Balance each of the following net ionic redox reactions.

a. Iron(II) ion plus permanganate ion produces iron(III) ion plus manganese(II) ion (acidic solution).

b. Iodine plus sulfur dioxide produces iodide ion plus sulfate ion (acidic solution).

c. Manganese(II) hydroxide plus nickel(IV) oxide produces manganese(III) oxide plus nickel(II) hydroxide (basic solution).

d. Chlorine disproportionates to produce chlorate ion and chloride ion (basic solution).

15.123 A solution is made by diluting 225 mL of a 0.245 M aluminum nitrate solution with water to a final volume of 0.750 L. Calculate the following:
  a. the molarities of the aluminum ion and nitrate ion in the original solution
  b. the molarities of the aluminum nitrate, aluminum ion, and nitrate ion in the diluted solution

15.124 A solution is made by diluting 315 mL of 0.115 M potassium phosphate solution with water to a final volume of 0.650 L. Calculate the following:
  a. the molarities of the potassium ion and the phosphate ion in the original solution
  b. the molarities of the potassium phosphate, potassium ion, and phosphate ion in the diluted solution

15.125 A 21.03 mL solution containing 2.341 g $Mg(NO_3)_2$ is mixed with a 22.51 mL solution containing 1.250 g Na_2CO_3. Calculate the molar concentration of the ions remaining in the solution after the reaction is complete. Assume volumes are additive.

15.126 A 26.55 mL solution containing 2.341 g $AgNO_3$ is mixed with a 13.72 mL solution containing 1.250 g NaCl. Calculate the molar concentration of the ions remaining in the solution after the reaction is complete. Assume volumes are additive.

15.127 The amount of $I_3^-(aq)$ in a solution can be determined by reacting it with a solution containing $S_2O_3^{2-}(aq)$.

$$I_3^-(aq) + 2\,S_2O_3^{2-}(aq) \longrightarrow 3\,I^-(aq) + S_4O_6^{2-}(aq)$$

Calculate the molarity of I_3^- in a solution, given that 43.2 mL of 0.300 M $S_2O_3^{2-}$ solution reacts with a 20.0 mL sample of the I_3^- solution.

15.128 Oxalic acid, $H_2C_2O_4$, reacts with dichromate ion, $Cr_2O_7^{2-}$, in an acidic solution as follows:

$$3\,H_2C_2O_4(aq) + Cr_2O_7^{2-}(aq) + 8\,H^+(aq) \longrightarrow$$
$$6\,CO_2(g) + 2\,Cr^{3+}(aq) + 7\,H_2O(l)$$

If 35.0 mL of an oxalic acid solution reacts completely with 25.0 mL of 0.0500 M $Cr_2O_7^{2-}$ solution, what is the molarity of the oxalic acid solution?

15.129 The amount of ozone, O_3, in polluted air can be determined in a two-step process. First the ozone is reacted with an acidic solution containing iodide ion.

$$O_3(g) + 2\,I^-(aq) + 2\,H^+(aq) \longrightarrow O_2(g) + I_2(s) + H_2O(l)$$

The iodine so produced is then reacted with thiosulfate, $S_2O_3^{2-}$, solution.

$$2\,S_2O_3^{2-}(aq) + I_2(s) \longrightarrow S_4O_6^{2-}(aq) + 2\,I^-(aq)$$

If 18.03 mL of a 0.00200 M $S_2O_3^{2-}$ solution completely reacts with the I_2 produced by a 28.09 g sample of polluted air, calculate the O_3 concentration, in ppm (m/m), in the sample of air.

15.130 The active ingredient in household bleach is the hypochlorite ion, ClO^-. The concentration of this ion in bleach can be determined using a two-step process. First, the bleach is reacted with an I^- (aq) solution.

$$ClO^-(aq) + 2\,I^-(aq) + 2\,H^+(aq) \longrightarrow$$
$$I_2(s) + Cl^-(aq) + H_2O(l)$$

The iodine so produced is then reacted with thiosulfate, $S_2O_3^{2-}$, solution.

$$I_2(s) + 2\,S_2O_3^{2-}(aq) \longrightarrow 2\,I^-(aq) + S_4O_6^{2-}(aq)$$

A 50.00 g sample of a certain household bleach is found to react completely with 42.5 mL of a 0.0150 M $S_2O_3^{2-}$ solution. Calculate the concentration, in mass percent, of ClO^- ion in the bleach.

Multiple-Choice Practice Test

Use this bank of 20 multiple-choice questions as a review of key concepts presented in this chapter. For many of the questions, there may be more than one correct answer (choice d) or no correct answer (choice e).

15.131 Which of the following statements concerning strong electrolytes is correct?
  a. All soluble acids are strong electrolytes.
  b. All soluble bases are strong electrolytes.
  c. All soluble salts are strong electrolytes.
  d. more than one correct response
  e. no correct response

15.132 Which of the following substances is a weak electrolyte?
  a. SO_2 **b.** HNO_3 **c.** NaCl
  d. more than one correct response
  e. no correct response

15.133 In which of the following pairs of substances would both members of the pair be written in molecular form in a ionic equation?
  a. HF and H_3PO_4
  b. NH_4Cl and NaCl
  c. NaOH and KOH
  d. more than one correct response
  e. no correct response

15.134 What are the spectator ions in the reaction

$$KOH + HNO_3 \longrightarrow KNO_3 + H_2O$$

  a. K^+ and H^+
  b. H^+ and OH^-
  c. K^+ and NO_3^-
  d. more than one correct response
  e. no correct response

15.135 In the following reaction, which species will be written in molecular form when the equation is converted to an ionic equation?

$$Mg(OH)_2(s) + 2\,HCl(aq) \longrightarrow$$
$$MgCl_2(aq) + 2\,H_2O(l)$$

   a. $Mg(OH)_2$
   b. HCl
   c. H_2O
   d. more than one correct response
   e. no correct response

15.136 Which of the following is the net ionic equation for the reaction

$$AgNO_3 + NaBr \longrightarrow AgBr + NaNO_3$$

   a. $Ag^+ + NaBr \longrightarrow AgBr + Na^+$
   b. $AgNO_3 + Br^- \longrightarrow AgBr + NO_3^-$
   c. $Na^+ + NO_3^- \longrightarrow NaNO_3$
   d. more than one correct response
   e. no correct response

15.137 The proper assignment of oxidation numbers to the elements in the compound Na_2SO_4 would be
   a. +1 for Na, +2 for S, and –2 for O.
   b. +2 for Na, +4 for S, and –2 for O.
   c. +2 for Na, +6 for S, and –8 for O.
   d. +1 for Na, +6 for S, and –2 for O.
   e. no correct response

15.138 The proper assignment of oxidation numbers to the elements in the polyatomic ion PO_4^{3-} would be
   a. +3 for P and –3 for O.   **b.** +3 for P and –2 for O.
   c. +5 for P and –8 for O.   **d.** +5 for P and –2 for O.
   e. no correct response

15.139 Which element is oxidized in the following redox reaction?

$$2\,H_2S + O_2 \longrightarrow 2\,H_2O + S$$

   a. sulfur in H_2S **b.** hydrogen in H_2S
   c. oxygen in O_2 **d.** hydrogen in H_2O
   e. no correct response

15.140 Which substance functions as the reducing agent in the following redox reaction?

$$2\,HNO_2 + 2\,HI \longrightarrow 2\,NO + I_2 + H_2O$$

   a. HNO_2 **b.** HI   **c.** NO   **d.** H_2O
   e. no correct response

15.141 In a redox reaction the substance oxidized
   a. contains an element that decreases in oxidation number.
   b. is also the reducing agent.
   c. always loses electrons.
   d. more than one correct response
   e. no correct response

15.142 Which of the following reactions is classified as a *nonredox* reaction?
   a. $2\,CuO \longrightarrow 2\,Cu + O_2$
   b. $NaOH + HCl \longrightarrow NaCl + H_2O$
   c. $CaCO_3 \longrightarrow CaO + CO_2$

   d. more than one correct response
   e. no correct response

15.143 In balancing the following redox equation using the change in oxidation number method, what are the values for the "bracket numbers" a and b?

$$\overset{\displaystyle b(a)}{\overbrace{Zn(s) + Cu^{2+}(aq) \longrightarrow Zn^{2+}(aq) + Cu(s)}}$$

   a. a = 1, b = 1   **b.** a = 1, b = 2
   c. a = 2, b = 1   **d.** a = 2, b = 2
   e. no correct response

15.144 Which of the following is a correctly balanced oxidation half-reaction?
   a. $I_2 + 2\,e^- \longrightarrow 2\,I^-$
   b. $H_2S \longrightarrow S + 2\,H^+ + 4\,e^-$
   c. $NO + 2\,H_2O \longrightarrow NO_3^- + 3\,e^- + 4\,H^+$
   d. more than one correct response
   e. no correct response

15.145 When the half-reaction

$$NO_3^- \longrightarrow NH_3 \quad \text{(basic solution)}$$

is correctly balanced, the
   a. OH^- and H_2O are both on the left side of the equation.
   b. OH^- and H_2O are both on the right side of the equation.
   c. OH^- is on the left side and H_2O on the right side of the equation.
   d. OH^- is on the right side and H_2O on the left side of the equation.
   e. no correct response

15.146 The balanced full-equation obtained by adding the following two balanced half-reactions together is

$$Zn \longrightarrow Zn^{2+} + 2e^-$$
$$Ag^+ + e^- \longrightarrow Ag$$

   a. $Zn + Ag^+ + e^- \longrightarrow Zn^{2+} + Ag$
   b. $Zn + 2\,Ag^+ \longrightarrow Zn^{2+} + 2\,Ag + 2\,e^-$
   c. $Zn + Ag^+ \longrightarrow Zn^{2+} + Ag$
   d. $Zn + 2\,Ag^+ \longrightarrow Zn^{2+} + 2\,Ag$
   e. no correct response

15.147 Which of the following is a disproportionation reaction?
   a. $H_2SO_3 \longrightarrow H_2O + SO_2$
   b. $3\,Br_2 + 3\,H_2O \longrightarrow HBrO_3 + 5\,HBr$
   c. $Mg + H_2SO_4 \longrightarrow MgSO_4 + H_2$
   d. more than one correct response
   e. no correct response

15.148 In balancing the equation for a disproportionation reaction using the oxidation number method, the substance undergoing disproportionation is

L
molecu

Mole

When
the rea
in orde
obviou
M
The rea
about,
Reactic
only a
and pr
reactio
type of

Activ

Not all
Someti
sion th
certain
cles m
kinetic
Each cl
I
the rea
underg
It
have it
activat
an exa
match
started

Collis

Even v
particle
sion th
lision,
orienta
wheth
A
chemic

I
cule. T
from N
shown
are sh

---

a. initially written twice on the reactant side of the equation.
b. initially written on both the reactant and product side of the equation.
c. always assigned an oxidation number of zero.
d. more than one correct response
e. no correct response

15.149 In stoichiometric calculations that involve ions, the molar mass value for the ion, on the amu scale, is

a. one amu unit less than that of a Mg atom.
b. two amu units less than that of a Mg atom.
c. one amu unit more than that of a Mg atom.
d. more than one correct response
e. no correct response

15.150 The molar concentration of NO_3^- ion in a 3.00 M $Al(NO_3)_3$ solution is

a. 1.00 M.   b. 3.00 M.
c. 6.00 M.   d. 9.00 M.
e. no correct response

FIGURE 16.3 Greater reactant surface area results in an increased reaction rate.

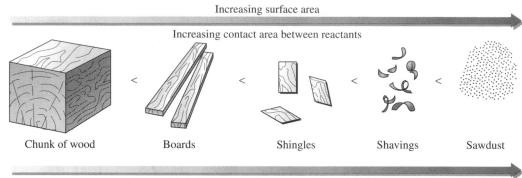

Increasing surface area

Increasing contact area between reactants

Chunk of wood   Boards   Shingles   Shavings   Sawdust

Increasing rate of reaction for the combustion reaction

When particle size becomes extremely small, reaction rates can be so fast that an explosion results. A lump of coal is difficult to ignite; coal dust ignites explosively. The spontaneous ignition of coal dust is a real threat to underground coal-mining operations. Grain dust (very finely divided grain particles) is a problem in grain-storage elevators; explosive ignition of the dust from an accidental spark is always a possibility. Figure 16.4 shows the destruction that can result from the accidental ignition of grain dust in a storage elevator.

Reactant Concentration

An increase in the concentration of a reactant causes an increase in the rate of the reaction. Combustible substances burn much more rapidly in pure oxygen than they do in air (21% oxygen). Increasing the concentration of a reactant means that there are more molecules of that reactant present in the reaction mixture and therefore there is a greater possibility for collisions between this reactant and other reactant particles. An analogy

FIGURE 16.4 Extremely rapid combustion of grain dust produced the explosive effect that destroyed this grain elevator. *Reuters*

to the reaction-rate–reactant-concentration relationship can be drawn from the game of billiards. The more billiard balls there are on the table, the greater the probability of a moving cue ball striking one of them.

The actual quantitative change in reaction rate as the concentration of reactants is increased varies with the specific reaction. The rate usually increases, but not to the same extent in all cases. Simply looking at the balanced equation for a reaction will not enable you to determine how changes in concentration will affect the reaction rate. This must be determined by actual experimentation. In some reactions the rate doubles with a doubling of concentration; however, this is not always the case.

Reaction Temperature

The effect of temperature on reaction rates can also be explained by using the molecular-collision concept. An increase in the temperature of a system results in an increase in the average kinetic energy of the reacting molecules. The increased molecular speed causes more collisions to take place in a given time. Also, since the average kinetic energy of the colliding molecules is greater, a larger fraction of the collisions will have sufficient kinetic energy to equal or exceed the activation energy.

We make use of our knowledge of the effect of temperature on chemical reactions on a regular basis in our daily lives. The chemical reaction of cooking takes place faster in a pressure cooker because of a higher cooking temperature (Sec. 11.15). On the other hand, foods are cooled or frozen to slow down the chemical reactions that result in the spoiling of food, the souring of milk, and the ripening of fruit.

A rough generalization for many common reactions is that the rate of the chemical reaction doubles for every 10°C increase in temperature in the temperature range we normally encounter.

Presence of Catalysts

A **catalyst** *is a substance that, when added to a reaction mixture, increases the rate of the reaction but is itself unchanged after the reaction is completed.* Catalysts can be classified into two categories: (1) *homogeneous catalysts* and (2) *heterogeneous catalysts*. A **homogeneous catalyst** *is a catalyst that is present in the same phase as the reactants in a reaction mixture.* Homogeneous catalysts are usually dispersed uniformly throughout the reaction mixture. A **heterogenous catalyst** *is a catalyst that exists in a phase different from the reactants in a reaction mixture.* Heterogeneous catalysts are usually solids.

Catalysts increase reaction rates by providing alternative reaction pathways with lower activation energies than the original uncatalyzed pathway. This lowering of activation energy effect is illustrated in Figure 16.5.

In homogeneous catalysis the alternative pathway involves the formation of an intermediate complex that contains the catalyst. This catalyst-containing intermediate

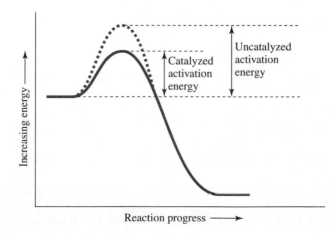

FIGURE 16.5 Catalysts lower the activation energy for chemical reactions. Reactions proceed more rapidly with the lowered activation energy.

FIGURE 16.6 A catalytic converter used in the exhaust system of an automobile. Such converters, which contain the metals platinum, palladium, and rhodium as catalysts, speed up reactions that convert air pollutants to less harmful products. For example, NO is converted to N_2 and CO is converted to CO_2. *Dorling Kindersley Media Library*

then breaks up to give the final products and regenerate the catalyst. The following equations, where C is the catalyst, illustrate this concept.

Uncatalyzed reaction: $X + Y \longrightarrow XY$

Catalyzed reaction: *Step 1*: $X + C \longrightarrow XC$

Step 2: $XC + Y \longrightarrow XY + C$

Catalysts that are solids are thought to provide a surface to which impacting reactant molecules are physically attracted and on which they are held with a particular orientation. Reactants so held are sufficiently close to, and thus favorably oriented toward, each other to allow the reaction to take place. The products of the reaction then leave the surface and make it available to catalyze other reactants.

Catalysts are used extensively in the chemical industry. Usually, very specific catalysts are used that accelerate one chemical reaction without influencing other possible competitive reactions. The small amounts of catalysts required, coupled with the fact that they are not used up, make the use of catalysts economically feasible in industrial processes. For example, a catalyst often makes it possible to avoid the high temperatures (costly) that would otherwise be necessary to cause a reaction with high activation energy to proceed.

Catalysts are a key element in the functioning of automobile emission control systems. In such systems, heterogeneous catalysts speed up reactions that convert air pollutants in the exhaust to less harmful products (see Fig. 16.6). For example, carbon monoxide is converted to carbon dioxide through reaction with the oxygen in air.

Catalysts are of extreme importance for the proper functioning of the human body and other biological systems. In the human body, catalysts called *enzymes*, which are proteins, cause many reactions to take place rapidly at body temperature and under mild conditions. These same reactions, uncatalyzed, proceed very slowly and then only under harsher conditions in a laboratory setting.

> Catalysts lower the activation energy for a reaction. Lowered activation energy increases the rate of the reaction as more reacting particles have the required activation energy.

> Biological catalysts, called *enzymes,* mediate nearly all the reactions that occur within a living organism.

16.4 CHEMICAL EQUILIBRIUM

Up to this point in our discussions of chemical reactions we have assumed that chemical reactions go to completion, that is, that reactions continue until one or more of the reactants is used up. This is usually not the case. Experiments show that in most chemical reactions the complete conversion of reactants to products does not occur regardless of the time allowed for the reactions to take place. The reason for this is that product molecules (provided they are not allowed to escape from the reaction mixture) begin to react with each other to again re-form the reactants. With time, a steady-state situation results where the rate of formation of products and the rate of re-formation of reactants are equal. At this point the concentrations of all reactants and all products remain constant; the reaction has reached a stage of *chemical equilibrium.* **Chemical equilibrium** *is the process wherein two opposing chemical reactions occur simultaneously at the same rate.* We have discussed equilibrium situations in previous chapters—see Sections 11.14 (vapor pressure), 13.2 (saturated solutions), and 14.3 (conjugate acids and bases). The first two of these previous equilibrium situations involved *physical* equilibrium (no chemical reaction) rather than *chemical* equilibrium. Conjugate acid–base relationships involve chemical equilibrium, a topic we now consider in detail.

The conditions that exist in a system in a state of chemical equilibrium can best be visualized by considering an actual chemical reaction. Suppose equal molar amounts of gaseous H_2 and I_2 are mixed together in a closed container and allowed to react.

$$H_2(g) + I_2(g) \longrightarrow 2\ HI(g)$$

Initially, no HI is present, so the only possible reaction that can occur is the one between H_2 and I_2. However, with time, as the HI concentration increases, some HI molecules collide with each other in a way that causes the reverse reaction to occur.

$$2\ HI(g) \longrightarrow H_2(g) + I_2(g)$$

The initial low concentration of HI makes this reverse action slow at first, but as the concentration of HI increases, so does the reaction rate. At the same time the reverse reaction rate is increasing, the forward reaction rate (production of HI) is decreasing as reactants are used up. Eventually, the concentrations of H_2, I_2, and HI in the reaction mixture reach a level at which the rates of the forward and reverse reactions become equal. At this point, a state of chemical equilibrium has been reached.

Figure 16.7 shows graphically the behavior of reaction rates and reaction concentrations with time for both the forward and reverse reactions in the H_2–I_2–HI system. Figure 16.7a shows that the forward and reverse reaction rates become equal as a result of the forward reaction rate decreasing (as reactants are used up) and the reverse reaction rate increasing (as product concentration increases). Figure 16.7b shows the important point that reactant and product concentrations are usually not equal at the point at which equilibrium is reached. Rates are equal, but concentrations are not. For the H_2–I_2–HI system, much more product HI is present than reactants H_2 and I_2 at equilibrium. In Figure 16.7b, note that the point at which equilibrium is established is the point where the two curves become straight lines.

At chemical equilibrium, forward and reverse reaction rates are equal. Reactant and product concentrations, although constant, do not have to be equal.

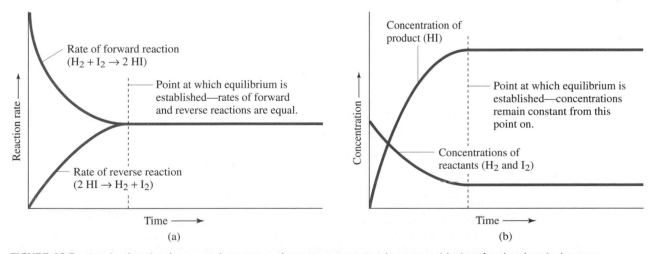

FIGURE 16.7 Graphs showing how reaction rates and reactant concentrations vary with time for the chemical system H_2–I_2–HI. (a) At equilibrium, rates of reaction are equal. (b) At equilibrium, concentrations of reactants and products remain constant but are not equal.

The equilibrium involving H_2, I_2, and HI could have been established just as easily by starting with pure HI and allowing it to change into H_2 and I_2 (the reverse reaction). The final position of equilibrium does not depend on the direction from which equilibrium is approached.

Instead of separate equations for the forward and reverse reactions for a system at equilibrium, it is normal procedure to represent the equilibrium by using a single equation and half-headed arrows pointing in both directions. Thus, the reactions between H_2 and I_2 and between two HI, at equilibrium, are written as

$$H_2(g) + I_2(g) \rightleftharpoons 2\,HI(g)$$

The term *reversible* is often used to describe a reaction like the one we have just discussed. A **reversible reaction** *is a chemical reaction in which the conversion of reactants to products (the forward reaction) and the conversion of products to reactants (the reverse reaction) occur simultaneously.* When the half-headed double arrow notation is used in a chemical equation, it means that the chemical reaction is reversible.

16.5 EQUILIBRIUM MIXTURE STOICHIOMETRY

Suppose that known amounts of reactants are placed in a reaction vessel and the system is allowed to reach chemical equilibrium (Sec. 16.4). To determine the composition of the resulting equilibrium mixture we need only to experimentally determine the equilibrium concentration of one of the substances in the mixture. With this one value and the substance amounts originally present, the concentrations of all other substances present in the equilibrium mixture can be calculated. Example 16.1 shows how such a calculation is carried out.

EXAMPLE 16.1 **Determining the Composition of an Equilibrium Mixture in Terms of Moles of Each Substance Present**

0.0930 mole of NO and 0.0652 mole of Br_2 are placed in a container and allowed to react until equilibrium is established.

$$2\,NO(g) + Br_2(g) \rightleftharpoons 2\,NOBr(g)$$

At equilibrium, 0.0612 mole of NOBr is present. What is the composition of the equilibrium mixture in terms of moles of each substance present?

SOLUTION

In solving this problem, we will deal with three quantities for each of the substances involved in the equilibrium: (1) starting amount of each substance, (2) amount that changes (undergoes reaction), and (3) equilibrium amount of each substance. The following table, the starting point for our calculation, summarizes the known (given) quantities in terms of these three parameters.

	2 NO(g)	+	Br_2(g)	$\rightleftharpoons$	2 NOBr(g)
Start	0.0930 mole		0.0652 mole		0 mole
Change	—		—		—
Equilibrium	—		—		0.0612 mole

Four of the nine "blanks" in the table have numbers in them. The key observation is that two of the three blanks for NOBr are known. In a problem of this type, any time two of the three key items (start, change, and equilibrium) are known for a substance, the third can be calculated by addition or subtraction. For NOBr, we started with zero amount and ended up with 0.0612 mole at equilibrium. Obviously, the amount of change for NOBr is +0.0612 mole, the amount of NOBr formed.

	2 NO(g)	+	Br$_2$(g)	⇌	2 NOBr(g)
Start	0.0930 mole		0.0652 mole		0 mole
Change	—		—		+0.0612 mole
Equilibrium	—		—		0.0612 mole

Once one of the change values is known, all other change quantities can quickly be calculated. The molar-change values are related to each other in the same manner as the coefficients in the equation are related to each other. Thus, we know that

1. the molar amount of NO that reacts is the same as the molar amount of NOBr produced, since these two substances have the same coefficients in the equation, and
2. the molar amount of Br$_2$ that changes (reacts) is one-half the molar amount of NOBr produced since the Br$_2$/NOBr coefficient ratio is 1:2.

Placing this information into the table gives

	2 NO(g)	+	Br$_2$(g)	⇌	2 NOBr(g)
Start	0.0930 mole		0.0652 mole		0 mole
Change	−0.0612 mole		−0.0306 mole		+0.0612 mole
Equilibrium	—		—		0.0612 mole

Note the minus signs placed in front of the NO and Br$_2$ change amounts. This is because these amounts are consumed (used up in the reaction). The plus sign in front of the NOBr change amount indicates a gain in the amount of this substance.

The last two blanks in the table are now easily determined through subtraction. For NO, 0.0930 mole (start) − 0.0612 mole (change) = 0.0318 mole (equilibrium). Similarly, for Br$_2$, we have 0.0652 mole − 0.0306 mole = 0.0346 mole. Our completed table is

	2 NO(g)	+	Br$_2$(g)	⇌	2 NOBr(g)
Start	0.0930 mole		0.0652 mole		0 mole
Change	−0.0612 mole		−0.0306 mole		+0.0612 mole
Equilibrium	0.0318 mole		0.0346 mole		0.0612 mole

The equilibrium mixture composition, which is the bottom line of the table, is

0.0318 mole NO;   0.0346 mole Br$_2$;   0.0612 mole NOBr

Answer Double Check:

Are the magnitudes of the equilibrium molar amounts reasonable? Yes. On the reactant side of the equation, the equilibrium molar amounts should be less than the starting amounts because some of the reactants are consumed in the reaction. Such is the case. On the product side of the equation, the equilibrium molar amount should be greater than the starting amount. Such is the case.

▶ **Practice Exercise 16.1**   Hydrogen and nitrogen react according to the following chemical equation.

$$N_2(g) + 3 H_2(g) \longrightarrow 2 NH_3(g)$$

When 0.87 mole of N$_2$ and 1.88 moles of H$_2$ are placed in an appropriate container and allowed to react until equilibrium is established, it is found that 0.52 mole of NH$_3$ has formed. What is the composition of the equilibrium mixture in terms of moles of each substance present?

Answers to all practice exercises in this chapter are found in the back-of-the-book answer section.

16.6 EQUILIBRIUM CONSTANTS

The concentrations of reactants and products are constant (not changing) in a system at chemical equilibrium (Sec. 16.4). The numerical values of these equilibrium concentrations can be used to calculate an *equilibrium constant*, a single number that describes the extent to which the chemical reaction of concern has occurred. An **equilibrium constant** *is a numerical value that characterizes the relationship between the concentrations of reactants and concentrations of products in a system that is at chemical equilibrium.*

The equilibrium constant for a chemical reaction is obtained by writing an *equilibrium expression* and then evaluating it numerically. To illustrate the calculation of an equilibrium constant, let us consider a general gas-phase reaction in which *a* moles of A and *b* moles of B react to produce *c* moles of C and *d* moles of D.

$$aA(g) + bB(g) \rightleftharpoons cC(g) + dD(g)$$

The equilibrium constant expression for this reaction is

$$K_{eq} = \frac{[C]^c[D]^d}{[A]^a[B]^b}$$

Note the following points about this general equilibrium constant expression:

1. The square brackets refer to molar (moles/liter) concentrations.
2. Product concentrations are always placed in the numerator of the equilibrium constant expression.
3. Reactant concentrations are always placed in the denominator of the equilibrium constant expression.
4. The coefficients in the balanced chemical equation for the equilibrium system determine the powers to which the concentrations are raised.
5. The abbreviation K_{eq} is used to denote an equilibrium constant.

An additional convention in writing equilibrium constant expressions, not apparent from the preceding equilibrium constant definition, is *only concentrations of gases and substances in solution are written in an equilibrium constant expression.* The reason for this convention is that other substances (pure solids and pure liquids) have constant concentrations. These constant concentrations are incorporated into the equilibrium constant itself. For example, pure water in the liquid state has a concentration of 55.5 moles/L. It does not matter whether we have 1.00, 50.0, or 750 mL of liquid water, the concentration will be the same. In the liquid state, pure water is pure water and it has only one concentration. Similar reasoning applies to other pure liquids and pure solids. All such substances have constant concentrations.

The concentrations of *pure liquids* and *pure solids*, which are constants, are never included in an equilibrium constant expression.

The only information we need to write an equilibrium constant expression is a balanced chemical equation that includes information about physical state. Using the preceding generalizations about equilibrium expressions, for the reaction

$$4\,NH_3(g) + 7\,O_2(g) \rightleftharpoons 4\,NO_2(g) + 6\,H_2O(g)$$

we write the equilibrium constant expression as

Coefficient of NO_2 ↘   ↙ Coefficient of H_2O

$$K_{eq} = \frac{[NO_2]^4[H_2O]^6}{[NH_3]^4[O_2]^7}$$

Coefficient of NH_3 ⌐⌐   ∟ Coefficient of O_2

| **EXAMPLE 16.2** | **Using Balanced Chemical Equations to Determine Equilibrium Constant Expressions** |

Write the equilibrium constant expression for each of the following reactions.

    a. $4\,NH_3(g) + 3\,O_2(g) \rightleftharpoons 2\,N_2(g) + 6\,H_2O(g)$
    b. $6\,Ca(s) + 2\,NH_3(g) \rightleftharpoons 3\,CaH_2(s) + Ca_3N_2(s)$
    c. $2\,Ag_2CO_3(s) \rightleftharpoons 4\,Ag(s) + 2\,CO_2(g) + O_2(g)$
    d. $NaCl(aq) + AgNO_3(aq) \rightleftharpoons AgCl(s) + NaNO_3(aq)$

SOLUTION

    a. All of the substances involved in this reaction are gases. Therefore, each reactant and product will appear in the equilibrium constant expression.

        The product concentrations, each raised to the power of its coefficient in the balanced equation, are placed in the numerator.

$$K_{eq} = \frac{[N_2]^2[H_2O]^6}{-} \qquad \text{Equation coefficients}$$

    The reactant concentrations, each raised to the power of its coefficient in the balanced equation, are placed in the denominator.

$$K_{eq} = \frac{[N_2]^2[H_2O]^6}{[NH_3]^4[O_2]^3}$$

    Note that H_2O as a gas (water vapor or steam) is included in an equilibrium constant expression. The concentration of a gas can vary. Water, as a liquid, is never included in equilibrium constant expressions.

    b. Three of the four substances involved in this reaction are solids and thus will not appear in the equilibrium constant expression. Since both products are solids, the numerator of the equilibrium constant expression is 1. The concentration of NH_3 raised to the second power is the only factor in the denominator because the other reactant is a solid.

$$K_{eq} = \frac{1}{[NH_3]^2}$$

    c. The reactant Ag_2CO_3 is a solid and thus will not appear in the equilibrium constant expression. Since Ag_2CO_3 is the only reactant, this means there will be no denominator in the equilibrium expression. Two of the three products are gases, and they appear in the numerator of the equilibrium constant expression.

$$K_{eq} = [CO_2]^2[O_2]$$

    d. All of the powers in this equilibrium constant expression are 1 because all of the coefficients in the balanced equation are ones.

$$K_{eq} = \frac{[NaNO_3]}{[NaCl][AgNO_3]}$$

    AgCl is not included in the equilibrium expression because it is a solid.

Answer Double Check:

Is the equilibrium constant expression correct in terms of the power to which the equilibrium concentrations are raised? Yes. In each case, the power present has the same numerical value as the coefficient for that substance in the chemical equation the equilibrium constant expression describes.

▶ **Practice Exercise 16.2** Write the equilibrium constant expression for each of the following chemical reactions.

a. $2\,NH_3(g) + H_2SO_4(l) \rightleftharpoons (NH_4)_2SO_4(s)$
b. $WCl_6(g) + 3\,H_2(g) \rightleftharpoons W(s) + 6\,HCl(g)$
c. $TiCl_3(s) \rightleftharpoons TiCl(s) + Cl_2(g)$
d. $HS^-(aq) + H^+(aq) \rightleftharpoons H_2S(aq)$

At a given temperature, the numerical value of the equilibrium constant for a reaction is obtained by substituting the experimentally determined equilibrium concentrations at that temperature into the equilibrium constant expression for the reaction, as is shown in Example 16.3.

EXAMPLE 16.3 Using Equilibrium Concentrations to Calculate the Value of an Equilibrium Constant

At a temperature of 927°C, the equilibrium molar concentrations for the reaction

$$CO(g) + 3\,H_2(g) \rightleftharpoons CH_4(g) + H_2O(g)$$

are

$$[CO] = 0.613,\ [H_2] = 1.839,\ [CH_4] = 0.387,\ \text{and}\ [H_2O] = 0.387$$

Calculate the value of the equilibrium constant for this reaction at 927°C.

SOLUTION

The general expression for the equilibrium constant is

$$K_{eq} = \frac{[CH_4][H_2O]}{[CO][H_2]^3}$$

Substituting the known equilibrium concentrations into this expression gives

$$K_{eq} = \frac{[0.387][0.387]}{[0.613][1.839]^3}$$

$$= 0.039284051 \quad \text{(calculator answer)}$$

$$= 0.0393 \quad\quad \textbf{(correct answer)}$$

Answer Double Check:

Were the powers to which the concentrations were raised incorporated into the mathematical calculation? Yes. In this problem the concentration (1.839 M) needed to be cubed, and it was.

▶ **Practice Exercise 16.3** At a given temperature, the equilibrium molar concentrations for the reaction

$$PCl_3(g) + Cl_2(g) \rightleftharpoons PCl_5(g)$$

are $[PCl_3] = 0.16$, and $[Cl_2] = 0.82$, and $[PCl_5] = 0.25$. Calculate the value of the equilibrium constant for this reaction at the given temperature.

There are many sets of equilibrium concentrations that will satisfy a given equilibrium constant expression at a given temperature. This is because, at any given temperature, an equilibrium may be reached from numerous different sets of starting

TABLE 16.1 Experimentally Determined Equilibrium Concentrations for the Chemical Reaction $H_2(g) + I_2(g) \rightleftharpoons 2\,HI(g)$

Experiment	Initial Concentration			Equilibrium Concentration			Calculated K_{eq} at 448°C $\dfrac{[HI]^2}{[H_2][I_2]}$
	$[H_2]$	$[I_2]$	$[HI]$	$[H_2]$	$[I_2]$	$[HI]$	
1	1.000	1.000	0	0.219	0.219	1.562	51
2	1.000	2.000	0	0.065	1.065	1.870	51
3	1.000	1.000	1.000	0.328	0.328	2.344	51
4	0	0	1.000	0.109	0.109	0.782	51

conditions. Illustrative of this is the data found in Table 16.1. Here, four sets of equilibrium concentrations are given that satisfy the equilibrium constant expression for the chemical reaction

$$H_2(g) + I_2(g) \rightleftharpoons 2\,HI(g)$$

The different sets of equilibrium concentrations result from different starting mixes of reactants and/or products that are allowed to reach equilibrium.

16.7 EQUILIBRIUM POSITION

Equilibrium position *is a qualitative indication of the relative amounts of reactants and products present when a chemical reaction reaches equilibrium.* The terms *far to the right, to the right, neither to the right nor the left, to the left,* and *far to the left* are used in describing equilibrium position. In equilibrium situations where the concentrations of products are greater than those of reactants, the equilibrium position is said to lie to the *right* because products are always listed on the right side of a chemical equation. Conversely, when reactants dominate at equilibrium, the equilibrium position lies to the *left*. The terminology *neither to the right nor the left* indicates that significant amounts of both reactants and products are present in an equilibrium mixture.

Equilibrium position can also be indicated by varying the length of the arrows in the half-headed double-arrow notation for a reversible reaction. The longer arrow indicates the direction of the predominant reaction. For example, the arrow notation in the equation

$$CO_2 + H_2O \rightleftharpoons H_2CO_3$$

indicates that the equilibrium position lies to the right.

The magnitude of the equilibrium constant for a reaction gives information about how far a reaction proceeds toward completion, that is, about where the equilibrium position lies. A large value of K_{eq} (greater than 10^3) means that the numerical value of the numerator is significantly greater than that of the denominator. In terms of reactants and products, this means that the concentrations of the products are greater than those of the reactants. The equilibrium position lies to the right.

Conversely, if the equilibrium constant is small (less than 10^{-3}), we have a situation where reactants will predominate over products in the reaction mixture. The equilibrium position is said to lie to the left in this situation.

For equilibrium conditions where K_{eq} has a value between 10^3 and 10^{-3}, appreciable concentrations of both products and reactants are present. The reaction described in Example 16.3 falls into this category.

16.8 TEMPERATURE DEPENDENCY OF EQUILIBRIUM CONSTANTS

When the numerical value of an equilibrium constant is given, the temperature at which the numerical value was determined must also be specified. This is because equilibrium constant values vary with temperature. A change in temperature changes molecular energies (Sec. 16.1), and molecular energies have a direct effect on the relative amounts of reactants and products present in an equilibrium mixture.

As an illustration of the effect that temperature has on equilibrium constant values, let us consider the equilibrium constant for the reaction

$$2\,H_2O(l) \rightleftharpoons H_3O^+(aq) + OH^-(aq)$$

The equilibrium constant expression for this reaction is

$$K_{eq} = [H_3O^+][OH^-]$$

(The concentration of water does not appear in the equilibrium constant expression as water is a pure liquid.)

This reaction and its accompanying equilibrium constant expression should seem familiar, as a detailed discussion about them occurred in Chapter 14 under the topic "Self-Ionization of Water" (Sec. 14.10). At that time, the terminology "ion product for water" was used instead of the yet-to-be-discussed terminology "equilibrium constant," and the value for the ion product for water expression was determined to be 1.00×10^{-14} at 24°C. The value 1.00×10^{-14}, which was designated as K_w, was used to determine hydronium and hydroxide ion concentrations in aqueous solutions (Sec. 14.10).

The value 1.00×10^{-14} is in reality an equilibrium constant and its value is thus dependent on temperature. An increase in temperature increases the probability of two water molecules interacting to produce ions. Hence ion concentrations increase, which increases the value of the ion product for water. Table 16.2 gives the ion product for water (equilibrium constant) value at temperatures above and below 24°C, the temperature for which the value 1.00×10^{-14} is valid.

TABLE 16.2 Ion Product for Water Value at Various Temperatures

Temperature	K_{eq} value
0°C	1.14×10^{-15}
10°C	2.92×10^{-15}
24°C	1.00×10^{-14}
25°C	1.01×10^{-14}
30°C	1.47×10^{-14}
40°C	2.92×10^{-14}
50°C	5.47×10^{-14}
60°C	9.61×10^{-14}

In most acid–base discussions, the temperature dependence of the ion product for water is usually ignored because the change in value with temperature is relatively small and because most laboratory acidity measurements are made at room temperature (roughly 24°C).

16.9 LE CHÂTELIER'S PRINCIPLE

A chemical system in a state of equilibrium remains in that state until it is disturbed by some change of condition. Disturbing an equilibrium has one of two results: Either the forward reaction speeds up (to produce additional products) or the reverse reaction speeds up (to produce additional reactants). Then with time, the forward and reverse reactions again become equal and a new equilibrium, not identical to the previous one, is established. If more products have been produced as a result of the disruption, the equilibrium position is said to have *shifted to the right*. Similarly, when the disruption causes more reactants to form, the equilibrium position has *shifted to the left*.

Qualitative predictions about the direction in which chemical equilibria shift can be made using a guideline (principle) introduced in 1888 by the French chemist Henri-Louis Le Châtelier (1850–1936)—see "The Human Side of Chemistry 16." **Le Châtelier's principle** *states that if a stress (change of conditions) is applied to a chemical system in equilibrium, the system will readjust (change the position of the equilibrium) in the direction that best reduces the stress imposed upon the system.* We will use this principle in considering how four types of changes affect equilibrium position. The changes are (1) concentration changes, (2) temperature changes, (3) pressure changes, and (4) addition of a catalyst.

The Human Side of Chemistry 16

Henri-Louis Le Châtelier (1850–1936)

Henri-Louis Le Châtelier (pronounced le-shot-lee-ay) was born in Paris in 1850. His father was inspector general of mines for France; his grandfather operated limekilns. Henri would visit his grandfather's kilns during vacations. Contacts with his father, grandfather, and their associates were a major factor in shaping his career, a career most noteworthy for his constant application of chemical principles in industrial settings.

After obtaining a formal education, he registered as a mining engineer. Later he obtained a degree in physical and chemical science. In 1887 he was appointed professor of industrial chemistry and metallurgy in the Écoles des Mines. Later came an appointment as professor of inorganic chemistry and still later an appointment in general chemistry.

Le Châtelier's interests were amazingly diverse. The chemistries of cement, of ceramics, and of glass occupied his attention for a time. Studies in combustion with the aim of preventing mine explosions were next. These led to studies of heat and its measurement. From his studies of heat came that for which he is best known, his principle of "stress and strain." This principle was first proposed in 1884; a simplified version was presented in 1888. Metallurgy also occupied his attention. In this area he was most concerned with the chemistry and metallurgy of iron and steel.

In later life he was a great national hero in France. His linking of science with industry (especially during World War I) was very important to France. In 1916, the president of the United States (Woodrow Wilson) engaged him as a consultant when the United States' National Research Council was established. He held positions on many commissions and boards that advised the French government on scientific and technical questions.

Concentration Changes

Adding or removing a *gaseous* reactant or *gaseous* product from a reaction mixture at equilibrium will always upset the equilibrium. Le Châtelier's principle predicts that the reaction will shift in the direction that will minimize the change in concentration caused by the addition or removal. If an additional amount of any *gaseous* reactant or product has been *added* to the system, the stress is relieved by shifting the equilibrium in the direction that *consumes* (uses up) some of the added reactant or product. Conversely, if a *gaseous* reactant or product is *removed* from an equilibrium system, the equilibrium will shift in a direction that *produces* more of the substance that was removed.

Let us consider the effect that selected concentration changes will have on the gaseous equilibrium.

$$N_2(g) + 3H_2(g) \rightleftharpoons 2NH_3(g)$$

Suppose some additional H_2 is added to the equilibrium mixture. The equilibrium will shift to the right; that is, the forward reaction rate will increase in order to use up additional H_2. Eventually a new equilibrium position will be reached. At this new position, the H_2 concentration will still be higher than it was before the addition; that is, not all of the added H_2 is consumed. In addition, the N_2 concentration will have decreased (some N_2 had to react with the H_2) and the NH_3 concentration will have increased as the product of the H_2 and N_2 reacting.

Removal of some NH_3 from this newly established equilibrium position will cause an additional shift to the right. The concentration of H_2 and N_2 will decrease as the system attempts to replace the NH_3 that was removed by producing more of it. Again, not all of the removed NH_3 will be replaced. When the new equilibrium position is achieved, the NH_3 concentration will be less than it was before the NH_3 removal.

Figure 16.8 shows graphically the effects that the H_2 addition just discussed has on the concentration of all substances present in the N_2–H_2–NH_3 equilibrium mixture.

Throughout this discussion of the effect of concentration changes on an equilibrium system, we have referred to the effect of adding or removing a *gaseous* species. If we

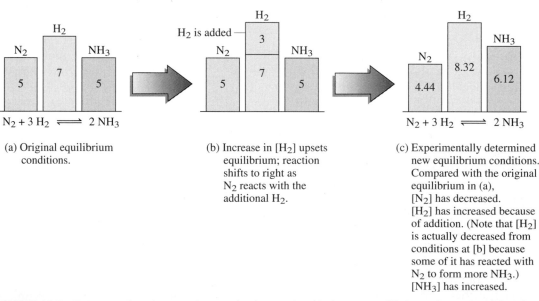

(a) Original equilibrium conditions.

(b) Increase in [H_2] upsets equilibrium; reaction shifts to right as N_2 reacts with the additional H_2.

(c) Experimentally determined new equilibrium conditions. Compared with the original equilibrium in (a), [N_2] has decreased. [H_2] has increased because of addition. (Note that [H_2] is actually decreased from conditions at [b] because some of it has reacted with N_2 to form more NH_3.) [NH_3] has increased.

FIGURE 16.8 Concentration changes that result when H_2 is added to an equilibrium mixture involving the system $N_2(g) + 3 H_2(g) \rightleftharpoons 2 NH_3(g)$.

add a pure solid or liquid to a gas-phase equilibrium system, there will be no shift in the equilibrium position; nothing happens. This is because there will be no change in concentration. The solid or liquid was 100% pure before the addition, and it is still 100% after the addition.

More specifically, consider the equilibrium system

$$CaCO_3(s) \rightleftharpoons CaO(s) + CO_2(g)$$

Adding or removing $CaCO_3$ or CaO from this system will cause no change in the equilibrium position. In general, *adding or removing a species disturbs an equilibrium system only if the concentration of that species appears in the expression for the equilibrium constant* (Sec. 16.6).

The concentrations of solids and liquids do not appear in equilibrium-constant expressions. The concentrations of gases and species in an *aqueous solution* do appear in equilibrium-constant expressions. The changing of the concentration of one of the two ions in the equilibrium

$$Pb^{2+}(aq) + 2\,Cl^-(aq) \rightleftharpoons PbCl_2(s)$$

would affect the equilibrium position.

Thousands of chemical equilibria simultaneously exist in biological systems. Many of them are interrelated. When the concentration of a single substance changes, many equilibria are affected.

Temperature Changes

Le Châtelier's principle can be used to predict the influence of temperature changes on an equilibrium, provided it is known whether the reaction of concern is endothermic or exothermic.

Consider the *exothermic reaction*

$$H_2(g) + F_2(g) \rightleftharpoons 2\,HF(g) + heat$$

Heat is produced when the reaction proceeds to the right. Thus, if we increase the temperature of an exothermic system at equilibrium, the system will shift to the left in an attempt to use up the added heat. When equilibrium is reestablished, the concentrations of H_2 and F_2 will be higher and the concentration of HF will have decreased. Lowering the temperature of an exothermic system at equilibrium will cause the reaction to shift to the right as the system attempts to replace the lost heat.

The behavior, with temperature change, of an equilibrium reaction mixture involving an *endothermic* reaction such as

$$heat + 2\,CO_2(g) \rightleftharpoons 2\,CO(g) + O_2(g)$$

is just the opposite of that of an exothermic reaction, since a shift to the left (rather than to the right) produces heat. Consequently, an increase in temperature will cause the equilibrium system to shift to the right (to use up the added heat), and a decrease in temperature will produce a shift to the left (to generate more heat).

The effect of temperature change on the position of an equilibrium depends on whether heat is a reactant or a product in the chemical reaction of concern.

Pressure Changes

Pressure changes affect systems at equilibrium only when gaseous substances are part of the equilibrium, and then only in cases where the chemical reaction is such that a change in the total number of moles of gaseous substances occurs. This latter point can be illustrated by considering the following two gas-phase reactions.

$$\underbrace{2\,H_2(g) + O_2(g)}_{\text{3 moles of gas}} \longrightarrow \underbrace{2\,H_2O(g)}_{\text{2 moles of gas}}$$

$$\underbrace{H_2(g) + Cl_2(g)}_{\text{2 moles of gas}} \longrightarrow \underbrace{2\,HCl(g)}_{\text{2 moles of gas}}$$

In the first reaction the total number of moles of gaseous reactants and products decreases as the reaction proceeds to the right, since 3 moles of reactants combine to give only 2 moles of products. In the second reaction there is no change in the total number of moles of gaseous substances present as the reaction proceeds, since 2 moles of reactants combine to give 2 moles of products. Thus, a pressure change will affect the position of equilibrium in the first reaction but not in the second reaction.

Pressure changes are usually brought about through volume changes. A pressure increase results from a volume decrease, and a pressure decrease from a volume increase (Sec. 12.3). The use of Le Châtelier's principle correctly predicts the direction of the equilibrium position shift resulting from a pressure change only when the pressure change is due to a volume change. It does not apply to pressure increases caused by the addition of a nonreactive (inert) gas to the reaction mixture. Such an addition has no effect on the equilibrium position. The partial pressure (Sec. 12.15) of each of the gases involved in the reaction remains the same.

According to Le Châtelier's principle, the stress of increased pressure is relieved by decreasing the number of moles of gaseous substances in the system. This is accomplished by the reaction shifting in the direction of the smaller number of moles, that is, to the side of the equation that contains the smaller number of moles of gaseous substances. For the reaction

$$2\,NO_2(g) + 7\,H_2(g) \rightleftharpoons 2\,NH_3(g) + 4\,H_2O(g)$$

an increase in pressure would shift the equilibrium position to the right, because there are 9 moles of gaseous reactants and only 6 moles of gaseous products. A stress of decreased pressure will result in an equilibrium system reacting in such a way as to produce more moles of gaseous substances.

> Increasing the pressure associated with an equilibrium system by adding an inert gas (a gas that is not a reactant or a product in the reaction) does not affect the position of the equilibrium.

Addition of a Catalyst

Catalysts do not change the position of equilibrium. This fact becomes clear when we remember that a catalyst functions by lowering the activation energy for a reaction (Sec. 16.3). The activation energy for the forward reaction is lowered, but so is the activation energy for the reverse reaction. Hence, a catalyst speeds up both the forward and reverse reactions and has no effect on the position of equilibrium. However, the lowered activation energy allows equilibrium to be established more quickly than if the catalyst were not present (see Fig. 16.9).

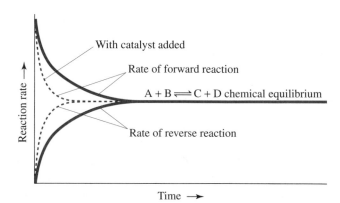

FIGURE 16.9 A catalyst decreases the time required for equilibrium to be reached. It does not, however, change the position of equilibrium; the amount of product produced remains the same.

| **EXAMPLE 16.4** | **Using Le Châtelier's Principle to Predict the Effects of Changes on the Equilibrium Position in an Equilibrium System** |

How will the gas-phase equilibrium

$$PCl_3(g) + Cl_2(g) \rightleftharpoons PCl_5(g) + heat$$

be affected by each of the following?

   a. removal of $PCl_5(g)$
   b. addition of $Cl_2(g)$
   c. temperature decrease
   d. an increase in the volume of the container (pressure decrease)

SOLUTION

   a. The equilibrium will shift to the right, according to Le Châtelier's principle, in an attempt to replenish the PCl_5 removed.
   b. The equilibrium will shift to the right in an attempt to use up the extra Cl_2 that has been placed in the system.
   c. Lowering the temperatures means that heat energy has been removed. The position of equilibrium will shift to the right in order to produce more heat to take the place of that removed.
   d. The system will shift to the left in an attempt to produce more moles of gaseous reactants; this will increase the pressure. In going to the left the reaction produces 2 moles of gaseous reactants for every 1 mole of gaseous product consumed.

▶ **Practice Exercise 16.4**   How will the gas-phase equilibrium

$$2 H_2O(g) + 2 SO_2(g) + heat \rightleftharpoons 2 H_2S(g) + 3 O_2(g)$$

be affected by each of the following?

   a. removal of $SO_2(g)$
   b. addition of $H_2O(g)$
   c. temperature decrease
   d. pressure decrease caused by an increase in the volume of the container

16.10 FORCING CHEMICAL REACTIONS TO COMPLETION

Reactions that would ordinarily reach a state of equilibrium can be forced to completion by using experimental conditions that place a "continual stress" on the potential equilibrium condition. Let us consider a few ways in which this "forcing" is done.

   Continuous removal of one or more products of a reaction will force the reaction to completion. To reach equilibrium, both reactants and products must be present. The removal of a product continually shifts the reaction to the right, that is, toward completion, according to Le Châtelier's principle. Eventually, one or more of the reactants is depleted, the sign of a completed reaction.

   Product removal is very easy to arrange in situations that involve gaseous products. If the reaction is run in an open container, the gaseous products automatically escape to the atmosphere as fast as they are produced. Such a reaction will never reach equilibrium and will continue until the limiting reactant is used up.

   Sometimes another chemical reaction is used to remove a product. Consider the situation of a saturated solution of NaCl.

$$NaCl(s) \rightleftharpoons Na^+(aq) + Cl^-(aq)$$

In such a solution, as much NaCl is dissolved as is possible. Adding $AgNO_3$ to the saturated solution will cause more NaCl to dissolve. The Ag^+ ions from the $AgNO_3$ react with the Cl^- ions in the saturated solution to form insoluble AgCl.

$$Ag^+(aq) + Cl^-(aq) \rightleftharpoons AgCl(s)$$

This removes Cl^- (one of the products in the original equilibrium) from the solution, thus upsetting the equilibrium. More NaCl will dissolve to compensate for the loss of the chloride ions (Le Châtelier's principle). Continued addition of $AgNO_3$ will eventually cause all of the NaCl to dissolve.

It is also possible to drive a reaction to completion by ensuring that an excess of one of the reactants is always present. The system will continually shift to the right (Le Châtelier's principle) to remove the stress caused by the excess reactant. Eventually other reactants will be depleted and the reaction will be completed. A procedure such as this is useful in a situation where one reactant is very expensive and others are much cheaper. To ensure that none of the expensive reactant goes unreacted, an excess of one of the less expensive reactants is used.

Summary

1. **Collision Theory** Collision theory is a set of statements that give the conditions that must be met before a chemical reaction will take place. The three basic tenets of collision theory are (1) reactant molecules must collide with each other before any reaction can occur; (2) colliding particles must possess a certain minimum energy, called the activation energy, if the collision is to result in reaction; and (3) colliding particles must come together in the proper orientation if the reaction is to occur.
2. **Endothermic and Exothermic Chemical Reactions** An endothermic chemical reaction requires an input of energy as the reaction occurs. This is because the energy required to break bonds in the reactants is greater than the energy released through bond formation in the products. An exothermic chemical reaction releases energy as the reaction occurs. Less energy is required to break reactant molecule bonds than is released through bond formation in product molecules in an exothermic reaction.
3. **Factors that Influence Chemical Reaction Rate** The rate of a chemical reaction is the rate at which reactants are consumed or products produced in a given time period. Four factors that affect reaction rates are (1) the physical nature of the reactants, (2) reactant concentrations, (3) reaction temperature, and (4) the presence of catalysts.
4. **Chemical Equilibrium** Chemical equilibrium is the condition in which two opposing chemical reactions (a forward reaction and a reverse reaction) occur simultaneously at the same rate. A state of chemical equilibrium is indicated in chemical equations by placing half-headed arrows pointing in both directions between reactants and products.
5. **Equilibrium Constant** An equilibrium constant is a numerical value that characterizes the relationship between the concentrations of reactants and products in a system at chemical equilibrium. The value of an equilibrium constant is obtained by writing an equilibrium expression and then numerically evaluating it. Equilibrium expressions can be obtained from the balanced chemical equation for a chemical reaction.
6. **Equilibrium Position** The relative amounts of reactants and products present in a system at equilibrium define the equilibrium position. The equilibrium position is toward the right when large amounts of products are present and is toward the left when large amounts of reactants are present.
7. **Le Châtelier's Principle** Le Châtelier's principle states that if a stress (change of conditions) is applied to a system in equilibrium, the system will readjust (change the position of the equilibrium) in the direction that best reduces the stress imposed on it. Stresses known to change equilibrium position include (1) concentration changes in reactants and/or products, (2) temperature changes, and (3) pressure changes. Catalysts do not change the position of an equilibrium.

Key Terms

The new terms defined in this chapter are

activation energy
   Sec. 16.1
catalyst *Sec. 16.3*
chemical equilibrium
   Sec. 16.4
collision theory *Sec. 16.1*

endothermic chemical
   reaction *Sec. 16.2*
equilibrium constant
   Sec. 16.6
equilibrium position
   Sec. 16.7

exothermic chemical
   reaction *Sec. 16.2*
heterogeneous catalyst
   Sec. 16.3
homogeneous catalyst
   Sec. 16.3

Le Châtelier's principle
   Sec. 16.9
rate of a chemical reac-
   tion *Sec. 16.3*
reversible reaction
   Sec. 16.4

Practice Problems

COLLISION THEORY (SEC. 16.1)

16.1 Why are reactions between substances in a solution usually faster than reactions between solid-
state reactants?

16.2 Why are gas-phase reactions usually faster than
solid-phase reactions?

16.3 Under similar concentration and temperature conditions, would a reaction with an activation
energy of 65 kJ/mole or one with an activation
energy of 45 kJ/mole proceed at a faster rate?
Explain your answer.

16.4 What is the relationship between activation energy
and the minimum combined kinetic energy reactant
particles must possess in order for their collision to
result in a reaction?

16.5 What two factors determine whether a collision
between two reactant molecules will result in
a reaction?

16.6 What happens to the reactants in an ineffective
molecular collision?

16.7 For the reaction

$$H_2 + Cl_2 \longrightarrow 2\,HCl$$

draw a sketch of a molecular orientation that is
highly favorable for an effective collision.

16.8 For the reaction

$$H_2 + Cl_2 \longrightarrow 2\,HCl$$

draw a sketch of a molecular orientation that is
unfavorable for an effective collision.

ENDOTHERMIC AND EXOTHERMIC CHEMICAL REACTIONS (SEC. 16.2)

16.9 Classify each of the following chemical reactions as
exothermic or *endothermic*.
   a. $H_2O_2 + heat \longrightarrow H_2 + O_2$
   b. $2\,CO + O_2 \longrightarrow 2\,CO_2 + heat$
   c. $2\,C_2H_6 + 7\,O_2 \longrightarrow 4\,CO_2 + 6\,H_2O + heat$
   d. $Fe_3O_4 + CO + heat \longrightarrow 3\,FeO + CO_2$

16.10 Classify each of the following chemical reactions as
exothermic or *endothermic*.
   a. $N_2 + 2\,O_2 + heat \longrightarrow 2\,NO_2$
   b. $N_2 + 3\,H_2 \longrightarrow 2\,NH_3 + heat$
   c. $2\,Fe + 3\,CO_2 + heat \longrightarrow Fe_2O_3 + 3\,CO$
   d. $4\,NH_3 + 5\,O_2 \longrightarrow 4\,NO + 6\,H_2O + heat$

16.11 For each of the chemical reactions in Problem 16.9
indicate whether the average energy of the reactant
molecules is (1) less than, (2) equal to, or (3) greater
than the average energy of the product molecules.

16.12 For each of the chemical reactions in Problem 16.10
indicate whether the average energy of the reactant
molecules is (1) less than, (2) equal to, or (3) greater
than the average energy of the product molecules.

16.13 Draw an energy diagram graph for a hypothetical
chemical reaction that is *exothermic* by 35 kJ/mole
and has an activation energy of 75 kJ/mole. Label
the following on the diagram.
   a. average energy of the reactants
   b. average energy of the products
   c. activation energy
   d. amount of energy liberated during the reaction

16.14 Draw an energy diagram graph for a hypothetical
chemical reaction that is *endothermic* by 35 kJ/mole
and has an activation energy of 75 kJ/mole. Label
the following on the diagram.
   a. average energy of the reactants
   b. average energy of the products
   c. activation energy
   d. amount of energy absorbed during the reaction

16.15 Reaction A occurs at room temperature and liberates 200 kJ of energy per mole of reactant. Reaction B
does not occur until a temperature of 150°C is
reached; it also liberates 200 kJ of energy per mole
of reactant. Draw an energy diagram graph for each
reaction, and indicate the similarities and differences between the two diagrams.

16.16 Reaction C occurs at room temperature and liberates
200 kJ of energy per mole of reactant. Reaction D also
occurs at room temperature but absorbs 200 kJ of
energy per mole of reactant. Draw an energy diagram

graph for each reaction, and indicate the similarities and differences between the two diagrams.

FACTORS THAT INFLUENCE CHEMICAL REACTION RATES (SEC. 16.3)

16.17 Using collision theory, indicate why each of the following factors influences the rate of a chemical reaction.
  a. temperature of reactants
  b. presence of a catalyst

16.18 Using collision theory, indicate why each of the following factors influences the rate of a chemical reaction.
  a. physical nature of reactants
  b. reactant concentrations

16.19 Will each of the changes listed increase or decrease the rate of the following chemical reaction?

$$2 H_2 + O_2 \longrightarrow 2 H_2O$$

  a. adding O_2 to the reaction mixture
  b. raising the temperature of the reaction mixture
  c. removing a catalyst present in the reaction mixture
  d. removing some H_2 from the reaction mixture

16.20 Will each of the changes listed increase or decrease the rate of the following chemical reaction?

$$2 CO + O_2 \longrightarrow 2 CO_2$$

  a. adding some CO to the reaction mixture
  b. lowering the temperature of the reaction mixture
  c. adding a catalyst to the reaction mixture
  d. removing some CO from the reaction mixture

16.21 Why will a spark cause coal dust in a mine to explode and yet not cause an explosion with charcoal in a barbeque?

16.22 Milk will sour in a couple of days when left at room temperature yet can remain unspoiled for 2 weeks when refrigerated. Explain why.

16.23 The characteristics of four reactions, each of which involves only two reactants, are as follows.

Reaction	Activation Energy	Temperature	Concentration of Reactants
1	low	low	1 mole/L of each
2	high	low	1 mole/L of each
3	low	high	1 mole/L of each
4	low	low	1 mole/L of 1st reactant; 4 moles/L of 2nd reactant

For each of the following pairs of the preceding reactions, indicate which reaction is faster. The rates to be compared are the rates when the two reactants are first mixed. Explain each of your answers.
a. 1 and 2 **b.** 1 and 3 **c.** 1 and 4 **d.** 2 and 3

16.24 The characteristics of four reactions, each of which involves only two reactants, are as follows.

Reaction	Activation Energy	Temperature	Concentration of Reactants
1	high	low	1 mole/L of each
2	high	high	1 mole/L of each
3	low	low	1 mole/L of 1st reactant; 4 moles/L of 2nd reactant
4	low	low	4 moles/L of each

For each of the following pairs of the preceding reactions, indicate which reaction is faster. The rates to be compared are the rates when the two reactants are first mixed. Explain each of your answers.
a. 1 and 2 **b.** 1 and 3 **c.** 1 and 4 **d.** 3 and 4

16.25 Draw an energy graph for an *endothermic* reaction where no catalyst is present. Then draw an energy diagram for the same reaction when a catalyst is present. Indicate the similarities and differences between the two diagrams.

16.26 Draw an energy graph for an *exothermic* reaction where no catalyst is present. Then draw an energy diagram for the same reaction when a catalyst is present. Indicate the similarities and differences between the two diagrams.

CHEMICAL EQUILIBRIUM (SEC. 16.4)

16.27 What condition must be met in order for a system to be in a state of chemical equilibrium?

16.28 What relationship exists between the rates of the forward and reverse reactions for a system in a state of chemical equilibrium?

16.29 Consider the following equilibrium system

$$CO(g) + H_2O(g) \rightleftharpoons CO_2(g) + H_2(g)$$

  a. Write the chemical equation for the forward reaction.
  b. Write the chemical equation for the reverse reaction.

16.30 Consider the following equilibrium system

$$CO(g) + 3 H_2(g) \rightleftharpoons CH_4(g) + H_2O(g)$$

  a. Write the chemical equation for the forward reaction.
  b. Write the chemical equation for the reverse reaction.

16.31 The following series of diagrams represent the reaction X $\rightleftharpoons$ Y followed over a period of time. The X molecules are light-colored and the Y molecules are dark-colored.

— Increasing time ⟶

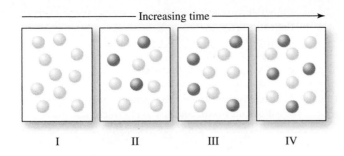

I   II   III   IV

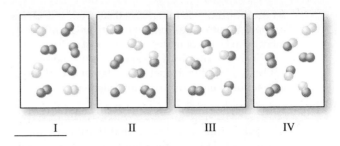

I   II   III   IV

At the end of the time period depicted, has the reaction system reached equilibrium? Justify your answer with a one-sentence explanation.

16.32 The following series of diagrams represent the reaction X ⇌ Y followed over a period of time. The X molecules are light-colored and the Y molecules are dark-colored.

— Increasing time ⟶

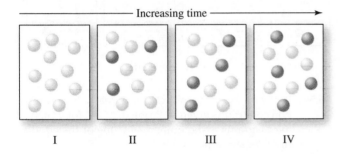

I   II   III   IV

At the end of the time period depicted, has the reaction system reached equilibrium? Justify your answer with a one-sentence explanation.

16.33 For the reaction $A_2 + 2B \longrightarrow 2AB$, diagram I depicts an initial reaction mixture where A_2 molecules are dark-colored and B atoms are light-colored. Which of the diagrams II through IV is a possible equilibrium state for the reaction system? There may be more than one correct answer.

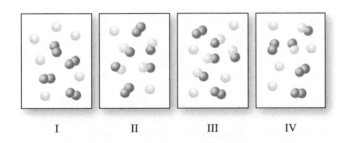

I   II   III   IV

16.34 For the reaction $A_2 + B_2 \longrightarrow 2AB$, diagram I depicts an initial reaction mixture where A_2 molecules are dark-colored and B_2 molecules are light-colored. Which of the diagrams II through IV is a possible equilibrium state for the reaction system? There may be more than one correct answer.

16.35 Sketch a graph showing how the concentration of reactant and product for a general reversible reaction involving one reactant and one product varies with time until equilibrium is reached. Assume that more product than reactant is present at equilibrium.

16.36 Sketch a graph showing how the rates of the forward and reverse reactions for a typical reversible chemical reaction vary with time.

EQUILIBRIUM MIXTURE STOICHIOMETRY (SEC. 16.5)

16.37 A 0.0200 mole sample of SO_3 is placed in a reaction container and allowed to decompose until equilibrium is established.

$$2\,SO_3(g) \rightleftharpoons 2\,SO_2(g) + O_2(g)$$

At equilibrium, 0.0029 mole of O_2 is present. What is the composition of the equilibrium mixture in terms of moles of each substance present?

16.38 A mixture of 0.100 mole of SO_2 and 0.100 mole of O_2 is placed in a reaction container and allowed to react until equilibrium is established.

$$2\,SO_2(g) + O_2(g) \rightleftharpoons 2\,SO_3(g)$$

At equilibrium, 0.0916 mole of SO_3 is present. What is the composition of the equilibrium mixture in terms of moles of each substance present?

16.39 A mixture of 0.296 mole of NH_3, 0.170 mole of N_2, and 0.095 mole of H_2 is allowed to reach equilibrium.

$$2\,NH_3(g) \rightleftharpoons N_2(g) + 3\,H_2(g)$$

At equilibrium, it is found that 0.268 mole of NH_3 is present. What is the composition of the equilibrium mixture in terms of moles of each substance present?

16.40 A mixture of 0.520 mole of NOCl, 0.010 mole of NO, and 0.053 mole of Cl_2 is allowed to reach equilibrium.

$$2\,NOCl(g) \rightleftharpoons 2\,NO(g) + Cl_2(g)$$

At equilibrium, it is found that 0.022 mole of NO is present. What is the composition of the equilibrium mixture in terms of moles of each substance present?

16.41 A 0.100 mole sample of CO and a 0.200 mole sample of O_2 are placed in a reaction container and allowed to react until equilibrium is established.

$$2\,CO(g) + O_2(g) \rightleftharpoons 2\,CO_2(g)$$

At equilibrium, it is found that 0.006 mole of CO has reacted. What is the composition of the equilibrium mixture in terms of moles of each substance present?

16.42 A 0.200 mole sample of N_2 and a 0.300 mole sample of O_2 are placed in a reaction container and allowed to react until equilibrium is established.

$$N_2(g) + 2 O_2(g) \rightleftharpoons 2 NO_2(g)$$

At equilibrium, it is found that 0.010 mole of N_2 has reacted. What is the composition of the equilibrium mixture in terms of moles of each substance present?

EQUILIBRIUM CONSTANTS (SEC. 16.6)

16.43 Write the expression for the equilibrium constant for each of the following chemical reactions.
a. $2 H_2(g) + C_2H_2(g) \longrightarrow C_2H_6(g)$
b. $2 SO_3(g) \longrightarrow 2 SO_2(g) + O_2(g)$
c. $3 Cl_2(g) + NH_3(g) \longrightarrow NCl_3(g) + 3 HCl(g)$
d. $N_2H_4(g) + 2 O_2(g) \longrightarrow 2 NO(g) + 2 H_2O(g)$

16.44 Write the expression for the equilibrium constant for each of the following chemical reactions.
a. $2 CH_4(g) \longrightarrow C_2H_6(g) + H_2(g)$
b. $2 PCl_3(g) + O_2(g) \longrightarrow 2 POCl_3(g)$
c. $SOCl_2(g) + H_2O(g) \longrightarrow SO_2(g) + 2 HCl(g)$
d. $CH_4(g) + 2 H_2S(g) \longrightarrow CS_2(g) + 4 H_2(g)$

16.45 Write the expression for the equilibrium constant for each of the following chemical reactions.
a. $CaCO_3(s) + SO_2(g) \longrightarrow CaSO_3(s) + CO_2(g)$
b. $2 FeBr_3(s) \longrightarrow 2 FeBr_2(s) + Br_2(g)$
c. $Mg(OH)_2(s) \longrightarrow MgO(s) + H_2O(g)$
d. $NaCl(aq) + AgNO_3(aq) \longrightarrow AgCl(s) + NaNO_3(aq)$

16.46 Write the expression for the equilibrium constant for each of the following chemical reactions.
a. $2 KClO_3(s) \longrightarrow 2 KCl(s) + 3 O_2(g)$
b. $2 C(s) + 2 H_2O(g) \longrightarrow CH_4(g) + CO_2(g)$
c. $3 CuO(s) + 2 NH_3(g) \longrightarrow 3 Cu(s) + N_2(g) + 3 H_2O(g)$
d. $BaCl_2(aq) + Na_2SO_4(aq) \longrightarrow 2 NaCl(aq) + BaSO_4(s)$

16.47 At a particular temperature, a hypothetical chemical system has the following equilibrium molar concentrations: A = 3.00, B = 2.00, and C = 5.00. Calculate the value of the equilibrium constant for the system if the reaction occurring were each of the following.
a. $A(g) \rightleftharpoons 2 B(g) + C(g)$
b. $A(g) + 3 B(g) \rightleftharpoons 2 C(g)$
c. $2 B(g) \rightleftharpoons A(g) + C(g)$
d. $4 C(g) + B(g) \rightleftharpoons 3 A(g)$

16.48 At a particular temperature, a hypothetical chemical system has the following equilibrium molar concentrations: A = 2.00, B = 4.00, and C = 3.00. Calculate the value of the equilibrium constant for

the system if the reaction occurring were each of the following.
a. $A(g) + 2 B(g) \rightleftharpoons C(g)$
b. $A(g) \rightleftharpoons B(g) + 3 C(g)$
c. $2 C(g) + B(g) \rightleftharpoons 2 A(g)$
d. $3 A(g) + 2 B(g) \rightleftharpoons 4 C(g)$

16.49 The following four diagrams represent gaseous equilibrium mixtures for the chemical reaction

$$A_2(g) + B_2(g) \rightleftharpoons 2 AB(g)$$

at four different temperatures. For which of the diagrams is the numerical value of the equilibrium constant the largest? (A atoms are light-colored and B atoms are dark-colored.)

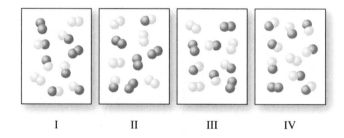

I   II   III   IV

16.50 Based on the diagrams, chemical reaction, and reaction conditions given in Problem 16.49, for which of the diagrams is the numerical value of the equilibrium constant the smallest?

16.51 The following four diagrams represent gaseous reaction mixtures for the chemical reaction

$$A_2(g) + B_2(g) \rightleftharpoons 2 AB(g)$$

If the numerical value of the equilibrium constant for the chemical reaction is 64, which of the diagrams represents the equilibrium mixture? (A atoms are light-colored and B atoms are dark-colored).

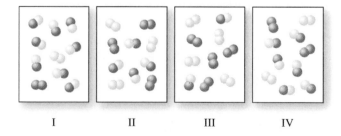

I   II   III   IV

16.52 Based on the diagrams, chemical reaction, and reaction conditions given in Problem 16.51, which of the diagrams represents the equilibrium mixture if the value of the equilibrium constant is 9.0?

16.53 The equilibrium constant for the reaction

$$CS_2(g) + 4 H_2(g) \rightleftharpoons CH_4(g) + 2 H_2S(g)$$

is 0.0280 at a particular temperature. The system at equilibrium has $[H_2S] = 1.43$, $[H_2] = 1.00$, and $[CH_4] = 0.00100$. What is $[CS_2]$?

16.54 The equilibrium constant for the reaction

$$CH_4(g) + 2\,H_2S(g) \rightleftharpoons CS_2(g) + 4\,H_2(g)$$

is 3.30×10^4 at a particular temperature. The system at equilibrium has $[CH_4] = 0.709$, $[H_2S] = 0.0100$, and $[H_2] = 2.34$. What is $[CS_2]$?

16.55 A 6.00 L vessel contains 0.0222 mole of PCl_3, 0.0189 mole of PCl_5, and 0.1044 mole of Cl_2 at 230°C in an equilibrium mixture. Calculate the value of K_{eq} for the reaction

$$PCl_3(g) + Cl_2(g) \rightleftharpoons PCl_5(g)$$

16.56 An 8.00 L vessel contains 0.650 mole of H_2, 0.275 mole of I_2, and 2.86 moles of HI at 491°C in an equilibrium mixture. Calculate the value of K_{eq} for the reaction

$$2\,HI(g) \rightleftharpoons H_2(g) + I_2(g)$$

16.57 For reactions with each of the following equilibrium constants, describe the position of equilibrium as (1) mostly products, (2) mostly reactants, or (3) significant amounts of both reactants and products.
  a. 10^{-10} at 25°C   **b.** 10^{30} at 25°C
  c. 10^9 at 127°C   **d.** 10^2 at 327°C

16.58 For reactions with each of the following equilibrium constants, describe the position of equilibrium as (1) mostly products, (2) mostly reactants, or (3) significant amounts of both reactants and products.
  a. 10^{25} at 25°C   **b.** 10^{-19} at 25°C
  c. 10^{-11} at 235°C   **d.** 10^{-1} at 1235°C

LE CHÂTELIER'S PRINCIPLE (SEC. 16.9)

16.59 For the chemical reaction

$$2\,NO(g) + 2\,CO(g) \rightleftharpoons N_2(g) + 2\,CO_2(g)$$

determine the direction that the equilibrium will be shifted by each of the following changes.
  a. increase in NO concentration
  b. increase in N_2 concentration
  c. decrease in CO_2 concentration
  d. decrease in CO concentration

16.60 For the chemical reaction

$$2\,HCl(g) + I_2(g) \rightleftharpoons 2\,HI(g) + Cl_2(g)$$

determine the direction that the equilibrium will be shifted by each of the following changes.
  a. increase in HCl concentration
  b. increase in Cl_2 concentration
  c. decrease in I_2 concentration
  d. decrease in HI concentration

16.61 For the reaction

$$2\,C_2H_2(g) + 5\,O_2(g) \rightleftharpoons 4\,CO_2(g) + 2\,H_2O(g) + heat$$

determine the direction that the equilibrium will be shifted by each of the following changes.
  a. increasing the concentration of C_2H_2
  b. decreasing the concentration of O_2
  c. increasing the temperature
  d. increasing the pressure by decreasing the volume of the container.

16.62 For the reaction

$$C_3H_8(g) + 5\,O_2(g) \rightleftharpoons 3\,CO_2(g) + 4\,H_2O(g) + heat$$

determine the direction that the equilibrium will be shifted by each of the following changes.
  a. increasing the concentration of CO_2
  b. decreasing the concentration of C_3H_8
  c. decreasing the temperature
  d. increasing the pressure by decreasing the volume of the container

16.63 Consider the following chemical system at equilibrium.

$$2\,H_2O(g) + 2\,Cl_2(g) + heat \rightleftharpoons 4\,HCl(g) + O_2(g)$$

For each of the following adjustments of conditions, indicate the effect on the position of equilibrium: shifts left, shifts right, no effect.
  a. heating the equilibrium mixture
  b. adding O_2 to the mixture
  c. increasing the pressure on the equilibrium mixture by adding an inert gas
  d. increasing the size of the reaction container

16.64 Consider the following chemical system at equilibrium.

$$2\,N_2(g) + 6\,H_2O(g) + heat \rightleftharpoons 4\,NH_3(g) + O_2(g)$$

For each of the following adjustments of conditions, indicate the effect on the position of equilibrium: shifts left, shifts right, no effect.
  a. adding N_2 to the mixture
  b. decreasing the size of the container holding the mixture
  c. adding a catalyst to the mixture
  d. refrigerating the warm equilibrium mixture

16.65 For which of the following reactions is product formation favored by high temperature?
  a. $N_2(g) + O_2(g) \rightleftharpoons 2\,NO(g) + heat$
  b. $N_2(g) + 3\,H_2(g) \rightleftharpoons 2\,NH_3(g) + heat$
  c. $CO(g) + 3\,H_2(g) + heat \rightleftharpoons CH_4(g) + H_2O(g)$
  d. $2\,H_2O(g) \rightleftharpoons 2\,H_2(g) + O_2(g) + heat$

16.66 For which of the reactions in Problem 16.65 is product formation favored by high pressure?

16.67 When the temperature of an equilibrium system for the following reaction is increased, the reaction shifts to the left. Is the forward reaction endothermic or exothermic?

$$CO(g) + H_2O(g) \rightleftharpoons CO_2(g) + H_2(g)$$

16.68 When the temperature of an equilibrium system for the following reaction is decreased, the reaction shifts to the left. Is the forward reaction endothermic or exothermic?

$$NO_2(g) + O_2(g) \rightleftharpoons NO(g) + O_3(g)$$

16.69 The following two diagrams represent the composition of an equilibrium mixture for the chemical reaction

$$A_2(g) + B_2(g) \rightleftharpoons 2\,AB(g)$$

at two different temperatures. Based on the diagrams, is the chemical reaction endothermic or exothermic? Explain your answer using Le Châtelier's principle. (A atoms are light-colored and B atoms are dark-colored in the diagrams.)

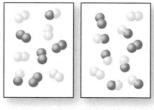

$T = 150°C$ $\qquad$ $T = 200°C$

16.70 The following two diagrams represent the composition of an equilibrium mixture for the chemical reaction

$$A_2(g) + B_2(g) \rightleftharpoons 2\,AB(g)$$

at two different temperatures. Based on the diagrams, is the chemical reaction endothermic or exothermic? Explain your answer using Le Châtelier's principle. (A atoms are light-colored and B atoms are dark-colored in the diagrams.)

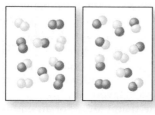

$T = 250°C$ $\qquad$ $T = 400°C$

ADDITIONAL PROBLEMS

16.71 Write a balanced chemical equation for a totally gaseous equilibrium system that would lead to the following expressions for the equilibrium constant.

a. $\dfrac{[NH_3]^2}{[N_2][H_2]^3}$ $\qquad$ **b.** $\dfrac{[N_2]^2[H_2O]^6}{[NH_3]^4[O_2]^3}$

c. $\dfrac{[N_2][O_2]}{[NO]^2}$ $\qquad$ **d.** $\dfrac{[NO]^2}{[N_2][O_2]}$

16.72 Write a balanced chemical equation for a totally gaseous equilibrium system that would lead to the following expressions for the equilibrium constant.

a. $\dfrac{[HCN]^2}{[H_2][C_2N_2]}$ $\qquad$ **b.** $\dfrac{[CH_4][H_2S]^2}{[CS_2][H_2]^4}$

c. $\dfrac{[NOBr]^2}{[NO]^2[Br_2]}$ $\qquad$ **d.** $\dfrac{[NO]^2[Br_2]}{[NOBr]^2}$

16.73 The following reaction at a certain temperature has an equilibrium-constant value of 25.9.

$$2\,CO_2(g) \rightleftharpoons 2\,CO(g) + O_2(g)$$

For each of the following compositions, decide whether the reaction mixture is at equilibrium. If it is not, decide which direction the reaction shifts to reach equilibrium.
a. $[CO_2] = 0.0300$, $[CO] = 0.350$, $[O_2] = 0.190$
b. $[CO_2] = 0.0600$, $[CO] = 0.700$, $[O_2] = 0.380$
c. $[CO_2] = 0.0280$, $[CO] = 0.356$, $[O_2] = 0.160$
d. $[CO_2] = 0.0100$, $[CO] = 0.330$, $[O_2] = 0.180$

16.74 The following reaction at a certain temperature has an equilibrium-constant value of 0.016.

$$2\,HI(g) \rightleftharpoons H_2(g) + I_2(g)$$

For each of the following compositions, decide whether the reaction mixture is at equilibrium. If it is not, decide which direction the reaction shifts to reach equilibrium.
a. $[HI] = 0.080$, $[H_2] = 0.010$, $[I_2] = 0.010$
b. $[HI] = 0.084$, $[H_2] = 0.012$, $[I_2] = 0.012$
c. $[HI] = 0.076$, $[H_2] = 0.012$, $[I_2] = 0.012$
d. $[HI] = 0.140$, $[H_2] = 0.031$, $[I_2] = 0.010$

16.75 At a given temperature, the equilibrium constant for the reaction

$$2\,NO(g) + Br_2(g) \rightleftharpoons 2\,NOBr(g)$$

is 2×10^3. What is the equilibrium constant, at the same temperature, for the following reaction?

$$2\,NOBr(g) \rightleftharpoons 2\,NO(g) + Br_2(g)$$

16.76 At a given temperature, the equilibrium constant for the reaction

$$CO(g) + H_2O(g) \rightleftharpoons CO_2(g) + H_2(g)$$

is 0.034. What is the equilibrium constant, at the same temperature, for the following reaction?

_____ $CO_2(g) + H_2(g) \rightleftharpoons CO(g) + H_2O(g)$

16.77 Which of the following changes would change the *value* of a system's equilibrium constant?
a. addition of a reactant or product
b. increase in the total pressure
c. decrease in the temperature
d. addition of an inert gas

16.78 Which of the following changes would change the *value* of a system's equilibrium constant?
a. removal of a reactant or product
b. decrease in the total pressure
c. increase in the temperature
d. addition of a catalyst

16.79 Predict the direction in which each of the following equilibria will shift if the pressure on the system is decreased by expansion.
a. $N_2O(g) + NO_2(g) \longrightarrow 3\,NO(g)$
b. $2\,HBr(g) \longrightarrow H_2(g) + Br_2(g)$
c. $SO_2(g) + Cl_2(g) \longrightarrow SO_2Cl_2(g)$
d. $C(s) + CO_2(g) \longrightarrow 2\,CO(g)$

16.80 Predict the direction in which each of the following equilibria will shift if the pressure on the system is increased by compression.
a. $PCl_5(g) \longrightarrow PCl_3(g) + Cl_2(g)$
b. $2\,NO(g) + Cl_2(g) \longrightarrow 2\,NOCl(g)$
c. $CS_2(g) + 4\,H_2(g) \longrightarrow CH_4(g) + 2\,H_2S(g)$
d. $Ni(s) + 4\,CO(g) \longrightarrow Ni(CO)_4(g)$

CUMULATIVE PROBLEMS

16.81 Given the following descriptions of reversible reactions, write the equilibrium-constant expression for each.
a. In a decomposition, reaction sulfur trioxide gas produces sulfur dioxide gas and oxygen gas.
b. Hydrogen gas reduces nitrogen dioxide gas to produce ammonia gas and steam.
c. Solid iron(II) oxide and carbon monoxide gas react to produce solid iron and carbon dioxide gas.
d. Solid magnesium carbonate decomposes to produce solid magnesium oxide and carbon dioxide gas.

16.82 Given the following descriptions of reversible reactions, write the equilibrium-constant expression for each.
a. In a synthesis reaction, gaseous hydrogen and gaseous bromine react to produce gaseous hydrogen bromide.
b. In a redox reaction, carbon disulfide gas reacts with hydrogen gas to produce methane gas and hydrogen sulfide gas.

c. Solid sodium carbonate reacts with sulfur dioxide gas and oxygen gas to produce solid sodium sulfate and carbon dioxide gas.
d. Chlorine gas reacts with liquid carbon disulfide to produce the liquids carbon tetrachloride and disulfur dichloride.

16.83 An equilibrium mixture at 900°C in a 1725 mL container involving the chemical system

$$CH_4(g) + 2\,H_2S(g) \rightleftharpoons CS_2(g) + 4\,H_2(g)$$

is found to contain 17.6 g CH_4, 50.8 g H_2S, 83.8 g CS_2, and 8.10 g H_2. Calculate the equilibrium constant for this reaction at the given temperature.

16.84 An equilibrium mixture at 472°C in a 1325 mL container involving the chemical system

$$N_2(g) + 3\,H_2(g) \rightleftharpoons 2\,NH_3(g)$$

is found to contain 4.23 g N_2, 0.915 g H_2, and 0.496 g NH_3. Calculate the equilibrium constant for this reaction at the given temperature.

16.85 For the chemical system

$$SbCl_5(g) \rightleftharpoons SbCl_3(g) + Cl_2(g)$$

it is found that an equilibrium mixture in a 1.00 L flask contains 2.48×10^{20} molecules of $SbCl_5$, 0.723 g of $SbCl_3$, and 0.00317 mole of Cl_2. Calculate the equilibrium constant for the reaction.

16.86 For the chemical system

$$PCl_5(g) \rightleftharpoons PCl_3(g) + Cl_2(g)$$

it is found that an equilibrium mixture in a 1.00 L flask contains 2.95×10^{20} molecules of PCl_5, 0.00451 mole of PCl_3, and 0.320 g of Cl_2. Calculate the equilibrium constant for the reaction.

16.87 Consider the following equilibrium situation at constant temperature.

$$N_2O_4(g) \rightleftharpoons 2\,NO_2(g)$$

An empty container is charged with pure $N_2O_4(g)$ until the pressure reaches 1.50 atm. The $N_2O_4(g)$ is then allowed to reach equilibrium with $NO_2(g)$, at which time the partial pressure of $N_2O_4(g)$ is 0.80 atm. What is the total pressure in atmospheres in the flask?

16.88 Consider the following equilibrium situation, at constant temperature.

$$2\,O_3(g) \rightleftharpoons 3\,O_2(g)$$

An empty container is charged with pure $O_3(g)$ until the pressure reaches 2.25 atm. The $O_3(g)$ is then allowed to reach equilibrium with $O_2(g)$, at which time the partial pressure of $O_3(g)$ is 0.11 atm. What is the total pressure, in atmospheres, in the flask?

16.89 At 750°C and 1.000 atm pressure, a gaseous mixture of carbon monoxide and carbon dioxide is in equilibrium with solid carbon.

$$C(s) + CO_2(g) \rightleftharpoons 2\,CO(g)$$

The gaseous mixture is 87.43% CO by mass.

Calculate the equilibrium constant for this reaction from the given information.

16.90 At 35°C and 1.00 atm pressure, a gaseous mixture of dinitrogen tetroxide and nitrogen dioxide at equilibrium.

$$N_2O_4(g) \rightleftharpoons 2\,NO_2(g)$$

The gaseous mixture is found to contain 32.7% by mass of N_2O_4.

Calculate the equilibrium constant for the reaction from the given information.

16.91 The equation for the self-ionization of water is

$$2\,H_2O \rightleftharpoons H_3O^+ + OH^-$$

Determine the direction that this equilibrium will be shifted by each of the following changes.
a. $[H_3O^+]$ increase
b. $[OH^-]$ decrease
c. pH increase by 2 units
d. pH decrease by 2 units

16.92 The equation for the self-ionization of water is

$$H_2O \rightleftharpoons H^+ + OH^-$$

Determine the direction that the equilibrium will be shifted by each of the following changes.
a. $[H^+]$ decrease
b. $[OH^-]$ increase
c. pH increase by 3 units
d. pH decrease by 1 unit

Multiple-Choice Practice Test

Use this bank of 20 multiple-choice questions as a review of key concepts presented in this chapter. For many of the questions, there may be more than one correct answer (choice d) or no correct answer (choice e).

16.93 Most chemical reactions are carried out in liquid solution or in the gaseous phase because in such situations
a. activation energies are higher.
b. products are less likely to decompose.
c. reactant collisions can occur more frequently.
d. more than one correct response
e. no correct response

16.94 For a collision between molecules to result in a reaction, the molecules must possess a certain minimum energy and
a. exchange electrons.
b. interact with a catalyst for at least 1 second.
c. have a favorable orientation relative to each other when they collide.
d. more than one correct response
e. no correct response

16.95 Which of the following statements about activation energy is correct?
a. It is the same at a given temperature for all chemical reactions.
b. It is the energy given off when two molecules collide.
c. It is low for reactions that take place rapidly.
d. more than one correct response
e. no correct response

16.96 Whether a chemical reaction is exothermic or endothermic is determined by
a. the magnitude of the activation energy.
b. the physical state of the reactants.
c. whether a catalyst is present.
d. more than one correct response
e. no correct response

16.97 Which of the following chemical reactions is endothermic?
a. $NH_3 + HBr \longrightarrow NH_4Br + heat$
b. $2\,NO_2 \longrightarrow N_2 + 2\,O_2 + heat$
c. $PCl_3 + Cl_2 \longrightarrow PCl_5 + heat$
d. more than one correct response
e. no correct response

16.98 Which of the following factors would most likely decrease the rate of a chemical reaction?
a. increase the state of subdivision of a reactant
b. decrease the concentration of a reactant
c. decrease the activation energy for the reaction
d. more than one correct response
e. no correct response

16.99 Increasing the temperature at which a chemical reaction occurs will
a. lower the activation energy, thus increasing the reaction rate.
b. increase collision frequency among reactants, thus increasing the reaction rate.
c. reduce collision energies, thus increasing reaction rate.
d. more than one correct response
e. no correct response

16.100 Which of the following statements concerning catalysts is correct?
a. They can be either solids, liquids, or gases.
b. They lower the activation energy for reactions.
c. They do not actively participate in the reaction.
d. more than one correct response
e. no correct response

16.101 Chemical equilibrium is reached in a chemical system when
  a. complete conversion of reactants to products has occurred.
  b. reactants are being consumed at the same rate they are being regenerated.
  c. reactant concentrations become equal to product concentrations.
  d. more than one correct response
  e. no correct response

16.102 A mixture of 2.00 moles of A and 2.00 moles of B is allowed to reach equilibrium, at which time 1.60 moles of A are present. How many moles of B are present at equilibrium if the reaction equations is

$$2\,A(g) + 3\,B(g) \rightleftharpoons 4\,C(g)$$

  a. 0.40 mole B   **b.** 1.40 moles B
  c. 1.60 moles B   **d.** 1.80 moles B
  e. no correct response

16.103 Which of the following statements is correct for an equilibrium-constant expression?
  a. Product concentrations are always placed in the numerator.
  b. Solid and liquid concentrations are always placed in the denominator.
  c. Concentrations for gaseous reactants and products are always squared.
  d. more than one correct response
  e. no correct response

16.104 Which one of the following will change the value of an equilibrium constant?
  a. changing the temperature
  b. adding other substances that do not react with the equilibrium mixture
  c. changing the concentrations of all of the reactants
  d. more than one correct response
  e. no correct response

16.105 What is the equilibrium constant value if equilibrium concentrations for the chemical system.

$$2\,NH_3(g) \rightleftharpoons N_2(g) + 3\,H_2(g)$$

are $[NH_3] = 0.40$, $[H_2] = 0.12$, and $[N_2] = 0.040$?
  a. 2.8×10^{-2}
  b. 4.3×10^{-4}
  c. 6.8×10^{-9}
  d. cannot be determined from the information given
  e. no correct response

16.106 When the position of an equilibrium is described as being "far to the left" it means that
  a. very few reactant molecules are present in the equilibrium mixture.
  b. the rate of the reverse reaction is greater than that of the forward reaction.

  c. many product molecules are present in the equilibrium mixture.
  d. more than one correct response
  e. no correct response

16.107 If, at equilibrium, nearly all of the reactants have been consumed, the equilibrium constant would be expected to have a
  a. very small numerical value.
  b. very large numerical value.
  c. numerical value slightly greater than 1.0.
  d. numerical value slightly less than 1.0.
  e. no correct response

16.108 Le Châtelier's principle states that
  a. only exothermic reactions can reach equilibrium.
  b. only reactions in which a catalyst is present can reach equilibrium.
  c. if a chemical reaction is disturbed, the temperature of the system will always increase.
  d. more than one correct response
  e. no correct response

16.109 Le Châtelier's principle can be used to predict the
  a. effect of a catalyst on the rate of a chemical reaction.
  b. effect of a concentration change on equilibrium position.
  c. temperature at which equilibrium will be established.
  d. more than one correct response
  e. no correct response

16.110 For a chemical reaction at equilibrium, which of the following would always decrease the concentrations of the products?
  a. an increase in the temperature
  b. a decrease in the pressure
  c. a decrease in a reactant concentration
  d. more than one correct response
  e. no correct response

16.111 According to Le Châtelier's principle, which of the following changes will shift the equilibrium position to the left for the reaction

$$N_2(g) + 3\,H_2(g) \rightleftharpoons 2\,NH_3(g) + heat$$

  a. increasing the concentration of N_2
  b. decreasing the concentration of NH_3
  c. increasing the temperature of the system
  d. more than one correct response
  e. no correct response

16.112 For which of the following equilibrium systems will the equilibrium position shift to the left when the pressure on the system is increased?
  a. $H_2(g) + Cl_2(g) \rightleftharpoons 2\,HCl(g)$
  b. $2\,SO_2(g) + O_2(g) \rightleftharpoons 2\,SO_3(g)$
  c. $2\,NH_3(g) \rightleftharpoons 3\,H_2(g) + N_2(g)$
  d. more than one correct response
  e. no correct response

GLOSSARY

accuracy An indicator of how close a measurement (or the average of multiple measurements) comes to a true or accepted value (2.3)

acid A hydrogen-containing molecular compound whose molecules yield hydrogen ions (H^+) when dissolved in water (8.6)

acid–base titration A procedure in which a measured volume of acid or base of known concentration is exactly reacted with a measured volume of a base or an acid of unknown concentration (14.14)

acidic hydrogen atom A hydrogen atom in an acid molecule that can be transferred to a base during an acid–base reaction (14.4)

acidic solution A solution in which the concentration of H_3O^+ ion is greater than that of OH^- ion (14.10)

activation energy The minimum combined kinetic energy that reactant particles must possess in order for their collision to result in a reaction (16.1)

actual yield The amount of product actually produced from a chemical reaction (10.11)

alkali metal A general name for any element in group IA of the periodic table, excluding hydrogen (6.2)

alkaline earth metal A general name for any element in group IIA of the periodic table (6.2)

alpha particle A particle having two protons and two neutrons, which is emitted by certain radioactive nuclei (5.8)

alpha-particle decay The radioactive decay process in which an alpha particle is emitted from an unstable nucleus (5.9)

amphiprotic substance A substance that can either lose or accept protons and thus can function as either an acid or a base (14.3)

anion A negatively charged ion (7.4)

applied scientific research Research whose major focus is the discovery of useful products and processes, which can be used to benefit humankind (1.2)

aqueous solution A solution in which water is the solvent (13.2)

area A measure of the extent of a surface (3.4)

Arrhenius acid A hydrogen-containing compound that in water produces hydrogen ions (H^+) (14.1)

Arrhenius base A hydroxide-ion-containing compound that in water produces hydroxide ions (OH^-) (14.1)

atom The smallest particle of an element that can exist and still have the properties of the element (4.10)

atomic mass The relative mass of an average atom of an element on a scale using the $^{12}_{6}C$ atom as the reference (5.4)

atomic number The number of protons in the nucleus of an atom (5.2)

atomic theory of matter A set of five statements that summarizes modern-day scientific thought about atoms (4.10)

aufbau diagram A listing of electron subshells arranged in the order in which electrons occupy them (6.7)

aufbau principle Electrons normally occupy electron subshells in order of increasing subshell energy (6.7)

Avogadro's law Equal volumes of different gases, measured at the same temperature and pressure, contain equal numbers of molecules (12.7); the volume of a gas is directly proportional to the number of moles of gas present if the temperature and pressure are kept constant (12.7)

Avogadro's number The name given to the numerical value 6.022×10^{23}, the number of particles in a mole (9.5)

balanced chemical equation A chemical equation that has the same number of atoms of each element involved in the reaction on each side of the equation (10.3)

barometer A device used for measuring atmospheric pressure (12.2)

basic scientific research Research whose major focus is the discovery of new fundamental information about humans and other living organisms and the universe in which they live (1.2)

basic solution A solution in which the concentration of OH^- ion is greater than that of H_3O^+ ion (14.10)

beta particle A particle, having charge and mass that are identical to those of an electron, which is emitted by certain radioactive nuclei (5.8)

beta-particle decay The radioactive decay process in which a beta particle is emitted from an unstable nucleus (5.9)

binary compound A compound in which only two elements are present (8.2)

binary ionic compound An ionic compound in which one element present is a metal and the other element present is a nonmetal (8.2)

binary molecular compound A molecular compound in which only two nonmetallic elements are present (8.5)

boiling A special form of evaporation in which conversion from the liquid to the vapor state occurs within the body of a liquid through bubble formation (11.15)

boiling point The temperature of a liquid at which the vapor pressure of the liquid becomes equal to the external (atmospheric) pressure exerted on the liquid (11.15)

bombardment reaction A nuclear reaction brought about by bombarding stable nuclei with small particles traveling at very high speeds (5.10)

bond length The distance between the nuclei of covalently bonded atoms (7.15)

bond polarity A measure of the degree of inequality in the sharing of electrons in a chemical bond (7.19)

bond strength A measure of the energy it takes to break a bond, that is, to separate bonded atoms to give neutral particles (7.15)

bonding electrons Pairs of valence electrons that are shared between atoms in a covalent bond (7.11)

Boyle's law The volume of a fixed mass of gas is inversely proportional to the pressure applied to the gas if the temperature is kept constant (12.3)

Brønsted–Lowry acid Any substance that can donate a proton (H^+) to some other substance (14.2)

Brønsted–Lowry base Any substance that can accept a proton (H^+) from some other substance (14.2)

buffer A solution that resists major changes in pH when small amounts of acid or base are added to it (14.13)

catalyst A substance that when added to a reaction mixture increases the rate of the reaction, but is itself unchanged after the reaction is completed (16.3)

cation A positively charged ion (7.4)

change of state A process in which a substance is transformed from one physical state to another physical state (11.8)

Charles's law The volume of a fixed mass of gas is directly proportional to its Kelvin temperature if the pressure is kept constant (12.4)

chemical bond The attractive force that holds two atoms together in a more complex unit (7.1)

chemical change A process in which a substance undergoes a change in chemical composition (4.4)

chemical equation A representation for a chemical reaction that uses chemical symbols and chemical formulas, instead of words, to describe the changes that occur in the chemical reaction (10.2)

chemical equilibrium A process wherein two opposing chemical reactions occur simultaneously at the same rate (16.4)

chemical formula A notation made up of the chemical symbols of the elements present in a compound and numerical subscripts (located to the right of each chemical symbol) that indicate the number of atoms of each element present in a structural unit of the compound (4.13)

chemical nomenclature The system of names used to distinguish compounds from each other and the rules needed to devise these names (8.1)

chemical periodicity The variation in properties of elements as a function of their positions in the periodic table (6.12)

chemical property A characteristic of a substance that describes the way the substance undergoes or resists change to form a new substance (4.3)

chemical reaction A process in which at least one new substance is produced as a result of chemical change (4.4)

chemical stoichiometry The study of the quantitative relationships among reactants and products in a chemical reaction (10.9)

chemical symbol A one- or two-letter designation for an element derived from the element's name (4.9)

chemistry The scientific discipline concerned with the characteristics, composition, and transformations of matter (4.1)

collision theory A set of statements that gives the conditions that must be met before a chemical reaction will take place (16.1)

combined gas law The product of the pressure and volume of a fixed amount of gas is directly proportional to its Kelvin temperature (12.6)

combustion analysis A method used to measure the amounts of carbon and hydrogen present in a combustible compound that contains these two elements (and perhaps other elements) when the compound is burned in pure O_2 (9.13)

combustion reaction A chemical reaction in which a substance reacts with oxygen (usually from air) that proceeds with evolution of heat and usually also a flame (10.6)

common name A name for a compound not based on IUPAC rules that does not convey information about the composition of the compound (8.5)

compound A pure substance that can be broken down into two or more simpler pure substances by chemical means (4.7)

compressibility A measure of the change in volume resulting from a pressure change (11.2)

concentrated solution A solution that contains a large amount of solute relative to the amount that could dissolve (13.2)

concentration The amount of solute present in a specified amount of solvent or a specified amount of solution (13.5)

condensed electron configuration An electron configuration in which the chemical symbol of the nearest noble gas element of lower atomic number is used to represent the electrons in the configuration up to that of the noble gas, and the remaining additional electrons are then appended to the chemical symbol of the noble gas (6.7)

conjugate acid The species formed when a proton (H^+ ion) is added to a Brønsted–Lowry base (14.3)

conjugate acid–base pair Two species, one an acid and one a base, that differ from each other through the loss or gain of a proton (H^+ ion) (14.3)

conjugate base The species that remains when a proton (H^+ ion) is removed from a Brønsted–Lowry acid (14.3)

conversion factor A ratio that specifies how one unit of measurement is related to another unit of measurement (3.6)

coordinate covalent bond A covalent bond in which both electrons of a shared electron pair come from one of the two atoms involved in the bond (7.14)

core electrons The inner-shell electrons of an atom that are not normally involved in determining the chemical properties of the atom (6.7)

covalent bond A chemical bond formed through the sharing of one or more pairs of electrons between two atoms (7.1); a chemical bond resulting from two nuclei attracting the same shared electrons (7.10)

cubic meter The SI system base unit of volume (3.4)

Dalton's law of partial pressures The total pressure exerted by a mixture of gases is the sum of the partial pressures of the individual gases (12.16)

daughter nuclide The nuclide that is produced in a radioactive decay process (5.9)

decomposition reaction A chemical reaction in which a single reactant is converted into two or more simpler substances (10.6)

density A ratio of the mass of an object to the volume occupied by the object (3.8)

diamagnetic atom An atom that has an electron arrangement containing one or more unpaired electrons (6.8)

diatomic molecule A molecule that contains two atoms (4.11)

dilute solution A solution that contains a small amount of solute relative to the amount that could dissolve (13.2)

dilution The process in which more solvent is added to a solution in order to lower its concentration (13.10)

dimensional analysis A general problem-solving method that uses the units associated with numbers as a guide in setting up the calculation (3.7)

dipole–dipole interaction An intermolecular force of attraction that occurs between polar molecules (11.16)

diprotic acid An acid that can transfer two H^+ ions (two protons) per molecule during an acid–base reaction (14.4)

disproportionation reaction A redox reaction in which some atoms of a single element in a reactant are oxidized and others are reduced (15.10)

dissociation The process whereby the already-existent ions in an ionic compound separate when the ionic compound is dissolved in a solvent (14.1)

distinguishing electron The last electron added to an element's electron configuration when the configuration is written following the aufbau principle (6.10)

double covalent bond A covalent bond in which two atoms share two pairs of valence electrons (7.12)

double-replacement reaction A chemical reaction in which two compounds exchange parts with each other and form two new compounds (10.6)

electrolyte A substance whose aqueous solution conducts electricity (15.2)

electron A subatomic particle that possesses a negative electrical charge (5.1)

electron capture A radioactive decay process in which an electron in a low-energy orbital, such as the 1s orbital, is pulled into an unstable nucleus, converting a proton to a neutron (5.11)

electron configuration A statement of how many electrons an atom has in each of its subshells (6.7)

electron orbital A region of space within an electron subshell where an electron with a specific energy is most likely to be found (6.6)

electron shell A region of space about a nucleus that contains electrons that have approximately the same energy and that spend most of their time approximately the same distance from the nucleus (6.4)

electron spin A property of an electron associated with the concept that an electron is spinning on its own axis (6.6)

electron subshell A region of space within an electron shell that contains electrons that have the same energy (6.5)

electronegativity A measure of the relative attraction that an atom has for the shared electrons in a bond (7.18)

electrostatic force An attractive force or repulsive force that occurs between charged particles (11.3)

element A pure substance that cannot be broken down into simpler pure substances by ordinary chemical means (4.7). A pure substance in which all atoms present have the same atomic number; that is, all atoms have the same number of protons (5.2)

empirical formula A chemical formula that gives the smallest whole-number ratio of atoms present in a formula unit of a compound (9.12)

endothermic change of state A change of state that requires the input (absorption) of heat energy (11.8)

endothermic chemical reaction A chemical reaction that requires the continuous input of energy as the reaction occurs (16.2)

equality conversion factor A ratio that converts one unit of a given measure to another unit of the same measure (3.8)

equation coefficient A number placed to the left of a chemical formula in a chemical equation that changes the amount, but not the identity, of a substance (10.3)

equilibrium constant A numerical value that characterizes the relationship between the concentrations of reactants and concentrations of products in a system that is in chemical equilibrium (16.6)

equilibrium position A qualitative indication of the relative amounts of reactants and products present when a chemical reaction reaches equilibrium (16.7)

equilibrium state A situation in which two opposite processes take place at equal rates (11.13)

equivalence conversion factor A ratio that converts one type of measure to a different type of measure (3.8)

evaporation A process in which molecules escape from a liquid phase to a gaseous phase (11.13)

exact number A number that has a value with no uncertainty in it; that is, it is known exactly (2.2)

exothermic change of state A change of state that requires heat energy to be given up (released) (11.8)

exothermic chemical reaction A chemical reaction in which energy is released as the reaction occurs (16.2)

experiment A well-defined, controlled procedure for obtaining information about a system under study (1.4)

exponent A number written as a superscript following another number that indicates how many times the first number is to be multiplied by itself (2.7)

extensive property A property that depends on the amount of substance present (4.3)

fixed-charge metal A metal that forms only one type of positive ion, which always has the same charge magnitude (8.2)

formula mass The sum of the atomic masses of all atoms present in one formula unit of a substance, expressed in atomic mass units (9.2)

formula unit The smallest whole-number repeating ratio of ions present in an ionic compound that results in charge neutrality (7.8)

gamma ray A form of high-energy radiation without mass or charge, that is emitted by radioactive nuclei (5.8)

gamma-ray emission The radioactive decay process in which gamma rays are emitted from an unstable nucleus (5.9)

gas The physical state characterized by both an indefinite shape and an indefinite volume (4.2); the physical state characterized by a complete dominance of kinetic energy (disruptive forces) over potential energy (cohesive forces) (11.6)

gas laws Generalizations that summarize in mathematical terms experimental observations about the relationships among the amount, pressure, temperature, and volume of a gas (12.2)

Gay-Lussac's law The pressure of a fixed mass of gas is directly proportional to its Kelvin temperature if the volume is kept constant (12.5)

Gay-Lussac's law of combining volumes The volumes of gases that participate in a chemical reaction, measured at the same temperature and pressure, are in the same ratio as the coefficients for these gases in the balanced equation for the reaction (12.11)

group A vertical column of elements in the periodic table (6.2)

half-life The time required for one-half of any given quantity of a radioactive substance to undergo decay (5.7)

half-reaction A chemical equation that describes one of the two parts of an overall oxidation–reduction reaction, either the oxidation part or the reduction part (15.9)

halogen A general name for any element in group VIIA of the periodic table (6.2)

heat capacity The amount of heat energy needed to raise the temperature of a given quantity of a substance in a specific physical state by 1°C (11.9)

heat of condensation Amount of heat energy evolved in the conversion of 1 gram of a gas to a liquid at the liquid's boiling point (11.11)

heat of fusion Amount of heat energy absorbed in the conversion of 1 gram of a solid to a liquid at the solid's melting point (11.11)

heat of solidification Amount of heat energy evolved in the conversion of 1 gram of a liquid to a solid at the liquid's freezing point (11.11)

heat of vaporization Amount of heat energy absorbed in the conversion of 1 gram of a liquid to a gas at the liquid's boiling point (11.11)

heteroatomic molecule A molecule in which two or more different kinds of atoms are present (4.11)

heterogeneous catalyst A catalyst that exists in a phase different from the reactants in a reaction mixture (16.3)

heterogeneous mixture A mixture that contains two or more visually distinguishable phases (parts), each of which has different properties (4.6)

homoatomic molecule A molecule in which all atoms present are of the same kind (4.11)

homogeneous catalyst A catalyst that is present in the same phase as the reactants in a reaction mixture (16.3)

homogeneous mixture A mixture that contains only one visually distinguishable phase (part), which has uniform properties throughout (4.6)

Hund's rule When electrons are placed in a set of orbitals of equal energy (the orbitals of a subshell), the order of filling for the orbitals is such that each orbital of the subshell receives an electron with the same spin before any orbital receives a second electron (of opposite spin) (6.8)

hydrogen bond An extra strong dipole–dipole interaction involving a hydrogen atom covalently bonded to a small, very electronegative atom (F, O, or N) and an unshared pair of electrons on another small, very electronegative atom (F, O, or N) (11.16)

hydrolysis The reaction of a substance with water to produce an H_3O^+ ion or OH^- ion or both (14.12)

ideal gas A gas that would obey gas laws exactly over all temperatures and pressures (12.8)

ideal gas law A gas law that describes the relationships among the four variables pressure, volume, molar amount, and temperature for a gaseous substance at a given set of conditions (12.9)

indicator A compound that exhibits different colors, depending on the pH of its surroundings (14.14)

inexact number A number that has a value with a degree of uncertainty in it (2.2)

inner-transition element An element found in the f area of the periodic table (6.11)

intensive property A property that is independent of the amount of substance present (4.3)

intermolecular force An attractive force that acts between a molecule and another molecule (11.16)

ion An atom (or group of atoms) that is electrically charged as the result of the loss or gain of electrons (7.4)

ion–dipole interaction An intermolecular attractive force between an ion and a polar molecule (11.16)

ion–ion interaction An intermolecular attractive force between oppositely charged ions present in liquid state (molten) ionic compounds (11.16)

ion product constant for water The numerical value (1.00×10^{-14}) associated with the product of the H_3O^+ ion and OH^- ion molar concentrations in water (14.10)

ionic bond A chemical bond formed through the transfer of one or more electrons from one atom or group of atoms to another (7.1); the chemical bond resulting from the attraction of positive and negative ions for each other (7.4)

ionic equation An equation in which the formulas of the predominant form of each compound in an aqueous solution are used; dissociated/ionized compounds are written as ions, and undissociated compounds are written in molecular form (15.3)

ionization The process whereby ions are produced from a molecular compound when it is dissolved in a solvent (14.1)

isobars Atoms that have the same mass number but different atomic numbers (5.3)

isoelectronic species Ions, or an atom and an ion, having the same number and configuration of electrons (7.5)

isotopes Atoms of an element that have the same number of protons and electrons but different numbers of neutrons (5.3)

joule The base unit for energy in the metric system (11.9)

kilogram The SI system base unit of mass (3.3)

kinetic energy The energy that matter possesses because of particle motion (11.3)

kinetic molecular theory of matter A set of five statements used to explain the physical behavior of the three states of matter (11.3)

law of conservation of mass Mass is neither created nor destroyed in any ordinary chemical reaction (10.1)

law of definite proportions In a pure compound the elements are always present in the same definite proportion by mass (9.1)

Le Châtelier's principle If a stress (change of conditions) is applied to a chemical system in equilibrium, the system will readjust (change the position of the equilibrium) in the direction that best reduces the stress imposed on the system (16.9)

Lewis structure A grouping of Lewis symbols that shows either the transfer of electrons or the sharing of electrons in chemical bonds (7.6)

Lewis symbol The chemical symbol of an element surrounded by dots equal in number to the number of valence electrons present in atoms of the element (7.2)

limiting reactant The reactant in a chemical reaction that is entirely consumed when the reaction goes to completion (10.10)

liquid The physical state characterized by both an indefinite shape and a definite volume (4.2); the physical state characterized by potential energy (cohesive forces) and kinetic energy (disruptive forces) of about the same magnitude (11.5)

liter A volume equal to that of a cube whose sides are 1 dm, or 10 cm (1 dm = 10 cm), in length (3.4)

London force A weak temporary dipole–dipole interaction that occurs between an atom or molecule (polar or nonpolar) and another atom or molecule (polar or nonpolar) (11.16)

manometer A device used to measure gas pressure in a laboratory (12.2)

mass A measure of the total quantity of matter in an object (3.3)

mass number The sum of the number of protons and the number of neutrons in the nucleus of an atom (5.2)

mass–volume percent concentration The mass of solute (in grams) divided by the total volume of solution (in milliliters) multiplied by 100 (13.6)

matter Anything that has mass and occupies space (4.1)

measurement The determination of the dimensions, capacity, quantity, or extent of something (2.1)

metal An element that has the characteristic properties of luster, thermal conductivity, electrical conductivity, and malleability (6.11)

metalloid An element with properties intermediate between those of metals and nonmetals (6.12)

meter The SI system base unit of length (3.2)

mixture A physical combination of two or more pure substances in which each substance retains its own chemical identity (4.5)

molality concentration A solution concentration unit that gives the number of moles of solute per kilogram of solvent (13.9)

molar mass A mass, in grams, of one mole of atoms, molecules, or formula units of a substance (9.6)

molar mass of a compound A mass, in grams, numerically equal to the formula mass of the compound (9.6)

molar mass of an element A mass, in grams, numerically equal to the atomic mass of the element, when the element is present in atomic form (9.6)

molar volume of a gas The volume occupied by one mole of a gas at a specified temperature and pressure (12.13)

molarity concentration A solution concentration unit that gives the number of moles of solute per liter of solution (13.8)

mole A counting unit based on the number 6.022×10^{23} (9.5); the amount of substance that contains as many particles (atoms, molecules, or formula units) as there are $^{12}_{6}C$ atoms in 12.000000 grams of $^{12}_{6}C$, which is 6.022×10^{23} (9.7)

mole fraction A dimensionless quantity that gives the ratio of the number of moles of a component in a mixture to the number of moles of all components present (12.16)

molecular formula A chemical formula that gives the actual number of atoms present in a formula unit of a compound (9.12)

molecular geometry A description of the three-dimensional arrangement of atoms within a molecule (7.17)

molecular polarity A measure of the degree of inequality in the attraction of bonding electrons to various locations within a molecule (7.20)

molecule A group of two or more atoms that functions as a unit because the atoms are tightly bound together (4.11)

monoatomic ion An ion formed from a single atom through loss or gain of electrons (7.9)

monoprotic acid An acid that can transfer only one H^+ ion (proton) per molecule during an acid–base reaction (14.4)

net ionic equation An ionic equation from which nonparticipating (spectator) species have been eliminated (15.3)

neutral solution A solution in which the concentrations of H_3O^+ and OH^- ions are equal (14.10)

neutralization The reaction between an acid and a hydroxide base to form a salt and water (14.7)

neutron A subatomic particle that is neutral, that is, has no charge (5.1)

noble gas A general name given to any element in group VIIIA of the periodic table (6.2)

noble gas element An element located in the far right column of the p area of the periodic table (6.11)

nonaqueous solution A solution in which a substance other than water is the solvent (13.2)

nonbonding electrons Pairs of valence electrons about an atom that are not involved in electron sharing (7.11)

nonelectrolyte A substance whose aqueous solution does not conduct electricity (15.2)

nonmetal An element characterized by the absence of the properties of luster, thermal conductivity, electrical conductivity, and malleability (6.11)

non-oxidation–reduction reaction A chemical reaction in which oxidation numbers do not change (15.6)

nonoxyacid A molecular compound composed of hydrogen and one or more nonmetals other than oxygen that produces H^+ ions in an aqueous solution (8.6)

nonpolar covalent bond A covalent bond in which there is equal sharing of bonding electrons between two atoms (7.19)

nonpolar molecule A molecule in which there is a symmetrical distribution of electronic charge (7.20)

normal boiling point The temperature of a liquid at which the liquid boils under a pressure of 760 mm Hg (11.15)

nuclear equation An equation in which the chemical symbols used represent atomic nuclei rather than atoms (5.9)

nucleon Any subatomic particle found in the nucleus of an atom (5.1)

nucleus The small, dense, positively charged center of an atom that contains an atom's protons and neutrons (5.1)

nuclide An atom with a specific atomic number and a specific mass number (5.6)

observation A statement that describes something we see, hear, smell, taste, or feel (1.4)

octet rule In forming compounds, atoms of elements lose, gain, or share electrons in such a way as to produce a noble gas electron configuration for each of the atoms involved (7.3)

orbital diagram A diagram that shows how many electrons an atom has in each of its occupied electron orbitals (6.8)

order of magnitude A single exponential value of the number 10 (2.7)

outer electrons The electrons in a condensed electron configuration given after the noble-gas core electrons (6.7)

oxidation The process whereby a substance in a chemical reaction loses one or more electrons (15.4)

oxidation number The charge that an atom appears to have when the electrons in each bond it is participating in are assigned to the more electronegative of the two atoms involved in the bond (15.5)

oxidation–reduction reaction A chemical reaction in which transfer of electrons between reactants occurs (15.4)

oxidizing agent The reactant in a redox reaction that causes oxidation by accepting electrons from the other reactant (15.4); the reactant in a redox reaction that contains the element that shows a decrease in oxidation number (15.5)

oxyacid A molecular compound composed of hydrogen, oxygen, and one or more other elements that produce H^+ ions in an aqueous solution (8.6)

paired electrons Two electrons of opposite spin present in the same orbital (6.8)

paramagnetic atom An atom that has an electron arrangement containing one or more unpaired electrons (6.8)

parent nuclide The nuclide that undergoes decay in a radioactive decay process (5.9)

part per billion concentration Concentration unit defined as number of parts of solute per billion parts of solution (13.7)

part per million concentration Concentration unit defined as number of parts of solute per million parts of solution (13.7)

partial pressure The pressure that a gas in a mixture would exert if it were present alone under the same conditions (12.16)

percent The number of items of a specified type in a group of 100 total items (3.10)

percent abundance The percent of atoms in a natural sample of a pure element that are a particular isotope of the element (5.3)

percent by mass concentration The mass of solute divided by the total mass of solution multiplied by 100 (13.6)

percent by mass of an element in a compound The number of grams of the element present in 100 grams of the compound (9.4)

percent by volume concentration The volume of solute divided by the total volume of solution multiplied by 100 (13.6)

percent composition of a compound The percent by mass of each element present in a compound (9.4)

percent error The ratio of the difference between a measured value and the accepted value for the measurement and the accepted value itself all multiplied by 100 (3.10)

percent ionic character of a bond A measure of how the actual charge separation (partial charge) in a bond that is due to electronegativity difference of the bonded atoms compares to the complete charge separation associated with ions (7.19)

percent purity The percent by mass of a specified substance in an impure sample of the substance (9.11)

percent yield The ratio of the actual yield of a product in a chemical reaction to its theoretical yield multiplied by 100 (10.11)

period A horizontal row of elements in the periodic table (6.2)

periodic law When elements are arranged in order of increasing atomic number, elements with similar chemical behavior occur at periodic (regularly recurring) intervals (6.1)

periodic table A tabular arrangement of the elements in order of increasing atomic number such that elements having similar chemical behavior are grouped in vertical columns (6.2)

pH The negative logarithm of the molar hydronium ion concentration (14.11)

pH scale A scale of small numbers that is used to specify molar hydronium ion concentration in an aqueous solution (14.11)

physical change A process in which a substance changes its physical appearance but not its chemical composition (4.4)

physical property A characteristic of a substance that can be observed without changing the substance into another substance (4.3)

polar covalent bond A covalent bond in which there is unequal sharing of bonding electrons between two atoms (7.19)

polar molecule A molecule in which there is an unsymmetrical distribution of electron charge (7.20)

polyatomic ion An ion formed from a group of atoms (held together by covalent bonds) through loss or gain of electrons (7.9)

polyprotic acid An acid that can transfer two or more H^+ ions (protons) per molecule during an acid–base reaction (14.4)

positron A particle with the same mass as an electron or a beta particle, but with a positive charge (5.11)

positron emission A radioactive decay process in which a positron is emitted from an unstable nucleus when a proton is converted to a neutron (5.11)

potential energy Stored energy that matter possesses as a result of its position, condition, and/or composition (11.3)

precision An indicator of how close a series of measurements of the same object are to each other (2.3)

pressure The force applied per unit area, that is, the total force on a surface divided by the area of that surface (12.2)

products Substances produced as a result of a chemical reaction (10.1)

properties The distinguishing characteristics of a substance, which are used in its identification and description (4.3)

proton A subatomic particle that possesses a positive electrical charge (5.1)

pure substance A single kind of matter that cannot be separated into other kinds of matter using physical means (4.5)

qualitative data Non-numerical data consisting of general observations about a system under study (1.4)

quantitative data Numerical data obtained by various measurements on a system under study (1.4)

quantized property A property that can have only certain values; that is, not all values are allowed (6.3)

radioactive atom An atom with an unstable nucleus from which radiation is spontaneously emitted (5.6)

radioactive decay The process whereby an unstable nucleus spontaneously gives off radiation (5.7)

radioactive decay series A sequence of nuclear reactions in which one radioactive nuclide decays to a second, which then decays to a third, and so on, until a stable nuclide is finally produced (5.13)

radioactivity The radiation spontaneously emitted from an unstable nuclide (5.6)

random error An error originating from uncontrollable variables in an experiment (2.3)

rate of a chemical reaction The rate at which reactants are consumed or products produced in a given time period in a chemical reaction (16.3)

reactants The starting substances that undergo change in a chemical reaction (10.1)

redox reaction A shortened designation for an oxidation–reduction reaction (15.4)

reducing agent The reactant in a redox reaction that causes reduction by providing electrons for the other reactant to accept (15.4); the reactant in a redox reaction that contains the element that shows a decrease in oxidation number (15.5)

reduction The process whereby a substance in a chemical reaction gains one or more electrons (15.4)

representative element An element located in the s area or the first five columns of the p area of the periodic table (6.11)

resonance structures Two or more Lewis structures for a molecule or polyatomic ion that have the same arrangement of atoms, contain the same number of electrons, and differ only in the location of the electrons (7.15)

reversible reaction A chemical reaction in which the conversion of reactants to products (the forward reaction) and the conversion of products to reactants (the reverse reaction) occur simultaneously (16.4)

rounding off Process of deleting unwanted (nonsignificant) digits from a calculated number (2.6)

salt An ionic compound containing a metal ion or polyatomic ion as the positive ion and a nonmetal ion or polyatomic ion (except hydroxide) as the negative ion (14.6)

saturated solution A solution that contains the maximum amount of solute that can be dissolved under the conditions at which the solution exists (13.2)

science The study in which humans attempt to organize and explain, in a systematic and logical manner, knowledge about themselves and their surroundings (1.1)

scientific discipline A branch of scientific knowledge limited in size and scope to make it more manageable (1.1)

scientific fact A reproducible piece of data about some phenomenon that is obtained from experimentation (1.4)

scientific hypothesis A model or statement that can be tested by experiment and offers an explanation for a scientific law (1.4)

scientific law A generalization that summarizes scientific facts about a natural phenomenon (1.4)

scientific method A set of procedures for acquiring knowledge and explaining phenomena (1.4)

scientific notation A numerical system in which numbers are expressed in the form $A \times 10^n$, where A is a number with a single nonzero digit to the left of the decimal point and n is a whole number (2.7)

scientific research The process of methodical investigation into a subject in order to discover new information about the subject (1.2)

scientific theory A scientific hypothesis that has been tested and validated over a long period of time (1.4)

semiconductor An element that does not conduct electrical current at room temperature but does so at higher temperatures (6.12)

SI base unit One of seven SI units of measurement from which all other SI measurement units can be derived (3.1)

SI-derived unit An SI unit derived by combining two or more SI base units (3.1)

SI system of units A particular choice of metric units that was adopted in 1960 as a standard for making metric system measurements (3.1)

significant figures Digits in any measurement that are known with certainty plus one digit that is uncertain (2.5)

single covalent bond A covalent bond in which two atoms share one pair of valence electrons (7.12)

single-replacement reaction A chemical reaction in which one element within a compound is replaced by another element (10.6)

solid The physical state characterized by both a definite shape and a definite volume (4.2); the physical state characterized by a dominance of potential energy (cohesive forces) or kinetic energy (disruptive forces) (11.4)

solubility The maximum amount of solute that will dissolve in a given amount of solvent under a fixed set of conditions (13.2)

solute A component of a solution present in a small amount relative to that of the solvent (13.1)

solution A homogeneous mixture of two or more substances, with each substance retaining its chemical identity (13.1)

solvent The component of a solution present in the greatest amount (13.1)

specific heat The amount of heat energy needed to raise the temperature of 1 gram of a substance in a specific physical state by 1°C (11.9)

stable nucleus A nucleus that does not easily undergo change (5.6)

standard molar volume of a gas The volume occupied by one mole of a gas, at STP conditions, that is 22.414 liters (12.13)

standard pressure for gases A pressure of 1 atm (760 mm Hg) (12.13)

standard temperature for gases A temperature of 0°C (273 K) (12.13)

STP conditions for gases The conditions of standard temperature and standard pressure (12.13)

strong acid An acid that, in an aqueous solution, transfers 100% or very nearly 100% of its acidic hydrogen atoms to water (14.5)

strong electrolyte A substance that ionizes/dissociates to a large extent (70%–100%) into ions in an aqueous solution (15.2)

subatomic particle A particle smaller than an atom that is a building block from which the atom is made (5.1)

surface tension A measure of the inward force on the surface of a liquid caused by unbalanced intermolecular forces (11.17)

supersaturated solution A solution that contains more dissolved solute than that needed for a saturated solution (13.2)

synthesis reaction A chemical reaction in which a single product is produced from two (or more) reactants (10.6)

systematic error An error originating from controllable variables in an experiment (2.3)

systematic name A name for a compound, based on IUPAC rules, that conveys information about the composition of the compound (8.5)

technology The application of scientific knowledge to the production of new products to improve human survival, comfort, and quality of life (1.2)

temperature A measure of the hotness or coldness of an object (3.11)

ternary compound A compound containing three different elements (8.4)

ternary ionic compound An ionic compound in which three elements are present, one element in a monoatomic ion and two other elements in a polyatomic ion (8.4)

ternary molecular compound A molecular compound that contains three different nonmetallic elements (8.6)

theoretical yield The maximum amount of a product that can be obtained from given amounts of reactants in a chemical reaction if no losses or inefficiencies of any kind occur (10.11)

thermal expansion A measure of the volume change resulting from a temperature change (11.2)

transition element An element located in the d area of the periodic table (6.11)

transmutation reaction A nuclear reaction in which a nuclide of one element is changed into a nuclide of another element (5.10)

triatomic molecule A molecule that contains three atoms (4.11)

triple covalent bond A covalent bond in which two atoms share three pairs of valence electrons (7.12)

triprotic acid An acid that can transfer three H^+ ions (protons) per molecule during an acid–base reaction (14.4)

unpaired electron A single electron in an orbital (6.8)

unsaturated solution A solution that contains less solute than the maximum amount that could dissolve (13.2)

unstable nucleus A nucleus that spontaneously undergoes change (5.6)

valence electron An electron in the outermost electron shell of a representative element or noble gas element (7.2)

vapor The gaseous state of a substance at a temperature and pressure at which the substance is normally a liquid or solid (11.13)

vapor pressure The pressure exerted by a vapor above a liquid when the liquid and vapor are in equilibrium (11.14)

variable-charge metal A metal that forms more than one type of positive ion, with the ion types differing in charge magnitude (8.2)

volatile substance A substance that readily evaporates at room temperature because of a high vapor pressure (11.14)

volume A measure of the amount of space occupied by an object (3.4)

VSEPR electron group A group of valence electrons present in a localized region about an atom in a molecule (7.17)

VSEPR theory A set of procedures for predicting the geometry of a molecule from the information contained in the molecule's Lewis structure (7.17)

weak acid An acid that, in an aqueous solution, transfers only a small percentage of its acidic hydrogen atoms to water (14.5)

weak electrolyte A substance that ionizes/dissociates to a small extent into ions in an aqueous solution (15.2)

weight A measure of the force exerted on an object by gravitational forces (3.3)

ANSWERS TO PRACTICE EXERCISES, SELECTED PRACTICE PROBLEMS, AND SELF-TEST QUESTIONS

Chapter 1

Practice Exercises

PE 1.1. (a) qualitative data (b) quantitative data (c) quantitative data **PE 1.2.** Scientific hypotheses 1 and 3 are disproved. **PE 1.3.** (a) scientific fact (b) scientific law (c) scientific hypothesis (d) scientific hypothesis

Practice Problems

1.1. (a) true (b) false (c) false (d) true **1.3.** (a) false (b) false (c) true (d) false **1.5.** (a) basic research (b) applied research (c) basic research (d) applied research **1.7.** c, b, e, a, and d **1.9.** (a) scientific hypothesis (b) nonscientific hypothesis (c) scientific hypothesis (d) nonscientific hypothesis **1.11.** (a) scientific hypothesis (b) scientific law (c) scientific fact (d) scientific hypothesis **1.13.** (a) false (b) false (c) true (d) true **1.15.** While a theory may not be an absolute answer, it is the best answer available. It may be supplanted only if repeated experimental evidence conclusively disproves it and a new theory is developed. **1.17.** (a) 4 is eliminated (b) 1 is eliminated (c) 1 and 4 are eliminated (d) 1 and 2 are eliminated **1.19.** (a) qualitative (b) quantitative (c) quantitative (d) qualitative **1.21.** The product of the pressure times the volume is a constant; or the pressure of the gas is inversely proportional to its volume; $P_1V_1 = P_2V_2$. **1.23.** Scientific laws are discovered by research. Researchers have no control over what the laws turn out to be. Societal laws are arbitrary conventions that can be and are changed by society when necessary. **1.25.** Publishing scientific data provides access to that data, enabling scientists to develop new theories based on a wide range of knowledge relating to a particular field. **1.27.** Conditions under which an experiment is conducted often affect the results of the experiment. If the experimental conditions are uncontrolled, the data from the experiment is not validated. **1.29.** A qualitative observation involves general nonnumerical information, and a quantitative observation involves numerical measurements.

Multiple-Choice Practice Test

1.31. a **1.32.** a **1.33.** d **1.34.** d **1.35.** e **1.36.** d **1.37.** b **1.38.** b **1.39.** c **1.40.** c

Chapter 2

Practice Exercises

PE 2.1. (a) Howard and Alice (b) Team B (c) The term *precision* does not apply to an individual count. (d) Team A **PE 2.2.** (a) ±0.01 quart (b) ±1 millimeter (c) ±0.1 degree (d) ±10 gram **PE 2.3.** (a) 4 (b) 2 (c) 2 (d) 2 **PE 2.4.** (a) 2, ±10,000 (b) 3, ±1000 (c) 5, ±10 (d) 6, ±1 **PE 2.5.** (a) 5 (b) 5 or 4 (100.31 or 99.87) (c) 3 or 2 (1.03 or 0.98) (d) 2 **PE 2.6.** (a) 25.1 (b) 25.7 (c) 0.000312 (d) 33,300 **PE 2.7.** (a) 2 (b) 2 (c) 1 (d) 3 **PE 2.8.** (a) 1270 (b) 970 (c) 73 (d) 7 **PE 2.9.** (a) 147 (b) 15.66 (c) 0.0200 (d) 430 **PE 2.10.** (a) 4.6 liters (b) 4.55 liters **PE 2.11.** (a) 6.6 $\times$ 10^{21} (b) 6 $\times$ 10^{-6} (c) 2.26 $\times$ 10^2 (d) 1.2 $\times$ 10^{10} **PE 2.12.** (a) 0.0012 (b) 16,000,000 (c) 160,000,000,000,000,000,000,000

(d) 0.00000000000013 **PE 2.13.** (a) ±100 (b) ±1 (c) ±0.01 (d) ±0.0001 **PE 2.14.** (a) 5.094 $\times$ 10^9 (b) 1.66 $\times$ 10^9 **PE 2.15.** 2.57 $\times$ 10^{-2} **PE 2.16.** (a) 4.26 $\times$ 10^4 (b) 5.8 $\times$ 10^{-18} **PE 2.17.** (a) 3.55 $\times$ 10^3 (b) 2.100 $\times$ 10^3 (c) 1.09 $\times$ 10^9

Practice Problems

2.1. (a) exact (b) exact (c) inexact (d) exact **2.3.** (a) exact (b) inexact (c) exact (d) exact **2.5.** The first 60 is an exact (defined) number, and the second 60 is a measured (inexact) number. **2.7.** Student A: low precision, low accuracy; Student B: high precision, high accuracy; Student C: high precision, low accuracy **2.9.** (a) 0.1 (b) 0.01 **2.11.** (a) 27 (b) 27.0 **2.13.** (a) ruler 4 (b) rulers 1 or 4 (c) ruler 2 (d) ruler 3 **2.15.** (a) ±0.001 (b) ±1 (c) ±0.0001 (d) ±0.1 **2.17.** The uncertainty in the first reading is ±0.1 second, and the uncertainty in the second reading is ±0.01 second. **2.19.** (a) 40,000–60,000 (b) 49,000–51,000 (c) 49,900–50,100 (d) 49,990–50,010 **2.21.** (a) 5 (b) 3 (c) 6 (d) 1 **2.23.** (a) 2 (b) 6 (c) 5 (d) 6 **2.25.** (a) 6 (b) 6 (c) 3 (d) 2 **2.27.** (a) 2, 0, 1, 0 (b) 2, 2, 2, 0 (c) 1, 0, 0, 2 (d) 0, 0, 0, 6 **2.29.** (a) the last 0 (b) the last 0 (c) the last 4 (d) the last 3 **2.31.** (a) ±0.01 (b) ±0.00000001 (c) ±100 (d) ±1,000,000 **2.33.** (a) yes (b) no (c) yes (d) yes **2.35.** (a) yes (b) no (c) yes (d) no **2.37.** (a) 23,000 (b) 23,$\overline{0}$00 (c) 23,000.0 (d) 23,000.000 **2.39.** (a) 3 (b) 4 (c) 2 (d) 5 **2.41.** four significant figures (0.9937, for example) or five significant figures (1.0022, for example) **2.43.** (a) 431.2 (b) 31.21 (c) 8.207 (d) 1.021 **2.45.** (a) 42.6 (b) 42.6 (c) 42.8 (d) 42.8 **2.47.** (a) 42,300 (b) 42,400 (c) 42,500 (d) 42,600 **2.49.** (a) 0.00033 (b) 0.012 (c) 0.20 (d) 0.36 **2.51.** (a) 42.3 (b) 42.0 (c) 42.6 (d) 42.3 **2.53.** (a) 0.351 (b) 653.9 (c) 22.556 (d) 0.2777 **2.55.** (a) 30,427.3 (b) 30,427 (c) 30,430 (d) 3$\overline{0}$,000 **2.57.** (a) 0.035 (b) 2.50 (c) 1,500,000 (d) 1$\overline{0}$0 **2.59.** (a) 0.12 (b) 120,000 (c) 12 (d) 0.00012 **2.61.** (a) 2 (b) 1 (c) 3 (d) 3 **2.63.** (a) 3 (b) 2 (c) 4 or more (d) 4 or more **2.65.** (a) 0.0299 (b) 140,000 (c) 1280 (d) 0.988 **2.67.** (a) 3.9 (b) 4.84 (c) 63 (d) 1 **2.69.** (a) tenths, ±0.1 (b) tenths, ±0.1 (c) units, ±1 (d) hundreds, ±100 **2.71.** (a) 162 (b) 9.3 (c) 1261 (d) 20.0 **2.73.** (a) 957.0 (b) 343 (c) 1200 (d) 132 **2.75.** (a) 2.7 (b) 2.9 (c) 0.2 (d) 46.4 **2.77.** (a) 267.3 (b) 260,000 (c) 201.3 (d) 3.8 **2.79.** (a) 170 (b) 123 (c) 131.2 (d) 0.18 **2.81.** (a) 168 (b) 123.3 (c) 131.19 (d) 0.185 **2.83.** (a) negative (b) positive (c) zero (d) positive **2.85.** (a) 2 (b) 4 (c) 3 (d) 1 **2.87.** (a) 4 (b) 5 (c) 4 (d) 6 **2.89.** (a) 2 (b) 5 (c) 2 (d) 7 **2.91.** (a) 4.732 $\times$ 10^2 (b) 1.234 $\times$ 10^{-3} (c) 2.3100 $\times$ 10^2 (d) 2.31 $\times$ 10^8 **2.93.** (a) 3.00 $\times$ 10^{-3} (b) 9.36 $\times$ 10^5 (c) 2.55 $\times$ 10^1 (d) 4.50 $\times$ 10^8 **2.95.** (a) 7 $\times$ 10^4 (b) 6.70 $\times$ 10^4 (c) 6.7000 $\times$ 10^4 (d) 6.700000 $\times$ 10^4 **2.97.** (a) 0.00230 (b) 4350 (c) 0.066500 (d) 111,000,000 **2.99.** (a) 3.42 $\times$ 10^6 (b) 2.36 $\times$ 10^{-3} (c) 3.2 $\times$ 10^2 (d) 1.2 $\times$ 10^{-4} **2.101.** (a) smaller (b) smaller (c) larger (d) larger **2.103.** (a) 10^1 (b) 10^3 (c) 10^3 (d) 10^{-3} **2.105.** (a) 3.65 $\times$ 10^5 (b) 3.6500 $\times$ 10^5 (c) 3.650000 $\times$ 10^5 (d) 3.65000000 $\times$ 10^5 **2.107.** (a) less than (b) the same (c) the same (d) greater than

2.109. (a) 10^8 (b) 10^{-8} (c) 10^2 (d) 10^{-2} **2.111.** (a) 2.992×10^8 (b) 9.1×10^2 (c) 2.7×10^{-9} (d) 2.7×10^{11} **2.113.** (a) 10^2 (b) 10^8 (c) 10^{-8} (d) 10^{-2} **2.115.** (a) 2.86×10^1 (b) 8.999×10^{17} (c) 3.49×10^{-2} (d) 1.111×10^{-18} **2.117.** (a) 10^1 (b) 10^{-1} (c) 10^{20} (d) 10^2 **2.119.** (a) 1.5×10^0 (b) 6.7×10^{-1} (c) 8.51×10^{-19} (d) 8×10^{15} **2.121.** (a) 4.42×10^3 (b) 9.30×10^{-2} (c) 9.683×10^5 (d) 1.919×10^4 **2.123.** (a) 7.713×10^7 (b) 8.253×10^7 (c) 8.307×10^7 (d) 8.313×10^7 **2.125.** (a) same (b) not the same (c) same (d) not the same **2.127.** (a) 3 (b) 4 (c) 4 (d) 4 **2.129.** (a) no (b) yes (c) yes (d) yes **2.131.** (a) 6.00×10^2 pounds (b) 6.000×10^2 pounds (c) 6.0×10^2 pounds (d) 6.00000×10^2 pounds **2.133.** (a) yes (b) no (c) no (d) no **2.135.** (a) 6.326×10^5 (b) 3.13×10^{-1} (c) 6.300×10^7 (d) 5.000×10^{-1} **2.137.** (a) 2 (b) 4 or more (c) 3 (d) 3 **2.139.** An exact number is a whole number; it cannot possess decimal digits. **2.141.** (a) 3.7×10^2 (b) 3.8×10^2 (c) 0.408 (d) 4.18 **2.143.** (a) 2.07×10^2, 243, 1.03×10^3 (b) 2.11×10^{-3}, 0.0023, 3.04×10^{-2} (c) 23,000, 9.67×10^4, 2.30×10^5 (d) 0.000014, 0.00013, 1.5×10^{-4}

Multiple-Choice Practice Test

2.145. d **2.146.** b **2.147.** d **2.148.** d **2.149.** b **2.150** c **2.151.** e **2.152.** a **2.153.** c **2.154.** d **2.155.** c **2.156.** b **2.157.** c **2.158.** c **2.159.** e **2.160.** e **2.161.** c **2.162.** b **2.163.** d **2.164.** e

Chapter 3

Practice Exercises

PE 3.1. (a) 10^3 (b) 10^{-2} (c) 10^9 (d) nano- (e) milli- (f) pico- **PE 3.2.** 1×10^{-3} m **PE 3.3.** 2.75×10^{-3} kg **PE 3.4.** 1.35×10^{24} cm^3 **PE 3.5.** 2790 miles **PE 3.6.** 227,000 g **PE 3.7.** 615 mm **PE 3.8.** (a) 450 cm^3 (b) 27 in.^{3} **PE 3.9.** 17,000 ft/qt **PE 3.10.** 0.53 g/cm^3 **PE 3.11.** (a) 59.9 g (b) 48.6 mL **PE 3.12.** 37.5 g **PE 3.13.** 1.8 hr **PE 3.14.** 1950 bushels **PE 3.15.** 54.8% **PE 3.16.** 11.7 lb **PE 3.17.** 7.00 **PE 3.18.** student 1, 34%; student 2, -6.8%; student 3, 23% **PE 3.19.** 41°C **PE 3.20.** 1391°F **PE 3.21.** (a) 21°C (b) 294 K

Practice Problems

3.1. (a) no (b) yes (c) no (d) yes **3.3.** (a) nano- (b) mega- (c) milli- (d) 10^3 (e) 10^{-2} (f) 10^{-6} **3.5.** (a) μg (b) km (c) cL (d) decimeter (e) milliliter (f) picogram **3.7.** (a) T, 10^{12} (b) nano-, 10^{-9} (c) G, 10^9 (d) micro-, μ **3.9.** (a) mass, kg (b) megameter, length (c) mass, ng (d) milliliter, volume **3.11.** (a) deci- (b) pico- (c) centi- (d) mega- **3.13.** (a) smaller by 10^2 (b) smaller by 10^3 (c) larger by 10^4 (d) smaller by 10^9 **3.15.** (a) length (b) area (c) volume (d) volume **3.17.** (a) 1 inch (b) 1 meter (c) 1 pound (d) 1 gallon **3.19.** (a) 20.4 cm^2 (b) 32 m^2 (c) 65.88 mm^2 (d) 8.2 mm^2 **3.21.** (a) 9.5 cm^3 (b) 1.4×10^2 cm^3 (c) 2.8×10^6 mm^3 (d) 3.7×10^2 cm^3 **3.23.** (a) equal (b) not equal (c) not equal (d) equal

3.25. (a) 24 hr = 1 day; $\dfrac{1 \text{ day}}{24 \text{ hr}}$; $\dfrac{24 \text{ hr}}{1 \text{ day}}$

(b) 60 sec = 1 min; $\dfrac{60 \text{ sec}}{1 \text{ min}}$; $\dfrac{1 \text{ min}}{60 \text{ sec}}$

(c) 10 decades = 1 century; $\dfrac{1 \text{ century}}{10 \text{ decades}}$; $\dfrac{10 \text{ decades}}{1 \text{ century}}$

(d) 365.25 days = 1 yr; $\dfrac{1 \text{ yr}}{365.25 \text{ days}}$; $\dfrac{365.25 \text{ days}}{1 \text{ yr}}$

3.27. (a) 2 pints = 1 quart (b) 3 feet = 1 yard (c) 2000 pounds = 1 ton (d) 36 inches = 1 yard

3.29. (a) $\dfrac{1 \text{ kL}}{10^3 \text{ L}}$; $\dfrac{10^3 \text{ L}}{1 \text{ kL}}$ (b) $\dfrac{1 \text{ mg}}{10^{-3} \text{ g}}$; $\dfrac{10^{-3} \text{ g}}{1 \text{ mg}}$

(c) $\dfrac{1 \text{ cm}}{10^{-2} \text{ m}}$; $\dfrac{10^{-2} \text{ m}}{1 \text{ cm}}$ (d) $\dfrac{1 \text{ }\mu\text{sec}}{10^{-6} \text{ sec}}$; $\dfrac{10^{-6} \text{ sec}}{1 \text{ }\mu\text{sec}}$

3.31. (a) 10^3 (b) 10^{-6} (c) 453.6 (d) 2.540 **3.33.** (a) four significant figures (b) exact (c) four significant figures (d) exact **3.35.** (a) exact (b) inexact (c) exact (d) exact **3.37.** (a) valid (b) not valid (c) valid (d) not valid **3.39.** (a) 3.70×10^2 dL (b) 3.70×10^{-2} m (c) 3.7×10^{-13} g (d) 3.7×10^5 L **3.41.** (a) 4.7×10^{10} mg (b) 5.00×10^{-7} cL (c) 6×10^{-7} dm (d) 3.7×10^{-14} km **3.43.** (a) 3.65×10^{-4} km^2 (b) 3.65×10^6 cm^2 (c) 3.65×10^4 dm^2 (d) 3.65×10^{-10} Mm^2 **3.45.** (a) 3.5×10^{10} mm^3 (b) 3.5×10^{37} pm^3 (c) 3.5×10^{-26} Gm^3 (d) 3.5×10^{-19} μm^3 **3.47.** (a) 6.0×10^{-4} m^2 (b) 7.2×10^{-9} m^3 (c) 2.5×10^{-9} dm^2 (d) 2.3×10^{34} nm^3 **3.49.** (a) 91.44 m (b) 9144 cm (c) 0.09144 km (d) 3.600×10^3 in. **3.51.** (a) 0.079 qt (b) 0.020 gal (c) 2.5 fl oz (d) 75 cm^3 **3.53.** (a) 6.0×10^{27} g (b) 6.0×10^{24} kg (c) 6.0×10^{36} ng (d) 2.1×10^{26} oz **3.55.** (a) 2.2×10^{-3} ft^3 (b) 8.0×10^{-5} yd^3 (c) 3.7 in.^{3} (d) 1.5×10^{-14} mi^3 **3.57.** (a) 5.2 cm^2 (b) 0.81 in.^{2} **3.59.** 51 ft^3 **3.61.** (a) 2.0×10^5 L/hr (b) 5.5×10^{-2} kL/sec (c) 3.3×10^4 dL/min (d) 4.8×10^9 mL/day **3.63.** (a) 5.7 μg/L (b) 5.7×10^{-9} g/mL (c) 5.7×10^{-9} mg/μL (d) 5.7×10^{-6} kg/kL **3.65.** (a) 0.789 g/mL (b) 7.18 g/mL (c) 0.916 g/mL (d) 1.49 g/mL **3.67.** (a) 37.7 g (b) 495 g (c) 59.5 g (d) 104 g **3.69.** (a) 17.1 mL (b) 0.912 mL (c) 13,600 mL (d) 17.5 mL **3.71.** 1.039 g/mL **3.73.** 61 lb **3.75.** 266 g **3.77.** (a) float, less dense than water (b) sink, more dense than water **3.79.** 36,300 g **3.81.** (a) equivalence (b) equality (c) equality (d) equivalence **3.83.** 5.1×10^2 mg antibiotic **3.85.** 138 lb **3.87.** (a) 5.29×10^{-6} g (b) 5.29×10^{-9} g (c) 5.01×10^{-6} g (d) 1.56×10^{-7} g **3.89.** (a) 1.3 min (b) 0.27 min (c) 0.15 min (d) 0.12 min **3.91.** 9.0 min **3.93.** (a) 109 bushels (b) $414.00 **3.95.** (a) $2.81 (b) $0.03 (c) $2110.10 (d) $131.94 **3.97.** (a) 17.6% (b) 29.4% (c) 70.6% (d) 52.9% **3.99.** (a) 95.05% copper (b) 4.95% zinc **3.101.** 22.3 g water **3.103.** 66.9 g salt **3.105.** 345 g **3.107.** three **3.109.** student 1 = -0.6%; student 2 = -0.8%; student 3 = 1.5% **3.111.** 35.4% **3.113.** (a) 32°F (b) 0°C (c) 273 K **3.115.** 36°F **3.117.** (a) 2466°F (b) 99.7°F (c) 19°F (d) 563°F **3.119.** (a) 733°C (b) 3.1°C (c) -22°C (d) 146°C **3.121.** (a) 548 K (b) 548.4 K (c) 548.88 K

(d) 408 K **3.123.** (a) 38°C (b) −361.1°F (c) 1077 K
(d) −321°F **3.125.** (14°F) therefore −10°C is a higher temperature than 10°F **3.127.** (−58°F) therefore −60°F is a lower temperature than 223 K **3.129.** (a) 2 (b) 3 (c) 5 (d) 4
3.131. (a) 6.301 km (b) 1.442 sec (c) 1.327 mg (d) 2.1 cL
3.133. (a) 10^3 (b) 10^{-12} (c) 10^{-9} (d) 10^{-1} **3.135.** 198 cm
3.137. (a) 14.1 L (b) 14.08 L (c) 14.081 L (d) 14.0814 L
3.139. 290,131,200 beats **3.141.** 497 g **3.143.** 661 mL
3.145. (a) 500 mg/dL, yes, a life-threatening situation
(b) 0.50 mg/dL, no, not a life-threatening situation
3.147. 0.19 g **3.149.** 1.7×10^{-2} mm
3.151. 2.2×10^7 worms **3.153.** −160°H
3.155. 122 bears/bag **3.157.** (a) 5.000×10^{-1} g/mL
(b) 5.0×10^{-1} g/mL (c) 5.0000×10^{-1} g/mL
(d) 5.000×10^{-1} g/mL **3.159.** (a) 8×10^3 cm^3
(b) 8.0×10^3 cm^3 (c) 8.00×10^3 cm^3 (d) 8.0×10^3 cm^3
3.161. (a) 3.256×10^3 g has the greatest precision.
(b) 3.34 mg has the greatest precision. (c) Each has the same precision. (d) 3.2500 g has the greatest precision.
3.163. (a) not equivalent, differ in significant figures
(b) equivalent (c) not equivalent (d) not equivalent

Multiple-Choice Practice Test

3.165. b **3.166.** e **3.167.** d **3.168.** c **3.169.** d **3.170.** a
3.171. c **3.172.** a **3.173.** c **3.174.** c **3.175.** d **3.176.** b
3.177. b **3.178.** b **3.179.** e **3.180.** b **3.181.** c **3.182.** e
3.183. e **3.184.** a

Chapter 4
Practice Exercises
PE 4.1. (a) chemical property (b) physical property
(c) chemical property (d) chemical property
PE 4.2. (a) chemical (b) physical (c) physical (d) physical
PE 4.3. first box, mixture; second box, compound
PE 4.4. (a) homogeneous mixture (b) homogeneous
mixture (c) element (d) compound
PE 4.5. (a) tetraatomic, heteroatomic, compound
(b) diatomic, homoatomic, element (c) tetraatomic, heteroatomic, compound (d) triatomic, heteroatomic, compound **PE 4.6.** (a) element (b) homogeneous mixture
(c) heterogeneous mixture (d) compound **PE 4.7.** (a) 2 H,
1 S, 4 O (b) 17 C, 20 H, 4 N, 6 O (c) 1 Ca, 4 H, 2 P, 8 O

Practice Problems
4.1. (a) shape (indefinite vs. definite) (b) indefinite shape
4.3. (a) does not take shape of container; definite volume
(b) takes shape of container; does not have a definite volume
(c) does not take shape of container; definite volume (d) takes
shape of container; definite volume **4.5.** (a) solid
(b) liquid (c) liquid (d) gas **4.7.** (a) chemical (b) chemical
(c) physical (d) physical **4.9.** (a) chemical (b) physical
(c) physical (d) physical **4.11.** (a) extensive (b) intensive
(c) intensive (d) intensive **4.13.** (a) differ in extensive
properties (amount) (b) differ in intensive properties (substance) (c) differ in intensive properties (temp) (d) differ in
extensive (amount) and intensive properties (temp)
4.15. (a) physical (b) physical (c) chemical (d) physical
4.17. (a) chemical (b) physical (c) chemical (d) physical
4.19. (a) physical (b) physical (c) chemical (d) chemical

4.21. (a) physical (b) physical (c) chemical (d) physical
4.23. (a) freezing (b) condensation (c) sublimation
(d) evaporation **4.25.** (a) heterogeneous mixture,
homogeneous mixture (b) homogeneous mixture, pure
substance **4.27.** (a) false (b) true (c) false (d) true
4.29. (a) heterogeneous mixture (b) homogeneous mixture
(c) homogeneous mixture (d) heterogeneous mixture
4.31. (a) homogeneous mixture, 1 phase (b) homogeneous
mixture, 1 phase (c) heterogeneous mixture, 2 phases
(d) heterogeneous mixture, 2 phases **4.33.** (a) chemically
homogeneous, physically homogeneous (b) chemically
heterogeneous, physically homogeneous (c) chemically
heterogeneous, physically heterogeneous
(d) chemically heterogeneous, physically heterogeneous
4.35. (a) compound (b) compound (c) no classification
possible (d) no classification possible **4.37.** (a) true
(b) false (c) false (d) false **4.39.** (a) A (no classification
possible), B (no classification possible), C (compound)
(b) D (compound), E (no classification possible), F (no
classification possible), G (no classification possible)
4.41. First box, mixture; second box, compound
4.43. (a) applies (b) applies (c) applies (d) does not apply
4.45. (a) false (b) true (c) false (d) true **4.47.** (a) true
(b) true (c) false (d) true **4.49.** (a) more abundant (b) less
abundant (c) less abundant (d) more abundant
4.51. (a) nitrogen (b) nickel (c) lead (d) tin (e) Al (f) Ne
(g) H (h) U **4.53.** (a) Na, S (b) Mg, Mn (c) Ca, Cd (d) As,
Ar **4.55.** (a) boron (b) barium (c) beryllium (d) bismuth
(e) berkelium (f) bromine **4.57.** (a) fluorine (b) zinc
(c) potassium (d) sulfur **4.59.** (a) iron, Fe (b) tin, Sn
(c) sodium, Na (d) gold, Au **4.61.** (a) Re-Be-C-Ca
(b) Ra-Y-Mo-Nd (c) Na-N-C-Y (d) Br-U-Ce or B-Ru-Ce
(e) S-H-Ar-O-N (f) Al-I-Ce **4.63.** (a) consistent
(b) consistent (c) not consistent (d) consistent
4.65. (a) heteratomic, compound (b) heteroatomic, compound (c) homoatomic, triatomic, element (d) heteroatomic,
compound **4.67.** (a) false, molecules must contain two or
more atoms (b) true (c) false, the structural unit for some
compounds is molecules and for other ions (d) false, the
diameter is approximately 10^{-10} meters **4.69.** (a) element
(b) mixture (c) mixture (d) mixture **4.71.** (a) AB_2 (b) BC_2
(c) B_2C_2 (d) A_3B_2C **4.73.** (a) compound (b) compound
(c) element (d) element **4.75.** (a) $C_{20}H_{30}$ (b) H_2SO_4
(c) HCN (d) $KMnO_4$ **4.77.** (a) same, both 3 (b) same, both
5 (c) fewer, 5 and 7 (d) more, 17 and 13 **4.79.** (a) NO is a
compound containing N and O; No is an element. (b) Cs_2 is
a diatomic molecule of the element Cs; CS_2 is a compound
containing C and S. (c) $CoBr_2$ is a compound containing Co
and Br; $COBr_2$ is a compound containing Co, O, and Br.
(d) H is an atom of the element H; H_2 is a diatomic molecule
of the element H. **4.81.** (a) H_3PO_4 (b) $SiCl_4$ (c) NO_2 (d) H_2O_2
4.83. (a) HCN (b) H_2SO_4 **4.85.** (a) colorless gas, odorless
gas, colorless liquid, boils at 43°C (b) toxic to humans, Ni
reacts with CO **4.87.** (a) compound (b) mixture
(c) compound (d) mixture **4.89.** (a) homogeneous mixture
(b) element (c) heterogeneous mixture (d) homogeneous
mixture **4.91.** (a) heterogeneous mixture (b) heterogeneous
mixture (c) homogeneous mixture (d) heterogeneous mixture

4.93. (a) and (c)   **4.95.** alphabetized: Bh, Bi, Bk, Cf, Co, Cs, Cu, Hf, Ho, Hs, In, Nb, Ni, No, Np, Os, Pb, Po, Pu, Sb, Sc, Si, Sn, Yb   **4.97.** (a) 2 (b) 3 (c) 3 (d) 6   **4.99.** (a) 4 (b) 2 (c) 4 (d) 6 (e) 54   **4.101.** (a) solid (b) state determination not possible (c) gas (d) state determination not possible   **4.103.** (a) 75.0% (b) 12.7% (c) 29.1% (d) 24.7%   **4.105.** (a) It is likely the students were all working with the same substances. (The densities are the same.) (b) No, density alone will not distinguish between elements and compounds.   **4.107.** compound   **4.109.** 1.90×10^{23} Au atoms   **4.111.** 2.84×10^9 miles

Multiple-Choice Practice Test
4.113. e   **4.114.** c   **4.115.** d   **4.116.** c   **4.117.** a   **4.118.** d   **4.119.** d   **4.120.** c   **4.121.** d   **4.122.** a   **4.123.** a   **4.124.** b   **4.125.** d   **4.126.** c   **4.127.** a   **4.128.** b   **4.129.** e   **4.130.** c   **4.131.** c   **4.132.** a

Chapter 5
Practice Exercises
PE 5.1. (a) 34 protons (b) 34 electrons (c) 46 neutrons (d) $^{80}_{34}\text{Se}$
PE 5.2. (a) 40 (b) 27 (c) 27 (d) +13   **PE 5.3.** (a) 28, 29, and 30
(b) 3.09%   **PE 5.4.** (a) isotopes (b) isobars (c) neither
(d) isobars   **PE 5.5.** Z = 12 splats, X = 6 splats, and Q = 3
splats   **PE 5.6.** 0.075 ppm   **PE 5.7.** 52.00 amu
PE 5.8. 0.575 g   **PE 5.9.** 83.4 days
PE 5.10. (a) $^{212}_{85}\text{At} \longrightarrow {}^{4}_{2}\alpha + {}^{208}_{83}\text{Bi}$
(b) $^{72}_{31}\text{Ga} \longrightarrow {}^{0}_{-1}\beta + {}^{72}_{32}\text{Ge}$
PE 5.11. (a) $^{91}_{42}\text{Mo} \longrightarrow {}^{0}_{1}\beta + {}^{91}_{41}\text{Nb}$ (b) $^{75}_{34}\text{Se} + {}^{0}_{-1}e \longrightarrow {}^{75}_{33}\text{As}$

Practice Problems
5.1. (a) electron (b) proton (c) proton, neutron (d) neutron
5.3. (a) false (b) false (c) false (d) true   **5.5.** 1837   **5.7.** (a) 50
(b) 47 (c) nickel (d) iodine   **5.9.** (a) 24, 53 (b) 44, 103 (c) 101,
256 (d) 16, 34   **5.11.** (a) 5 (b) 8 (c) 13 (d) 18   **5.13.** (a) 11 (b) 16
(c) 27 (d) 40   **5.15.** (a) 11 (b) 16 (c) 27 (d) 40   **5.17.** (a) +5
(b) +8 (c) +13 (d) +18   **5.19.** (a) $^{11}_{5}\text{B}$ (b) $^{16}_{8}\text{O}$ (c) $^{27}_{13}\text{Al}$ (d) $^{40}_{18}\text{Ar}$
5.21. (a) atomic number (b) atomic number and mass
number (c) mass number (d) atomic number and
mass number   **5.23.** (a) nitrogen (N) (b) aluminum (Al)
(c) barium (Ba) (d) gold (Au)   **5.25.** (a) $^{60}_{28}\text{Ni}$, 28, 32
(b) $^{37}_{18}\text{Ar}$, 18, 19 (c) $^{90}_{38}\text{Sr}$, 38, 38 (d) 92, 235, 92, 143
5.27. (a) 27 protons, 27 electrons, 32 neutrons (b) 45 protons,
45 electrons, 58 neutrons (c) 69 protons, 69 electrons,
100 neutrons (d) 9 protons, 9 electrons, 10 neutrons
5.29. (a) 34 (b) 23 (c) 23 (d) +11   **5.31.** (a) $^{31}_{15}\text{P}$ (b) $^{18}_{8}\text{O}$ (c) $^{54}_{24}\text{Cr}$
(d) $^{197}_{79}\text{Au}$ **5.33.** (a) 24 (b) 13 (c) 25 (d) 12   **5.35.** (a) 43 (b) 43
(c) 60 (d) 30   **5.37.** (a) same total number of subatomic parti-
cles, 60 (b) same number of neutrons, 16 (c) same number of
neutrons, 12 (d) same number of electrons, 3
5.39. (a) scandium (b) chlorine (c) strontium (d) arsenic
5.41. (a) no (b) yes (c) no (d) yes   **5.43.** $^{54}_{26}\text{Fe}$, $^{56}_{26}\text{Fe}$, $^{57}_{26}\text{Fe}$, $^{58}_{26}\text{Fe}$
5.45. $^{96}_{40}\text{Zr}$, $^{94}_{40}\text{Zr}$, $^{92}_{40}\text{Zr}$, $^{91}_{40}\text{Zr}$, $^{90}_{40}\text{Zr}$ **5.47.** (a) 54, 56, 57, and 58
(b) 2.19%   **5.49.** (a) not isotopes (b) not isotopes (c) not iso-
topes   **5.51.** (a) not the same (b) same (c) not the same
(d) same   **5.53.** (a) isotopes (b) isobars (c) neither
(d) isotopes   **5.55.** (a) isotopes (b) isobars (c) neither
(d) isotopes   **5.57.** (a) $^{31}_{15}\text{P}$, $^{33}_{15}\text{P}$, $^{35}_{15}\text{P}$ (b) $^{31}_{15}\text{P}$, $^{32}_{15}\text{P}$, $^{33}_{15}\text{P}$
(c) $^{31}_{15}\text{P}$, $^{32}_{15}\text{P}$, $^{34}_{15}\text{P}$ (d) $^{30}_{15}\text{P}$, $^{32}_{15}\text{P}$, $^{34}_{15}\text{P}$ **5.59.** $^{40}_{18}\text{Ar}$, $^{40}_{19}\text{K}$, $^{40}_{20}\text{Ca}$
5.61. (a) 59 (b) 40 (c) $^{18}_{9}\text{F}$ (d) $^{18}_{8}\text{O}$ **5.63.** (a) 55.85 amu

(b) 14.01 amu (c) calcium (d) iodine   **5.65.** Values on this
hypothetical relative mass scale are
Q = 8.00 bebs, X = 4.00 bebs, and Z = 2.00 bebs.
5.67. (a) Z = 9 amu, X = 27 amu, Q = 108 amu (b) Z is Be,
X is Al, Q is Ag   **5.69.** 224.8 lb   **5.71.** 35.46 amu
5.73. 47.88 amu   **5.75.** (a) ^{14}N (b) ^{51}V (c) ^{121}Sb (d) ^{193}Ir
5.77. (a) possible (b) possible (c) possible (d) impossible
5.79. (a) 75%, 25% (b) 49 amu, 75% (c) 80%, 20% (d) 49 amu,
20%   **5.81.** (a) 0.415 (b) 0.415 (c) 0.849 (d) 1.12
5.83. (a) iron, Fe (b) strontium, Sr (c) beryllium, Be
(d) helium, He   **5.85.** The number 12.0000 amu applies only
to ^{12}C isotope. The number 12.011 amu applies to naturally
occurring carbon, a mixture of ^{12}C, ^{13}C, and ^{14}C.
5.87. (a) true (b) false (c) false (d) false   **5.89.** spontaneous
emission of radiation   **5.91.** 209   **5.93.** no difference
5.95. (a) $^{9}_{5}\text{B}$ or boron-9 (b) $^{44}_{19}\text{K}$ or potassium-44 (c) $^{96}_{45}\text{Rh}$ or
rhodium-96 (d) $^{182}_{73}\text{Ta}$ or tantalum-182   **5.97.** (a) $^{14}_{7}\text{N}$
(b) $^{197}_{79}\text{Au}$ (c) rubidium-92 (d) tin-121   **5.99.** (a) 1/4
(b) 1/32 (c) 1/8 (d) 1/64   **5.101.** (a) 4 (b) 6 (c) 5 (d) 9
5.103. (a) 2.0 g (b) 0.50 g (c) 0.062 g (d) 10.0 g
5.105. (a) 7.50 g (b) 9.84 g (c) 9.98 g (d) 10.0 g
5.107. (a) 16 hr (b) 32 hr (c) 4.0×10^{-1} hr (d) 56 hr
5.109. (a) $^{0}_{2}\alpha$ (b) $^{0}_{-1}\beta$ (c) $^{0}_{0}\gamma$ **5.111.** two protons and two
neutrons   **5.113.** (a) $^{200}_{84}\text{Po} \longrightarrow {}^{4}_{2}\alpha + {}^{196}_{82}\text{Pb}$
(b) $^{244}_{96}\text{Cm} \longrightarrow {}^{4}_{2}\alpha + {}^{240}_{94}\text{Pu}$ (c) $^{240}_{96}\text{Cm} \longrightarrow {}^{4}_{2}\alpha + {}^{236}_{94}\text{Pu}$
(d) $^{238}_{92}\text{U} \longrightarrow {}^{4}_{2}\alpha + {}^{234}_{90}\text{Th}$ **5.115.** (a) $^{10}_{4}\text{Be} \longrightarrow {}^{0}_{-1}\beta + {}^{10}_{5}\text{B}$
(b) $^{77}_{32}\text{Ge} \longrightarrow {}^{0}_{-1}\beta + {}^{77}_{33}\text{As}$ (c) $^{60}_{26}\text{Fe} \longrightarrow {}^{0}_{-1}\beta + {}^{60}_{27}\text{Co}$
(d) $^{25}_{11}\text{Na} \longrightarrow {}^{0}_{-1}\beta + {}^{24}_{12}\text{Mg}$ **5.117.** (a) decreases by 4
(b) does not change (c) does not change   **5.119.** (a) alpha
particle decay (b) beta particle decay   **5.121.** (a) $^{0}_{-1}\beta$ (b) $^{125}_{52}\text{Te}$
(c) $^{4}_{2}\alpha$ (d) $^{229}_{90}\text{Th}$ **5.123.** (a) $^{199}_{79}\text{Au} \longrightarrow {}^{0}_{-1}\beta + {}^{199}_{80}\text{Hg}$
(b) $^{120}_{48}\text{Cd} \longrightarrow {}^{0}_{-1}\beta + {}^{120}_{49}\text{In}$ (c) $^{152}_{67}\text{Ho} \longrightarrow {}^{4}_{2}\alpha + {}^{148}_{65}\text{Tb}$
(d) $^{226}_{88}\text{Ra} \longrightarrow {}^{4}_{2}\alpha + {}^{222}_{86}\text{Rn}$ **5.125.** (a) $^{4}_{2}\alpha$ (b) $^{2}_{1}\text{H}$ (c) $^{81}_{34}\text{Se}$
(d) $^{9}_{4}\text{Be}$ **5.127.** (a) $^{9}_{4}\text{Be} + {}^{4}_{2}\alpha \longrightarrow {}^{12}_{6}\text{C} + {}^{1}_{0}\text{n}$
(b) $^{58}_{28}\text{Ni} + {}^{1}_{1}\text{H} \longrightarrow {}^{55}_{27}\text{Co} + {}^{4}_{2}\alpha$
(c) $^{113}_{48}\text{Cd} + {}^{1}_{0}\text{n} \longrightarrow {}^{114}_{48}\text{Cd} + {}^{0}_{0}\gamma$
(d) $^{27}_{13}\text{Al} + {}^{4}_{2}\alpha \longrightarrow {}^{30}_{15}\text{P} + {}^{1}_{0}\text{n}$
5.129. nine   **5.131.** Mass number does not change; atomic
number decreases by one.   **5.133.** $^{0}_{-1}\beta$ emission
5.135. (a) $^{29}_{15}\text{P} \longrightarrow {}^{0}_{1}\beta + {}^{29}_{14}\text{Si}$ (b) $^{112}_{51}\text{Sb} \longrightarrow {}^{0}_{1}\beta + {}^{112}_{50}\text{Sn}$
(c) $^{46}_{23}\text{V} \longrightarrow {}^{0}_{1}\beta + {}^{46}_{22}\text{Ti}$ (d) $^{132}_{58}\text{Ce} \longrightarrow {}^{0}_{1}\beta + {}^{132}_{57}\text{La}$
5.137. (a) $^{76}_{36}\text{Kr} + {}^{0}_{-1}e \longrightarrow {}^{76}_{35}\text{Br}$ (b) $^{122}_{54}\text{Xe} + {}^{0}_{-1}e \longrightarrow {}^{122}_{53}\text{I}$
(c) $^{100}_{46}\text{Pd} + {}^{0}_{-1}e \longrightarrow {}^{100}_{45}\text{Rh}$ (d) $^{175}_{73}\text{Ta} + {}^{0}_{-1}e \longrightarrow {}^{175}_{72}\text{Hf}$
5.139. (a) $^{0}_{1}\beta$ (b) $^{0}_{-1}e$ (c) $^{103}_{47}\text{Ag}$ (d) $^{133}_{55}\text{Cs}$ **5.141.** beta decay
5.143. (a) 1.32 before; 1.24 after (b) 1.46 before; 1.47 after
(c) 1.39 before; 1.43 after (d) 1.18 before; 1.23 after
5.145. (a) $^{87}_{36}\text{Kr}$, beta-particle emission; $^{74}_{36}\text{Kr}$, positron
emission; n/p ratio is higher for $^{87}_{36}\text{Kr}$ (b) $^{84}_{34}\text{Se}$, beta-particle
emission; $^{63}_{33}\text{As}$, positron emission; n/p ratio is higher for
$^{84}_{34}\text{Se}$ (c) $^{74}_{31}\text{Ga}$, beta-particle emission; $^{64}_{31}\text{Ga}$, positron emission; n/p ratio is higher for $^{74}_{31}\text{Ga}$ (d) $^{99}_{41}\text{Nb}$, beta-particle
emission; $^{99}_{46}\text{Pd}$, positron emission; n/p ratio is higher for
$^{99}_{41}\text{Nb}$ **5.147.** stable; a decay series always ends with a
stable nuclide   **5.149.** $^{234}_{91}\text{Pa}$ **5.151.** (1) $^{220}_{86}\text{Rn} \rightarrow {}^{216}_{84}\text{Po} + {}^{4}_{2}\alpha$
(2) $^{216}_{84}\text{Po} \rightarrow {}^{212}_{82}\text{Pb} + {}^{4}_{2}\alpha$ (3) $^{212}_{82}\text{Pb} \rightarrow {}^{212}_{83}\text{Bi} + {}^{0}_{-1}\beta$
(4) $^{212}_{83}\text{Bi} \rightarrow {}^{212}_{84}\text{Po} + {}^{0}_{-1}\beta$ **5.153.** (a) false (b) false (c) true

(d) true **5.155.** (a) $^{37}_{18}$Ar, $^{39}_{19}$K, $^{42}_{20}$Ca, $^{44}_{21}$Sc, $^{43}_{22}$Ti
(b) $^{44}_{21}$Sc, $^{42}_{20}$Ca, $^{43}_{22}$Ti, $^{39}_{19}$K, $^{37}_{18}$Ar (c) $^{37}_{18}$Ar, $^{39}_{19}$K, $^{42}_{20}$Ca, $^{44}_{21}$Sc, $^{43}_{22}$Ti
(d) $^{44}_{21}$Sc, $^{43}_{22}$Ti, $^{42}_{20}$Ca, $^{39}_{19}$K, $^{37}_{18}$Ar **5.157.** (a) $^{8}_{5}$B (b) $^{12}_{5}$B (c) $^{15}_{5}$B
(d) $^{16}_{5}$B **5.159.** (a) 29 and 29 (b) 29 and 29 (c) 34 and 36
5.161. (a) 13 (b) 27 (c) 14 (d) 27.0 amu **5.163.** (a) chromium
(b) thorium (c) chromium (d) zirconium **5.165.** (a) 36.756
new amu (b) 67.117 new amu **5.167.** (a) radon-221
(b) potassium-47 (c) krypton-78 (d) argon-37
5.169. (a) $^{228}_{88}$Ra, $^{228}_{89}$Ac, $^{228}_{90}$Th (b) $^{228}_{90}$Th, $^{224}_{88}$Ra, $^{220}_{86}$Rn
5.171. phosphorus-28, positron emission; phosphorus-34,
beta-particle emission **5.173.** A, negligible amount (zero); B,
approximately 0.250 mole; C, negligible amount (zero); D,
approximately 0.750 mole **5.175.** (a) 7.0 days (b) 3.80 g Q
5.177. (a) isobars (b) neither (c) isotopes (d) isotopes
5.179. 82 **5.181.** 672 protons **5.183.** 2.4 miles

Multiple-Choice Practice Test
5.185. e **5.186.** d **5.187.** e **5.188.** c **5.189.** c **5.190.** d
5.191. c **5.192.** c **5.193.** e **5.194.** c **5.195.** e **5.196.** c
5.197. c **5.198.** b **5.199.** b **5.200.** b **5.201.** d **5.202.** b
5.203. a **5.204.** a

Chapter 6
Practice Exercises
PE 6.1. (a) S (b) Be (c) Cl (d) Xe **PE 6.2.** (a) 2 (b) 1 (c) 10 (d) 2
PE 6.3. (a) $1s^22s^22p^63s^23p^64s^1$ (b) $1s^22s^22p^63s^23p^64s^23d^8$
PE 6.4. (a) [He]$2s^22p^5$ (b) [Ne] $3s^23p^2$ (c) [Ar]$4s^2$
(d) [Kr]$5s^24d^{10}$
PE 6.5.

Answer:

PE 6.6. (a) p for Kr, s for Ba (b) p^6 for Kr, s^2 for Ba (c) 4 for
Kr, 6 for Ba **PE 6.7.** $1s^22s^22p^63s^23p^64s^23d^{10}4p^65s^24d^{10}5p^2$
PE 6.8. (a) Ar (b) C (c) Ne (d) Be

Practice Problems
6.1. (a) Al (b) Be (c) Sn (d) K **6.3.** (a) same group (b) same
period (c) neither same group nor same period (d) same
period **6.5.** (a) $_{19}$K, $_{37}$Rb (b) $_{15}$P, $_{33}$As (c) $_9$F, $_{53}$I
(d) $_{11}$Na, $_{55}$Cs **6.7.** (a) group (b) periodic law (c) periodic
law (d) period **6.9.** (a) bromine (b) lithium (c) argon
(d) strontium **6.11.** (a) 3 (b) 4 (c) 4 (d) 4 **6.13.** (a) 3 (b) 1
(c) 2 (d) 2 **6.15.** (a) 2 (b) 10 (c) 6 (d) 14 **6.17.** (a) 2 (b) 2
(c) 2 (d) 2 **6.19.** (a) 1 (b) 1 (c) 5 (d) 3 **6.21.** (a) $3d$ subshell
(b) $2p$ subshell (c) $3p$ subshell (d) third shell **6.23.** (a) true
(b) false (c) true (d) true (e) false **6.25.** (a) spherical
(b) dumbbell (c) cloverleaf (d) spherical **6.27.** (a) allowed
(b) allowed (c) not allowed (d) not allowed **6.29.** (a) $3p$
(b) $5s$ (c) $4d$ (d) $4p$ **6.31.** (a) $3s$ (b) $3p$ (c) $4f$ (d) $3d$
6.33. (a) $1s, 2s, 3s, 4s,$ (b) $4s, 4p, 4d, 4f$ (c) $2p, 6s, 4f, 5d$ (d) $3p, 4s,$
$3d, 5p$ **6.35.** (a) $1s^22s^22p^63s^23p^1$ (b) $1s^22s^22p^3$ (c) $1s^22s^22p^6$
(d) $1s^22s^22p^63s^23p^3$ **6.37.** (a) $1s^22s^22p^63s^23p^64s^23d^6$
(b) $1s^22s^22p^63s^23p^64s^23d^{10}4p^65s^1$
(c) $1s^22s^22p^63s^23p^64s^23d^{10}4p^65s^24d^{10}5p^5$
(d) $1s^22s^22p^63s^23p^64s^23d^{10}4p^65s^24d^{10}5p^66s^24f^{14}5d^{10}6p^6$
6.39. (a) $_{10}$Ne (b) $_{19}$K (c) $_{22}$Ti (d) $_{30}$Zn **6.41.** (a) [Ne]$3s^23p^1$
(b) [Ar]$4s^1$ (c) [Ar]$4s^23d^{10}$ (d) [Kr]$5s^24d^{10}5p^2$
6.43. (a) [He]$2s^22p^4$ (b) [Ne]$3s^2$ (c) [Ar]$4s^2$ (d) [Ar]$4s^23d^{10}4p^5$

6.45. (a) magnesium (b) chlorine (c) tellurium (d) barium
6.47. (a) 10 (b) 10 (c) 36 (d) 54 **6.49.** (a) 2 (b) 7 (c) 16 (d) 2
6.51. (a) (b)
(c)
(d)

6.53. (a)
(b)
(c)
(d)

6.55. (a)
(b) (c)
(d)

6.57. (a) (b) (c)
(d)

6.59. (a) 2 (b) 1 (c) 0 (d) 2 **6.61.** (a) paramagnetic
(b) paramagnetic (c) diamagnetic (d) paramagnetic
6.63. (a) no (b) yes (c) no (d) yes **6.65.** (a) p area (b) d area
(c) d area (d) s area **6.67.** (a) p subshell (b) d subshell
(c) d subshell (d) s subshell **6.69.** (a) $3p$ subshell
(b) $3d$ subshell (c) $4d$ subshell (d) $6s$ subshell **6.71.** (a) 3
(b) 3 (c) 2 (d) 2 **6.73.** (a) Al (b) Li (c) La (d) Sc **6.75.**
(a) Kr (b) Li (c) K (d) Lu **6.77.** (a) $1s^22s^22p^63s^23p^64s^23d^{10}4p^3$
(b) $1s^22s^22p^63s^23p^64s^23d^{10}4p^65s^24d^{10}5p^3$
(c) $1s^22s^22p^63s^23p^64s^23d^{10}4p^65s^1$
(d) $1s^22s^22p^63s^23p^64s^23d^{10}4p^65s^24d^{10}$ **6.79.** (a) $1s^22s^22p^63s^2$
(b) $1s^22s^22p^63s^23p^64s^2$ (c) $1s^22s^22p^63s^23p^64s^23d^{10}$
(d) $1s^22s^22p^63s^23p^64s^23d^{10}4p^65s^24d^{10}5p^2$ **6.81.** (a) 2 (b) 8
(c) 10 (d) 10 **6.83.** (a) 4 (b) 9 (c) 2 (d) 0 **6.85.** (a) metallic
(b) nonmetallic (c) metallic (d) nonmetallic **6.87.** (a) no
(b) no (c) yes (d) yes **6.89.** (a) more metals (b) more metals
(c) more nonmetals (d) more nonmetals **6.91.** (a) Li (b) K
(c) Fe (d) Hg **6.93.** (a) metal (b) nonmetal (c) good
conductor (d) good conductor **6.95.** (a) transition element (b) representative element (c) noble gas (d) transition

element   **6.97.** (a) $_1$H (b) $_2$He (c) $_3$Li (d) $_{58}$Ce   **6.99.** (a) 4 (b) 7 (c) 6 (d) 1   **6.101.** (a) 1 (b) 1 (c) 3 (d) 2   **6.103.** (a) false (b) true (c) false   **6.105.** (a) Ge (b) B (c) Po (d) Te   **6.107.** (a) Rb (b) Ti (c) Se (d) Ba   **6.109.** (a) F (b) P (c) Zn (d) Cl   **6.111.** (a) Bi (b) Be (c) K (d) Ne   **6.113.** (a) B (b) Ga (c) K (d) Rb   **6.115.** (a) N (b) Ra (c) Kr (d) Li   **6.117.** (a) The $2s$ subshell can hold a maximum of two electrons. (b) The $2p$ subshell must be completely filled (six electrons) before the $3s$ is filled. (c) The $2p$ subshell is filled after the $2s$ subshell. (d) The $3p$ and $4s$ subshells fill after the $3s$.   **6.119.** (a) period 3, group IA (b) period 3, group IIIA (c) period 4, group IIIB (d) period 4, group VIIA   **6.121.** (a) 1 (b) 1 (c) 1 (d) 1   **6.123.** (a) paramagnetic (b) paramagnetic (c) paramagnetic (d) paramagnetic   **6.125.** (a) $x = 6, y = 2$ (b) $x = 2, y = 1$ (c) $x = 2, y = 10$ (d) $x = 2, y = 6$ **6.127.** (a) [Ne]$3s^2$ and [Kr]$5s^2$ (b) [Ar]$4s^23d^3$ and [Xe]$6s^24f^{14}5d^3$ (c) [He]$2s^22p^5$ and [Ar]$4s^23d^{10}4p^5$ (d) [Ar]$4s^23d^{10}4p^1$ and [Xe]$6s^24f^{14}5d^{10}6p^1$ **6.129.** (a) [Ar]$4s^1$ and [Ar]$4s^24p^1$ (b) [Kr]$5s^24d^2$ and [Kr]$5s^24d^4$ (c) [Ne]$3s^23p^4$ and [Ne]$3s^23p^6$ (d) [Kr]$5s^24d^{10}$ and [Kr]$5s^24d^{10}5p^2$ **6.131.** (a) 6 (Be, Mg, Ca, Sr, Ba, Ra) (b) 2 (N, P) (c) 2 (V, Nb) (d) 2 (Br, I)   **6.133.** (a) $1s^22s^22p^63s^23p^2$ (b) $1s^22s^22p^2$ (c) $1s^22s^22p^3$ (d) $1s^22s^22p^3$ **6.135.** (a) Be (b) Be (c) Ne (d) Ar   **6.137.** (a) 2 (b) 4 (c) 1 (d) 2   **6.139.** (a) IVA, VIA (b) VB, middle column of VIIIB (c) IVB, last column of VIIIB (d) IA   **6.141.** (a) Po (b) Cr (c) elements 88–116 and 118 (d) elements 12–18   **6.143.** (a) F (b) Ag   **6.145.** (a) O, Li, He, B, K (b) O, He, B (the nonmetals) (c) O, Li, He, B, Sr, K (all of them) (d) K   **6.147.** the same   **6.149.** 54   **6.151.** +38   **6.153.** (a) 22 (b) 72.7% (c) 54.5% (d) 50.0%   **6.155.** P, S, Cl, Se, Br

Multiple-Choice Practice Test

6.157. b   **6.158.** a   **6.159.** a   **6.160.** d   **6.161.** e   **6.162.** d   **6.163.** c   **6.164.** c   **6.165.** d   **6.166.** e   **6.167.** b   **6.168.** c   **6.169.** a   **6.170.** d   **6.171.** d   **6.172.** e   **6.173.** c   **6.174.** b   **6.175.** c   **6.176.** c

Chapter 7
Practice Exercises

PE 7.1. (a) 4 (b) 5 (c) 1   **PE 7.2.** (a) Na K Rb (b)
PE 7.3. (a) Mg^{2+} (b) Cl^- **PE 7.4.** (a) 11 protons, $\cdot\ddot{N}:$ $\cdot\ddot{O}:$ $\cdot\ddot{F}:$
trons (b) 34 protons, 36 electrons   **PE 7.5.** (a) two electron gain, O^{2-} (b) one electron gain, F^- (c) three electron loss, Al^{3+} (d) two electron loss, Ca^{2+}

PE 7.6. (a)

PE 7.7. (a) Na_2S (b) MgF_2 (c) Ca_3N_2 **PE 7.8.** (a) Na_2SO_4 (b) KNO_3 (c) NaOH (d) $(NH_4)_2S$

PE 7.9. (a)

PE 7.10. (a) SiF_4 (b) OCl_2

PE 7.11.

PE 7.12.

PE 7.13.

PE 7.14. (a) trigonal pyramidal (b) trigonal pyramidal (c) tetrahedral about each C atom   **PE 7.15.** (a) F (b) O (c) Li (d) N   **PE 7.16.** (a) polar covalent, partial positive charge on N, partial negative charge on F (b) polar covalent, partial positive charge on Si, partial negative charge on Cl (c) ionic (d) polar covalent, partial positive charge on Al, partial negative charge on P   **PE 7.17.** (a) polar (b) polar (c) polar (d) nonpolar

Practice Problems

7.1. (a) 5 (b) 2 (c) 2 (d) 6   **7.3.** (a) 4 (b) 3 (c) 6 (d) 2   **7.5.** (a) more (b) same number (c) more (d) fewer   **7.7.** (a) 1 (b) 1 (c) 2 (d) 1   **7.9.** (a) IA, 1 (b) VIIIA, 8 (c) IIA, 2 (d) VIIA, 7   **7.11.** (a) C (b) F (c) Mg (d) P   **7.13.** (a) C (b) Si (c) Cl (d) Ba   **7.15.** (a) B (b) C (c) F (d) Li   **7.17.** (a) incorrect (b) correct (c) incorrect (d) correct   **7.19.** (a) fewer (b) same number (c) more (d) same number   **7.21.** (a) Li^+ (b) P^{3-} (c) Br^- (d) Ba^{2+} **7.23.** (a) Be, 2e (b) I, I^- (c) Al^{3+}, 13p (d) S, 18e, 16p   **7.25.** (a) neutral (b) positively charged (c) positively charged (d) negatively charged   **7.27.** (a) Al^{3+} (b) O^{2-} (c) Mg^{2+} (d) Be^{2+} **7.29.** (a) anion (b) cation (c) not an ion (d) anion   **7.31.** (a) negative ion (b) positive ion (c) positive ion (d) negative ion   **7.33.** (a) anion (b) cation (c) cation (d) anion   **7.35.** (a) 3e gained (b) 2e lost (c) 2e lost (d) 1e gained   **7.37.** (a) loss (b) loss (c) loss (d) gain   **7.39.** (a) $1s^22s^22p^6$ (b) $1s^22s^22p^6$ (c) $1s^22s^22p^63s^23p^6$ (d) $1s^22s^22p^63s^23p^6$ **7.41.** (a) [Ne] (b) [Ne] (c) [Ar] (d) [Ar]   **7.43.** (a) 4 (b) 3 (c) 1 (d) 2   **7.45.** (a) IA (b) VIA (c) VA (d) IVA   **7.47.** (a) He (b) Xe (c) Kr (d) Xe   **7.49.** (a) Na^+ (b) F^- (c) O^{2-} (d) Mg^{2+} **7.51.** (a) yes (b) no (c) yes (d) no

7.53. (a)

(b)

(c)

(d)

7.55. (a) Lithium has one valence electron (lose one); nitrogen has five valence electrons (gain three).

(b) Mg has two valence electrons (lose two); O has six valence electrons (gain two).

$$Mg\cdot\ \ddot{O}: \longrightarrow Mg^{2+}\ \ :\ddot{O}:^{2-}$$

(c) Cl has seven valence electrons (gain one); barium has two valence electrons (lose two).

$$Ba \quad \overset{\cdot\ddot{C}l:}{\underset{\cdot\ddot{C}l:}{}} \longrightarrow Ba^{2+} \quad \begin{matrix}:\ddot{C}l:^{-}\\ :\ddot{C}l:^{-}\end{matrix}$$

(d) F has seven valence electrons (gain one); K has one valence electron (lose one).

$$K\cdot\ \ \ddot{F}: \longrightarrow K^{+}\ \ :\ddot{F}:^{-}$$

7.57. (a) $CaCl_2$ (b) BeO (c) AlN (d) K_2S **7.59.** (a) CaF_2, CaO, Ca_3N_2, Ca_2C (b) AlF_3, Al_2O_3, AlN, Al_4C_3 (c) AgF, Ag_2O, Ag_3N, Ag_4C (d) ZnF_2, ZnO, Zn_3N_2, Zn_2C
7.61. (a) Mg^{2+}, S^{2-} (b) Al^{3+}, N^{3-} (c) 2 Na^+, O^{2-}
(d) 3 Ca^{2+}, 2 N^{3-} **7.63.** (a) X_2Z (b) XZ_3 (c) X_3Z (d) ZX_2
7.65. (a) Be_3N_2 (b) NaBr (c) SrF_2 (d) Al_2S_3
7.67. (a) $Mg(CN)_2$ (b) $CaSO_4$ (c) $Al(OH)_3$ (d) NH_4NO_3
7.69. (a) $AlPO_4$ (b) $Al_2(CO_3)_3$ (c) $Al(ClO_3)_3$ (d) $Al(C_2H_3O_2)_3$
7.71. (a) KOH, KCN, KNO_3, K_2SO_4 (b) $Mg(OH)_2$, $Mg(CN)_2$, $Mg(NO_3)_2$, $MgSO_4$ (c) $Al(OH)_3$, $Al(CN)_3$, $Al(NO_3)_3$, $Al_2(SO_4)_3$
(d) NH_4OH, NH_4CN, NH_4NO_3, $(NH_4)_2SO_4$
7.73. (a) $:\ddot{I}:\ddot{I}:$ (b) $:\ddot{C}l:\ddot{F}:$ (c) $H:\ddot{S}:H$ (d) $:\ddot{F}:\ddot{P}:\ddot{F}:$ with $:\ddot{F}:$ below

7.75. (a) $H:\ddot{B}r:$ (b) $:\ddot{F}:\ddot{O}:\ddot{F}:$ (c) $:\ddot{C}l:\ddot{N}:\ddot{C}l:$ with $:\ddot{C}l:$ below (d) $:\ddot{I}:$ with $:\ddot{I}:\ddot{S}i:\ddot{I}:$ and $:\ddot{I}:$ below

7.77. (a) three bonding pairs, two nonbonding pairs (b) six bonding pairs, zero nonbonding pairs (c) four bonding pairs, four nonbonding pairs (d) seven bonding pairs, two nonbonding pairs   **7.79.** (a) one triple (b) four single, one double (c) two double (d) two triple, one single
7.81. (a) yes (b) yes (c) no, C has to form four bonds (d) yes   **7.83.** a bond in which both electrons of the shared pair come from one of the two atoms   **7.85.** (a) the N—O bond (b) none (c) the O—Cl bond (d) the two O—Br bonds   **7.87.** Resonance structures are two or more Lewis structures for a molecule or ion that have the same arrangement of atoms, contain the same number of electrons, and differ only in the location of the electrons.
7.89.

$$\left[:\ddot{O}=N-\ddot{O}:\right]^{-} \longleftrightarrow \left[:\ddot{O}-N=\ddot{O}:\right]^{-}$$
with $:\ddot{O}:$ below N in both

7.91. (a) 24 (b) 26 (c) 24 (d) 32   **7.93.** (a) 24 (b) 26 (c) 24 (d) 32
7.95. (a) 1b, 3nb (b) 3b, 2nb (c) 1b, 6nb (d) 1b, 3nb
7.97. (a)

H, H on top; H—C—C—H; H, H on bottom

(b) H—$\ddot{N}$—$\ddot{N}$—H with H, H below each N

(c) :$\ddot{F}$—C—$\ddot{F}$: with H above and H below C

(d) H, :$\ddot{C}l$: on top; H—C—C—$\ddot{C}l$:; H, :$\ddot{C}l$: on bottom

7.99. (a)

$$\left[H-\underset{\overset{|}{H}}{\overset{\overset{H}{|}}{N}}-H\right]^{+}$$

(b)

$$\left[H-\underset{\overset{|}{H}}{\overset{\overset{H}{|}}{Be}}-H\right]^{2-}$$

(c)

$$\left[:\ddot{O}-\underset{\overset{|}{\underset{\ddot{O}:}{}}}{Cl}-\ddot{O}:\right]^{-}$$

(d)

$$\left[:\ddot{O}-\underset{\overset{|}{\underset{\ddot{O}:}{}}}{I}-\ddot{O}:\right]^{-}$$

7.101. (a) :$\ddot{C}l$—C=C—H with :$\ddot{C}l$: and H below the carbons

(b) H—C—C≡N: with H above and H below the first C

(c) H—C≡C—C—H with H above and H below the last C

(d) :$\ddot{F}$—$\ddot{N}$=$\ddot{N}$—$\ddot{F}$:

7.103. Follow the steps used in Problems 7.97–7.102.
(a)

H—C—N=$\ddot{O}$ with H, H on C and :$\ddot{O}$: below N ⟷ H—C—N—$\ddot{O}$: with H, H on C and :O: double bonded below N

(b) :N≡N—$\ddot{O}$: ⟷ :$\ddot{N}$=N=$\ddot{O}$: ⟷ :$\ddot{N}$—N≡O:

(c)

$$\left[\ddot{O}=C-\ddot{O}:\right]^{2-} \longleftrightarrow \left[:\ddot{O}-C-\ddot{O}:\right]^{2-} \longleftrightarrow \left[:\ddot{O}-C=\ddot{O}\right]^{2-}$$
with $:\ddot{O}:$ / $:O:$ below C

(d) $\left[:\ddot{S}-C≡N:\right]^{-} \longleftrightarrow \left[\ddot{S}=C=\ddot{N}\right]^{-} \longleftrightarrow \left[:S≡C-\ddot{N}:\right]^{-}$

7.105. Total electrons in structure = 32; total valence electrons = 4(6) + 7 = 31; 31 − 32 = −1; $n = 1$.   **7.107.** (a) angular (b) linear (c) angular (d) linear (diatomic molecule)   **7.109.** (a) The electron arrangement about the central atom involves two bonding locations and two nonbonding locations. This arrangement produces an angular geometry. (b) The electron arrangement about the central atom involves two bonding locations and zero nonbonding locations. This arrangement produces a linear geometry. (c) The electron arrangement about the central atom involves two bonding locations and zero nonbonding locations. This arrangement produces a linear geometry. (d) The electron arrangement about the central atom involves two bonding locations and one nonbonding location. This arrangement produces an angular geometry.   **7.111.** (a) The electron arrangement about the central atom involves three bonding locations and one nonbonding location. This arrangement produces a trigonal pyramidal geometry. (b) The electron arrangement about the central atom involves two bonding locations and zero nonbonding locations. This arrangement produces a linear geometry. (c) The electron arrangement about the central atom involves two bonding locations and one nonbonding location. This arrangement produces an angular geometry. (d) The electron arrangement about the central atom involves four bonding locations and zero nonbonding locations. This arrangement produces a tetrahedral geometry. **7.113.** (a) The electron arrangement about the central atom involves four bonding locations and zero nonbonding locations. This arrangement produces a tetrahedral geometry. (b) The electron arrangement about the central atom involves two bonding locations and zero nonbonding locations. This arrangement produces a linear geometry. (c) The electron arrangement about the central atom involves three bonding locations and one nonbonding location. This arrangement produces a trigonal pyramidal geometry. (d) The electron arrangement about the central atom involves three bonding locations and one nonbonding location. This arrangement produces a trigonal pyramidal geometry. **7.115.** (a) Each central atom has three bonding locations and one nonbonding location, giving a trigonal pyramid for each center. The centers are joined by one axis of each trigonal pyramid. (b) The C has four bonding locations and zero nonbonding locations; tetrahedral. The O has two bonding locations and two nonbonding locations; angular. **7.117.** increases across a period; decreases down a group **7.119.** (a) O (b) Be (c) C (d) Ca **7.121.** (a) F, O, B, Li (b) F, Cl, S, P (c) N, P, As, Sb (d) F, Si, Mg, Fr **7.123.** (a) Br, N, Cl, O, F (b) Na, Rb, K, Cs, Fr, Ba, Ra (c) F, O, N, Cl (d) 0.5 unit **7.125.** (a) true (b) false (c) false (d) false **7.127.** (a) S (b) Br (c) N (d) N **7.129.** (a) H—O has the greatest polarity (b) O—Al has the greatest polarity (c) B—N has the greatest polarity (d) Al—Cl has the greatest polarity **7.131.** (a) H—O (b) O—Al (c) B—N (d) Al—Cl **7.133.** (a) H—O (b) O—Al (c) B—N (d) Al—Cl **7.135.** (a) H—O (b) O—Al (c) B—N (d) Al—Cl **7.137.** (a) polar covalent (b) ionic (c) polar covalent (d) nonpolar covalent **7.139.** (a) nonpolar covalent (b) polar covalent (c) could be polar covalent or ionic (d) ionic **7.141.** (a) more covalent character (b) more covalent character (c) more ionic character (d) more ionic character **7.143.** (a) more covalent character (b) more ionic character (c) more covalent character (d) more ionic character **7.145.** (a) nonpolar (b) polar (c) polar (d) polar **7.147.** (a) nonpolar (b) polar (c) polar (d) nonpolar **7.149.** (b and c). In (b) both are polar molecules. In (c) both are nonpolar molecules. In (a) and (b) one is polar and the other is nonpolar. **7.151.** (a) 3 (b) 2 (c) 2 (d) 2 **7.153.** (a) Mg: $1s^2 2s^2 2p^6 3s^2$; Mg^{2+}: $1s^2 2s^2 2p^6$ (b) F: $1s^2 2s^2 2p^5$; F^-: $1s^2 2s^2 2p^6$ (c) N: $1s^2 2s^2 2p^3$; N^{3-}: $1s^2 2s^2 2p^6$ (d) Ca^{2+}: $1s^2 2s^2 2p^6 3s^2 3p^6$; S^{2-}: $1s^2 2s^2 2p^6 3s^2 3p^6$ **7.155.** (a) nonisoelectronic cations (b) nonisoelectronic anions (c) isoelectronic cations (d) nonisoelectronic anions **7.157.** (a) ionic (b) molecular (c) ionic (d) molecular **7.159.** (a) molecule (b) formula unit (c) formula unit (d) molecule **7.161.** (a) monoatomic only (b) both monoatomic and polyatomic (c) polyatomic only (d) monoatomic only **7.163.** (a) yes (b) yes (c) no (d) no **7.165.** (a) O (b) N (c) O (d) F **7.167.** (a) correct number of electron dots, but improperly placed (b) not enough electron dots (c) correct number of electron dots, but improperly placed (d) correct number of electron dots, but improperly placed **7.169.** BA, CA, DB, and DA **7.171.** (a) same, both single (b) different, single and triple (c) different, triple and

double (d) different, triple and double **7.173.** (a) x = Cl (b) x = Cl
7.175.

$$Ca^{2+} \left[\begin{array}{c} :\ddot{O}: \\ | \\ :\ddot{O}—S—\ddot{O}: \\ | \\ :\ddot{O}: \end{array} \right]^{2-} ; \left[\begin{array}{c} H \\ | \\ H—N—H \\ | \\ H \end{array} \right]^{+} \left[\begin{array}{c} :\ddot{O}—N—\ddot{O}: \\ \| \\ \ddot{O}: \end{array} \right]^{-}$$

Note that other resonance structures are possible for the NO_3^- ion.
7.177. (a) tetrahedral electron-pair geometry; tetrahedral molecular geometry (b) tetrahedral electron-pair geometry; trigonal pyramidal molecular geometry (c) linear electron-pair geometry; linear molecular geometry (d) tetrahedral electron-pair geometry; tetrahedral molecular geometry **7.179.** (a) 109.5° because the electron-group arrangement about the oxygen atom is tetrahedral (b) 120° because the electron-group arrangement about the carbon atom is trigonal planar **7.181.** (a) 2 (b) 0 (c) 0 (d) 0 **7.183.** (a) strontium (b) nitrogen (c) calcium (d) gallium **7.185.** A = Al, D = N, formula = AlN **7.187.** D = Ca, A = S, formula = CaS **7.189.** A = O and D = F. The Lewis structure of the compound is $:\ddot{F}:\ddot{O}:\ddot{F}:$ **7.191.** A = H and D = O; x is 2 and y is 1; the formula is H_2O. **7.193.** A = Be and D = F; the Lewis structure is

$$\left[\begin{array}{c} :\ddot{F}: \\ :\ddot{F}:Be:\ddot{F}: \\ :\ddot{F}: \end{array} \right]^{2-}$$

Multiple-Choice Practice Test
7.195. a **7.196.** d **7.197.** c **7.198.** b **7.199.** c **7.200.** c
7.201. b **7.202.** e **7.203.** c **7.204.** d **7.205.** c **7.206.** c
7.207. d **7.208.** d **7.209.** b **7.210.** c **7.211.** c **7.212.** b
7.213. d **7.214.** b

Chapter 8
Practice Exercises
PE 8.1. (a) molecular (b) ionic (c) ionic (d) molecular
PE 8.2. (a) magnesium chloride (b) potassium nitride (c) aluminum oxide (d) barium sulfide **PE 8.3.** (a) CaS (b) KF (c) Mg_3P_2 (d) K_2O **PE 8.4.** (a) lead(II) sulfide (b) lead(IV) sulfide (c) cobalt(II) chloride (d) copper(I) fluoride **PE 8.5.** (a) FeS (b) Fe_2S_3 (c) $PbCl_4$ (d) $PbCl_2$
PE 8.6. (a) ferric oxide (b) aurous sulfide (c) copper(II) bromide (d) lead(IV) chloride **PE 8.7.** (a) calcium carbonate (b) aluminum sulfate (c) magnesium nitrate (d) ammonium chloride **PE 8.8.** (a) NH_4Br (b) Ag_3PO_4 (c) Cu_3PO_4 (d) $Mg(ClO_4)_2$ **PE 8.9.** (a) BrI (b) SO_3 (c) NH_3 (d) NO
PE 8.10. (a) sulfur dioxide (b) silicon tetrabromide (c) dichlorine monoxide (d) dinitrogen trioxide
PE 8.11. (a) nitrous acid (b) phosphorous acid (c) hydrocyanic acid (d) chloric acid **PE 8.12.** (a) HNO_3 (b) H_2SO_3 **PE 8.13.** (a) disulfur monoxide (b) beryllium phosphate (c) carbonic acid (d) calcium sulfide
PE 8.14. (a) $AgNO_3$ (b) $SnCl_4$ (c) N_2O_5 (d) K_2CO_3

Practice Problems
8.1. (a) molecular (b) ionic (c) ionic (d) molecular
8.3. (a) ionic (b) ionic (c) ionic (d) ionic **8.5.** (a) yes (b) yes (c) no, both molecular (d) no, both ionic **8.7.** (a) binary

(b) not binary (c) binary (d) binary **8.9.** (a) not ionic (b) ionic (c) ionic (d) not ionic **8.11.** (a) not binary ionic (b) not binary ionic (c) binary ionic (d) not binary ionic **8.13.** (a) fixed charge (b) variable charge (c) fixed charge (d) variable charge **8.15.** (a) and (d) **8.17.** (a) variable charge (b) variable charge (c) fixed charge (d) fixed charge **8.19.** (a) magnesium ion (b) potassium ion (c) zinc ion (d) silver ion **8.21.** (a) copper(II) ion (b) copper(I) ion (c) cobalt(III) ion (d) cobalt(II) ion **8.23.** (a) bromide ion (b) nitride ion (c) sulfide ion (d) chloride ion **8.25.** (a) Zn^{2+} (b) Pb^{2+} (c) Ca^{2+} (d) N^{3-} **8.27.** (a) magnesium oxide (b) lithium sulfide (c) silver chloride (d) zinc bromide **8.29.** (a) fixed-charge (b) variable-charge (c) fixed-charge (d) variable-charge **8.31.** (a) not needed (b) needed (c) not needed (d) needed **8.33.** (a) +2 (b) +2 (c) +3 (d) +2 **8.35.** (a) iron(II) oxide (b) nickel(II) chloride (c) gold(III) oxide (d) cobalt(II) nitride **8.37.** (a) iron(II) bromide, iron(III) bromide (b) copper(I) sulfide, copper(II) sulfide (c) tin(II) sulfide, tin(IV) sulfide (d) nickel(II) oxide, nickel(III) oxide **8.39.** (a) aluminum oxide (b) cobalt(III) fluoride (c) silver nitride (d) barium sulfide **8.41.** (a) plumbic oxide (b) auric chloride (c) iron(III) iodide (d) tin(II) bromide **8.43.** In b and c both names denote the same compound **8.45.** (a) Li_2S (b) ZnS (c) Al_2S_3 (d) Ag_2S **8.47.** (a) CuS (b) Cu_3N_2 (c) SnO (d) SnO_2 **8.49.** (a) Al_2O_3, aluminum oxide (b) Pb^{2+}, Br^-, lead(II) bromide (c) Fe^{2+}, FeS (d) Zn^{2+}, Br^-, $ZnBr_2$ **8.51.** (a) OH^- (b) NH_4^+ (c) NO_3^- (d) ClO_4^- **8.53.** (a) peroxide (b) thiosulfate (c) oxalate (d) chlorate **8.55.** (a) SO_4^{2-}, SO_3^{2-} (b) PO_4^{3-}, HPO_4^{2-} (c) OH^-, O_2^{2-} (d) CrO_4^{2-}, $Cr_2O_7^{2-}$ **8.57.** (a) +1 (b) +2 (c) +2 (d) +1 **8.59.** (a) fixed-charge (b) fixed-charge (c) variable-charge (d) variable-charge **8.61.** (a) zinc sulfate (b) barium hydroxide (c) iron(III) nitrate (d) copper(II) carbonate **8.63.** (a) iron(III) carbonate, iron(II) carbonate (b) gold(I) sulfate, gold(III) sulfate (c) tin(II) hydroxide, tin(IV) hydroxide (d) chromium(III) acetate, chromium(II) acetate **8.65.** (a) Ag_2CO_3 (b) $AuNO_3$ (c) $Fe_2(SO_4)_3$ (d) $CuCN$ **8.67.** (a) −2 (b) −1 (c) −2 (d) −1 **8.69.** (a) $(NH_4)_2SO_4$ (b) NH_4CN (c) Na_2O_2 (d) KN_3 **8.71.** (a) four different elements present (b) ternary ionic (c) binary ionic (d) binary ionic **8.73.** (a) $Al_2(SO_4)_3$, aluminum sulfate (b) Cu^{2+}, CN^-, copper(II) cyanide (c) Fe^{2+}, $Fe(OH)_2$ (d) Zn^{2+}, N_3^-, $Zn(N_3)_2$ **8.75.** The less electronegative element is written first. (a) CO (b) Cl_2O (c) N_2O (d) HI **8.77.** (a) 7 (b) 5 (c) 3 (d) 10 **8.79.** (a) tetraphosphorus decoxide (b) sulfur tetrafluoride (c) carbon tetrabromide (d) chlorine dioxide **8.81.** (a) SO (b) S_2O (c) SO_2 (d) S_7O_2 **8.83.** (a) hydrogen sulfide (b) hydrogen fluoride (c) ammonia (d) methane **8.85.** (a) PH_3 (b) HBr (c) C_2H_6 (d) H_2Te **8.87.** A polyatomic ion (SO_3^{2-}) is present **8.89.** (a) no (b) yes (c) yes (d) no **8.91.** (a) NO_3^- (b) I^- (c) ClO^- (d) PO_3^{3-} **8.93.** (a) HCN (b) H_2SO_4 (c) HNO_2 (d) H_3BO_3 **8.95.** (a) nitric acid (b) hydroiodic acid (c) hypochlorous acid (d) phosphorous acid **8.97.** (a) hydrocyanic acid (b) sulfuric acid (c) nitrous acid (d) boric acid **8.99.** (a) HCl (b) H_2CO_3 (c) $HClO_3$ (d) H_2SO_4 **8.101.** (a) H^+, HNO_2, nitrous acid (b) H^+, PO_4^{3-}, phosphoric acid (c) CN^-, HCN (d) H^+, ClO^-, $HClO$ **8.103.** (a) arsenous acid (b) periodic acid

(c) hypophosphorous acid (d) bromous acid **8.105.** (a) hydrogen bromide (b) hydrocyanic acid (c) hydrogen sulfide (d) hydroiodic acid **8.107.** selenium dibromide **8.109.** (a) hydrogen fluoride (b) magnesium sulfide (c) lithium nitride (d) sulfur dibromide **8.111.** (a) no (b) no (c) yes (d) no **8.113.** (a) yes (b) no (c) no (d) yes **8.115.** (a) sulfate (b) perchlorate (c) peroxide (d) dichromate **8.117.** (a) Ca_3N_2 (b) $Ca(NO_3)_2$ (c) $Ca(NO_2)_2$ (d) $Ca(CN)_2$ **8.119.** (a) K_3P (b) K_3PO_4 (c) K_2HPO_4 (d) KH_2PO_4 **8.121.** (a) N_2O, CO_2 (b) NO_2, SO_2 (c) SF_2, SCl_2 (d) N_2O_3 **8.123.** (a) $CaCO_3$, HNO_2 (b) $NaClO_4$, $NaClO_3$ (c) $HClO_2$, $HClO$ (d) Li_2CO_3, Li_3PO_4 **8.125.** (a) sodium nitrate (b) aluminum sulfide, magnesium nitride, beryllium phosphide (c) iron(III) oxide (d) gold(I) chlorate **8.127.** (a) $Ni_2(SO_4)_3$ (b) Ni_2O_3 (c) $Ni_2(C_2O_4)_3$ (d) $Ni(NO_3)_3$ **8.129.** The superoxide ion is O_2^-. The nitronium ion is NO_2^+. The formula of nitronium superoxide is NO_2O_2. **8.131.** (a) ternary molecular (b) binary molecular (c) binary ionic (d) binary molecular **8.133.** (a) metal (b) nonmetal (c) X_3Y_2 (d) beryllium phosphide **8.135.** (a) sodium sulfate (b) sodium oxide (c) sodium nitride (d) sodium bromide **8.137.** (a) magnesium chloride, $MgCl_2$ (b) oxygen difluoride, OF_2 **8.139.** (a) $x = 4$; silicon tetrachloride (b) $x = 2$; magnesium chloride (c) $x = 3$; potassium nitride (d) $x = 3$; nitrogen trichloride **8.141.** beryllium bromate **8.143.** aluminum nitride **8.145.** carbon dioxide **8.147.** beryllium cyanide

Multiple-Choice Practice Test
8.149. d **8.150.** a **8.151.** d **8.152.** c **8.153.** b **8.154.** c **8.155.** d **8.156.** c **8.157.** a **8.158.** d **8.159.** a **8.160.** b **8.161.** b **8.162.** c **8.163.** d **8.164.** e **8.165.** d **8.166.** a **8.167.** c **8.168.** d

Chapter 9
Practice Exercises
PE 9.1. The mass percents of oxygen in the two samples are the same: 72.72%. **PE 9.2.** (a) 101.96 amu (b) 124.12 amu **PE 9.3.** (a) 176.14 amu (b) 376.41 amu **PE 9.4.** 39.99% C, 6.73% H, and 53.28% O **PE 9.5.** 44.77% C, 7.52% H, and 47.71% O **PE 9.6.** (a) 1.93×10^{24} H_2O_2 molecules (b) 3.23×10^{23} Cu atoms **PE 9.7.** (a) 84.6 g CO (b) 169 g Fe **PE 9.8.** 151 g/mole **PE 9.9.** 3.818×10^{-23} g **PE 9.10.** (a) 4.086 moles H atoms, 2.043 moles O atoms (b) 4.086 moles H atoms, 4.086 moles O atoms **PE 9.11.** 3.952×10^{22} molecules C_3H_8S **PE 9.12.** 1.60×10^{-13} g CH_4 **PE 9.13.** (a) 3.271×10^{-22} g Au (b) 5.324×10^{-23} g N_2H_4 **PE 9.14.** 9.38 g C **PE 9.15.** 9.49×10^{22} H atoms **PE 9.16.** (a) 38.93 g H_3PO_4 (b) 1.07 g impurities **PE 9.17.** 1.91×10^{23} molecules $C_2H_6O_2$ **PE 9.18.** (a) NO_2 (b) C_3H_7 (c) CH_2 (d) NH_3 **PE 9.19.** $C_{13}H_{17}ClNO$ **PE 9.20.** Fe_2O_3 **PE 9.21.** C_5H_4 **PE 9.22.** C_2H_6O **PE 9.23.** (a) N_2O_4 (b) SO_2 **PE 9.24.** (a) C_3H_8 (b) C_3H_8

Practice Problems
9.1. 42.9% C, 57.1% O **9.3.** According to the law of definite proportions, the percentage A in each sample should be the same and the percentage D should also be the same. Sample I: % A = 57.81%, % D = 42.2%

Sample II: % A = 57.81%, % D = 42.2%   **9.5.** The mass ratio between X and Q present in the sample will be the same for samples of the same compound. Experiments 1 and 3 produced the same compound.   **9.7.** 100.0 g SO_2

9.9. (a) 156.71 amu (b) 151.92 amu (c) 342.34 amu (d) 354.48 amu   **9.11.** (a) $C_2H_4(OH)_2$ = 62.08 amu (b) $(H_2N)_2CO$ = 60.07 amu (c) $Mg_3(Si_2O_5)_2(OH)_2$ = 379.31 amu (d) $Al_2Si_2O_5(OH)_2$ = 224.16 amu   **9.13.** (a) 62.08 amu (b) 90.09 amu   **9.15.** $y = 4$ **9.17.** 16.1 amu   **9.19.** 44.1 amu **9.21.** N_2O **9.23.** (a) 4.90% H, 17.48% B, 77.62% O (b) 83.88% C, 16.12% H (c) 71.616% C, 5.521% H, 6.962% N, 15.90% O (d) 70.275% C, 8.059% H, 9.429% Cl, 3.726% N, 8.511% O **9.25.** (a) 79.89% Cu, 20.1% O (b) 32.37% Na, 22.57% S, 45.06% O (c) 25.9% N, 74.08% O (d) 33.4% S, 66.6% O **9.27.** (a) no (b) no (c) yes (d) yes   **9.29.** They match. **9.31.** C_2H_2 = 92.24% C, 7.76% H; C_6H_6 = 92.24% C, 7.76% H. Both compounds have the same ratio of carbon to hydrogen. Thus, the %C and %H will be the same in each. **9.33.** 363 g N   **9.35.** (a) 6.02×10^{23} Ag atoms (b) 6.02×10^{23} H_2O molecules (c) 6.02×10^{23} $NaNO_3$ formula units (d) 6.02×10^{23} SO_4^{2-} ions   **9.37.** (a) 1.93×10^{24} Si atoms (b) 2.2×10^{23} Si atoms (c) 7.29×10^{23} Si atoms (d) 1.01×10^{25} Si atoms   **9.39.** (a) 9.03×10^{23} molecules CO_2 (b) 3.01×10^{23} molecules NH_3 (c) 1.40×10^{24} molecules PF_3 (d) 6.715×10^{23} molecules N_2H_4 **9.41.** 1.19×10^{-15} mole people   **9.43.** (a) 40.08 g (b) 28.09 g (c) 58.93 g (d) 107.9 g   **9.45.** (a) 58.44 g (b) 78.05 g (c) 85.00 g (d) 163.94 g   **9.47.** (a) 58.44 g (b) 78.05 g (c) 85.00 g (d) 163.9 g   **9.49.** (a) 180.5 g (b) 233.6 g (c) 434.0 g (d) 753.9 g **9.51.** (a) 44.01 g (b) 44.02 g (c) 44.11 g   **9.53.** (a) 2.00 moles Cu (b) 1.00 mole Br (c) 1.50 moles N_2O (d) 4.87 moles B_2H_6 **9.55.** (a) 18.02 g/mole (b) 39.41 g/mole (c) 149.6 g/mole (d) 79.81 g/mole   **9.57.** compound A **9.59.** (a) 3.155×10^{-23} g (b) 8.635×10^{-23} g (c) 1.971×10^{-23} g (d) 3.271×10^{-22} g   **9.61.** The element with an atomic mass of 30.97 amu is phosphorus. **9.63.** 6.022×10^{23} atoms Al   **9.65.** 3 moles Na/1 mole Na_3PO_4; 1 mole P/1 mole Na_3PO_4; 4 moles O/1 mole Na_3PO_4; 3 moles Na/ 1 mole P; 3 moles Na/4 moles O; 1 mole P/4 moles O   **9.67.** (a) 4 moles N/1 mole $S_4N_4Cl_2$ (b) 2 moles Cl/1 mole $S_4N_4Cl_2$ (c) 10 moles atoms/1 mole $S_4N_4Cl_2$ (d) 2 moles Cl/4 moles S   **9.69.** (a) Both compounds contain the same moles of S. (b) They contain different moles of S. (c) Both compounds contain the same moles of S. (d) Both compounds contain the same moles of S.   **9.71.** (a) $NaAuBr_4$ = 12.0 moles of atoms (b) $C_2H_2Cl_4$ = 8 moles of atoms (c) $Ba(NO_3)_2$ = 27.0 moles of atoms (d) NH_4CN = 8.40 moles of atoms **9.73.** (a) 1.54×10^{24} atoms (b) 5.70×10^{23} atoms (c) 3.46×10^{23} atoms (d) 1.58×10^{23} atoms **9.75.** (a) 5.02×10^{23} molecules (b) 3.42×10^{23} molecules (c) 1.98×10^{23} molecules (d) 1.39×10^{23} molecules **9.77.** (a) greater than (b) less than (c) less than (d) less than   **9.79.** (a) 17.14 g P (b) 18.82 g PH_3 (c) 1.250×10^{-19} g P (d) 1.373×10^{-19} g PH_3 **9.81.** (a) 3.818×10^{-23} g Na (b) 4.037×10^{-23} g Mg (c) 9.655×10^{-23} g C_4H_{10}

(d) 1.297×10^{-22} g C_6H_6 **9.83.** (a) 60.66 g Cl (b) 903.5 g Cl (c) 9.72 g Cl (d) 17.0 g Cl   **9.85.** (a) 1.71×10^{23} atoms P (b) 3.38×10^{23} atoms P (c) 9.82×10^{22} atoms P (d) 1.23×10^{23} atoms P   **9.87.** (a) 370 g O (b) 13.04 g O (c) 29 g O (d) 215 g O   **9.89.** (a) 0.09251 mole $B(OH)_3$ (b) 1.393×10^{22} molecules B_4H_{10} (c) 1.696 g $C_2B_4H_6$ (d) 5.571×10^{22} atoms B   **9.91.** (a) 7.64 g NH_3 (b) 2.55 g NH_3 (c) 1.91 g NH_3 (d) 7.64 g NH_3 **9.93.** (a) 22.9 g SiH_4 (b) 185 g SiH_4 (c) 36.9 g SiH_4 (d) 25.0 g SiH_4 **9.95.** (a) 4.213 moles atoms (b) 1.005×10^{24} atoms C (c) 2.544 g O (d) 4.787×10^{22} molecules $C_{21}H_{30}O_2$ **9.97.** (a) 9.03×10^{23} atoms (b) 4.52×10^{23} molecules **9.99.** (a) 310 g Fe_2S_3 (b) 15 g impurities   **9.101.** 99.29% **9.103.** 20.0 g sample   **9.105.** 3.063×10^{23} atoms Cu **9.107.** 10.9 g Cr   **9.109.** (a) CH (b) NS (c) C_2H_6O (d) BNH_2 **9.111.** (a) not the same (b) not the same (c) same (d) same   **9.113.** (a) $C_6H_8O_7$, $C_6H_8O_7$ (b) $C_6H_{12}O_6$, CH_2O **9.115.** (a) Na_2S (b) $KMnO_4$ (c) H_2SO_4 (d) $C_2H_3O_5N$ **9.117.** (a) 3 to 5 (b) 2 to 5 (c) 6 to 7 (d) 4 to 7 to 6   **9.119.** (a) P_2O_5 (b) Mg_3N_2 (c) $Na_2S_2O_3$ (d) $Mg_2P_2O_7$ **9.121.** CuS, Cu_2S **9.123.** C_2H_6S **9.125.** BeO **9.127.** (a) CH_4 (b) CH (c) C_2H_5 (d) CH_3 **9.129.** CH_2 **9.131.** C_5H_6O **9.133.** (a) C_3H_6 (b) $Na_2S_4O_6$ (c) $C_3H_6O_2$ (d) $C_5H_5N_5$ **9.135.** (a) X_3Y_9 (b) XY_3 (c) X_2Y_6 (d) XY_3 **9.137.** (a) $C_4H_6O_2$ (b) $C_6H_9O_3$ (c) $C_{12}H_{18}O_6$ (d) $C_4H_6O_2$ **9.139.** (a) $C_8H_8O_2$ (b) $C_8H_8O_2$ **9.141.** (a) $C_3H_6O_3$ (b) $C_3H_6O_3$ **9.143.** (a) gold (b) sulfur (c) Cl_2 molecules (d) Ne   **9.145.** (a) P_4 (b) 1.00 mole Na (c) Cu (d) Be **9.147.** (a) false (b) true (c) false (d) false   **9.149.** 2.837 g K and 1.163 g S   **9.151.** 37.8 g B   **9.153.** 4.48 g $C_6H_{12}O_6$ **9.155.** 288 g O   **9.157.** (a) 8.37% H (b) 64.3% H (c) 64.3% H **9.159.** (a) Na_3AlF_6 (b) SO_2 (c) $BaCO_3$ (d) HClO **9.161.** HO, H_2O_2 **9.163.** 2.43×10^{23} atoms C **9.165.** 5.013 g sample burned   **9.167.** (a) CH_4S (b) 7.221 g **9.169.** 40.0% NaF, 40.0% $NaNO_3$, and 20.0% Na_2SO_4 **9.171.** 1.364% C, 6.365% N, 9.53% H, 82.74% O   **9.173.** $x = 3$ **9.175.** 40.08 g/mole   **9.177.** $C_3H_8O_2$ **9.179.** (a) 142 (b) 141.9 (c) 141.94 (d) 141.943   **9.181.** (a) 120.4 amu (b) 148.3 amu (c) 223.4 amu (d) 115.0 amu **9.183.** (a) 80.03% S (b) 39.07% S (c) 24.26% S (d) 94.07% S **9.185.** 11.3 g/cm^3 **9.187.** 84.2 L CO_2 **9.189.** 69.4 mL water **9.191.** 7×10^{17} atoms ^{40}K   **9.193.** (a) 1.40×10^{24} formula units $Al_2(SO_4)_3$ (b) 2.81×10^{24} Al^{+3} ions (c) 4.21×10^{24} SO_4^{2-} ions (d) 7.02×10^{24} ions **9.195.** 32.7 g KCl   **9.197.** 4.52×10^{24} Cl^- ions **9.199.** 6.82×10^{26} atoms Ni   **9.201.** 81.0 mL solution

Multiple-Choice Practice Test
9.203. e   **9.204.** c   **9.205.** d   **9.206.** d   **9.207.** c   **9.208.** c **9.209.** b   **9.210.** b   **9.211.** d   **9.212.** d   **9.213.** d   **9.214.** c **9.215.** a   **9.216.** c   **9.217.** d   **9.218.** e   **9.219.** d   **9.220.** e **9.221.** a   **9.222.** c

Chapter 10
Practice Exercises
PE 10.1. (a) 3 (b) 4 (c) 14 (d) 18
PE 10.2. $SiO_2 + 3 C \longrightarrow SiC = 2 CO$
PE 10.3. $4 HNO_3 \longrightarrow 4 NO_2 + O_2 + 2 H_2O$
PE 10.4. $N_2 + 3 H_2 \longrightarrow 2 NH_3$

PE 10.5. (a) decomposition (b) synthesis
(c) double-replacement (d) single-replacement
PE 10.6. $C_7H_8 + 9\,O_2 \longrightarrow 7\,CO_2 + 4\,H_2O$
PE 10.7. (a) 3.75 moles H_2O (b) 9.38 moles O_2
PE 10.8. 156.23 g (reactants) = 156.23 g (products)
PE 10.9. 1.52 g SO_2 **PE 10.10.** 1.978 g I_2
PE 10.11. 0.6250 mole N_2O **PE 10.12.** $8.859 \times 10^{21}\,O_2$
molecules   **PE 10.13.** Nuts are the limiting reactant.
PE 10.14. C_6H_{12} is the limiting reactant.   **PE 10.15.** 84.18 g
$CHCl_3$ **PE 10.16.** (a) 52.5 g $NaClO$ (b) 82.3% yield
PE 10.17. 2.03 moles SO_2 **PE 10.18.** 76.5 g H_2SO_4
PE 10.19. $2\,SO_2 + O_2 + 2\,H_2SO_4 \longrightarrow 2\,H_2S_2O_7$

Practice Problems

10.1. (a) not consistent with (b) consistent with   **10.3.** 44.01
10.5. 2.2 g CO_2 **10.7.** diagram III   **10.9.** (a) appropriate
(b) appropriate (c) appropriate (d) not appropriate
10.11. (a) solid, liquid, aqueous, gas (b) solid, aqueous,
solid, gas, liquid   **10.13.** (a) balanced (b) not balanced
(c) balanced (d) not balanced   **10.15.** (a) 2 Cu, 2 O
(b) 4 H, 2 O (c) 1 Ba, 2 Cl, 2 Na, 1 S (d) 6 C, 16 H, 20 O
10.17. (a) $2\,N_2 + 3\,O_2 \longrightarrow 2\,N_2O_3$
(b) $4\,NH_3 + 6\,NO \longrightarrow 5\,N_2 + 6\,H_2O$
(c) $CS_2 + 3\,O_2 \longrightarrow CO_2 + 2\,SO_2$
(d) $Mg + 2\,HBr \longrightarrow MgBr_2 + H_2$
10.19. (a) $3\,PbO + 2\,NH_3 \longrightarrow 3\,Pb + N_2 + 3\,H_2O$
(b) $2\,NaHCO_3 + H_2SO_4 \longrightarrow Na_2SO_4 + 2\,CO_2 + 2\,H_2O$
(c) $TiO_2 + C + 2\,Cl_2 \longrightarrow TiCl_4 + CO_2$
(d) $2\,NBr_3 + 3\,NaOH \longrightarrow N_2 + 3\,NaBr + 3\,HBrO$
10.21. (a) $C_5H_{12} + 13\,O_2 \longrightarrow 5\,CO_2 + 6\,H_2O$
(b) $2\,C_5H_{10} + 15\,O_2 \longrightarrow 10\,CO_2 + 10\,H_2O$
(c) $C_5H_8 + 7\,O_2 \longrightarrow 5\,CO_2 + 4\,H_2O$
(d) $C_5H_{10}O + 7\,O_2 \longrightarrow 5\,CO_2 + 5\,H_2O$
10.23. (a) $Ca(OH)_2 + 2\,HNO_3 \longrightarrow Ca(NO_3)_2 + 2\,H_2O$
(b) $BaCl_2 + (NH_4)_2SO_4 \longrightarrow BaSO_4 + 2\,NH_4Cl$
(c) $2\,Fe(OH)_3 + 3\,H_2SO_4 \longrightarrow Fe_2(SO_4)_3 + 6\,H_2O$
(d) $Na_3PO_4 + 3\,AgNO_3 \longrightarrow 3\,NaNO_3 + Ag_3PO_4$
10.25. (a) $AgNO_3 + KCl \longrightarrow AgCl + KNO_3$ (b)
$CS_2 + 3\,O_2 \longrightarrow CO_2 + 2\,SO_2$ (c) $2\,H_2 + O_2 \longrightarrow 2\,H_2O$
(d) $2\,Ag_2CO_3 \longrightarrow 4\,Ag + 2\,CO_2 + O_2$
10.27. (a) $6\,A_2 + 2\,B_2 \longrightarrow 4\,A_3B$
(b) $3\,A_2 + 3\,B_2 \longrightarrow 6\,AB$ **10.29.** diagram III
10.31. (a) synthesis (b) synthesis (c) double-replacement
(d) single-replacement   **10.33.** (a) synthesis (b) synthesis
10.35. (a) $Zn + Cu(NO_3)_2 \longrightarrow Zn(NO_3)_2 + Cu$
(b) $2\,Ca + O_2 \longrightarrow 2\,CaO$
(c) $K_2SO_4 + Ba(NO_3)_2 \longrightarrow BaSO_4 + 2\,KNO_3$
(d) $2\,Ag_2O \longrightarrow 4\,Ag + O_2$
10.37. (a) $Rb_2CO_3 \longrightarrow Rb_2O + CO_2$
(b) $SrCO_3 \longrightarrow SrO + CO_2$
(c) $Al_2(CO_3)_3 \longrightarrow Al_2O_3 + 3\,CO_2$
(d) $Cu_2CO_3 \longrightarrow Cu_2O + CO_2$
10.39. (a) $2\,C_3H_6 + 9\,O_2 \longrightarrow 6\,CO_2 + 6\,H_2O$
(b) $C_2H_4 + 3\,O_2 \longrightarrow 2\,CO_2 + 2\,H_2O$
(c) $C_7H_{16} + 11\,O_2 \longrightarrow 7\,CO_2 + 8\,H_2O$
(d) $C_8H_{16} + 12\,O_2 \longrightarrow 8\,CO_2 + 8\,H_2O$
10.41. (a) $CH_2O + O_2 \longrightarrow CO_2 + H_2O$
(b) $2\,C_5H_{12}O + 15\,O_2 \longrightarrow 10\,CO_2 + 12\,H_2O$

(c) $C_5H_{10}O + 7\,O_2 \longrightarrow 5\,CO_2 + 5\,H_2O$
(d) $2\,C_5H_{10}O_2 + 13\,O_2 \longrightarrow 10\,CO_2 + 10\,H_2O$
10.43. (a) $4\,C_2H_7N + 19\,O_2 \longrightarrow 8\,CO_2 + 14\,H_2O + 4\,NO_2$
(b) $CH_4S + 3\,O_2 \longrightarrow CO_2 + 2\,H_2O + SO_2$
10.45. (a) synthesis, single-replacement, combustion
(b) decomposition, single-replacement (c) synthesis, decomposition, single-replacement, double-replacement, combustion (d) synthesis, decomposition, single-replacement,
double-replacement, combustion   **10.47.** (a) 0.50 mole
(b) 1.0 mole (c) 2.5 moles (d) 3.5 moles   **10.49.** 4 moles
NH_3/3 moles O_2; 3 moles O_2/4 moles NH_3; 4 moles
NH_3/2 moles N_2; 2 moles N_2/4 moles NH_3; 4 moles
NH_3/6 moles H_2O; 6 moles H_2O/4 moles NH_3; 3 moles
O_2/2 moles N_2; 2 moles N_2/3 moles O_2; 3 moles
O_2/6 moles H_2O; 6 moles H_2O/3 moles O_2; 2 moles
N_2/6 moles H_2O; 6 moles H_2O/2 moles N_2
10.51. (a) 3 moles H_2O/6 moles ClO_2
(b) 1 mole HCl/5 moles $HClO_3$
(c) 1 mole HCl/3 moles H_2O
(d) 6 moles ClO_2/5 moles $HClO_3$ **10.53.** (a) 2.00 moles
NaN_3 (b) 9.00 moles CO (c) 6.00 moles NH_2Cl
(d) 2.00 moles $C_3H_5O_9N_3$ **10.55.** (a) 1.42 moles H_2O_2
(b) 0.473 mole CS_2 (c) 0.710 mole Mg (d) 4.26 moles HCl
10.57. (a) 6.12 moles products (b) 4.38 moles products
(c) 2.62 moles products (d) 4.81 moles products
10.59. (a) 3.0 moles Ag_2CO_3 (b) 8.0 moles Ag_2CO_3
(c) 4.8 moles Ag_2CO_3 (d) 4.29 moles Ag_2CO_3
10.61. (a) 5.67 g C (b) 3.26 g B
10.63. (a) 452.23 g reactants = 452.23 g products
(b) 173.01 g reactants = 173.01 g products
(c) 166.07 g reactants = 166.07 g products
(d) 1230.71 g reactants = 1230.71 g products
10.65. (a) 336 g O_2 (b) 112 g O_2 (c) 112 g O_2 (d) 56.0 g O_2
10.67. (a) 126 g HNO_3 (b) 126 g HNO_3 (c) 31.5 g HNO_3
(d) 2.10×10^2 g HNO_3 **10.69.** (a) 3.983 g HNO_3
(b) 0.6134 g H_2O (c) 1.054 g NaOH (d) 7.579 g $AgC_2H_3O_2$
10.71. (a) 30.0 g SiO_2 (b) 76.7 g CO (c) 66.8 g SiC (d) 20.7 g C
10.73. (a) 216 g LiOH (b) 239 g LiOH (c) 26.6 g LiOH
(d) 6.48 g LiOH   **10.75.** (a) 0.2977 mole Na_2SiO_3 (b) 79.95 g
HF (c) 9.867×10^{21} molecules H_2SiF_6 (d) 65.57 g HF
10.77. 26.6 g Al   **10.79.** 65.68 g Cr and 134.3 g Cl_2
10.81. 284 bolts   **10.83.** 213 kits   **10.85.** (a) NH_3 (b) Be
(c) NH_3 (d) Be   **10.87.** (a) 3.6 g NH_3 (b) 36.5 g NH_3 (c) 45.0 g
NH_3 (d) 67 g NH_3 **10.89.** (a) 4.4 g H_2 (b) 3.5 g H_2 (c) 13.0 g
N_2 (d) 1 g N_2 **10.91.** (a) A_2 (b) A_2 **10.93.** 175 $CoCl_3$ formula units   **10.95.** 0 g Fe_3O_4, 9.6 g O_2
10.97. 2.57 g SF_4, 3.21 g S_2Cl_2, 5.56 g NaCl   **10.99.** 30.8%
10.101. (a) 172.2 g (b) 125.6 g (c) 72.94%   **10.103.** diagram III
10.105. 96.6%   **10.107.** 131 g   **10.109.** 31.8 g CO_2
10.111. 34%   **10.113.** 54.6 g SO_2 **10.115.** 293 g O_2
10.117. (a) 4.00 moles $NaClO_3$ (b) 22.2 g $NaClO_3$
10.119. (a) 4.00 moles HNO_3 (b) 9.00 g HNO_3
10.121. 6.69 g $AgNO_3$
10.123. $2\,NaClO_3 + 3\,S \longrightarrow 2\,NaCl + 3\,SO_2$
10.125. $2\,N_2 + 2\,H_2O + 5\,O_2 \longrightarrow 4\,HNO_3$
10.127. (a) 6.00 moles O_2 (b) 9.00 moles H_2O
(c) 1.00 mole H_3PO_3 (d) 1.50 moles Cl_2

10.129. 8.34 g N_2, 21.4 g H_2O, 45.2 g Cr_2O_3
10.131. 41.53 g H_2S **10.133.** 80.1 g NO **10.135.** 44.4%
10.137. 75.4% **10.139.** 38.3%
10.141. (a) $Zn + 2\,AgNO_3 \longrightarrow Zn(NO_3)_2 + 2\,Ag$
(b) $HCl + NaOH \longrightarrow NaCl + H_2O$
(c) $PCl_3 + Cl_2 \longrightarrow PCl_5$ (d) $2\,Cu + O_2 \longrightarrow 2\,CuO$
10.143. C_3H_6 **10.145.** 43%
10.147. The formula of the copper oxide is Cu_2O;
$2\,Cu_2S + 3\,O_2 \longrightarrow 2\,Cu_2O + 2\,SO_2$.
10.149. Molecular formula = C_6H_6;
$2\,C_6H_6 + 15\,O_2 \longrightarrow 12\,CO_2 + 6\,H_2O$
10.151. (a) 260.88 g BeF_2 (b) 0.26082 kg BeF_2
(c) 2.608×10^8 g BeF_2 (d) 0.5750 lb BeF_2
10.153. 2.040 moles chlorine-containing products
10.155. 7.91 moles of electrons **10.157.** 8.139×10^{24} positive ions **10.159.** 87.3 mL solution
10.161. 0.0788 g HNO_3 **10.163.** 0.16 ton $CaSO_3$

Multiple-Choice Practice Test
10.165. c **10.166.** c **10.167.** e **10.168.** e **10.169.** d
10.170. d **10.171.** b **10.172.** e **10.173.** a **10.174.** d
10.175. a **10.176.** b **10.177.** d **10.178.** b **10.179.** c
10.180. c **10.181.** d **10.182.** b **10.183.** e **10.184.** c

Chapter 11
Practice Exercises
PE 11.1. (a) 83,800 joules (b) 2.00×10^4 calories (c) 20.0 kilocalories **PE 11.2.** 499 J **PE 11.3.** 147°C
PE 11.4. 2.42 J/g°C **PE 11.5.** 5.0 g wood **PE 11.6.** 7180 J
PE 11.7. 294 J/g **PE 11.8.** −259,000 J **PE 11.9.** −48,400 J
PE 11.10. (a) no (b) no (c) yes

Practice Problems
11.1. (a) gaseous (b) liquid (c) gaseous (d) gaseous
11.3. (a) both gases (b) liquid and solid (c) both gases
(d) both solids **11.5.** (a) 3 (b) 2 **11.7.** (a) potential
(b) kinetic (c) potential (d) potential **11.9.** (a) Average
velocity increases with increased temperature, and vice
versa. (b) potential (attractive) (c) direct; higher temperature, higher disruptive forces (d) all three **11.11.** (a) solid
(b) liquid (c) solid (d) solid or liquid **11.13.** (a) The predominant cohesive forces in the solid hold the particles in
essentially fixed positions. (b) The gas particles are widely
separated (disruptive forces). The solid and liquid particles have very little space between them (cohesive forces).
The space between the particles can be decreased greatly
in gases, but not in solids or liquids. (c) The cohesive
forces are dominant enough that changing the temperature has only a small effect on the space between particles.
(d) The disruptive forces in a gas are so dominant that
each particle can act independently of the others.
11.15. (a) liquid (b) liquid and gas (c) solid
11.17. (a) different (b) different (c) same (d) same
11.19. (a) different (b) different (c) different (d) same
11.21. (a) not opposite (b) not opposite (c) not opposite
(d) not opposite **11.23.** (a) 2.0 calories (b) 1.0 kilocalorie
(c) 100 Calories (d) 1000 kilocalories **11.25.** (a) 2.29×10^6 J
(b) 547 kcal (c) 5.47×10^5 cal (d) 547 Cal **11.27.**
(a) 153 J (b) 178 J (c) 4.0×10^2 J (d) 968 J **11.29.** (a) 63.9°C
(b) 22.5°C (c) 89.5°C (d) −31.6°C **11.31.** (a) 11 g (b) 13 g

(c) 2.4 g (d) 2.0 g **11.33.** (a) 0.33 J/g°C (b) 0.39 J/g°C (c)
0.48 J/g°C (d) 0.62 J/g°C **11.35.** (a) 15.5 J/°C (b) 36.3 J/°C
(c) 167 J/°C (d) 35 J/°C **11.37.** 80.0 g copper
11.39. 0.382 J/g°C **11.41.** 47 g **11.43.** (a) heat of solidification (b) heat of condensation (c) heat of fusion (d) heat
of vaporization **11.45.** (a) $−1.96 \times 10^4$ J (b) $−1.13 \times 10^5$ J
(c) $−1.02 \times 10^4$ J (d) $−1.13 \times 10^5$ J **11.47.** 3075 J released
11.49. 5.125 times as much **11.51.** 41.5 moles
11.53. copper **11.55.** 7.30×10^2 J
11.57.

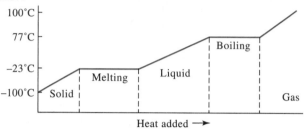

11.59. (a) 1600 J (b) 27,400 J (c) 41,100 J (d) 1.61×10^5 J
11.61. (a) 8690 J (b) 9080 J (c) 1.510×10^4 J (d) 32,200 J
11.63. 2.15×10^4 J **11.65.** (a) boiling point (b) vapor
pressure (c) boiling (d) boiling point **11.67.** (a) Increasing
the temperature increases the average kinetic energy of the
particles, enabling more particles to evaporate. (b) The
boiling point is lower; reactions occur more slowly at
lower temperatures. (c) The boiling point is higher; reactions occur more rapidly at higher temperatures. (d) The
particles leaving have higher than average kinetic energy,
reducing the average energy of particles of the liquid as
they leave. **11.69.** (a) increase (b) no change (c) increase
(d) no change **11.71.** (a) no change (b) decrease (c) no
change (d) no change **11.73.** (a) increase (b) no change
(c) no change (d) no change **11.75.** B must have lower
cohesive forces between particles than A to make B evaporate faster. **11.77.** At the same temperature, the substance
with the higher vapor pressure is more volatile. Thus, CS_2
is more volatile. **11.79.** Polar molecules must be present.
11.81. The stronger the intermolecular forces, the higher
the boiling point. **11.83.** (a) London forces (b) London
forces (c) dipole–dipole interactions (d) London forces
11.85. (a) no (b) yes (c) yes (d) no **11.87.** (a) F_2, larger
mass (b) HF, hydrogen bonding (c) CO, greater polarity
(d) O_2, larger mass **11.89.** (a) SiH_4 (b) $SiCl_4$ (c) $GeBr_4$
(d) C_2H_4 **11.91.** The water must be heated to a higher
temperature before the effects of hydrogen bonding are
overcome. **11.93.** At 4°C, the distance between water
molecules maximizes because of "balance" between
random motion and hydrogen-bonding effects. Below this
temperature, random motion decreases and hydrogen-
bonding causes the molecules to move farther apart, causing a density decrease. **11.95.** Hydrogen bonding
between water molecules causes ice to have an "open"
structure that is less dense than liquid water.
11.97. Hydrogen bonding causes water to have higher
than normal heats of vaporization and condensation.
Thus larger amounts of heat are absorbed during evaporation and released during condensation, which produces

a temperature-moderating effect. **11.99.** A measure of the inward force on the surface of a liquid caused by unbalanced intermolecular forces. **11.101.** (a) $SnCl_4$ (b) SnI_4 (c) SnI_4 (d) SnI_4 **11.103.** Vaporizing 50.0 g water **11.105.** 9.52 g **11.107.** 38.3°C **11.109.** 0.421 J/g°C **11.111.** 1.4×10^3 J/g **11.113.** 6.5×10^2 J **11.115.** 25 kJ **11.117.** The unknown is likely a mixture of A and B. **11.119.** 0.775 g/mL **11.121.** 30.1 g/mole **11.123.** 9.45 kJ **11.125.** 2.38×10^3 J

Multiple-Choice Practice Test
11.127. a **11.128.** c **11.129.** a **11.130.** e **11.131.** a **11.132.** e **11.133.** c **11.134.** d **11.135.** d **11.136.** d **11.137.** c **11.138.** a **11.139.** d **11.140.** d **11.141.** b **11.142.** e **11.143.** d **11.144.** d **11.145.** b **11.146.** d

Chapter 12
Practice Exercises
PE 12.1. (a) 0.8872 atm (b) 26.55 in. Hg (c) 13.02 psi
PE 12.2. 1.31 L Cl_2 **PE 12.3.** 46.0 mL **PE 12.4.** 907 mL
PE 12.5. 184°C **PE 12.6.** $V_1 = V_2 \times \dfrac{P_2}{P_1} \times \dfrac{T_1}{T_2}$
PE 12.7. 3.50 L Xe **PE 12.8.** −41°C **PE 12.9.** 37.5 L
PE 12.10. 80.0 L **PE 12.11.** 8.69 L HCl **PE 12.12.** 112°C
PE 12.13. 32.0 g/mole **PE 12.14.** 4.47 g/L
PE 12.15. 36.4 g/mole **PE 12.16.** 5.54 L SO_2 **PE 12.17.** I_2
PE 12.18. (a) 25.7 L N_2 (b) 17.0 L N_2 **PE 12.19.** (a) 30.4 L F_2 (b) 30.4 L Ne (c) 30.4 L CO_2 (d) 30.4 L NH_3
PE 12.20. (a) 2.45 g/L SO_2 (b) 2.859 g/L SO_2
PE 12.21. 9.06 L $COCl_2$ **PE 12.22.** (a) 2.50 g NH_3 (b) 4.98 g PH_3 **PE 12.23.** 92.3 L Cl_2 **PE 12.24.** 20.1 g $CuCO_3$
PE 12.25. 27.6 L CO_2 **PE 12.26.** 9.39 L **PE 12.27.** 1.00 atm for each gas **PE 12.28.** $X_N = 0.0275$, $X_O = 0.0120$, $X_{He} = 0.962$ **PE 12.29.** $P_{CO} = 6.25$ atm, $P_{Ne} = 2.50$ atm, $P_{Ar} = 1.25$ atm **PE 12.30.** 2.4 atm **PE 12.31.** 662 mm Hg
PE 12.32. (a) 31.6% Ne (b) 31.6% Ne (c) 31.6% Ne

Practice Problems
12.1. (a) 4710 mm Hg (b) 186 in. Hg (c) 91.0 psi (d) 471 cm Hg
12.3. (a) smaller than (b) smaller than (c) equal to (d) greater than **12.5.** 999 mm Hg **12.7.** (a) increase (b) increase (c) decrease (d) increase **12.9.** (a) 1.34 L (b) 0.381 L (c) 6.71 L (d) 56.1 L **12.11.** (a) 2.0×10^2 mL (b) 9.4 mL (c) 1400 mL (d) 550 mL **12.13.** diagram II **12.15.** 1.94×10^3 mm Hg
12.17. (a) decrease (b) increase (c) decrease (d) increase
12.19. (a) 6.43 L (b) 8.05 L (c) 4.03 L (d) 18.8 L
12.21. (a) 22.2 mL (b) 9.40×10^2 mL (c) 4330 mL (d) 1100 mL
12.23. diagram II **12.25.** −125°C **12.27.** (a) decrease (b) increase (c) decrease (d) increase **12.29.** (a) 1.03 atm (b) 0.946 atm (c) 0.800 atm (d) 0.461 atm **12.31.** (a) 1.24 atm (b) 1.07 atm (c) 1.30 atm (d) 30.8 atm **12.33.** 3.1 atm
12.35. (a) $T_2 = T_1 \times \dfrac{P_2}{P_1} \times \dfrac{V_2}{V_1}$ (b) $\dfrac{V_2}{P_1} = \dfrac{V_1}{P_2} \times \dfrac{T_2}{T_1}$
12.37. (a) 0.871 mL (b) 0.483 mL (c) 0.231 mL (d) 0.112 mL
12.39. (a) 5.90 L (b) 3.70×10^3 mL (c) 2.11 atm (d) −171°C
12.41. (a) 235 mm Hg (b) 727°C **12.43.** (a) 2.48×10^{3}°C (b) −196.7°C (c) 186°C (d) 33°C **12.45.** (a) 48.0 L (b) 55.7 L (c) 1.30×10^2 L (d) 12.0 L **12.47.** (a) 2.58 moles (b) 3.43 moles (c) 4.29 moles (d) 5.15 moles
12.49. diagram III **12.51.** (a) 31.0 L (b) 38.0 L (c) 24.0 L

(d) 17.0 L **12.53.** (a) 0.677 L (b) 0.785 L (c) 1.83 L (d) 0.169 L
12.55. (a) 10.9 L (b) 10.9 L (c) 10.9 L (d) 10.9 L
12.57. (a) 34.9 L (b) 52.3 L (c) 43.6 L (d) 105 L **12.59.** 0.540 L
12.61. (a) 0.450 mole O_2 (b) 2.32 moles CH_4 (c) 100.0 g N_2O (d) 100 g CO **12.63.** 2.6 L **12.65.** (a) 26.9 L (b) 22.1 L (c) 5.24 L (d) 24.5 L **12.67.** 0.900 mole **12.69.** −2°C
12.71. (a) 1.2 atm (b) 4.89 atm (c) 0.044 atm (d) 0.175 atm
12.73. 0.0824 atm L/mole K **12.75.** 36.1 L **12.77.** (a) 29.5 g (b) 3.19 g (c) 31.5 g (d) 0.363 g **12.79.** 44 g Cl_2 added
12.81. 28.0 g **12.83.** 103 g **12.85.** 90.1 g
12.87. CO gas **12.89.** (a) 1.24 g/L (b) 1.68 g/L (c) 0.840 g/L (d) 0.996 g/L **12.91.** (a) 0.937 atm (b) 0.200 atm (c) 0.482 atm (d) 0.597 atm **12.93.** CO gas
12.95. 25.01 g/mole **12.97.** 78°C
12.99. $4\,NH_3(g) + 5\,O_2(g) \longrightarrow 4\,NO(g) + 6\,H_2O(g)$
12.101. (a) 0.433 L C_3H_8 (b) 0.325 L C_3H_8
12.103. 0.11 L C_3H_8 **12.105.** (a) 1.50 L H_2 (b) 5.50 L H_2 (c) 1.00 L N_2 (d) 1.00 L H_2
12.107. (a) 22.5 L CO_2, 90.0 L H_2 (b) 33.0 L CO_2, 132 L H_2 (c) 64.0 L CO_2, 256 L H_2 (d) 16.0 L CO_2, 64.0 L H_2
12.109. (a) 20.9 L (b) 21.7 L (c) 32.5 L (d) 16.2 L
12.111. (a) 12.1 L (b) 12.1 L (c) 12.1 L (d) 12.1 L
12.113. (a) 0.509 atm (b) 0.565 atm (c) 1.02 atm (d) 1.49 atm
12.115. (a) 22.41 L (b) 22.41 L (c) 22.41 L (d) 22.41 L
12.117. (a) 0.9901 g/L (b) 0.7604 g/L (c) 0.7301 g/L (d) 0.4869 g/L **12.119.** (a) 2.16 g/L (b) 1.47 g/L (c) 0.7162 g/L (d) 1.342 g/L **12.121.** (a) O_3 (b) PH_3 (c) CO_2 (d) F_2 **12.123.** (a) 44.1 g/mole (b) 28.0 g/mole (c) 16.0 g/mole (d) 71.0 g/mole **12.125.** (a) 42.2 g (b) 46.6 g (c) 84.7 g (d) 36.0 g **12.127.** (a) 41.9 g (b) 46.2 g (c) 84.0 g (d) 35.7 g **12.129.** (a) 28.0 L (b) 28.0 L (c) 28.0 L (d) 28.0 L **12.131.** (a) 18.9 L (b) 18.9 L (c) 18.9 L (d) 18.9 L
12.133. (a) NH_3 (b) O_2 (c) NO_2 (d) NO **12.135.** 18.7 L NO_2
12.137. 61.1 g NO **12.139.** 12.2 L **12.141.** 259 L
12.143. 78.0 L **12.145.** 0.891 L CO_2 **12.147.** 7.39 L
12.149. (a) 5.51 L (b) 6.11 L (c) 76.8 L (d) 197 L
12.151. 8.01 atm **12.153.** 9.0 atm for He, 5.0 atm for Ne, 15.0 atm for Ar **12.155.** (a) 167 mm Hg (b) 354 mm Hg (c) 235 mm Hg (d) 86 mm Hg **12.157.** (a) 0.018 atm (b) 0.030 atm (c) 0.012 atm **12.159.** (a) 0.407 mole fraction CO, 0.259 mole fraction CO_2, 0.334 mole fraction H_2S (b) 0.700 atm P_{CO}, 0.445 atm P_{CO_2}, 0.574 atm P_{H_2S}
12.161. (a) 4.8 atm (b) 4.8 atm (c) 4.8 atm (d) 0.015 mole O_2
12.163. (a) 0.60 atm (b) 0.40 atm (c) 0.400 atm (d) 0.400 atm
12.165. 5.2 atm **12.167.** (a) 5.10 atm (b) 10.3 atm
12.169. 0.455 atm for He, 0.227 atm for Ar, 0.318 atm for Xe
12.171. 1.28 atm **12.173.** $P_{N_2} = 213$ mm Hg, $P_{H_2} = 639$ mm Hg **12.175.** (a) 726 mm Hg (b) 617 mm Hg (c) 722 mm Hg (d) 690 mm Hg **12.177.** (a) 0.0909 (b) 72.7% Ar (c) 18.2% Ne (d) 9.09% He **12.179.** (a) 25% O_2 (b) 37% Ar (c) 37% Ne (d) 0.25 atm **12.181.** 2.69×10^{22} molecules CO_2 **12.183.** 3.48 L NH_3 **12.185.** 0.14 g O_2
12.187. 43.8 mL **12.189.** 25% **12.191.** 50.0%
12.193. (a) 1.14:1 (b) 1.14:1 **12.195.** $P_T = 1200.0$ mm Hg
12.197. 1.3 atm **12.199.** 2.5 atm **12.201.** 1240 mm Hg
12.203. 75 amu **12.205.** $x = 3$ **12.207.**
(a) $X_{HCl} = 0.216$, $X_{H_2} = 0.260$, $X_{He} = 0.524$

(b) P_{HCl} = 0.259 atm, P_{H_2} = 0.312 atm, P_{He} = 0.629 atm
(c) 10.50 g/mole (d) 0.4685 g/L **12.209.** C_4H_{10}
12.211. 120.0 L products total **12.213.** 0.0868 m³ HCl
12.215. 2.0 atm

Multiple-Choice Practice Test
12.217. e **12.218.** b **12.219.** d **12.220.** b **12.221.** b
12.222. b **12.223.** d **12.224.** a **12.225.** e **12.226.** b
12.227. d **12.228.** c **12.229.** d **12.230.** d **12.231.** c
12.232. a **12.233.** b **12.234.** a **12.235.** d **12.236.** b

Chapter 13
Practice Exercises
PE 13.1. (a) soluble (b) soluble (c) soluble (d) soluble
(e) insoluble **PE 13.2.** 4.51% (m/m) **PE 13.3.** 36.0%
(v/v) **PE 13.4.** 250.6 g NH_4Cl **PE 13.5.** 27.3 g $NiSO_4$
PE 13.6. 56.5 mL HNO_3 solution
PE 13.7. (a) 1.25×10^4 ppm (m/m)
(b) 1.25×10^7 ppb (m/m) **PE 13.8.** 1.5×10^{-4} g Cu
PE 13.9. 0.868 M **PE 13.10.** 17.1 g $C_{12}H_{22}O_{11}$
PE 13.11. 294 mL H_2O_2 solution **PE 13.12.** 18.0 M H_2SO_4
PE 13.13. 47.8 g H_2O **PE 13.14.** 16.1 m **PE 13.15.** 270.3 g
$C_6H_{12}O_6$ **PE 13.16.** 11.8 M H_2O_2 **PE 13.17.** 0.76 m
$NaHCO_3$ **PE 13.18.** 1.64 M HNO_3 **PE 13.19.** 475 mL
PE 13.20. 0.390 M **PE 13.21.** 0.200 L Na_2S solution
PE 13.22. 932 g AgCl **PE 13.23.** 54 g Ag_3PO_4
PE 13.24. 2.80 L CO_2

Practice Problems
13.1. (a) true (b) true (c) true (d) false **13.3.** (a) solute
(sodium chloride), solvent (water) (b) solute (sucrose),
solvent (water) (c) solute (water), solvent (ethyl alcohol)
(d) solute (ethyl alcohol), solvent (methyl alcohol)
13.5. (a) slightly soluble (b) soluble (c) very soluble
(d) very soluble **13.7.** (a) supersaturated (b) saturated
(c) unsaturated (d) saturated **13.9.** (a) saturated
(b) unsaturated (c) unsaturated (d) supersaturated
13.11. (a) dilute (b) concentrated (c) concentrated
(d) concentrated **13.13.** (a) second (b) first (c) second
(d) second **13.15.** (a) none (b) 44 g **13.17.** (a) hydrated
ion (b) hydrated ion (c) oxygen atom (d) hydrogen atom
13.19. (a) decrease (b) increase (c) increase (d) increase
13.21. (a) ethanol (b) carbon tetrachloride (c) ethanol
(d) ethanol **13.23.** (a) generally soluble (b) generally soluble (c) generally insoluble (d) generally soluble
13.25. (a) carbonate rule (b) sulfide rule (c) ammonium
rule (d) sulfate rule **13.27.** (a) soluble (b) soluble with
exceptions (c) soluble (d) insoluble with exceptions
13.29. (a) like (b) like (c) unlike (d) like **13.31.** (a) no,
$MgSO_4$ soluble (b) yes (c) no, both soluble (d) no, both
soluble **13.33.** (a) 7.20% (m/m) (b) 2.77% (m/m)
(c) 19.4% (m/m) (d) 9.09% (m/m) **13.35.** (a) 26.3% (m/m)
(b) 28.5% (m/m) (c) 55.0% (m/m) (d) 90.5% (m/m)
13.37. (a) 2.50 g (b) 4.38 g (c) 12.50 g (d) 10.3 g
13.39. (a) 5.18 g H_2O (b) 51.8 g H_2O (c) 56.9 g H_2O (d) 2.3 g
H_2O **13.41.** (a) 9.50×10^2 g H_2O (b) 9.50×10^2 g H_2O
(c) 9.50×10^2 g H_2O (d) 9.50×10^2 g H_2O
13.43. (a) 79.1% ethyl alcohol (b) 38.1% ethyl alcohol
(c) 20.9% ethyl alcohol (d) 5.14% ethyl alcohol

13.45. (a) 36.06% methyl alcohol (b) 66.72% H_2O
13.47. 2.81 mL **13.49.** 1.40 gal H_2O **13.51.** (a) 0.0867%
(m/v) (b) 0.467% (m/v) (c) 2.25% (m/v) (d) 6.29% (m/v)
13.53. (a) 7.27% (m/v) (b) 0.857% (m/v) (c) 0.857% (m/v)
(d) 68.7% (m/v) **13.55.** (a) 83.3 mL (b) 117 mL (c) 3750 mL
(d) 3.75 mL **13.57.** (a) 11.4 g Na_3PO_4
(b) 3.75×10^3 g Na_3PO_4 **13.59.** 24% **13.61.** (a) 1.79 ppm
(b) 170 ppm (c) 0.312 ppm (d) 2270 ppm **13.63.** (a) 1790 ppb
(b) 170,000 ppb (c) 312 ppb (d) 2,270,000 ppb **13.65.** yes
13.67. 11,500 L air **13.69.** (a) 1.6×10^{-6} g (b) 3.1×10^{-3} g
(c) 13 g (d) 27 g **13.71.** (a) 4.0 M (b) 3.81 M (c) 1.06 M
(d) 0.40 M **13.73.** (a) 29.8 g (b) 0.852 g (c) 5.33×10^4 g
(d) 337 g **13.75.** (a) 84.8 mL (b) 567 mL (c) 1.80×10^3 mL
(d) 1.52 mL **13.77.** (a) diagram IV (b) diagrams I and III
13.79. (a) 0.631 L (b) 0.530 L (c) 0.669 L (d) 0.327 L
13.81. 22.72 M **13.83.** 17.96% **13.85.** 5.49 M
13.87. (a) 0.0357 m (b) 6.92 m (c) 0.00800 m (d) 0.449 m
13.89. (a) 11.2 g (b) 5.62 g (c) 29.8 g (d) 48.0 g
13.91. (a) 1940 g (b) 680 g (c) 1.20×10^2 g (d) 73 g
13.93. 6.3 m **13.95.** 0.584 m **13.97.** 0.762 M
13.99. 1.191 g/mL **13.101.** 1.47 m **13.103.** (a) 0.250 M
(b) 0.115 M (c) 0.0444 M (d) 0.00690 M **13.105.** (a) 0.553 M
(b) 0.604 M (c) 0.793 M (d) 2.24 M **13.107.** (a) 43.3 mL
(b) 3600 mL (c) 85.2 mL (d) 0.15 mL **13.109.** diagram II
13.111. (a) 380 mL H_2O (b) 30.0 mL H_2O (c) 3040 mL H_2O
(d) 1.4×10^5 mL H_2O **13.113.** (a) 3.3 M (b) 2.5 M (c) 8.86 M
(d) 0.00909 M **13.115.** (a) 0.468 M (b) 6.00 M (c) 4.75 M
(d) 5.99 M **13.117.** 8.7 L **13.119.** 1.00 L **13.121.** 17.3 g S
13.123. 610 mL **13.125.** 0.0632 M **13.127.** 1.7 L NO
13.129. 0.0510 M **13.131.** (a) $(NH_4)_3PO_4$ (b) $Ca(OH)_2$
(c) $AgNO_3$ (d) CaS, $Ca(NO_3)_2$, $Ca(C_2H_3O_2)_2$ **13.133.** 80 g
13.135. (a) 17.8% (m/v) (b) 1.05 M **13.137.** 6.72 g Na_2SO_4
13.139. 1245 g H_2O **13.141.** 3.74 mg solute/kg solution
13.143. 0.171 M **13.145.** 8.06 mL **13.147.** 340 g H_2O
13.149. 298 g solution, 266 mL solution
13.151. (a) 11.9% (m/v) (b) 12.2% (m/m) (c) 15.1% (v/v)
13.153. 1.01 g/mL **13.155.** (a) AgCl is insoluble.
(b) $Ba_3(PO_4)_2$ is insoluble. (c) $PbSO_4$ is insoluble. (d) $BaSO_4$
and CuS are insoluble. **13.157.** 117 L **13.159.** 2.32 g AgCl
13.161. 289 g $BaCrO_4$ **13.163.** 37.1% (m/m) **13.165.** zinc
13.167. 0.0833 M

Multiple-Choice Practice Test
13.169. b **13.170.** b **13.171.** d **13.172.** e **13.173.** b
13.174. b **13.175.** b **13.176.** b **13.177.** d **13.178.** d
13.179. c **13.180.** c **13.181.** c **13.182.** b **13.183.** d
13.184. d **13.185.** b **13.186.** c **13.187.** b **13.188.** d

Chapter 14
Practice Exercises
PE 14.1. $HClO_2/ClO_2^-$ and H_3O^+/H_2O
PE 14.2. (a) HCO_3^- (b) HCO_3^-
PE 14.3. (a) $FeSO_4 + 2 NaCl \longrightarrow FeCl_2 + Na_sSO_4$
(b) $2 Al(NO_3)_3 + 3 K_2S \longrightarrow Al_2S_3 + 6 KNO_3$
(c) $2 H_3PO_4 + 3 Na_2CO_3 \longrightarrow 2 Na_3PO_4 + 3 H_2CO_3$
PE 14.4. 3.00×10^{-12} M OH^- **PE 14.5.** (a) 5.0 (b) 10.0
(c) 6.0 **PE 14.6.** 7.03 **PE 14.7.** 2.5×10^{-8} M
PE 14.8. (a) acidic (b) neutral (c) basic (d) basic

PE 14.9. (a) CN^-; $CN^- + H_2O \longrightarrow HCN + OH^-$
(b) NH_4^+; $NH_4^+ + H_2O \longrightarrow NH_3 + H_3O^+$
PE 14.10. (a) no (b) no (c) no (d) yes
PE 14.11. (a) $H_3O^+ + F^- \longrightarrow HF + H_2O$
(b) $H_2CO_3 + OH^- \longrightarrow HCO_3^- + H_2O$
PE 14.12. 0.0181 M $Ba(OH)_2$

Practice Problems

14.1. (a) H^+ ion (b) dissociation   **14.3.** (a) Arrhenius acid
(b) Arrhenius acid   **14.5.** (a) ionization (b) dissociation
(c) dissociation (d) ionization   **14.7.** (a) $HBr \longrightarrow H^+ + Br^-$
(b) $HClO_2 \longrightarrow H^+ + ClO_2^-$ (c) $LiOH \longrightarrow Li^+ + OH^-$
(d) $Ba(OH)_2 \longrightarrow Ba^{2+} + 2\,OH^-$ **14.9.** (a) acid: HBr, base:
H_2O (b) acid: H_2O, base: N_3^- (c) acid: H_2S, base: H_2O
(d) acid: HS^-, base: H_2O
14.11. (a) $HOCl + NH_3 \longrightarrow NH_4^+ + OCl^-$
(b) $H_2CO_3 + H_2O \longrightarrow H_3O^+ + HCO_3^-$
(c) $H_2O + H_2O \longrightarrow H_3O^+ + OH^-$
(d) $HC_2O_4^- + H_2O \longrightarrow H_3O^+ + C_2O_4^{2-}$
14.13. (a) NO_2^- (b) $H_2PO_4^-$ (c) PO_4^{3-} (d) NH_3
14.15. (a) $HClO_2$ (b) H_2SO_4 (c) HSO_3^- (d) PH_4^+
14.17. (a) $H_2C_2O_4$ and $HC_2O_4^-$; $HClO$ and ClO^- (b) HSO_4^-
and SO_4^{2-}; H_3O^+ and H_2O (c) $H_2PO_4^-$ and HPO_4^{2-}; NH_4^+
and NH_3 (d) H_2CO_3 and HCO_3^-; H_2O and OH^-
14.19. (a) yes (b) no (c) no (d) yes
14.21. (a) (1) $HS^- + H_3O^+ \longrightarrow H_2S + H_2O$
(2) $HS^- + OH^- \longrightarrow H_2O + S^{2-}$
(b) (1) $HPO_4^{2-} + H_3O^+ \longrightarrow H_2PO_4^- + H_2O$
(2) $HPO_4^{2-} + OH^- \longrightarrow H_2O + PO_4^{3-}$
(c) (1) $HCO_3^- + H_3O^+ \longrightarrow H_2CO_3 + H_2O$
(2) $HCO_3^- + OH^- \longrightarrow H_2O + CO_3^{2-}$
(d) (1) $H_2PO_3^- + H_3O^+ \longrightarrow H_3PO_3 + H_2O$
(2) $H_2PO_3^- + OH^- \longrightarrow H_2O + HPO_3^{2-}$
14.23. (a) monoprotic (b) monoprotic (c) monoprotic
(d) diprotic   **14.25.** (a) 1 (b) 1 (c) 1 (d) 2
14.27. (a) 0 (b) 5 (c) 3 (d) 2
14.29. (a) $H_2C_2O_4 + H_2O \longrightarrow H_3O^+ + HC_2O_4^-$
$HC_2O_4^- + H_2O \longrightarrow H_3O^+ + C_2O_4^{2-}$
(b) $H_2C_3H_2O_4 + H_2O \longrightarrow H_3O^+ + HC_3H_2O_4^-$
$HC_3H_2O_4^- + H_2O \longrightarrow H_3O^+ + C_3H_2O_4^{2-}$
14.31. to indicate that only one of the six H atoms are acidic
14.33. Monoprotic. Hydrogens attached to carbon atoms are
not acidic.   **14.35.** (a) strong (b) weak (c) weak (d) weak
14.37. (a) stronger (b) stronger (c) weaker (d) weaker
14.39. (a) strong, strong (b) strong, strong (c) weak, strong
(d) weak, strong   **14.41.** (a) HNO_3 (b) $HClO_4$ (c) H_3PO_4
(d) HF   **14.43.** Weak acid pertains to extent of dissociation and dilute acid pertains to concentration.   **14.45.** acid
IV   **14.47.** (a) salt (b) acid (c) salt (d) salt
14.49. (a) calcium sulfate (b) lithium carbonate (c) sodium
bromide (d) aluminum sulfide   **14.51.** (a) insoluble
(b) soluble (c) soluble (d) insoluble   **14.53.** (a) 2 (b) 2 (c) 3
(d) 5   **14.55.** (a) $NaNO_3 \longrightarrow Na^+ + NO_3^-$
(b) $CuCO_3 \longrightarrow Cu^{2+} + CO_3^{2-}$ (c) $BaCl_2 \longrightarrow Ba^{2+} + 2\,Cl^-$
(d) $Al_2(SO_4)_3 \longrightarrow 2\,Al^{3+} + 3\,SO_4^{2-}$
14.57. (a) yes (b) yes (c) no (d) yes
14.59. (a) $Ni + 2\,HCl \longrightarrow NiCl_2 + H_2$
(b) $Ca + 2\,H_2O \longrightarrow Ca(OH)_2 + H_2$

(c) $Mg + 2\,HCl \longrightarrow MgCl_2 + H_2$
(d) $Zn + 2\,H_2O \longrightarrow Zn(OH)_2 + H_2$ **14.61.** (a) no
(b) yes (c) yes (d) no   **14.63.** (a) 2 to 1 (b) 1 to 1 (c) 1 to 1
(d) 1 to 3
14.65. (a) $2\,HBr + Sr(OH)_2 \longrightarrow SrBr_2 + 2\,H_2O$
(b) $HCN + LiOH \longrightarrow LiCN + H_2O$
(c) $H_2SO_4 + Mg(OH)_2 \longrightarrow MgSO_4 + 2\,H_2O$
(d) $H_3PO_4 + 3\,KOH \longrightarrow K_3PO_4 + 3\,H_2O$
14.67. (a) $HC_3H_5O_3 + NaOH \longrightarrow NaC_3H_5O_3 + H_2O$
(b) $H_2C_4H_4O_5 + 2\,NaOH \longrightarrow Na_2C_4H_4O_5 + 2\,H_2O$
(c) $H_2C_4H_4O_4 + 2\,NaOH \longrightarrow Na_2C_4H_4O_4 + 2\,H_2O$
(d) $H_3C_6H_5O_7 + 3NaOH \longrightarrow Na_3C_6H_5O_7 + 3\,H_2O$
14.69. (a) $NaNO_3$ (b) KCN (c) Li_2SO_4 (d) $BaCO_3$
14.71. (a) H_2S, NaOH (b) HNO_3, KOH (c) $HClO_3$, $Al(OH)_3$
(d) HBr, $Ca(OH)_2$
14.73. (a) $Zn + 2\,HCl \longrightarrow ZnCl_2 + H_2$
(b) $HCl + NaOH \longrightarrow NaCl + H_2O$
(c) $2\,HCl + Na_2CO_3 \longrightarrow 2\,NaCl + CO_2 + H_2O$
(d) $HCl + NaHCO_3 \longrightarrow NaCl + CO_2 + H_2O$
14.75. (a) no (b) yes (c) no (d) yes
14.77. (a) $Fe + CuSO_4 \longrightarrow FeSO_4 + Cu$
(b) $Sn + 2\,AgNO_3 \longrightarrow Sn(NO_3)_2 + 2\,Ag$
(c) $Zn + NiCl_2 \longrightarrow ZnCl_2 + Ni$
(d) $Cr + Pb(C_2H_3O_2)_2 \longrightarrow Cr(C_2H_3O_2)_2 + Pb$
14.79. (a) An insoluble salt is formed. (b) An insoluble
salt is formed. (c) An insoluble salt is formed; a weak acid
is formed. (d) A gas is evolved.
14.81. (a) $2\,Al(NO_3)_3 + 3\,(NH_4)_2S \longrightarrow 6\,NH_4NO_3 + Al_2S_3$
(b) $2\,HCl + Ba(OH)_2 \longrightarrow BaCl_2 + 2\,H_2O$ (c) no reaction
(d) no reaction   **14.83.** (a) 2.9×10^{-12} M (b) 2.1×10^{-9} M
(c) 9.1×10^{-7} M (d) 1.1×10^{-5} M
14.85. (a) 2.4×10^{-11} M (b) 1.7×10^{-10} M
(c) 2.9×10^{-6} M (d) 1.4×10^{-4} M
14.87. (a) 1.8×10^{-13} M (b) 1.1×10^{-10} M
(c) 4.3×10^{-8} M (d) 1.5×10^{-3} M   **14.89.** (a) acidic
(b) acidic (c) acidic (d) basic   **14.91.** (a) 3.0×10^{-12} M,
basic (b) 1.5×10^{-7} M, basic (c) 1.4×10^{-7} M, acidic
(d) 2.2×10^{-9} M, acidic   **14.93.** (a) pH = 4.0
(b) pH = 11.0 (c) pH = 5.0 (d) pH = 9.0   **14.95.** (a) 2.2
(b) 3.2 (c) 7.5 (d) 9.2   **14.97.** (a) 2.5 (b) 2.52 (c) 2.523
(d) 2.5229   **14.99.** (a) both acidic (b) acidic, basic (c) both
basic (d) acidic, neutral   **14.101.** (a) 1×10^{-4} M
(b) 6×10^{-5} M (c) 3×10^{-5} M (d) 2×10^{-5} M
14.103. (a) 3.7×10^{-3} M (b) 1.9×10^{-6} M
(c) 1.9×10^{-8} M (d) 1.78×10^{-9} M
14.105. (a) 2.7×10^{-12} M (b) 5.2×10^{-9} M
(c) 5.4×10^{-7} M (d) 5.62×10^{-6} M
14.107. (a) 1.4×10^{-5} M, 9.14, basic (b) 5.0×10^{-6} M,
2.0×10^{-9} M, acidic (c) 1.4×10^{-5} M, 4.85, acidic
(d) 5.9×10^{-9} M, 1.7×10^{-6} M, basic
14.109. (a) Solution A (b) Solution B   **14.111.** (a) 4.199
(b) 3.900 (c) 3.500 (d) 1.500   **14.113.** (a) 2.20 (b) 0.70
(c) 4.38 (d) 10.37   **14.115.** (a) PO_4^{3-} (b) CN^- (c) NH_4^+
(d) none   **14.117.** (a) neutral (b) basic (c) acidic (d) neutral
14.119. (a) $NH_4^+ + H_2O \longrightarrow H_3O^+ + NH_3$
(b) $C_2H_3O_2^- + H_2O \longrightarrow OH^- + HC_2H_3O_2$
(c) $F^- + H_2O \longrightarrow OH^- + HF$

(d) $CN^- + H_2O \longrightarrow OH^- + HCN$ **14.121.** (a) no
(b) yes (c) no (d) yes **14.123.** (a) HCN/CN^-
(b) $H_3PO_4/H_2PO_4^-$ (c) H_2CO_3/HCO_3^- (d) HCO_3^-/CO_3^{2-}
14.125. All four represent buffers.
14.127. (a) $HF + OH^- \longrightarrow F^- + H_2O$
(b) $PO_4^{3-} + H_3O^+ \longrightarrow HPO_4^{2-} + H_2O$
(c) $CO_3^{2-} + H_3O^+ \longrightarrow HCO_3^- + H_2O$
(d) $H_3PO_4 + OH^- \longrightarrow H_2PO_4^- + H_2O$
14.129. (a) 20.0 mL (b) 40.0 mL (c) 40.0 mL (d) 80.0 mL
14.131. (a) 20.0 mL (b) 40.0 mL (c) 1.00×10^2 mL
(d) 1.50×10^2 mL **14.133.** (a) 0.0705 M (b) 0.176 M
(c) 0.0587 M (d) 0.352 M **14.135.** (a) strong (b) weak
(c) weak (d) strong **14.137.** 200 **14.139.** 6.76
14.141. (a) $A > B > D > C$ (b) $C < D < B < A$
(c) $C > D > B > A$ (d) $A < B < D < C$
14.143. (a) 1.6×10^{-7} M (b) 1.6×10^{-7} M (c) 6.8 (d) neutral
14.145. (a) no, higher pH (b) no, higher pH (c) yes (d) yes
14.147. pH = 1.5 **14.149.** $Ba(OH)_2$, $LiCN$, K_2SO_4, NH_4Br,
$HClO_4$ **14.151.** (a) HCN/CN^- (b) HF/F^-
14.153. $H_3PO_4/H_2PO_4^-$ and $H_2PO_4^-/HPO_4^{2-}$
14.155. 0.0061 mole $Ca(OH)_2$ **14.157.** (a) 2.00×10^{-14} M
(b) 0.500 M (c) 13.699 **14.159.** (a) the iodide ion, I^- (b) the
hydrogen phosphate ion, HPO_4^{2-} (c) the hydroxide ion, OH^-
(d) the hydronium ion, H_3O^+ **14.161.** (a) 0.48 (b) 14.312
(c) 1.22 (d) 0.46 **14.163.** (a) 0.10 M (b) 0.0010 M (c) 1.00
(d) 3.00 **14.165.** 2.3×10^{16} H_3O^+ ions
14.167. 1.82×10^{23} ions **14.169.** 0.67 L HCl

Multiple-Choice Practice Test
14.171. a **14.172.** b **14.173.** b **14.174.** c **14.175.** b
14.176. b **14.177.** c **14.178.** c **14.179.** b **14.180.** e
14.181. a **14.182.** a **14.183.** a **14.184.** d **14.185.** d
14.186. c **14.187.** e **14.188.** e **14.189.** b **14.190.** c

Chapter 15
Practice Exercises
PE 15.1. $Pb^{2+} + SO_4^{2-} \longrightarrow PbSO_4$
PE 15.2. $2 H_3PO_4 + 3 Cu^{2+} \longrightarrow Cu_3(PO_4)_2 + 6 H^+$
PE 15.3. $H^+ + OH^- \longrightarrow H_2O$
PE 15.4. (a) +1 for H, +6 for S, −2 for O
(b) +4 for C, −2 for S (c) +1 for K, +5 for Cl, −2 for O
(d) −3 for N, +1 for H **PE 15.5.** (a) Fe(0), Cu^{2+}(+2),
Cu(0), and Fe^{2+}(+2); Fe is the reducing agent, Cu^{2+} is the
oxidizing agent (b) SO_2(+4,−2), O_2(0), and SO_3(+6,−2);
SO_2 is the reducing agent, O_2 is the oxidizing agent
(c) PH_3(−3, +1), NO_2(+4,−2), H_3PO_4(+1,+5,−2), and
N_2(0); PH_3 is the reducing agent, NO_2 is the oxidizing agent
PE 15.6. (a) nonredox decomposition (b) redox decomposi-
tion (c) redox single-replacement (d) redox combustion
PE 15.7. $3 PbO + 2 NH_3 \longrightarrow N_2 + 3 H_2O + 3 Pb$
PE 15.8.
$4 Sb + 4 NO_3^- + 4 H^+ \longrightarrow Sb_4O_6 + 4 NO + 2 H_2O$
PE 15.9. $SeO_3^{2-} + Cl_2 + 2 OH^- \longrightarrow$
$SeO_4^{2-} + 2 Cl^- + H_2O$
PE 15.10. $2 CrO_4^{2-} + 3 N_2O + 10 H^+ \longrightarrow$
$2 Cr^{3+} + 6 NO + 5 H_2O$
PE 15.11. $2 MnO_4^- + 3 SO_3^{2-} + H_2O \longrightarrow$
$2 MnO_2 + 3 SO_4^{2-} + 2 OH^-$

PE 15.12. $MnO_2 + NO_2^- + 2H^+ \longrightarrow Mn^{2+} + NO_3^- + H_2O$
PE 15.13. (a) and (b)
$2 NO_2 + 2 OH^- \longrightarrow NO_3^- + NO_2^- + H_2O$
PE 15.14. 99 g K^+ **PE 15.15.** 167 mL $(NH_4)_3PO_4$ solution
PE 15.16. 658 mL SeO_3^{2-} solution

Practice Problems
15.1. (a) weak electrolyte (b) strong electrolyte (c) strong
electrolyte (d) weak electrolyte **15.3.** (a) both
(b) molecular form (c) ionic form (d) both **15.5.** (a) 2 (b) 3
(c) 2 (d) 2 **15.7.** (a) $NaCl \longrightarrow Na^+ + Cl^-$
(b) $Mg(NO_3)_2 \longrightarrow Mg^{2+} + 2 NO_3^-$
(c) $NH_4CN \longrightarrow NH_4^+ + CN^-$
(d) $HClO_4 \longrightarrow H^+ + ClO_4^-$ **15.9.** diagram III
15.11. (a) yes (b) no (c) yes (d) no **15.13.** (a) yes (b) yes
(c) yes (d) no **15.15.** (a) molecular (b) net ionic (c) ionic
(d) net ionic **15.17.** (a) $Pb^{2+} + 2 Br^- \longrightarrow PbBr_2$
(b) $Fe^{3+} + 3 OH^- \longrightarrow Fe(OH)_3$
(c) $Zn + 2 H^+ \longrightarrow Zn^{2+} + H_2$
(d) $H_2S + 2 OH^- \longrightarrow S^{2-} + 2H_2O$
15.19. (a) $2 Ag^+ + Pb \longrightarrow 2 Ag + Pb^{2+}$
(b) $Cl_2 + 2 Br^- \longrightarrow 2 Cl^- + Br_2$
(c) $2 Al^{3+} + 3 S^{2-} \longrightarrow Al_2S_3$ (d) everything cancels
15.21. (a) loss of electrons (b) increase in oxidation
number **15.23.** (a) An oxidizing agent gains electrons
from another substance. (b) An oxidizing agent contains
the atom that shows an oxidation number decrease. (c) An
oxidizing agent is itself reduced. **15.25.** (a) oxidized
(b) decrease (c) reducing (d) loses **15.27.** (a) 0 (b) 0 (c) 0
(d) 0 **15.29.** (a) +3 (b) +4 (c) +4 (d) +2 **15.31.** (a) +2
(b) −1 (c) +5 (d) +4 **15.33.** (a) −1 (b) +1 (c) −1 (d) +1
15.35. (a) +1 for H, +5 for P, −2 for O (b) +2 for Ba,
+6 for Cr, −2 for O (c) −3 for N, +1 for H, +7 for Cl,
−2 for O (d) +1 for H, +5 for P, −2 for O
15.37. H_2S (−2), S_8 (0), H_2SO_3 (+4), SO_3 (+6)
15.39. (a) +2 (b) +3 (c) +2 (d) +2 **15.41.** (a) −2 (b) +2
(c) −1 (d) −2 **15.43.** (a) −1 (b) +1 (c) +1 (d) −1
15.45. (a) H_2 oxidized, N_2 reduced (b) I^- oxidized, Cl_2
reduced (c) Fe oxidized, Sb reduced (d) S oxidized,
N reduced **15.47.** (a) N_2 is oxidizing agent, H_2 is reducing agent. (b) Cl_2 is oxidizing agent, I^- is reducing agent.
(c) Sb_2O_3 is oxidizing agent, Fe is reducing agent.
(d) HNO_3 is oxidizing agent, H_2SO_3 is reducing agent.
15.49. (a) SO_2 (b) HNO_3 (c) HNO_3 (d) SO_2
15.51. (a) The metal is oxidized and the nonmetal is
reduced. (b) The metal is oxidized and the nonmetal
is reduced. (c) The metal is oxidized and the nonmetal is
reduced. (d) The metal is oxidized and the nonmetal is
reduced. **15.53.** (a) redox (b) nonredox (c) redox (d) redox
15.55. (a) redox, synthesis (b) redox, single-replacement
(c) non-redox, decomposition (d) non-redox, double-
replacement **15.57.** (a) redox (b) redox (c) can't classify
(d) redox **15.59.** (a) $2 Cr + 6 HCl \longrightarrow 2 CrCl_3 + 3 H_2$
(b) $2 Cr_2O_3 + 3 C \longrightarrow 4 Cr + 3 CO_2$
(c) $SO_2 + NO_2 \longrightarrow SO_3 + NO$
(d) $BaSO_4 + 4 C \longrightarrow BaS + 4 CO$
15.61. (a) $Br_2 + 2 H_2O + SO_2 \longrightarrow 2 HBr + H_2SO_4$

(b) $3 H_2S + 2 HNO_3 \longrightarrow 3 S + 2 NO + 4 H_2O$
(c) $SnSO_4 + 2 FeSO_4 \longrightarrow Sn + Fe_2(SO_4)_3$
(d) $Na_2TeO_3 + 4 NaI + 6 HCl \longrightarrow$
$6 NaCl + Te + 3H_2O + 2 I_2$
15.63. (a) $I_2 + 5 Cl_2 + 6 H_2O \longrightarrow 2 HIO_3 + 10 Cl^- + 10 H^+$
(b) $8 MnO_4^- + 5 AsH_3 + 24 H^+ \longrightarrow$
$5 H_3AsO_4 + 8 Mn^{2+} + 12 H_2O$
(c) $2 Br^- + SO_4^{2-} + 4 H^+ \longrightarrow Br_2 + SO_2 + 2 H_2O$
(d) $Au + 4 Cl^- + 3 NO_3^- + 6 H^+ \longrightarrow$
$AuCl_4^- + 3 NO_2 + 3 H_2O$
15.65. (a) $8 OH^- + S^{2-} + 4 Cl_2 \longrightarrow$
$SO_4^{2-} + 8 Cl^- + 4 H_2O$
(b) $3 SO_3^{2-} + 2 CrO_4^{2-} + 5 H_2O \longrightarrow$
$2 Cr(OH)_4^- + 3 SO_4^{2-} + 2 OH^-$
(c)
$2 MnO_4^- + 3 IO_3^- + H_2O \longrightarrow 2 MnO_2 + 3 IO_4^- + 2 OH^-$
(d)
$I_2 + 7 Cl_2 + 18 OH^- \longrightarrow 2 H_3IO_6^{2-} + 14 Cl^- + 6 H_2O$
15.67. (a) reduction (b) oxidation (c) oxidation (d) reduction
15.69. (a) $Te \longrightarrow TeO_2$ (b) $Fe^{2+} \longrightarrow Fe^{3+}$
(c) $CN^- \longrightarrow CNO^-$ (d) $Cl^- \longrightarrow Cl_2$
15.71. (a) $MnO_2 + 4 H^+ + e^- \longrightarrow Mn^{3+} + 2 H_2O$
(b) $H_3MnO_4 + 5 H^+ + 5 e^- \longrightarrow Mn + 4 H_2O$
(c) $MnO_4^- + 8 H^+ + 5 e^- \longrightarrow Mn^{2+} + 4 H_2O$
(d) $MnO_4^- + 4 H^+ + 3 e^- \longrightarrow MnO_2 + 2 H_2O$
15.73. (a) $SeO_4^{2-} + 4 H_2O + 6 e^- \longrightarrow Se + 8 OH^-$
(b) $Se^{2-} + 6 OH^- \longrightarrow SeO_3^{2-} + 3 H_2O + 6 e^-$
(c) $SeO_4^{2-} + H_2O + 2 e^- \longrightarrow SeO_3^{2-} + 2 OH^-$
(d) $Se + 6 OH^- \longrightarrow SeO_3^{2-} + 3 H_2O + 4 e^-$
15.75. (a) $Zn + Cu^{2+} \longrightarrow Cu + Zn^{2+}$
(b) $Br_2 + 2 I^- \longrightarrow 2 Br^- + I_2$
(c) $S_2O_3^{2-} + 4 Cl_2 + 5 H_2O \longrightarrow 2 HSO_4^- + 8 Cl^- + 8 H^+$
(d) $6 Zn + As_2O_3 + 12 H^+ \longrightarrow 2 AsH_3 + 6 Zn^{2+} + 3 H_2O$
15.77.
(a) $I_2 + 5 Cl_2 + 6 H_2O \longrightarrow 2 HIO_3 + 10 Cl^- + 10 H^+$
(b) $8 MnO_4^- + 5 AsH_3 + 24 H^+ \longrightarrow$
$5 H_3AsO_4 + 8 Mn^{2+} + 12 H_2O$
(c) $2 Br^- + SO_4^{2-} + 4 H^+ \longrightarrow Br_2 + SO_2 + 2 H_2O$
(d) $Au + 4 Cl^- + 3 NO_3^- + 6 H^+ \longrightarrow$
$AuCl_4^- + 3 NO_2 + 3 H_2O$
15.79. (a) $2 NH_3 + ClO^- \longrightarrow N_2H_4 + Cl^- + H_2O$
(b)
$Cr(OH)_2 + 2 BrO^- + 2 OH^- \longrightarrow CrO_4^{2-} + 2 Br^- + 2 H_2O$
(c) $2 CrO_2^- + 3 H_2O_2 + 2 OH^- \longrightarrow 2 CrO_4^{2-} + 4 H_2O$
(d)
$2 Bi(OH)_3 + 3 Sn(OH)_3^- + 3 OH^- \longrightarrow 3 Sn(OH)_6^{2-} + 2 Bi$
15.81.
(a) $S^{2-} + 4 Cl_2 + 8 OH^- \longrightarrow SO_4^{2-} + 8 Cl^- + 4 H_2O$
(b) $3 SO_3^{2-} + 2 CrO_4^{2-} + 5 H_2O \longrightarrow$
$2 Cr(OH)_4^- + 3 SO_4^{2-} + 2 OH^-$
(c) $2 MnO_4^- + 3 IO_3^- + H_2O \longrightarrow 2 MnO_2 + 3 IO_4^- + 2 OH^-$
(d) $I_2 + 7 Cl_2 + 18 OH^- \longrightarrow 2 H_3IO_6^{2-} + 14 Cl^- + 6 H_2O$
15.83. (a) $3 HNO_2 \longrightarrow 2 NO + NO_3^- + H^+ + H_2O$
(b) $ClO^- + Cl^- + 2 H^+ \longrightarrow Cl_2 + H_2O$
(c) $3 S + 6 OH^- \longrightarrow 2 S^{2-} + SO_3^{2-} + 3 H_2O$
(d) $3 Br_2 + 6 OH^- \longrightarrow BrO_3^- + 5 Br^- + 3 H_2O$
15.85. (a) $3 HNO_2 \longrightarrow 2 NO + NO_3^- + H^+ + H_2O$
(b) $ClO^- + Cl^- + 2 H^+ \longrightarrow Cl_2 + H_2O$

(c) $3 S + 6 OH^- \longrightarrow 2 S^{2-} + SO_3^{2-} + 3 H_2O$
(d) $3 Br_2 + 6 OH^- \longrightarrow BrO_3^- + 5 Br^- + 3 H_2O$
15.87. (a) 35.45 g/mole (b) 18.05 g/mole (c) 216.00 g/mole
(d) 118.96 g/mole **15.89.** (a) 40.4 g (b) 40.4 g (c) 20.2 g
(d) 13.5 g **15.91.** (a) 1270 mL (b) 634 mL (c) 1110 mL
(d) 8.00×10^2 mL **15.93.** 265 mL **15.95.** (a) 0.20 M Na^+,
0.20 M Cl^- (b) 0.40 M K^+, 0.20 M SO_4^{2-} (c) 0.20 M Al^{3+},
0.60 M NO_3^- (d) 0.20 M Mg^{2+}, 0.40 M Cl^-
15.97. (a) 9.89×10^{22} ions (b) 3.32×10^{22} ions
(c) 9.91×10^{23} ions (d) 8.59×10^{23} ions **15.99.** (a) 563 mL
(b) 430 mL (c) 4540 mL (d) 896 mL **15.101.** (a) 0.287 M
(b) 0.0267 M (c) 0.0866 M **15.103.** N_2O, NO,
N_2O_3, NO_2, N_2O_5 **15.105.** (a) The oxidation number is
already at its minimum value and cannot go any lower.
(b) The oxidation number is already at its maximum
value and cannot go any higher. (c) The oxidation
number is at an intermediate value and can be either
increased or decreased. (d) The oxidation number is
already at its maximum value and cannot go any higher.
15.107. (a) +2, +1 (b) +3 in both (c) +1 in both (d) +3 in
both **15.109.** (a) two reduction half-reactions (b) one reduction and one oxidation half-reaction (c) two oxidation half-
reactions (d) one reduction and one oxidation half-reaction
15.111. $4 MnO_4^- + 5 PH_3 + 12 H^+ \longrightarrow$
$5 H_3PO_2 + 4 Mn^{2+} + 6 H_2O$;
$2 SO_4^{2-} + PH_3 + 4 H^+ \longrightarrow H_3PO_2 + 2 SO_2 + 2 H_2O$;
$3 MnO_4^- + 5 As + 3 H_2O + 9 H^+ \longrightarrow 5 H_3AsO_3 + 3 Mn^{2+}$;
$3 SO_4^{2-} + 2 As + 6 H^+ \longrightarrow 2 H_3AsO_3 + 3 SO_2$
15.113. $Zn \longrightarrow Zn^{2+} + 2 e^-$; $NO_3^- + 10 H^+ + 8 e^- \longrightarrow$
$NH_4^+ + 3 H_2O$
15.115. $2 KMnO_4 + 5 K_2C_2O_4 + 16 HCl \longrightarrow$
$2 MnCl_2 + 10 CO_2 + 8 H_2O + 12 KCl$ **15.117.** (a) redox,
HNO_3 is the oxidizing agent (b) acid–base, H_2S is the acid
(c) redox, H_2O_2 is the oxidizing agent (d) acid–base, H_2SO_4
is the acid **15.119.** (a) $Sn^{2+} + 2 Fe^{2+} \longrightarrow Sn + 2 Fe^{3+}$
(b) $PH_3 + 2 NO_2 \longrightarrow H_3PO_4 + N_2$
(c) $S + 3 H_2O + 2 Pb^{2+} \longrightarrow 2 Pb + H_2SO_3 + 4 H^+$
(d) $4 Zn + 10 H^+ + NO_3^- \longrightarrow 4 Zn^{2+} + NH_4^+ + 3 H_2O$
15.121. (a)
$3 H_2S + Cr_2O_7^{2-} + 8 H^+ \longrightarrow 2 Cr^{3+} + 3 S + 7 H_2O$
(b) $5 ClO_3^- + 3 I_2 + 3 H_2O \longrightarrow 6 IO_3^- + 5 Cl^- + 6 H^+$
(c) $S^{2-} + 4 Br_2 + 8 OH^- \longrightarrow SO_4^{2-} + 8 Br^- + 4 H_2O$
(d) $2 NO_2 + 2 OH^- \longrightarrow NO_3^- + NO_2^- + H_2O$
15.123. (a) 0.245 M Al^{3+}, 0.735 M NO_3^-
(b) 0.0735 M $Al(NO_3)_3$, 0.0735 M Al^{3+}, 0.220 M NO_3^-
15.125. 0.0916 M Mg^{2+}, 0.5418 M Na^+, 0.7251 M NO_3^-
15.127. 0.324 M I_3^- **15.129.** 30.8 ppm (m/m) O_3

Multiple-Choice Practice Test
15.131. c **15.132.** e **15.133.** a **15.134.** c **15.135.** d
15.136. e **15.137.** d **15.138.** d **15.139.** a **15.140.** b
15.141. d **15.142.** d **15.143.** c **15.144.** c **15.145.** d
15.146. d **15.147.** b **15.148.** a **15.149.** e **15.150.** d

Chapter 16
Practice Exercises
PE 16.1. 0.61 mole N_2, 1.10 moles H_2, 0.52 mole NH_3
PE 16.2. (a) $1/[NH_3]^2$ (b) $[HCl]^6/[WCl_6][H_2]^3$ (c) $[Cl_2]$
(d) $[H_2S]/[HS^-][H^+]$ **PE 16.3.** $K_{eq} = 1.9$

PE 16.4. (a) shifts to the left (b) shifts to the right (c) shifts to the left (d) shifts to the right

Practice Problems

16.1. Contact between molecules occurs with more ease in a solution. **16.3.** 45 kJ/mole; the lower activation energy is easier to overcome. **16.5.** molecular orientation at time of collision and molecular energy

16.7.

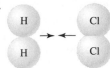

16.9. (a) endothermic (b) exothermic (c) exothermic (d) endothermic **16.11.** (a) less than (b) greater than (c) greater than (d) less than

16.13.

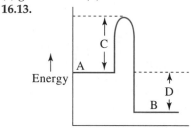

Reaction progress

(a) The average energy of the reactants is shown as A. (b) The average energy of the products is shown as B. (c) The activation energy is shown as C. (d) The energy liberated in the reaction is shown as D, or A − B.

16.15.

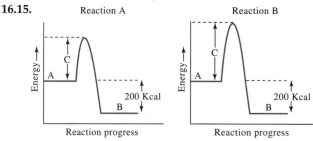

The reactions are exothermic to the same extent, and the energy difference between reactants and products is the same. The reactions differ in activation energy (C).

16.17. (a) With increased temperature, molecules move faster and collide more often. (b) A catalyst allows for an alternate pathway, which requires less activation energy. **16.19.** (a) increase (b) increase (c) decrease (d) decrease **16.21.** Surface area is much greater for the coal dust. **16.23.** (a) 1 (b) 3 (c) 4 (d) 3

16.25.

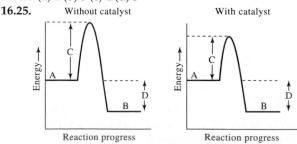

The energy difference between reactants and products (D) is the same. The activation energy (C) differs. **16.27.** The forward and reverse reaction rates must be equal.

16.29. (a) $CO(g) + H_2O(g) \longrightarrow CO_2(g) + H_2(g)$
(b) $CO_2(g) + H_2(g) \longrightarrow CO(g) + H_2O(g)$ **16.31.** yes, diagrams III and IV have the same composition

16.33. diagrams II and IV

16.35.

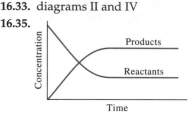

16.37.
$SO_3 = 0.0142$ mole, $SO_2 = 0.0058$ mole, $O_2 = 0.0029$ mole
16.39.
$NH_3 = 0.268$ mole, $N_2 = 0.184$ mole, $H_2 = 0.137$ mole
16.41. 0.094 mole CO, 0.197 mole O_2, 0.006 mole CO_2

16.43. (a) $\dfrac{[C_2H_6]}{[C_2H_2][H_2]^2}$ (b) $\dfrac{[SO_2]^2[O_2]}{[SO_3]^2}$ (c) $\dfrac{[NCl_3][HCl]^3}{[Cl_2]^3[NH_3]}$
(d) $\dfrac{[NO]^2[H_2O]^2}{[N_2H_4][O_2]^2}$ **16.45.** (a) $\dfrac{[CO_2]}{[SO_2]}$ (b) $[Br_2]$ (c) $[H_2O]$
(d) $\dfrac{[NaNO_3]}{[NaCl][AgNO_3]}$ **16.47.** (a) 6.67 (b) 1.04 (c) 3.75

(d) 0.0216 **16.49.** diagram IV **16.51.** diagram I
16.53. 0.0730 M **16.55.** 48.9 **16.57.** (a) mostly reactants (b) mostly products (c) mostly products (d) significant amounts of both reactants and products **16.59.** (a) right (b) left (c) right (d) left **16.61.** (a) to the right (b) to the left (c) to the left (d) to the right **16.63.** (a) shifts right (b) shifts left (c) no effect (d) shifts right **16.65.** (a) no (b) no (c) yes (d) no **16.67.** exothermic **16.69.** endothermic; reaction shifts to the right (more AB)
16.71. (a) $N_2 + 3 H_2 \rightleftharpoons 2 NH_3$
(b) $4 NH_3 + 3 O_2 \rightleftharpoons 2 N_2 + 6 H_2O$
(c) $2 NO \rightleftharpoons N_2 + O_2$ (d) $N_2 + O_2 \rightleftharpoons 2 NO$
16.73. (a) at equilibrium (b) shifts to the left (c) at equilibrium (d) shifts to the left **16.75.** 0.0005 **16.77.** (a) no (b) no (c) yes (d) no **16.79.** (a) right (b) does not shift (c) left
(d) does not shift **16.81.** (a) $K_{eq} = \dfrac{[SO_2]^2[O_2]}{[SO_3]^2}$
(b) $K_{eq} = \dfrac{[NH_3]^2[H_2O]^4}{[H_2]^7[NO_2]^2}$ (c) $K_{eq} = \dfrac{[CO_2]}{[CO]}$ (d) $K_{eq} = [CO_2]$

16.83. $K_{eq} = 38.9$ **16.85.** 0.0244 **16.87.** 2.20 atm
16.89. 0.1192 **16.91.** (a) shift to the left (b) shift to the right (c) shift to the left (d) shift to the left

Multiple-Choice Practice Test
16.93. c **16.94.** c **16.95.** c **16.96.** e **16.97.** e **16.98.** b
16.99. b **16.100.** d **16.101.** b **16.102.** b **16.103.** a
16.104. a **16.105.** b **16.106.** e **16.107.** b **16.108.** e
16.109. b **16.110.** c **16.111.** c **16.112.** c

INDEX

(A boldfaced term is defined on the indicated page.)

A

Accuracy, 17
contrasted with precision, 17–19
Acid(s), 322
Arrhenius, 615
Brønsted–Lowry, 616–18
characteristics of, early known, 614
classification of, based on negative ion
formed, 322–23
diprotic, 620, 622
ionization of, 615
monoprotic, 620, 622
net ionic equations and, 663–64
neutralization of, 627–28
nomenclature of, 321–26
nonoxy, 324–25
oxy, 325
polyprotic, 621
reaction with bases, 627–28
reaction with carbonates and
bicarbonates, 628
reaction with metals, 626–27
reaction with salts, 628–30
strengths of, 622–24
strong, 622–23
titration of, 648–50
triprotic, 621–22
weak, 622–24
Acid–base neutralization, 627–28
Acid–base theory
Arrhenius, concepts of, 615
Brønsted–Lowry, amphoteric substances
and, 620
Brønsted–Lowry, concepts of, 616–18
Brønsted–Lowry, conjugate pairs and,
618–20
Acid–base titration(s), 648
concentration determination and, 649–50
equipment needed for, 648–49
indicator use in, 649
Acidic hydrogen atom(s), 621
Acidic solution, 635
defined in terms of hydronium ion
concentration, 635
defined in terms of pH value, 637
Activation energy, 709
energy diagrams showing, 710
Actual yield, 426
calculations involving concept of, 426–29

Air
composition of, table, 547
Alkali metal(s), 201
location of, within periodic table, 201
Alkaline earth metal(s), 201
location of, within periodic table, 201
Alpha particle(s), 172
emission of, equations for, 173
notation for, 172
Alpha particle decay, 173
examples of, 173–75
general equation for, 173
neutron-to-proton ratio and, 180–81
Amphiprotic substance(s), 620
examples of, 620
Anion(s), 247
nomenclature for monoatomic, 306
Applied scientific research, 2
role of, 2
Aqueous solution(s), 570
Area, 62
formulas for calculation of, 62–63
units for, 62
Arrhenius, Svante August
Human Side of Chemistry feature
about, 616
Arrhenius acid, 615
ionization of, 615
Arrhenius base, 615
dissociation of, 615
Atom(s), 126
atomic numbers for, 149–51
building blocks for matter, 126–27
charge neutrality and, 148
diamagnetic, 218
electron configurations for, 210–15
electron orbital diagrams for, 216–18
ion formation from, 246–49
Lewis symbols for, 244–45
limit of chemical subdivision and, 130
mass numbers for, 150–51
mass of, 127
nuclear and extranuclear regions of,
147–48
nucleus of, 147–48, 165–66
paramagnetic, 218
radii for, 229
relative mass scale for, 158
size of, 127

size relationships among components
parts of, 148–49
subatomic particles, arrangement
within, 147–48
Atomic mass(es), 156
calculation of, procedure for, 161
informational value of, 156
irregularities in sequence of, 202–03
relative mass scale concept and, 157–58
units for, 158
values of, changes in, 162
values of, synthetic elements and, 162
values of, table of, inside front cover
values of, uncertainty associated
with, 162
weighted average concept and, 159–60
Atomic mass unit
relationship to grams unit, 358–59
relative scale for, 158
Atomic number(s), 149
electrons and, 150
for elements, listing of, inside front cover
general symbol for, 150
identifying characteristic for elements,
149–50
informational value of, 149–50
protons and, 149–50
use of, with elemental symbols, 151–52
Atomic radii
values, listing of, 229
values, periodic trends in, 229–30
Atomic theory of matter, 126
concepts associated with, 126–27
formulation of, 125–26
isotopes and, 153–54
law of conservation of mass and, 396–97
law of definite proportions and, 341–42
Aufbau diagram, 210–11
electron configurations and, 210–13
Aufbau principle, 210
electron configurations and, 210–13
Avogadro, Lorenzo Romano Amedeo
Carlo
Human Side of Chemistry feature
about, 349
Avogadro's law, 514
combined with the combined gas law,
516–17
mathematical statement of, 515
use of, in calculations, 516

W

Y

Mathematical Meanings of Metric System Prefixes

Prefix	Meaning	Prefix	Meaning
Tera (T)	10^{12}	Pico (p)	10^{-12}
Giga (G)	10^{9}	Nano (n)	10^{-9}
Mega (M)	10^{6}	Micro (μ)	10^{-6}
Kilo (k)	10^{3}	Milli (m)	10^{-3}
Hecto (h)	10^{2}	Centi (c)	10^{-2}
Deca (da)	10^{1}	Deci (d)	10^{-1}

Common Fixed-Charge Metallic Cations and Nonmetallic Anions

Cation	Name	Anion	Name
Li^+	lithium ion	F^-	fluoride ion
Na^+	sodium ion	Cl^-	chloride ion
K^+	potassium ion	Br^-	bromide ion
Rb^+	rubidium ion	I^-	iodide ion
Cs^+	cesium ion	O^{2-}	oxide ion
Be^{2+}	beryllium ion	S^{2-}	sulfide ion
Mg^{2+}	magnesium ion	N^{3-}	nitride ion
Ca^{2+}	calcium ion	P^{3-}	phosphide ion
Sr^{2+}	strontium ion	C^{4-}	carbide ion
Ba^{2+}	barium ion		
Ag^+	silver ion		
Zn^{2+}	zinc ion		
Cd^{2+}	cadmium ion		
Al^{3+}	aluminum ion		
Ga^{3+}	gallium ion		

Common Variable-Charge Metallic Cations

Cation	IUPAC Name	Older Name
Cu^+	copper(I) ion	cuprous ion
Cu^{2+}	copper(II) ion	cupric ion
Fe^{2+}	iron(II) ion	ferrous ion
Fe^{3+}	iron(III) ion	ferric ion
Sn^{2+}	tin(II) ion	stannous ion
Sn^{4+}	tin(IV) ion	stannic ion
Pb^{2+}	lead(II) ion	plumbous ion
Pb^{4+}	lead(IV) ion	plumbic ion
Au^+	gold(I) ion	aurous ion
Au^{3+}	gold(III) ion	auric ion